Organic Chemistry

Organic Chemistry

Concepts and Applications

Allan D. Headley

Texas A&M University
Commerce, Texas, USA

This edition first published 2020
© 2020 John Wiley & Sons, Inc.

All rights reserved. No part of this publication may be reproduced, stored in a retrieval system, or transmitted, in any form or by any means, electronic, mechanical, photocopying, recording or otherwise, except as permitted by law. Advice on how to obtain permission to reuse material from this title is available at http://www.wiley.com/go/permissions.

The right of Allan D. Headley to be identified as the authors of this work has been asserted in accordance with law.

Registered Office
John Wiley & Sons, Inc., 111 River Street, Hoboken, NJ 07030, USA

Editorial Office
111 River Street, Hoboken, NJ 07030, USA

For details of our global editorial offices, customer services, and more information about Wiley products visit us at www.wiley.com.

Wiley also publishes its books in a variety of electronic formats and by print-on-demand. Some content that appears in standard print versions of this book may not be available in other formats.

Limit of Liability/Disclaimer of Warranty
In view of ongoing research, equipment modifications, changes in governmental regulations, and the constant flow of information relating to the use of experimental reagents, equipment, and devices, the reader is urged to review and evaluate the information provided in the package insert or instructions for each chemical, piece of equipment, reagent, or device for, among other things, any changes in the instructions or indication of usage and for added warnings and precautions. While the publisher and authors have used their best efforts in preparing this work, they make no representations or warranties with respect to the accuracy or completeness of the contents of this work and specifically disclaim all warranties, including without limitation any implied warranties of merchantability or fitness for a particular purpose. No warranty may be created or extended by sales representatives, written sales materials or promotional statements for this work. The fact that an organization, website, or product is referred to in this work as a citation and/or potential source of further information does not mean that the publisher and authors endorse the information or services the organization, website, or product may provide or recommendations it may make. This work is sold with the understanding that the publisher is not engaged in rendering professional services. The advice and strategies contained herein may not be suitable for your situation. You should consult with a specialist where appropriate. Further, readers should be aware that websites listed in this work may have changed or disappeared between when this work was written and when it is read. Neither the publisher nor authors shall be liable for any loss of profit or any other commercial damages, including but not limited to special, incidental, consequential, or other damages.

Library of Congress Cataloging-in-Publication Data

Names: Headley, Allan D., 1955– author.
Title: Organic chemistry : concepts and applications / Allan D. Headley
   (Texas A&M University).
Description: First edition. | Hoboken, NJ : Wiley, 2020. | Includes
   bibliographical references and index. |
Identifiers: LCCN 2019018485 (print) | LCCN 2019020628 (ebook) |
   ISBN 9781119504627 (Adobe PDF) | ISBN 9781119504672 (ePub) |
   ISBN 9781119504580 (pbk.)
Subjects: LCSH: Chemistry, Organic–Textbooks.
Classification: LCC QD251.3 (ebook) | LCC QD251.3 .H43 2020 (print) | DDC
   547–dc23
LC record available at https://lccn.loc.gov/2019018485

Cover Design: Wiley
Cover Images: Background © Sean Nel/Shutterstock, Chemical images courtesy of Allan D. Headley

Set in 10/12pt Warnock by SPi Global, Pondicherry, India

Printed in the United States of America

Contents

Preface   *xvii*
About the Campanion Website   *xxiii*

1   Bonding and Structure of Organic Compounds   *1*
1.1   Introduction   *1*
1.2   Electronic Structure of Atoms   *4*
1.2.1   Orbitals   *4*
1.2.2   Electronic Configuration of Atoms   *6*
1.2.3   Lewis Dot Structures of Atoms   *8*
1.3   Chemical Bonds   *9*
1.3.1   Ionic Bonds   *9*
1.3.2   Covalent Bonds   *9*
1.3.3   Shapes of Molecules   *12*
1.3.4   Bond Polarity and Polar Molecules   *12*
1.3.5   Formal Charges   *14*
1.3.6   Resonance   *15*
1.4   Chemical Formulas   *18*
1.4.1   Line-Angle Representations of Molecules   *18*
1.5   The Covalent Bond   *20*
1.5.1   The Single Bond to Hydrogen   *20*
1.5.2   The Single Bond to Carbon   *21*
1.5.3   The Single Bond to Heteroatoms   *22*
1.5.4   The Carbon–Carbon Double Bond   *23*
1.5.5   The Carbon–Heteroatom Double Bond   *25*
1.5.6   The Carbon–Carbon Triple Bond   *26*
1.5.7   The Carbon–Heteroatom Triple Bond   *27*
1.6   Bonding – Concept Summary and Applications   *28*
1.7   Intermolecular Attractions   *29*
1.7.1   Dipole–Dipole Intermolecular Attractions   *29*
1.7.2   Intermolecular Hydrogen Bond   *30*
1.7.3   Intermolecular London Force Attractions   *31*
1.8   Intermolecular Molecular Interactions – Concept Summary and Applications   *31*
           End of Chapter Problems   *34*

2   Carbon Functional Groups and Organic Nomenclature   *39*
2.1   Introduction   *39*
2.2   Functional Groups   *39*

2.3	Saturated Hydrocarbons *41*
2.3.1	Classification of the Carbons of Saturated Hydrocarbons *44*
2.4	Organic Nomenclature *45*
2.5	Structure and Nomenclature of Alkanes *45*
2.5.1	Nomenclature of Straight Chain Alkanes *45*
2.5.2	Nomenclature of Branched Alkanes *46*
2.5.3	Nomenclature of Compounds that Contain Heteroatoms *49*
2.5.4	Common Names of Alkanes *50*
2.5.5	Nomenclature of Cyclic Alkanes *51*
2.5.6	Nomenclature of Branched Cyclic Alkanes *51*
2.5.7	Nomenclature of Bicyclic Compounds *52*
2.6	Unsaturated Hydrocarbons *54*
2.7	Structure and Nomenclature of Alkenes *56*
2.7.1	Nomenclature of Branched Alkenes *56*
2.7.2	Nomenclature of Polyenes *57*
2.7.3	Nomenclature of Cyclic Alkenes *58*
2.8	Structure and Nomenclature of Substituted Benzenes *58*
2.8.1	Nomenclature of Disubstituted Benzenes *59*
2.9	Structure and Nomenclature of Alkynes *60*
	End of Chapter Problems *61*

3   Heteroatomic Functional Groups and Organic Nomenclature   *63*

3.1	Properties and Structure of Alcohols, Phenols, and Thiols *63*
3.1.1	Types of Alcohols *65*
3.2	Nomenclature of Alcohols *66*
3.2.1	Nomenclature of Difunctional Alcohols *67*
3.2.2	Nomenclature of Cyclic Alcohols *67*
3.2.3	Nomenclature of Substituted Phenols *68*
3.3	Nomenclature of Thiols *68*
3.4	Structure and Properties of Aldehydes and Ketones *69*
3.5	Nomenclature of Aldehydes *70*
3.5.1	Nomenclature of Difunctional Aldehydes *70*
3.6	Nomenclature of Ketones *71*
3.6.1	Nomenclature of Difunctional Ketones *71*
3.6.2	Nomenclature of Cyclic Ketones *72*
3.7	Structure and Properties of Carboxylic Acids *73*
3.8	Nomenclature of Carboxylic Acids *75*
3.8.1	Nomenclature of Difunctional Carboxylic Acids *76*
3.8.2	Nomenclature of Cyclic Carboxylic Acids *76*
3.9	Structure and Properties of Esters *78*
3.9.1	Nomenclature of Esters *79*
3.9.2	Nomenclature of Cyclic Esters *80*
3.10	Structure and Properties of Acid Chlorides *82*
3.10.1	Nomenclature of Acid Chlorides *82*
3.10.2	Nomenclature of Difunctional Acid Chlorides *83*
3.11	Structure and Properties of Anhydrides *83*
3.11.1	Nomenclature of Anhydrides *84*
3.12	Structure and Properties of Amines *84*
3.12.1	Nomenclature of Amines *86*

3.12.2	Nomenclature of Difunctional Amines 88
3.13	Structure and Properties of Amides 88
3.13.1	Nomenclature of Amides 89
3.14	Structure and Properties of Nitriles 90
3.14.1	Nomenclature of Nitriles 90
3.15	Structure and Properties of Ethers 91
3.15.1	Nomenclature of Ethers 93
3.15.2	Nomenclature of Oxiranes 93
3.16	An Overview of Spectroscopy and the Relationship to Functional Groups 94
3.16.1	Infrared Spectroscopy 95
	End of Chapter Problems 99

4 Alkanes, Cycloalkanes, and Alkenes: Isomers, Conformations, and Stabilities 103

4.1	Introduction 103
4.2	Structural Isomers 103
4.3	Conformational Isomers of Alkanes 104
4.3.1	Dashed/Wedge Representation of Isomers 104
4.3.2	Newman Representation of Conformers 105
4.3.3	Relative Energies of Conformers 107
4.4	Conformational Isomers of Cycloalkanes 108
4.4.1	Isomers of Cyclopropane 108
4.4.2	Conformational Isomers of Cyclobutane 109
4.4.3	Conformational Isomers of Cyclopentane 109
4.4.4	Conformational Isomers of Cyclohexane 110
4.4.5	Conformational Isomers of Monosubstituted Cyclohexane 112
4.4.6	Conformational Isomers of Disubstituted Cyclohexane 113
4.5	Geometric Isomers 114
4.5.1	IUPAC Nomenclature of Alkene Geometric Stereoisomers 116
4.6	Stability of Alkanes 119
4.7	Stability of Alkenes 121
4.8	Stability of Alkynes 122
	End of Chapter Problems 123

5 Stereochemistry 125

5.1	Introduction 125
5.2	Chiral Stereoisomers 126
5.2.1	Determination of Enantiomerism 127
5.3	Significance of Chirality 129
5.3.1	Molecular Chirality and Biological Action 130
5.4	Nomenclature of the Absolute Configuration of Chiral Molecules 131
5.5	Properties of Stereogenic Compounds 133
5.6	Compounds with More Than One Stereogenic Carbon 134
5.6.1	Cyclic Compounds with More Than One Stereogenic Center 136
5.7	Resolution of Enantiomers 137
	End of Chapter Problems 140

6 An Overview of the Reactions of Organic Chemistry 145

| 6.1 | Introduction 145 |
| 6.2 | Acid–Base Reactions 145 |

6.2.1	Acids	*146*
6.2.2	Bases	*147*
6.3	Addition Reactions	*149*
6.4	Reduction Reactions	*150*
6.5	Oxidation Reactions	*153*
6.6	Elimination Reactions	*154*
6.7	Substitution Reactions	*156*
6.8	Pericyclic Reactions	*158*
6.9	Catalytic Coupling Reactions	*158*
	End of Chapter Problems	*159*

7 Acid–Base Reactions in Organic Chemistry *165*

7.1	Introduction	*165*
7.2	Lewis Acids and Bases	*165*
7.3	Relative Strengths of Acids and Conjugate Bases	*166*
7.4	Predicting the Relative Strengths of Acids and Bases	*169*
7.5	Factors That Affect Acid and Base Strengths	*170*
7.5.1	Electronegativity	*171*
7.5.2	Type of Hybridized Orbitals	*171*
7.5.3	Resonance	*172*
7.5.4	Polarizability/Atom Size	*174*
7.5.5	Inductive Effect	*175*
7.6	Applications of Acid–Bases Reactions in Organic Chemistry	*176*
	End of Chapter Problems	*180*

8 Addition Reactions Involving Alkenes and Alkynes *183*

8.1	Introduction	*183*
8.2	The Mechanism for Addition Reactions Involving Alkenes	*183*
8.3	Addition of Hydrogen Halide to Alkenes (Hydrohalogenation of Alkenes)	*185*
8.3.1	Addition Reactions to Symmetrical Alkenes	*185*
8.3.2	Addition Reactions to Unsymmetrical Alkenes	*186*
8.3.3	Predicting the Major Addition Product	*187*
8.3.4	Predicting the Stereochemistry of Addition Reaction Products	*190*
8.3.5	Predicting the Major Addition Product – Markovnikov Rule	*190*
8.3.6	Unexpected Hydrohalogenation Products	*191*
8.3.7	Anti-Markovnikov Addition to Alkenes	*192*
8.4	Addition of Halogens to Alkenes (Halogenation of Alkenes)	*196*
8.5	Addition of Halogens and Water to Alkenes (Halohydrin Formation)	*198*
8.6	Addition of Water to Alkenes (Hydration of Alkenes)	*199*
8.6.1	Hydration by Oxymercuration–Demercuration	*203*
8.6.2	Hydration by Hydroboration-Oxidation	*204*
8.7	Addition of Carbenes to Alkenes	*207*
8.7.1	Structure of Carbenes	*207*
8.7.2	Reactions of Carbenes	*207*
8.8	The Mechanism for Addition Reactions Involving Alkynes	*209*
8.8.1	Addition of Bromine to Alkynes	*209*
8.8.2	Addition of Hydrogen Halide to Alkynes	*210*

8.8.3	Addition of Water to Alkynes	*211*
8.9	Applications of Addition Reactions to Synthesis	*213*
	End of Chapter Problems	*214*

9 Addition Reactions Involving Carbonyls and Nitriles *223*

9.1	Introduction	*223*
9.2	Mechanism for Addition Reactions Involving Carbonyl Compounds	*223*
9.3	Addition of HCN to Carbonyl Compounds	*224*
9.4	Addition of Water to Carbonyl Compounds	*226*
9.4.1	Reactivity of Carbonyl Compounds Toward Hydration	*227*
9.5	Addition of Alcohols to Carbonyl Compounds	*230*
9.5.1	Ketals and Acetals as Protection Groups	*234*
9.6	Addition of Ylides to Carbonyl Compounds (The Wittig Reaction)	*235*
9.6.1	Synthesis of Phosphorous Ylides	*236*
9.7	Addition of Enolates to Carbonyl Compounds	*237*
9.8	Addition of Amines to Carbonyl Compounds	*240*
9.9	Mechanism for Addition Reactions Involving Imines	*241*
9.9.1	Addition of Water to Imines	*242*
9.10	Mechanism for Addition Reactions Involving Nitriles	*242*
9.10.1	Addition of Water to Nitriles	*243*
9.11	Applications of Addition Reactions to Synthesis	*244*
	End of Chapter Problems	*246*

10 Reduction Reactions in Organic Chemistry *251*

10.1	Introduction	*251*
10.2	Reducing Agents of Organic Chemistry	*252*
10.2.1	Metal Hydrides	*252*
10.2.2	Organometallic Compounds	*253*
10.2.3	Dissolving Metals	*254*
10.2.4	Hydrogen in the Presence of a Catalyst	*254*
10.3	Reduction of C=O and C=S Containing Compounds	*255*
10.3.1	Reduction Using $NaBH_4$ and $LiAlH_4$	*255*
10.3.2	Reduction Using Organometallic Reagents	*257*
10.3.3	Reduction Using Acetylides	*259*
10.3.4	Reduction Using Metals	*260*
10.3.5	Reduction Using Hydrogen with a Catalyst	*261*
10.3.6	The Wolff Kishner Reduction	*261*
10.4	Reduction of Imines	*263*
10.4.1	Reduction Using $NaBH_4$ and $LiAlH_4$	*263*
10.4.2	Reduction Using Hydrogen with a Catalyst	*265*
10.5	Reduction of Oxiranes	*266*
10.6	Reduction of Aromatic Compounds, Alkynes, and Alkenes	*268*
10.6.1	Reduction Using Dissolving Metals	*268*
10.6.2	Reduction Using Catalytic Hydrogenation	*269*
	End of Chapter Problems	*272*

11 Oxidation Reactions in Organic Chemistry *275*

11.1	Introduction	*275*
11.2	Oxidation	*275*

11.3	Oxidation of Alcohols and Aldehydes	279
11.3.1	Oxidation Using Potassium Permanganate ($KMnO_4$)	280
11.3.2	Oxidation Using Chromic Acid (H_2CrO_4)	281
11.3.3	Swern Oxidation	283
11.3.4	Dess-Martin Oxidation	284
11.3.5	Oxidation Using Pyridinium Chlorochromate	285
11.3.6	Oxidation Using Silver Ions	286
11.3.7	Oxidation Using Nitrous Acid	286
11.3.8	Oxidation Using Periodic acid	287
11.4	Oxidation of Alkenes Without Bond Cleavage	288
11.4.1	Epoxidation of Alkenes	288
11.4.1.1	Reactions of Epoxides	289
11.4.2	Oxidation of Alkenes with $KMnO_4$	291
11.4.3	Oxidation of Alkenes with OsO_4	292
11.5	Oxidation of Alkenes with Bond Cleavage	293
11.5.1	Oxidation of Alkenes with $KMnO_4$ at Elevated Temperatures	293
11.5.2	Ozonolysis of Alkenes	295
11.6	Applications of Oxidation Reactions of Alkenes	296
11.7	Oxidation of Alkynes	299
11.8	Oxidation of Aromatic Compounds	300
11.9	Autooxidation of Ethers and Alkenes	301
11.10	Applications of Oxidation Reactions to Synthesis	302
	End of Chapter Problems	304
12	**Elimination Reactions of Organic Chemistry**	**309**
12.1	Introduction	309
12.2	Mechanisms of Elimination Reactions	309
12.2.1	Elimination Bimolecular (E2) Reaction Mechanism	310
12.2.2	Elimination Unimolecular (E1) Reaction Mechanism	314
12.2.3	Elimination Unimolecular – Conjugate Base (E1cB) Reaction Mechanism	315
12.3	Elimination of Hydrogen and Halide (Dehydrohalogenation)	316
12.4	Elimination of Water (Dehydration)	319
12.4.1	Dehydration Products	319
12.4.2	Carbocation Rearrangement	321
12.4.3	Pinacol Rearrangement	322
12.5	Applications of Elimination Reactions to Synthesis	323
	End of Chapter Problems	326
13	**Spectroscopy Revisited, A More Detailed Examination**	**331**
13.1	Introduction	331
13.2	The Electromagnetic Spectrum	331
13.2.1	Types of Spectroscopy Used in Organic Chemistry	333
13.3	UV-Vis Spectroscopy and Conjugated Systems	334
13.4	Infrared Spectroscopy	337
13.5	Mass Spectrometry	343
13.6	Nuclear Magnetic Resonance (NMR) Spectroscopy	346
13.6.1	Theory of Nuclear Magnetic Resonance Spectroscopy	347
13.6.2	The NMR Spectrometer	348
13.6.3	Magnetic Shielding	349

13.6.4	The Chemical Shift, the Scale of the NMR Spectroscopy	350
13.6.5	Significance of Different Signals and Area Under Each Signal	351
13.6.6	Splitting of Signals	353
13.6.7	Carbon-13 NMR (^{13}C NMR)	363
13.6.8	Carbon-13 Chemical Shifts and Coupling	363
	End of Chapter Problems	367

14 Free Radical Substitution Reactions Involving Alkanes 369

14.1	Introduction	369
14.2	Types of Alkanes and Alkyl Halides	371
14.2.1	Classifications of Hydrocarbons	371
14.2.2	Bond Dissociation Energies of Hydrocarbons	373
14.2.3	Structure and Stability of Radicals	374
14.3	Chlorination of Alkanes	376
14.3.1	Mechanism for the Chlorination of Methane	377
14.3.2	Chlorination of Other Alkanes	379
14.4	Bromination of Alkanes	380
14.4.1	Bromination of Propane and Other Alkanes	380
14.5	Applications of Free Radical Substitution Reactions	386
14.6	Free Radical Inhibitors	388
14.7	Environmental Impact of Organohalides and Free Radicals	389
	End of Chapter Problems	391

15 Nucleophilic Substitution Reactions at sp^3 Carbons 393

15.1	Introduction	393
15.2	The Electrophile	393
15.3	The Leaving Group	394
15.3.1	Converting Amines to Good Leaving Groups	395
15.3.2	Converting the OH of Alcohols to a Good Leaving Group in an Acidic Medium	395
15.3.3	Converting the OH of Alcohols to a Good Leaving Group Using Phosphorous Tribromide	396
15.3.4	Converting the OH of Alcohols to a Good Leaving Group Using Thionyl Chloride	396
15.3.5	Converting the OH of Alcohols to a Good Leaving Group Using Sulfonyl Chlorides	396
15.4	The Nucleophile	397
15.5	Nucleophilic Substitution Reactions	397
15.5.1	Mechanisms of Nucleophilic Substitution Reactions	399
15.6	Bimolecular Substitution Reaction Mechanism (S_N2 Mechanism)	400
15.6.1	The Electrophile of S_N2 Reactions	400
15.6.2	The Nucleophile of S_N2 Reactions	402
15.6.3	The Solvents of S_N2 Reactions	403
15.6.4	Stereochemistry of the Products of S_N2 Reactions	404
15.6.5	Intramolecular S_N2 Reactions	405
15.7	Unimolecular Substitution Reaction Mechanism (S_N1 Mechanism)	406
15.7.1	The Nucleophile and Solvents of S_N1 Reactions	407
15.7.2	Stereochemistry of the Products of S_N1 Reactions	408

15.7.3	The Electrophile of SN1 Reactions *409*
15.8	Applications of Nucleophilic Substitution Reactions – Synthesis *414*
15.8.1	Synthesis of Ethers *415*
15.8.2	Synthesis of Nitriles *416*
15.8.3	Synthesis of Silyl Ethers *416*
15.8.4	Synthesis of Alkynes *418*
15.8.5	Synthesis of α-Substituted Carbonyl Compounds *419*
	End of Chapter Problems *420*

16 Nucleophilic Substitution Reactions at Acyl Carbons *425*

16.1	Introduction *425*
16.2	Mechanism for Acyl Substitution *426*
16.2.1	The Leaving Group of Acyl Substitution Reactions *427*
16.2.2	Reactivity of Electrophiles of Acyl Substitution Reactions *427*
16.2.3	Nucleophiles of Acyl Substitution Reactions *428*
16.3	Substitution Reactions Involving Acid Chlorides *428*
16.3.1	Substitution Reactions Involving Acid Chlorides and Water *429*
16.3.2	Substitution Reactions Involving Acid Chlorides and Alcohols *430*
16.3.3	Substitution Reactions Involving Acid Chlorides and Ammonia and Amines *431*
16.3.4	Substitution Reactions Involving Acid Chlorides and Carboxylate Salts *432*
16.3.5	Substitution Reactions Involving Acid Chlorides and Soft Organometallic Reagents *433*
16.3.6	Substitution Reactions of Acid Chlorides with Hard Organometallic Reagents *433*
16.3.7	Substitution Reactions of Acid Chlorides with Soft Metal Hydrides Reagents *434*
16.3.8	Substitution Reactions of Acid Chlorides with Hard Metal Hydrides Reagents *435*
16.4	Substitution Reactions Involving Anhydrides *436*
16.4.1	Substitution Reactions of Anhydrides with Water *437*
16.4.2	Substitution Reactions of Anhydrides with Alcohols *438*
16.4.3	Substitution Reactions of Anhydrides with Ammonia and Amines *439*
16.4.4	Substitution Reactions of Anhydrides with Carboxylate Salts *439*
16.4.5	Substitution Reactions of Anhydrides with Soft Organometallic Reagents *440*
16.4.6	Substitution Reactions of Anhydrides with Hard Organometallic Reagents *440*
16.4.7	Substitution Reactions of Anhydrides with Soft Metallic Hydrides *441*
16.4.8	Substitution Reactions of Anhydrides with Hard Metallic Hydrides *441*
16.5	Substitution Reactions Involving Esters *442*
16.5.1	Substitution Reactions of Esters with Water *444*
16.5.2	Substitution Reactions of Esters with Alcohols *445*
16.5.3	Substitution Reactions of Esters with Ammonia and Amines *446*
16.5.4	Substitution Reactions of Esters with Soft Organometallic Reagents *447*
16.5.5	Substitution Reactions of Esters with Hard Organometallic Reagents *447*
16.5.6	Substitution Reactions of Esters with Soft and Hard Metallic Hydrides *448*
16.5.7	Substitution Reactions of Esters with Enolates of Esters *449*
16.6	Substitution Reactions Involving Amides *451*
16.6.1	Substitution Reactions of Amides with Water *452*
16.6.2	Substitution Reactions of Amides with Hard Metallic Hydrides *453*
16.7	Substitution Reactions Involving Carboxylic Acids *454*
16.7.1	Substitution Reactions of Carboxylic Acids with Alcohols *455*
16.7.2	Substitution Reactions of Carboxylic Acid with Ammonia and Amines *456*

16.7.3	Substitution Reactions of Carboxylic Acids with Hard Metallic Hydrides	*457*
16.8	Substitution Reactions Involving Oxalyl Chloride	*458*
16.9	Substitution Reactions Involving Sulfur Containing Compounds	*458*
16.10	Applications of Acyl Substitution Reactions	*460*
16.10.1	Preparation of Esters	*460*
16.10.2	Preparations of Amides	*461*
	End of Chapter Problems	*462*

17 Aromaticity and Aromatic Substitution Reactions *467*

17.1	Introduction	*467*
17.2	Structure and Properties of Benzene	*468*
17.3	Nomenclature of Substituted Benzene	*470*
17.3.1	Nomenclature of Monosubstituted Benzenes	*470*
17.3.2	Nomenclature of Di-Substituted Benzenes	*471*
17.4	Stability of Benzene	*473*
17.5	Characteristics of Aromatic Compounds	*475*
17.5.1	Carbocyclic Compounds and Ions	*475*
17.5.2	Polycyclic Compounds	*476*
17.5.3	Heterocyclic Compounds	*477*
17.6	Electrophilic Aromatic Substitution Reactions of Benzene	*478*
17.6.1	Substitution Reactions Involving Nitronium Ion	*479*
17.6.2	Substitution Reactions Involving the Halogen Cation	*480*
17.6.3	Substitution Reactions Involving Carbocations	*481*
17.6.4	Substitution Reactions Involving Acyl Cations	*483*
17.6.5	Substitution Reactions Involving Sulfonium Ion	*484*
17.7	Electrophilic Aromatic Substitution Reactions of Substituted Benzene	*484*
17.7.1	Electron Activators for Electrophilic Aromatic Substitution Reactions	*485*
17.7.2	Electron Deactivators for Electrophilic Aromatic Substitution Reactions	*488*
17.7.3	Substitution Involving Disubstituted Benzenes	*490*
17.8	Applications – Synthesis of Substituted Benzene Compounds	*491*
17.9	Electrophilic Substitution Reactions of Polycyclic Aromatic Compounds	*494*
17.10	Electrophilic Substitution Reactions of Pyrrole	*496*
17.11	Electrophilic Substitution Reactions of Pyridine	*497*
17.12	Nucleophilic Aromatic Substitution	*499*
17.12.1	Nucleophilic Aromatic Substitution Involving Substituted Benzene	*499*
17.12.2	Nucleophilic Aromatic Substitution Involving Substituted Pyridine	*502*
	End of Chapter Problems	*504*

18 Conjugated Systems and Pericyclic Reactions *511*

18.1	Conjugated Systems	*511*
18.1.1	Stability of Conjugated Alkenes	*511*
18.2	Pericyclic Reactions	*513*
18.2.1	Cycloaddition Reactions	*513*
18.2.1.1	Cycloaddition Reactions [2+2]	*514*
18.2.1.2	Cycloaddition Reactions [4+2]	*516*
18.2.2	Electrocyclic Reactions	*519*
18.2.3	Sigmatropic Reactions	*521*
	End of Chapter Problems	*522*

19	**Catalytic Carbon–Carbon Coupling Reactions** *525*	
19.1	Introduction *525*	
19.2	Reactions of Transition Metal Complexes *525*	
19.2.1	Oxidative Addition Reactions *526*	
19.2.2	Transmetallation Reactions *526*	
19.2.3	Ligand Migration Insertion Reactions *527*	
19.2.4	β-Elimination Reactions *527*	
19.2.5	Reductive Elimination Reactions *527*	
19.3	Palladium-Catalyzed Coupling Reactions *528*	
19.3.1	The Heck Reaction *528*	
19.3.2	The Suzuki Reaction *531*	
19.3.3	The Stille Coupling Reaction *533*	
19.3.4	The Negishi Coupling Reaction *534*	
	End of Chapter Problems *535*	
20	**Synthetic Polymers and Biopolymers** *537*	
20.1	Introduction *537*	
20.2	Cationic Polymerization of Alkenes *537*	
20.2.1	Cationic Polymerization of Isobutene *538*	
20.2.2	Cationic Polymerization of Styrene *538*	
20.3	Anionic Polymerization of Alkenes *540*	
20.3.1	Anionic Polymerization of Vinylidene Cyanide *540*	
20.4	Free Radical Polymerization of Alkenes *540*	
20.4.1	Free Radical Polymerization of Isobutylene *541*	
20.5	Copolymerization of Alkenes *542*	
20.5.1	Cationic Copolymerization *542*	
20.5.2	Epoxy Resin Copolymers *543*	
20.6	Properties of Polymers *543*	
20.6.1	Solubility of Polymers *544*	
20.6.2	Thermal Properties of Polymers *544*	
20.7	Biopolymers *544*	
20.8	Amino Acids, Monomers of Peptides and Proteins *545*	
20.9	Acid–Base Properties of Amino Acids *547*	
20.10	Synthesis of α-Amino Acids *547*	
20.10.1	Synthesis of α-Amino Acids Using the Strecker Synthesis *547*	
20.10.2	Synthesis of α-Amino Acids Using Reductive Amination *548*	
20.10.3	Synthesis of α-Amino Acids Using Hell Volhard Zelinsky Reaction *548*	
20.10.4	Synthesis of α-Amino Acids Using the Gabriel Malolic Ester Synthesis *549*	
20.11	Reactions of α-Amino Acids *550*	
20.11.1	Protection–Deprotection of the Amino Functionality *550*	
20.11.2	Reactions of the Carboxylic Acid Functionality *551*	
20.11.3	Reaction of α-Amino Acids to Form Dipeptides *552*	
20.11.4	Reaction of α-Amino Acids With Ninhydrin *554*	
20.12	Primary Structure and Properties of Peptides *556*	
20.12.1	Identification of Amino Acids of Peptides *556*	
20.12.2	Identification of the Amino Acid Sequence *556*	
20.13	Secondary Structure of Proteins *558*	
20.14	Monosaccharides, Monomers of Carbohydrates *559*	
20.15	Reactions of Monosaccharides *560*	

20.15.1	Hemiacetal Formation Involving Monosaccharides	*560*
20.15.2	Base-catalyzed Epimerization of Monosaccharides	*562*
20.15.3	Enediol Rearrangement of Monosaccharides	*563*
20.15.4	Oxidation of Monosaccharides with Silver Ions	*563*
20.15.5	Oxidation of Monosaccharides with Nitric Acid	*563*
20.15.6	Oxidation of Monosaccharides with Periodic Acid	*564*
20.15.7	Reduction of Monosaccharides	*565*
20.15.8	Ester Formation of Monosaccharides	*565*
20.15.9	Ether Formation of Monosaccharides	*565*
20.15.10	Intermolecular Acetal Formation Involving Monosaccharides	*565*
20.16	Disaccharides and Polysaccharides	*566*
20.17	N-Glycosides and Amino Sugars	*567*
20.18	Lipids	*568*
20.19	Properties and Reactions of Waxes	*569*
20.20	Properties and Reactions of Triglycerides	*569*
20.20.1	Saponification (Hydrolysis) of Triglycerides	*570*
20.20.2	Reduction of Triglycerides	*571*
20.20.3	Transesterification of Triglycerides	*571*
20.21	Properties and Reactions of Phospholipids	*572*
20.22	Structure and Properties of Steroids, Prostaglandins, and Terpenes	*572*
	End of Chapter Problems	*573*

Index *577*

Preface

About This Book

This book is written from the students' perspective. Addressing the questions that students of organic chemistry typically have, the errors they typically make, along with some fundamental misconceptions that they typically formulate, are all the focus of this textbook. A major difference between this textbook and the majority of other textbooks is with the presentation of the information. The objective of this textbook is to develop the student's ability to think critically and creatively and equally important to improve the problem-solving skills of students. The content information is presented in such a way to assist students develop these skills. These are skills critically needed for students of science as they prepare for today's workforce. This approach also gives students the assurance that their opinions and thoughts are valued. As a result, students will become confident as they master the subject material. With this approach, students will quickly realize that it is in their best interest to develop these skills instead of relying on memorization as they approach this course and other science courses. The development of these skills will eventually prepare students to become better scientists. The problems in each chapter and at the end-of-chapter problems are designed to get students to solve problems by using their critical thinking skills.

For the majority of textbooks, the vast amount of organic chemistry information is dealt with primarily by categorizing the information into functional group categories. Thus, each of the approximately 20 chapters of a typical organic chemistry textbook is basically an exhaustive study of compounds with the different functional groups found in organic chemistry. This approach does not lend itself to aid students understand and master the vast content information of organic chemistry; this approach only presents large categories of information for students to handle. As a result, some students tend to rely on memorization instead of developing a scientific approach to handle all the information presented. In this textbook, the vast amount of organic chemistry information is not presented by functional group categories, but instead by reaction types; this approach presents much fewer categories of information for students to handle. In this textbook, the content information is divided into eight general categories based on reaction types, and not functional groups. An overview of the eight reaction types that are covered in the textbook is covered in Chapter 6. Since the majority of these types of reactions are the basic reactions covered in general chemistry, this approach provides a much better method to bridge the gap between general chemistry and organic chemistry. For example, there is a chapter that covers oxidation, a concept covered in general chemistry, but in this textbook, the concept of oxidation is applied to organic molecules that have different functional groups. Thus, after students have learned the concept of oxidation, they will be better prepared to apply that concept to a wide variety of organic molecules. The first part of the textbook covers relevant concepts of chemistry and the later sections deal with the applications of the concepts learned to the reactions of a wide cross section of molecules with different functional groups, hence the title of the textbook – *Organic Chemistry: Concepts and Applications*.

The first chapter covers the description of the atom and molecules; the next two chapters give a basic description of functional groups and the nomenclature of organic molecules so that students can readily recognize different types of molecules and learn the language of organic chemistry encountered in later chapters. The philosophy is that once students are able to recognize different functional groups, they will be better able to predict and communicate the various outcomes of different reactions encountered in organic chemistry. As a result, students will be able to apply their creative thinking skills to solve various problems encountered in this course. Since students are taught early in the textbook how to recognize the different reaction types, they will not only recognize the connection with general chemistry and organic chemistry but also how to apply the knowledge gained from general chemistry to new concepts that will be learned in organic chemistry.

Another aspect that this textbook covers is the importance and relevance of organic chemistry to our environment, the pharmaceutical and chemical industries, and biological and physical sciences. For example, in the study of the properties and the types of reactions that alkanes undergo, students will recognize the relevance of using different types of reactions to convert fossil and petroleum products into important compounds, such as polymers, pharmaceutical products, everyday household chemicals, insecticides, and herbicides. Also, the importance and significance of reactive intermediates including radicals are discussed. As a result, throughout the textbook, there are various "*Did you Know*?" sections. In these sections, students are shown the importance and the relevance of the content material being covered to the environment; often times, this is information that students may not have realized or know. There is a supplemental package that accompanies this text that includes multiple-choice questions similar to those of most national standardized tests and there are answers and detailed explanations for the questions. This supplemental package is included since most students who take organic chemistry eventually take an aptitude test for professional schools, including the Medical College Admissions Test (MCAT) for medical school, Dental Aptitude Test (DAT) for dental school, Pharmacy College Admission Test (PCAT) for pharmacy school, or the GRE subject test for most graduate programs. Organic chemistry makes up a large percentage of these exams since students' critical, analytical, and creative skills are needed to be successful in organic chemistry and these programs.

In summary, this textbook offers a new approach to not only teach organic chemistry but also as a guide to assist students to become better scientists by developing their critical, analytical, and creative thinking skills. These skills will prepare students for today's job market, which relies heavily on the creative application of knowledge.

To the Student of Organic Chemistry

Chemistry is all around us and plays a very important role in just about every aspect of our everyday lives. Our society benefits from chemistry, especially organic chemistry, in many ways. A large percentage of just about everything around us is derived through a process that involves chemistry. For example, a large percentage of the clothes that we wear are synthetic polymers; the plastic containers for milk, water, and other liquids are made from polymers, which are different types of polymers from the kind that are used to make some of the clothes that we wear. So, it is important to understand and learn how chemistry can be used to benefit our everyday lives, and how chemists can utilize chemistry to improve the quality of our lives and solve various problems. In order to succeed in this course, you must have a positive attitude about chemistry. The same is true for any of your other courses and anything that you want to succeed at in life. Can you imagine an athlete who wants to be the best at his or her sport keeps saying that they just do not like the game or thinks that the game that they are playing is extremely difficult and that they will never master that particular game! I am of the impression that such an individual will not be very successful at that particular sport. As a result, this cannot be the approach to succeed at mastering something that needs to be mastered. A very positive approach must be taken in order to be successful in organic chemistry. One way of achieving the goal of benefiting the maximum from organic chemistry is to become involved in chemistry; get to know, understand, and appreciate its benefits to society. This approach will require constant and persistent work on this subject. Develop a schedule for study and try to study consistently for at least five to six hours per week. Depending on your background in chemistry, some students may require a bit more time. Most people who succeed at a particular discipline have to put aside a large percentage of time to practice and perfect their skills. Each member of the football team must practice regularly so that the team can be the best in the conference and the nation. We can learn something from their approach to achieve success – they set aside time to practice regularly. Whether the discipline is baseball, football, cheerleading, or chemistry, success appears to come from disciplined and consistent hard work. Like anything that we do in life that we are successful at, we must dedicate time in order to achieve perfection. An important aspect of time dedicated toward mastering organic chemistry is to attend classes and taking good notes. Just hearing the subject being discussed goes a long way. As you start to master the subject, you will require less time to understand the different topics of organic chemistry and you will be able to spend more time analyzing and applying the concepts learned.

There are strategies that have been proven to be useful in order to be successful in organic chemistry. It may sound simple, but the first strategy to succeed in organic chemistry is to *attend lectures* and it is important to attend each and every lecture. *Read ahead* of the lecture material that will be discussed. Sometimes, you may not fully understand the materials that you read, but the main point is to get familiar with the material so that when you get to lecture, you will have already seen some of the materials and understanding it then will be much easier. *Practice, practice, practice*! Work the problems at the end of the chapter and those in the chapter – do not just work problems to get the answers that are in the solutions manual, but spend most of your time understanding the concept of each problem. The problems in this textbook are designed to apply your understanding of specific concepts to solve a wide variety of problems. The problems are not designed to determine how well you have memorized the information and can reproduce it. Remember that the solutions that are found in the solutions manual are not always the only solutions; there are typically other reasonable possibilities. If your answer is different from the one shown in the solutions manual, you should use your critical thinking skills to determine why the difference before coming to a final conclusion. In working your problems, you should be able to formulate a very similar question by changing a few words or structures of molecules of the problem to get another problem that can test the same concept. You will have to think through possible solutions. It is best to work a few problems and understand the concepts involved than to work lots of problems and not fully understand the concepts or principles. In solving problems, make sure that you "work" through the problems and not just look at the problem and then look at the solutions manual for the "answer." It is always a good practice to go over your graded exams. Some instructors offer regrades that allows students to challenge possible solutions and grading errors. Take advantage of this opportunity since it serves to reinforce your thinking ability and confidence, plus it may get you a few extra points on an exam!

It is impossible to learn chemistry and master the subject without *getting questions*. Scientists are curious individuals and are constantly seeking explanations for different observations. A good test of how well you are doing in this course is to determine how many questions come to you as the different topics are covered. If you read the textbook and attend lectures and have not developed a question or become curious about something, such as why does this happen, etc., you should try to carry out a deeper analysis of the topic that you are studying. The type of questions that should cross your mind should be of the curious type, the "what if" question is one that demonstrates curiosity. The next aspect of being a good scientist is to get your questions answered. Seek to get answers to your questions by first thinking through the concepts instead of just checking the solutions manual for the answers, or just getting an answer from someone without a discussion. With this approach, you have not utilized your critical and analytical thinking skills by just getting an answer. A major aspect of our work as scientists is centered on our ability to critically analyze information and formulate reasonable explanations. If you still need to get additional explanations for your questions, start seeking individuals who can assist. Most professors have posted office hours – use them. Some schools have help sessions or other forms of tutorials – capitalize on these opportunities. Some universities are very fortunate to have graduate students or tutorial study groups; these are tremendous resources to assist in getting your questions answered. Some students find it very helpful to form study groups. This approach is very helpful since you will learn from your peers. Peer-led team learning environments are typically found in the workplace, the team approach is very useful in finding solutions to various problems. Remember that it is extremely difficult for you to succeed in this course by just working alone; this course is also intended to assist students to become good at working in teams. Molecular models and molecular modeling computer programs will play an important role in helping you to better visualize and understand most of the concepts that will be discussed in this course. There are lots of computer programs that will assist in the visualization of the actual three-dimensional structures of molecules; some give good descriptions of the arrangements of electrons about atoms and molecules. Also, become very familiar with the periodic table and the meaning of each number on the table and the approximate location of each atom on the periodic table. This knowledge will become very useful in analyzing various properties of atoms and molecules.

There are many benefits to taking a course such as organic chemistry. Most of the principles and reactions that will be discussed in this course may not be remembered in years to come, but students will develop a more scientific mind from the various exercises, including the exams and discussions encountered throughout the course. Critical thinking, combined with a scientific approach developed in this course, is the key to being successful at your chosen profession and will be invaluable as you continue to prepare for your profession. From this course, you will not only gain knowledge of the basic principles of organic chemistry, but another major benefit, which is of equal importance, is the development and constant utilization of the critical and analytical thinking skills, which will be invaluable to assist you in solving work and life's everyday challenges. Most science students are required to take organic chemistry in order to assist in the development of better critical thinking skills. You will discover that if you take the scientific approach to learn organic chemistry, you will not have to memorize your way through this course. Instead, you will have the ability to apply the concepts learned to solve various problems and be better prepared to analyze and evaluate new information, and eventually be able to create new knowledge.

In summary, the ultimate goal of a course of this type is for students to be able to evaluate information learned and eventually to be able to generate new knowledge to benefit the society. Today's society is often described as a knowledge-based society because of the need to have creative thinkers find innovative avenues to apply new knowledge learned. You will need to be disciplined, be ready to work hard and consistently, and not be afraid to think. This approach keeps research, innovation, and new discoveries alive. At the end of the semester, you should reflect on your accomplishments over the semester and determine if you have made any change in the way you think or approach problems and if you have become a better scientist. If you have, then you have had a very successful semester of organic chemistry!

To the Instructor

We have all heard the comment from some students of organic chemistry that there is a major disconnect between their general chemistry course and organic chemistry. One of the goals of this textbook is to address that disconnect. In this book, concepts that are learned in general chemistry are constantly being reinforced and are used as the foundation for students to gain a better understanding of concepts that are discussed in organic chemistry. Fundamental concepts are introduced early so that students can get a clear understanding of a topic that is being introduced. This approach is important so that when specific topics are re-introduced throughout the textbook, students will be comfortable in applying the concepts learned to solve different problems.

In this book, students will find only relevant material throughout the text. Some textbooks try to introduce very advanced topics, and students at this level do not have a deep enough understanding of concepts involved to fully appreciate such advanced topics. As a result, students find such topics very confusing and often times serve as a distraction from the important topic being discussed. Information in this textbook is designed to stimulate students' critical thinking skills and to get students to apply these skills to find possible solutions to various problems. It is also designed to get students to fully develop the scientific method and to reach conclusions based on the scientific process. In this textbook, each concept is presented in a timely manner so that students are constantly building on their knowledge – most on the principles learned in general chemistry. Problems are carefully designed so that students have the opportunity to apply their critical thinking skills to determine possible solutions to problems encountered. As a result, there is no unique solution to most problems, but a discussion is given for each problem with possible solutions in the solutions manual. This approach makes students aware that there are sometimes not just one unique answer to some questions. This approach also serves to build students' confidence in making decisions about possible solutions. In this textbook, whenever a new topic is introduced, it is done so by reintroducing and building on the fundamental principles learned in general chemistry. As a result, this is a perfect textbook to bridge the gap between the courses of general chemistry and organic chemistry.

About the Companion Website

This book is accompanied by a companion website:

www.wiley.com/go/Headley_OrganicChemistry

The website includes:

- Solution manual
- MCQs

1

Bonding and Structure of Organic Compounds

1.1 Introduction

The word "organic" was first used to describe compounds that were derived from plants or animals, but this term was later used to describe compounds that contain mostly carbon and hydrogen atoms. Today, the term organic is loosely used to describe food that is produced without the use of pesticides, hormones, antibiotics, or fertilizers.

In organic chemistry, we will carry out a detailed study of the composition, properties, and reactions of compounds that contain primarily carbon and hydrogen atoms, also known as organic compounds. Even though many organic compounds contain only carbon and hydrogen atoms, a large percentage contains other atoms, such as oxygen, nitrogen, sulfur, as well as halogens; these atoms are referred to as heteroatoms. Atoms other than carbon and hydrogen that are present in organic compounds are called *heteroatoms*.

Prior to the start of the nineteenth century, chemists were familiar with inorganic compounds; for example, it was known that ammonium cyanate, an inorganic compound, could be easily made by the exchange reaction shown in Reaction (1-1).

$$NH_4Cl + AgOCN \longrightarrow NH_4OCN + AgCl \quad (1\text{-}1)$$
Ammonium chloride, Silver cyanate, Ammonium cyanate, Silver chloride

Even though organic compounds were known, similar reactions that could be used for their synthesis were not known. Instead, organic compounds were obtained primarily from natural sources, such as extraction from plants and other natural sources. As early as 1828, a medical doctor, Friedrich Wöhler, synthesized urea, a known organic compound. The synthesis of urea was accomplished by heating ammonium cyanate (an inorganic compound), as shown by the reaction in Reaction (1-2).

$$NH_4OCN \xrightarrow{\text{Heat}} H_2N-\underset{\underset{O}{\parallel}}{C}-NH_2 \quad (1\text{-}2)$$
Ammonium cyanate, an inogranic compound → Urea, an organic compound

This was a major discovery that initiated the era of organic chemistry. For the first time, an organic compound could be synthesized and these types of compounds did not have to be obtained naturally. In the early 1800s, just about all compounds that were used for different reasons, mostly medical, were obtained from natural sources. Today, a large percentage of organic compounds, including urea, which is a major component of fertilizer, adhesives, and resins, are synthesized and are not obtained naturally.

Organic Chemistry: Concepts and Applications, First Edition. Allan D. Headley.
© 2020 John Wiley & Sons, Inc. Published 2020 by John Wiley & Sons, Inc.
Companion website: www.wiley.com/go/Headley_OrganicChemistry

1 Bonding and Structure of Organic Compounds

> **DID YOU KNOW?**
>
> Fruits and vegetables that are produced without the use of pesticides or fertilizers are described as "organic."

Organic chemistry is that branch of science that deals with the synthesis and properties of compounds that contain primarily carbon and hydrogen atoms. As mentioned earlier, many other compounds that also contain heteroatoms are also considered organic. It is truly remarkable that the millions of known organic compounds, with new ones being constantly synthesized, all contain only carbons, hydrogens, and just a few heteroatoms! Today, most of the organic compounds that are synthesized are not made from inorganic compounds, but from simpler organic compounds. Some known everyday organic compounds that are made from simple starting organic compounds are shown below.

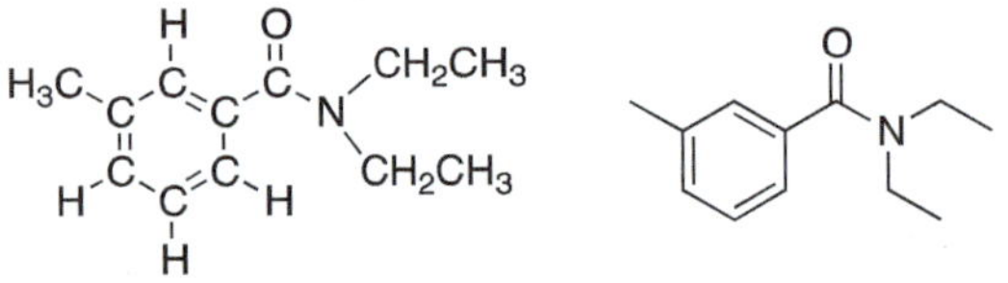

Two different representations of
N,N-diethyl-3-methylbenzamide (DEET)
Insect repellent

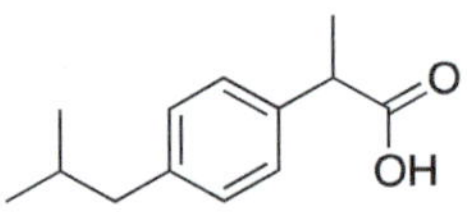

Ibuprofen (pain killer)

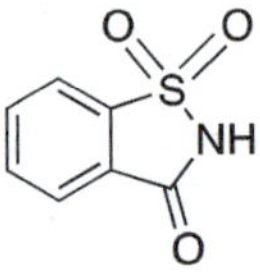

Saccharin
Artificial sweetener

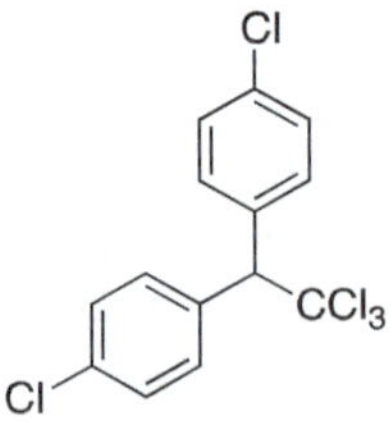

Dichlorodiphenyltrichloroethane (DDT)
Insecticide

You may be wondering where are the carbon and hydrogen atoms in these compounds since they are not shown in the structure, except for the first structure, which shows two representations of *N,N*-diethyl-3-methylbenzamide (DEET). You will learn later in this chapter that at each intersection in the structure, there are carbon and hydrogen atoms or just carbon atoms. DEET contains carbon, hydrogen, oxygen, and nitrogen atoms. Ibuprofen, a painkiller, contains carbon, hydrogen, and oxygen atoms. The artificial sweetener, saccharin, contains carbon, hydrogen, oxygen, nitrogen, and sulfur atoms. Dichlorodiphenyltrichloroethane (DDT), which is used as an insecticide, contains carbon, hydrogen, and chlorine atoms. As mentioned earlier, these compounds are still considered organic even though they contain heteroatoms and not just carbon and hydrogen atoms. Today, a variety of useful organic compounds, like those shown above, are made from simple starting compounds and they are not obtained from natural sources. Various drugs, pesticides, herbicides, plastic bottles, and various household cleaners are examples of compounds that are synthesized from simple starting compounds. A specific branch of chemistry that deals with the synthesis of such compounds from simple starting compounds is called *organic synthesis*. Today's pharmaceutical industries routinely synthesize important drugs to cure various diseases, but there are many factors that must be considered before a decision is made to synthesize a particular drug or to isolate it from nature as was typically done in the 1800s as pointed out earlier. It is extremely expensive to develop a particular drug, and isolation from natural sources has environmental impacts that must be considered.

> **DID YOU KNOW?**
>
> Taxol is an organic compound that is used for the cure of cancer, specifically cervical and breast cancer. 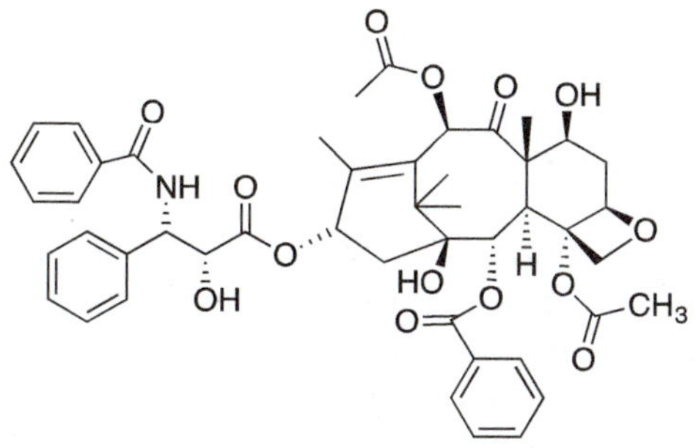
>
> Taxol
>
> When taxol was first discovered as a possible cure for cancer, its primary source was from the dried inner bark of the Pacific Yew tree. Taxol is only about 0.02% by weight of the Pacific Yew tree. This means that in order to produce approximately 1 kg of the drug, it would require at least 5000 kg of bark from approximately 2500 trees. Thus, a requirement of about 2 g of taxol for each treatment would result in the reduction or loss of a number of these trees. Since the Pacific Yew tree is a slow-growing evergreen tree, it is obvious that obtaining taxol for medicinal purpose would have a tremendous environmental impact and perhaps lead to the extinction of Pacific Yew trees. Fortunately, organic chemistry allows taxol to be synthesized from simpler starting compounds, but its synthesis involves a time-consuming and costly process.

1.2 Electronic Structure of Atoms

Before we actually examine how atoms are bonded together to form different organic molecules, we need to review our understanding and concept of atoms, and more specifically, at the electronic level. There are four elements that we will encounter frequently throughout our study of organic chemistry: hydrogen (H), carbon (C), nitrogen (N), and oxygen (O). You should locate these elements on the periodic table before continuing. Note carefully their location on the periodic table in relationship to each other and the various numbers that are associated with each atom. First, we will review the electronic structure of these atoms. It is extremely important that students and instructors as well as other scientists across the world visualize the structure of each atom and compounds similarly in order to effectively communicate concepts and explain various scientific observations. Studying organic chemistry is like learning a new language; if we are to learn a new language, we will have to learn the basics, such as the alphabet and symbols in order to make sure that there is a universal understanding of the language so that it can be utilized effectively to communicate with others across the world.

Thus, the first part of this course focuses on gaining a universal concept and visualization of the three-dimensional description of atoms and molecules. It is extremely important to stress the three-dimensional visualization since our understanding and explanations of different observations will depend on our three-dimensional concept of atoms and molecules. Reactions of molecules take place in a three-dimensional world and hence it is important that we visualize molecules from that perspective. The use of molecular model sets will be very helpful to better visualize atoms and molecules in three dimensions. Once students gain a good understanding of the three-dimensional world of atoms and molecules, it becomes much easier to communicate chemistry concepts on a two-dimensional paper.

1.2.1 Orbitals

The simplest description of the atom is that it consists of neutrons, protons, and electrons, and this simple description is enough for us to appreciate and understand most of the concepts of organic chemistry. The question now becomes, where are these electrons, protons, and neutrons in the atom? We know from general chemistry that the nucleus contains the neutrons, which are neutral, and also the protons, which are positively charged, but where are the electrons? Electrons are not randomly distributed throughout the atom, but they are located in specific regions of the atom. Thus, our first task is to get a picture of the structure of the atom and then try to visualize the location of the electrons. A comparison that can be used to gain a visual description of how electrons populate the atom is to make the comparison of students populating a large dorm of a college or university. In a dorm, there are many rooms and different types of rooms, i.e. the bedroom, bathroom, living room, and so on. Once a dorm is built, the next task is to get students to occupy the dorm. It is very important to ensure that as students start to occupy the dorm, the dorm is occupied properly based on rules established by the university. For example, one of these rules could be that students must occupy the first floor before occupying the second and third floors, and so on.

Regarding the structure of the atom, there are no rooms, bathrooms, and so on, like you would find in a dormitory; but in addition to the nucleus, there are regions outside the nucleus called orbitals. The nucleus is a spherical tiny space, which contains the protons and neutrons and is located in the center of the atom. Orbitals are another region of the atom, which are located outside the nucleus, and an orbital is the region where electrons occupy. There are various types of orbitals like there are various types of rooms in a dormitory, and we will discuss each type later in the chapter. In order to fully appreciate how electrons are distributed within the atom, we need to get a good concept of orbitals and their location around the nucleus. The nucleus of an atom is very small and is approximately 10^{-14} m in diameter, and the diameter of an entire atom, including the electrons, is approximately 10^{-10} m. Although it is impossible to determine the exact location of the electrons in an atom, it is possible to gain a good approximation of the region in space where the electrons are most likely to be found. There is a greater probability of finding the electrons closer to the nucleus than further from the nucleus (remember that the nucleus contains the positively charged protons). The electrons are not randomly distributed in the space outside the nucleus, but they are most likely to be found in specific regions of space called ***orbitals***. A comparison could be made with students in the dormitory; sometimes, it is almost impossible to tell exactly where in the dormitory a specific student is located, but around 2:00 in the morning, we could say that there is a good probability of finding the student in the room sleeping!

Orbitals have different sizes and shapes, similar to dormitory rooms that have different sizes and shapes. For atoms, numbers are used to represent orbitals of different energies, and letters are used to represent orbitals of different shapes. The **s** orbitals are spherical, and there are **s** orbitals that are larger than other **s** orbitals. Principal quantum numbers are used to differentiate between **s** orbitals of different sizes and energy. A small principal quantum number that is associated with the letter **s** suggests that the **s** orbital is smaller and close to the nucleus, and hence low in energy, compared to another **s** orbital that has a larger principal number, which would be larger and higher in energy and hence further from the nucleus. Thus, the **1s** orbital is small, close to the nucleus, and low in energy, compared to the **2s** orbital, which is larger, further from the nucleus, and higher in energy as illustrated in Figure 1.1.

The **s** orbitals are not the only type of orbitals that are present in an atom, there are also other types of orbitals, and the **p** orbitals are of another type. The shape of **p** orbitals is different from the spherical shape of the **s** orbitals. The shape of the **p** orbitals is often described as dumbbell shaped or as that of an hourglass, as shown in Figure 1.2.

Thus, for these orbitals, the probability of finding the electrons is not in a spherical region of space as that for the **s** orbitals, but only in the three-dimensional region that is outlined by the geometry of the shape of the orbitals. As shown in Figure 1.2, some **p** orbitals are small and close to the nucleus and other **p** orbitals are larger and further from the nucleus. Principal quantum numbers are used also to describe the relative size of different **p** orbitals, relative

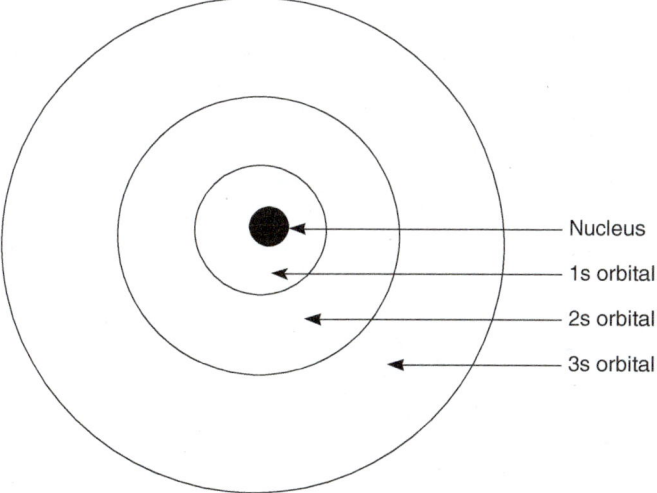

Figure 1.1 A slice through an atom showing **s** orbitals with different principal quantum numbers and different sizes. Remember that this is a two-dimensional representation of a three-dimensional atom; this is actually a slice through a sphere.

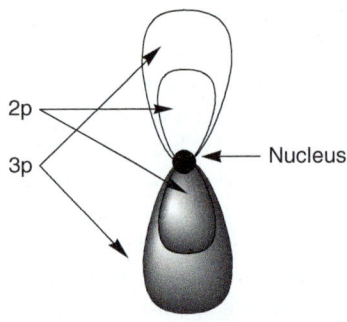

Figure 1.2 A slice through an atom showing **p** orbitals with different principal quantum numbers and different sizes. Remember that this is a two-dimensional representation of a three-dimensional atom.

energy, and distance from the nucleus. Thus, a **2p** orbital is smaller than a **3p** orbital; note that there is no **1p** orbital.

In reality, the representation shown in Figure 1.2 for different **p** orbitals is only a partial description of **p** orbitals. For each **p** orbital that has the same principal quantum number, there are actually three equivalent **p** orbitals. All three equivalent **p** orbitals have exactly the same size and shape, but they are arranged in three different directions in the three-dimensional space. To differentiate between the three equivalent (or degenerate) orbitals, the subscripts, x, y, and z are used to indicate the direction in which they point in space. That is, for the 2p orbital shown in Figure 1.2, there are a total of three equivalent **2p** orbitals as shown in Figure 1.3, $2p_x$, $2p_y$, and $2p_z$, and they all point in three different directions in the three-dimensional space. Similarly, for the **3p** orbitals, there are a total of three 3p orbitals: $3p_x$, $3p_y$, and $3p_z$ p orbitals.

Problem 1.1

List the following orbitals in order of their relative size:

2s, $2p_y$, 1s, $2p_z$, 3s

In summary, Figure 1.4 gives an illustration of the orbitals around the nucleus of an atom.

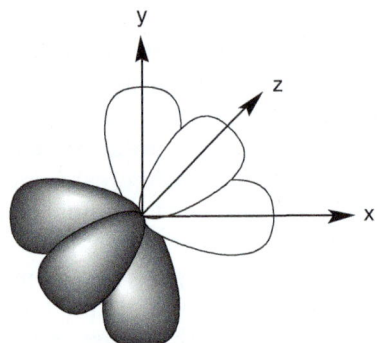

Figure 1.3 The three equivalent **p** orbitals all point in three different directions based on the x, y, and z planes.

1.2.2 Electronic Configuration of Atoms

Now that we have a good visual of the three-dimensional structure of the atom, we need to now concentrate on populating the atom with neutrons, protons, and electrons. As mentioned earlier, the nucleus has the neutrons and protons, and the electrons of an atom are located in regions outside the nucleus and they are not just randomly distributed, but they are in the different orbitals as described in the previous section. Let us start by looking at the simplest atom, the hydrogen atom. Try to locate this atom on the periodic table and identify the numbers associated with this atom. There are essentially two numbers associated with each atom on the periodic table as shown in Figure 1.5 for the hydrogen atom.

The first number is 1 (an integer and no units), which is the atomic number and indicates that there is only one electron and hence one proton. The other number is 1.0079 amu (atomic

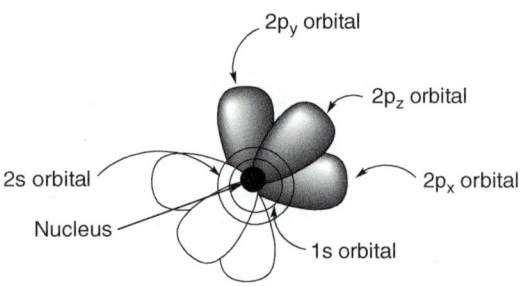

Figure 1.4 Two-dimensional illustration of the atom showing the nucleus, the 1s, 2s, $2p_x$, $2p_y$, and $2p_z$ orbitals of an atom.

mass unit), which gives the average atomic mass of the atom based on the natural abundance of the different isotopes of the hydrogen atom. You will recall from your general chemistry course that isotopes have different number of neutrons. Based on our description of the orbitals, the one electron of hydrogen will be found in the orbital that is lowest in energy and closest to the nucleus. This orbital would be the **1s** orbital, and not the **2s** or the **2p** orbitals, which are further from the nucleus and higher in energy.

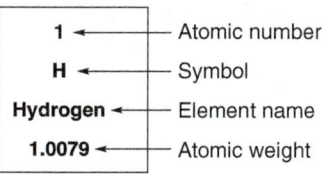

Figure 1.5 The hydrogen atom as seen on the periodic table.

Let us examine the next and most encountered atom in organic chemistry, the carbon atom. Locate this atom on the periodic table and note its location relative to other atoms. Also note the different numbers that are associated with carbon, which are shown in Figure 1.6.

You will notice the integer 6 (which indicates the number of electrons and hence the number of protons); there is the other number, 12.011, which is the atomic weight and represents the average mass of the isotopes of carbon that exist in natural abundance. Like that of the hydrogen atom, the num-

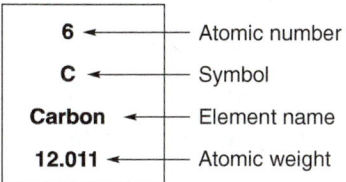

Figure 1.6 The carbon atom as seen on the periodic table.

ber that is of most importance to us is the integer, which indicates the number of electrons that are present in the atom. Thus, based on the information obtained from the periodic table, a carbon atom has six electrons, but where are these six electrons? Are they all in the **1s** orbital or are they all in the **2s**, or **2p** orbitals, or are they distributed randomly in these orbitals? Getting back to our original comparison of students populating the dorm, students are not randomly distributed in the various compartments of the dorm, but there are rules that must be followed to assign students to the different rooms. As mentioned earlier for the hydrogen atom, electrons much prefer to be closer to the nucleus since it is positively charged due to the presence of the protons. Thus, the electrons always occupy orbitals lower in energy (closer to the nucleus) first before occupying orbitals higher in energy; you will recall that this observation was described in your general chemistry course as the ***Aufbau Principle***. Much like when students move into the dorm, they occupy the first floor first, and when that floor is full, then they start occupying the second floor, and so on. You will also recall from your general chemistry course that a maximum of *two electrons* can occupy any one orbital and the electron spin must be paired (***Pauli's Exclusion Principle***). This principle is comparable to a rule for occupying the dormitory rooms, only two students to a room. Thus, of the six electrons of a carbon atom, only a maximum of two electrons can occupy any one orbital. Thus, the **1s** orbital will have two electrons and the remaining electrons must occupy the other orbitals. The magnitude of the principal quantum number associated with an orbital indicates the relative energy of that orbital. Thus, the orbital next in energy to the **1s** orbital is the **2s** orbital. In order to determine the next set of orbitals, we will have to look at the principal quantum number. There are three **2p** orbitals, and as a result, these orbitals are next in energy since they have the same principal quantum number. The next set of orbitals would then be the **3s**, followed by the **3p** orbitals. For the **s** and **p** orbitals with the same principal quantum number, the **s** orbital is lower in energy compared to the **p** orbital. Even though they both have the same principal quantum number, the **s** orbital is more compact, compared to the more diffused **p** orbital. Thus, for a carbon atom, two electrons would be in the **1s** orbital, two would be in the **2s** orbital, and the rest would occupy the **2p** orbitals and not the **3s** orbital.

As you will note from Figure 1.4, it is very cumbersome to utilize a two-dimensional artistic representation to effectively illustrate all the features of the atom, including the relative shapes, directions, and energies of orbitals. This representation is complicated even more when we try to show the electrons in the orbitals. A much easier representation that is often used to show the orbitals and the electrons that are contained in an atom is called the **electronic configuration**. With this representation, numbers, letters, subscripts, and superscripts are used to represent the relative energy (principal quantum number), shape, orientation, and number of electrons in each orbital, respectively. A superscript is used to represent the number of electrons in each orbital and a subscript is used to represent the orientation of each orbital in space. Subscripts are used in association with the **p** orbitals and not the **s** orbital since the **s** orbitals are spherical and not directional like the **p** orbitals. The **electronic configurations** for the atoms that will be encountered frequently in organic chemistry are shown below.

Hydrogen (H): $1s^1 2s^0 2p_x^0 2p_y^0 2p_z^0 3s^0$ (or just: $1s^1$)

Carbon (C): $1s^2 2s^2 2p_x^1 2p_y^1 2p_z^0 3s^0$ (or just: $1s^2 2s^2 2p_x^1 2p_y^1$)

Nitrogen (N): $1s^2 2s^2 2p_x^1 2p_y^1 2p_z^1 3s^0$ (or just: $1s^2 2s^2 2p_x^1 2p_y^1 2p_z^1$)

Oxygen (O): $1s^2 2s^2 2p_x^2 2p_y^1 2p_z^1 3s^0$ (or just: $1s^2 2s^2 2p_x^2 2p_y^1 2p_z^1$)

Note that for orbitals of equal energy (**degenerate orbitals**), such as the $2p_x$, $2p_y$, and $2p_z$, electrons occupy separate orbitals unless there are more than one electron in each orbital; you will recall this observation as **Hund's rule** from your general chemistry course.

Problem 1.2

With the aid of your periodic table, give the electronic configuration for each of the following atoms: F, B, and Na.

1.2.3 Lewis Dot Structures of Atoms

Compounds are made of atoms held together by different bonds, which are formed by the electrons of the atoms. For these bonds, not all the electrons are involved, but typically electrons furthest from the nucleus and in the orbitals of highest energies. These electrons are classified into a category described as the *valence electrons*. By definition, valence electrons are the electrons of the outer shell of the atom, or electrons that are in orbitals with the same and highest principal quantum number. For example, there are four valence electrons for carbon, those are the electrons of the 2s and 2p orbitals; the two electrons that are in the 1s orbital are known as the *core electrons*. Note that even though there are two different types of orbitals, the s and the p, the principal quantum number for both is the same, which is 2 and hence classified as the valence shell. It is possible to determine the valence electrons for each atom from the periodic table, based on the number above the column of a particular atom. For our study in organic chemistry, the number of valence electrons will be of extreme importance when we examine bonding and reactions.

An American chemist, Gilbert N. Lewis (1875–1946) was the first to devise a method to easily show the number of valence electrons associated with different atoms by using dots; hence, these structures are known as the *Lewis dot structures*. In the Lewis dot structure, dots are arranged around an atom and the number of dots reflects the number of valence electrons in the atom as shown in Figure 1.7.

Note that for the atoms shown in Figure 1.7, only the valence electrons as shown in the square brackets are used for the Lewis dot structure.

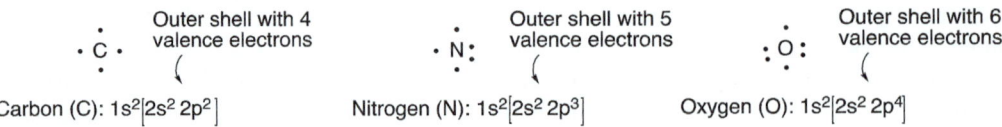

Figure 1.7 Lewis dot structures for selected atoms.

Problem 1.3

With the aid of your periodic table, give the electronic configuration and Lewis dot structures for the following atoms:

Li, Be, F, Ne, Na, and Mg.

1.3 Chemical Bonds

Atoms combine with each other to form molecules, and the atoms in a molecule are held together by chemical bonds, which involve the valence electrons. There are two types of chemical bonds – *ionic bonds* and **covalent bonds**.

1.3.1 Ionic Bonds

For molecules that contain ionic bonds, the atoms involved acquire the electronic configuration of the nearest noble gas by either gaining or losing valence electrons. As a result, each atom of an ionic bond acquires a formal charge. Atoms that lose electrons acquire a positive formal charge and become cations, and atoms that gain electrons acquire a negative formal charge and become anions. The attraction that results between these two oppositely charged species (cation and anion) is called an *ionic bond*. Figure 1.8 shows the ionic bond that results between Li and F.

Note that in Figure 1.8, the lithium cation, which is formed after the loss of one electron, acquires the electronic configuration of the noble gas, helium (He). Similarly, fluorine acquires the electronic configuration of the nearest noble gas, neon (Ne), by gaining an electron. You will discover that ionic bonds are formed between atoms of the extreme columns of the periodic table. For example, the bond that is most likely formed between potassium and chlorine will be an ionic bond; similarly for sodium and bromine.

Problem 1.4

Using a formulism as that shown in Figure 1.8, illustrate how ionic bonds are formed between the following pairs of atoms.

a) Na and Cl
b) Li and Cl
c) Mg and F

1.3.2 Covalent Bonds

A covalent bond in a molecule results from the sharing of valence electrons of the atoms of that molecule. Among organic molecules, the covalent bond is the most common and this is the type of bond that you will encounter the most throughout this course. The valence electrons that are used to make a covalent bond are called *bonding electrons*, and valence electrons that are not involved in a covalent bond are called *nonbonding electrons*, or *unshared electrons*. The arrangements

Lithium (Li): $1s^2\ 2s^1\ 2p^0$ $\xrightarrow{\text{Loss of one electron}}$ Lithium cation (Li$^+$): $1s^2\ 2s^0\ 2p^0$ (Li$^+$) $\Big\}$ Li$^+$ F$^-$

Fluorine (F): $1s^2\ 2s^2\ 2p^5$ $\xrightarrow{\text{Gain of one electron}}$ Fluorine anion (F$^-$): $1s^2\ 2s^2\ 2p^6$ (F$^-$)

Figure 1.8 The formation of an ionic bond between lithium and fluorine.

of the atoms and electrons (both bonding and nonbonding electrons) in molecules can also be represented using **Lewis dot structures**. As mentioned earlier, in the Lewis dot structure, only the valence electrons of an atom are considered, and recall that the ***valence*** electrons for a particular atom are the number of electrons that are contained in the orbitals that have the highest principal quantum number (outer shell). Thus, for carbon, the number of valence electrons is four (4) because there are a total of four (4) electrons in the 2s and 2p orbitals.

In the Lewis dot structure of a covalent molecule, the valence electrons are represented by dots, with the bonding electrons located between the atoms, illustrating the covalent bond, and the nonbonding electrons located around the atoms of the molecule. The total number of valence electrons for a molecule is determined by adding the valence electrons of each atom of the molecule. In drawing Lewis dot structures, the **octet rule** is obeyed. The octet rule states that there must be a total of eight electrons (bonding and nonbonding) around each atom in a molecule. There are exceptions to this rule, however, and hydrogen is one exception that will be encountered frequently throughout this course. The number of electrons that are associated with a hydrogen atom in a molecule will not be eight, but two (2). Other atoms, such as boron (B), beryllium (Be), and aluminum (Al), are also exceptions, and they will be discussed later.

In order to draw the Lewis dot structure, the atoms of the molecule are typically arranged in the most symmetrical manner, and the first atom of the chemical formula is typically the central atom. The valence electrons are distributed so that each atom has an octet of electrons, except hydrogen, which has a duet of electrons. Since organic chemistry is the chemistry of carbon-containing compounds, it is extremely important that we fully understand how to draw the Lewis dot structure of compounds that contain carbons. Thus, we will start by looking at the simplest organic molecule, methane (CH_4), the Lewis dot structure can be determined based on the number of valence electrons as shown below:

Atom	# valence electrons in each atom	Total # valence electrons
Carbon (C)	4	4
Hydrogen (H)	1	(4 × 1) 4

Total valence electrons for CH_4 = 8 electrons.

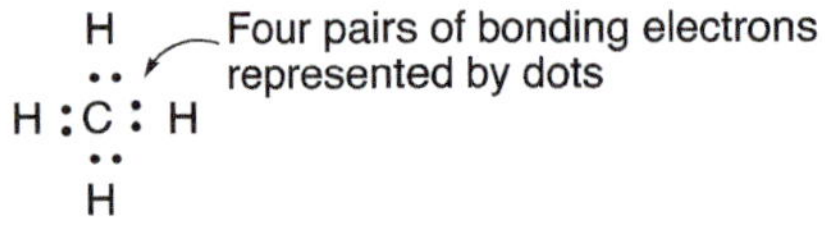

$$H:\overset{..}{\underset{..}{C}}:H \quad \text{Four pairs of bonding electrons represented by dots}$$

Lewis dot structure of CH_4

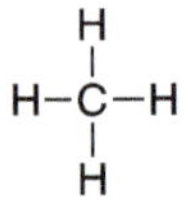

$$H-\underset{|}{\overset{|}{C}}-H$$

Lewis structure of CH_4

Note that in the Lewis dot structure of methane, the bonding electrons are represented by dots. Another representation of the same structure is to use a single line to represent a pair of bonding electrons that are shared between any two atoms. The representation of using a line to represent bonding electrons is very important since that will be the representation used throughout this course.

Another example of using the Lewis dot structure is shown below for carbon dioxide molecule (CO_2).

Atom	# valence electrons in each atom	Total # valence electrons
Carbon (C)	4	4
Oxygen (O)	6	(6 × 2) 12

Total valence electrons for CO_2 = 16 electrons

These 16 valence electrons must be distributed among the three atoms. One possibility is to try and distribute these 16 electrons so that each of the oxygen atoms gets an octet of electrons, as shown below.

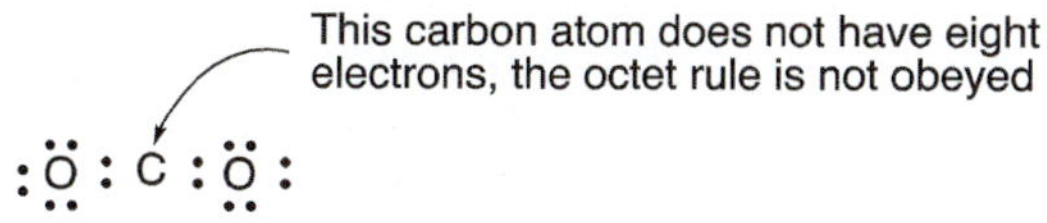

$$:\ddot{\underset{..}{O}}:C:\ddot{\underset{..}{O}}:$$

This carbon atom does not have eight electrons, the octet rule is not obeyed

Incorrect Lewis structure for CO_2

However, you will quickly realize that for this Lewis dot structure, the octet rule is not obeyed because the carbon does not have an octet of electrons, even though the oxygen atoms have an octet of electrons. As a result, this is not a correct Lewis dot structure for CO_2, but it can be transformed into the correct Lewis dot structure by using two pairs of unshared electrons from the oxygen atoms to share with the carbon and oxygen atoms as shown below.

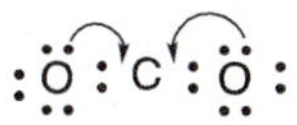

$$:\ddot{\underset{..}{O}}:\!:C\!:\!:\ddot{\underset{..}{O}}:$$

The correct Lewis structure of CO_2 can be gained by using two pairs of unshared electrons from the oxygen atoms to share with the carbon and oxygen atoms

Note carefully that the curved arrow formulism is used to indicate the movement of electrons from the oxygen atom to form bonding electrons for the carbon and oxygen atoms. This method of using an arrow to show electron movement will be used routinely throughout this course. A double-barbed arrow is used to indicate the movement of two electrons (or a pair of electrons), and a single-barbed arrow is used to indicate the movement of one electron. Thus, the correct Lewis dot structure for carbon dioxide (CO_2) is shown in Figure 1.9.

Note that the most symmetrical arrangement of the atoms results in a linear structure, in which the central atom, C, is the first atom of the chemical formula. Also, note that the bonds to carbon are double bonds. Thus, there are two double bonds in the carbon dioxide molecule and each oxygen atom has two pairs of unshared electrons.

Problem 1.5

Draw the Lewis dot structure for the following molecules and clearly show all nonbonding electrons.

a) H_2O b) NH_3 c) CS_2 d) SiH_4 e) CH_2O f) CH_2O_2

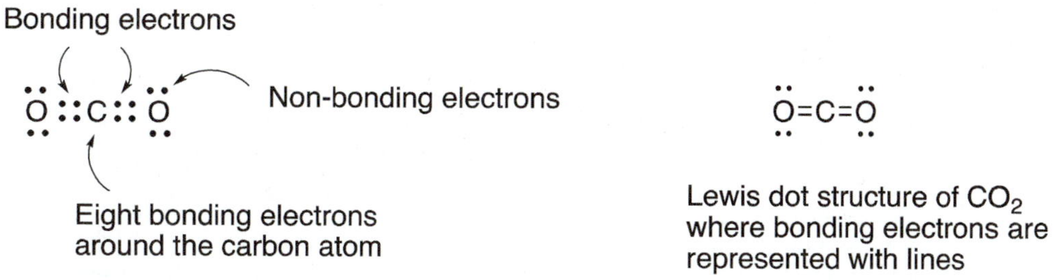

Bonding electrons

$$\ddot{\underset{..}{O}}::C::\ddot{\underset{..}{O}}$$

Non-bonding electrons

Eight bonding electrons around the carbon atom

$$\ddot{\underset{..}{O}}=C=\ddot{\underset{..}{O}}$$

Lewis dot structure of CO_2 where bonding electrons are represented with lines

Figure 1.9 Correct Lewis dot structure for carbon dioxide (CO_2).

1.3.3 Shapes of Molecules

You saw from the example in the previous section that the geometry of CO_2 is linear. This geometry results since electrons are negatively charged and are located at opposite ends of the molecule. Remember that like charges repel and opposite charges attract. Thus, the bonding electrons and the nonbonding electrons on oxygen are located at opposite ends of the molecule. For CH_4, the geometry is different since there are now four pairs of bonding electrons, and as a result, the four hydrogens are located at opposite ends around the central carbon atom. This geometry is different from that of CO_2 and is called a *tetrahedral geometry*. Remember that we should always be visualizing molecules in three dimensions. Valence shell electron pair repulsion (VSEPR) theory can be used to explain the geometry of CO_2, CH_4, and other molecules. This theory is based on the fact that electrons (bonding and nonbonding) have the same charge (negative), and as a result, will repel each other in a molecule. Thus, electrons (both bonding and nonbonding) of a molecule will be located at opposite ends of a molecule, which will result in different geometries around the central atom of different molecules. Figure 1.10 shows the geometries of some common molecules.

If there are only two electrons (one pair of electrons) between any two atoms in a molecule, the covalent bond is called a ***single bond***. If there are four electrons (two pairs of electrons) between any two atoms as is the case with carbon and oxygen atoms of CO_2, the bond is described as a ***double bond***, and if there are six electrons (three pairs of electrons), the bond is called a ***triple bond***.

Problem 1.6

Give the Lewis dot structure for the following molecules and predict the geometry about each atom of your structure that is bonded to at least two other atoms: NF_3, H_2S, CH_4, CS_2, CH_2O, CH_3OH, CH_3N.

1.3.4 Bond Polarity and Polar Molecules

Because there are typically different types of atoms in a molecule, the bonding electrons between two different atoms in the molecule are not equally distributed within the bond. You will recall from your previous course in general chemistry that electronegativity is defined as the tendency of an atom to attract electrons toward itself and that the most electronegative atom is fluorine. The relative electronegativities of atoms can be determined from the periodic table. Electronegativity increases in going from left to right across a particular row of the periodic

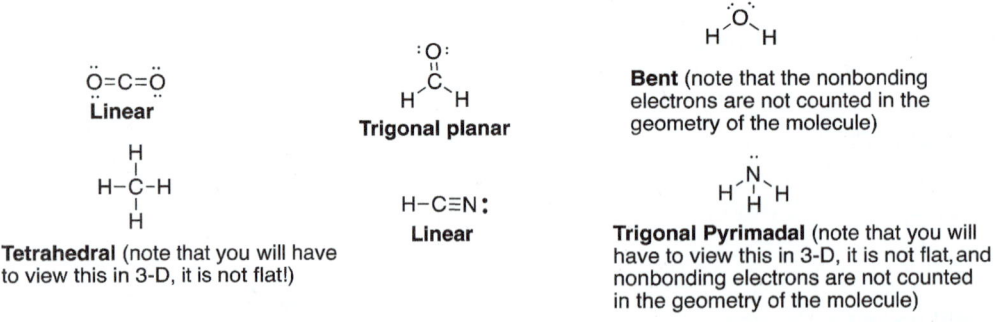

Figure 1.10 Examples of common organic molecules with different geometries as predicted by the VSEPR theory. Note that they also contain different types of covalent bonds, i.e. single, double, and triple covalent bonds.

table and also increases in going up a particular group (column) on the periodic table. This means that electronegativity of different atoms increases in going diagonally toward fluorine on the periodic table. Thus, by using the periodic table, it is possible to determine relative electronegativities of any two atoms without knowledge of the actual electronegativity values.

Problem 1.7

Of the following pairs of atoms, look at the periodic table and determine which is more electronegative.

a) K and Br b) Cl and Br c) N and C d) Mg and C.

For some molecules, two atoms of a covalent bond (single, double, or triple) may have different electronegativities. Both atoms do not equally share the bonding electrons of such a bond. The more electronegative atom attracts the bonding electrons closer to itself, compared to the less electronegative atom. The relative electronegativities can be readily determined from the periodic table. Remember, the most electronegative atoms are located at the top right of the periodic table. Covalent bonds that have atoms of different electronegativities are called *polar covalent bonds*. On the other hand, if the electronegativities of both atoms in a covalent bond are the same, then the bonding electrons are shared equally and the covalent bond is called a *nonpolar covalent bond*. Thus, covalent bonds that involve atoms of equal electronegativity are classified as nonpolar covalent bonds.

Problem 1.8

Classify the following covalent bonds as polar or nonpolar.

a) O—H b) H—H c) HN=NH d) C—Mg e) C—O f) C—Cl

There are different ways to show the polarities of polar covalent bonds. The most commonly used representation in organic chemistry is the use of partial charges, δ^+ or δ^- (δ is the Greek letter delta, and the representations are referred to as delta negative and delta positive). The most electronegative atom gets the δ^- symbol, and the least electronegative atom is assigned the δ^+ symbol. The unequal distribution of bonding electrons for the polar bonding of H—Cl is shown in Figure 1.11. Also shown in Figure 1.11 is another representation of the distribution of the electrons of the polar covalent bond of H—Cl in which an arrow (with a cross) is used. For this representation, the head of the arrow points in the direction of the most electronegative atom and away from the least electronegative atom, as shown in Figure 1.11. Note that the very electronegative chlorine gets the δ^- and the head of the arrow points to the chlorine atom. The most commonly used representation in organic chemistry is the use of partial charges, δ^+ or δ^-.

More examples in which these representations are used to show the polarities of polar covalent bonds are shown in Figure 1.12.

Problem 1.9

i) Give the Lewis dot structures of the molecules shown below and use the δ^+ and δ^- representations to indicate the polarity of the covalent bonds.

HF, HCN, NH_3 and CS_2.

ii) Which of the following molecules have nonpolar covalent bonds?

H_2, Cl_2, CH_3Cl, CO_2, COS, and H_2O.

1 Bonding and Structure of Organic Compounds

$$\overset{\delta+}{H}\text{---}\overset{\delta-}{Cl} \qquad H\overset{\longmapsto}{\text{---}}Cl$$

Figure 1.11 Two different representations of polar covalent bonds.

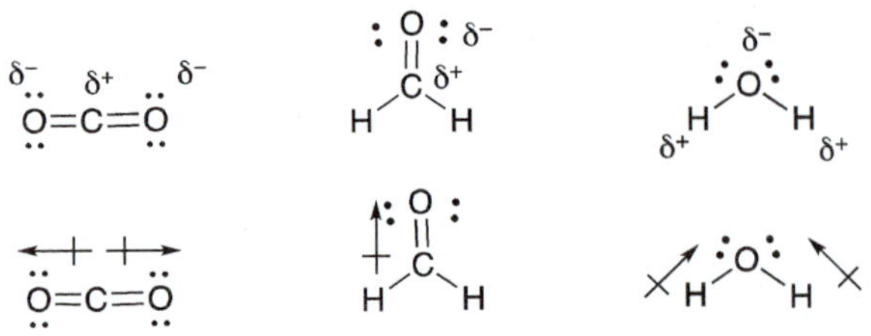

[Structures showing bond polarities for CO₂, H₂CO (formaldehyde), and H₂O with δ+ and δ− labels, as well as dipole arrow representations]

Figure 1.12 Examples of selected molecules in which different representations are used to show bond polarities.

Molecules that have at least one polar covalent bond may have a net molecular polarity based on the individual bond polarities; for example, hydrochloric acid is a polar molecule because there is a net dipole in the direction of the chlorine atom. For molecules with more than one polar bond, a carefully examination of the geometry of the molecule, along with the bond polarities, must be carried out in order to determine if the molecule is polar or not polar. The dipole moment of molecules is determined by the sum of the bond polarities and the three-dimensional geometry of the molecule. For example, water as shown in Figure 1.12 is a polar molecule since the bond dipoles do not cancel each other, but instead reinforce each other. On the other hand, carbon dioxide, as shown in Figure 1.12, is linear and since the two atoms bonded to the central carbon are the same (oxygen atoms) and bonded to the central carbon atom at a bond angle of 180°, carbon dioxide is not a polar molecule. The **dipole moment** is a measure of the overall polarity of a molecule, and the Greek letter μ is used to represent dipole moments of molecules. A close analysis of molecules must be carried out in order to determine if they are polar molecules or not. An analysis of the polarity of each covalent bond, along with an analysis of the three-dimension geometry of the molecule, must be carried out. For example, CO_2 has two polar C=O double bonds, but the molecule is linear and the two bond dipoles of equal magnitude point in opposite directions. As a result, the bond dipoles cancel, and the net dipole of the molecule is zero. On the other hand, H_2O, which is a bent molecule, has a dipole moment greater than zero ($\mu > 0$).

Problem 1.10

Draw the Lewis dot structures of each of the following molecules; determine the geometries and which have dipole moments of zero or nonzero.

a) CH_4 b) $CHCl_3$ c) H_2O d) CS_2 e) NH_3 f) CH_2Cl_2 g) COS h) BF_3

1.3.5 Formal Charges

The atoms of some molecules have charges, even though the overall charge of the molecule may be neutral. For most molecules that will be encountered in organic chemistry, the charge on most of the atoms is zero, but for some molecules, atoms may acquire a formal charge of +1 or −1 or even higher charges. The classic example of a molecule that is neutral, but has atoms with formal charges, is nitric acid and the Lewis dot structure is shown in Figure 1.13.

Name	Formula	Lewis dot structure
Nitric acid	HNO_3	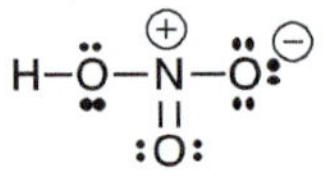 [Lewis structure of H−O−N(+)(=O)−O(−)]

Figure 1.13 Lewis dot structure of nitric acid showing the formal charges.

1.3 Chemical Bonds

Shown below are different equations that are used to calculate the formal charge for a specific atom in a molecule.

$$\text{Formal Charge (FC)} = \left(\#\text{valence }e^- \text{ in neutral atom}\right) - \left(\#\text{of unshared }e^-\right) - \left(\tfrac{1}{2}\text{ shared }e^-\right)$$

OR

$$\text{Formal Charge (FC)} = [\text{Group number}] - [\text{nonbonding electrons}] - \tfrac{1}{2}[\text{shared electrons}]$$

OR

$$\text{Formal Charge (FC)} = \left(\#\text{valence }e^- \text{ in neutral atom}\right) - \left(\#\text{nonbonding }e^-\right) - \left(\tfrac{1}{2}\text{ bonding }e^-\right)$$

In order to determine the formal charge of an atom in a molecule, the Lewis dot structure of the molecule must first be determined. For example, the formal charge of nitrogen in the Lewis dot structure of nitric acid (HNO_3) shown in Figure 1.13 is +1. This value can be obtained by first drawing the Lewis dot structure of nitric acid, as shown below; then by using one of the above equations the formal charge can be calculated, as shown in Eq. (1-3).

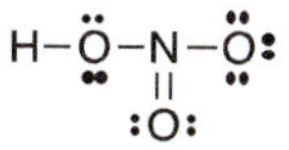

$$\text{Formal Charge (Nitrogen)} = 5 - 0 - \tfrac{1}{2}(8) = +1 \tag{1-3}$$

Similarly, the formal charge of one of the oxygen atoms (oxygen at the top right) as shown in Figure 1.13 is –1 and below is shown the calculation to obtain its formal charge as shown in Eq. (1-4).

$$\text{Formal Charge (Oxygen at the top right)} = 6 - 6 - \tfrac{1}{2}(2) = -1 \tag{1-4}$$

The formal charge calculation shows that the formal charge of the other oxygen atom (oxygen at the bottom) is zero as shown in Eq. (1-5).

$$\text{Formal Charge (Oxygen at the bottom)} = 6 - 4 - \tfrac{1}{2}(4) = 0 \tag{1-5}$$

Problem 1.11

Utilize one of the equations above to calculate the formal charge for the central atom of each of the following species.

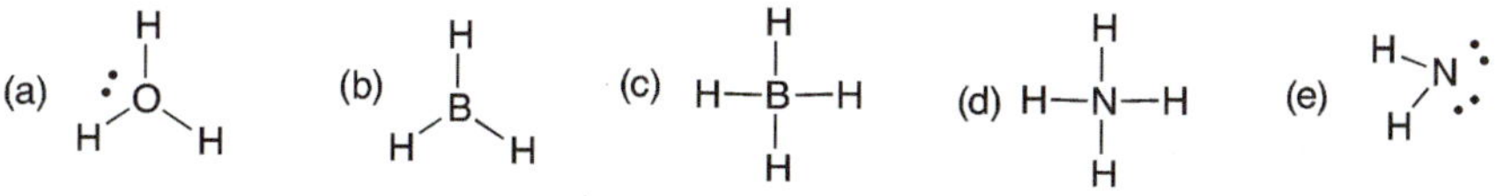

1.3.6 Resonance

For some molecules or ions, it is possible to draw more than one Lewis dot structures that meet the requirements of the octet rule. For example, nitric acid has two equivalent Lewis dot structures. Such equivalent Lewis dot structures are also called **resonance structures**, and the double-headed arrow is used to show this relationship between resonance structures as shown in Figure 1.14.

Since electrons are in constant motion in specific regions of space, resonance structures result from the movement of electrons (typically of unshared electrons or electrons of a double

1 Bonding and Structure of Organic Compounds

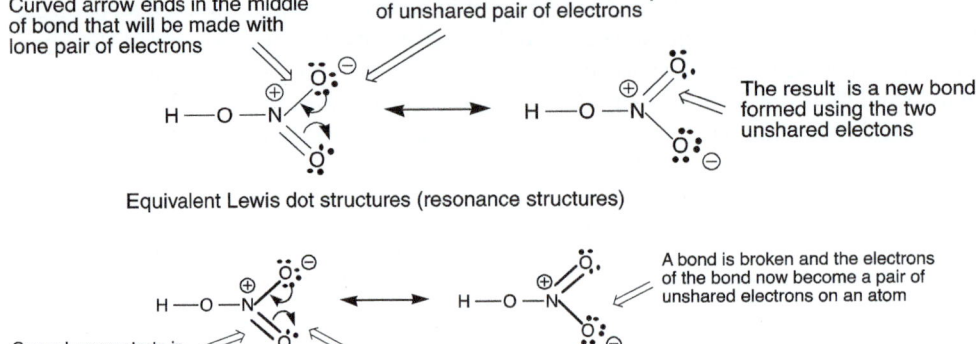

Figure 1.14 Equivalent Lewis dot structures (resonance structures) for nitric acid.

or triple bond) across bonds or atoms within the molecule. Resonance structures (or equivalent Lewis dot structures) are possible only for molecules or ions that have double and triple bonds or at least one lone pair of electrons that are adjacent to each other. Equivalent Lewis dot structures can be achieved by a method called *curved-arrow formulism* in which arrows are used to show the movements of electrons, as shown below.

Note that the curved arrow starts with one pair of electrons and ends in the middle of the bond that will be made with these two electrons. The other curved arrow movement starts in the middle of the double bond and shows how the electrons are moved onto the electronegative oxygen to give another resonance structure. Note that a double-headed arrow is used and each barb of the arrow represents an electron. There will be the need in later chapters to move just one electron at a time and then a single-barbed curved arrow is used to indicate the movement of just one electron. This curved-arrow formulism will be used frequently throughout this course to obtain different resonance structures and to show the movement of electrons in general.

Nitric acid is just one example where resonance structures are possible, there are many other molecules and ions that will be encountered throughout the course. In this section, different types of molecules and ions that have possible resonance structures will be discussed. The first example is shown in the resonance structures given in 1-6.

$$\text{Major contributor} \quad \longleftrightarrow \quad \text{Minor contributor}$$
Octet rule for all atoms   No octet rule for carbon   (1-6)

Note carefully that a curved double-headed arrow is used in the resonance structure to the left to indicate the electron movement of two electrons of the double-bonded electrons to the more electronegative oxygen. A close examination of the resonance structures shown above reveals that the atoms of the resonance structure of the left all have an octet of electrons;

whereas, for the resonance structure on the right, the carbon atom does not have an octet of electrons. As a result, the resonance structure on the left is described as contributing more to the overall picture of the molecule and is a major contributor to the overall structure of the molecule. The other resonance structure is considered to be a minor resonance contributor to the overall resonance structure of the molecule shown.

For resonance structures in which all the atoms have an octet of electrons, there may be other factors to consider in determining the major and minor resonance contributors. For the anion shown below, all the atoms have an octet of electrons, but in the resonance structure to the left, there is a negative formal charge on carbon, which is not as electronegative as oxygen. On the other hand, the resonance structure on the right has the negative formal charge on the more electronegative oxygen atom. Thus, the resonance structure to the left is the minor resonance contributor, and the structure to the right is a major resonance contributor.

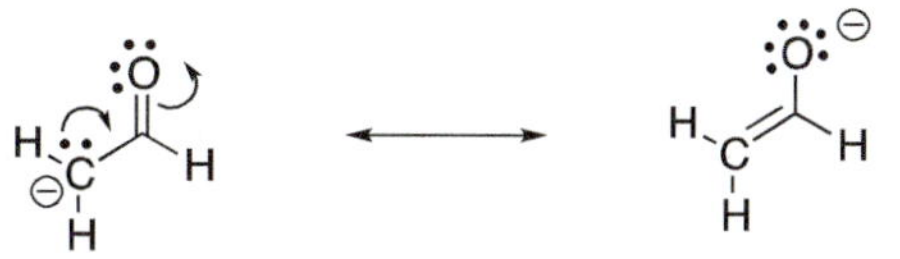

Minor resonance contributor   Major resonance contributor

In determining major resonance contributors, the charge separation must also be considered. For the resonance structures shown below, the resonance structure to the left is neutral with no charge separation; whereas, the structure to the right, even though it is neutral, there is a charge separation. As a result, the resonance structure to the left is a major resonance contributor, since there is minimum charge separation, actually zero, and the resonance structure to the right is minor, since there is a charge separation.

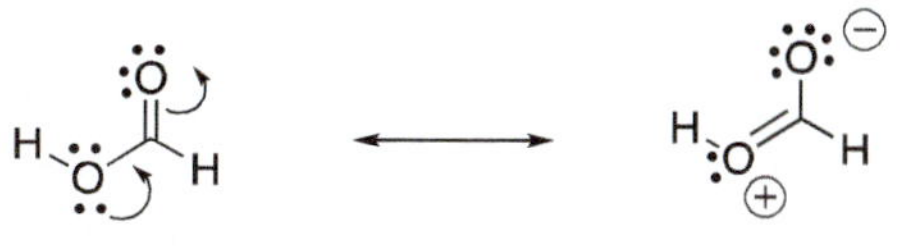

Major resonance contributor,   Minor resonance contributor
minimum charge separation   due to charge separation

Problem 1.12

i) Use *curved arrow formulism* to show electron movement. Show how the resonance structure on the left is transformed to the resonance structures on the right.

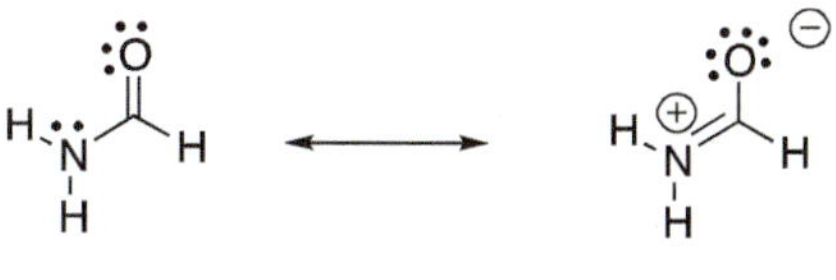

ii) Give another resonance structure for each of the following species. Use *curved arrow formulism* to indicate electron movement and include formal charges where appropriate.

(a) 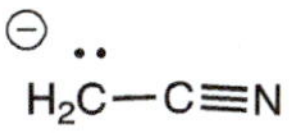 $H_2C-C\equiv N$ with $\ominus$ on C

(b) 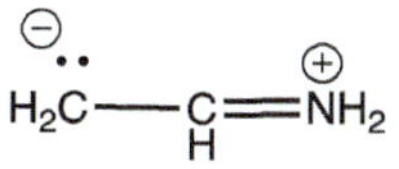 $H_2C-\underset{H}{C}=NH_2$ with $\ominus$ on left C and $\oplus$ on N

(c) $H_2C-\underset{H}{C}=CH_2$ with $\ominus$ on left C

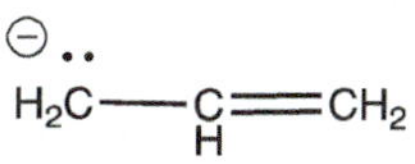

(d) 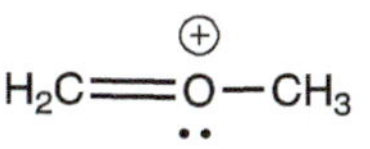 $H_2C=\overset{\oplus}{O}-CH_3$

(e) 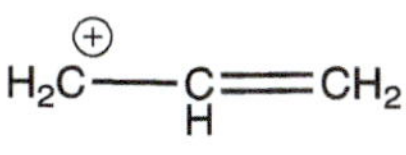 $H_2C-\underset{H}{\overset{\oplus}{C}}=CH_2$

1.4 Chemical Formulas

Condensed formulas show the types and fixed ratio of atoms that are contained in molecules. A number is used as a subscript to represent the same number of atoms in a molecule. The subscript is written to the right of the atoms in the condensed formula. Given condensed formulas for compounds, the Lewis dot structures give fairly good descriptions of the arrangements of the atoms, covalently bonded and nonbonded electrons in molecules and ions, but ***molecular formulas*** or ***structural formulas*** are used frequently to represent organic molecules. The Lewis dot structure of a molecule or ion can be transformed easily into structural formulas. For structural formulas, lines represent the bonding electrons of the Lewis dot structures, and dots are still used for the nonbonded electrons. Examples of both representations are shown below.

Lewis dot structure	Structural formula	Condensed formula
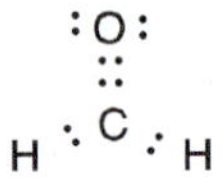 (H₂C=O with dots)	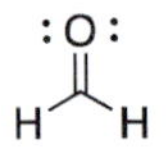 (H₂C=O with lone pairs on O)	CH_2O
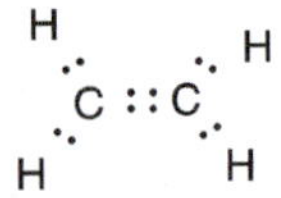 (H₂C::CH₂ with dots)	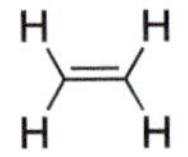 (H₂C=CH₂)	C_2H_4

1.4.1 Line-Angle Representations of Molecules

Throughout the course, there is yet another representation of molecular structures that will be used and is known as the line-angle representation. In this representation, lines that intersect each other at an angle of 120° are used. Each intersection represents a carbon along with the appropriate number of hydrogens, but the hydrogens are not shown. For the line-angle representation, the start of a line represents a carbon with three hydrogens; at an intersection of two lines, there are two hydrogens; at an intersection of three lines, there is one hydrogen; and at the intersection of four lines, there are no hydrogens. Thus, for propane, shown below are the various representations, but as pointed out, the line-angle representation will be routinely used throughout this course.

Molecular formula	Structural formula	Condensed formula	Line-angle formula
C_3H_8	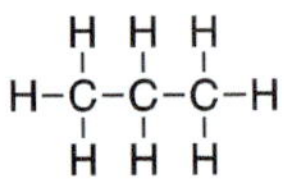 H-C(H₂)-C(H₂)-C(H₂)-H	$CH_3CH_2CH_3$	 ⌃

Shown in Figures 1.15 and 1.16 are examples of a more detailed description of line-angle representations.

Shown below is another example in which different representations are used.

Structural formula	Condensed formula	Line-angle formula
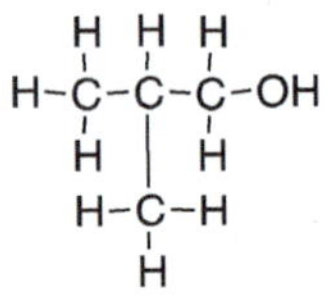 (isobutanol structural)	$CH_3CH(CH_3)CH_2OH$	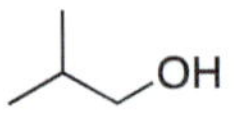 (isobutanol line-angle with OH)

1.4 Chemical Formulas | 19

Figure 1.15 A detailed description of CH$_3$CH$_2$CH$_3$ using a line-angle representation.

Figure 1.16 A detailed description of CH$_3$CH(CH$_3$)CH$_3$ using a line-angle representation.

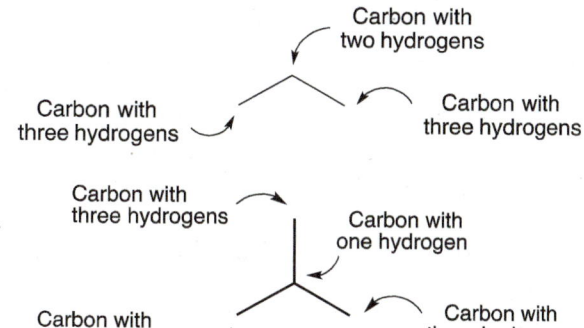

We will be using the line-angle representation mostly in this course and you should be very familiar with drawing this representation for different organic molecules. Shown below is a more detailed description of the line-angle representation for CH$_3$CH(CH$_3$)CH$_2$COH.

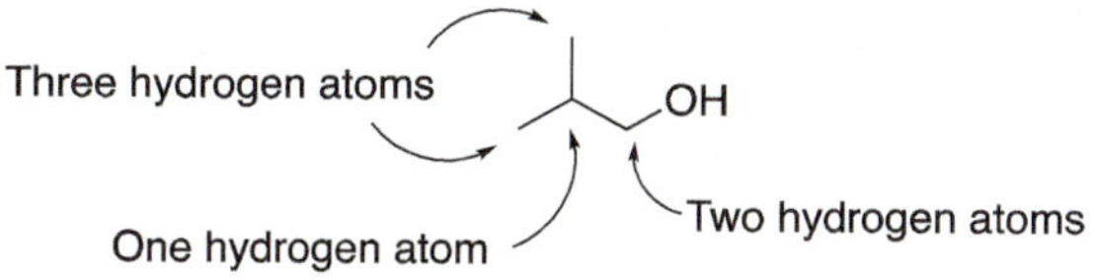

Three hydrogen atoms
One hydrogen atom
Two hydrogen atoms
OH

Problem 1.13

Give the structural formulas (line-angle representation) for each of the following molecules.

a) CH$_3$CH$_2$CH$_2$CH$_3$
b) CH$_3$C(CH$_3$)$_2$CH$_2$CH$_3$
c) CH$_3$CH$_2$CH(CH$_3$)CH(CH$_3$)CH$_3$

Critical analytical thinkers should be able to not only convert condensed formulas to line angle but also the reverse, from line-angle to condensed. Problem 1.14 is designed to build on the ability of thinking in reverse.

Problem 1.14

Give the condensed formulas for the following molecules.

(a) 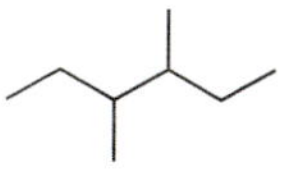   (b)

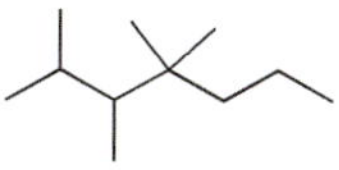

For molecules that have double bonds, the same line-angle representation is used, shown below are examples of acyclic and cyclic alkene molecules.

Molecular formula	Structural formula	Condensed formula	Line-angle formula
C$_3$H$_6$	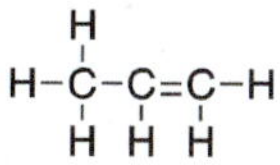 H-C-C=C-H (with H's)	CH$_3$CH=CH$_2$	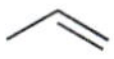
C$_6$H$_{10}$	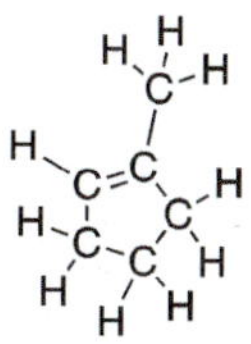 (cyclopentene with CH)	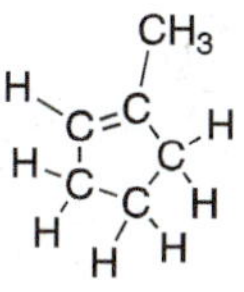 (cyclopentene with CH$_3$)	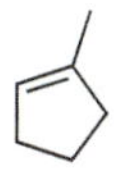

Note that the number of hydrogens is reduced by one for each carbon of the double bond, compared to the structures shown earlier. Also, the number of hydrogens is reduced by two for a cyclic compound, compared to acyclic compounds discussed earlier.

Owing to the linear geometry of the carbon atoms about triple bonds, the line-angle representation of a triple bond is drawn as linear as shown in the example below.

Molecular formula	Structural formula	Condensed formula	Line-angle formula
C_3H_4	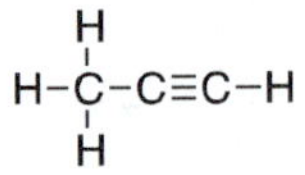 H–C(H)(H)–C≡C–H	$CH_3C≡CH$	 —≡

Problem 1.15

Give the structural formulas (line-angle representations) for each of the following molecules shown below.

(a) $CH_3CH_2C(CH_3)_2CH=C(CH_3)CH_2CH_3$

(b) 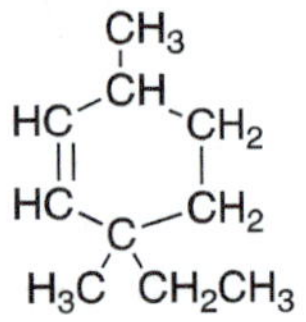
```
         CH3
         |
         CH
        /   \
    HC       CH2
    ||        |
    HC       CH2
       \   /
        C
       / \
     H3C  CH2CH3
```

(c) $CH_3CH_2CH(CH_3)C≡CCH_3$

1.5 The Covalent Bond

The VSEPR theory has worked very well to assist in determining the geometry of molecules, but there is another model that can be used to not only assist in determining the geometry of molecules but can also be used to explain important properties of molecules, such as why a hydrogen that is bonded to a triple bonded carbon is more acidic than one that is bonded to a double-bonded carbon or even a single-bonded carbon. In this model, the *molecular orbital (MO) theory*, atomic orbitals of different atoms form molecular orbitals, which are generated by mixing atomic orbitals.

1.5.1 The Single Bond to Hydrogen

The hydrogen molecule (H_2) is the simplest molecule; it contains only a single covalent bond (two electrons, one from each hydrogen atom) and no nonbonding electrons. As a result, it is an ideal molecule to be used to illustrate how atomic orbitals are mixed to form molecular orbitals. To form a hydrogen molecule, the 1s orbital of each atom interacts to form two new orbitals, a molecular orbital, referred to as a sigma (σ) orbital and another molecular orbital, which is higher in energy than the σ orbital and is called a sigma star orbital (σ*). Since there can only be a maximum of two electrons in any of these two orbitals, the two electrons available (one from each hydrogen atom) will occupy the lowest molecular orbital, creating a covalent bond, a sigma (σ) bond. This bond is known as a sigma (σ) bond since the bonding electrons occupy the sigma (σ) molecular orbital. An illustration of the molecular orbitals and the bonding formed in the H_2 molecule is shown in Figure 1.17.

In this illustration, short horizontal lines are used to represent atomic and molecular orbitals. The short horizontal lines in the middle of the diagram represent molecular orbitals, and the short horizontal lines at the outer ends of the diagram represent atomic orbitals. Note also that the molecular and atomic orbitals have different energy levels. Molecular orbitals that contain the bonding electrons are lower in energy (more stable) than either of the atomic orbitals that contributed to the molecular orbital. As pointed out earlier, there is another type of molecular orbital that is generated from the mixing of the two atomic orbitals. This orbital is higher in energy than either of the atomic orbitals or the sigma molecular orbital; this orbital is called an anti-bonding molecular orbital. If electrons are placed in this orbital, the molecule is not stable.

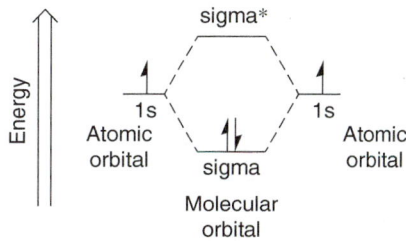

Figure 1.17 Molecular orbital diagram of hydrogen molecule.

1.5.2 The Single Bond to Carbon

This theory (molecular orbital theory) can be used also to explain the bonding and geometry of methane (CH_4) and other organic molecules. It is known from the VSEPR theory, molecular modeling and other experimental data that the H—C—H bond angles of methane are 109.5° and that the geometry of methane is tetrahedral, and each C—H bond length is 1.10Å. To account for this observation, four equivalent molecular orbitals from the carbon atom to each of the hydrogen atoms of methane must be used; and in each of these molecular orbitals, there must be two electrons – one from carbon and one from each hydrogen to form the four equivalent covalent bonds. In order to create four equivalent orbitals from the carbon atom, there has to be mixing (hybridization) of four atomic orbitals. Atomic orbitals that contain the valence electrons are the most important orbitals to be considered for covalent bonds. As a result, the 2s and all three of the 2p orbitals of carbon are the most important orbitals to be considered for the four new molecular orbitals. In Figure 1.18, the hybridization (mixing) of the 2s and 2p orbitals of carbon to create four new atomic orbitals is shown.

The new hybridized orbitals that are created are called sp^3 orbitals. The name is derived from the orbitals that are used to generate these orbitals: <u>one</u> 2s orbital and the <u>three</u> 2p orbitals, hence the name s^1p^3, or just simply sp^3. These orbitals are higher in energy than either the 1s or the 2s orbitals, but lower in energy than the 2p orbitals, as illustrated in Figure 1.18. Also, the mixing of the orbitals results in hybridized orbitals that have 75% **p** character and 25% **s**

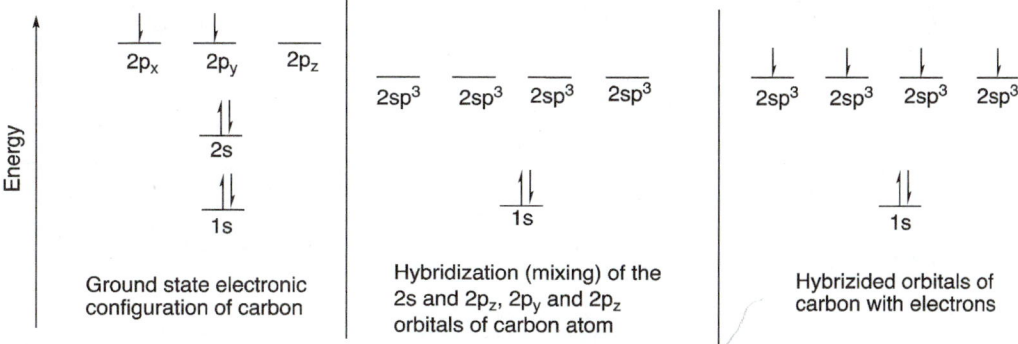

Figure 1.18 Hybridization of the 2s and 2p orbitals of carbon to form four equivalent $2sp^3$ orbitals.

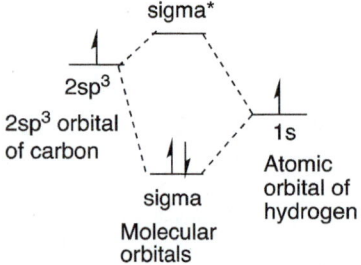

Figure 1.19 Molecular orbital diagram of one of the bonds of methane, note that one electron is from hydrogen and the other is from one of the sp³ hybridized orbitals of carbon.

character since these are essentially the contributions from these orbitals. The significance of this percentage mixing of atomic orbitals will be important when the bond lengths and electronegativities of atoms in molecules that have different hybridized orbitals are discussed.

Each C—H bond of methane can be represented by a similar molecular orbital diagram as that used for hydrogen (Figure 1.17), but in this case, four molecular orbitals are needed since there are four covalent bonds in methane. Figure 1.19 gives the illustration of just one of those molecular orbitals of methane in which only one of the sp³ hybridized orbitals of the four shown in Figure 1.18 is used to bond with one hydrogen atom.

Based on this molecular orbital model, there are four equivalent sigma bonds from carbon to four hydrogen atoms, and the result is a molecule with a tetrahedral geometrical arrangement, as shown in Figure 1.20. Note that the geometry is the same as that predicted by the VSEPR theory.

1.5.3 The Single Bond to Heteroatoms

A similar approach using the molecular orbital theory can be used to explain the experimental observations of ammonia (NH_3). It is known that the bond angles of ammonia are slightly less than that of methane, actually 107°. The molecular orbital picture shown in Figure 1.21 best explains this observation in which sp³ hybridized orbitals from the nitrogen atom are used to bond to the hydrogen atoms from nitrogen and the unshared electrons are in a sp³ hybridized molecular orbital. Since these electrons are nonbonding and hence occupy more space than

Figure 1.20 Methane showing the tetrahedral arrangement of the four equivalent bonds from carbon to four hydrogen atoms.

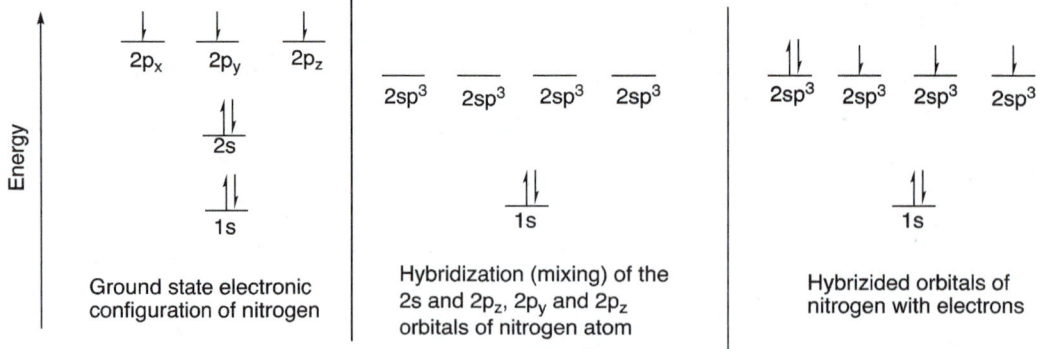

Figure 1.21 Hybridization of the 2s and 2p orbitals of nitrogen to form four equivalent 2sp³ orbitals. Note that the number of electrons is different from that of carbon; there are a total of seven electrons in nitrogen and a total of six electrons in carbon.

bonding electrons, they will slightly compress the N—H bond angles, compared to that of methane.

The orbitals of nitrogen are sp^3 hybridized, with bond angles close to that of methane, but slightly less as explained above. The model for ammonia is shown in Figure 1.22.

The molecular orbital theory can be used to explain the bonding and other properties of water (H$_2$O). From the VSEPR theory, there are two bonds from oxygen to two hydrogen atoms in water and there are also two pairs of nonbonding electron. Similar to ammonia, there must be molecular orbitals available for not only the bonding electrons but also nonbonding electrons. As a result, the orbitals of water are sp^3, in which two are used to bond to hydrogens and the other two contain two pairs of nonbonding electrons.

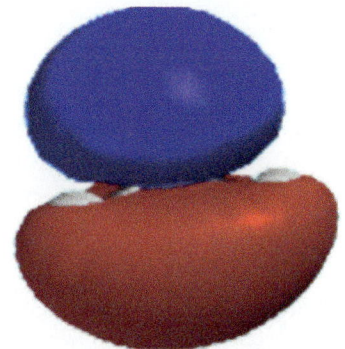

Figure 1.22 Ammonia showing the trigonal pyramidal arrangement of the three equivalent bonds from nitrogen to three hydrogen atoms and the fourth orbital, which contains the unshared pair of electrons (shown in blue).

Problem 1.16

a) The oxygen atom of water (H$_2$O) also has four sp^3 hybridized orbitals, two to the hydrogen atoms and two that contain two lone pairs of nonbonding electrons. Use a figure similar to that in Figure 1.21 to represent the orbitals and electrons in water.
b) Explain why the observed bond angle of water is 104.5° and not 109.5°, similar to methane (CH$_4$) or ammonia (NH$_3$), which is 107°.

1.5.4 The Carbon–Carbon Double Bond

The expanded structure of ethylene, C$_2$H$_4$, is shown below.

$$\underset{H}{\overset{H}{\diagdown}}C=C\underset{H}{\overset{H}{\diagup}}$$

Ethylene (ethene)

Based on the VSEPR theory, ethylene is a trigonal planar (flat) molecule, with bond angles about each carbon atom of 120°. Since the bond angles about the carbon for CH$_4$ are 109.5°, it is obvious that sp^3 hybridized orbitals cannot be used to explain the bonding and geometry of ethylene. For ethylene, there are three atoms bonded to each carbon – two bonds to two hydrogen atoms and the other to another carbon atom; note that for ethylene, there are no nonbonding electrons. Thus, for each carbon atom, three equivalent orbitals are needed to bond to three atoms. The creation of three equivalent orbitals from each carbon atom requires the hybridization (mixing) of the 2s and 2p orbitals that must be considered; but in this case, only the hybridization of the <u>one</u> 2s and only <u>two</u> of the 2p orbitals are needed to create three new orbitals. The three new orbitals that are created are called s^{1}p^2 or simply, sp^2 orbitals. Figure 1.23 gives an illustration of the hybridization needed to form three equivalent sp^2 orbitals from one of the carbon atoms of ethylene.

Note that not all three 2p orbitals are used as in the case to create four sp^3 hybridized orbitals for CH$_4$. For ethylene, there is one 2p orbital that is left unhybridized. Also note that the three hybridized orbitals of ethylene have 66.6% p character and 33.3% s character; hence, there is more s character for the carbon atom of molecules that utilize sp^2 orbitals, compared to molecules, such as methane, that use sp^3 hybridized orbitals. Also, note that when the

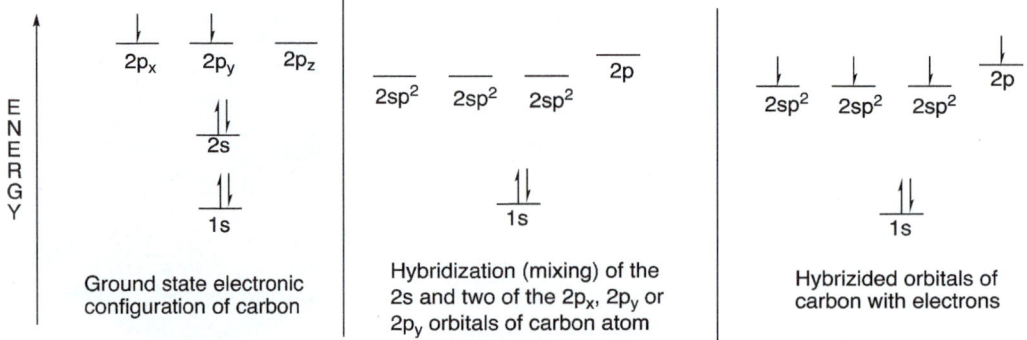

Figure 1.23 Hybridization of the 2s and two 2p orbitals of carbon to form three equivalent $2sp^2$ orbitals and one 2p orbital. Note that since the 2p orbitals are degenerate, there are several options to get three equivalent $2sp^2$ orbitals.

electrons are placed in the orbitals, they are distributed across all three hybridized orbitals and the 2p orbital before pairing occurs. Based on *Hunds rule*, the expectation is that since the 2p orbital is a bit higher in energy than the $2sp^2$ orbitals that the electrons would pair before occupying the vacant 2p orbital. In this case, however, the energy difference between the $2sp^2$ and 2p orbitals is extremely small and the energy needed to pair the electrons is a slightly more, and as a result, the electrons occupy separate orbitals before pairing as shown in Figure 1.23.

Thus, for each carbon atom of ethylene, there are three sigma bonds, one is bonded to each of the two hydrogen atoms and the third to another sp^2 hybridized carbon as shown in Figure 1.24. The carbon–carbon bond length is shorter than a sp^3 carbon–carbon length; the C—C bond length of CH_3CH_3 is 1.54 Å and that of ethylene is 1.33 Å. This observation can be explained due to the different types of hybridized orbitals that are used to make the molecular orbitals. The sp^2 orbital has more s character and electrons that are in those orbitals are closer to the nucleus, compared to the sp^3 orbital, which has less **s** character resulting in the electrons in the sp^3 orbital being further from the nucleus. Figure 1.24 gives a graphical representation of ethylene.

As shown in Figure 1.24, each carbon atom in ethylene has an unhybridized 2p orbital, which contains one electron. There is a second type of bond, which is created between the p atomic orbitals of the adjacent carbon atoms as illustrated in Figure 1.25. This new molecular orbital that is created is called the pi (π) molecular orbital and the bond that is formed is called a *pi (π) bond* and is different from a sigma bond in that it is created using electrons from the p orbitals.

Another representation of this type of bonding is shown in Figure 1.26, in which the atomic orbitals of each p orbital form a molecular orbital, which is low in energy, and the other molecular orbital, which is higher in energy, is known as an antibonding or pi* (π^*) molecular orbital.

Since the p orbitals have an hourglass shape as described earlier in the chapter, a single pi (π) bond has electron distribution above and below the plane of the sigma bond. Figure 1.27 gives

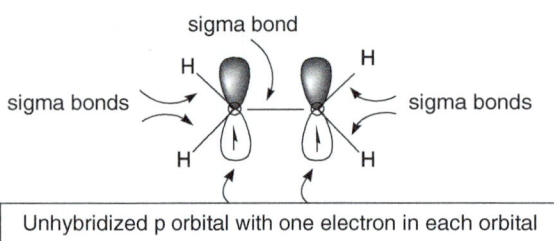

Figure 1.24 Graphical representation of sp^2 sigma bonds and unhybridized p orbital of ethylene.

Figure 1.25 Graphical illustration of the pi (π) bond of ethylene.

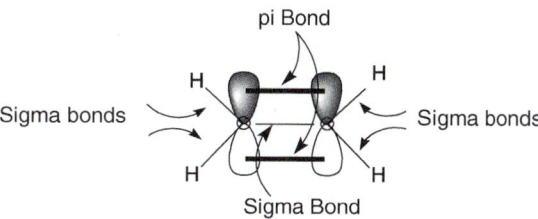

Figure 1.26 Molecular orbital diagram of the pi (π) bond of ethylene

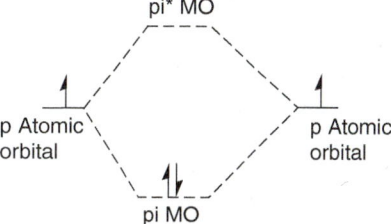

another illustration of ethylene showing the pi (π) bond, where the electron density is above and below the plane of the molecule.

This newly created bond, the pi (π) bond, in combination with the sigma (σ) bond, result in a carbon–carbon double bond. This carbon–carbon double bond has four electrons, but in different molecular orbitals, two are in the σ molecular orbital and two in the π molecular orbital. Figure 1.28 gives another representation of the relative energies of these two orbitals, a sigma (σ) and pi (π). Note that the antibonding orbitals, σ* and π*, have no electrons.

Problem 1.17

a) What type of orbitals would you expect for the bonding from the central atom in BF_3?
b) Give a diagram similar to that in Figure 1.17 for BF_3.
c) Based on the location of Al in the periodic table, predict the geometry that would exist for $AlCl_3$.

1.5.5 The Carbon–Heteroatom Double Bond

It is possible for carbon to form a double bond to heteroatoms, such as nitrogen or oxygen. It is known that the geometry around the carbon of methyleneimine and formaldehyde is 120 °C and that these molecules are flat and described as trigonal planar; the Lewis dot structures of these molecules are shown below.

Figure 1.27 Graphical illustration of the pi (π) orbital of ethylene showing the pi (π) bond, where the electron density is above and below the plane of the molecule.

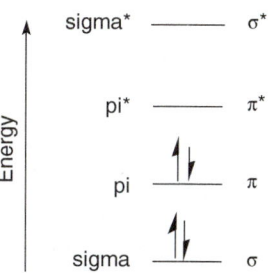

Figure 1.28 Relative energies of the molecular orbitals of ethylene

H₂C=N̈H   H₂C=Ö̈
Methyleneimine   Formaldehyde

For methyleneimine, the carbon atom is bonded to two hydrogen atoms and a nitrogen atom; as a result, there should be three sp^2 orbitals from carbon to these atoms. For the nitrogen atom, which is bonded to a sp^2 carbon, a hydrogen has one pair of nonbonded electrons; thus, three equivalent orbitals must be created: one to form a sigma bond to carbon, one to hydrogen, and another orbital for the unshared pair of electrons. As a result, the orbitals of the nitrogen atom are sp^2 hybridized orbitals. Similarly, the carbon atom of formaldehyde is bonded to three atoms (two hydrogen atoms and an oxygen atom); thus, the carbon atom uses sp^2 hybridized orbitals to form sigma bonds to these atoms. However, for the oxygen atom, which is bonded to an sp^2 carbon and has two pairs of nonbonded electrons, three equivalent orbitals must be created: one to form a sigma bond to carbon and the other two orbitals for the two unshared pairs of electrons. As a result, the orbitals of the oxygen atom are sp^2. For these two molecules, there are two electrons in each of the unhybridized p orbitals. These electrons form a pi (π) bond similar to that described for ethylene.

Problem 1.18

Formic acid has two oxygen atoms that use different types of orbitals for bonding. (i) For each oxygen atom, determine the type of orbitals used for bonding? (ii) Determine the type of orbitals used for bonding around the carbon atom?

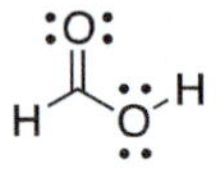

:O:
‖
H—C—O—H
  ̈ ̈

Formic acid

1.5.6 The Carbon–Carbon Triple Bond

The Lewis dot structure of acetylene, C_2H_2, is: H:C:::C:H and is also shown below.

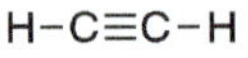

H−C≡C−H
Acetylene

Based on the VSEPR theory, acetylene is a linear molecule. It is known that the bond angle around each carbon is 180° and that the C—H bond length is 1.061 Å, which is shorter than either the C—H bond length of ethylene or ethane. These observations imply that sp^2 or sp^3 molecular orbitals cannot be used to describe the geometry and bond lengths of acetylene. Since there are two atoms bonded to each carbon atom (one to hydrogen and the other to another carbon atom), two equivalent orbitals are needed. In order to create two equivalent orbitals from one of the carbon atoms to a hydrogen atom and to another carbon atom, there needs to be hybridization (mixing) of <u>one</u> 2s orbital and only <u>one</u> of the 2p orbitals to create s^1p^1 or simply sp orbitals. The hybridization to create an sp orbital is illustrated in Figure 1.29.

In this case, the orbitals are 50% s in character and 50% p in character. Thus, these orbitals have the most s character, compared to the sp^3 or the sp^2 orbitals. Since acetylene has two similar carbon atoms, there must be two sp hybridized carbon atoms. For each carbon, one is bonded to a hydrogen atom and to the other sp hybridized carbon resulting in two sigma bonds, one to another carbon and one to a hydrogen atom, as illustrated in Figure 1.30.

The two unhybridized atomic 2p orbitals on each carbon atom of acetylene are perpendicular to each other, i.e. 90°; another term that is often used to describe the geometric relationship between these two orbitals is that the orbitals are *orthogonal* to each other. The electrons in these orbitals create two new pi (π) bonds that are orthogonal to each other, as shown in Figure 1.31.

1.5 The Covalent Bond

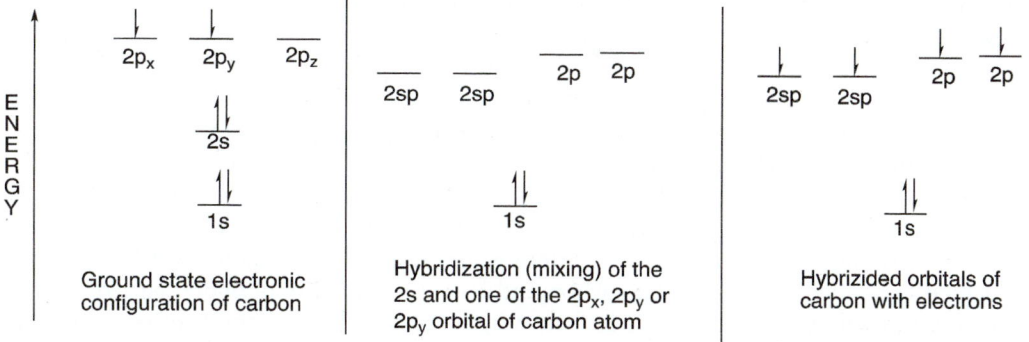

Figure 1.29 Hybridization of the 2s and one 2p orbital of carbon to form two equivalent *2sp* orbitals. Note that since the 2p orbitals are degenerate, there are several options to get two equivalent 2sp orbitals.

Figure 1.30 Representation of the sigma bonds and two unhybridized 2p orbitals (each with one electron) of acetylene.

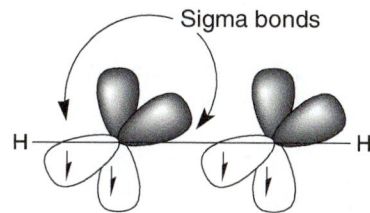

Figure 1.31 Representation of the sigma bonds and two perpendicular pi (π) bonds of acetylene.

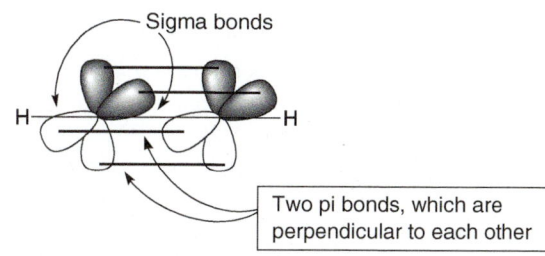

The combination of two pi bonds and one sigma bond that is created between the carbon atoms is called a carbon–carbon triple bond – two pi bonds and one sigma bond.

Since the geometry of acetylene is similar to that of carbon dioxide, with bond angles of 180° about the central carbon atom of carbon dioxide, the orbitals that are used for the carbon atom of carbon dioxide are also *sp*.

Problem 1.19

i) What type of orbitals would you expect for the bonding in $BeCl_2$?
ii) Give a diagram similar to that in Figure 1.29 for this molecule.

1.5.7 The Carbon–Heteroatom Triple Bond

It is possible to form a similar triple bond from carbon to nitrogen or oxygen. The structures of examples of molecules, in which a triple bond is formed from carbon to nitrogen atom and from carbon to oxygen atom, are shown below.

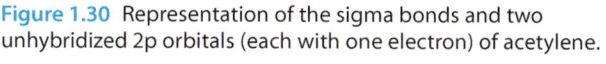

H·C≡N: :C≡O:
Hydrogen cyanide   Carbon monoxide

Note that the case of carbon monoxide is unique in that there are formal charges on the atoms as shown above. For carbon monoxide there is one carbon atom that is bonded to an oxygen atom by a triple bond, which is a very strong bond. This triple bond consists of two covalent bonds as well as one dative covalent bond, also known as a coordinate covalent bond. A dative covalent bond is a covalent bond in which one of the atoms provides a lone pair of electrons to a make a bond. For carbon monoxide, oxygen is the atom that provides a pair of electrons to form this bond. We will not encounter many of these types of bonds throughout this course of organic chemistry. For hydrogen cyanide, the carbon atom is bonded to hydrogen and nitrogen; as a result, sp orbitals must be used to form bonds from carbon to these atoms. For the nitrogen atom, which is bonded to only the sp carbon and has one pair of nonbonded electrons, two equivalent orbitals must be created: one to form a sigma bond to carbon and another orbital for the unshared pair of electrons. As a result, the orbitals of the nitrogen atom are sp.

1.6 Bonding – Concept Summary and Applications

Table 1.1 gives a summary of the differences between bonds formed to different hybridized orbitals of carbon.

An observation that can be made from Table 1.1 is that the carbon–carbon triple bond is the shortest bond of the three types of bonds involving hybridized orbitals. Recall that the carbon of the carbon–carbon triple bond uses sp orbitals and that they have 50% s character and 50% p character, compared to 25% s and 75% p for the carbon of methane. Since the s orbitals are closer to the nucleus, which is positive due to the presence of the protons, the attractions of the bonding electrons in the sp orbitals are stronger than those of methane, which contains only 25% s character. The carbon–carbon triple bond is also the strongest bond since its breakage requires the most energy ($200\,\text{kJ}\,\text{mol}^{-1}$), compared to methane or ethane.

Table 1.2 gives a more detailed summary of the bonding features of carbon to other common heteroatoms typically found in most organic chemistry. First, note the number of bonds that are possible to each atom. With carbon, even though there are different possible hybridization states, there are still only four bonds possible to each carbon. For nitrogen, even though it may have the same hybridization state as carbon, the number of bonds involving neutral nitrogen is different; there will always be three bonds to nitrogen if the formal charge on the nitrogen is zero. Likewise, oxygen, which has four hybridized orbitals, has just two bonds to the oxygen atom if the formal charge on oxygen is zero.

Table 1.1 Comparisons of the three types of orbitals found in most organic compounds.

Hybrid orbitals	Angle around carbon	Geometry	Example	C—C bond strength (kJ mol^{-1})	C—H bond length (Å)	C—C bond length (Å)
sp^3	109.5°	Tetrahedral	Ethane (CH$_3$CH$_3$)	90	1.09	1.54
sp^2	120°	Trigonal planar	Ethene (CH$_2$=CH$_2$)	146	1.08	1.33
sp	180°	Linear	Ethyne (H:C:::C:H)	200	1.06	1.20

Table 1.2 Number of bonds that are possible from a neutral atom in molecules.

Atom	Hybridized orbitals	Number of bonds	Geometry	Bonds to other atoms	Example	Specific example
C	sp^3	4	Tetrahedral	4	$-\overset{\shortmid}{\underset{\shortmid}{C}}-$	$H-\overset{H}{\underset{H}{C}}-H$
C	sp^2	4	Trigonal planar	3	$-\overset{\shortmid}{C}=$	$\overset{H}{\underset{H}{>}}C=C\overset{H}{\underset{H}{<}}$
C	sp	4	Linear	2	$-C\equiv$	$H-C\equiv C-H$
N	sp^3	3	Trigonal Pyramidal	3	$-\overset{..}{\underset{\shortmid}{N}}-$	$H-\overset{..}{\underset{H}{N}}-H$
N	sp^2	3	Bent	2	$-\overset{..}{N}=$	$\overset{..}{\underset{H}{N}}=C\overset{H}{\underset{H}{<}}$
N	sp	3	—	1	$:N\equiv$	$:N\equiv C-H$
O	sp^3	2	Bent	2	$-\overset{..}{\underset{..}{O}}-$	$H-\overset{..}{\underset{..}{O}}-H$
O	sp^2	2	—	1	$\overset{..}{\underset{..}{O}}=$	$\overset{..}{\underset{..}{O}}=C\overset{H}{\underset{H}{<}}$

1.7 Intermolecular Attractions

In this section, a visual description of the various interactions that exist among molecules will be examined. The types of attractions that exist among molecules (intermolecular attractions) are very weak compared to the bonding found within organic molecules, i.e. the ionic or covalent bond.

1.7.1 Dipole–Dipole Intermolecular Attractions

A very common intermolecular interaction encountered among molecules is **dipole–dipole attraction**. As we have seen in the earlier section of this chapter, some molecules are polar molecules, and different regions of the molecule have partial charges. Chloromethane (CH_3Cl) is an example, it has four covalent bonds, three from the central carbon to hydrogens and one to chlorine. Since this molecule has a carbon–chlorine bond, this covalent bond is a polar covalent bond and hence based on the geometry, is a polar molecule, i.e. it has a dipole moment greater than zero. For chloromethane, as shown in Figure 1.32, the chlorine bears a partial negative charge (since it is electronegative), and the carbon bears a partial positive charge (since it is more electropositive than chlorine). If two of these molecules are in close contact with each other, there will be an attraction between them since opposite charges are attracted to each other. This type of attraction is called a **dipole–dipole** attraction as shown in Figure 1.32.

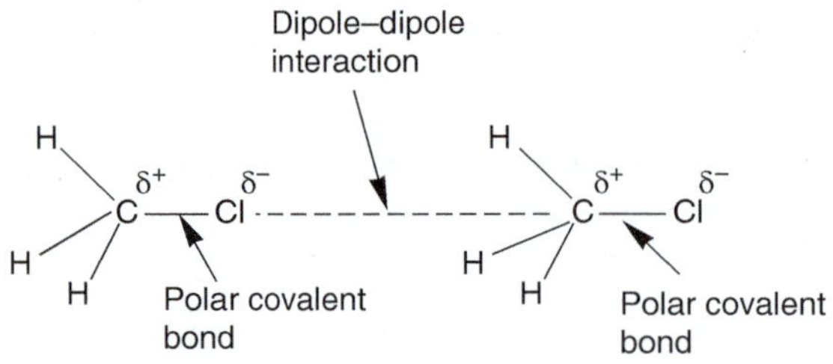

Figure 1.32

Figure 1.32 Illustration of the dipole–dipole attraction between two molecules of CH_3–Cl molecules.

Problem 1.20

i) Using an illustration as shown in Figure 1.32, show the intermolecular attractions between each of the following molecules.

(a) 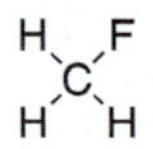 CH_2F_2 (H,H,F,H on C) (b) 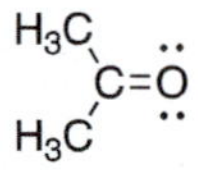 $(H_3C)_2C=O$ (c) $H_3C-C≡N:$

ii) For compounds (b) and (c) above, what factors should be considered to determine which will form a stronger intermolecular attraction?

1.7.2 Intermolecular Hydrogen Bond

If hydrogen is bonded to an electronegative atom in a molecule such as nitrogen or oxygen, a very polar covalent bond is created due to the difference in electronegativity between hydrogen and oxygen or nitrogen. As shown in Figure 1.33, there is a similar intermolecular attraction between two methanol molecules as that shown for chloromethane in Figure 1.32; in this case, however, the very electropositive hydrogen is involved.

Since the magnitude of charge separation between the electronegative oxygen and hydrogen is very large compared to other electropositive atoms, this type of intermolecular interaction is called a **hydrogen bond**. Thus, the hydrogen bond is a specific type of dipole–dipole interaction, which specifically involves hydrogen bonded to an electronegative atom.

Problem 1.21

i) Using an illustration as shown in Figure 1.33, show the hydrogen bond intermolecular attractions between each of the following molecules.

(a) H_3C-NH_2 (b) 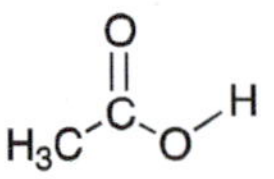 $H_3C-C(=O)-O-H$ (c) $H_3C-C(=O)-N(H)-H$

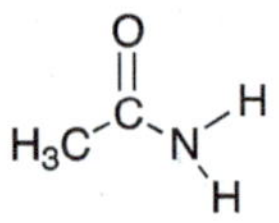

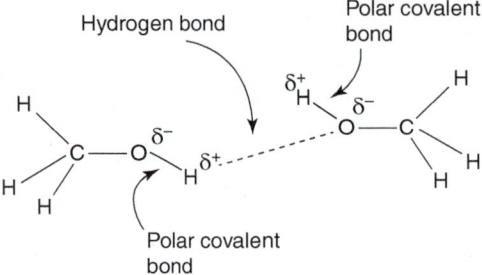

Figure 1.33 Hydrogen bond between two methanol molecules.

ii) Would you expect intermolecular hydrogen bonding to exist for the following molecule? Explain your answer.

```
    H   ..  H
    |   ..  |
H – C – O – C – H
    |   ..  |
    H       H
```

1.7.3 Intermolecular London Force Attractions

Nonpolar molecules are also attracted to each other by very weak attractive forces called *London forces*; these attractive forces are also known as *Van der Waals* attraction. This type attraction is based on the induced dipole that is possible in nonpolar molecules based on an instantaneous shift of electrons. As a result of such instantaneous shifts, it is possible for nonpolar molecules to become instantaneously polarized with partial negative and positive charges distributed at different regions of the molecule. These partial charges can induce a similar charge distribution in another molecule, and as a result, an attraction exists between two fairly nonpolar molecules, as illustrated in Figure 1.34. Such molecules are often described as polarizable molecules and are typically large nonpolar molecules, compared to smaller molecules. Thus, this type of intermolecular attraction would be greater for pentane compared to ethane. In general, large polarizable molecules will result in stronger interactions, compared to smaller, compact less polarizable molecules.

Problem 1.22

Of the following pairs of molecules, determine which molecule would have the stronger Van der Waals intermolecular attraction. Explain your answer

(a)
```
    H H                H H H
    | |                | | |
H – C – C – H   and  H – C – C – C – H
    | |                | | |
    H H                H H H
```

(b)
```
    H H H H H H                              H
    | | | | | |                              |
H – C – C – C – C – C – C – H              H – C – H
    | | | | | |                  and        H   |   H H
    H H H H H H                              | | | |
                                         H – C – C – C – C – H
                                             |   |   | |
                                             H   |   H H
                                             H – C – H
                                                 |
                                                 H
```

1.8 Intermolecular Molecular Interactions – Concept Summary and Applications

The types of intermolecular interactions discussed in the previous section, especially hydrogen bonds, drastically alter the expected physical properties of some compounds. The physical properties of any compound are measurable and observable characteristics of that compound.

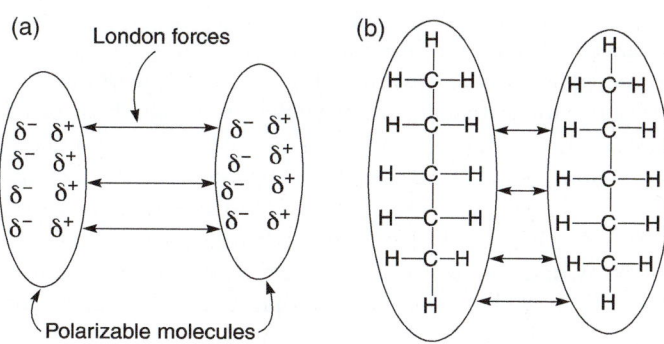

Figure 1.34 Graphical illustration of the attraction between two nonpolar molecules (a) and illustration of these attractions of pentane (b).

Sets of characteristics are unique for a specific compound, and these characteristic properties are often used to identify unknown compounds. The boiling and melting points of compounds are measurable properties and that a liquid is colorless is an observable property. Knowledge of the physical properties of compounds can give an indication of certain structural features of the molecule. For example, ethanol has a boiling point of 78.5 °C, whereas dimethylether has a boiling point of −23.6 °C, even though they both have the same molecular formula and molecular weight and the same type of atoms.

```
    H H  H                         H     H
    | | ..                         |  .. |
  H-C-C-O:                     H - C - O - C - H
    | |  ..                        |  .. |
    H H                            H     H
```
Ethanol, boiling point 78.5 °C   Dimethylether, boiling point −23.6 °C

The boiling point of a liquid is the temperature at which the vapor phase of the liquid is the same as that of the atmospheric pressure. At that temperature, the molecules move freely from the liquid phase to the vapor phase. Since the two compounds mentioned above have the same molecular mass, it is expected that the same amount of energy would be required to vaporize both, and hence they should have the same boiling point. The boiling point of a liquid has a direct relationship with possible intermolecular attractions and the types of attractive forces that exist. If the intermolecular attractions for one of these compounds are very strong, the individual molecules are not as free to escape from the liquid phase to the vapor phase, and thus more energy must be supplied to overcome these intermolecular attractions before boiling of that liquid can take place. This increase in energy input results in higher boiling points for liquids that have strong intermolecular attractions. The type of intermolecular attractions that are encountered in ethanol is hydrogen bonding, and this type of interaction is not possible for dimethylether since all the hydrogens are bonded to carbons and the partial positive charge is not as great, compared to those of ethanol.

Problem 1.23

Butaneamine ($CH_3CH_2CH_2NH_2$) has approximately the same molecular weight as propanol ($CH_3CH_2CH_2OH$), yet the boiling point of butaneamine is 48 °C and that of propanol is 97 °C. Explain this difference in boiling points.

London forces, even though much weaker than hydrogen bonding and dipole–dipole attractions, also affect the expected boiling points of liquids. For example, the boiling points of most alkanes are usually very low, compared to other organic compounds of similar molecular weights that contain heteroatoms. In addition, the boiling points of highly branched alkanes are typically lower than alkanes of similar molecular weights that are not highly branched as shown in Figure 1.35.

Since hydrocarbons are nonpolar molecules, the type of intermolecular attractions that are expected are *London forces*. For branched hydrocarbons, there is less surface area and hence the attractive forces that are created will be less than those of straight chain hydrocarbons that have greater surface areas as shown in Figure 1.34. The boiling point of pentane is 36.0 °C, whereas the boiling point of 2,2-dimethylpropane is 4.4 °C, a compound that has the same molecular formula, but a lower boiling point. Intermolecular attractions can be used to explain this observation. For branched alkanes, the electrons are close to each other, compared to a straight chain alkane, where the electrons are more dispersed throughout the molecule. As a result, the surface area of a branched alkane is less than that of a straight chain alkane. Thus, the electrons of a

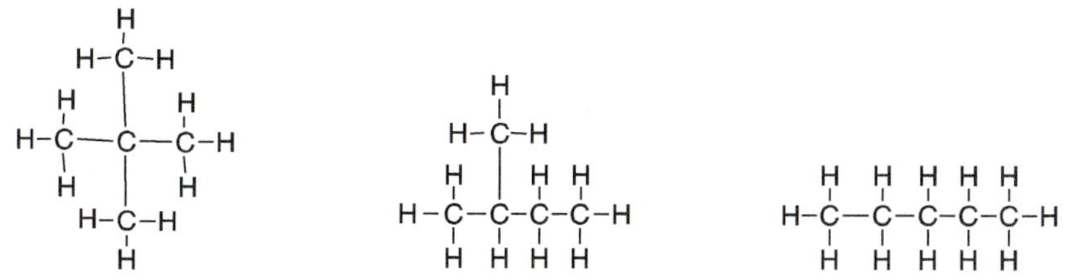

1.8 Intermolecular Molecular Interactions – Concept Summary and Applications

2,2-Dimethylpropane
Boiling point = 4.4 °C

2-Methylbutane
Boiling point = 29.9 °C

Pentane
Boiling point = 36.0 °C

Figure 1.35 Boiling points of different alkanes with the same molecular weight and same kinds of atoms (carbon and hydrogen atoms).

branched alkane are less polarizable; whereas for the less branched alkanes, the electrons are more polarizable. Due to the increased *Van der Waals* attractions for the more polarizable molecules, straight chain alkanes have higher boiling points than highly branched alkanes. 2,2-Dimethylpropane has less surface area and more compact and hence a lower boiling point than pentane, which has a greater surface area and is less compact. In general, isomers/molecules with larger surface areas have higher boiling points than those with less surface area. Since alkanes are nonpolar molecules, the attractions between the different molecules are minimal, resulting in low boiling points in general, compared to more polar molecules.

Problem 1.24

Arrange the following compounds in the order of boiling points, i.e. the lowest boiling liquid first:

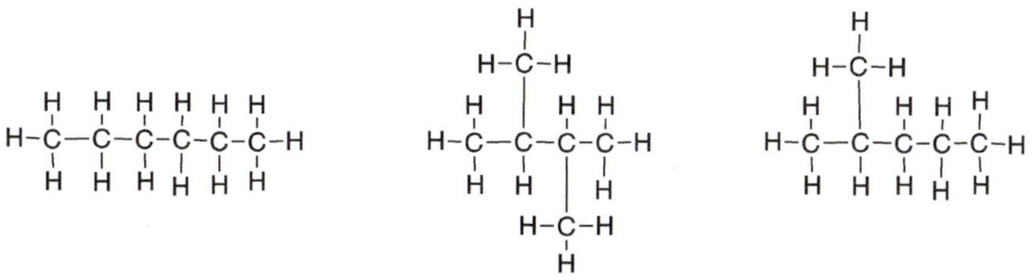

Hexane   **2,3-Dimethylbutane**   **2-Methylpentane**

The solubility of compounds in different solvents is also dictated to a large extent by the intermolecular attractions that are created between the solute and solvent molecules. As a result, different compounds have different solubilities in water, which is a polar solvent. Water has two hydrogens bonded to the very electronegative oxygen. Solvents that have such hydrogens carry a partial positive charge on the hydrogens and these solvents are called polar-protic solvents. Methanol (CH_3OH) and ethanol (CH_3CH_2OH) are examples of *polar-protic solvents*. Based on potential intermolecular interactions, it is possible to predict the relative solubilities of different solutes in water. Glucose (structure shown below) is very soluble in water, whereas cyclohexane (structure shown below), which has a similar appearance, is not soluble in water. In a solution of glucose and water, intermolecular attractions via hydrogen bonds are possible, whereas for cyclohexane, all the hydrogens are bonded to carbons and a similar type of solvent solute intermolecular attraction as that for glucose–water does not exist.

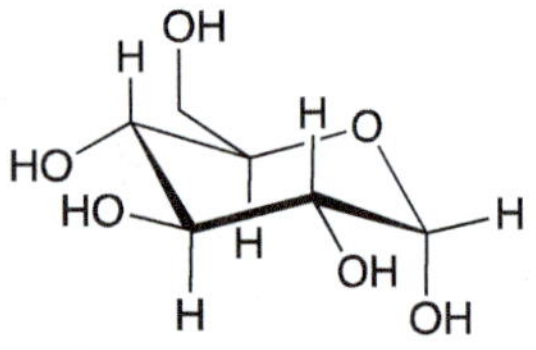

Glucose

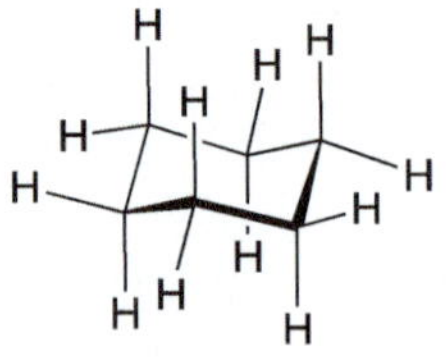

Cyclohexane

Problem 1.25

Draw structures to demonstrate the hydrogen bonds that are possible with water and glucose (structure shown above).

End of Chapter Problems

1.26 Write the ground state electronic configuration for each of the following atoms.

a) Carbon
b) Oxygen
c) Nitrogen
d) Fluorine
e) Sodium
f) Lithium
g) Neon
h) Beryllium
i) Helium
j) Boron

1.27 How many valence electrons are there in each of the following atoms?

a) Carbon
b) Oxygen
c) Nitrogen
d) Hydrogen
e) Fluorine
f) Boron
g) Lithium
h) Sodium
i) Beryllium

1.28 Give Lewis dot structures for the following molecules.

a) C_2H_6O
b) CH_2Cl_2
c) C_2H_7N
d) C_3H_6
e) CH_3OH
f) CH_3Cl
g) CH_3SH
h) CS_2
i) H_2O_2
j) N_2H_2

1.29 Determine which of the following molecules is(are) polar or nonpolar?

a) O_2
b) Cl_2
c) N_2
d) H_2O
e) CO_2
f) H_2O_2
g) HCl
h) NH_3
i) H_2S
j) CO
k) CH_2Cl_2
l) BF_3
m) CF_4

1.30 List the following atoms in terms of increasing electronegative, i.e. the least electronegative first: Li; C; O; N; Na

1.31 Draw the Lewis dot structures for the following compounds, showing appropriate formal charges.

a) NH_4NCl
b) $NaBH_4$
c) CH_3NO_2
d) CH_4
e) HCN
f) CH_3N

1.32 For the molecule shown below, indicate: (a) the type of hybridized orbitals for each atom (except hydrogen); (b) the geometry around each atom; and (c) the number of unshared pair(s) of electrons on each atom.

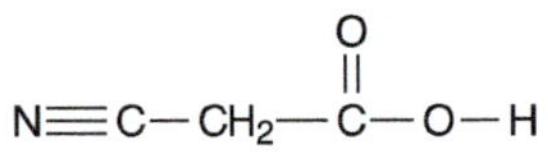

$$N\equiv C-CH_2-\overset{\overset{\displaystyle O}{\|}}{C}-O-H$$

End of Chapter Problems

1.33 What is the geometry about the central atom of the following?

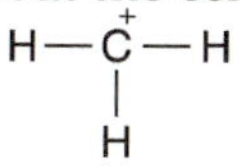

$$H-\overset{+}{\underset{H}{C}}-H$$

a) BH_3 b) Methyl carbocation

1.34 Which of the following compounds would you expect to have a covalent bond(s)? Explain your answer.
NaCl, KCl, CCl_4, CaO

1.35 Draw Lewis dot structures of each of the following molecules and use δ^+ and δ^- where possible to show the polarity of each covalent bond in the following molecules.

a) CO_2 b) N_2 c) HCl   d) H_2O e) CH_3OH
f) CH_3F g) NH_3 h) CH_2O

1.36 Write Lewis dot structures for each of the following ions.

a) OH^- b) CO_3^{2-} c) HCO_3^- d) NH_2^- e) CH_3O^-

1.37 Give Lewis dot structures of the following and determine the formal charge on the nitrogen and the oxygen atoms.

a) CH_3NH_3 b) HNO_3 c) CH_3NH_2 d) H_3O^+

1.38 What is the formal charge on the central atom of each of the species shown below?

$$F-\overset{\overset{F}{|}}{\underset{F}{N:}} \quad\quad H-\overset{\overset{H}{|}}{\underset{H}{\overset{..}{O}}} \quad\quad H-\overset{\overset{H}{|}}{\underset{H}{Al}}-H \quad\quad H-\overset{\overset{H}{|}}{\underset{\underset{H}{|}}{Al}}-H$$

1.39 What are the molecular formulas for the following compounds?

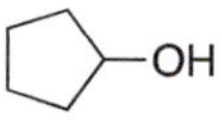

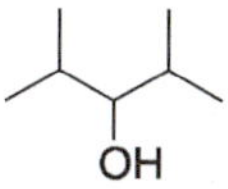

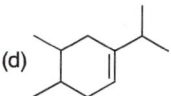

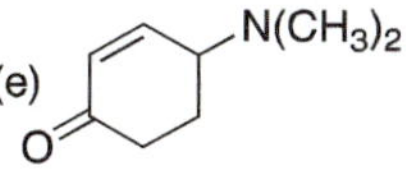

1.40 For the following molecules, give the approximate bond angles about each of the atoms indicated by an arrow and determine the type of hybridized orbitals used in bonding to the other atoms.

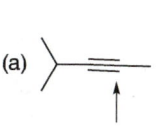

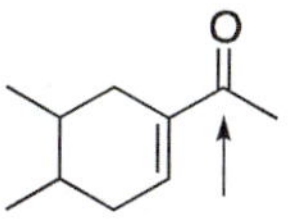

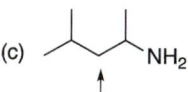

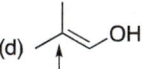

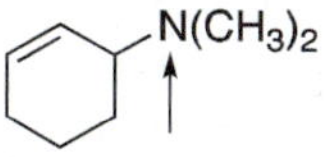

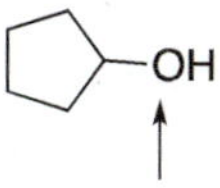

1 Bonding and Structure of Organic Compounds

1.41 Below are the resonance structures of an ion. Determine which is the major resonance contributor and explain your answer.

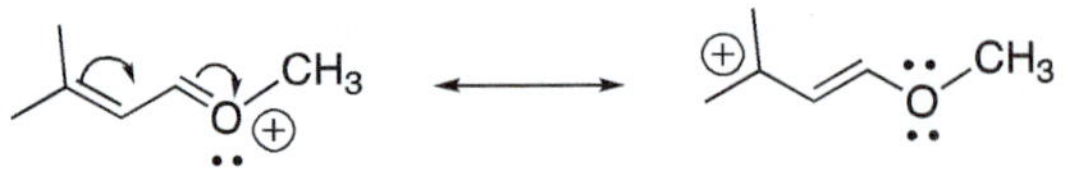

1.42 For the molecules shown below, draw another resonance structure. Use curved arrows to indicate electron movement and include formal charges where appropriate.

(a) 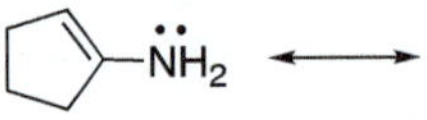 cyclopentenyl–NH₂ ⟷

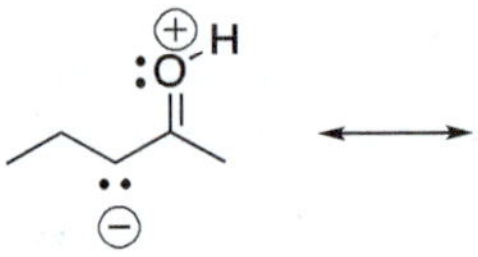

(b) 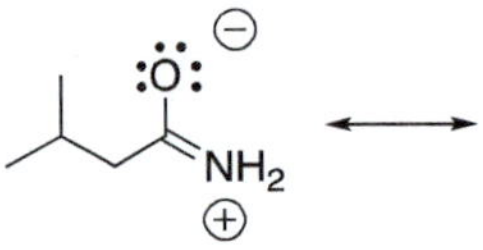 (structure with :Ö:⁻ and NH₂⁺) ⟷

(c) H—N≡N⁺=N:⁻ ⟷

(d) H₂C=O⁻ / H—C—H ⟷

(e) H₂C=CH—C≡N: ⟷

(f) cyclohexyl-C(=O)-NH₂ ⟷

(g) (structure with ⁺OH and O⁻) ⟷

(h) (nitro-substituted benzene with –OCH₃) ⟷
(Delocalize electrons from the OCH₃ group)

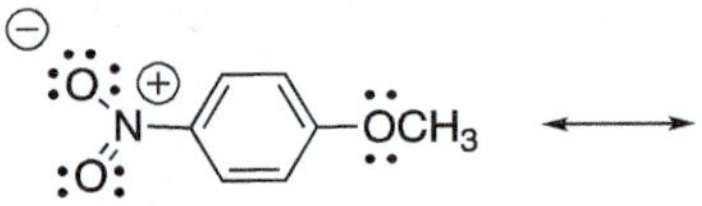

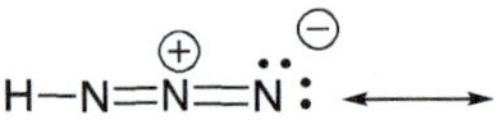

(i)  phenyl-C≡N: ⟷

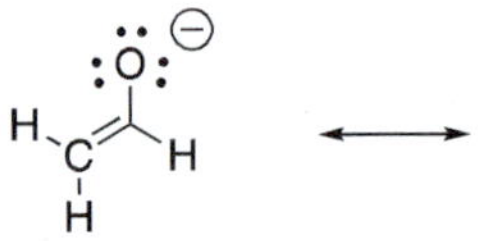

(j) 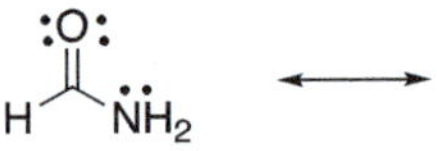 H-C(=O)-NH₂ ⟷

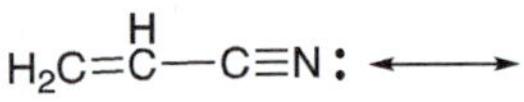

(k) 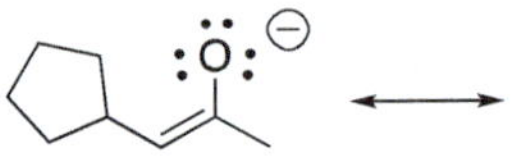 cyclopentyl-CH=C(O⁻)- ⟷

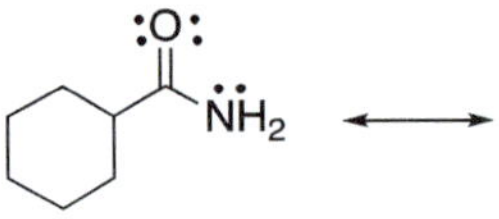

1.43 Using dotted lines to indicate intermolecular interactions, indicate the hydrogen bond formed for each of the following molecules.

a) $CH_3CH_2NH_2$ b) CH_3SH c) HF d) CH_3CO_2H

1.44 Consider the molecular formulas shown below, then answer the following questions.

C_2H_6O CH_2Cl_2 C_2H_7N C_3H_6

a) For each molecule, draw Lewis dot structure(s). [Note: there may be more than one structure for each molecule.]
b) Using the partial charge notation (δ^+ or δ^-), indicate the polarity of each covalent bond for each isomer that you have drawn.
c) Is the molecule that you have drawn capable of forming hydrogen bonds to water?
d) Use dotted lines (...) to indicate the hydrogen bonded structure that would exist between two molecules that you have drawn in question b.

1.45 Even though the molecular weight of methanamine (CH₃NH₂) is approximately the same as that of methanol (CH₃OH), the boiling point of methanol is greater than that of methanamine (65.0 °C vs. −6.3 °C). Briefly explain this observation.

1.46 The molecular weight of CH₃SH (methanethiol) is greater than that of methanol (CH₃OH), yet the boiling point of methanethiol is 6 °C, compared to the boiling point of methanol which is 65.0 °C. Briefly explain this observation.

1.47 Which of the following pairs of molecules has a higher boiling point? Explain your answer.

(a) 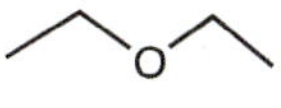 CH₃CH₂—O—CH₂CH₃ and CH₃CH₂CH₂CH₂OH

(b) H₃C—N—CH₃ and CH₃CH₂CH₂NH₂
          |
          CH₃

1.48 For each of the following pairs of compounds, which has a higher boiling point? Explain your answer.

(a)  CH₃CH₂CH₂CH₂CH₂CH₃ and (CH₃)₃CCH(CH₃)₂

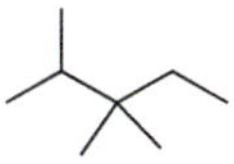

(b)  CH₃CH₂CH₂CH₂CH₃ and (CH₃)₄C

(c)  CH₃CH₂CH₂CH₂CH₃ and (CH₃)₃CCH₂CH₃

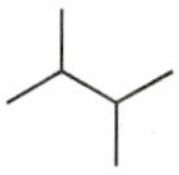

1.49 Which of the following compounds have dipole moments greater than zero?

a) CH_2Cl_2 b) BF_3 c) CF_4 d) H_2O e) NH_3

2

Carbon Functional Groups and Organic Nomenclature

2.1 Introduction

There are practically millions of different organic compounds that exist today and research labs and various industries, especially the pharmaceutical industry, are constantly synthesizing new ones. As scientists, it is very difficult, if not impossible to study all these compounds individually. One approach to better understand the chemistry of the numerous compounds is to classify them into different groups and then study the properties and reactions of the various groups. Organic compounds can be classified into various groups based on specific atomic arrangements that are common within molecules. The groups that are used to classify organic molecules are known as *functional groups*. The physical and chemical properties of compounds that contain a particular functional group tend to be similar. For example, molecules that contain the —OH group are known as alcohols. Rubbing alcohol (isopropanol) is one example of an alcohol and another is ethyl alcohol, which is used as an organic solvent; methanol is another example of an alcohol, which is used as a fuel. Molecules that contain the —NH_2 group are classified as amines. Amino acids are a well-known group of molecules that contain the amine and carboxylic acid functionalities; amino acids are the essential building blocks of protein and life. A wide variety of medicines contain the amine functionality. Morphine and demerol that are commonly used as analgesics are amines, and novocain is an amine that is commonly used as an anesthetic. Many insecticides are amines, and amines are also widely used in the tanning industry and the manufacture of dyes. Ketones and aldehydes are two other functional groups that are found in many compounds that are commonly used; ketones and aldehydes are some of the key components of perfumes and they are commonly used solvents. For example, acetone is the key component in nail polish remover. Shown in Figure 2.1 are well-known molecules in which functional groups have been highlighted.

2.2 Functional Groups

The functional groups shown in Figure 2.1 all have basically the same heteroatoms, but for each functional group, the atomic arrangements are different. A *functional group* can be defined also as a specific arrangement of a particular set of atoms in regions of molecules. In later chapters, we will see that reactions of organic molecules take place at the functional groups. Thus, it is important to be able to recognize functional groups. The heteroatoms that comprise functional groups are mostly oxygen, nitrogen, and halogens, but many contain other heteroatoms, such as phosphorous, sulfur, and silicon. There are few functional groups, however, that contain only carbon atoms that are either sp^2 or sp hybridized; specifically, the carbon–carbon double bond and the carbon–carbon triple bond. Table 2.1 lists the common functional groups that we will encounter in organic chemistry.

Organic Chemistry: Concepts and Applications, First Edition. Allan D. Headley.
© 2020 John Wiley & Sons, Inc. Published 2020 by John Wiley & Sons, Inc.
Companion website: www.wiley.com/go/Headley_OrganicChemistry

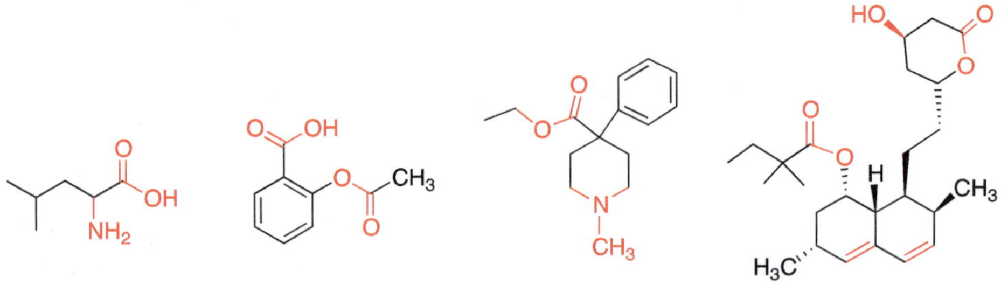

2 Carbon Functional Groups and Organic Nomenclature

Leucine (an amino acid)

Aspirin (pain medication)

Demerol (pain medication)

Zocor (lipid-lowering agent)

Figure 2.1 Examples of common compounds that have different functional groups, functional groups are highlighted in red.

Table 2.1 Common functional groups of organic molecules.

Structural feature	Name of functional group	
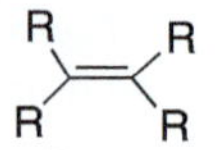 R₂C=CR₂	Alkene	Functional groups that contain only carbon atoms
R−C≡C−R	Alkyne	
 (benzene ring)	Benzene	
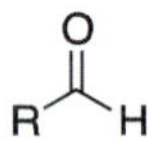 R−C(=O)−H	Aldehyde	Functional groups that contain oxygen atoms
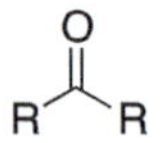 R−C(=O)−R	Ketone	
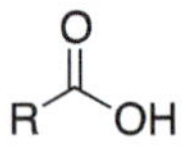 R−C(=O)−OH	Carboxylic acid	
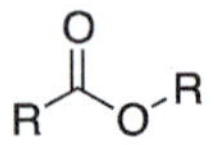 R−C(=O)−O−R	Ester	
R−OH	Alcohol	
R−O−R	Ether	
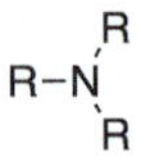 R−NR₂	Amine	Functional groups that contain nitrogen atoms
R−C≡N	Nitrile	
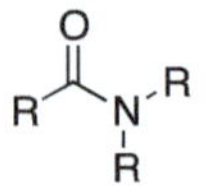 R−C(=O)−NR₂	Amide	Functional group that contains oxygen and nitrogen atoms

R represents a group of atoms, typically alkyl groups, bonded to the functional group.

In Figure 2.2, the functional groups of some of the molecules shown in Figure 2.1 are identified.

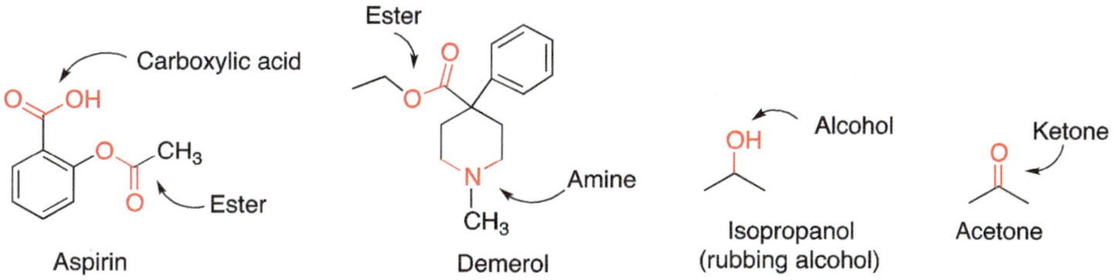

Figure 2.2 Commonly known compounds with functional groups identified.

Problem 2.1

i) Give the name of the functional group in each of the following molecules.

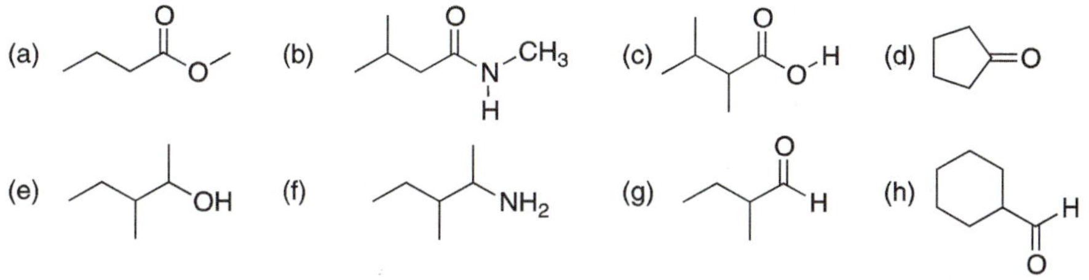

ii) Utilizing structures similar to those shown in part (i) of this question, give structural formulas for different compounds that contain no more than four carbons and contain the following functional groups.

a) alcohol   b) ester   c) aldehyde
d) alkene   e) carboxylic acid   f) ketone

2.3 Saturated Hydrocarbons

Throughout our course of organic chemistry, we will encounter numerous compounds that contain only carbon and hydrogen atoms, these compounds are called **hydrocarbons**, and a study of these compounds will be the emphasis of this chapter. First, we will concentrate on a category of hydrocarbons, called **saturated hydrocarbons or alkanes**, sometimes referred to as aliphatic hydrocarbons, which are compounds that contain only carbon and hydrogen atoms, and the carbon atoms are all sp^3 hybridized and hence have four single bonds to each carbon atom. The term aliphatic comes from the Greek word *aleiphar*, which pertains to fats or oils. Saturated hydrocarbons can be acyclic or cyclic. Acyclic saturated hydrocarbons have the chemical formula C_nH_{2n+2}, where n represents an integer, such as 1, 2, 3, and so on. On the other hand, saturated hydrocarbons that are in the form of a ring, or have a ring as part of the molecule, all have the general formula C_nH_{2n}, where n represents an integer. Figure 2.3 gives examples of different representations of acyclic hydrocarbons. Note that there are other possible compounds with the same molecular formula, but just one example of each is given.

Figure 2.4 gives examples of cyclic hydrocarbons. Note that the general formula that represents cycloalkanes is C_nH_{2n} only if there is one ring in the compound, but if there are additional rings, such as the case of the last compound in Figure 2.4, then the formula is slightly different.

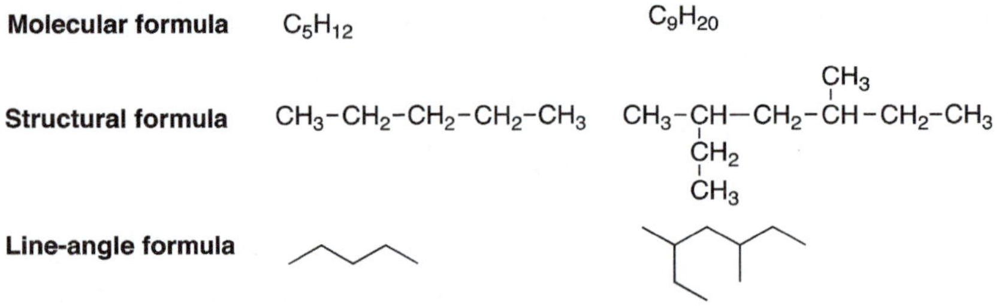

Molecular formula	C$_5$H$_{12}$	C$_9$H$_{20}$
Structural formula	CH$_3$-CH$_2$-CH$_2$-CH$_2$-CH$_3$	CH$_3$-CH(CH$_2$CH$_3$)-CH$_2$-CH(CH$_3$)-CH$_2$-CH$_3$
Line-angle formula	(zigzag)	(branched structure)

Figure 2.3 Examples of acyclic saturated hydrocarbon compounds, also known as alkanes.

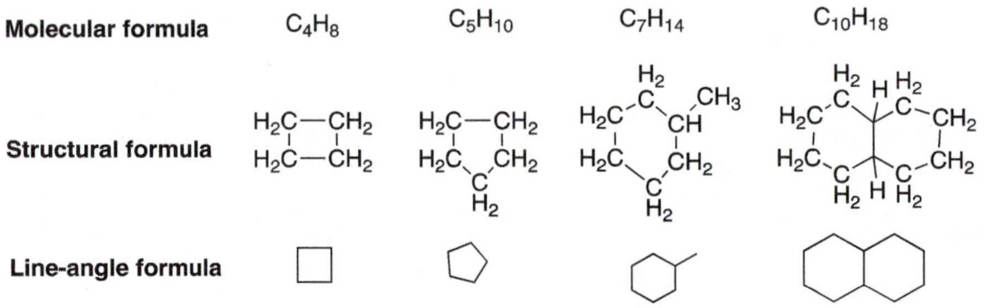

Molecular formula	C$_4$H$_8$	C$_5$H$_{10}$	C$_7$H$_{14}$	C$_{10}$H$_{18}$
Structural formula	(cyclobutane)	(cyclopentane)	(methylcyclohexane)	(decalin)
Line-angle formula	□	⬠	⬡	(fused bicyclic)

Figure 2.4 Examples of cyclic saturated hydrocarbon compounds (cycloalkanes).

If there are two rings, the formula becomes C_nH_{2n-2}. Therefore, it is possible to distinguish between an acyclic alkane and a cyclic and further a bicyclic alkane just based on the formula.

Problem 2.2

Determine which of the following saturated hydrocarbons are cyclic or acyclic. For hydrocarbons that are identified as cyclic, determine the number of rings present.

a) $C_{12}H_{24}$ b) $C_{12}H_{22}$ c) $C_{12}H_{20}$ d) $C_{10}H_{22}$ e) $C_{10}H_{20}$ f) $C_{10}H_{16}$

Alkanes are nonpolar molecules, so they are not miscible in water, but they are miscible in nonpolar, or weakly polar organic solvents. Alkanes are less dense than water; most alkanes have densities of approximately $0.7\,g\,ml^{-1}$, compared to water, which has a density of $1.0\,g\,ml^{-1}$. Thus, a mixture of water and alkanes, such as water and gasoline or oil, readily separates into two phases, with the alkane being the top layer. Alkanes that have four or less carbons (CH_4, C_2H_6, C_3H_8, and C_4H_{10}) are gases at room temperature. However, under low enough temperatures and pressures, these gases can be liquefied and can be stored and transported. These alkanes burn very cleanly and hence are used as fuels for heating. Owing to the possibility of leakage, these gaseous hydrocarbons are not used widely as fuels in vehicles. Alkanes that have five to eight carbons (C_5H_{12}, C_6H_{14}, C_7H_{16}, C_8H_{18}) are liquid at room temperature and are the major components of gasoline. These liquids are very volatile and hence make very effective motor fuel. Other alkanes (C_9 to about C_{16}) are higher boiling liquids and are used mostly as kerosene and diesel fuel. Table 2.2 shows the boiling point ranges of the different categories of alkanes, along with typical uses.

Alkanes are obtained mostly from petroleum, which was first obtained from oil wells around the middle 1800s. The majority of alkanes that are used in industry are derived from petroleum and petroleum by-products. Petroleum, also called crude oil, comes from the remains of prehistoric plants, and it is pumped from wells. The principal components of crude oil are alkanes and

Table 2.2 Various uses and properties of different categories of alkanes.

Number of carbons	Uses	Boiling point range
C_1–C_4	Typically called natural gas and used as fuel for cooking	Below 20 °C (volatile gases)
C_5–C_7	Industrial and research lab solvents	20–100 °C (volatile liquids)
C_7–C_{12}	Gasoline for vehicles and other combustion engines	20–200 °C (volatile liquids)
C_{12}–C_{18}	Kerosene	200–300 °C
C_{12} and larger	Diesel oil and heating homes	200–400 °C
C_{20} and larger	Grease and other lubricants	> 400 °C (nonvolatile liquids)
C_{20} and larger	Asphalt and tar	> 400 °C (nonvolatile solids)

another type of compounds called aromatics (which will be discussed in Chapter 17). Crude oil also contains other compounds, such as sulfur and nitrogen. After the crude oil is pumped from the ground (as shown in Figure 2.5), a process called refining takes place, in which the alkanes are separated into fractions based on their boiling points, also called fractional distillation. The distillation of petroleum gives different fractions based on the number of carbon atoms. Another process called cracking is used to convert some of the larger alkanes to smaller more useful alkanes, such as those of gasoline. Cracking involves heating the alkanes in the presence of a catalyst, such as SiO_2 or Al_2O_3. Natural gas, which is about 70% CH_4 (methane) and 15% C_2H_6 (ethane) depending on the source of the petroleum, is collected and stored from the petroleum production process. Natural gas is typically used to heat buildings and generate electricity.

Figure 2.5 The process of pumping petroleum, also called crude oil, from the ground.

Hydrocarbons are not only used as fuels but they are also found in nature. The simplest alkanes (C_1–C_4) are considered to be asphyxiants (in that they will not support respiration with insufficient oxygen.). Inhalation of higher molecular weight alkanes may cause central nervous system depression, dizziness, and loss of coordination. Also, higher molecular weight alkanes will dissolve fatty substances around the nerve. As a result, extreme care must be exercised in the industrial workplace to minimize the exposure to workers. For some insects, hydrocarbons provide a waterproof layer to prevent desiccation. Some alkanes also serve as pheromones for some insects and in some cases act as recognition cues to protect insects from parasites and other undesired invasions.

> **DID YOU KNOW?**
>
> The saturated hydrocarbon shown below is a sex pheromone used by female tiger moths. 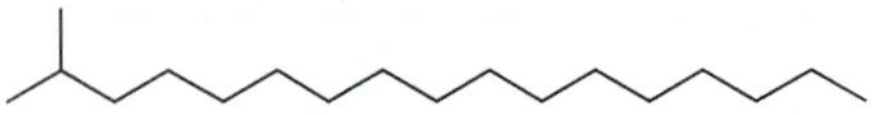
>
> 2-Methylheptadecane

2.3.1 Classification of the Carbons of Saturated Hydrocarbons

Owing to the numerous different possible saturated hydrocarbons, a system of classification has been devised. This classification is based on the number of alkyl groups that are bonded to the carbons. For example, if a carbon in a hydrocarbon is bonded to one alkyl group, that carbon is classified as a primary carbon, or sometimes the symbol 1° is used. Since carbons of saturated hydrocarbon have four single bonds, the classifications are primary (1°), secondary (2°), and tertiary (3°), as illustrated below.

One alkyl group, **R** and three hydrogens	Two alkyl groups, **R** and two hydrogens	Three alkyl groups, **R** and one hydrogen
H–C–H with R above and H below	R–C–H with R above and H below	R–C–H with R above and R below
A primary carbon (1°)	A secondary carbon (2°)	A tertiary carbon (3°)

This method of classification will be used throughout our course to describe different organic compounds that contain these types of carbons.

Problem 2.3

Label the starred carbon atoms in the following compounds as primary (1°), secondary (2°), or tertiary (3°).

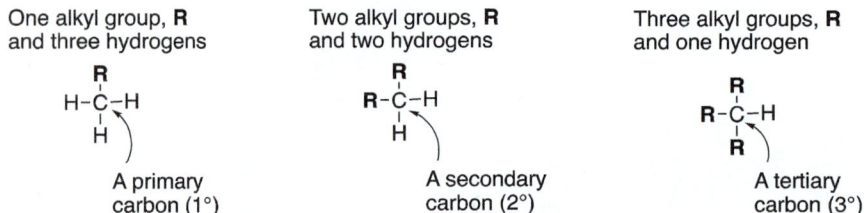

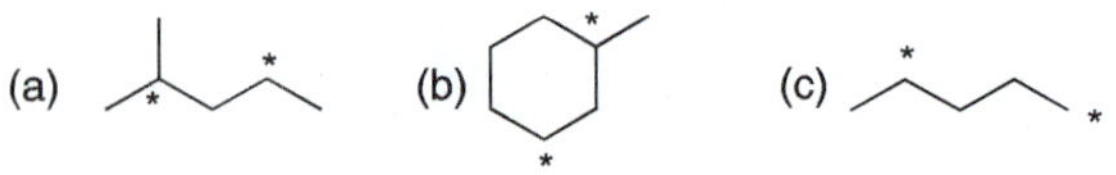

2.4 Organic Nomenclature

Around the middle of the nineteenth century, many compounds were synthesized or isolated from natural sources and there had to be a systematic way of communicating the names and the structures of these compounds within the scientific community. The system of naming chemical compounds used at that time was not based on a scientific method. For example, barbituric acid, which is the parent compound for barbiturate drugs, it is believed that the name was derived from a woman's name, Barbara. Adolph von Baeyer, a German chemist, synthesized a new compound from urea and malonic acid and he did not have a name for this new compound, so he decided to name it after St. Barbara's Day because he was celebrating both the discovery of this new compound and Saint Barbara's birthday that day.

> **DID YOU KNOW?**
>
> Phenobarbital, which is a derivative of barbituric acid, is used to treat insomnia and is also used as a sedative to relieve the symptoms of anxiety or tension, and to control certain types of seizures. It is believed that the parent name of this class of compounds known as barbiturates was derived from a woman's name, Barbara.
>
> Barbituric acid   Phenobarbital

During that period, a lot of trivial names were used, and other chemists could not readily determine the exact structures of the compounds, unless they had prior knowledge of that system of naming compounds. The International Union of Pure and Applied Chemists (IUPAC) decided on a set of rules that should be used by the scientific community for the naming of compounds in general, not only organic compounds but also inorganic compounds as well. As a result, there are specific rules that were established for the naming of compounds.

2.5 Structure and Nomenclature of Alkanes

Alkanes are the first set of organic compounds that will be examined and named using the IUPAC rules.

2.5.1 Nomenclature of Straight Chain Alkanes

Table 2.3 shows the basic vocabulary of the IUPAC nomenclature most commonly encountered in the naming of organic compounds. These names are often referred to as the root or parent names of organic compounds.

In naming organic compounds, the IUPAC name of the molecule is based of the number of carbons that are connected in a continuous chain. Of course, not all hydrocarbons have simple structures as those shown in Table 2.3.

2 Carbon Functional Groups and Organic Nomenclature

Table 2.3 The IUPAC names of straight chain hydrocarbons.

Number of carbons	Structure	Name
1	CH_4	Methane
2	CH_3CH_3	Ethane
3	$CH_3CH_2CH_3$	Propane
4	$CH_3(CH_2)_2CH_3$	Butane
5	$CH_3(CH_2)_3CH_3$	Pentane
6	$CH_3(CH_2)_4CH_3$	Hexane
7	$CH_3(CH_2)_5CH_3$	Heptane
8	$CH_3(CH_2)_6CH_3$	Octane
9	$CH_3(CH_2)_7CH_3$	Nonane
10	$CH_3(CH_2)_8CH_3$	Decane

2.5.2 Nomenclature of Branched Alkanes

Most organic compounds encountered in organic chemistry have branches attached to a main continuous chain and they must have unique names that reflect their structures. For example, isooctane is a highly branched alkane, which is used in the petroleum industry as a fuel additive. The structure of this compound is shown below, but isooctane is not the IUPAC name, but a common name.

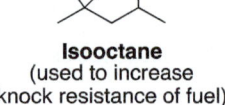

Isooctane
(used to increase knock resistance of fuel)

In naming organic molecules with branches, the following rules must be followed. The application of these rules to determine the IUPAC name of a compound is shown below.

$$CH_3-CH_2-\underset{\underset{CH_3}{|}}{CH}-CH_2-CH_2-CH_3$$

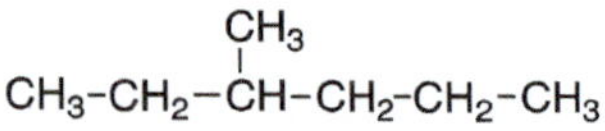

Rule #1 Find the longest continuous chain. Note that the longest continuous chain may not always be the horizontal chain of carbon atoms.

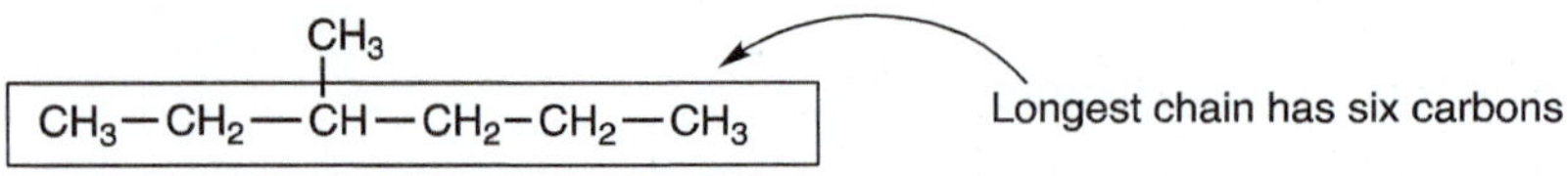

$$\boxed{CH_3-CH_2-\underset{\underset{CH_3}{|}}{CH}-CH_2-CH_2-CH_3}$$ Longest chain has six carbons

Rule #2 Based on the number of carbons in the longest chain, use Table 2.3 to determine the parent or root name. Based on the number of carbons that are present in the compound, a root or parent name is assigned such as shown below.

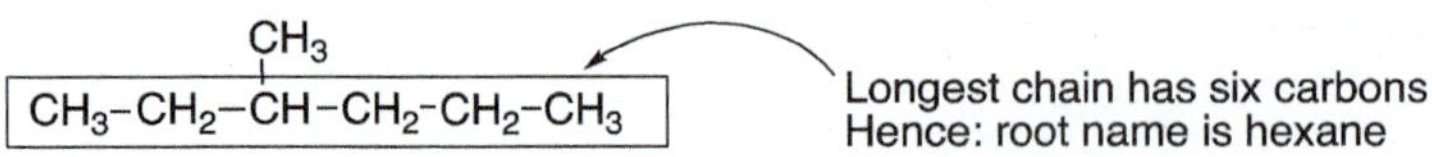

$$\boxed{CH_3-CH_2-\underset{\underset{CH_3}{|}}{CH}-CH_2-CH_2-CH_3}$$ Longest chain has six carbons
Hence: root name is hexane

Rule #3 In order to determine the positions of branches, *assign numbers to each carbon of that longest continuous chain*. Start the assignment so that the end closest to a branch gets the number 1, as shown below.

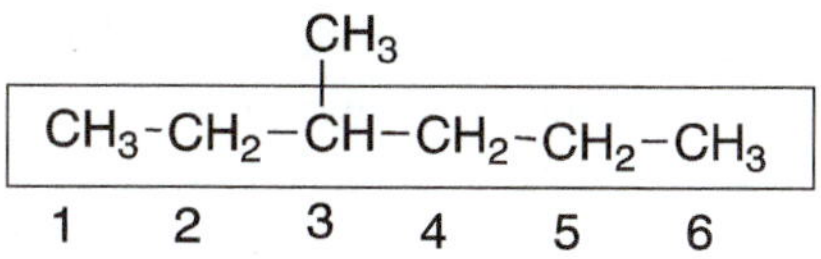

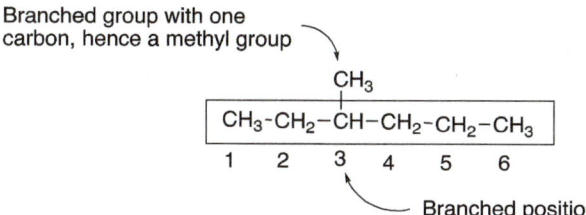

Rule #4 Identify the branched position by the number assigned in rule 3.

Rule #5 Name the branch based on the number of carbons of that branch. Determine the number of carbons present in the branch and use the names in Table 2.3 to derive a name. Derive the name of the branch by converting the corresponding alkane from Table 2.3 to alkyl. That is, if one carbon is present, methane becomes methyl; if two carbons are present, ethyl is used, etc.

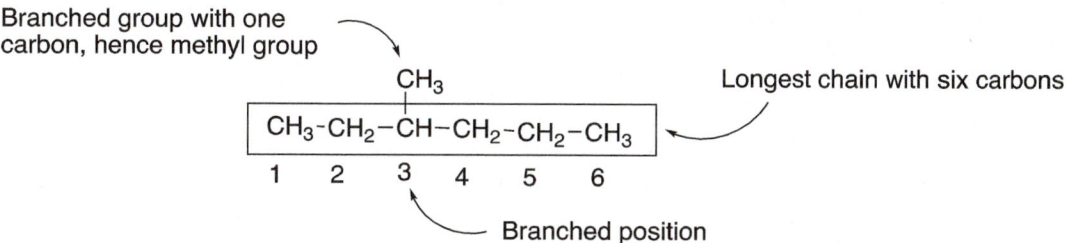

Rule #6 Write the complete IUPAC name so that the root name is at the end, the name of the branch is next to the root name, and the number that indicates the position of the branch is next to the name of the branch and separated by a hyphen. Remember that the parent or root name is at the end of the IUPAC name as shown in the example below.

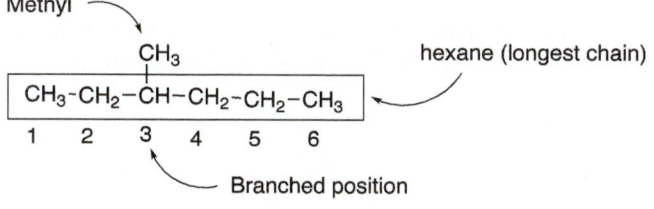

IUPAC name: 3-Methylhexane

Note that the number is separated from a letter by a hyphen.

Problem 2.4

i) Give the IUPAC names for the following compounds.

(a) 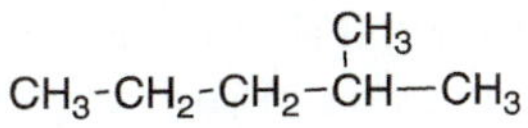 $CH_3-CH_2-CH_2-\underset{\underset{CH_3}{|}}{CH}-CH_3$ (b) 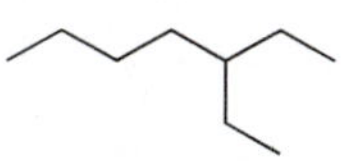 (c)

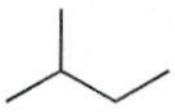

ii) Give the line-angle formula for the following compounds.

   a) 3-Methylhexane   b) 3-Ethylhexane   c) 2-Methylheptane

Rule #7 If there are two or more of the same branches that are present in a molecule, the prefix di, tri, tetra, etc., along with the corresponding numbers are used as shown in the example below.

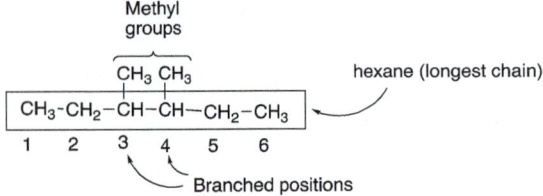

IUPAC name: 3,4-Dimethylhexane

Note that a comma is used to separate numbers, and a hyphen is used to separate a number from a letter. If there is a choice, start at the end with the most branches so that numbers used in the prefix are the lowest combination, as shown in the example below.

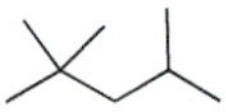

IUPAC name: 2,2,4-Trimethylpentane
and not 2,4,4-Trimethylpentane

Problem 2.5

i) Give the IUPAC names for the following compounds.

(a) 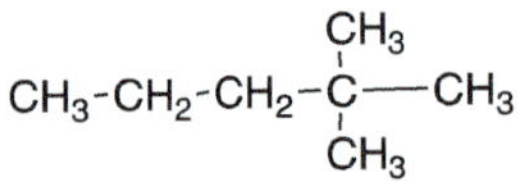 $CH_3-CH_2-CH_2-\underset{\underset{CH_3}{|}}{\overset{\overset{CH_3}{|}}{C}}-CH_3$ (b) 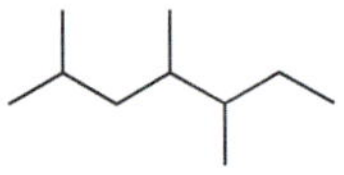   (c) 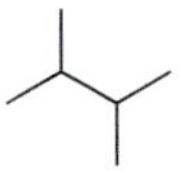   (d)

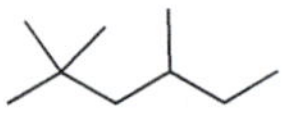

ii) Give the line-angle formula for the following compounds.

   a) 2,3-Dimethylhexane   b) 3,3-Dimethyloctane   c) 2,2,4-Trimethylheptane

If there are different types of alkyl branches, identify each branch by their location on the longest chain and list the branches in alphabetical order as part of the name as shown below.

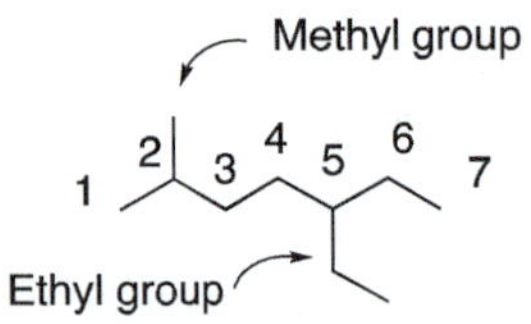

IUPAC name: 5-Ethyl-2-methylheptane
Incorrect IUPAC name: 2-Methyl-5-ethylpentane

Note that the IUPAC name is 5-ethyl-2-methylheptane and not 2-methyl-5-ethylheptane. The groups bonded to the main longest chain must be listed in alphabetical order. As a result, ethyl, which starts with an "e" comes before methyl, which starts with "m". It should be pointed out that the prefixes, di, tri, tetra, etc., that are used to describe the number of similar groups do not play a role in the alphabetical ordering of the names. That is, if there is a dimethyl in the name, the "m" of the methyl and not the "d" of the dimethyl is used to determine the alphabetical order. Similarly, for trimethyl and tetramethyl, the methyl will dictate which comes first in the nomenclature.

Correct IUPAC name: 4-Ethyl-2,2-dimethylhexane
Incorrect IUPAC name: 2,2-Dimethyl-4-ethylhexane

Problem 2.6

i) Give the IUPAC names for the following compounds.

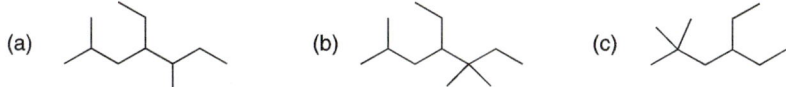

ii) Give the line-angle formula for the following compounds.

a) 3-Ethyl-2-methylheptane b) 4-Ethyl-2,2,3-trimethyloctane

2.5.3 Nomenclature of Compounds that Contain Heteroatoms

As pointed out earlier, some organic compounds have heteroatoms; that is, some organic compounds have carbons, hydrogens, and other types of atoms, such as oxygen, nitrogen, or halogens present. In naming such compounds using the IUPAC naming system, rules 1 through 7 apply, except the heteroatomic substituent is named as shown in Table 2.4.

For example, if a compound has a chlorine atom bonded to the longest chain, *chloro* is used in the IUPAC name of such a compound, as shown in the example below.

$$CH_3 - \underset{1}{CH} - \underset{2}{\underset{|}{CH}} - \underset{3}{\underset{Cl}{CH}} - \underset{4}{CH_2} - \underset{5}{CH_2} - \underset{6}{CH_3}$$

2,3-Dichlorohexane

Table 2.4 IUPAC names for selected heteroatomic groups.

Group	Name
Cl	Chloro
Br	Bromo
I	Iodo
F	Fluoro
NO$_2$	Nitro

For compounds that have different substituents, including heteroatomic and alkyl substituents, the groups are listed in an alphabetical order as shown in the compound below.

```
    Cl   CH₃
    |    |
CH₃-CH—CH—CH₂-CH₂-CH₃
 1   2   3   4   5   6
   2-Chloro-3-methylhexane
```

The longest chain contains six carbons and hence the root name is hexane; there are two different types of substituents, an alkyl substituent and a chloro substituent. Since chloro starts with the letter "c" and comes before the letter "m" for the methyl group in the alphabet, the chloro group is listed first for the IUPAC name. Note that numbers are separated from letters by a hyphen.

Problem 2.7

i) Give the IUPAC names for the following compounds.

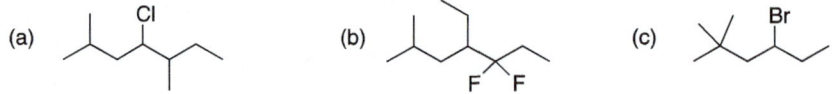

ii) Give the line-angle formulas for the following compounds.

    a) 3-Chloro-2,2-dimethylpentane   b) 2,2-Dibromo-4,4-dimethylheptane

2.5.4 Common Names of Alkanes

Throughout this course, common names of some organic compounds are used and examples are shown in Figure 2.6. In this text, however, IUPAC names will be used routinely.

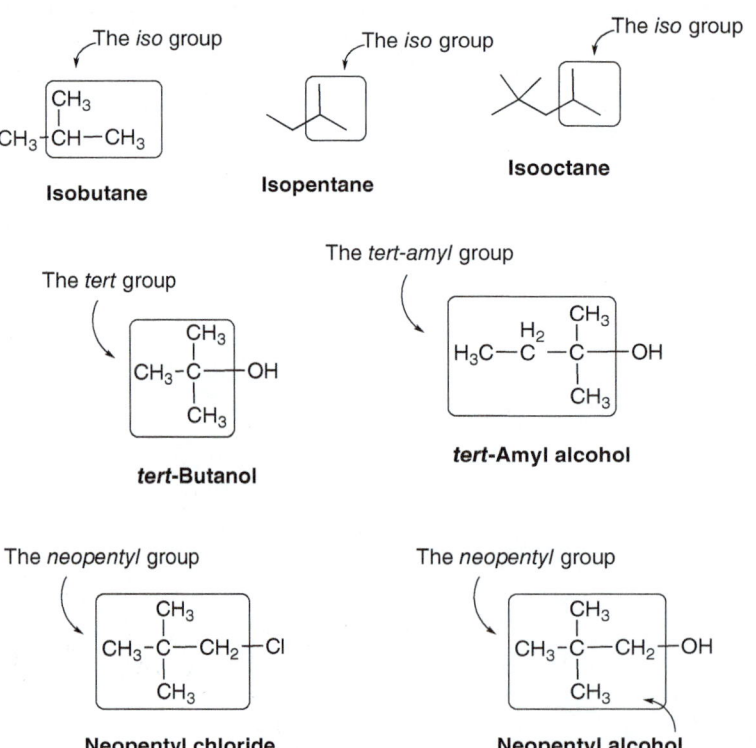

Figure 2.6 Examples of using common names for some common compounds.

2.5.5 Nomenclature of Cyclic Alkanes

For cyclic saturated hydrocarbons, Table 2.5 shows the names of the most common cyclic compounds that will be encountered throughout this course. As is obvious, there are no cyclic compounds for compounds that contain one or two carbons.

It is obvious that the prefix **cyclo** is added to the names of the straight chain alkane given in Table 2.3 to determine the names of cycloalkanes. Thus, if the cyclic molecule has six carbons, its root name is cyclohexane.

Problem 2.8

Give the molecular formulas for (a) cyclononane and (b) cyclodecane.

2.5.6 Nomenclature of Branched Cyclic Alkanes

Most cycloalkanes that will be encountered throughout this course contain the basic structural features as shown in Table 2.5, but they will have branches. In naming cyclic compounds, the root or parent name is that of the cyclic structure shown in Table 2.5, but the numbering system is a little different from that of a straight chain, the position that gets the number 1 is the carbon that has a branch.

Rule #8 In naming cyclic compounds, start numbering at the carbon that has a branch and continue in the direction that has the most branches. Below is an example of a compound with two numbering possibilities.

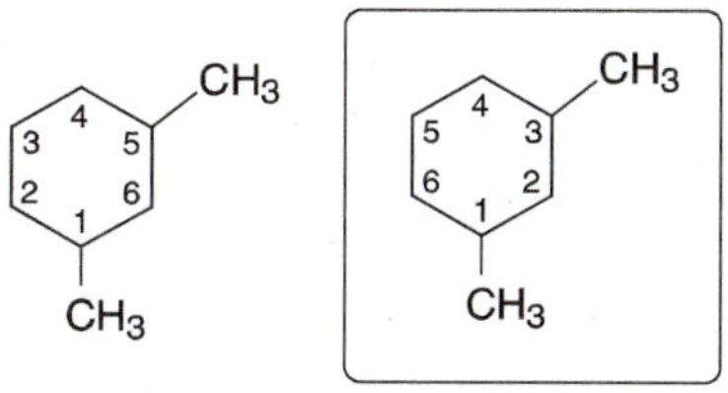

Table 2.5 The IUPAC names of cyclic hydrocarbons.

Number of carbons	Structure	Name
3	 △	Cyclopropane
4	□	Cyclobutane
5	 ⬠	Cyclopentane
6	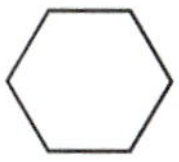 ⬡	Cyclohexane
7	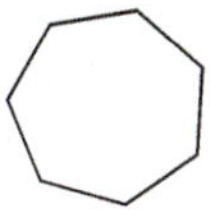	Cycloheptane
8	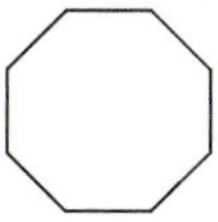	Cyclooctane

Note that the numbering system used for the molecule on the right gives the branches on the ring the numbers 1 and 3, compared to the numbering system for the molecule on the left, which gives the numbers 1 and 5. For IUPAC names of cyclic molecules with branches, the lower number combination is always used. Thus, the correct IUPAC name of the molecule shown above is 1,3-dimethylcyclohexane.

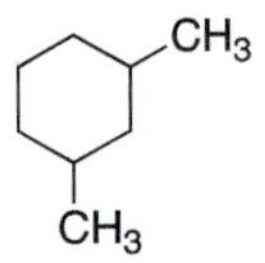

Correct IUPAC name: 1,3-Dimethylcyclohexane
Incorrect IUPAC name: 1,5-Dimethylcyclohexane

Rules 2 through 7 also apply in the naming of cyclic compounds that have more than two branches or heteroatomic substituents.

Problem 2.9

i) Give the IUPAC names for the compounds shown below.

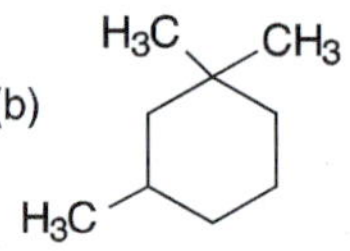

ii) Give the structures of the compounds shown below.

   a) 1,1-Dichloro-2-methylcyclohexane   b) 1-Chloro-3-methylcyclopentane

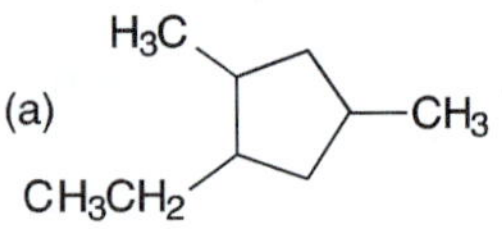

2.5.7 Nomenclature of Bicyclic Compounds

Some alkanes have two rings that are fused at different points. Compounds that have the rings bonded at the same carbon are the simplest and are called spiro compounds, spiranes, or spiratanes, which is derived from a Latin term, which means twisted or coiled. Examples of spiranes are shown below.

Examples of spiro compounds (spiratanes)

This next section concentrates on compounds in which two rings are bonded at different carbons and not at the same carbon as discussed for spiranes. Such compounds are called bicyclic compounds and examples are shown below.

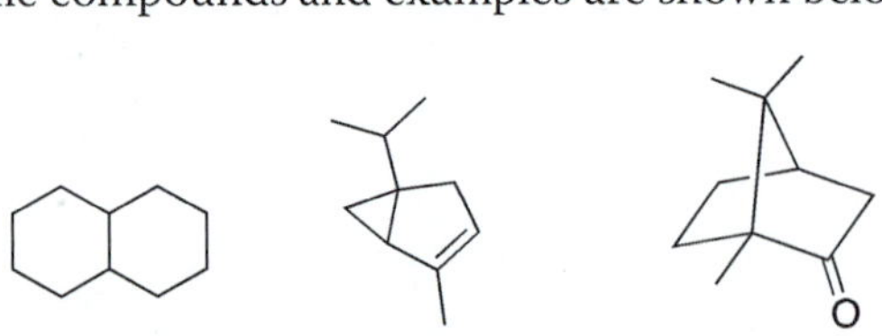

Decalin   α-Thujene   Camphor
Specific examples of bicyclic compounds

Decalin is a bicyclic compound, which is used as an industrial solvent in the synthesis of resin and also used as fuel additive, and this type of bicyclic framework is found in many important organic compounds. Another type of bicyclic compound frequently found in many natural products is shown in the structure of α-thujene, which is found in nature and is an essential oil of some plants. Camphor contains another type of bicyclic system, and it is frequently found in many natural products. Camphor has a distinct odor and is found in the wood of the camphor laurel and kapur trees, which are found primarily in different regions of Asia.

In naming bicyclic compounds, the total number of carbons must first be determined, which will become the root name of the compound. Next, the points where the rings are bonded must be identified (also known as the bridgehead carbons), as shown below.

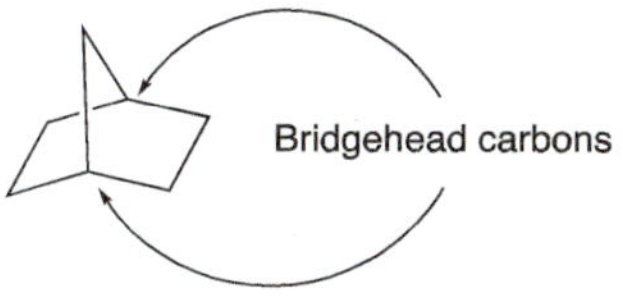

Seven-carbon bicyclic compound 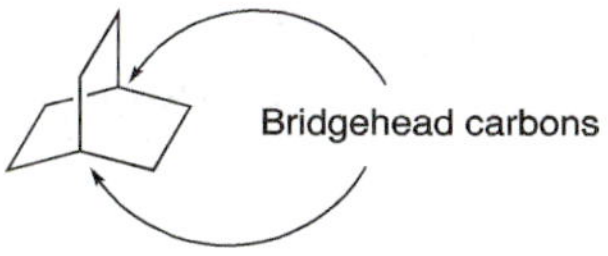   Eight-carbon bicyclic compound

Problem 2.10

For the compounds shown below that are used in the treatment of different forms of cancer, circle the carbons of bicyclic systems.

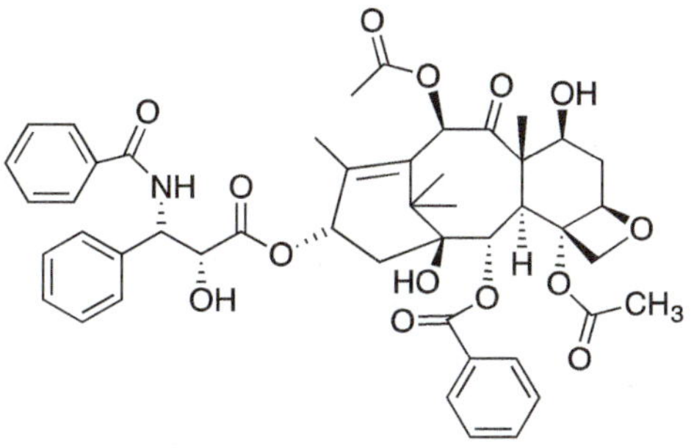

Taxol 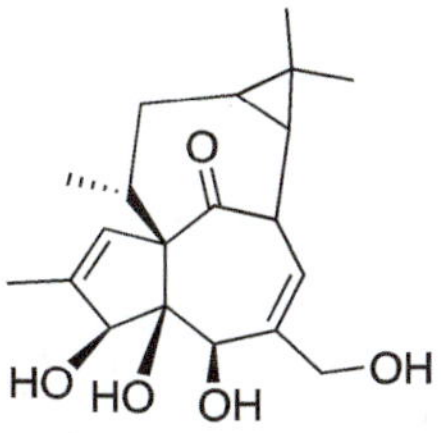   Ingenol

Once the points of fusion are identified, number the carbons starting at a point of fusion as shown below. Note that number 1 is a bridgehead carbon, and the numbering continues in the direction so that an entire ring is numbered. Then, include the carbons of the other ring, carbons 7 and 8. Also, note that only carbons 1 and 4 are the bridgehead carbons, and carbons 2 and 3; 5 and 6; and 7 and 8 are not.

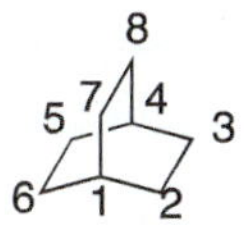

For the IUPAC name of the above molecule, the root is octane since there are eight carbons in the molecule. Next, the numbers of carbons that are not bridgehead carbons, but bonded to the bridgehead carbons, are indicated as shown in the name of the molecules below.

2 Carbon Functional Groups and Organic Nomenclature

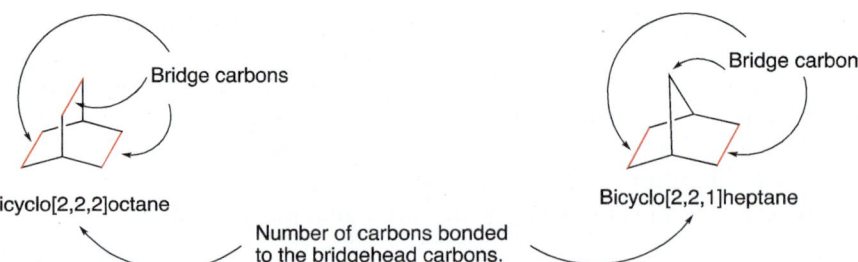

Bicyclo[2,2,2]octane

Bicyclo[2,2,1]heptane

Number of carbons bonded to the bridgehead carbons.

Problem 2.11

Give the IUPAC names for the bicyclic molecules shown below.

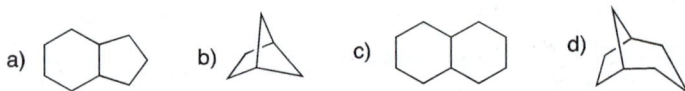

If there are substituents present, the molecule must be numbered so that the lowest number combination, which describes the positions of the substituents, is achieved. The starting point in numbering a bicyclic compound is always at a point of a bridgehead carbon. Shown below are some examples.

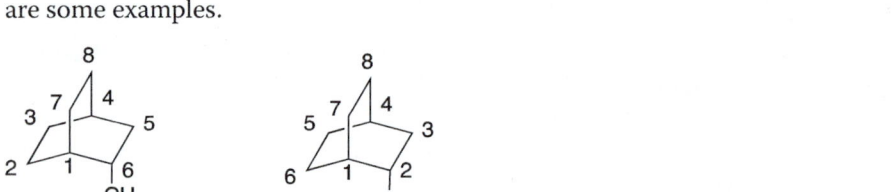

Incorrect numbering Correct numbering

IUPAC name: 2-Methylbicyclo[2,2,2]octane

Problem 2.12

Give the IUPAC names for the bicyclic molecules shown below.

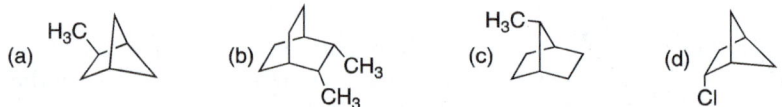

2.6 Unsaturated Hydrocarbons

In this section, compounds that contain at least two sp^2 or sp carbon atoms will be examined. These compounds are also known as unsaturated hydrocarbon compounds and contain the alkene and alkyne functional groups. Examples are shown below.

Ethylene Benzene Acetylene

Examples of unsaturated hydrocarbon compounds

One of the most popular alkenes is ethylene, which is produced naturally in the ripening of fruits. Ethylene is a sweet-smelling gas, which is an asphyxiant, anesthetic to animals, and phytotoxic to plants. Compounds that contain the alkene functionality are typically synthesized from petroleum products; fossil products, which were pointed out earlier, are mostly hydrocarbons. Saturated hydrocarbons under catalytic conditions and appropriate temperatures can be converted into alkenes.

Benzene is one compound of a very large category of compounds, which are described as aromatic compounds and will be studied in more details in Chapter 17. Benzene was isolated from compressed illuminating gas in 1825 by Michael Faraday of the Royal Institution. In 1834, Eilhardt Mitscherlich of the University of Berlin synthesized benzene by heating benzoic acid with calcium oxide, and he also showed that the molecular formula of benzene is C_6H_6. Today, benzene is obtained primarily from the distillation of petroleum, and it is used as the starting material for many useful chemicals in our society. Benzene is also a solvent that is sometimes used in chemical labs, both industrial and academic research labs.

Acetylene is the simplest alkyne. It has just one carbon–carbon triple bond, two carbons, and two hydrogens. It is easily made by heating lime and coke to produce calcium carbide, and the addition of water to calcium carbide yields acetylene as shown below in Reactions (2-1) and (2-2).

$$\text{Coke + Lime} \xrightarrow{2500\,°C} \text{Calcium carbide (CaC}_2\text{) + Carbon monoxide (CO)} \quad (2\text{-}1)$$

$$\text{CaC}_2 + 2H_2O \longrightarrow \underset{\text{Acetylene}}{H-C\equiv C-H} + \underset{\text{Calcium hydroxide}}{Ca(OH)_2} \quad (2\text{-}2)$$

Acetylene is one of the most useful of the alkyne family of compounds. It is used in combination with pure oxygen to produce a very high temperature when it burns; as a result, it is used as the primary fuel for oxyacetylene welding torches in industry to cut and weld steel. When acetylene is burned in the presence of oxygen, a very hot blue flame is produced; temperatures as high as 3000 °C can be generated making it an ideal fuel for this purpose. Propyne, which is also known as methylacetylene, is sometimes used as a substitute for acetylene as fuel for welding torches and even has been tested as a possible rocket fuel. Another major use of alkynes is that they are often used as starting materials in the synthesis of polymers. Even though there is a triple bond in alkynes, it is a nonpolar bond. As a result, alkynes are nonpolar molecules and are immiscible in polar solvents, such as water. They have relatively low boiling points, but the boiling points of alkynes with high molecular weights can be relatively high.

Since the carbons of the alkyne functionality are sp hybridized, which means that they contain 50% s character, these carbons are relatively electronegative, compared to carbons that are sp^2 (33% s character) and sp^3 (25% s character). Remember that the s orbital is closer to the nucleus, compared to the more diffused p orbitals. As a result, hydrogens that are bonded to alkynes are relatively acidic, compared to alkenes and alkanes and can be removed by a strong enough base as shown in the Reaction (2-3).

$$\underset{\text{Acidic hydrogen}}{R-C\equiv C-H} + \underset{\text{Strong base}}{NaNH_2} \longrightarrow \underset{\text{Alkyne salt}}{R-C\equiv C{:}^-\ Na^+} + NH_3 \quad (2\text{-}3)$$

As we will see in later chapters, the salts of alkynes are useful synthetic intermediate of larger useful compounds.

2.7 Structure and Nomenclature of Alkenes

2.7.1 Nomenclature of Branched Alkenes

For the nomenclature of alkenes, there are only minor changes to rules 1 through 3 that were used for the nomenclature of alkanes.

Rule #1 (in the naming of alkenes): Determine the longest continuous chain that contains the alkene functionality. Note that there may be other chains that are longer than the one that has the carbon–carbon double bond, but the longest chain that contains the carbon–carbon double bond is used to derive the names of alkenes.

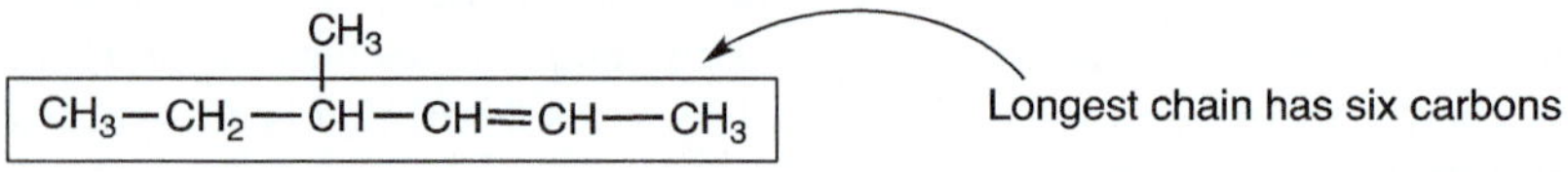

$$CH_3-CH_2-\underset{\underset{CH_3}{|}}{CH}-CH=CH-CH_3 \qquad \text{Longest chain has six carbons}$$

Rule #2 Assign numbers to each carbon of this longest continuous chain that contains the double bond. Start the assignment so that the end closest to the double bond gets number 1.

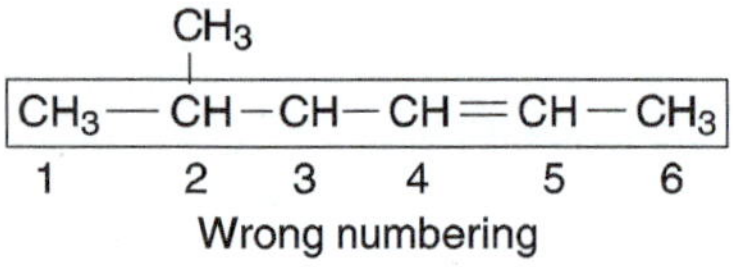

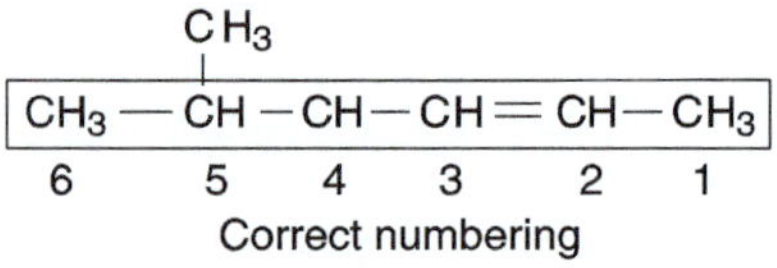

$$\underset{\underset{\text{Wrong numbering}}{1\ \ \ 2\ \ \ 3\ \ \ 4\ \ \ 5\ \ \ 6}}{CH_3-\underset{\underset{CH_3}{|}}{CH}-CH-CH=CH-CH_3} \qquad \underset{\underset{\text{Correct numbering}}{6\ \ \ 5\ \ \ 4\ \ \ 3\ \ \ 2\ \ \ 1}}{CH_3-\underset{\underset{CH_3}{|}}{CH}-CH-CH=CH-CH_3}$$

Rule #3 Use Table 2.3 to determine the name of the parent or root name and specify the position of the double bond with a number. Based on the numbering of carbons present in the compound, a root or parent name is derived by changing the **-ane** of the corresponding alkane to **-ene**. Thus, a compound that has six carbons and contains a carbon–carbon double bond is a hexene, and a number is used to indicate the position of the double bond. The rest of the rules are the same as those discussed for the naming of alkanes. The IUPAC name for an alkene is shown below.

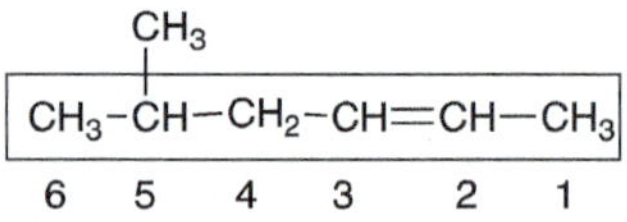

$$\underset{6\ \ \ 5\ \ \ 4\ \ \ 3\ \ \ 2\ \ \ 1}{CH_3-\underset{\underset{CH_3}{|}}{CH}-CH_2-CH=CH-CH_3}$$

5-Methyl-2-hexene (not 2-methyl-4-hexene)

For alkenes that have the carbon–carbon double bond in position 1, the use of 1 is optional.

$CH_3CH_2CH_2CH=CH_2$

1-Pentene (or Pentene)

An isomer of pentene, however, that has the double bond in a position other than number 1, the position of the double bond must be specified so that the position of the double bond gets the lowest possible number.

$CH_3CH=CHCH_2CH_3$

2-Pentene (not 3-Pentene)

Problem 2.13

i) Give the line-angle structures of the following molecules

   a) 2-Methyl-2-pentene
   b) 2,5-Dimethyl-3-hexene
   c) 2,3-Dimethyl-2-butene
   d) 3,4,4-Trimethyl-1-pentene

ii) Give IUPAC names for the following molecules.

(a) 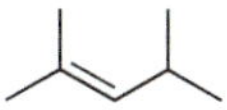   (b) 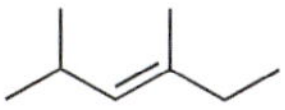   (c) 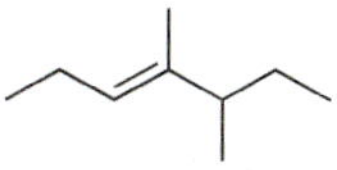   (d)

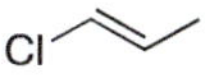

2.7.2 Nomenclature of Polyenes

Compounds that have two double bonds are named as **dienes**, and the position of each double bond must be specified with appropriate numbers as shown in the example below.

$$CH_2 = CHCH_2CH = CHCH_3$$
1,4-hexadiene (not 2,5-hexadiene)

A great majority of organic compounds found in nature have alternating double bonds and such compounds are referred to as having conjugated double bonds. Isoprene is one of the simplest conjugated systems, but conjugated systems that have extended conjugation are typically highly colored compounds. Examples of conjugate polyenes, along with their IUPAC names, are shown below.

2-Methyl-1,3-butadiene (isoprene)   1,3-Butadiene   1,3-Cyclohexadiene   1,3-Cyclopentadiene

2,3-Dimethyl-1,3-butadiene   1,3,5-Hexatriene   2,4-Dimethyl-1,3,5-hexatriene

As shown in the examples above, numbers are used to describe the locations of the double bonds of polyenes.

Problem 2.14

i) Give the structures of the following molecules.

   a) 2,4-Dimethyl-1,4-pentadiene
   b) 4-Chloro-2,6-dimethyl-1,3,5-heptatriene

ii) Give the IUPAC names for the following molecules.

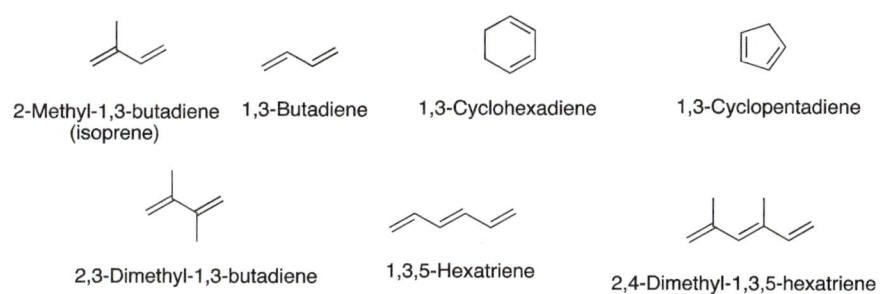

(a) 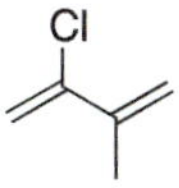   (b)

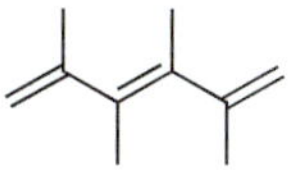

2.7.3 Nomenclature of Cyclic Alkenes

Cyclic compounds that have a double bond are named as a **cycloalkenes**. In naming cycloalkenes, one of the carbons of the double bond is assigned position 1 and the other carbon of the double bond is assigned position 2. In other words, number in the direction of the double bond as shown in the example below.

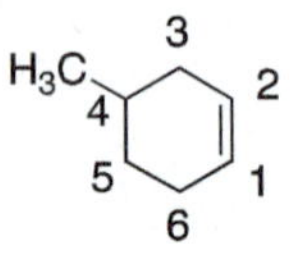

Correct: must number is the direction of the double bond

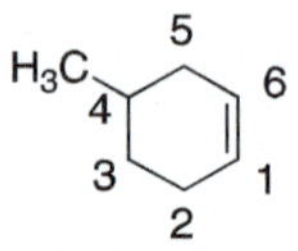

Incorrect: must number is the direction of the double bond

For cycloalkenes that have branches, number the cycloalkene in the direction of the double bond and also in the direction so that the branch gets the lowest possible number as shown in the example below.

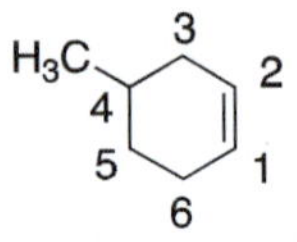

Correct: must number is the direction of the double bond and towards the closest branch

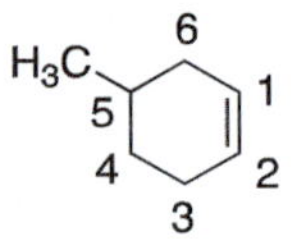

Incorrect: must number is the direction of the double bond and towards the closest branch

The correct IUPAC name for the compound above is:

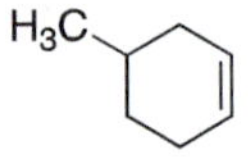

IUPAC name: 4-Methyl-1-cyclohexene
(or 4-Methylcyclohexene)

The number 1 for the double bond is optional, and often not used, in naming organic compounds.

Problem 2.15

i) Give the IUPAC names for the following cyclic alkenes.

(a) 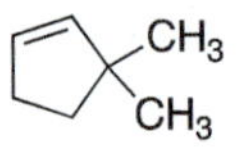   (b)

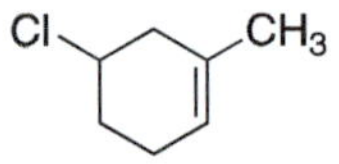

ii) Give line-angle structures for the following cyclic alkenes.

a) 1,2-Dimethylcyclohexene
b) 2-Bromo-3-methylcyclopentene

2.8 Structure and Nomenclature of Substituted Benzenes

Benzene, which has the chemical formula C_6H_6, is a unique compound and Chapter 17 is dedicated to the study of benzene. Figure 2.7 shows different representations of the structure of benzene.

2.8 Structure and Nomenclature of Substituted Benzenes

Figure 2.7. Representations of benzene.

Substituted benzene results from the substitution of a hydrogen for a group that we have discussed earlier and examples are shown below.

Methylbenzene   Nitrobenzene   Chlorobenzene

2.8.1 Nomenclature of Disubstituted Benzenes

If there are two substituents bonded to the benzene ring, the exact relationship between the substituents on the ring must be specified. The numbering system is similar to that of cyclohexane in that the name should have the lowest possible number combination, as shown in the example below.

1,3-Dibromobenzene   1-Bromo-3-nitrobenzene   1-Methyl-2-nitrobenzene

Note that when there are more than one substituents on the benzene ring, they are listed in an alphabetical order.

Problem 2.16

i) Give the structures of each of the following compounds.

  a) 1,2-Dibromobenzene
  b) 3-Chlorophenol
  c) 1-Fluoro-4-nitrobenzene.

ii) Give the names for each of the following compounds.

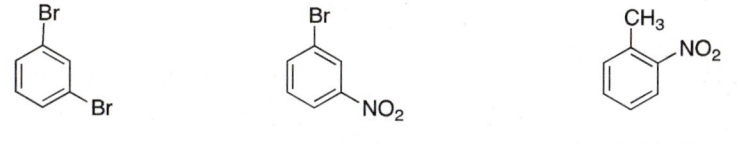

If benzene is a substituent, it is named as the phenyl group (—C$_6$H$_5$) in the nomenclature of the molecule, as shown in the example below.

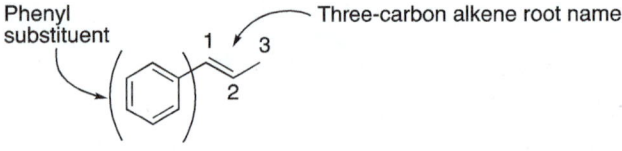

IUPAC name: 1-Phenyl-1-propene

2.9 Structure and Nomenclature of Alkynes

For the nomenclature of alkynes, there is only a minor change in rules that were discussed in Section 2.7 for the nomenclature of alkenes.

Rule #1 Find the longest continuous chain that contains the alkyne functionality.

$$CH_3-CH_2-CH_2-C\equiv C-CH_3 \quad \text{Longest chain has six carbons}$$

Rule #2 Assign numbers to each carbon of this longest continuous chain that contains the triple bond. Start the assignment so that the end closest to the triple bond has number 1.

$$\underset{6\ \ \ \ 5\ \ \ \ \ 4\ \ \ \ 3\ \ \ 2\ \ \ \ 1}{CH_3-CH_2-CH_2-C\equiv C-CH_3} \qquad \underset{1\ \ \ \ 2\ \ \ \ \ 3\ \ \ \ 4\ \ \ 5\ \ \ \ 6}{CH_3-CH_2-CH_2-C\equiv C-CH_3}$$

Correct   Incorrect

Rule #3 Use Table 2.3 to determine the name of the parent or root name and specify the position of the triple bond with a number. Based on the number of carbons that are present in the compound, a root or parent name is assigned by changing the **-ane** of the corresponding alkane to **-yne**. Thus, a compound that has six carbons and has a carbon–carbon triple bond is named as a hexyne, and the position of the triple bond is specified with a number as illustrated in the name of the molecule below.

5-Methyl-2-hexyne (not 2-Methyl-4-hexyne)

The rest of the rules are the same as those discussed for the naming of alkanes and alkenes. For compounds that have two triple bonds, the name of the compound is a **diyne**, and numbers are used to indicate the positions of the triple bonds as shown in the example below.

H—≡—CH$_2$—≡—CH$_3$
1,4-Hexadiyne

If there are other functional groups present, such as shown in the molecule below, which has both an alkene and an alkyne functionality identify the longest chain that contains both functionalities.

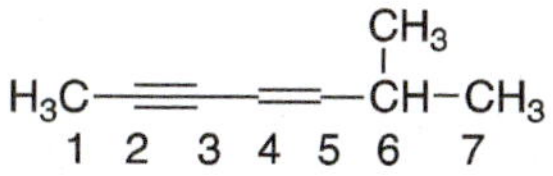

$$H_3C-\!\!\equiv\!\!=\!\!-\overset{\overset{\displaystyle CH_3}{|}}{CH}-CH_3$$
$$1\ 2\ 3\ 4\ 5\ 6\ \ 7$$

Start numbering at the end closest to either the double or triple bond, in this case the triple bond. For this molecule, the alkyne will become the main root, heptyne, and more specific, 2-heptyne with the alkene functionality in position #4. The combined root name becomes: 4-hepten-2-yne (also can be written as hep-4-en-2-yne). Since there is a methyl group in position #6, the complete name is shown below.

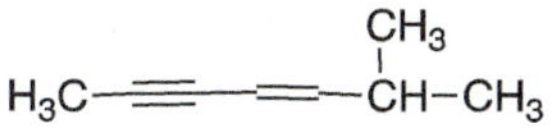

$$H_3C-\!\!\equiv\!\!=\!\!-\overset{\overset{\displaystyle CH_3}{|}}{CH}-CH_3$$

6-Methyl-4-hepten-2-yne

If the double and triple bonds are equidistant from the ends, the alkene functionality will become the root name to determine the numbering.

Problem 2.17

i) Give structures of the following molecules.

  a) 5-Chloro-4-methyl-2-hexyne
  b) 5,5-Dimethyl-3-heptyne
  c) 4-Bromo-5-chloro-2-octyne
  d) 5,5-Dichloro-2-heptyne

ii) Give the IUPAC names for the compounds shown below

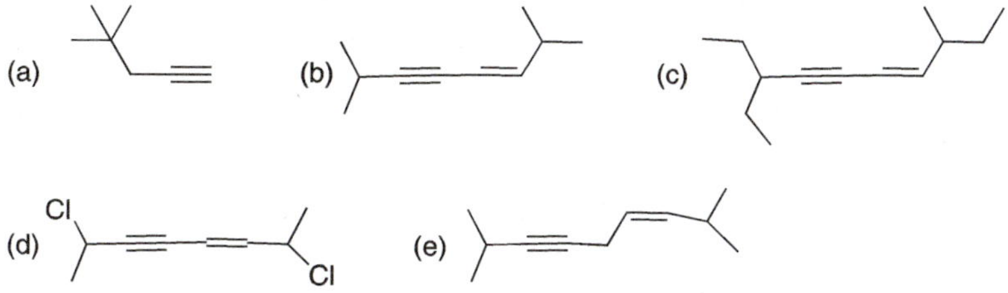

End of Chapter Problems

2.18 Give line-angle structural formulas for each of the following compounds.

  a) 3-Ethyl-2-methylhexane
  b) 2,4-Dichlorohexane
  c) 2,2,4,4-Tetramethylpentane
  d) Cyclopentylcyclopentane
  e) 2,3-Dimethylbutane
  f) Bromo-3-propylcyclohexane
  g) 3,6-Dimethyldecane
  h) 4-Ethyl-2,2,7-trimethyloctane
  i) 4-Ethyl-2,4-dimethyl-2-hexene
  j) 4-Methyl-2-pentyne

2.19 Give the IUPAC names for each of the following molecules.

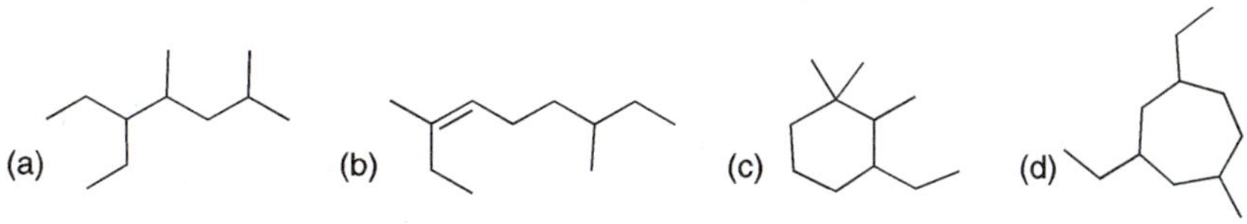

(e), 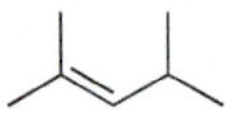 (f), 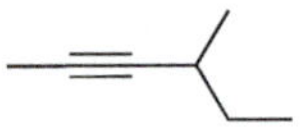 (g), 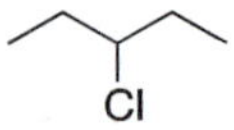 (h) H₃C−C≡C−CH₃

(i), 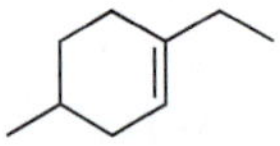 (j), 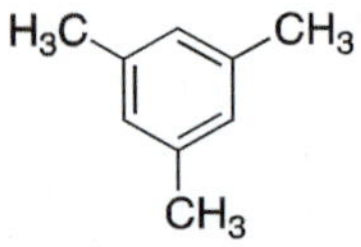 (j), 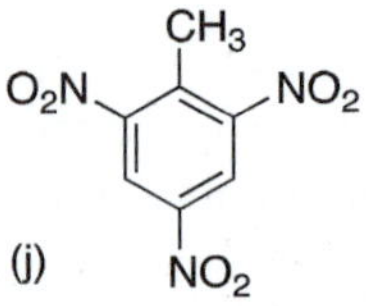 (k)

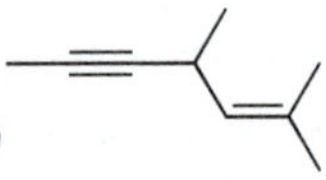

2.20 Draw and name *all* the different molecules that have the same formula of C_7H_{16}. Also give the IUPAC names for each of these different compounds.

2.21 For each of the following molecular formulas, write condensed structural formulas that have the indicated functional groups.

a) C_5H_8 (cyclic alkene)   b) C_6H_{10} (alkene)   c) C_6H_{10} (alkyne)

2.22 Give a complete IUPAC name for each of the following compounds.

(a), 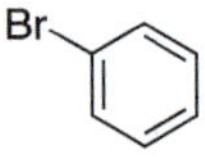 (b), 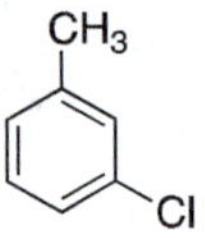 (c), 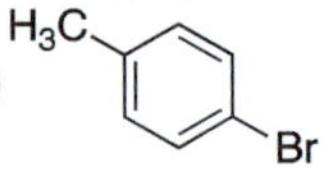 (d)

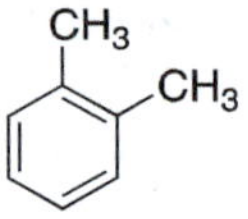

2.23 Give structural formulas for the following compounds.

a) 1,2-Dimethylbenzene
b) Iodobenzene
c) 1,4-Dichlorobenzene
d) 1,3-Dimethylbenzene
e) 3-Nitrotoluene
f) 2-Chlorophenol

3

Heteroatomic Functional Groups and Organic Nomenclature

3.1 Properties and Structure of Alcohols, Phenols, and Thiols

Alcohols have the —OH functional group, and alcohols are very important compounds in organic chemistry and today's economy. Alcohols that have three or less carbons are some of the most common alcohols. Methanol, which has one carbon, is often used as an alternate fuel for the internal combustion engines. It is a desirable alternative fuel since its combustion is very efficient. It is typically combined directly with gasoline but can be used directly as is the case with some racing cars. Methanol can typically be oxidized to form formaldehyde and formic acid, which are compounds that are used widely in industry. Ethanol has two carbons and is also used as a blend in combination with gasoline for fuel. It is also the alcohol of alcoholic beverages. Since pure ethanol is a strong dehydrating agent, it cannot be consumed directly. Alcoholic liquids, such as whiskey, are mixtures of ethanol and water, and the typical mixture is in the range of 45–50% in ethanol/water by volume. The term proof is an old terminology, which refers to the alcoholic content of an alcoholic mixture and is defined as twice the percentage of alcohol by volume; thus, for example, a 50% alcohol solution, is also referred to as 100 proof. A major health disadvantage in the use of ethanol is that it is oxidized rapidly and can result in cirrhosis of the liver. On the other hand, methanol, which has a very similar structure to ethanol, is toxic to the body; it damages the optic nerve, which results in blindness, so it should never be consumed. Ethylene glycol, another alcohol that has two —OH groups, is the main component of antifreeze/coolant. Owing to the low vapor pressure, very high boiling point (197 °C), and very low freezing point (−13 °C) of ethylene glycol, it remains a liquid even in the most severe winter and summer weather. Interestingly, propylene glycol, another alcohol that has two —OH groups, is used in small amounts in the food and cosmetic industries, whereas ethylene glycol is poisonous. Another important alcohol is isopropanol, which is also known as rubbing alcohol. Isopropanol has many household and personal care uses, and it is also used in industry as a solvent.

CH$_3$OH	CH$_3$CH$_2$OH	Isopropanol	Ethylene glycol	Propylene glycol
Methanol	Ethanol			

Phenolic compounds are a unique category of alcohols and will be studied in greater details in Chapter 17, aromatic compounds. Phenolic compounds were first used as antiseptics for

Organic Chemistry: Concepts and Applications, First Edition. Allan D. Headley.
© 2020 John Wiley & Sons, Inc. Published 2020 by John Wiley & Sons, Inc.
Companion website: www.wiley.com/go/Headley_OrganicChemistry

surgery, but they were determined to be poisonous because these compounds also kill cells and are easily absorbed through the skin and cause damage to key organs including the kidney, spleen, and pancreas. As a result, extreme care must be exercised in the use of phenolic compounds. Today, most phenols and derivatives of phenol are used as antiseptics and disinfectants; o-phenylphenol and hexachlorophene are two that are commonly used, which are shown below along with other examples of phenolic compounds.

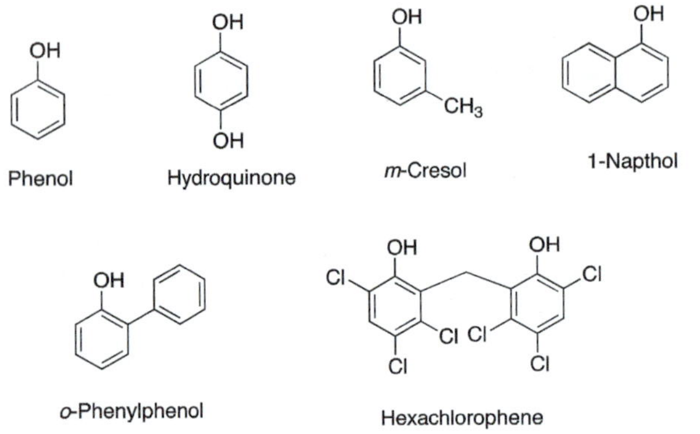

Examples of phenolic compounds

Phenol and phenol derivatives are also found in natural products. For example, thymol and vanillin are the constituents of thyme and vanilla beans are phenolic compounds. BHT (butylated hydroxytoluene) is a food additive that prolongs shelf life and protects against oxidation, and it is also a phenolic compound.

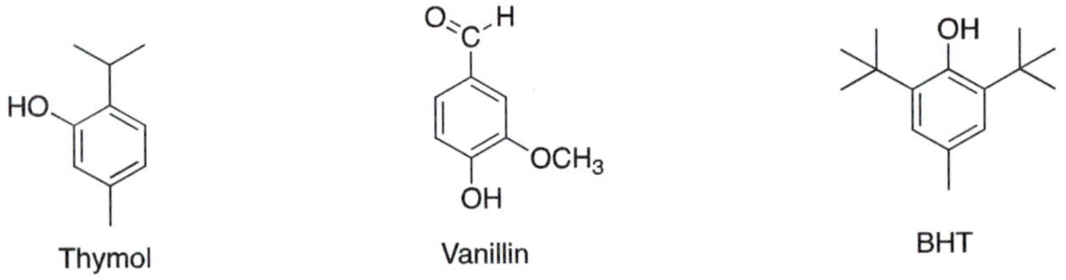

Alcohols are often converted, through different types of reactions, to other important compounds that are essential everyday compounds, such as polymers and pharmaceutical compounds. The alcohol functionality is also found in many biological compounds, such as glucose and even an amino acid, serine. The alcohol functionality is found in many pharmaceutical products, such as formoterol, which is one of the drugs used in the management of asthma and chronic obstructive pulmonary disease (COPD).

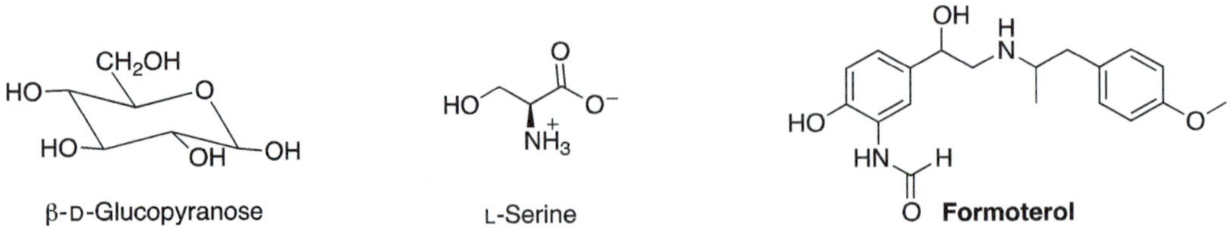

Thiols, also known as mercaptans, contain the —SH functional group, and low molecular weight thiols, such as methanethiol, which has one carbon, have a distinct pungent, very disagreeable characteristic odor, often described as that of rotten eggs. As a result, lower molecular weights thiols are used as additive odorants to natural gas. Since natural gas in its pure form is odorless, the addition of thiols assists in the detection of natural gas, especially in situations where there are leakages. The SH functionality is also found in one of the naturally occurring amino acids, cysteine.

$$\text{R-SH}$$
Structural feature of thiols

$$H_3C\text{-}SH$$
Methanethiol

$$H_2N\text{-}CH(CH_2\text{-}SH)\text{-}C(=O)\text{-}OH$$
Cysteine

Thiols have a common feature in that a hydrogen atom is bonded to the fairly electronegative atom, sulfur, resulting in the formation of hydrogen bonds among thiol molecules. Similar hydrogen bonds are possible for alcohols, but for alcohols, the hydrogen atom is bonded to the oxygen atom, which is more electronegative, compared to sulfur. As a result, hydrogen bonds that are formed among alcohol molecules are stronger that those formed with thiol molecules. Hence, a thiol, with a comparable molecular weight as an alcohol, has a lower boiling point than the corresponding alcohol.

3.1.1 Types of Alcohols

There are a large number of alcohols. As a result, alcohols are often classified based on the type of carbon that contains the OH functionality. If the —OH functionality is bonded to a carbon that has only one other alkyl group (R), the alcohol is classified as a primary alcohol (1°). If the —OH is bonded to a carbon that has two alkyl groups, the alcohol is a secondary alcohol (2°), and if the —OH group is bonded to a carbon that has three alkyl groups, the alcohol is classified as a tertiary alcohol (3°).

Methanol, a specific alcohol

A primary alcohol (1°)

A secondary alcohol (2°)

A tertiary alcohol (3°)

Isopropanol, also known as rubbing alcohol, has two methyl groups bonded to the carbon that has the —OH functionality, and as a result, it is classified as a secondary alcohol. A similar classification as that described above for alcohols can be carried out for thiols.

Problem 3.1
Classify the following alcohols as primary (1°), secondary (2°), or tertiary (3°).

(a), (b), (c)

3.2 Nomenclature of Alcohols

Alcohols are named as *alkanols* in the IUPAC system of nomenclature. For the alkanes that are used to derive the root names for alcohols, the **ane** is changed to **-anol** (or the **e** of the alkane is changed to **ol**). In naming alkanols, the longest continuous chain that contains the —OH functionality is identified and used as the root name for the alkanol. In numbering alkanols, start at the end closest to the OH functionality and use the number of the carbon to which the —OH group is attached to specify the location of the functionality in the IUPAC name, as shown in the example below.

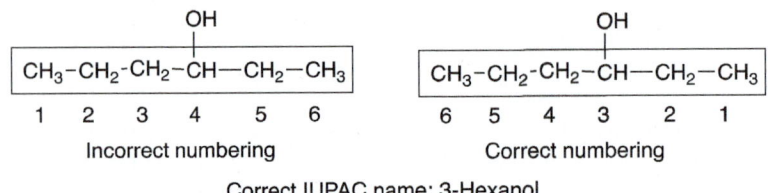

OH
$CH_3-CH_2-CH_2-CH-CH_2-CH_3$
1 2 3 4 5 6
OH
$CH_3-CH_2-CH_2-CH-CH_2-CH_3$
6 5 4 3 2 1

Incorrect numbering   Correct numbering

Correct IUPAC name: 3-Hexanol
Incorrect IUPAC name: 4-Hexanol

In naming alcohols that have branches, start the numbering from the end closest to the alcohol functionality and not at the branch as in the case in numbering alkanes. An example of the correct and incorrect method of numbering the longest chain of a branched alcohol is shown below.

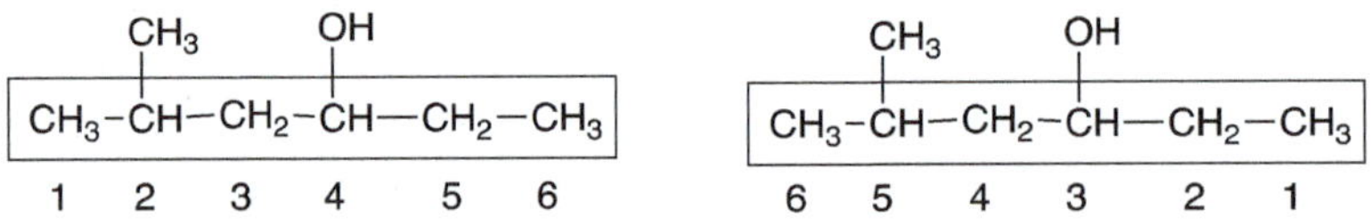

CH_3 OH
$CH_3-CH-CH_2-CH-CH_2-CH_3$
1 2 3 4 5 6
CH_3 OH
$CH_3-CH-CH_2-CH-CH_2-CH_3$
6 5 4 3 2 1

Incorrect numbering   Correct numbering

Correct IUPAC name: 5-Methyl-3-hexanol
Incorrect IUPAC name: 2-Methyl-4-hexanol

Note that in the above incorrect example, even though the branch is in position 2 and the —OH is in position 4, a branch does not take priority over the —OH functionality. That is, the —OH functionality must always get the lowest number and not the location of branching. In naming an alcohol, the position of the OH functionality must be specified by a number, and in the above example, the number 3 is used to specify the location of the OH functionality.

Problem 3.2

i) Give the IUPAC names for the following alcohols.

(a)
OH

(b)
OH

(c)
OH

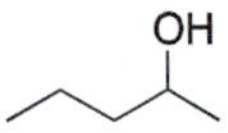

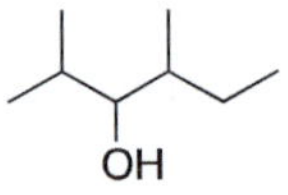

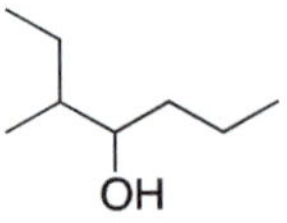

ii) Give the line-angle structures of the following molecules.

a) 2,2-Dimethyl-3-pentanol
b) 2,3,4-Trimethyl-2-hexanol

3.2.1 Nomenclature of Difunctional Alcohols

A source of confusion in the nomenclature of alkanols arises typically when the OH functionality is present in the same molecule that has another functionality, such as an alkene. The OH functionality is sometimes named as a hydroxy substituent if additional functional groups are present in the molecule, but typically the functionalities are combined as shown in the example below.

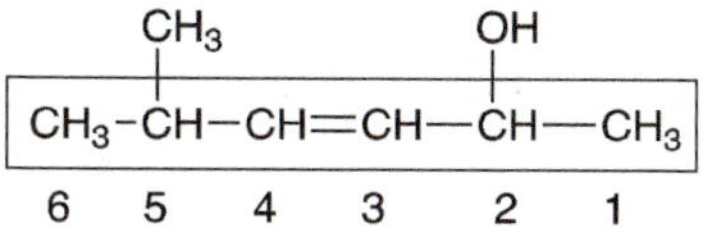

```
        CH3          OH
         |            |
CH3─CH─CH═CH─CH─CH3
 6    5   4   3   2   1
```

Correct IUPAC name: 5-Methyl-3-hexen-2-ol
Incorrect IUPAC name: 2-Methyl-3-hexen-5-ol

Note that in this case, the numbering system is based on the longest chain that contains both the alkene and OH functionalities. Also, note that the numbering is determined based on the alcohol functionality having, the lowest number. The alcohol functionality has priority over the alkene functionality.

Problem 3.3

i) Give the line-angle structure of 1-pentyn-3-ol.
ii) Give the IUPAC names for the compounds shown below.

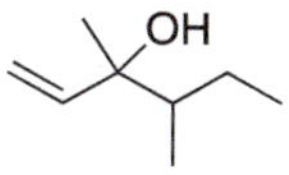

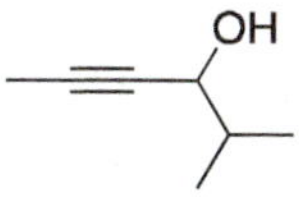

3.2.2 Nomenclature of Cyclic Alcohols

In naming cyclic alcohols, the root name becomes *cycloalkanol*. In numbering cyclic alcohols, the carbon that has the —OH functionality gets #1, and the numbering proceeds in the direction of the branches to give the lowest possible number combination, an example of naming a substituted cycloalkanol is shown below.

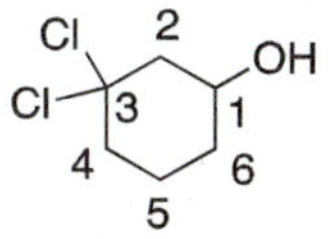

Correct numbering
3,3-Dichlorocyclohexanol

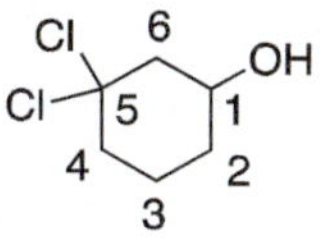

Incorrect numbering
5,5-Dichlorocyclohexanol
(Incorrect IUPAC name)

Problem 3.4

i) Give the IUPAC name of the molecule shown below.

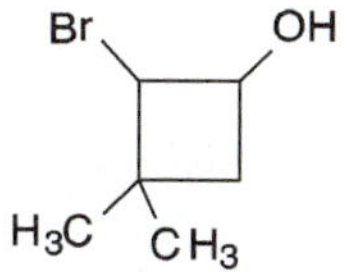

ii) Give the structure of 2,3-dichloro-4-methylcyclobutanol

3.2.3 Nomenclature of Substituted Phenols

The structure of phenol was given earlier in the chapter, and substituted phenols are named such that phenol is the root name as shown in the examples below. Note that the carbon that has the OH group gets #1 and the numbering is similar to other cyclic structures in that the number combination of the name is the lowest.

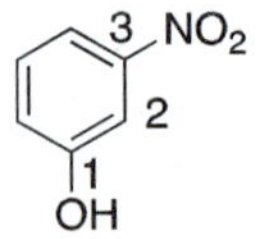

3-Nitrophenol 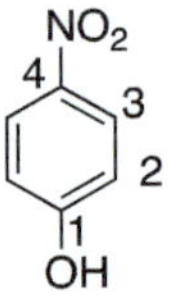   4-Nitrophenol 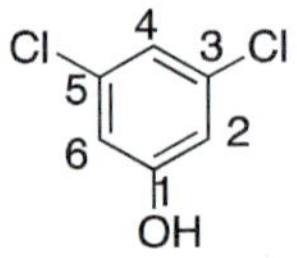   3,5-Dichlorophenol

If there are more than two different groups on the phenol ring, they are listed alphabetically.

Problem 3.5

i) Give the names for the following substituted phenols.

(a), (b), 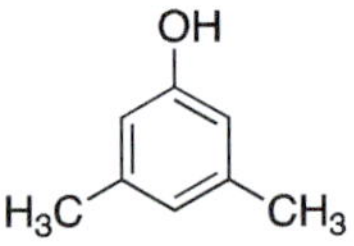 (c)

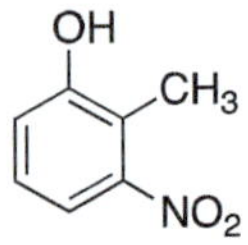

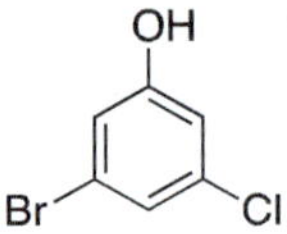

ii) Give the structures that correspond to the following molecules.

a) 2,4-Dibromophenol
b) 2,4,6-Trimethylphenol
c) 2-Bromo-4-chlorophenol

3.3 Nomenclature of Thiols

Thiols are organic compounds that have the —SH functionality and they are named as **thiols**. The **-ane** of the corresponding alkane is unchanged, but thiol is added to alkane to form one word, ***alkanethiol***. In naming thiols, the longest continuous chain that contains the —SH functionality forms the root name as shown below.

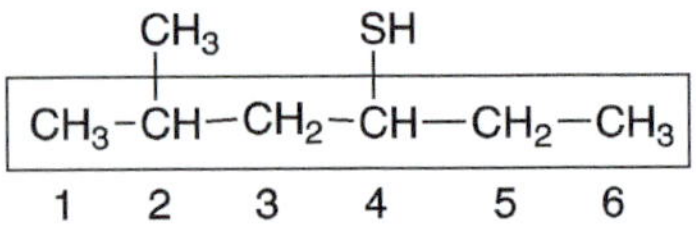

Incorrect numbering 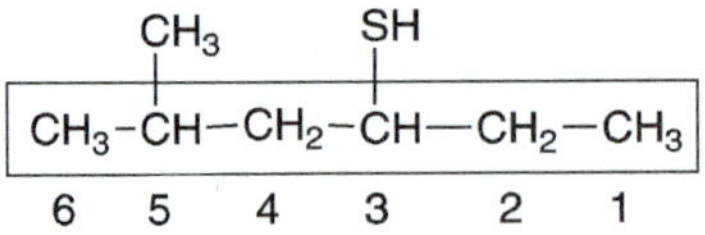   Correct numbering

Correct IUPAC name: 5-Methyl-3-hexanethiol
Incorrect IUPAC name: 2-Methyl-4-hexanethiol

The thiol (—SH) functionality takes priority over branching, which is similar to that in the naming of alcohols. The position of the —SH functionality must be specified with a number. In the above example, the number 3 is used to specify the position of the —SH functionality.

Problem 3.6

i) Is the IUPAC name shown for the compound shown below correct?

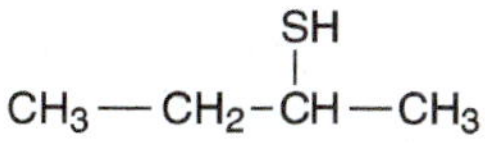

$$CH_3-CH_2-CH(SH)-CH_3$$

2-Butanethiol

ii) Give line-angle structures for the following molecules.

a) 3,3-Dimethyl-2-pentanethiol
b) 4-Methyl-3-penten-2-thiol
c) 3-Methylcyclohexanethiol

3.4 Structure and Properties of Aldehydes and Ketones

Aldehydes and ketones have the carbon–oxygen double bond, also known as the carbonyl group, as a common feature. The type of groups that are bonded to the carbon atom of the carbon–oxygen double bond determine whether these types of compounds are ketones or aldehydes. If two alkyl groups are bonded to the carbonyl carbon, as shown below, the functional group is a ketone. If there is one alkyl group, along with a hydrogen atom bonded to the carbonyl group, the functional group is an aldehyde.

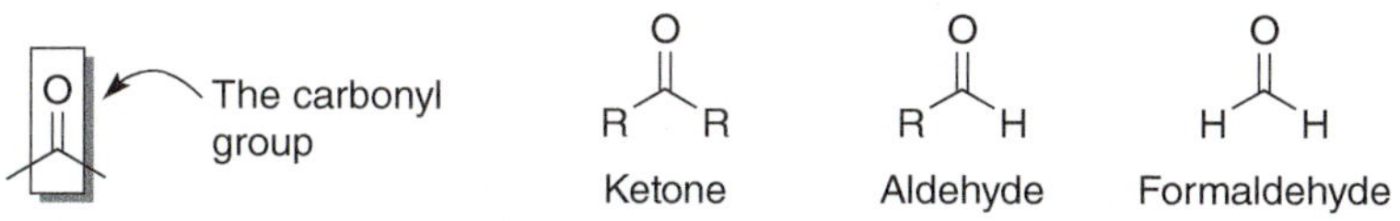

The simplest aldehyde is formaldehyde in which there are two hydrogen atoms bonded to the carbon of the carbonyl group. Formaldehyde is commonly used in industry and has a pungent suffocating odor. It is typically sold as formalin, which is a mixture in aqueous solution (about 37–50%). It is easily oxidized to formic acid. Aldehydes that contain large alkyl groups are less toxic and do not penetrate the respiratory track as formaldehyde. Acetone, which is a ketone, dissolves fats and skin. A large number of compounds that are used in the flavoring industry are aldehydes or ketones. Some commonly used aldehydes and ketones are shown below.

Vanillin (vanilla flavor)

(R)-Carvone (spearmint flavor)

Cinnamaldehyde (cinnamon flavor)

Benzaldehyde (almond flavor)

Problem 3.7

For the compounds given in the example above, determine which are aldehydes and ketones.

3.5 Nomenclature of Aldehydes

The systematic IUPAC naming of aldehydes is similar to the naming of organic molecules discussed previously in that the root names are based on the longest chain that contains the functional group, the aldehyde functionality in this case. Compounds that have the aldehyde functionality, —CHO, are named as **alkanals**. The **e** of the corresponding alkane is changed to **al**. Thus, a compound that has five carbons and has the aldehyde functionality, the IUPAC name is pentanal. In naming aldehydes, the longest continuous chain that contains the —CHO functionality forms the root name of the alkanal. Since the —CHO functionality is a terminal functionality, the carbon of the carbonyl group is assigned number 1. Thus, the position of the —CHO functionality is not specified since it is always #1. Shown below is an example for the correct numbering of aldehydes.

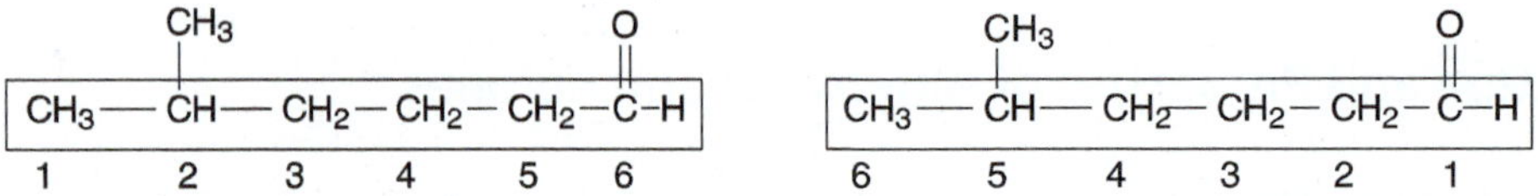

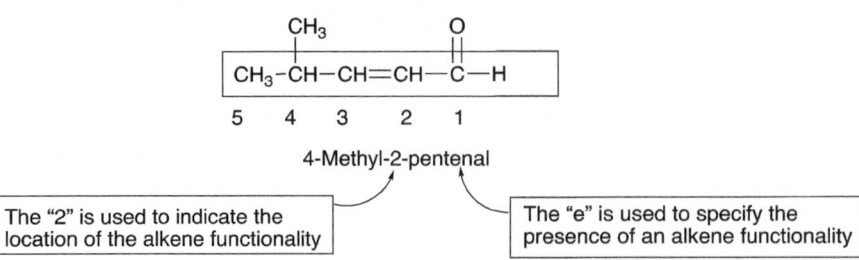

Note that for the above incorrect example, even though the branch is in position 2 and the —CHO functionality is in position 6, the branch does not take priority over the —CHO functionality. That is, the —CHO functionality must always be assigned the #1 position.

Problem 3.8

i) Give the IUPAC names for the compounds shown below.

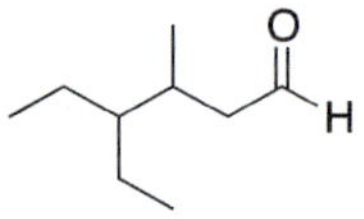

ii) Give structures for each of the following molecules.

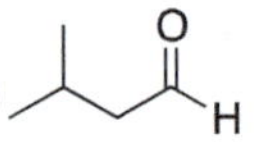

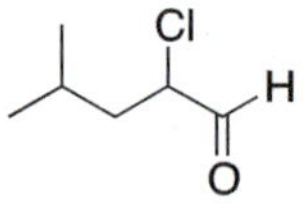

a) 2,3-Dimethylhexanal
b) 2-Chloro-3-methylbutanal
c) 2,2-Dimethylbutanal

3.5.1 Nomenclature of Difunctional Aldehydes

It is possible to have the —CHO functionality present in a molecule that has another functionality, such as an —OH, alkene, or alkyne functionalities. For such a molecule, the aldehyde group will always get priority and gets assigned position #1. If there is only one other functionality present, the position of the other functionality is specified based on the numbering system of the root name given to the alkanal. An example is shown below for the nomenclature of a molecule that contains the aldehyde and alkene functionalities.

4-Methyl-2-pentenal

The "2" is used to indicate the location of the alkene functionality

The "e" is used to specify the presence of an alkene functionality

3.6 Nomenclature of Ketones

Note that the "e" is used to specify the alkene functionality and that the number that directly precedes gives the position of the alkene functionality. If there are other functionalities, such as an —OH, it is specified as a hydroxyl group.

Problem 3.9

i) Give the IUPAC names for the compounds shown below.

(a) CH₃—CH(OH)—C(=O)—H (b) H₂C=CH—CH(CH₃)—C(=O)—H (c) H₂C=C(CH₃)—CH₂—C(=O)—H

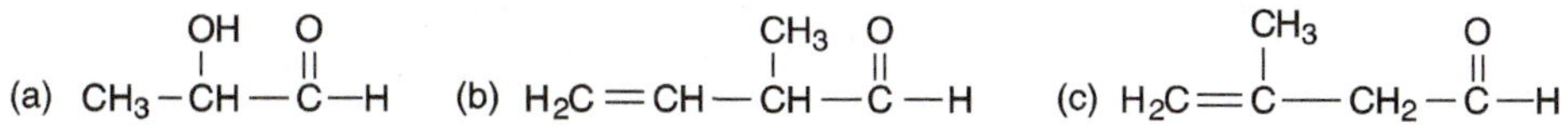

ii) Give the structures of the compounds shown below.

a) 3-Ethyl-2-hydroxyhexanal
b) 2-Propenal
c) 3,7-Dimethyl-2,6-octadienal (Geranial)
d) 3-Phenyl-2-propenal (Cinnamaldehyde) [note that the phenyl substituent is —C₆H₅]

3.6 Nomenclature of Ketones

Compounds that have the ketone functionality are named as **alkanones**. The **-ane** of the corresponding alkane root name is changed to **-anone** (or the **e** of the corresponding alkane root name is changed to **one**). In naming alkanones, the longest continuous chain that contains the carbonyl functionality is identified and based on the number of carbons present, the root name of the alkanone is determined. In naming ketones, the carbon that gets number 1 is the carbon that is closest to the carbonyl functionality and a number must be used to specify the position of the carbonyl functionality of the ketone, as shown in the example below.

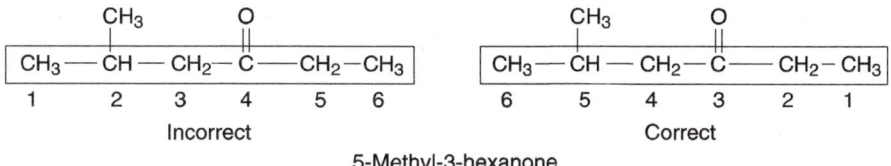

5-Methyl-3-hexanone

Note that in the above incorrect example, even though the branch is in position 2 and the ketone functionality is in position 4, the branch does not take priority over the carbonyl functionality for the assignment of numbers in naming ketones. That is, the carbonyl functionality must always get the lowest number and not the location of the branch. In the above correct example, the number 3 is used to specify the position of the carbonyl functionality.

3.6.1 Nomenclature of Difunctional Ketones

A source of confusion in the nomenclature of ketones comes when there is another functionality present in the same molecule. If the other functionality is an alkene, for example, the position of the alkene functionality is specified based on the numbering system of the root name given to the alkanone, as shown in the example below.

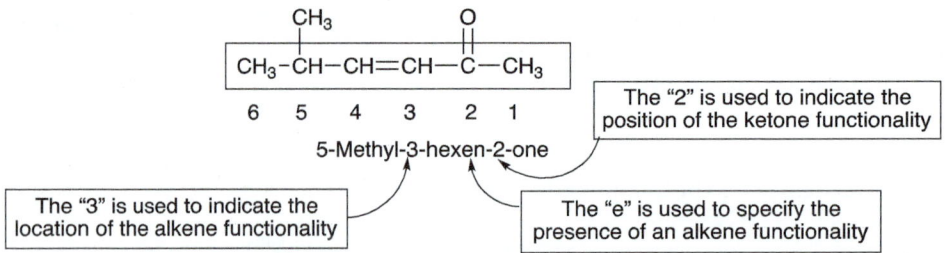

If an —OH functionality is present in the same molecule that contains a ketone functionality, a number is used to indicate the presence of the —OH functionality and the molecule is named as a hydroxy alkanone (and not as an alkanol). For the compound below, which has three functionalities, note that the numbering is based on the presence of the *ketone*, and not on the presence of the —OH or alkene functionalities. The ketone takes priority over the branching, an alkene, or an alcohol functionality.

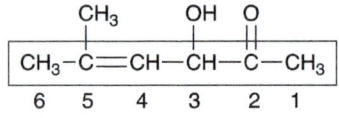
3-Hydroxy-5-methyl-4-hexen-2-one

Problem 3.10

i) Give the structures of the compounds shown below.

  a) 3-Hydroxy-2-pentanone
  b) 4-Methyl-2-pentanone
  c) 3,3-Dimethyl-2-butanone
  d) 5-Methyl-3-hexen-2-one
  e) 3-Pentyn-2-one

ii) Give IUPAC names for the compounds shown below.

(a)   (b)   (c)

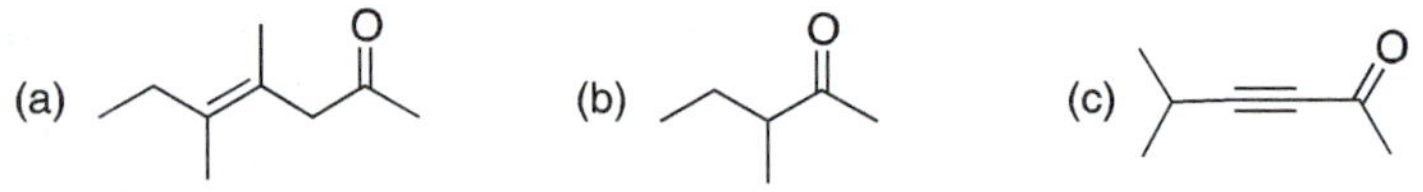

3.6.2 Nomenclature of Cyclic Ketones

In naming cyclic alkanones, the root name is derived from the corresponding cyclic alkane and the cycloalkane becomes **cycloalkanone**. In numbering cycloalkanones, the carbon that has the carbonyl functionality gets #1. Examples are shown below.

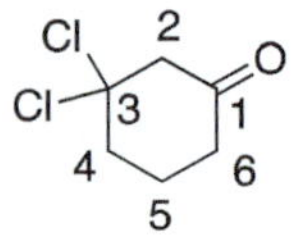

Correct numbering
3,3-Dichlorocyclohexanone

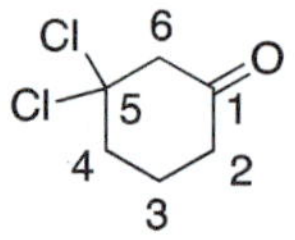

Incorrect numbering
5,5-Dichlorocyclohexanone
(Incorrect IUPAC name)

Note that the numbering from position #1 proceeds in the direction of the closest branch or functionality. Shown below is an example of a cyclic ketone, which also contains an alkene functionality.

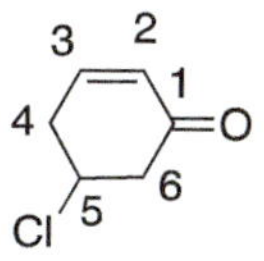

Correct numbering
5-Chloro-2-cyclohexenone

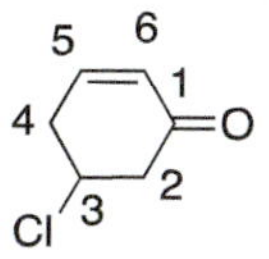

Incorrect numbering
3-Chloro-5-cyclohexenone
(incorrect IUPAC name)

Problem 3.11

i) Give line-angle structures of the following molecules.

  a) 2,3-Dichloro-4-methylcyclopentanone
  b) 5-Methyl-3-hexanone
  c) CH₃C(CH₃)₂CH₂COCH₂CH₃

ii) Give the IUPAC name of the molecule shown below.

(a) 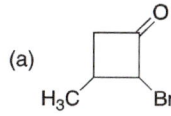 (b) 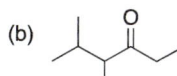 (c)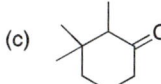

3.7 Structure and Properties of Carboxylic Acids

As the name of these compounds suggests, these are acids and they have the functional feature —COOH; shown below are different ways of writing the carboxylic acid functional group.

R–COOH   R–CO₂H

Various representations of the carboxylic acid functionality

The simplest carboxylic acid is formic acid, in which a hydrogen atom is bonded to the carbonyl carbon of the carboxylic acid functionality. Formic acid is fairly acid and can cause damage to tissues. Formic acid is the main constituent of ant venom, which causes severe irritation when one is stung by an ant. Other carboxylic acids are very important compounds in our everyday lives, and some common ones are shown below.

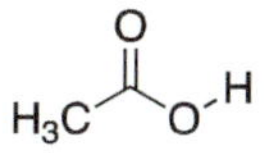

Acetic acid, found in vinegar and most salad dressings

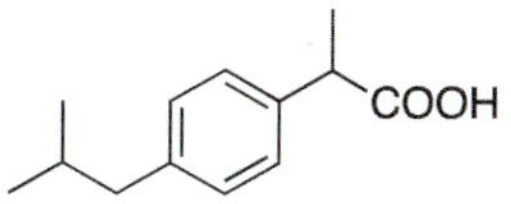

Ibuprofen (a pain killer)

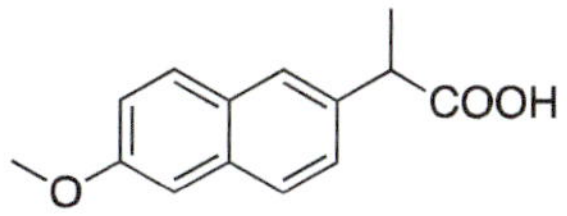

Naproxen (used in arthritis treatment)

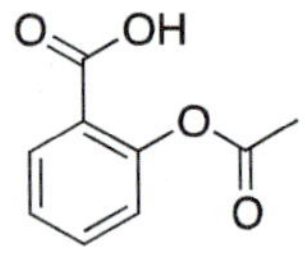

Acetylsalicylic acid
(aspirin, an analgesic)

> **DID YOU KNOW?**
>
> Formic acid is the main constituent of ant venom, which causes severe irritation when one is stung by an ant.
>
>

Carboxylic acids have an acidic hydrogen that is bonded to the electronegative oxygen of the carboxylic acid functionality, and as a result, it will be involved in hydrogen bonding. Thus, carboxylic acids are typically high boiling liquids, compared to other liquids of comparable molecular weights that do not have intermolecular hydrogen bonds, as shown below.

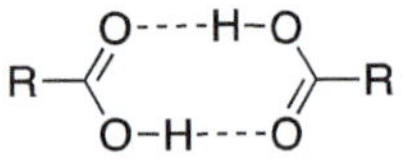

Intermolecular hydrogen bonds
 involving carboxylic acids

Owing to the type of hydrogen bond shown above, the boiling point of acetic acid is higher than expected, compared to other type compounds of similar molecular weights, the boiling point of acetic acid is 118 °C, and it is soluble in polar solvents such as water.

Carboxylic acids that have long chains are typically classified as saturated fatty acids or unsaturated fatty acids. Unsaturated fatty acids can be broken further into cis and trans fatty acids as shown below.

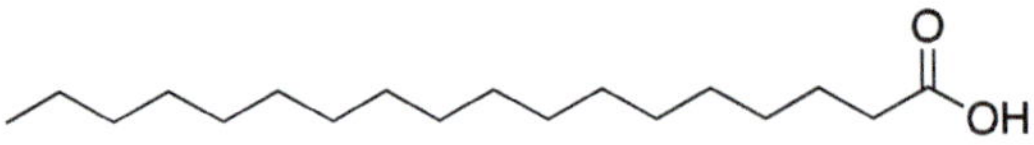

Saturated fatty acid

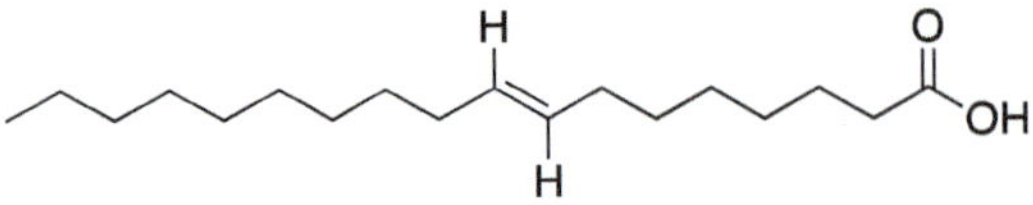

trans-Unsaturated fatty acid

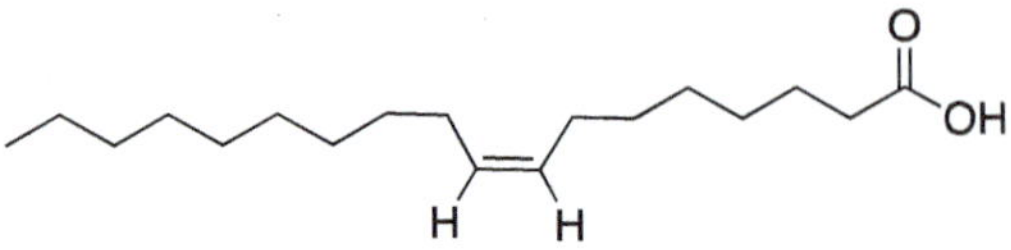

cis-Unsaturated fatty acid

Unsaturated fatty acids have lower melting points than saturated fatty acids with the same number of carbons owing to the structured arrangement of the molecules. For example, the melting point of stearic acid ($C_{18}H_{36}O_2$) is 69.6 °C, but the melting point of oleic acid ($C_{18}H_{34}O_2$), which contains one cis-double bond, is 13.4 °C. Polyunsaturated fatty acids have even lower melting points.

3.8 Nomenclature of Carboxylic Acids

Compounds that have the —COOH functionality are named as **alkanoic acids**. The **-e** of the corresponding root alkane name is changed to **-oic** followed by the addition of the word **acid**. In naming carboxylic acids, the longest continuous chain that contains the —COOH functionality forms the root name of the alkanoic acid as shown below.

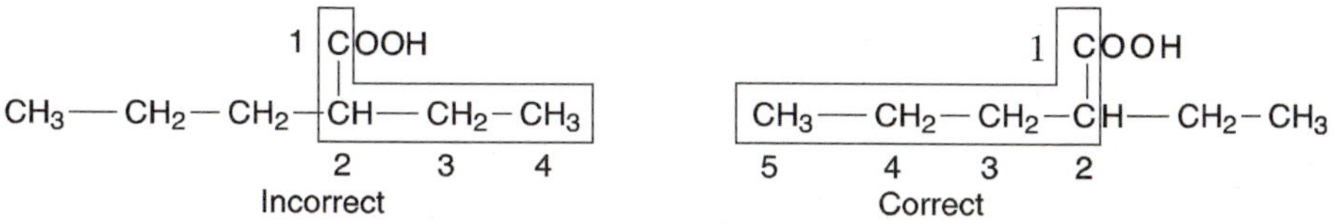

$$CH_3-CH_2-CH_2-\underset{2}{CH}-\underset{3}{CH_2}-\underset{4}{CH_3}$$
with $\overset{1}{COOH}$ branch
Incorrect

$$CH_3-\underset{5}{}CH_2-\underset{4}{CH_2}-\underset{3}{CH}-\underset{2}{CH_2}-CH_3$$
with $\overset{1}{COOH}$ branch
Correct

2-Ethylpentanoic acid

Since the carboxylic acid functionality is a terminal functionality, the carbon of the COOH is always assigned #1. Therefore, the position of the —COOH functionality does not have to be specified by a number. The position of other groups (substituents) that are bonded to the root chain must be specified by a number based on the numbering system dictated by the —COOH position, an example is shown below.

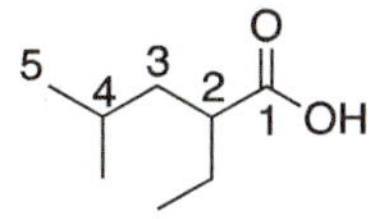

2-Ethyl-4-methylpentanoic acid

Problem 3.12

i) Give the line-angle structures of the following molecules.

    a) 2-Chloro-3-ethylhexanoic acid
    b) 2-Chloro-3,3-dimethylpentanoic acid
    c) 2-Ethylbutanoic acid

ii) Give the IUPAC name for the compound shown below.

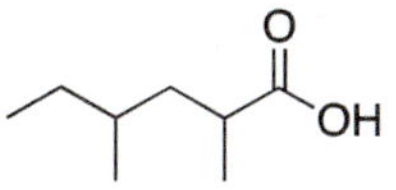

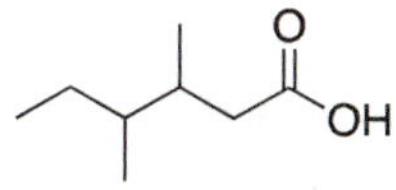

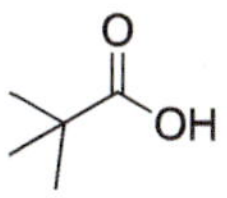

Over the years, trivial names have been used for carboxylic acids; one type of trivial name is based on the use of letters of the Greek alphabet. Letters of the Greek alphabet are used to indicate the position of substituents on the chain of the carboxylic acids. For this system, the first carbon that is next to the carbonyl carbon of the reference carboxylic acid functionality is assigned the Greek letter α and a substituent that is bonded to this carbon is described as being in the α position. The carbon next to that first carbon of the longest chain, or the carbon that is two carbons away from the carbonyl carbon, is assigned the Greek letter β. Likewise, the carbon that is in the third position from the carbonyl carbon of the reference carboxylic acid functionality is assigned the Greek letter γ, as shown below.

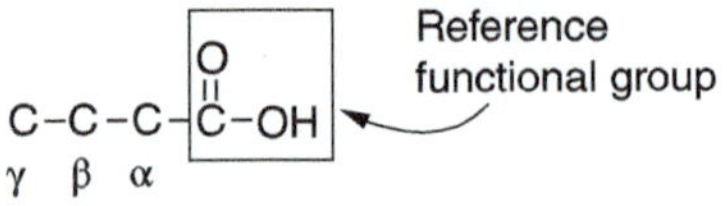

Substituents that are bonded to these carbons are assigned the corresponding Greek letter to indicate the position of the substituent as shown in the examples below.

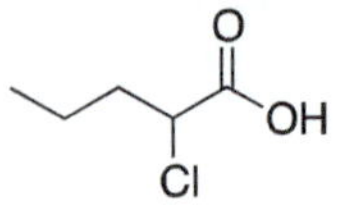

α-Chloropentanoic acid 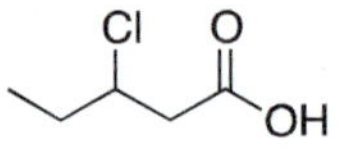   β-Chloropentanoic acid

3.8.1 Nomenclature of Difunctional Carboxylic Acids

A large percentage of carboxylic acids contains not only the —COOH functionality, but other functionalities, such as an alcohol (—OH), alkene, or alkyne functionalities. For these molecules, the carboxylic acid group will always get priority and is assigned #1. In naming such polyfunctional molecules, the position of another functionality is specified based on the numbering system dictated by the location of the carboxylic acid functionality, an example is shown below. Note that the number 2 indicates the position of the alkene, which signifies the letter "e" in "hexenoic" of the name.

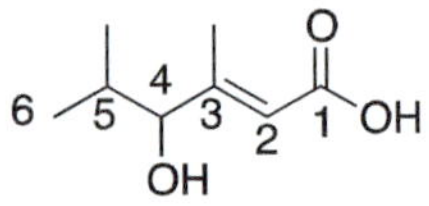

4-Hydroxy-3,5-dimethyl-2-hexenoic acid

Problem 3.13

i) Give the line-angle structure of the following molecules.

    a) 3-Hydroxypentenoic acid
    b) 4-Hydroxy-2-pentenoic acid
    c) 4-Chloro-2-hexynoic acid
    d) 5-Chloro-2,2-dimethyl-3-hexynoic acid

ii) Give the IUPAC name for the compounds shown below.

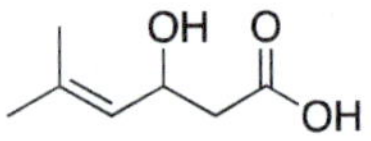

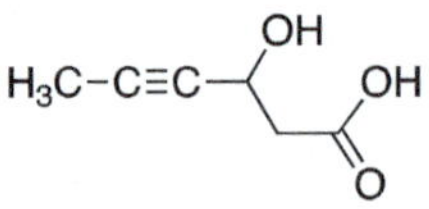

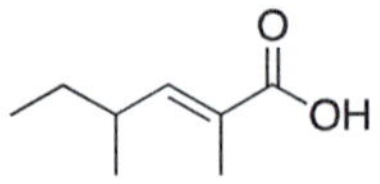

3.8.2 Nomenclature of Cyclic Carboxylic Acids

The carboxylic acid functionality is a terminal functionality and hence cannot be in a ring; that is, this functionality must be bonded to a ring. In naming cyclic carboxylic acids, carboxylic acid becomes the root name and the cyclic structure becomes the substituent. The IUPAC

3.8 Nomenclature of Carboxylic Acids

name then becomes **cycloalkanecarboxylic acid** (note, two words). The general nomenclature system is illustrated below.

IUPAC name: Cyclopentanecarboxylic acid

Note that number 1 is not included in the name to indicate the position of the carboxylic acid functionality. In numbering cyclic carboxylic acids that have substituents, the carbon that has the —COOH functionality gets #1 and the numbering proceeds in the direction so that the lowest number combination results. Examples of naming a cyclic carboxylic acid are given below.

IUPAC name: 2,3-Dimethylcyclopentanecarboxylic acid
(and not 4,5-Dimethylcyclopentanecarboxylic acid)

IUPAC name: 3,3-Dichlorocyclohexanecarboxylic acid
(and not 5,5-Dichlorocyclohexanecarboxylic acid)

Problem 3.14

i) Give the line-angle structures for the following compounds.

   a) 2,3-Dichloro-4-methylcyclopentanecarboxylic acid
   b) 1-Methylcyclohexanecarboxylic acid
   c) 2,2-Dimethylcyclobutanecarboxylic acid

ii) Give the IUPAC names of the molecules shown below.

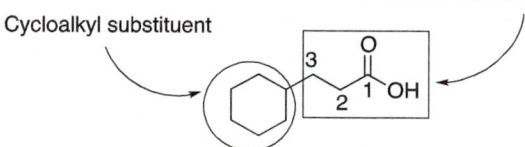

Note that if the carboxylic acid functionality is not bonded directly to a cyclic structure, but to a carbon of the longest continuous linear region of the molecule, the cyclic structure is then named as a cycloalkyl group, as shown in the example below.

IUPAC name: 3-Cyclohexylpropanoic acid

If there are two carboxylic acid functionalities present, then both carboxylic groups are indicated by numbers as shown in the example below.

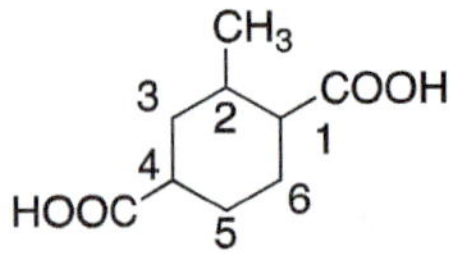

2-Methylcyclohexane-1,4-dicarboxylic acid

As usual, always ensure that the number combination is always the lowest. Thus, the IUPAC name in the above example is not 6-methylcyclohexane-1,4-dicarboxylic acid.

If there is a different functionality present in a cyclic carboxylic acid molecule, a system similar to that described in the previous section regarding the nomenclature of molecules with difunctional groups is used, as shown in the example below. Note that the "e" is used to indicate the presence of an alkene functionality and the number that precedes gives the position of the double bond in the molecule.

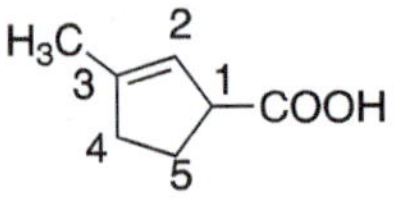

3-Methyl-2-cyclopentenecarboxylic acid 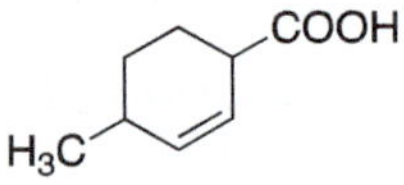   4-Methyl-2-cyclohexenecarboxylic acid

At this point, it is worth establishing a priority of functional groups in the naming of polyfunctional molecules. The functionality of highest priority will be used to determine the root name of the molecule. The priority trend of some functional groups is shown below.

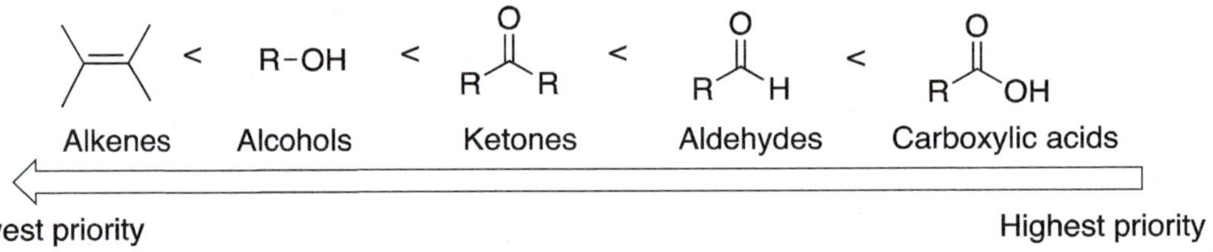

Alkenes < Alcohols < Ketones < Aldehydes < Carboxylic acids

Lowest priority   Highest priority

Hence, if a molecule contains a carboxylic acid functionality and an alcohol functionality, the carboxylic acid functionality is used to form the root name and the molecule is not named as an alkanol, but as a carboxylic acid as shown in the example below.

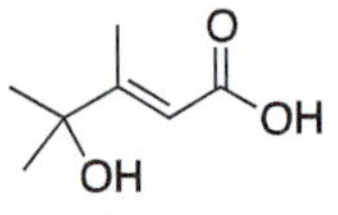

4-Hydroxy-3,4-dimethyl-2-pentenoic acid

3.9 Structure and Properties of Esters

Esters have the functional group as shown below, in which an alkyl group is bonded to the carbon of the carbon–oxygen double bond and another alkyl group is bonded to oxygen as shown below.

3.9 Structure and Properties of Esters

Various representations of the ester functionality

Molecules that contain the ester functionality are also known as esters and they are typically sweet-smelling compounds, and the sweet aroma of most fruits is due to the presence of esters. For example, isoamyl acetate is the ester responsible for the banana aroma of ripened bananas.

> **DID YOU KNOW?**
>
> An ester, known as isoamyl acetate, is responsible for the sweet-smelling aroma of ripened bananas.

Esters are also good solvents and can even dissolve body tissue. Diethylhexyl phthalate (DEHP), which is an ester, is used as plasticizer to impart flexibility to polyvinyl chloride (PVC) plastics. Aspirin (acetylsalicilic acid), which was mentioned earlier as an analgesic, also contains an ester functionality.

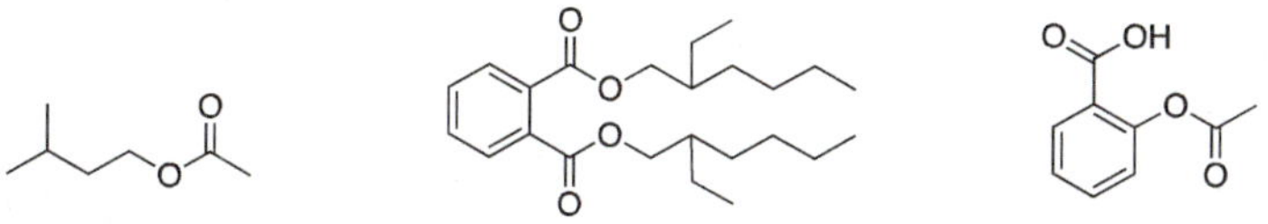

Isoamyl acetate
(Responsible for the fruity aroma in bananas)

Diethylhexyl phthalate (DEHP)

Acetylsalicylic acid
(aspirin, an analgesic)

Problem 3.15

For the molecules shown above, identify and circle the ester functional group.

3.9.1 Nomenclature of Esters

In naming esters, first identify the ester functionality and name the group bonded to the oxygen of the ester group as an alkyl group. The group that is bonded to the carbon of the carbonyl group is named as an *alkanoate*; the format of the combined IUPAC name is shown below. Note that there are two words in the name of esters.

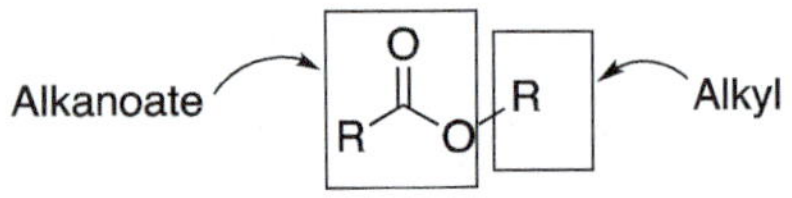

Alkyl alkanoate

A specific example of the nomenclature of an ester is shown below.

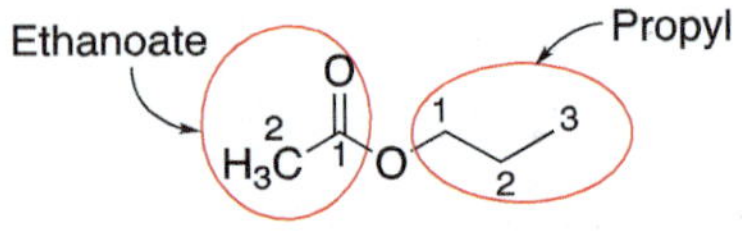

IUPAC name: Propyl ethanoate

If there are substituents of either of both groups of an ester, the groups are assigned numbers based on the numbering starting from the groups bonded to the ester functionality, as illustrated in the example below.

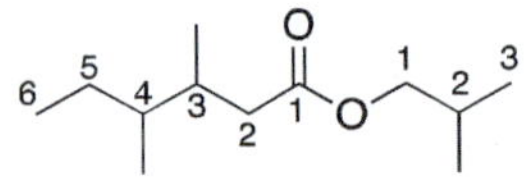

2-Methylpropyl 3,4-dimethylhexanoate

Problem 3.16

i) Give line-angle structures for the following molecules.

  a) Ethyl propanoate
  b) Methyl butanoate
  c) Butyl butanoate

ii) Give IUPAC names for the molecules shown below.

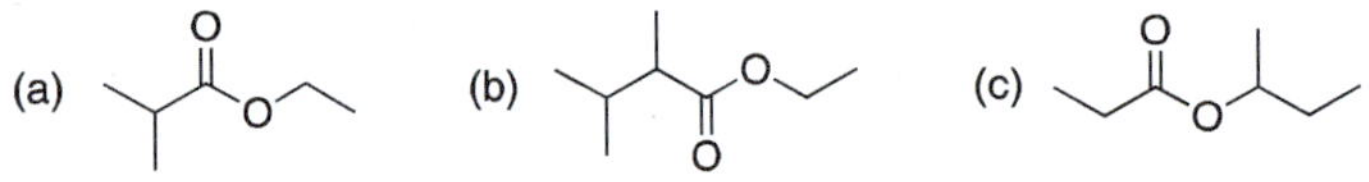

3.9.2 Nomenclature of Cyclic Esters

Cyclic esters are also known as lactones. The IUPAC names of cyclic esters are based on the following general system: 2-oxacycloalkanone, as shown below. Note that this naming system is essentially based on an alkanone (which is assigned position #1, and the oxygen, which is named as an "oxa" gets position #2.

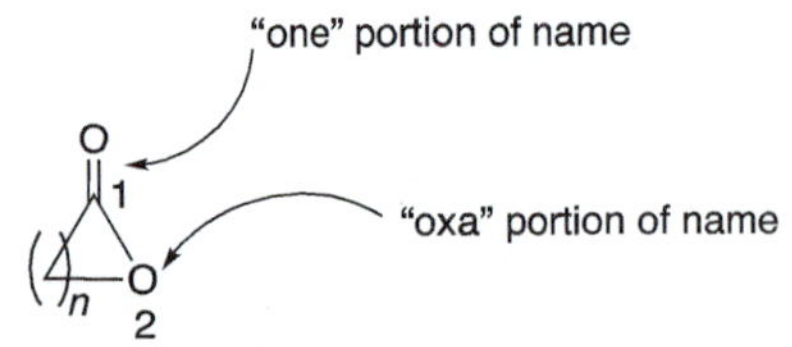

"one" portion of name

"oxa" portion of name

2-Oxacycloalkanone
where n represents an integer

3.9 Structure and Properties of Esters

Thus, if *n* in the above example is 4, a six-membered ring is formed and named as 2-oxacyclohexanone. A more detailed illustration is shown below.

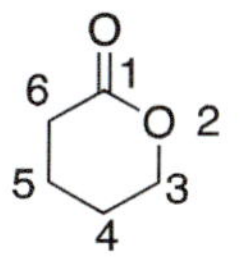

2-Oxacyclohexanone

The oxygen of the ester is represented by the "2-oxa"

The carbonyl group of the ester is represented by the "one"

If there are substituents on the cyclic structure, it is identified based on the numbering system above; note that the numbering starts at the carbonyl carbon and proceeds in the direction of the oxygen of the ester. Examples giving the IUPAC names of different substituted lactones are shown below.

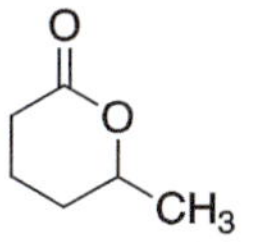

3-Methyl-2-oxacyclohexanone 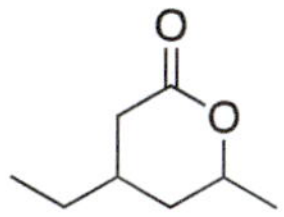   5-Ethyl-3-methyl-2-oxacyclohexanone 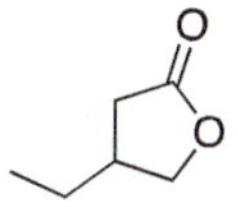   4-Ethyl-2-oxacyclopentanone

Common names are sometimes used for lactones. A very commonly used naming system for lactones is derived from the common names of the corresponding carboxylic acid in which Greek letters are assigned to the carbons adjacent to the carbonyl carbon of carboxylic acids. For the common names of lactones, the Greek letter is used to indicate the carbon that is bonded to the oxa oxygen in the cyclic structure, hence indicating the number of carbons that are in the lactone, examples are shown below.

β-Propiolactone   β-Butyrolactone   γ-Butyrolactone   δ-Valerolactone

The first lactone has three carbons and is labeled as a β since the second carbon is bonded to the oxa oxygen. Likewise, the second is labeled β even though it has four carbons, but it is the second carbon that is bonded to the oxa oxygen.

Problem 3.17

i) Give the IUPAC names for the following molecules.

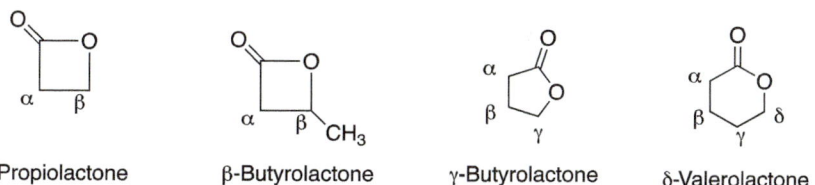

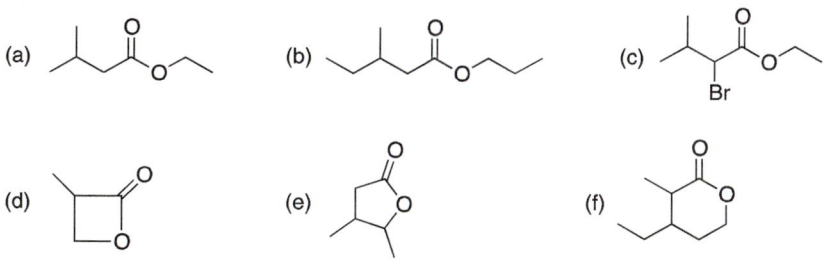

ii) Give structures for the following molecule.

a) Methyl butanoate
b) Ethyl-2-methylpentanoate
c) Methyl 2,3-dimethylhexenoate
d) Ethyl 4-chlorobutanoate
e) 2-Oxacyclopentanone
f) 3-Methyl-2-oxacyclohexanone
g) 4,4-Dimethyl-2-oxacyclopentanone
h) 3-Ethyl-4-propyl-2-oxacyclohexanone

3.10 Structure and Properties of Acid Chlorides

Compounds that have the —COCl functionality are commonly called acid chlorides or acyl chlorides. They are very reactive compounds and are used primarily for the synthesis of other compounds. The reactivity of acid chlorides is due primarily to the presence of the very electronegative chlorine, which is bonded to a carbon atom of the carbonyl group, which produces a very polarized functional group. Acid chlorides are highly reactive with water, and if inhaled causes severe medical problems owing to a reaction with moisture of the respiratory track.

3.10.1 Nomenclature of Acid Chlorides

Acyl chlorides are named based on the longest continuous chain that has the acid chloride functionality, these compound are named as **alkanoyl chlorides** in the IUPAC system. Note that the **e** of the corresponding alkane is changed to **oyl** and the word chloride is added. The IUPAC nomenclature is based on the longest chain that contains this functional group, which is changed from the alkane to *alkanoyl chloride*. The acid chloride functional group, along with the nomenclature of selected examples, is shown below.

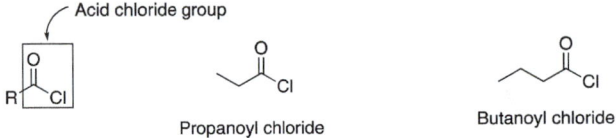

In naming acid chlorides, identify the longest chain that contains the acid chloride functional group, the carbonyl carbon is assigned #1.

Incorrect (not the longest chain)   Correct (longest chain)

The correct root name for the molecule above is pentanoyl chloride (and not butanoyl chloride). If that longest chain has groups (substituents) attached, the position of each substituent must be specified based on the numbering of the longest continuous chain. More examples of nomenclature are shown below.

2-Ethyl-4-methylpentanoyl chloride   5-Chloro-2-ethylpentanoyl chloride

Since the alkanoyl chloride is a terminal functionality, it is not necessary to indicate its position by the #1.

3.10.2 Nomenclature of Difunctional Acid Chlorides

It is possible to have the —COCl functionality present in a molecule that has another functionality, such as an OH, alkene, or alkyne. For these molecules, the alkanoyl chloride functionality will always get priority and the carbonyl carbon is assigned #1. The position of the other functionality is specified based on the numbering system dictated by the location of the alkanoyl chloride functionality. An example of the nomenclature of an acid chloride, which also contains an alkene functionality is shown below.

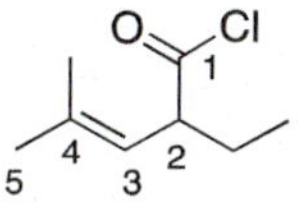

2-Ethyl-4-methyl-3-pentenoyl chloride

The "3" indicates the alkene functionality

The "e" indicates the alkene functionality

Problem 3.18

i) Give IUPAC names for the compounds shown below.

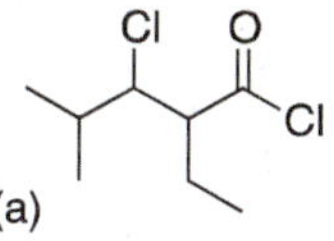

(a) 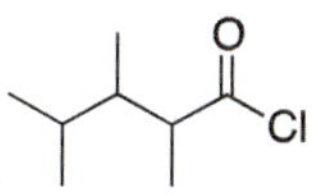   (b)    (c) 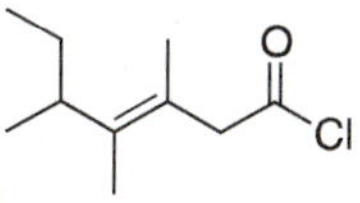   (d)

ii) Give line-angle structure for the following molecules.

  a) 2-Pentenoyl chloride
  b) 3-Methyl-4-nitrohexanoyl chloride
  c) 3-Chloro-3-butenoyl chloride
  d) 4-Methyl-2-pentynoyl chloride

3.11 Structure and Properties of Anhydrides

Anhydrides are compounds that typically have a strong odor, they are also lachrymators, corrosive to the eyes, can cause blisters, and they are very reactive with water. Owing to the reactivity of these compounds, they are typically used as intermediates for the synthesis of other compounds. Commonly used anhydrides are shown below. Note that they are symmetrical around a central oxygen atom and the two carbonyl groups; this feature is typical for most common anhydrides, but there are others where this is not the case.

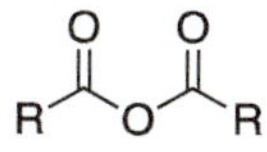

Structural feature of anhydrides

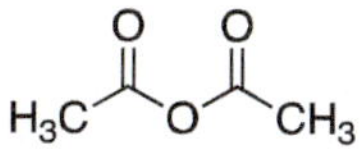

Acetic anhydride

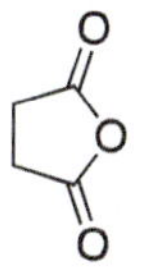

Succinic anhydride (a cyclic anhydride)

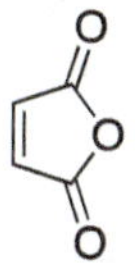

Maleic anhydride (a cyclic anhydride with an alkene functionality)

3.11.1 Nomenclature of Anhydrides

The root nomenclature of anhydrides is *alkanoic alkonoic anhydride* of which the carbonyl carbon is a part of the chain included in the alkanoic group, as demonstrated below.

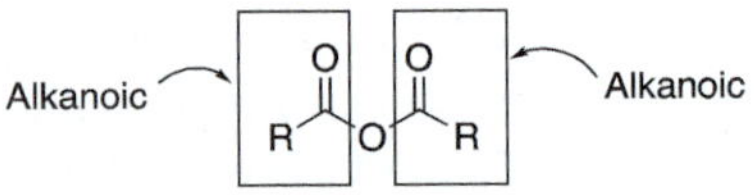

Alkanoic alkanoic anhydride

Based on this system, the name for the anhydride shown below is ethanoic propanoic anhydride. Since "E" comes before "P" in the alphabet, the ethanoic portion comes first in the nomenclature of this molecule.

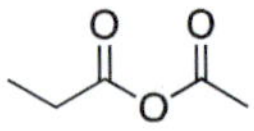

Ethanoic propanoic anhydride

Problem 3.19

i) Give the IUPAC names for the molecules shown below.

(a)   (b)   (c)

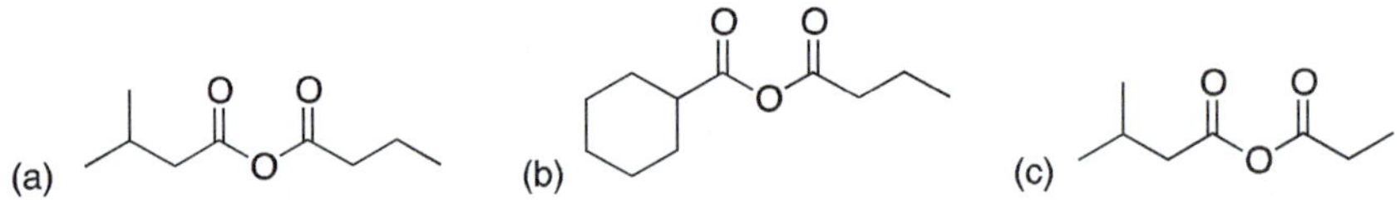

ii) Give line-angle structures for the molecules shown below.

   a) Butanoic ethanoic anhydride
   b) 3-Choropentanoic propanoic anhydride
   c) Cyclopentylmethanoic pentanoic anhydride

3.12 Structure and Properties of Amines

Most of the organic compounds examined thus far all have primarily carbon, hydrogen, and oxygen atoms. In organic chemistry and biological chemistry, the nitrogen atom is another commonly encountered element. Amines are organic compounds that contain nitrogen; most have a distinct odor, and they are readily soluble in water due to their polarity. Some amines are toxic and cause cancer of the liver and other organs, but a large percentage of amines have pharmaceutical benefits. The amine functionality is found in some very useful everyday drugs, which are used routinely in medicine, some are shown below.

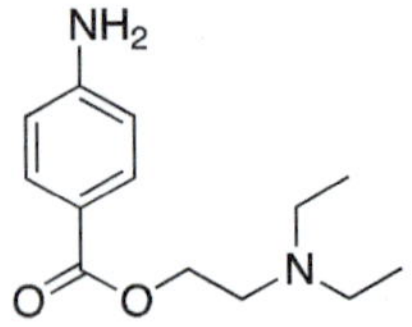

Novocaine, an anesthetic

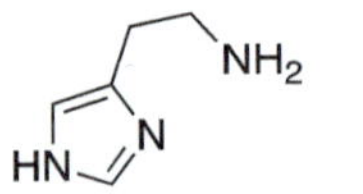

Histamine, used to dilate blood vessels

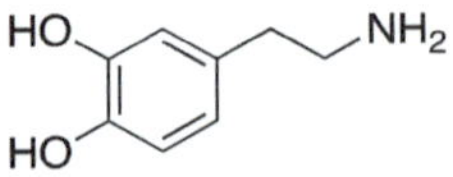

Dopamine a neurotransmitter

3.12 Structure and Properties of Amines

Amphetamine and metamphetamine are stimulants of the central nervous system.

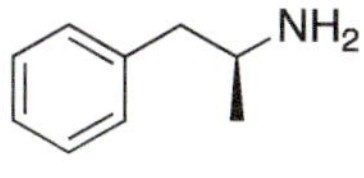

Amphetamine

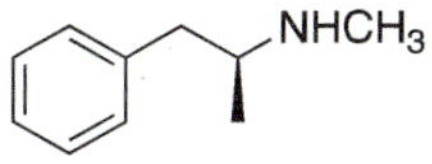

Methamphetamine

Nitrogen is also found in proteins, nucleic acids, and other natural occurring molecules, which will be discussed in more detail in Chapter 20.

There are a large number of organic molecules that are amines and they are routinely classified based on the number of alkyl groups or hydrogens that are bonded to the nitrogen. Ammonia is a specific amine in that it has only hydrogens bonded to the nitrogen. Primary amines have two hydrogen atoms bonded to the nitrogen and one alkyl group. Secondary amines have one hydrogen and two alkyl groups bonded to the nitrogen atom. Tertiary amines do not have a hydrogen bonded to the nitrogen, but have three alkyl groups bonded to the nitrogen atom. Thus, the number of alkyl groups bonded to the nitrogen atom is used to classify amines as primary (1°), secondary (2°), or tertiary (3°). The general representations of these different classifications are shown below.

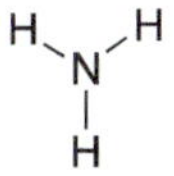

Ammonia 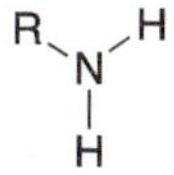   Primary amine (1°) 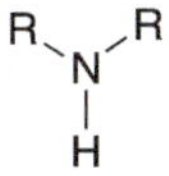   Secondary amine (2°) 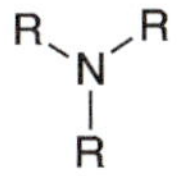   Tertiary amine (3°)

A fourth category exists if there are four alkyl groups bonded to the nitrogen; this category is classified as an ammonium salts, since it is ionic.

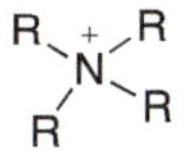

Quarternary ammonium ion

Shown below are examples of primary (1°) amines.

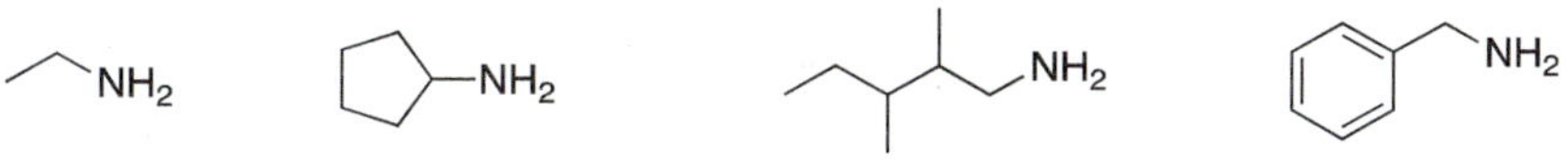

Shown below are examples of secondary (2°) amines.

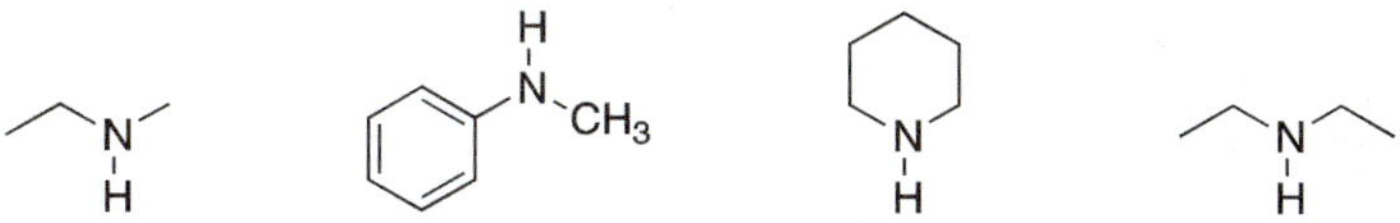

Shown below are examples of tertiary (3°) amines.

Shown below are examples of ammonium ion salts.

Problem 3.20

Classify the following amines as primary (1°), secondary (2°), or tertiary (3°).

(a) 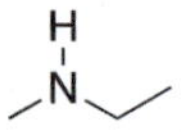 H-N (b) H₃C–NH₂ (c) 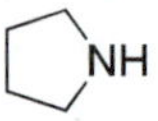 ⬠NH

(d) 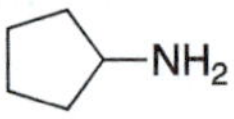 ⬠–NH₂ (e) 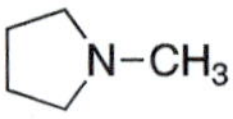 ⬠N–CH₃ (f) (CH₃)₃N

3.12.1 Nomenclature of Amines

In naming amines using the IUPAC system of nomenclature, the root name of the corresponding alkane that has the amine functionality is changed to alkanamine, i.e. the **e** of the alkane is changed to **amine**. In naming primary amines, identify the longest chain that contains the amine functional group and start numbering at the end closest to the amine group, as shown in the example below.

NH₂		NH₂	
CH₃—CH₂—CH₂—CH—CH₂—CH₃		CH₃—CH₂—CH₂—CH—CH₂—CH₃	
6 5 4 3 2 1		1 2 3 4 5 6	
Correct	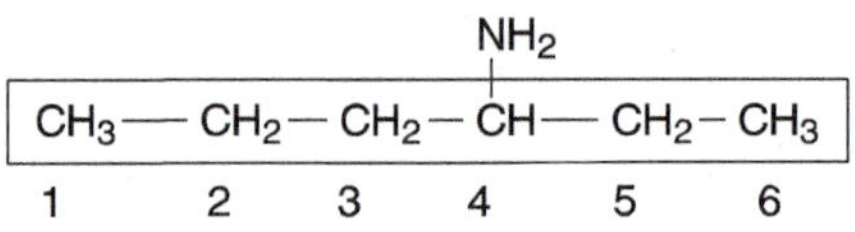	Not correct	

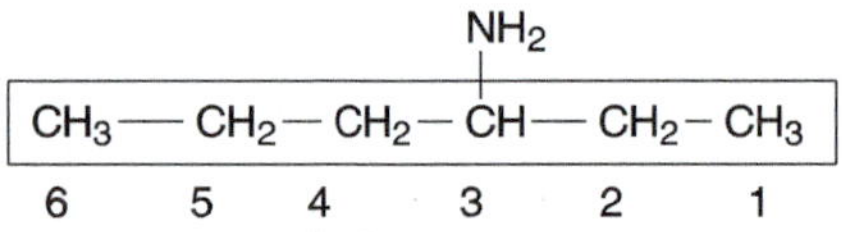

Next, the position of the amine functionality must be indicated with a number, as shown in the example below.

NH₂		NH₂	
CH₃—CH₂—CH₂—CH—CH₂—CH₃		CH₃—CH₂—CH₂—CH—CH₂—CH₃	
6 5 4 3 2 1		1 2 3 4 5 6	
3-Hexanamine (Correct)	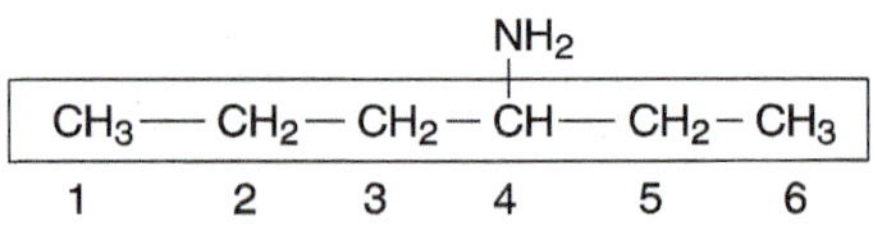	4-Hexanamine (Incorrect)	

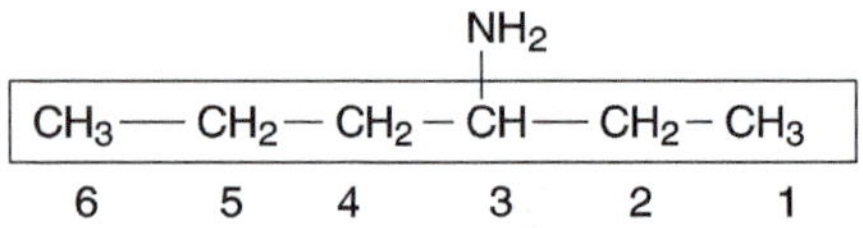

The locations of branches (substituents) are indicated by the numbers that are from the numbering system determined based on the presence of the amine functionality, as shown in the example below.

CH₃ NH₂		CH₃ CH₃ NH₂	
CH₃–CH–CH₂–CH–CH₂–CH₃		CH₃–CH–CH–CH–CH₂–CH₃	
6 5 4 3 2 1		6 5 4 3 2 1	
5-Methyl-3-hexanamine	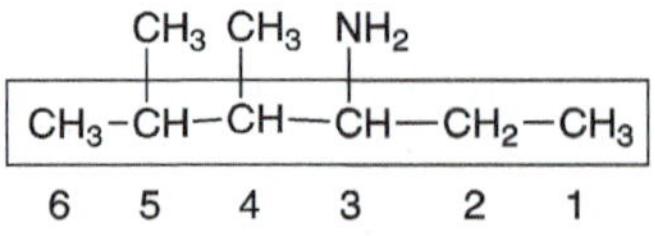	4,5-Dimethyl-3-hexanamine	

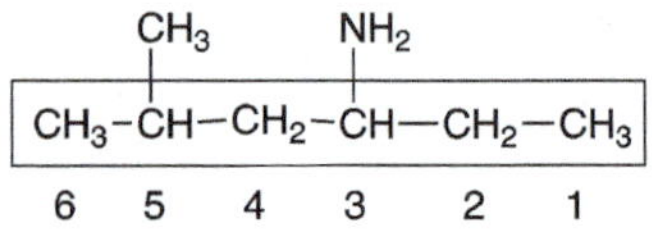

3.12 Structure and Properties of Amines

Problem 3.21

i) Give the IUPAC names for the compounds shown below.

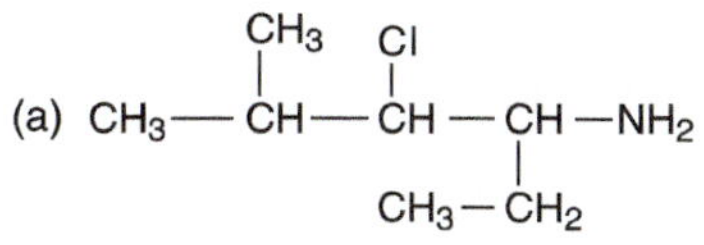

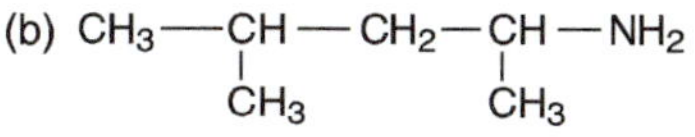

(a) CH₃—CH(CH₃)—CH(Cl)—CH(CH₃—CH₂)—NH₂

(b) CH₃—CH(CH₃)—CH₂—CH(CH₃)—NH₂

ii) Give line-angle structures for the following molecules.

a) 3-Methyl-2-pentanamine
b) 2,4-Dimethyl-2-hexamine

In naming (2°) or tertiary (3°) amines, each alkyl group that is bonded to the amine nitrogen is named and their presence is signified by the prefix *N*. If the amine is secondary with only one alkyl group bonded to the nitrogen, then only one *N* is used to indicate its presence. On the other hand, if the amine is tertiary with two groups bonded, then both groups are identified and their presence is indicated by two *N*s separated by a comma, as shown in the examples below.

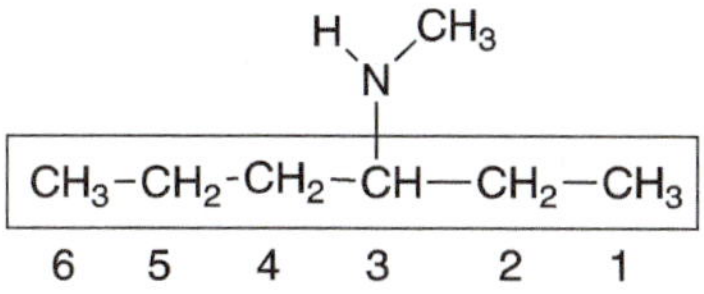

CH₃-CH₂-CH₂-CH(N(H)(CH₃))-CH₂-CH₃
6   5   4   3   2   1
N-Methyl-3-hexanamine

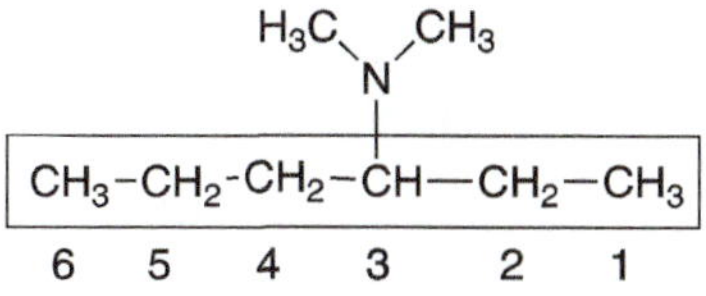

CH₃-CH₂-CH₂-CH(N(CH₃)(CH₃))-CH₂-CH₃
6   5   4   3   2   1
N,*N*-Dimethyl-3-hexanamine

Problem 3.22

i) Give the IUPAC names for the compounds shown below.

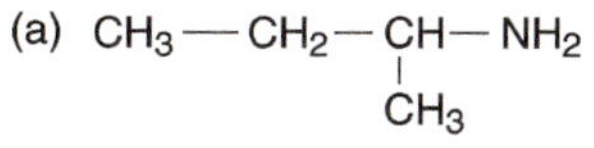

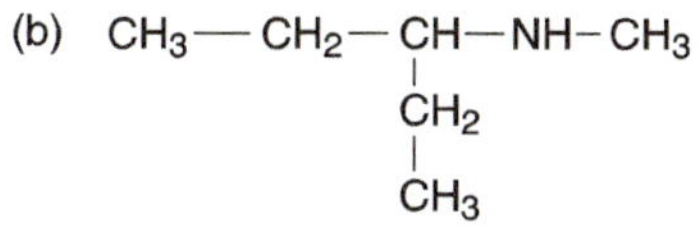

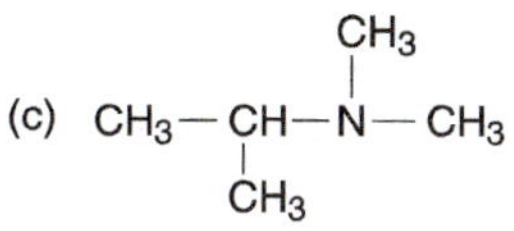

(a) CH₃—CH₂—CH(CH₃)—NH₂

(b) CH₃—CH₂—CH(CH₂—CH₃)—NH—CH₃

(c) CH₃—CH(CH₃)—N(CH₃)—CH₃

ii) Give line-angle structures for the following molecules.

a) *N*,*N*-Dimethyl-3-hexanamine
b) *N*-Ethyl-3-heptanamine
c) *N*,*N*-Dimethyl-2-propanamine

For amines that have cyclic structures, identify the cyclic system and if the nitrogen is not a part of the ring, name the amine as a derivative of the cyclic system; examples are shown below.

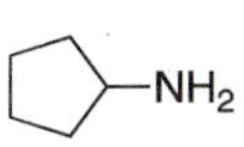

Cyclopentanamine 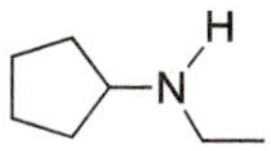   *N*-Ethylcyclopentanamine 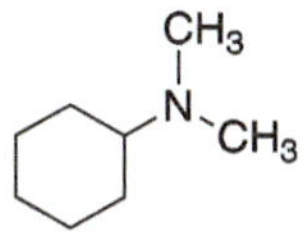   *N*,*N*-Dimethylcyclohexanamine

If the amine group is bonded to the phenyl ring (—C₆H₅), this compound has a special name, aniline. If groups are bonded to the phenyl ring, these compounds are named as substituted anilines, as shown in the examples below.

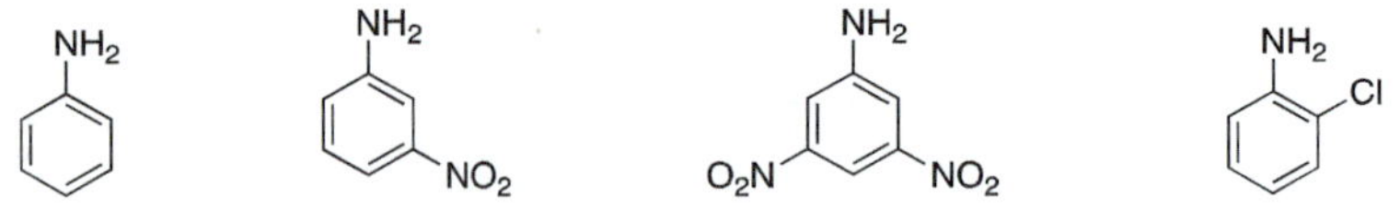

Aniline   3-Nitroaniline   3,5-Dinitroaniline   2-Chloroaniline

If the nitrogen is part of the ring system, then they are typically named as a derivative of a common name, such as shown below:

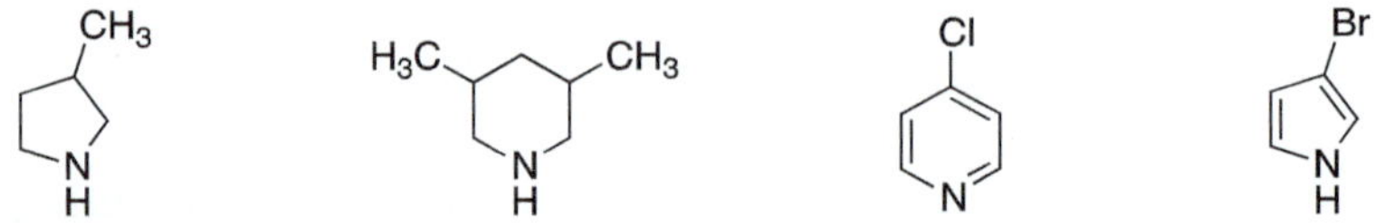

3-Methylpyrrolidine   3,5-Dimethylpiperdine   4-Chloropyridine   3-Bromopyrrole

3.12.2 Nomenclature of Difunctional Amines

For most difunctional molecules that contain the amine functionality, the amino group is named as a substituent, based on the root name of another functional group that is present in the molecule. Examples are shown below.

3-Aminopentanoic acid   3-*N,N*-Dimethylaminopentanol

Functional groups that are of low priority, such as the alkene or alkyne functionalities, are often incorporated in the nomenclature using the alkanamine system as shown in the example below.

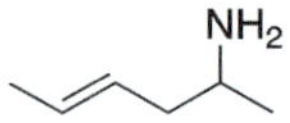

4-Hexen-2-amine

Problem 3.23

i) Give the IUPAC names for the compounds shown below.

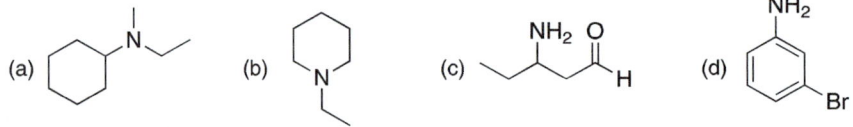

ii) Give line-angle structures for the following molecules.

    a) *N,N*-Dimethylcyclopentanamine
    b) 4-Amino-2-pentanone
    c) 2-Fluoroaniline

3.13 Structure and Properties of Amides

The amide functional group is found in various molecules, not only in organic chemistry but also in biochemistry and biology. Proteins and peptides are important molecules to life, which are amides and some very useful polymers are polyamides. The amide functionality consists of a carbonyl group that is bonded to an sp^3 nitrogen. It is possible to have two hydrogens bonded to the nitrogen of the amide functionality; one hydrogen atom and one alkyl group; or two alkyl groups. Amides can be described as primary, secondary, or tertiary based on the number of hydrogens or alkyl groups bonded to the nitrogen as illustrated below.

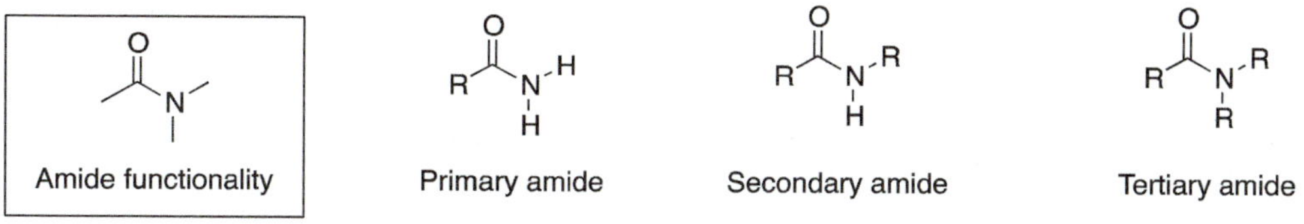

Shown below is the amide bond as part of a dipeptide.

A dipeptide

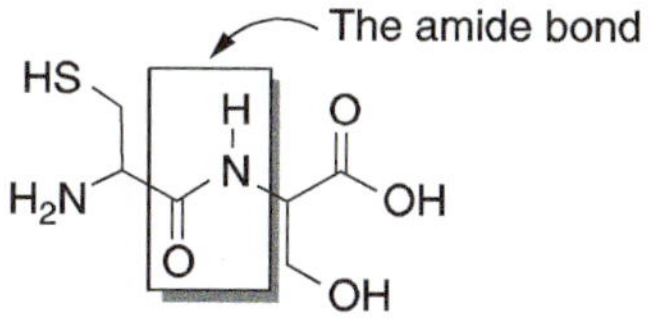

3.13.1 Nomenclature of Amides

In naming primary amides, identify the longest chain that contains the amide functional group and assign the carbonyl carbon #1, as illustrated below.

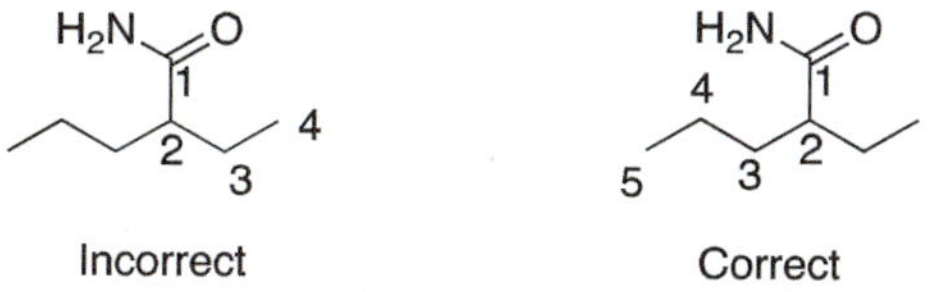

The root name of an amide is based on the number of carbons of the corresponding alkane – the alkane is changed to ***alkanamide***. Thus, the root name for the molecule above is pentanamide and not butanamide. Since the amide functionality is a terminal functionality, it is not necessary to indicate its position by #1. The position of any group (substituent) must be specified by a number based on the numbering system used to determine the longest chain. Thus, the complete IUPAC name of the primary amide above is shown below, along with an example of the IUPAC name of another primary amide.

2-Ethyl-4-methylpentanamide

5-Chloro-2-ethyl-4-methylpentanamide

In naming secondary and tertiary amides, the substituents of the nitrogen are named and specified by the prefix, *N*; examples are shown below.

N-Methylpentanamide

N,*N*-Dimethylpentanamide

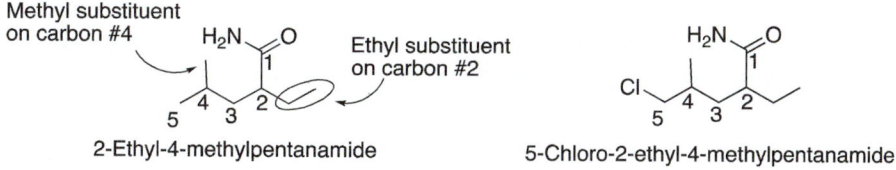

If there are substituents on the longest chain, they are specified based on the numbers assigned to derive the root name of the molecule. Examples are shown below.

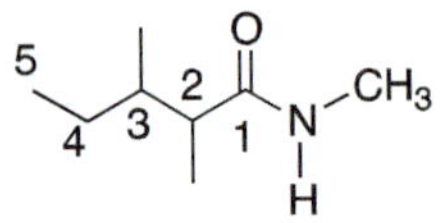

N-Methyl-2,3-dimethylpentanamide 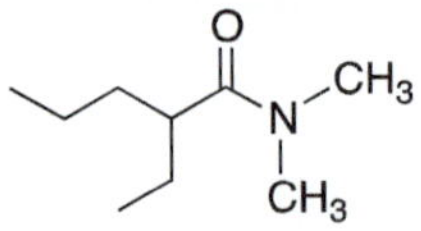   N,N-Dimethyl-2-ethylpentanamide

Problem 3.24

i) Give IUPAC names of the following amides.

(a)   (b)

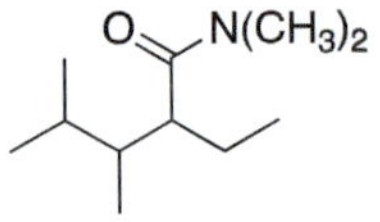

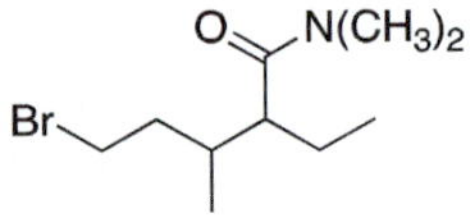

ii) Give the structures for the molecules below.

a) N,N-Diethyl-3-methylpentanamide
b) 2,2,3-Trimethylhexanamide

3.14 Structure and Properties of Nitriles

The nitriles functionality (—CN) contains an sp hybridized carbon, which is bonded to an sp hybridized nitrogen and to another group, typically an alkyl group. Nitrile rubber is a synthetic copolymer made from acrylonitrile and butadiene. Nitrile rubber is used to make protective gloves, hoses, and seals since it is highly resistant to chemicals and moisture. Methyl cyanoaclylate, another type of nitrile, is an integral ingredient of many types of glues that are used for everyday household needs. Nitriles have the functional feature shown below.

R—C≡N   H$_2$C═CH—C≡N   N≡C—CH$_2$—C(=O)—OCH$_3$

Nitrile   Acrylonitrile   Methyl cyanoacrylate

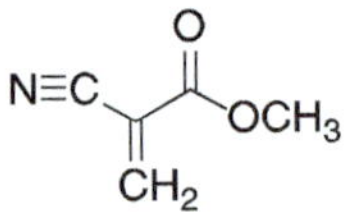

3.14.1 Nomenclature of Nitriles

In naming nitriles, the longest continuous chain is named as an **alkanenitrile.** For the nitrile shown below, the longest continuous chain is named as an alkanenitrile, note that the carbon of the nitrile gets #1.

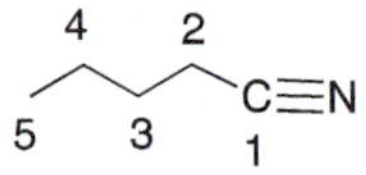

Pentanenitrile

Substituents (groups) that are bonded to the longest chain of the nitrile are assigned numbers based on the numbering system of that root name.

Problem 3.25

Give the IUPAC names for each of the following compounds.

(a) 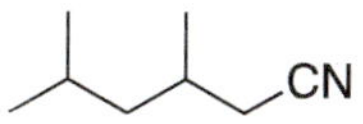 [structure: 3-methylpentanenitrile-like chain with CN]

(b) 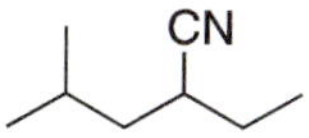 [structure with CN on middle carbon]

(c) 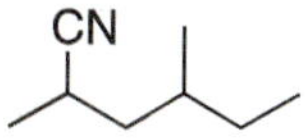 [structure with CN on middle carbon]

3.15 Structure and Properties of Ethers

Ethers have the functional group R—O—R, where R represents alkyl saturated groups or unsaturated groups, including the phenyl group. These alkyl groups may be the same or different. Ethers have very strong C—O—C bonds, and as a result, their toxicity effects are low, but they can depress the central nervous system and have been used as anesthetic for surgery. Diethyl ether was first used as an anesthetic in 1846. It is a powerful anesthetic and not very toxic to patients, but some patients had symptoms of nausea and vomiting after surgery. Another ether that was widely used for patients as well as in veterinary medicine is isoflurane. Structures of common ethers are shown below.

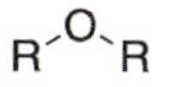

| R—O—R | Diethyl ether | Isoflurane | Ethyl methyl ether | Ethyl phenyl ether |

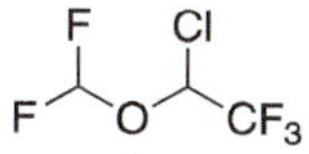

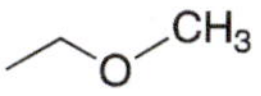

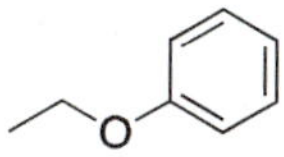

Structural feature of ethers

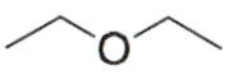

Ethers are relatively unreactive compounds and as a result are used as common organic solvents in the lab. Owing to the low boiling points of ethers that have low molecular weights, they are typically flammable and explosive liquids. Diethylether, commonly referred to as ether, is used as a starter fluid for diesel and gasoline engines. Methyl *tert*-butyl ether (MTBE) is a common organic solvent, which is also used as an octane-boosting additive for gasoline engines.

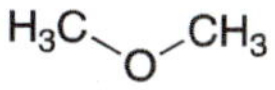

Dimethyll ether (boiling point = –25 °C)

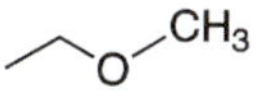

Ethyl methyl ether (boiling point = 8 °C)

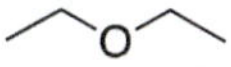

Diethyl ether (boiling point = 34 °C)

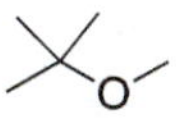

MTBE Methyl *tert*-butyl ether (boiling point = 55 °C)

Cyclic ethers are also possible. The smallest cyclic ether, also called an epoxide or oxirane, is one that contains an oxygen atom and two carbons to form a three member-ring molecule. These molecules are very reactive and are used frequently as reagents of organic synthesis. One of the reasons for the reactivity of these molecules is due to the release of the ring strain found in these molecules. Carbons, which are sp^3 hybridized, would prefer to have bond angles of approximately 109.5°, but the bond angles in oxiranes are approximately 60°.

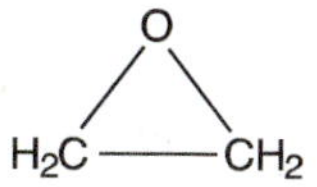

Epoxide (oxirane)

Larger cyclic ethers include tetrahydrofuran (THF) and 1,4-dioxane, which are used as common solvents in industry and the organic research labs. Many sugar molecules have the pyran structure and are usually referred to as a pyranose ring, as we will see later when we study monosaccharides and carbohydrates in Chapter 20. Shown below are examples of common cyclic ethers.

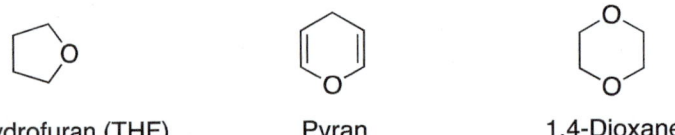

Tetrahydrofuran (THF)　　　　Pyran　　　　1,4-Dioxane

Crown ethers are large polycyclic ethers, owing to their size, they are often referred to as macromolecules.

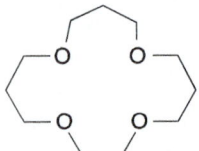

A crown ether

Depending on the number of oxygen atoms, the size of the cavity in the middle of crown ethers can be different. Crown ethers form stable complexes with different metal ions, such as Na^+, K^+, or Hg^{2+}. Owing to different sizes of the cavity in the middle of these compounds, they are often used to extract specific metal ions that can fit in the cavity.

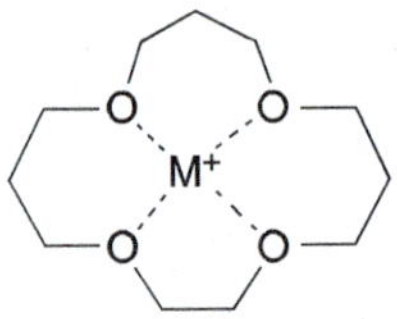

A crown ether with a metal ion in its cavity

12-Crown-4 is shown below, and the size of the cavity is just the right for the lithium cation, as shown below. It is labeled as 12-crown-4 since there are 12 atoms in the complete structure and 4 oxygen atoms that are involved in the interaction with the lithium cation.

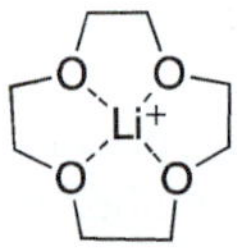

12-Crown-4 ether with a lithium cation (Li^+)

For larger cations, such as sodium and potassium, larger crown ethers would be needed. For the sodium cation, 15-crown-5 is used, and for potassium, which is even larger, would require 18-crown-6.

Problem 3.26

Give the structures of the crown ether that would be necessary for sodium cation ion (15-crown-5).

Epoxides are also important intermediates in biosynthesis of compounds, such as steroid hormones, which is illustrated below.

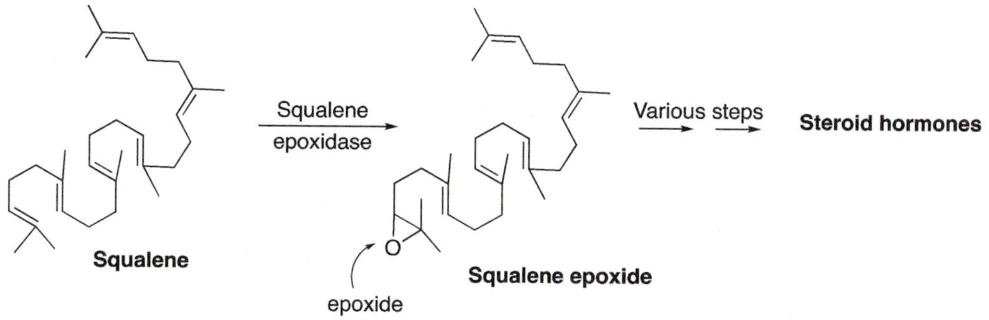

3.15 Structure and Properties of Ethers

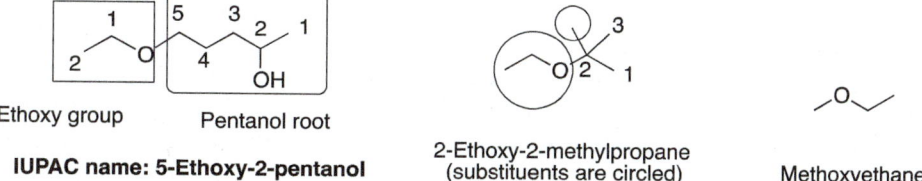

trans-4,5-Epoxy-(E)-2-decenal is the compound in human blood that gives it its characteristic metallic odor. The metallic odor of *trans*-4,5-epoxy-(E)-2-decenal is used by predators to locate blood or prey.

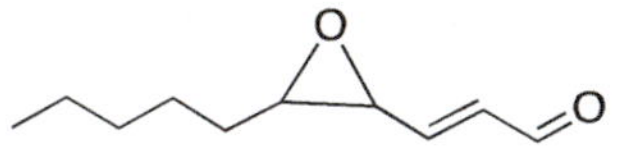

trans-4,5-Epoxy-(E)-2-decenal

3.15.1 Nomenclature of Ethers

The common method of naming ethers is to name the two groups on either side of the ether functionality followed by the word ether. So CH_3-O-CH_3 is dimethyl ether and $CH_3OCH_2CH_3$ is ethyl methyl ether. Note that since the letter "e" comes before the letter "m" in the alphabet, ethyl comes before methyl in the name of this compound. However, for the IUPAC system, the longest chain that does not include the RO is named as an alkane and the other portion is named as an alkoxy substituent. Thus, $CH_3OCH_2CH_3$ is methoxyethane. Other examples are shown below.

Ethoxy group   Pentanol root

IUPAC name: 5-Ethoxy-2-pentanol

2-Ethoxy-2-methylpropane
(substituents are circled)

Methoxyethane

3.15.2 Nomenclature of Oxiranes

In naming epoxides, or three member-ring ethers, a numbering system must be established and the numbering starts at the oxygen and proceeds in the direction that contains a group, or the group that has the highest priority. Shown below is the numbering of a substituted epoxide.

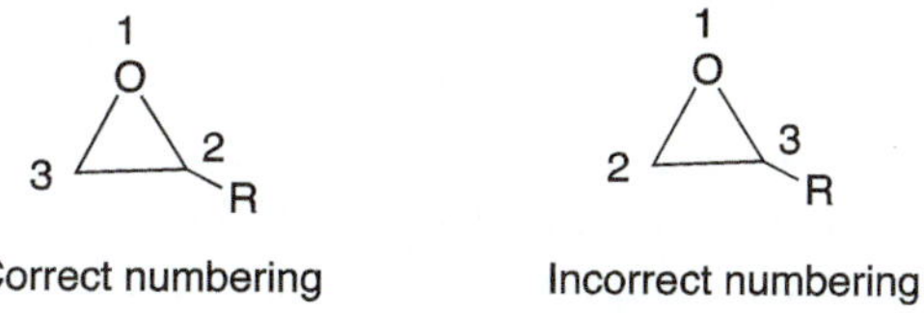

Correct numbering   Incorrect numbering

Numbering of oxiranes

In naming epoxides, identify the group on the carbon and name it as a substituent of oxirane. If there are two groups on the same carbon or on different carbons, numbers are used to indicate the locations of these groups. That is, if there are two groups on carbons 2 and 3 of an oxirane, it is named as a 2,3-disubstituted oxirane. If two groups are on the same carbon of an oxirane, it is named as 2,2-disubstituted oxirane. Examples are shown below.

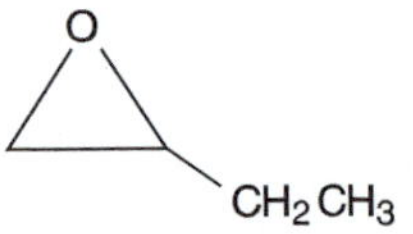

2-Ethyloxirane 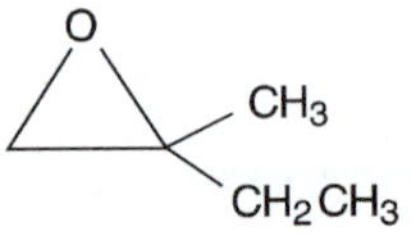   2-Ethyl-2-methyloxirane 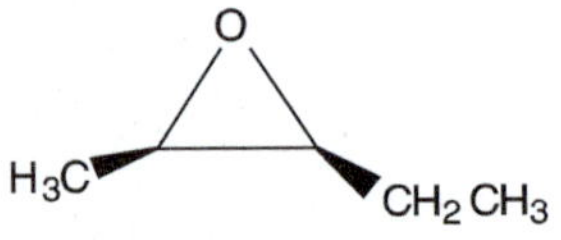   cis-2-Ethyl-3-methyloxirane

Since the three-member oxirane ring is a rigid system, groups can be bonded to either side of the of epoxide ring, resulting in different isomers. The cis and trans notations are used to name such isomers.

Problem 3.27

Give the structures of the following molecules.

a) 2-Methoxybutane
b) Butyl ethyl ether
c) *trans*-2-Ethyl-3-methyloxirane

3.16 An Overview of Spectroscopy and the Relationship to Functional Groups

Everyday, new organic compounds are being identified and isolated from natural sources, and new compounds are also constantly being synthesized in the research laboratories. As we know, some of these compounds from natural sources, such as plants, have healing and medicinal benefits, and as we have pointed out in Chapter 1, if they are to be reproduced in large quantities for pharmaceutical use, we have to know their structures, and especially the type of functional groups present. In addition, for such compounds to be synthesized, chemists need to make sure that the newly synthesized compounds contain the correct functional groups and that they are actually the intended compounds. Thus, an integral aspect of an organic chemist's work is the determination of the actual structure of newly discovered and synthesized compounds. For less complicated compounds, it is possible to make structure determination based on physical methods. Since we know that the physical properties of compounds are different and the probability of any two compounds with the exact same set of physical properties is very unlikely, once the physical properties of an unknown compound are known, a comparison can be made with known compounds to determine its identity by matching the physical properties of the known with the unknown. This method is not the best method for larger more complex organic molecules that have many functional groups, such as those of compounds obtained from natural sources and compounds synthesized in a research lab. As a result, chemists have resorted to another method to determine the type of functional groups that are present in unknown compounds. Spectroscopic methods can be used to assist in the determination of the presence of different functional groups

3.16.1 Infrared Spectroscopy

and eventually the structures of unknown compounds. We will cover spectroscopy in much greater details in Chapter 13; in this section, we will briefly cover one type of spectroscopy, infrared (IR) spectroscopy, that is integral for the identification of functional groups on unknown molecules.

3.16.1 Infrared Spectroscopy

Once molecules are allowed to interact with energy of a specified frequency, the molecules reach excited states owing to the energy absorbed. Once this energy is absorbed, there are different outcomes. For example, when molecules are excited by energy in the IR region of the electromagnetic spectrum, the bonds of the molecules vibrate and bend in a predictable manner. The bonds in a molecule can be described as a spring holding together the two atoms of a bond and that there is a constant vibration of this spring. The energy required to vibrate a strong bond is less than the energy required to vibrate a weaker bond. Thus, a scan of energy of the IR region of the electromagnetic spectrum will cause different bonds of a molecule to vibrate at different frequencies. This type of scan is known as a spectrum and if a comparison of the IR spectrum of an unknown compound is made against that of known compounds, an indication of the type of bonds, and hence functional groups, that are present in a molecule can be achieved. The IR spectrum of water is given in Figure 3.1.

There are important features of an IR spectrum that should be pointed out. The x-axis is given in wavenumbers (cm^{-1}) and reflects the energy supplied to the molecule in the IR region of the electromagnetic spectrum. For the IR spectrum shown in Figure 3.1, the energy was scanned from around $4000\ cm^{-1}$ to around $900\ cm^{-1}$. The y-axis (transmittance) shows regions of the scan where the molecule absorbed energy and specific vibrational modes were excited as described earlier. Note that there is a noticeable signal (also known as band) at $3400\ cm^{-1}$. This signal is due to the O—H stretch vibration brought about by the energy in that region of the electromagnetic spectrum. All molecules that contain an O—H bond have a signal in this region of the IR spectrum. Figure 3.2 shows the IR spectrum of methanol;

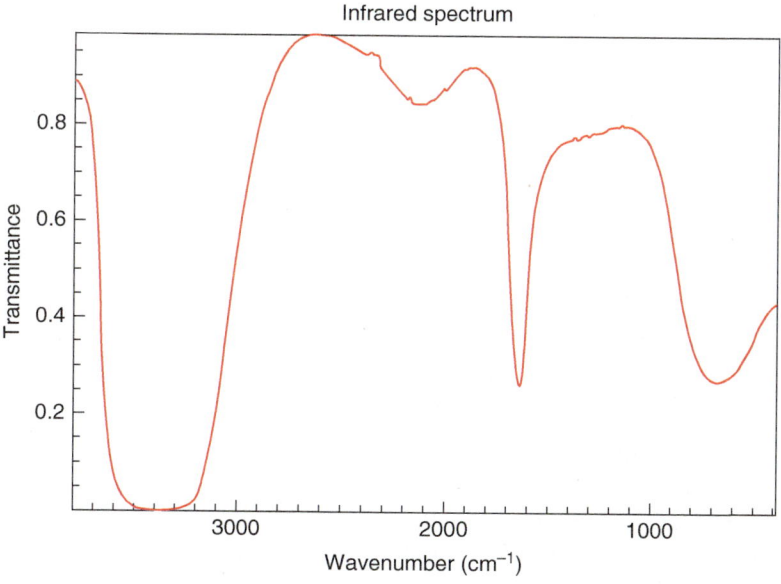

Figure 3.1 Infrared spectrum of water (H—O—H). *Source:* with permission from NIST.

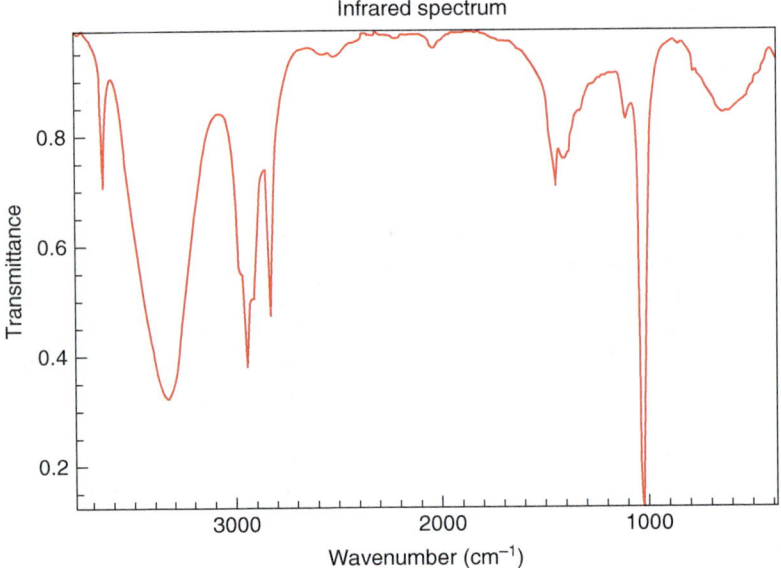

Figure 3.2 Infrared spectrum of methanol (CH$_3$—OH). *Source:* with permission from NIST.

note that there is a signal that occurs in the region of 3300 cm^{-1}, which indicates the presence of an O—H group.

Problem 3.28

Shown below are two IR spectra, one is for tetrachloromethane (CCl$_4$) and the other for 2-propanol. Predict which belongs to each compound.

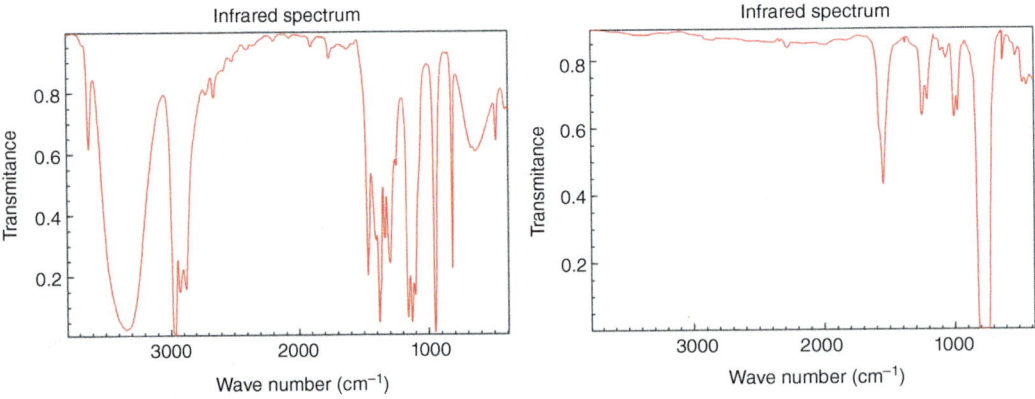

This type of analysis can be used to identify the different types of bonds and functional groups present in a molecule. Table 3.1 shows the regions of the IR spectrum where different vibrational modes of different functional groups occur. By comparison of this information with that obtained from the IR spectrum of an unknown compound, the presence of specific functional groups can be determined.

Table 3.1 Selected IR frequencies of common functional groups.

Bond	Functional group	Type vibration	Frequency (cm^{-1})	Intensity[a]
C—H	Alkane	Stretch	3000–2850	s
	CH$_3$	Bend	1450 and 1375	m
	–CH$_2$–	Bend	1465	m
	Alkenes	Stretch	3100–3000	m
		Out-of-plane bend		
	Aromatics	Stretch	3150–3050	s
		Out-of-plane bend	900–690	s
		Stretch		w
C≡C	Alkyne	Stretch	~3300	s
C=O	Aldehyde		1740–1720	s
	Ketone		1725–1705	s
	Carboxylic acid		1725–1700	s
	Ester		1750–1730	s
	Amide		1670–1640	s
	Anhydride		1810–1740	s
	Acid chloride		1800	s
C—O	Alcohols, ethers, esters, carboxylic acids, anhydrides		1300–1000	s
O—H	Alcohols, phenols		3650–3600 (free) 3650–3600 (H-bonded)	m
—COOH	Carboxylic acid		3400–2400	m
N—H	Primary and secondary amines and amides	Stretch	2500–3100	m
		Bend	1640–1550	m-s
C—N	Amines		1350–1000	m-s
C=N	Imines and oximes		1690–1640	m-s
C≡N	Nitriles		2260–2240	m
X=C=Y	Allenes, ketenes, isocyanates, isothiocyanates		2270–1950	m-s
N=O	Nitro (—NO$_2$)		1550–1350	s
S—H	Thiols		2250	w
S=O	Sulfoxides		1050	s
	Sulfones, sulfonyl chlorides, sulfates, sulfonamides		1375–1300	s
C—X	X = Fluorine		1400–1000	s
	X = Chlorine		800–600	s
	X = Bromine, iodine		<667	s

[a] Intensity of the signal.

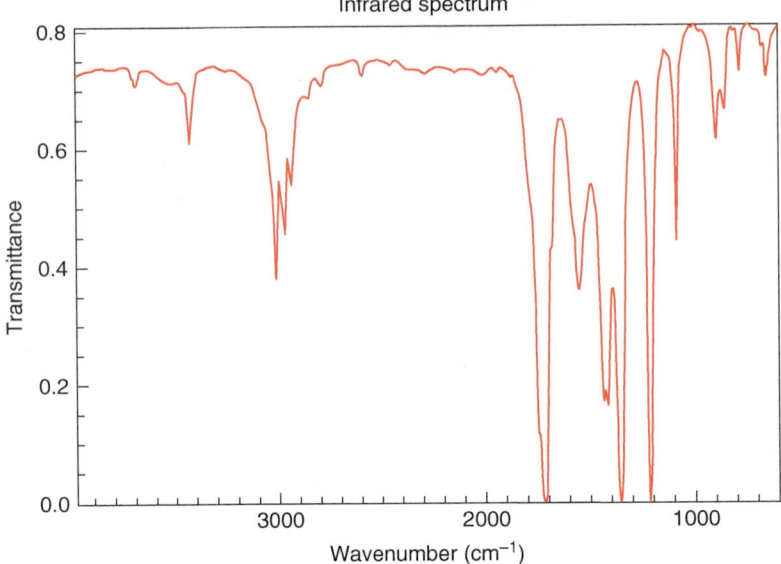

Figure 3.3 Infrared spectrum of 2-propanone (CH$_3$COCH$_3$). *Source:* with permission from NIST.

Shown in Figure 3.3 is the IR spectrum for 2-propanone. A close examination of the spectrum reveals that there is no signal around 3400 cm^{-1}, which indicates that this molecule does not have an O—H functionality. However, there is a prominent band at 1700 cm^{-1}. Table 3.1 shows that a band in the region of 1725–1705 cm^{-1} corresponds to the carbonyl of a ketone. Thus, this is confirmation that the molecule has a carbonyl group and more than likely in the form of a ketone.

Problem 3.29

Shown below are two IR spectra, one is for benzyl alcohol (C$_6$H$_5$CH$_2$OH) and the other is for acetophonone (C$_6$H$_5$COCH$_3$), identify which belongs to each compound.

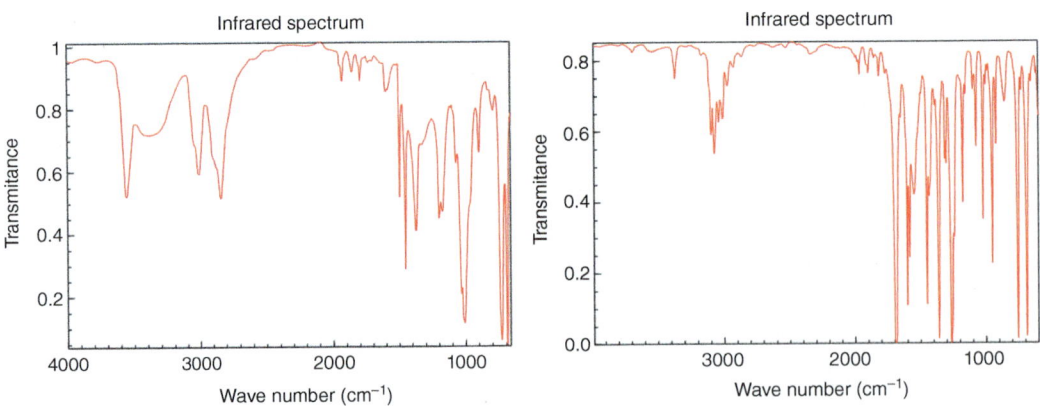

Thus, IR spectroscopy is a major tool that gives chemists the ability to determine the presence of different functional groups in different molecules.

End of Chapter Problems

3.30 Give the structure of a molecule that has each of the following functional groups.

a) alcohol   b) carboxylic acid   c) alkyne   d) alkene   e) amine

3.31 Determine if the names given below for each of the molecules given are correct. If not, give the correct IUPAC name.

(a) 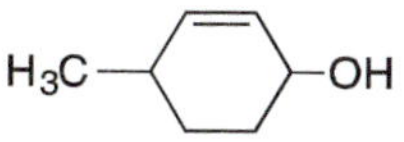 H₃C—⟨cyclohexene⟩—OH
4-Methyl-2-cyclohexenol

(b) 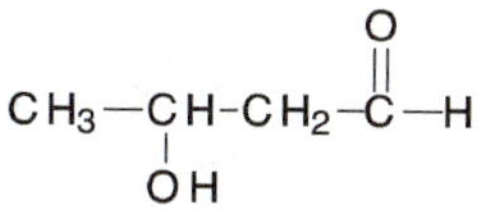 CH₃—CH(OH)-CH₂-C(=O)—H
3-Hydroxybutanol

(c) 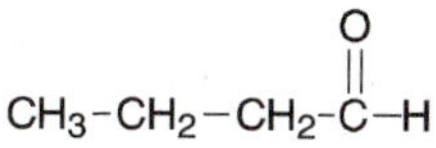 CH₃-CH₂-CH₂-C(=O)-H
Butanol

(d) 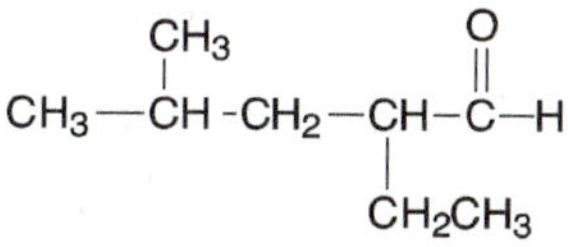 CH₃—CH(CH₃)-CH₂—CH(CH₂CH₃)-C(=O)—H
2-Ethyl-4-methylpentanol

(e) 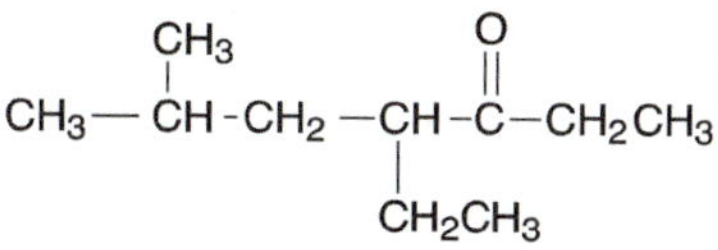 CH₃—CH(CH₃)-CH₂—CH(CH₂CH₃)-C(=O)-CH₂CH₃
4-Ethyl-6-methyl-3-heptanone

(f) 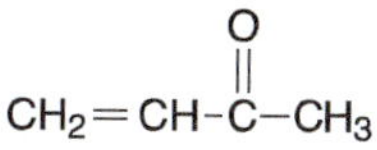 CH₂=CH-C(=O)-CH₃
3-Penten-2-one

3.32 Give the name of the functional group found in each of the following molecules.

(a)  alkyne structure
(b) 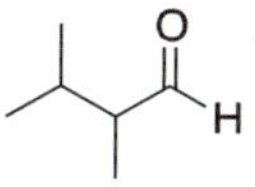 aldehyde structure
(c) 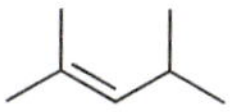 alkene structure
(d) 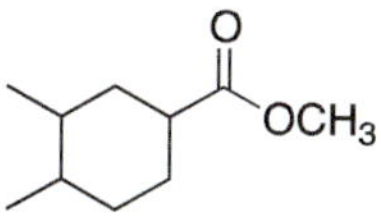 ester with OCH₃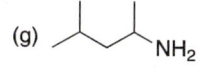
(e) 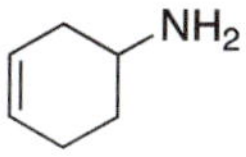 cyclohexene with NH₂
(f) 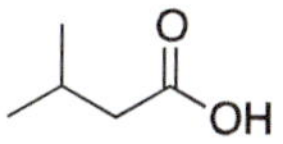 carboxylic acid
(g) amine with NH₂
(h) 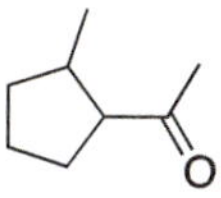 ketone on cyclopentane
(i) 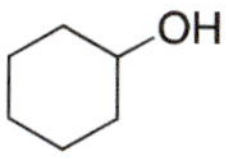 cyclohexanol OH

3.33 Give <u>line-angle formulas</u> for the following molecules.

i) Ethyl ethanoate
ii) 2-Chlorohexanal
iii) 6-Methyl-4-hepten-2-one
iv) 3-Chlorobutanoic acid
v) 3-Hexanol
vi) Methyl-2-hexanethiol
vii) 4-Methyl-2-pentanamine
viii) 4-Bromo-3-methyl-2-pentanone
ix) 3-Bromobutanal

3.34 There are four different alcohols with the molecular formula $C_4H_{10}O$. Give the line-angle formula for each and indicate which are primary, secondary, or tertiary.

3.35 Give IUPAC names for the following molecules.

i) 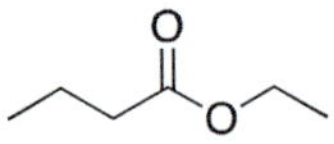 [structure] ii) 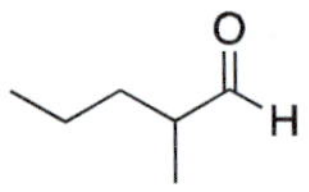 [structure]

iii) 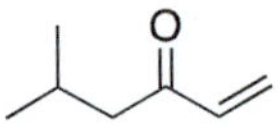 [structure] iv) 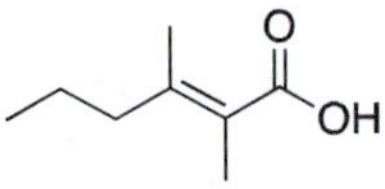 [structure]

v) 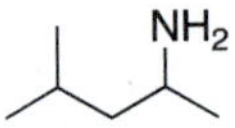 [structure] vi) 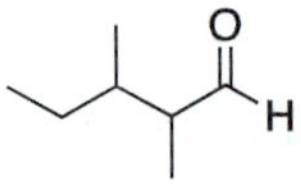 [structure]

vii) 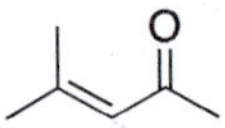 [structure] viii) [structure]

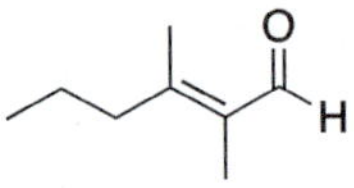

ix) 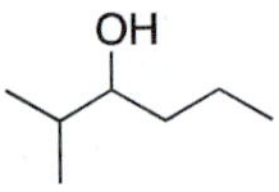 [structure] x) 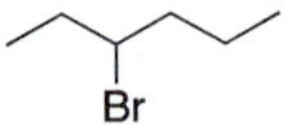 [structure]

3.36 Shown below is the structure of ethambutol, which is used to treat tuberculosis. Identify the functional groups present in ethambutol.

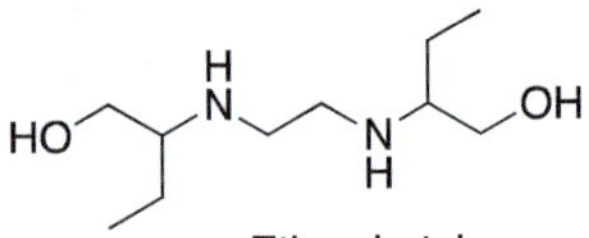

Ethambutol

3.37 Shown below is the lipid-lowering drug, simvastatin. Identify the functional groups present in simvastatin.

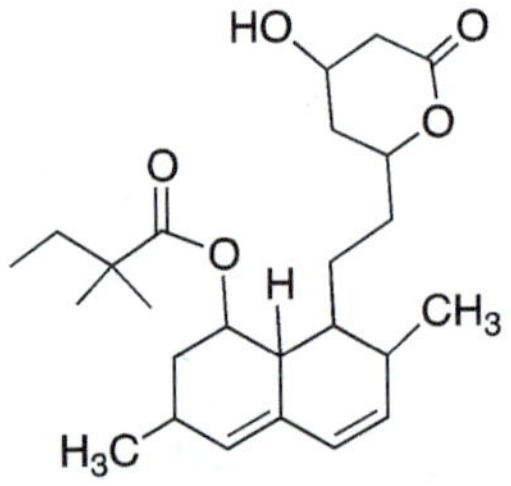

Simvastatin (lipid-lowering agent)

3.38 Naltrexone is used to treat narcotic addition, identify all the functional groups present in naltrexone.

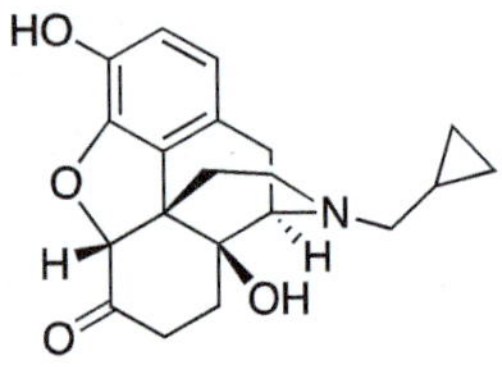

Naltrexone

3.39 The drug acebutolol is used in the treatment of various heart diseases. Identify the functional groups present in acebutolol.

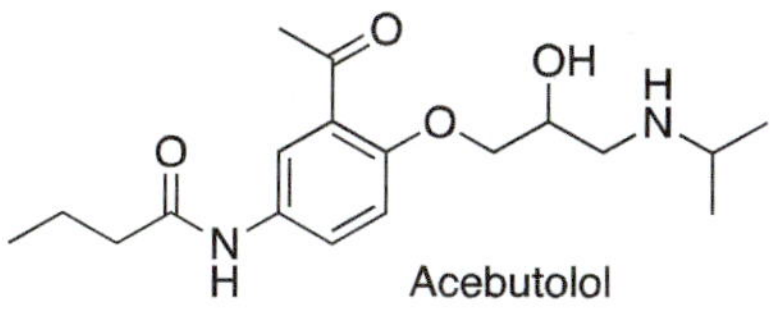

Acebutolol

3.40 An unknown compound shows a strong IR band at 1710 cm^{-1} and has a molecular formula of $C_5H_{10}O$, it was shown to be an acyclic compound. What is a most likely structure of this unknown compound?

3.41 For each of the following molecular formulas shown below, give a line-angle structure that contains only the indicated functionality.

a) $C_5H_8O_2$ (carboxylic acid)
b) $C_6H_{10}O$ (ketone)
c) C_6H_{10} (alkyne)
d) $C_5H_{10}O$ (aldehyde)
e) C_3H_5N (nitrile)
f) $C_4H_{10}O$ (ether)

g) C_3H_9N (2° amine)
h) $C_4H_{11}N$ (3° amine)
i) $C_4H_8O_2$ (ester)
j) $C_5H_{10}O_2$ (carboxylic acid)
k) $C_4H_{10}O$ (alcohol)
l) $C_5H_{11}NO$ (amide)

3.42 Label each alcohol shown below as primary (1°), secondary (2°), or tertiary (3°).

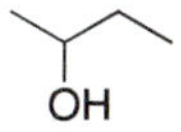

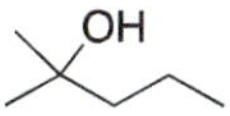

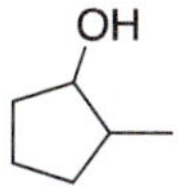

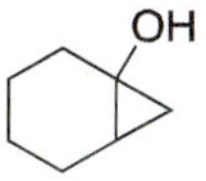

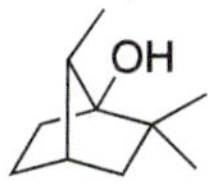

3.43 Give the structures of all alcohol isomers with the molecular formula $C_5H_{12}O$, name them and indicate which are primary, secondary, or tertiary.

3.44 Give the IUPAC names for each of the following molecules.

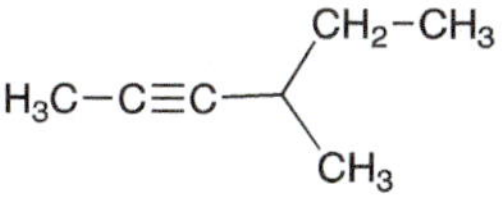

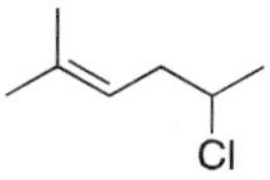

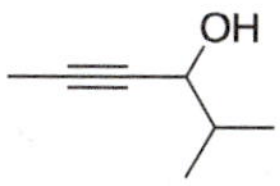

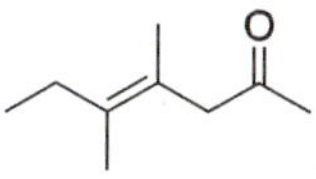

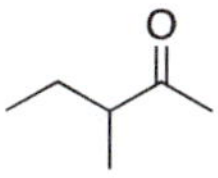

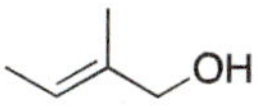

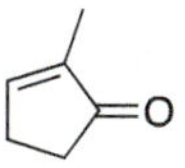

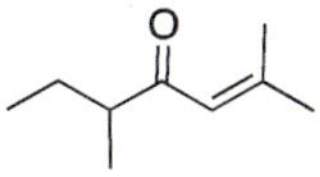

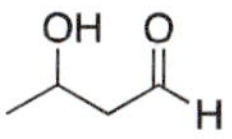

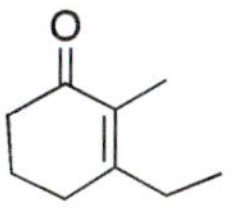

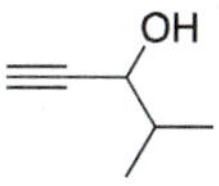

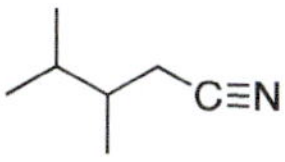

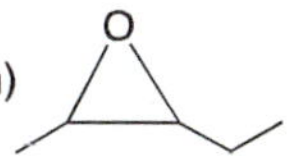

3.45 Give <u>line-angle structures</u> for the following compounds.
   a) 1-Pentyn-3-ol
   b) 2-Chloro-3-methylbutanal
   c) 4-Methyl-2-pentanone
   d) 5-Methyl-3-hexen-2-one
   e) 2-Chlorohexanal
   f) 6-Methyl-4-hepten-2-one
   g) 3-Hexanol
   h) 2-Chloro-2-hexenal
   i) 3-Bromo-4-methylcyclopentanone
   j) 3-Chloro-2-butenoic acid
   k) 2-Methyl-2-hexenal
   l) Ethyl ethanoate
   m) Propyl butanoate
   n) 3-Bromo-4-methylcyclohexanone
   o) 3-Hydroxybutanal
   p) 3-Chloro-2-butenoic acid
   q) *trans*-2,3-Diethyloxirane
   r) 3-Methylpentanenitrile

3.46 Shown below are the structures of four compounds with different functional groups.

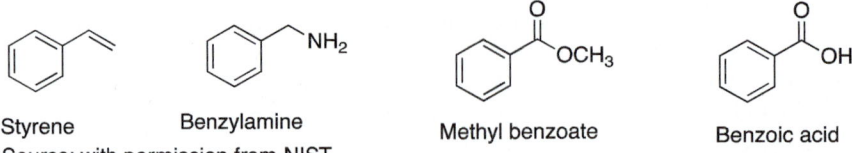

Styrene Benzylamine Methyl benzoate Benzoic acid
Source: with permission from NIST.

The IR spectra of the compounds given above are shown below. Utilizing the information contained in Table 3.1, match the spectra with the compounds given.

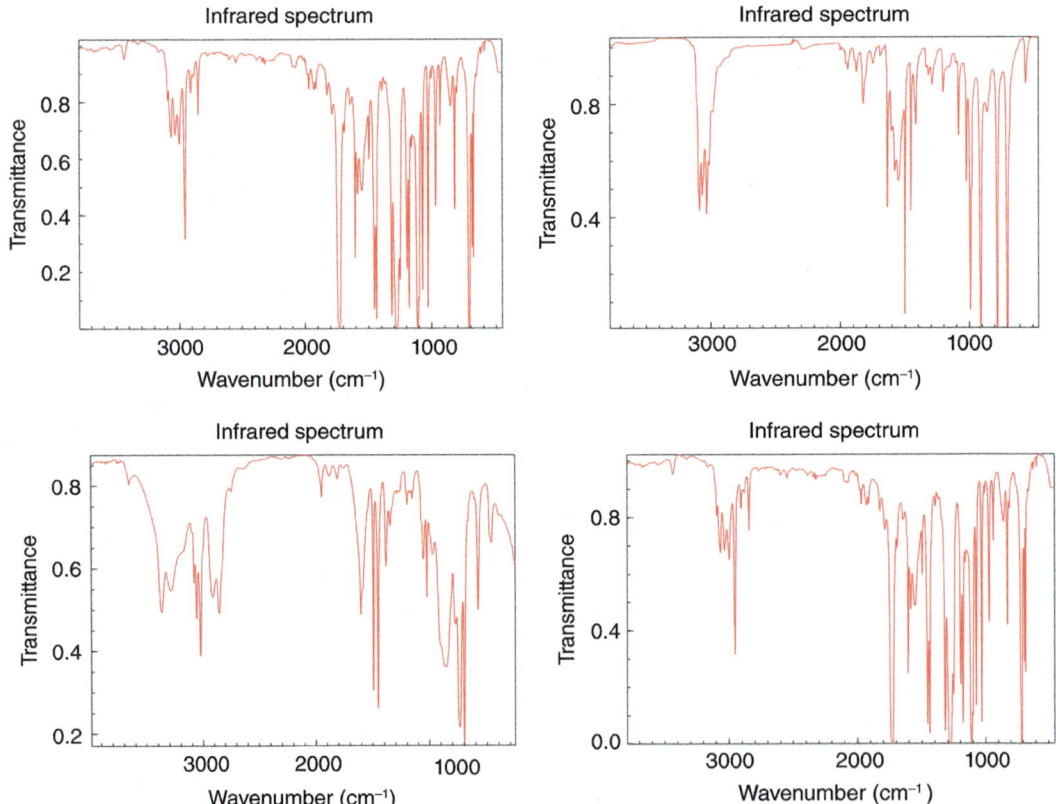

Utilizing the information contained in Table 3.1, match the spectra with the compounds given.

4

Alkanes, Cycloalkanes, and Alkenes: Isomers, Conformations, and Stabilities

4.1 Introduction

In the previous chapters, the properties and nomenclature of alkanes and alkenes were covered. In this chapter, there will be a closer examination of the three-dimensional arrangements of the atoms of these types of molecules and how different arrangements affect their stabilities and properties. The first concept that should be reviewed is that of isomers. By definition, isomers are compounds that have the same number and type of atoms but are arranged differently, resulting in different molecules. As a result, these compounds, although having the same molecular formulas, are different and hence have different properties. As the number of atoms of a molecule increases, so does the number of possible isomers (possible atomic arrangements), compared to molecules with fewer atoms. Structural isomers are one form of isomers; there are other types of isomers that are categorized as geometric isomers, conformational isomers, and stereoisomers. In this chapter, each type will be covered, and owing to the importance of stereoisomers, that category of isomers will be covered in the next chapter.

4.2 Structural Isomers

The simplest alkane that has structural isomers is butane since it is possible to have different bonding arrangements of the atoms. For butane, there are two different possible arrangements of the atoms that result in structurally different molecules with the same C_4H_{10} formula as shown below.

$$CH_3-CH_2-CH_2-CH_3 \qquad CH_3-\underset{\underset{\displaystyle CH_3}{|}}{CH}-CH_3$$

Two structural isomers that have the chemical formula C_4H_{10}

As mentioned earlier, the number of structural isomers increases as the number of atoms in a molecule increases. Molecules with many atoms have many more structural isomers, compared to molecules with a fewer number of atoms. Structural isomers are different molecules since the bonding arrangements are different for the compounds. The two compounds shown above with molecular formula C_4H_{10} are known as **structural isomers** or **constitutional isomers**. Since structural isomers are different compounds, they have different properties, such as density and melting points.

Organic Chemistry: Concepts and Applications, First Edition. Allan D. Headley.
© 2020 John Wiley & Sons, Inc. Published 2020 by John Wiley & Sons, Inc.
Companion website: www.wiley.com/go/Headley_OrganicChemistry

Problem 4.1

i) Give structural formulas for all possible structural isomers of C_5H_{12}.
ii) Give IUPAC names for the isomers of question (i) above.

4.3 Conformational Isomers of Alkanes

At room temperature, rotation about a carbon–carbon single bond of molecules occurs freely. For some molecules that have bulky groups bonded to adjacent carbons of a carbon–carbon single covalent bond, rotation is a bit more difficult compared to molecules that have smaller groups bonded to similar adjacent carbon atoms. Owing to the restricted rotation that results around the carbon–carbon bond of such a molecule that has bulky groups, the bulky groups spend more time in specific regions of the molecule, typically as far away as possible from each other, compared to being closer to each other resulting in different conformations of the molecules. The terms *conformers* and *rotamers* are used to describe these types of isomers. By definition, conformers are isomers with the same atom connectivity but have different arrangements of specific groups about a single bond, which comes about due to rotation about a carbon–carbon single bond. The term isomeric conformer is used to describe conformers in which groups around a carbon–carbon single bond are in different locations. The illustration of this three-dimensional concept on two-dimensional paper can be challenging and specific representations must be used to convey the orientation of the different conformational isomers. There are two types of representations that are typically used to represent conformers in organic chemistry: *dashed/wedge* and the *Newman projection*.

4.3.1 Dashed/Wedge Representation of Isomers

For the dashed/wedge representation, imagine that there are four groups bonded to a central carbon and that the molecule is oriented in such a way that two groups, along with the central atom are in the same plane, i.e. the plane of the paper. Solid lines are used to represent the bonds from the central atom to the two groups in the plane. With this orientation of molecules, one bond will be in front of the paper, while the other will be in the back. A dashed line is used to represent the group that is behind the plane of the paper and a wedged line represents the group that is in front of the plane of the paper. The dashed/wedge representation is illustrated in Figure 4.1.

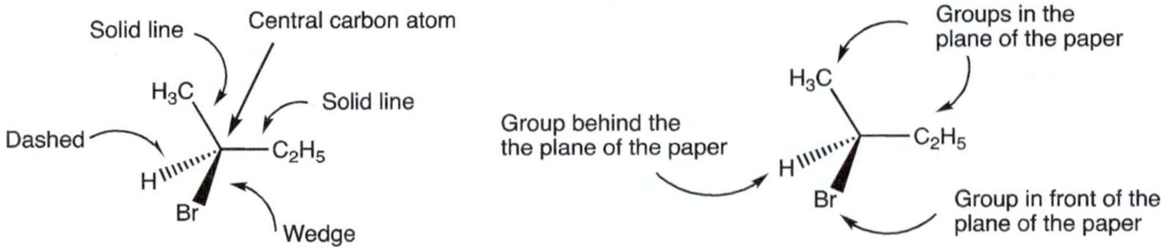

Figure 4.1 Illustration of the dashed-wedge representation of 2-bromobutane.

Problem 4.2

Draw a dashed-wedge representation of the following molecules.

a) 2-Chlorobutane (use carbon #2 as the central atom)
b) 2-Bromo-2-chloropentane (use carbon #2 as the central atom)

The dashed/wedge representation can be used to show not only the three-dimensional arrangement of atoms about one central carbon atom but also the three-dimensional arrangement about two carbon atoms in a molecule. A careful examination of the molecule shown below, 3,4-dimethylhexane, reveals that there are many possible representations of this molecule in three-dimensional space using the dashed-wedge representation.

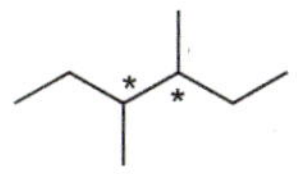

3,4-Dimethylhexane

A dashed-wedge representation is shown in Figure 4.2, in which the groups around the starred carbons are represented by solid, dashed, and wedged lines.

The use of a model will assist tremendously to effectively visualize the three-dimensional arrangements of the groups of the conformer shown in Figure 4.2.

The conformer shown in Figure 4.2 is only one representation of many possible conformers. Other conformers can be achieved by rotation about the starred carbon–carbon single bond as shown in Figure 4.3.

Problem 4.3

Draw dashed-wedge representations of the molecule shown in Figure 4.2 after rotation about the carbon–carbon bond by 60° and 120°, respectively.

4.3.2 Newman Representation of Conformers

The Newman projection is another representation that is used to represent the arrangements of the atoms or groups of atoms of a molecule in three-dimensional space. For the

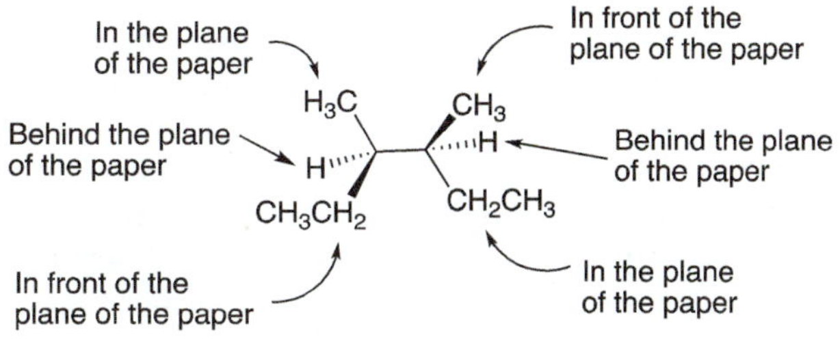

Figure 4.2 A dashed-wedge representation for 3,4-dimethylhexane illustrating the three-dimensional arrangement about the two central carbons that are starred in the molecule above.

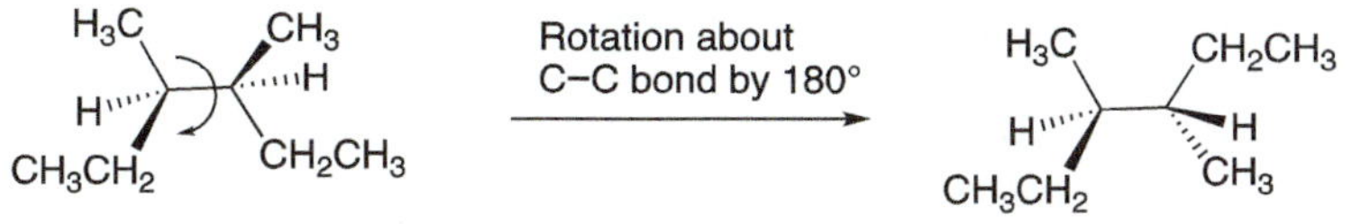

Figure 4.3 Dashed/wedge conformer that results after the rotation about the C$_3$ and C$_4$ single bond by 180° of the conformer of 3,4-dimethylhexane shown in Figure 4.2.

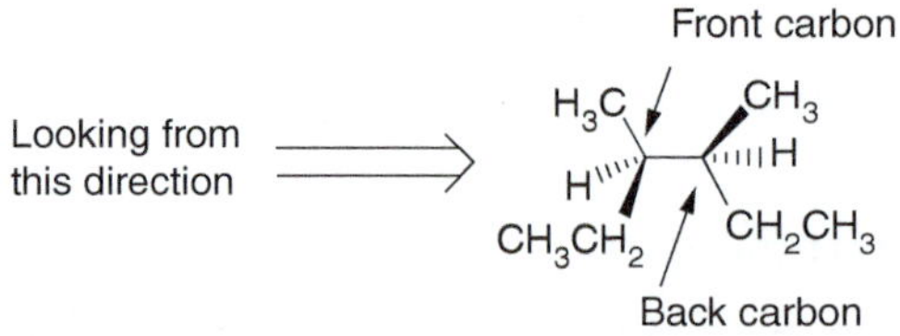

Figure 4.4 Example of the orientation of molecules for the Newman projection.

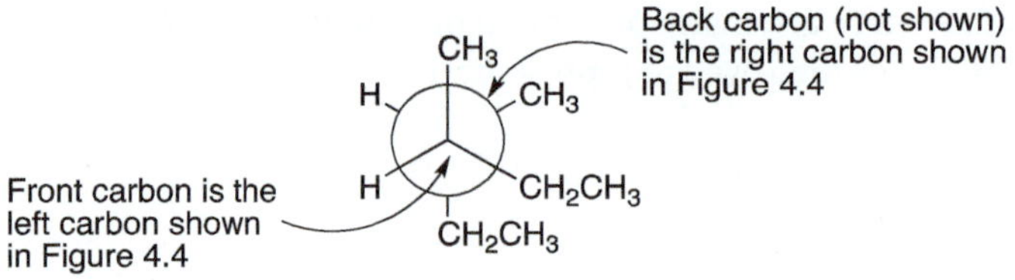

Figure 4.5 The Newman projection of 3,4-dimethylhexane shown in Figure 4.4

Newman projection, the molecule is viewed by looking directly down a carbon–carbon single bond. That is, one carbon will be in front (close to the viewer) and another is in the back (further from the viewer). This is illustrated in Figure 4.4, in which the viewer is looking down the carbons indicated in the figure.

The projection that the viewer sees from the perspective shown in Figure 4.4 is the Newman projection and is shown in Figure 4.5. The Newman projection is best visualized utilizing a model to fully appreciate especially this three-dimensional representation.

Problem 4.4

Give a dashed-wedged representation and its Newman representation of one isomer of $CH_3CH(Cl)CH(Cl)CH_3$ (note that carbons 2 and 3 will be the carbons that will be the central atoms).

The energy difference between conformational isomers is approximately $3 \, \text{kcal} \, \text{mol}^{-1}$ depending on the size of the groups around the bonds of the molecule. If the groups that are bonded to each other via a single covalent bond are very large however, then rotation about that carbon–carbon single bond is more difficult, compared to a molecule with smaller groups. Owing to the difficulty encountered when two bulky groups pass each other, rotation is restricted, and as a result, the energy difference between conformers is greater, compared to a molecule with smaller groups. For some molecules that have extremely large groups bonded to a single covalent bond, two conformers can be separated from each other, but only at low temperatures.

There are specific names that are used to describe different conformational isomers of molecules. The ***anti***-conformer is the conformer that results if the two bulkiest groups of two adjacent carbons are exactly opposite to each other, or trans to each other. If the largest groups are directly behind each other, this conformer is called an ***eclipsed*** conformer. If the two bulkiest groups are not directly behind each other, the ***staggered*** conformer results. A specific staggered conformer in which the two largest groups are close to each other is called a ***gauche*** conformer. These concepts are illustrated using the Newman projections in Figure 4.6.

Problem 4.5

Using the Newman representation, give the anti, eclipsed, and gauche conformers of 2,3-dichlorobutane.

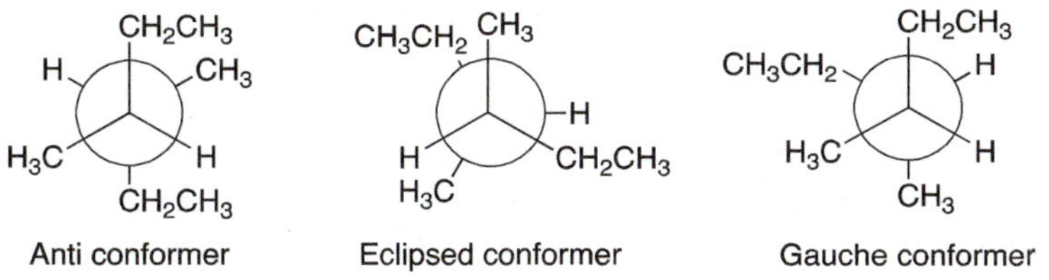

4.3 Conformational Isomers of Alkanes

Figure 4.6 Newman projections of different conformers of 2,3-dimethylhexane.

4.3.3 Relative Energies of Conformers

Another method that is often used to demonstrate the different stabilities of the conformers of a molecule is based on their relative energies. Conformers that are most stable will have the largest group furthest from each other (the anti conformer). Thus, the most stable conformer of 1,2-dibromoethane is the anti conformer. Conformers that have large groups or atoms close to each other are the least stable conformer. Therefore, the least stable conformer of 1,2-dibromoethane is the eclipsed conformer. The relative energies for the various conformers of 1,2-dibromoethane can be represented on an energy diagram as shown in Figure 4.7.

Note that in Figure 4.7 the conformer that is highest in energy is the eclipsed conformer in which both large bromine atoms are directly behind each other (the first conformer at 0°). By rotating about the C_1–C_2 bond keeping the front carbon stationary and rotating only around the back carbon by 60° results in another conformer, a staggered conformer. For the conformer that results after a 60° rotation, the bromine atoms are not directly behind each other, but beside each other. This conformer is more stable than the eclipsed conformer. As a result, the staggered conformer is lower in energy than the eclipsed conformer. The rotation about the same carbon–carbon bond by another 60° results in another eclipsed conformer. For this eclipsed conformer, however, the bromine atoms are not directly behind each other, but a bromine atom and a hydrogen atom are directly behind each other. Since the hydrogen atom is smaller than the bromine atom, there is less interaction between these atoms, compared to

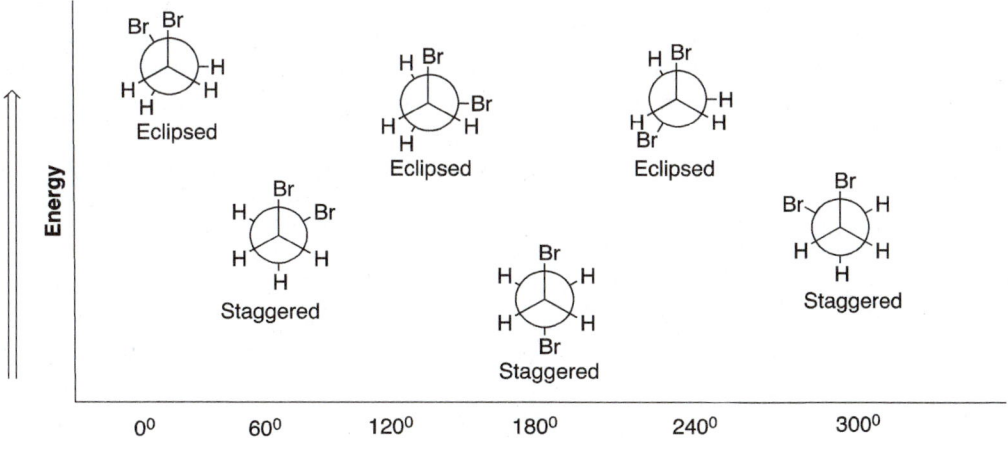

Figure 4.7 Relative energies of the various conformers of 1,2-dibromoethane obtained by rotation about the carbon–carbon bond by 60°.

an eclipsed conformer where the large bromine atoms are directly behind each other. As a result, this conformer is less stable than the second conformer, but more stable than the first eclipsed conformer. The next conformer, the fourth conformer, is the most stable conformer and, as a result, is shown in the lowest position on the energy diagram. For this conformer, the large bromine atoms are the furthest from each other; hence, a staggered, and specifically an anti conformer results, since the large bromine atoms are anti across from each other. The rotation of this anti conformer by another 60° results in another eclipsed conformer (the fifth conformer). This conformer is very similar to the third conformer in that the bromine and the hydrogen atoms are directly behind each other, thus they are similar in energy. Rotation of the C_1—C_2 bond by another 60° results in the sixth conformer, which is another staggered conformer, and is similar to the second conformer, which is also known as a gauche conformer (since the two largest groups are beside each other).

Problem 4.6

Give the Newman representations of the most and least stable conformers of the following molecules.

a) 3,4-Dimethylhexane (consider rotation about C_3 and C_4).
b) 3-Methyl-2-pentanol (consider rotation about C_2 and C_3).

4.4 Conformational Isomers of Cycloalkanes

Free rotation about carbon–carbon bonds in cycloalkanes is not possible. There is the possibility of very limited internal rotation about any carbon–carbon bond in a cyclic molecule, however. Of course, for alkanes with small rings, internal bond rotation is very limited, compared to cycloalkanes with larger rings. In this section, we will examine the various conformations brought about by the limited internal rotation of carbon–carbon single bonds of cycloalkanes.

4.4.1 Isomers of Cyclopropane

For cycloalkanes that are smaller than the cyclopentane and cyclohexane molecules, there is very limited internal rotation about any carbon–carbon single bond. As a result, groups that are bonded to any of the carbons of cyclopropane are either on the same side of the ring or on opposite sides of the ring. The different location of groups relative to the fixed ring structure results in different isomers, called *geometric isomers*. Figure 4.8 shows two different representations of

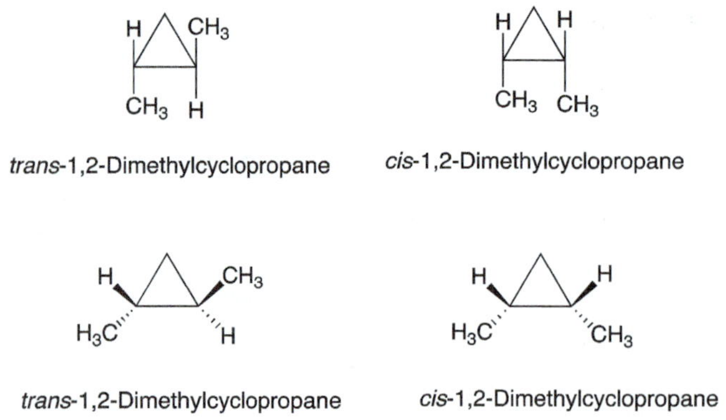

Figure 4.8 Representations of trans and cis arrangements of 1,2-dimethylcyclopropane.

trans-1,2-Dimethylcyclopropane

cis-1,2-Dimethylcyclopropane

trans-1,2-Dimethylcyclopropane

cis-1,2-Dimethylcyclopropane

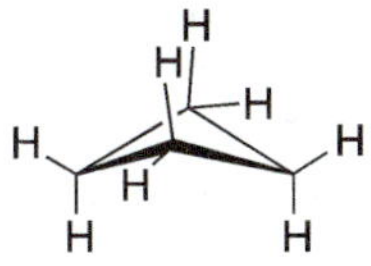

Cyclobutane

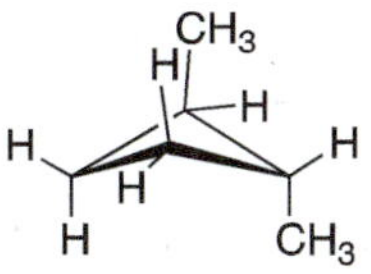

trans-Dimethylcyclobutane

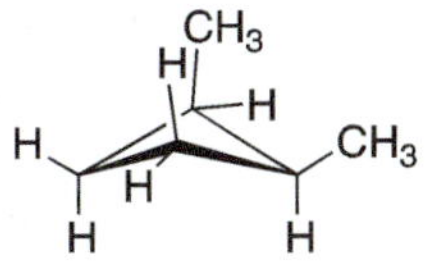

cis-Dimethylcyclobutane

Figure 4.9 Representations of cyclobutane and *trans* and *cis*-dimethylcyclobutane.

the arrangements of groups about the rigid cyclopropane ring. If two groups are on the same side of the ring, the relationship is a cis and the isomer is called a cis isomer. If the groups are on opposite sides of the rigid ring structure, the isomer that results is called a trans isomer.

4.4.2 Conformational Isomers of Cyclobutane

Limited internal rotation about the carbon–carbon bonds of cyclobutane is just slightly greater than that of cyclopropane. The cyclobutane ring is considered to be rigid and as a result, hydrogens and substituents bonded to carbons of cyclobutane can be either on the same side of the fairly rigid cyclobutane ring or on opposite sides as shown in Figure 4.9.

Compared to cyclopropane, the cyclobutane ring is not as flat and sometimes described as puckered as illustrated in Figure 4.9.

Problem 4.7

i) Draw the cis and trans isomers of 1,2-dibromocyclobutane.
ii) Label the following as cis or trans 1,2-dibromocyclobutane.

(a)
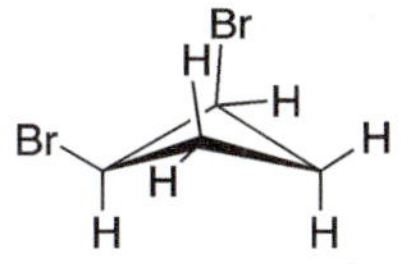

(b)
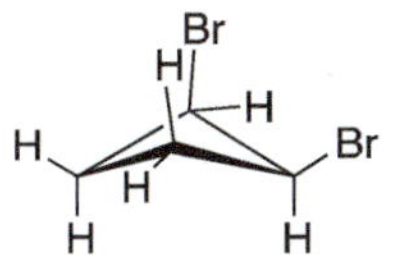

(c)
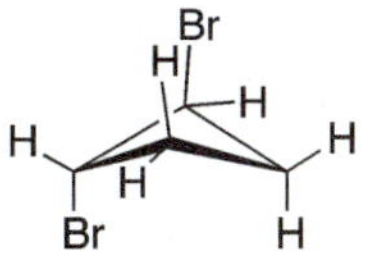

iii) Draw the cis and trans isomers of 1,2-dimethylcyclobutane.

4.4.3 Conformational Isomers of Cyclopentane

The increased size of cyclopentane ring, compared to cyclobutane and cyclopropane rings, introduces greater internal rotation about the carbon–carbon single bonds. For the cyclopentane ring, the hydrogens can be described based on their orientation relative to the plane of the cyclopentane ring. A close examination of the hydrogens bonded to the cyclopentane ring reveals that five hydrogens (one of each carbon) appear to be almost in the same plane as the ring, and as a result, they are described as *equatorial* hydrogens. The other five hydrogen atoms (one on each carbon) appear to be either above or below the plane of the cylopentane ring and they are described as *axial*. This observation is illustrated in Figure 4.10, where H_a represents axial hydrogens and H_e represents equatorial hydrogens.

As a result, the relationship between any two substituents on the cyclopentane ring can be described in two ways: cis and trans or based on their location as *axial* or *equatorial*. An example of the two conformers of 1,2-dimethylcyclopentane is shown in Figure 4.11.

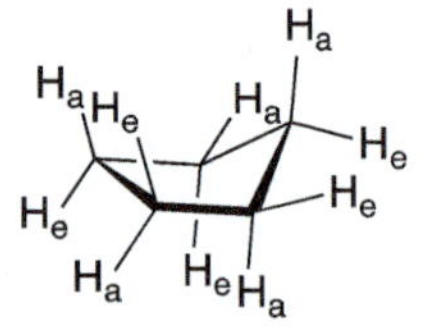

Figure 4.10 Cyclopentane showing axial hydrogens (H_a) and equatorial hydrogens (H_e).

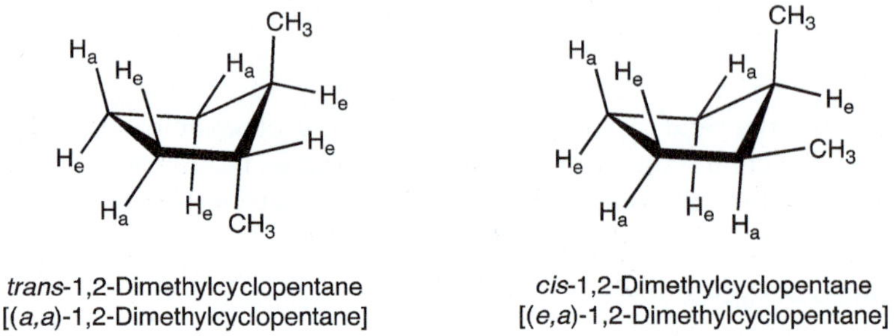

trans-1,2-Dimethylcyclopentane
[(a,a)-1,2-Dimethylcyclopentane]

cis-1,2-Dimethylcyclopentane
[(e,a)-1,2-Dimethylcyclopentane]

Figure 4.11 *cis* and *trans*-Dimethylcyclopentanes.

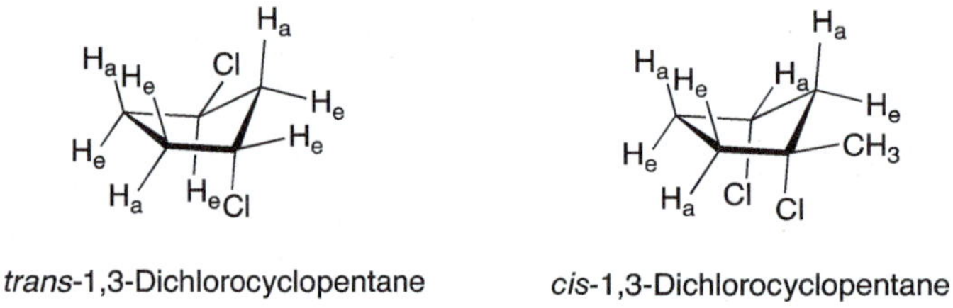

trans-1,3-Dichlorocyclopentane

cis-1,3-Dichlorocyclopentane

Figure 4.12 Different conformations of *cis* and *trans*-1,3-dichlorocyclopentane.

The possibility of different conformers of 1,3-disubstituted cyclopentane also exists, and Figure 4.12 shows two conformers of 1,3-dichlorocyclopentane, the cis and trans conformers.

Problem 4.8

Draw the trans conformers of the following molecules.

a) 1,2-Dibromocyclopentane
b) 1,3-Dibromocyclopentane

Since there is greater internal carbon–carbon rotation within cyclopentane, compared to that of cyclopropane and cyclobutane, the difference between the cis and trans conformers of cyclopentane is not as obvious, compared to cis and trans conformers of disubstituted cyclopropane and cyclobutane.

4.4.4 Conformational Isomers of Cyclohexane

There is more internal carbon–carbon bond rotation within cyclohexane than cyclopentane, which leads to a more obvious difference in the conformers, compared to cyclic structures discussed earlier. One of the first observations about cyclohexane is that it can acquire one of two different conformations based on greater internal bond rotation around the carbon–carbon single bonds, of which two are shown in Figure 4.13. The conformer shown on the left looks like

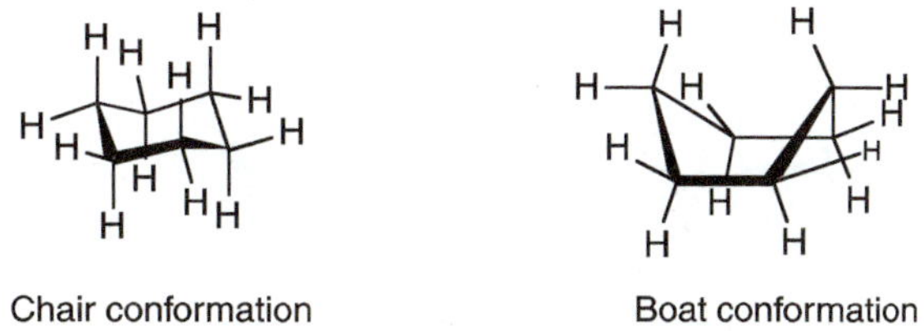

Figure 4.13 Chair and boat conformations of cyclohexane.

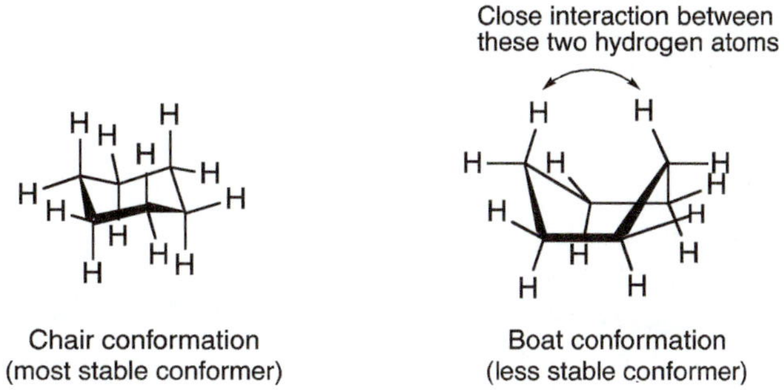

Figure 4.14 Illustration of the rationale for the relative stabilities of the chair and boat conformers of cyclohexane.

a chair, hence it is called the chair conformer. The other conformer on the right looks like a boat, and hence it is called the boat conformer.

Of these two conformers, the chair conformer is the most stable conformer. The boat conformer is less stable because the hydrogens are closer to each other, especially those in the 1,4 positions as illustrated in Figure 4.14.

Since the chair conformer is the most stable conformer, it will be the conformer that will be analyzed in details relative to the boat conformer, which is less stable. A close examination of the chair conformer of cyclohexane reveals that there are six hydrogens (one on each carbon) that are essentially in the same plane as the six carbon atoms; as a result, they are described as *equatorial (e)*. Likewise, there are six hydrogens (one on each carbon) that are above and below the plane of the cyclohexane ring and these hydrogens are described as *axial (a)*. In fact, three hydrogens of these hydrogens are above the plane and three are below the plane. This observation is illustrated in Figure 4.15.

At room temperature, the cyclohexane molecule is in equilibrium with both conformations, but the equilibrium favors the stable chair conformer, compared to the boat conformer. That is, the molecule is constantly flipping from one chair conformation to another chair through the boat conformation, as shown in Figure 4.16. The energy required to "flip" from one cyclohexane conformer to another conformer through the boat conformer is very small. Thus, a flip from one chair conformation to another is accomplished easily at room temperature.

Figure 4.15 Representation of the different hydrogens of cyclohexane chair conformation. H_e represents hydrogens in the equatorial position, and H_a represents hydrogens in the axial position.

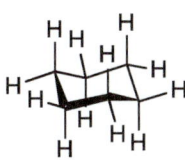

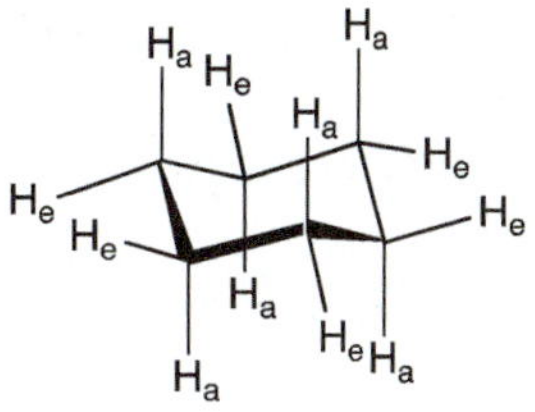

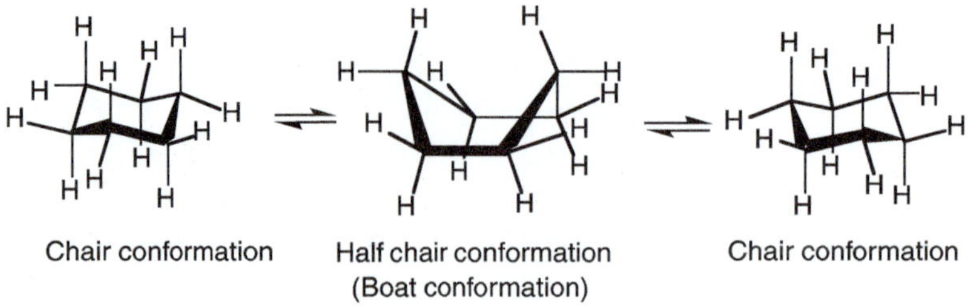

Figure 4.16 Conformational changes of cyclohexane from one chair conformer to another through the boat conformation.

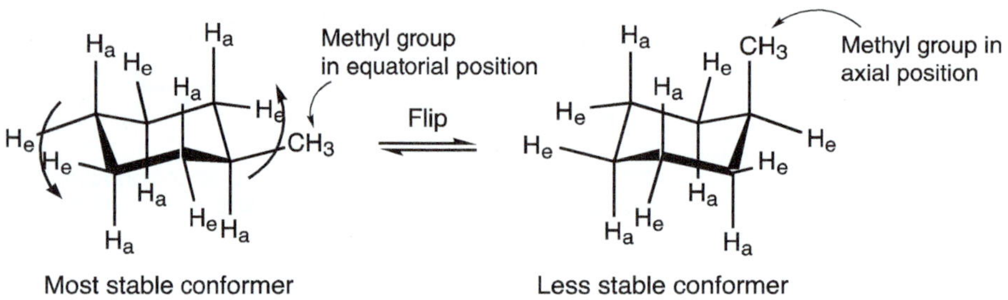

Figure 4.17 Two chair conformers of methylcyclohexane.

4.4.5 Conformational Isomers of Monosubstituted Cyclohexane

Whenever there is a ring flip for cyclohexane, there is no noticeable change between both conformers. If one of the hydrogens of cyclohexane is substituted for a methyl group, to produce methylcyclohexane, then the methyl group can occupy either the axial or the equatorial position, and there is a noticeable difference in the location of the methyl group between both chair conformers when there is a ring flip. Consider the ring flip of methylcyclohexane as shown in Figure 4.17

For the chair conformer to the left, the methyl group is in the equatorial position and after the ring flip to form the other chair conformation, the methyl group is in the axial position (conformer on the right). A close examination of these two conformers reveals that the larger methyl group (compared to the hydrogens) would much prefer to occupy the equatorial position since there is more space in the equatorial plane to accommodate a large group, compared to the axial plane. For these conformers, the chair conformer to the left is more stable than the conformer shown on the right. The greater stability of the conformation on the left comes from the methyl group being in the equatorial position, where there is more space, compared to the axial. The other conformer (the one to the right) is the least stable conformer since the methyl group is in the axial position where there are more steric interactions with the other axial hydrogens and the methyl group.

If larger groups such as the 1,1-dimethylethyl [—C(CH$_3$)$_3$, *tert*-butyl] are placed on the cyclohexane ring, the energy difference between the two conformers would be greater, compared to the energy difference for methylcyclohexane. As a result, the conformer which has the *tert*-butyl group in the equatorial position would be the preferred conformer, compared to the conformer in which the group is in the axial position.

Problem 4.9

a) Draw the most stable conformer of chlorocyclohexane.
b) Draw the least stable conformer of bromocyclocyclohexane.

4.4.6 Conformational Isomers of Disubstituted Cyclohexane

Disubstituted cyclohexanes, such as 1,4-dimethylcyclohexane, have more isomers and hence conformational isomers, compared to monosubstituted cyclohexane, such as methylcyclohexane, as we have seen in the previous section. Figure 4.18 shows the possible chair conformers of 1,4-dimethylcyclohexane.

Note that since there are now two groups on the cyclohexane ring, the possibility of cis and trans isomers exists for each conformer. Hence, there are four possible isomers for a disubstituted cyclohexane, compared to two conformers for a monosubstituted cyclohexane, as discussed in the previous section. For the first two conformers shown in Figure 4.18, notice that the two methyl substituents are on opposite sides of the plane of the cyclohexane ring, hence they are trans isomers. The methyl groups of the first conformer of this equilibrium are at the opposite ends of the cyclohexane ring and for the other conformer of this equilibrium, the methyl groups are above and below the plane of the cyclohexane ring. As a result, they are described a trans conformers. Thus, if the relationship between two groups in the 1,4 positions of the cyclohexane ring is in the equatorial-equatorial locations, or axial-axial locations, the isomers are trans isomers.

For the second equilibrium shown in Figure 4.18, notice that the two methyl substituents are not on the opposite sides of the plane of the cyclohexane ring. This is a bit more difficult to visualize, but in the first case, they are axial-equatorial, which gives a cis relationship; for the second conformer, the groups are also axial-equatorial, which also gives a cis relationship. As a result, these two conformers are described cis conformers.

Problem 4.10

a) Figure 4.18 shows four conformers of 1,4-dimethylcyclohexane, which is the least stable conformer and explain your answer?
b) Draw most stable conformer of 1,4-dichlorocyclohexane.

Another representation of the conformers shown in Figure 4.18 is shown in Figure 4.19, in which the dashed-wedge representation is used. Note that with this type of representation, the

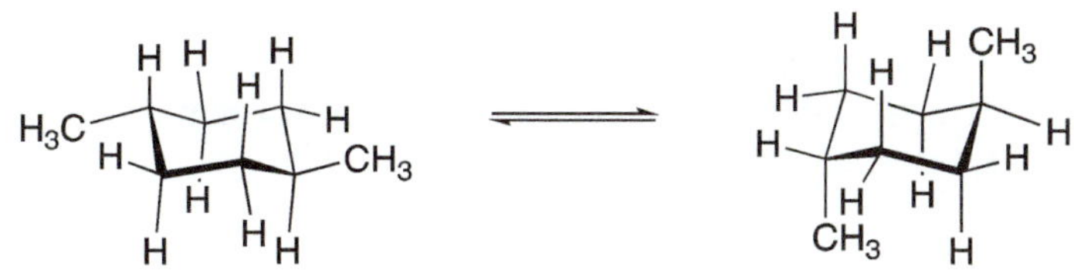

trans-1,4-Dimethylcyclohexane   *trans*-1,4-Dimethylcyclohexane

cis-1,4-Dimethylcyclohexane   *cis*-1,4-Dimethylcyclohexane

Figure 4.18 Possible conformers of 1,4-dimethylcyclohexane.

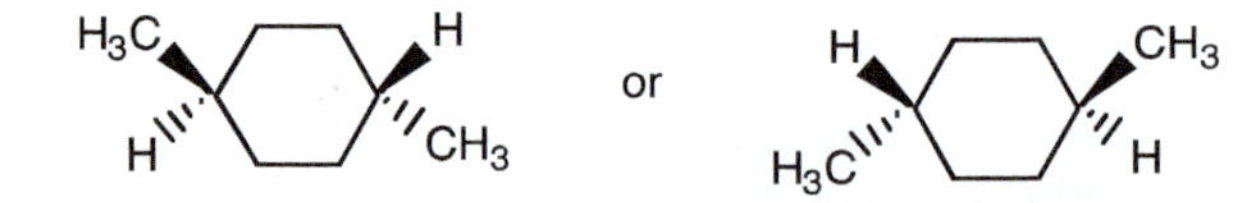

trans-1,4-Dimethylcyclohexane   or   *trans*-1,4-Dimethylcyclohexane

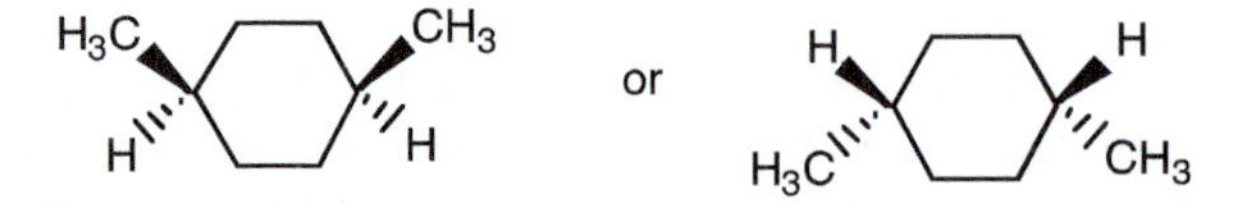

cis-1,4-Dimethylcyclohexane   or   *cis*-1,4-Dimethylcyclohexane

Figure 4.19 Possible conformers of 1,4-dimethylcyclohexane in which the dashed-wedge representation is used.

Table 4.1 Conformers of cis and trans isomers of disubstituted cyclohexanes.

Relationship	Cis	Trans
1,2	a,e or e,a	e,e or a,a
1,3	a,a or e,e	a,e or e,a
1,4	a,e or e,a	a,a or e,e

axial and equatorial positions cannot be shown. This representation just gives the relationship between the groups bonded to the ring.

For other disubstituted cyclohexanes, such as 1,3-dimethylcyclohexane and 1,2-dimethylcyclohexane, a similar analysis can be carried out. For these compounds, it is possible to have different cis and trans conformers similar to those shown in Figure 4.18. Table 4.1 gives a summary of the types of isomers, based on the arrangements of the groups in the equatorial or axial positions of the cyclohexane ring. At this point, a model set would be very helpful to better visualize the relationships for the different conformers.

Problem 4.11

a) Draw a chair conformer of *trans*-1,2-dimethylcyclohexane.
b) Draw the chair conformer of the <u>most</u> stable conformer of 1,2-dimethylcyclohexane.
c) Draw a chair conformer of *cis*-1,3-dimethylcyclohexane.

4.5 Geometric Isomers

For compounds that have a double bond, there can be different arrangements in space of the atoms or groups bonded to the carbons of the double bond. There is no rotation about a carbon–carbon double bond like that of a carbon–carbon single bond. Due to the lack of rotation about a carbon–carbon double bond, it is impossible to have conformational isomers, but it is possible to have different groups on the same side of the double bond, or on opposite side of the carbon–carbon double bond. *Geometric isomers* are molecules that have different spatial atomic arrangements based on a rigid *framework* in the molecule, such as a double bond or a rigid cyclic structure as demonstrated with the cyclopropane ring, which was discussed earlier in the chapter and illustrated in Figure 4.8. The terms cis and trans are used to describe the spatial arrangement of two atoms or groups about a rigid system, such as a double bond. The

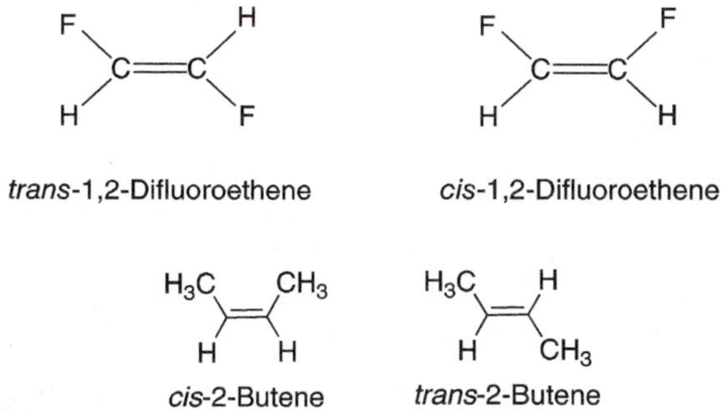

Figure 4.20 Different isomers (geometric isomers) of 1,2-difluoroethene and 2-butene.

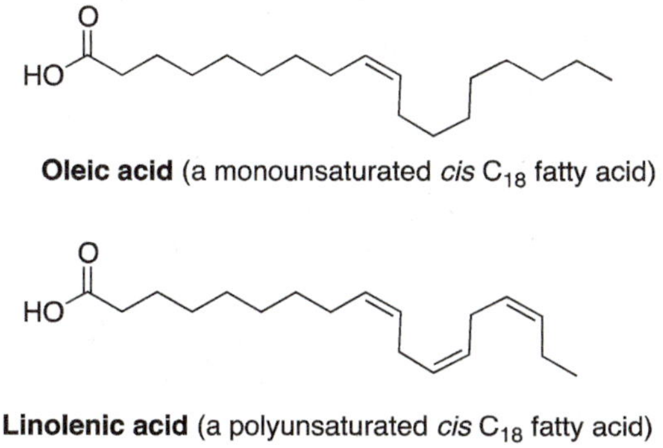

trans-1,2-Difluoroethene

cis-1,2-Difluoroethene

cis-2-Butene

trans-2-Butene

Figure 4.21 Examples of fatty acids (unsaturated fatty acids with cis double bonds).

Oleic acid (a monounsaturated *cis* C_{18} fatty acid)

Linolenic acid (a polyunsaturated *cis* C_{18} fatty acid)

cis geometric stereoisomer has the same atoms or groups on the same side of the double bond, and the trans stereoisomer has the same atoms or groups on the opposite side of the double bond. 1,2-Difluoroethene has two different geometric isomers, which have the same atom-to-atom connectivity, yet the dipole moments (μ) of these molecules are different; the same is true for cis and *trans* 2-butene as shown in Figure 4.20.

The term unsaturation is sometimes used to describe the presence of the carbon–carbon double bonds in molecules. For example, unsaturated fats have at least one double bond (site of unsaturation), and as a result, it is possible to have geometric isomers about the double bonds of unsaturated fats, as shown in the examples in Figure 4.21. Due to the presence of trans and cis double bonds in fatty acids, the melting points of these compounds are typically lower than their unsaturated counterpart. Unsaturated fatty acids, like the ones shown in Figure 4.21, cannot fit into a nicely defined crystal structure, compared to saturated fatty acids; as a result, they are typically liquids at room temperature, compared to saturated fatty acids that are typically solids at room temperature.

Another example of the differences in properties brought about due to different geometric isomers is that of fumaric acid and maleic acid. As shown below, these compounds are geometric isomers and the body recognizes one as essential for health, whereas the other is toxic; the only difference between both molecules is the orientation of the groups in space across a rigid double bond, they are geometric isomers of each other.

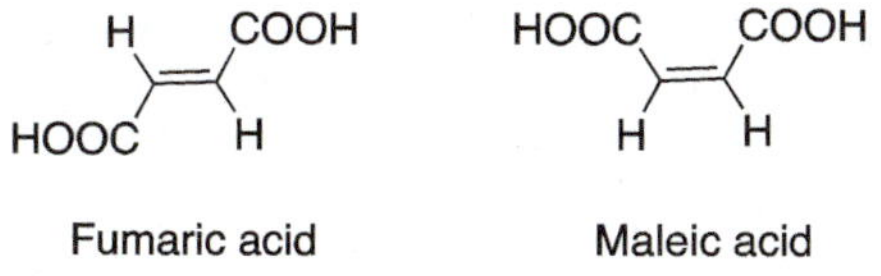

Fumaric acid

Maleic acid

4.5.1 IUPAC Nomenclature of Alkene Geometric Stereoisomers

Owing to the difference in properties of geometric isomers arising from orientations of groups about the double bond, it is essential to be able to appropriately identify different geometric isomers. Since geometric stereoisomers are obviously different molecules; as a result, they must have different names. Cis and trans are common descriptions of such molecules, but another system, the Z and E symbolism is used in the IUPAC system of nomenclature. The designations Z and E are derived from the German terms zusammen, which means together and entgegen, which means opposite. Hence the Z and E are used to designate the relationship of the groups of highest priority about the carbon–carbon double bond. The priority of a group is determined based on the atomic number of the atoms that are directly bonded to the carbons of the alkene double bond. If an atom that is bonded directly to the carbon of the double bond has a larger atomic number than another atom that is bonded to the same carbon of the double bond, the atom with the larger atomic number is assigned the highest priority. If the atoms or groups of highest priority are on the same side of the double bond, the letter Z is assigned to describe that arrangement. The letter E is assigned if the groups of highest priority are on opposite sides of the double bond. The E and Z system of naming is typically used for alkenes, and the trans and cis descriptions are typically used for cyclic compounds that have different geometric stereoisomers.

An example of the use of the E and Z system is demonstrated in naming the geometric isomers of 1,2-difluoroethene, which is shown in Figure 4.22. In this figure, the fluorine atom has the highest atomic number, compared to the hydrogen atom; hence, fluorine gets the higher priority. Thus, for the geometric isomer on the left, the groups of highest priority are on opposite side of the double bond and the designation for that geometric isomer is the (E) isomer. The geometric isomer on the right has the groups of highest priority of the same side and is designated the (Z) geometric isomer.

Shown below is the application of the E and Z system for naming the geometric isomers of 2-butene; the highest priority group (or atom) is assigned #1.

For more complex molecules where the groups bonded to the carbon–carbon double bond are completely different, a careful examination of the atomic numbers of each atom bonded to the carbons of the double bond must be carried out. Shown in the examples below are more complex molecules, in which a methyl group, a hydrogen, and two different halogens are bonded to the carbons of the double bond. For the first example, the carbon on the left of the

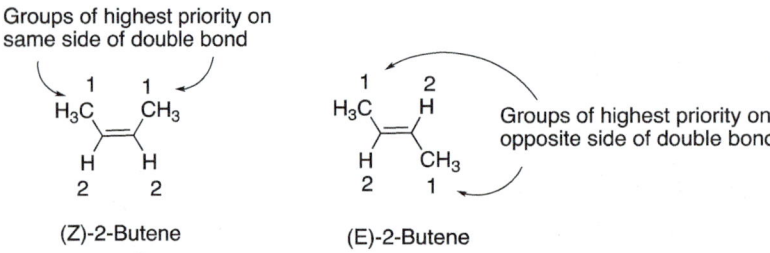

Figure 4.22 E and Z designation in the IUPAC nomenclature system of 1,2-difluoroethene. Atoms with the largest atomic numbers are on opposite side of the double bond. Atoms with the largest atomic numbers are on same side of the double bond.

double bond has two groups bonded, a bromine atom and a chlorine atom. Since the atomic number of the bromine is greater than that of chlorine, bromine is assigned the highest priority and gets #1 and chlorine gets #2. For the other carbon of the double bond, there is a methyl group and a hydrogen bonded to that carbon. Since the carbon, which is directly bonded to the carbon of the double bond, has a larger atomic number than that of hydrogen, the methyl group has the highest priority and gets assigned #1 and hydrogen #2.

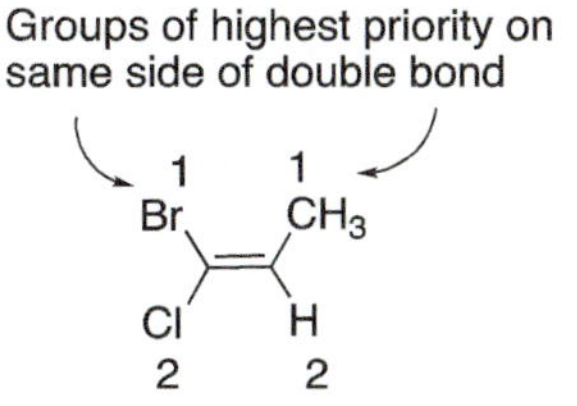

(Z)-1-Bromo-1-chloro-1-propene

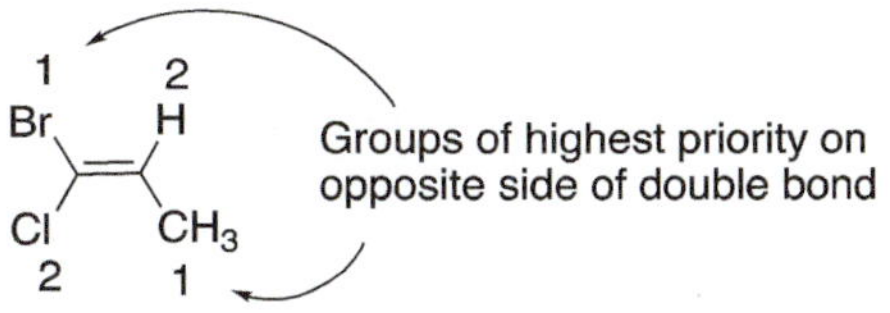

(E)-1-Bromo-1-chloro-1-propene

For this molecule, the groups of highest priority are on the same side, that is, both #1s are on the same side of the double bond. As a result, the geometric isomer is assigned a Z, hence the name shown above. For the other molecule to the right, a similar analysis can be carried out. The carbon on the left of the double bond has a bromine and a chlorine that are bonded directly to the alkene carbon. Hence, the assignment shown is based on the atomic numbers of bromine and chlorine. The same analysis for the right carbon of the double bond gives the methyl group the highest priority and it is assigned #1 and hydrogen #2. Based on this arrangement, the molecule is named as the E geometric isomer.

Problem 4.12

Give the structure of the following molecules.

a) (E)-3,4-Dimethyl-3-heptene
b) (Z)-3-Chloro-4-methyl-4-octene

For most of the molecules that will be encountered in organic chemistry, the determination of the atom, or group, of highest priority is not as obvious as that of the examples discussed so far. As a result, a set of rules must be used to determine the highest priority for the more complex groups.

1) Determine the atomic numbers of the atoms that are directly bonded to the carbon of each of the carbons of the double bond and assign priorities. As mentioned earlier, the atom with the highest atomic number gets priority #1.
2) If there are two isotopes, the isotope with the higher atomic mass is given the higher priority number.
3) If there are two of the same atoms bonded to a carbon of the double bond, consider the atomic number of the next atom bonded to those atoms and use atomic number of the adjacent atom to determine the priority of that group. For example, —CH_2CH_3 gets a higher priority, compared to —CH_3.

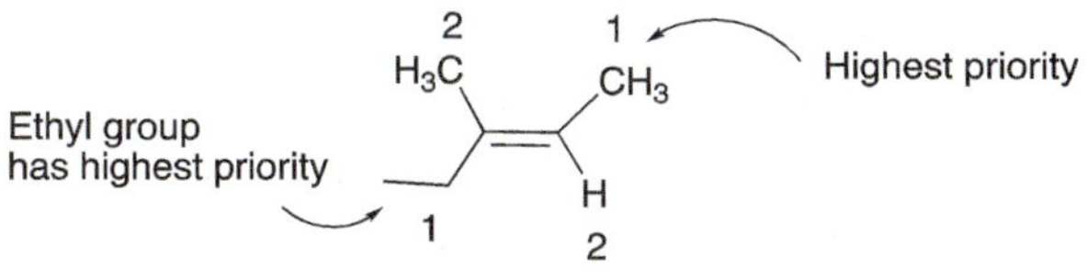

E-3-Methyl-2-pentene

Difunctional molecules discussed in the previous chapter, including molecules that contain the alkene and alcohol functionalities can also exist as geometric isomers as shown in the example below.

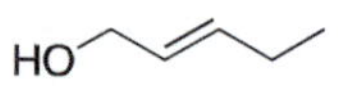

(E)-2-Penten-1-ol 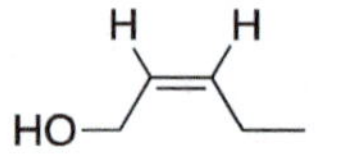   (Z)-2-Penten-1-ol

4) For groups that have multiple bonds, the scheme below for the priority assignment is used.

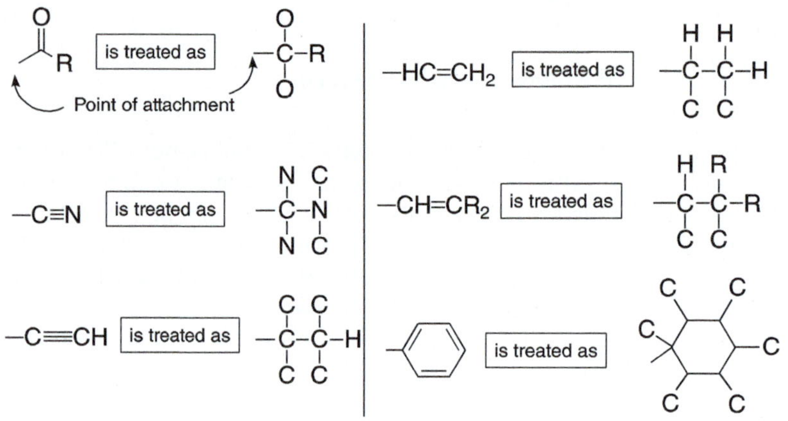

Problem 4.13

a) Of the following pairs of groups, determine which has the highest priority.
   a) $-CH_3$ and $-CH(CH_3)_2$
   b) $-CH=CH_2$ and $-CH_2CH_3$
   c) $-CHO$ and $-OH$
   d) NH_2 and Br.

b) Give the structures and complete IUPAC names (including E and Z description) for the two geometric stereoisomers of 3-methyl-2-pentene.

c) Give the IUPAC for the compounds shown below.

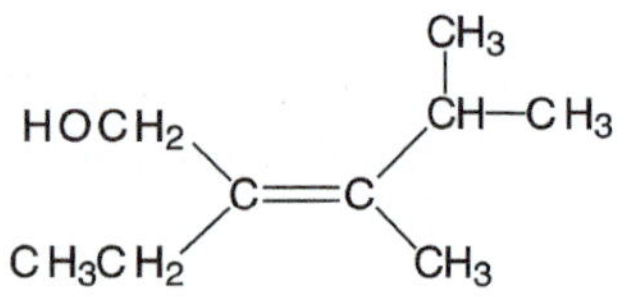

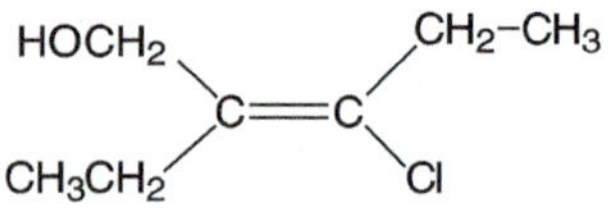

d) Give the IUPAC names for the molecules shown below.

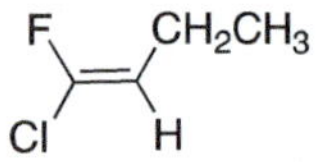

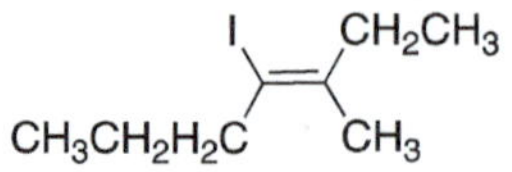

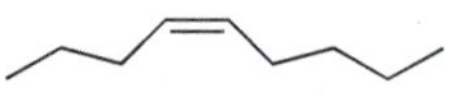

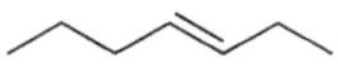

e) Give the line-angle structures for the following molecules.
   a) (Z)-1-Chloropropene.
   b) (E)-3-Methyl-2-hexene.

4.6 Stability of Alkanes

Since it is possible to have different structural isomers of alkanes, a question that chemists typically probe is the relative stability of different isomers of compounds. That is, of two isomers, which is the most stable? For various reasons, chemists are constantly concerned about the stability of compounds; stable compounds are typically not as reactive as compounds that are not stable. Chemists have developed a method to determine the relative stability of compounds, which is determined based on the heat liberated from specific reactions. A basic concept learned in general chemistry is that an exothermic reaction gives off heat and an endothermic reaction must be heated in order for the reaction to occur. The energy profile for an exothermic reaction is shown in Figure 4.23.

Compounds that are low on the energy profile diagram are more stable than compounds higher on an energy profile diagram. Thus, in the diagram, as shown in Figure 4.23, as a reaction proceeds from reactants to products, the reactants are less stable than the products and the difference in heat (ΔH) for the reaction gives an idea of the stability of the reactants, relative to the products. A comparison of two reactions that give the same products can be carried out to reveal the relative stabilities of the reactants, as shown in Figure 4.24.

From the energy profile shown in Figure 4.24, the reactants of the reaction with the larger ΔH are less stable than the reactants of the reaction with the smaller ΔH. As a result, this type of analysis will give us an idea of the relative stabilities of the reactants of two different reactions that give the same products.

The reaction of alkanes with oxygen to produce heat, also known as combustion, is an important type of exothermic reaction. As a result, alkanes make exceptionally good fuels for heating homes, cooking, and as a fuel for the heat engines of vehicles. The combustion reactions of two isomers of octane are shown in Figure 4.25, along with the amount of heat given off. The energy profile of the type shown in Figure 4.24 can be used to analyze the stabilities of these two isomers of octane. Note that both reactions give the same products, but the energy for each reaction is different since different isomers are used. As a result, it can be concluded that 2,2,3,3-tetramethylbutane is more stable than octane since the ΔH is less for its combustion, compared to octane.

Figure 4.23 Energy profile for an exothermic reaction.

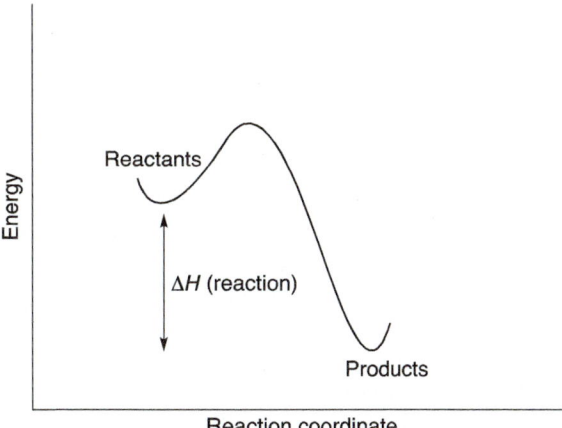

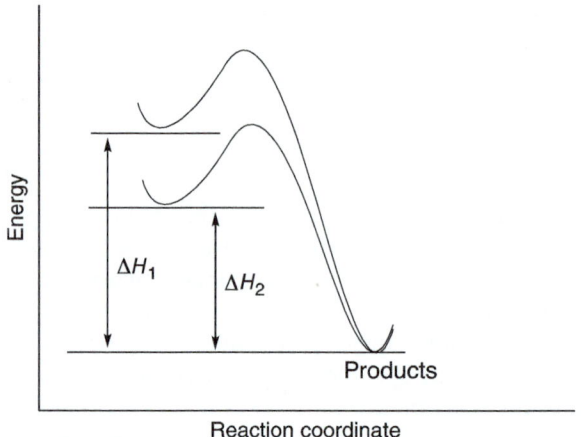

Figure 4.24 Reaction profile for the reactions of two different reactants that give the same products.

$$C_8H_{18} + 25/2\, O_2 \longrightarrow 8\,CO_2 + 9\,H_2O \quad \Delta H = -1307.5 \text{ kcal mol}^{-1}$$
Octane

$$C_8H_{18} + 25/2\, O_2 \longrightarrow 8\,CO_2 + 9\,H_2O \quad \Delta H = -1303.6 \text{ kcal mol}^{-1}$$
2,2,3,3-Tetramethylbutane

Figure 4.25 Combustion reactions of two isomers of octane showing the amounts of heat liberated.

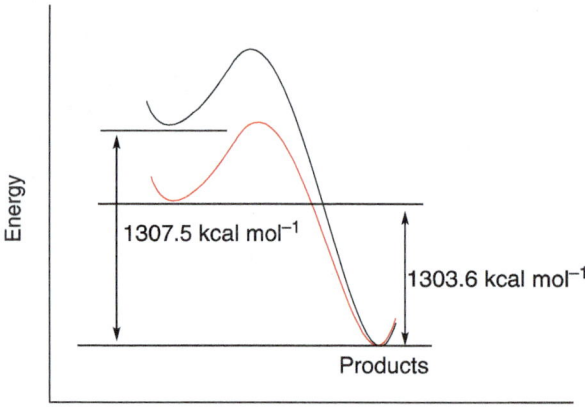

Figure 4.26 Reaction profile for the combustion reactions of octane (shown in black) and 2,2,3,3-tetramethylbutane (shown in red). Note that the more branched isomer is the most stable isomer.

Figure 4.26 gives the energy profile for both reactions. Note that the more branched isomer liberates the least amount of heat, hence the more stable isomer, compared to an isomer that is not as branched or to the straight chain isomer.

Figure 4.27 shows four different alkanes, along with the heat liberated upon combustion of each alkane. You will notice that as the number of carbons and hydrogens increases, the amount of heat that is liberated also increases. Approximately 157 kcal mol^{-1} of heat is liberated for each CH_2 added to a straight chain alkane. Thus, it is possible to estimate the amount of heat liberated for a straight chain alkane. Based on knowledge of the heats of combustion, one can

CH₄ + 2O₂ → CO₂ + 2H₂O   ΔH = −213 kcal mol⁻¹
Methane

C₃H₈ + 5O₂ → 3CO₂ + 4H₂O   ΔH = −531 kcal mol⁻¹
Propane

C₄H₁₀ + 13/2 O₂ → 4CO₂ + 5H₂O   ΔH = −689 kcal mol⁻¹
Butane

C₅H₁₂ + 8O₂ → 5CO₂ + 6H₂O   ΔH = −839 kcal mol⁻¹
Pentane

Figure 4.27 Combustion of different alkanes, along with the amount of heat liberated.

predict the relative stabilities of isomers of alkanes and also estimate the amount of heat liberated for the combustion of straight chain alkanes.

4.7 Stability of Alkenes

The same type of analysis can be used to determine the relative stability of different structural isomers of alkenes. Consider the hydrogenation of the two alkenes shown in Reactions (4-1) and (4-2). Since both reactions give the same product (2-methylbutane), a comparison of relative energies of the reactants can be carried out by a similar approach as demonstrated with the combustion of alkanes. For the alkene that has the terminal double bond (Reaction 4-1), more heat is released to give the same product, compared to that of the same reaction for the alkene that has an internal double bond (Reaction 4-2).

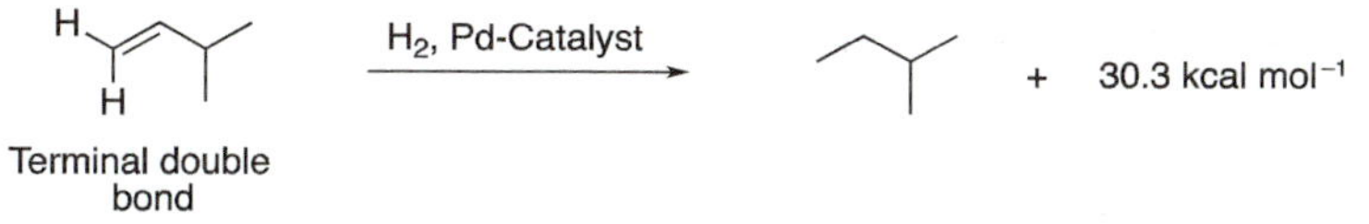

Terminal double bond →(H₂, Pd-Catalyst)→ product + 30.3 kcal mol⁻¹   (4-1)

Internal double bond →(H₂, Pd-Catalyst)→ product + 26.9 kcal/mol⁻¹ 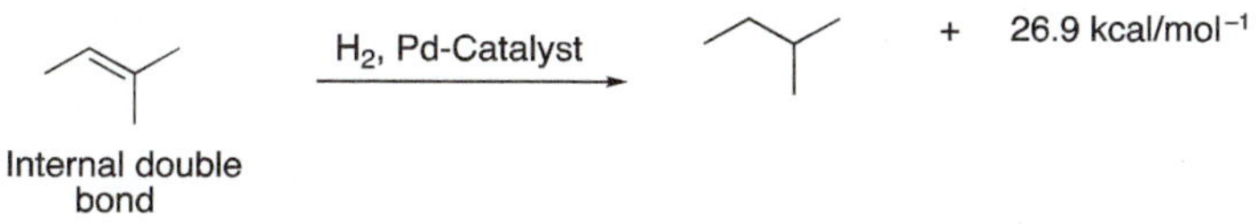   (4-2)

This concept can be represented using an energy profile diagram as shown in Figure 4.28.

Thus, it can be concluded that 2-methyl-2-butene is more stable than 3-methyl-1-butene, and an alkene that has a double bond in a terminal position is less stable than a similar alkene that has the double bond internally.

A similar analysis can be carried out to determine the relative stabilities of cis and trans geometric isomers of alkenes as shown in Reactions (4-3) and (4-4).

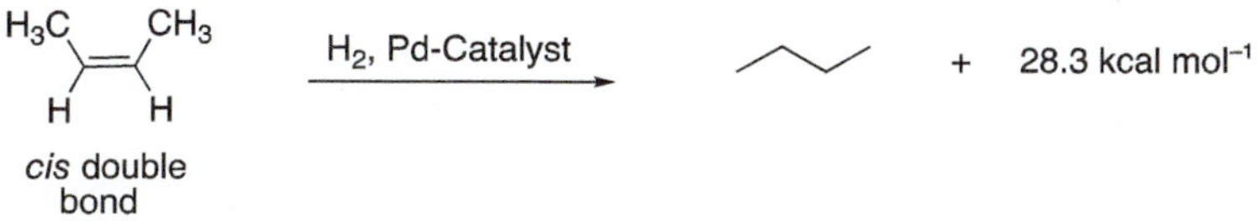

cis double bond →(H₂, Pd-Catalyst)→ product + 28.3 kcal mol⁻¹   (4-3)

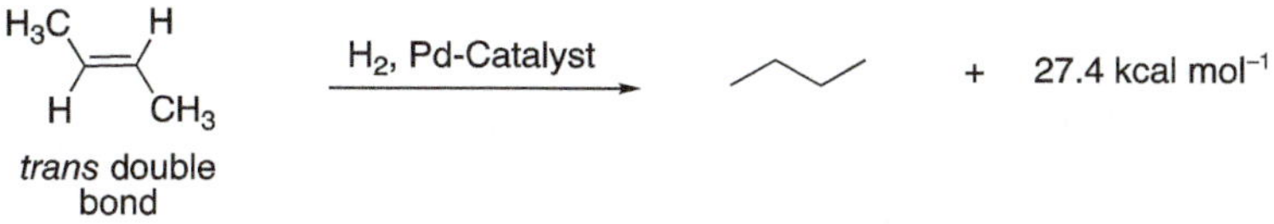

trans double bond →(H₂, Pd-Catalyst)→ product + 27.4 kcal mol⁻¹   (4-4)

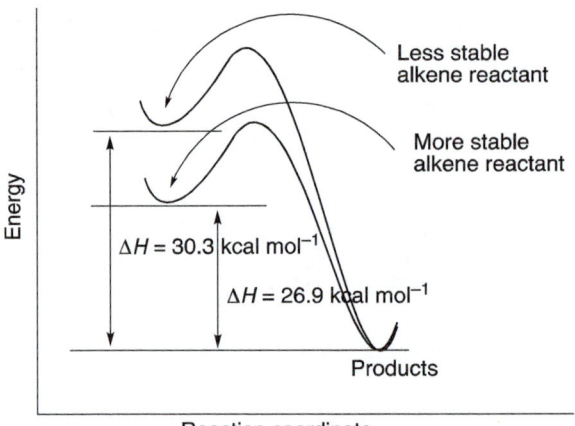

Figure 4.28 Reaction profile for the catalytic hydrogenation of 3-methyl-1-butene and 2-methyl-2-butene.

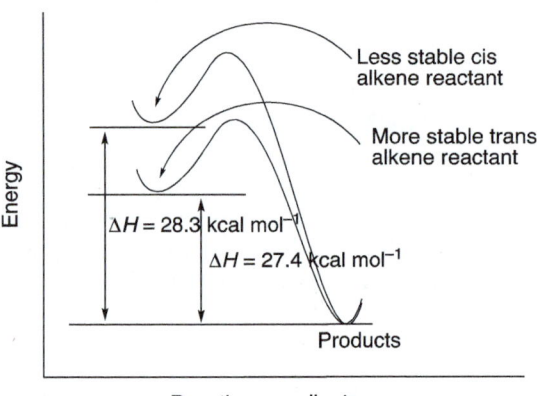

Figure 4.29 Reaction profile for the catalytic hydrogenation of *cis*-2-butene and *trans*-2-butene.

Note that for these reactions, more heat is liberated for the cis butene (Reaction 4-3) and less heat is liberated from trans butene (Reaction 4-4). The energy profile for these reactions is shown in Figure 4.29.

Based on these results, it can be concluded that (E)-2-butene is more stable than (Z)-2 butene and that trans alkenes are more stable that cis alkenes.

4.8 Stability of Alkynes

There are no geometric isomers of alkynes, but structural isomers are possible in terms of the location of the triple bond. The triple bond can be internal or terminal, and Reactions (4-5) and (4-6) give the hydrogenation of a terminal and internal alkyne.

$$H-C\equiv C-\diagup\diagdown\quad +\quad H_2\quad \longrightarrow\quad \diagup\diagdown\diagup\diagdown\quad \Delta H = -69.2 \text{ kcal mol}^{-1} \quad (4\text{-}5)$$

$$\diagdown C\equiv C\diagdown\quad +\quad H_2\quad \longrightarrow\quad \diagup\diagdown\diagup\diagdown\quad \Delta H = -65.2 \text{ kcal mol}^{-1} \quad (4\text{-}6)$$

The energy profile for Reactions (4-5) and (4-6) is shown in Figure 4.30.

Figure 4.30 Reaction profile for the catalytic hydrogenation of 1-hexyne and 3-hexyne.

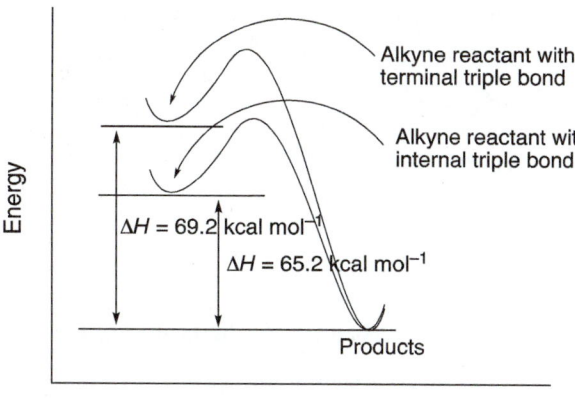

Since the least amount of heat is liberated from 1-hexyne, compared to 3-hexyne, it can be concluded that 3-hexyne is more stable than 1-hexyne, and a general conclusion can be drawn from this observation that alkynes that have internal triple bonds are more stable than alkynes with terminal triple bonds.

Problem 4.14

Of the following pair of isomers, determine which is more stable.

a) (E)-3-hexene and (Z)-3-hexene
b) 1-pentene and 2-pentene
c) 2,3-dimethylcyclohexene and 1,2-dimethylcyclohexene

End of Chapter Problems

4.15 Give a dashed-wedged representation for the following molecules.
   a) 2-Butanol (dashed-wedged representation about C_2)
   b) Cyclopentanol (dashed-wedged representation about C_1)
   c) 4-Bromo-2-pentanol (dashed-wedged representations about C_2 and C_4)

4.16 Answer the following questions that relate to 1,2-dibromoethane.
   a) Give the Newman projection and dashed/wedge representation of the least stable conformer of 1,2-dibromoethane.
   b) Consider the most stable conformer of 1,2-dibromoethane, give the Newman projection of the conformer that results after rotation about the carbon–carbon bond by 60°.
   c) Consider the most stable conformer of 1,2-dibromoethane, give the dashed-wedge representation of the conformer that results after rotation about the carbon–carbon bond by 180° has occurred.
   d) Using the Newman projection, draw all possible staggered conformations of 1,2-dibromoethane.
   e) Using the dashed-wedged representation, draw all possible eclipsed conformations of 1,2-dibromoethane.

4.17 Consider carefully the conformers of molecules A and B shown.

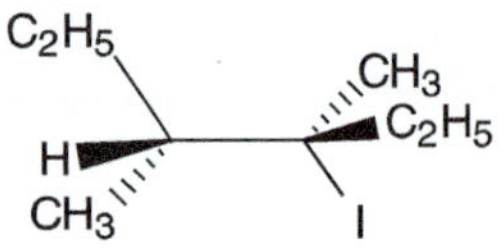

Assume that the trend in size for the groups of molecules **A** and **B** is:
I > C$_2$H$_5$ > CH$_3$ > H, i.e. the largest is the iodine atom.

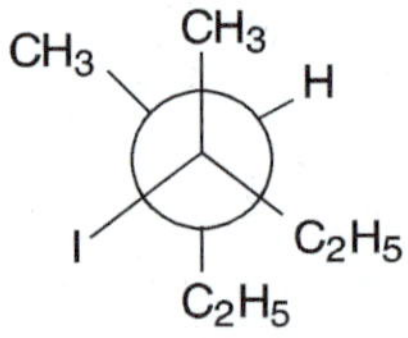

a) Draw the Newman projection for the most stable conformer of molecule **A**.
b) Draw the dashed-wedge representation of the least stable conformer of molecule **B**.
c) Draw the Newman projection that results after rotation about the middle C—C bond of molecule **A** by 180°.

4.18

a) Draw a chair conformation of the molecule shown below.

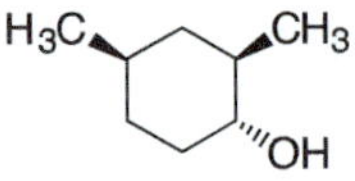

b) Now convert your structure to the other chair conformer.
c) Which of the two conformers that you have drawn above is the most stable? Explain your answer.

4.19 Answer the following questions that relate to dimethylcyclohexane.
a) Draw the chair conformer of the least stable conformer of 1,4-dimethylcyclohexane
b) Draw the chair conformer of the most stable conformer of 1,4-dimethylcyclohexane
c) Draw the chair conformer of the least stable conformer of 1,2-dimethylcyclohexane
d) Draw a trans conformer of 1,4-dimethylcyclohexane
e) Draw a cis conformer of 1,3-dimethylcyclohexane

4.20 Give complete IUPAC names, including E or Z assignment, for the following molecules.

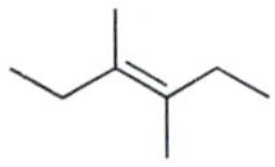

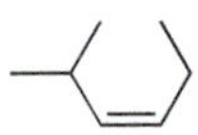

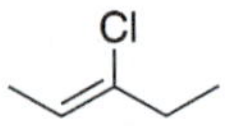

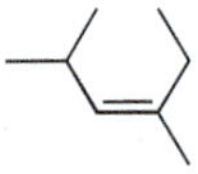

5

Stereochemistry

5.1 Introduction

In a three-dimensional world, our analysis of atoms and molecules must be from that perspective. In this chapter, a detailed examination of the three-dimensional arrangement of atoms in a molecule will be carried out. The study of differences in isomers brought about by differences in location of atoms or groups in molecules in three-dimensional space is called stereochemistry. Geometric and conformational stereoisomers have already been examined in the previous chapter. Geometric isomers differ from each other based on the arrangement of the atoms or groups across a rigid plane, such as a double bond or rigid cyclic ring in the molecule. For example, *trans*-1,2-dichloroethene is different from *cis*-1,2-dichloroethene since the chlorine atoms are on different sides of the rigid double bond.

trans-1,2-Dichloroethene   *cis*-1,2-Dichloroethene

On the other hand, conformational isomers differ from each in that groups or atoms in the molecule are in different locations due to rotation about a single bond, typically a carbon–carbon single bond. Thus, the anti-conformer of 1,2-dichloroethane is different from the *gauche* conformer. These conformers are different only in the spatial arrangement of the groups around the carbon–carbon bond; and as a result, their relative energies are different.

Anti conformer of
1,2-dichloroethane

Gauch conformer of
1,2-dichloroethane

For some stereoisomers, it is extremely difficult to visualize differences that result in three-dimensional space for molecules brought about by different spatial arrangements of the atoms or groups, unless a model set is used. **Stereochemistry** is the study of molecules that have the same atom–atom connectivity, but the spatial arrangement among the groups or atoms is different, resulting in different molecules, known as stereoisomers. Just as a pair of shoes looks similar, but the spatial arrangements in three-dimensional space are different. Try putting on

Organic Chemistry: Concepts and Applications, First Edition. Allan D. Headley.
© 2020 John Wiley & Sons, Inc. Published 2020 by John Wiley & Sons, Inc.
Companion website: www.wiley.com/go/Headley_OrganicChemistry

the left foot of the shoe on your right foot, it just will not fit since it is different from the right foot of a pair of shoes. Just as the left foot pair of shoes and its right foot can be described as different, we will see in this chapter that the same comparison can be made for some molecules.

Being able to recognize stereoisomers is an essential technique in that it requires students to be able to analyze molecules in the three-dimensional space and carry out various manipulations mentally in order to make important conclusions about molecules. This technique is an important analytical thinking tool since most of today's highly technical activities are typically carried out while viewing the procedure on a 2-D computer screen. A large percentage of surgeries today are carried out using techniques in which various surgical manipulations are viewed and controlled by viewing three-dimensional procedures on a computer 2-D screen.

5.2 Chiral Stereoisomers

As pointed out earlier, some objects and molecules, even though they look very much alike, are not really the same. For example, objects that are mirror images of each other look very much alike, but are they really the same? It is very important to be able to determine subtle differences in the stereochemistry of molecules. A practical way to better understand this concept is to imagine a pair of gloves. They look pretty much the same, but they are really different. The left hand of a pair of gloves will not fit the right hand and vice versa. Thus, one evidence that each glove in a pair of gloves is different from the other is that they are not interchangeable on the right hand and left hand. One test that can be used to determine if two molecules, which look very much alike, are really the same or different is to determine if they are *superimposable* on each other. For molecules that are superimposable, there is a complete overlay of all atoms in three-dimensional space. If two molecules are superimposable, then they are the same, but if they are not superimposable, then they are not the same – they are different molecules. Molecules in which their mirror images are not superimposable on each other are called **chiral** or **stereogenic** molecules. L and D α-amino acids are very similar in appearance, but they are really different molecules since they are not superimposable on each other. Shown in Figure 5.1 are the L and D forms of the α-amino acid, alanine.

Since the mirror image of a stereogenic compound is not the same as the molecule being reflected in the mirror, a stereogenic (or chiral) molecule and its mirror image are really two different compounds, and hence they are isomers, and to be specific, they are stereoisomers. The relationship between such stereoisomers is a special one and is referred to as **enantiomerism**, and the compounds are called *enantiomers*. *Enantiomers* are stereoisomers in which the mirror images of each other are not superimposable on themselves. This concept of enantiomers is illustrated in Figure 5.2.

The molecules shown in Figure 5.2 differ from each other only in the spatial arrangements of the four different groups around the central carbon. This central carbon of compounds that have four different groups is known as a **stereogenic carbon**. Thus, a *stereogenic* carbon is a carbon that is bonded to four different groups. An examination of a slightly different molecule, which has the same two atoms or groups bonded to a carbon and two different atoms or groups, gives a totally different outcome, as shown in the molecules in Figure 5.3.

H_3C H
H_2N $COOH$
L-Alanine, an amino acid

H CH_3
H_2N $COOH$
D-Alanine, an amino acid

Figure 5.1 L-Alanine and D-alanine are different molecules; they are mirror images of each other and are not superimposable on each other.

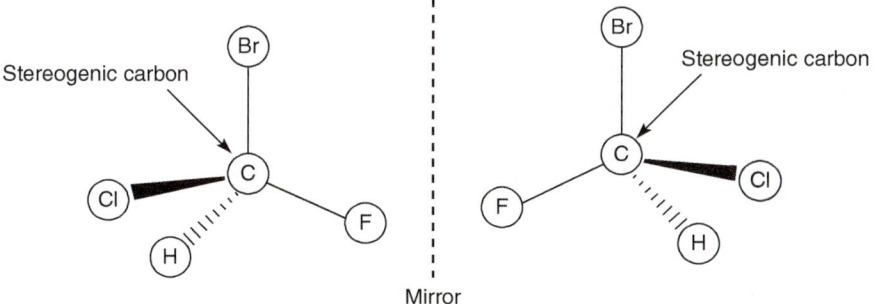

Figure 5.2 Nonsuperimposable mirror images, also known as enantiomers.

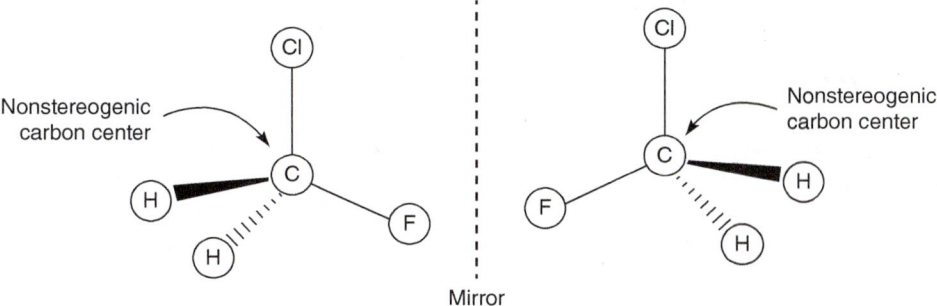

Figure 5.3 Superimposable mirror images, or same molecules.

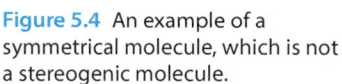

Figure 5.4 An example of a symmetrical molecule, which is not a stereogenic molecule.

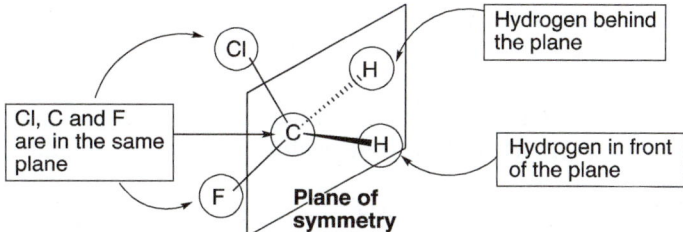

Even though these compounds are mirror images of each other, they are superimposable on each other and hence they are not enantiomers, but instead they are the same molecules. Note that for these molecules there is a plane of reflection through the carbon, chlorine, and fluorine atoms, as shown in Figure 5.4. That is, the two hydrogen atoms reflect across that plane. Such molecules are referred to as ***symmetric*** molecules or have a plane of symmetry. Such molecules are achiral or not stereogenic.

5.2.1 Determination of Enantiomerism

The use of an imaginary mirror to assist in the determination of enantiomerism is a very tedious exercise. Some simple observations of the molecules in both examples given above can be made to decide if molecules are stereogenic or not. Molecules that are ***asymmetric*** or molecules that do not have a plane of symmetry (as the molecules shown in Figure 5.2) are stereogenic and will exhibit enantiomerism. If a molecule has a carbon atom that has four different groups

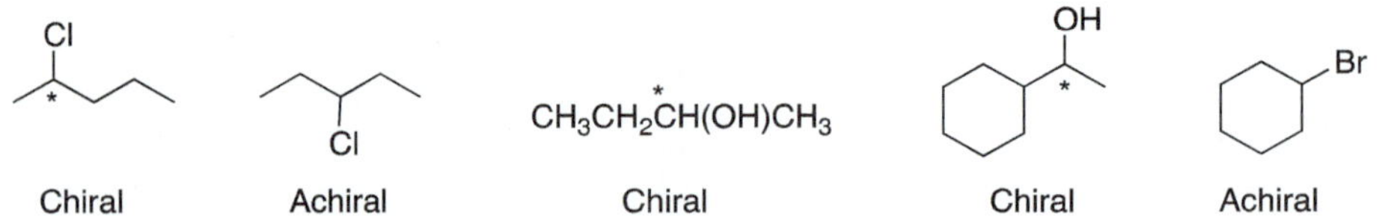

5 Stereochemistry

Figure 5.5 Examples of chiral and achiral molecules, an asterisk indicates the stereogenic carbon.

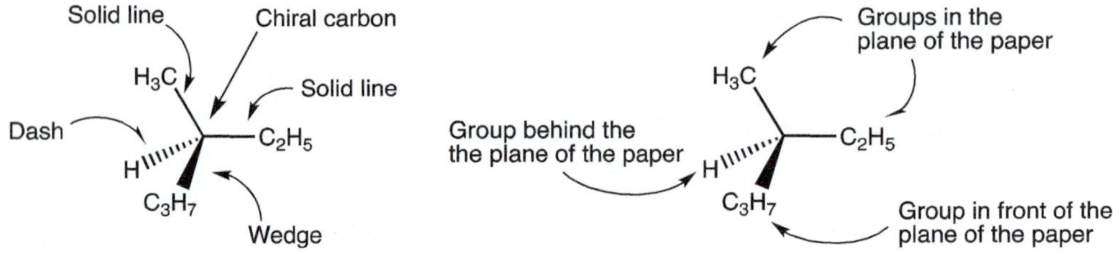

Figure 5.6 Illustration of the dashed-wedge representation for stereogenic molecules.

bonded to that carbon, then that molecule is stereogenic. These simple observations will assist in a quick determination of enantiomerism in molecules. Figure 5.5 gives examples of molecules that are stereogenic because they contain one atom (the starred atom) that has four different groups that are bonded to that carbon. Note that for the achiral molecules, there is not a carbon present that has four different groups.

The dashed-wedge representation can also be used to show the stereochemistry of the 3-D arrangement of atoms or groups about a central carbon, and this type of representation is typically used to show the 3D stereochemistry about stereogenic carbons. Figure 5.6 gives examples in which dashed-wedge representations are used to show the stereochemistry about stereogenic carbons.

Problem 5.1

Molecules that are asymmetric or have a carbon that has four different groups are chiral. Utilize these two criteria to determine which of the following molecules are chiral?

The **Fischer Projection** is another representation that is typically used to represent a molecule in the three-dimensional space, and this type of representation is very useful to visualize the three-dimensional arrangements of the atoms in a molecule. Emil Fischer, a Nobel-prize winner, determined a method to visualize the configuration of an isomer that has four groups bonded to a central carbon. Compounds that are represented by the Fischer projection are viewed in such a way that the groups that are in front of the plane of the paper are located in a horizontal plane and the groups that are in a vertical plane are away from the viewer (in the back of the plane of the paper). The dashed-wedge representation can be used to illustrate the orientation of the Fischer projection and is demonstrated using 1-bromo-1-chloroethane shown in Figure 5.7.

The Fischer projection and dashed-wedge representation for 2-bromobutane are given in Figure 5.8.

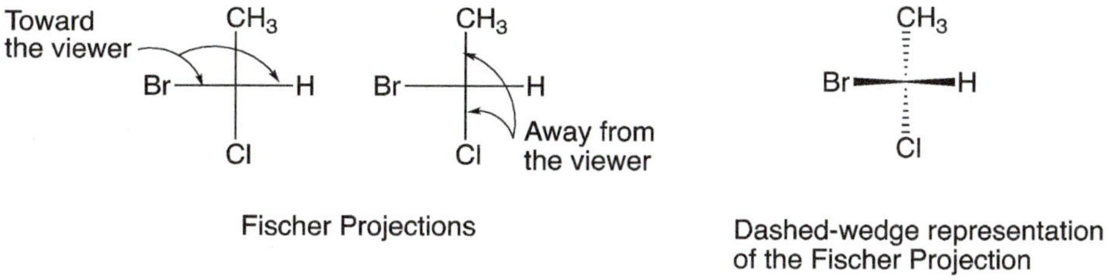

Figure 5.7 The Fischer projection of a specific isomer of 1-bromo-1-chloroethane.

Figure 5.8 Dashed/wedge and Fischer projection of 2-bromobutane.

Problem 5.2

i) Draw a Fischer projection of the following molecules.

  a) 2-Chlorobutane (use C_2 as the central carbon).
  b) 2-Pentanol (use C_2 as the central carbon).
  c) 3-Hexanethiol (use C_3 as the central carbon).
  d) 3-Methyl-3-hexanol use (C_3 as the central carbon).

ii) Give the Fischer projection for each of the molecules shown below. Note that each molecule is represented by a specific dashed-wedge representation.

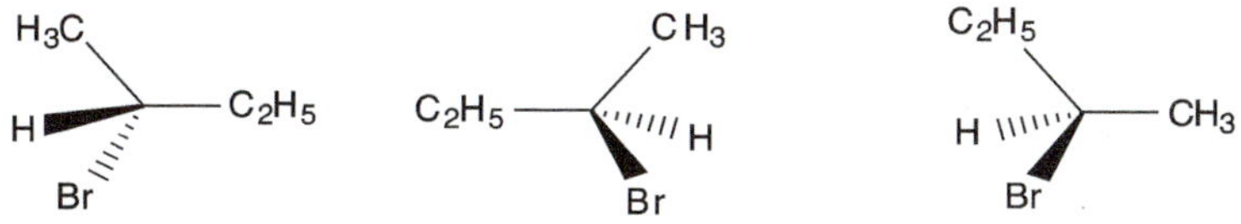

5.3 Significance of Chirality

Chirality (enantiomerism) in molecules plays a key role in nature and technology. The effective functioning of the human body, for example, depends on the chiral recognition of different molecules. Many biological molecules are chiral, and the survival of living systems depends on interactions with a specific enantiomer. For example, one enantiomeric form of amino acids is not toxic to the body, yet their enantiomeric stereoisomers are toxic to the body. Just as a right-hand glove cannot fit on the left hand and vice versa; typically, only one enantiomer of a drug can fit in the receptor/drug binding site for effective biological activity. This concept is illustrated in Figure 5.9. In order for the drug to show the desired pharmacological effect, the portions B, C, and D of the drug must fit in the corresponding sockets b, c, and d of the drug-binding site (receptor). As shown in the figure, the interactions of one of the enantiomers correspond ideally to the receptor, resulting in desired biological effects. The other enantiomer also has the same groups A, B, C, and D, but is inactive since it is not able to fit into the receptor.

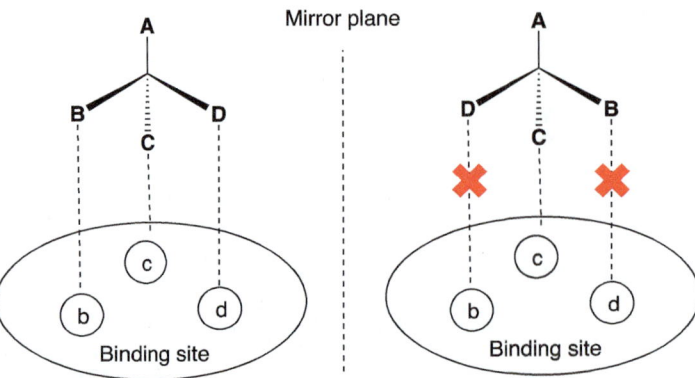

Figure 5.9 Hypothetical interaction between two enantiomers of a chiral drug and a binding site.

5.3.1 Molecular Chirality and Biological Action

In 1848, the French chemist, Louis Pasteur discovered molecular chirality in tartaric acid. In 1858, he showed that one enantiomer (the dextro form) of the ammonium tartrate was fermented by *Penicillium glaucum*, while its enantiomer (the levs form) was unaffected.

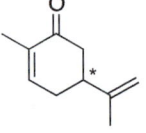

Tartaric acid

One enantiomer of ammonium tartrate is fermented by *Penicillium glaucum*

The other enantiomer is unaffected by *Penicillium glaucum*

Since then, scientists have recognized the significance of stereoisomerism in relationship to physiological activity. A practical importance of chirality in biological system is that of carvone and the effect of the different enantiomers on the olfactory receptors. As shown below, different odors are detected from the two enantiomers of carvone, an asterisk indicates the stereogenic carbon.

Carvone

One enantiomer has an odor of caraway seeds

The other enantiomer has an odor of spearmint

Problem 5.3

Atorvastatin and simvastatin are the active ingredients in the lipid-lowering drugs, Lipitor and Zocor, respectively. Identify chiral carbons that are present in atorvastatin and simvastatin.

Atorvastatin

Simvastatin

The urgency of the importance of stereochemistry and chirality, especially in the pharmaceutical industry, lies in the history of the thalidomide tragedy. Thalidomide was a drug first used in the market in 1957 in Germany. It was prescribed as a sedative and was claimed to cure anxiety, insomnia, gastritis, and tension. Later, in the 1960s, it was widely used against nausea for morning sickness for pregnant women. Shortly after the drug was introduced in the market, many infants were born with malformation of limbs. The research led to the discovery of two enantiomers of thalidomide and found that the one enantiomer has the desired effect in treating morning sickness, whereas the other enantiomer is teratogenic.

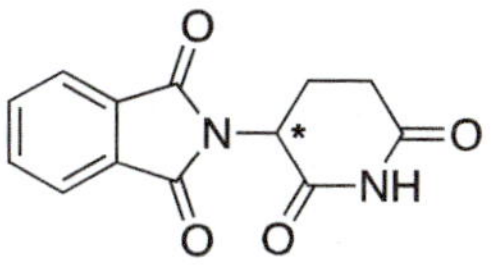

Thalidomide

One enantiomer is a sedative and used in the cure of morning sickness in pregnant women

The other enantiomer is a teratogenic and causes severe deformity in babies

Problem 5.4

Identify the chiral centers and give a dashed-wedge representation around the chiral carbon for each of the molecules shown below.

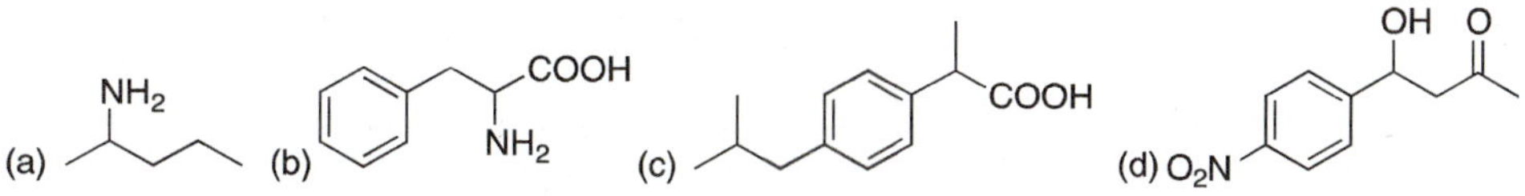

5.4 Nomenclature of the Absolute Configuration of Chiral Molecules

Since enantiomers are different molecules, they must have different names. The IUPAC nomenclature system that has been used so far gives exactly the same name for enantiomers. Since enantiomers are different compounds, they should have different names even if they are slightly different. Three chemists, R.S. Cahn, C. K. Ingold, and V. Prelog, were the first to devise a system to allow chemists to differentiate between enantiomers by using an absolute assignment. In the Cahn-Ingold-Prelog system, a set of rules were developed using the designations R (from the Latin rectus, which means right handed) and S (from the Latin sinister, which means left handed). The R, S system was developed to assign an absolute configuration for the stereogenic centers of chiral molecules.

Rules for R and S Assignments

1) Assign priority numbers to each atom bonded to the stereogenic carbon based on the atomic number of the atom that is directly bonded to the chiral carbon. This system of priority assignment has been mentioned before in the section on geometric isomers of alkenes (Section 4.5.1).
2) Orient the molecule in such a way that the group of lowest priority is away from view, i.e. in the back.
3) Trace the path from the group of highest priority to the group of lowest priority. That is, #1 to #2 then to #3, exclude #4 since it is away from view.
4) If that path traced is clockwise, assign the absolute configuration of **R**; and if the path is counterclockwise, assign the absolute configuration of **S**.

Some students find it much easier to use the Fischer projection and to orient the molecule such that the group of lowest priority is in one of the vertical positions. Remember that the groups of the vertical positions are in the back when using the Fischer projection. With the molecule in the Fischer projection and if the group (or atom) of lowest priority is in one of the vertical positions, trace the priority of the other groups from #1 to #2, then #3 and if that path traced is clockwise, assign the absolute configuration of **R**; and if the path is counterclockwise, assign the absolute configuration of **S**. If the group of lowest priority is not in one of the vertical positions, but in one of the horizontal positions, trace the priority of the other groups from #1 to #2, then #3 and if that path traced is clockwise assign the opposite configuration, **S**. The same is done if the counterclockwise is obtained. Since there are only two possibilities for the assignment, R or S, the orientation obtained is opposite to that if the group of lowest priority were in the back. Shown below is an illustration of this method.

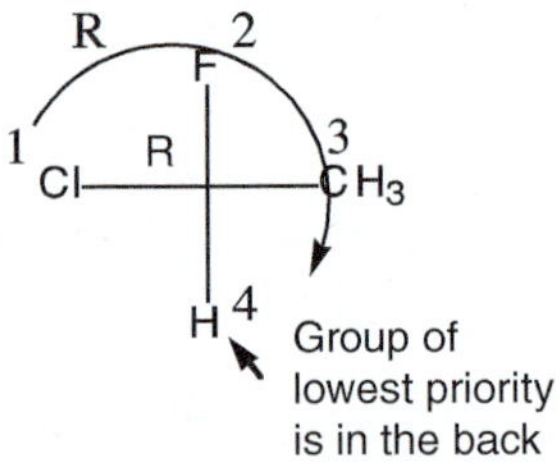

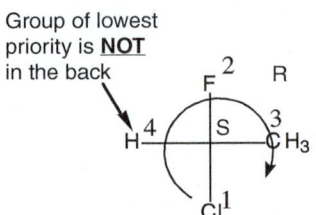

Since the group of lowest priority is in the back, the absolute configuration of the chiral carbon is the same as that shown by the direction of the priority groups

Since the group of lowest priority is **NOT** in the back, the absolute configuration of the chiral carbon is **OPPOSITE** to that which is shown by the direction of the priority of the groups

Thus, the complete IUPAC names for stereogenic compounds should include the absolute configuration, R or S. The complete IUPAC names of the stereogenic molecules of 1-bromo-1-chloroethane are shown below.

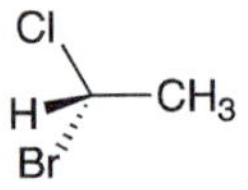

(S)-1-Bromo-1-chloroethane 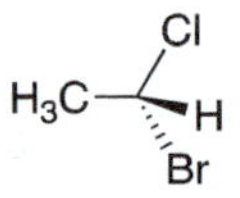   (R)-1-Bromo-1-chloroethane

Problem 5.5

Give the absolute configuration for each of the following molecules; that is, assign R or S absolute configuration.

(a), 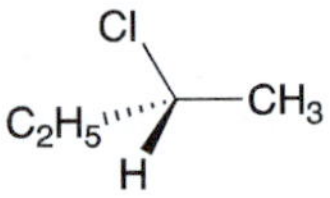 (b), 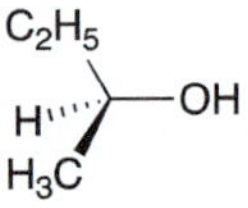 (c), 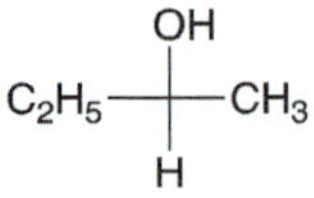 (d)

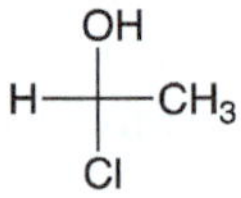

It is important to know that enantiomers are really different compounds and that each plays important roles in biological systems. The discovery that there are enantiomers of thalidomide and that one enantiomer is the compound for desired biological activity and that the other enantiomer led to serious medical consequences led chemists to emphasize the importance of the purity of enantiomers, especially when used for drugs. A large percentage of the pharmaceutical drugs on the market today are chiral and are sold as single enantiomers since slight

Table 5.1 Examples of different enantiomers of drugs having different physiological actions, the star indicates the chiral center.

Drug	R-Enantiomer	S-Enantiomer	Structure
Ibuprofen	Slow acting	Fast acting	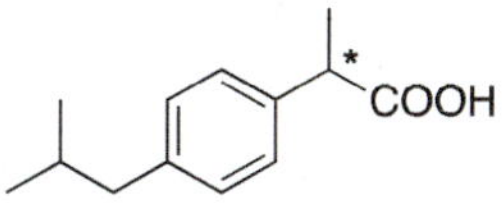
Naproxen	No analgesic effect	Arthritis treatment	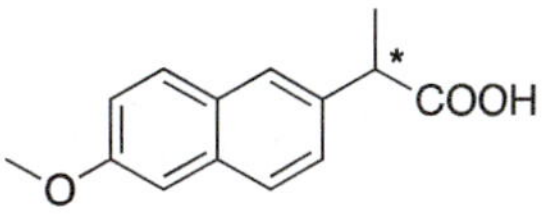
Citalopram	Inactive	Antidepressant	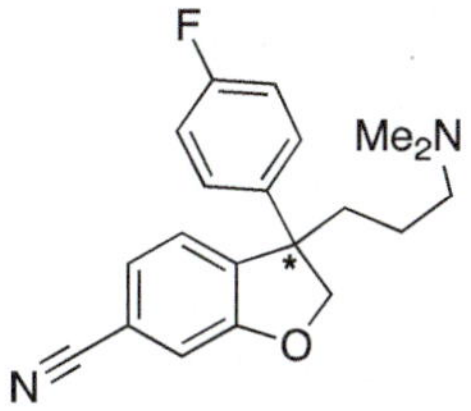
Dopa	Biologically inactive	Anti-Parkinson	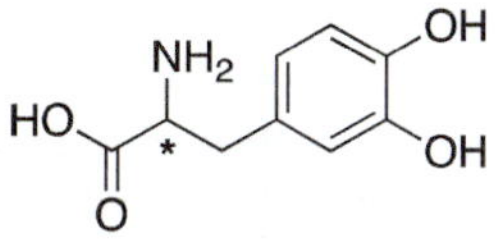
Ethambutol	R,R enantiomer: Causes optical neuritis, eventually results in blindness	S,S enantiomer: Used to treat tuberculosis	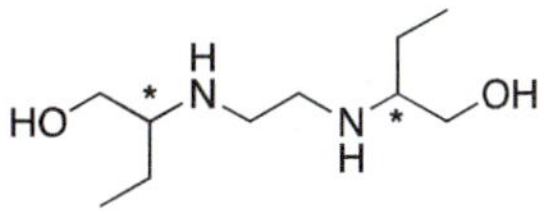
Penicillamine	Toxic	Antiarthritic	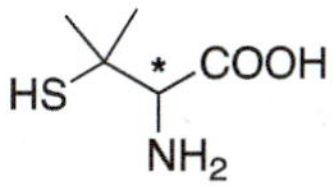
Clopidogrel (Plavix)	Less active enantiomer for the inhibition of platelet aggregation	More active enantiomer for the inhibition of platelet aggregation	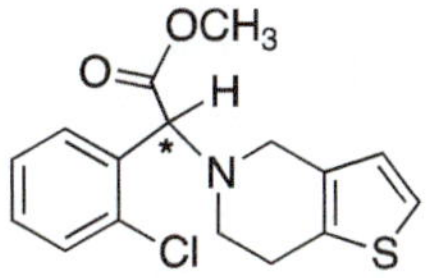

contamination of the enantiomer can cause serious medical consequences. Thus, it is imperative that the pharmaceutical industry exercise extreme care not to have contamination of enantiomers in drugs that can cause harmful side effects. Table 5.1 shows the structures of some popular drugs on the market today and the importance of specific enantiomers.

5.5 Properties of Stereogenic Compounds

Since enantiomers are different molecules, they must have different properties. For a pair of enantiomers, there is only one physical property difference observed and that is their interaction with polarized light. Ordinary light can be thought of as electromagnetic radiation, which consists of electric and magnetic field vectors, which oscillate in all directions. Polarized light, on the other hand, has vectors in only one plane; vectors in the other planes have been filtered out, this light is also known as monochromatic polarized light. When ordinary light is passed through a prism, such as a calcite crystal ($CaCO_3$) prism, the light that results is polarized light. If polarized light is passed through a solution of an enantiomer of a stereogenic compound, the plane of the polarized light is rotated to the right or to the left depending on the absolute configuration of the enantiomer. Based on this property, enantiomers are also referred to as optically active compounds. This property, specific rotation [α], is measured using a polarimeter and is defined by Eq. (5-1).

$$\text{Specific rotation} = [\alpha] = \text{observed rotation}(°) / \text{length}(dm) \times \text{concentration}(g/ml^{-1}) \quad (5\text{-}1)$$

One enantiomer of an optically active compounds will rotate the plane of polarized light in one direction and its enantiomer will rotate the plane of polarized light by an equal amount, but in the opposite direction. Optically active compounds that rotate the plane of polarized light to the right (or clockwise) are called **dextrorotatory** (*d*) compounds, whereas compounds that rotate the plane of polarized light to the left (or counter clockwise) are called **levorotatory** (*l*) compounds. Another method that is often used to differentiate between the enantiomers of optically active compounds is the use of a positive sign (+) for dextrorotatory compounds and a negative sign (−) for levorotatory compounds. This system was an arbitrary assignment by Emil Fischer, a German organic chemist. He arbitrarily assigned the glyceraldehyde enantiomer that rotates the plane of polarized light to the right with a (+) sign and called it the *d* isomer. Optically active compounds are assigned + or − (or *d* and *l*) based on experimental results, and as a result, these experimental assignments do not necessarily correlate with the R and S absolute assignments discussed earlier. Thus, there is no direct relationship between R and S and *l* and *d* or R and S and the direction that enantiomers rotate the plane of polarized light. The experimentally determined specific rotation values of glyceraldehyde are shown below.

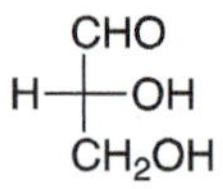

```
       CHO                        CHO
   H ──┼── OH                 HO ──┼── H
       CH₂OH                      CH₂OH
```

(+) or (*d*) Glyceraldehyde   (−) or (*l*) Glyceraldehyde
    [α] = +8.7   [α] = −8.7

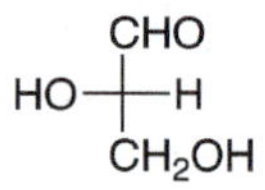

The specific rotation for an equal mixture of enantiomers will be zero since for every molecule that rotates the plane of polarized light in one direction, its enantiomer will rotate it in the opposite direction. A mixture of equal amounts of enantiomers is called a **racemic mixture**. Racemic mixtures are *optically inactive*.

5.6 Compounds with More Than One Stereogenic Carbon

It is possible for an optically active molecule to have more than one stereogenic carbon. For each stereogenic carbon in such a molecule, it is possible to assign absolute configuration about that carbon. As a result, there are many different stereoisomers that are possible. The relationship between such stereoisomers may or may not be enantiomerism. For a stereoisomer that has more than one stereogenic carbon, the possible combinations of R and S stereoisomers for the stereogenic carbons are 2*n*, where n represents the number of stereogenic carbons. Thus, for a compound with two stereogenic carbons, there is a maximum of four different stereoisomers. Shown in Figure 5.10 are the four stereoisomers of 2-bromo-3-chlorobutane, which has two stereogenic carbons.

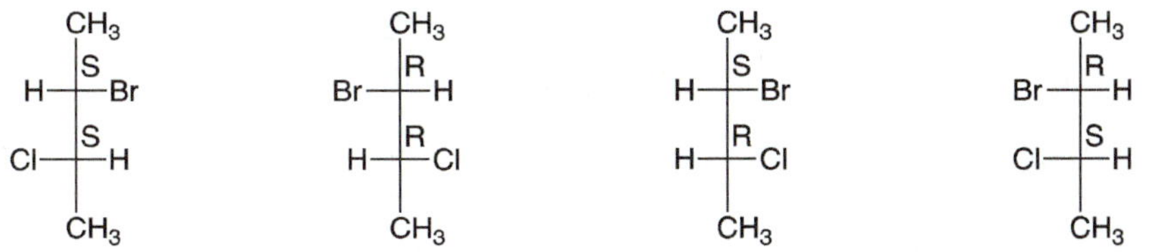

5.6 Compounds with More Than One Stereogenic Carbon

```
     CH₃              CH₃              CH₃              CH₃
      |S               |R               |S               |R
  H---+---Br       Br--+---H        H---+---Br       Br--+---H
      |S               |R               |R               |S
  Cl--+---H        H---+---Cl       H---+---Cl       Cl--+---H
      |                |                |                |
     CH₃              CH₃              CH₃              CH₃
 (2S, 3S)-2-Bromo- (2R, 3R)-2-Bromo- (2S, 3R)-2-Bromo- (2R, 3S)-2-Bromo-
 3-chlorobutane    3-chlorobutane    3-chlorobutane    3-chlorobutane
```

Figure 5.10 Fischer projections of different stereoisomers of 2-bromo-3-chlorobutane.

The first and second molecules are enantiomers of each other, and the third and fourth molecules are enantiomers of each other. The first and third molecules and the first and fourth molecules are not enantiomers of each other, however. Since the only difference between these molecules is the difference in configuration about one carbon, these molecules are different molecules and the relationship is not enantiomerism, but different stereoisomers and this special relationship is called **diastereomerism**, and the molecules are called **diastereomers**. Thus, (2S,3S)-2-bromo-3-chlorobutane and (2S,3R)-2-bromo-3-chlorobutane are diastereomers. Likewise, (2S,3S)-2-bromo-3-chlorobutane and (2R,3S)-2-bromo-3-chlorobutane are diastereomers. The other diastereomers are (2R,3R)-2-bromo-3-chlorobutane and (2R,3S)-2-bromo-3-chlorobutane. **Diastereomers** *are stereoisomers that are not mirror images of each other and are not enantiomers of each other.* The physical properties of diastereomers, such as the melting points and boiling points, are different from each other.

Problem 5.6

For 2-bromo-3-fluorobutane, (a) give a Fischer projection of this molecule, (b) assign R and S absolute configuration about each stereogenic carbon, (c) give the Fischer projection of the diastereomers of the molecule given in part (a) of this question.

The same analysis of stereoisomers can be carried out also using the dashed-wedge representation instead of the Fischer projection. Figure 5.11 shows the dashed-wedge representation for the stereoisomers shown in Figure 5.10.

Note that the Fischer projections shown in Figure 5.10 are the least stable since they are all eclipsed conformational isomers. Figure 5.12 gives the most stable anti conformers, which is accomplished by rotation about the C_2–C_3 bond by 180°.

For some molecules with more than one stereogenic carbon, not all stereoisomers that result from different spatial arrangements of the atoms or groups about the stereogenic carbons result in different stereoisomers. Consider the stereoisomers of 2,3-dibromobutane, which are shown below.

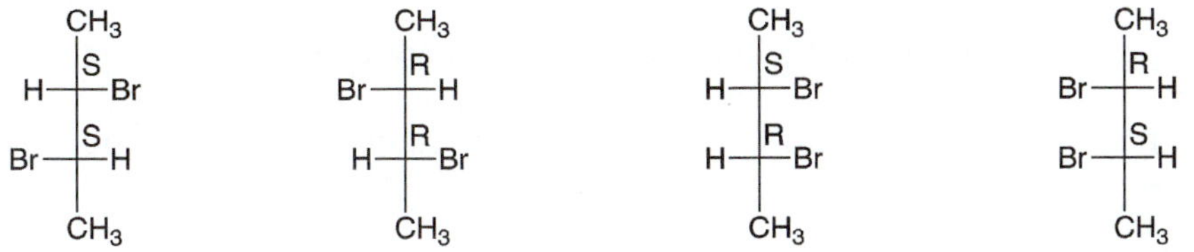

```
     CH₃              CH₃              CH₃              CH₃
      |S               |R               |S               |R
  H---+---Br       Br--+---H        H---+---Br       Br--+---H
      |S               |R               |R               |S
  Br--+---H        H---+---Br       H---+---Br       Br--+---H
      |                |                |                |
     CH₃              CH₃              CH₃              CH₃
 (2S, 3S)-2,3-     (2R, 3R)-2,3-     (2S, 3R)-2,3-     (2R, 3S)-2,3-
 Dibromobutane     Dibromobutane     Dibromobutane     Dibromobutane
```

For these stereoisomers, the first and second molecules are enantiomers; however, the third and fourth molecules are not enantiomers, but identical molecules. Making this assignment is

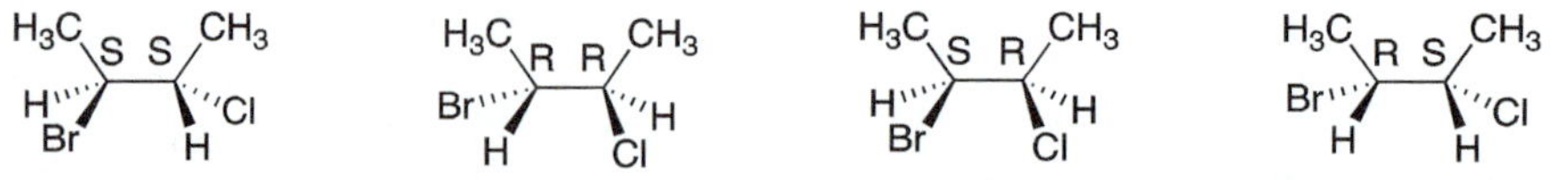

H₃C⟍S S⟋CH₃
H⟋   ⟍Cl
Br   H
(2S, 3S)-2-Bromo-
3-chlorobutane

H₃C⟍R R⟋CH₃
Br⟋   ⟍H
H   Cl
(2R, 3R)-2-Bromo-
3-chlorobutane

H₃C⟍S R⟋CH₃
H⟋   ⟍H
Br   Cl
(2S, 3R)-2-Bromo-3-
chlorobutane

H₃C⟍R S⟋CH₃
Br⟋   ⟍Cl
H   H
(2R, 3S)-2-Bromo-
3-chlorobutane

Figure 5.11 Dashed-wedge representations of the stereoisomers shown in Figure 5.10.

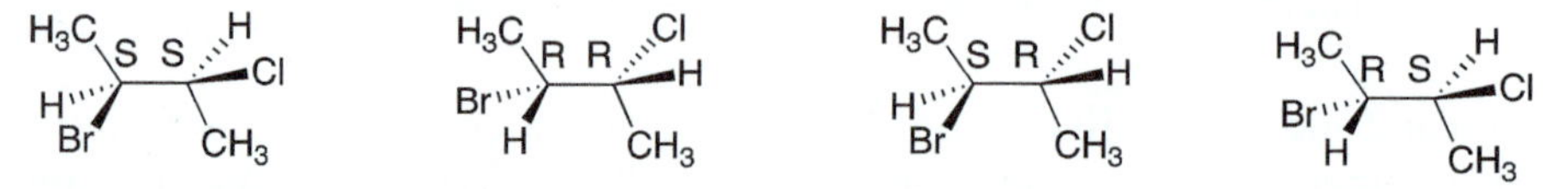

(2S, 3S)-2-Bromo-3-chlorobutane

(2R, 3R)-2-Bromo-3-chlorobutane

(2S, 3R)-2-Bromo-3-chlorobutane

(2R, 3S)-2-Bromo-3-chlorobutane

Figure 5.12 Dashed-wedge representation of the stereoisomers shown in Figure 5.11, but a rotation about the central carbon–carbon bond by 180° gives the most stable conformers.

tricky since the third and fourth molecules are mirror images of each other. Close examination of these two molecules reveals that they are superimposable on each other, hence the same. Another method that can be used to demonstrate that they are the same is to determine if there is a plane of symmetry in the molecule. As demonstrated earlier in the chapter, if there is a plane of symmetry, the mirror image of that molecule is not an enantiomer, but the same. For the third and fourth molecules, there is a plane of symmetry, which is a horizontal plane between carbons 2 and 3, as demonstrated below.

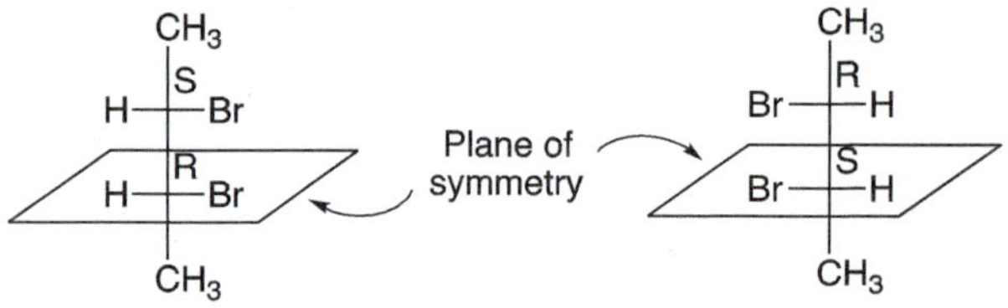

(2S, 3R)-2,3-Dibromo-butane

(2R, 3S)-2,3-Dibromo-butane

Stereoisomers that have stereogenic carbons, and also a plane of symmetry, are called ***meso compounds***. ***Meso compounds*** are stereoisomers that contain stereogenic carbons, but their mirror images are superimposable on each other.

Problem 5.7

i) Give a Fischer projection of (2S,3R)-2,3-butanediol; is this compound a meso compound or optically active?
ii) Give a Fischer projection of (2S,3S)-2,3-butanediol; is this compound a meso compound or optically active?

5.6.1 Cyclic Compounds with More Than One Stereogenic Center

It is possible to have cyclic compounds that have stereogenic carbons, such as carvone and thalidomide, as discussed earlier in the chapter, and the R and S stereoisomers are shown below.

Figure 5.13 Symmetrical analysis of isomers of 1,2-dimethylcyclopropane.

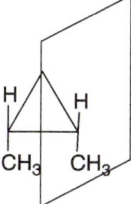

cis-1,2-Dimethylcyclopropane has a plane of symmetry

trans-1,2-Dimethylcyclopropane does not have a plane of symmetry

(S)-Carvone (R)-Carvone (S)-Thalidomide (R)-Thalidomide

It is possible to have cyclic compounds with two chiral carbons and for such compounds to be symmetric or asymmetric as discussed in the previous section. Shown in Figure 5.13 are the structures of 1,2-dimethylcyclopropane. Note that in the structure to the left, where the two methyl groups are on the same side of the plane of the cyclopropane ring (the cis stereoisomer), there is a plane of symmetry going through carbons 1 and 2. Hence, its mirror image is superimposable on itself and symmetric making the molecule achiral. On the other hand, the stereoisomer to the right (the trans stereoisomer) does not have a plane of symmetry and hence this stereoisomer is asymmetric and its mirror image is not the same; hence, it is a chiral molecule.

Problem 5.8

i) Which of the following molecules have a plane of symmetry and hence are achiral?

ii) Give the dashed/wedge structures of (1S,2R)-1,2-dichlorocyclopropane and (1S,2S)-1,2-dichlorocyclopropane and determine if these compounds are meso compounds or not.

5.7 Resolution of Enantiomers

As demonstrated earlier in the chapter, pure enantiomeric compounds are typically needed for specific biological activities; purity of different enantiomeric compounds is of special importance to the pharmaceutical industry. Most of the reactions that will be discussed in later chapters give a mixture of enantiomers. Thus, it is very important to have experimental methods, which can be used to separate a mixture of enantiomers. For a mixture of enantiomers, one

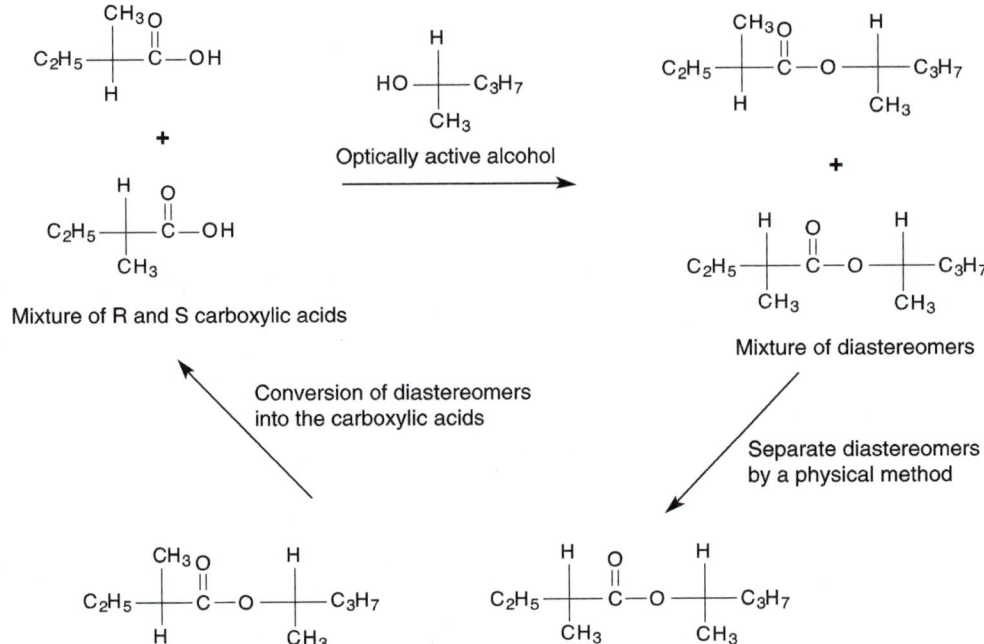

Figure 5.14 An experimental method for the separation of a mixture of enantiomers of 2-methylbutanoic acid.

strategy that is used is to convert both enantiomers of the mixture into other compounds, such as diastereomers, which have different properties. A mixture of diastereomers can be separated based on differences in their physical properties. Once the separation has occurred, each diastereomer can be converted back to the original pure enantiomer. Figure 5.14 shows a reaction scheme, which can be used to separate enantiomers of 2-methylbutanoic acid.

The first reaction involves the conversion of the enantiomeric mixture of carboxylic acids into the diastereomers in which an optically active alcohol, (R)-2-pentanol, is used. Since the products of this reaction are diastereomers, they can be separated by a physical method such as chromatography. Once the separation has been accomplished, each diastereomer can be converted back into the carboxylic acid. In this case, the reaction is a hydrolysis reaction, which converts the ester into the original reactants, the carboxylic acid and alcohol, and the carboxylic acid is purified from the alcohol, typically by crystallization or distillation depending on the boiling and melting points of the acid. Shown in Figure 5.15 is a schematic illustration of the experimental method used for this purification.

An alternative method of separating enantiomeric mixtures is to pass the mixture of enantiomers through a column that has optically active packing materials. With the appropriate choice of optically active column, there is an attraction between the enantiomers and the optically active packing material of the column, resulting in a diastereomeric complex. Since the properties, including polarity of these diastereomeric complexes, are different, they will move through the column at different rates and hence separated. This technique is typically done using chiral columns with a liquid passing through at a high pressure, this technique is known as high-pressure liquid chromatography (HPLC). Chiral packing materials typically include polysaccharides.

In organic synthesis, one of the major goals is the synthesis of optically pure compounds. As mentioned earlier, most reactions give a mixture of enantiomers, but if specific catalysts

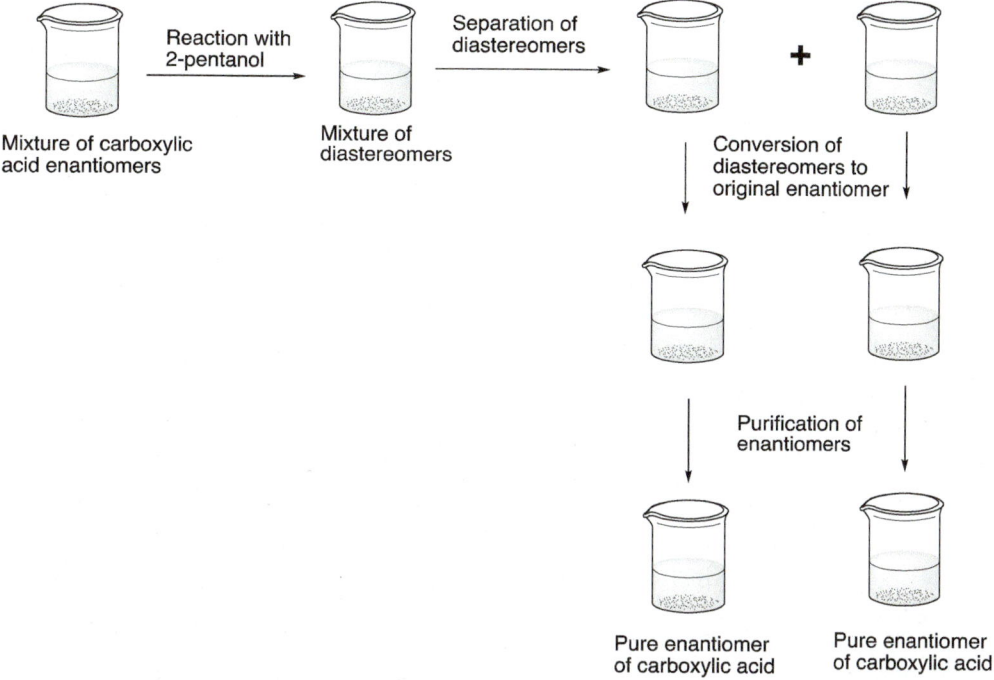

Figure 5.15 Illustration of purification of carboxylic acid by the scheme shown in Figure 5.14.

are used, a higher percentage of one enantiomer can be achieved over the other enantiomer. To determine the effectiveness of such reactions, the term percent enantiomeric (%ee) excess is used and is defined by Eq. (5-2).

$$\%ee = \text{enantiomer1} - \text{enantiomer2} / \text{enantiomer1} + \text{enantiomer2} \times 100\% \qquad (5\text{-}2)$$

For the reaction below, a special catalyst was used to achieve a percent enantiomeric excess (%ee) of 76% of one enantiomer over the other.

For some reactions, special catalysts can give up to 99%ee, which is extremely important in the pharmaceutical industry since as pointed out in the introduction section, one enantiomer can be poisonous to the body, whereas the other is active. Another example is in the use of Baker's yeast for the following conversion.

For this reaction, an 84%ee can be obtained.

End of Chapter Problems

5.9 Arrange the following groups in decreasing order of priorities (highest first).

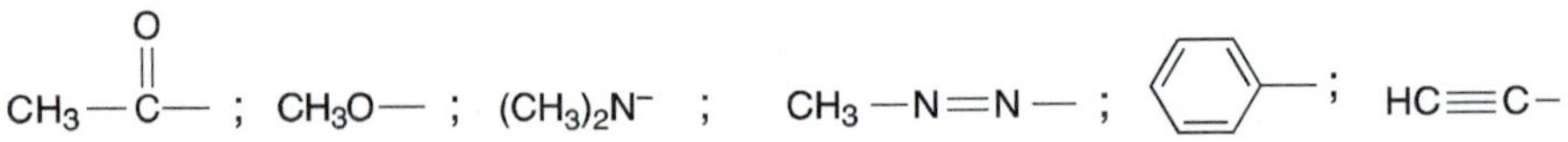

$$CH_3-\overset{O}{\underset{\|}{C}}- \quad ; \quad CH_3O- \quad ; \quad (CH_3)_2N- \quad ; \quad CH_3-N=N- \quad ; \quad C_6H_5- \quad ; \quad HC\equiv C-$$

5.10 Using (E)-(Z) designation, give the IUPAC names for the following compounds.

(i) 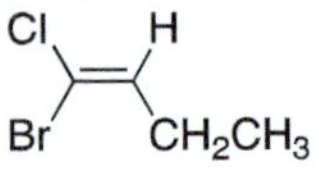 Cl, H / Br, CH₂CH₃ on C=C

(ii) 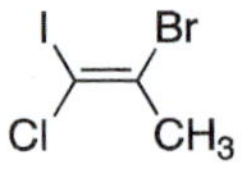 I, Br / Cl, CH₃ on C=C

(iii) 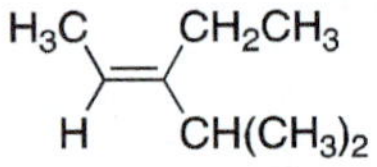 H₃C, CH₂CH₃ / H, CH(CH₃)₂ on C=C

(iv) 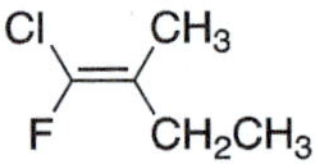 Cl, CH₃ / F, CH₂CH₃ on C=C

5.11 Give the complete name (including E, Z prefix) of the following molecule.

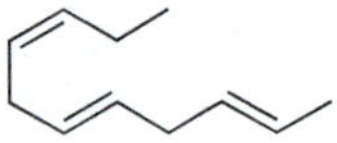

Undecane is the root name for alkane with 11 carbons

5.12 The structures of two naturally occurring amino acids, L-alanine and L-phenylalanine, are shown below. What is the absolute configuration of the stereogenic center in each?

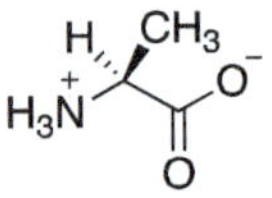

L-Alanine 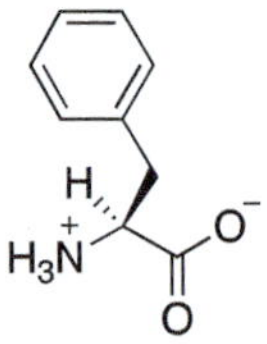   L-Phenylananine

5.13 What is the relationship between the following compounds, i.e. same, different, enantiomers, diastereomers, etc.?

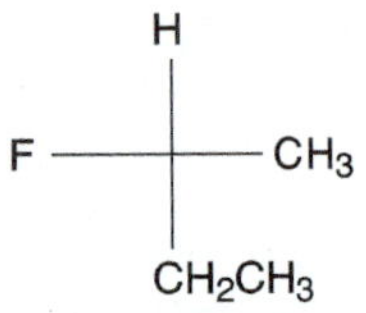

H (up), F (left), CH₃ (right), CH₂CH₃ (down)   and   F (up), H (left), CH₃ (right), CH₂CH₃ (down)

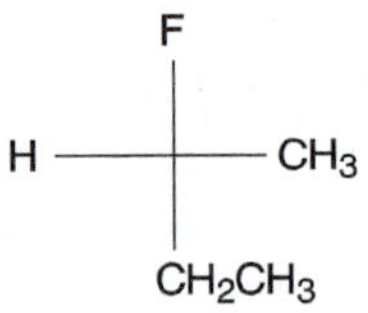

5.14 Define the following: (a) stereogenic center; (b) enantiomers; (c) racemic mixture; (d) diastereomers; (e) meso compound.

5.15 Label each of the following compounds as optical active or optical inactive; explain each answer.

a) *cis*-1,3-Dimethylcyclohexane
b) *trans*-1,3-Dimethylcyclohexane
c) *cis*-1-Chloro-3-methylcyclohexane
d) *trans*-1-Chloro-3-methylcyclohexane
e) *cis,cis*-1,3,5-Trimethylcyclohexane
f) *trans,trans*-1,3,5-Trimethylcyclohexane
g) *cis,cis*-1-Chloro-3,5-dimethylcyclohexane
h) *trans,cis*-1-Chloro-3,5-dimethylcyclohexane

5.16 Draw dashed/wedge three-dimensional structures of two different stereoisomers of each of the following compounds and determine the absolute configuration (R or S) around each chiral carbon.

a) CH3CHClCHClCH3
b) CH3CH(OH)CH2CH(OH)CH3
c) CH2ClCHFCHFCH2Cl
d) CH3CH(OH)CH2CHClCH3
e) CH3CHBrCHFCH3

5.17

i) Draw a Fischer projection for (1S,2R)-1,2-dibromo-1,2-dichloroethane. Is this compound optically active? Explain your answer.
ii) Draw a Fischer projection for (1S,2S)-1,2-dibromo-1,2-dichloroethane. Is this compound optically active or optically inactive? Explain your answer.

5.18 Draw the structure of the most stable conformer of the following molecules:

a) *trans*-1,3-Dichlorocyclohexane
b) *cis*-2-Chlorocyclohexanol (assume that —Cl is larger than —OH)

5.19

a) How many stereogenic centers are found in vitamin E structure shown below?

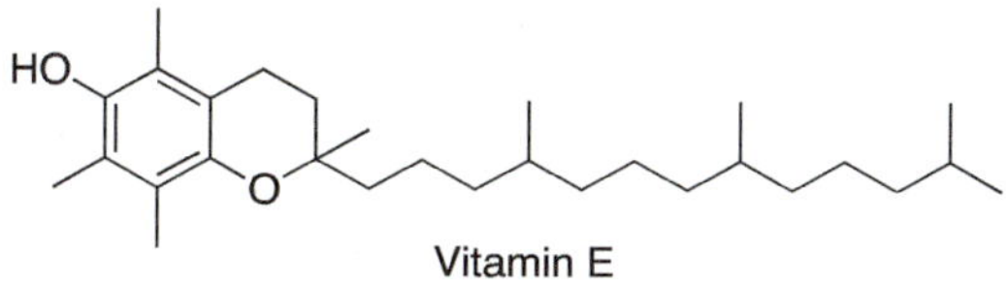

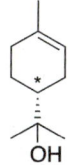

Vitamin E

b) Use dashed-wedge representation to assign each of the chiral carbons identified with the S configuration.

5.20 Shown below is the structure of α-terpineol, with its stereogenic center starred. What is the configuration, (R) or (S), at the starred C atom?

α-Terpineol

5 Stereochemistry

5.21 Consider carefully the following molecules (A, B, and C) shown below:

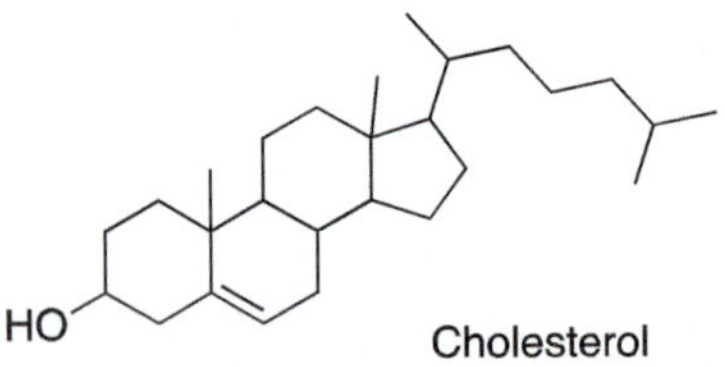

Cholesterol

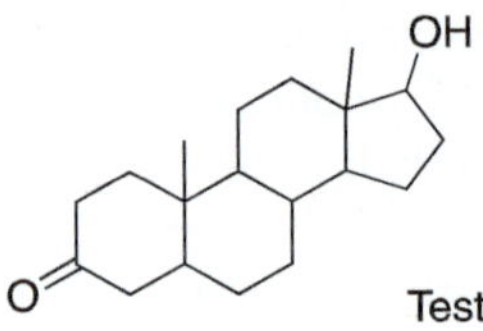

Testosterone

i) Give <u>complete</u> IUPAC names (including stereochemistry) for the molecules A, B, and C.
ii) What is the relationship between the following pairs of molecules: A and B; A and C; B and C, i.e. same, enantiomers, diastereomers, etc.?

5.22 Identify the chiral carbons in the molecules shown below.

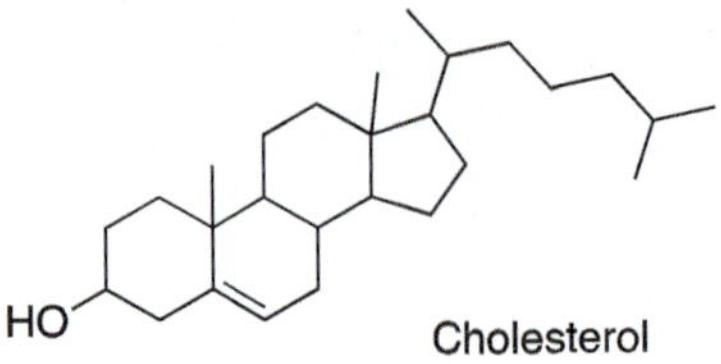

Cholesterol

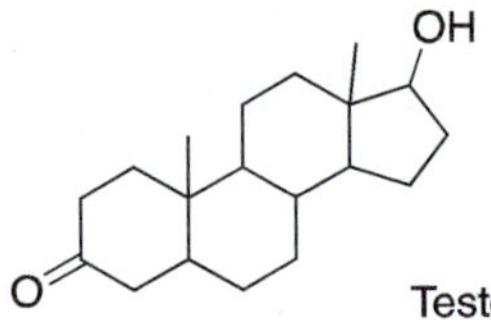

Testosterone

5.23 What is the absolute configuration, (R) or (S), at the starred C atoms for the molecules below?

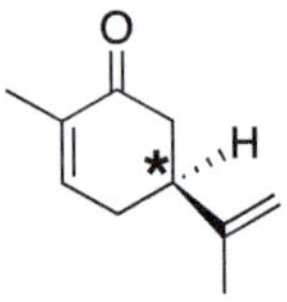

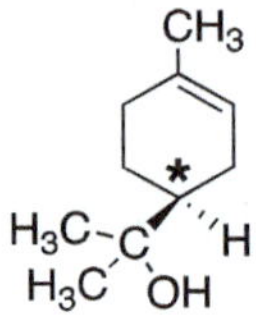

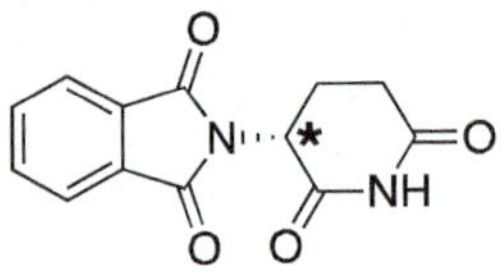

5.24 Analyze carefully the structures shown below, **I** through **VI**, and answer the following questions. Possible choices include enantiomers, structural isomers, same molecules, diastereomers, or meso compounds.

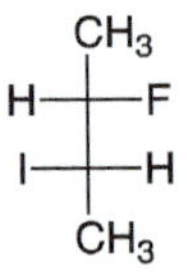

I 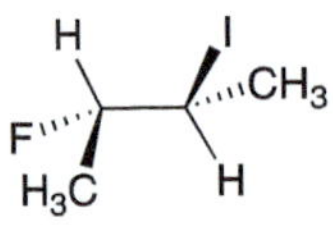   II 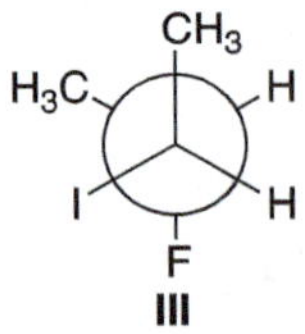   III 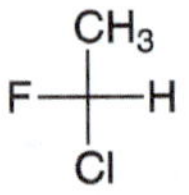   IV 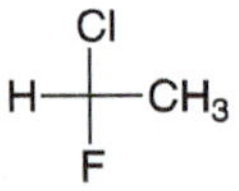   V 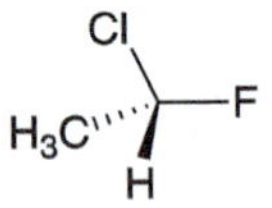   VI

a) What is the relationship between molecule (V) and molecule (VI)?
b) What is the relationship between molecule (I) and molecule (II)?
c) What is the relationship between molecule (II) and molecule (III)?
d) What is the relationship between molecule (IV) and molecule (VI)?
e) What is the relationship between molecule (IV) and molecule (V)?
f) What is the relationship between molecule (I) and molecule (III)?

5.25 Shown below is the structure of ethambutol, which is used to treat tuberculosis. There are two chiral centers in this molecule. The S,S isomer is the active form, whereas the R,R causes optical neuritis, and eventually results in blindness.

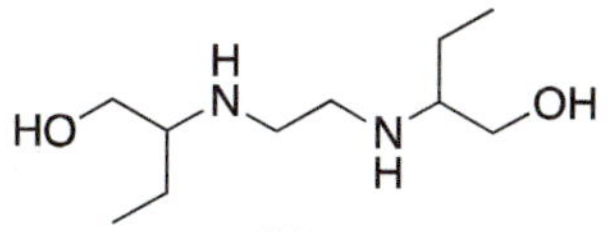

Ethambutol

a) For the molecule above, clearly circle the two chiral (stereogenic) centers.
b) Draw the **S,S** enantiomer of ethambutol and clearly show the stereochemistry around each chiral carbon by using dashed-wedge representation.

5.26 Shown below is the structure of taxol, a drug that is used to cure cervical cancer. It has many stereogenic centers; identify the stereogenic centers in this molecule.

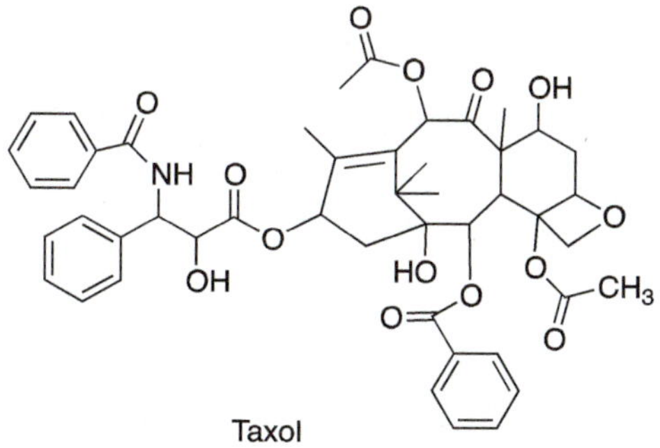

Taxol

6

An Overview of the Reactions of Organic Chemistry

6.1 Introduction

Throughout our course in organic chemistry, some of the reactions studied in your general chemistry course, along with some new types of reactions, will be covered. There are numerous reactions that organic molecules undergo, but these reactions can be divided into a few general reaction categories: (i) acid–base, (ii) addition, (iii) reduction, (iv) oxidation, (v) elimination, (vi) substitution (vii) pericyclic, and (viii) catalytic coupling reactions. If the fundamental concepts of these reaction types are understood, it will be easier to recognize these types of reactions in later chapters and students will be able to predict the outcomes of a wide cross section of organic reactions. With this approach of fully understanding concepts and their applications, students are better able to solve problems instead of having to rely on memorization to solve the various problems that will be encountered throughout the course. In this chapter, a brief examination of the underlying concepts of these reaction types will be carried out so that when more complex reactions are encountered, students will be better able to appreciate the versatility of organic reactions and to use the knowledge gained to predict expected and unexpected outcomes of reactions. The different categories of reactions that will be covered in this course will involve the functional groups covered in Chapters 3 and 4. In this chapter, a brief overview of basic concepts of these reaction types and their applications will be covered. In addition, specific features of these reaction types will be easily recognized, and students will be able to apply basic concepts to predict possible products and reactants that these reactions undergo. In the remaining chapters of the book, more specific details and applications of various reactions will be studied.

6.2 Acid–Base Reactions

Acids and bases are encountered in just about all aspects of our everyday life; the different salad dressings that are used regularly to make salads taste tangy contain a small amount of an acid, acetic acid. The tart taste of lemons, grapefruits, and other fruits is due to the presence of small amounts of different acids, of which the most commonly known is citric acid. Oxalic acid is used in the bleaching of wood pulp. Benzoic acid is commonly used in cosmetics and dyes, and the salts of these acids are most commonly included in food as a preservative. Some of the everyday kitchen cleaners contain different acids or bases. The liquid that is used to unclog most drains contains a very strong base. Examples of common acids are shown in Figure 6.1.

Organic Chemistry: Concepts and Applications, First Edition. Allan D. Headley.
© 2020 John Wiley & Sons, Inc. Published 2020 by John Wiley & Sons, Inc.
Companion website: www.wiley.com/go/Headley_OrganicChemistry

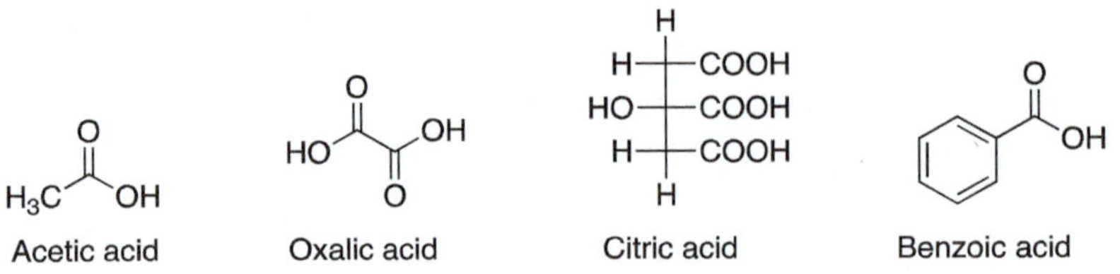

Figure 6.1 Examples of common organic acids.

Acids react with bases to produce salts and water, and the reaction of benzoic acid and sodium hydroxide is shown in Reaction (6-1).

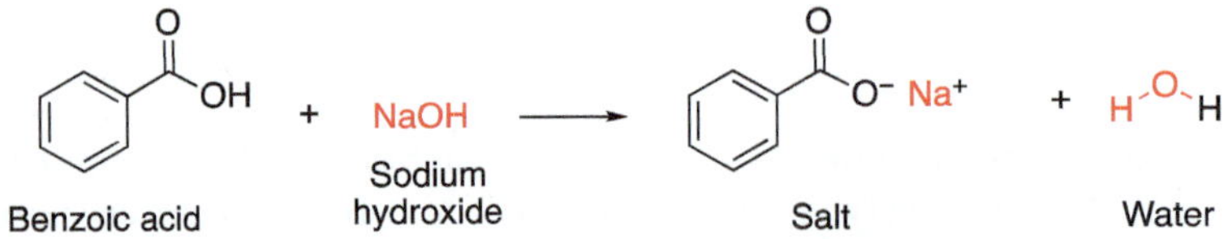

$$\text{Benzoic acid} + \text{NaOH (Sodium hydroxide)} \longrightarrow \text{Salt} + \text{Water} \quad (6\text{-}1)$$

Acid–base reactions that will be encountered in organic chemistry include bases that are much stronger that those encountered in general chemistry. Bases that were encountered in the general chemistry lab included sodium hydroxide and potassium hydroxide, but in organic chemistry, much stronger bases such as sodium amide and potassium methoxide will be encountered. Similarly, some of the acids in organic chemistry are much stronger than those encountered in general chemistry, but the reaction principles remain the same. In the next sections, a review of the definition and concepts of organic acids and bases is carried out.

6.2.1 Acids

One of the earliest definitions of an acid was coined by Arrhenius, which states that an acid is a substance that when added to water increases the concentration of protons (H^+ ions). On the other hand, a base is any substance that when added to water increases the hydroxide ions (OH^-) concentration. Over the years, as chemists gained a better understanding of the role of acids and bases in nature and chemical reactions, other definitions have developed. In order to understand the importance of acids and bases and how they work, other definitions of acids and bases have been more useful. Another important chemical definition of acids and bases that is used frequently in organic chemistry is that of **Brønsted–Lowry**, named after a Danish chemist, Johannes Brønsted and an English chemist, Thomas Lowry, who coined another definition of acids and bases. A Brønsted–Lowry acid is any substance that will donate a proton (H^+). A proton is a hydrogen atom without its electron. Recall from Chapter 1 that a hydrogen atom has only one electron and if this electron were removed, a proton is the result. Hydrochloric acid (HCl) is an acid because it can undergo a heterolytic cleavage of the polar covalent bond to give a proton (H^+) and the chloride anion (Cl^-). If HCl is placed in water, it will donate its proton to the water to form the protonated water and a Cl^- anion, also known as the conjugate base of the acid, HCl. The reaction is shown in Reaction (6-2).

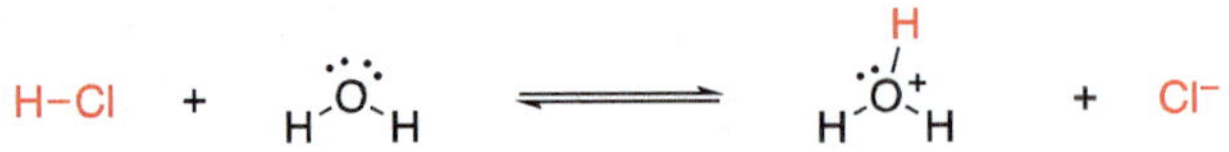

$$H\text{-}Cl + H_2O \rightleftharpoons H_3O^+ + Cl^- \quad (6\text{-}2)$$

Thus, HCl is a proton donor and it can donate its proton to water. In the above example, a H_2O molecule accepted the proton from the HCl.

Problem 6.1

i) Which of the following are Brønsted–Lowry Acids?
   HI, H_2O, NaCN, $AlCl_3$, and KCl.
ii) Write a reaction similar to the one given in Reaction (6-2) for each of the acids identified above with water.

Based on the Brønsted–Lowry definition of an acid, any molecule that has hydrogen atoms is a potential acid. Since organic compounds have mostly carbons and hydrogens, most are potential acids, but to varying degrees. Some organic molecules are fairly acidic, whereas most are extremely weak acids and will require extremely strong bases to abstract a proton.

6.2.2 Bases

A **Brønsted–Lowry base** is a proton acceptor (H^+ acceptor). Thus, water in the presence of HCl will accept the proton from HCl to become H_3O^+; hence, water in this case is a base and that reaction is shown in Reaction (6-2). Sodium hydroxide is a base since it will accept a proton from an acid such as HCl, as shown in Reaction (6-3).

$$\text{H–Cl} + \text{NaOH} \rightleftharpoons \text{H–Ö–H} + \text{Na}^+\text{Cl}^- \qquad (6\text{-}3)$$
Acid   Base   Conjugate acid   Conjugate base

In this case, water is the conjugate acid of sodium hydroxide. With this knowledge of acids and bases, it becomes clear that for the reaction given in Reaction 6.1, which involves benzoic acid and sodium hydroxide, benzoic acid is the proton donor and sodium hydroxide is the proton acceptor, hence the base. Another application of this concept is shown in Reaction (6-4).

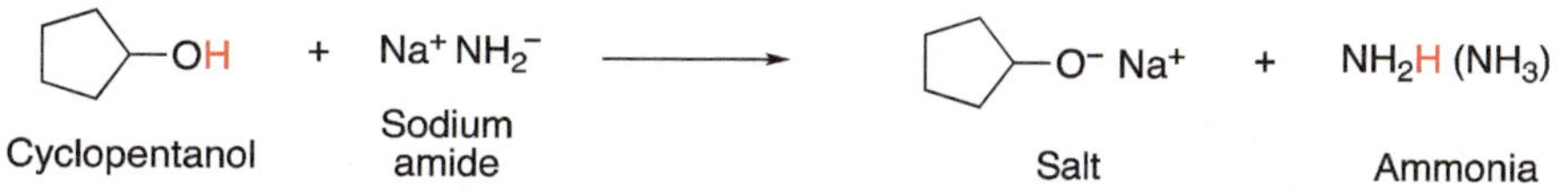

$$\text{Cyclopentanol–OH} + \text{Na}^+\text{NH}_2^- \longrightarrow \text{Cyclopentyl–O}^-\text{Na}^+ + \text{NH}_2\text{H (NH}_3\text{)} \qquad (6\text{-}4)$$
Cyclopentanol   Sodium amide   Salt   Ammonia

For the above reaction shown in Reaction (6-4), cyclopentanol is the acid (the acidic hydrogen is shown in red), in which the proton goes to the sodium amide, which is the base. Note in this case, the product is a salt and ammonia. Thus, for acid–base reactions that will be encountered in organic chemistry, each reaction has to be carefully analyzed by applying the concepts and definitions of acids and bases. Students should also be able to readily identify an acid and a base and, of course, be able to predict the products of an acid–base reaction.

Problem 6.2

i) Identify the acid and base in the reactants and the conjugate acid and conjugate base in the products for each of the following reactions.

(a) 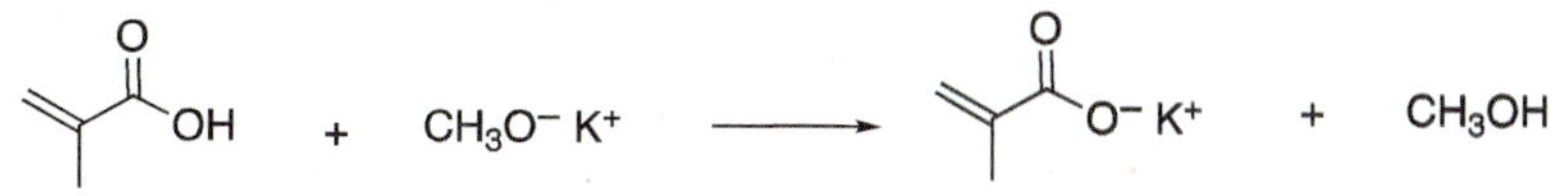 $CH_3C(=O)OH + CH_3O^- K^+ \longrightarrow CH_3C(=O)O^- K^+ + CH_3OH$

(b)  $C_6H_5\text{–OH} + CH_3CH_2CH_2CH_2^- \, Li^+ \longrightarrow CH_3CH_2CH_2CH_3 + C_6H_5\text{–O}^- Li^+$

ii) Predict the products of the reactions of the acid–base pair shown below.

(a) PhC(=O)OH + NaNH$_2$ ⟶

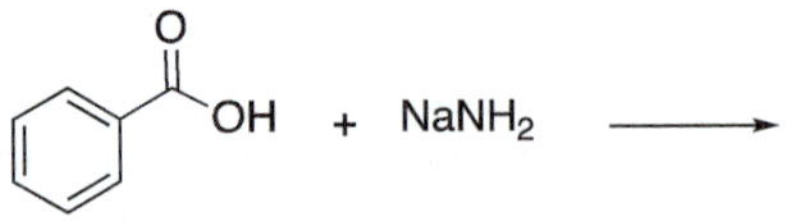

(b) CH$_3$C(=O)OH + CH$_3$O$^-$K$^+$ ⟶

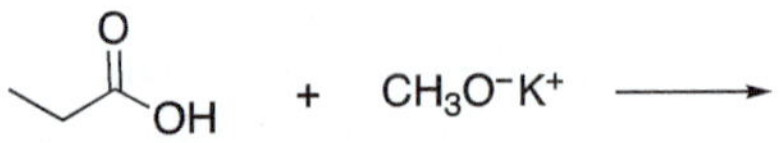

There is yet another definition of acids and bases and it is that of G. N. Lewis of the University of California, Berkeley. In 1923, he defined a **Lewis acid** as an electron pair acceptor. In order for a molecule to be able to accept a pair of electrons, there must be an available empty orbital in which these electrons must be placed. The proton (H$^+$) is the simplest, which is a hydrogen atom without its electron, which means that it has an empty orbital. On the other hand, a **Lewis base** is defined as an electron pair donor. Thus, a Lewis base must have at least one unshared pair of electrons. Lewis bases include molecules, which have functional groups such as amines (RNH$_2$) and alcohols (ROH) since they have at least one unshared pair of electrons. Bases that we have seen before, such as NaOH and CH$_3$O$^-$ K$^+$, are considered bases under the Lewis definition since they have at least one unshared pair of electrons on the oxygen, but this definition now includes a larger number of molecules; the same is true for acids. The reactions of a Lewis base (ammonia) with a proton and the reaction of ammonia with BF$_3$, an electron pair acceptor, are shown in Reactions (6-5) and (6-6).

$$H^+ + :NH_3 \rightleftharpoons H-\overset{+}{N}H_3 \qquad (6\text{-}5)$$

Proton   Ammonia   Ammonium
(Lewis acid) (Lewis base)   cation

$$BF_3 + :NH_3 \rightleftharpoons F_3\overset{-}{B}-\overset{+}{N}H_3 \qquad (6\text{-}6)$$

Boron trifluoride   Ammonia
(Lewis acid)   (Lewis base)

Note that in these reactions, the lone pair of electrons from the ammonia (shown in red) forms the single bond in the product to the proton, which has an empty orbital. Also, note that since nitrogen in the product has four bonds, it acquires a formal charge of positive one (+1). The same is true for the reaction of ammonia with boron trifluoride. The lone pair of electrons of ammonia bonds to the empty p orbital of boron trifluoride to form a new ionic complex shown in Reaction (6-6); boron of boron trifluoride is sp^2 hybridized, with an empty p orbital.

Problem 6.3

Identify the Lewis acid and Lewis base in each of the following reactions.

(a) AlCl$_3$ + :NH$_3$ ⟶ Cl$_3$$\overset{-}{Al}$—$\overset{+}{N}H_3$

(b) AlCl$_3$ + (CH$_3$)$_2$C=$\overset{..}{\overset{..}{O}}$: ⟶ (CH$_3$)$_2$C=$\overset{+}{\underset{..}{O}}$—$\overset{\ominus}{A}lCl_3$

(c) H$^+$ + (CH$_3$)$_2$C=$\overset{..}{\overset{..}{O}}$: ⟶ (CH$_3$)$_2$C=$\overset{+}{\underset{..}{O}}$—H

6.3 Addition Reactions

As the name suggests, these reactions involve the addition of a reactant to another reactant to produce the product. The reaction shown in Reaction (6-7) involves the addition of bromine molecule across a carbon–carbon double bond, an alkene functional group, to form an addition product. Hence, this specific type of addition reaction is called **brominaton** reaction of an alkene.

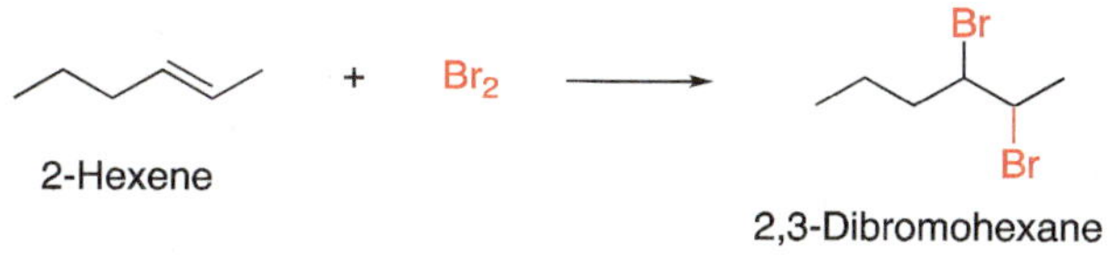

$$\text{2-Hexene} + \text{Br}_2 \longrightarrow \text{2,3-Dibromohexane} \tag{6-7}$$

Another example of an addition reaction is given in Reaction (6-8) in which chlorine is added across a carbon–carbon double bond of cyclohexene.

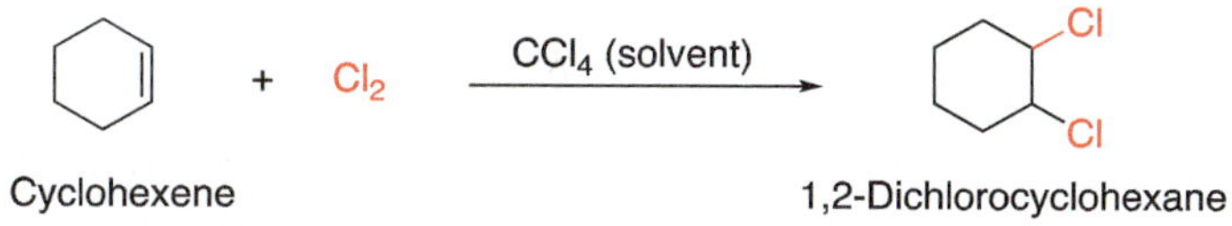

$$\text{Cyclohexene} + \text{Cl}_2 \xrightarrow{\text{CCl}_4 \text{ (solvent)}} \text{1,2-Dichlorocyclohexane} \tag{6-8}$$

The addition of a molecule such as HBr can also take place to an alkene, as shown in the reaction in Reaction (6-9).

$$\text{cyclohexene} + \text{HBr} \longrightarrow \text{bromocyclohexane} \tag{6-9}$$

The reaction in Reaction (6-10) involves the addition of HCl to a nonsymmetrical alkene to form two possible products.

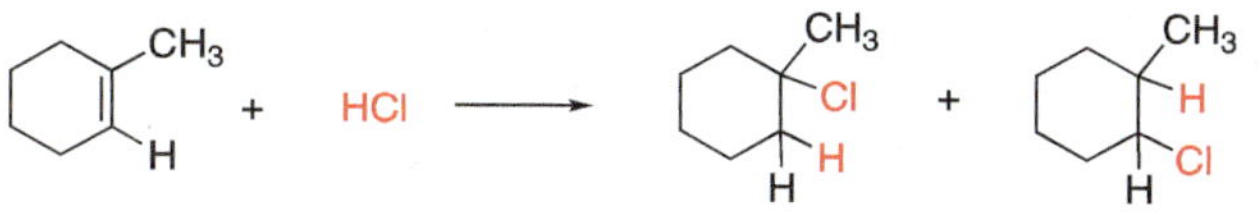

$$\tag{6-10}$$

Since the addition can take place in two ways, there are two different products. Since this reaction involves the addition of hydrochloric acid (HCl) across the double bond, this reaction is called **hydrochlorination**. Reaction (6-11) involves another example of the addition of HCl across the carbon–carbon double bond.

$$\tag{6-11}$$

Note that for the reaction shown in Reaction (6-11), there are two possible products since there are two possible addition modes, similar to the addition reaction of HCl to an alkene shown in Reaction (6-10). For addition reactions, there are two important observations: (i) an alkene functionality is transformed into single bonds and (ii) one reactant molecule adds to another molecule at the site of the functional group to form the addition product(s). More details regarding these and other similar addition reaction will be discussed in Chapter 8.

Problem 6.4

Predict all possible products for the following addition reactions.

(a) 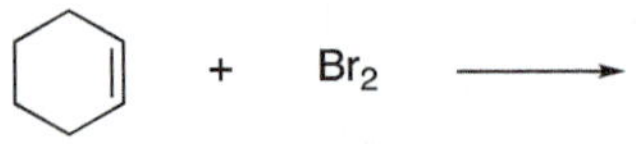 cyclohexene + Br$_2$ ⟶

(b) 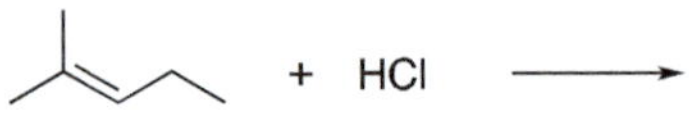 2-methyl-2-butene + HCl ⟶

(c)  methylenecyclopentane + HBr ⟶

6.4 Reduction Reactions

Reduction is typically defined as the supply of electrons to a molecule and also the loss of oxygen from a molecule. The former definition is most often seen in organic chemistry. Shown below in Reaction (6-12) is a typical reduction reaction from general chemistry in which Zn metal is the reducing agent and supplies two electrons to the Cu^{2+} ions, the species that is being reduced.

$$Zn(s) + Cu^{2+}(aq) \longrightarrow Zn^{2+}(aq) + Cu(s) \quad (6\text{-}12)$$

Reducing agent: Zn(s); Species being reduced: Cu^{2+}(aq)

Most of the reducing agents that we will encounter in organic chemistry will not be Zn, but compounds that contain hydrogen or even carbon. As you can imagine, for hydrogen to be a reducing agent, it must be able to supply electrons to the species to be reduced. Organic reducing reagents that have at least one electron pair to donate to the compound being reduced typically include LiAlH$_4$ and NaBH$_4$, these reagents will supply the hydride anion (H:$^-$) to the species to be reduced. The structures of these reducing agents are shown below.

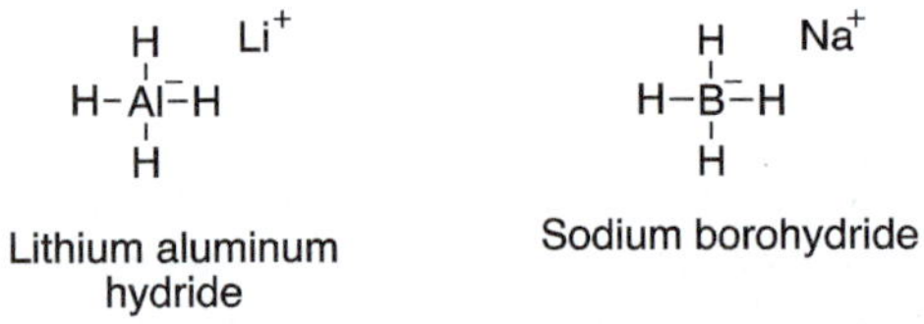

Lithium aluminum hydride (H–Al(H)(H)–H with Li$^+$)

Sodium borohydride (H–B(H)(H)–H with Na$^+$)

Note that the species being reduced has to be able to accept the electrons. In the case of Reaction (6-12), the copper ion has a positive charge, which means that it can readily accept electrons from the reducing agent. For organic reactions, the species that will be reduced are typically not copper or other metal ions, but organic compounds that typically have polar functional groups, such as carbonyls and nitriles. Such groups are readily reduced since the reducing agent can supply the electrons to the partially positive region of the polar covalent bond; an example of a reduction reaction, in which LiAlH$_4$ is used as the reducing reagent for the reduction of a ketone is shown in Reaction (6-13).

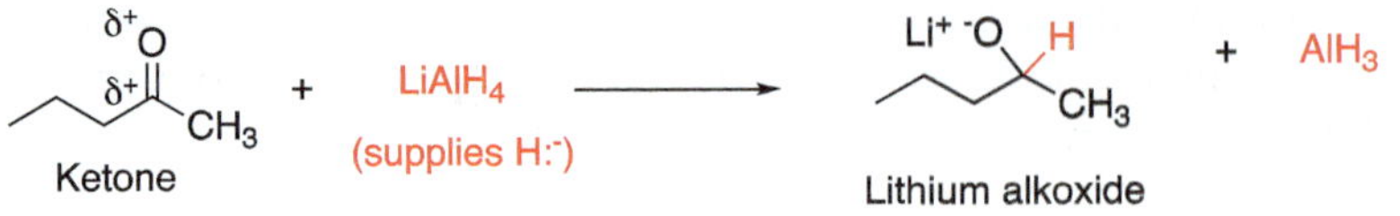

Ketone + LiAlH$_4$ (supplies H:$^-$) ⟶ Lithium alkoxide + AlH$_3$ $\quad$ (6-13)

6.4 Reduction Reactions

For this reaction, the reducing agent introduces a hydride ion (H:⁻) to the electropositive carbon of the carbon–oxygen polar double bond to produce the organic salt (lithium alkoxide). Since organic salts, such as the product of the reaction shown above, are rarely used in organic chemistry, such salts are neutralized in an acidic solution to form another organic chemistry functional group, in this case, an alcohol. The neutralization, which is essentially an acid–base reaction, is shown in Reaction (6-14). This type of neutralization reaction is sometimes referred to as acidic workup.

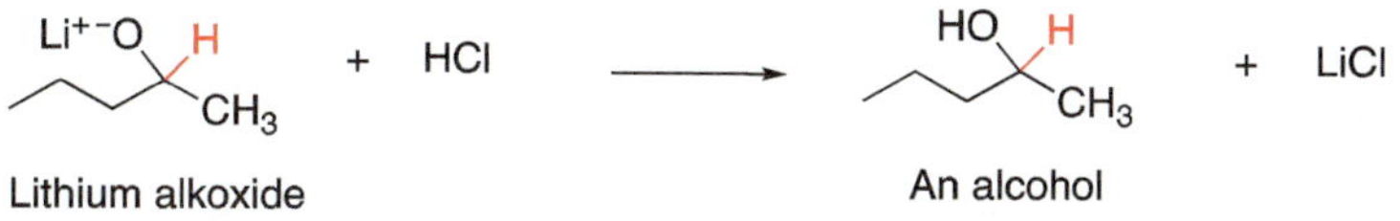

$$\text{Lithium alkoxide} + \text{HCl} \longrightarrow \text{An alcohol} + \text{LiCl} \tag{6-14}$$

For most organic reactions where there is a sequence of reactions to produce a specific organic product, such as the ones in Reactions (6-13) and (6-14) to produce an alcohol from a ketone using a reducing agent, the reactions are combined in a specific manner. The main organic reactant is written on the left of the reaction, and other reagents, catalysts, solvents, and reaction conditions, such as temperature, are written on the arrow. Also, different steps of the reaction sequence are shown using numbers on the arrow. Thus, an alternate way of writing Reactions (6-13) and (6-14) is to use the format shown in Reaction (6-15).

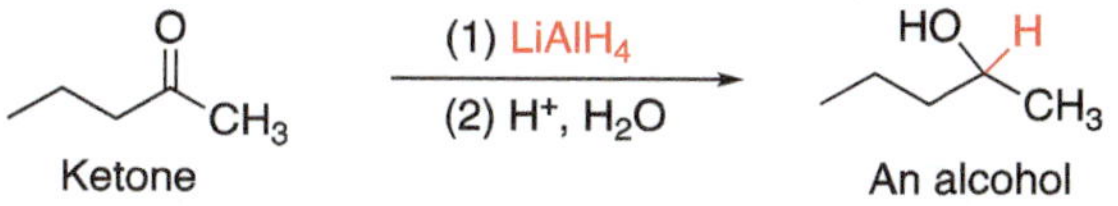

$$\text{Ketone} \xrightarrow[\text{(2) H}^+\text{, H}_2\text{O}]{\text{(1) LiAlH}_4} \text{An alcohol} \tag{6-15}$$

Note that by using this format, the reaction is not balanced as is typically done in general chemistry and that only the main organic product is typically shown and other reagents, such as the inorganic lithium chloride salt, are not included. Another method that is often used to show the different reactions in a sequence of reactions is shown below for the reactions of Reactions (6-13) and (6-14), in which the two steps are shown adjacent to each other as shown in Reaction (6-16).

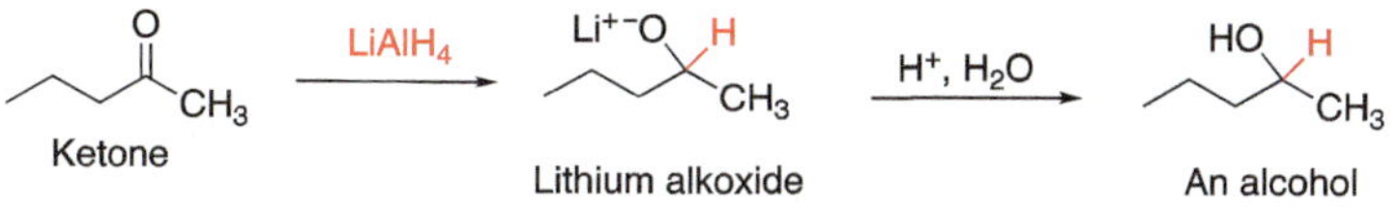

$$\text{Ketone} \xrightarrow{\text{LiAlH}_4} \text{Lithium alkoxide} \xrightarrow{\text{H}^+\text{, H}_2\text{O}} \text{An alcohol} \tag{6-16}$$

Based on the concepts discussed above, it is possible to predict the reduction products if given the reactant and reducing agent, as given in Reaction (6-17).

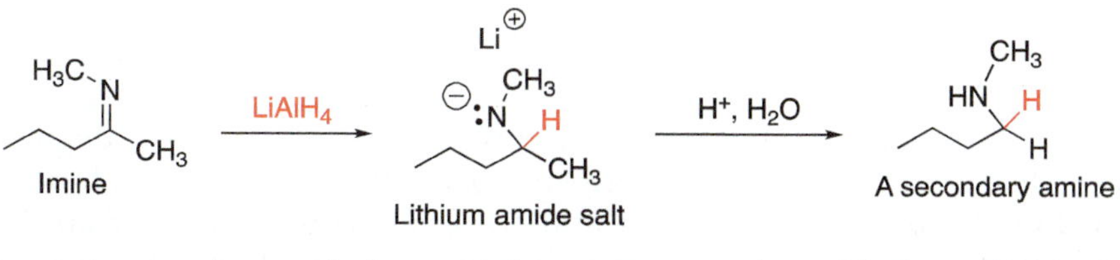

$$\text{Imine} \xrightarrow{\text{LiAlH}_4} \text{Lithium amide salt} \xrightarrow{\text{H}^+\text{, H}_2\text{O}} \text{A secondary amine} \tag{6-17}$$

This reaction involves an imine organic reactant reacting with the reducing agent LiAlH$_4$, which delivers a hydride ion (H:⁻) to the electropositive carbon of the polar carbon–nitrogen double bond to form a salt (lithium amide). The second reaction is an acidic neutralization reaction of the salt to form, in this case, an amine.

At this point, just about any reduction reaction involving LiAlH$_4$ or NaBH$_4$ can be examined and a prediction made about possible organic products. Problem 6.5 is designed to apply this concept.

Problem 6.5

Give the reduction organic products for the following reactions.

(a) 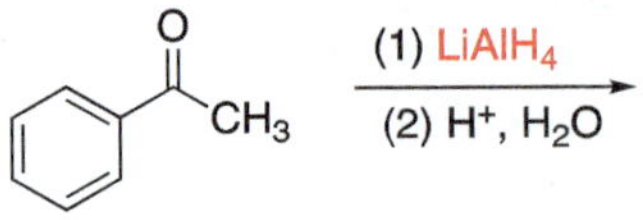 PhC(O)CH₃ $\xrightarrow{\text{(1) LiAlH}_4}{\text{(2) H}^+, \text{H}_2\text{O}}$

(b) 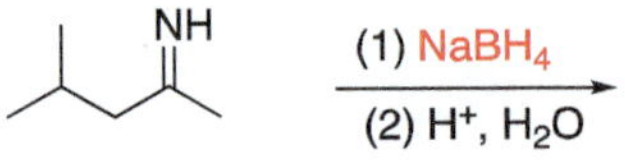 (CH₃)₂CHC(=NH)CH₃ $\xrightarrow{\text{(1) NaBH}_4}{\text{(2) H}^+, \text{H}_2\text{O}}$

(c) 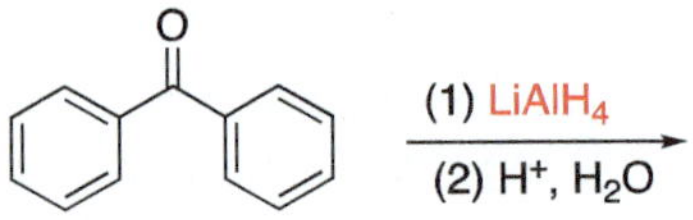 PhC(O)Ph $\xrightarrow{\text{(1) LiAlH}_4}{\text{(2) H}^+, \text{H}_2\text{O}}$

For functional groups that do not have polar covalent bonds as discussed above, such as the alkene and alkyne, reduction is much more difficult, and as a result, a catalyst is typically required. For the reduction (or addition of hydrogen molecule) across a carbon–carbon double bond that is shown in Reaction (6-18), a catalyst is required.

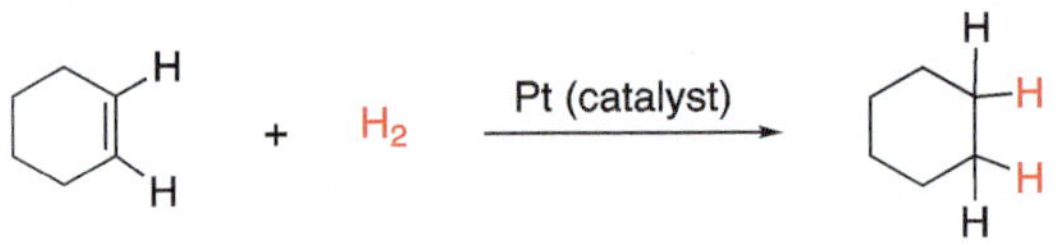

cyclohexene + H₂ $\xrightarrow{\text{Pt (catalyst)}}$ cyclohexane (6-18)

Reduction of alkenes using hydrogen and a catalyst is very important in organic chemistry since the products of these reactions are saturated compounds. Another term that is often used for this specific type of reduction reaction is hydrogenation since a hydrogen molecule is added across a carbon–carbon double bond. Based on the outcome of the reduction of alkenes with hydrogen in the presence of a catalyst, it should not be difficult to predict the outcome of the reduction of ketones under the same reaction conditions, as shown in Reaction (6-19).

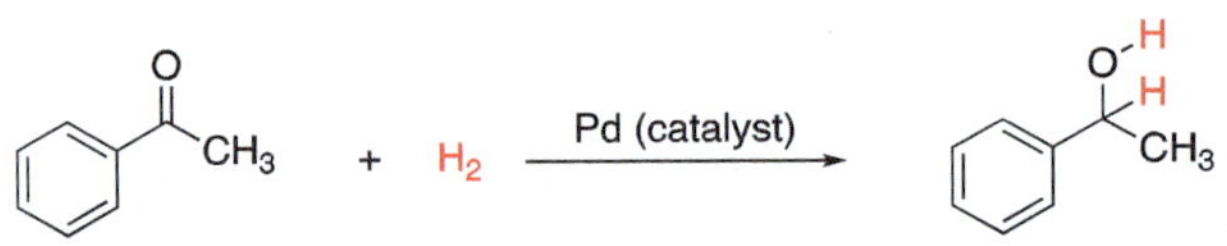

PhC(O)CH₃ + H₂ $\xrightarrow{\text{Pd (catalyst)}}$ PhCH(OH)CH₃ (6-19)

Note that even though the reactant has a polar covalent carbon–carbon double bond, reduction is possible using hydrogen in the presence of a catalyst. In this case, however, the product is an alcohol. Note that the product of this reaction is the same as if LiAlH₄ were used as the reducing agent, followed by a neutralization reaction. It is not surprising that the same organic product should result since for both types of reactions, H–H is added across the double bond. These examples give an idea of the versatility of the reactions of organic chemistry and especially the choice of different reagents available to accomplish a particular transformation.

Problem 6.6

Give the reduction products for the following reactions (note the format of writing the second reaction!).

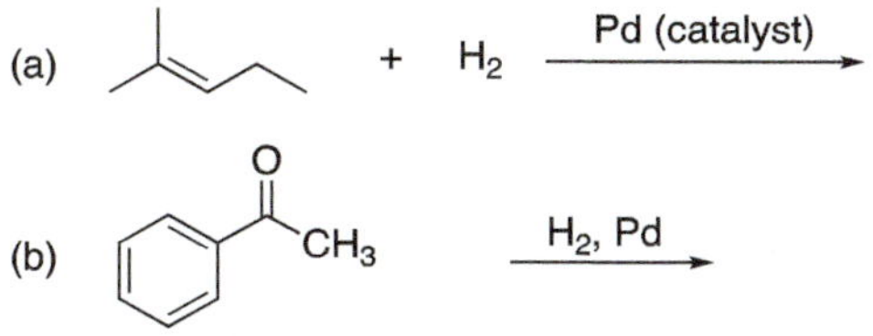

(a) CH₃-CH=CH-CH₃ (2-butene structure) + H₂ →[Pd (catalyst)]

(b) Phenyl-C(=O)-CH₃ →[H₂, Pd]

6.5 Oxidation Reactions

Based on one of the definitions that was learned in general chemistry, oxidation is the supply of oxygen and/or the removal of hydrogen and electrons from molecules. Oxidation reactions are the opposite of reduction reactions, which are discussed in Section 6.4. Oxidizing agents typically are reagents that have oxygen atoms. Thus, the following molecules are good oxidizing agents: $Na_2Cr_2O_7$, $KMnO_4$, OsO_4, and O_3 (ozone) since they all contain oxygen. In organic chemistry, we will frequently encounter oxidation reactions in which there is also a removal of a hydrogen atom from molecules. Shown below are examples of oxidation reactions, in which oxygen is added (or a change in oxidation state of oxygen is achieved) in going from the reactant to the organic product, and at the same time, a hydrogen atom is removed as shown in the reactions in Reactions (6-20)–(6-22).

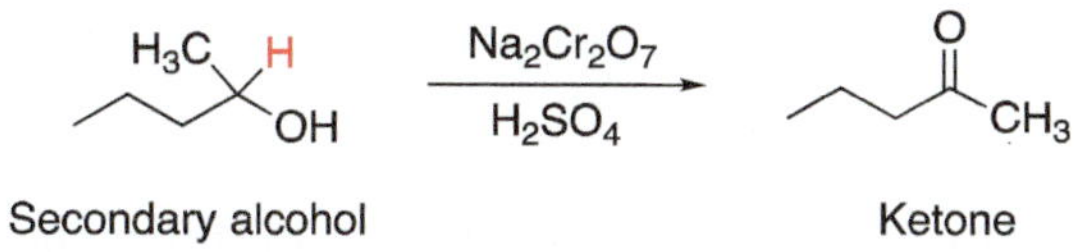

Secondary alcohol →[$Na_2Cr_2O_7$ / H_2SO_4] Ketone (6-20)

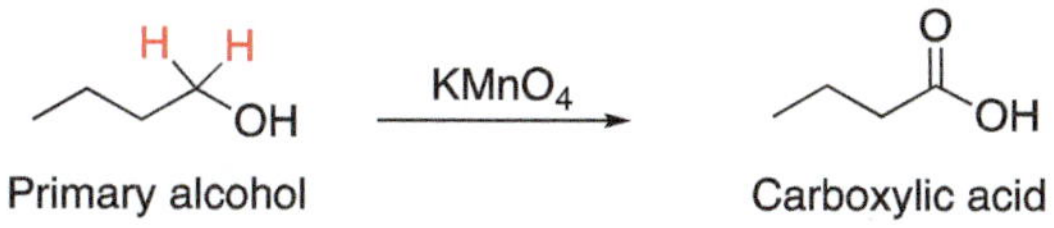

Primary alcohol →[$KMnO_4$] Carboxylic acid (6-21)

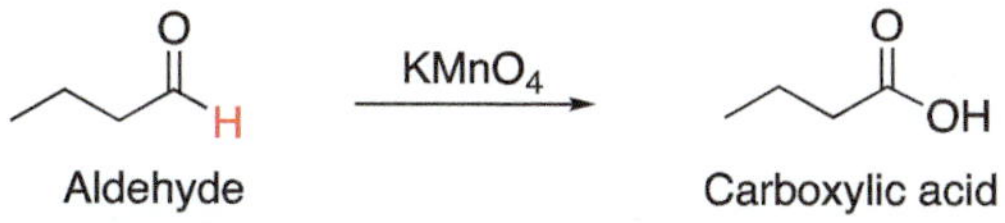

Aldehyde →[$KMnO_4$] Carboxylic acid (6-22)

Note that in the oxidation of alcohols, the hydrogen that is bonded directly to the carbon that has the –OH group (indicated in red) is lost. Thus, based on this observation, a tertiary alcohol is not expected to undergo oxidation.

Tertiary alcohol →[$Na_2Cr_2O_7$ / H_2SO_4] **No reaction** since there is not a hydrogen atom bonded to the carbon that has the alcohol functional group as shown in Reactions (6-20) and (6-21).

Also notice that the hydrogen that is removed during the oxidation of an aldehyde is the hydrogen that is bonded to the carbonyl carbon. Thus, a ketone is not expected to be oxidized since it does not have a hydrogen atom bonded to the carbonyl carbon, as shown below.

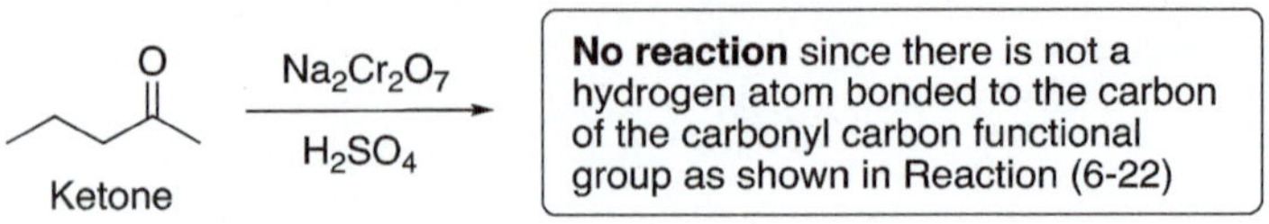

Ketone $\xrightarrow[H_2SO_4]{Na_2Cr_2O_7}$ **No reaction** since there is not a hydrogen atom bonded to the carbon of the carbonyl carbon functional group as shown in Reaction (6-22)

A very practical application of oxidation reactions is to detect the presence of alcohol for drivers suspected to be intoxicated, also known as driving while intoxicated of alcohol (DWI). In this test, $Na_2Cr_2O_7$ is used as an oxidizing agent, which has an orange color, and after it oxidizes ethanol to acetic acid, it changes to green, the reaction is given in Reaction (6-23).

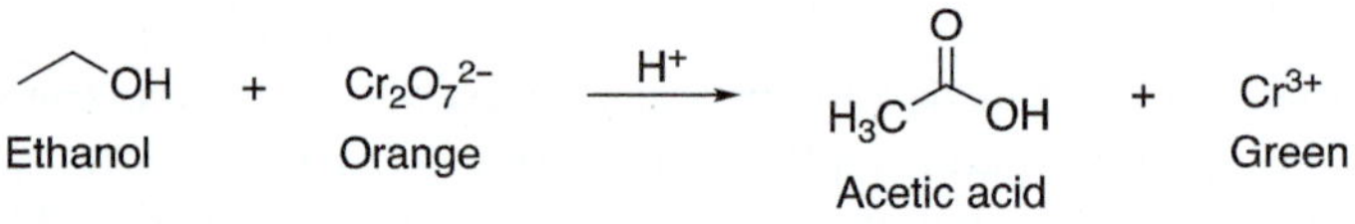

Ethanol (OH) + $Cr_2O_7^{2-}$ (Orange) $\xrightarrow{H^+}$ Acetic acid ($H_3C-COOH$) + Cr^{3+} (Green) (6-23)

Problem 6.7

i) Give the oxidation products for the following reactions.

(a) 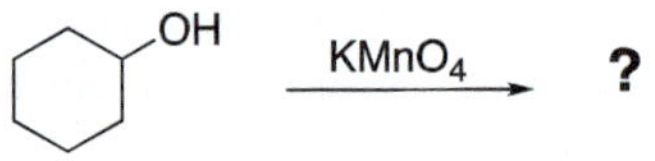 cyclohexanol $\xrightarrow{KMnO_4}$?

(b)  2-methylpropan-1-ol (or similar secondary alcohol with OH) $\xrightarrow{KMnO_4}$?

ii) Give the organic reactant that is needed to give the following oxidation products shown for the reactions below.

(a) 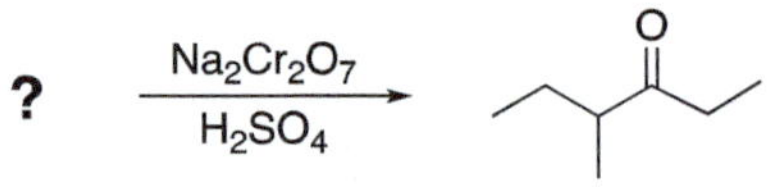 ? $\xrightarrow[H_2SO_4]{Na_2Cr_2O_7}$ (methyl ethyl ketone with methyl branch)

(b) 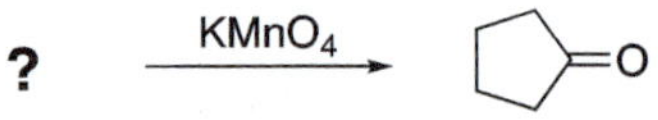 ? $\xrightarrow{KMnO_4}$ cyclopentanone

Note that the last question of Problem 6.7 is intended to get students to think critically about these reactions by reversing the thinking process and coming up with possibilities for reactants to give a desired organic product.

6.6 Elimination Reactions

Elimination reactions are the opposite of the addition reaction discussed in Section 6.3. Most of these reactions typically involve the removal of atoms or groups of atoms from a carbon–carbon single bond to form a double bond at the site of the removal of the groups. An example of an elimination reaction is shown in Reaction (6-24).

6.6 Elimination Reactions

$$\text{3-Pentanol} \xrightarrow[\text{Heat}]{\text{Acid catalyst}} \text{2-Pentene} + H_2O \qquad (6\text{-}24)$$

Note that for this reaction, water is the molecule that is eliminated to give only one organic product, which is an alkene. If on the other hand, the reactant was not a symmetrical alcohol, then there would have been two possible alkenes as shown in Reaction (6-25).

$$\text{3-Hexanol} \xrightarrow[\text{Heat}]{\text{Acid catalyst}} \text{3-Hexene} + \text{2-Hexene} + H_2O \qquad (6\text{-}25)$$

Most elimination reactions typically require more severe reaction conditions, compared to addition reactions. Addition reactions typically proceed in the presence of a small amount of a catalyst, such as dilute sulfuric acid. Hence, the reaction of the type shown in Reactions (6-24) and (6-25) is described as acid-catalyzed *dehydration* since it involves the removal of water.

An elimination reaction can also occur in which the molecule, HCl, is eliminated from a reactant in the presence of heat and base, such as KOH, to facilitate the removal; an example is shown in Reaction (6-26).

$$\text{3-Chloropentane} \xrightarrow{\text{KOH, heat}} \text{2-Pentene} + \text{HOH} + \text{KCl} \qquad (6\text{-}26)$$

Note that for the reaction given in Reaction (6-26), an alkene is also formed as the product. Also, if the reactant is a symmetrical chloroalkane, there is only one alkene as product, but if the reactant chloroalkane is unsymmetrical, there will be two possible organic products. Elimination reactions of this type in which there is a removal of hydrogen and a halogen, such as the chloride from reactants, are known as *dehydrohalogenation* reactions. Also, note that if one of the adjacent carbons to the carbon that has the chlorine atom does not have a hydrogen, elimination will not occur at that site, as shown in Reaction (6-27) involving 3,3-dimethyl-4-chlorohexane.

$$\text{3,3-Dimethyl-4-chlorohexane} \xrightarrow{\text{KOH, heat}} \text{4,4-Dimethyl-2-hexene (only organic product)} + \text{HOH} + \text{KCl} \qquad (6\text{-}27)$$

Since one of the adjacent carbons to the carbon that has the chlorine atom has two methyl groups and no hydrogen, elimination does not occur at that carbon. Similarly, a reaction will not occur if there is not a hydrogen on either of carbons adjacent to the carbon that has the chlorine atom as shown for the reactant of 1,1,3,3-tetramethyl-2-chlorocyclohexane given in Reaction (6-28).

$$\text{1,1,3,3-Tetramethyl-2-chlorocyclohexane} \xrightarrow{\text{KOH, heat}} \text{No reaction} \qquad (6\text{-}28)$$

Problem 6.8

i) Give the elimination products for the following reactions.

(a) 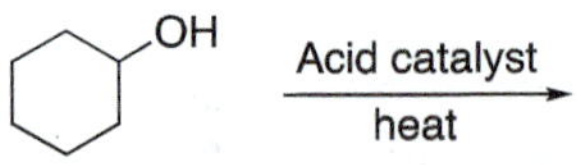 cyclohexanol $\xrightarrow{\text{Acid catalyst, heat}}$

(b) 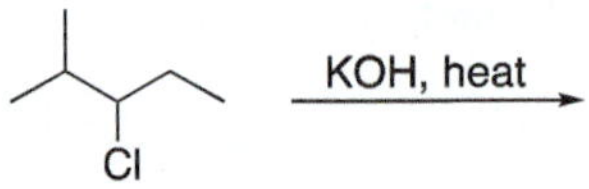 2-chloro-3-methylbutane $\xrightarrow{\text{KOH, heat}}$

ii) There are three possible elimination products for the following reaction, what are they?

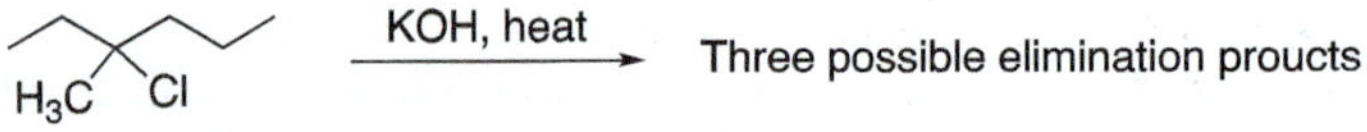

2-chloro-2-methylpentane (H₃C, Cl labeled) $\xrightarrow{\text{KOH, heat}}$ Three possible elimination products

It is possible for elimination reactions to occur from a carbon–heteroatom bond, as shown in Reaction (6-29).

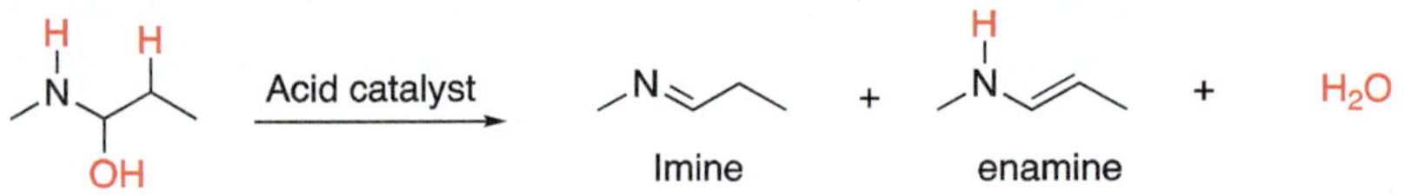

$$\text{R-NH-CH(OH)-R'} \xrightarrow{\text{Acid catalyst}} \text{R-N=CH-R'} \; (\text{Imine}) \;+\; \text{R-NH-CH=CH-R'} \; (\text{enamine}) \;+\; H_2O \quad (6\text{-}29)$$

The above reaction is very similar to the ones discussed earlier in this section where the elimination of water occurred to form an alkene. In this case, the elimination of water occurred, but in addition to the formation of an alkene, another organic product is also formed which has a carbon–nitrogen double bond and this functional group is called an imine.

Problem 6.9

Give the elimination products for the following reactions.

(a) 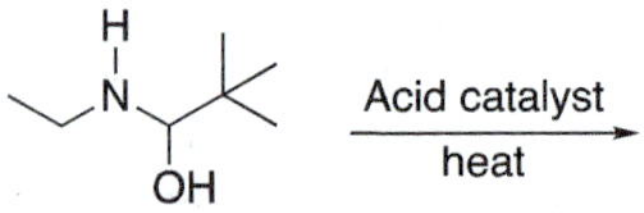 [structure with NH and OH] $\xrightarrow{\text{Acid catalyst, heat}}$

(b) 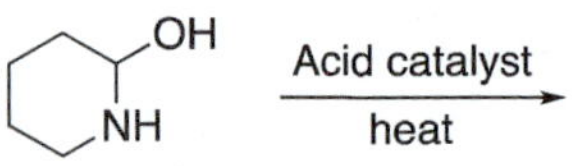 2-hydroxypiperidine (cyclohexane with OH and NH) $\xrightarrow{\text{Acid catalyst, heat}}$

For elimination reactions, it is important to recognize the functional group, which is typically an alcohol or alkyl halide, and recognize that the elimination reaction will occur from either of the adjacent carbons where there is a hydrogen to form an alkene functional group in the organic product.

6.7 Substitution Reactions

A substitution reaction is simply a reaction in which one atom (or a group of atoms) in the reactant is substituted by another atom or group of atoms to form a product. We have seen this type of reaction in our general chemistry course, and an example is shown in Reaction (6-30).

$$Ag^{2+}(NO_3)_2^- + 2Na^+Cl^- \rightarrow 2Na^+NO_3^- + Ag^{2+}Cl_2^- \quad (6\text{-}30)$$

For this reaction, the NO_3^- anion (of silver nitrate) in the reactant is substituted for the chloride (the anion) from the sodium chloride in the reactant to form the product. Thus, for this substitution reaction, the nitrate anion replaces the chloride anion to form the product. Hence, this type of reaction is called a substitution reaction. Substitution reactions in organic chemistry are similar except the substitution takes place at a carbon atom. Since organic chemistry is the study of organic compounds that contain mostly carbon and hydrogen, most of the reactions that will be encountered throughout this course do not contain transition elements, but the atoms mentioned in Chapter 2 (functional groups). An example of a substitution reaction of an organic molecule is shown in Reaction (6-31). Note that the sodium is not involved in the substitution reaction, just the iodide and the bromine atom. The sodium ion is referred to as a spectator ion in this case.

$$CH_3CH_2CH_2CH_2Br + NaI \longrightarrow CH_3CH_2CH_2CH_2I + NaBr \quad (6\text{-}31)$$

A similar analysis can be carried out for the reaction given in Reaction (6-31) as the reaction given in Reaction (6-30). For the substitution reaction involving 1-bromobutane and sodium iodide (Reaction 6-31), the iodide anion of sodium iodide replaces the bromine of 1-bromobutane to form the product 1-iodobutane. Since in organic chemistry, the emphasis is on the organic compounds in the reactant and the product, the other substitution product, sodium iodide, is considered to be inorganic and sometimes not included in the overall reaction. Other examples of substitution reactions are shown in Reactions (6-32) and (6-33).

$$CH_3CH_2CH_3 + Br_2 \xrightarrow{\text{Light}} CH_3CHBrCH_3 + HBr \quad (6\text{-}32)$$

$$\text{cyclopentane} \xrightarrow{Br_2,\ \text{light}} \text{bromocyclopentane} + HBr \quad (6\text{-}33)$$

Note that for these reactions, a hydrogen atom in the reactant is substituted for a halogen (in this case, bromine) to form the organic product. This category of reactions (substitution reactions) can involve many different organic compounds, and as a result, substitution reactions form a major category of reactions in organic chemistry. At this point, students should be able to recognize a substitution reaction, compared to the other types of reactions covered so far and also be able to recognize the atom (or group of atoms) that is involved in the substitution and to predict the organic product. Problem 6.10 is designed to have students identify the atom (or group of atoms) involved in the substitution reactions.

Problem 6.10

For the substitution reactions below, identify the atom (or group of atoms) in the reactants that are involved in the substitution reactions.

(a) $CH_3CH_2CH_2Br + NaCN \longrightarrow CH_3CH_2CH_2CN + NaBr$

(b) cyclohexane + $Br_2 \xrightarrow{\text{Light}}$ bromocyclohexane + HBr

(c) $CH_3CH_2CH_2I + NaSH \longrightarrow CH_3CH_2CH_2SH + NaI$

For the organic substitution reactions discussed thus far, substitution takes place at an sp^3 carbon, but it is possible to have substitution reactions take place at an sp^2 carbon, these types of reactions will be studied in detail in Chapters 16 and 17.

6.8 Pericyclic Reactions

Pericyclic reactions are a lot different from the reactions discussed thus far. The reactants of pericyclic reactions are typically conjugated alkenes, and throughout the course of pericyclic reactions, the pi (π) electrons of the reactants are transformed to form new bonds. Typically, heat or light is required for pericyclic reactions. The most noted pericyclic reaction is given in Reaction (6-34) and is also known as the Diels–Alder reaction, named after the German chemists, Otto Diels and Kurt Alder, who discovered this reaction. They discovered the reaction in 1938 and received the Nobel Prize for their contribution to the advancement of chemistry in 1950. For the pericyclic reaction given in Reaction (6-34), arrows are used to show the movement of the pi (π) electrons to form new bonds of a six-member ring.

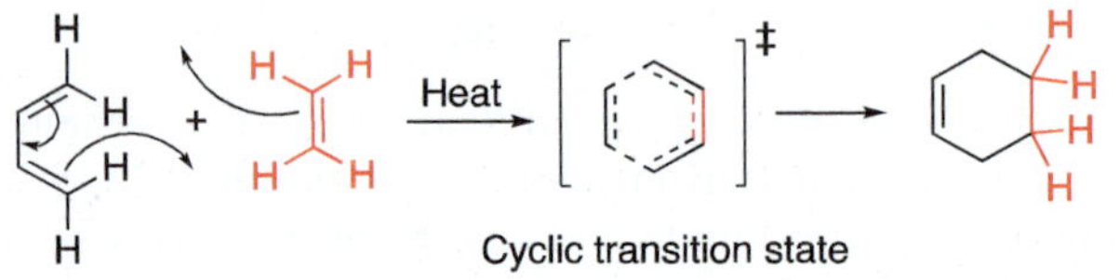

(6-34)

Cyclic transition state

Problem 6.11 is designed to help students recognize the reactants and to predict possible products for this type of pericyclic reactions based on your analysis of the above reaction.

Problem 6.11

Give the product for the following pericyclic reactions.

(a)

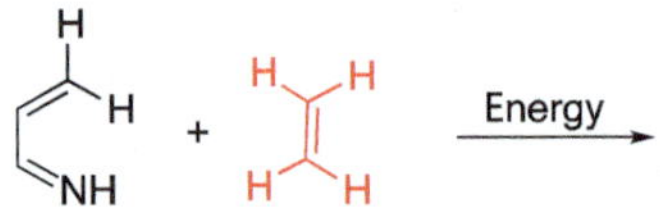

(b)

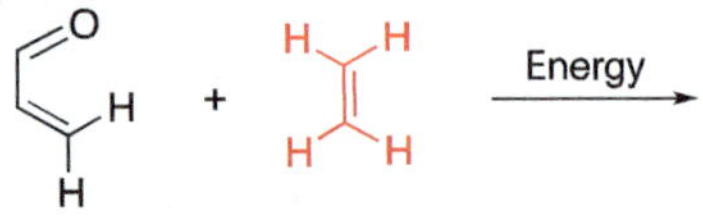

Pericyclic reactions will be covered in a lot more detail in Chapter 18, but at this point, students should be able to identify the uniqueness of pericyclic reactions, compared to the other reactions covered thus far and the special feature of conjugated double bond requirement for these reactions.

6.9 Catalytic Coupling Reactions

Catalytic coupling reactions are very important reactions in organic chemistry since these reactions involve the coupling of carbons to form new carbon–carbon bonds in the presence of a transition metal catalyst. An example of a famous reaction of this type is shown in Reaction (6-35), which is also known as the Heck reaction in which the palladium is used to catalyze this reaction.

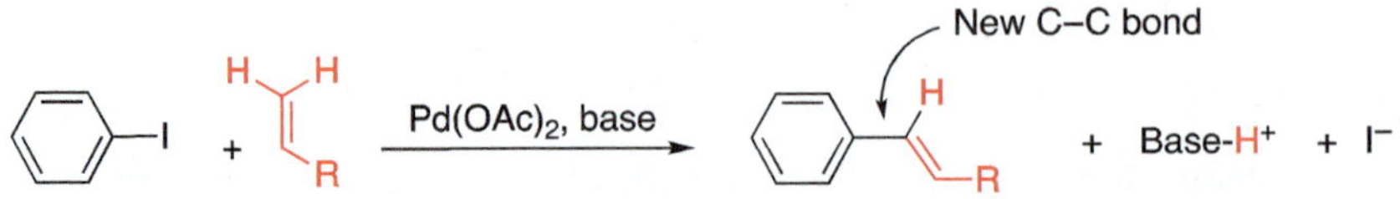

(6-35)

The Heck reaction is named after Richard F. Heck (1931–2015), an American chemist who worked at the University of Delaware, USA. He shared the Nobel Prize in Chemistry in 2010 with Ei-ichi Negish and Akira Suzuki, two Japanese chemists who discovered this type of coupling reaction. The Heck reaction involves the carbon–carbon bond-forming reaction in the presence of the catalyst, palladium. These catalytic coupling reactions are of extreme importance in the formation of new carbon–carbon bonds, which are sometimes challenging type of bonds to make in organic chemistry. More details on these reactions will be discussed in Chapter 19, but at this point, students should be able to recognize the features of these types of catalytic coupling reactions, which are carried out using sp^2 carbons and students should be able to predict the coupled organic product. Problem 6.12 is designed to get students to recognize the features of a catalytic coupling reaction.

Problem 6.12

i) Give the product for the following catalytic coupling reactions.

(a) 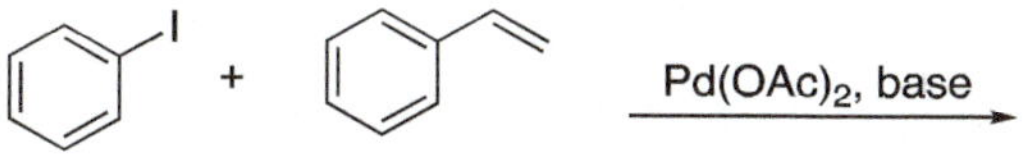 PhI + PhCH=CH$_2$ $\xrightarrow{\text{Pd(OAc)}_2,\ \text{base}}$

(b) 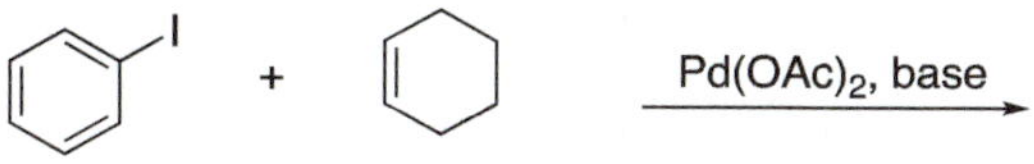 PhI + cyclohexane $\xrightarrow{\text{Pd(OAc)}_2,\ \text{base}}$

ii) Give the reactants for the following catalytic coupling reactions.

(a) 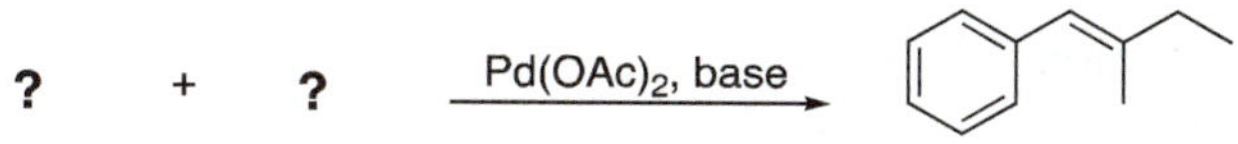 ? + ? $\xrightarrow{\text{Pd(OAc)}_2,\ \text{base}}$ PhCH=C(CH$_3$)CH$_3$

(b) 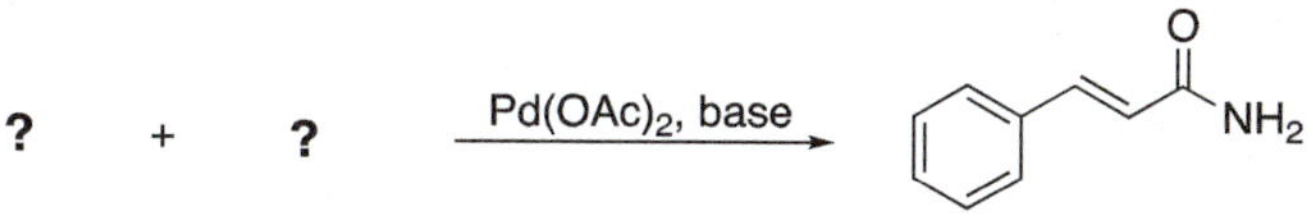 ? + ? $\xrightarrow{\text{Pd(OAc)}_2,\ \text{base}}$ PhCH=CHC(O)NH$_2$

With this overview of the different types of important reactions of organic chemistry, students can better appreciate the versatility of the reactions of organic chemistry in producing an extremely wide variety of organic products. In later chapters, we will discuss in detail, the mechanisms for these types of reactions and how these types of reactions can be used to produce desired organic products. This aspect of organic chemistry is called organic synthesis and is routinely used in industry and research labs.

End of Chapter Problems

6.13 Give the products of the following acid–base reactions.

(a) 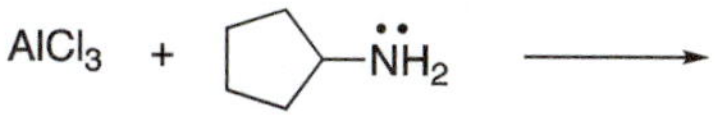 AlCl$_3$ + cyclopentyl-NH$_2$ $\longrightarrow$

(b) 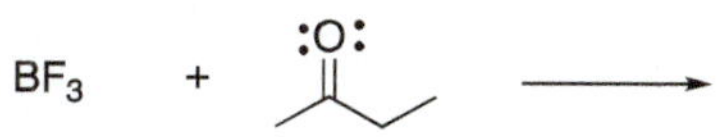 BF$_3$ + acetone $\longrightarrow$

6 An Overview of the Reactions of Organic Chemistry

6.14 For the following addition reactions, predict the all possible organic products needed to complete each reaction.

(a) 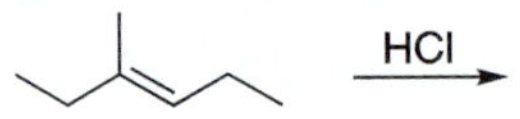 2-methyl-2-butene + HCl →

(b) 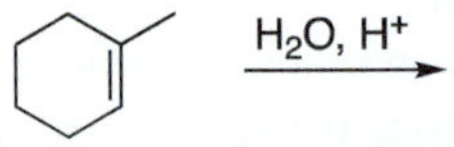 cyclohexene + H$_2$O, H$^+$ →

6.15 Predict the products for the following substitution reactions.

(a) 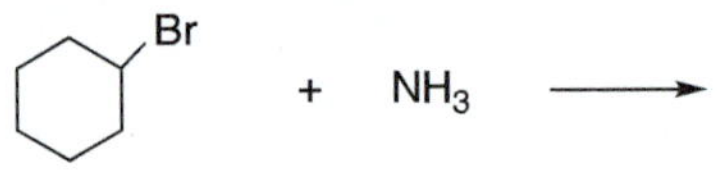 bromocyclohexane + NH$_3$ →

(b) 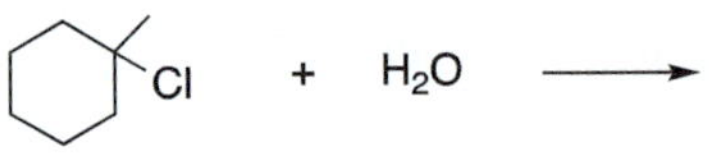 chlorocyclohexane + H$_2$O →

(c) 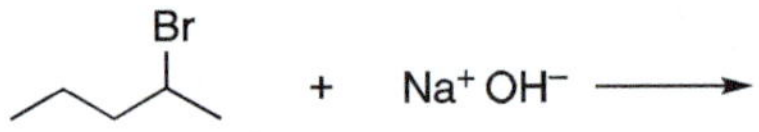 2-bromobutane + Na$^+$ OH$^-$ →

(d) 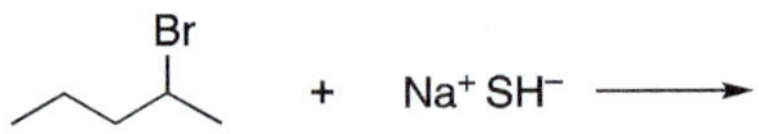 2-bromobutane + Na$^+$ SH$^-$ →

6.16 Predict all possible products for the following elimination reactions.

(a) 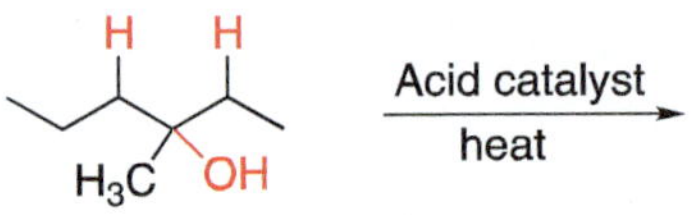 2-methyl-2-butanol, Acid catalyst, heat →

(b) 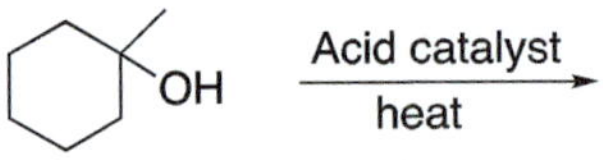 1-methylcyclohexanol, Acid catalyst, heat →

6.17 Predict products for the following oxidation reactions.

(a) 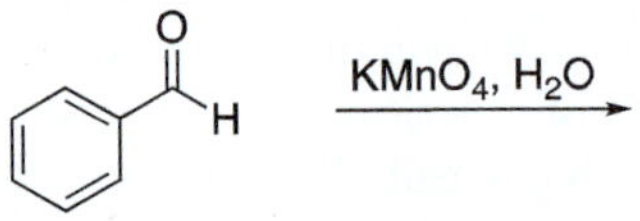 benzaldehyde + KMnO$_4$, H$_2$O →

(b) 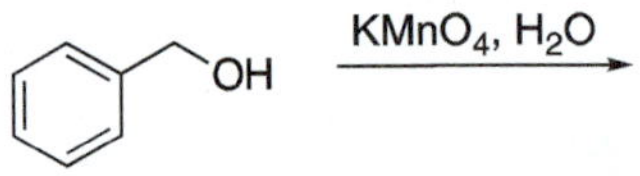 benzyl alcohol + KMnO$_4$, H$_2$O →

6.18 Predict products for the following reduction reactions.

(a) 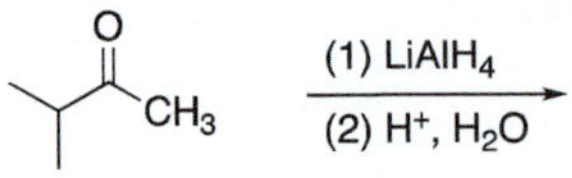 methyl isopropyl ketone, (1) LiAlH$_4$, (2) H$^+$, H$_2$O →

(b) 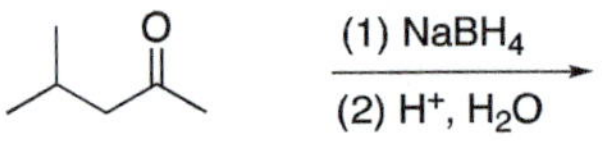 butan-2-one (or similar ketone), (1) NaBH$_4$, (2) H$^+$, H$_2$O →

(c) 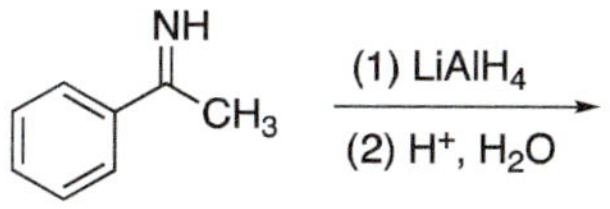 [structure: phenyl-C(=NH)-CH₃] $\xrightarrow{\text{(1) LiAlH}_4 \quad \text{(2) H}^+, \text{H}_2\text{O}}$

6.19 Predict the products of the following Diels–Alder pericyclic reactions.

(a)  cyclopentadiene + ethylene $\xrightarrow{\text{Energy}}$

(b) 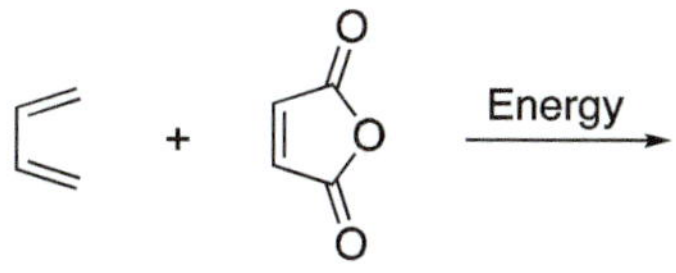 1,3-butadiene + maleic anhydride $\xrightarrow{\text{Energy}}$

6.20 Predict products for the following catalytic coupling reactions.

(a) 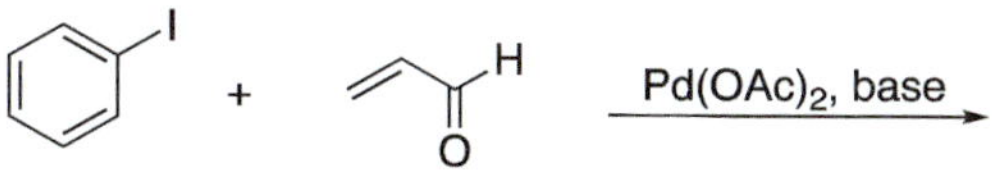 iodobenzene + acrolein (CH₂=CH-CHO) $\xrightarrow{\text{Pd(OAc)}_2, \text{ base}}$

(b) 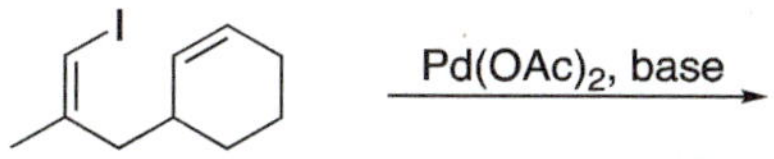 [vinyl iodide tethered to cyclohexene] $\xrightarrow{\text{Pd(OAc)}_2, \text{ base}}$

6.21 What types are the following reactions, i.e. substitution, elimination, or oxidation? Note that only the major organic product is shown for each reaction, other inorganic products are not shown.

(a) 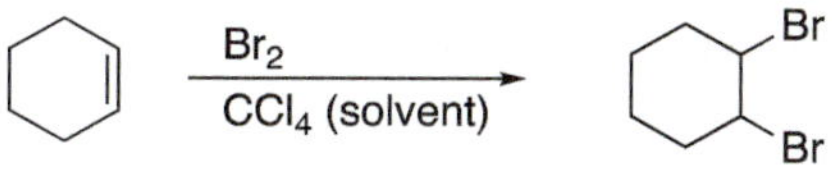 cyclohexene $\xrightarrow{\text{Br}_2 \\ \text{CCl}_4 \text{ (solvent)}}$ trans-1,2-dibromocyclohexane

(b) 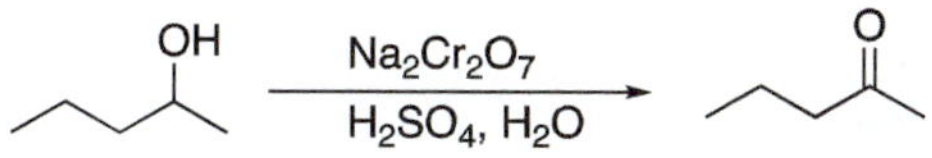 2-butanol $\xrightarrow{\text{Na}_2\text{Cr}_2\text{O}_7 \\ \text{H}_2\text{SO}_4, \text{H}_2\text{O}}$ 2-butanone

(c) 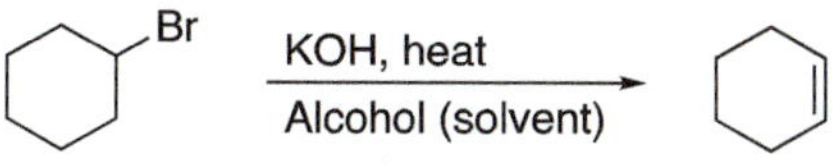 bromocyclohexane $\xrightarrow{\text{KOH, heat} \\ \text{Alcohol (solvent)}}$ cyclohexene

(d) 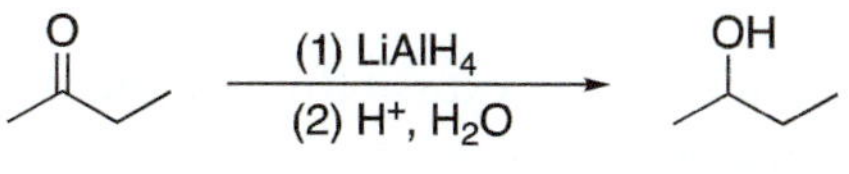 [ketone] $\xrightarrow{\text{(1) LiAlH}_4 \\ \text{(2) H}^+, \text{H}_2\text{O}}$ [alcohol, OH]

(e) 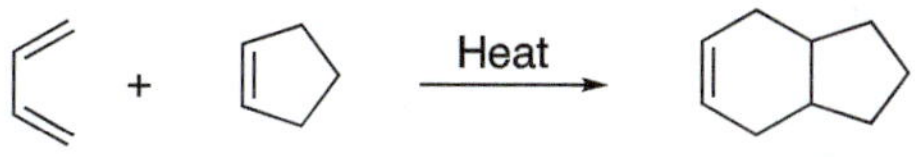 1,3-butadiene + cyclopentadiene $\xrightarrow{\text{Heat}}$ bicyclic product

(f) 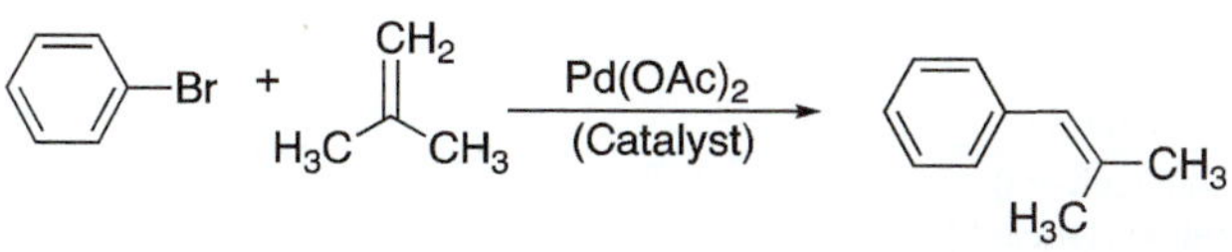 bromobenzene + isobutylene (H₃C-C(CH₃)=CH₂) $\xrightarrow{\text{Pd(OAc)}_2 \\ \text{(Catalyst)}}$ PhCH=C(CH₃)-CH₃ (with H₃C)

(g) PhCOCH$_3$ $\xrightarrow{\text{NaOCH}_3, \text{CH}_3\text{OH}}$ PhCOCH$_2^-$ Na$^+$

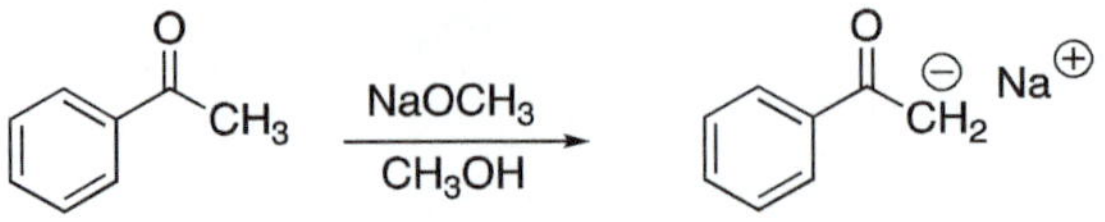

(h) cyclohexane $\xrightarrow{\text{Cl}_2, \text{light}}$ chlorocyclohexane + HCl

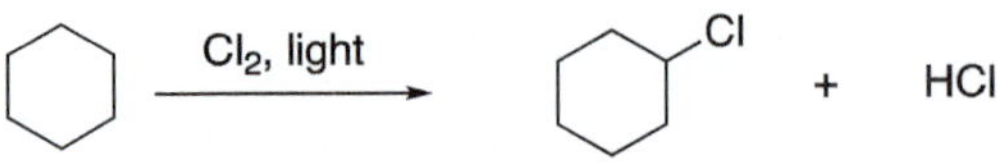

6.22 In the space provided, complete the following reactions by supplying the structure of the reactant or an organic product. The types of reactions are shown below the arrow.

(i) _____ $\xrightarrow[\text{light}]{\text{Cl}_2}$ chlorocyclohexane + HCl

Cyclic saturated hydrocarbon   (Substitution reaction)

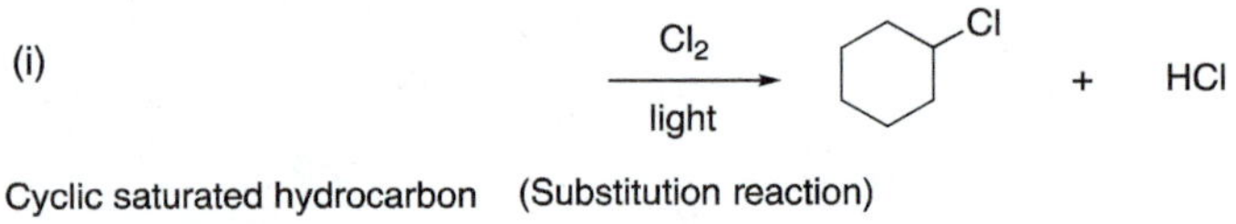

(ii) _____ $\xrightarrow[\text{H}_2\text{SO}_4]{\text{K}_2\text{Cr}_2\text{O}_7}$ CH$_3$CH$_2$CH$_2$COOH

(alcohol or aldehyde)   (Oxidation)

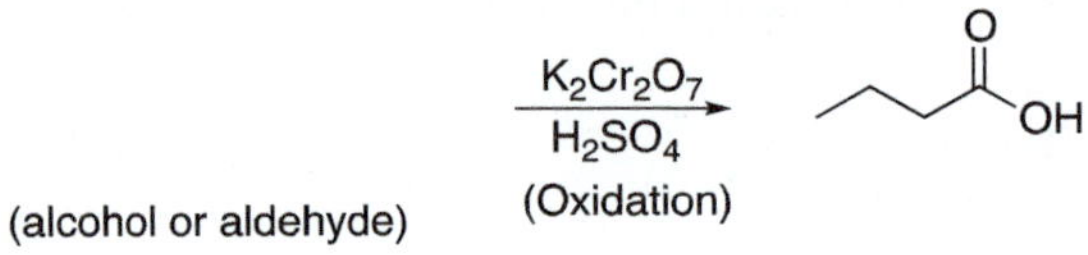

(iii) CH$_3$CH$_2$CH$_2$CHO $\xrightarrow{\text{K}_2\text{Cr}_2\text{O}_7/\text{H}_2\text{SO}_4}$ _____

(Oxidation)   Carboxylic acid

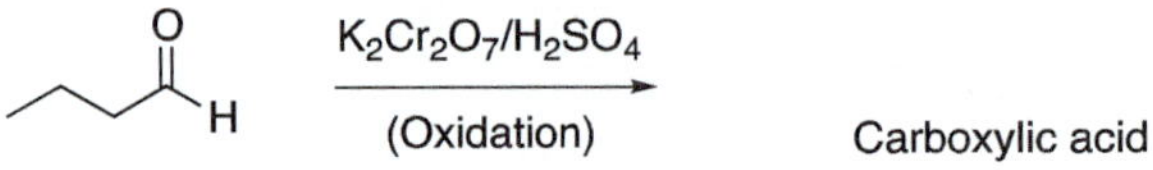

(iv) 2-pentanol (CH$_3$CH(OH)CH$_2$CH$_2$CH$_3$) $\xrightarrow{\text{K}_2\text{Cr}_2\text{O}_7/\text{H}_2\text{SO}_4}$ _____

(oxidation)

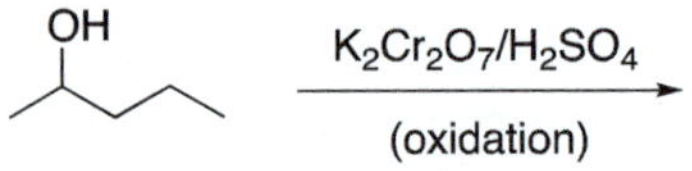

(v) _____ $\xrightarrow[\text{(Oxidation)}]{\text{K}_2\text{Cr}_2\text{O}_7 / \text{H}_2\text{SO}_4}$ cyclohexane-COOH

Alcohol

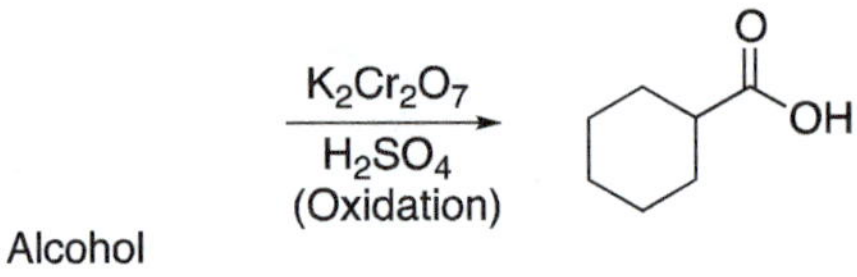

(vi) PhI + CH$_2$=CHCHO $\xrightarrow[\text{(catalytic coupling)}]{\text{Pd(OAc)}_2, \text{ base}}$ _____

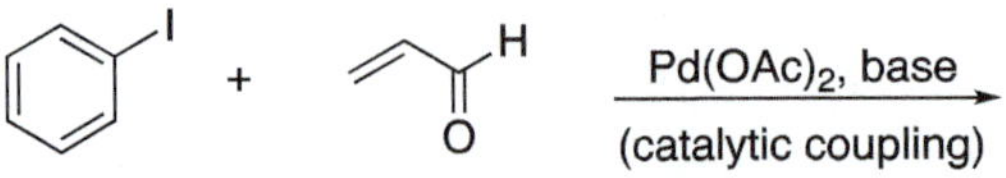

(vii) cyclopentadiene + CH$_2$=CHCN $\xrightarrow[\text{(pericyclic reaction)}]{\text{energy}}$ _____

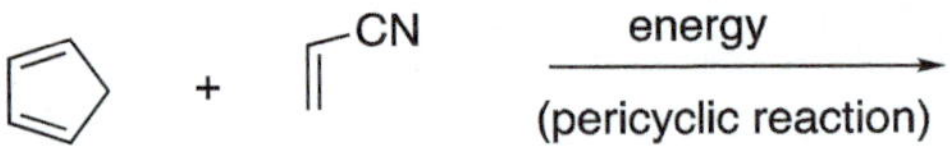

6.23 Complete the following reactions by supplying the structure of the reactant or an organic product. The types of the reactions are shown below the arrow.

(a) Alkene $\xrightarrow[\text{(Addition reaction)}]{\text{Br}_2, \text{CCl}_4 \text{ (solvent)}}$ CH$_3$CH(Br)CH(Br)CH$_3$

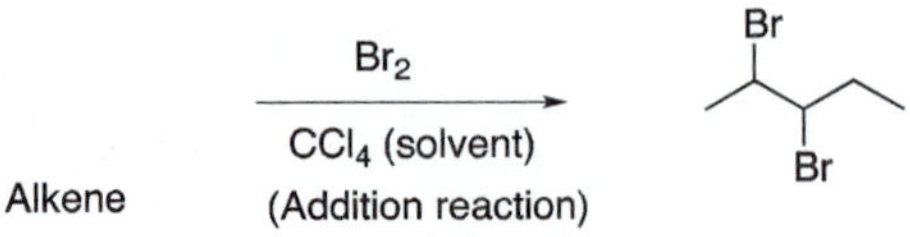

(b) Alkyl halide $\xrightarrow[\text{(Elimination Reaction)}]{\text{KOH, heat, Alcohol (solvent)}}$ CH$_3$CH=CHCH$_3$ + KCl + H$_2$O

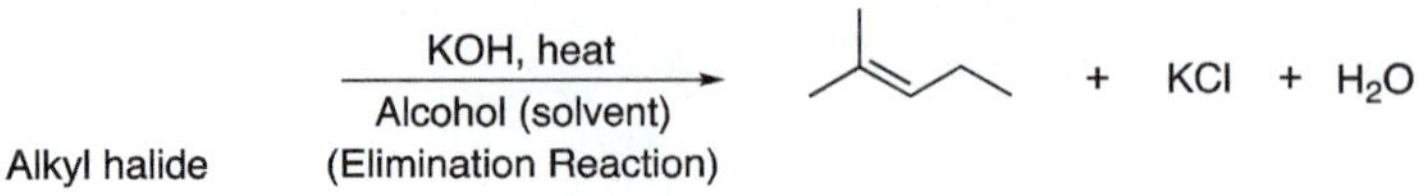

End of Chapter Problems

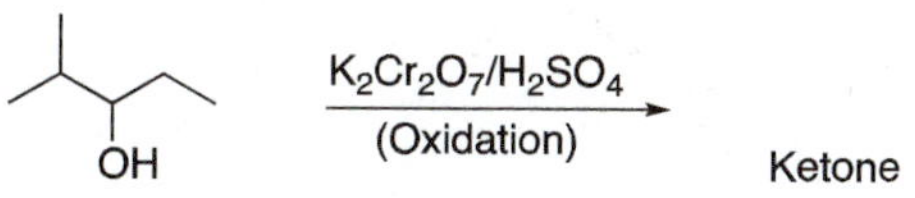

(c) (CH₃)₂CH-CH(OH)-CH₂CH₃ →[K₂Cr₂O₇/H₂SO₄]{(Oxidation)} Ketone

(d) Alcohol →[K₂Cr₂O₇/H₂SO₄]{(Oxidation)} cyclohexanone

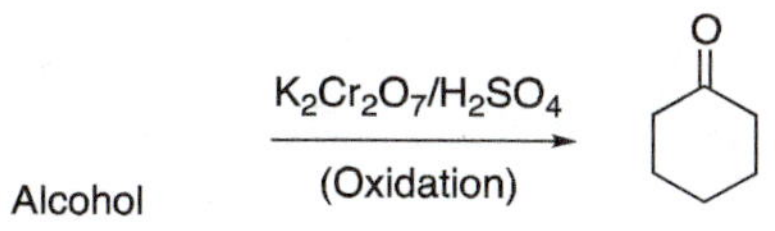

(e) C₆H₅-NH₂ + H-Cl →[(Acid-base reaction)] Salt

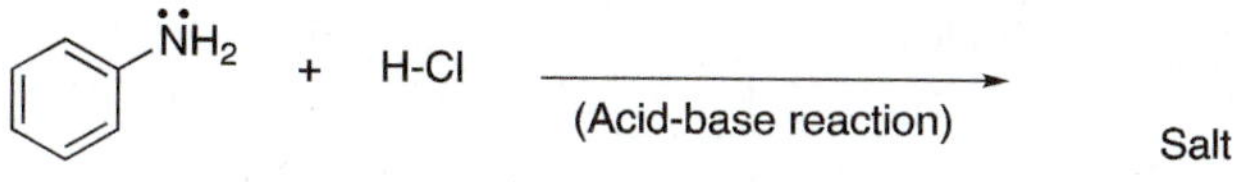

(f) (CH₃)₂N-CH(CH₃)-CH₂CH₃ (dimethylamine on sec-butyl) + H-Br →[(Acid-base reaction)] Salt

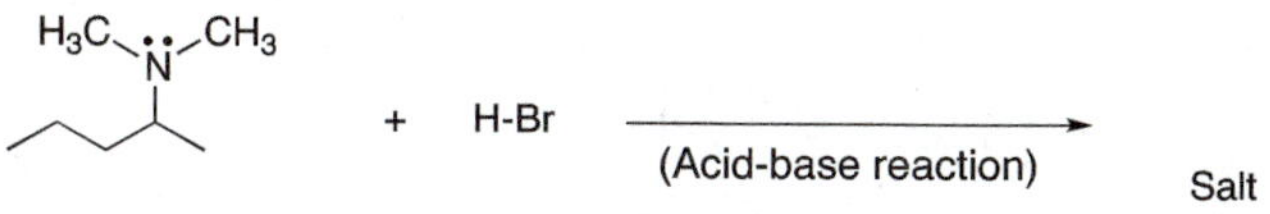

(g) (CH₃)₂CH-CO-CH₃ →[(1) NaBH₄ / (2) H₂O]{(Reduction)} (An alcohol)

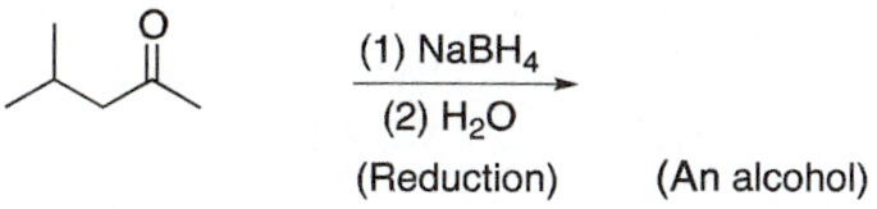

(h) Cyclopentanecarboxylic acid + NaOH (base) →[Acid-base reaction] Cyclopentyl-C(=O)-O⁻Na⁺ + H₂O
(Carboxylic acid) (Carboxylate salt)

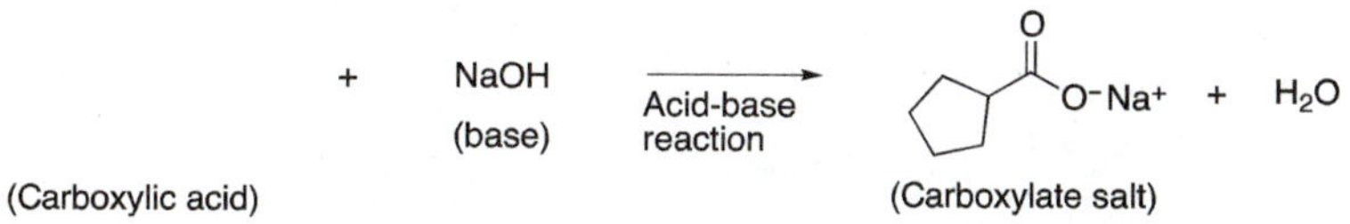

(i) Methylcyclopentene →[Cl₂ / CCl₄ (solvent) / (Addition reaction)] Cycloalkyl halides

(j) Alcohol →[K₂Cr₂O₇/H₂SO₄]{(Oxidation)} 2-Methyl-3-pentanone

(k) Alkane →[Br₂ / Light]{(Substitution reaction)} 2-Bromobutane + HBr

(l) Cycloalkyl halide →[KOH, heat / Alcohol (solvent) / (Elimination Reaction)] methylcyclopentene + KCl + H₂O

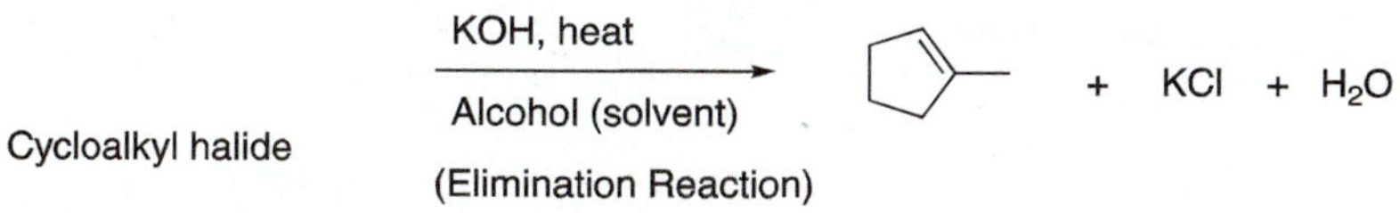

(m) 2-Methylcyclohexanone →[(1) NaBH₄ / (2) H₂O]{(Reduction)} Alcohol

(n) 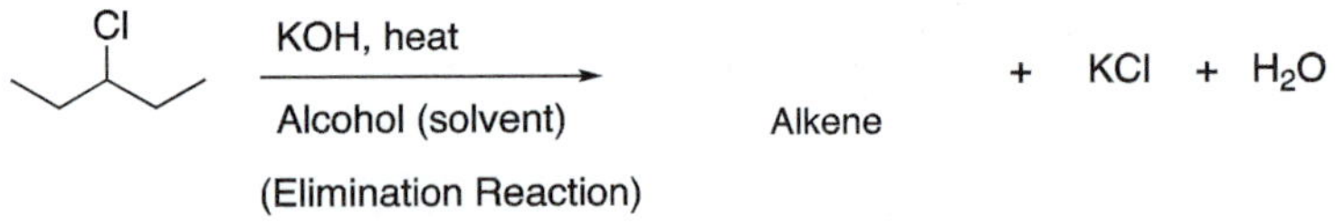 CH₃CH(Cl)CH₂CH₃ →[KOH, heat / Alcohol (solvent)] Alkene + KCl + H₂O
(Elimination Reaction)

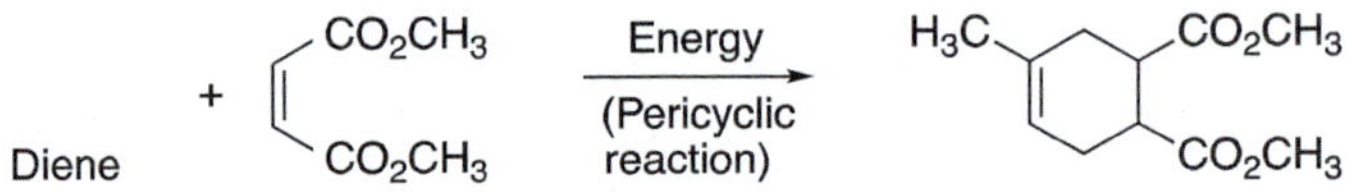

(o) 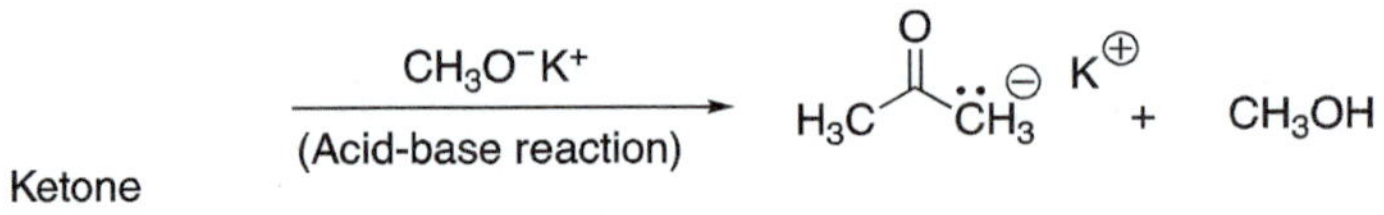 Ketone →[CH₃O⁻K⁺ / (Acid-base reaction)] H₃C-C(=O)-CH₂⁻ K⁺ + CH₃OH

(p) 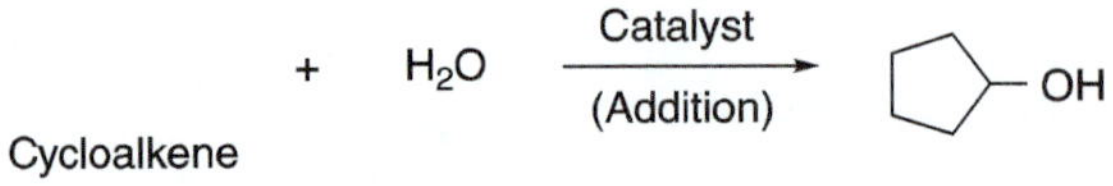 Cycloalkene + H₂O →[Catalyst / (Addition)] cyclopentanol (−OH)

(q) 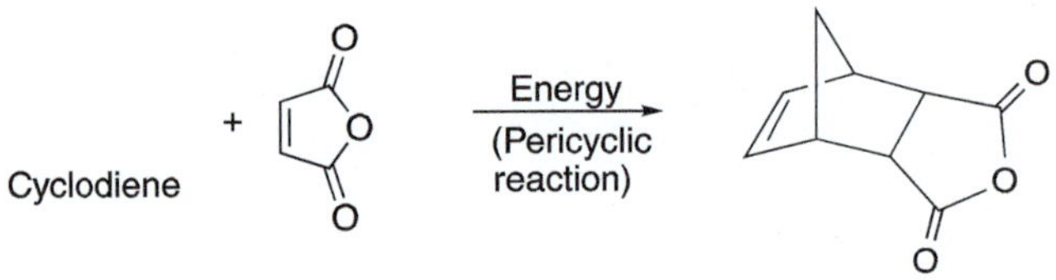 Cyclodiene + maleic anhydride →[Energy / (Pericyclic reaction)] norbornene-fused anhydride

(r) Diene + dimethyl maleate →[Energy / (Pericyclic reaction)] methylcyclohexene dicarboxylate

(s) Cyclopentanol →[K₂Cr₂O₇ / H₂SO₄ / (Oxidation)] Ketone

(t) 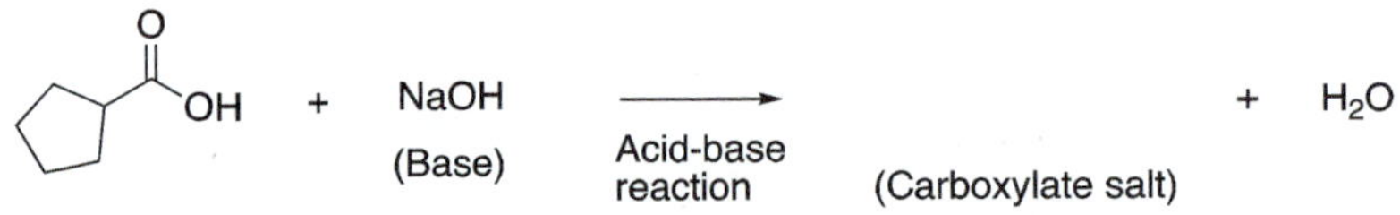 Cyclopentanecarboxylic acid + NaOH (Base) →[Acid-base reaction] (Carboxylate salt) + H₂O

(u) 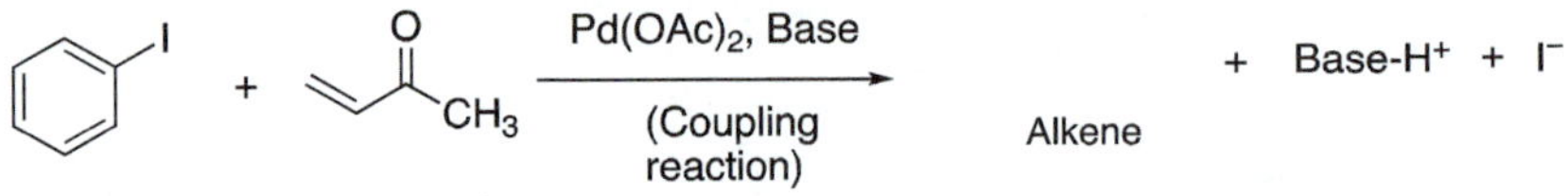 Iodobenzene + CH₂=CH-C(=O)CH₃ →[Pd(OAc)₂, Base / (Coupling reaction)] Alkene + Base-H⁺ + I⁻

(v) 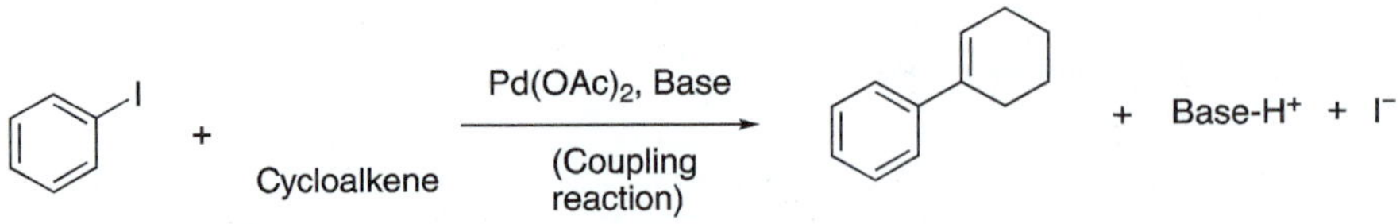 Iodobenzene + Cycloalkene →[Pd(OAc)₂, Base / (Coupling reaction)] phenylcyclohexene + Base-H⁺ + I⁻

(w) 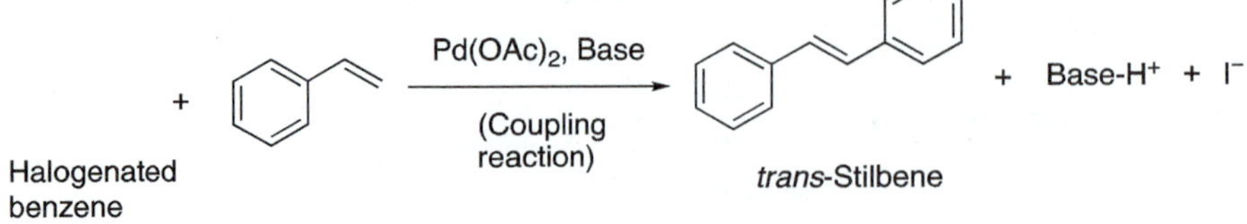 Halogenated benzene + styrene →[Pd(OAc)₂, Base / (Coupling reaction)] trans-Stilbene + Base-H⁺ + I⁻

7

Acid–Base Reactions in Organic Chemistry

7.1 Introduction

The acid–base reactions that will be encountered in organic chemistry will involve extremely strong acids and bases. Since organic compounds consist of mostly carbons and hydrogens, they are typically very weak acids, and as a result, very strong bases are typically required to deprotonate most organic compounds. Some bases that will be encountered in this course are extremely strong, and as a result, water is typically not the solvent of choice for most acid–base reactions, but aprotic solvents such as diethyl ether and tetrahydrofuran (THF) are typically the solvents of choice. An extremely strong base has the capability of explosively abstracting a proton from water if it were used as a solvent for organic acid–base reactions. As a result, extreme care must be exercised in the choice of a solvent for most organic reactions. In this chapter, a more detailed examination of acid–base reactions will be carried out by applying the concepts learned in the previous chapter to analyze a wide range of acid–base reactions and students will be better able to determine appropriate reaction conditions.

7.2 Lewis Acids and Bases

The Arrhenius, Brønsted-Lowry, and the Lewis definitions of acids and bases were covered in the previous chapter. In organic chemistry, the Lewis definition is more frequently used, compared to the other definitions. A ***Lewis acid*** is an electron pair acceptor. In order for a molecule to be an acid and be able to accept a pair of electrons, there must be a vacant orbital in which these electrons must be placed. Of course, the simplest Lewis acid is a proton (H^+) since it has a vacant orbital. On the other hand, a Lewis base is an electron pair donor. Thus, Lewis bases have at least one unshared pair of electrons. A Lewis acid–base reaction involves an electron pair from the base being transferred to the vacant orbital of the acid, as illustrated in Reaction (7-1) for the reaction of methylamine (a Lewis base) with BF_3, a Lewis acid.

$$F-B(F)(F) \;+\; CH_3\ddot{N}H_2 \longrightarrow {}^-BF_3-\overset{+}{N}HCH_3 \qquad (7\text{-}1)$$

Electron pair acceptor Electron pair donor New bond

Organic Chemistry: Concepts and Applications, First Edition. Allan D. Headley.
© 2020 John Wiley & Sons, Inc. Published 2020 by John Wiley & Sons, Inc.
Companion website: www.wiley.com/go/Headley_OrganicChemistry

For the reaction given in Reaction (7-1), the lone pair of electrons on the methylamine forms a bond to BF_3 using the vacant p orbital of BF_3 to produce the product shown. Note, for the product for Reaction (7-1), the boron has acquired a formal negative charge due to the gain of a pair of electrons from the methylamine. On the other hand, the nitrogen has a positive charge since it has given up a pair of electrons to form a bond to boron.

A carbocation, which is sp^2 hybridized and has a vacant p orbital, is also a Lewis acid and, as a result, can undergo a reaction with a Lewis base, such as the chloride anion since the chloride anion has four unshared pair of electrons. The reaction of a carbocation and the chloride anion is shown in Reaction (7-2). Acid–base reactions of this type will be encountered frequently throughout organic chemistry.

$$\text{Lewis acid (carbocation with empty p orbital)} + \text{Lewis base (Anion with four lone pairs of electrons)} \longrightarrow \text{Lewis acid-base product} \qquad (7\text{-}2)$$

For the acid–base reaction given in Reaction (7-2), both reactants have formal charges of +1 and −1, respectively, and as a result, the product is neutral. The new bond is a result of using one pair of electrons from the chloride anion to form a bond using the vacant p orbital on the carbocation.

Problem 7.1

Which of the following are bases or acids? Explain your answer.

HI, H_2O, NaCN, $AlCl_3$, and KCl.

Ralph G. Pearson categorized acids and bases as hard and soft acids and bases (HSAB). Hard Lewis bases have small ionic radii and are typically highly solvated in a solvent; they are also relatively electronegative and nonpolarizable, NH_2^- and OH^- are two examples of hard Lewis bases. On the other hand, soft Lewis bases have large ionic radii, they are relatively large and polarizable, typically not highly solvated; the anions of P, S, and Cl are soft Lewis bases. Hard Lewis acids have small ionic radii and are highly solvated; the proton, H^+ is considered to be the hardest acid and will be encountered frequently throughout organic chemistry. An example of a soft acid is Cu^{2+}. Typically, hard acids bind strongly with hard bases and soft acids bind with soft bases.

7.3 Relative Strengths of Acids and Conjugate Bases

The degree to which acids dissociate in water (and other solvents) is different for each acid. Each acid dissociates based on the equilibrium established in a particular solvent. The magnitude of such an equilibrium constant (K_{eq}) reflects the relative distribution of the dissociated species to the undissociated acid in solution. It is known that nitric acid cannot be ingested; on the other hand, acetic acid (in the form of vinegar in salad dressings) can be ingested! Nitric acid is a much stronger acid than acetic acid and, as a result, is a very dangerous corrosive acid.

7.3 Relative Strengths of Acids and Conjugate Bases

The relative strengths of acids are typically measured in water. Since water is a base, an acid–base chemical reaction takes place between the acid (HA) and water (the base) to form protonated water, H_3O^+, and A^-, which are also known as the conjugate acid and base, respectively, as shown in Reaction (7-3) for an acid HA.

$$H-A + H_2O \underset{}{\overset{K_{eq}}{\rightleftharpoons}} H_3O^+ + A^- \quad (7\text{-}3)$$

Acid — Base — Conjugate acid — Conjugate base

At equilibrium, all the species, H_3O^+, A^-, HA, and H_2O, exist in solution, and the concentrations of all the species are used to calculate the equilibrium constant for a particular acid, HA. The expression for the equilibrium constant is shown in Eq. (7-4), where K_{eq} is the equilibrium constant and the concentrations for the various species are shown in square brackets.

$$K_{eq} = [A^-][H_3O^+]/[H_2O][HA] \quad (7\text{-}4)$$

Since water is the solvent making its concentration very high, compared to the other species in solution, its concentration can be considered to be a constant. Thus, the expression in Eq. (7-4) can be modified as shown in Eq. (7-5) and even further to Eq. (7-6) if $[H_2O]K$ made equal to a new constant, K_a, the acidity equilibrium constant in water as a solvent.

$$[H_2O]K_{eq} = [H_3O^+]/[A^-][HA] \quad (7\text{-}5)$$

$$K_a = [H_3O^+]/[A^-][HA], \quad \text{where} \quad [H_2O]K_{eq} = K_a \quad (7\text{-}6)$$

The K_a for HCl has been determined experimentally to be 1×10^7, and for acetic acid, the K_a is 1.75×10^{-5}. Since 1×10^7 is a much larger number than 1.75×10^{-5}, HCl is a stronger acid producing more protons (H^+) in water, compared to acetic acid. Another way of stating this observation is that for an equal concentration of both HCl and acetic acid, the concentration of the protons in a solution of HCl is much greater than that of a solution of acetic acid.

Since these K_a values tend to be either very large or very small and very cumbersome to use, an easier method was devised to compress or expand these numbers, and at the same time convey the same meaning of the relative strengths of acids. The negative log of K_a or pK_a values are routinely used to reflect the relative strength of acids, and the relationship between K_a and pK_a is shown in Eq. (7-7).

$$pK_a = -\log(K_a) \quad (7\text{-}7)$$

Thus, the pK_a for HCl is -7 and for acetic acid is 4.75.

Problem 7.2

i) Determine the pK_a of the acids that have the following K_a values? Which of the acids with the given K_a values would be the weakest?

a) $K_a = 1.3 \times 10^{-5}$
b) $K_a = 1.3 \times 10^5$
c) $K_a = 1.3 \times 10^{-10}$
d) $K_a = 2.3 \times 10^{25}$
e) $K_a = 7.9 \times 10^{-12}$

ii) Given the pK_a values shown below, determine the K_a values? Which of the acids with the given pK_a values would be the strongest?

a) pK_a = 35
b) pK_a = 12
c) pK_a = 4.6
d) pK_a = 24
e) pK_a = −14

A pK_a table gives the pK_a values of common acids used in organic chemistry and the pK_a values of some acids are shown in Table 7.1.

An acid with a pK_a value that is more positive than the pK_a value of another acid means that the acid with the more positive pK_a value is the weaker acid. For example, if two acids have pK_a values of 3.75 and 8.34, respectively, the acid with a pK_a value of 3.75 is a stronger acid than the acid with a pK_a value of 8.34. Also, notice that the pK_a for some of these acids are outside the range of the acidity of water, this means that their pK_a values are not measurable in water, but in other solvents. For example, NH_3 is a weak acid, pK_a = 38, but its conjugate base is extremely strong and will deprotonate water.

The pK_a table is very helpful in predicting the acidity of compounds and equally important, it is also very helpful in predicting the relative strength of conjugate bases. Consider the two acids, HBr and HCN with the acidity in water is shown in Reactions (7-8) and (7-9).

Table 7.1 pK_a for various acids of organic chemistry.

Acid	Conjugate base	pK_a
HI	I^-	−9.5
HBr	Br^-	−9.0
HCl	Cl^-	−7.0
H_3O^+	H_2O	−1.7
CH_3COOH	CH_3COO^-	4.8
H_2S	SH^-	7.0
HCN	CN^-	9.2
NH_4^+	NH_3	9.4
CH_3SH	CH_3S^-	10.0
R-CHCOR	Enolate	~19
CH_3OH	CH_3O^-	15.2
H_2O	OH^-	15.7
CH_3CH_2CN	CH_3CHCN^-	25
$CH_3C{\equiv}CH$ (alkynes)	$CH_3C{\equiv}C^-$	25
H_2	H^-	35
NH_3	NH_2^-	38
C_6H_6	$C_6H_5^-$	43
$CH_3CH{=}CH_2$	$CH_3CH{=}CH^-$	44
CH_4	CH_3^-	50
CH_3CH_3	$CH_3CH_2^-$	50

$$\underbrace{\underset{\substack{\text{Acid}\\ \text{p}K_a = -9.0}}{\text{H–Br}} + \underset{\text{Base}}{\text{H}_2\text{O}}}_{\text{Conjugate acid-base pair}} \rightleftharpoons \underset{\substack{\text{Conjugate}\\ \text{acid}}}{\text{H}_3\text{O}^+} + \underset{\substack{\text{Conjugate}\\ \text{base}}}{\text{Br}^-} \quad (7\text{-}8)$$

$$\underset{\substack{\text{Acid}\\ \text{p}K_a = 9.2}}{\text{H–CN}} + \underset{\text{Base}}{\text{H}_2\text{O}} \rightleftharpoons \underset{\substack{\text{Conjugate}\\ \text{acid}}}{\text{H}_3\text{O}^+} + \underset{\substack{\text{Conjugate}\\ \text{base}}}{\text{CN}^-} \quad (7\text{-}9)$$

It is known that the conjugate base of a strong acid is weak and as a result, a determination can be made whether a conjugate base of an acid is strong or weak based by just knowing the pK_a value of an acid. Note that the conjugate bases shown in Reactions (7-8) and (7-9) are negatively charged. A stable negative charge has the ability to accommodate that charge. Thus, very electronegative or polarizable species will be better able to accommodate a charge, compared to less electronegative or polarizable species. For the reaction given in Reaction (7-8), the Br⁻ anion is a large atom and hence good at stabilizing the negative charge of the conjugate base. As a result, it is a weak base and its conjugate acid is a strong acid. On the other hand, the CN⁻ anion of Reaction (7-9) is not as polarizable as the bromide anion and is considered to be a harder base as described by the Pearson hard-soft base concept. As a result, the cyanide anion (CN⁻) is a stronger conjugate base than the bromide anion, which makes the conjugate acid, HCN, a weaker acid than HBr. This conclusion that Br⁻ is a weak conjugate base and hence the conjugate acid, HBr is a strong acid is confirmed from the pK_a values in the pK_a table. Similarly, the conjugate base CN⁻ is a strong conjugate base, which makes its conjugate acid, HCN a weak acid. This concept will be utilized throughout this course to predict the relative strengths of conjugate bases and acids.

Problem 7.3

i) Give the structure of the conjugate base of each of the following acids.

  a) H_2O
  b) CH_3OH
  c) HCN
  d) CH_3SH
  e) H_2S

ii) Utilize Table 7.1 to determine which of the following pairs of acids is stronger and give the structures of the conjugate bases of each acid.

  a) CH_3COOH and CH_3OH
  b) NH_3 and CH_3OH
  c) CH_4 and CH_3OH

7.4 Predicting the Relative Strengths of Acids and Bases

The pK_a table is very helpful in predicting the acidity of individual compounds, but if an acid were to be mixed with a base, the question now becomes would the equilibrium favor the right or the left? To fully appreciate this concept, consider mixing hydrochloric acid (HCl) with water (a base); the equilibrium is shown in Reaction (7-10), along with the pK_a values of the acids involved.

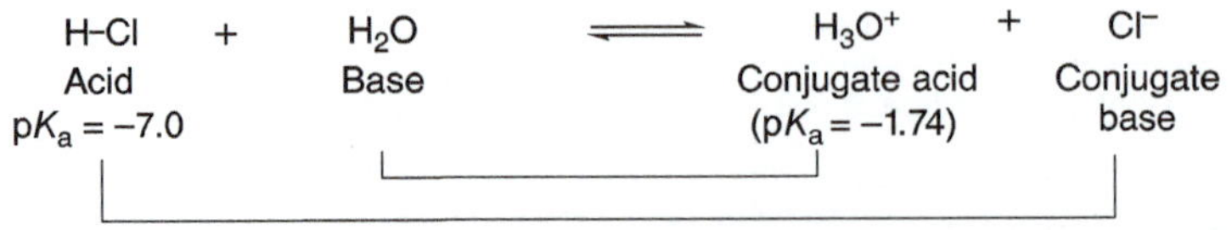

$$\text{H-Cl} + \text{H}_2\text{O} \rightleftharpoons \text{H}_3\text{O}^+ + \text{Cl}^- \quad (7\text{-}10)$$

Acid (p$K_a = -7.0$) + Base → Conjugate acid (p$K_a = -1.74$) + Conjugate base

Based on these pK_a values of the acids of this equilibrium, one has a pK_a value more negative (more acidic) than the other acid, which is less acidic. The position of the equilibrium is always towards the weaker acid. Thus, in this case, HCl is a stronger acid, compared to the conjugate acid of water, and as a result, the equilibrium will favor the right (the weaker acid), as represented in Reaction (7-11).

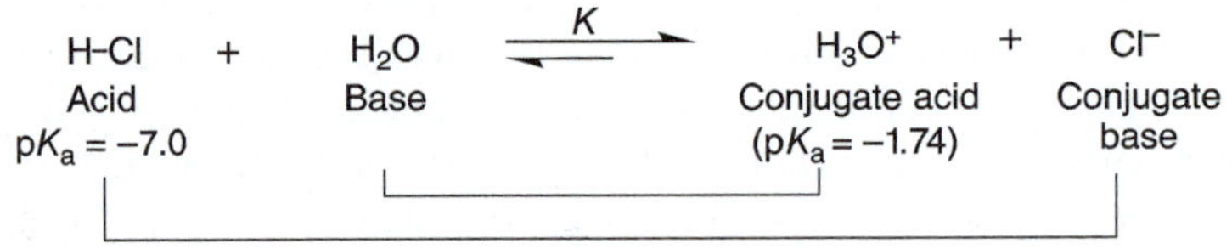

$$\text{H-Cl} + \text{H}_2\text{O} \xrightarrow{K} \text{H}_3\text{O}^+ + \text{Cl}^- \quad (7\text{-}11)$$

Acid (p$K_a = -7.0$) + Base → Conjugate acid (p$K_a = -1.74$) + Conjugate base

Problem 7.4

Predict the position of the following equilibria (pK_a values are shown below the acids).

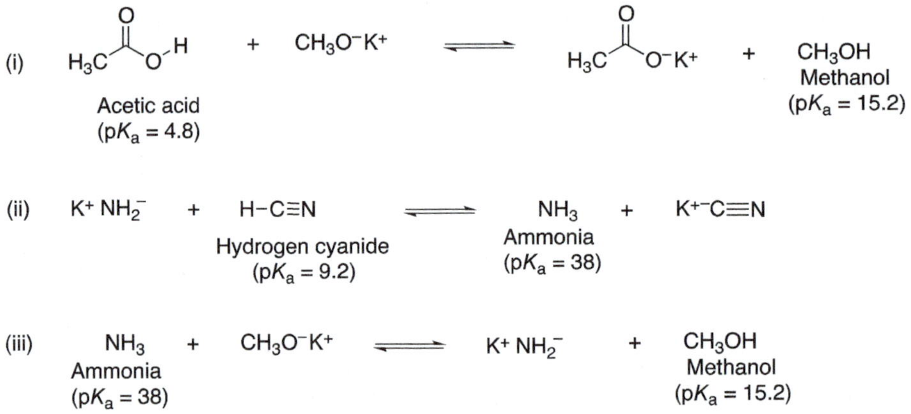

(i) Acetic acid $\text{H}_3\text{C-C(=O)-O-H}$ (p$K_a = 4.8$) + $\text{CH}_3\text{O}^-\text{K}^+$ ⇌ $\text{H}_3\text{C-C(=O)-O}^-\text{K}^+$ + CH_3OH Methanol (p$K_a = 15.2$)

(ii) K^+NH_2^- + H-C≡N Hydrogen cyanide (p$K_a = 9.2$) ⇌ NH_3 Ammonia (p$K_a = 38$) + $\text{K}^+{}^-\text{C}\equiv\text{N}$

(iii) NH_3 Ammonia (p$K_a = 38$) + $\text{CH}_3\text{O}^-\text{K}^+$ ⇌ K^+NH_2^- + CH_3OH Methanol (p$K_a = 15.2$)

This concept will be used continuously throughout this course since a large percentage of organic reactions involves the use of an acid or a base. As a result, a basic knowledge of the position of equilibria is needed in order to decide on the best acid (or base) to use for various reactions.

7.5 Factors That Affect Acid and Base Strengths

It is possible to predict the relative acidity of compounds without knowledge of the actual pK_a values. For most of the reactions that will be examined in organic chemistry, knowledge of the acidity or basicity of the reactants will assist in the prediction of products and the type of reaction needed to produce desired products. There are various factors that affect the acidity of compounds and they do so in a predictable manner, and each is discussed in the next sections.

7.5.1 Electronegativity

By definition, electronegativity is the ability of an atom to attract electrons toward itself. Thus, the electronegativity of the atom, which is bonded to a hydrogen atom of an acid, plays an important role in the relative acidity of that compound. For example, the relative acidities of ammonia and methane can be determined from examination of the electronegativity of the atoms to which the acidic hydrogen is directly bonded. The dissociation of these compounds is shown in Reactions (7-12) and (7-13).

$$H_3C-H \longrightarrow H_3C:^{\ominus} + H^{\oplus} \qquad (7\text{-}12)$$

$$H_2N-H \longrightarrow H_2N:^{\ominus} + H^{\oplus} \qquad (7\text{-}13)$$

Since nitrogen is a more electronegative atom than carbon, the nitrogen can accommodate the negative charge of the conjugate base much better than the less electronegative carbon atom. Thus, the conjugate base, NH_2^-, is a weaker base compared to the conjugate base CH_3^-. As a result, methane is a weaker acid than ammonia. This information can be confirmed from the pK_a table, which gives the pK_a values of methane and ammonia as 50 and 38, respectively. The ability of the conjugate base to stabilize its negative charge is very important. As mentioned before, the most electronegative atoms are the most appropriate atoms to accommodate a negative charge.

Problem 7.5

For the Brønsted–Lowry pairs of acids shown below, give the structures of the conjugate bases? Without using the pK_a table, determine which of the following pairs of acids is more acidic?

i) CH_4 and CH_3OH
ii) NH_3 and CH_3OH
iii) CH_4 and NH_3

7.5.2 Type of Hybridized Orbitals

As seen in Chapter 1, hybridized orbitals of carbon result from the mixture of s and p orbitals: sp^3 orbitals are 75% p character and 25% s character; sp^2 orbitals are 66% p character and 33% s character; and sp orbitals are 50% p character and 50% s character. Since the s orbitals are closer to the nucleus, compared to the more diffused p orbitals, hybridized orbitals that have more s character are closer to the nucleus than orbitals that have less s character. Thus, alkynes are more acidic than alkenes and alkanes as shown by Reactions (7-14)–(7-16).

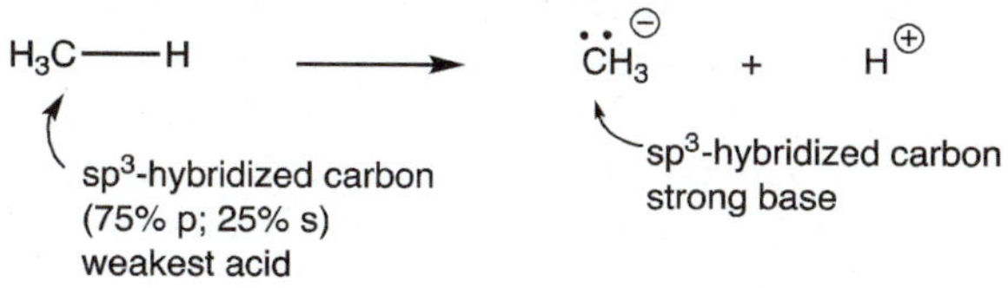

$$H_3C-H \longrightarrow \overset{..}{C}H_3^{\ominus} + H^{\oplus} \qquad (7\text{-}14)$$

sp³-hybridized carbon
(75% p; 25% s)
weakest acid

sp³-hybridized carbon
strong base

$$H_2C=CH-H \longrightarrow H_2C=\overset{..}{C}H^{\ominus} + H^{\oplus} \qquad (7\text{-}15)$$

sp²-hybridized carbon
(66% p; 33% s)
weak acid

sp²-hybridized carbon
weaker base

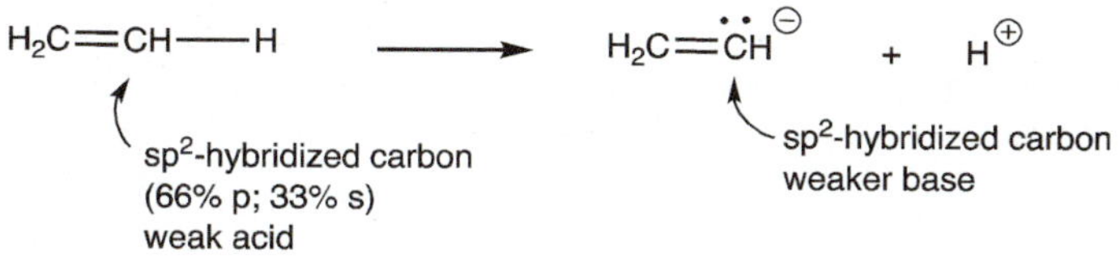

$$HC\equiv C-H \longrightarrow HC\equiv C:^{\ominus} + H^{\oplus} \qquad (7\text{-}16)$$

sp-hybridized carbon (50% p; 50% s) strongest acid

sp-hybridized carbon weakest base

Since the conjugate base of an alkyne has a negative charge on the more electronegative sp-hybridized carbon, it is a weaker base than that of the conjugate base of either an alkene or an alkane. Hence, alkynes are stronger acids than either alkenes or alkanes, and alkenes are stronger acids than alkanes. From the pK_a table, the pK_a values of most terminal alkynes are around 25, which make them acidic, even though weak acids and they can be involved in an acid–base reaction with an appropriate base, as shown in Reaction (7-17).

$$H_3C-C\equiv C-H + NaNH_2 \longrightarrow H_3C-C\equiv C^- Na^+ + NH_3 \qquad (7\text{-}17)$$

Propyne ($pK_a = 25$)

Ammonia ($pK_a = 38$)

As shown in the reaction in Reaction (7-17), in order to remove the proton from propyne, a strong enough base is needed to ensure that the equilibrium lies to the right. Sodium amide ($NaNH_2$) is a strong enough base to remove the proton from an alkyne since the pK_a of the conjugate acid of $NaNH_2$ is 38.

Problem 7.6

Using the pK_a table, predict appropriate bases that can be used to complete the following acid–base reactions.

(a) [cyclohexene with H] —Base?→ [cyclohexenyl anion]

(b) [isopropyl alkyne with H] —Base?→ [isopropyl acetylide anion]

7.5.3 Resonance

As mentioned in Chapter 1, resonance structures are equivalent Lewis dot structures and represent the movement of nonbonded electrons or pi (π) electrons across a number of adjacent atoms. Resonance structures for charged species also indicate the degree of possible charge delocalization within an ion, and hence the degree of stabilization of an ion. The greater the number of resonance structures there are for an ion, the greater is the stability of that ion. Hence, a conjugate base, which is typically negatively charged, that has a number of resonance structures is more stable than another similar conjugate base, which does not have as many resonance structures. The conjugate base of phenol has many more resonance structures, compared to the conjugate base of cyclohexanol, as shown in Reactions (7-18) and (7-19).

[Phenol ionization showing resonance-stabilized phenoxide anion with four resonance structures] (7-18)

Resonance stabilized anion (conjugate base)

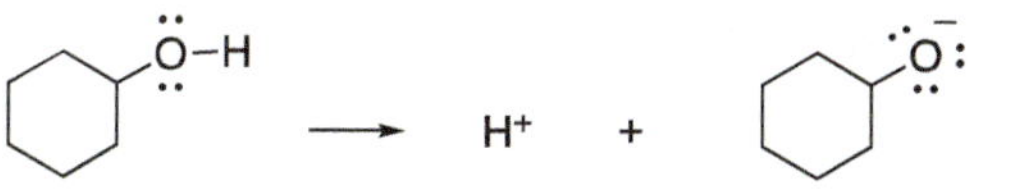

7.5 Factors That Affect Acid and Base Strengths

$$C_6H_{11}OH \longrightarrow H^+ + C_6H_{11}O^- \quad \text{(Localized anion, conjugate base)} \tag{7-19}$$

Since the conjugate base of phenol is resonance stabilized, it is more stable than the conjugate base of cyclohexanol, which is not resonance stabilized. Thus, phenol is a stronger acid (resulting in a weaker conjugate base) than cyclohexanol, which has a stronger conjugate base, as shown in the equilibrium reaction given in Reaction (7-20), which is to the right.

$$\text{PhOH} \; (pK_a = 10) + \text{C}_6\text{H}_{11}\text{O}^-\text{Na}^+ \rightleftharpoons \text{PhO}^-\text{Na}^+ + \text{C}_6\text{H}_{11}\text{OH} \; (pK_a = 16) \tag{7-20}$$

Problem 7.7

Of the pairs of molecules shown below, (a) remove the proton from the most acidic molecule and draw another resonance of the conjugate base; (b) which is the stronger acid?

(a) 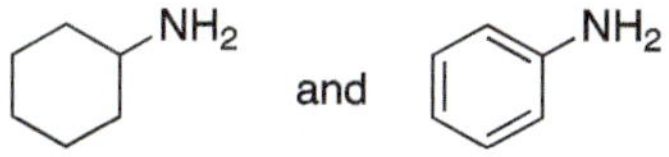 cyclohexyl-NH$_2$ and phenyl-NH$_2$ (b) 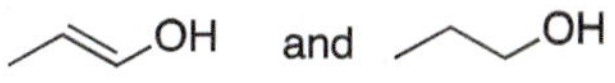 CH$_2$=CH-CH$_2$-OH and CH$_3$-CH$_2$-CH$_2$-OH

As shown in Reaction (7-20) and Table 7.1, the pK_a values of aliphatic alcohols are approximately 15–16, and as a result, very strong bases are required to abstract the acidic hydrogen bonded to the oxygen. One of the strongest bases in organic chemistry is sodium hydride, which can react with alcohols to form the conjugate alkoxide ion of the alcohol as shown in Reaction (7-21).

$$\underset{\text{Cyclohexanol } (pK_a = 16)}{\text{C}_6\text{H}_{11}\text{O-H}} + \underset{\text{Sodium hydride}}{\text{NaH}} \rightleftharpoons \underset{\text{Localized anion (conjugate base)}}{\text{C}_6\text{H}_{11}\text{O}^-\text{Li}^+} + \underset{\text{Hydrogen } (pK_a = 35)}{\text{H}_2} \tag{7-21}$$

Compared to the hydrogens of alkanes, the hydrogens bonded to the α-carbon of carbonyl compounds are fairly acidic. The reason for the higher acidity (lower pK_a values) of the hydrogens bonded to the α-carbon of carbonyl compounds is due to the stability of the conjugate base that results upon deprotonation as shown in Reaction (7-22).

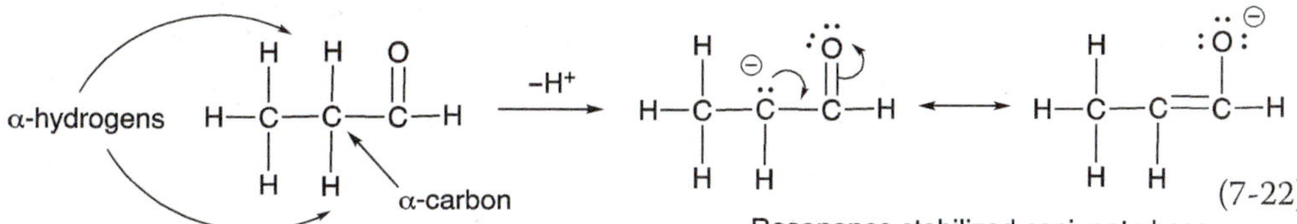

$$\text{α-hydrogens: H-C(H)(H)-C(H)(H)-C(=O)-H} \xrightarrow{-H^+} \text{H-C(H)-C(H)^--C(=O)-H} \longleftrightarrow \text{H-C(H)-C(H)=C(O^-)-H} \tag{7-22}$$

Resonance stabilized conjugate base

As shown in the reaction in Reaction (7-22), the conjugate base is resonance stabilized and hence fairly stable resulting in pK_a values of most ketones and aldehydes of approximately 20 as shown in Table 7.1. The conjugate bases of carbonyl compounds are called enolates since the protonation of one of the resonance structures results in enols. The removal of the α-hydrogen requires a strong enough base in which its conjugate acid's pK_a would have to be greater than 20. The conjugate base of water (the hydroxide anion) is often used to deprotonate the α-hydrogen of carbonyl compounds even though the pK_a of the conjugate acid (water) is around 15. Since the pK_a values are fairly close, an equilibrium exists and slightly favors the neutral carbonyl compound, but a mixture of the acids and conjugate bases will exist as shown in Reaction (7-23). The use of a stronger base would favor the enolate formation.

$$H_3C-\underset{Ketone\ (pK_a=20)}{\underset{\|}{\overset{:O:}{\overset{\|}{C}}}}-CH_3\ +\ NaOH\ \rightleftharpoons\ H_2C-\underset{Resonance\ stabilized\ conjugate\ base}{\overset{:\overset{..}{O}:^{\ominus}}{\overset{\|}{C}}}-CH_3\ \longleftrightarrow\ H_2C=\underset{}{\overset{:\overset{..}{O}:^{\ominus}\ Na^{\oplus}}{\overset{|}{C}}}-CH_3\ +\ \underset{Water\ (pK_a=15)}{H_2O}$$

(7-23)

Problem 7.8

Give the structure of the conjugate bases that would result from the following acid–base reactions.

(a) cyclohexanone + Base ⟶ ? + Base-H⁺

(b) $H_3C\overset{O}{\overset{\|}{C}}CH_3$ + Base ⟶ ? + Base-H⁺

(c) 2-methylcyclopentanone + Base ⟶ ? + Base-H⁺

(d) 2,4-pentanedione + Base ⟶ ? + Base-H⁺

7.5.4 Polarizability/Atom Size

The size of the atom, which bears the negative charge of a conjugate base, is also an important factor in the determination of the stability of the conjugate base, and hence, the acidity of the conjugate acid. A large atom can best accommodate a negative charge compared to a smaller atom. This concept, also known as hard and soft bases, was discussed earlier in the chapter. For the two equilibria shown in Reaction (7-24) and (7-25), note that iodine is the larger atom, compared to chlorine, and as a result, the iodide anion can accommodate the negative charge of the conjugate base much better than the smaller chloride anion.

$$\underset{Acid}{H-I}\ +\ \underset{Base}{H_2O}\ \overset{K}{\rightleftharpoons}\ \underset{Conjugate\ acid}{H_3O^+}\ +\ \underset{Conjugate\ base}{I^-} \qquad (7\text{-}24)$$

$$\underset{Acid}{H-Cl}\ +\ \underset{Base}{H_2O}\ \overset{K}{\rightleftharpoons}\ \underset{Conjugate\ acid}{H_3O^+}\ +\ \underset{Conjugate\ base}{Cl^-} \qquad (7\text{-}25)$$

The ability of a large atom to accommodate a charge is based on the fact that it has more electrons that can be polarized (or moved about) to accommodate a negative charge.

Problem 7.9

Of the following pairs of acids, which is the stronger acid? Explain your answer.

a) HBr and HCI
b) HI and HCl

7.5.5 Inductive Effect

If the conjugate base of an acid contains an electronegative atom, then the negatively charged conjugate base can be stabilized by an electronegative atom that is close to the negative charge. From the periodic table, electronegativity of the atoms increases across a period from left to top right and decreases down a family or group. Thus, the carboxylate salt of a conjugate carboxylic acid that has a fluorine atom bonded in the α-position is more stable than the carboxylate salt that has a less electronegative chlorine atom bonded to the same α-carbon. As a result, a carboxylic acid that has a very electronegative atom bonded to the α-position is more acidic than another similar carboxylic acid that has a less electronegative atom bonded in the same α-position, this concept is illustrated below.

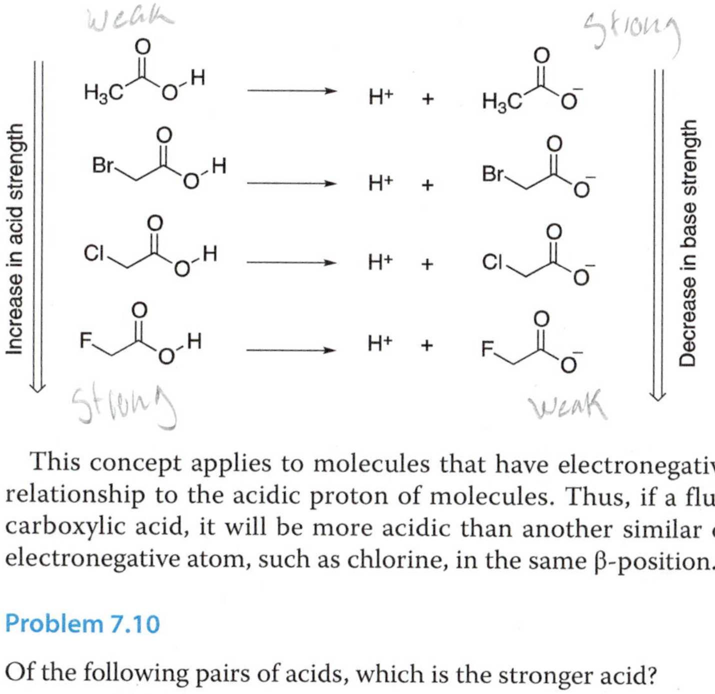

This concept applies to molecules that have electronegative atoms in other positions in relationship to the acidic proton of molecules. Thus, if a fluorine is in the β-position of a carboxylic acid, it will be more acidic than another similar carboxylic acid that has a less electronegative atom, such as chlorine, in the same β-position.

Problem 7.10

Of the following pairs of acids, which is the stronger acid?

(a) 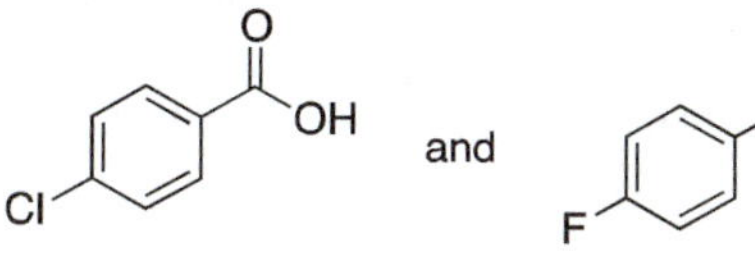 4-chlorobenzoic acid and 4-fluorobenzoic acid
(b) 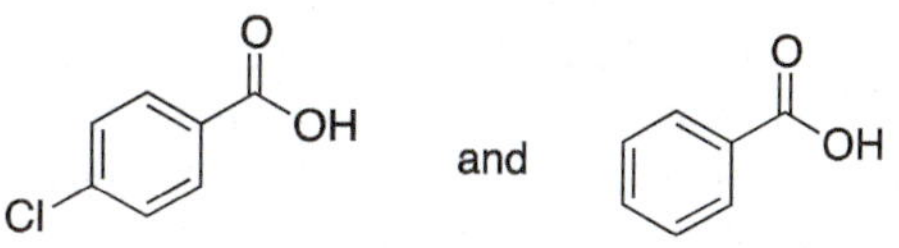 4-chlorobenzoic acid and benzoic acid

Of course, if the same electronegative atom resides at different locations on different conjugate bases, the more stable (weaker) conjugate base is the one that has the electronegative atom closest to the negative charge of the conjugate base. Thus, 2-fluorobutanoic acid is a stronger acid than 4-fluorobutanoic acid.

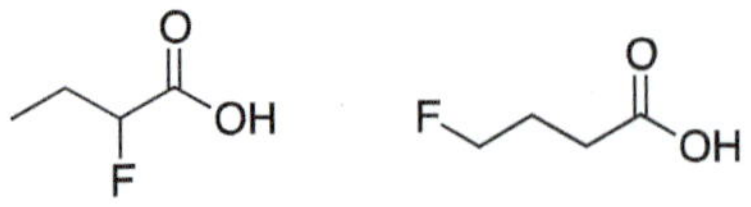

[Structure: 2-fluorobutanoic acid] **Stronger acid**   [Structure: 4-fluorobutanoic acid] **Weaker acid**

The same is true for substituted benzoic acids, 2-fluorobenzoic acid is a stronger acid than 4-fluorobenzoic acid.

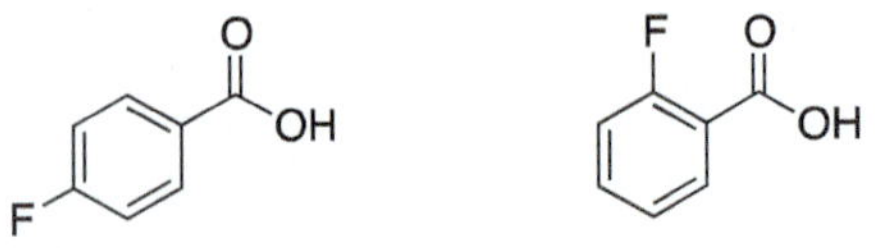

[Structure: 4-fluorobenzoic acid] **Weaker acid**   [Structure: 2-fluorobenzoic acid] **Stronger acid**

Problem 7.11

Of the two pairs of acids shown below, determine which is the stronger acid.

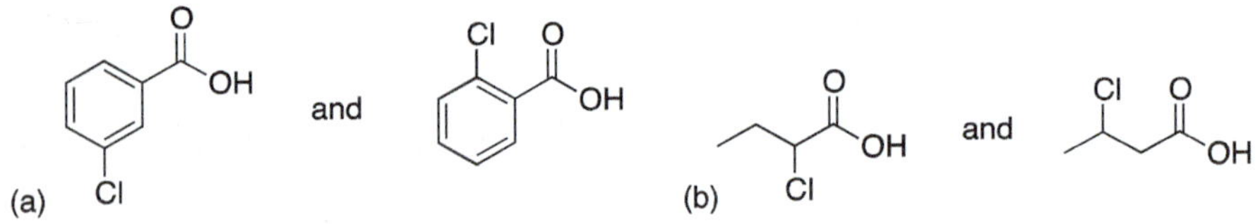

(a) [3-chlorobenzoic acid] and [2-chlorobenzoic acid] (b) [2-chlorobutanoic acid] and [3-chlorobutanoic acid]

7.6 Applications of Acid–Bases Reactions in Organic Chemistry

Some of the strong bases encountered in general chemistry include sodium hydroxide, potassium hydroxide, sodium carbonate, and so on, but in organic chemistry, much stronger bases are required to remove acidic hydrogens of some organic compounds. Based on the information contained in the pK_a table, it is possible to predict strong conjugate bases, and it is obvious that one of the strongest organic bases would be the conjugate base of an alkane, since it is one of the weakest organic acids. Reaction (7-26) shows the conjugate base that results from the deprotonation of methane.

$$CH_3\text{-}H \;\rightleftharpoons\; {:}CH_3^{\ominus} \;+\; H^+ \tag{7-26}$$

Weak acid, pK_a = 50   Very strong conjugate base

It turns out that there is not a base strong enough to carry out the deprotonation as shown above to generate the conjugate base of an alkane. As a result, the conjugate base has to be synthesized indirectly. Victor Grignard, a French chemist (1871–1935), was one of the first chemists to recognize that compounds of the type shown in Reaction (7-27), could be synthesized by reacting an alkyl halide with magnesium. He received the Nobel prize in chemistry, along with Paul Sabatier, for this discovery.

$$H_3C\text{-}H_2C\text{-}Cl \;+\; Mg \;\xrightarrow{\text{Diethylether}}\; H_3C\text{-}H_2\overset{\delta-}{C}\text{-}\overset{\delta+}{MgCl} \tag{7-27}$$

Chloroethane   Organomagnesium

A close inspection of the product of this reaction, an organomagnesium, shows that the carbon is bonded to magnesium, which is a polar covalent bond in which the bonding electrons reside mostly on the carbon since magnesium is more electropositive. Reagents of this type are essentially the conjugate bases of alkanes, and this procedure can be used to synthesize a wide variety of organomagnesium (or very strong conjugate bases of alkanes). Since Victor Grignard was the first to discover this type of reaction, this type of reaction of an alkyl halide with magnesium is known as the Grignard reaction and the product is the organomagnesium, which is also known as a Grignard reagent. The reaction of a Grignard reagent with phenol is shown in Reaction (7-28).

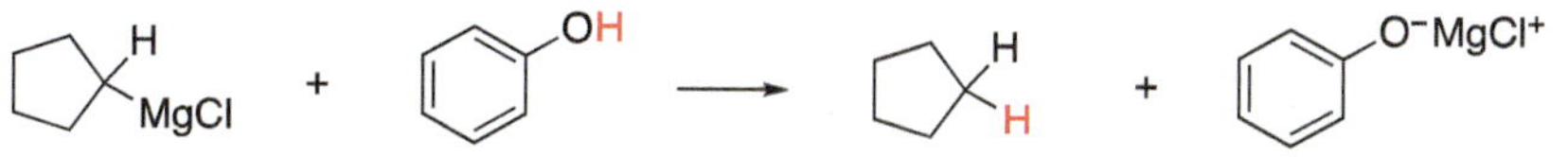

$$\text{cyclopentyl-MgCl} + \text{PhOH} \longrightarrow \text{cyclopentane} + \text{PhO}^-\text{MgCl}^+ \quad (7\text{-}28)$$

Since this discovery, many different organometallic compounds have been synthesized using different metals. Shown below are some commonly used organometallic reagents, including an example of the Grignard reagent, these are all extremely strong bases.

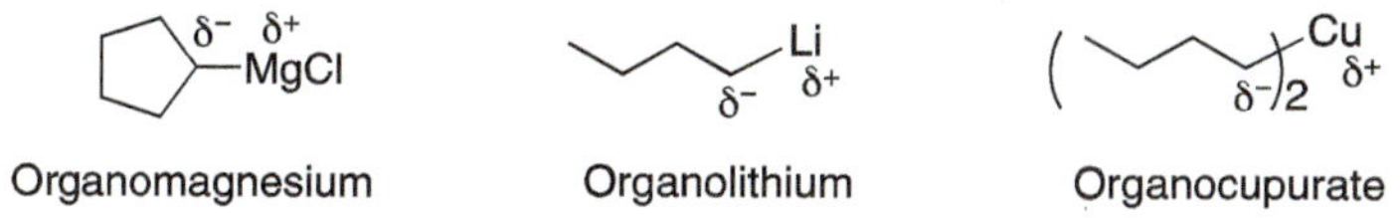

Organomagnesium   Organolithium   Organocupurate

Problem 7.12

Give the products for each of the following acid–base reactions.

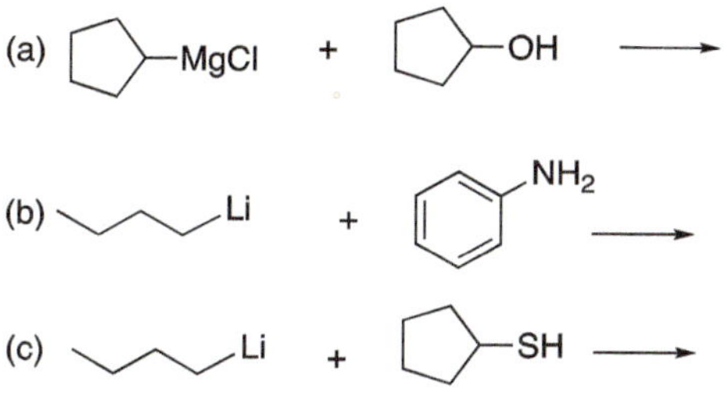

(a) cyclopentyl–MgCl + cyclopentyl–OH ⟶

(b) butyl–Li + PhNH$_2$ ⟶

(c) butyl–Li + cyclopentyl–SH ⟶

The conjugate base of ammonia is also a very strong base that is commonly used in organic chemistry, and its structure is shown in Reaction (7-29).

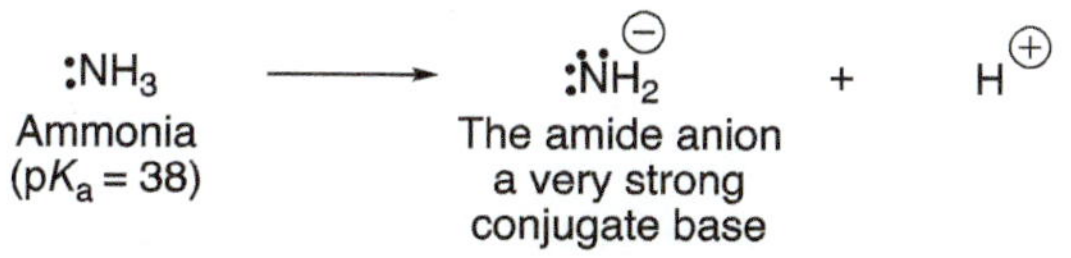

$$:NH_3 \longrightarrow :NH_2^{\ominus} + H^{\oplus} \quad (7\text{-}29)$$

Ammonia (pK$_a$ = 38) → The amide anion, a very strong conjugate base

Since the conjugate base is an anion and there is a charge on the nitrogen atom, it is known as a nitrogen base; shown below are two commonly used nitrogen bases in organic chemistry.

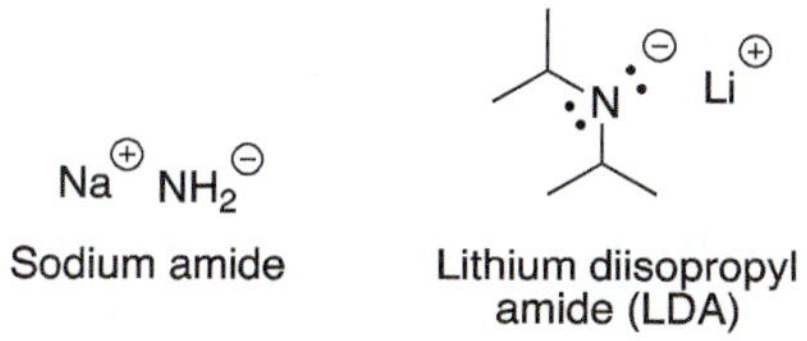

Na$^\oplus$ NH$_2^\ominus$
Sodium amide

Lithium diisopropyl amide (LDA)

Another category of strong bases that are used in organic chemistry is the conjugate bases of hydrogen, the hydride anion, as shown in Reaction (7-30).

$$H_2 \longrightarrow :H^{\ominus} + H^{\oplus} \tag{7-30}$$

Hydrogen (pK_a = 35) — The hydride anion, a very strong conjugate base

Shown below are examples of hydrogen bases that are routinely used in organic chemistry.

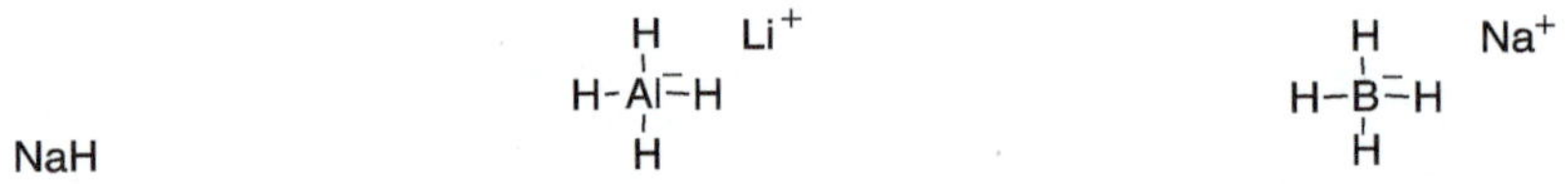

NaH — Sodium hydride

Lithium aluminum hydride (LiAlH$_4$)

Sodium borohydride (NaBH$_4$)

Shown in Reaction (7-31) is an example of a reaction in which LiAlH$_4$ is used as a base.

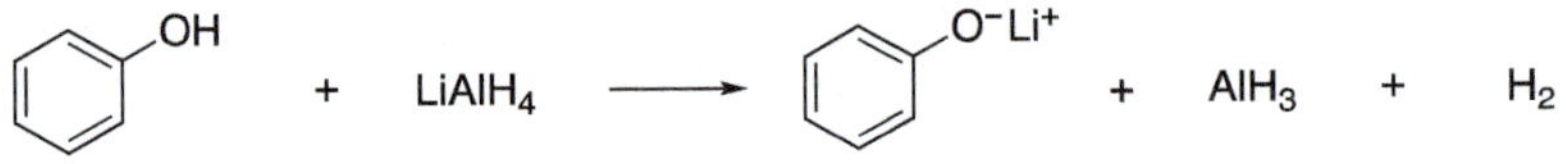

$$Ph\text{-}OH + LiAlH_4 \longrightarrow Ph\text{-}O^-Li^+ + AlH_3 + H_2 \tag{7-31}$$

Problem 7.13

Give the products for each of the following acid–base reactions.

(a) cyclohexyl-SH + LiAlH$_4$ ⟶

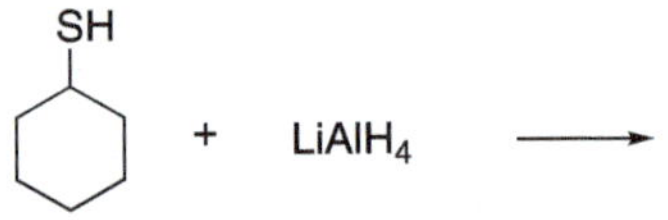

(b) PhCH$_2$– (benzyl) + LiAlH$_4$ ⟶

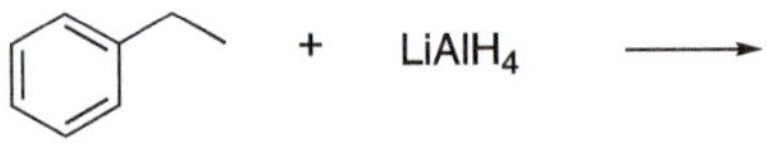

(c) H$_3$C–C≡CH + NaH ⟶

Another category of strong bases that are commonly used in organic chemistry are the conjugate bases of alcohols, as shown in Reaction (7-32).

$$RO\text{-}H \longrightarrow RO:^{\ominus} + H^{\oplus} \tag{7-32}$$

Alcohol (pK_a = 35) — The alkoxide anion, a very strong conjugate base

Shown below are some oxygen bases that are routinely used in organic chemistry.

CH$_3$O$^\ominus$ K$^\oplus$
Potassium methoxide
(MeO$^-$K$^+$)

(CH$_3$)$_3$C-O$^\ominus$ K$^\oplus$
Potassium *tert*-butoxide
(*t*-BuO$^-$K$^+$)

Using the hard–soft description mentioned in the previous chapter, potassium methoxide is relatively small in size, hence a hard base; on the other hand, potassium *tert*-butoxide is relatively large and classified as a softer base, an example of an acid–base reaction in which an oxygen base is used is shown in Reaction (7-33).

$$\text{PhCH}_2\text{H} + t\text{-BuO}^-\text{K}^+ \longrightarrow \text{PhCH}^-\text{H} \; \text{K}^+ + t\text{-BuOH} \tag{7-33}$$

Problem 7.14

Give the products for each of the following acid–base reactions.

(a) $H_3C-C{\equiv}CH \; + \; MeO^-K^+ \longrightarrow$

(b) cyclopentane-1,3-dione $+ \; t\text{-BuO}^-K^+ \longrightarrow$

(c) cyclohexanone $+ \; MeO^-K^+ \longrightarrow$

Knowledge of pK_a values and acid strengths is necessary for choosing appropriate bases for the deprotonation of just about any organic molecule. It is very important that a strong enough base is used for appropriate deprotonation. Extremely strong bases are often very expensive and very reactive. At times, it may be necessary to use a cost-effective weaker base, which can accomplish the same deprotonation task. Curved arrows are typically used to show how the unshared pair of electrons of the base are used to abstract the proton from an organic acid. An example of the curved arrow formulism is shown in Reaction (7-34), in which ammonia abstracts a proton from acetic acid.

$$\underset{\substack{\text{Acetic acid}\\(pK_a = 4.8)}}{H_3C-COOH} + \underset{\text{Ammonia}}{:NH_3} \longrightarrow \underset{\text{Acetate anion}}{H_3C-COO^-} + \underset{\substack{\text{Ammonium}\\\text{cation}\\(pK_a = 9.4)}}{H-\overset{+}{N}H_3} \tag{7-34}$$

Note that in this notation, a curved arrow with two barbs indicates the movement of two electrons from the ammonia that abstracts the proton from acetic acid. Also, note that the electrons from the ammonia (in the reactant) form a single bond with the proton (H⁺) that was abstracted to form the ammonium ion in the product, which has a formal charge of positive one (+1). Also, note that since acetic acid has lost a proton (H⁺), but keeps the bonding electron, it acquires a formal charge of negative one (−1) to form the acetate ion in the product.

Since the double bond of an alkene consists of a pair of π (pi) electrons, alkenes can be considered as Lewis bases and can react with an acid, such as HCl, as shown in Reaction (7-35).

$$\underset{\text{Lewis base}}{\text{cyclohexene}} + \underset{\text{Acid}}{H-Cl} \longrightarrow \underset{\text{Conjugate acid}}{\text{cyclohexyl cation}} + \underset{\text{Conjugate base}}{Cl^-} \tag{7-35}$$

The conjugate acid (also known as a carbocation) that is shown in the reaction of Reaction (7-35) is sp² hybridized and has an empty p orbital; it can accept a pair of electrons, and hence, a Lewis acid. As a result, carbocations can react with a base in an acid–base reaction as shown in the reaction of Reaction (7-36).

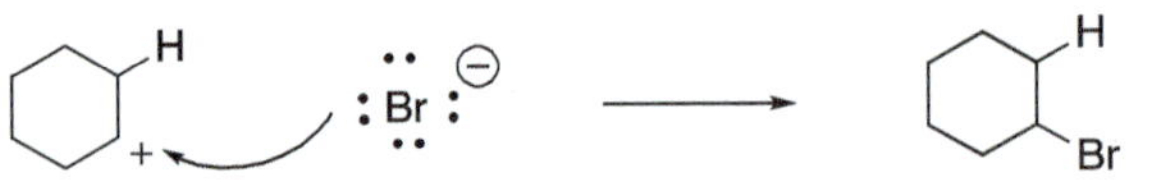

$$\text{Lewis acid} + \text{Lewis base} \longrightarrow \text{Product} \quad (7\text{-}36)$$

Another example in the use of an appropriate base for an acid–base reaction includes the deprotonation of the α-hydrogen of carbonyl compounds, which is shown in the reaction in Reaction (7-37).

$$\text{Cyclohexanone} \ (pK_a \sim 19) + \text{Lithium diisopropyl amide (LDA)} \longrightarrow \text{Enolate} + \text{Diisopropylamine} \ (pK_a \sim 36) \quad (7\text{-}37)$$

The introduction of a second carbonyl group adjacent to the α-carbon makes the α-hydrogens more acidic, compared to the carbonyl compound given in Reaction (7-37). As a result, a weaker base can be used to remove the proton to form the conjugate base as shown in Reaction (7-38).

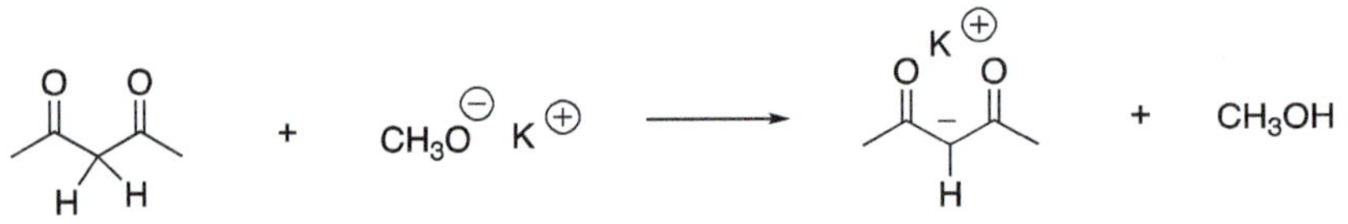

$$\beta\text{-Ketopentanone} \ (pK_a \sim 9) + \text{Potassium methoxide (MeO}^-\text{K}^+\text{)} \longrightarrow \text{Enolate} + \text{Methanol} \ (pK_a \sim 16) \quad (7\text{-}38)$$

Esters are another category of compounds that contain at least one α-hydrogen and are acidic compounds. With an appropriate base, deprotonation of an α-hydrogen from esters can occur as shown in Reaction (7-39).

$$\text{Methyl ethanoate} \ (pK_a = 19) + \text{Potassium methoxide (MeO}^-\text{K}^+\text{)} \longrightarrow \text{Ester enolate} + \text{Methanol} \ (pK_a = 16) \quad (7\text{-}39)$$

The conjugate base products of the reactions given in the examples above are very important compounds for the synthesis of larger organic molecules, and they will be used in later chapters.

End of Chapter Problems

7.15 Arrange the bases shown below in terms of base strengths, i.e. the most basic first (pK_a values of the conjugate acids are shown in parenthesis).

Cl^- (−7); CH_3O^- (15.5); F^- (3.2); CN^- (9.1)

7.16 For the acids shown below of which the pK_a values are given in parenthesis, answer the following questions.

HCN (9.31); CH$_3$CH$_2$OH (15.9); HCl (−7); CH$_3$COOH (4.75); CH$_3$NO$_2$ (10.21)

a) Arrange the acids in order of decreasing acidity, i.e. most acidic first.
b) Give the structure of the conjugate base for each acid.
c) Arrange the conjugate bases that you have given in the previous question in order of decreasing basicity, i.e. most basic conjugate base first.

7.17 Of the following acids, of which pK_a values of the acids are shown in parenthesis, determine which has the weakest conjugate base? Explain your answer.

HF (2.55); H$_2$O (15.9); HCN (9.31); HI (−9.5)

7.18 Without the aid of a pK_a table, determine which of the following two acids is the strongest acid: ClCH$_2$COOH and FCH$_2$COOH, explain your answer.

7.19 Of the pairs of acids shown below, which is the strongest acid, briefly explain your answer?

(i) 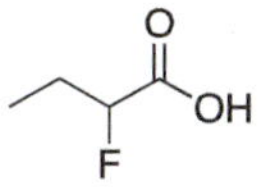 [structure] and [structure]
(ii) 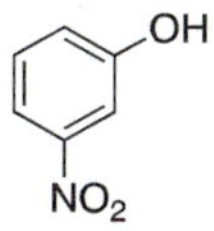 [structure] and [structure]
(iii)  [structure] and [structure]
(iv) 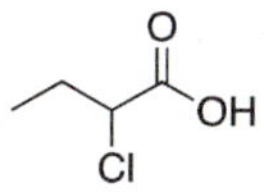 [structure] and [structure]

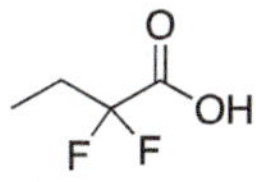

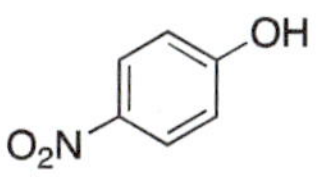

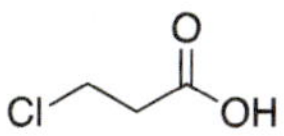

7.20 Consider the molecules shown below, then answer the following questions.

(i)  [structure] (ii) 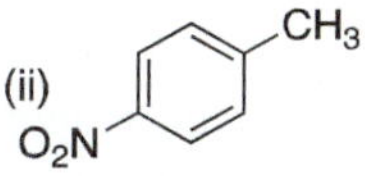 [structure] (iii) 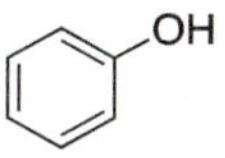 [structure] (iv) 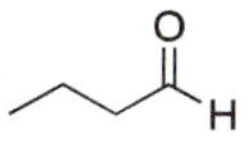 [structure] (v) H$_3$C−C≡N

a) For each molecule, remove the most acidic hydrogen and draw the structure of the resulting conjugate base.
b) For each of the conjugate base that you have drawn, give as many resonance structures as you can.
c) Of the resonance structures that you have drawn in the above question, determine which contributes the most to the overall structure of each conjugate base.

7.21 For each of the following sets of compounds, indicate which is the weaker base and give a brief explanation for your choice.

a) CH$_3$COO$^−$; ClCH$_2$COO$^−$; Cl$_2$CHCOO$^−$; CCl$_3$COO$^−$
b) Cl$^−$; CN$^−$; CH$_3$O$^−$; CH$_3^−$
c) OH$^−$; CN$^−$; Br$^−$; NH$_2^−$

7.22 For the reactions shown below, which is the Lewis acid and which is the Lewis base?

:NH$_3$ + AlCl$_3$ ⟶ H$_3$N$^{(+)}$−AlCl$_3^{(−)}$

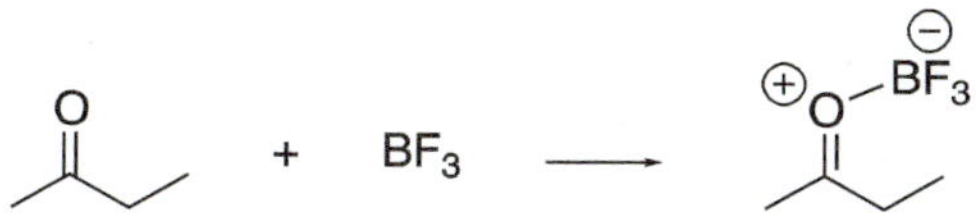

[structure] + BF$_3$ ⟶ [structure]

7.23 Naringenin is a flavonoid, which has many hydrogens. The structure of Naringenin is shown below. Identify all hydrogens and determine their relative acidities.

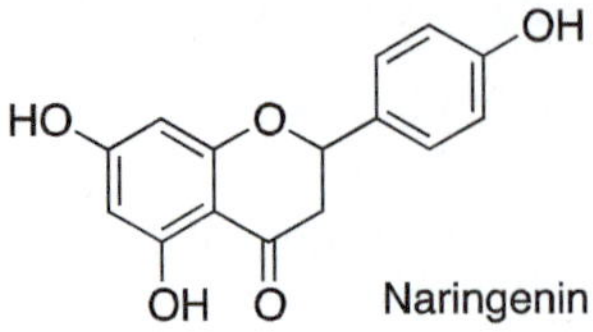

Naringenin

7.24 Complete the following reactions for supplying the products, the solvent for each reaction is shown below the arrow.

(a) [indane-1,3-dione] $\xrightarrow[\text{Methanol}]{\text{CH}_3\text{O}^{\ominus} \text{ K}^{\oplus}}$

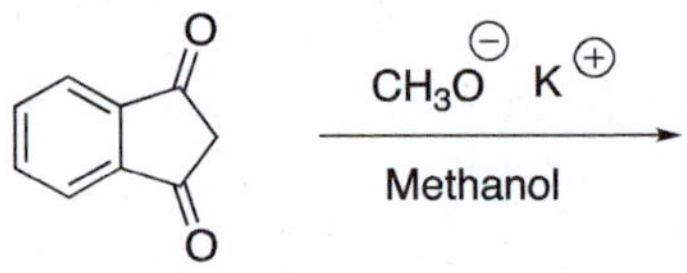

(b) [ethyl acetoacetate-like diester] $\xrightarrow[\text{Ethanol}]{\text{EtO}^-\text{K}^+}$

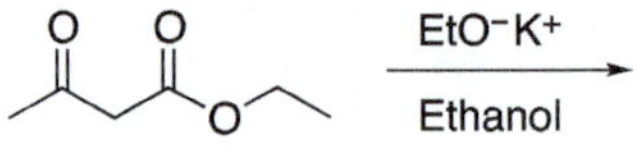

(c) [substituted cyclohexanone with ester side chain] $\xrightarrow[\text{Ethanol}]{\text{EtO}^-\text{K}^+}$

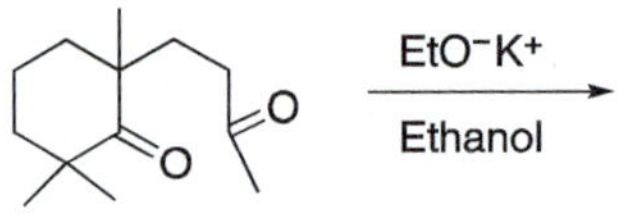

7.25 Organometallic reagents, such as the Grignard reagent, are extremely strong bases, and extreme care must be used in the selection of an appropriate solvent when using these reagents. For example, protic solvents, such as alcohols, are not good solvents since they are readily deprotonated in the presence of an organometallic reagent. Tetrahydrofuran is a good solvent when using these reagents, yet a very similar reagent, 2,2,5,5-tetramethyltetrahydrofuran, is a poor solvent. Explain this observation.

Tetrahydrdofuran 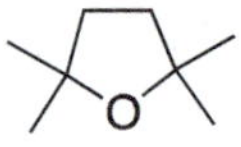   2,2,5,5-Tetramethyltetrahydrofuran

8

Addition Reactions Involving Alkenes and Alkynes

8.1 Introduction

The word addition when used in chemistry implies the addition of atom(s), ion(s), or molecule(s) to a reactant molecule to form a new molecule as product or an intermediate, which eventually goes on to form a final product. Two examples of addition reactions that were briefly mentioned in Chapter 6 are shown in Reactions (8-1) and (8-2).

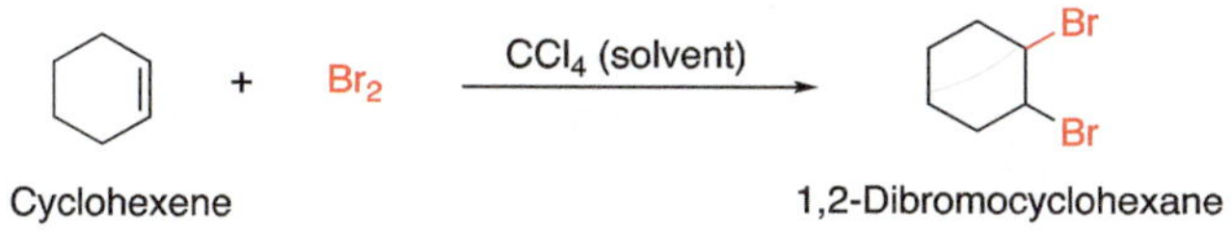

$$\text{Cyclohexene} + \text{Br}_2 \xrightarrow{\text{CCl}_4 \text{ (solvent)}} \text{1,2-Dibromocyclohexane} \quad (8\text{-}1)$$

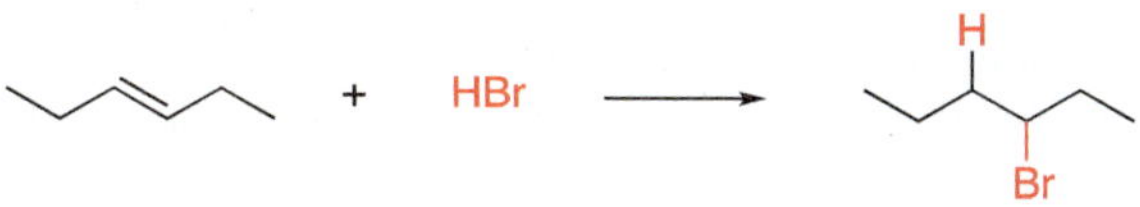

$$\text{CH}_3\text{CH}=\text{CHCH}_2\text{CH}_3 + \text{HBr} \longrightarrow \text{CH}_3\text{CH}_2\text{CH(Br)CH}_2\text{CH}_3 \quad (8\text{-}2)$$

In this chapter, a much deeper analysis of this type of reaction will be carried out. Also, a detailed mechanism of how addition reactions occur will be carried out, which will assist in the prediction of possible addition products. The concept of acids and bases will be essential in understanding and explaining some of the observations of addition reactions.

8.2 The Mechanism for Addition Reactions Involving Alkenes

Reaction (8-1), (8-2), and the reactions covered in Chapter 6 do not just magically proceed from reactants to products. For most reactions, there are different steps before the products are actually formed; the reactants react to form intermediates and go through various steps before forming the products. This step-by-step description of reactions is also known as the reaction mechanism. Most of the addition reactions that will be examined in this course, take place by going through a number of different steps before the final product is finally formed. In this chapter, the addition of different molecules to alkene will be examined. As shown in Chapter 2, alkenes have the carbon–carbon double bond, which has a pair of pi (π) electrons that can be used to react with a Lewis acid as shown in Reaction (8-3).

Organic Chemistry: Concepts and Applications, First Edition. Allan D. Headley.
© 2020 John Wiley & Sons, Inc. Published 2020 by John Wiley & Sons, Inc.
Companion website: www.wiley.com/go/Headley_OrganicChemistry

$$\underset{\underset{\text{(alkene)}}{\text{Lewis base}}}{\underset{R}{\overset{R}{>}}\!\!=\!\!\underset{R}{\overset{H}{<}}} + \underset{\underset{\text{(electrophile)}}{\text{Lewis acid}}}{E^+} \longrightarrow \underset{\text{Carbocation}}{E\!-\!\underset{R}{\overset{R}{>}}\!\!\overset{+}{-}\!\!\underset{R}{\overset{H}{<}}} + \underset{\text{Carbocation}}{\underset{R}{\overset{R}{>}}\!\!\overset{+}{-}\!\!\underset{R}{\overset{H}{<}}\!-\!E} \qquad (8\text{-}3)$$

New covalent bond

The curved arrow formulism is often used to show electron movements from the Lewis base to the Lewis acid as shown in Reaction (8-3). For these addition types of reactions, there are some specific names that are used to represent the Lewis base and Lewis acids. The carbon–carbon double bond of alkenes is electron rich due to the presence of the pi (π) electrons, and as a result, alkenes will act as a source of electrons in a reaction. This double bond, which is a Lewis base, as shown in the reaction in Reaction (8-3), is also known as a nucleophile. A *nucleophilic* species has at least one pair of electrons available to form a new covalent bond with a Lewis acid; thus, alkenes are nucleophilic. Another term that is often used for the Lewis acid of these reactions is electrophile. An *electrophile*, as the name suggests, is an electron-loving species, which will react with the electron pair of a nucleophile to form a new covalent bond as shown in Reaction (8-3). The proton is easy to identify as an electrophile since it has a vacant orbital to accept electrons and form a new covalent bond with a nucleophile. Hence, Lewis bases and acids that were discussed in the previous chapter are also nucleophiles and electrophiles, respectively. Note that in Reaction (8-3), the pi (π) electrons of the carbon–carbon double bond can form a bond to either side of the initial double bond giving rise to two different intermediate carbocations.

Whenever these reactions are carried out in the lab, the electrophile is a neutral molecule, and not just a cation as shown in Reaction (8-3). The counter ion for an electrophile is an anion, which is also a nucleophile. Thus, the reaction of a neutral electrophile with an alkene will take place in two steps. First is the addition of the electrophile to the nucleophilic double bond to form an electrophile carbocation, and the second is the addition of the nucleophilic counter anion of the electrophile to the electrophilic carbocation intermediate to form a neutral product. The first step of the reaction in this type of reaction mechanism is shown in the reaction in Reaction (8-4).

$$\underset{\text{Alkene}}{\underset{R}{\overset{R}{>}}\!\!=\!\!\underset{R}{\overset{H}{<}}} + \underset{\text{Electrophile}}{E^+\,\ddot{N}u^-} \longrightarrow \underset{\text{Carbocation}}{E\!-\!\underset{R}{\overset{R}{>}}\!\!\overset{+}{-}\!\!\underset{R}{\overset{H}{<}}} + \underset{\text{Carbocation}}{\underset{R}{\overset{R}{>}}\!\!\overset{+}{-}\!\!\underset{R}{\overset{H}{<}}\!-\!E} + \ddot{N}u^- \qquad (8\text{-}4)$$

Note that since the alkene is unsymmetrical, there are two possible carbocations formed, and as a result, there will be two possible products, as shown in the reactions in Reactions (8-5) and (8-6).

$$\underset{\text{Electrophile}}{E\!-\!\underset{R}{\overset{R}{>}}\!\!\overset{+}{-}\!\!\underset{R}{\overset{H}{<}}} + \underset{\text{Nucleophile}}{:Nu^-} \longrightarrow \underset{\text{Final addition product}}{E\!-\!\underset{R}{\overset{R}{>}}\!\!-\!\!\underset{R}{\overset{H}{<}}\!-\!Nu} \qquad (8\text{-}5)$$

$$\underset{\text{Electrophile}}{\underset{R}{\overset{R}{>}}\!\!\overset{+}{-}\!\!\underset{R}{\overset{H}{<}}\!-\!E} + \underset{\text{Nucleophile}}{:Nu^-} \longrightarrow \underset{\text{Final addition product}}{Nu\!-\!\underset{R}{\overset{R}{>}}\!\!-\!\!\underset{R}{\overset{H}{<}}\!-\!E} \qquad (8\text{-}6)$$

Figure 8.1 Energy profile for the hypothetical electrophilic addition reaction.

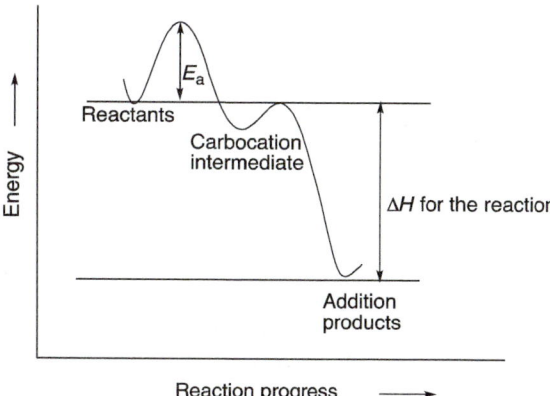

This type of reaction in which electrophiles (typically along with its counter anion) are added to the carbon–carbon double bond of an alkene is called an **electrophilic addition** reaction. The overall general reaction is shown in Reaction (8-7).

$$\underset{R}{\overset{R}{>}}=\underset{R}{\overset{H}{<}} + E^+ : Nu^- \longrightarrow Nu \rightarrow \underset{R}{\overset{R}{>}}\underset{R}{\overset{H}{<}} E + E \rightarrow \underset{R}{\overset{R}{>}}\underset{R}{\overset{H}{<}} Nu \qquad (8\text{-}7)$$

The two steps of the reaction mechanism shown in Reactions (8-5) and (8-6) are called the elementary steps of the reaction mechanism, and these steps are shown in the energy profile diagram given in Figure 8.1. Note that in the energy profile diagram, the relative energies of the reactants, intermediates, and products are shown for each of the elementary steps and that the intermediates are always shown in the middle of the reaction profile diagram.

Energy profile diagrams will be used frequently throughout this course to analyze various types of reactions.

8.3 Addition of Hydrogen Halide to Alkenes (Hydrohalogenation of Alkenes)

8.3.1 Addition Reactions to Symmetrical Alkenes

In this section, the knowledge gained in the previous section will be applied to real molecules to not only predict the organic products, but the major organic product. The first reaction that will be examined is the addition of hydrochloric acid (H-Cl) to cyclohexene as shown in Reaction (8-8).

Recall that the H-Cl bond is a polar covalent bond and due to chlorine being more electronegative, compared to hydrogen, the bonding electrons are closer to the chlorine, compared to the hydrogen. Hence, the hydrogen becomes the electrophile and the counter chloride anion is a nucleophile. Thus, the first step of the reaction mechanism will involve the addition of the electrophilic H⁺ to the carbon–carbon double bond of cyclohexene as shown in the elementary step of Reaction (8-9) to form a carbocation.

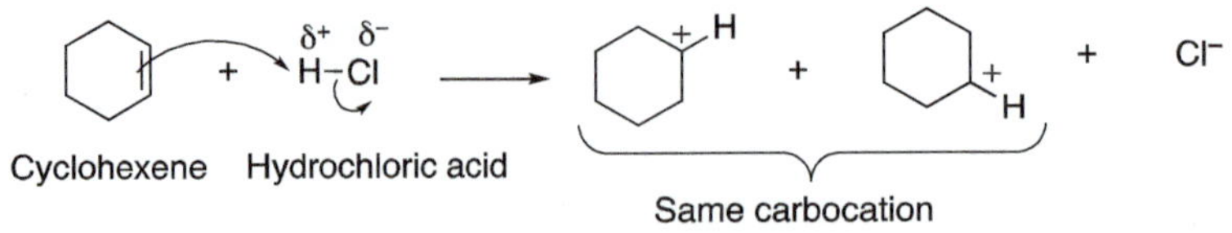

$$\text{Cyclohexene} + \overset{\delta+}{H}\overset{\delta-}{-}Cl \longrightarrow \underbrace{\text{carbocation}^+ + \text{carbocation}^+}_{\text{Same carbocation}} + Cl^- \tag{8-9}$$

Note that since the alkene is symmetrical, the same carbocation results after the addition of the proton to either carbon of the carbon–carbon double bond. As a result, only one organic product will be formed from the second step in the mechanism, where the nucleophilic chloride anion adds to the carbocation, as shown in Reactions (8-10) and (8-11).

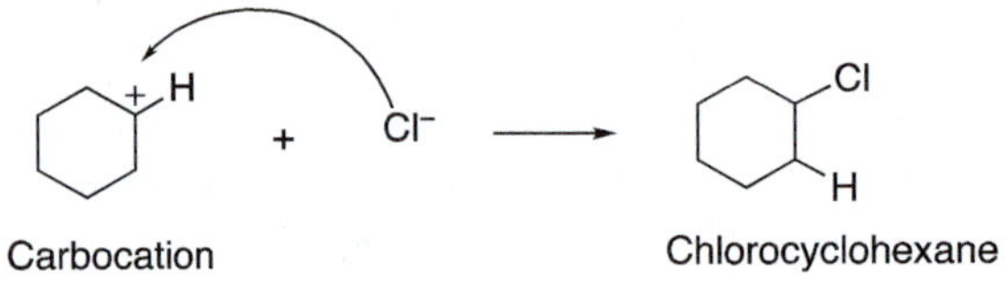

$$\text{Carbocation} + Cl^- \longrightarrow \text{Chlorocyclohexane} \tag{8-10}$$

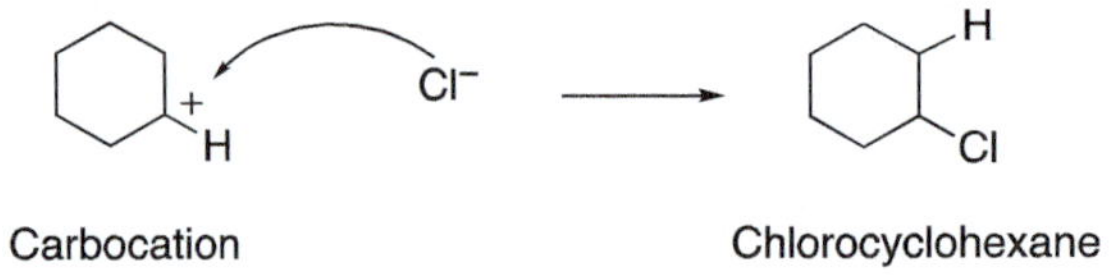

$$\text{Carbocation} + Cl^- \longrightarrow \text{Chlorocyclohexane} \tag{8-11}$$

Problem 8.1

i) Give the products for the reactions shown below:

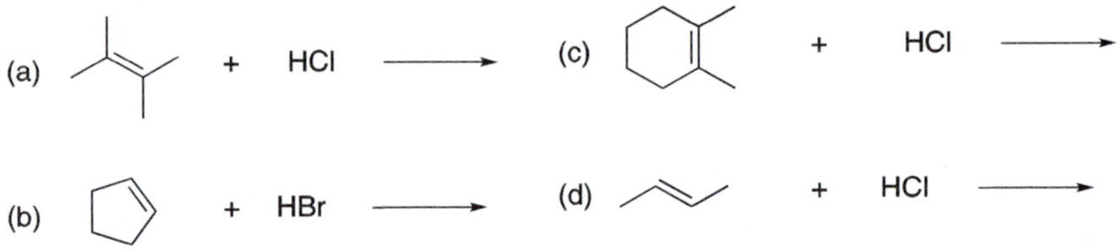

(a) (CH₃)₂C=C(CH₃)₂ + HCl ⟶

(b) cyclopentene + HBr ⟶

(c) 1-methylcyclohexene + HCl ⟶

(d) propene + HCl ⟶

ii) Are there two different or one addition product that results from each of the above reactions? Explain your answer.

8.3.2 Addition Reactions to Unsymmetrical Alkenes

Based on the information presented in the previous section, if the alkene is symmetrical, only one product results from the addition of HCl or similar molecules. On the other hand, if the alkene is not symmetrical, that is there are different groups bonded to the carbons of the double bond, then two possible carbocations result, as shown in Reaction (8-4). The addition of the nucleophilic anion to these carbocations results in two different electrophilic addition products, as shown in Reaction (8-12).

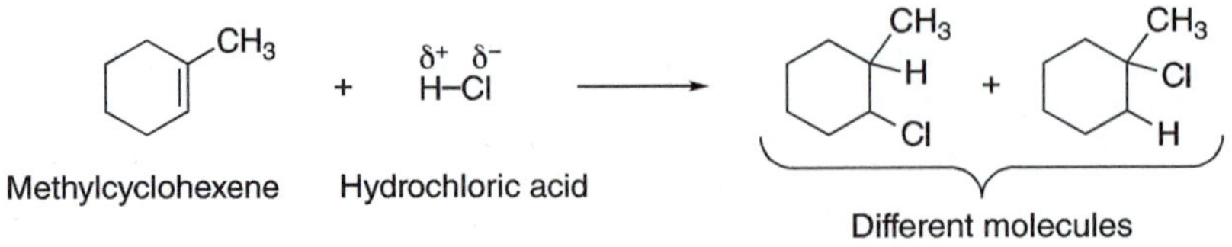

$$\text{Methylcyclohexene} + \overset{\delta+}{H}\overset{\delta-}{-}Cl \longrightarrow \underbrace{\text{product with CH}_3, H, Cl + \text{product with CH}_3, Cl, H}_{\text{Different molecules}} \tag{8-12}$$

One of these products is the major organic product and the method to determine which is the major product will be discussed in the next section.

Problem 8.2

Give the two possible products for the reactions shown below:

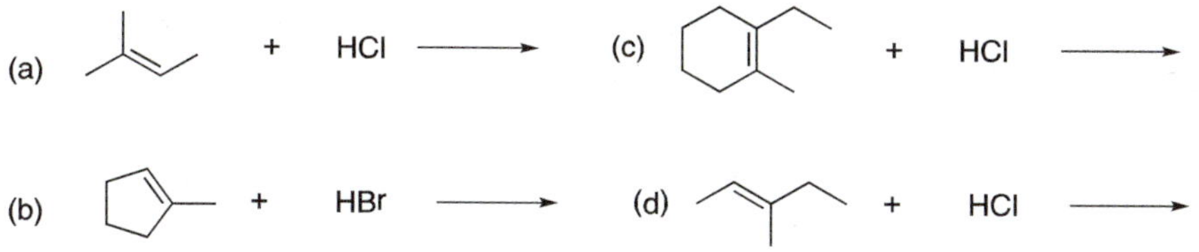

(a) (CH₃)₂C=CH₂ (isobutylene structure) + HCl ⟶

(b) cyclopentene + HBr ⟶

(c) 1-ethylcyclohexene + HCl ⟶

(d) 2-methyl-2-pentene type alkene + HCl ⟶

8.3.3 Predicting the Major Addition Product

For the reaction involving methylcyclohexene and hydrochloric acid (Reaction 8-12), one of the products shown is the major product and the other is the minor product. In order to be able to predict the major organic product, a close examination of the reaction mechanism must be carried out. It should be remembered that in the first step of the mechanism, carbocations are the intermediates that result from the initial addition of the electrophile (H^+) to the nucleophilic double bond (Reactions 8-4 and 8-9). For this reaction however, one of these carbocations is a more stable carbocation than the other and will be formed in greater abundance compared to the less stable carbocation. Thus, a systematic method must be established to determine the relative stabilities of carbocations.

A carbocation that has three alkyl groups bonded to the carbon of the carbocation is more stable than one that has two alkyl groups bonded to the carbon of the carbocation. A carbocation that has three alkyl groups bonded to the carbon of the carbocation is called a tertiary carbocation (3°), whereas a carbocation that has two alkyl groups bonded to the carbon of the carbocation is called a secondary carbocation (2°). Alkyl groups, which are electron rich, can donate electrons to help stabilize the positive charge of a carbocation. Thus, a tertiary carbocation, which has three alkyl groups, is more stable than a secondary carbocation, which has two alkyl groups. Based on this trend, it can be readily concluded that a carbocation that has only one alkyl group bonded to the carbon of the carbocation is called a primary (1°) and is not as stable as a secondary carbocation or a tertiary carbocation. The methyl carbocation is therefore the least stable carbocation. Figure 8.2 gives illustrations of the structure of carbocations and how adjacent alkyl groups can stabilize different types of carbocations.

Thus, the order of the stability of different carbocations is as shown below:

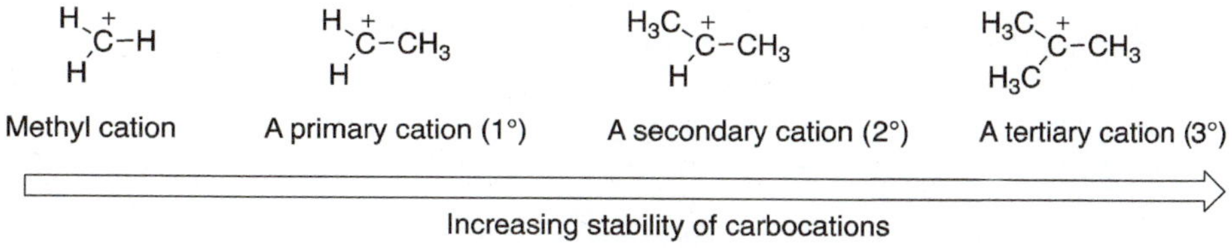

| Methyl cation | A primary cation (1°) | A secondary cation (2°) | A tertiary cation (3°) |

Increasing stability of carbocations ⟶

A different representation of the order of the stability of carbocations is shown below.

$$\text{Tertiary (3°)} > \text{Secondary (2°)} > \text{Primary (1°)} > \text{Methyl (CH}_3\text{)}$$

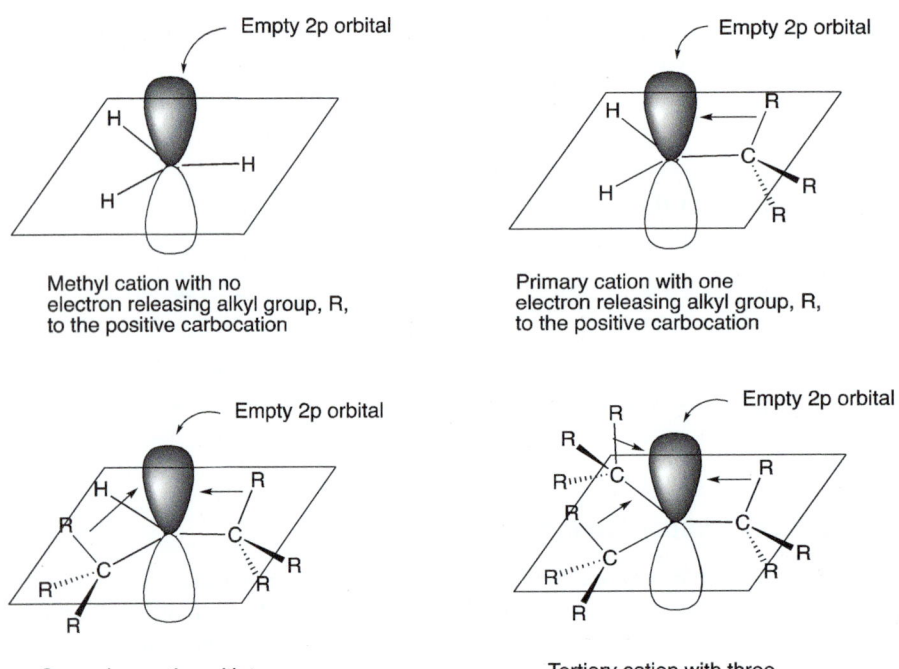

Figure 8.2 Relative stabilities of different types of carbocations.

The addition of an electrophile, such as the proton, to the carbon–carbon double bond of methylcyclohexene results in two possible carbocations, which are shown in Reaction (8-13).

$$\text{Methyl cyclohexene} + \text{H-Cl}\ (\delta^+\ \delta^-) \xrightarrow{-Cl^-} \text{Tertiary carbocation} + \text{Secondary carbocation} \tag{8-13}$$

In the final step of the mechanism, the nucleophilic chloride anion adds to the carbocation to form the products, but since the tertiary carbocation is more stable and formed in greater abundance, compared to the secondary carbocation, the tertiary carbocation will lead to the major organic product, as shown in Reactions (8-14) and (8-15).

$$\text{Tertiary carbocation} + Cl^- \longrightarrow \text{Major product} \tag{8-14}$$

$$\text{Secondary carbocation} + Cl^- \longrightarrow \text{Minor product} \tag{8-15}$$

The overall reaction is shown in Reaction (8-16), and the Figure 8.3 shows the energy profile for the two steps in the addition reaction of hydrochloric acid to methylcyclohexene to give two intermediates, a more stable tertiary carbocation (lower in energy), compared to the secondary carbocation, which is higher in energy.

Figure 8.3 Energy profile for the addition reaction of methylcyclohexene with hydrochloric acid.

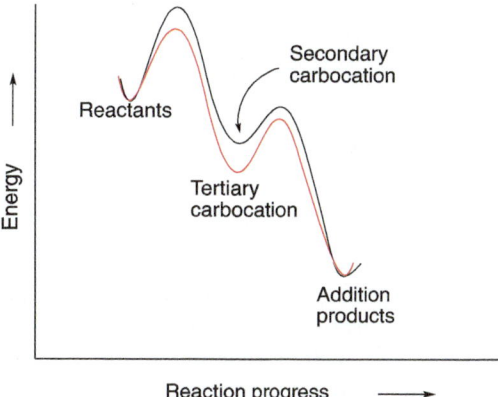

$$\text{Methylcyclohexene} + \text{H-Cl} \longrightarrow \text{Major product} + \text{Minor product} \tag{8-16}$$

This analysis of determining the relative stability of the intermediate carbocation in order to predict the major addition product can be applied to similar addition reactions involving different alkenes. The addition of hydrochloric acid to propene results in two products as shown in Reaction (8-17). During the course of the reaction, a primary and a secondary carbocation are formed, but the major product results via the secondary and more stable carbocation.

$$\text{Propene} + \text{H-Cl} \longrightarrow \text{2-Chloropropane (major product)} + \text{Chloropropane (minor product)} \tag{8-17}$$

The first step of the mechanism involves the addition of the electrophilic proton of the hydrochloric acid to the nucleophilic double bond of the alkene to form two possible carbocations as shown in Reaction (8-18).

$$H_3C\text{-}CH=CH_2 \xrightarrow{H-Cl} H_3C\text{-}\overset{+}{C}H\text{-}CH_3 \text{ (Secondary carbocation)} + H_3C\text{-}CH_2\text{-}\overset{+}{C}H_2 \text{ (Primary carbocation)} + Cl^- \tag{8-18}$$

As pointed out earlier, a primary carbocation is one that has only one alkyl group that is bonded to the carbon of the carbocation. As a result, it is less stable than either a secondary or a tertiary carbocation. Owing to the greater stability of the secondary carbocation, compared to the primary carbocation, it will be in greater abundance and hence lead to the major organic product after the addition to the nucleophilic chloride anion, as shown in Reaction (8-19) with the primary carbocation leading to the minor product, as shown in Reaction (8-20).

$$H_3C\text{-}\overset{+}{C}H\text{-}CH_3 \text{ (Secondary carbocation)} + Cl^- \longrightarrow H_3C\text{-}CHCl\text{-}CH_3 \text{ (Major organic product)} \tag{8-19}$$

$$H_3C-\overset{H}{\underset{H}{\overset{+}{C}}}-CH_2 + Cl^- \longrightarrow H_3C-\overset{H}{\underset{H}{C}}-CH_2\cdot Cl \qquad (8\text{-}20)$$

Primary carbocation   Minor organic product

8.3.4 Predicting the Stereochemistry of Addition Reaction Products

Since the carbon of carbocations is sp^2 hybridized, there is a vacant p orbital, which is orthogonal to the plane of the sigma (σ) bonds. As a result, the nucleophile can attack from either side of the flat carbocation as illustrated in Figure 8.4.

Thus, there will be an equal mixture of enantiomers making the product racemic as shown in Reaction (8-21) if a stereogenic carbon is generated in the product.

$$\text{CH}_2=\text{CHCH}_3 \xrightarrow{\text{HCl}} \underset{\substack{\text{(S)-2-Chlorobutane}\\(50\%)}}{\overset{H_3C}{\underset{Cl}{\overset{|}{\underset{|}{C}}}}{-}CH_2CH_3} + \underset{\substack{\text{(R)-2-Chlorobutane}\\(50\%)}}{\overset{Cl}{\underset{H_3C}{\overset{|}{\underset{|}{C}}}}{-}CH_2CH_3} \qquad (8\text{-}21)$$

8.3.5 Predicting the Major Addition Product – Markovnikov Rule

Markovnikov, a Russian Chemist, was one of the first scientists to predict the major organic product for addition reactions involving the addition of H^+Nu^- across a double bond. He predicted that the major product of these addition reactions is the one in which H^+ adds to the carbon of the alkene double bond that has the most hydrogens. This observation is now known as ***Markovnikov's Rule***. For the reaction shown below, the hydrogen of HCl adds to the carbon with two hydrogens instead of the carbon with only one hydrogen to give the major product.

$$H_3C-CH=CH_2 \xrightarrow{\text{HCl}} \underset{\text{Minor product}}{H_3C-\underset{H}{\overset{|}{C}H}-\underset{Cl}{\overset{|}{C}H_2}} + \underset{\text{Major product}}{H_3C-\underset{Cl}{\overset{|}{C}H}-\underset{H}{\overset{|}{C}H_2}}$$

This carbon has one hydrogen, H will **not** add here

This carbon has two hydrogens, H will add here

For electrophilic addition reactions, the major product results from the addition of the electrophile to the carbon that has the most hydrogens and the nucleophile adds to the carbon

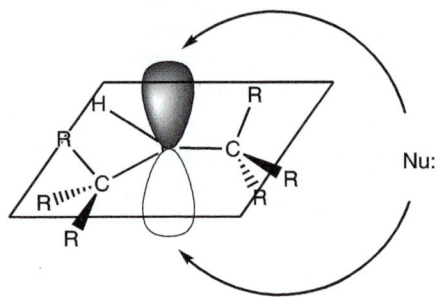

Figure 8.4 Flat carbocation showing equal probability of attack of nucleophile for either side.

8.3 Addition of Hydrogen Halide to Alkenes (Hydrohalogenation of Alkenes)

that has the least hydrogens. As a result, these reactions are known *regiospecific* reactions since the electrophile and the nucleophile add to specific regions of the reactant molecule.

Problem 8.3

Using Markovnikov's rule, predict the major product that results from the addition of HCl to the following alkenes.

(a) 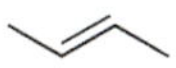 [structure] (b)  [structure] (c) 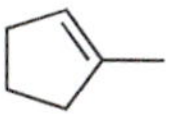 [structure]

8.3.6 Unexpected Hydrohalogenation Products

For a large percentage of the reactions that will be encountered in organic chemistry, there are sometimes products that are obtained that are unexpected. To better understand the formation of unexpected products, it is important to know mechanistic possibilities for the reactions. For electrophilic addition reactions, often times, the occurrence of unexpected organic products is due to the rearrangement of carbocation intermediates to form more stable carbocations. Consider Reaction (8-22).

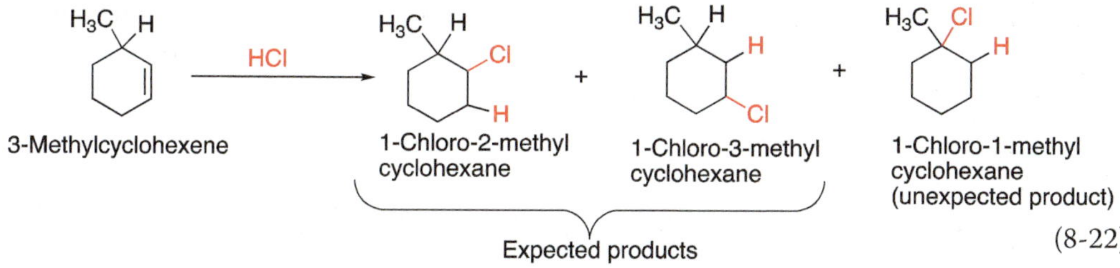

3-Methylcyclohexene + HCl → 1-Chloro-2-methyl cyclohexane + 1-Chloro-3-methyl cyclohexane + 1-Chloro-1-methyl cyclohexane (unexpected product)

Expected products

(8-22)

The first step of the mechanism for this reaction involves the addition of the electrophilic proton to the nucleophilic alkene, as shown in Reaction (8-23).

(8-23)

The secondary carbocation that is formed can be converted to a tertiary carbocation by a hydride shift as shown in Reaction (8-24), which eventually leads to the unexpected organic product.

2° carbocation —1,2-hydride migration→ 3° carbocation → Unexpected hydro halogenation product

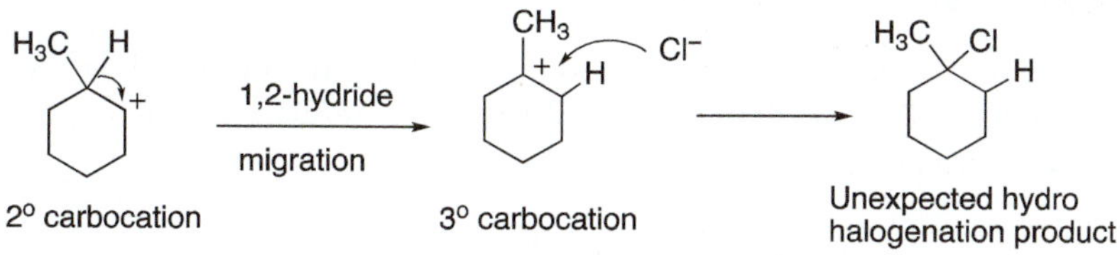

(8-24)

For the addition reaction of hydrochloric acid to 3,3-dimethylcyclohexene, an unexpected product is obtained, as shown in Reaction (8-25).

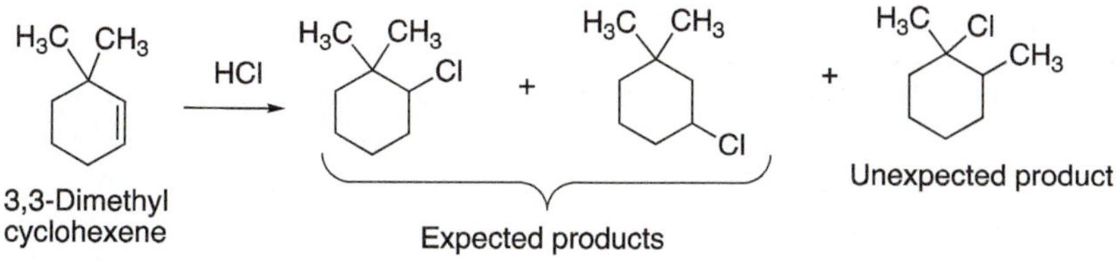

3,3-Dimethyl cyclohexene + HCl → Expected products + Unexpected product

(8-25)

The first step of the mechanism for this reaction is similar to Reaction (8-23) and is shown in Reaction (8-26).

$$\text{1,1-dimethylcyclohexene} + H-Cl \longrightarrow \text{2° carbocation} + Cl^- \tag{8-26}$$

In this case, the secondary carbocation that is formed can rearrange to form a tertiary carbocation, but in this case, a methyl group with its bonding electrons (methide anion) migrates as shown in Reaction (8-27) to form a tertiary carbocation, which eventually leads to the unexpected organic product.

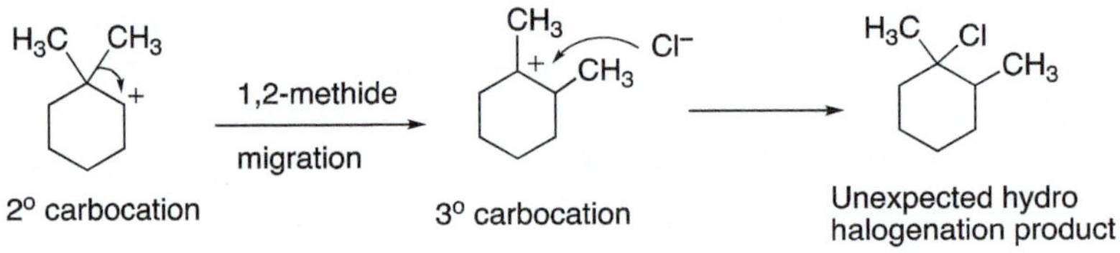

$$\text{2° carbocation} \xrightarrow[\text{migration}]{\text{1,2-methide}} \text{3° carbocation} \xrightarrow{Cl^-} \text{Unexpected hydro halogenation product} \tag{8-27}$$

For some carbocations, the methyl group is not the only group that can migrate, but sometimes an adjacent sigma bond in order to create a more stable cation as needed to explain the unexpected products shown in Problem 8.4.

Problem 8.4

Using arrows to represent electron movement, provide a reasonable mechanism to explain the following reactions.

(a) [structure] $\xrightarrow{HBr}$ [structure]

(b) [structure] $\xrightarrow{HCl}$ [structure]

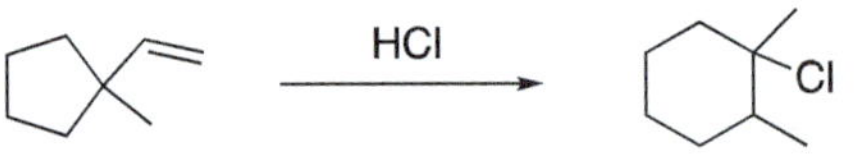

8.3.7 Anti-Markovnikov Addition to Alkenes

A challenge that chemists are often faced with is to obtain the anti-Markovnikov product as the major organic product. That is, the addition of the electrophile and nucleophile in an anti-Markovnikov manner, as shown in Reaction (8-28).

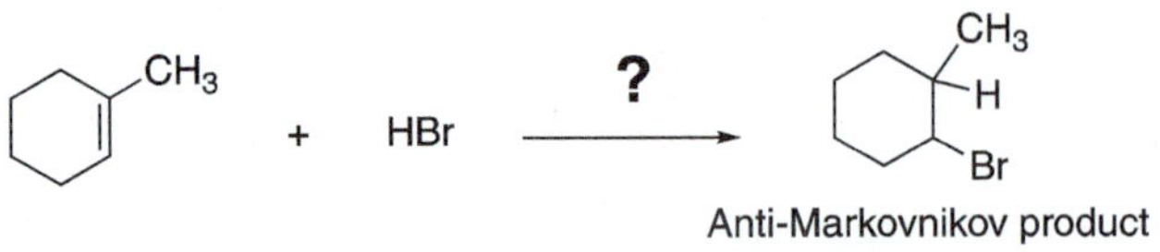

$$\text{1-methylcyclohexene} + HBr \xrightarrow{?} \text{Anti-Markovnikov product} \tag{8-28}$$

In order to determine if Reaction (8-28) is possible, the mechanism by which it would most likely occur will have to be examined. In order to fully understand how best to take advantage of the chemistry that will make this occur, possible mechanisms by which reactions occur must be examined. It is obvious that just adding HBr will not work since the major product would be the Markovnikov product owing to the most stable carbocation intermediate formed during the course of the reaction. As a result, a different approach must be used and the mechanism for this reaction would have to be different from the regular mechanism discussed earlier. It is known that H-Br, in the presence of peroxide and heat, generates the bromine radical and the use of radical intermediates, instead of carbocation intermediates, would produce a different mechanism for the addition of H-Br to a carbon–carbon double bond.

Radicals are very important intermediates in chemistry; radicals are reactive intermediates and are typically generated by the homolytic cleavage of a sigma bond. One of the most common ways of generating free radicals is the homolytic breakage of the oxygen–oxygen bond of peroxides. In the presence of energy, such as heat or light, an oxygen–oxygen bond will break homolytically to produce radicals as shown in Reaction (8-29).

$$RO-OR \xrightarrow{\text{Heat}} 2RO\cdot \text{ (Radicals)} \quad (8\text{-}29)$$

(Peroxide)

Examples of common peroxides that are used in organic chemistry to generate radicals are shown in Reactions (8-30) and (8-31).

$$C_2H_5O-OC_2H_5 \xrightarrow{134\text{-}185\,°C} 2\ C_2H_5O\cdot \quad (8\text{-}30)$$

$$H_3C-C(O)-O-O-C(O)-CH_3 \xrightarrow{55\text{-}85\,°C} 2\ H_3C-C(O)-O\cdot \quad (8\text{-}31)$$

Resonance stabilized radical

Note that the radical generated in Reaction (8-30) requires a fairly high temperature, whereas less energy is required for the generation of the more resonance-stabilized radical shown in Reaction (8-31). Another type of compound that is often used to generate radicals is azoalkane, such as the one shown in Reaction (8-32).

$$(CH_3)_3C-N=N-C(CH_3)_3 \xrightarrow{40\text{-}70\,°C} (CH_3)_3C\cdot + N\equiv N\ (N_2)\uparrow \quad (8\text{-}32)$$

Azoalkane → Tertiary radical + Stable nitrogen gas

The energy requirement for the generation of the radical of Reaction (8-32) is the least of the three reactions. Since this reaction liberates nitrogen as a gas, the reaction is entropy favored. Reactions that generate radicals as those given above are typically referred to as radical initiation reactions or steps.

Since radicals are reactive intermediates, they provide an alternative mechanism by which H-Br can add in a different manner across a carbon–carbon double bond. The bromine radical can be generated from the elementary reaction as shown in Reaction (8-33) in which the radical abstracts a hydrogen radical from HBr to generate another radical. This step is also known as radical propagation step since radicals are produced in a reaction with another radical.

$$RO\cdot\ +\ H-Br\ \longrightarrow\ RO-H\ +\ Br\cdot \quad (8\text{-}33)$$

Once bromine radicals are produced, even though it is fairly stable owing to its size, it will add to an unsymmetrical alkene to generate two carbon radicals as shown in Reactions (8-34) and (8-35).

$$\text{(structure)} + \text{Br}\cdot \longrightarrow \text{(structure with Br)} \qquad (8\text{-}34)$$

$$\text{(structure)} + \text{Br}\cdot \longrightarrow \text{(structure with Br)} \qquad (8\text{-}35)$$

Since there are two carbon radicals that are generated, the more stable will be formed in greater abundance and lead to the major product. Carbon radicals are similar to carbocations except that carbon radicals have one electron in the adjacent unhybridized p orbital, compared to carbocations, which have no electrons in the p orbital. As a result, a similar analysis as that carried out to determine the stability of carbocations can be used to determine the stability of radicals.

A carbon radical that has three alkyl groups bonded to the carbon of the radical is more stable than the one that has two groups bonded to the carbon of the radical. A radical that has three alkyl groups bonded to the carbon of the radical is called a tertiary radical (3°), whereas a radical that has two alkyl groups bonded to the carbon of the radical is called a secondary radical (2°). Alkyl groups, which are electron rich, can donate electrons to help stabilize the single electron of a radical. Thus, a tertiary radical, that has three alkyl groups bonded to the carbon of the radical, is more stable than a secondary radical, which has two alkyl groups bonded to the radical carbon. A radical that has only one alkyl group bonded to the carbon of the radical is called a primary (1°) and is not as stable as the secondary or the tertiary radicals. Figure 8.5 gives an illustration of the stability of radicals. The methyl radical is therefore the least stable radical and a tertiary radical is the most stable.

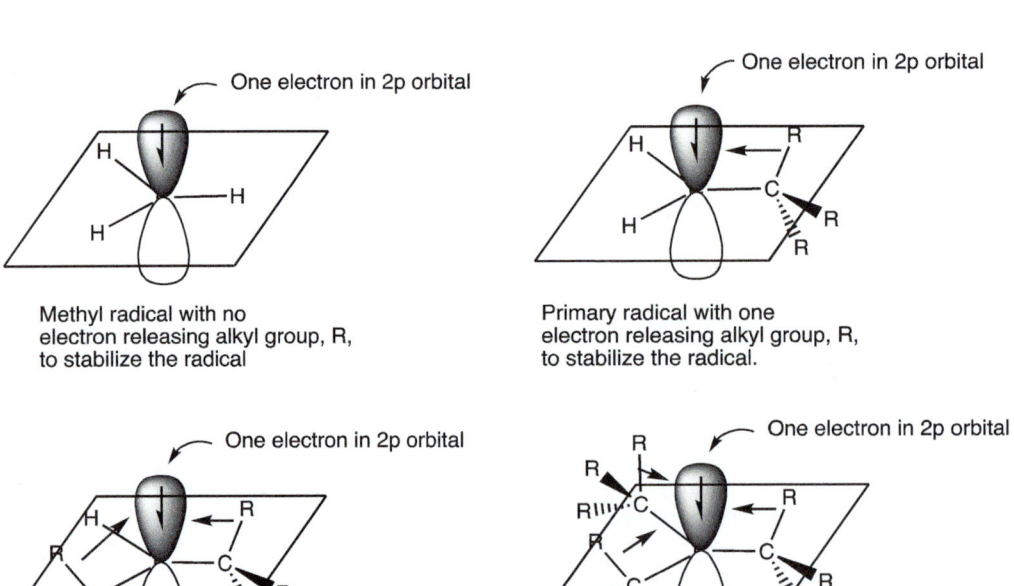

Figure 8.5 Relative stabilities of different types of radicals.

8.3 Addition of Hydrogen Halide to Alkenes (Hydrohalogenation of Alkenes)

Thus, the order of stability of different radicals is as shown below:

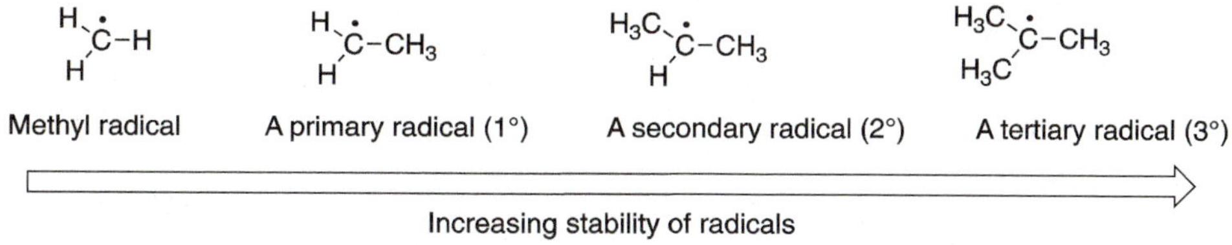

| Methyl radical | A primary radical (1°) | A secondary radical (2°) | A tertiary radical (3°) |

Increasing stability of radicals →

A different representation of the order of the stability of radicals is shown below.

$$\text{Tertiary (3°)} > \text{Secondary (2°)} > \text{Primary (1°)} > \text{Methyl (CH}_3\text{)}$$

Problem 8.5

Consider each of the following pairs of radicals, then determine which radical is the most stable.

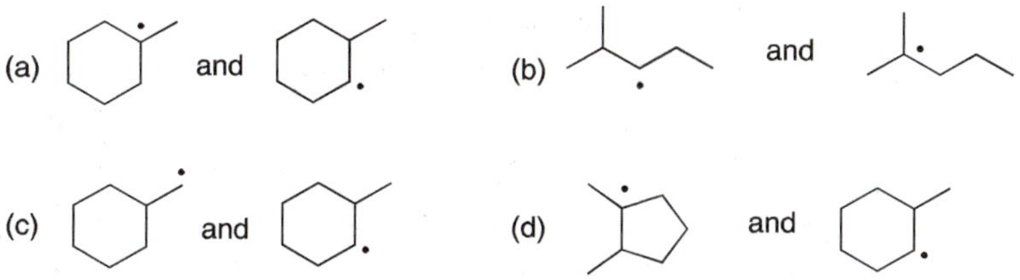

(a), (b), (c), (d)

Based on the stability trend for radicals, the radicals produced in Reactions (8-34) and (8-35) can be summarized as shown in Reaction (8-36), one radical is tertiary and more stable and the other radical is a secondary and less stable.

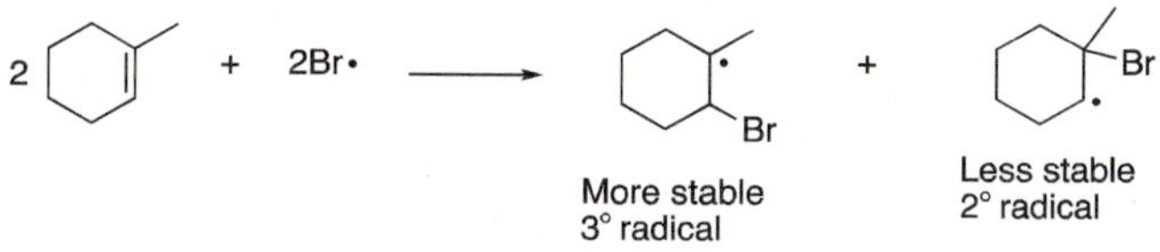

$$2\,\text{cyclohexene} + 2\text{Br}\cdot \longrightarrow \text{More stable 3° radical} + \text{Less stable 2° radical} \tag{8-36}$$

Typically, radicals are not as prone to rearrangement as carbocations and there are no rearranged unexpected products for these reactions. The products will be obtained from these two radicals, but the major product will result from the more stable radical intermediate as shown in Reactions (8-37) and (8-38).

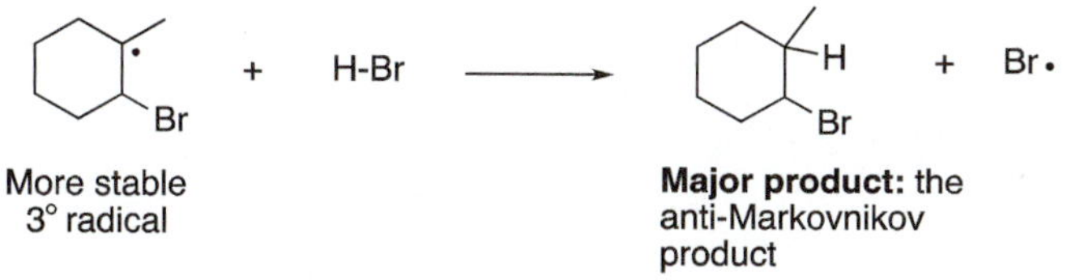

More stable 3° radical + H–Br ⟶ **Major product:** the anti-Markovnikov product + Br• (8-37)

Less stable 2° radical + H–Br ⟶ **Minor product** + Br• (8-38)

For these reactions, the radical reacts with H-Br to produce the organic products and another bromine radical. Note that the major product will be the one produced in Reaction (8-37) since the tertiary radical is more stable than the secondary radical.

In a final step where radicals react to form a neutral molecule, also known as the termination steps, bromine radicals react with another radical to form either Br$_2$ or ROBr as shown in Reactions (8-39) and (8-40).

$$Br\cdot + Br\cdot \rightarrow Br_2 \quad (8\text{-}39)$$

$$RO\cdot + Br\cdot \rightarrow ROBr \quad (8\text{-}40)$$

There are other possibilities for the termination of radicals involving coupling of the carbon radicals, but the reaction conditions can be controlled to produce the desired product. Thus, the overall reaction for the hydrobromination of methylcyclohexene in the presence of peroxide is given in Reaction (8-41). Note that it is a *regiospecific* reaction since for the major product, the bromine adds to the carbon with the most hydrogens, and the hydrogen adds to the carbon with the least number of hydrogens.

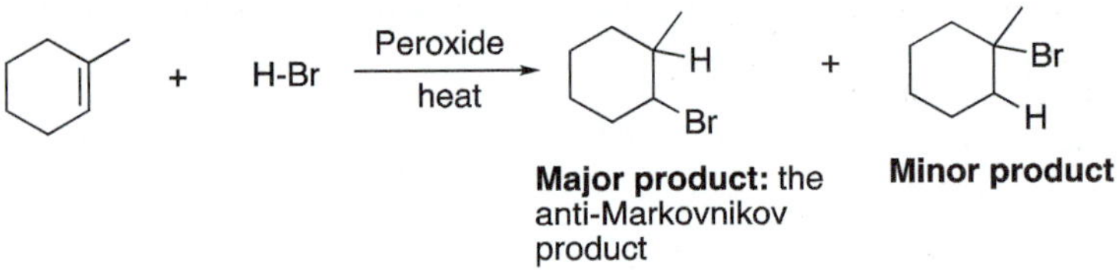

Gaining different outcomes of these addition reactions (Markovnikov and anti-Markovnikov reactions) by a slight change in reaction conditions demonstrates the importance of understanding the mechanisms of reactions. Chemists can manipulate reaction conditions to get a high percentage of a major organic product or to get a totally different organic product. Reaction (8-42) gives a summary of this concept that slight changes in reaction conditions can lead to different products.

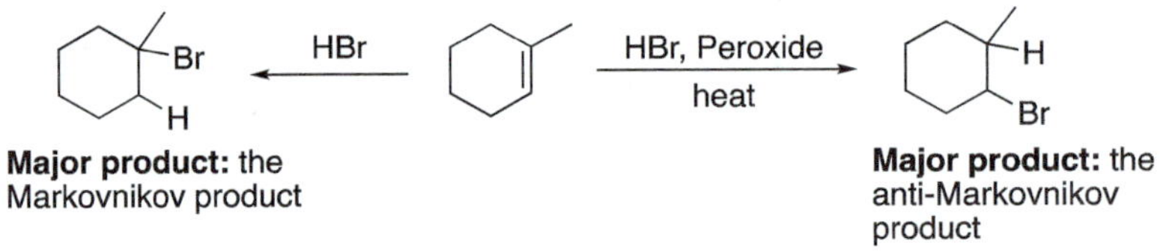

Problem 8.6

Give the major product that results from the addition of HBr to the following alkenes in the presence of peroxides and heat.

(a) 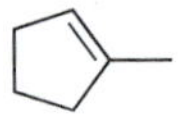 (b) 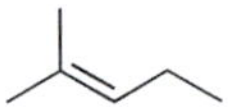 (c) 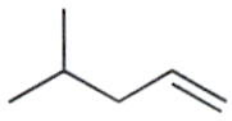 (d)

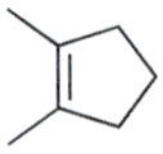

8.4 Addition of Halogens to Alkenes (Halogenation of Alkenes)

Chlorine and bromine readily add across the carbon–carbon double bonds of alkenes. The reaction of alkenes with bromine is often used as a test for the presence of carbon–carbon double bond functionality in compounds. The reddish bromine solution turns colorless in the presence of alkenes due to the reaction of both reagents. Since bromine (Br$_2$) is a large polarizable molecule (molecular weight is 160 amu), it is possible that at any instant the molecule is polarized. That is, a partial positive charge resides at one end of this large molecule and a partial negative charge resides at the other end. Since the pi (π) electrons of the double bond of alkenes are nucleophilic, a reaction of the nucleophilic alkene and the partial positive end of the polarized bromine molecule will occur as shown in Reaction (8-43) to form a bromonium ion and a bromide anion.

$$(8\text{-}43)$$

The bromonium ion intermediate is relatively stable since the positive charge is distributed throughout the three atoms, including the very large bromine atom, which is located just above the carbon–carbon bond. Owing to the unusual stability of the bromonium ion, it will not rearrange like some carbocations. Even though the bromonium ion is relatively stable, it is reactive and will react with a nucleophile since it is electrophilic. Since the bromide anion that was generated in Reaction (8-43) is also nucleophilic, it is possible for it to attack the bromonium ion to give a neutral product, as shown in Reaction (8-44).

$$(8\text{-}44)$$

There is one very important observation that should be made about this last step of the reaction mechanism; the nucleophilic bromide anion attacks the electrophilic bromonium ion opposite to the side that has the bromine of the bromonium ion. The very large bromine atom occupies most of the space of one side of the bromonium ion, and as a result, the nucleophilic bromide anion adds to the opposite side. Hence, the stereochemistry of the product that results is a **trans** product. As pointed out earlier, the bromonium ion does not rearrange, and as a result, there are typically no unexpected products observed for the bromination of alkene addition reactions. Shown in Reaction (8-45) is the addition of bromine to a 4-methyl-2-pentene.

$$(8\text{-}45)$$

The trans-addition product is given in Reaction (8-45) showing two different representations, the dashed-wedge and Newman projection. This reaction represents an ideal application of the use of the stereochemistry concepts and representations that were learned in earlier chapters. A similar type of reaction of alkenes with chlorine is expected, but since chlorine is not as large as the bromine molecule, the chloronium ion is not as stable as the bromonium ion. As a result, the stereochemical outcome (trans isomer) is not as pronounced, compared to the reaction of alkenes with the bromine. Since a specific stereoisomer is formed from either a trans or cis alkene, this type of reaction is described as a stereospecific reaction. A ***stereospecific reaction*** is one in which the stereochemistry of the reactant completely dictates the stereochemistry of the product.

Problem 8.7

Give the major product that results from the addition of bromine to the following alkenes, indicate stereochemistry where appropriate.

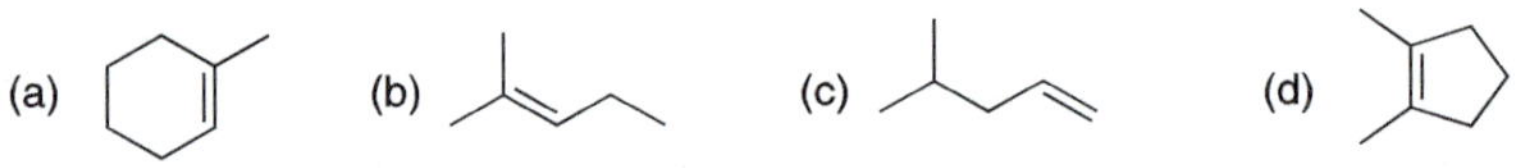

(a) [cyclohexene structure]   (b) [2-methyl-2-butene structure]   (c) [3-methyl-1-butene structure]   (d) [methylenecyclopentane structure]

8.5 Addition of Halogens and Water to Alkenes (Halohydrin Formation)

An alkene in the presence of a bromine and water will result in the addition of Br-OH across the double bond, and it is observed from experiments that this type of reaction occurs in a *stereospecific* manner, as shown in Reaction (8-46) involving an alkene in the presence of bromine and water to give the trans product, which is known as a bromohydrin.

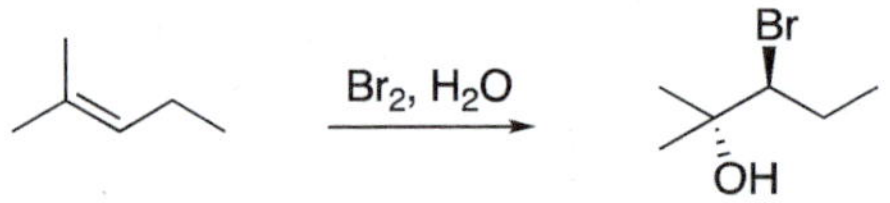

$$\text{alkene} \xrightarrow{Br_2, H_2O} \text{bromohydrin} \tag{8-46}$$

In the first step of the mechanism for Reaction (8-46), the nucleophilic alkene attacks the bromine molecule as shown in Reaction (8-47) to form the bromonium ion and a bromide anion.

$$\text{alkene} + Br\!-\!Br \longrightarrow \text{bromonium ion} + :Br:^{\ominus} \tag{8-47}$$

In the next step of the mechanism, the more nucleophilic water attacks the bromonium electrophilic ion instead of the bromide anion as shown in Reaction (8-48).

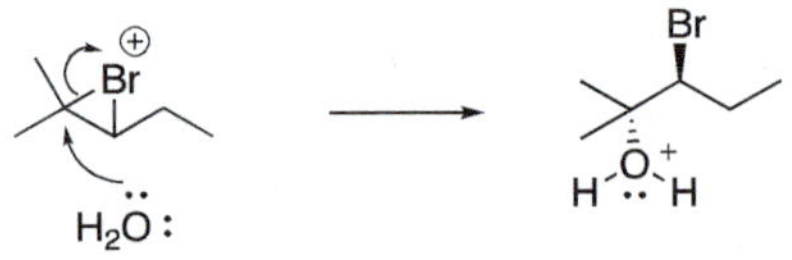

$$\text{bromonium ion} + H_2O: \longrightarrow \text{protonated bromohydrin} \tag{8-48}$$

Note that the water attached the more tertiary-like carbon of the bromonium ion. The final step of the mechanism involves the loss of a proton, which is abstracted by the bromide anion as shown in Reaction (8-49).

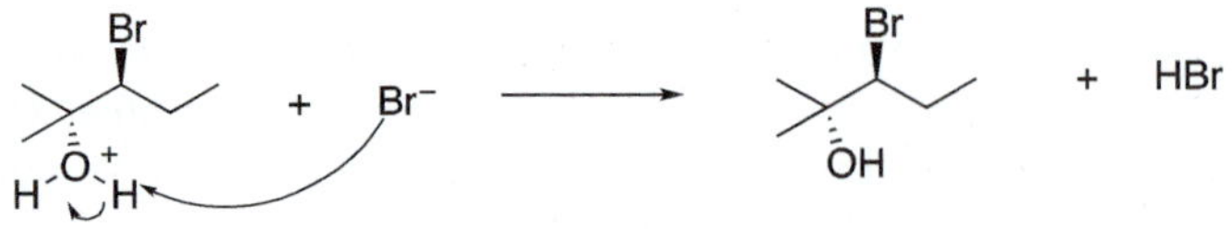

$$\text{protonated bromohydrin} + Br^- \longrightarrow \text{bromohydrin} + HBr \tag{8-49}$$

For the reactions that lead to the formation of bromohydrins, rearrangement is typically not observed owing to the stability of the bromonium ion. These reactions are stereospecific resulting in trans-additions since attack of the bridged intermediate by water takes from the opposite side of the bromonium ion as shown in Reaction (8-50).

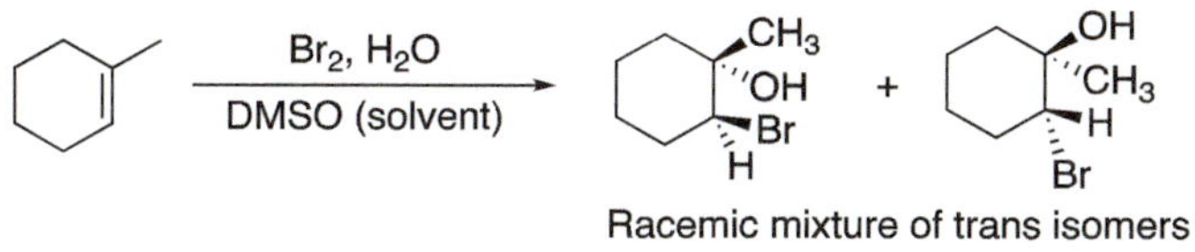

8.6 Addition of Water to Alkenes (Hydration of Alkenes)

$$\text{cyclohexene} \xrightarrow[\text{DMSO (solvent)}]{Br_2,\ H_2O} \underset{\text{Racemic mixture of trans isomers}}{\text{trans-2-bromo-1-methylcyclohexanol products}} \quad (8\text{-}50)$$

These reactions are also *regiospecific* since the water will bond to the carbon of the bromonium ion that is more carbocationic-like. That is, if carbocations were to be formed, the order of stability of carbocations will dictate and water will prefer to bond to a tertiary type more stable carbon, compared to a secondary type carbon. This concept is illustrated in Reaction (8-51), and the bromohydrin shown will not form.

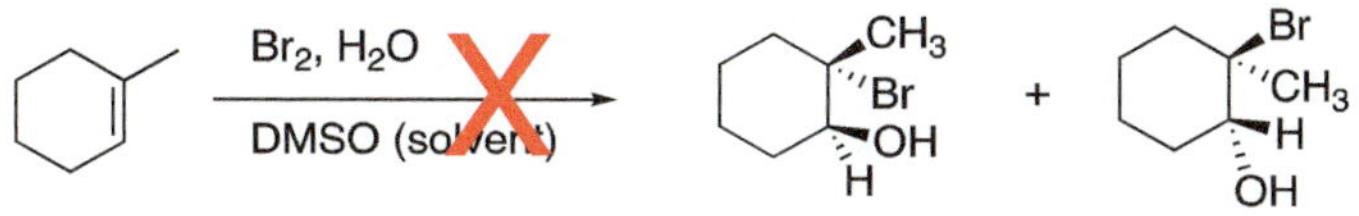

$$\text{1-methylcyclohexene} \xrightarrow[\text{DMSO (solvent)}]{Br_2,\ H_2O} \times \text{(incorrect regiochemistry products)} \quad (8\text{-}51)$$

Problem 8.8

Give the major organic product that results from the addition of bromine and water in the presence of dimethylsulfoxide (DMSO) as solvent to the alkenes shown below. Show stereochemistry where appropriate.

(a) cyclopentene   (b) 2-methyl-2-butene   (c) 1-butene   (d) 1-methylcyclopentene

8.6 Addition of Water to Alkenes (Hydration of Alkenes)

As demonstrated earlier, addition reactions involving H-Cl across the double bond involve the formation of a carbocation intermediate. The mechanism of those addition reactions can be used to explain the outcome of the addition reaction of H-OH across a double bond. Water is a polar molecule in which the hydrogen is partially positive and the OH group is partially negative. Thus, the partially positive hydrogen would act as an electrophile and the OH group as the nucleophile for an addition reaction to the carbon–carbon double bonds of alkenes, as illustrated in Reaction (8-52).

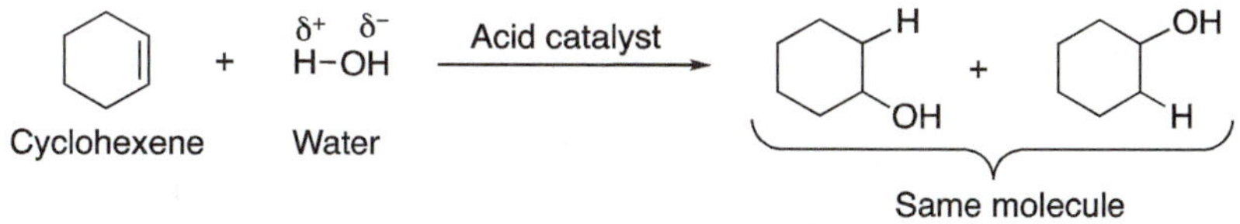

$$\text{Cyclohexene} + \overset{\delta^+\ \delta^-}{H\text{-}OH} \xrightarrow{\text{Acid catalyst}} \underbrace{\text{cyclohexanol} + \text{cyclohexanol}}_{\text{Same molecule}} \quad (8\text{-}52)$$

It turns out that water is a very stable molecule, and as a result, a catalyst is needed for this type of addition reaction to proceed. The catalyst in this case is an acid (H⁺), and the first step of the reaction mechanism is shown in Reaction (8-53) in which the electrophilic proton of the catalyst adds to the nucleophilic alkene to form a carbocation.

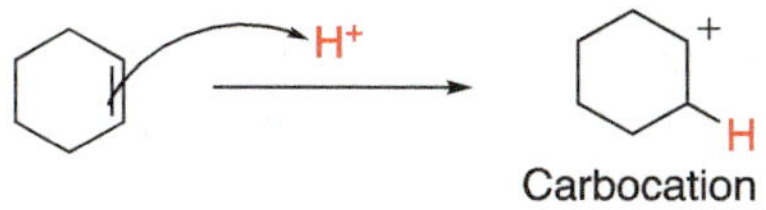

$$\text{cyclohexene} \xrightarrow{H^+} \text{Carbocation} \quad (8\text{-}53)$$

In the next step of the reaction mechanism, the nucleophilic water reacts with the electrophilic carbocation, as shown in Reaction (8-54).

$$\text{(8-54)}$$

Notice that since one pair of the electrons from water is used to bond with the electrophilic carbocation forming a new covalent bond, the oxygen acquires a formal charge of positive one (+1). The newly formed intermediate carries a positive charge and in the next step of the reaction mechanism deprotonation occurs to produce a neutral organic product and regenerates a proton, which is essentially the catalyst, as shown in Reaction (8-55).

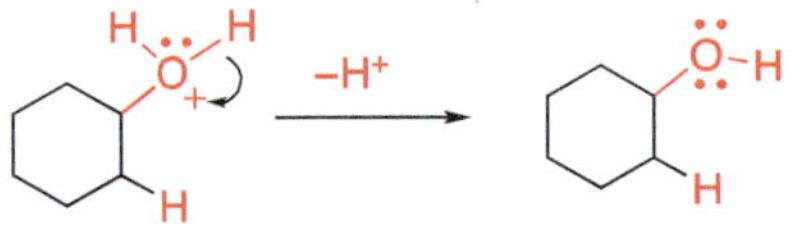

$$\text{(8-55)}$$

Hence, the catalyst proton is used in the first step of the reaction, but is regenerated in the last step of the reaction mechanism. The overall reaction mechanism is shown in Reaction (8-56).

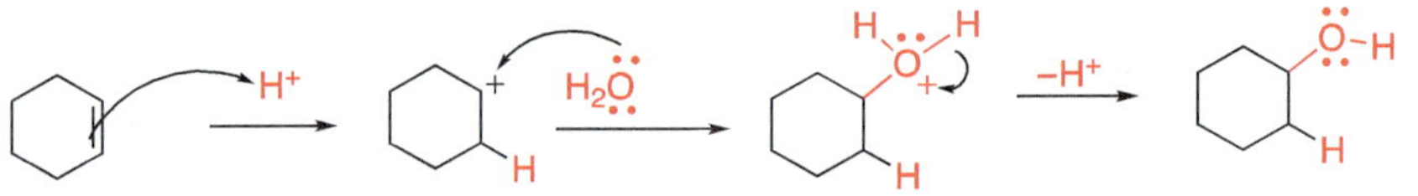

$$\text{(8-56)}$$

A challenge that is encountered for the addition to unsymmetrical alkenes is to predict the major addition product. A basic concept for these reactions is that the reaction will proceed via the more stable cation to give the major product; thus, the key is to determine the most stable carbocation. For the addition of water to methylcyclohexene, the reaction mechanism is shown in Reaction (8-57), in which the more stable tertiary carbocation is formed.

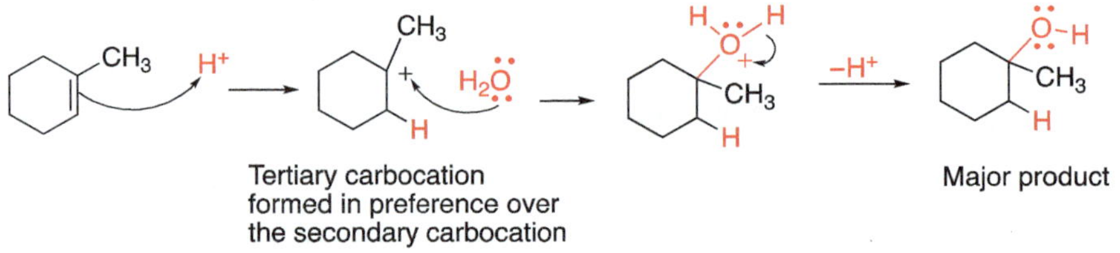

$$\text{(8-57)}$$

Tertiary carbocation formed in preference over the secondary carbocation

Major product

Problem 8.9

Give the major organic product that results from addition of water in the presence of an acid catalyst to the alkenes shown below.

(a) 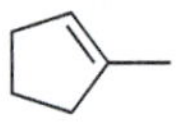   (b) 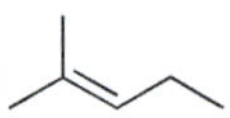   (c) 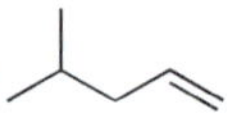   (d)

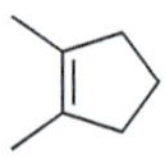

Another challenge that exists in determining the products for the catalytic addition of water across the double bond of an alkene is that rearrangement of the intermediate carbocation is possible giving rise to unexpected products, as in the example in Reaction (8-58).

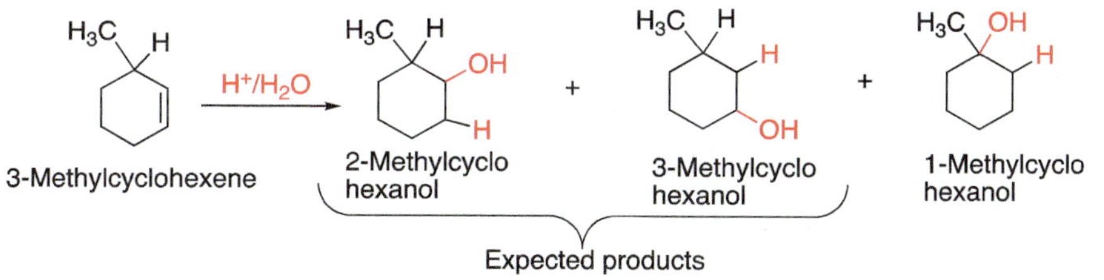

8.6 Addition of Water to Alkenes (Hydration of Alkenes)

[Reaction 8-58: 3-Methylcyclohexene + H⁺/H₂O → 2-Methylcyclohexanol + 3-Methylcyclohexanol + 1-Methylcyclohexanol (Expected products)] (8-58)

1-Methylcyclohexanol is an unexpected product and a closer look at the reaction mechanism will help explain its formation. In the first step of the reaction mechanism, the pi (π) electrons react with the catalyst (H⁺) to form a secondary carbocation as shown in Reaction (8-59).

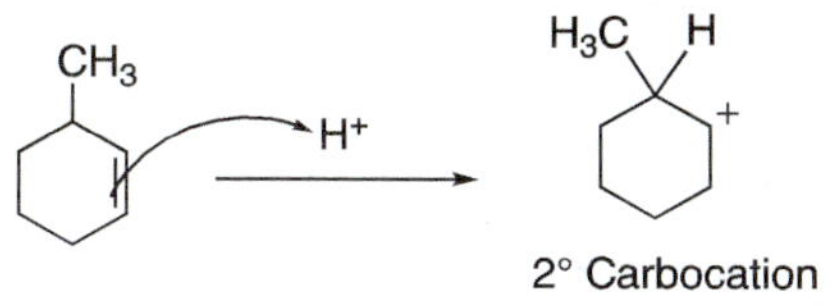

[Reaction 8-59: 3-Methylcyclohexene + H⁺ → 2° Carbocation] (8-59)

Since the carbocation formed is a secondary carbocation and can rearrange, it will rearrange to give a more stable tertiary carbocation. Such a rearrangement is possible by the migration of a hydride ion (H⁻) from the adjacent carbon to form the more stable tertiary carbocation as shown in Reaction (8-60).

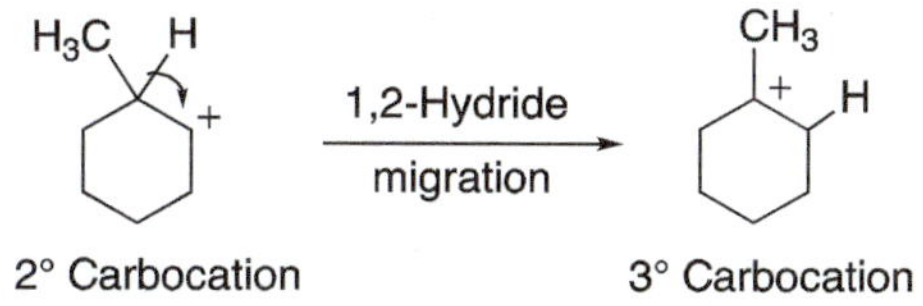

[Reaction 8-60: 2° Carbocation → (1,2-Hydride migration) → 3° Carbocation] (8-60)

Note that the migrating hydride anion moves with its bonding electrons and that explains the charge left behind after the migration has occurred as a positive charge on the carbon, resulting in the formation of a stable tertiary carbocation. In the next step of the reaction mechanism, the nucleophilic water bonds to the carbocation to form a protonated alcohol intermediate, as shown in Reaction (8-61).

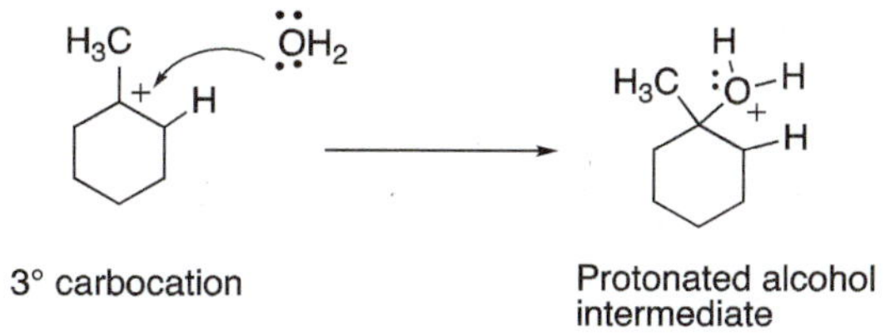

[Reaction 8-61: 3° carbocation + :ÖH₂ → Protonated alcohol intermediate] (8-61)

The last step of the reaction mechanism involves the loss of a proton to form the final product as shown in Reaction (8-62).

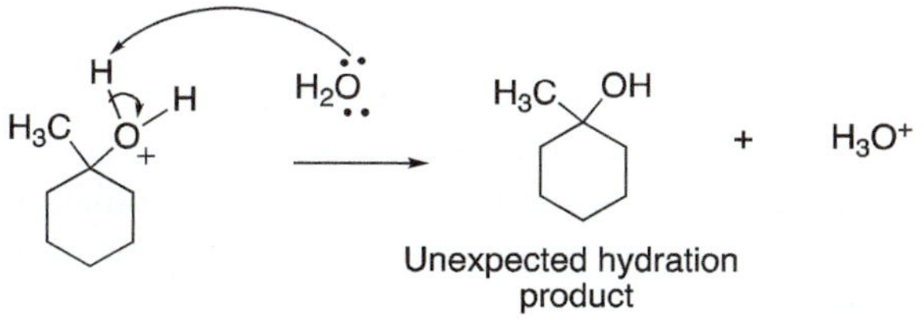

[Reaction 8-62: Protonated alcohol + H₂O → 1-Methylcyclohexanol (Unexpected hydration product) + H₃O⁺] (8-62)

One of the products of the reaction of 3,3-dimethylcyclohexene with water in the presence of an acid catalyst is an unexpected hydration product as shown in Reaction (8-63).

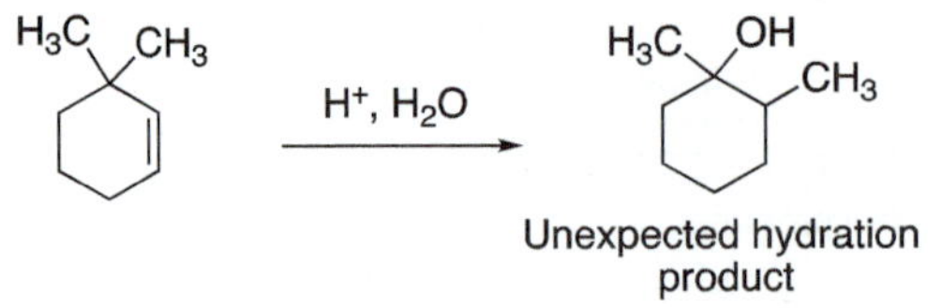

$$\text{3,3-dimethylcyclohexene} \xrightarrow{H^+, H_2O} \text{Unexpected hydration product} \quad (8\text{-}63)$$

A close examination of the reaction mechanism reveals that there is a migration of a methyl group with its electron (methide ion) to form a more stable carbocation as shown in Reaction (8-64).

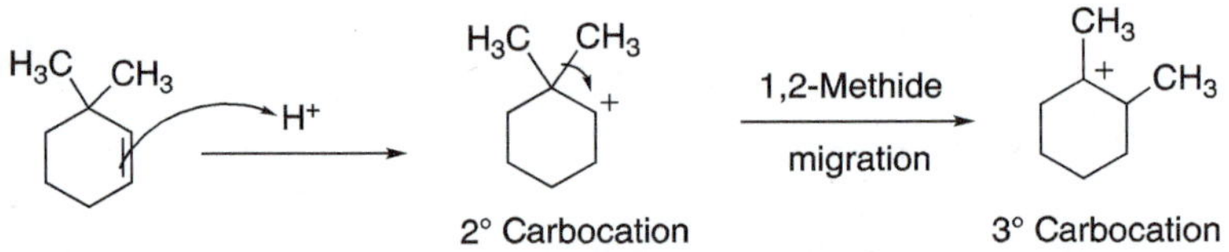

$$\xrightarrow{H^+} 2° \text{ Carbocation} \xrightarrow{1,2\text{-Methide migration}} 3° \text{ Carbocation} \quad (8\text{-}64)$$

In the final steps of the reaction mechanism, water reacts with the tertiary carbocation to form a protonated alcohol, which is deprotonated in a final step to form the unexpected alcohol as shown in Reaction (8-65).

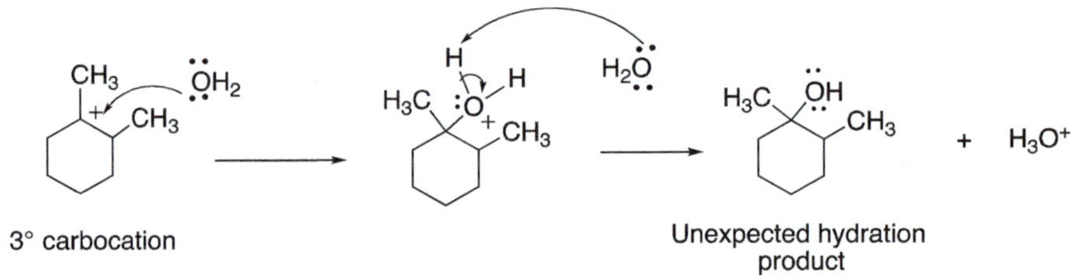

$$3° \text{ carbocation} \longrightarrow \longrightarrow \text{Unexpected hydration product} + H_3O^+ \quad (8\text{-}65)$$

Problem 8.10

Propose a reasonable mechanism to explain the formation of the rearranged products shown for the following hydration reactions.

(a) $\text{CH}_2=\text{CHCH}(\text{CH}_3)_2 \xrightarrow{H^+/H_2O} (\text{CH}_3)_3\text{C-CH}(OH)\text{CH}_3$

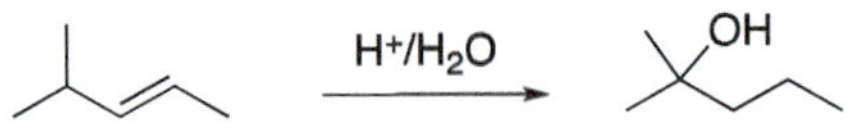

(b) methylenecyclopentane $\xrightarrow{H^+/H_2O}$ 1-methylcyclohexan-1-ol

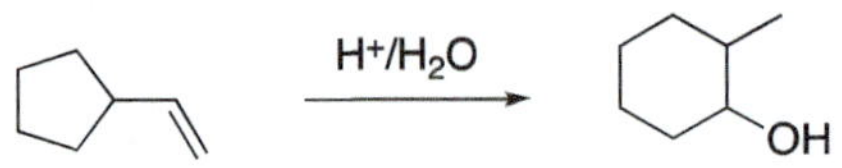

For the hydration of alkenes, it is obvious that it is sometimes difficult to predict the Markovnikov addition product since rearrangement often times occurs. As a result, there is a challenge if the Markovnikov addition product is the desired product for a particular addition reaction. In the next section, specific strategies to achieve the goal of obtaining one product as the major organic product will be discussed.

8.6.1 Hydration by Oxymercuration–Demercuration

As shown in the previous section, a major problem encountered in trying to achieve the synthesis of a specific hydration product is that rearrangement sometimes occurs, which often gives unexpected products. One way to get around possible rearrangement is to examine the mechanism of another route to add water across a double bond. Mercury ions can add across a carbon–carbon double bond to form a bridged relatively stable intermediate, similar to the bromonium ion that was encountered in the bromination of alkenes. Recall that due to the size of the bromine atom, the charge of the carbocation is distributed across three atoms: the two carbons of the alkene and bromine, see Reaction (8-43). Owing to the stability of the bromonium ion, no rearrangement occurs. The addition of an electrophilic mercury (in the form of mercury acetate), which is fairly large and polarizable, to a double bond leads to a similar stable bridged mercury cationic intermediate as shown in Reaction (8-66).

$$\text{AcO-Hg-OAc} + \text{alkene} \longrightarrow \text{bridged mercury cation} + \text{OAc}^- \qquad (8\text{-}66)$$

This bridged mercury cationic intermediate is fairly stable due to the size of the large polarizable mercury and the positive charge that is distributed throughout the three atoms of the bridged intermediate. One carbon of this bridged mercury cationic intermediate, however, has a greater positive character than the other carbon, and the carbon that has the greater number of alkyl substitution bonded to that carbon is more carbocationic-like, compared to the other carbon with less alkyl groups. In other words, the carbon that is more tertiary-like carries a greater positive character, compared to another carbon that is secondary or primary. Thus, an attack by a nucleophile, such as water, will occur at the carbon that bears the most alkyl groups, and the intermediate that results will have the mercury acetate bonded to the adjacent carbon, as shown in Reaction (8-67).

$$(8\text{-}67)$$

In a second reaction, the mercury acetate can be removed by a reduction reaction with a strong reducing agent such as $NaBH_4$, as shown in Reaction (8-68).

$$(8\text{-}68)$$

Thus, the overall reaction that results from the addition of water across a double bond, by introducing mercury acetate, and then removing the mercury salt is shown in Reaction (8-69).

$$\underset{H}{\overset{R}{\underset{R}{\diagup}}}\!\!=\!\!\underset{H}{\overset{H}{\diagdown}} \quad \xrightarrow{\text{Hg(OAc)}_2/\text{H}_2\text{O}} \quad \underset{R}{\overset{R}{\diagdown}}\!\!\underset{\underset{\text{HgOAc}}{|}}{\overset{\text{OH}}{\underset{\text{CH}_2}{\diagup}}} \quad \xrightarrow{\text{NaBH}_4,\ \text{OH}^-} \quad \underset{R}{\overset{R}{\diagdown}}\!\!\underset{\text{CH}_3}{\overset{\text{OH}}{\diagup}} \qquad (8\text{-}69)$$

Since the strategy used in this type of reaction is to introduce a mercury salt to stabilize the intermediate formed, then removing it, this reaction is often referred to as oxymercuration–demercuration. Note that the result of this addition reaction is a Markovnikov addition of water to a double bond and that there is no rearrangement for this type of reaction. A specific example oxymecuration–demercuration (hydration) is shown in Reaction (8-70).

2-Methyl-2-pentene $\xrightarrow[\text{(2) NaBH}_4,\ \text{OH}^-]{\text{(1) Hg(OAc)}_2/\text{H}_2\text{O}}$ 2-Methyl-2-pentanol (major organic product) (8-70)

Note that in the reaction mechanism, the attack of the water on the bridged mercury ion takes place to give the trans-product, but stereochemistry at the carbon that has the mercury acetate is lost during the reduction step of the reaction sequence. As a result, these reactions are not considered to be stereospecific, and racemic mixture of the enantiomers will result as the products as shown in Reaction (8-71).

Pentene $\xrightarrow[\text{(2) NaBH}_4]{\text{(1) Hg(OAc)}_2,\ \text{H}_2\text{O}}$ (S)-2-Pentanol + (R)-2-Pentanol (8-71)

Problem 8.11

Give the major organic products for the following hydration reactions.

(a) [1-methylcyclohexene] $\xrightarrow[\text{2) NaBH}_4,\ \text{OH}^-]{\text{1) Hg(OAc)}_2/\text{H}_2\text{O}}$

(b) [2-methyl-2-butene] $\xrightarrow[\text{2) NaBH}_4,\ \text{OH}^-]{\text{1) Hg(OAc)}_2/\text{H}_2\text{O}}$

Thus, oxymercuration–demercuration reactions give no rearrangement and the addition takes place in a regiospecific manner by the Markovnikov addition rule and these reactions are not considered to be stereospecific.

8.6.2 Hydration by Hydroboration-Oxidation

A question that comes to mind at this point is it possible to add water to alkenes in an anti-Markovnikov manner? In this case, a different strategy will have to be used, which of course cannot involve the formation of carbocation intermediates since carbocation intermediates result in Markovnikov addition. A different approach will have to be taken and this involves the use of borane (BH_3), which exists as a dimer as shown in Reaction (8-72).

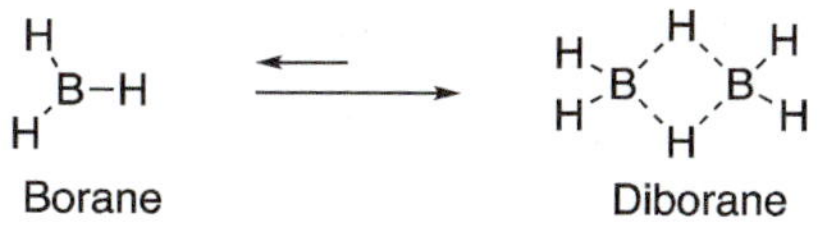

8.6 Addition of Water to Alkenes (Hydration of Alkenes)

$$\text{Borane} \rightleftharpoons \text{Diborane} \tag{8-72}$$

This molecule is actually a Lewis acid, with boron being sp² hybridized and having a vacant p orbital, it can accept electrons into the vacant p orbital. If another molecule of borane is close, the bonding electrons from the B—H bonds attempt to share its bonding electrons with the empty p orbital on the boron and the result is the dimer and the bonds are shown as partial B—H bonds (as shown in Reaction 8-72). As a result, the above equilibrium involving the diborane and its monomer, as shown above, lies to the right, but in the presence of a Lewis base solvent, such as tetrahydrofuran (THF), it forms a stable complex as shown in Reaction (8-73).

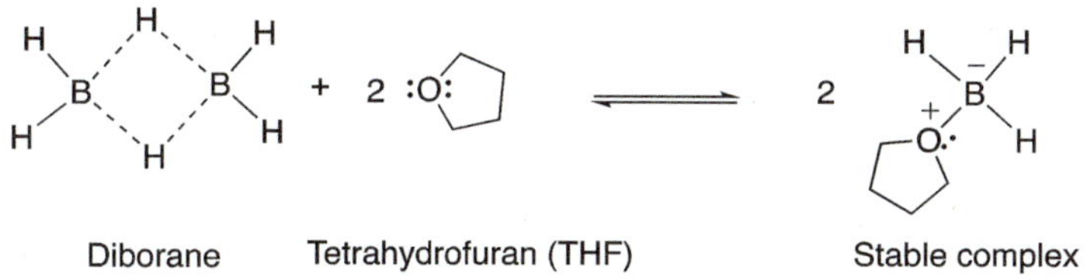

$$\text{Diborane} + 2\,\text{Tetrahydrofuran (THF)} \rightleftharpoons 2\,\text{Stable complex} \tag{8-73}$$

Since the BH₃ is electron deficient, it will add across the double bond of the alkene. As shown in the Reaction (8-74), BH₃ adds across the double bond to form the product shown.

$$\tag{8-74}$$

Note that the borane (BH₂-H) adds to the same side of the double bond, resulting in an addition that is described as a **cis** addition. For an unsymmetrical alkene, there are two ways that the cis addition of borane to the double bond can take place to give two different cis products as shown in Reaction (8-75).

$$\tag{8-75}$$

More stable due to less steric interaction

Less stable due to more steric interaction

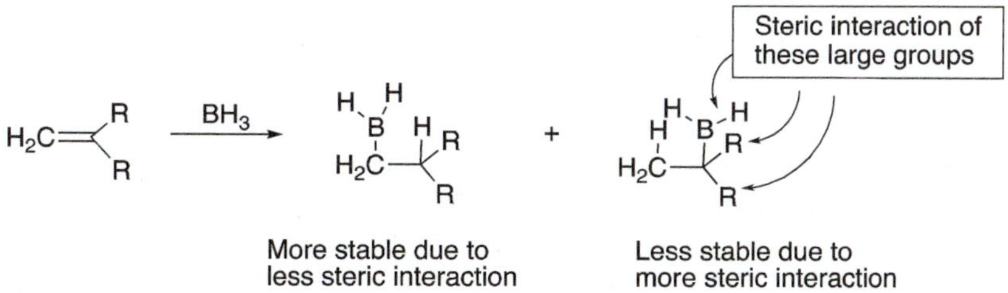

Steric interaction of these large groups

The stability of the products is a factor in determining which product is the major product. As shown in Reaction (8-75), the first product is more stable than the second product due to less steric interaction between the borane group and the two hydrogens on the adjacent carbon. For the second product, there is steric interaction between the borane and the alkyl groups on the adjacent carbon. Similar steric interactions played an important role in the stability of conformers. As demonstrated in Chapter 4, large bulky groups prefer not to be close to each other. Thus, due to steric interactions, the first product is more stable, and hence the major product, compared to the second product, which is less stable. Shown in Reaction (8-76) is a reaction involving the addition of borane to 2-methyl-2-pentene showing the major and minor products.

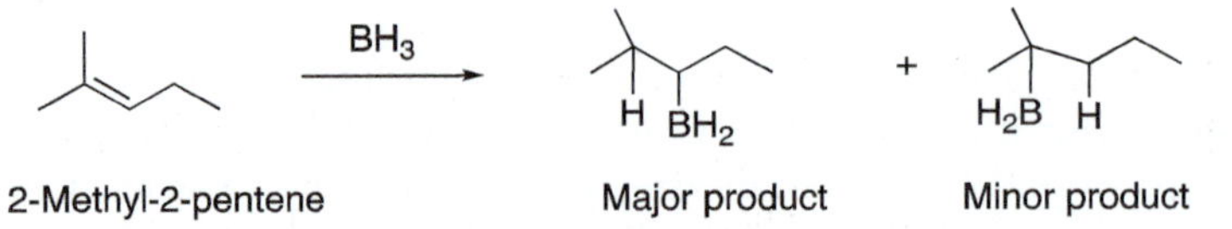

$$\text{2-Methyl-2-pentene} \xrightarrow{BH_3} \text{Major product} + \text{Minor product} \tag{8-76}$$

The initial intent in considering these types of reactions is the addition H-OH across the carbon–carbon double bond to give the anti-Markovnikov product. The products shown in the above reaction can be converted to alcohols by oxidation of the borane functionality of the molecules, as shown in Reaction (8-77), in which hydrogen peroxide (H_2O_2) is used as the oxidizing agent.

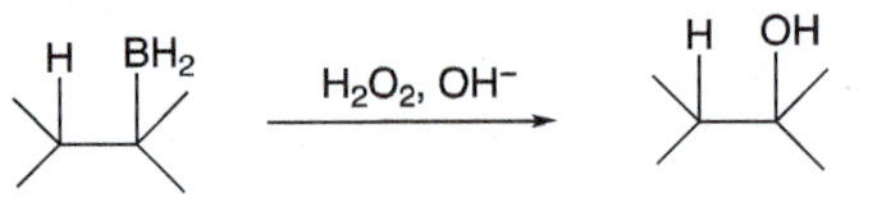

$$\xrightarrow{H_2O_2, OH^-} \tag{8-77}$$

It is always important to know how reactions work (the mechanism) so that if similar reactions are encountered later in this course, they will not present a challenge to predict possible products. In the first step of the mechanism for the oxidation reaction, the peroxide anion is produced in the presence of hydrogen peroxide and a base, such as sodium hydroxide as shown in Reaction (8-78).

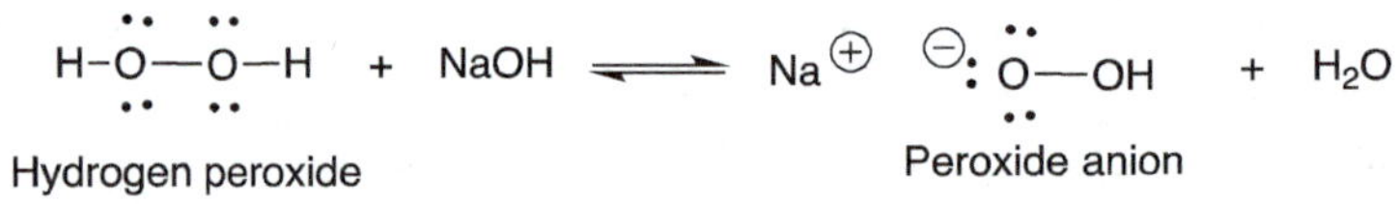

$$H\text{-}\ddot{O}\text{--}\ddot{O}\text{-}H + NaOH \rightleftharpoons Na^{\oplus} \;\; {}^{\ominus}{:}\ddot{O}\text{--}OH + H_2O \tag{8-78}$$

Hydrogen peroxide · · · · · · · · · · · · · · · · · · Peroxide anion

In a second step, the peroxide anion reacts with the vacant p orbital of the borane, followed by migration of the alkyl group to the electron-deficient oxygen as shown in Reaction (8-79) to eventually give the final product.

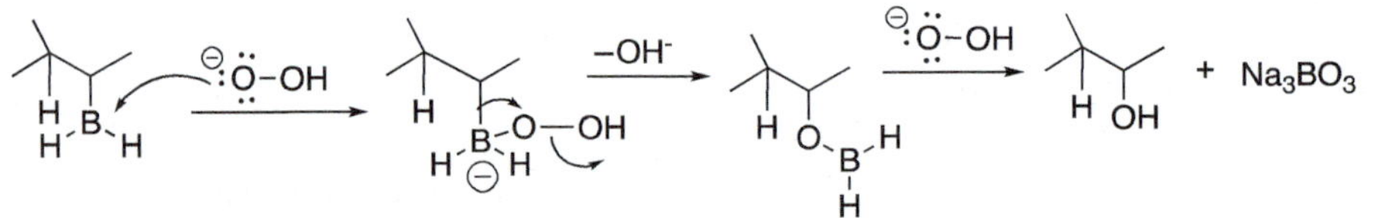

$$\cdots \xrightarrow{:\ddot{O}-OH} \cdots \xrightarrow{-OH^-} \cdots \xrightarrow{:\ddot{O}-OH} \cdots + Na_3BO_3 \tag{8-79}$$

Reaction (8-80) shows the overall two-step reaction sequence in which an alkene is converted to an alcohol by the addition of essential H-OH across the double bond, but in an anti-Markovnikov manner using the hydroboration–oxidation reaction.

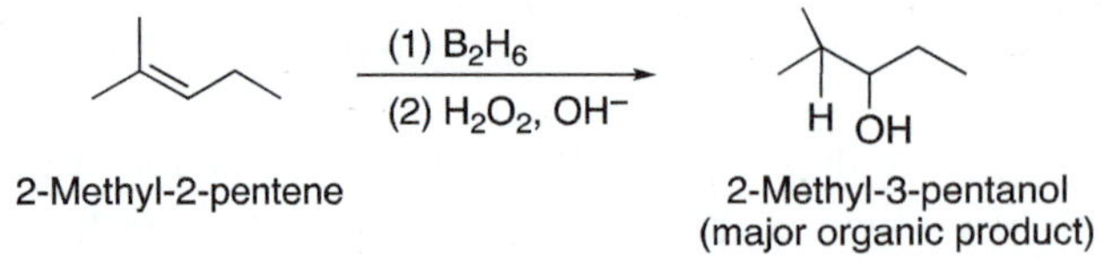

$$\text{2-Methyl-2-pentene} \xrightarrow[\text{(2) } H_2O_2,\; OH^-]{\text{(1) } B_2H_6} \text{2-Methyl-3-pentanol (major organic product)} \tag{8-80}$$

For the hydration (addition of H-OH) of alkenes in which the hydroboration–oxidation reaction is used, there is no rearrangement. The addition takes place in a regiospecific manner by the anti-Markovnikov addition, and the reaction is stereospecific to give the cis addition.

Problem 8.12

Give the major organic product of the reactions shown below.

(a) [cyclohexene] → (1) B₂H₆ (2) H₂O₂, OH⁻

(b) [2-methyl-2-pentene] → (1) B₂H₆ (2) H₂O₂, OH⁻

8.7 Addition of Carbenes to Alkenes

So far, the different reactive intermediates that have been discussed include radicals and carbocations. Another type of reactive intermediate is the carbene. Carbenes were discovered in 1959, they are neutral carbon species and have two bonds to carbon, instead of the typical four, which is expected for neutral carbon molecules. It was also observed that the most stable carbene contained a pair of electrons in a single orbital. In order to explain these observations, a systematic consistent scientific explanation must be developed to account for the structure of carbenes.

8.7.1 Structure of Carbenes

It was observed that there are two types of carbenes: one is described as a singlet carbene and the other as a triplet carbene. It is obvious that either of the traditional sp^3, sp^2, and sp hybridized orbitals could not provide an adequate description for this newly found species. The model that best explains these experimental results is one in which the carbon is sp^2 hybridized, but the gap between the p and the three sp^2 orbitals is larger than that found in the sp^2 hybridized orbitals and p orbitals of the carbons for alkenes. That is, ΔG between the three equivalent sp^2 orbitals and the p orbital is large and the electrons prefer to pair instead of going in separate orbitals. This type of carbene is called a singlet carbene (Figure 8.6).

8.7.2 Reactions of Carbenes

When considering the reaction of the very reactive carbene intermediate, there are two aspects that are very important: first, it has two electrons in a sp^2 sigma (σ) orbital, which makes it nucleophilic. The second aspect is that it has an empty p orbital, which means that it is also electrophilic. Thus, it is possible for carbenes to have a very favorable reaction with the nucleophilic double bond to produce an electrophilic carbon on the alkene, which can then react with

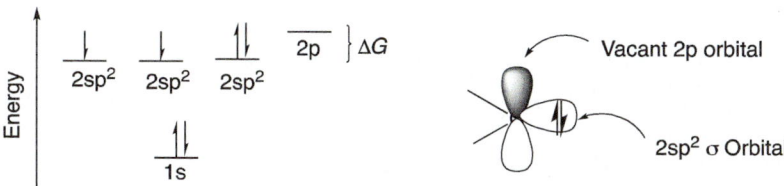

Figure 8.6 Proposed model for the electronic configuration of singlet carbenes.

the nucleophilic two electrons in the sigma (σ) orbital of the carbene to produce a neutral product as shown in Reaction (8-81) to produce a cyclopropane ring. As a result, this type of reaction is often referred to as cyclopropanation of alkenes and is a concerted reaction and does not occur in a step-wise manner as described for some of the other reactions studied.

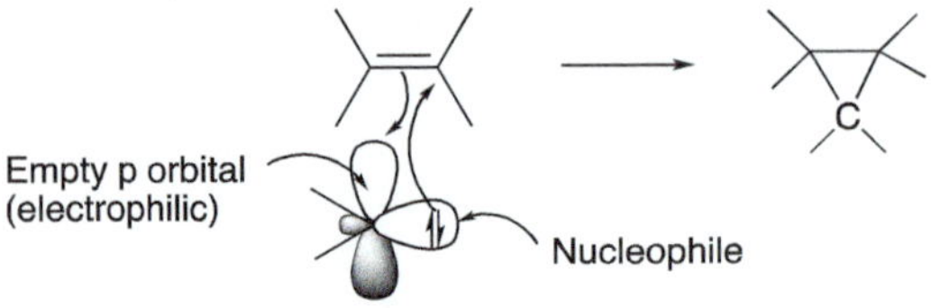

(8-81)

The Simmons–Smith reaction is commonly used to generate carbenes for the cyclopropanation of alkenes. The main reactant intermediate of the Simmons–Smith reaction is a "carbinoid" species since a carbene is not actually formed, as illustrated in Reaction (8-82).

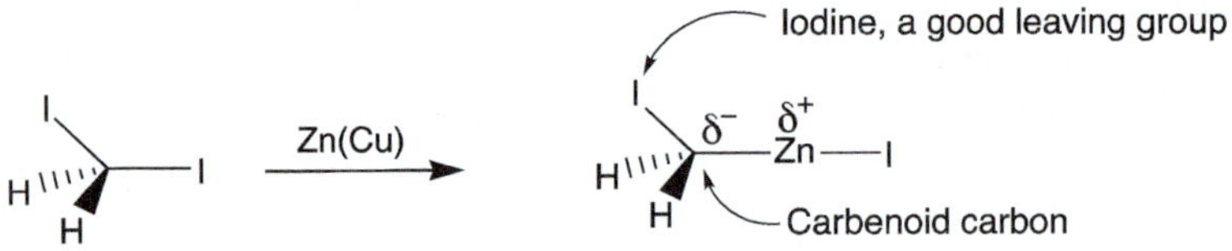

(8-82)

The initial product of the Simmons–Smith reaction is classified as carbenoid, and not a carbene. This species contains a very good leaving group, iodine that is bonded to a carbon. The same carbon is also bonded to a zinc, which means that the bonding electrons of the polar carbon–zinc bond are closer to the more electronegative carbon than zinc, hence the carbon of this compound is classified as carbene-like hence, "carbenoid." Reaction (8-83) gives an example of cyclopropanation using the Simmons–Smith reaction.

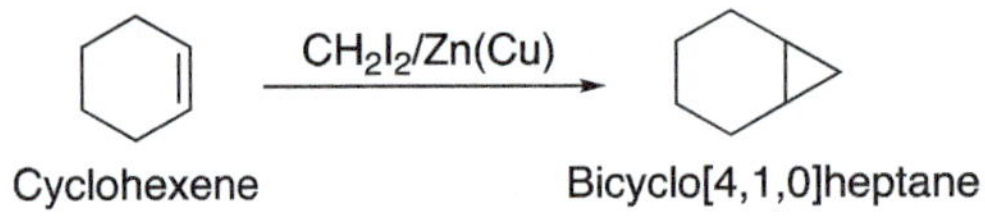

Cyclohexene → $CH_2I_2/Zn(Cu)$ → Bicyclo[4,1,0]heptane (8-83)

Problem 8.13

Give the major organic product of the reactions shown below.

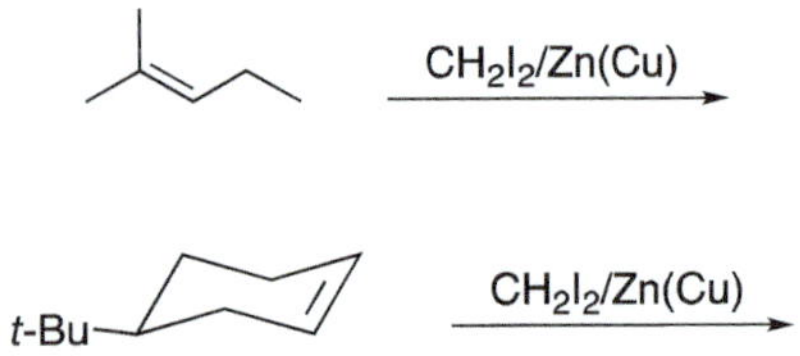

$CH_2I_2/Zn(Cu)$

t-Bu— $CH_2I_2/Zn(Cu)$

Another method for the generation of carbenes is accomplished by exposing diazomethane to energy in the form of heat or light, as shown in Reaction (8-84).

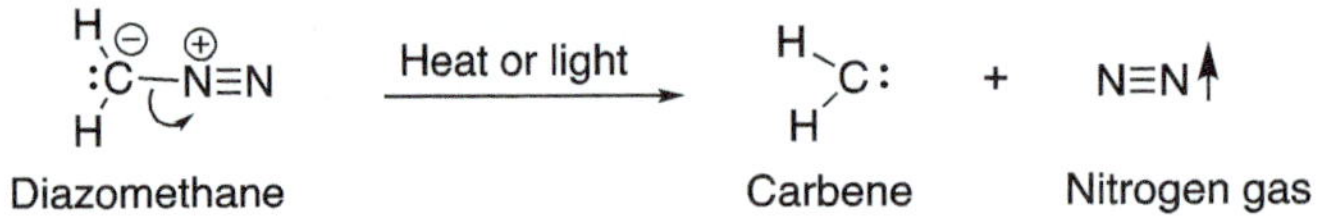

Diazomethane — Heat or light → Carbene + Nitrogen gas (8-84)

Diazomethane is very reactive since it liberates a gas (nitrogen) and hence an increase in entropy when exposed to energy to break the C—N bond to produce a carbene and nitrogen.

8.8 The Mechanism for Addition Reactions Involving Alkynes

Owing to the presence of pi (π) bonds in alkynes, addition reactions involving alkynes are similar to those encountered with alkenes. A major difference, however, is that there are two pairs of pi (π) electrons; thus, alkynes can react with two moles of the electrophile and nucleophile as shown in Reactions (8-85) and (8-86).

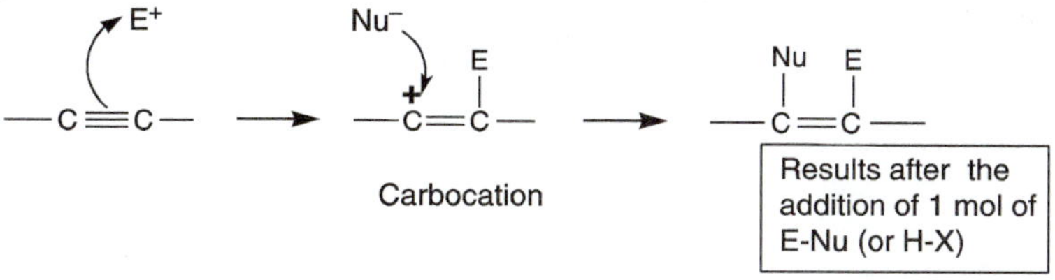

(8-85)

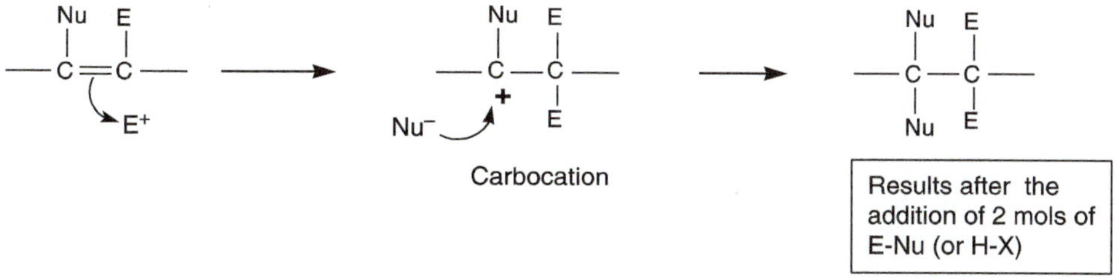

(8-86)

8.8.1 Addition of Bromine to Alkynes

As mentioned earlier, owing to the polarizability of the very large bromine molecule, it is possible for it to add to the nucleophilic alkyne. Shown in Reaction (8-87) is the first step in the mechanism for addition of bromine to 2-butyne to form a bromonium ion.

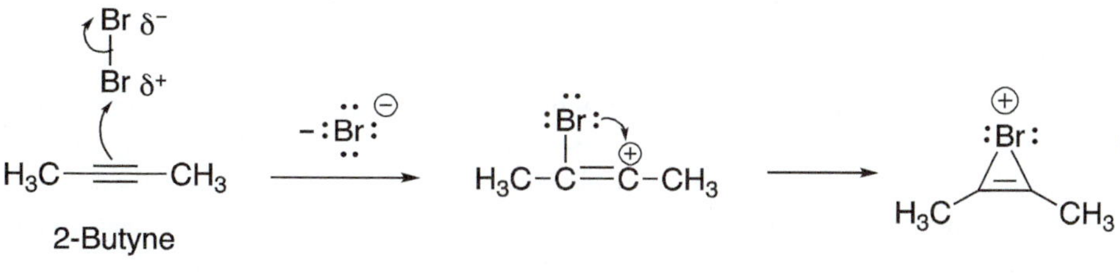

(8-87)

Note the stabilization of the alkene bromonium ion due to the large polarizable bromine atom. In the final step of the reaction, the nucleophilic bromide anion adds to the bromonium ion to give the trans-dibromide product alkene as shown in Reaction (8-88).

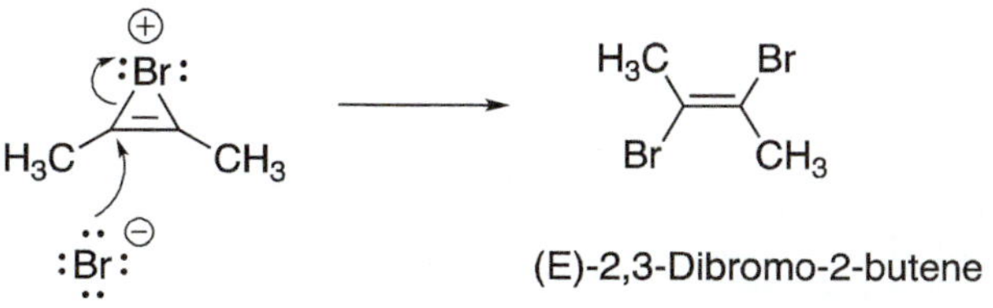

(8-88)

(E)-2,3-Dibromo-2-butene

In the presence of excess bromine, an additional reaction will take place with the resulting product of Reaction (8-88) as shown in Reaction (8-89).

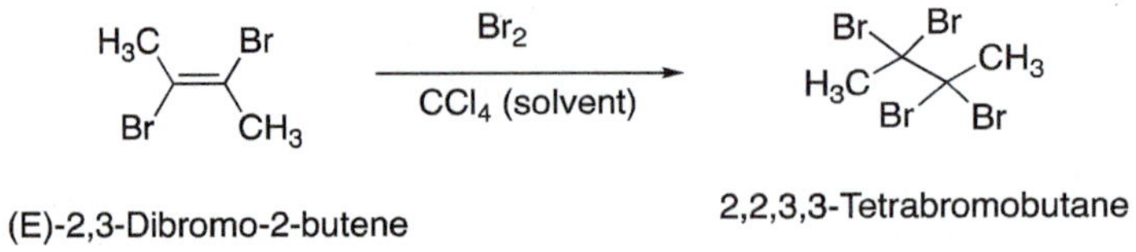

$$\text{(E)-2,3-Dibromo-2-butene} \xrightarrow{\text{Br}_2, \text{CCl}_4 \text{ (solvent)}} \text{2,2,3,3-Tetrabromobutane} \tag{8-89}$$

Problem 8.14

Give the major organic product of the reactions shown below.

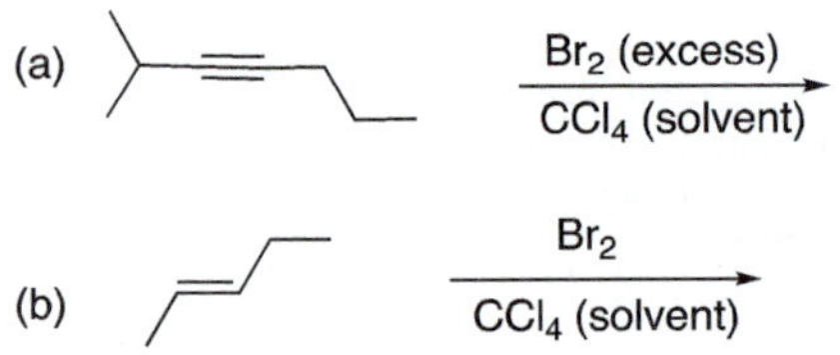

(a) (CH₃)₂CH−C≡C−CH₂CH₃ $\xrightarrow{\text{Br}_2 \text{ (excess)}}_{\text{CCl}_4 \text{ (solvent)}}$

(b) (alkene) $\xrightarrow{\text{Br}_2}_{\text{CCl}_4 \text{ (solvent)}}$

8.8.2 Addition of Hydrogen Halide to Alkynes

The mechanism for the addition of HBr or HCl to alkynes is similar to the addition reaction of these reagents to alkenes, except that the possibility exists for two moles to be added since alkynes have two sets of pi (π) electrons. Reaction (8-90) shows the addition of hydrobromic acid to propyne.

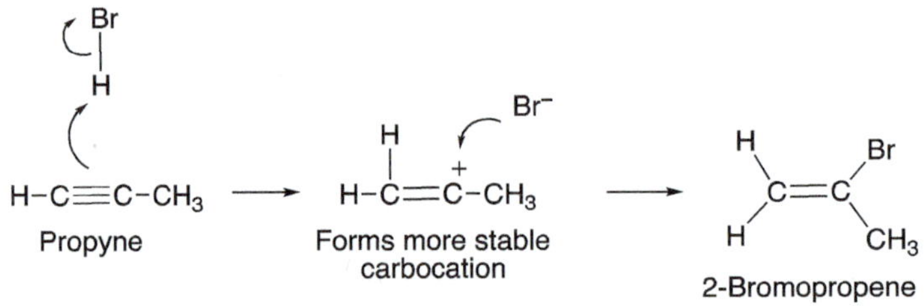

$$\text{H-C}\equiv\text{C-CH}_3 \longrightarrow \text{H-C}=\overset{+}{\text{C}}\text{-CH}_3 \longrightarrow \text{2-Bromopropene} \tag{8-90}$$

Propyne → Forms more stable carbocation → 2-Bromopropene

Note that in the first step of the reaction mechanism, the addition is regiospecific and follows Markovnikov addition, in which the hydrogen adds to the carbon of the triple bond with the greatest number of hydrogens, in this case the terminal carbon, which has only one hydrogen. As expected, in the presence of excess HBr, another addition reaction will take place to the resulting alkene, as shown in Reaction (8-91).

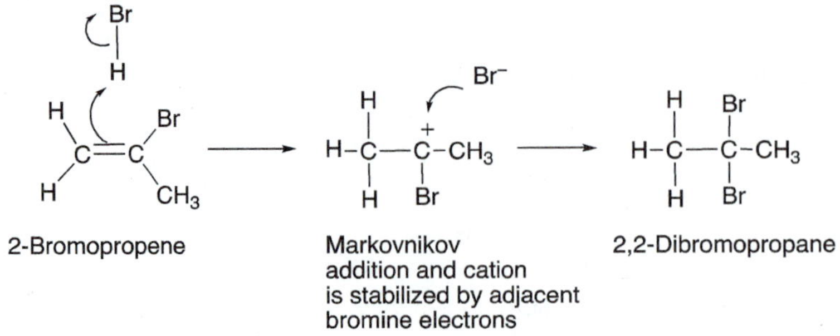

$$\text{2-Bromopropene} \longrightarrow \text{Markovnikov addition and cation is stabilized by adjacent bromine electrons} \longrightarrow \text{2,2-Dibromopropane} \tag{8-91}$$

8.8 The Mechanism for Addition Reactions Involving Alkynes

Note that the addition of the second mole of HBr to the alkene also follows the Markovnikov addition. If the same reaction were carried out involving an alkyne with an internal triple bond, there is a mixture of addition products.

Problem 8.15

Give the products that result after the addition of excess HCl to 2-pentyne.

8.8.3 Addition of Water to Alkynes

Once a good understanding of the mechanism of different addition reactions is gained, that knowledge can be applied to understanding of the same type of reactions involving a wide range of different molecules. The addition of H-OH (H_2O) to alkynes follows a very similar mechanism as that of the addition of water to alkenes. Recall that the addition of water to alkenes requires a catalyst, such as a proton, which is an extremely good electrophile to get the reaction started. For the addition of water to alkynes, the electrophilic catalyst that is typically used is Hg^{2+} in the form of the mercury salt $HgSO_4$. In the first step of the mechanism, the nucleophilic alkyne adds to the electrophilic mercury cation, which is then attacked by water to yield the mercury enol as shown in the Reaction (8-92).

$$R-C\equiv C-H \xrightarrow[H_3O^+]{HgSO_4} \cdots \quad (8\text{-}92)$$

In the next step of the mechanism, the nucleophilic double bond reacts with a proton to generate a stabilized carbocation, which is stabilized through resonance by the electrons on the adjacent oxygen, as shown in Reaction (8-93).

$$\cdots \quad (8\text{-}93)$$

In the next step of the reaction mechanism, the mercury ion is lost to form an enol as shown in Reaction (8-94).

$$\cdots \xrightarrow{-Hg^{2+}} \text{Enol} \xrightleftharpoons[\text{tautomerism}]{K_T} \text{Keto} \quad (8\text{-}94)$$

Enols are not stable and they quickly undergo a migration of the double bond and hydrogen to form the more stable carbonyl-containing compound. The carbonyl compound is more stable than the enol since the carbon–oxygen double bond is shorter and stronger than the carbon–carbon double bond of the enol. The equilibrium involving these two species is called a **keto-enol equilibrium or keto-enol tautomerism**. The mechanism for the keto-enol tautomerism in an acidic medium is shown in Reaction (8-95).

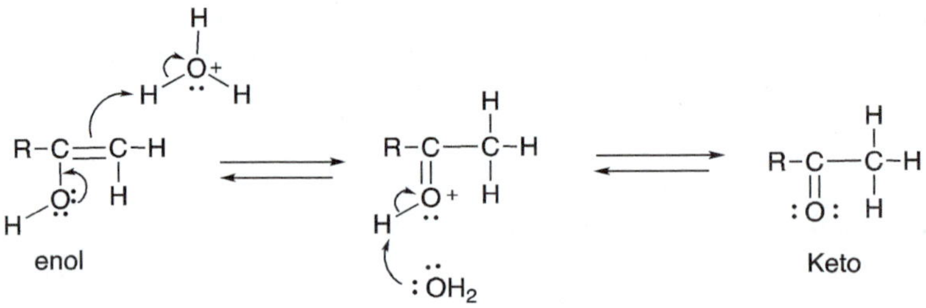

8 Addition Reactions Involving Alkenes and Alkynes

$$\text{enol} \rightleftharpoons \cdots \rightleftharpoons \text{Keto} \tag{8-95}$$

Thus, for the addition of water to terminal alkynes in the presence of an electrophilic catalyst, such as Hg^{2+}, a methyl ketone is the major organic product since the reaction is regiospecific. For the addition of water to symmetrical alkynes in the presence of an electrophilic catalyst, such as Hg^{2+}, there is only one product as the major organic product. On the other hand, if the alkyne is unsymmetrical, the possibility exists for the formation of two different organic ketones as shown in Reaction (8-96).

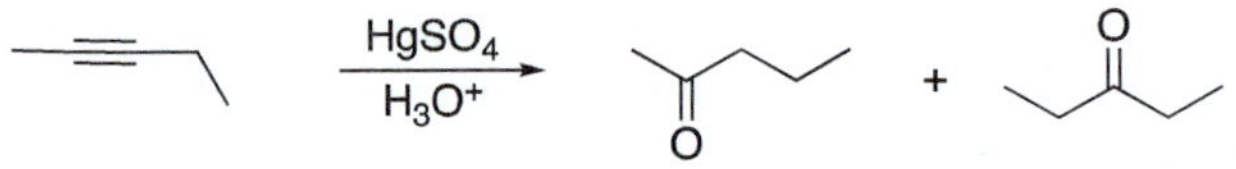

$$\xrightarrow{HgSO_4 / H_3O^+} \quad + \quad \tag{8-96}$$

It is possible to have a regiospecific hydration reaction to a terminal alkyne which is the opposite of that discussed above. Consider the reaction of terminal alkyne shown in Reaction (8-97) with a much bulkier borane than the one used earlier for the hydroboration–oxidation of alkenes. For this reaction, the very bulky substituted disiamylborane[bis(1,2-dimethylpropyl)borane], abbreviated as sia$_2$BH, is used and will ensure an addition to the alkyne in an anti-Markovnikov manner for the same reason explained previously in Section 8.6.2.

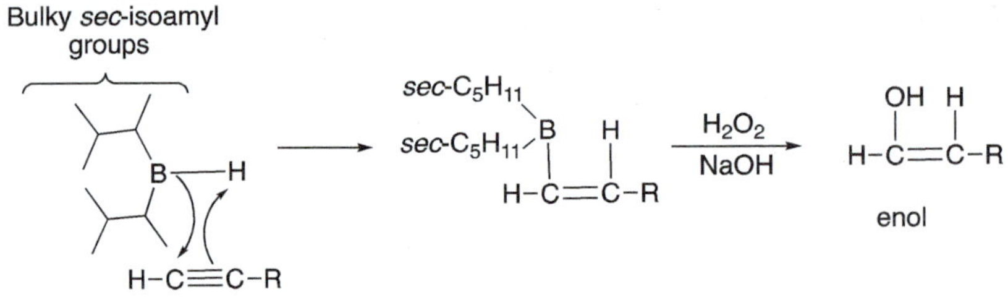

$$\xrightarrow{} \xrightarrow{H_2O_2 / NaOH} \tag{8-97}$$

The final product of the hydroboration reaction (after oxidation with H_2O_2 in the presence of NaOH) is an enol. As mentioned earlier, the keto-enol equilibrium favors the carbonyl form, which is the final product as shown in Reaction (8-98).

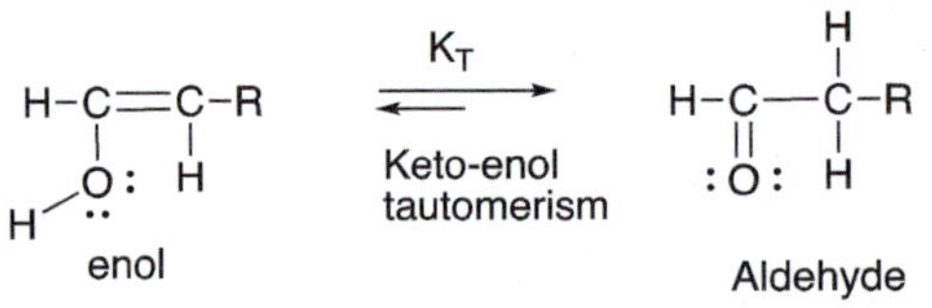

$$\text{enol} \xrightleftharpoons[\text{tautomerism}]{K_T \text{ Keto-enol}} \text{Aldehyde} \tag{8-98}$$

Thus, for a reaction involving a terminal alkyne and this very bulky diborane, disiamylborane [bis(1,2-dimethylpropyl)borane], followed by oxidation, the final product is an aldehyde as shown in Reaction (8-99).

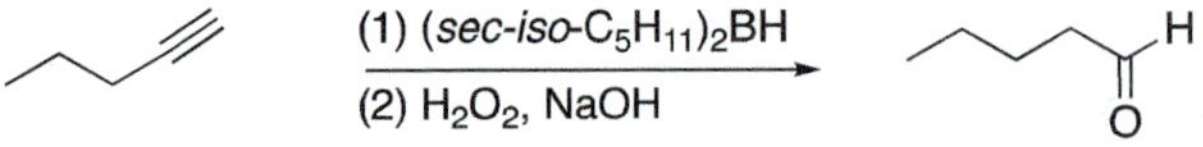

$$\xrightarrow[\text{(2) } H_2O_2, \text{ NaOH}]{\text{(1) } (sec\text{-}iso\text{-}C_5H_{11})_2BH} \tag{8-99}$$

Note that these reactions using disiamylborane, followed by oxidation result in reactions that are regiospecific and anti-Markovnikov.

Problem 8.16

For the reactions given below, provide either the major organic product or an appropriate reactant.

(a) (CH₃)₂CHC≡CH → (1) (sec-iso-C₅H₁₁)₂BH (2) H₂O₂, NaOH

(b) ? → (1) (sec-iso-C₅H₁₁)₂BH (2) H₂O₂, NaOH → CH₃CH(CH₃)CH₂CHO

8.9 Applications of Addition Reactions to Synthesis

In this section, methods will be developed to make strategic choices in the use of addition reactions, along with some of the other reactions covered in Chapter 6 in order to accomplish specific transformations and the synthesis of specific target products. An important aspect of organic chemistry is the application of various reaction types to synthesize specific target molecules, which are typically larger molecules and with new functional groups. This area of organic chemistry is called synthetic organic chemistry. For the addition reaction involving nonsymmetrical alkenes, there is typically a mixture of products, but if specific reactions are used, one product can be obtained as the major product. As a result, a careful analysis of the functional groups of a target molecule must be carried out in order to determine the type of addition reaction that must be used in order to synthesize the target molecule. Consider the transformation shown in Reaction (8-100).

1-Methylcyclohexene —?→ 2-methylcyclohexanone   (8-100)

Starting reagent   Target molecule

Your first task is to make some observations about the target molecule. One observation is that it has a ketone functionality, and one of the reactions that was covered in Chapter 6 to make ketones is the oxidation of secondary alcohols to give ketones. Thus, it is possible to oxidize the corresponding alcohol to obtain the target ketone, as shown in Reaction (8-101).

2-Methylcyclohexanol —Oxidation (KMnO₄)→ Target molecule   (8-101)

The next task is to recognize a type of reaction that can be used to make 2-methylcyclohexanol. An anti-Markovnikov addition of water to an alkene will accomplish this transformation as shown in Reaction (8-102).

1-Methylcyclohexene —(1) B₂H₆ (2) H₂O₂, OH⁻→ 2-Methylcyclohexanol   (8-102)

Note that for the Reaction (8-102), a strategic choice must be made in using the hydroboration–oxidation reaction and not mercuration–demercuration. Another choice would be the addition of water in a catalytic amount of acid, but there are two potential problems; first, the major product would be the Markovnikov addition and second, there is the possibility of rearrangement to give different products. Thus, the sequence of reactions that would be used for the transformation shown in Reaction (8-100) is shown in the reaction sequence given in Reaction (8-103).

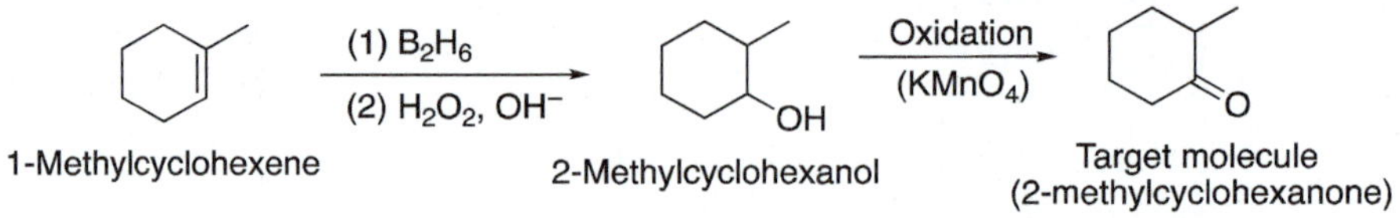

$$\text{1-Methylcyclohexene} \xrightarrow[\text{(2) } H_2O_2,\ OH^-]{\text{(1) } B_2H_6} \text{2-Methylcyclohexanol} \xrightarrow{\text{Oxidation} \atop (KMnO_4)} \text{Target molecule (2-methylcyclohexanone)} \quad (8\text{-}103)$$

Problem 8.17

Determine the appropriate starting alkene and reaction conditions for the synthesis of the compounds shown below.

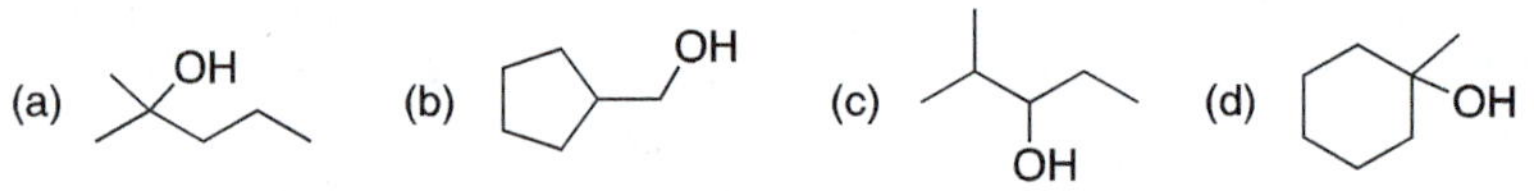

(a), (b), (c), (d) [structures shown]

The same analysis can be applied for the synthesis of other similar compounds, and Problem 8.18 is designed to apply that approach to devise appropriate synthetic routes for the synthesis of different target molecules.

Problem 8.18

Show how to carry out the following transformations, clearly show all steps in the synthesis.

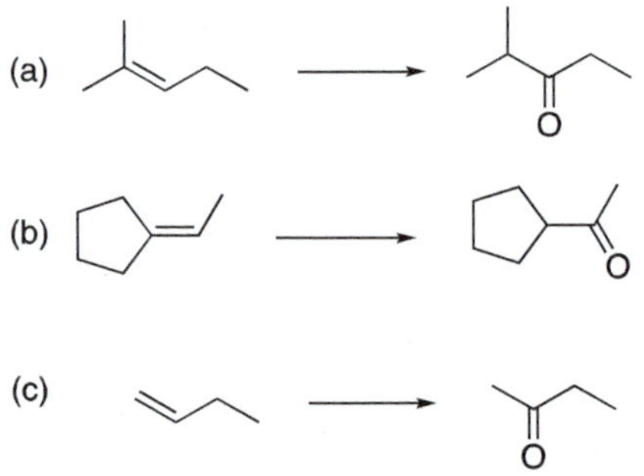

(a), (b), (c) [structures shown]

End of Chapter Problems

8.19 Give the major organic product for the reaction of 3-methyl-2-pentene with each of the following reagents:

a) (i) $Hg(OAc)_2/H_2O$, (ii) $NaBH_4$
b) (i) BH_3, (ii) H_2O_2, OH^-
c) CH_2I_2, Zn/Cu
d) H_2O, Br_2, DMSO (solvent)

8.20 Give a mechanism to explain the reaction shown below:

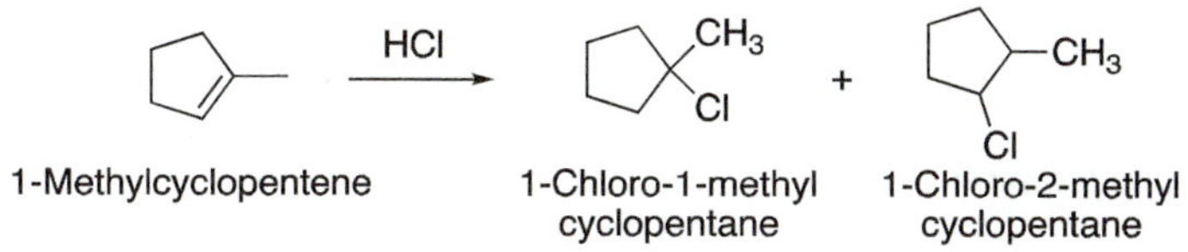

1-Methylcyclopentene →(HCl) 1-Chloro-1-methylcyclopentane + 1-Chloro-2-methylcyclopentane

8.21 Give the products for each of the following addition reactions.

(i) CH₃CH=CHCH₃ + H-Cl → ?

(ii) (CH₃)₂C=C(CH₃)₂ + Br₂ / CCl₄ (solvent) → ?

(iii) methylenecyclopentane + H-Cl → ?

(iv) cyclohexene + HBr → ?

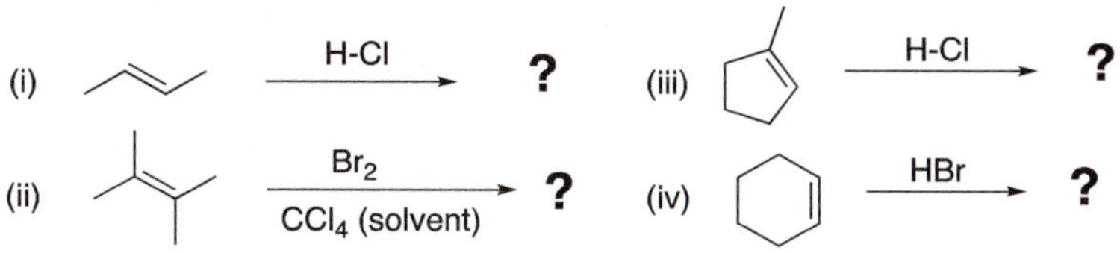

8.22 Give the reactant for each of the following addition reactions.

(i) ? + Br₂, H₂O / DMSO (solvent) → (CH₃)₂C(OH)CH(Br)CH₃

(ii) ? + (1) BH₃ (2) H₂O₂, NaOH → (CH₃)₂CHCH(OH)CH₃

(iii) ? + (1) Hg(OAc)₂, H₂O (2) NaBH₄ → 1-methylcyclopentanol

(iv) ? + HBr → 1-bromo-1-methylcyclohexane

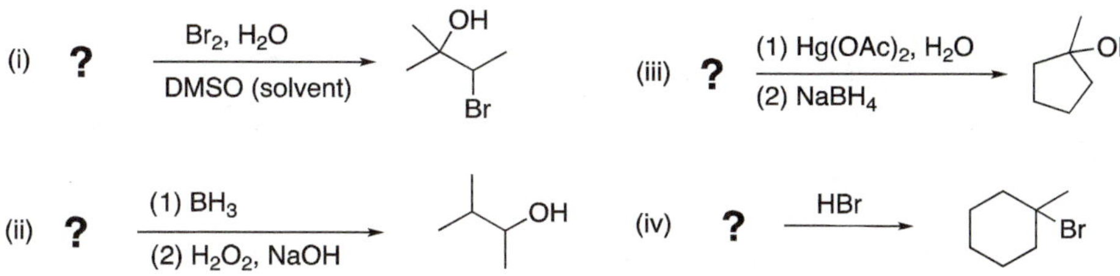

8.23 Complete the following reactions by giving the <u>structures</u> of the reactant, missing reagents, or <u>major organic product.</u> Include appropriate stereochemistry in the products where applicable.

(a) (Z)-Pentene →(Br₂, CCl₄) [Newman projection with H, Br, CH₂CH₃ on front; H, Br, CH₃ on back]

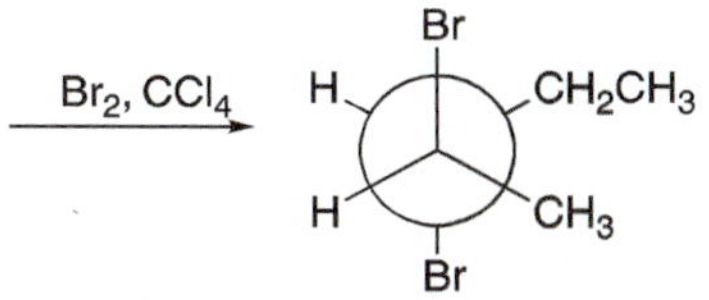

(b) 1-methylcyclohexene →(Br₂, H₂O / DMSO (solvent))

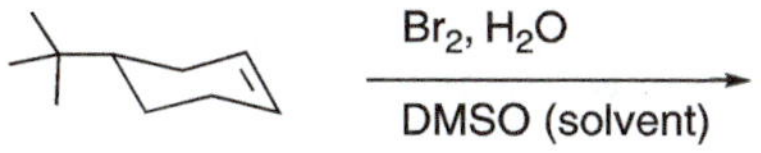

(c) (CH₃)₂C=CHCH₂CH₃ → → (CH₃)₂C(OH)CH(Br)CH₂CH₃... → product with OH and Br

(d) IUPAC name of reactant →(1) Hg(OAc)₂, H₂O (2) NaBH₄ → 1-methylcyclopentanol with CH₃ and OH

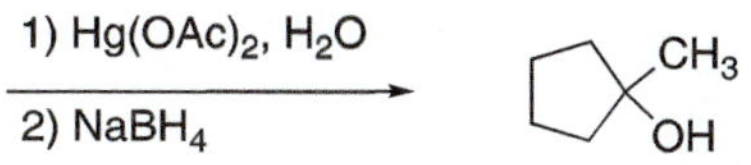

(e) 2-Methyl-1-pentene → → product with OH and Br

(f)  [alkene] → [product with OH and Br]

(g) 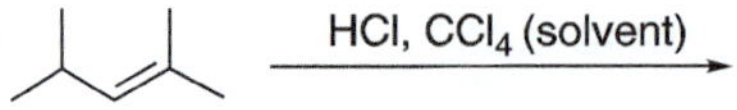 [alkene] —HCl, CCl₄ (solvent)→

(h)  [alkene] —HBr→ [cyclohexane with Br]

(i)  [methylenecyclohexane] —HCl→

(j)  [alkene] → [tertiary bromide]

(k)  —HBr→ [bromocyclopentane derivative]

(l) 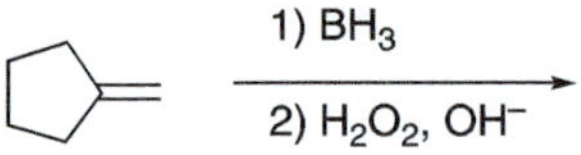 [methylenecyclopentane] —1) BH₃; 2) H₂O₂, OH⁻→

(m) 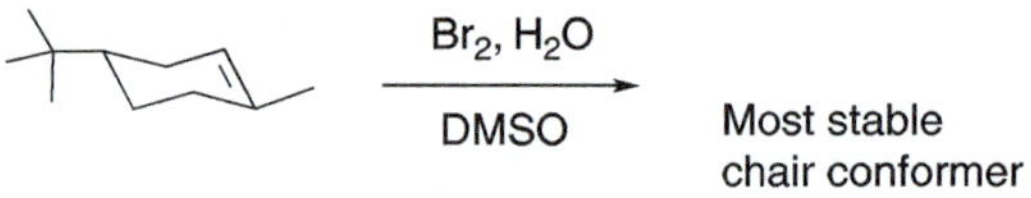 [substituted cyclohexene] —Br₂, H₂O / DMSO→ Most stable chair conformer

(n) 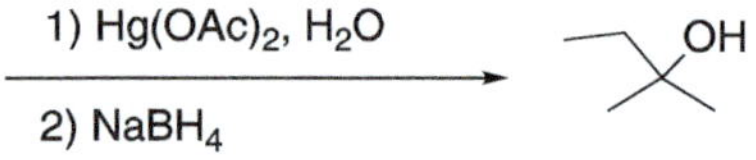 —1) Hg(OAc)₂, H₂O; 2) NaBH₄→ [tertiary alcohol]

(o) 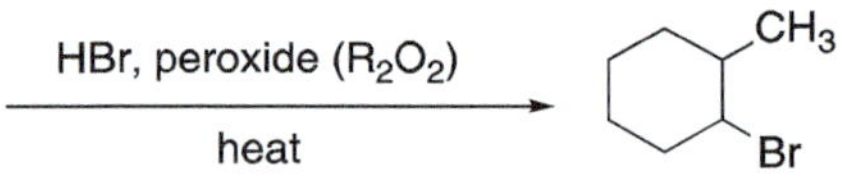 —HBr, peroxide (R₂O₂), heat→ [cyclohexane with CH₃ and Br]

(p) 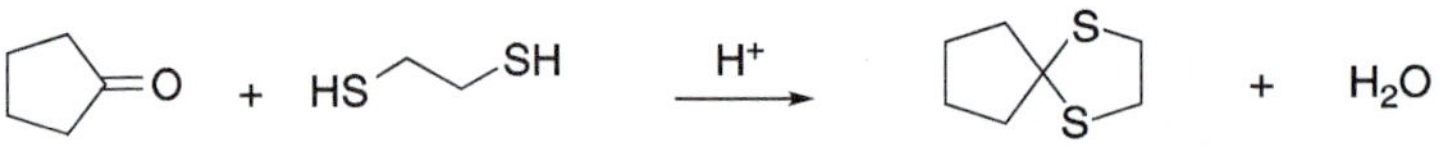 [cyclopentanone] + HS−CH₂CH₂−SH —H⁺→ [dithiolane spiro] + H₂O

(q) 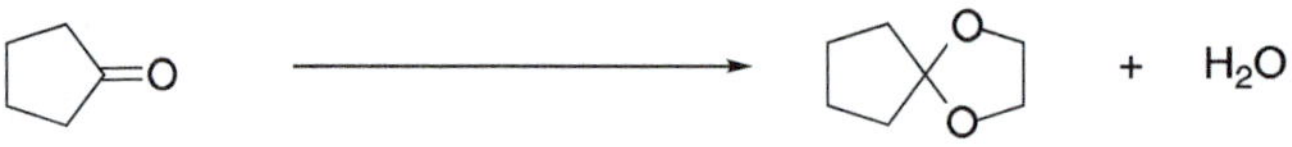 [cyclopentanone] → [dioxolane spiro] + H₂O

(r) 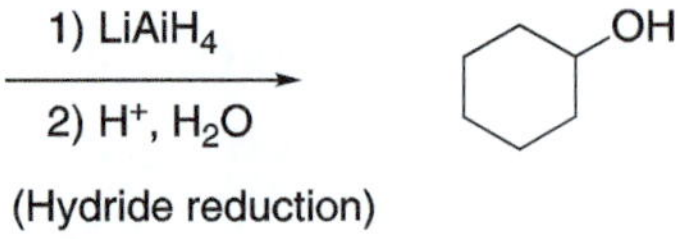 —1) LiAlH₄; 2) H⁺, H₂O→ [cyclohexanol]
(Hydride reduction)

(s) 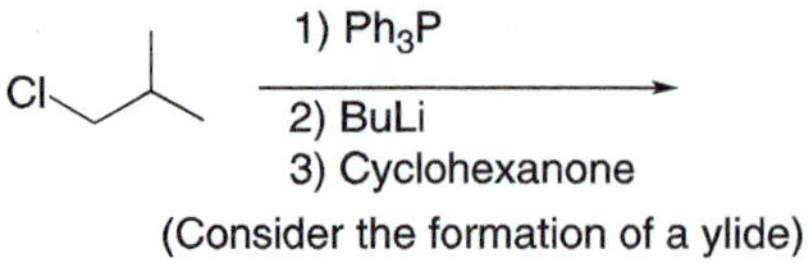 Cl−CH₂CH(CH₃)₂ —1) Ph₃P; 2) BuLi; 3) Cyclohexanone→
(Consider the formation of a ylide)

(t) 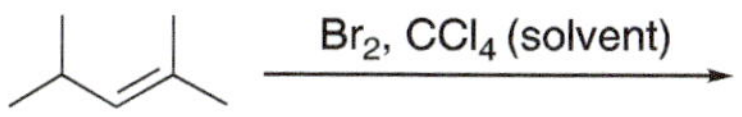 [2-methyl-2-butene] — Br₂, CCl₄ (solvent) →

(u)  [2-methyl-1-pentene] → [product shown: tertiary OH, primary Br on adjacent carbons]

(v)  [methylenecyclopentane] — HCl → [1-chloro-1-methylcyclopentane]

(w) 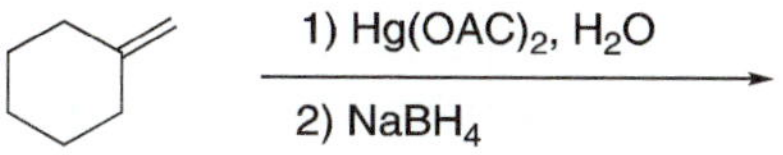 [methylenecyclohexane] — 1) Hg(OAc)₂, H₂O; 2) NaBH₄ →

(x) 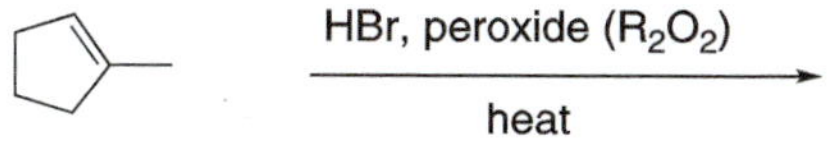 [cyclopentene] — HBr, peroxide (R₂O₂), heat →

(y) 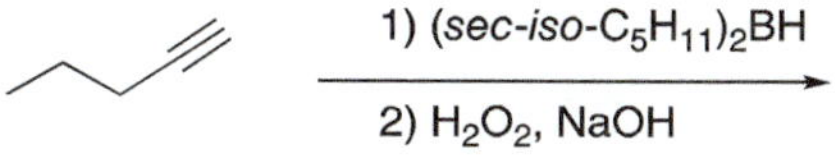 [1-butyne] — 1) (sec-iso-C₅H₁₁)₂BH; 2) H₂O₂, NaOH →

8.24 Provide the reagent(s) necessary to carry out that lettered transformation.

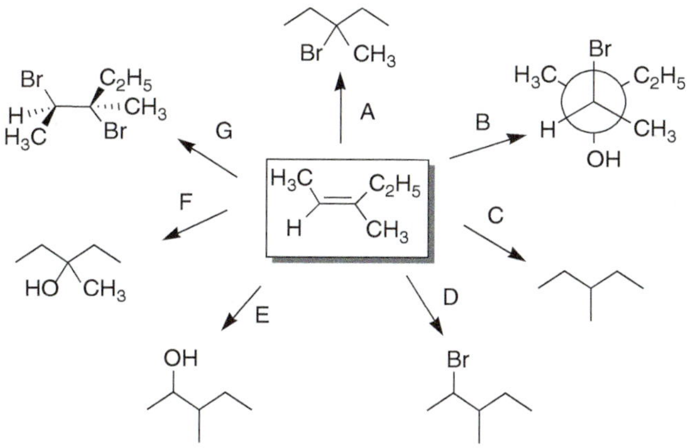

[Central alkene: (E)-3-methyl-2-pentene with substituents H₃C, C₂H₅, H, CH₃; surrounded by products A–G reached via arrows labeled A, B, C, D, E, F, G]

- A: dibromide (Br, CH₃ shown at top)
- B: bromohydrin (Br and OH with H₃C, C₂H₅, CH₃)
- C: 2-methylbutane (hydrogenation product, shown at right)
- D: secondary bromide (3-bromo-2-methylbutane-like structure)
- E: alcohol (OH on secondary carbon)
- F: tertiary alcohol (HO, CH₃)
- G: vicinal dibromide (Br, H, H₃C, C₂H₅, CH₃, Br — anti addition)

8.25 Using arrows to indicate electron movement, give a step-by-step description of a mechanism to explain the product shown for the reaction below. Clearly show the structures of all intermediates, along with the appropriate formal charge in your mechanism (hint: note that there are two alkene groups, which means two nucleophiles present in the molecule).

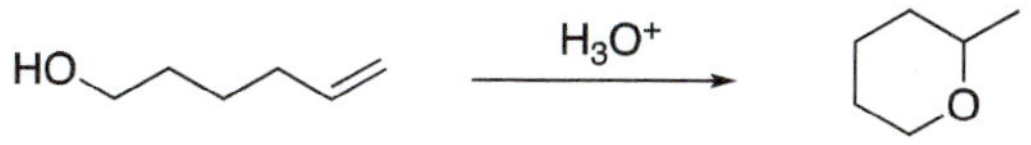

HO–CH₂CH₂CH₂CH₂–CH=CH₂ —H₃O⁺→ [2-methyltetrahydropyran]

8.26 Using arrows to indicate electron movement, give a step-by-step description of a mechanism to explain the product shown for the reactions below. Clearly show the structures of all intermediates, along with the appropriate formal charge in your mechanism (hint: note that there are two alkene groups, which means two nucleophiles present in the molecule).

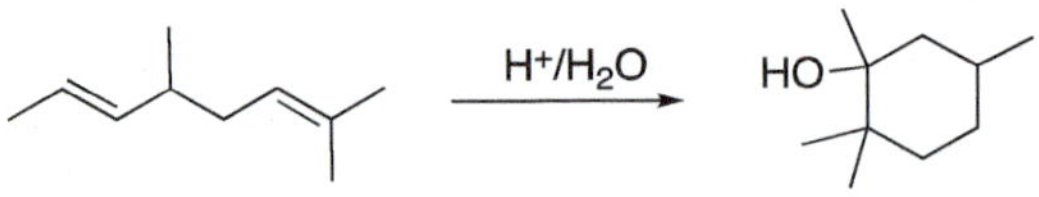

8.27 The following cyclization was observed when trying to carry out an addition of Br₂ to the unsaturated compound shown below. Using arrows to indicate electron movement, give a step-by-step description of a mechanism to explain the products shown for the reaction below. Clearly show the structures of all intermediates, along with the appropriate formal charge in your mechanism.

8.28 The following cyclization was observed when trying to carry out an addition of Br₂ to the unsaturated compound shown below. Using arrows to indicate electron movement, give a step-by-step description of a mechanism to explain the unexpected product shown for the reaction below. Clearly show the structures of all intermediates, along with the appropriate formal charge in your mechanism (hint: the first step in the mechanism is given).

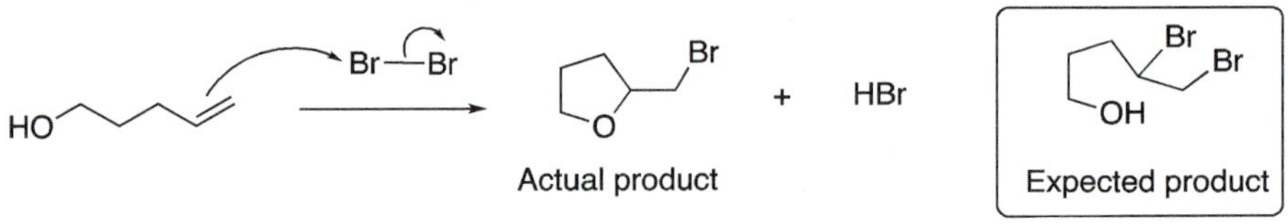

8.29 The routine addition of HCl across the double bond of a vinylcyclohexane gave a small amount of an unexpected rearranged product. Using arrows to indicate electron movement, give a step-by-step description of a mechanism to explain the product shown for the reaction below. Clearly show the structures of all intermediates, along with the appropriate formal charge in your mechanism.

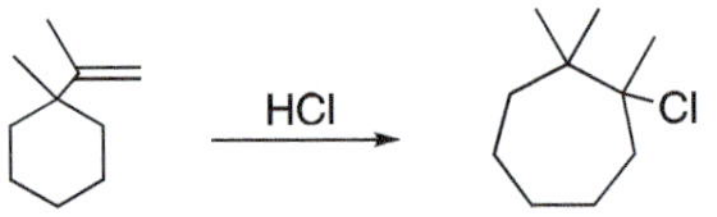

8.30 Rearrangement of carbocation intermediates often occurs for addition reactions. With this information in mind, explain by appropriate *mechanisms* (i.e. step-by-step descriptions and curved arrows to indicate electron movements) the formation of the organic product shown for the following reaction (hint: the first step in the mechanism is shown).

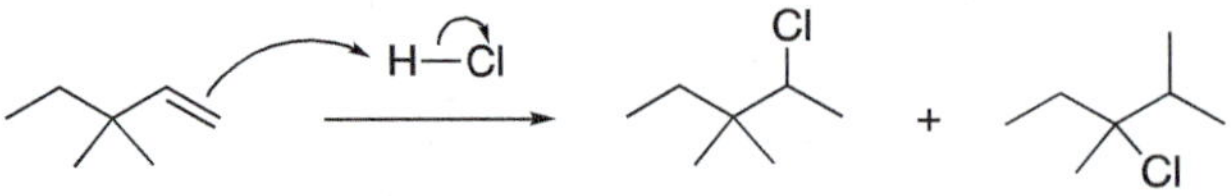

8.31 The following cyclization was observed in the oxymercuration–demercuration of the unsaturated compound shown below. Provide a reasonable mechanism to explain the formation of the products shown.

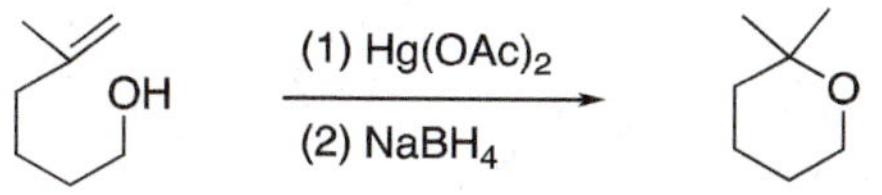

8.32 Explain by an appropriate *mechanism* (i.e. step-by-step descriptions and curved arrows to indicate electron movements) the formation of the organic product shown for the following reaction (hint: the first step in the mechanism is shown).

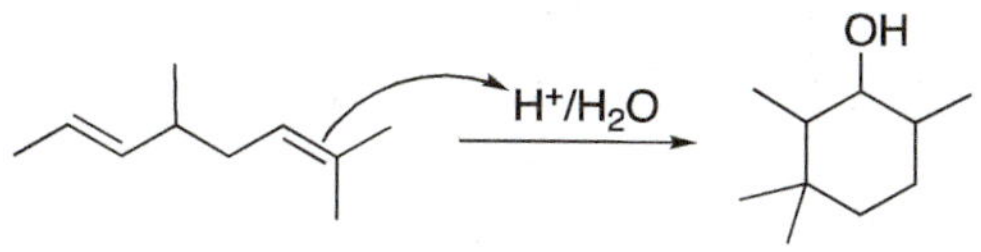

8.33 Using arrows to indicate electron movement, give a step-by-step description of a mechanism to explain the reaction shown below. Clearly show the structures of all intermediates, along with the appropriate formal charge in your mechanism.

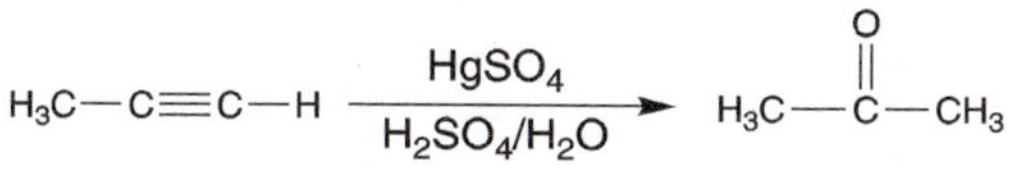

$$H_3C-C{\equiv}C-H \xrightarrow[H_2SO_4/H_2O]{HgSO_4} H_3C-\overset{\overset{O}{\|}}{C}-CH_3$$

8.34 Determine an appropriate starting alkene and reaction conditions for the synthesis of the compounds shown below.

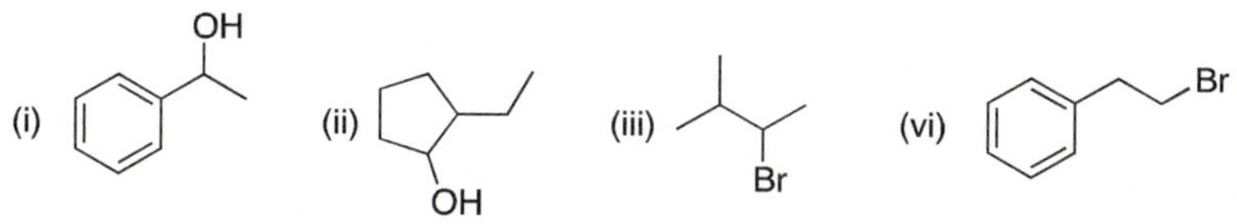

8.35 Give the structures for the molecules missing in the sequence of reactions shown below. The reagents used to carry out each transformation are represented by numbers and are shown below the reaction scheme.

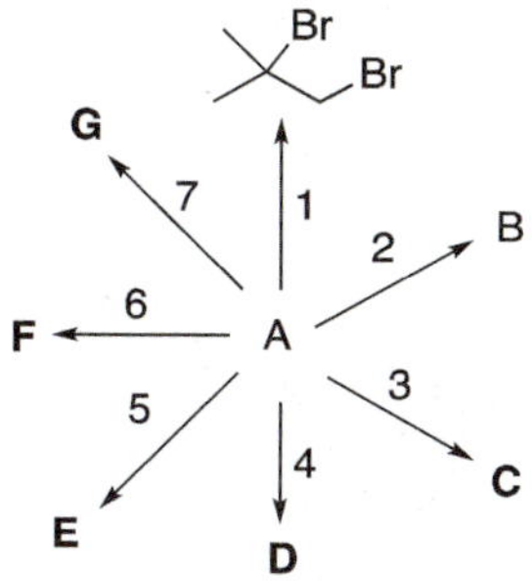

1. Br_2/CCl_4; **2.** HBr, peroxides; **3.** Br_2/H_2O; **4.** HBr (no peroxides); **5.** (i) BH_3, (ii) H_2O_2, OH^-; **6.** (i) $Hg(OAc)_2/H_2O$, (ii) $NaBH_4$; **7.** H_2/Pd.

8.36 For the reaction scheme shown below, provide the reagent(s) necessary to carry out each lettered transformation: (Note: some transformations may require more than one step and these steps must be indicated clearly).

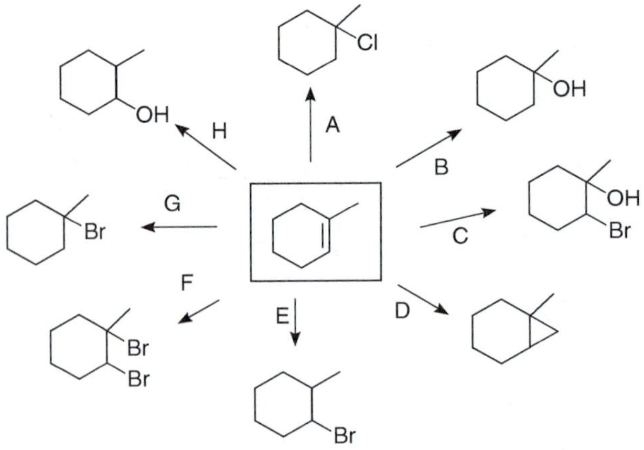

8.37 Supply the necessary reagents and reaction conditions to complete the reaction scheme shown below.

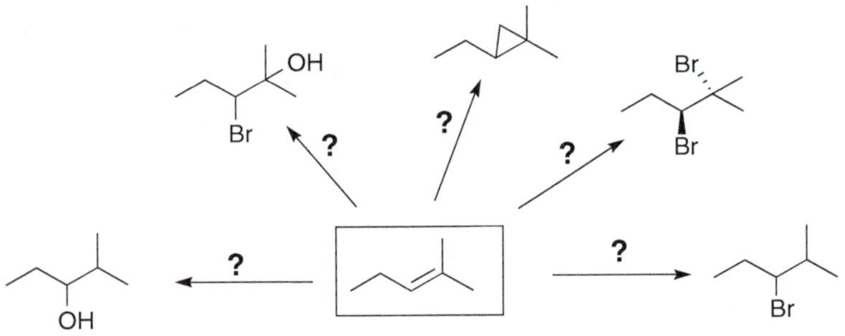

8.38 The routine addition of HBr across the double bond of the compound shown below gave an unexpected rearranged compound as one of the products. Propose a step-by-step mechanism for the formation of this product.

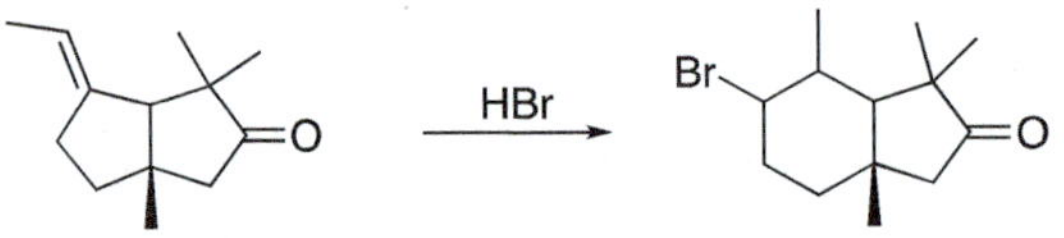

8.39 Due to the presence of pi (π) bonds, alkenes are nucleophilic. With this information in mind, propose a step-by-step mechanism using curved arrows to indicate electron movements for the reaction shown below (hint: in the first step a carbocation is formed, which is attacked by the other nucleophilic double bond to form a ring before being attacked by the bromide anion).

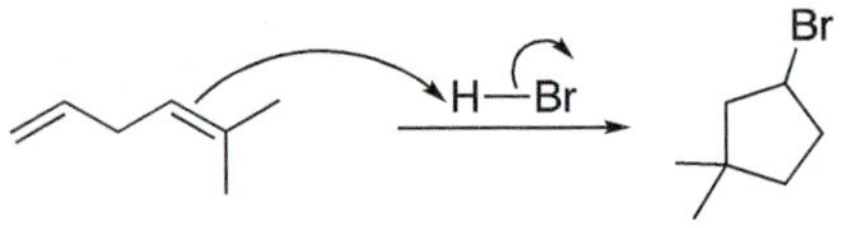

8.40 Show how to accomplish the following transformations (synthesis). All reagents, reaction conditions, and appropriate solvents must be **clearly** shown.

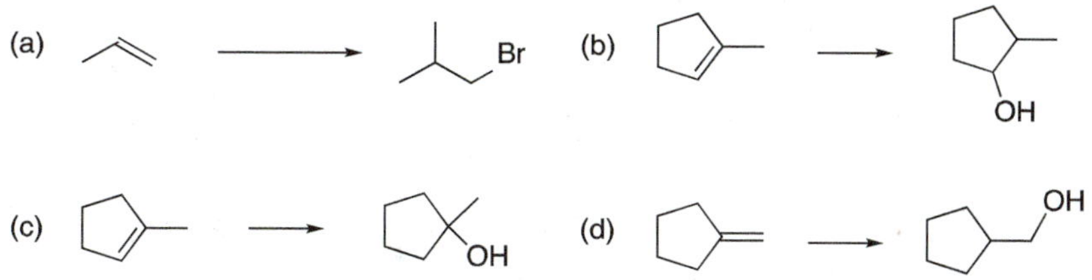

(a) CH₂=CHCH₃ ⟶ (CH₃)₂CHCH₂Br

(b) cyclopentene ⟶ cyclopentanol

(c) cyclopentene ⟶ 1,2-dimethylcyclopentane-OH (cyclopentanol with substituents as shown)

(d) methylenecyclopentane ⟶ cyclopentyl-CH₂OH

8.41 For the addition of HBr to 2-methylpropene, the major product that was isolated is 2-bromo-2-methylpropane, and only a trace of 1-bromo-2-methylpropane obtained. On the other hand, for the same addition reaction, but in the presence of peroxide and heat, the major product is 1-bromo-2-methylpropane and the minor product is 2-bromo-2-methylpropane. Write appropriate mechanisms to explain the formation of both products under the two different reaction conditions, i.e. no peroxide and with peroxide and heat, respectively (hint: peroxides (ROOR) readily dissociate to form radicals).

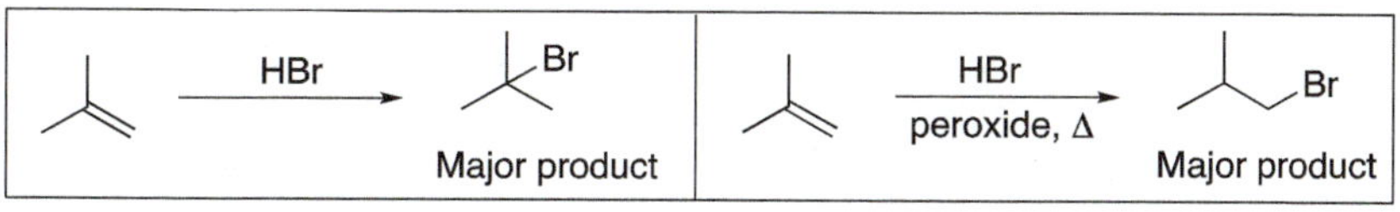

(CH₃)₂C=CH₂ —HBr→ (CH₃)₃CBr   Major product

(CH₃)₂C=CH₂ —HBr, peroxide, Δ→ (CH₃)₂CHCH₂Br   Major product

9

Addition Reactions Involving Carbonyls and Nitriles

9.1 Introduction

In this chapter, we will continue our discussion of addition reactions, but we will be examining the addition of an electrophile and nucleophile to carbon–oxygen and carbon–nitrogen multiple bonds, instead of carbon–carbon multiple bonds. As we will see, these reactions are very important, especially in biology. For example, a very stable form of glucose, comes about due to an addition reaction involving an intramolecular addition of an alcohol to the carbon–oxygen double bond of an aldehyde functional group.

9.2 Mechanism for Addition Reactions Involving Carbonyl Compounds

Owing to the polarity of the carbonyl bond, which is present in aldehydes and ketones, the dipole moment of these compounds is greater than zero. For example, the dipole moment of acetone (2-propanone) is 2.85 debyes. Shown in (9-1) are the two resonance structures of the carbonyl functionality, note the charge separation in one of the resonance structures (the right structure), hence a minor contributor to the overall structure of the molecule; it does however illustrate that the carbon of the carbonyl group is electropositive and the oxygen is nucleophilic.

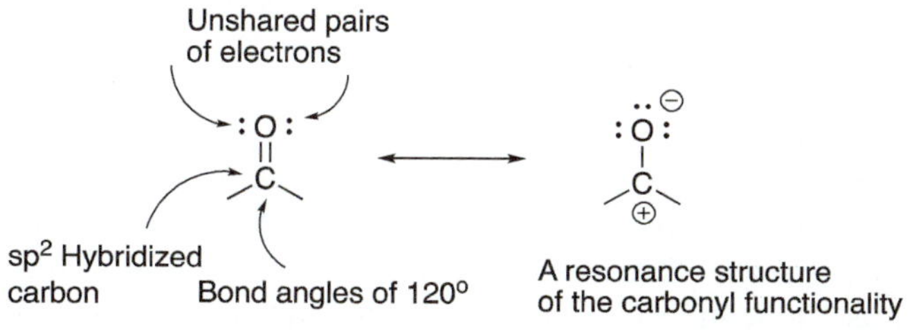

(9-1)

One observation that is readily made about compounds that contain the carbon–oxygen double bond functionality is that the pi (π) electrons are not equally distributed about the double bond, compared to an alkene. As a result, it is much easier to predict the addition of an electrophile or nucleophile to these compounds, compared to that of an alkene or alkyne. The addition of an electrophile to a carbonyl group is shown in Reaction (9-2).

Organic Chemistry: Concepts and Applications, First Edition. Allan D. Headley.
© 2020 John Wiley & Sons, Inc. Published 2020 by John Wiley & Sons, Inc.
Companion website: www.wiley.com/go/Headley_OrganicChemistry

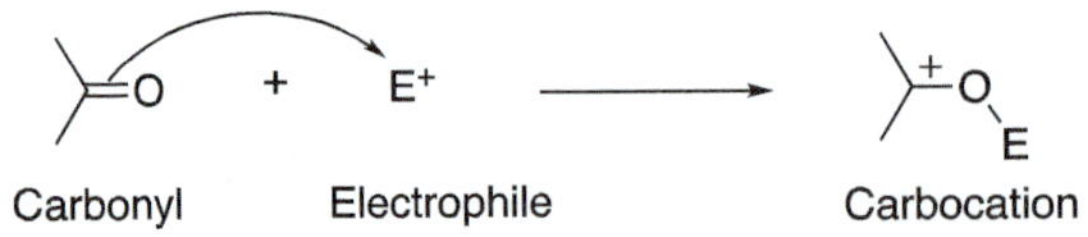

$$\text{Carbonyl} + E^+ \longrightarrow \text{Carbocation} \tag{9-2}$$

As you can imagine, in another step, a nucleophile will bond to the electrophilic carbon, as shown in Reaction (9-3) to form an addition product.

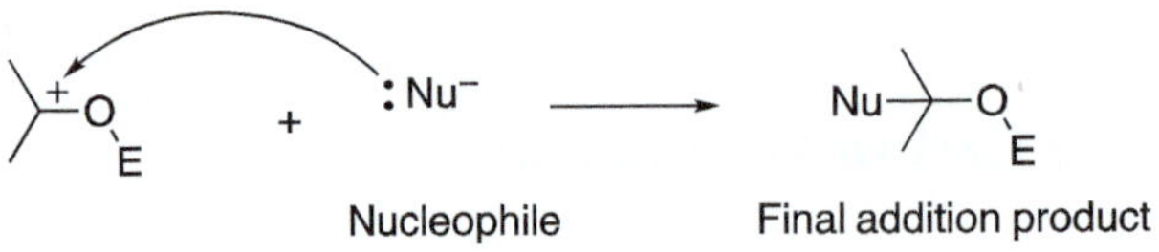

$$\text{(Carbocation)} + :Nu^- \longrightarrow \text{Final addition product} \tag{9-3}$$

Based on the predictability of the addition of an electrophile and nucleophile to the carbonyl functionality, these reactions are considered *regiospecific*. Of course, it is possible for the sequence of the addition steps to take place differently, and the first step involves the nucleophile adding to the electrophilic carbon of the carbonyl group, as shown in Reaction (9-4).

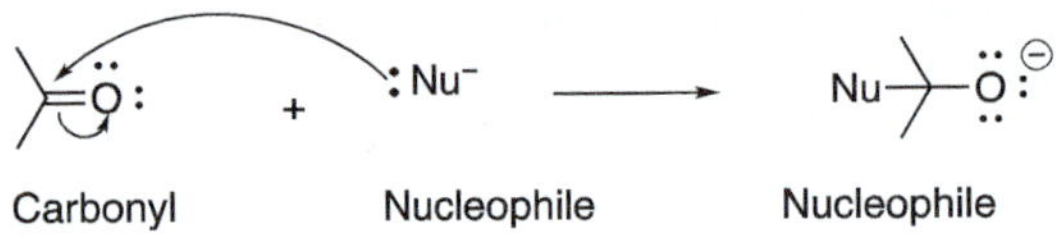

$$\text{Carbonyl} + :Nu^- \longrightarrow \text{Nucleophile} \tag{9-4}$$

In the next step, the electrophile will add to the nucleophilic intermediate that was created by the addition of the nucleophile, as shown in Reaction (9-5) to give a final neutral product.

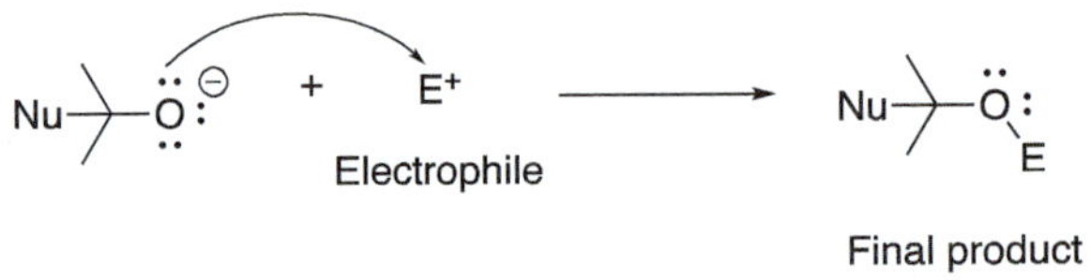

$$\text{(Nucleophile)} + E^+ \longrightarrow \text{Final product} \tag{9-5}$$

The overall hypothetical addition reaction involving the addition of a nucleophile and an electrophile to a carbonyl functionality is shown in Reaction (9-6).

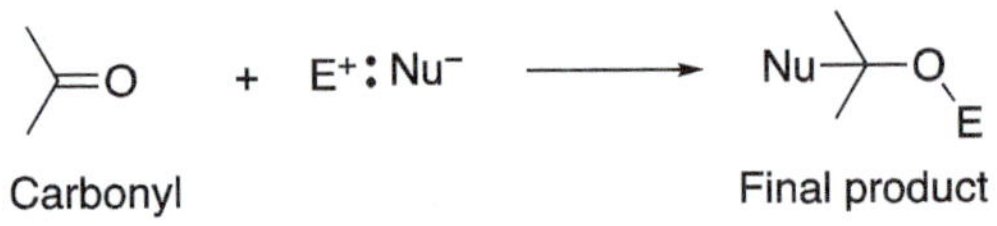

$$\text{Carbonyl} + E^+ : Nu^- \longrightarrow \text{Final product} \tag{9-6}$$

As we have seen from previous sections, it is possible to dictate a reaction path (or mechanism) by varying the reaction conditions. If this addition reaction were carried out in the presence of excess electrophile, such as an acidic medium, the electrophilic would add first. On the other hand, if carried out in the presence of excess nucleophile, such as a basic medium, the nucleophile would add first. Thus, the sequence of the addition pathway can be dictated based on the reaction conditions.

9.3   Addition of HCN to Carbonyl Compounds

Owing to the polarity of the covalent bond of hydrogen cyanide (HCN), it will add across the carbon–oxygen double bond in a *regiospecific* manner as shown in Reaction (9-7) to form a new type of compound called a cyanohydrin.

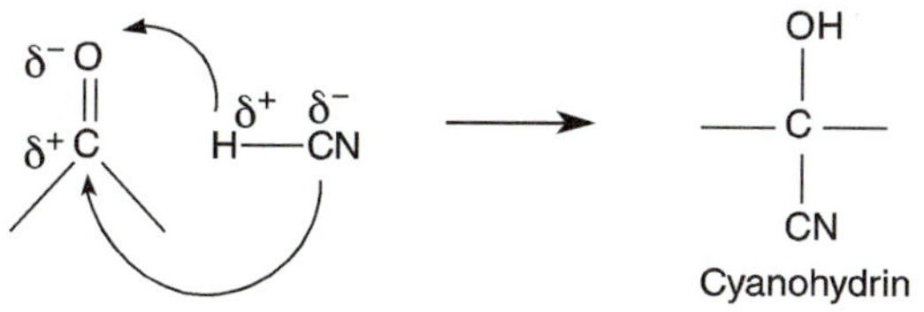

9.3 Addition of HCN to Carbonyl Compounds

$$\overset{\delta^-}{O} \parallel \overset{\delta^+}{C} + \overset{\delta^+}{H}-\overset{\delta^-}{CN} \longrightarrow \underset{\underset{CN}{|}}{\overset{\overset{OH}{|}}{C}} \quad \text{Cyanohydrin} \quad (9\text{-}7)$$

HCN is not very acidic ($pK_a = 9.2$), and it has been shown that reactions of the type as shown in Reaction (9-7) are best carried out in a slightly basic medium and typically catalyzed by KCN or NaCN. In the presence of a carbonyl compound, the CN⁻ will add first to the electrophilic carbon of the carbonyl bond as shown in Reaction (9-8).

$$N\equiv C:^{\ominus} + \;\;\;\;\;\; C=O: \longrightarrow N\equiv C-C-O:^{\ominus} \quad (9\text{-}8)$$

In a second step, the basic oxygen abstracts a proton from HCN to give a product called a cyanohydrin and in the process the CN⁻ catalyst is regenerated as shown in Reaction (9-9).

$$N\equiv C-C-\overset{..}{\underset{..}{O}}:^{\ominus} + H-C\equiv N \longrightarrow N\equiv C-C-OH + N\equiv C:^{\ominus} \quad (9\text{-}9)$$

Cyanohydrin

An example of the addition of HCN to benzaldehyde in the presence of NaCN leading to the formation of the cyanohydrin is shown in Reaction (9-10).

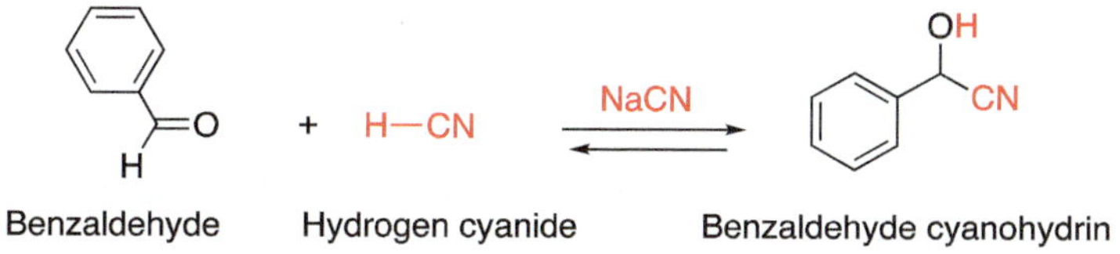

PhCHO + H—CN $\xrightarrow{\text{NaCN}}$ Ph-CH(OH)-CN   (9-10)

Benzaldehyde   Hydrogen cyanide   Benzaldehyde cyanohydrin

Another example of the addition of HCN to ketone is shown in Reaction (9-11), in which HCN is added to 2-pentanone in the presence of NaCN.

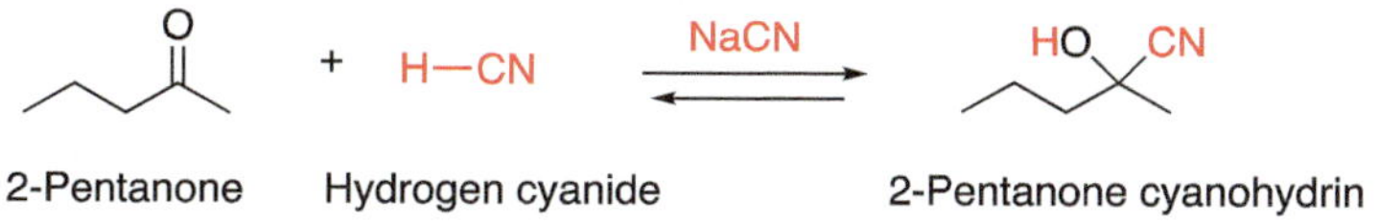

2-Pentanone + H—CN $\xrightarrow{\text{NaCN}}$ 2-Pentanone cyanohydrin   (9-11)

Problem 9.1

Complete the following addition reactions by supplying the major organic product, reactants, or the reaction conditions (appropriate catalyst) to give the products shown.

(a) 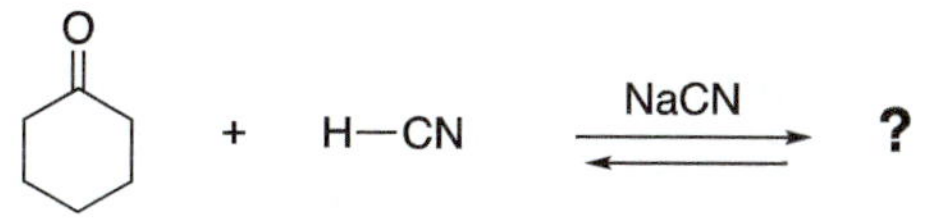 cyclohexanone + H—CN $\xrightarrow{\text{NaCN}}$?

(b) 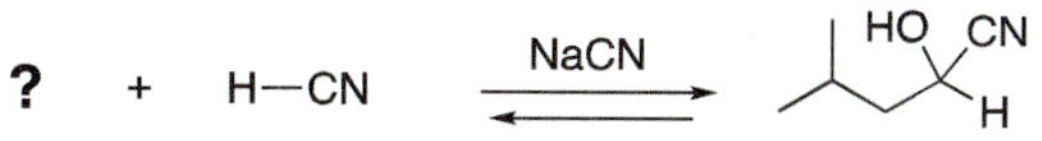 ? + H—CN $\xrightarrow{\text{NaCN}}$ (CH₃)₂CH-C(OH)(CN)H

(c) CH₃COCH₂CH₃ + ? ⇌ (CH₃)(CH₂CH₃)C(OH)(CN)

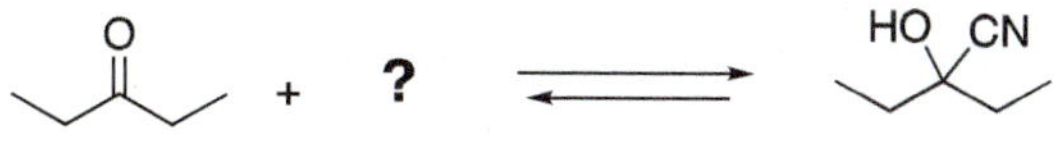

At this point, it may be a concern about the importance of this type of reaction, and specifically cyanohydrins? These types of reactions can serve to produce intermediates for the synthesis of other important compounds. For the synthesis of α-amino acids, the cyanohydrin can serve as an intermediate as shown in Reaction (9-12) in the synthesis of alanine; the details for the last few reactions will be covered in later chapters.

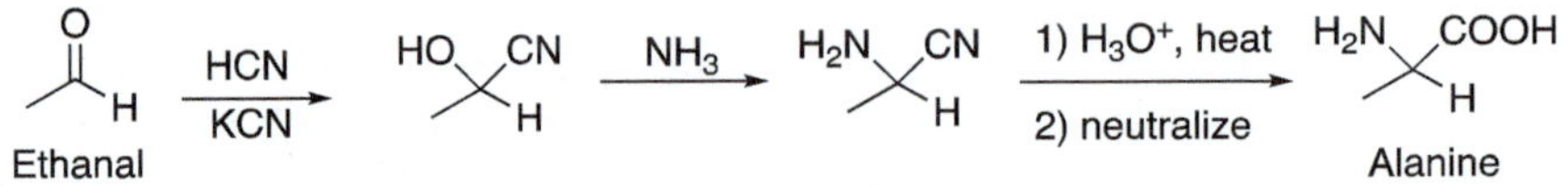

$$\text{Ethanal} \xrightarrow{\text{HCN, KCN}} \text{HO-CH(CN)-CH}_3 \xrightarrow{\text{NH}_3} \text{H}_2\text{N-CH(CN)-CH}_3 \xrightarrow[\text{2) neutralize}]{\text{1) H}_3\text{O}^+, \text{heat}} \text{Alanine} \quad (9\text{-}12)$$

Problem 9.2

Using Reaction (9-12) as a model, show how the cyanohydrin intermediate can be used to synthesize the following amino acids. This problem is designed to get students to start thinking in reverse. That is, you will have to look at the product in the above example and work backward to figure out the starting aldehyde!

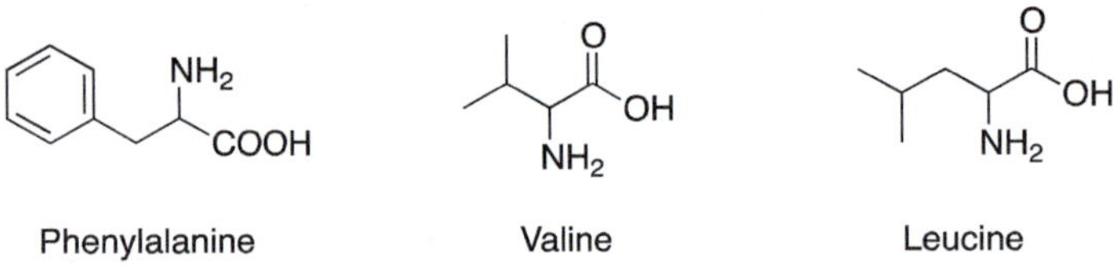

Phenylalanine   Valine   Leucine

9.4 Addition of Water to Carbonyl Compounds

In a similar manner, the addition of water to the carbon–oxygen double bond can be easily predicted. The general reaction showing the regiospecificity of the addition is given in Reaction (9-13).

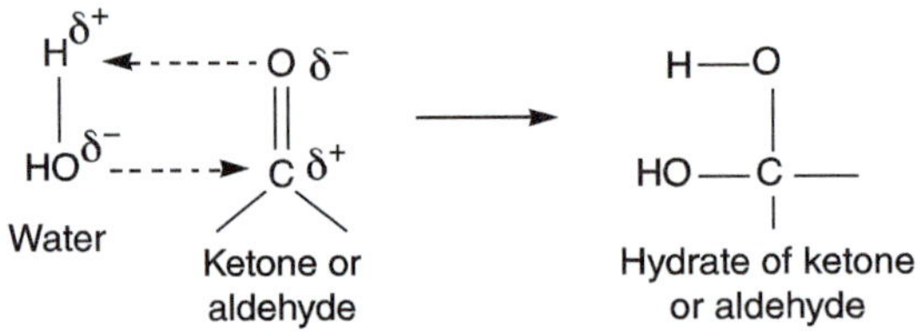

$$\text{Water} + \text{Ketone or aldehyde} \longrightarrow \text{Hydrate of ketone or aldehyde} \quad (9\text{-}13)$$

As mentioned earlier, this type of addition can take place under either acidic or basic conditions. Let us look at a specific example of the addition of water and the mechanism for the formation of the product, in this case, the **hydrate**. The acid-catalyzed addition of water to acetone (2-proponone) is shown in Reaction (9-14).

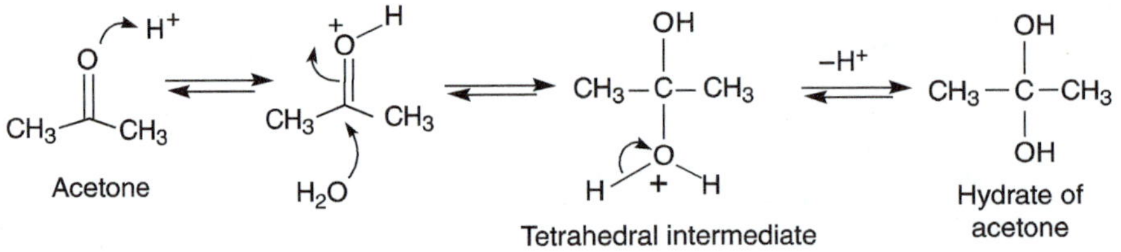

$$\text{Acetone} \xrightarrow{\text{H}^+} \xrightarrow{\text{H}_2\text{O}} \text{Tetrahedral intermediate} \xrightarrow{-\text{H}^+} \text{Hydrate of acetone} \quad (9\text{-}14)$$

Since an acidic medium has an excess of protons, the proton is added in a first step to form a protonated ketone. In the next step, the nucleophilic water adds to the electrophilic carbon of the carbonyl carbon to form a tetrahedral intermediate in which there is a formal charge of positive one (+1) on the oxygen since one pair of electrons are used from the water to form a covalent bond to the carbon atom of the protonated carbonyl compound. In the next step of the mechanism, a proton is lost to generate the neutral hydrate of acetone. Thus, the proton is a catalyst since it is used in the first step of the mechanism and regenerated in the last step.

The exact same product of Reaction (9-14) can be obtained if the conditions were basic instead of acidic. In an aqueous basic medium, there is an excess of the nucleophilic hydroxide anions, which will bond to electrophilic carbonyl carbon in the first step of the mechanism as shown in Reaction (9-15) to form a tetrahedral intermediate.

$$\underset{\underset{OH^-}{}}{\overset{O}{\underset{\|}{CH_3-C-CH_3}}} \rightleftharpoons \underset{\text{Tetrahedral intermediate}}{\overset{O^-}{\underset{OH}{CH_3-\overset{|}{\underset{|}{C}}-CH_3}}} \xrightarrow{-OH^-} \underset{\text{Hydrate of acetone}}{\overset{OH}{\underset{OH}{CH_3-\overset{|}{\underset{|}{C}}-CH_3}}} \quad (9\text{-}15)$$

The initially formed tetrahedral intermediate is basic and will abstract a proton from water to form the hydrate and regenerate the catalyst, OH^-. Thus, this reaction is often described as base catalyzed.

Problem 9.3

Give the hydrates that would be formed from each of the following ketones or aldehydes.

(a) cyclopentanone   (b) propanal   (c) 2,2-difluoropropanal   (d) acetaldehyde

9.4.1 Reactivity of Carbonyl Compounds Toward Hydration

Depending on the groups that are bonded to the carbonyl carbon, the electrophilicity of the carbonyl carbon varies. Owing to the greater electronegativity of fluorine, compared to that of hydrogen, the positive character of the carbon of difluoroformaldehyde is greater than that of the carbonyl carbon of formaldehyde. As a result, a nucleophilic attack will be much favored for the difluoroformaldehyde, compared to formaldehyde. Thus, the equilibrium constants for the hydration of these two compounds are different. The equilibrium favors the hydrate of difluoroformaldehyde (Reaction 9-16), compared to equilibrium for the hydrate of formaldehyde (9-17). As a result, difluoroformaldehyde is more reactive than formaldehyde toward hydration.

Very electrophilic carbon due to the fluorine atoms

$$F_2C=O + H_2O \xrightleftharpoons[]{H^+} \underset{FF}{\overset{HOOH}{>\!\!<}} \quad (9\text{-}16)$$

Very reactive toward hydration

Less electrophilic carbon since only hydrogens

$$H_2C=O + H_2O \xrightleftharpoons[]{H^+} \underset{HH}{\overset{HOOH}{>\!\!<}} \quad (9\text{-}17)$$

Less reactive toward hydration

Carbonyl compounds that have electronegative atoms and that are close to the carbonyl carbon are more reactive than comparable compounds that have electronegative groups further from the carbonyl carbon. Reactions (9-18) and (9-19) are examples that illustrate this concept.

$$\text{2-Fluorobutanal} + H_2O \xrightleftharpoons{H^+} \text{product} \tag{9-18}$$

$$\text{4-Fluorobutanal} + H_2O \xrightleftharpoons{H^+} \text{product} \tag{9-19}$$

Since the fluorine in 2-fluorobutanal is closer to the carbonyl carbon, that carbon is more electrophilic than the carbonyl carbon of 4-fluorobutanal (Reaction 9-19) in which the fluorine is further from the carbonyl carbon. The same analysis can be applied to determine the reactivity of carbonyl compounds that have groups of different electronegativities that are within the same proximity from the carbonyl carbon as shown in Reactions (9-20) and (9-21).

$$\text{2-Fluorobutanal} + H_2O \xrightleftharpoons{H^+} \text{product} \tag{9-20}$$

$$\text{2-Methylbutanal} + H_2O \xrightleftharpoons{H^+} \text{product} \tag{9-21}$$

Due to the presence of a highly electronegative fluorine adjacent to the carbonyl compound in 2-fluorobutanal, the carbonyl carbon is very electrophilic, compared to the carbonyl carbon in 2-methylbutanal, which contains the less electronegative methyl group in the same position. Thus, the equilibrium in Reaction (9-20) lies further to the right, compared to the equilibrium in Reaction (9-21). An extension of this analysis would imply that 2,2-difluorobutanal is more reactive than 2-fluorobutanal.

The same type of analysis can be carried out to determine the relative reactivity of carbonyl compounds that have crowding around the carbonyl carbon. A more crowded carbonyl compound is less reactive than one that is not crowded. Let us consider the acid-catalyzed hydration of 2,2,4,4-tetramethyl-3-pentanone. The first step of the mechanism is shown in Reaction (9-22) where the carbonyl carbon is protonated.

$$\text{2,2,4,4-Tetramethyl-3-pentanone} + H^+ \rightleftharpoons \text{protonated carbonyl} \tag{9-22}$$

In the next step of the reaction mechanism, the nucleophilic water attacks the protonated carbonyl to form a tetrahedral intermediate, which leads to the last step of the mechanism which involves the regeneration of the proton catalyst. These steps are shown in Reaction (9-23).

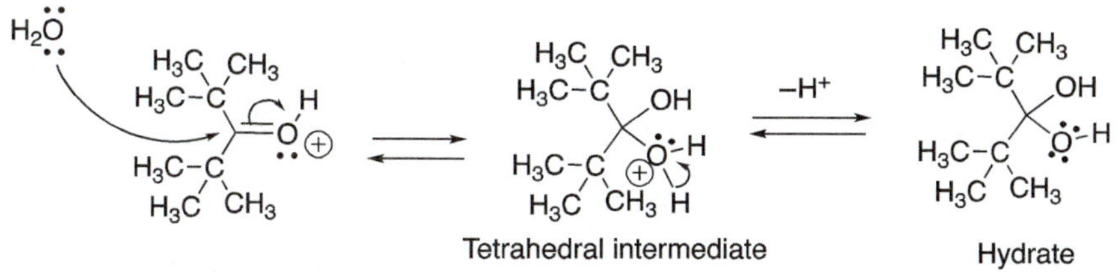

$$\text{(9-23)}$$

It becomes obvious that the tetrahedral intermediate that is formed is very crowded, hence not as stable as one that is not as crowded. You will recall that the ideal bond angle about an sp^3 carbon is 109.5°. As a result, the tetrahedral intermediate is very strained and not very stable, compared to another tetrahedral intermediate that is not as crowded. Thus, a very crowded carbonyl compound is not as reactive as another that is not as crowded. The hydration of 2,2,4,4-tetramethyl-3-pentanone (Reaction 9-24) and propanone (Reaction 9-25) illustrates this concept.

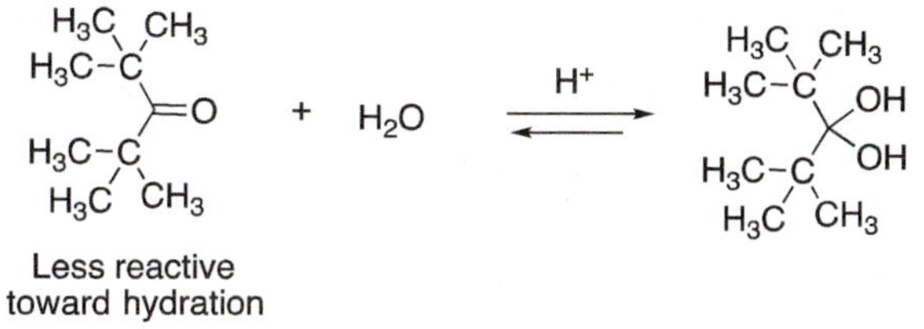

$$\text{(9-24)}$$

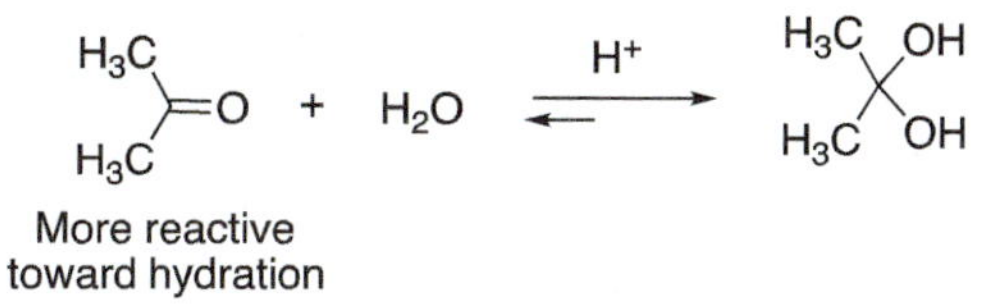

Less reactive toward hydration

$$\text{(9-25)}$$

More reactive toward hydration

Problem 9.4

i) Determine which of each of the following pairs of carbonyl compounds is more reactive toward hydration.

(a) and   (b) and

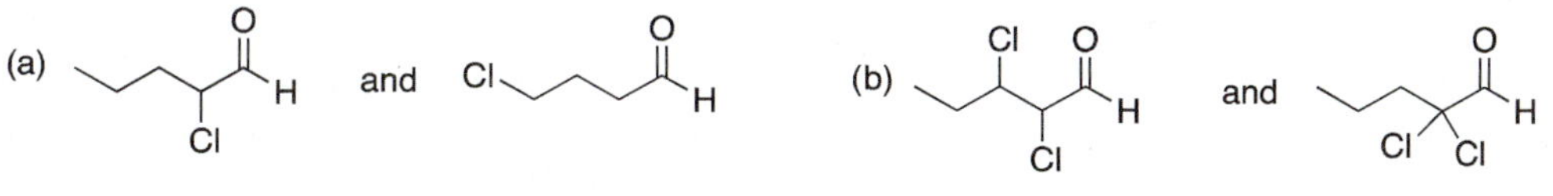

ii) Determine which of each of the following pairs of ketones is more reactive toward hydration.

(a) and   (b) and

(c) and   (d) and

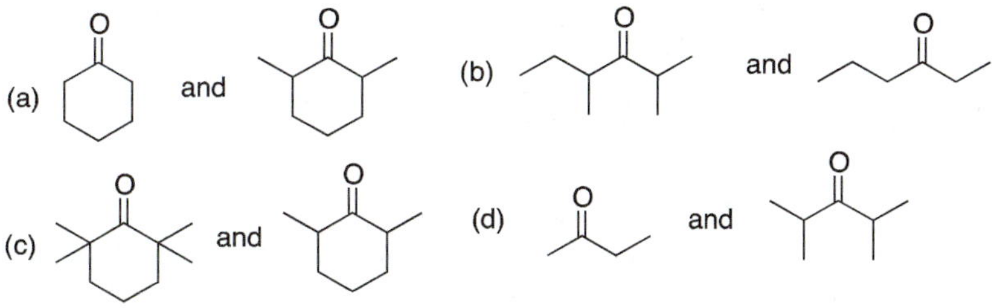

9.5 Addition of Alcohols to Carbonyl Compounds

The mechanism for the addition of alcohols to the carbonyl functionality of aldehydes and ketones is similar to that of the addition of water to carbonyl compounds. Under acidic conditions, the reaction of an aldehyde with one mole of an alcohol gives a **hemiacetal**. The general reaction mechanism for the acid-catalyzed addition of an alcohol to an aldehyde is shown in Reaction (9-26).

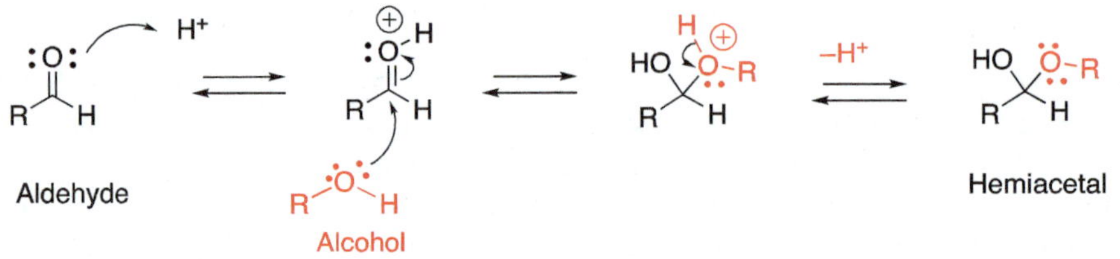

(9-26)

Note that these reactions are reversible and the equilibrium can be shifted in any direction by increasing the concentration of either the alcohol or the aldehyde in the presence of an acid. Reaction (9-27) shows the acid-catalyzed reaction of propanol with butanal.

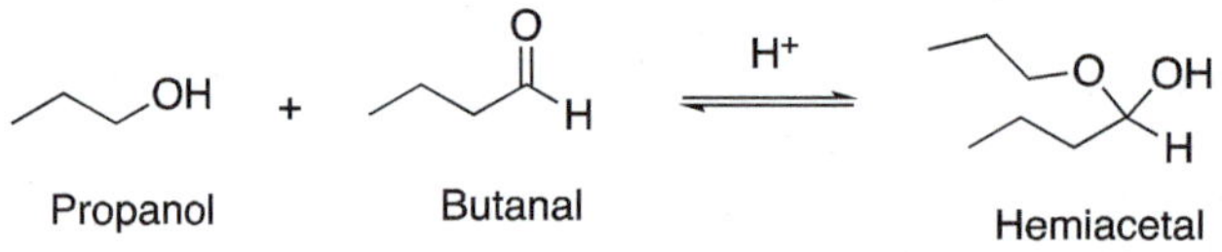

(9-27)

It is possible to have intramolecular hemiacetal formation in which the alcohol and aldehyde functionalities of a difunctional molecule react in an acidic medium, as shown in Reaction (9-28).

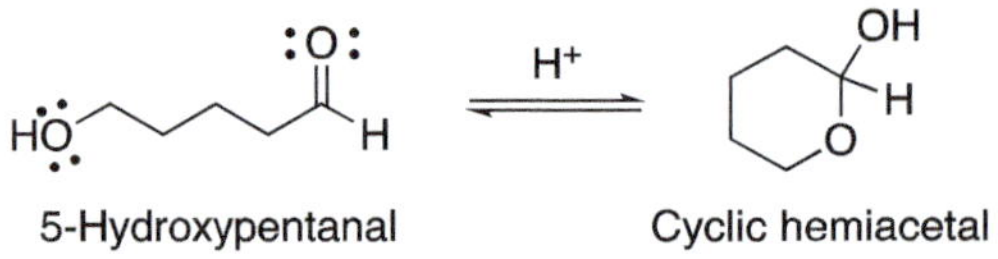

(9-28)

The mechanism for the intramolecular acid-catalyzed addition reaction is similar to that of the intermolecular reaction in that the carbonyl oxygen is first protonated. The next step in the mechanism involves the intramolecular attack of the alcohol functionality on the electrophilic carbonyl carbon to form the protonated cyclic hemiacetal. The last step involves the regeneration of the proton catalyst and the cyclic hemiacetal as outlined in Reaction (9-29).

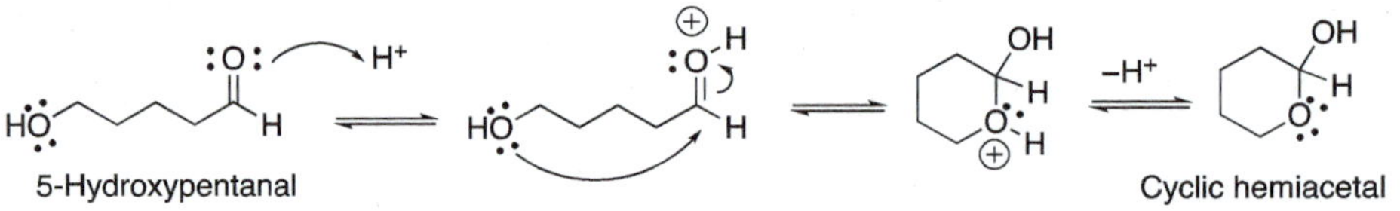

(9-29)

Problem 9.5

Complete each of the following reactions by giving the structure of the hemiacetal product or the appropriate aldehyde and alcohol.

9.5 Addition of Alcohols to Carbonyl Compounds

(a) 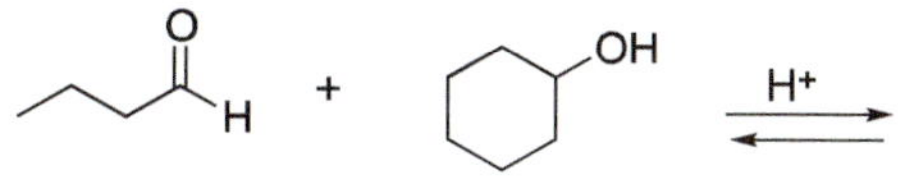 CH₃CH₂CH₂C(=O)H + cyclohexanol $\xrightarrow{H^+}$

(b) 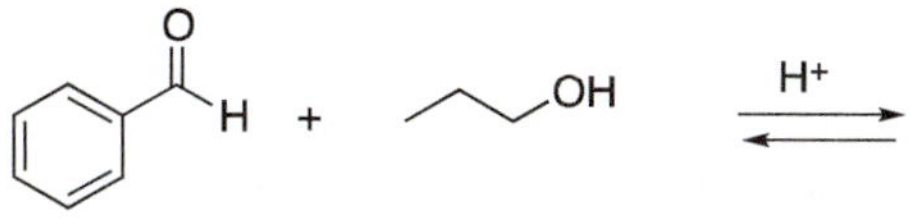 PhC(=O)H + CH₃CH₂CH₂OH $\xrightarrow{H^+}$

(c) 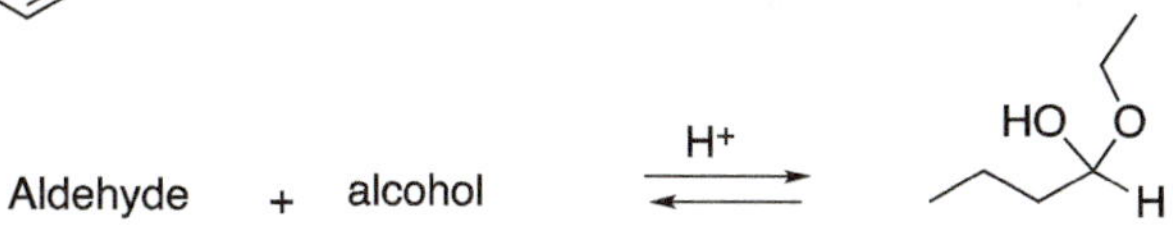 Aldehyde + alcohol $\xrightarrow{H^+}$ [hemiacetal product shown with HO and OEt groups]

(d) 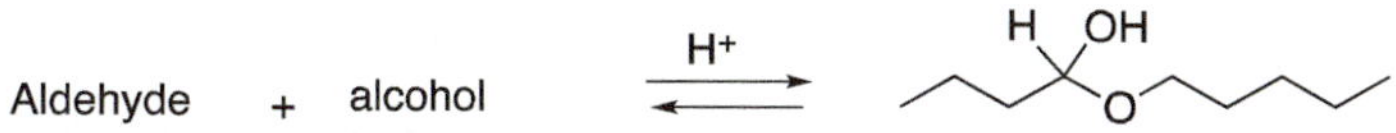 Aldehyde + alcohol $\xrightarrow{H^+}$ [acetal product shown with H, OH and OR groups]

An important intramolecular hemiacetal formation exists in biological science. Glucose and similar monosaccharaides contain both an aldehyde and alcohol functionalities, and in solution, these compounds form cyclic hemiacetals as shown in Reaction (9-30).

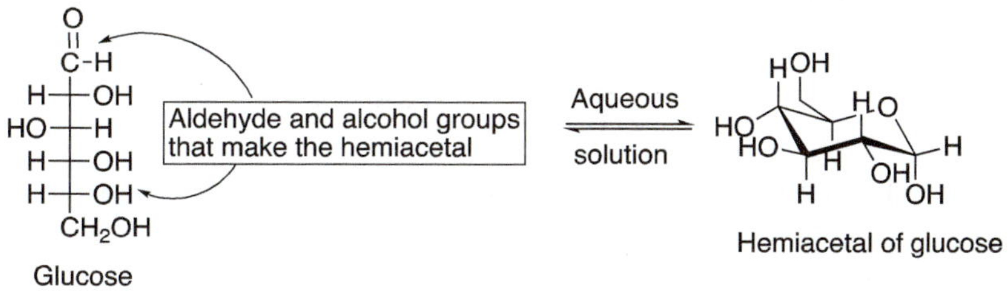

Glucose (Fischer projection with C-H aldehyde, H—OH, HO—H, H—OH, H—OH, CH₂OH) — "Aldehyde and alcohol groups that make the hemiacetal" — Aqueous solution ⇌ Hemiacetal of glucose (chair form) (9-30)

Most monosaccharides exist as hemiacetals or hemiketals, which we will be covering in Chapter 20.

Problem 9.6

The open-chain structure of mannose is shown below. Give the most stable chair conformation of the six-member cyclic hemiacetal for mannose that would exist in aqueous solution.

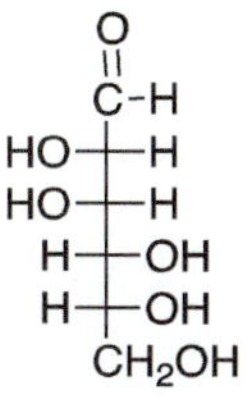

Mannose (Fischer projection: C-H aldehyde, HO—H, HO—H, H—OH, H—OH, CH₂OH)

Hemiacetals, in the presence of an alcohol and in the presence of an acid catalyst, react to form acetals. In the first step of the reaction mechanism leading to the acetal formation, the hemiacetal is protonated as shown in Reaction (9-31) to produce a resonance-stabilized carbocation after loss of water.

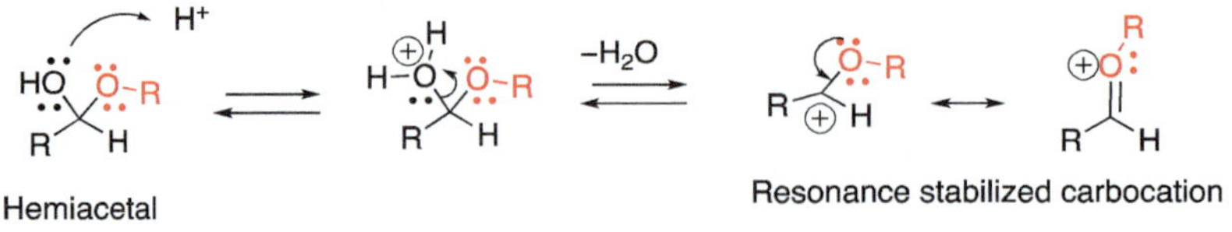

9 Addition Reactions Involving Carbonyls and Nitriles

[Reaction 9-31: Hemiacetal → protonated hemiacetal → resonance stabilized carbocation, with loss of H₂O] (9-31)

In the next step of the mechanism, the newly developed carbocation reacts with a molecule of the nucleophilic alcohol to produce a protonated acetal, which loses a proton to form the acetal, as shown in Reaction (9-32).

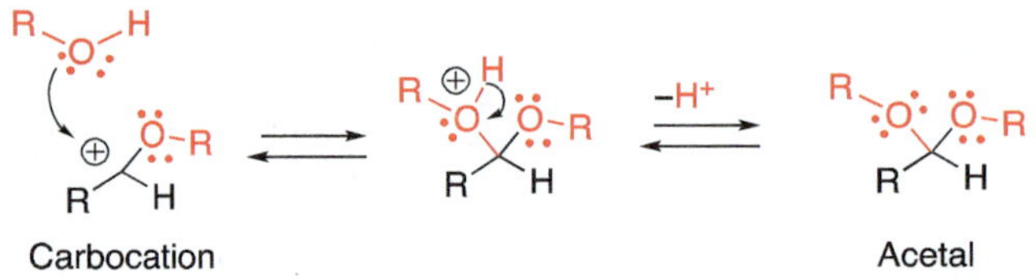

[Reaction 9-32: Carbocation + ROH → protonated acetal → Acetal, with loss of H⁺] (9-32)

Thus, the overall reaction of an aldehyde in the presence of excess alcohol and in an acidic medium results in the acetal that is reflected in the reaction of butanal with an excess of methanol in an acidic medium as shown in Reaction (9-33).

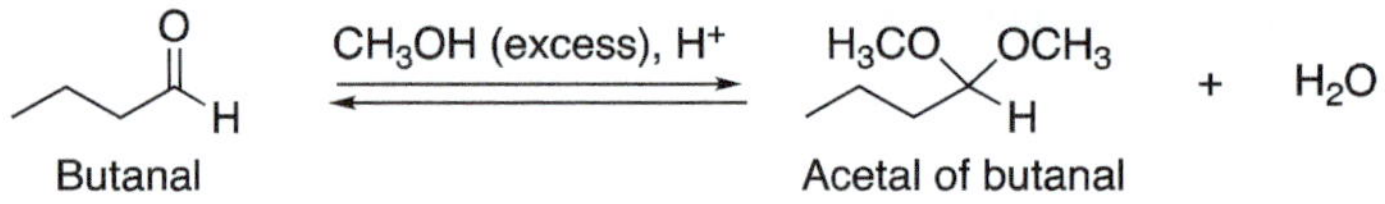

Butanal + CH$_3$OH (excess), H⁺ ⇌ Acetal of butanal (H$_3$CO, OCH$_3$) + H$_2$O   (9-33)

Problem 9.7

Complete each of the following reactions by giving the structure of the acetal product or the appropriate aldehyde and alcohol.

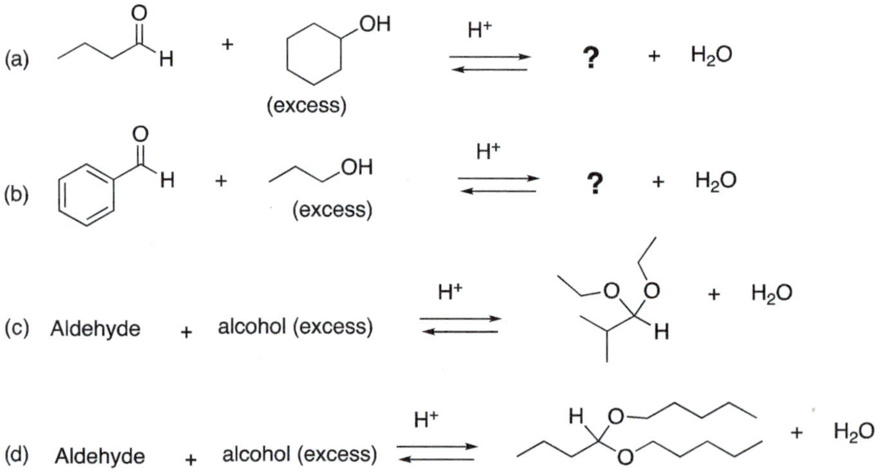

(a) Butanal + cyclohexanol (excess), H⁺ ⇌ ? + H$_2$O

(b) Benzaldehyde + propanol (excess), H⁺ ⇌ ? + H$_2$O

(c) Aldehyde + alcohol (excess), H⁺ ⇌ [acetal structure shown] + H$_2$O

(d) Aldehyde + alcohol (excess), H⁺ ⇌ [acetal structure shown] + H$_2$O

Similar addition reactions as those shown above involving aldehydes to form hemiacetals and acetals are possible for ketones. In this case, however, the products are known as **hemiketals** and **ketals**. The hemiketal formation of propanone and methanol is shown in Reaction (9-34).

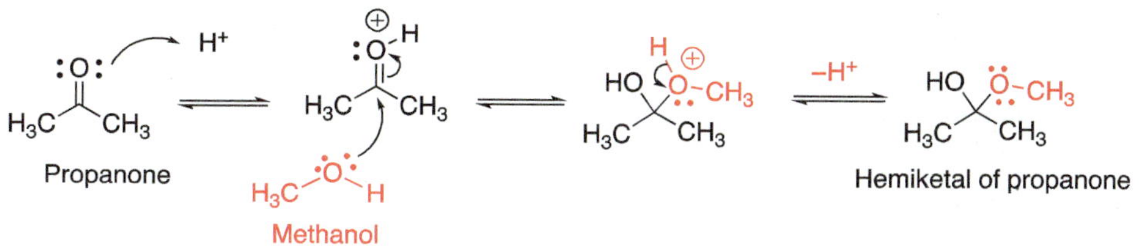

9.5 Addition of Alcohols to Carbonyl Compounds

$$\text{(9-34)}$$

Propanone + Methanol ⇌ ⇌ ⇌ Hemiketal of propanone

Hemiketals in the presence of an excess of an alcohol will react further to form ketals as shown below in the reaction of the hemiketal of propanone in the presence of another mole of methanol. In the first step of the mechanism, the hemiketal is protonated and water leaves to create a fairly stable resonance-stabilized carbocation as shown in Reaction (9-35).

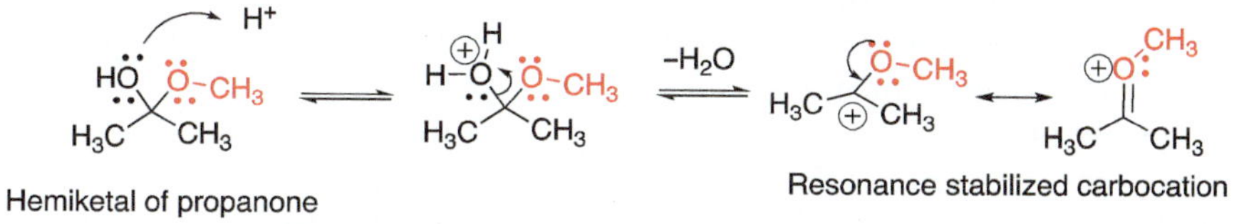

$$\text{(9-35)}$$

Hemiketal of propanone ⇌ ⇌ Resonance stabilized carbocation

In the next step of the reaction mechanism, the carbocation reacts with another mole of methanol to form a protonated ketal, and after the loss of a proton in the final step, the ketal is formed as shown in Reaction (9-36).

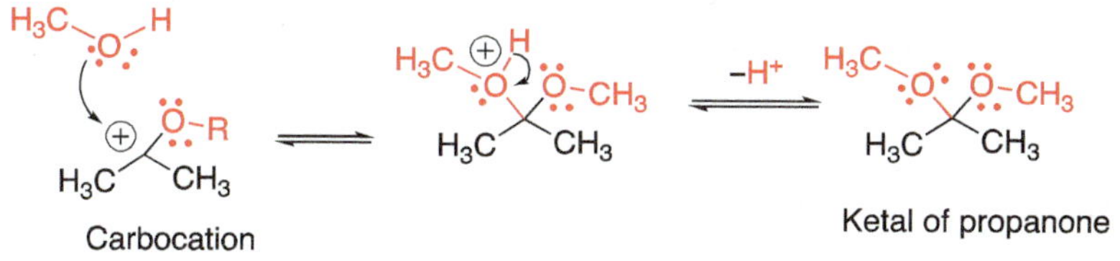

$$\text{(9-36)}$$

Carbocation ⇌ ⇌ Ketal of propanone

Problem 9.8

i) Give the ketal or acetal that would result from the reaction of each of the carbonyl compounds shown below in an acidic medium and in the presence of excess methanol.

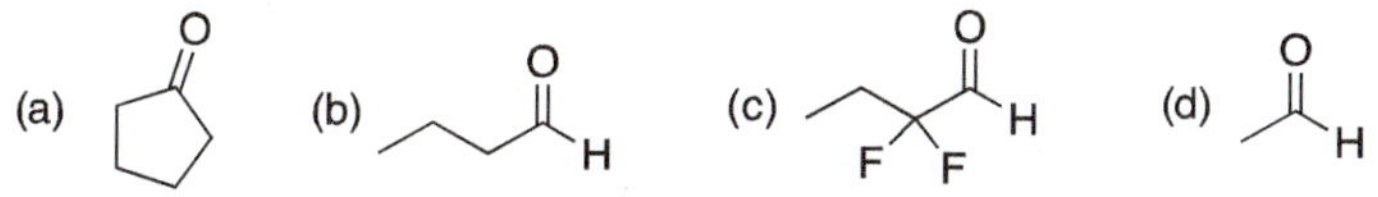

(a) cyclopentanone (b) propanal (c) 2,2-difluoropropanal (d) acetaldehyde

ii) Which of the above molecules would you expect to be the most reactive? Explain your answer.

It is possible to use a diol instead of an excess of alcohol to react with a carbonyl compound to produce a ketal or acetal. 1,2-Ethanediol is one such molecule that has two alcohol functionalities and readily reacts with aldehydes or ketones in the presence of an acid to form cyclic acetals or ketals as shown in the example in Reaction (9-37).

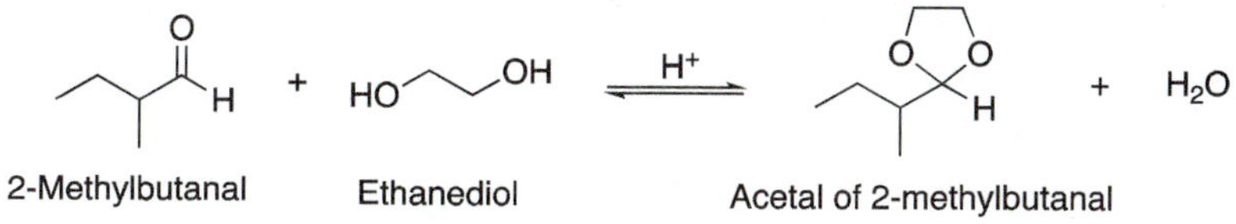

$$\text{(9-37)}$$

2-Methylbutanal + Ethanediol ⇌ Acetal of 2-methylbutanal + H_2O

The importance of these types of reactions will be discussed in the next section.

Problem 9.9

Using arrow-pushing formulism to indicate electron movement, propose a reasonable mechanism for the reaction shown below.

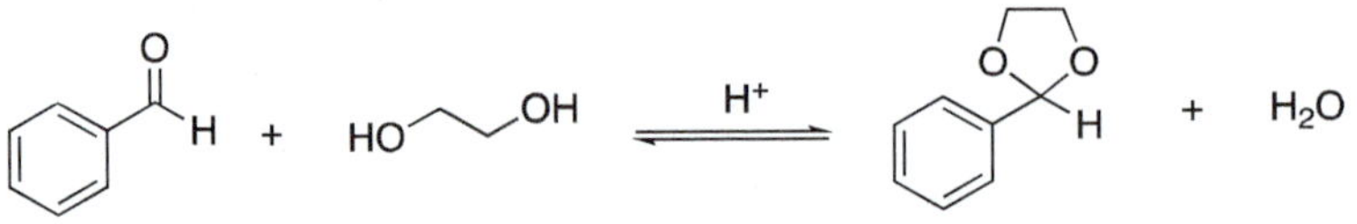

PhCHO + HOCH₂CH₂OH ⇌ (H⁺) Ph-CH(OCH₂CH₂O) + H₂O

9.5.1 Ketals and Acetals as Protection Groups

For molecules that have more than one of the functional groups listed in Chapters 3 and 4, extreme care must be exercised if it is desired to have just one functional group that undergoes a specific reaction and not another functional group. For polyfunctional molecules, many times the reaction conditions that are suited for one functional group are also suitable for a reaction to occur not only at the desired functional group but also at other functional groups. We have seen in this chapter that alkenes and carbonyl compounds can undergo addition reaction. Thus, if a molecule has both alkene and carbonyl functionalities present, it becomes very difficult to carry out a reaction involving just one functionality without affecting another. A strategy that is often used in organic chemistry is to protect one of the functionalities from the reaction that can take place at another functional group. Once the reaction is complete, the protection group is removed to generate the original functional group. Since the addition reactions of alcohols to the carbonyl functionality form ketal and acetals are reversible reactions, alcohols are ideal to act as protecting groups for carbonyl groups. Thus, for a difunctional molecule that has a carbonyl functionality and another functionality, the carbonyl functionality can be converted to the ketal functionality, then a reaction can be carried out on another functionality on the same molecule. Once that reaction is complete, the protecting acetal or ketal can be removed to regenerate the original carbonyl functionality. The challenge is illustrated in the desired transformation shown in Reaction (9-38).

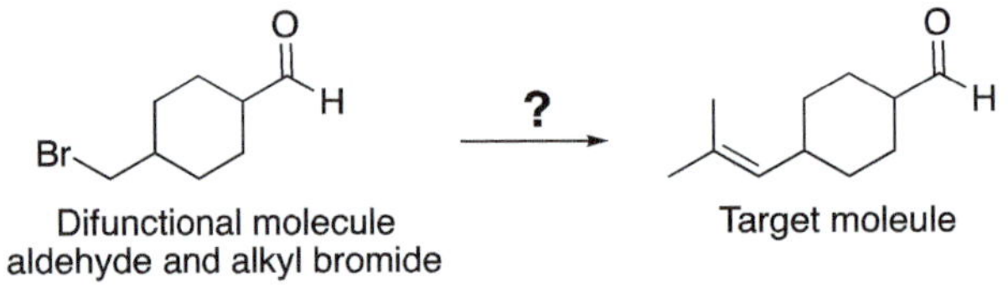

Difunctional molecule → Target molecule (9-38)
aldehyde and alkyl bromide

In order to carry out the transformation shown in Reaction (9-38), the reaction needs to take place at the alkylbromide functional end and not at the carbonyl group. As we will see in the next section, there is a reaction that can transform an alkyl bromide to an alkene, but it involves the reaction with another molecule that has a carbonyl functionality. Thus, in order to carry out this transformation, the aldehyde functionality must be first protected before carrying out the reaction at the alkylbromide functionality. Once the reaction is complete, the protecting group is removed to generate the desired product. The strategy is illustrated in the sequence of reactions shown in Reaction (9-39).

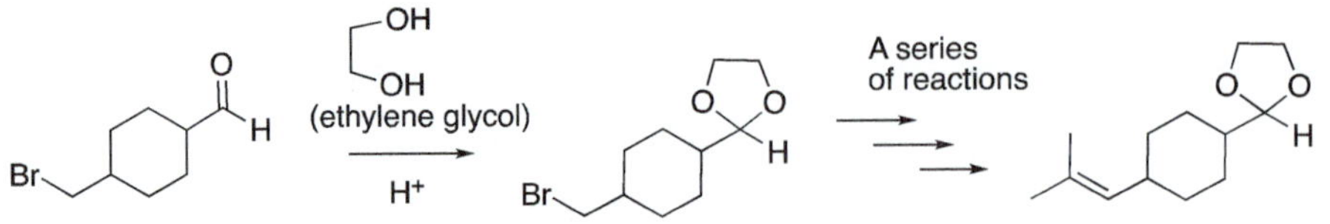

(9-39)

The first reaction to be carried out is the protection of the aldehyde to form an acetal. Once protected, specific reactions can occur involving the alkylbromide functionality to give the protected target molecule. The last reaction in this sequence of reactions involves the conversion of the protected target molecule to get the target molecule, as shown in Reaction (9-40).

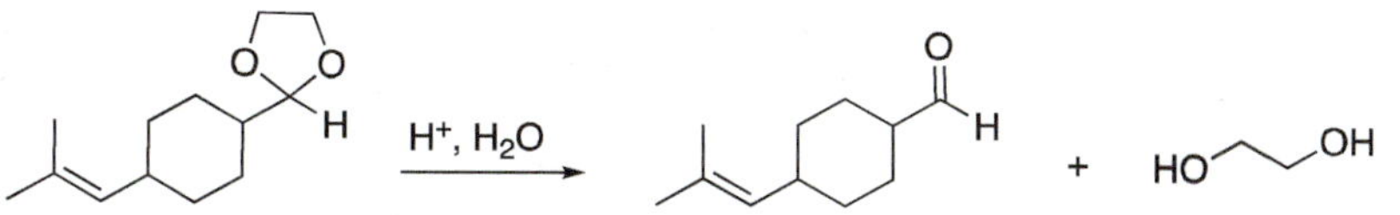

$$\text{(protected target molecule)} \xrightarrow{H^+, H_2O} \text{Target molecule} + HO\text{-}CH_2CH_2\text{-}OH \quad (9\text{-}40)$$

As you can imagine, the reaction of carbonyl compounds and dithiols should be very similar to that of diols. In fact, 1,2-ethanedithiol is sometimes used as a protecting group for carbonyl compounds. The reaction involving a 1,2-ethanedithiol and cyclohexanone to produce 1,3-dithiolanes is shown in Reaction (9-41), but ketals and acetals are used more often as protecting groups in organic synthesis.

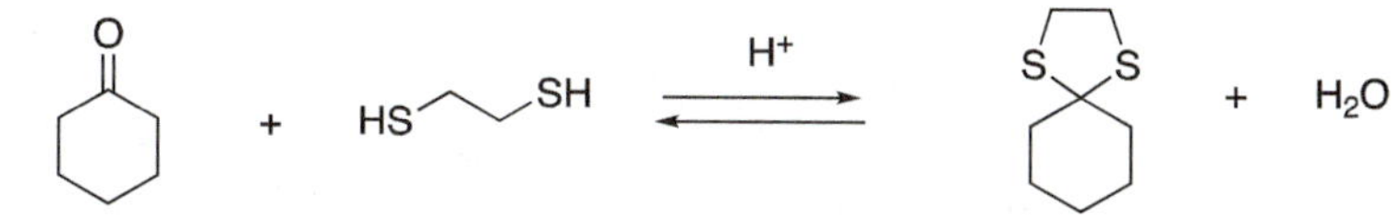

$$\text{Cyclohexanone} + \text{HS-CH}_2\text{CH}_2\text{-SH} \underset{}{\overset{H^+}{\rightleftharpoons}} \text{1,3-Dithiolane} + H_2O \quad (9\text{-}41)$$

Cyclohexanone   1,2-Ethanedithiol   1,3-Dithiolane

Problem 9.10

i) Give the ketal or acetal that would result from the reaction of each of the carbonyl compounds shown below with ethylene glycol (1,2-ethanediol) in the presence of an acid.

(a) 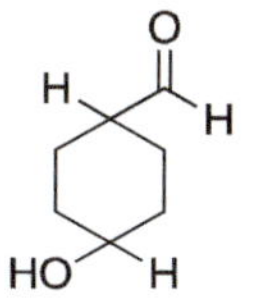 *[structure: cyclohexane with CHO at one position and with H and OH substituents]*   (b) 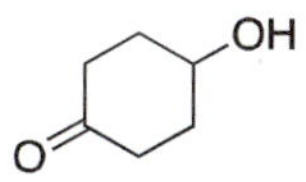 *[structure: cyclohexanone with OH substituent]*

ii) Give the 1,3-dithiolane that would result from the reaction of each of the carbonyl compounds shown above with 1,2-ethanedithiol in the presence of an acid.

9.6 Addition of Ylides to Carbonyl Compounds (The Wittig Reaction)

First, let us define ylides; ylides are neutral compounds that have a negatively charged carbon atom directly bonded to a positively charged atom, typically sulfur, phosphorus, or nitrogen. Two resonance structures for a phosphorous ylide are shown in (9-42).

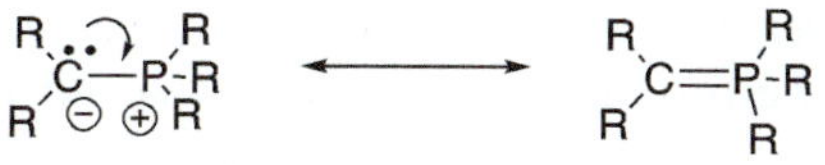

$$\underset{R}{\overset{R}{C}}\text{-}\underset{R}{\overset{R}{P}}\text{-R} \quad \longleftrightarrow \quad \underset{R}{\overset{R}{C}}=\underset{R}{\overset{R}{P}}\text{-R} \quad (9\text{-}42)$$

One observation that you will make about ylides is that they are both nucleophilic and electrophilic; a similar observation was made for carbenes that were encountered in the previous chapter. As a result, ylides can undergo an addition reaction with carbonyl compounds in a regiospecific manner, as shown in Reaction (9-43) to give oxaphosphetanes, which readily decompose to form an alkene and phosphine oxide. The phosphine oxide is relatively stable due to the very favorable and strong phosphorus–oxygen double bond.

$$
\underset{\substack{\text{Reaction of carbonyl}\\\text{with a phosphorous ylide}}}{\overset{\text{C=O}}{\underset{H_3C}{\overset{H_3C}{\diagdown}}}\overset{+}{\underset{Ph}{\overset{-}{C}\text{-}\overset{+}{P}\text{-}Ph}}} \longrightarrow \underset{\text{Oxaphosphetane}}{\overset{\text{C-O}}{\underset{H_3C}{\overset{H_3C}{\diagdown}}}\overset{|}{\underset{Ph}{\overset{|}{C}\text{-}\overset{|}{P}\text{-}Ph}}} \longrightarrow \underset{\text{Alkene}}{\overset{H_3C}{\underset{H_3C}{\diagdown}}C=C} \; + \; \underset{\substack{\text{Triphenylphosphine}\\\text{oxide}}}{O=\overset{Ph}{\underset{Ph}{P}\text{-}Ph}} \quad (9\text{-}43)
$$

When a phosphorous ylide, which contains the phenyl (Ph, C_6H_5) groups, is used for Reaction (9-43), the reaction is known as the Wittig reaction. An example of the Wittig reaction is shown in Reaction (9-44).

$$\text{cyclohexanone} \xrightarrow{Ph_3\overset{+}{P}\text{-}\overset{-}{C}H_2} \text{methylenecyclohexane} + O=\overset{Ph}{\underset{Ph}{P}\text{-}Ph} \text{ (Triphenylphosphine oxide)} \quad (9\text{-}44)$$

You will recognize that the reaction is a very efficient strategy to convert a carbonyl functionality to an alkene functionality. This type of reaction was discovered in 1954 by Georg Wittig, a German chemist, who shared the Nobel Prize in 1979 with Herbert C. Brown for his discovery.

Problem 9.11

Give the structure of the ylide (Wittig reagent) necessary to carry out the following transformations.

(a) cyclohexanone ⟶ ethylidenecyclohexane

(b) butan-2-one ⟶ 2-methylpent-2-ene (shown structure)

9.6.1 Synthesis of Phosphorous Ylides

The synthesis of phosphorous ylides is fairly straightforward. First, a substitution reaction is carried out involving triphenylphosphine and alkylbromide in which the bromide is substituted for the triphenylphosphine to form a bromide salt product as shown in the substitution reaction in Reaction (9-45). We have seen a similar reaction in Chapter 6 when we were introduced to substitution reactions. For this reaction, the bromine in the reactant is substituted with the PPh_3 molecule to form the product.

$$\underset{Ph}{\overset{Ph}{Ph\text{-}P:}} \; + \; \underset{R}{\overset{R\;H}{\diagup\text{-}Br}} \longrightarrow \underset{Ph}{\overset{Ph}{Ph\text{-}\overset{+}{P}}}\overset{Br^-}{\underset{R}{\diagup\text{-}\overset{H}{R}}} \quad (9\text{-}45)$$

A second reaction involves an acid-base reaction in which the bromide salt of the product of Reaction (9-45) reacts with strong base, typically butyl lithium, to remove a proton adjacent to the phosphorous to generate the ylide as shown in Reaction (9-46).

$$\text{Ph}_3\text{P}^+\text{CHR}(R)\, \text{Br}^- + \text{CH}_3\text{CH}_2\text{CH}_2\text{CH}_2^-\text{Li}^+ \longrightarrow \text{Ph}_3\text{P}^+\text{-CR}^-(R) + \text{CH}_3\text{CH}_2\text{CH}_2\text{CH}_3 + \text{LiBr} \quad (9\text{-}46)$$

Butyl lithium, a strong base → Ylide + Butane

You will recall from Chapter 7 on acid–base reactions that the butyl lithium is an extremely strong base since its conjugate acid (butane) is an extremely weak acid ($pK_a \sim 50$). One important observation that is made for the synthesis of Wittig ylides that are used for these reactions is that its synthesis depends on having a hydrogen adjacent to the phosphorous atom to be abstracted to form the ylide. Thus, the starting alkyl bromide cannot be tertiary, but methyl, primary or secondary. Reaction (9-47) shows the synthesis of the ylide that is needed to synthesize the alkene shown in Reaction (9-48).

$$\text{Ph}_3\text{P:} + \text{iPr-Br} \longrightarrow \text{Ylide salt} \xrightarrow{\text{BuLi}} \text{Ylide} \quad (9\text{-}47)$$

$$\text{cyclohexanone} + \text{Ylide} \longrightarrow \text{alkylidenecyclohexane} + \text{O=PPh}_3 \quad (9\text{-}48)$$

Problem 9.12

i) Determine the ylides and carbonyl compounds necessary to synthesize the following compounds.

(a) cyclopentylidene-ethane (b) 2-methyl-3-pentene

ii) Show how to synthesize the ylides of the above question starting from an appropriate alkylbromide and triphenyl phosphine.

9.7 Addition of Enolates to Carbonyl Compounds

As we have seen in Chapter 7 on acids and bases, enolates are created by the abstraction of an alpha-proton of carbonyl compounds to give a resonance-stabilized enolate conjugate base as shown in Reaction (9-49).

$$\underset{\text{H}}{\overset{\text{O}}{\|}}\text{CH}_3 \xrightleftharpoons[\text{(base)}]{\text{KOH}} \underset{\text{H}}{\overset{\text{O}^-\text{K}^+}{\underset{}{}}}\!=\!\text{CH}_2 \longleftrightarrow \underset{\text{H}}{\overset{\text{O}}{\|}}\text{CH}_2^-\; \text{K}^+ \quad (9\text{-}49)$$

Resonance-stabilized enonate

Enolates are nucleophiles and they react with carbonyl compounds in a similar manner as any of the nucleophiles mentioned earlier in this chapter. Shown in Reaction (9-50) is the addition of the enolate of ethanal to another molecule of ethanal.

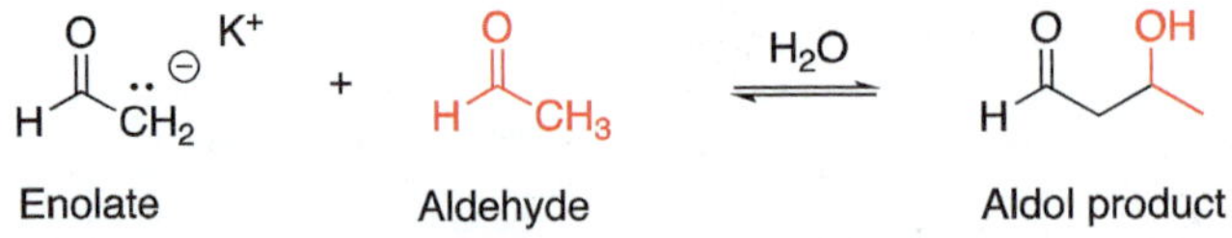

$$\text{Enolate} + \text{Aldehyde} \xrightleftharpoons{H_2O} \text{Aldol product} \tag{9-50}$$

Reactions of this type in which an enolate of a carbonyl compound reacts with another mole of the carbonyl compound in the presence of a base are known as aldol condensation reactions and they are very useful reactions for the synthesis of new carbon–carbon bonds to make larger molecules as shown in Reaction (9-51).

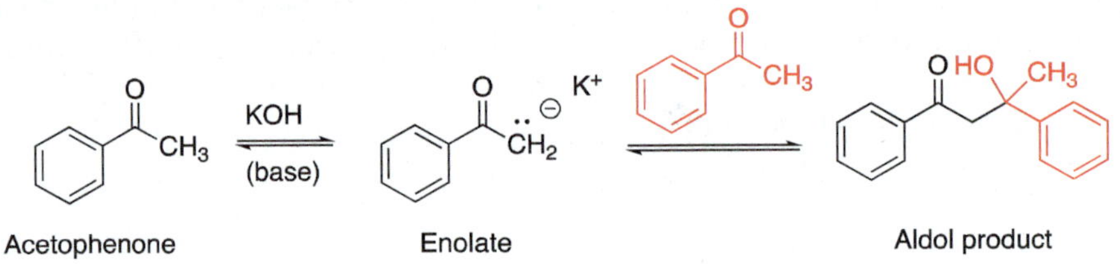

$$\text{Acetophenone} \xrightleftharpoons{\text{KOH (base)}} \text{Enolate} \rightleftharpoons \text{Aldol product} \tag{9-51}$$

For such aldol adducts, the elimination of water is possible to give a very stable unsaturated α,β-unsaturated compound as shown in Reaction (9-52).

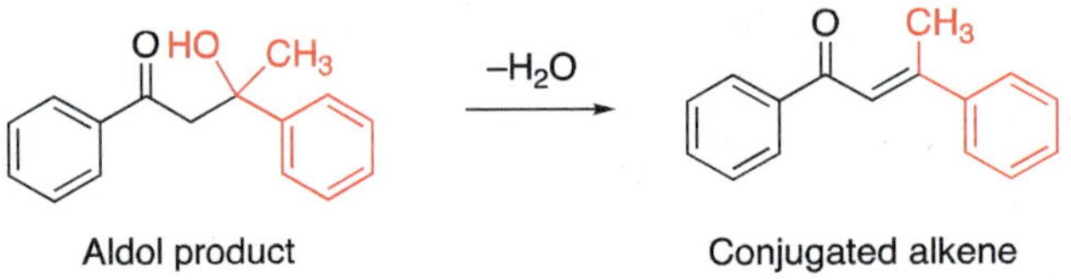

$$\text{Aldol product} \xrightarrow{-H_2O} \text{Conjugated alkene} \tag{9-52}$$

For carbonyl compounds that have only one type of α-hydrogen(s), it is possible for its enolate to react with another carbonyl compound that does not have α-hydrogen(s) as shown in the Reaction (9-53).

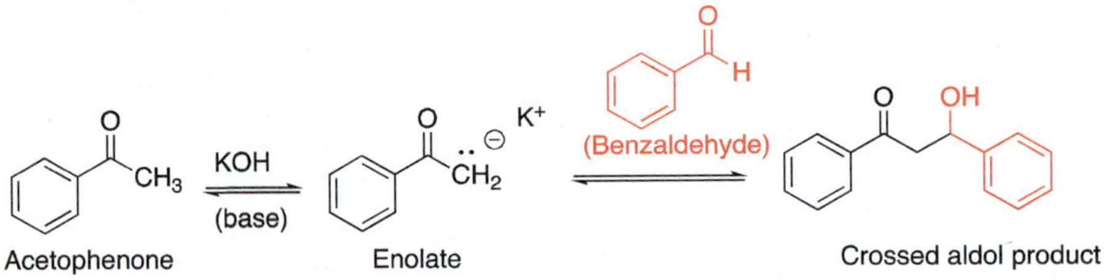

$$\text{Acetophenone} \xrightleftharpoons{\text{KOH (base)}} \text{Enolate} + \text{(Benzaldehyde)} \rightleftharpoons \text{Crossed aldol product} \tag{9-53}$$

Such aldol condensation reactions are known as crossed-aldol condensation reactions. It is possible to have an intramolecular aldol condensation as shown in Reaction (9-54).

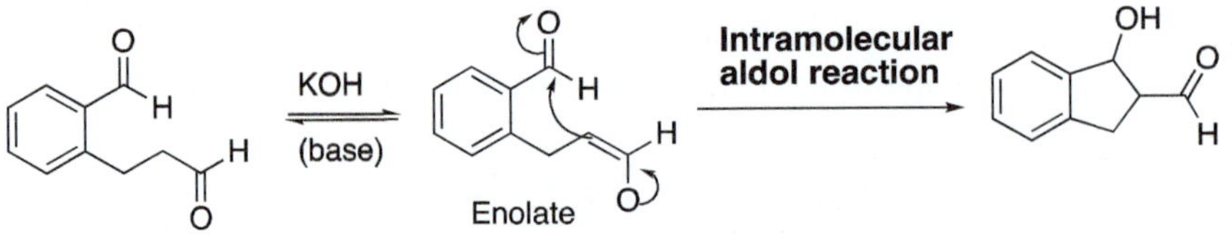

$$\xrightleftharpoons{\text{KOH (base)}} \text{Enolate} \xrightarrow{\text{Intramolecular aldol reaction}} \tag{9-54}$$

As demonstrated above, it is possible to have loss of water to give the α,β-unsaturated carbonyl compound as shown in Reaction (9-55).

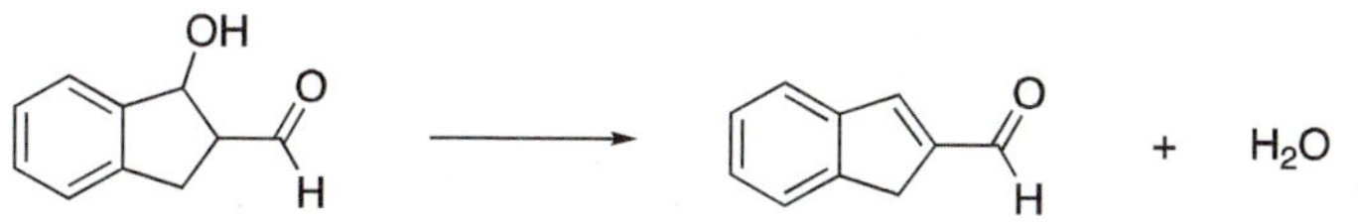

9.7 Addition of Enolates to Carbonyl Compounds

[Reaction 9-55: dehydration of β-hydroxy aldehyde to α,β-unsaturated aldehyde + H₂O] (9-55)

The addition of an enolate to an α,β-unsaturated carbonyl takes place in a 1,4 manner, instead of a 1,2 addition as discussed thus far. This type of addition is also known as a Michael addition as shown in Reaction (9-56).

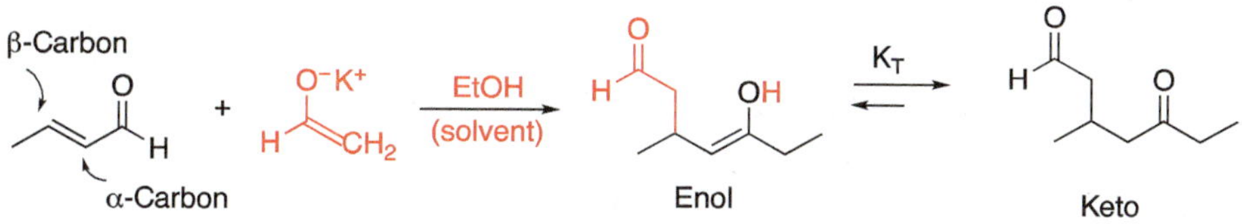

[Reaction 9-56: β-Carbon / α-Carbon of α,β-unsaturated carbonyl + enolate (O⁻K⁺, CH₂) in EtOH (solvent) → Enol ⇌K_T Keto] (9-56)

The final step involves a keto-enol tautomerization, which favors the formation of the keto form as shown in Reaction (9-56).

The combination of the Michael addition with the aldol reaction to form a cyclic system is known as the Robinson annulation, which also means ring-formation reaction. The first step of a Robinson annulation, which is a Michael addition reaction, is shown in Reaction (9-57).

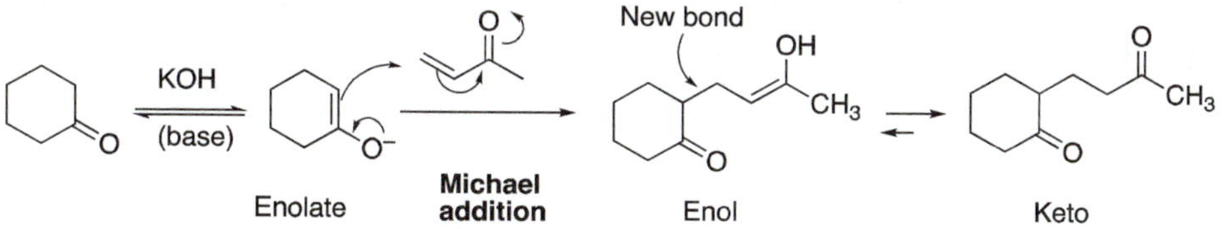

[Reaction 9-57: cyclohexanone + KOH (base) → Enolate → Michael addition with methyl vinyl ketone → Enol → Keto, showing new bond] (9-57)

In the next step of the reaction, an enolate is formed from the ketone in the presence of a base as shown in Reaction (9-58).

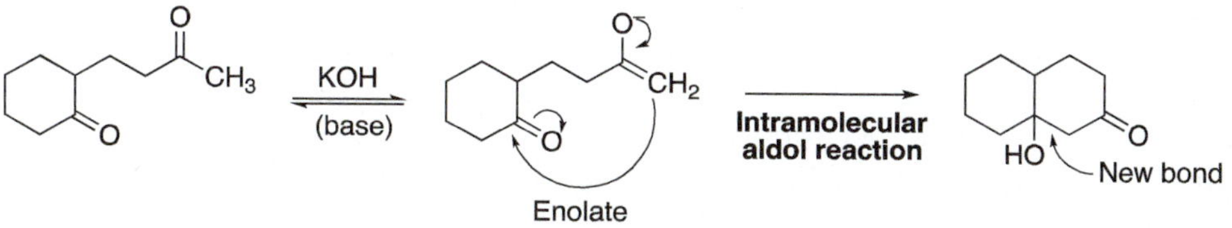

[Reaction 9-58: Michael adduct + KOH (base) → Enolate → Intramolecular aldol reaction → bicyclic β-hydroxy ketone with new bond] (9-58)

Note that there are other possible ring formations, but the six-member ring is the most stable. In the final step of the reaction, loss of water occurs to form the α,β-unsaturated compound as shown in Reaction (9-59).

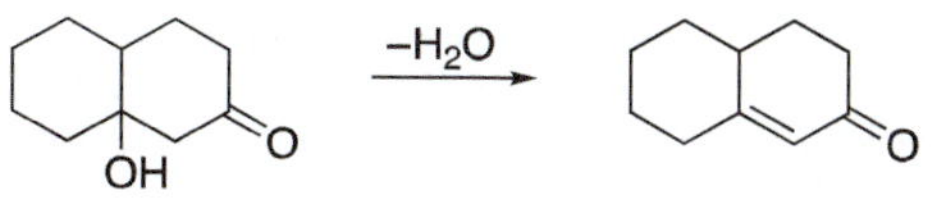

[Reaction 9-59: β-hydroxy decalone −H₂O → octalone] (9-59)

Another example of the Robinson annulation reaction is shown in Reaction (9-60).

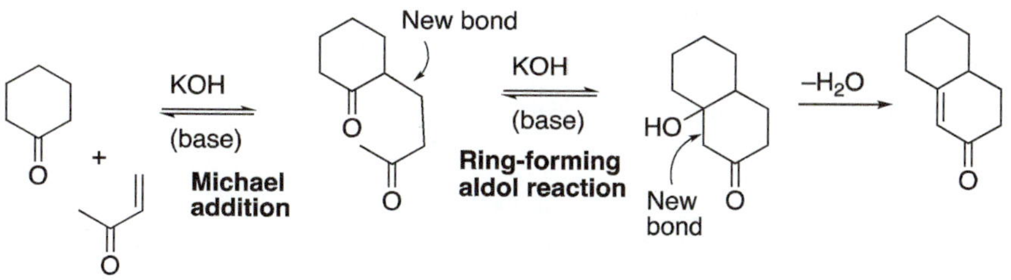

[Reaction 9-60: cyclohexanone + methyl vinyl ketone, KOH (base), Michael addition → diketone (new bond) → KOH (base), Ring-forming aldol reaction → bicyclic β-hydroxy ketone (new bond) → −H₂O → octalone] (9-60)

9.8 Addition of Amines to Carbonyl Compounds

In this section, we will examine the addition of ammonia and other types of amines to carbonyl compounds. Reaction (9-61) shows the addition of ammonia to 2-butanone.

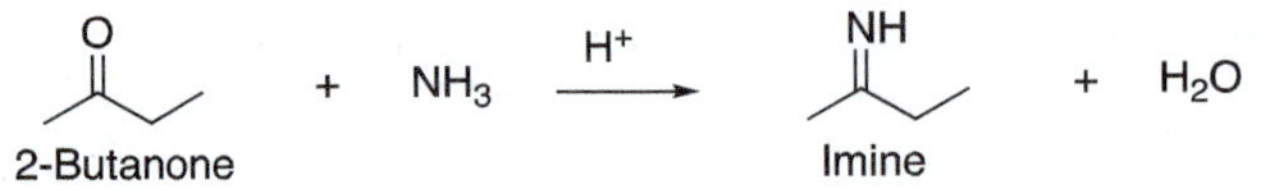

$$\text{2-Butanone} + NH_3 \xrightarrow{H^+} \text{Imine} + H_2O \tag{9-61}$$

Reaction (9-62) gives the reaction of 2-butanone with methanamine, which is a primary amine.

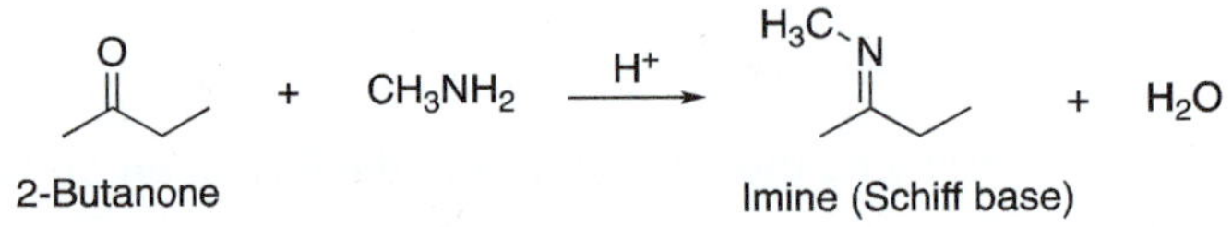

$$\text{2-Butanone} + CH_3NH_2 \xrightarrow{H^+} \text{Imine (Schiff base)} + H_2O \tag{9-62}$$

Imines that have an alkyl group bonded to the nitrogen of the carbon–nitrogen double bond are also known as a Schiff base, named after Hugo Schiff, an Italian chemist, who discovered these types of reactions. The mechanism for the formation of imines from the addition of ammonia and amines to carbonyl compounds involves a combination of two reactions introduced in Chapter 6, addition–elimination reactions. In the first step of the reaction mechanism for Reaction (9-62), the carbonyl group is protonated since the medium is acidic, followed by the addition of methylamine as shown in Reaction (9-63).

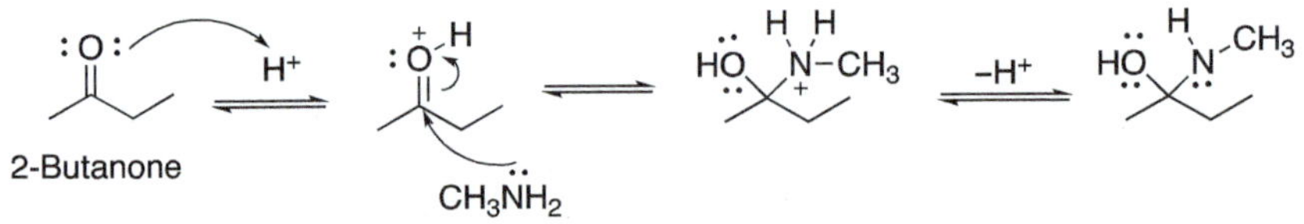

$$(9\text{-}63)$$

Once the addition of methanamine to 2-butanone occurs, there is a loss of a proton. In the next step in the sequence of reactions, the protonation of the OH group occurs converting it to a good leaving group, H_2O, followed by the loss of water and the formation of a resonance-stabilized carbocation, which is stabilized from the lone-pair of electrons from the adjacent nitrogen, as shown in the reaction steps in Reaction (9-64).

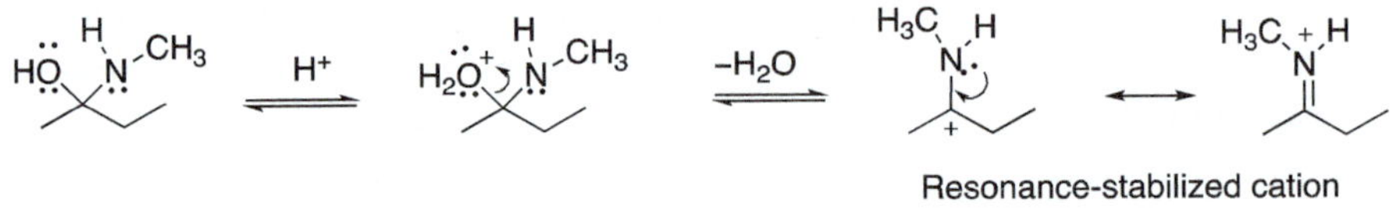

$$(9\text{-}64)$$

Resonance-stabilized cation

In the last step of the mechanism, a proton is lost to form the imine, as shown in Reaction (9-65).

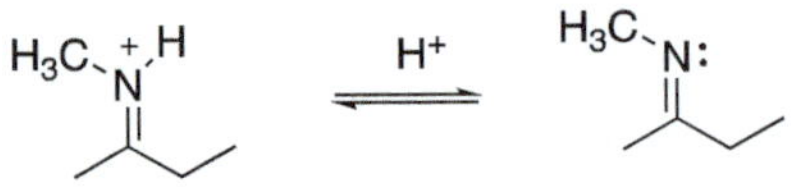

$$(9\text{-}65)$$

Other addition reactions involving different types of amines are shown in Reactions (9-66)–(9-68).

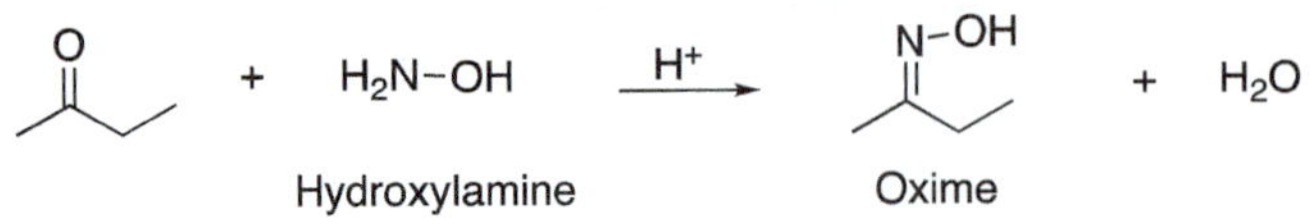

$$\text{2-Butanone} + H_2N\text{–}OH \xrightarrow{H^+} \text{Oxime} + H_2O \tag{9-66}$$

Hydroxylamine

$$\text{CH}_3\text{COCH}_2\text{CH}_3 + \text{H}_2\text{N-NH}_2 \xrightarrow{\text{H}^+} \text{(CH}_3\text{)(CH}_2\text{CH}_3\text{)C=N-NH}_2 + \text{H}_2\text{O} \qquad (9\text{-}67)$$

Hydrazine → Hydrozone

$$\text{CH}_3\text{COCH}_2\text{CH}_3 + \text{H}_2\text{N-NH-Ph} \xrightarrow{\text{H}^+} \text{(CH}_3\text{)(CH}_2\text{CH}_3\text{)C=N-NH-Ph} + \text{H}_2\text{O} \qquad (9\text{-}68)$$

Phenylhydrazine → Phenylhydrazone

Problem 9.13

Give a mechanism for the addition–elimination (of water) reactions for the reaction shown in Reaction (9-66).

9.9 Mechanism for Addition Reactions Involving Imines

Imines are very important functional group in organic chemistry; most are used as ligands for the extraction of metals. The polarity of the imine functionality is very similar to that of the carbonyl functionality. The knowledge gained and used for the addition reactions of the previous section can be applied to predict the outcome of just about any addition reaction with imines. As you can imagine, the addition of an electrophile to the imine is regiospecific and will add to the nitrogen since it is more electronegative than carbon as shown in Reaction (9-69).

$$\underset{\delta^+ \; \delta^-}{\text{C=NR}} + \text{E}^+ \longrightarrow \underset{\text{Carbocation}}{\text{C}^+\text{-N(R)(E)}} \qquad (9\text{-}69)$$

Imine   Electrophile

The next step of the addition reaction mechanism involves the nucleophile adding to the electrophilic carbon as shown in Reaction (9-70).

$$\text{C}^+\text{-N(R)(E)} + :\text{Nu}^- \longrightarrow \text{Nu-C-N(R)(E)} \qquad (9\text{-}70)$$

Nucleophile   Final addition product

Thus, the overall hypothetical reaction is shown in Reaction (9-71), note again, it a regiospecific reaction with the electrophile adding to the nitrogen and the nucleophile adding to the electrophilic carbon.

$$\text{C=NR} + :\text{Nu}^-\text{E}^+ \longrightarrow \text{Nu-C-N(R)(E)} \qquad (9\text{-}71)$$

Imine   Addition product

Of course, it is possible for the sequence of the addition steps to take place in a different sequence in which the first step involves the nucleophile adding to the electrophilic carbon of the imine, as shown in Reaction (9-72).

$$\text{C=NR} + :\text{Nu}^- \longrightarrow \text{Nu-C-NR}^{\ominus} \qquad (9\text{-}72)$$

Imine   Nucleophile   Iminium anion

In the next step, the electrophile will add to the iminium intermediate that was created by the addition of the nucleophile, as shown in Reaction (9-73) to give a final neutral product.

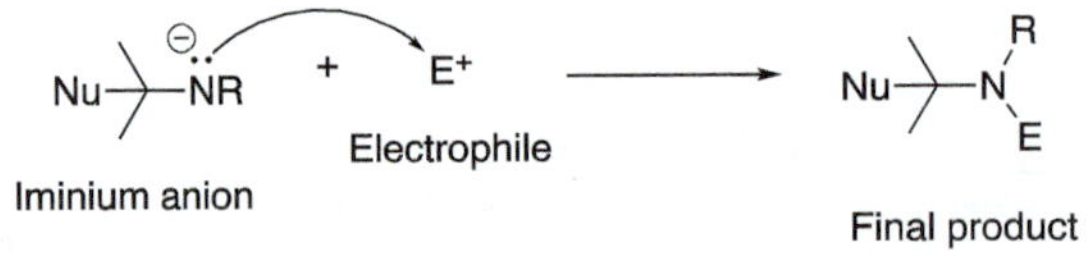

$$\text{Nu}-\overset{\ominus}{\text{NR}} \;+\; \text{E}^+ \;\longrightarrow\; \text{Nu}-\underset{\text{E}}{\overset{R}{\text{N}}} \quad\quad (9\text{-}73)$$

Iminium anion   Electrophile   Final product

The overall hypothetical addition reaction is the same as given in Reaction (9-71). The next section will cover the addition of water to imines.

9.9.1 Addition of Water to Imines

In this section, we will discuss the addition of water to imines, which is accomplished in the presence of an acid catalyst. The addition of water to imine functionalities is a reversible reaction and an example is shown in Reaction (9-74).

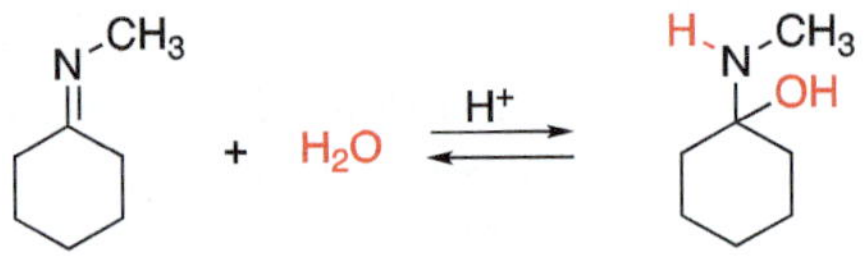

$$\text{cyclohexanone N-methylimine} \;+\; H_2O \;\underset{}{\overset{H^+}{\rightleftharpoons}}\; \text{N-methyl-1-aminocyclohexanol} \quad\quad (9\text{-}74)$$

The mechanism for the addition of water to the imine functionality is shown in Reaction (9-75).

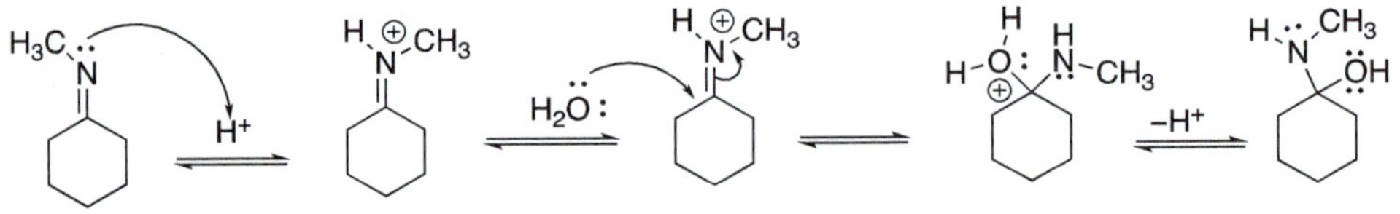

$$(9\text{-}75)$$

You will notice that the product of the addition reaction of water results in an intermediate that was formed by the addition of amines to carbonyl compound (Reaction 9-63). As a result, it is possible to convert an imine to a carbonyl compound as shown in Reaction (9-76).

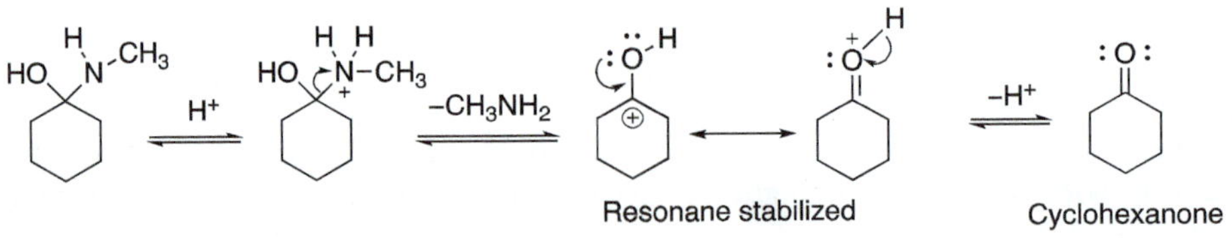

$$(9\text{-}76)$$

Resonane stabilized   Cyclohexanone

As we will see in later chapters, these reactions are very important for the synthesis of molecules that have these functionalities.

9.10 Mechanism for Addition Reactions Involving Nitriles

Owing to the polarity of the nitrile bond, as shown below, its reaction with an electrophile and a nucleophile is similar to that of carbonyl and imine functionalities. In the case of nitriles, there are two sets of pi (π) electrons, which are not equally distributed between the carbon–nitrogen triple bond, but are closer to the more electronegative nitrogen. As a result, addition reactions involving this functionality are regiospecific as shown in Reaction (9-77).

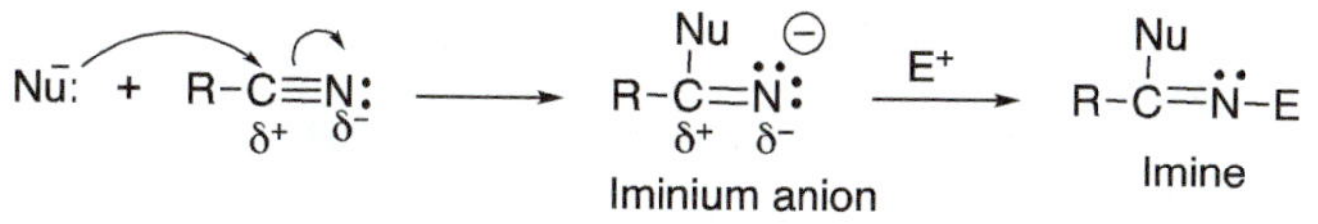

9.10 Mechanism for Addition Reactions Involving Nitriles

$$\text{Nu:}^- + \text{R-C}\equiv\text{N:} \longrightarrow \underset{\text{Iminium anion}}{\text{R-C=N:}^{\ominus}} \overset{\text{E}^+}{\longrightarrow} \underset{\text{Imine}}{\text{R-C=N-E}} \quad (9\text{-}77)$$

You will notice that the product of an initial addition reaction to the nitrile functionality generates another functionality (iminium anion) that we have seen before in the previous section. As shown in the previous section, the iminium anion can undergo a similar addition reaction, owing to the presence of its remaining double bond, as shown in Reaction (9-78).

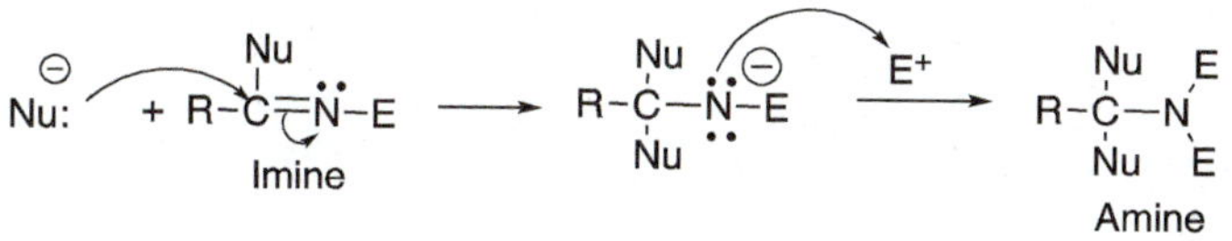

$$\text{Nu:}^- + \underset{\text{Imine}}{\text{R-C=N-E}} \longrightarrow \underset{\text{Nu}}{\text{R-C-N-E}^{\ominus}} \overset{\text{E}^+}{\longrightarrow} \underset{\text{Amine}}{\text{R-C-N}} \quad (9\text{-}78)$$

Of course, it is possible for the sequence of the addition steps to take place in a different sequence in which the first step involves the addition of the electrophile to the carbon of the nitrile, followed by the addition of the nucleophile as shown in Reaction (9-79).

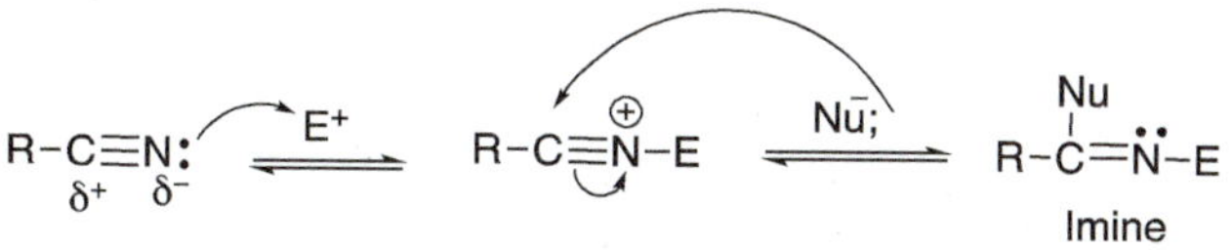

$$\text{R-C}\equiv\text{N:} \overset{\text{E}^+}{\rightleftharpoons} \text{R-C}\equiv\overset{\oplus}{\text{N}}\text{-E} \overset{\text{Nu:}^-}{\rightleftharpoons} \underset{\text{Imine}}{\text{R-C=N-E}} \quad (9\text{-}79)$$

In the next step of the reaction, another mole of electrophile is added to the imine and eventually the final product, the amine, is formed as shown in Reaction (9-80).

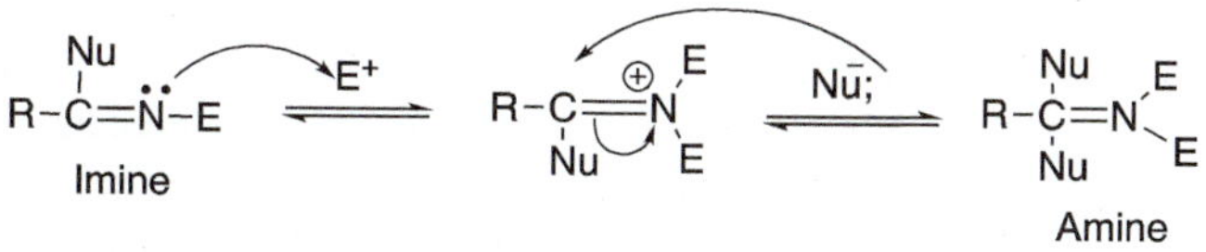

$$\underset{\text{Imine}}{\text{R-C=N-E}} \overset{\text{E}^+}{\rightleftharpoons} \text{R-C=N}^{\oplus}\underset{\text{Nu}}{^{\text{E}}_{\text{E}}} \overset{\text{Nu:}^-}{\rightleftharpoons} \underset{\text{Amine}}{\text{R-C=N}} \quad (9\text{-}80)$$

9.10.1 Addition of Water to Nitriles

Since nitriles can undergo two successive addition reactions. Nitriles are compounds that are often used as starting materials for the synthesis of a wide variety of other organic compounds. One of the most common transformations that nitriles undergo is the addition of water (hydrolysis) to form the corresponding amide or carboxylic acid. The first reaction that we will examine is the addition of water to cyanobenzene in the presence of an acid catalyst to give an amide functionality. Since the nitrile is a very strong triple bond, these hydrolysis reactions are typically carried out at elevated temperatures and in the presence of an acid catalyst as shown in Reaction (9-81).

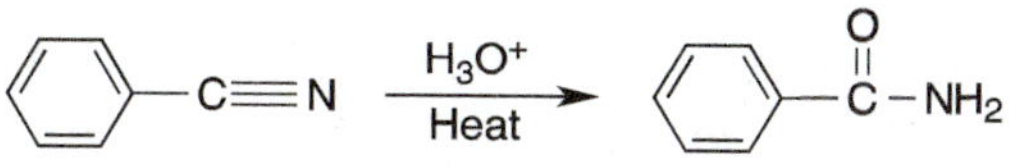

$$\text{Ph-C}\equiv\text{N} \xrightarrow[\text{Heat}]{\text{H}_3\text{O}^+} \text{Ph-C(=O)-NH}_2 \quad (9\text{-}81)$$

The mechanism for this reaction is shown below. In the initial step, the electrophilic proton adds to the nucleophilic nitrogen of the nitrile triple bond, followed by an attack of the nucleophilic water molecule on the electrophilic carbon of the triple bond as shown in Reaction (9-82).

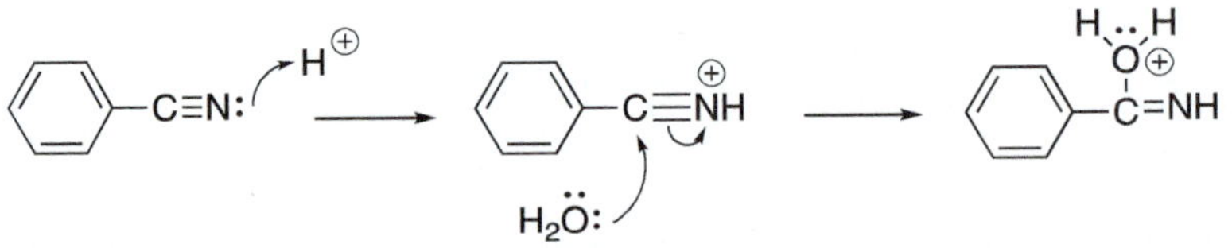

244 | 9 Addition Reactions Involving Carbonyls and Nitriles

$$\text{Ph-C≡N:} \xrightarrow{H^+} \text{Ph-C≡NH}^+ \xrightarrow{H_2\ddot{O}:} \text{Ph-C(OH}_2^+\text{)=NH} \quad (9\text{-}82)$$

After the loss of proton and subsequent protonation of the nitrogen, hydrolysis of the imine (as shown in the previous section) gives the amide as shown in Reaction (9-83).

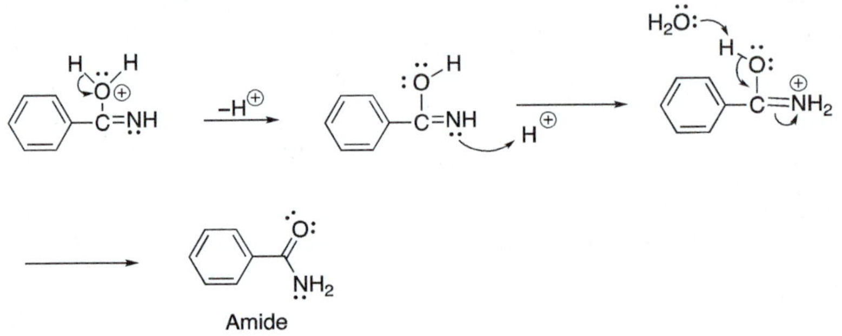

$$\text{Ph-C(OH}_2^+\text{)=NH} \xrightarrow{-H^+} \text{Ph-C(OH)=NH} \xrightarrow{H^+, H_2O} \text{Ph-C(OH)(OH}_2^+\text{)-NH}_2 \longrightarrow \text{Ph-C(=O)-NH}_2 \quad (9\text{-}83)$$

Amide

Problem 9.14
Complete the following reactions by giving the amide product that results.

(a) $(CH_3)_2CHCH_2CH_2\text{-CN} \xrightarrow[\text{Heat}]{H^+, H_2O}$

(b) $PhCH_2\text{-CN} \xrightarrow[\text{Heat}]{H^+, H_2O}$

If the hydrolysis is carried out under much more severe conditions, such as prolonged heating and in very strong acidic conditions, the amide that is initially formed is readily transformed into a carboxylic acid. The conversion involved is a combination of addition reaction and a substitution reaction, and we will cover the mechanism for the substitution reaction in Chapter 16, but the overall reaction is shown in Reaction (9-84).

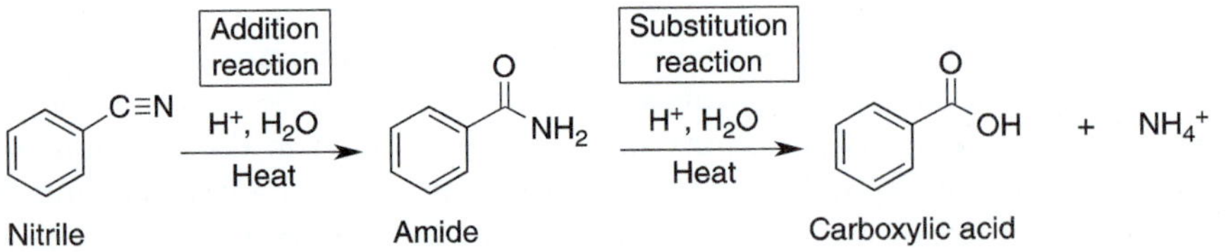

$$\text{Ph-C≡N} \xrightarrow[\text{Heat}]{\substack{\text{Addition reaction}\\ H^+, H_2O}} \text{Ph-C(=O)-NH}_2 \xrightarrow[\text{Heat}]{\substack{\text{Substitution reaction}\\ H^+, H_2O}} \text{Ph-C(=O)-OH} + NH_4^+ \quad (9\text{-}84)$$

Nitrile   Amide   Carboxylic acid

9.11 Applications of Addition Reactions to Synthesis

In this section, we will examine how best to make strategic choices in the use of specific reactions to accomplish transformations and the best combination of reactions if more than one reaction is required to accomplish a transformation. As pointed out in the previous chapter, an important aspect of organic chemistry is the synthesis of specific target molecules utilizing the reaction types that we have studied so far. Let us first consider the type of reactions needed to carry out the following transformation shown in Reaction (9-85).

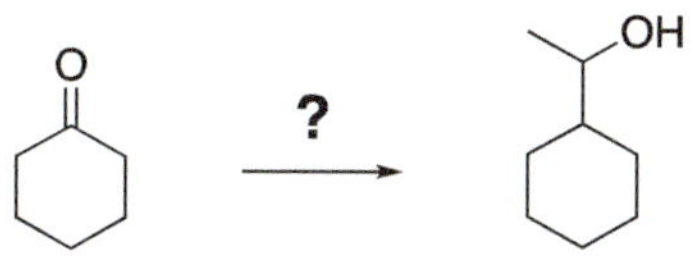

9.11 Applications of Addition Reactions to Synthesis

$$\text{cyclohexanone} \xrightarrow{?} \text{1-(cyclohexyl)methanol with OH} \quad (9\text{-}85)$$

Initial inspection of the reaction shows that the oxygen of the carbonyl functionality in the reactant has been replaced by a carbon. We have covered an addition reaction that can convert a carbonyl functionality to an alkene and that is the Wittig reaction. We have also covered a reaction that can convert an alkene into alcohols via a specific hydration reaction. Now that we have established a strategy to carry out the transformation, the next task is to select an appropriate sequence of reactions and reaction conditions to accomplish the transformation. Working backward, the last reaction needed for this transformation is shown in Reaction (9-86).

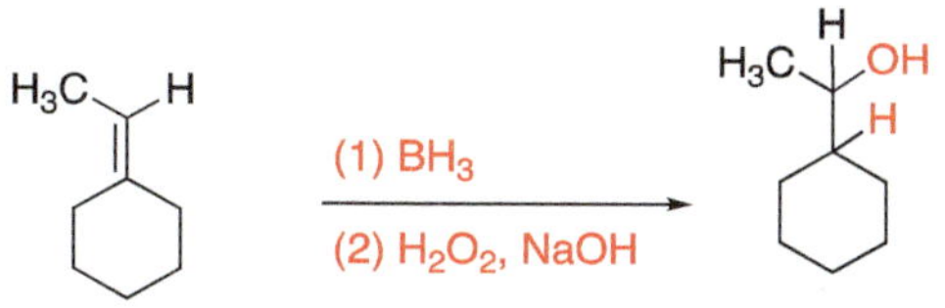

$$\text{ethylidenecyclohexane} \xrightarrow[\text{(2) } H_2O_2,\ NaOH]{\text{(1) } BH_3} \text{alcohol product} \quad (9\text{-}86)$$

The synthesis of the alkene can be accomplished from a carbonyl compound as shown below. Note that you will have to make the appropriate selection of Wittig reagent with appropriate groups bonded to the reagent so that it gives the desired product. In this case, the carbon should be bonded to a hydrogen and a methyl group, as shown in red in Reaction (9-87).

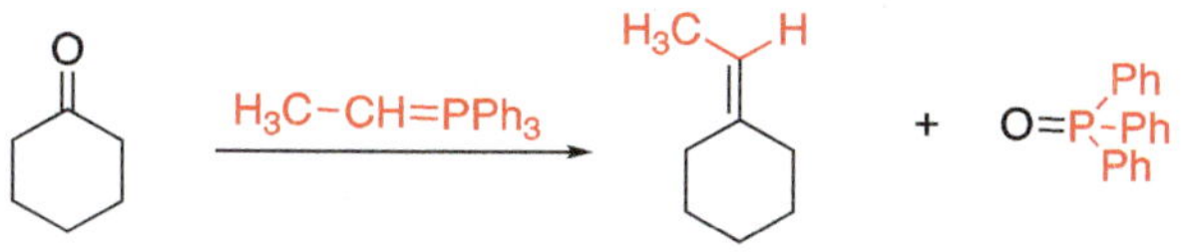

$$\text{cyclohexanone} \xrightarrow{H_3C-CH=PPh_3} \text{ethylidenecyclohexane} + O=PPh_3 \quad (9\text{-}87)$$

Thus, the overall sequence of reactions needed for this transformation can be summarized as shown in Reaction (9-88). Note that the inorganic products are typically not written since we are more concerned about obtaining an organic molecule.

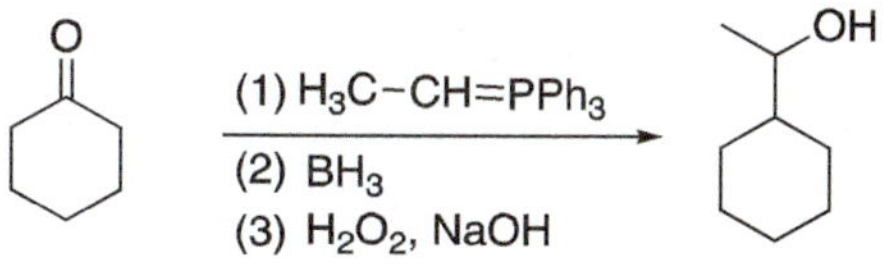

$$\text{cyclohexanone} \xrightarrow[\substack{\text{(2) } BH_3 \\ \text{(3) } H_2O_2,\ NaOH}]{\text{(1) } H_3C-CH=PPh_3} \text{alcohol product} \quad (9\text{-}88)$$

Problem 9.15

Show how to carry out the transformations shown below. Clearly show all reagents needed and steps in any sequence of reactions carried out. Note that these transformations are essentially the same as shown in the example, except that instead of using the ketone as a starting compound, Wittig reagents are the starting reagents.

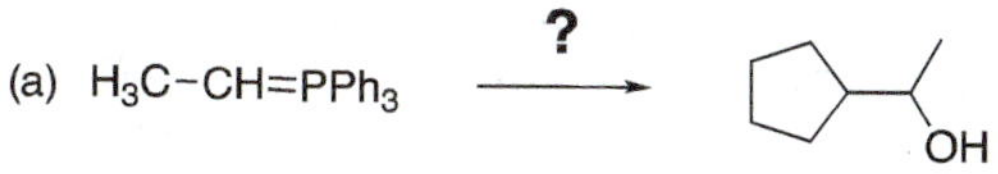

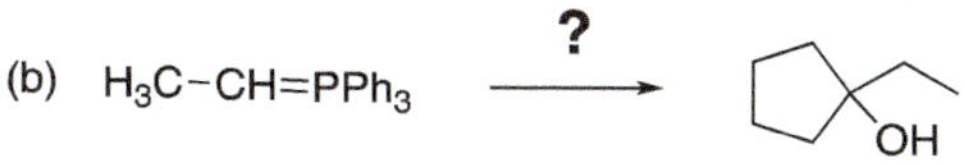

(a) $H_3C-CH=PPh_3 \xrightarrow{?}$ cyclopentyl-CH(CH_3)OH

(b) $H_3C-CH=PPh_3 \xrightarrow{?}$ cyclopentyl-C(CH_3)_2OH

End of Chapter Problems

9.16 Show how to carry out the following transformations. For each step in your synthesis, clearly show all reagents and reaction conditions.

(a) [imine] → [secondary amine]

(b) [cyclopentyl-C≡N] → [cyclopentyl-C(=O)-NH₂]

(c) [benzaldehyde] → [hemiacetal/acetal with HO and O-propyl]

(d) [cyclopentanone] → [cyclic ketal]

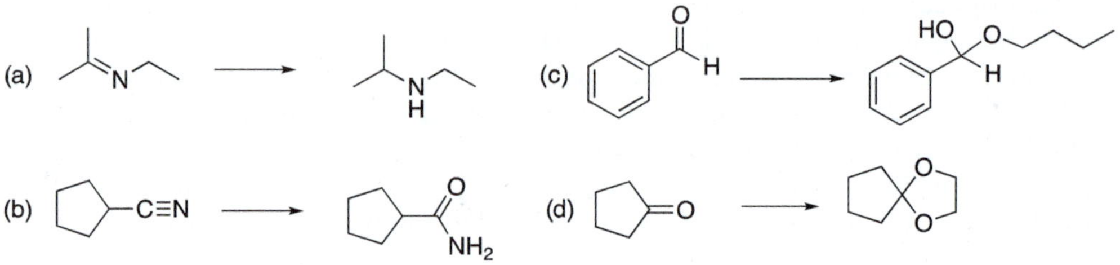

9.17 Propose a reasonable mechanism for the hydrolysis of the acetal shown below (hint: the first step in the mechanism is given).

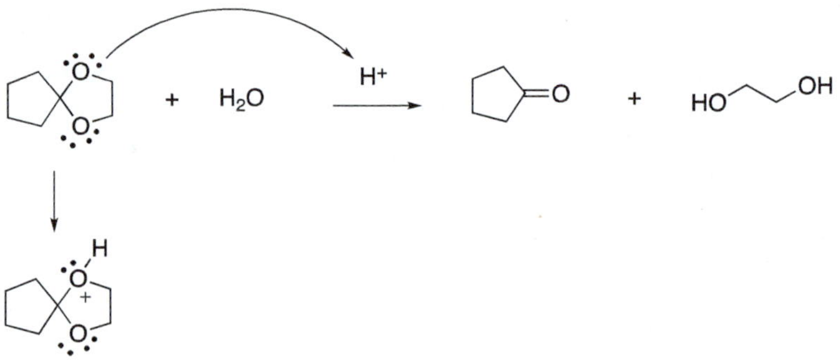

[spiro acetal] + H₂O —H⁺→ [cyclopentanone] + HO–CH₂CH₂–OH

↓

[protonated acetal intermediate]

9.18 Show how to carry out the following transformation, clearly show all reagents and steps in your reaction sequence if your synthesis is a multistep synthesis (hint: you may have to protect one of these functional groups while carrying out a reaction on the other functional group).

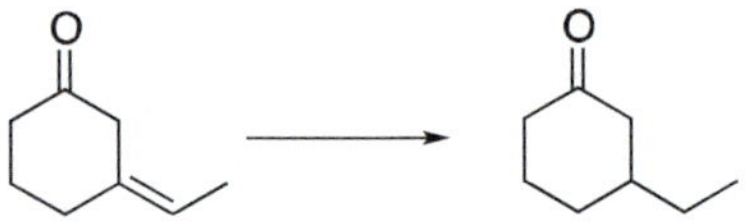

[cyclohexanone with exocyclic alkene] → [cyclohexanone with ethyl group]

9.19 Three different alkenes (**A, B,** and **C**) yield 2-methylbutane when they are hydrogenated in the presence of a metal catalyst. Give the structures and IUPAC names of the three compounds.

9.20 Show how to accomplish the following transformation, clearly show all reagents and reaction conditions necessary to carry out each step in your synthesis (hint: consider using an appropriate hydration reaction followed by an oxidation reaction to form the ketone).

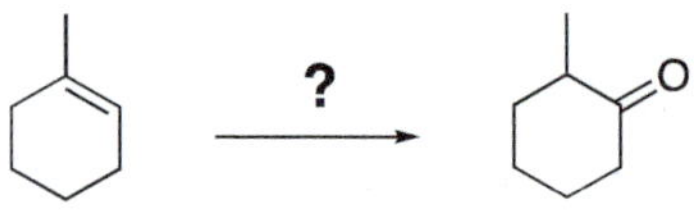

[methylcyclohexene] —?→ [methylcyclohexanone]

9.21 Determine which of each of the following pairs of ketones is more reactive toward hydration?

(a) 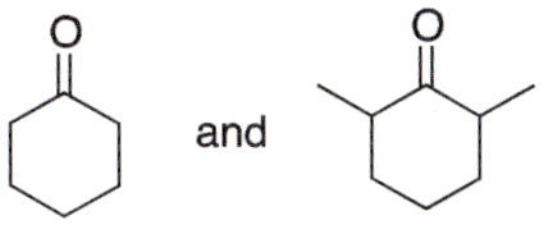 [cyclohexanone] and [2,6-dimethylcyclohexanone] (b) 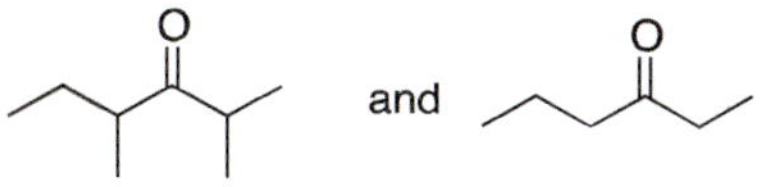 [2,4-dimethyl-3-pentanone] and [2-methyl-3-pentanone]

(c) 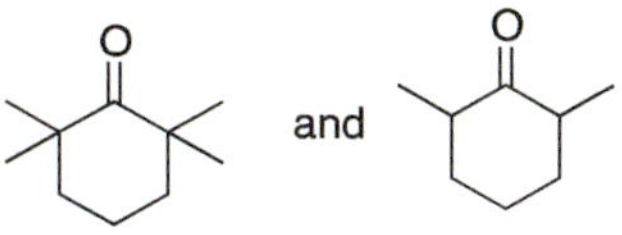 [2,2,6,6-tetramethylcyclohexanone] and [2,6-dimethylcyclohexanone] (d) 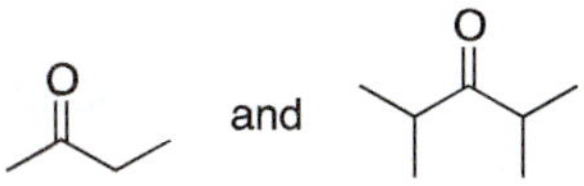 [2-butanone-like structure] and [2,2-dimethyl-3-pentanone]

9.22 Propose a reasonable mechanism for the hydrolysis of the nitrile shown below to form an amide (hint: the first step in the mechanism is shown).

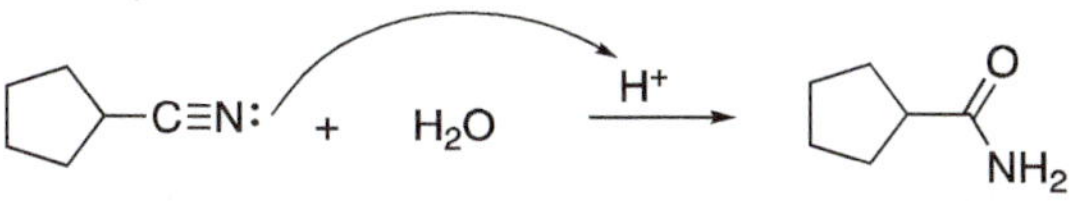

$$\text{cyclopentyl-C}\equiv\text{N:} + H_2O \xrightarrow{H^+} \text{cyclopentyl-C(=O)-NH}_2$$

9.23 Propose reasonable mechanisms for the following reaction (hint: the first steps for each mechanism are shown).

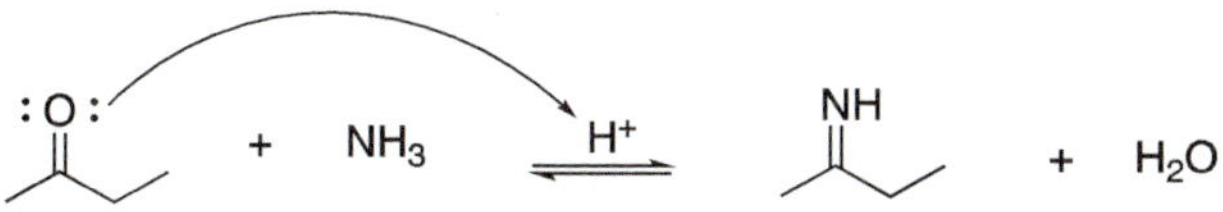

$$\text{ketone} + NH_3 \xrightleftharpoons{H^+} \text{imine} + H_2O$$

9.24 Show how to carry out the following transformations. For the steps in your syntheses, <u>clearly</u> show all reagents and reaction conditions (hint: note that the products of these reactions have more carbons than the reactants, additional carbons are shown in red; think of adding extra carbons via a Wittig reaction to form an alkene, then an addition reaction, shown in green).

(a) 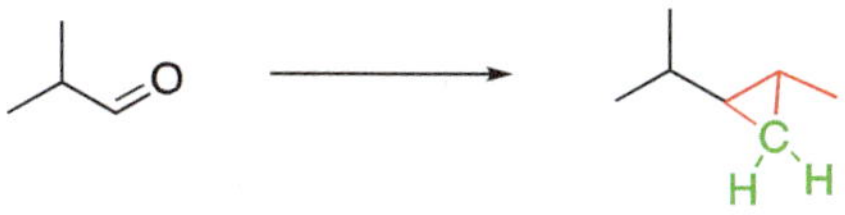 [acetone] ⟶ [product with added C,H,H in green]

(b) 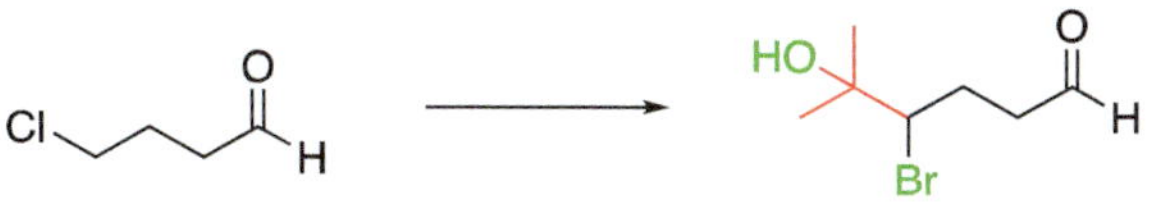 Cl-CH₂CH₂-C(=O)H ⟶ HO-C(CH₃)(Br)-CH₂CH₂-C(=O)H

9.25 Show how to accomplish the following transformations (synthesis). All reagents, reaction conditions, and appropriate solvents must be clearly shown.

(a)  cyclopentanone ⟶ cyclopentyl-CH(OH)

(b) 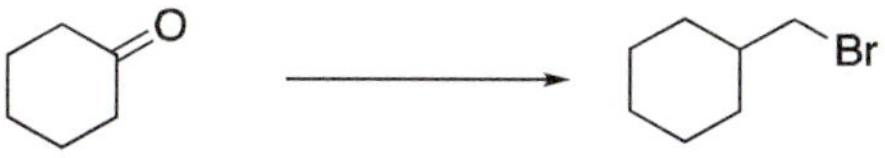 cyclohexanone ⟶ cyclohexyl-CH₂Br

(c)  [structure: Br-CH2-CH2-CHO → (CH3)2CH-CH(OH)-CH2-CHO]

(d)  [structure: Br-CH2CH2CH2-C(O)-CH3 → cyclopropane derivative with -CH2-C(O)-CH3]

9.26 The cyclopropane group is typically found in insecticides. Molecule **2**, shown below, is a long-lasting inhibitor for some enzymes. On the arrow, give appropriate reagents that can be used to carry out the transformation shown.

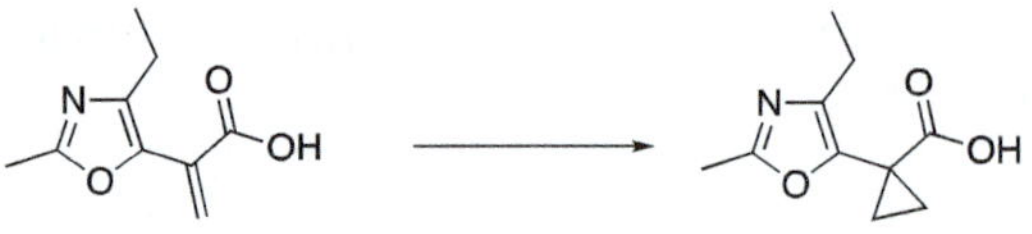

[structures of compounds **1** and **2**]

9.27 Ketones can be converted to 1,3-dithiolanes as shown in the reaction below. Using arrow-pushing formulism to indicate electron movement, propose a reasonable mechanism for the reaction shown below, which is carried out in an acidic medium (hint: the first step in the mechanism is given, the next step involves the nucleophilic sulfur adding to the electrophilic carbon).

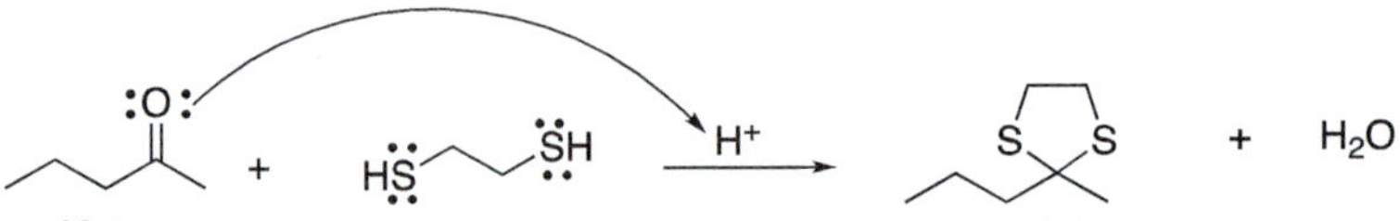

[Ketone + HS-CH2CH2-SH, H+ → 1,3-Dithiolane + H2O]

9.28 Provide structures for the lettered compounds for each of the reaction scheme shown below.

(a) [Br-CH2CH2CH2-CHO →(EG, H+)→ **A** →(PPh3)→ **B** →(BuLi)→ **C**; **C** + acetone-like ketone →(H+, H2O)→ **D**; **D** ... →(CH2I2, Zn/Cu)→ **E**]

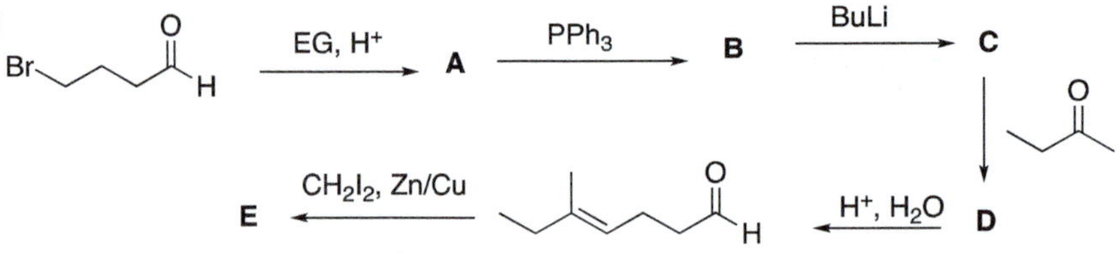

(b) [Br-CH2CH2-CHO →(EG, H+)→ **A** →(PPh3)→ **B** →(BuLi)→ **C**; **C** + cyclopentanone →(H+, H2O)→ **D**; **D** →(Br2, H2O, DMSO solvent)→ **E**]

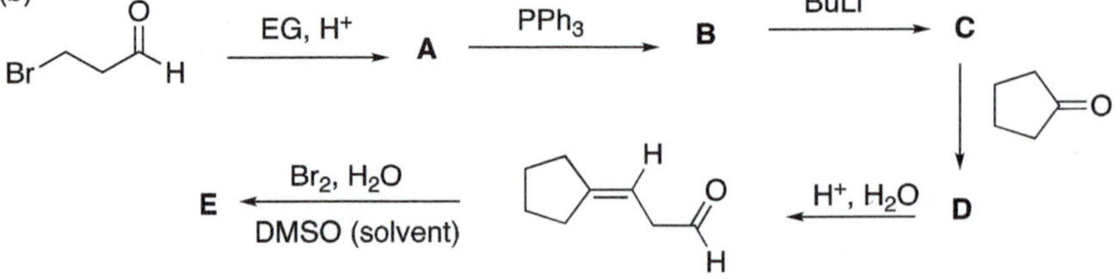

9.29 Using arrow-pushing formulism to indicate electron movement, provide a reasonable mechanism for the following reaction; hint the first step is shown.

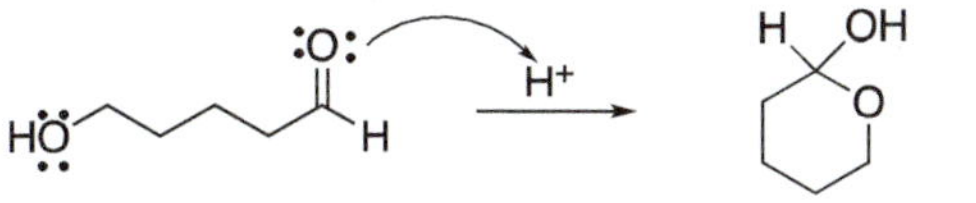

Cyclic hemiacetal

9.30 Using arrow-pushing formulism to indicate electron movement, provide a reasonable mechanism for the following reaction.

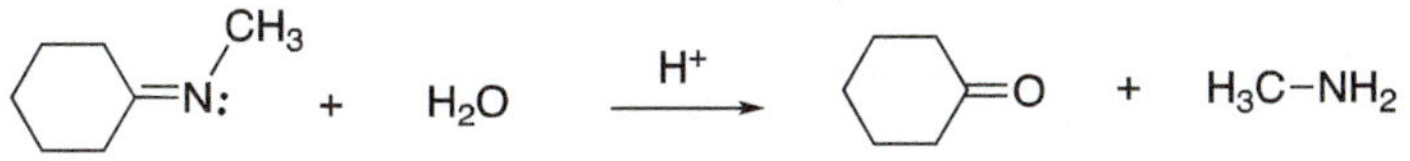

9.31 Show how to carry out the following transformations. For the steps in your syntheses, <u>clearly</u> show all reagents and reaction conditions.

(a) (hint: consider using a Wittig reaction)

(b) (hint: consider using a protecting group and a Wittig reaction)

(c) (hint: consider an oxidation reaction followed by a Grignard reaction)

(d) (hint: consider using an oxirane in your synthesis)

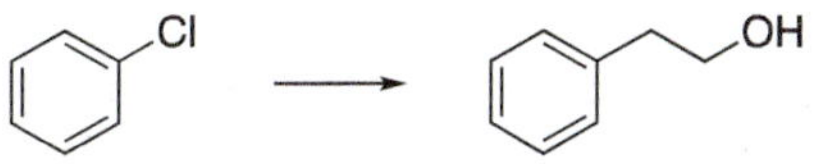

10

Reduction Reactions in Organic Chemistry

10.1 Introduction

As mentioned in Chapter 6 in the overview of the reactions of organic chemistry, reduction is the supply of electrons and or the removal of oxygen, and this process typically is accompanied by the supply of hydrogen. In organic chemistry, this type of reaction is very common in order to synthesize many useful compounds. First, it is important to recognize compounds that can be reduced. Such compounds are typically compounds that are electrophilic and can accept electrons from reducing agents. We have seen from previous chapters, various types of electrophiles and some are shown below, in which the electrophilic atom is indicated with a partial positive (δ^+).

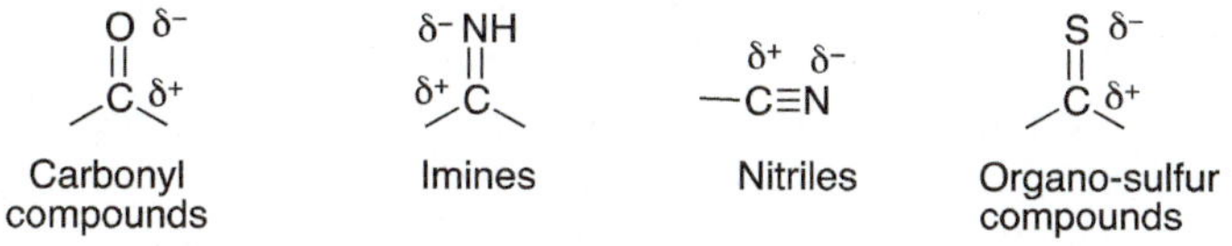

- Carbonyl compounds
- Imines
- Nitriles
- Organo-sulfur compounds

It is also important to be able to recognize possible reducing agents. As mentioned above, reducing agents are reagents that have the ability to supply electrons to another molecule. You will recall that this definition is also the same definition as a Lewis base and also a nucleophile. For the reactions of this chapter, however, the difference is that the reducing agents have much stronger potential to supply electrons, compared to the nucleophiles that we have discussed thus for, such as water, alcohols, and amines. The reduction potentials of reducing agents are much greater than typical Lewis bases or nucleophiles. Owing to the similarity of the structures of Lewis bases, nucleophiles and reducing agents, the symbol Nu:$^-$, along with its counter cation, is typically used to represent a reducing agent. Reaction (10-1) is a hypothetical reduction reaction of a polar covalent double bond with a reducing agent.

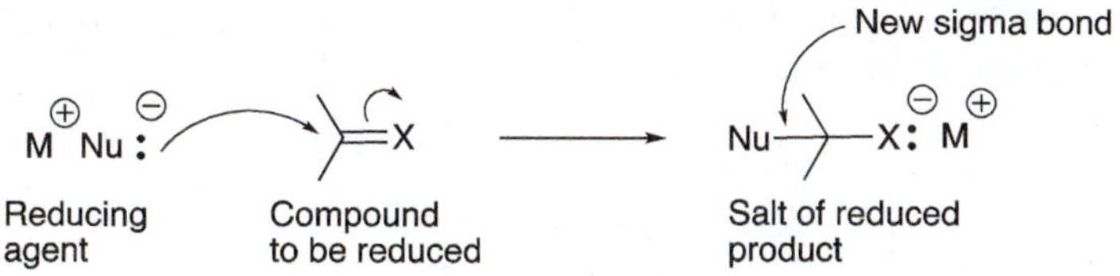

$$M^+ \; Nu{:}^- \quad + \quad \underset{\text{Compound to be reduced}}{\overset{}{\diagup}\hspace{-0.5em}=\hspace{-0.5em}X} \quad \longrightarrow \quad \underset{\text{Salt of reduced product}}{Nu\hspace{-0.2em}\diagup\hspace{-0.5em}-\hspace{-0.2em}X{:}^- \; M^+} \qquad (10\text{-}1)$$

Note that reduction takes place because the reducing agent, Nu:$^-$, uses its unshared pair of electrons to form a new sigma bond in the product. Also, notice that this reaction is a regiospecific addition reaction in that the reducing agent bonds to the electrophilic carbon of the

Organic Chemistry: Concepts and Applications, First Edition. Allan D. Headley.
© 2020 John Wiley & Sons, Inc. Published 2020 by John Wiley & Sons, Inc.
Companion website: www.wiley.com/go/Headley_OrganicChemistry

molecule that is being reduced. As you can imagine, the reduction of compounds where there are no polar bonds, such as alkenes and alkynes, is a bit more difficult and requires extremely strong reducing agents and severe reaction conditions, and often a catalyst is used for the reduction of the molecules as shown below.

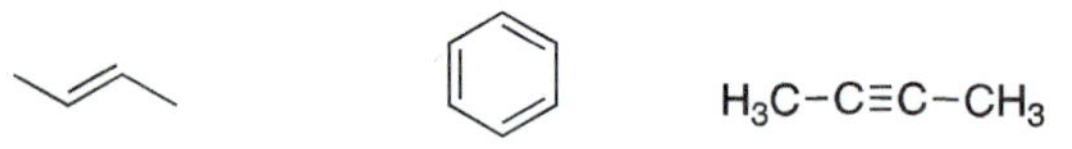

2-Butene   Benzene $H_3C-C\equiv C-CH_3$ Butyne

Appropriate reducing agents and reaction conditions that are suited for the reduction of these types of molecules will be covered later in this chapter. In first sections of this chapter, we will examine the various reducing agents that are commonly used in organic chemistry and also discuss the mechanism for the reduction of different functional groups.

10.2 Reducing Agents of Organic Chemistry

Our next task is to identify compounds that are potential reducing agents; that is, agents that can readily supply electrons to another compound. In organic chemistry, electrons are typically delivered via a specific atom of the reducing agent, and as you can imagine, such an atom is not very electronegative. In fact, some of the best reducing agents contain very electropositive atoms, such as hydrogen or carbon atoms. In a reduction reaction, electrons are delivered from these very electropositive atoms to the molecule being reduced. If a hydrogen atom has an unshared pair of electrons, it is called a *hydride* anion, and the counter cation for hydride anions are typically metal cations, such as lithium, sodium, or potassium. The same is true for carbon atom with a pair of electrons, these carbon anions are typically bonded to metal cations, and as a result, they are referred to as organometallic compounds. It is not surprising that since reducing agents contain at least a pair of electrons that they are also bases. In fact, the reducing agents that will be discussed in this chapter are some of the strongest bases of organic chemistry. In this section, we will discuss each type of hydrogen and carbon reducing agents that will be encountered throughout our course of organic chemistry.

10.2.1 Metal Hydrides

Metal hydrides are extremely strong reducing agents, and they deliver the electrons via the hydride anion to form a new covalent bond to the molecule being reduced. One of the strongest metal hydrides reducing agents as mentioned above is a hydrogen atom that has a pair of electrons and has a counter cation, such as sodium or potassium. Another type contains the hydride anion bonded to aluminum or boron. Examples of metal hydrides that are typically used in organic chemistry as reducing agents are shown below.

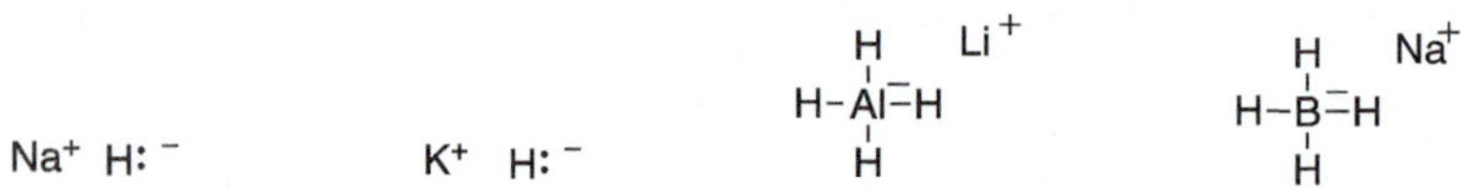

Na^+ H:$^-$ K^+ H:$^-$ Li^+ [H-Al(H)$_3$] Na^+ [H-B(H)$_3$]

Sodium hydride   Potassium hydride   Lithium aluminum hydride   Sodium borohydride

Notice that these compounds are ionic, and the actual reducing hydride anion is bonded to an electropositive atom: sodium, potassium, aluminum, and boron, respectively. For lithium aluminum hydride and sodium borohydride, the central atoms in the neutral molecules typically have three bonds instead of four bonds as shown for the compounds above. As a result, lithium aluminum hydride and sodium borohydride react readily delivering a hydride anion ($H:^-$) to molecules to be reduced resulting in the formation of a trivalent aluminum and boron products, along with the reduced organic product. Reactions in which these reducing agents are typically used are carried out in non-protic solvents, such diethyl ether or tetrahydrofuran (THF) and not in protic solvents, such as water or alcohols. You will recall that protic solvents have an electrophile in the form of a potential proton (H^+), which will react violently with the hydride anion (an extremely strong base) to produce hydrogen (H_2) gas. We will use these reducing agents frequently throughout our course in organic chemistry.

10.2.2 Organometallic Compounds

The next set of reducing agents that we will examine are those that contain carbon, and some common organometallic compounds are shown below.

$\overset{\delta-\ \delta+}{R-Li}$ $\overset{\delta-\ \delta+}{R-MgX}$ $\overset{\delta-\ \delta+}{R_2-CuLi}$

Organolithium   Organomagnesium   Organocuprate
                 (Grignard reagent)

The first is a carboanion, which is bonded to the electropositive lithium atom. Since lithium is very electropositive, you can imagine that the carbon–lithium bond is very polar, in fact almost ionic, and as a result, the carboanion is very reactive making organolithium compounds strong reducing agents. Organolithium reducing agents are readily made by the reaction of an alkyl halide with lithium, as shown in Reaction (10-2).

$$R-X \ + \ 2\,Li \ \xrightarrow[\text{(solvent)}]{\text{Diethyl ether}} \ R\text{-}Li \ + \ Li^+\ X^- \tag{10-2}$$

R = alkyl group
X = Cl, Br, or I   (Organolithium)

The other reducing agent is a carboanion bonded to magnesium, which is not as electropositive as lithium, but these compounds are still very reactive in delivering electrons, making organomagnesium compounds very reactive reducing agents. François Auguste Victor Grignard, who was born in Cherbourg, France, in 1871, discovered these types of compounds and in 1912 shared the Nobel Prize in Chemistry for this discovery. As a result, these types of reducing agents are often referred to as *Grignard reagents*. They are readily synthesized by the reaction of an alkyl halide with magnesium metal in a non-protic solvent, typically diethyl ether, an example for the synthesis of a Grignard reagent is shown in Reaction (10-3).

$$CH_3CH_2CH_2-Cl \ + \ Mg \ \xrightarrow{\text{Ether}} \ CH_3CH_2CH_2-Mg-Cl \tag{10-3}$$

1-Chloropropane   Propylmagnesium chloride

The last type of organometallic reagent shown is the organocuperate (R_2CuLi). As you will see from the periodic table, copper is not as electropositive as either lithium or magnesium, and as a result, the carboanion bond to the copper is polar, but not as polar as that of the other organometallic compounds discussed thus far. As a result, organocuperates are reducing agents, but not as strong as organolithium or the Grignard reagent. Organocuperates are readily made from organolithium compounds, as shown in Reaction (10-4).

$$\underset{\substack{\text{Organolithium} \\ \text{R = alkyl group}}}{\text{2 R-Li}} + \underset{\substack{\text{X = Cl} \\ \text{Br or I}}}{\text{CuX}} \xrightarrow[\text{(solvent)}]{\text{Diethyl ether}} \underset{\text{Organocuperate}}{\text{R}_2\text{CuLi}} + \text{Li}^+ \text{X}^- \quad (10\text{-}4)$$

As you can imagine, over the years, a wide variety of organometallic compounds, such as organocadmium (R_2Cd), have been developed and used as effective reducing reagents in organic chemistry.

Problem 10.1

Starting with 1-bromopentane, show how to synthesize the following organometallic reagents, include appropriate solvents in your synthesis.

(a) ⌇⌇⌇Li   (b) ⌇⌇⌇MgBr   (c) (⌇⌇⌇)$_2$CuLi

Another carbanion that we will encounter is the acetylide anion, and it can be synthesized as shown in Reaction (10-5)

$$\text{R-C} \equiv \text{C-H} + \text{NaNH}_2 \longrightarrow \text{R-C} \equiv \text{C:}^{\ominus} \text{Na}^{\oplus} + \text{NH}_3 \quad (10\text{-}5)$$

You will recall from Table 7.1 that the hydrogen that is bonded to a carbon of a triple bond is acidic, $pK_a \sim 25$. As a result, a strong base, such as $NaNH_2$, can abstract that proton creating the conjugate acetylide anion base. Note that NH_2 anion is a strong base since its conjugate acid is a very weak acid, $pK_a = 38$.

10.2.3 Dissolving Metals

You will recall from your general chemistry course that the elements to the extreme left of the periodic table are very reactive primarily because they want to achieve the electronic configuration of the nearest noble gas, which can be accomplished by these elements giving up electrons. As a result, these elements are extremely good reducing agents. The most reactive of these elements are the ones at the top left of the periodic table, K and Na. The loss of an electron by sodium atom to produce sodium cation is shown in Reaction (10-6).

$$\text{Sodium (Na): 1s}^2\, 2s^2\, 2p^6\, 3s^1 \xrightarrow{\text{Loss of one electron}} \text{Sodium cation (Na}^+\text{): 1s}^2\, 2s^2\, 2p^6\, 3s^0 \quad (10\text{-}6)$$

These elements in an appropriate solvent will act as very strong reducing agents as we will see in the reactions to be discussed in the next sections.

10.2.4 Hydrogen in the Presence of a Catalyst

Hydrogen is fairly unreactive and is typically not considered to be a strong reducing agent, but in the presence of a catalyst, such as platinum or palladium and under extreme conditions such as elevated temperatures and pressures, reduction using hydrogen gas can take place. An example showing the use of hydrogen to reduce a molecule that has a carbon–oxygen double bond is shown in Reaction (10-7). A wider range of these types of reactions will be covered in the later part of the chapter.

$$\text{cyclohexanone} \xrightarrow[\text{High pressure (~65 atm)}]{\text{H}_2,\ \text{Ni, heat}} \text{cyclohexanol} \quad (10\text{-}7)$$

10.3 Reduction of C=O and C=S Containing Compounds

In this section, the reducing agents discussed in the previous section will be applied to specific type of compounds. As pointed out earlier, the C=O and C=S functionalities have polarized bonds in which the electrons are closer to the oxygen or sulfur compared to the carbon. As a result, the carbon in the C=O and C=S bonds is partially positive, and this region of these molecules is electrophilic and can accept electrons from a reducing agent. In this section, we will utilize the reducing reagents discussed in the previous section to reduce various compounds that have the C=O and C=S functionalities to produce new compounds with different functional groups.

10.3.1 Reduction Using $NaBH_4$ and $LiAlH_4$

The reduction of compounds that have the C=O functionality by $LiAlH_4$ gives alkoxide salts. A condensed mechanism for the reaction is shown in Reaction (10-8).

$$H-\overset{H}{\underset{H}{Al}}-H \;\; Li^+ \;\; + \;\; \underset{\text{A ketone or aldehyde}}{\overset{O \,\delta^-}{\underset{}{\overset{\|}{C}}}\delta^+} \longrightarrow \underset{\text{An alkoxide salt}}{-\overset{O^- \; Li^+}{\underset{H}{\overset{|}{C}}}-} \;\; + \;\; AlH_3 \tag{10-8}$$

Note that it is the hydride anion (a hydrogen atom with two electrons) that is transferred from the reducing agent, lithium aluminum hydride, to the carbon of the carbonyl functionality (shown in red). In another step, the alkoxide salt, which is strongly basic, can be converted into an organic molecule by the abstraction of a proton from an acid to give an alcohol; shown in Reaction (10-9) is the acid–base reaction of the basic alkoxide salt with dilute HCl to give an alcohol.

$$\underset{\text{An alkoxide salt}}{-\overset{O^- \; Li^+}{\underset{H}{\overset{|}{C}}}-} \;\; \xrightarrow{\text{Dil. HCl}} \;\; \underset{\text{An alcohol}}{-\overset{OH}{\underset{H}{\overset{|}{C}}}-} \;\; + \;\; LiCl \tag{10-9}$$

Note that for the two steps of the reaction sequence given in Reactions (10-8) and (10-9), a carbonyl compound is transformed into a different functionality, an alcohol. The sequence of reactions can be summarized in the reaction format shown in Reaction (10-10).

$$\underset{\text{Ketone or aldehyde}}{\overset{O}{\underset{}{\overset{\|}{C}}}} \;\; \xrightarrow[\text{(2) } H^+, H_2O]{\text{(1) } LiAlH_4} \;\; \underset{\text{An alcohol}}{-\overset{OH}{\underset{H}{\overset{|}{C}}}-} \tag{10-10}$$

Note that in the second step, that a generic symbol for a dilute inorganic acid (H^+) is used; this symbol is typically used to represent one of the general inorganic acids: HCl, H_2SO_4, or HNO_3, and this step is sometimes referred to as a hydrolysis step or acidic workup. Also, note that the inorganic salt is not typically included as a product; typically, just the organic products are shown for organic reactions. A similar reaction can be carried out using $NaBH_4$, and that type of reaction (including the hydrolysis) is shown in Reaction (10-11).

$$\underset{\substack{H\\|\\H-\overset{|}{\underset{|}{B}}-H\\|\\H}}{}\text{Li}^+ \quad + \quad \underset{\text{A ketone or aldehyde}}{\overset{O\,\delta-}{\underset{}{\underset{|}{\overset{||}{C}}\,\delta+}}} \longrightarrow \underset{\text{An alkoxide salt}}{\overset{O^-\;\text{Li}^+}{\underset{H}{\underset{|}{\overset{|}{-C-}}}}} \overset{H^+,\,H_2O}{\longrightarrow} \underset{\text{An alcohol}}{\overset{OH}{\underset{H}{\underset{|}{\overset{|}{-C-}}}}} \quad (10\text{-}11)$$

You will readily notice that from these reactions, different types of alcohols result depending on the structure of the starting carbonyl compound. As you saw from Chapter 3, an aldehyde has an alkyl group and a hydrogen bonded to the carbonyl carbon. Whereas, a ketone has two alkyl groups bonded to the carbon of the carbonyl carbon. Formaldehyde is a specific carbonyl compound, which has two hydrogens bonded to the carbonyl carbon.

Formaldehyde	Aldehyde	Ketone
H—C(=O)—H	R—C(=O)—H	R—C(=O)—R

The reduction of formaldehyde with reducing agents that supply the hydride anion (H:⁻) gives methanol as a product; the reduction of aldehydes gives primary alcohols as the products; and the reduction of ketones gives secondary alcohols as the products. Examples of the reduction of these types of compounds are shown in Reactions (10-12)–(10-14).

Formaldehyde $\xrightarrow{\text{(1) LiAlH}_4,\;\text{(2) H}^+,\,H_2O}$ Methanol (10-12)

An aldehyde (ethanal) $\xrightarrow{\text{(1) LiAlH}_4,\;\text{(2) H}^+,\,H_2O}$ A primary alcohol (ethanol) (10-13)

A ketone (propanone) $\xrightarrow{\text{(1) LiAlH}_4,\;\text{(2) H}^+,\,H_2O}$ A secondary alcohol (2-propanol) (10-14)

It is not surprising due to the similarity of the structure of C=S functionality to the carbonyl functionality that reactions involving these compounds with a hydride as the reducing agent are very similar and predictable. An example of the reduction of a C=S functionality to produce the thiol functionality is shown in Reaction (10-15).

$$H_3C-\underset{}{\overset{S}{\underset{}{\overset{||}{C}}}}-CH_3 \xrightarrow{\text{(1) LiAlH}_4,\;\text{(2) H}^+,\,H_2O} \underset{H_3C}{\overset{HS}{\diagdown}}\underset{CH_3}{\overset{H}{\diagup}} \quad (10\text{-}15)$$

Problem 10.2

i) Give the major organic products of the reduction of each of the following compounds with LiAlH₄, followed by hydrolysis.

(a) cyclopentanone (b) propanal (c) 3,3-difluoro-butanal (d) acetaldehyde (H₃C—CHO)

ii) Complete the following reactions by supplying an appropriate reactant or the major organic product.

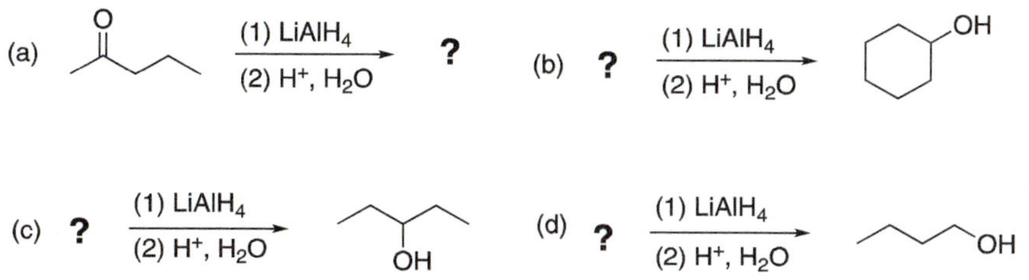

(a) CH₃COCH₂CH₃ $\xrightarrow{\text{(1) LiAlH}_4}{\text{(2) H}^+, \text{H}_2\text{O}}$?

(b) ? $\xrightarrow{\text{(1) LiAlH}_4}{\text{(2) H}^+, \text{H}_2\text{O}}$ cyclohexanol

(c) ? $\xrightarrow{\text{(1) LiAlH}_4}{\text{(2) H}^+, \text{H}_2\text{O}}$ (CH₃)₂CHOH (isopropanol)

(d) ? $\xrightarrow{\text{(1) LiAlH}_4}{\text{(2) H}^+, \text{H}_2\text{O}}$ CH₃CH₂CH₂CH₂OH (1-butanol)

10.3.2 Reduction Using Organometallic Reagents

We learned in Chapter 7 that the conjugate bases of very weak acids are extremely strong Lewis bases; the conjugate bases of alkanes are known as carboanions, and as mentioned earlier, they exist in the form of neutral organometallic compounds. As a result, these compounds are also extremely good reducing agents and react with the electrophilic carbons of C=O and C=S bonds, similar to that of the reaction of the hydride anion with these types of compounds, but the type of organic products produced from these reactions are different from that of the reaction with metal hydrides. For these reactions, a new carbon–carbon bond is formed, instead of a C—H bond, as observed from the reaction involving metal hydrides. One of the most popular organometallic compounds is the Grignard reagent (RMgX), where X represents a halogen, usually chlorine or bromine. The mechanism for the reaction of Grignard reagents with the C=O functionality is very similar to that involving the reaction of the hydride ion, in this case, the nucleophilic carbon of the alkyl group adds to the carbonyl carbon and forms a new carbon–carbon bond, as shown in the generic mechanism of Reaction (10-16).

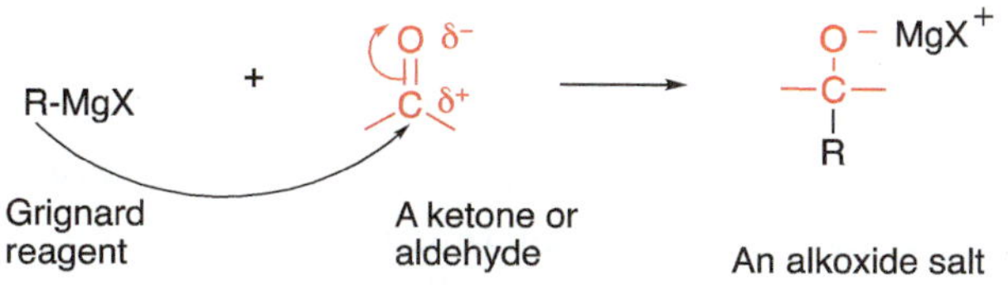

$$\text{R-MgX} + \underset{\text{A ketone or aldehyde}}{\overset{\delta-}{\underset{\delta+}{C=O}}} \longrightarrow \underset{\text{An alkoxide salt}}{\overset{O^- \text{MgX}^+}{\underset{R}{-C-}}} \qquad (10\text{-}16)$$

As pointed out in the previous section, a second step that involves an acid–base reaction (hydrolysis) of the alkoxide salt will convert the basic salt to an alcohol, as shown in Reaction (10-17).

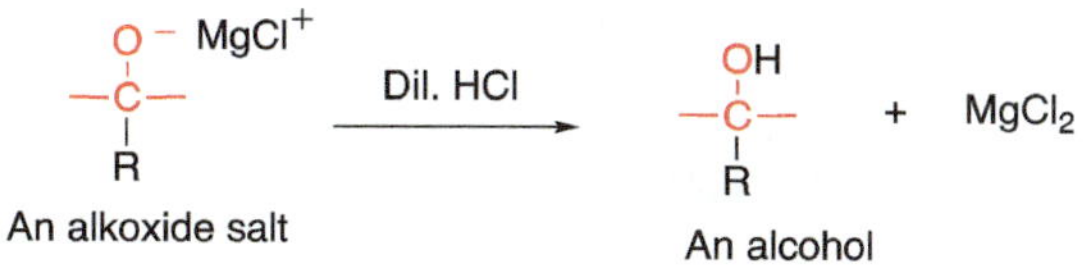

$$\underset{\text{An alkoxide salt}}{\overset{O^- \text{MgCl}^+}{\underset{R}{-C-}}} \xrightarrow{\text{Dil. HCl}} \underset{\text{An alcohol}}{\overset{OH}{\underset{R}{-C-}}} + \text{MgCl}_2 \qquad (10\text{-}17)$$

You will notice that this sequence of reactions transforms a carbonyl functionality to an alcohol, with a new carbon–carbon bond in the product. The reaction of formaldehyde, which has two hydrogens bonded to the carbonyl functionality, with a Grignard reagent gives a primary alcohol after hydrolysis of the alkoxide salt. The reaction of a Grignard reagent with aldehydes gives secondary alcohols after hydrolysis of the alkoxide salts, and the reaction of a Grignard reagent with ketones gives tertiary alcohols after hydrolysis of the alkoxide salts as shown in Reactions (10-18)–(10-20).

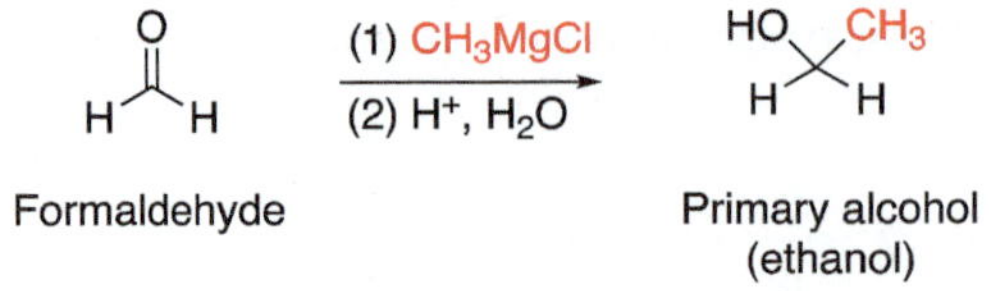

$$\underset{\text{Formaldehyde}}{\overset{O}{\underset{H}{\overset{\|}{C}}}_H} \xrightarrow[\text{(2) H}^+\text{, H}_2\text{O}]{\text{(1) CH}_3\text{MgCl}} \underset{\underset{\text{(ethanol)}}{\text{Primary alcohol}}}{\overset{HO\ \ CH_3}{\underset{H\ \ \ H}{C}}} \quad (10\text{-}18)$$

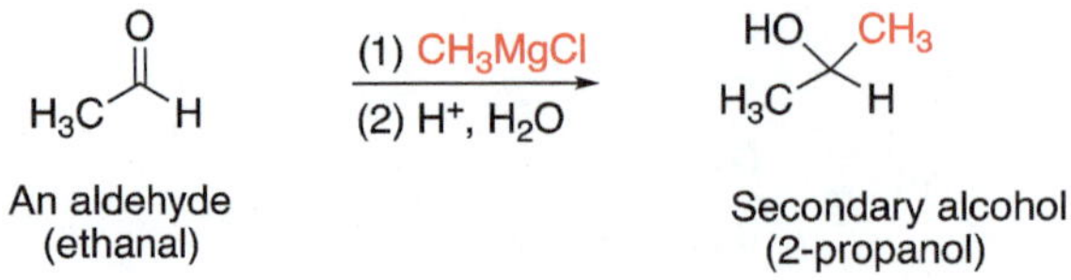

$$\underset{\underset{\text{(ethanal)}}{\text{An aldehyde}}}{\overset{O}{\underset{H_3C}{\overset{\|}{C}}}_H} \xrightarrow[\text{(2) H}^+\text{, H}_2\text{O}]{\text{(1) CH}_3\text{MgCl}} \underset{\underset{\text{(2-propanol)}}{\text{Secondary alcohol}}}{\overset{HO\ \ CH_3}{\underset{H_3C\ \ \ H}{C}}} \quad (10\text{-}19)$$

$$\underset{\underset{\text{(propanone)}}{\text{A ketone}}}{\overset{O}{\underset{H_3C\ \ CH_3}{\overset{\|}{C}}}} \xrightarrow[\text{(2) H}^+\text{, H}_2\text{O}]{\text{(1) CH}_3\text{MgCl}} \underset{\underset{\text{(2-methyl-2-propanol)}}{\text{Tertiary alcohol}}}{\overset{HO\ \ CH_3}{\underset{H_3C\ \ \ CH_3}{C}}} \quad (10\text{-}20)$$

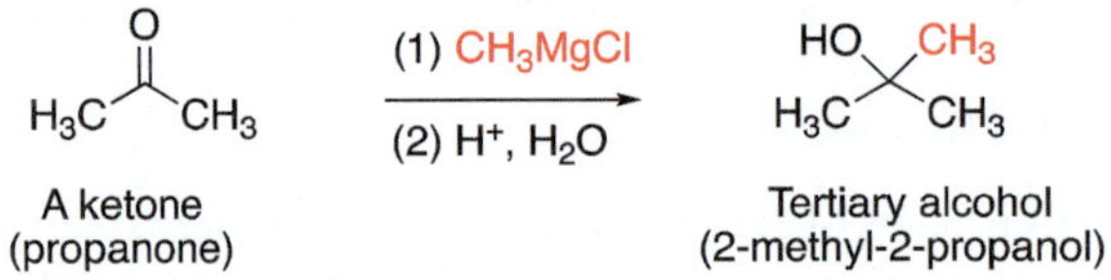

An example of the reduction of a C=S functionality to produce the thiol functionality using methyl Grignard is shown in Reaction (10-21); in this case, the product is not an alcohol, but a thiol.

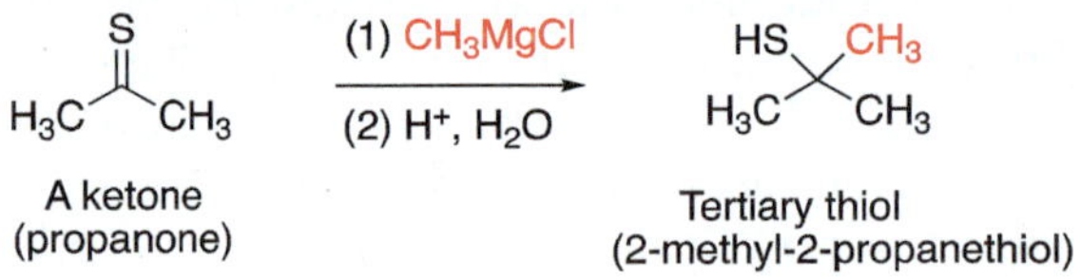

$$\underset{\underset{\text{(propanone)}}{\text{A ketone}}}{\overset{S}{\underset{H_3C\ \ CH_3}{\overset{\|}{C}}}} \xrightarrow[\text{(2) H}^+\text{, H}_2\text{O}]{\text{(1) CH}_3\text{MgCl}} \underset{\underset{\text{(2-methyl-2-propanethiol)}}{\text{Tertiary thiol}}}{\overset{HS\ \ CH_3}{\underset{H_3C\ \ \ CH_3}{C}}} \quad (10\text{-}21)$$

Problem 10.3

i) Show how you could use a compound that contains a C=O or C=S functionality and an appropriate Grignard reagent to synthesize the molecules shown below.

(a) (CH₃)₂CH–C(CH₃)₂–OH   (b) cyclopentyl-C(CH₃)₂–SH   (c) (CH₃)₃C–SH   (d) cyclopentyl-C(CH₃)(CH₂CH₃)–SH

ii) Tertiary alcohols can be synthesized by the reaction of a Grignard reagent (RMgX) with an appropriate ketone. Outline three different syntheses of 3-methyl-3-hexanol, but utilizing different Grignard reagents and ketones in each of your synthesis.

An important reduction reaction using organometallic reagents is its reaction with carbon dioxide, as shown in Reaction (10-22).

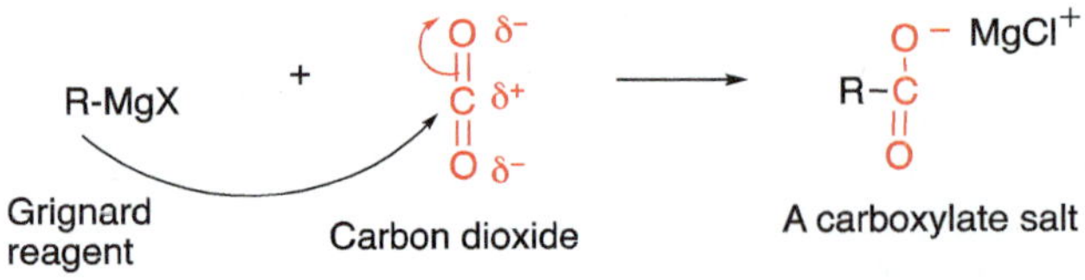

$$\underset{\underset{\text{reagent}}{\text{Grignard}}}{\text{R-MgX}} + \underset{\text{Carbon dioxide}}{\overset{O\ \delta-}{\underset{O\ \delta-}{\overset{\|}{C}\ \delta+}}} \longrightarrow \underset{\text{A carboxylate salt}}{\overset{O^-\ MgCl^+}{\underset{O}{\overset{|}{\underset{\|}{R-C}}}}} \quad (10\text{-}22)$$

As we have seen, these types of basic salts can be converted into a nonionic organic molecule in the presence of a weak acid, as shown in the reaction given in Reaction (10-23). In this case, the final organic products are carboxylic acids.

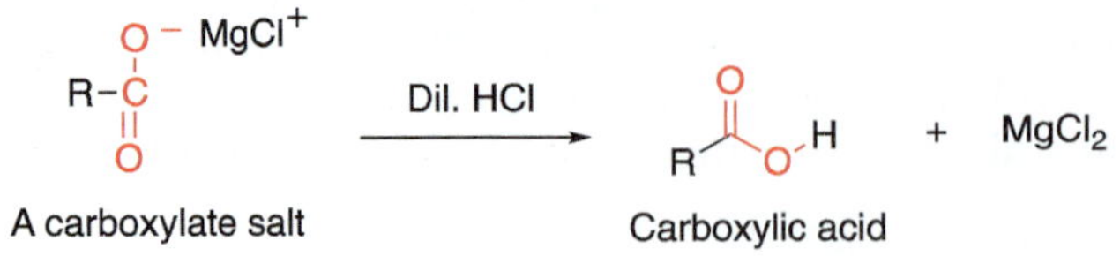

$$\underset{\text{A carboxylate salt}}{\overset{O^-\ MgCl^+}{\underset{O}{\overset{|}{\underset{\|}{R-C}}}}} \xrightarrow{\text{Dil. HCl}} \underset{\text{Carboxylic acid}}{\overset{O}{\underset{R\ \ \ O-H}{\overset{\|}{C}}}} + MgCl_2 \quad (10\text{-}23)$$

Note that this transformation using a Grignard reagent to make carboxylic acid increases the number of carbons in the carboxylic acid product.

Problem 10.4

Show how to carry out the following transformations, your synthesis should involve a Grignard reagent and carbon dioxide.

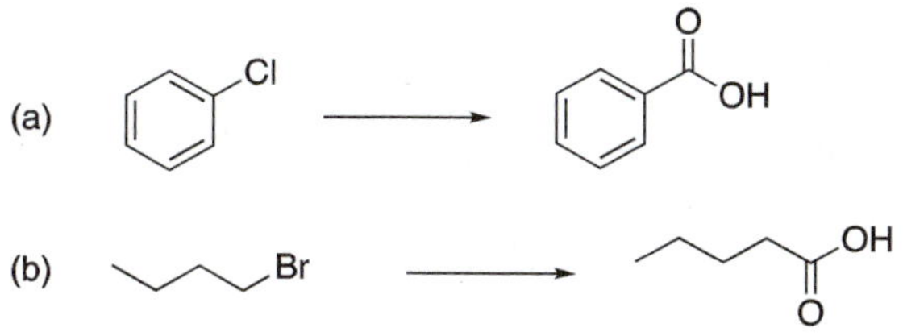

(a) C₆H₅–Cl ⟶ C₆H₅–COOH

(b) CH₃CH₂CH₂CH₂–Br ⟶ CH₃CH₂CH₂CH₂–COOH

10.3.3 Reduction Using Acetylides

We have mentioned earlier in Chapter 7 that terminal alkynes are relatively acidic, and the acidic hydrogen can be deprotonated with an appropriate base. The acetylide anion that results is not only a conjugate Lewis base but is also a potential reducing agent. As you can imagine, it is not as strong a reducing agent, compared to the reducing agents covered earlier since the carbon of acetylides are more electronegative than that of alkyl carbons, but nonetheless, the acetylide anion can give its pair of nonbonding electrons to an electrophile to form a new carbon–carbon bond. Examples are shown in Reactions (10-24)–(10-26) and note that the products of these reactions are not only alcohols but also difunctional containing an alcohol and an alkyne functionality.

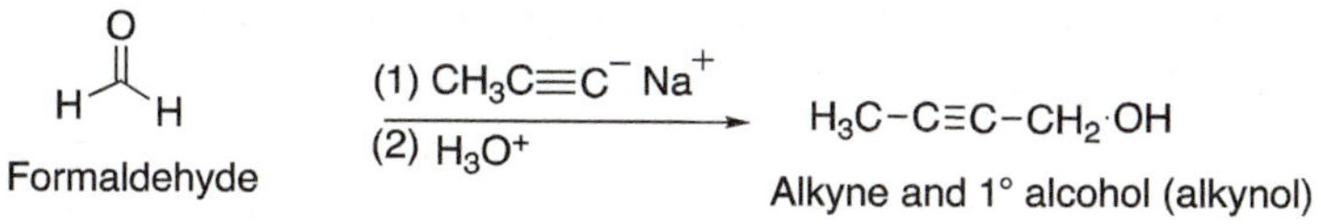

$$\text{H-CHO (Formaldehyde)} \xrightarrow[\text{(2) H}_3\text{O}^+]{\text{(1) CH}_3\text{C}\equiv\text{C}^-\text{Na}^+} \text{H}_3\text{C-C}\equiv\text{C-CH}_2\text{OH} \quad \text{Alkyne and 1° alcohol (alkynol)} \quad (10\text{-}24)$$

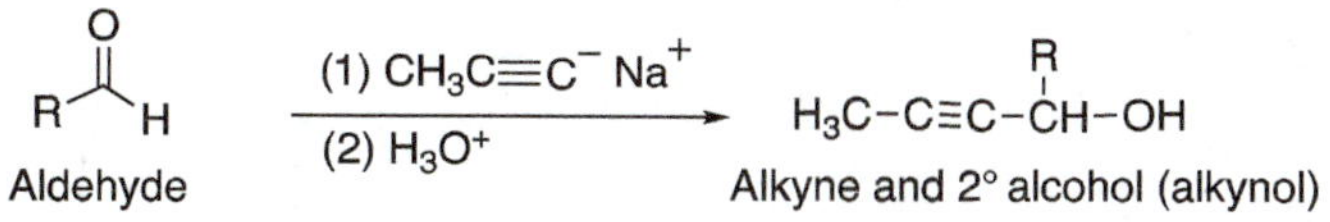

$$\text{R-CHO (Aldehyde)} \xrightarrow[\text{(2) H}_3\text{O}^+]{\text{(1) CH}_3\text{C}\equiv\text{C}^-\text{Na}^+} \text{H}_3\text{C-C}\equiv\text{C-CH(R)-OH} \quad \text{Alkyne and 2° alcohol (alkynol)} \quad (10\text{-}25)$$

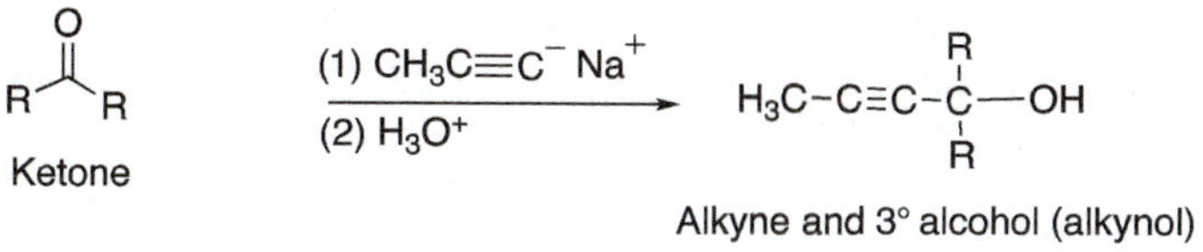

$$\text{R-CO-R (Ketone)} \xrightarrow[\text{(2) H}_3\text{O}^+]{\text{(1) CH}_3\text{C}\equiv\text{C}^-\text{Na}^+} \text{H}_3\text{C-C}\equiv\text{C-C(R)(R)-OH} \quad \text{Alkyne and 3° alcohol (alkynol)} \quad (10\text{-}26)$$

At this point, you are beginning to realize that with much creativity, you can devise the synthesis of a wide variety of different compounds by using various reaction types and carefully selecting specific reactants and specific reducing agents. The result is that new molecules with different functional groups are produced.

Problem 10.5

Complete the following reactions by supplying an appropriate reactant or the major organic product.

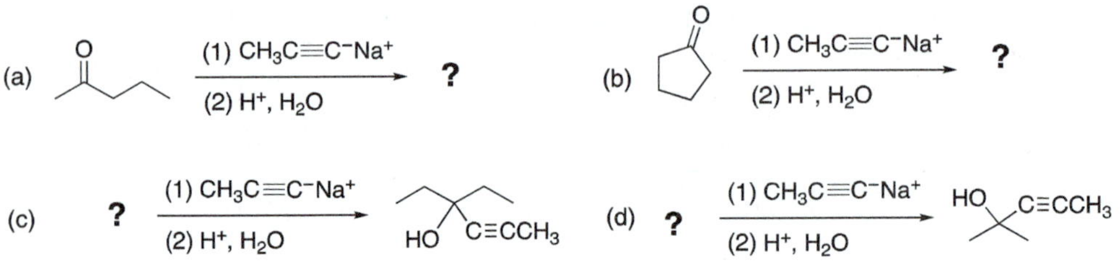

(a) $CH_3COCH_2CH_3 \xrightarrow{\text{(1) } CH_3C\equiv C^-Na^+}_{\text{(2) } H^+, H_2O}$?

(b) cyclopentanone $\xrightarrow{\text{(1) } CH_3C\equiv C^-Na^+}_{\text{(2) } H^+, H_2O}$?

(c) ? $\xrightarrow{\text{(1) } CH_3C\equiv C^-Na^+}_{\text{(2) } H^+, H_2O}$ (CH$_3$)$_2$C(OH)C≡CCH$_3$

(d) ? $\xrightarrow{\text{(1) } CH_3C\equiv C^-Na^+}_{\text{(2) } H^+, H_2O}$ HO-C(CH$_3$)(CH$_2$CH$_3$)-C≡CCH$_3$

10.3.4 Reduction Using Metals

This type of reduction is very important since a carbonyl functional group can be reduced to a methylene unit using transition metals, such as zinc. One of these types of reactions is shown in Reaction (10-27) and is often referred to as the Clemmensen reduction.

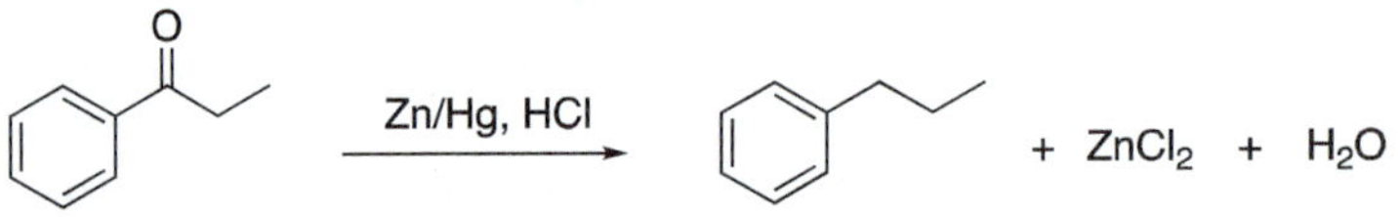

$$\text{PhCOCH}_2\text{CH}_3 \xrightarrow{\text{Zn/Hg, HCl}} \text{PhCH}_2\text{CH}_2\text{CH}_3 + ZnCl_2 + H_2O \qquad (10\text{-}27)$$

This reaction is named after a Danish chemist, Erik Christian Clemmensen (1876–1941). In this reaction, an alloy of zinc/mercury amalgam is used as the reducing agent. In the first step of the mechanism, the carbonyl oxygen is protonated in the acidic medium, which makes the carbonyl carbon even more electrophilic. The next step involves the delivery of an electron pair from zinc to the protonated carbonyl compound, and in another step of the reaction mechanism, the chloride anion reacts with the reduced intermediate as shown in Reaction (10-28).

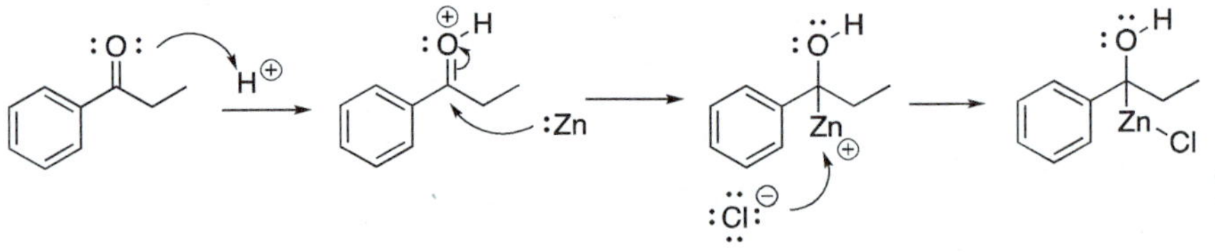

$$(10\text{-}28)$$

In the next step of the reaction mechanism, water is lost in the acidic medium to form a cationic intermediate, to which another mole of zinc delivers another pair of electrons as shown in Reaction (10-29).

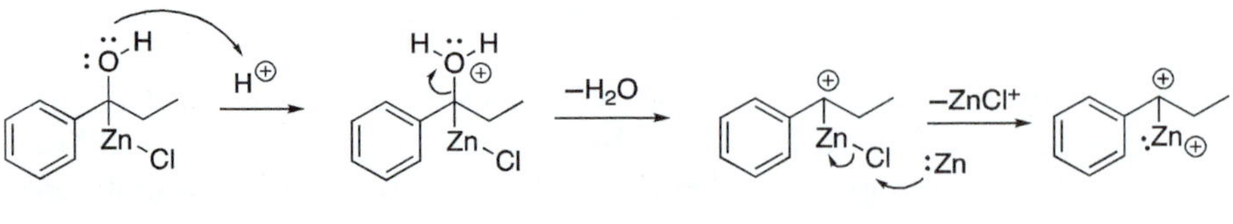

$$(10\text{-}29)$$

In the next step, the electrons are delivered to the carbocation, converting it into a carboanion shown in Reaction (10-30).

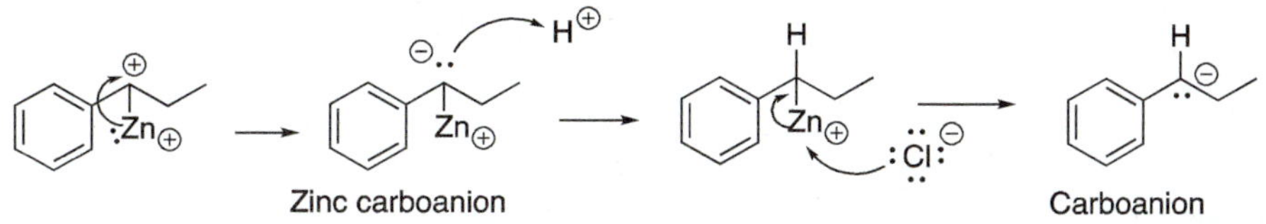

$$\text{(10-30)}$$

Zinc carboanion   Carboanion

In the last step of the reaction mechanism, the carboanion reacts with the proton of the acidic medium to give the reduced product as shown in Reaction (10-31).

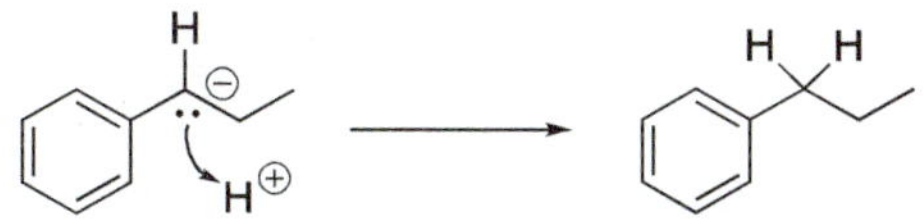

$$\text{(10-31)}$$

10.3.5 Reduction Using Hydrogen with a Catalyst

In general chemistry, we learned that reduction is typically accompanied by the gain of hydrogen. Hydrogen, in the presence of a catalyst, is routinely used in organic chemistry to accomplish reduction of carbon–carbon double and triple bonds. The reduction of carbon–oxygen double bonds and carbon–nitrogen double and triple bonds can be accomplished by hydrogenation, but more severe conditions, such as high temperatures and pressures in the presence of an appropriate catalyst, are required. The reduction of a molecule that has a carbon–oxygen double bond using hydrogen in the presence of a catalyst is shown in Reaction (10-32).

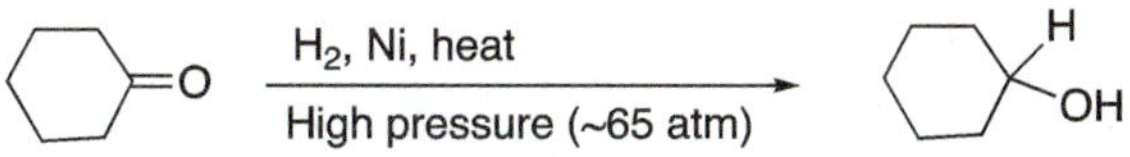

$$\text{(10-32)}$$

Problem 10.6

Give the reactant or the organic product to complete the reactions shown below.

(a) 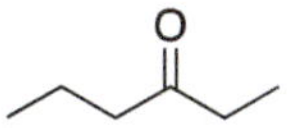 [ketone] $\xrightarrow[\text{High pressure (~65 atm)}]{H_2, \text{ Ni, heat}}$?

(b) ? $\xrightarrow[\text{High pressure (~65 atm)}]{H_2, \text{ Ni, heat}}$ [alcohol with OH]

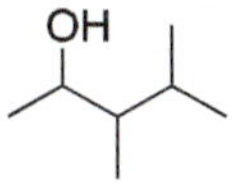

(c) 2-Methylcyclopentanone $\xrightarrow[\text{High pressure (~65 atm)}]{H_2, \text{ Ni, heat}}$?

(d) ? $\xrightarrow[\text{High pressure (~65 atm)}]{H_2, \text{ Ni, heat}}$ 3,4-Dimethyl-2-hexanol

10.3.6 The Wolff Kishner Reduction

The addition of hydrazine to a carbonyl functionality gives a product that we have seen in the previous chapter, called a hydrazone (Reaction 10-33).

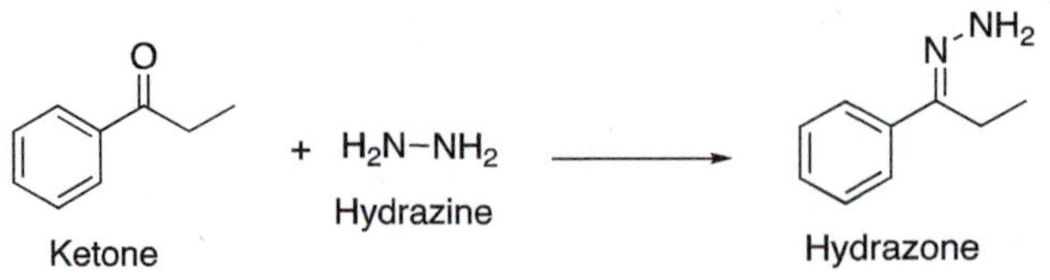

10 Reduction Reactions in Organic Chemistry

[Ketone] + H₂N–NH₂ (Hydrazine) ⟶ [Hydrazone]   (10-33)

In the presence of a strong base, such as KOH, the hydrazone is reduced and in the process liberates nitrogen gas as shown in Reaction (10-34).

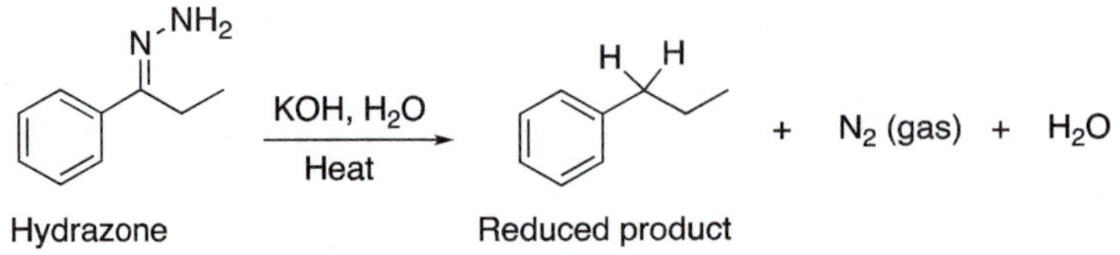

[Hydrazone] —KOH, H₂O, Heat→ [Reduced product] + N_2 (gas) + H_2O (10-34)

A driving force for this reaction to proceed to the right is the liberation of nitrogen gas (an increase in entropy). The mechanism for this reaction involves the abstraction of a proton bonded to the nitrogen as shown below to generate a carboanion, which abstracts a proton from the solvent, water as shown in Reaction (10-35).

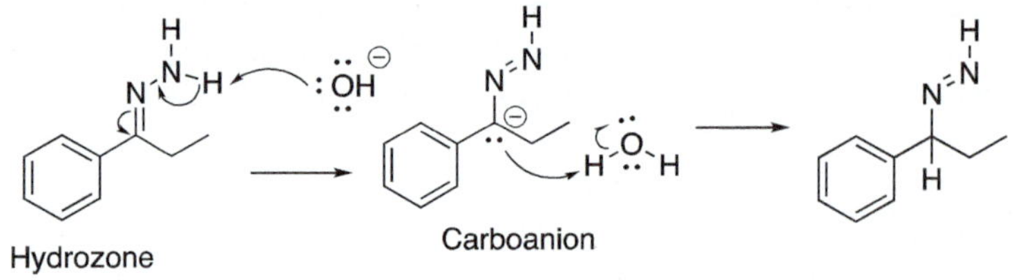

[Hydrozone] ⟶ [Carboanion] ⟶ [product]   (10-35)

In the next step of the mechanism, another mole of the hydroxide anion abstracts another proton from the hydrogen bonded to the nitrogen, releasing nitrogen gas and another carboanion, which abstracts a proton from the solvent, water to give the reduced product as shown in Reaction (10-36).

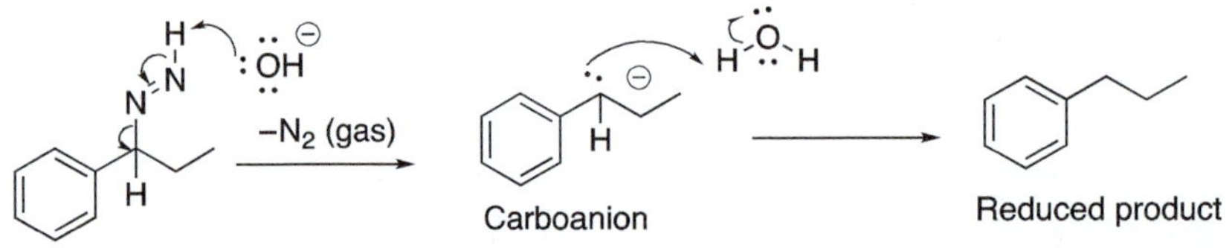

[structure] —$-N_2$ (gas)→ [Carboanion] ⟶ [Reduced product]   (10-36)

This type of reaction was discovered in the early 1900s by Nikolai Kishner and Ludwig Wolff and now known as the Wolff-Kishner reaction. For the Wolff-Kishner reaction, note that this reduction takes place in a very basic medium, compared to the Clemmensen reduction, which takes place is an acidic medium. Reaction (10-35) gives an example of the Wolff-Kishner reduction reaction.

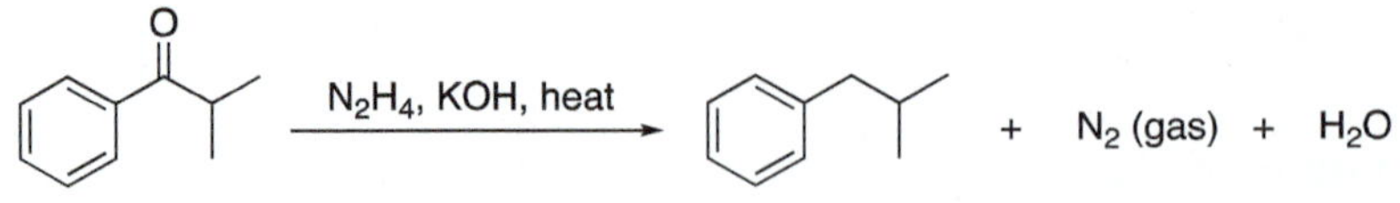

[ketone] —N_2H_4, KOH, heat→ [product] + N_2 (gas) + H_2O (10-37)

At the end of this chapter, we will demonstrate the need for carrying out reduction under different conditions, such as acidic versus basic condition, and an appropriate reduction must be selected in order to accomplish the synthesis of specific target molecules.

Problem 10.7

Complete the following reactions by supplying an appropriate reactant or the major organic product.

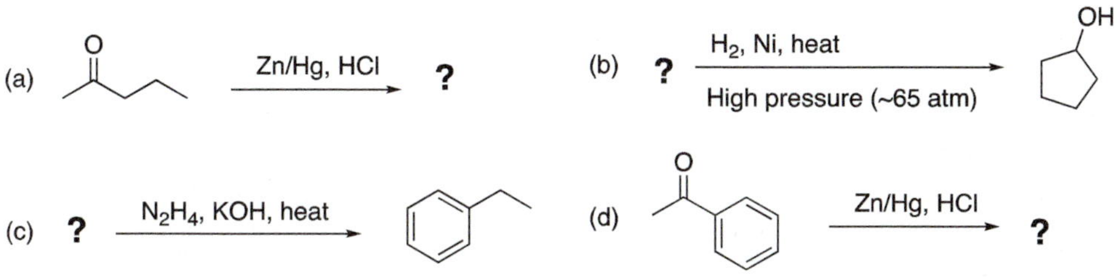

(a) pentan-2-one (CH₃-CO-CH₂-CH₂-CH₃) $\xrightarrow{\text{Zn/Hg, HCl}}$?

(b) ? $\xrightarrow[\text{High pressure (~65 atm)}]{\text{H}_2\text{, Ni, heat}}$ cyclopentanol

(c) ? $\xrightarrow{\text{N}_2\text{H}_4\text{, KOH, heat}}$ ethylbenzene

(d) acetophenone $\xrightarrow{\text{Zn/Hg, HCl}}$?

10.4 Reduction of Imines

Close inspection of the imine functionality reveals that it should undergo similar reactions as those of the carbonyl functionality since they are both polarized double bonds. As a result, imines can be reduced in a similar manner as carbon–oxygen double bonds to form the corresponding amine.

10.4.1 Reduction Using NaBH₄ and LiAlH₄

The reduction of compounds that have C=N—R functionality by LiAlH₄ gives amide salts. A condensed mechanism for the reaction is shown in Reaction (10-38).

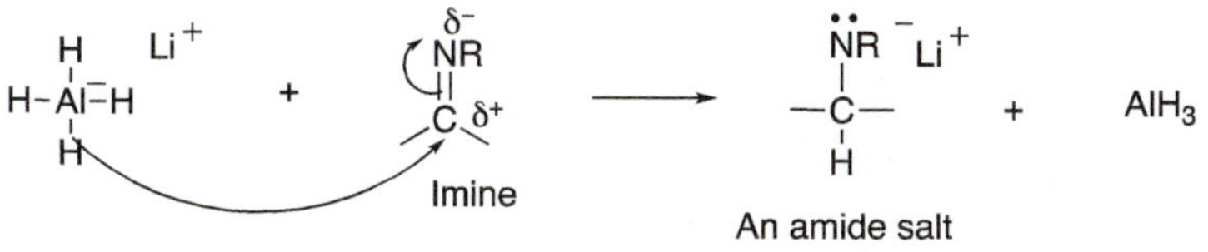

$$\text{H-AlH}_3^- \text{Li}^+ \;+\; \overset{\delta-}{\text{NR}}=\overset{\delta+}{\text{C}} \;\longrightarrow\; \underset{\text{An amide salt}}{\overset{\ddot{\text{N}}\text{R}^-\;\text{Li}^+}{\underset{\text{H}}{-\text{C}-}}} \;+\; \text{AlH}_3 \qquad (10\text{-}38)$$

In the presence of an acidic aqueous solution, amide salts, which are strongly basic, are neutralized by an acid–base reaction to give secondary amines if R is an alkyl group. In the case where R is hydrogen, the product is a primary amine. These reactions are shown in Reactions (10-39) and (10-40).

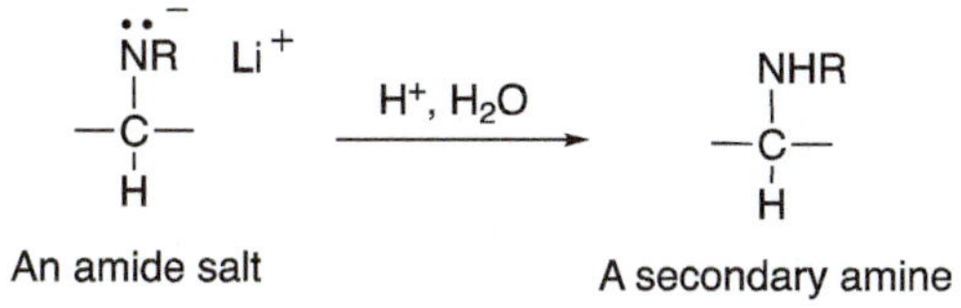

$$\underset{\text{An amide salt}}{\overset{\ddot{\text{N}}\text{R}^-\;\text{Li}^+}{\underset{\text{H}}{-\text{C}-}}} \xrightarrow{\text{H}^+\text{, H}_2\text{O}} \underset{\text{A secondary amine}}{\overset{\text{NHR}}{\underset{\text{H}}{-\text{C}-}}} \qquad (10\text{-}39)$$

$$\underset{\text{An amide salt}}{\overset{\ddot{\text{N}}\text{H}^-\;\text{Li}^+}{\underset{\text{H}}{-\text{C}-}}} \xrightarrow{\text{H}^+\text{, H}_2\text{O}} \underset{\text{A primary amine}}{\overset{\text{NH}_2}{\underset{\text{H}}{-\text{C}-}}} \qquad (10\text{-}40)$$

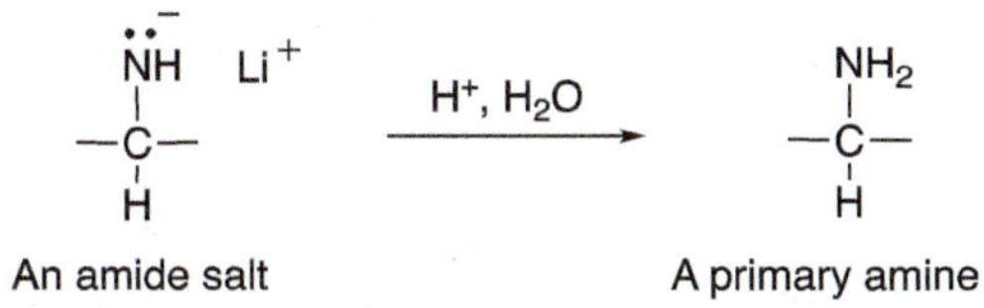

The overall reaction showing the two-reaction sequence, which includes the acidic workup, is shown in Reaction (10-41).

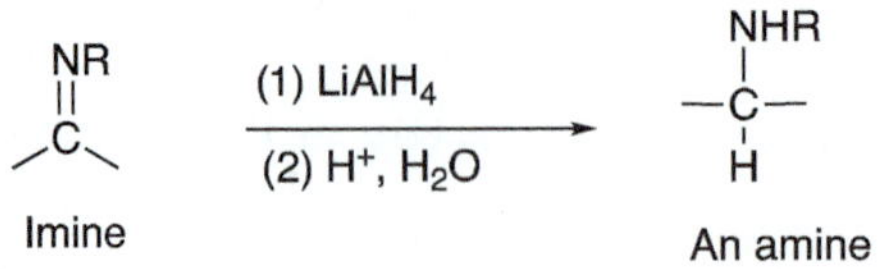

10 Reduction Reactions in Organic Chemistry

$$\underset{\text{Imine}}{\overset{NR}{\underset{C}{\|}}} \xrightarrow[\text{(2) H}^+,\text{ H}_2\text{O}]{\text{(1) LiAlH}_4} \underset{\text{An amine}}{\overset{NHR}{\underset{H}{-C-}}} \tag{10-41}$$

A similar type of reaction is expected using NaBH$_4$, as shown in Reaction (10-42).

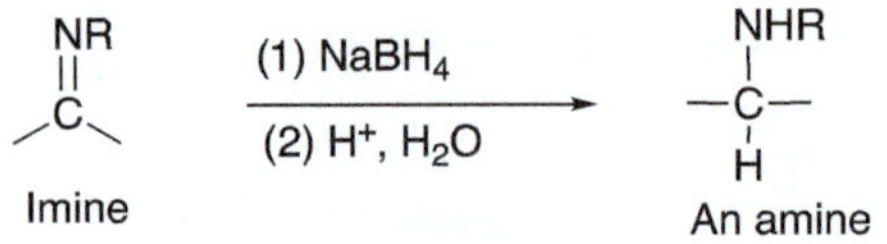

$$\underset{\text{Imine}}{\overset{NR}{\underset{C}{\|}}} \xrightarrow[\text{(2) H}^+,\text{ H}_2\text{O}]{\text{(1) NaBH}_4} \underset{\text{An amine}}{\overset{NHR}{\underset{H}{-C-}}} \tag{10-42}$$

Reaction (10-43) shows an example of the reduction of the *N*-methyl-2-butaneimine.

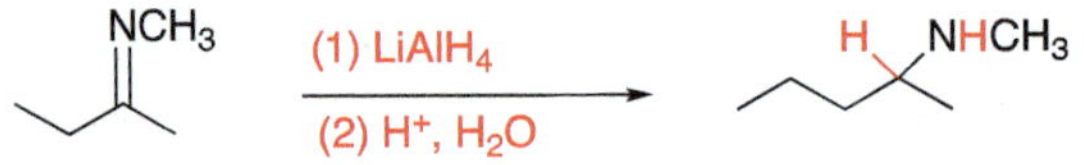

$$\tag{10-43}$$

You will readily notice that different secondary amines result from these reduction reactions depending on the structure of the starting imine compound. Thus, the imines shown below will result in different amines upon reduction.

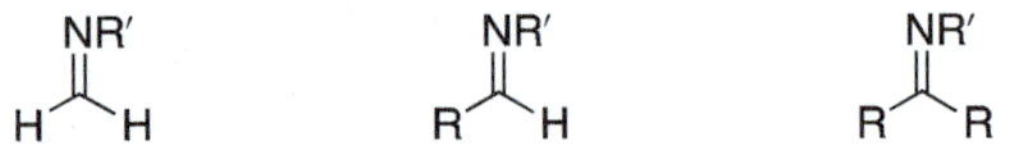

Reactions (10-44) through (10-46) show the reduction of different imines with lithium aluminum hydride, followed by hydrolysis.

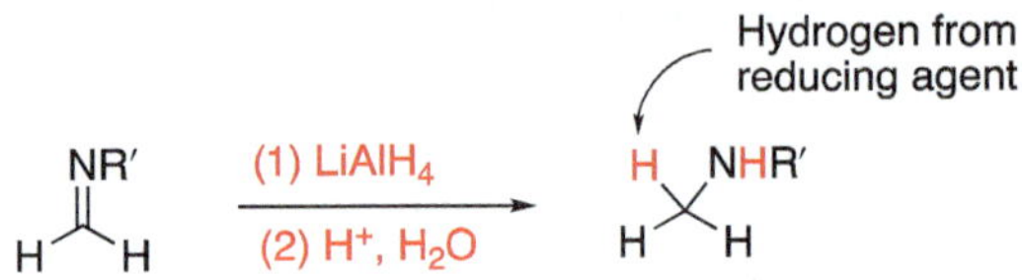

$$\tag{10-44}$$

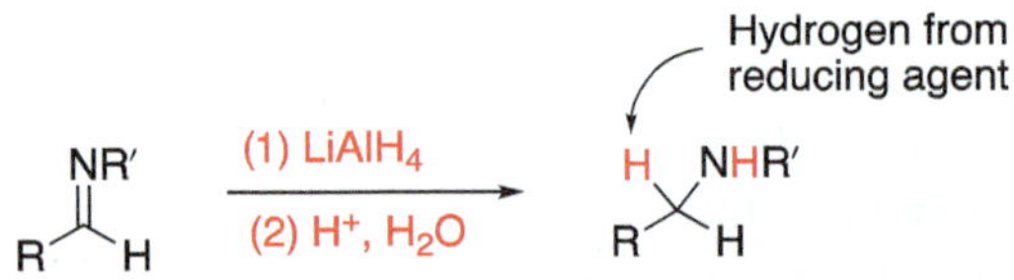

$$\tag{10-45}$$

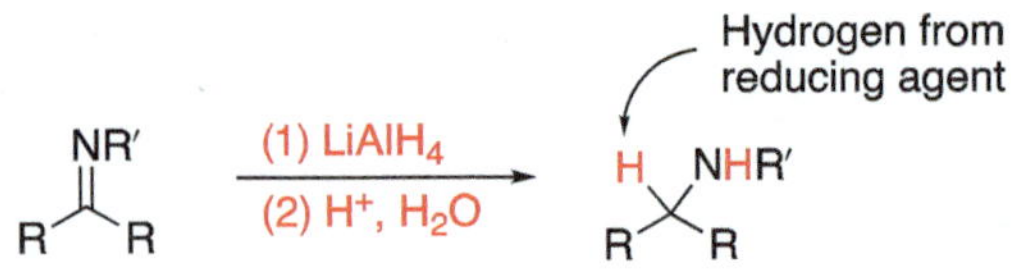

$$\tag{10-46}$$

Problem 10.8

i) Give the major organic products of the reaction of each of the following compounds with LiAlH$_4$, followed by hydrolysis.

(a) cyclopentanone N-H imine (=NH) (b) CH$_3$-N=C(CH$_3$)CH$_2$CH$_3$ (c) CH$_3$-N=CH-CH$_2$CH$_2$CH$_3$ (d) cyclohexanone N-methylimine

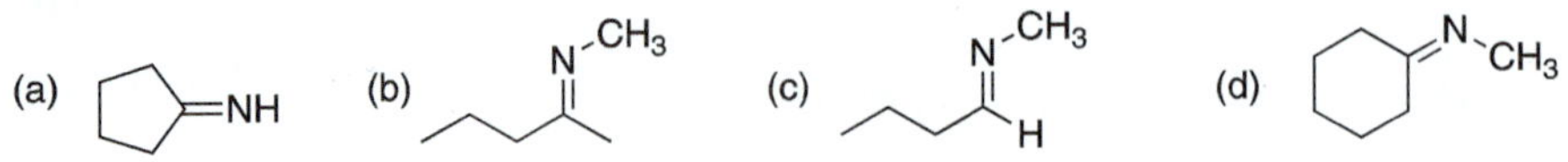

ii) Complete the following reactions by supplying an appropriate reactant or the major organic products.

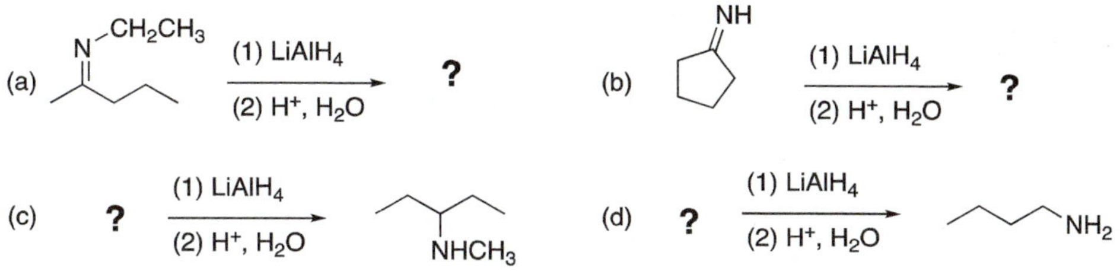

(a) CH₃CH(=N-CH₂CH₃)CH₂CH₃ (1) LiAlH₄ (2) H⁺, H₂O → ?

(b) cyclopentanone imine (=NH) (1) LiAlH₄ (2) H⁺, H₂O → ?

(c) ? (1) LiAlH₄ (2) H⁺, H₂O → CH₃CH₂CH(NHCH₃)CH₂CH₃ (product shown with NHCH₃ group)

(d) ? (1) LiAlH₄ (2) H⁺, H₂O → CH₃CH₂CH₂CH₂NH₂ (propyl–NH₂ shown)

10.4.2 Reduction Using Hydrogen with a Catalyst

The reduction of imines as discussed in the previous section can be accomplished using hydrogen, an appropriate catalyst and under the right reaction conditions, such as elevated temperature and pressure. Reaction (10-47) shows the reduction of an imine to form the amine in the presence of hydrogen and a catalyst.

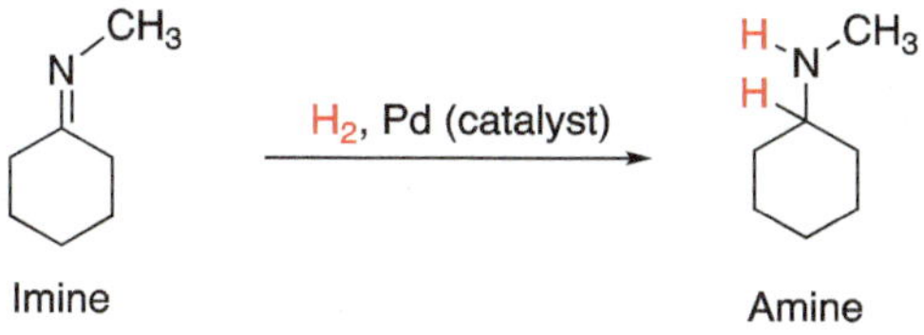

Imine (cyclohexylidene-N-methyl) — H₂, Pd (catalyst) → Amine (N-methylcyclohexylamine, with added H's shown in red)

(10-47)

Problem 10.9

Give the reactant or the organic product to complete the reactions shown below.

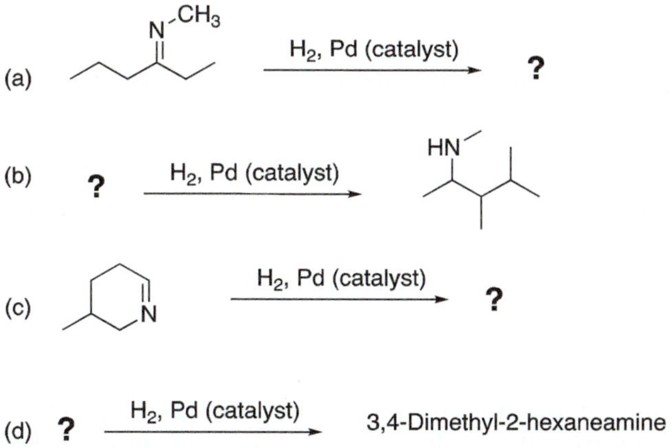

(a) CH₃CH₂CH₂C(=N-CH₃)CH₃ — H₂, Pd (catalyst) → ?

(b) ? — H₂, Pd (catalyst) → (CH₃)₂CHCH(NH-iPr)CH₃ (secondary amine shown, HN-iPr)

(c) (2-methyl-piperidine-type cyclic imine) — H₂, Pd (catalyst) → ?

(d) ? — H₂, Pd (catalyst) → 3,4-Dimethyl-2-hexaneamine

Using reactions that we have already covered thus far, we can transform a carbonyl-containing compound into an amine with strategic selection of appropriate reactions. Reactions (10-48) and (10-49) show how benzaldehyde can be transformed into benzylamine. You will recall from Section 9.8 that the addition of ammonia and primary amines to ketones and aldehydes results in imines as the products as shown in Reaction (10-48).

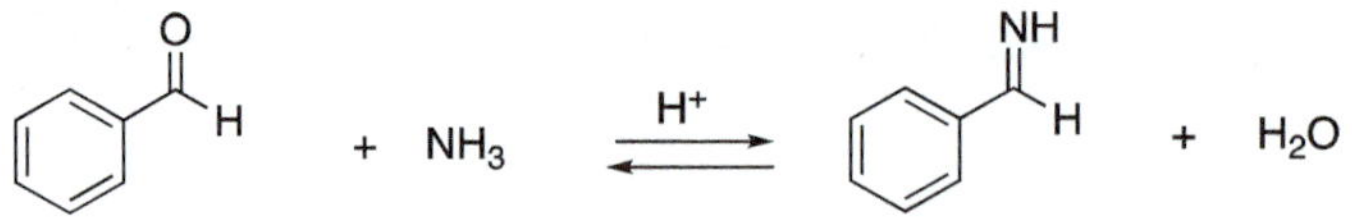

266 | 10 Reduction Reactions in Organic Chemistry

$$\text{Benzaldehyde} \; (PhCHO) + NH_3 \underset{}{\overset{H^+}{\rightleftharpoons}} \text{An imine} \; (PhCH=NH) + H_2O \tag{10-48}$$

In another reaction, the imine can be reduced to form a primary amine, as shown in Reaction (10-49).

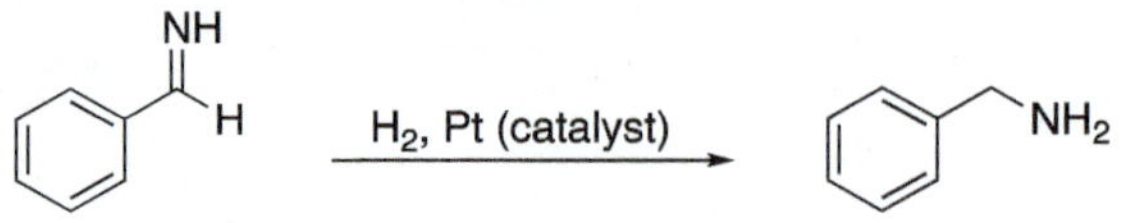

$$\text{An imine} \; (PhCH=NH) \xrightarrow{H_2,\; Pt\; (catalyst)} \text{Benzylamine} \; (PhCH_2NH_2) \tag{10-49}$$

Reaction (10-48) shows the formation of the imine from the corresponding benzaldehyde, and Reaction (10-49) shows the reduction reaction. This combination of reactions, which essentially converts a carbonyl compound to an amine, is called *reductive amination*. Thus, the above two reactions are used to accomplish the transformation shown in Reaction (10-50).

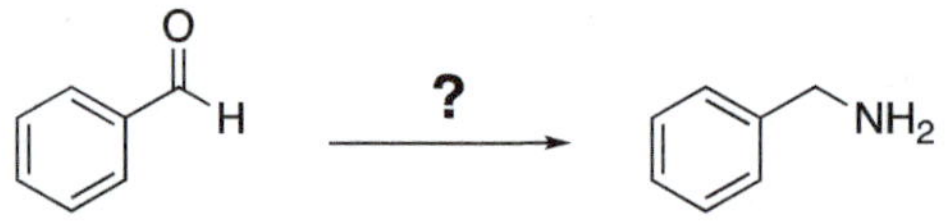

$$\text{Benzaldehyde} \; (PhCHO) \xrightarrow{?} \text{Benzylamine} \; (PhCH_2NH_2) \tag{10-50}$$

Students should be able to examine the transformation shown in Reaction (10-50) and critically think through the type of reactions and specific reactions that could be used to accomplish the transformation shown. Problem 10.10 is designed to get students to apply that critical thinking process to carry out the transformations.

Problem 10.10

Starting with any carbonyl-containing compound, show how reductive amination can be used to synthesize the following compounds.

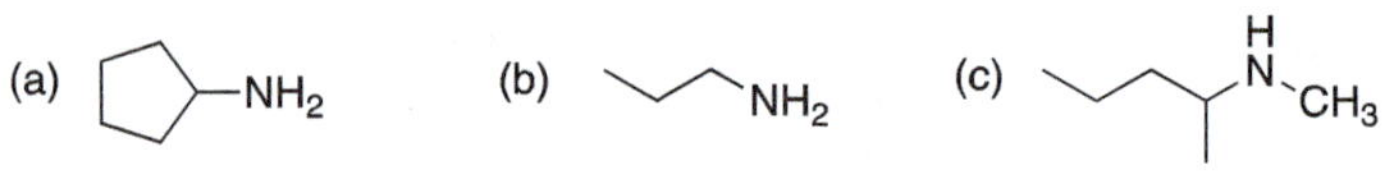

(a) cyclopentyl-NH$_2$ (b) CH$_3$CH$_2$CH$_2$NH$_2$ (c) CH$_3$CH$_2$CH$_2$CH(NHCH$_3$)—

10.5 Reduction of Oxiranes

Oxiranes are cyclic ethers, and as you observe from the structure below, they are very similar to carbonyl compounds in that there are polar bonds, in which the carbons bonded to the oxygen are partially positive and the oxygen is partially negative.

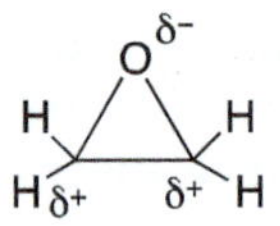

Oxirane (with $O^{\delta-}$ and H's on carbons labeled $\delta+$)

10.5 Reduction of Oxiranes

Based on the partial charges of oxiranes, reducing agents will attack either of the partially positive carbons to generate an alkoxide anion salt as shown in Reaction (10-51), which is similar to the alkoxide anion salt that we saw in the section with carbonyl compounds.

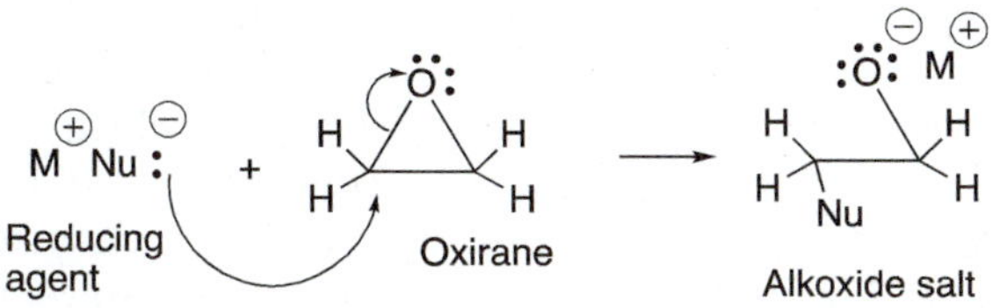

$$\overset{\oplus}{M} \overset{\ominus}{Nu:} + \text{Oxirane} \longrightarrow \text{Alkoxide salt} \tag{10-51}$$

In a separate step, the alkoxide salt can be hydrolyzed under acidic conditions to produce an alcohol as shown in Reaction (10-52).

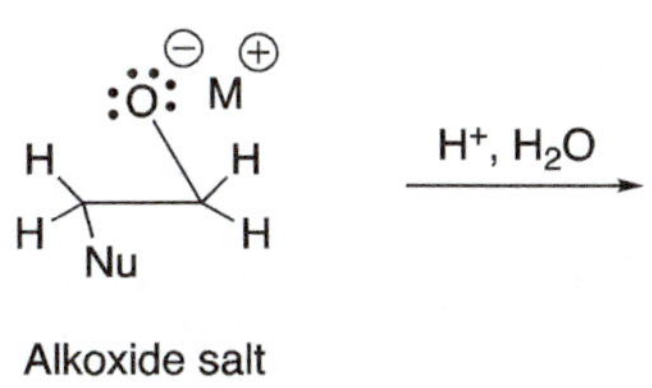

$$\text{Alkoxide salt} \xrightarrow{H^+, H_2O} \text{Alcohol} \tag{10-52}$$

Note that via this transformation, two carbons are added to the reducing agent, such as a Grignard reagent, to form the alcohol product. A summary of this concept is shown in Reaction (10-53).

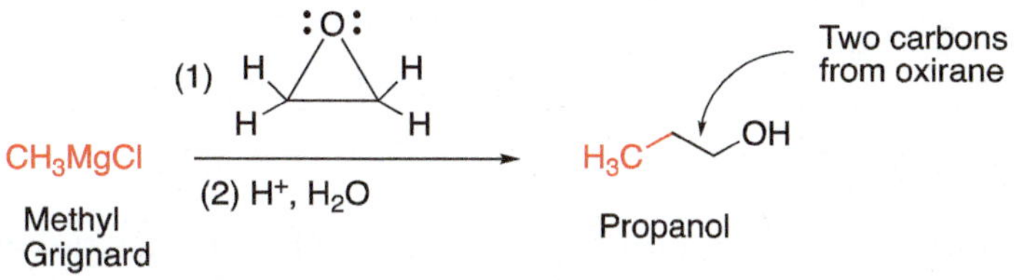

$$CH_3MgCl \xrightarrow[\text{(2) }H^+, H_2O]{\text{(1) oxirane}} H_3C\diagdown OH \text{ (Propanol)} \tag{10-53}$$

Methyl Grignard → Propanol (Two carbons from oxirane)

Thus, it is possible to use methyl chloride as a starting compound to synthesize propanol by this method as shown in Reaction (10-54).

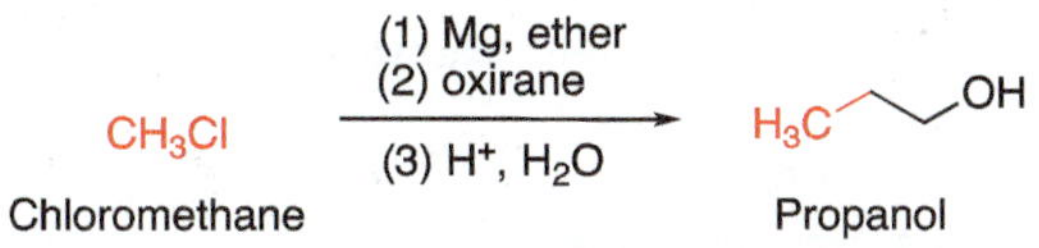

$$CH_3Cl \xrightarrow[\text{(3) }H^+, H_2O]{\text{(1) Mg, ether; (2) oxirane}} H_3C\diagdown OH \tag{10-54}$$

Chloromethane → Propanol

Students should be able to apply analytical thinking skills to determine the type of reactions needed for the above transformation and equally important the specific reactions needed. Problem 10.11 is designed to help students apply their analytical thinking skills to determine possible starting compounds and possible reactions in order to synthesize the compounds shown.

Problem 10.11

Show how you could use ethylene oxide and appropriate organohalides to synthesize the alcohols shown below.

(a) (CH₃)₂CHCH₂CH₂OH (b) PhCH₂CH₂OH

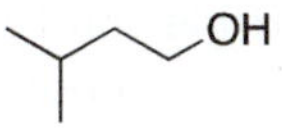

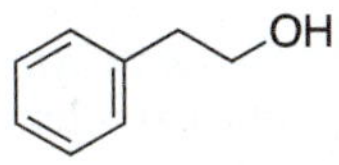

10.6 Reduction of Aromatic Compounds, Alkynes, and Alkenes

10.6.1 Reduction Using Dissolving Metals

We will be studying the chemistry of some very stable molecules in later chapters, these molecules are known as aromatic compounds, of which benzene is the most common, its structure is shown below. As you will see, there are six pi (π) electrons in conjugation, which makes it very difficult to add more electrons through a reduction process. These are not impossible reactions, but can be accomplished in the presence of very strong reducing agents, such as reducing metals, as shown in Reaction (10-55). For this reaction, ammonia is the solvent.

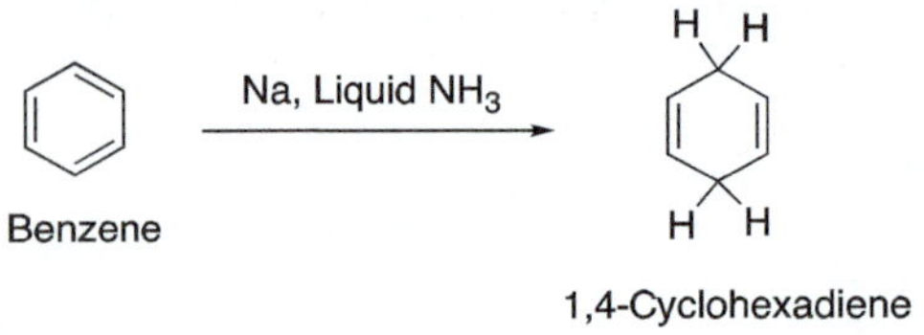

$$\text{Benzene} \xrightarrow{\text{Na, Liquid NH}_3} \text{1,4-Cyclohexadiene} \tag{10-55}$$

This type of reaction is often referred to as the Birch reduction. In the first step of the mechanism, the very strong reducing agent, sodium metal, supplies one electron to the conjugated system to form a radical anion, which then abstracts a proton from the solvent, ammonia as shown in Reaction (10-56) to form a radical.

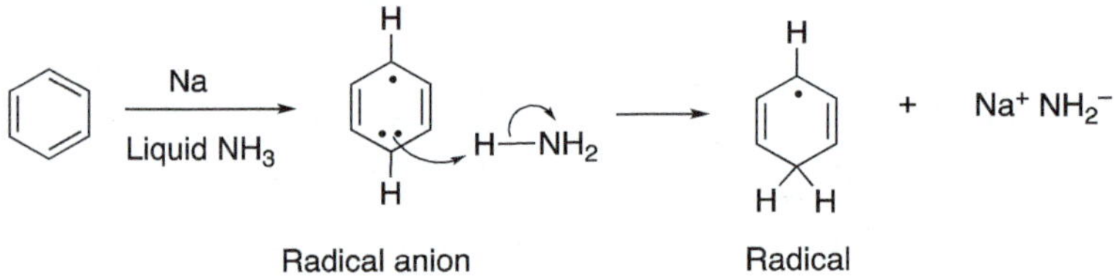

$$\text{Benzene} \xrightarrow{\text{Na, Liquid NH}_3} \text{Radical anion} \xrightarrow{\text{H—NH}_2} \text{Radical} + \text{Na}^+ \text{NH}_2^- \tag{10-56}$$

Sodium then delivers another electron to form an anion, which abstracts another proton from the solvent, ammonia to form the final reduced product as shown in Reaction (10-57).

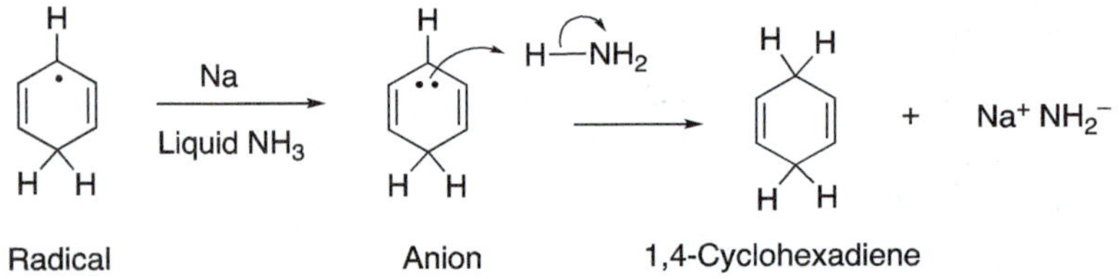

$$\text{Radical} \xrightarrow{\text{Na, Liquid NH}_3} \text{Anion} \xrightarrow{\text{H—NH}_2} \text{1,4-Cyclohexadiene} + \text{Na}^+ \text{NH}_2^- \tag{10-57}$$

The same reaction condition as given above can be used to reduce alkynes to alkenes, specifically trans-alkenes, which is given in Reaction (10-58).

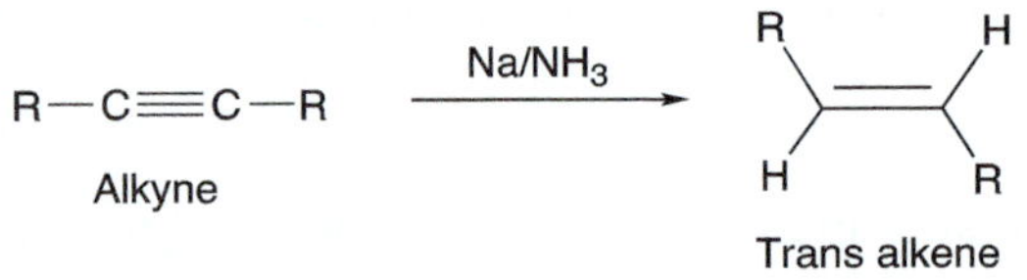

$$R-C\equiv C-R \xrightarrow{\text{Na/NH}_3} \text{Trans alkene} \tag{10-58}$$

The general reaction mechanism for Reaction (10-56) is similar to the reduction of benzene above, in that the reducing metal supplies an electron to the pi (π) system of the triple bond in the initial step to form a trans radical anion, which abstracts a proton from the solvent, ammonia, as shown in Reaction (10-59). As you will recall from the valence shell electron pair repulsion

(VSEPR) theory, electrons repel each other, and as a result, the three electrons of the radical anion will be in a trans arrangement as shown in Reaction (10-59). This concept is the key to the formation of the trans-product.

$$R\text{—}\!\!\equiv\!\!\text{—}R \xrightarrow{Na} \underset{\text{Trans radical anion}}{\begin{array}{c}R\\ \diagup\!\!=\!\!\diagdown\\ R\end{array}} \xrightarrow{H\text{—}NH_2} \begin{array}{c}R\quad H\\ \diagup\!\!=\!\!\diagdown\\ R\end{array} + Na^+\, NH_2^- \qquad (10\text{-}59)$$

Alkyne

In the second step of the mechanism, another electron is delivered to the radical to form an anion, which then abstracts a proton from the solvent, as shown in Reaction (10-60).

$$\begin{array}{c}R\quad H\\ \diagup\!\!=\!\!\diagdown\\ R\end{array} \xrightarrow{Na} \begin{array}{c}R\quad H\\ \diagup\!\!=\!\!\diagdown\\ R\end{array} \xrightarrow{H_2N\text{—}H} \underset{\text{Trans-alkene}}{\begin{array}{c}R\quad H\\ \diagup\!\!=\!\!\diagdown\\ H\quad R\end{array}} + Na^+\, NH_2^- \qquad (10\text{-}60)$$

10.6.2 Reduction Using Catalytic Hydrogenation

Another important reduction reaction involving the carbon–carbon multiple bonds is one where a hydrogen molecule adds across the double bond of alkenes in the presence of a catalyst. As we have demonstrated from the reductions of carbonyl compounds and imines with hydrogen, hydrogen is a very stable and unreactive molecule, and as a result, a catalyst and high temperatures and pressures are necessary since the activation barrier for such reactions is very high. These reduction reactions transform unsaturated compounds into saturated compounds and are widely used in industry. Two examples are shown in Reactions (10-61) and (10-62).

1,2-Dimethylcyclohexene $\xrightarrow{H_2,\ Pt\ (catalyst)}$ 1,2-Dimethylcyclohexane (10-61)

(Z)-2-Hexene + H_2 $\xrightarrow{Pd\ (catalyst)}$ Hexane (10-62)

An important observation that should be made for these reactions is that the hydrogen adds to the same side of the double bonds, cis-addition. Figure 10.1 shows how a catalyst is used to

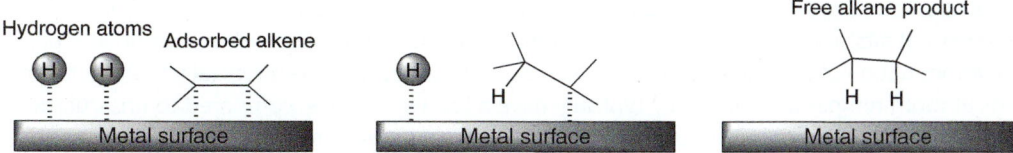

Figure 10.1 Steps for the cis hydrogenation of an alkene in the presence of a catalyst.

reduce the energy of activation required to break the hydrogen–hydrogen bond, which has to take place in order for the addition reaction to occur.

Based on the mechanism shown in Figure 10.1, the addition of the hydrogen atoms across the carbon–carbon double is a **cis**-addition, which means that both hydrogen atoms add to the same side of the double bond. These reactions are described as stereospecific reactions, and Reaction (10-63) is another example of the reduction of an unsymmetric alkene showing the stereochemistry of the products.

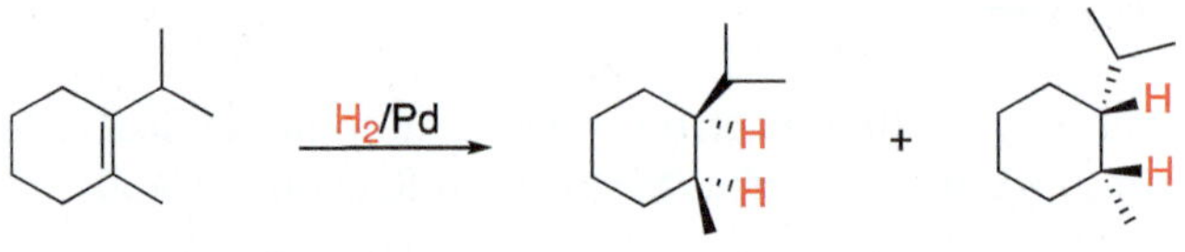

(10-63)

Cis addition of hydrogen to an unsymmetric alkene

For the hydrogenation of alkenes, rearrangement is not observed since as shown in Figure 10.1, the reactants are attached to the surface of the catalyst, and as a result, rearrangement does take place.

Problem 10.12

Give the major product that results from hydrogenation of the following alkenes, show stereochemistry where appropriate.

(a) 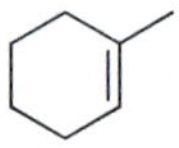   (b) 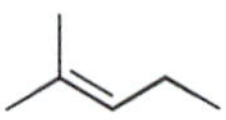   (c) 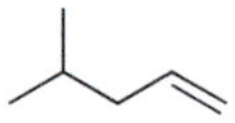   (d)

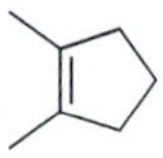

DID YOU KNOW?

In today's health conscious society, we often hear of saturated and unsaturated fats and that unsaturated fats are better for us. Unsaturated fats are classified as unsaturated due to the presence of double bonds. Unsaturated fats can be converted to saturated fats by a hydrogenation process. The example of hydrogenation of oleic acid is shown below.

$$CH_3(CH_2)_7CH=CH(CH_2)_7COOH \xrightarrow{H_2,\ catalyst} CH_3(CH_2)_7\overset{H}{\underset{|}{C}}H-\overset{H}{\underset{|}{C}}H(CH_2)_7COOH$$

Oleic acid (unsaturated)   Stearic acid (saturated)

Fats can be categorized into the following categories: saturated (e.g. butter, lard, coconut oil); monounsaturated (e.g. olive or canola oils); and polyunsaturated (e.g. omega-6 oils, including sunflower and safflower oil, and omega-3 oils, including fish and flaxseed oils). Hydrogenated fats are unnatural fats, and they have been considered to be unhealthy fats. Hydrogenation and partial hydrogenation of unsaturated fats are carried out on unsaturated fats typically to make them solids at room temperature and they typically have a longer shelf life, compared to unsaturated fats. They are marketed as butters and margarines. Any polyunsaturated oil can be hydrogenated, but the more common ones that are hydrogenated include cottonseed, palm, soy, and corn oils.

Different oils, butter, and margarines

Now that we have seen that it is possible to add hydrogen in a cis-manner across a double bond, the question that comes to mind is if it is possible to add just one mole of hydrogen across a triple bond to get a cis-alkene? In the presence of a catalyst, such as Pd, hydrogen will add across the triple bond to first give the alkene and eventually the alkane in the presence of excess hydrogen, as shown in Reaction (10-64).

$$R-C\equiv C-R \xrightarrow{H_2 \text{ (excess)/Pd}} \underset{\text{Alkane}}{R-CH_2-CH_2-R} \tag{10-64}$$

By using specialized catalysts, it is possible to add just one mole of hydrogen across the triple bond of alkynes to form the corresponding alkene. For these reactions, the addition of hydrogen takes place on the same side of the double bond, cis-addition.

$$\underset{\text{2-Butyne}}{H_3C-\!\!\!\equiv\!\!\!-CH_3} \xrightarrow[\text{(Lindlar's catalyst)}]{H_2 \text{, Deactivated Pd}} \underset{(Z)\text{-2-Butene}}{\overset{H\quad H}{\underset{H_3C\quad CH_3}{\diagdown\!\!=\!\!\diagup}}} \tag{10-65}$$

The mechanism is similar to that shown for the addition of hydrogen to alkenes, which was shown earlier, except the metal surface in this case is deactivated allowing only one mole of hydrogen to add.

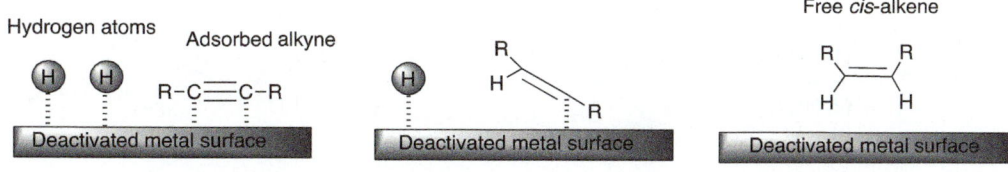

One deactivated catalyst that is used to accomplish this task is called the Lindlar's catalyst, which is Pd/BaSO₄ in the presence of quinone.

Problem 10.13

Complete the following reactions by supplying either the major organic product or the reaction conditions (appropriate catalyst) to give the products shown.

(a) 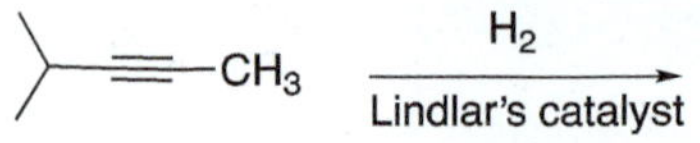 (CH₃)₂CHC≡C-CH₃ $\xrightarrow{\text{H}_2}$ Lindlar's catalyst

(b) 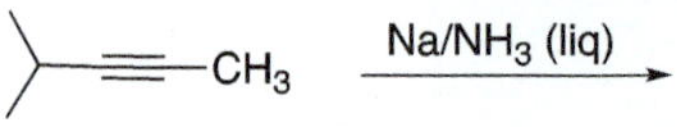 (CH₃)₂CHC≡C-CH₃ $\xrightarrow{\text{Na/NH}_3 \text{ (liq)}}$

(c)

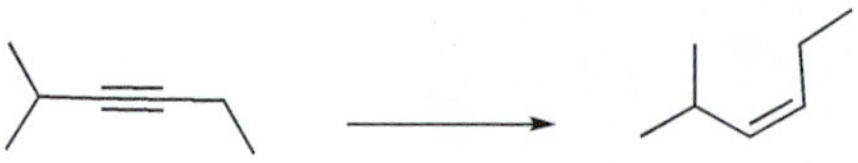

(d)

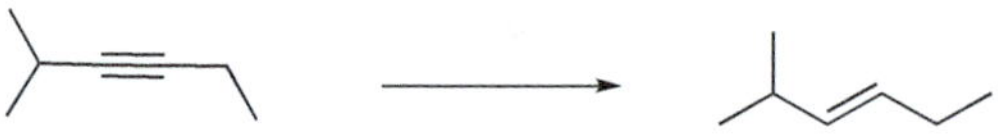

In summary, the addition of hydrogen across a triple bond involving Lindlar's catalyst is a stereospecific reaction in that it gives the *syn*-product (syn-addition) to form (Z) alkenes. The reaction using dissolving metal is also stereospecific in that it gives the trans product (*anti*-addition) to form (E) alkenes.

End of Chapter Problems

10.14 Provide the major organic product for each of the following reactions.

(a) 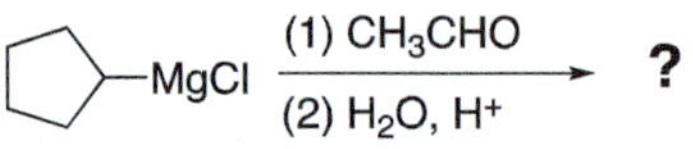 cyclopentyl-MgCl $\xrightarrow[\text{(2) H}_2\text{O, H}^+]{\text{(1) CH}_3\text{CHO}}$?

(b) 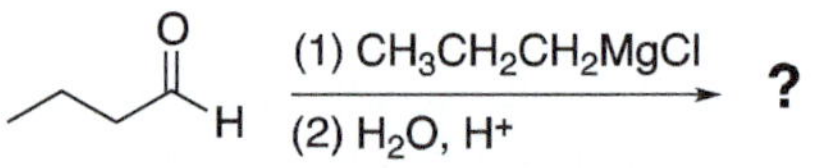 [ketone] $\xrightarrow[\text{(2) H}_2\text{O, H}^+]{\text{(1) CH}_3\text{CH}_2\text{CH}_2\text{MgCl}}$?

(c) 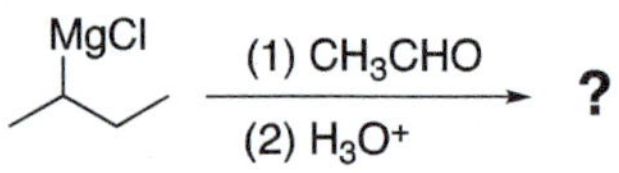 isobutyl-MgCl $\xrightarrow[\text{(2) H}_3\text{O}^+]{\text{(1) CH}_3\text{CHO}}$?

(d) 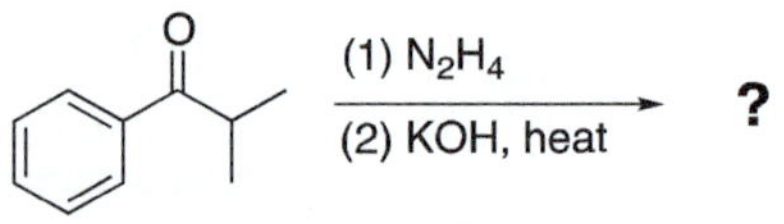 [phenyl isopropyl ketone] $\xrightarrow[\text{(2) KOH, heat}]{\text{(1) N}_2\text{H}_4}$?

10.15 Provide the organic reactant to complete each of the following reactions.

(a) 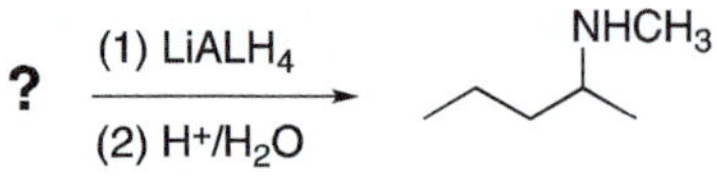 ? $\xrightarrow[\text{(2) H}^+/\text{H}_2\text{O}]{\text{(1) LiAlH}_4}$ [product with NHCH₃]

(b) 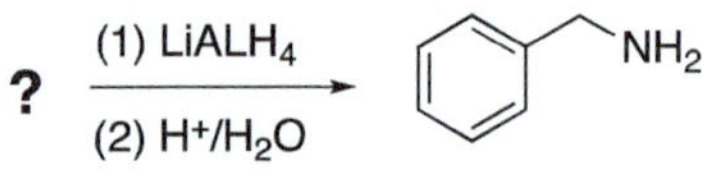 ? $\xrightarrow[\text{(2) H}^+/\text{H}_2\text{O}]{\text{(1) LiAlH}_4}$ [benzyl-NH₂]

(c) 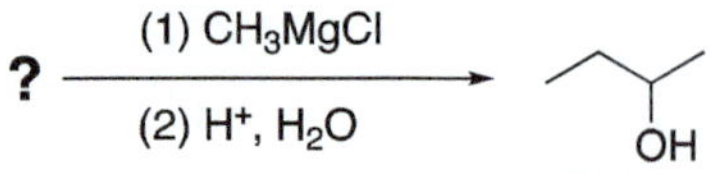 ? $\xrightarrow[\text{(2) H}^+, \text{H}_2\text{O}]{\text{(1) CH}_3\text{MgCl}}$ [product with OH]

(d) 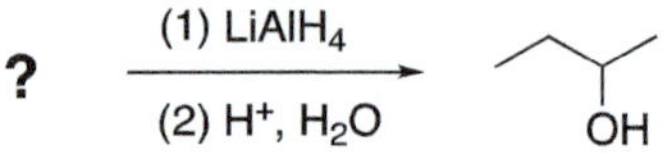 ? $\xrightarrow[\text{(2) H}^+, \text{H}_2\text{O}]{\text{(1) LiAlH}_4}$ [product with OH]

10.16 Show how you could use a Grignard reagent and a C=O containing compound to make each of the following compounds.

(a) 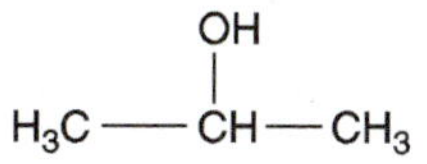 $H_3C-\underset{\underset{H}{|}}{\overset{\overset{OH}{|}}{C}}-CH_3$ (b) 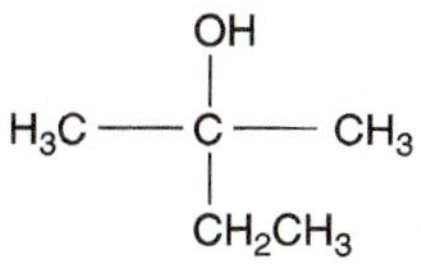 $H_3C-\underset{\underset{CH_2CH_3}{|}}{\overset{\overset{OH}{|}}{C}}-CH_3$ (c) $CH_3CH_2CH_3OH$

10.17 Show how to carry out the following transformation.

cyclopentanone $\longrightarrow$ cyclopentyl-NHCH$_3$

10.18 Show how to carry out the following transformations. For each step in your synthesis, <u>clearly</u> show all reagents and reaction conditions.

(a)  propanal $\longrightarrow$ 3-Hexanol

(b)  cyclopentyl-Cl $\longrightarrow$ cyclopentyl-C(OH)(ethyl)

(c) cyclohexene oxide $\longrightarrow$ trans-2-methylcyclohexanol

10.19 Show how you could use ethylene oxide (oxirane) and an alkyne to synthesize each of the alcohols shown below.

(a) 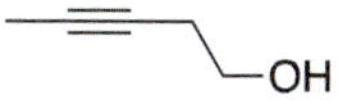 HC≡C–CH$_2$–CH$_2$–OH (b) 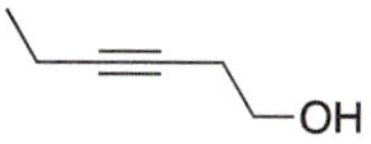 CH$_3$–C≡C–CH$_2$–CH$_2$–OH

10.20 Show how to carry out the following transformations.

cis-4-methyl-2-pentene $\xleftarrow{?}$ 4-methyl-2-pentyne $\xrightarrow{?}$ trans-4-methyl-2-pentene

10.21 Muscalure is the sex attractant of the common housefly, it is *cis*-tricos-9-ene and its structure is shown below. Starting with any alkyne, devise a synthesis for this compound.

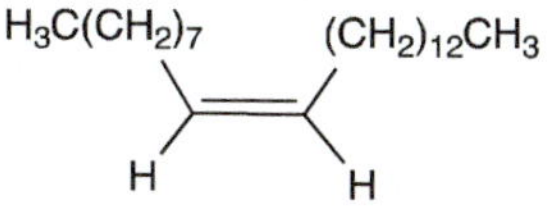

$H_3C(CH_2)_7$ $(CH_2)_{12}CH_3$
$\quad\quad\quad C=C$
$\quad\quad H \quad\quad\quad H$

cis-Tricos-9-ene, "muscalure"

10.22 Tertiary alcohols can be synthesized by the reaction of a Grignard reagent (RMgX) with an appropriate ketone as shown in the reaction below for the synthesis of 3-methyl-3-hexanol.

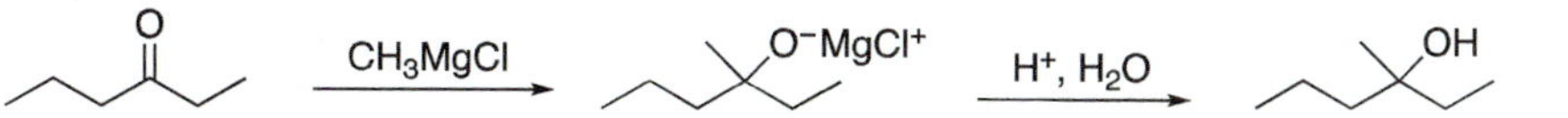

3-Hexanone $\xrightarrow{CH_3MgCl}$ intermediate (O$^-$MgCl$^+$) $\xrightarrow{H^+, H_2O}$ 3-Methyl-3-hexanol (Tertiary alcohol)

Utilizing a similar strategy as given in the example above, outline the synthesis of the same molecule, 3-methyl-3-hexanol, but utilizing a different Grignard reagent and ketone.

10.23 Show how to carry out the following transformation. For each step in your synthesis, clearly show all reagents and reaction conditions.

(CH₃)₂CHCH₂Br ⟶ (CH₃)₂CHCH₂C(O)OH

10.24 Show how to carry out the following transformations. For each step in your synthesis, clearly show all reagents and reaction conditions.

(a)  (CH₃)₂CHCHO ⟶ (CH₃)₂CHCH₂NHCH₃

(b)  cyclopentanone ⟶ cyclopentyl-NHCH₃

(c)  (CH₃)₂CHCH₂Br ⟶ (CH₃)₂CHCH₂CH₂OH

(d)  cyclohexyl-Br ⟶ cyclohexyl-CH₂CH₂OH

11

Oxidation Reactions in Organic Chemistry

11.1 Introduction

Oxidation is the opposite of reduction, which was covered in the previous chapter. Oxidation involves the removal of electrons and or hydrogen atoms and the supply of oxygen to another molecule. Oxidation reactions are very important reactions in organic chemistry, and many of these reactions are used in synthetic organic chemistry to transform the functionalities of molecules into other functionalities. Oxidation reactions are also important in that the combustion of some compounds releases large amounts of energy in the form of heat. Before we proceed, however, we need to be able to identify oxidizing agents. One of the easiest oxidizing agents to recognize is oxygen. We are all familiar with rust, which is a form of oxidation, in which oxygen oxidizes metals. There are other oxidizing agents, and most that are used in organic chemistry typically contain a number of oxygen atoms. Two common oxidizing agents of organic chemistry are shown below.

Chromic acid (H_2CrO_4)   Potassium permanganate ($KMnO_4$)

We will see in this chapter that these oxidizing agents are some of the strongest oxidizing agents of organic chemistry, and sometimes salts of chromic acid are used in combination with sulfuric acid to make an even stronger oxidizing agent. In this chapter, we will apply the concept of oxidation to a number of different types of functional groups to obtain molecules with different functionalities. We will also utilize a combination of various reaction types, including oxidation reaction, to strategically synthesize new target molecules.

11.2 Oxidation

Combustion is the reaction of compounds with oxygen under suitable reaction conditions. Reaction (11-1) shows the combustion of methane, and Reaction (11-2) shows the combustion of butane.

$$CH_4 + 2O_2 \rightarrow CO_2 + 2H_2O + \text{heat}$$
Methane   (11-1)

Organic Chemistry: Concepts and Applications, First Edition. Allan D. Headley.
© 2020 John Wiley & Sons, Inc. Published 2020 by John Wiley & Sons, Inc.
Companion website: www.wiley.com/go/Headley_OrganicChemistry

$$CH_3CH_2CH_2CH_3 + 13/2 O_2 \rightarrow 4CO_2 + 5H_2O + \text{heat} \quad (11\text{-}2)$$
Butane

Complete combustion of alkanes yields carbon dioxide and water as the products. These reactions are very important, during the winter, and the combustion of natural gas is used to heat our homes, classrooms, and workplaces. Natural gas is 60–90% (methane) and the other 10–40% consists of C_2H_6 and C_3H_8, N_2, CO_2. A great deal of natural gas is found in the Texas Panhandle and Oklahoma of the United States. Petroleum, also called crude oil, is a mixture of aliphatic and aromatic compounds and a small percentage of sulfur and nitrogen compounds. There are as many as 500 different compounds present in petroleum or crude oil. This crude product from the ground is not very useful since it consists of so many different compounds. However, refining – the process of distillation – separates the different components of crude oil into various fractions. One fraction is gasoline, which as we know is used as the fuel for the combustion engine. This fraction when used as fuel, however, causes knocking in engines, and to prevent knocking, sometimes other alkanes and aromatic compounds are added to the fuel. It was observed that 2,2,4-trimethylpentane (isooctane) could be used as an antiknocking compound for gasoline engines. As a result, it is used to rate fuels. Examples of the amount of heat liberated from the combustion of different alkanes are given in Reactions (11-3)–(11-6).

$$\text{Cyclopropane} + O_2 \rightarrow CO_2 + H_2O \quad \Delta H° = -498.8 \text{ kcal mol}^{-1} \quad (11\text{-}3)$$

$$\text{Cyclopentane} + O_2 \rightarrow CO_2 + H_2O \quad \Delta H° = -793.5 \text{ kcal mol}^{-1} \quad (11\text{-}4)$$

$$\text{Propane} + O_2 \rightarrow CO_2 + H_2O \quad \Delta H° = -531.1 \text{ kcal mol}^{-1} \quad (11\text{-}5)$$

$$\text{Butane} + O_2 \rightarrow CO_2 + H_2O \quad \Delta H° = -688.4 \text{ kcal mol}^{-1} \quad (11\text{-}6)$$

The negative sign indicates that heat is released from the combustion, i.e. these are all exothermic reactions.

Problem 11.1
Give a balanced equation for the combustion of hexane.

Oxidation reactions involving oxygen are easy to recognize since oxygen is the oxidizing agent. Another definition of oxidation is that there is a loss of hydrogen. For the reaction given in Reaction (11-1) in which methane is oxidized, methane (CH_4) loses hydrogen atoms to become CO_2, and hence meets the requirement of being oxidized since it loses hydrogen atoms in going from reactant to product.

The other definition of oxidation states that there is a loss of electrons. In order to determine a loss of electrons, we must determine the oxidation state of the carbon of the reactant and also in the product. Let us consider the oxidation of methane (Reaction 11-1). In order to determine the oxidation state of carbon in the reactant (methane), we will have to start by looking at the oxidation state of elemental carbon, and how many valence electrons are associated with this atom. You will recall that there are four valence electrons associated with elemental carbon and since it is an element, it is assigned an oxidation state of zero. Since carbon is bonded to other atoms, typically by covalent bonds, we will have to take into consideration the electrons from the other bonding atoms. In the case of methane, they are four hydrogen atoms. Shown in Table 11.1 is the calculation for the oxidation of the carbon of methane and carbon dioxide. For methane, there are eight electrons that are shared in the bonding to methane and they are used in the calculation and shown in the second column, last line. For the product, CO_2 shown in

Table 11.1 Calculating the oxidation states for methane and carbon dioxide.

Molecule	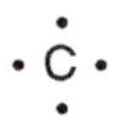 $\cdot \overset{\cdot\cdot}{\underset{\cdot\cdot}{C}} \cdot$	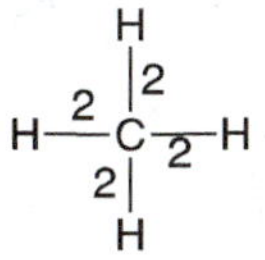 $H\overset{2}{-}\overset{\overset{H}{\mid}2}{\underset{\underset{H}{\mid}}{C}}\overset{2}{-}H$	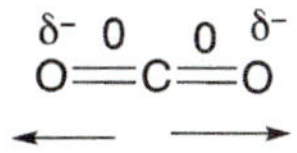 $\overset{\delta^-}{O}\overset{0}{=}\overset{0}{C}\overset{\delta^-}{=}\overset{}{O}$
Number of electrons used for calculation	4	8	0
Oxidation state	0	4 − 8 = −4	4 − 0 = +4

Reaction 11-1, the number of electrons considered for the calculation is zero since these bonding electrons are closely associated with the electronegative oxygen, compared to the carbon. Thus, the oxidation state of the carbon in carbon dioxide is positive four (+4), as shown in the calculation in Table 11.1 (last column, last line).

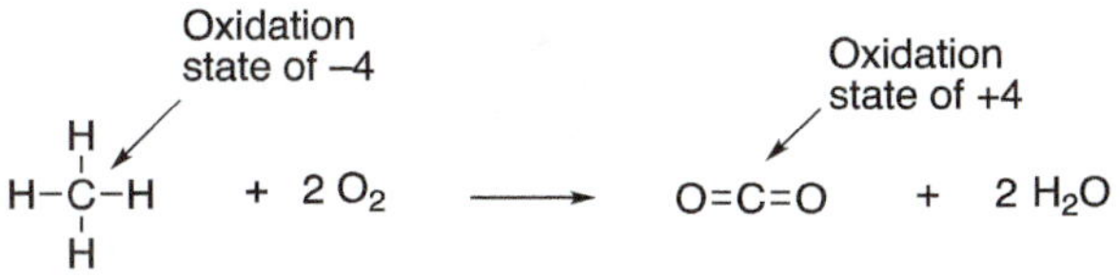

$$\underset{\text{Oxidation state of −4}}{H-\underset{\underset{H}{\mid}}{\overset{\overset{H}{\mid}}{C}}-H} + 2\,O_2 \longrightarrow \underset{\text{Oxidation state of +4}}{O=C=O} + 2\,H_2O$$

Therefore, for the combustion reaction shown in Reaction (11-1), the carbon of methane has an oxidation state of −4 and is transformed into carbon dioxide via this reaction, with an oxidation state of +4. This involves a loss of electrons, hence oxidation. We will see more examples in the next section when we look at the reactions of more complex molecules, but the same concept of determining oxidation states can be applied.

Let us now consider a molecule, where all four hydrogens of methane are substituted with alkyl groups, such as methyl groups. The bonding electrons are now equally shared since both atoms are carbon atoms with the same electronegativity. Thus, there are four electrons associated with the central carbon in this case and not eight as was the case with methane. As a result, the oxidation state of 2,2-dimethylpropane is zero (0) as shown in the calculation in Table 11.2, second column, last row. The other example shown in the table has one hydrogen atom and three alkyl groups bonded to the central carbon. As shown in the last column, there are now five electrons associate with this calculation giving an oxidation state of negative one (−1) as shown in the last column, last row.

Thus, for the oxidation of ethane, the oxidation state of the carbon shown below is −3; in the product, the oxidation state of the carbon of carbon dioxide is +4. Since the carbon goes from negative to positive, ethane has been oxidized.

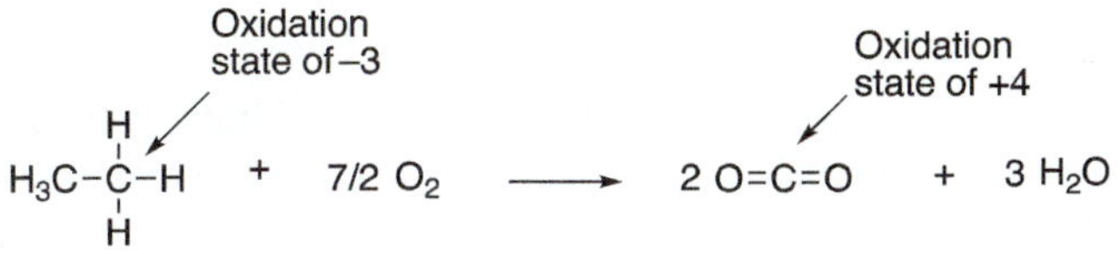

$$\underset{\text{Oxidation state of −3}}{H_3C-\underset{\underset{H}{\mid}}{\overset{\overset{H}{\mid}}{C}}-H} + 7/2\,O_2 \longrightarrow \underset{\text{Oxidation state of +4}}{2\,O=C=O} + 3\,H_2O$$

Let us now consider the oxidation states of carbons of molecules that are bonded to heteroatoms, such as oxygen, which are more electronegative than carbon. As a result, the covalent bonding electrons will all be associated with the heteroatom and not with carbon, and the method of counting bonding electrons is the same as with carbon dioxide. Table 11.3 gives examples of secondary and tertiary alcohols and carboxylic acids.

Table 11.2 Calculation the oxidation states for various types of hydrocarbons.

Molecule	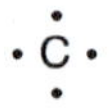 ·C·	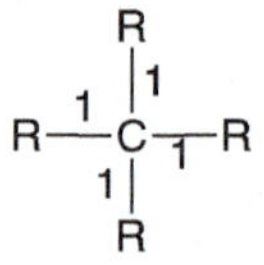 R—C(R)(R)—R	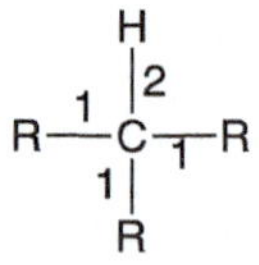 R—C(H)(R)—R
Number of electrons used for calculation	4	4	5
Oxidation state	0	4 − 4 = 0	4 − 5 = −1

Table 11.3 Calculation the oxidation states for molecules with oxygen.

Molecule	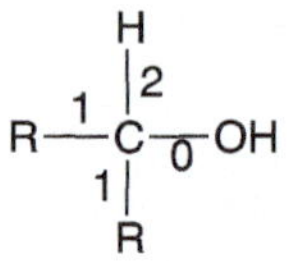 R—C(H)(R)—OH	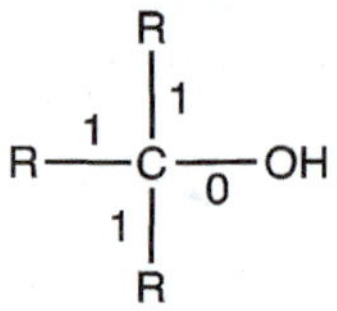 R—C(R)(R)—OH	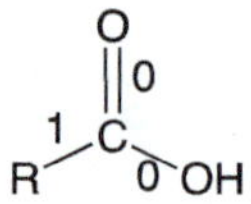 R—C(=O)—OH
Number of electrons used for calculation	4	3	1
Oxidation state	4 − 4 = 0	4 − 3 = +1	1 − 4 = +3

Reactions (11-7)–(11-9) give examples of oxidation reactions, in which the oxidation states of the carbons in both the reactants and products are shown.

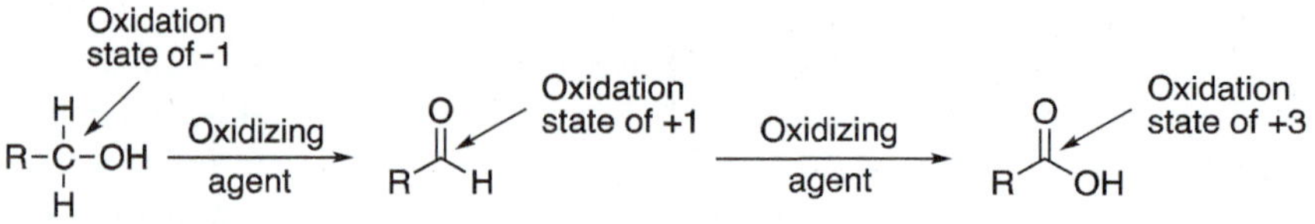

$$\underset{\text{Oxidation state of }-1}{R\text{-}CH_2\text{-}OH} \xrightarrow{\text{Oxidizing agent}} \underset{\text{Oxidation state of }+1}{R\text{-}CHO} \xrightarrow{\text{Oxidizing agent}} \underset{\text{Oxidation state of }+3}{R\text{-}COOH} \quad (11\text{-}7)$$

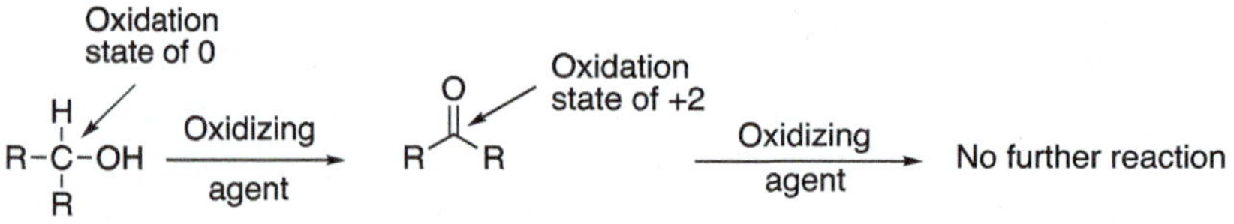

$$\underset{\text{Oxidation state of }0}{R\text{-}CH(R)\text{-}OH} \xrightarrow{\text{Oxidizing agent}} \underset{\text{Oxidation state of }+2}{R\text{-}CO\text{-}R} \xrightarrow{\text{Oxidizing agent}} \text{No further reaction} \quad (11\text{-}8)$$

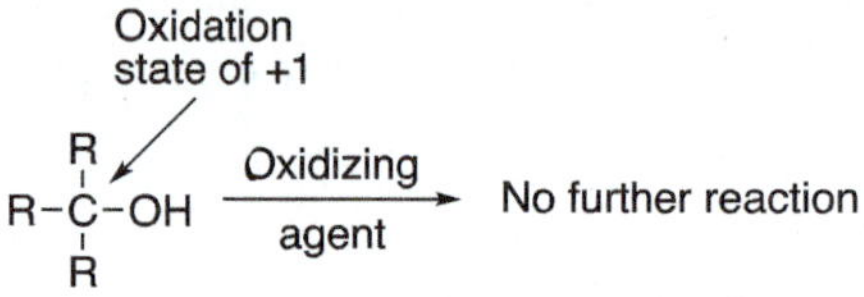

$$\underset{\text{Oxidation state of }+1}{R\text{-}C(R)(R)\text{-}OH} \xrightarrow{\text{Oxidizing agent}} \text{No further reaction} \quad (11\text{-}9)$$

For Reaction (11-7), note that the reactant is a primary alcohol, and it is oxidized initially to an aldehyde, which can be oxidized further to the carboxylic acid. Note that since the aldehyde has one hydrogen atom bonded to the carbon of the carbonyl, it can be oxidized further by losing that hydrogen and of course a further loss of electrons. For Reaction (11-8), the reactant is a secondary alcohol and is oxidized to a ketone. Note that the ketone cannot be oxidized further since it does not have a hydrogen bonded to the carbonyl carbon. The same is true for tertiary alcohols; for Reaction (11-9), there is no reaction since a tertiary alcohol cannot be oxidized further since it does not have a hydrogen atom bonded to the carbon of the alcohol functionality.

Problem 11.2

i) Give the oxidation state of the starred carbons for each of the following molecules.

(a) H–C*(H)(H)–C₂H₅ (b) H–C*(CH₃)(CH₃)–CH₃ (c) H₃C–C*(CH₃)(CH₃)–CH₃ (d) (CH₃)₂CH–CH₂–CH₃ (starred)

ii) Give the oxidation state of the starred carbons for each of the following molecules.

(a) H–C*(OH)(H)–CH₃ (b) H–C(=O)–CH₂–CH₃ (starred on carbonyl) (c) CH₃CH₂–C*(=O)–OH (d) (CH₃)₂CH*–CH(OH)–CH₃

iii) For the reactions below, determine the oxidation state of the starred carbons of the reactants and products and determine if an oxidation occurred.

(a) CH₃CH₂–C*H(OH)–CH₃ $\xrightarrow{\text{KMnO}_4,\ \text{KOH, H}_2\text{O}}$ CH₃CH₂–C*(=O)–CH₃

(b) CH₃CH₂–C*H₂–CH₂OH $\xrightarrow{\text{Na}_2\text{Cr}_2\text{O}_7,\ \text{H}_2\text{SO}_4}$ CH₃CH₂–C*H₂–C(=O)OH

(c) Ph–C*H(OH)–CH₃ $\xrightarrow{\text{KMnO}_4,\ \text{KOH, H}_2\text{O}}$ Ph–C*(=O)–OH

(d) Ph–C*H₂–OH $\xrightarrow{\text{CrO}_3,\ \text{pyridine}}$ Ph–C*(=O)–H

11.3 Oxidation of Alcohols and Aldehydes

Some of the oxidizing agents that are typically used in organic chemistry include H_2CrO_4/H_2SO_4, $KMnO_4/KOH$, CrO_3/pyridine, and CrO_3/H_2SO_4. The last reagent, CrO_3/H_2SO_4, is a very common oxidizing agent and is often referred to as the **Jones reagent**. Note that all these agents have oxygen and at least three oxygen atoms. An oxidizing agent, however, does not necessarily need to have oxygen atoms, but most that are used in organic chemistry have oxygen atoms, and hence is a very good indication that a particular agent is an oxidizing agent. In this section, we will examine the oxidation of molecules that have specific functional groups, alcohols and aldehydes. The first set of molecules that will be examined are alcohols. Alcohols that have at least one hydrogen atom bonded to the carbon of the alcohol functionality (–OH) can be oxidized to different functionalities, which include aldehydes, ketones, and even carboxylic acids. That is, methanol, primary and secondary alcohols can be oxidized but not a tertiary alcohol since tertiary alcohols have three alkyl groups bonding to the carbon of the alcohol functionality and no hydrogen. Aldehydes that are the products of an alcohol oxidation can be oxidized further to give another molecule with a different functionality, carboxylic acids.

Aldehydes have a hydrogen atom bonded to the carbonyl carbon functionality, which makes it possible to be oxidized further.

Consider the oxidation reaction shown in Reaction (11-10), a close examination shows that a hydrogen atom is lost in going from the reactant (alcohol) to the organic product (ketone). Note that it is not just any hydrogen from the molecule that is lost, but the hydrogen that is bonded to the carbon of the alcohol functionality, along with the hydrogen from the alcohol.

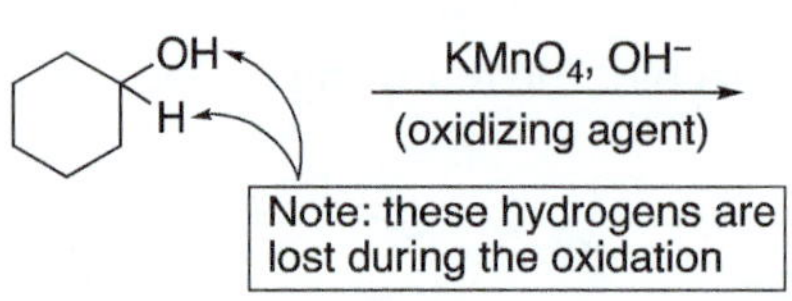

$$\text{cyclohexanol} \xrightarrow{\text{KMnO}_4, \text{OH}^- \text{ (oxidizing agent)}} \text{cyclohexanone} \tag{11-10}$$

Note: these hydrogens are lost during the oxidation

In the next section, we will examine the mechanism for oxidation reactions and account for the loss of the hydrogens and change of oxidation state of the carbon of the functional group being oxidized.

11.3.1 Oxidation Using Potassium Permanganate (KMnO$_4$)

An examination of the reaction mechanism will show how the loss of the hydrogen atoms occurs during the oxidation. During the first step of the mechanism, the nucleophilic oxygen of the alcohol functionality adds to the very electrophilic manganese of KMnO$_4$ followed by the loss of water as shown in Reaction (11-11).

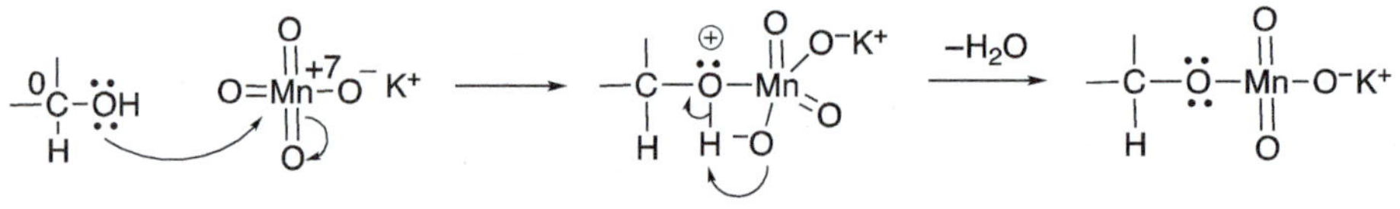

$$\tag{11-11}$$

In the next step of the mechanism, the hydrogen that is bonded to the alcohol carbon is abstracted by the base to release the oxidized carbonyl compound as shown in Reaction (11-12). Note that the oxidation state of the carbon is changed from 0 to +2.

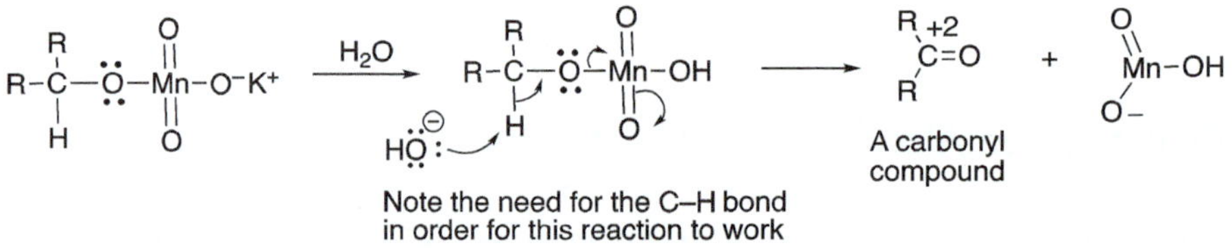

$$\tag{11-12}$$

Note the need for the C–H bond in order for this reaction to work

A carbonyl compound

Aldehydes can be oxidized to carboxylic acids by the same reagent, KMnO$_4$, in a basic solution. The first step of the mechanism is shown in Reaction (11-13).

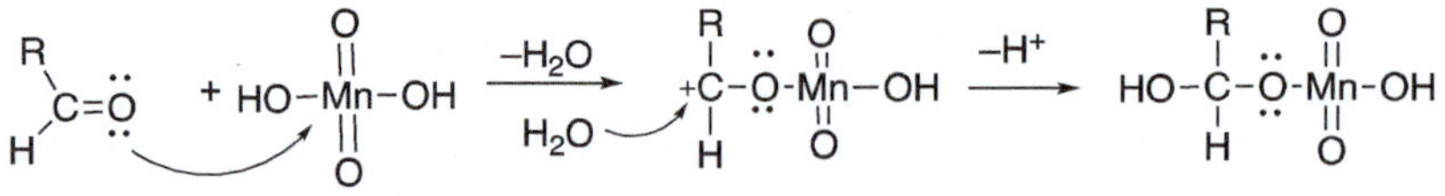

$$\tag{11-13}$$

In the final step of the reaction mechanism, the hydrogen is lost to give the oxidized product, as shown in Reaction (11-14).

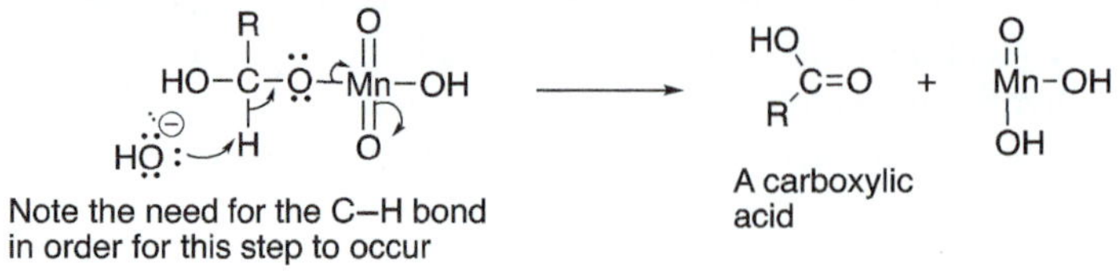

$$\tag{11-14}$$

Note the need for the C–H bond in order for this step to occur

A carboxylic acid

11.3 Oxidation of Alcohols and Aldehydes

As you can imagine, the oxidation of a primary alcohol gives the aldehyde first, which is oxidized to the carboxylic acid as shown in Reaction (11-15).

Butanol $\xrightarrow{\text{KMnO}_4, \text{Heat}}$ [Butanal (not isolated)] $\xrightarrow{\text{KMnO}_4, \text{Heat}}$ Butanoic acid (11-15)

Problem 11.3

Give the oxidation products of the reaction of the following molecules with KMnO$_4$ in an aqueous solution of KOH.

(a) cyclopentanol (b) cyclohexylmethanol (c) 1-cyclohexylethanol (d) 1-methylcyclohexanol

11.3.2 Oxidation Using Chromic Acid (H$_2$CrO$_4$)

A very similar oxidation as discussed in the previous section can be accomplished using chromic acid. Even in an acidic medium, there is a similar abstraction of the hydrogen bonded to the carbon that has the alcohol functionality as shown in Reaction (11-16).

$$R-\underset{H}{\underset{|}{\overset{R}{\overset{|}{C}}}}-\ddot{O}H + HO-\underset{O}{\overset{O}{\overset{||}{Cr}}}-OH \xrightarrow[H_2O]{-H_2O} R-\underset{H}{\underset{|}{\overset{R}{\overset{|}{C}}}}-\ddot{O}-\underset{O}{\overset{O}{\overset{||}{Cr}}}-OH \longrightarrow \underset{R}{\overset{R}{C}}=O + \underset{O^-}{\overset{O}{\overset{||}{Cr}}}-OH$$

A carbonyl compound, a ketone (11-16)

Note the need for the C–H bond in order for this reaction to work

Since there is a hydrogen atom bonded to the carbonyl carbon of aldehydes, these molecules can be oxidized further to give carboxylic acids as shown in the mechanism in Reaction (11-17). The first step is similar to that of the oxidation of alcohols, but instead of the alcohol oxygen attacking the electropositive chromium atom, it is the carbonyl oxygen. A carbocation is created, which is readily attacked by the nucleophilic water to form the intermediate shown in Reaction (11-17).

$$\underset{H}{\overset{R}{C}}=\ddot{O} + HO-\overset{O}{\overset{||}{Cr}}-OH \xrightarrow[H_2O]{-H_2O} \underset{H}{\overset{R}{\overset{+}{C}}}-\ddot{O}-\overset{O}{\overset{||}{Cr}}-OH \xrightarrow{-H^+} HO-\underset{H}{\overset{R}{C}}-\ddot{O}-\overset{O}{\overset{||}{Cr}}-OH$$ (11-17)

In the next step of the mechanism, water abstracts the proton bonded to the carbonyl carbon to form the carboxylic acid and reduced chromium salt as shown in Reaction (11-18).

$$HO-\underset{H}{\overset{R}{C}}-\ddot{O}-\overset{O}{\overset{||}{Cr}}-OH \longrightarrow \underset{R}{\overset{HO}{C}}=O + \overset{O}{\overset{||}{Cr}}-OH$$ (11-18)

Note the need for the C–H bond in order for this reaction to work

A carboxylic acid

An example of the overall reaction is shown in Reaction (11-19).

$$\text{Butanal} \xrightarrow[\text{H}_2\text{SO}_4]{\text{H}_2\text{Cr}_2\text{O}_7} \text{Butanoic acid} \tag{11-19}$$

As a result, the oxidation of primary alcohols with a strong oxidizing agent, such as chromic acid or potassium permanganate results in the carboxylic acid, and not the aldehyde. Even though the aldehyde is formed, it is readily oxidized further into the carboxylic acid as shown in Reaction (11-20).

$$\text{Butanol} \xrightarrow[\text{H}_2\text{SO}_4]{\text{H}_2\text{Cr}_2\text{O}_7} [\text{Butanal is formed but is readily oxidized further}] \xrightarrow[\text{H}_2\text{SO}_4]{\text{H}_2\text{Cr}_2\text{O}_7} \text{Butanoic acid} \tag{11-20}$$

Since tertiary alcohols do not have a hydrogen atom bonded to the carbon of the alcohol functionality, they cannot be oxidized, an example is shown in Reaction (11-21).

$$\text{3-Methyl-3-hexanol} \xrightarrow[\text{H}_2\text{SO}_4]{\text{H}_2\text{Cr}_2\text{O}_7} \text{No reaction} \tag{11-21}$$

Problem 11.4

Give the oxidized organic product that results from the reaction of each of the following with H_2CrO_4.

(a) cyclohexanol (b) propan-1-ol (c) butan-2-ol (d) 1-methylcyclopentan-1-ol

A practical application of this oxidation reaction is that ethanol can be oxidized to acetic acid using a dichromate salt. Dichromate salts are reddish orange in color and once the oxidation has occurred, it is transformed into a green dichromate chromium (III) ion. The reaction is shown in Reaction (11-22). The breath analyzer test, which is used to determine the presence of alcohol, relies on this color change for a positive result based on the reaction given in Reaction (11-22).

$$\underset{\text{Ethanol}}{\text{CH}_3\text{CH}_2\text{OH}} + \underset{\substack{\text{Dichromate}\\\text{(reddish orange)}}}{\text{Cr}_2\text{O}_7^{-2}} \xrightarrow[\text{H}_2\text{O}]{\text{H}_2\text{SO}_4} \underset{\text{Acetic acid}}{\text{CH}_3\text{COOH}} + \underset{\substack{\text{Chromium (III) ion}\\\text{(green)}}}{\text{Cr}^{+3}} \tag{11-22}$$

Another variation of this oxidizing agent in which a chromium salt is used is the Jones reagent, which is named after its discoverer by Sir Ewart Jones, an English chemist. The Jones reagent consists of a solution of chromium trioxide in dilute sulfuric acid and acetone and is used to oxidize alcohols to ketones and carboxylic acids, an example is shown in Reaction (11-23).

$$\text{3-Hexanol} \xrightarrow[\substack{\text{aq. H}_2\text{SO}_4\\\text{acetone}}]{\text{CrO}_3} \text{3-Hexanone} \tag{11-23}$$

11.3.3 Swern Oxidation

Chromium salts are typically toxic, and extreme care must be exercised in their use. As a result, an alternate set of reagents that is used for oxidation is oxalyl chloride and dimethyl sulfoxide (DMSO); this oxidation was developed by Daniel Swern, an American Chemist, and is now known as the Swern oxidation. The general reaction is shown in Reaction (11-24).

$$R-\underset{H}{\underset{|}{\overset{R}{\overset{|}{C}}}}-OH \xrightarrow[(2)\ Et_3N]{(1)\ ClC(O)C(O)Cl\ (oxalyl\ chloride)\ and\ (CH_3)_2S=O\ (DMSO)} \underset{R}{\overset{R}{>}}C=O \qquad (11\text{-}24)$$

Note again that in order for the oxidation to take place, there must be a hydrogen atom bonded to the carbon of the alcohol group. For the first step of the mechanism, the nucleophilic oxygen of the DMSO attacks the electrophilic carbon of oxalyl chloride to displace a chloride anion followed by the attack of the chloride anion on the electrophilic sulfur atom, as shown in Reaction (11-25).

$$\text{Oxalyl chloride (COCl)}_2 + \text{Dimethyl sulfoxide (DMSO)} \xrightarrow{-Cl^-} \text{intermediate} \longrightarrow \text{chlorosulfonium intermediate} \qquad (11\text{-}25)$$

In the next step of the reaction, the intermediate falls apart to release two gases, carbon dioxide and carbon monoxide, and a chlorosulfonium salt, as shown in Reaction (11-26).

$$\text{intermediate} \longrightarrow O=C=O + C=O + \underset{Cl}{\overset{H_3C-\overset{+}{S}-CH_3}{|}} \ Cl^- \qquad (11\text{-}26)$$

Carbon dioxide   Carbon monoxide   Chlorosulfonium salt

In the next step of the mechanism, the chlorosulfonium salt reacts with the alcohol, as shown in Reaction (11-27) to form an alkylsulfonium salt.

$$H_3C-\overset{+}{S}-CH_3\ (Cl)\ Cl^- + HO-\underset{H}{\overset{R}{\underset{|}{C}}}-R \xrightarrow{-HCl} \underset{H_3C-\overset{+}{S}-CH_3}{\overset{H\diagdown\ _{R}}{\underset{|}{\overset{C-R}{\underset{O}{|}}}}}\ Cl^- \qquad (11\text{-}27)$$

Alkylsulfonium salt

In the next step of the reaction mechanism, the alkylsulfonium salt reacts with the base, triethylamine, to form a ylide, as shown in Reaction (11-28). You will recall that ylides are neutral compounds, in which there is a formal charge of negative charge on a carbon and on the adjacent atom, typically phosphorous, sulfur or nitrogen, there is a formal positive charge.

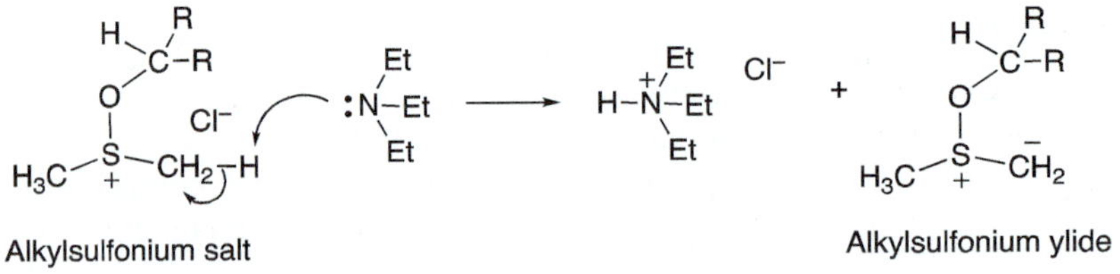

$$\text{(11-28)}$$

In the next step of the mechanism, the ylide breaks apart to form the oxidized carbonyl compound, as shown in Reaction (11-29).

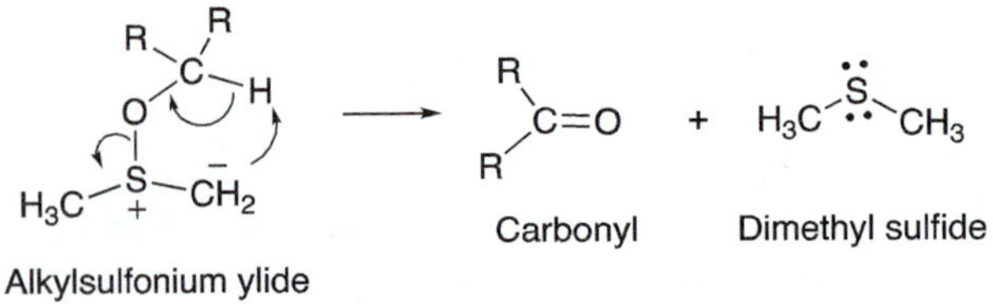

$$\text{(11-29)}$$

11.3.4 Dess-Martin Oxidation

Another environmentally-friendly method for the oxidation of alcohols is with the use of hypervalent iodine compounds. This reaction, also known as the Des-Martin periodinane (DMP) reaction, derived its name from two American chemists, Daniel Benjamin Dess and James Cullen Martin, who developed this reaction. This reaction is similar to the Swern oxidation in that primary alcohols are oxidized to aldehydes and secondary alcohols are oxidized to ketones; tertiary alcohols are not oxidized as expected. An example of the reaction is shown in Reaction (11-30).

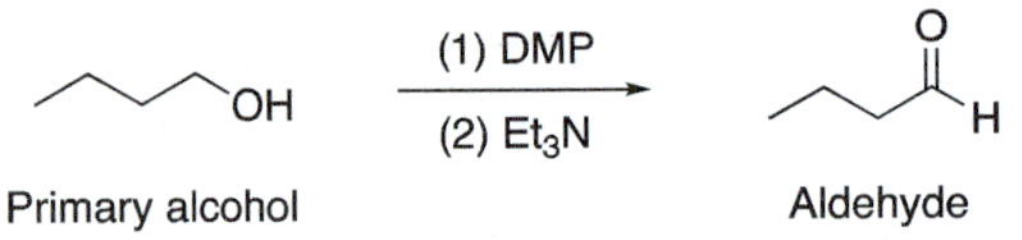

$$\text{(11-30)}$$

The mechanism involves an attack of the alcohol on the electrophilic iodine of the periodinane compound displacing an acetate anion as shown in Reaction (11-31).

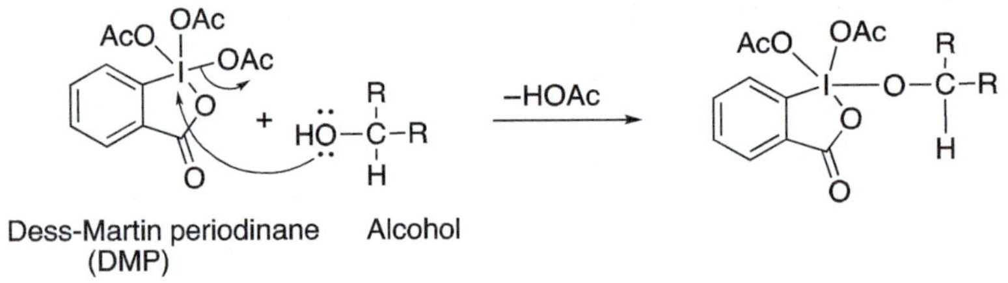

$$\text{(11-31)}$$

In the next step of the reaction mechanism, the amine base abstracts a proton as shown in Reaction (11-32) to give the oxidized carbonyl compound.

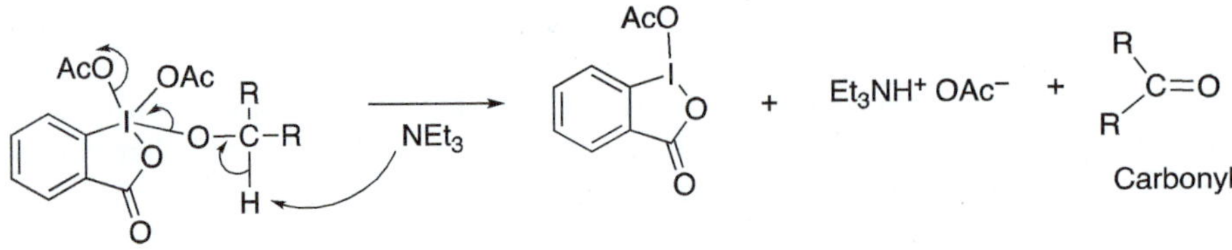

$$\text{(11-32)}$$

11.3.5 Oxidation Using Pyridinium Chlorochromate

As we have seen from the oxidizing agents discussed thus far, some will oxidize primary alcohols to the aldehyde and will not proceed further to give the carboxylic acid, while other stronger oxidizing agents will oxidize primary alcohols to carboxylic acids. Another mild oxidizing agent results from mixing CrO_3 in with HCl and pyridine to give pyridine chlorochromate as shown in Reaction (11-33).

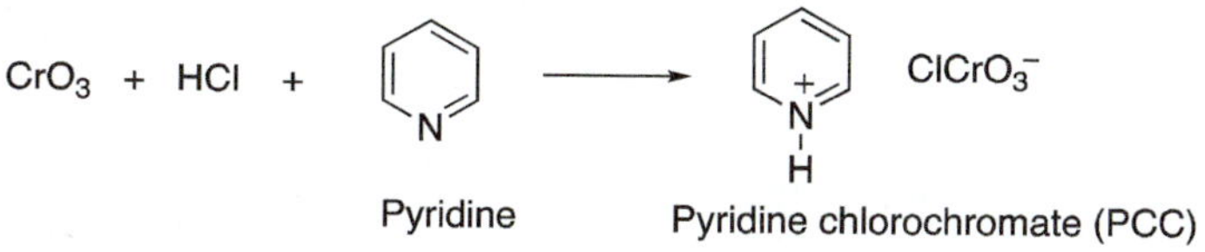

$$CrO_3 + HCl + \text{Pyridine} \longrightarrow \text{Pyridine chlorochromate (PCC)} \quad ClCrO_3^- \tag{11-33}$$

This much milder oxidizing agent is also referred to as PCC and will oxidize primary alcohols just to the aldehyde and not further to give the carboxylic acid, as shown in Reaction (11-34).

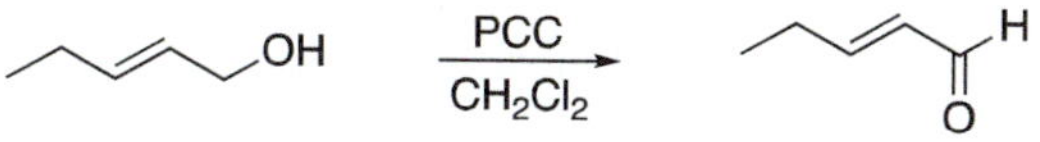

$$\text{R-OH} \xrightarrow[CH_2Cl_2]{PCC} \text{R-CHO} \tag{11-34}$$

Since it is such a mild oxidizing agent, it will not oxidize other functional groups, such as double bonds, as we will see later in the chapter that stronger oxidizing agents will react with double bonds.

Problem 11.5

i) Give the oxidized products for the following reactions.

(a) 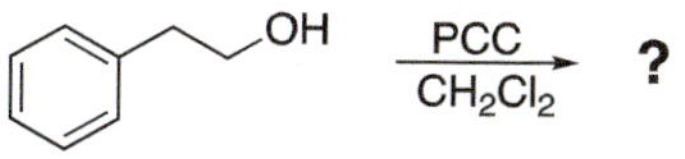 PhCH$_2$CH$_2$OH $\xrightarrow[CH_2Cl_2]{PCC}$?

(b) 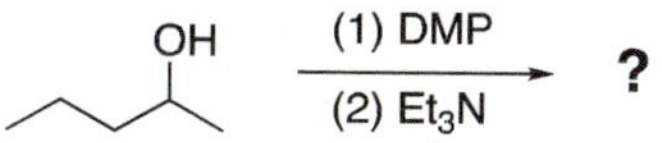 sec-butanol $\xrightarrow[(2)\ Et_3N]{(1)\ DMP}$?

(c) 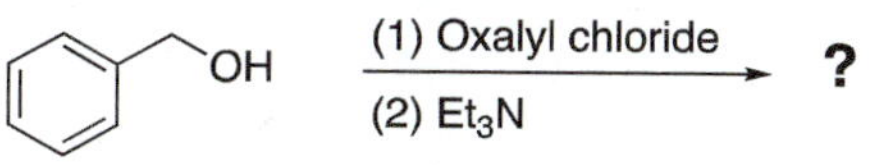 PhCH$_2$OH $\xrightarrow[(2)\ Et_3N]{(1)\ Oxalyl\ chloride}$?

ii) Give the organic reactants to complete the following reactions.

(a) 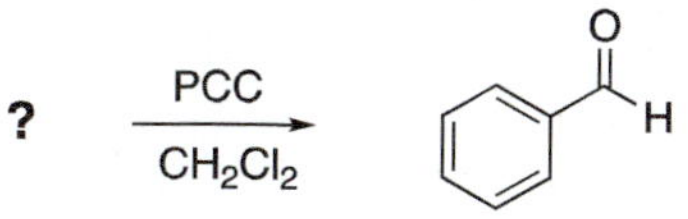 ? $\xrightarrow[CH_2Cl_2]{PCC}$ PhCHO

(b)  ? $\xrightarrow[(2)\ Et_3N]{(1)\ DMP}$ diethyl ketone

(c) 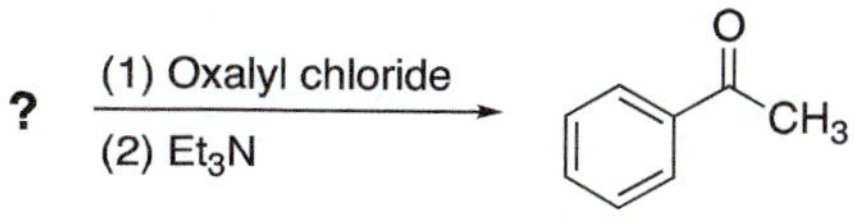 ? $\xrightarrow[(2)\ Et_3N]{(1)\ Oxalyl\ chloride}$ PhCOCH$_3$

11.3.6 Oxidation Using Silver Ions

The oxidation of aldehydes by silver ions gives an interesting result (a silver mirror), and as a result, this reaction is used as a test for the presence of aldehydes. This test is called the **Tollens Test** for aldehydes. The general reaction is shown in Reaction (11-35).

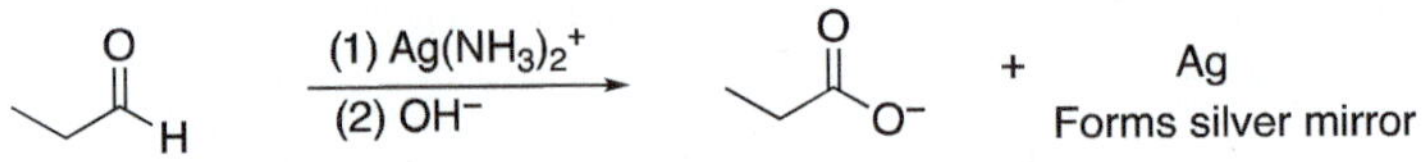

$$\text{CH}_3\text{CH}_2\text{CHO} \xrightarrow[\text{(2) OH}^-]{\text{(1) Ag(NH}_3)_2^+} \text{CH}_3\text{CH}_2\text{COO}^- + \underset{\text{Forms silver mirror}}{\text{Ag}} \quad (11\text{-}35)$$

Problem 11.6

i) Which of the following compounds would give a positive Tollens test?

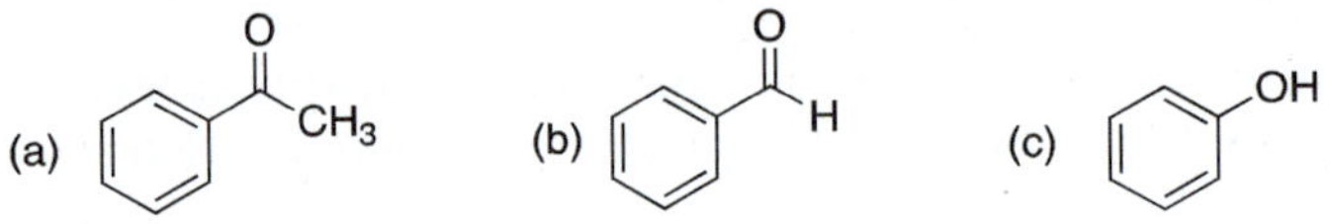

(a) PhCOCH$_3$ (b) PhCHO   (c) PhOH

ii) Give the organic reactants that could be used to give the following compounds by an oxidation reaction.

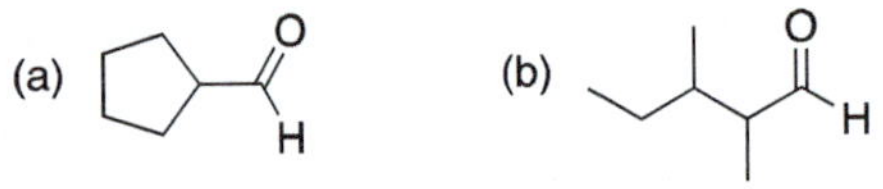

(a) cyclopentyl-CHO   (b) (CH$_3$)$_2$CHCH(CH$_3$)CHO

11.3.7 Oxidation Using Nitrous Acid

Nitrous acid is an oxidizing agent and is typically used in organic chemistry to oxidize amines to azide salts, as shown in Reaction (11-36).

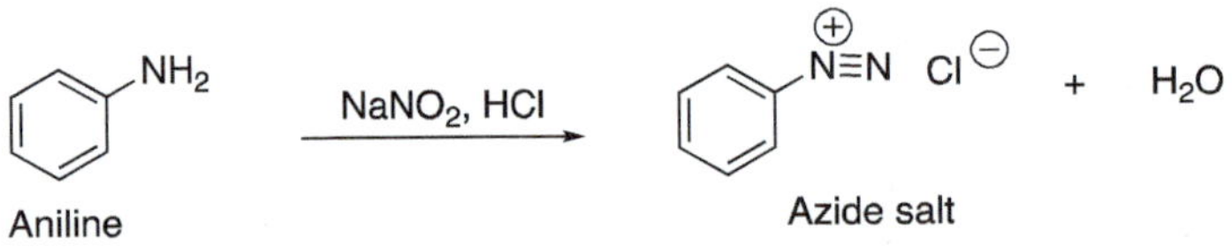

$$\underset{\text{Aniline}}{\text{PhNH}_2} \xrightarrow{\text{NaNO}_2, \text{HCl}} \underset{\text{Azide salt}}{\text{Ph–N}\overset{+}{\equiv}\text{N Cl}^-} + \text{H}_2\text{O} \quad (11\text{-}36)$$

The first step of the mechanism for this reaction involves an attack of the nucleophilic amine functionality on the electrophilic nitrogen of the NO$_2$ group. In the next step, one of the —OH groups is protonated and leaves as water, as shown below.

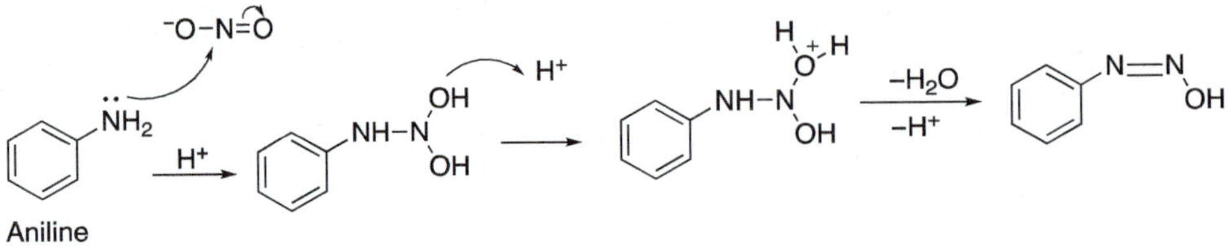

Ph–NH$_2$ + $^-$O–N=O $\xrightarrow{\text{H}^+}$ Ph–NH–N(OH)$_2$ $\xrightarrow{\text{H}^+}$ Ph–NH–N(OH)(OH$_2^+$) $\xrightarrow[-\text{H}^+]{-\text{H}_2\text{O}}$ Ph–N=N–OH

Aniline

In the next step, protonation of the other —OH group occurs and it is converted to a good leaving group and leaves as water resulting in the azide salt as shown below.

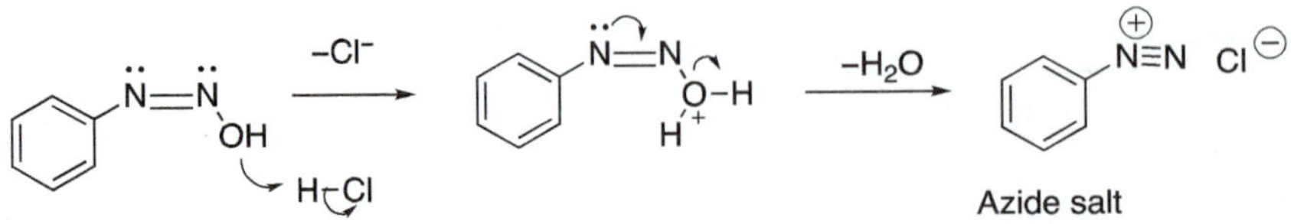

Ph–N=N–OH $\xrightarrow[-\text{Cl}^-]{\text{HCl}}$ Ph–N=N–O(H)–H$^+$ $\xrightarrow{-\text{H}_2\text{O}}$ Ph–N$^+$≡N Cl$^-$

Azide salt

11.3 Oxidation of Alcohols and Aldehydes

These reactions will be discussed in Chapter 17 on aromaticity, but these reactions are typically used in the preparation of azo dyes. Azo dyes are brightly colored compounds, which are used as dyes for various textiles, particularly cotton, silk, wool, and some synthetic fabrics.

11.3.8 Oxidation Using Periodic Acid

Based on the number of oxygen atoms that are present in periodic acid ($HIO_4 \cdot 2H_2O$), it is not surprising that it is a strong oxidizing agent. This reagent is used in organic chemistry to cleave 1,2-diols via an oxidation reaction as shown in Reaction (11-37).

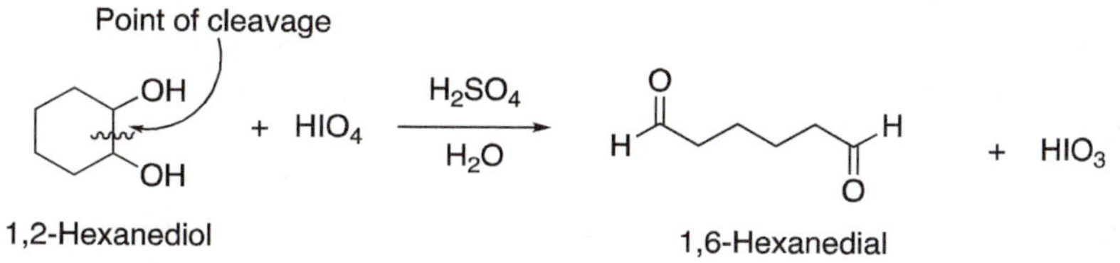

$$\text{1,2-Hexanediol} + HIO_4 \xrightarrow[H_2O]{H_2SO_4} \text{1,6-Hexanedial} + HIO_3 \quad (11\text{-}37)$$

Note that the product for this reaction is a dialdehyde. If a methyl group is bonded to one of the carbons that contains the alcohol functionality, that functionality will be converted into a ketone, as shown in the Reaction (11-38).

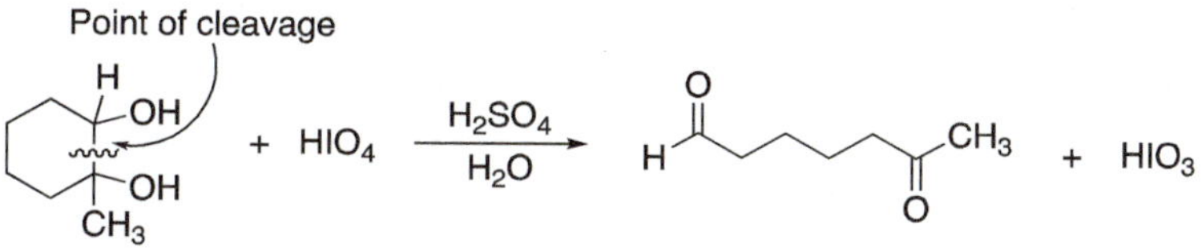

$$\text{(diol with CH}_3\text{)} + HIO_4 \xrightarrow[H_2O]{H_2SO_4} \text{(keto-aldehyde)} + HIO_3 \quad (11\text{-}38)$$

Examination of the mechanism for the reaction will explain the products formed. In the first step, one of the alcohol functionalities bonds to the very electrophilic iodine, after an attack of the other alcohol oxygen and loss of water, a cyclic periodate intermediate is formed as shown in Reaction (11-39).

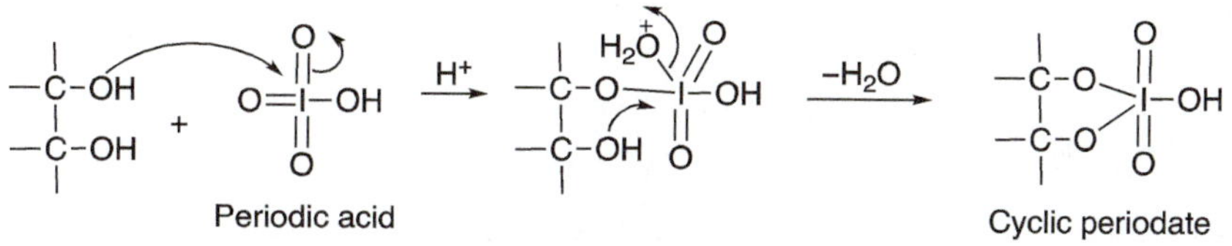

$$\text{diol} + \text{Periodic acid} \xrightarrow{H^+} \text{intermediate} \xrightarrow{-H_2O} \text{Cyclic periodate} \quad (11\text{-}39)$$

In the next step of the mechanism, the cyclic periodate decomposes with a carbon–carbon bond cleavage to form two carbonyl compounds (or a single compound if the initial diol is cyclic) and iodic acid, as shown in Reaction (11-40).

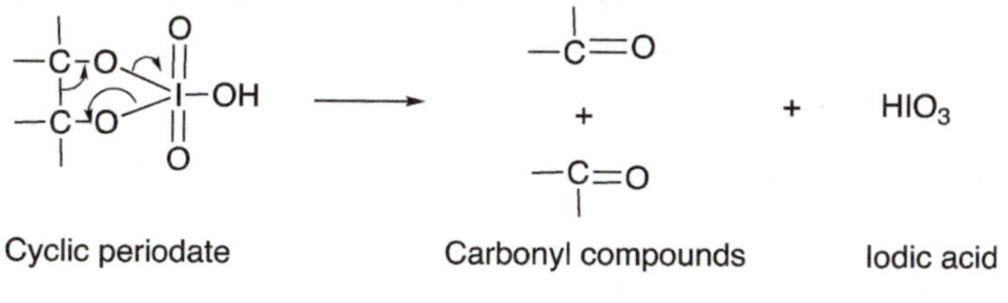

$$\text{Cyclic periodate} \longrightarrow \text{Carbonyl compounds} + HIO_3 \quad (11\text{-}40)$$

Problem 11.7

Give the organic product(s) that results from the reaction of each of the following molecules with HIO_4. If the reaction is not like to occur, indicate no reaction.

(a) structure with OH groups
(b) structure with OH groups
(c) structure with OH groups

11.4 Oxidation of Alkenes Without Bond Cleavage

There are two types of oxidation reactions involving alkenes, the first is oxidation reaction without cleavage of the double bond and the second is oxidation with cleavage of the double bond. In this section, we will examine oxidation of alkenes without cleavage of the double bond, and in the next section, we will examine oxidation of alkenes with cleavage of the double bond. Reaction (11-41) shows the general oxidation reaction of alkenes without bond cleavage to form an epoxide. Reaction (11-42) also shows oxidation of the double bond without cleavage but forms a 1,2-diol. As you can imagine, the oxidizing agents and reaction conditions for these reactions are different to enable different types of products.

$$\text{Alkene} \xrightarrow{\text{Oxidizing agent}} \text{Epoxide} \quad (11\text{-}41)$$

$$\text{Alkene} \xrightarrow{\text{Oxidizing agent}} \text{1,2-Diol} \quad (11\text{-}42)$$

11.4.1 Epoxidation of Alkenes

The epoxidation of ethylene to form ethylene oxide is one of the most important reactions of industrial importance. Each year, millions of pounds of ethylene oxide are produced by the reaction of ethylene with oxygen in the presence of silver catalyst, as shown in Reaction (11-43).

$$\text{Ethylene} + O_2 \xrightarrow{\text{Ag (catalyst)}} \text{Ethylene oxide} \quad (11\text{-}43)$$

Peroxyacids, also known as peracids, are oxidizing agents. One that is highly used in industry for the bleaching of pulp is peroxysulfuric acid, but a more commonly used peracid is hydrogen peroxide. Hydrogen peroxide can be used to oxidize organic carboxylic acids to peroxyacids, which are commonly used in organic chemistry as an oxidizing reagent, the reaction that shows the synthesis of peroxyacids from carboxylic acids is shown in Reaction (11-44).

$$\underset{\text{Carboxylic acid}}{RCO_2H} + \underset{\text{Hydrogen peroxide}}{H_2O_2} \longrightarrow \underset{\text{Peroxyacid}}{RCO_3H} + \underset{\text{Water}}{H_2O} \quad (11\text{-}44)$$

Compared to organic carboxylic acids, peroxyacids have an additional oxygen, and peroxyacids are known to be mild oxidizing agents. Alkenes are oxidized by peracids (RCOOOH) to give oxiranes. The solvent used for these reactions is usually a very inert solvent, such as $CHCl_3$ or CH_2Cl_2. The peroxyacid that is typically used to oxidize the large majority of

11.4 Oxidation of Alkenes Without Bond Cleavage

alkenes is *meta*-chloroperoxybenzoic acid (MCPBA), and Reaction (11-45) gives an example of this oxidation reaction.

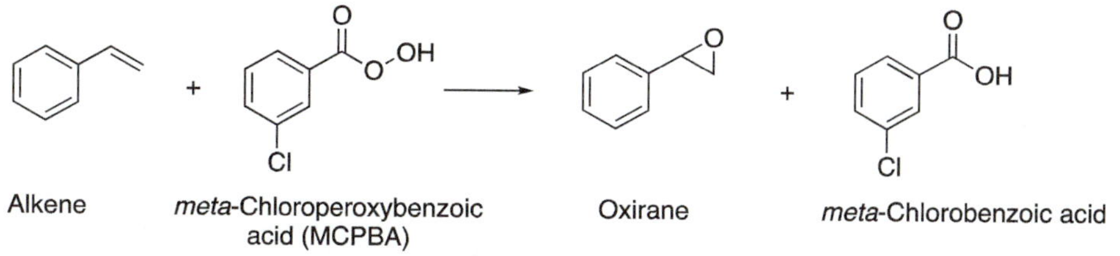

styrene + MCPBA → styrene oxide (oxirane) + *meta*-chlorobenzoic acid (11-45)

The mechanism for epoxidation of alkenes with a peracid is a concerted one in which bonds are broken and formed simultaneously as shown in Reaction (11-46).

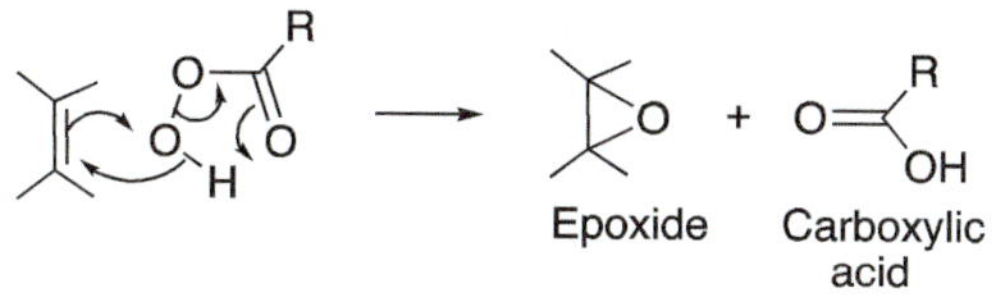

alkene + peracid → Epoxide + Carboxylic acid (11-46)

Problem 11.8

i) Complete the following reactions by supplying the major organic products.

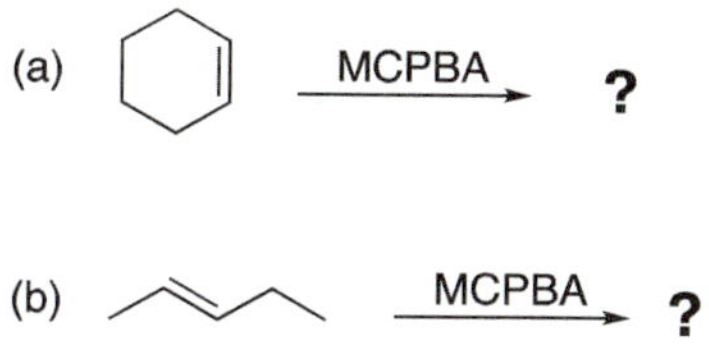

(a) cyclohexene $\xrightarrow{\text{MCPBA}}$?

(b) 2-butene $\xrightarrow{\text{MCPBA}}$?

ii) Give the reactants needed to complete the following reactions.

(a) ? $\xrightarrow{\text{MCPBA}}$ 2,2-dimethyl epoxide

(b) ? $\xrightarrow{\text{MCPBA}}$ cyclopentene oxide

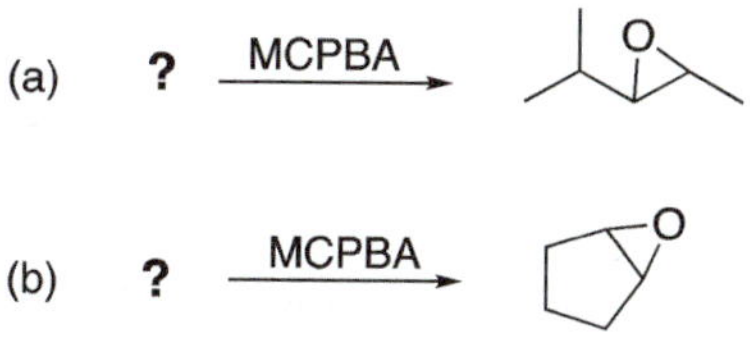

11.4.1.1 Reactions of Epoxides

Epoxides are very reactive, and we have seen the reactions of epoxides with different reducing agents in the previous chapter; epoxides can be easily reduced to undergo ring opening with a reducing agent, such as an organometallic reagent or metal hydride. Epoxides can also undergo ring opening reactions under aqueous acidic or basic conditions to produce *trans*-diols. The acid catalyzed opening of epoxides in water is shown in Reaction (11-47).

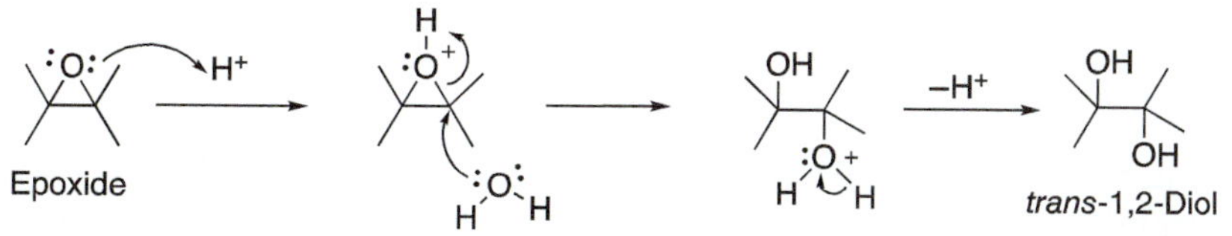

Epoxide + H⁺ → protonated epoxide → ring-opened intermediate → *trans*-1,2-Diol (11-47)

Note that attack of the nucleophilic water occurs from the opposite side of the epoxide, resulting in a *trans*-diol as product. An important ring opening reaction of ethylene oxide in the presence of water is to produce ethylene glycol, which is the main component of anti-freeze as shown in Reaction (11-48).

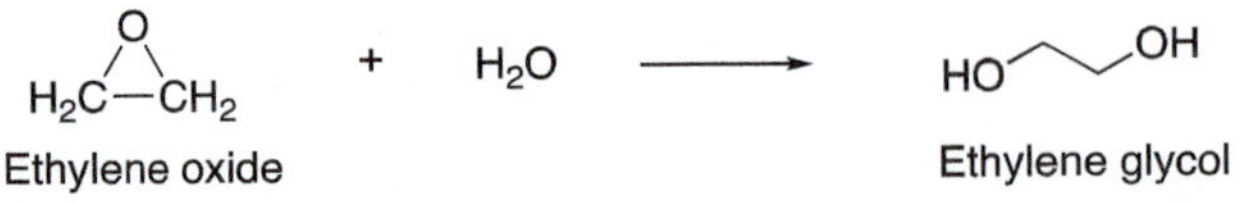

$$\text{H}_2\text{C}-\text{CH}_2\text{(epoxide)} + \text{H}_2\text{O} \longrightarrow \text{HO-CH}_2\text{-CH}_2\text{-OH} \quad (11\text{-}48)$$

Ethylene oxide → Ethylene glycol

You will notice from the above mechanism that the nucleophilic water can react with any of the electrophilic carbons of the protonated epoxide to produce the final product. In the case of a ring opening reaction in which HCl is used, the chloride anion is the nucleophile, instead of water, and the attack can take place on either electrophilic carbon atoms of the protonated epoxide. In the case of a symmetrical epoxide, only one product will result, but in the case of an unsymmetrical oxirane, there are two possible products, as shown in Reaction (11-49).

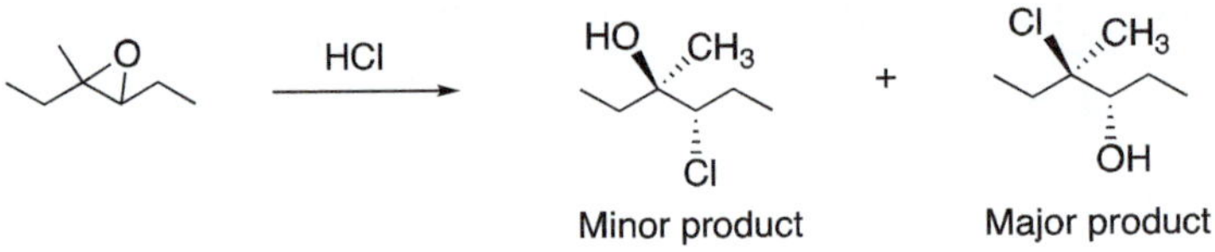

(11-49)

Minor product + Major product

We will have to examine the mechanism in order to determine a rationale for the formation of the product shown above as the major product. Under acidic conditions, the carbon that contains the most groups, i.e. the one with the least hydrogens, is the carbon that is usually attacked by the nucleophile. The carbon with the most surrounding alkyl groups in an acidic medium is more carbocationic-like and we know that the order of stability is that a tertiary carbocation is more stable than a secondary. This concept is illustrated below for the resonance structures shown in Reaction (11-50).

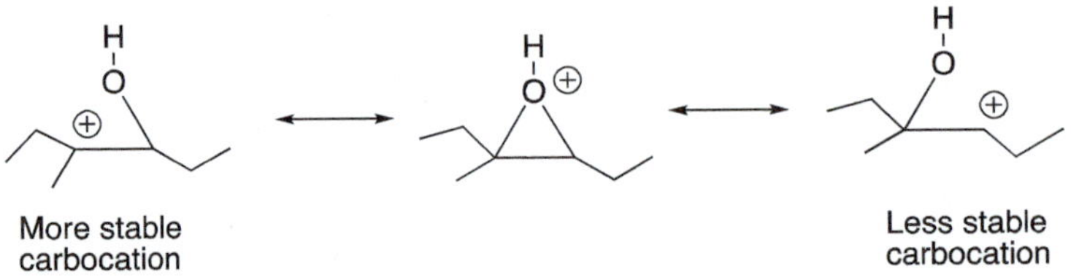

(11-50)

More stable carbocation ↔ ↔ Less stable carbocation

Based on this concept, the chloride anion will attack the carbon that has the methyl and ethyl groups, compared to the other electrophilic carbon that has one hydrogen and an ethyl group.

Problem 11.9

Give the final organic product for each of the following sequence of reactions. Show appropriate stereochemistry of the final product.

(a) 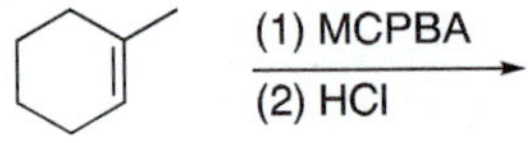 cyclohexene (1) MCPBA (2) HCl

(b) 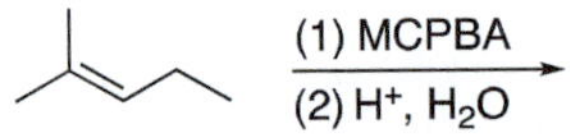 (1) MCPBA (2) H⁺, H₂O

The ring opening of oxiranes in a basic medium involves the attack of the nucleophile on one of the electrophilic carbons, followed by protonation in a second step to give the final *trans*-diol, as shown below for the reaction using water as the nucleophile.

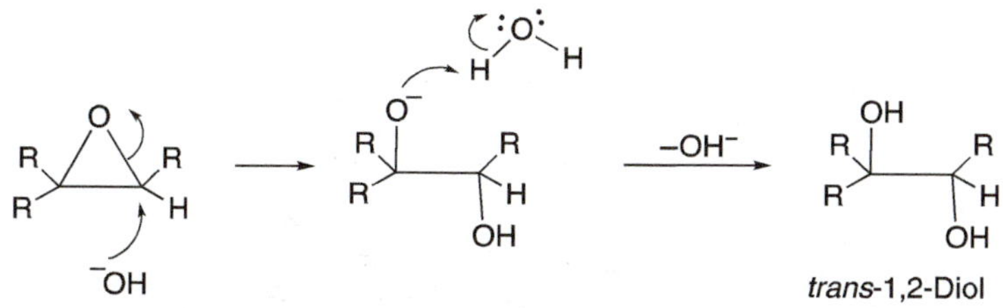

11.4 Oxidation of Alkenes Without Bond Cleavage

$$\text{(11-51)}$$

For the ring-opening reactions in basic media, the attack of the nucleophile on an unsymmetrical oxirane occurs at the electrophilic carbon that is less crowded; that is, the carbon with the least number of alkyl groups, as shown in Reaction (11-52).

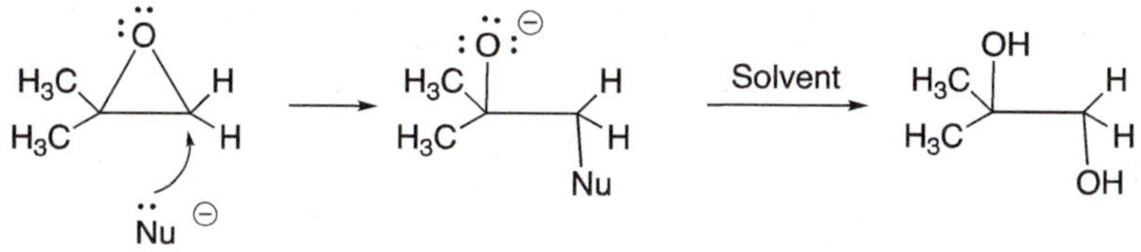

$$\text{(11-52)}$$

As we have seen from the previous chapter, the Grignard reagent and other reducing agents react with epoxides in this manner.

Problem 11.10

Give the final organic products for each of the following sequence of reactions. Show appropriate stereochemistry of the final product.

(a) cyclohexene
(1) MCPBA
(2) CH_3MgCl
(3) H^+, H_2O

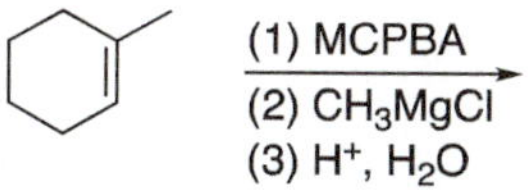

(b) 2-methyl-2-butene
(1) MCPBA
(2) CH_3MgCl
(3) H^+, H_2O

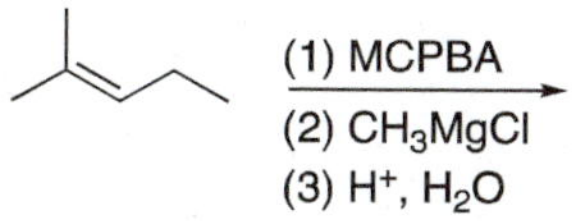

11.4.2 Oxidation of Alkenes with $KMnO_4$

The reaction of an alkene with $KMnO_4$ at room temperature proceeds without cleavage of the double bond, even though $KMnO_4$ is a strong oxidizing agent. Note from Reaction (11-53) that the addition is a *cis*-addition and not a *trans*-addition as observed using an aqueous ring opening of an epoxide.

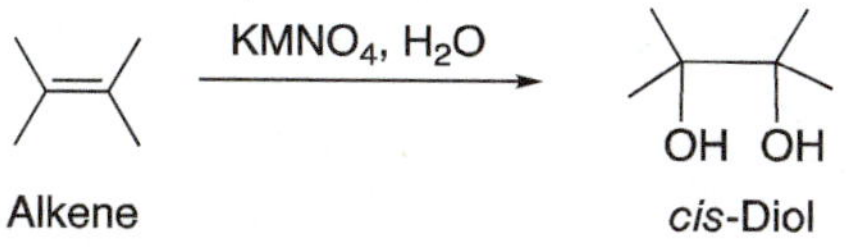

Alkene $\xrightarrow{KMNO_4, H_2O}$ *cis*-Diol $\qquad$ (11-53)

In order to understand the stereochemical outcome of this type of reaction, we will have to examine the reaction mechanism. In the first step of the reaction mechanism, the $KMnO_4$ adds to the alkene double bond as shown in Reaction (11-54).

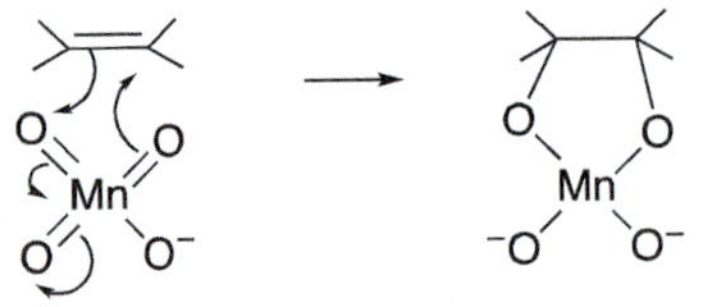

$$（11\text{-}54）$$

In the next step of the reaction mechanism, water hydrolyzes the Mn complex to give the cis-diol and MnO_2, as shown in Reaction (11-55).

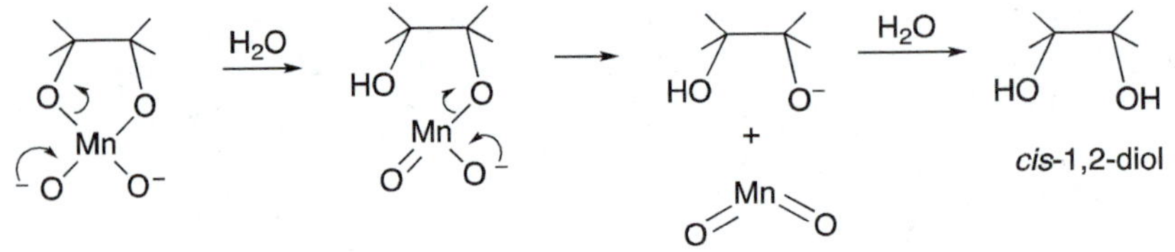

$$（11\text{-}55）$$

Note that the *syn*-addition results because both oxygen atoms of the diol are from the oxidizing reagent, $KMnO_4$. Based on the above observations, you should be able to predict the organic product of any oxidation reaction involving $KMnO_4$ and any alkene at room temperature. An example of an oxidation reaction of cyclohexene is shown in Reaction (11-56). Note the stereochemistry of the OH groups in the product.

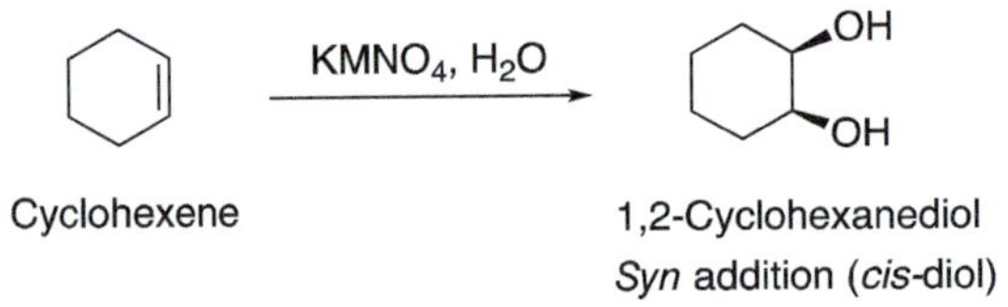

$$（11\text{-}56）$$

Cyclohexene → 1,2-Cyclohexanediol
Syn addition (*cis*-diol)

11.4.3 Oxidation of Alkenes with OsO_4

The reaction of an alkene with osmium tetroxide (OsO_4) at room temperature proceeds without cleavage of the double bond. Like $KMnO_4$, OsO_4 is a strong oxidizing agent, and as you can suspect from the similar structure of both oxidizing reagents, the addition is also a *cis*-diol addition as shown in Reaction (11-57).

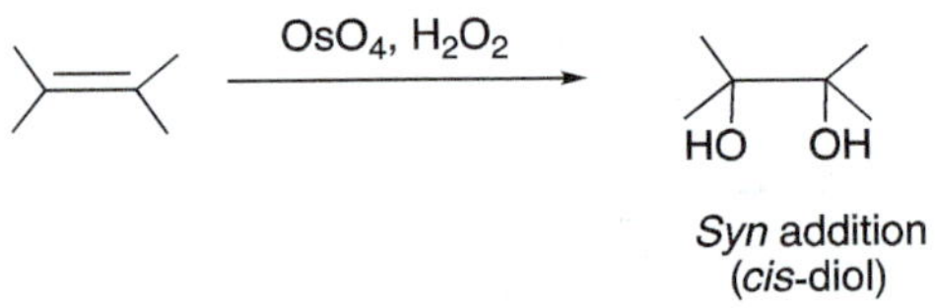

$$（11\text{-}57）$$

Syn addition
(*cis*-diol)

A compressed mechanism to show the cis-addition for this reaction is shown in Reaction (11-58).

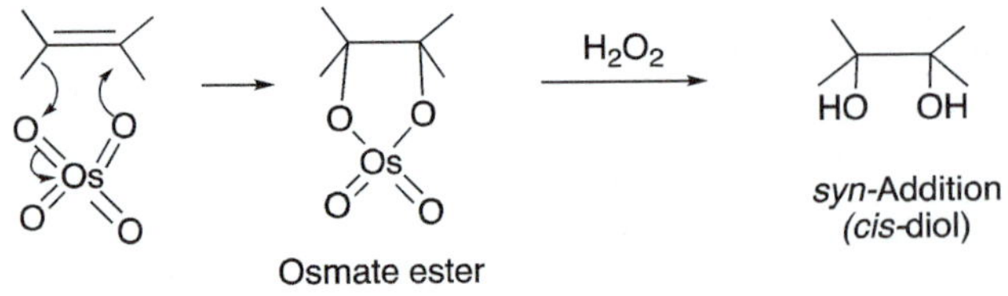

$$（11\text{-}58）$$

Osmate ester

syn-Addition
(*cis*-diol)

Osmium tetroxide in the presence of hydrogen peroxide can be used to give the same 1,2-*cis* diol from cyclohexene in which $KMnO_4$ was used, Reaction (11-56). Reaction (11-59) shows the oxidation of cyclohexene with osmium tetroxide.

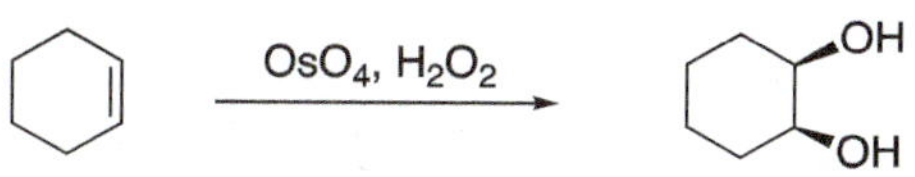

$$\text{Cyclohexene} \xrightarrow{\text{OsO}_4, \text{H}_2\text{O}_2} \text{1,2-Cyclohexanediol} \quad \textit{Syn addition (cis-diol)} \tag{11-59}$$

For the oxidation of alkene, with KMnO₄ (room temperature), or with OsO₄ in the presence of hydrogen peroxide the reaction proceeds without cleavage of the double bond and the product is a *syn*-addition of two OH groups to form a *cis*-diol.

Problem 11.11

i) Give the final organic products for each of the following reactions. Show appropriate stereochemistry of the final product.

(a) 1-methylcyclohexene $\xrightarrow[\text{Room temp.}]{\text{KMnO}_4, \text{OH}^-}$ (b) 2-methyl-2-butene $\xrightarrow[\text{Room temp.}]{\text{KMnO}_4, \text{OH}^-}$

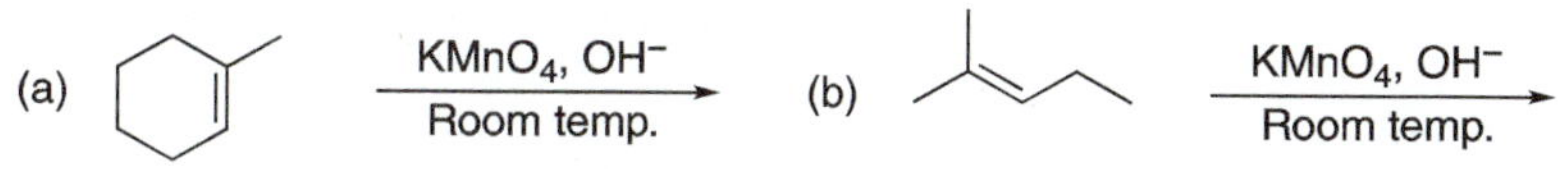

ii) Give the final organic products for each of the following reactions. Show appropriate stereochemistry of the final product.

(a) 1-methylcyclopentene $\xrightarrow{\text{OsO}_4, \text{H}_2\text{O}_2}$? (b) trans-2-pentene $\xrightarrow{\text{OsO}_4, \text{H}_2\text{O}_2}$?

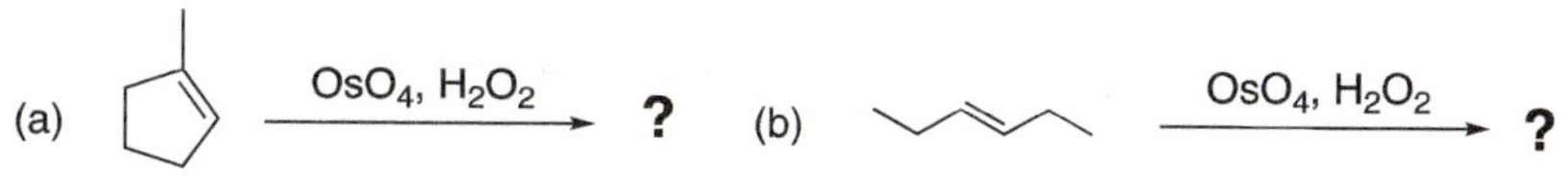

11.5 Oxidation of Alkenes with Bond Cleavage

There are two types of oxidation reactions that proceed with cleavage of the double bond; the first occurs when using KMnO₄, the same oxidizing reagent used in the previous section, but if the reaction is carried out at elevated temperatures, bond cleavage occurs. The second oxidation reaction of alkenes that occurs with bond cleavage is by using the oxidizing reagent ozone (O₃), followed by reduction using $\text{Zn}^{2+}/\text{H}_3\text{O}^+$ or $(\text{CH}_3)_2\text{S}$. As a result, this type of reaction is called ozonolysis. Reaction (11-60) gives a general representation of oxidation reactions that occur with bond cleavage.

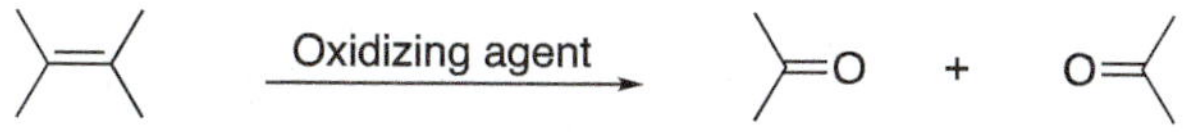

$$\text{Alkene} \xrightarrow{\text{Oxidizing agent}} \text{Carbonyl compounds} \tag{11-60}$$

As we have seen from previous reactions, the products that are actually obtained depend to a large degree on the oxidizing reagent used and reaction conditions. In this section, we will examine different oxidizing agents and reaction conditions necessary to accomplish the specific type of reactions given in Reaction (11-60). Each type of reaction will be examined in the next section, along with their mechanisms.

11.5.1 Oxidation of Alkenes with KMnO₄ at Elevated Temperatures

The general oxidation of alkenes with KMnO₄ in the presence of heat is shown in Reaction (11-61).

$$\text{Alkene} \xrightarrow{\text{KMnO}_4,\ \text{heat}} \text{Carbonyl compounds} \quad (11\text{-}61)$$

A specific example of the oxidation of an alkene using a strong oxidizing reagent at elevated temperature is shown in Reaction (11-62).

$$\text{3-Ethyl-2-methyl-2-pentene} \xrightarrow{\text{KMnO}_4,\ \text{heat}} \text{Propanone} + \text{3-Pentanone} \quad (11\text{-}62)$$

Note that if one of the carbonyl products obtained contains a hydrogen bonded to the carbonyl carbon, further oxidation occurs in the presence of the strong oxidizing reagent, KMnO$_4$, to produce the carboxylic acid as illustrated in Reaction (11-63).

$$\text{Alkene} \xrightarrow{\text{KMnO}_4\ \text{Heat}} \text{Ketone} + [\text{Aldehyde (not isolated)}] \xrightarrow{\text{KMnO}_4\ \text{Heat}} \text{Carboxylic acid} \quad (11\text{-}63)$$

A specific example of the oxidation of an alkene that has one hydrogen bonded to the carbon–carbon double bond and a methyl group bonded to the other carbon of the double bond is shown in Reaction (11-64).

$$\text{3-Methyl-3-hexene} \xrightarrow{\text{KMnO}_4\ \text{Heat}} \text{2-Butanone} + \text{Propanoic acid} \quad (11\text{-}64)$$

The outcome of the oxidation of this unsymmetrical alkene is that two different products are produced, one a carboxylic acid and the other a ketone. Reaction (11-65) shows the oxidation of an alkene that has one hydrogen bonded to each of the carbons of the carbon–carbon double bond. The aldehyde that is initially produced is readily oxidized into the carboxylic acid.

$$\text{3-Hexene} \xrightarrow{\text{KMnO}_4\ \text{Heat}} \text{Propanoic acid (2 moles)} \quad (11\text{-}65)$$

Reaction (11-66) shows the oxidation of a cyclic alkene in which the carbon–carbon double bond has a methyl and hydrogen bonded to each of the carbons. The result of this oxidation is an open chain difunctional molecule.

$$\text{methylcyclohexene} \xrightarrow{\text{KMnO}_4\ \text{Heat}} \text{H}_3\text{C-CO-CH}_2\text{CH}_2\text{CH}_2\text{-COOH} \quad (11\text{-}66)$$

An examination of the reaction mechanism gives the rationale for the products formed. The first step of the reaction mechanism is given in Section 11.4.2, Reaction (11-54), but in the presence of heat, the intermediate decomposes as shown in Reaction (11-67).

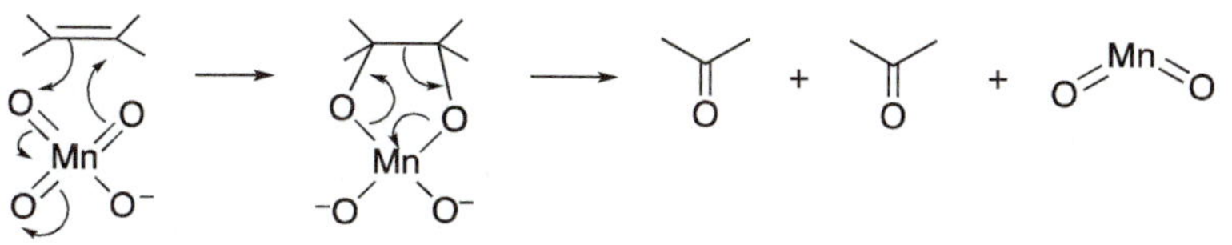

Problem 11.12

Determine the reactant alkene needed to complete the following reactions.

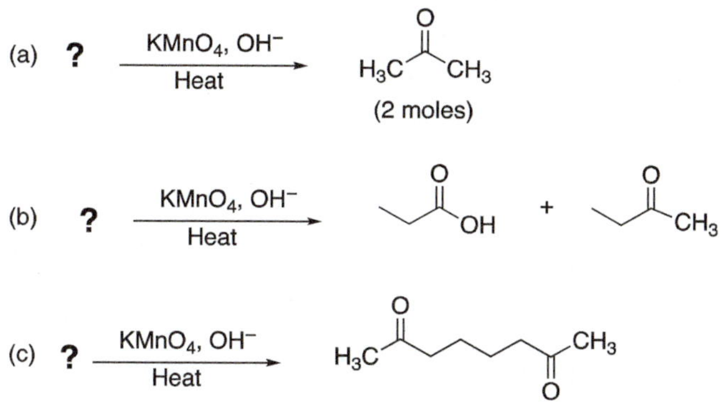

(a) ? $\xrightarrow{\text{KMnO}_4,\ \text{OH}^-,\ \text{Heat}}$ H₃C–C(=O)–CH₃ (2 moles)

(b) ? $\xrightarrow{\text{KMnO}_4,\ \text{OH}^-,\ \text{Heat}}$ CH₃CH₂C(=O)OH + CH₃CH₂C(=O)CH₃

(c) ? $\xrightarrow{\text{KMnO}_4,\ \text{OH}^-,\ \text{Heat}}$ H₃C–C(=O)–(CH₂)₃–C(=O)–CH₃

11.5.2 Ozonolysis of Alkenes

The oxidation of the carbon–carbon double bond with ozone (O_3), followed by a reaction with dimethyl sulfide, results in the cleavage of the double bond. The reaction conditions for this oxidation reaction are mild enough that if an aldehyde is produced as a product, it is not oxidized further to form the carboxylic acidic as we have seen using $KMnO_4$ and heat. The general reaction is shown below.

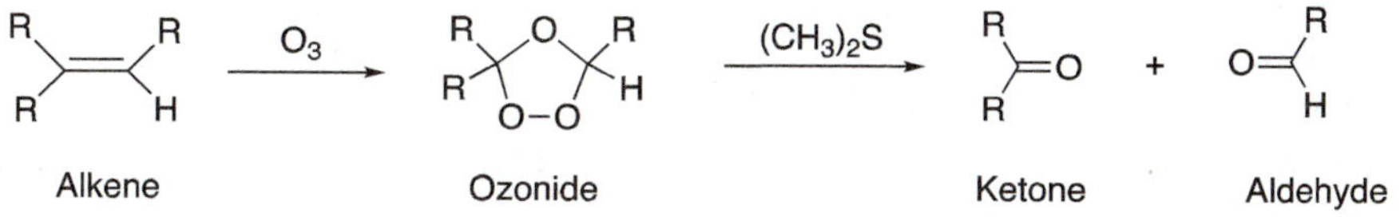

Alkene $\xrightarrow{O_3}$ Ozonide $\xrightarrow{(CH_3)_2S}$ Ketone + Aldehyde (11-68)

The mechanism for this reaction is shown below; this first step involves an addition of the ozone to the alkene, followed by the formation of an ozonide.

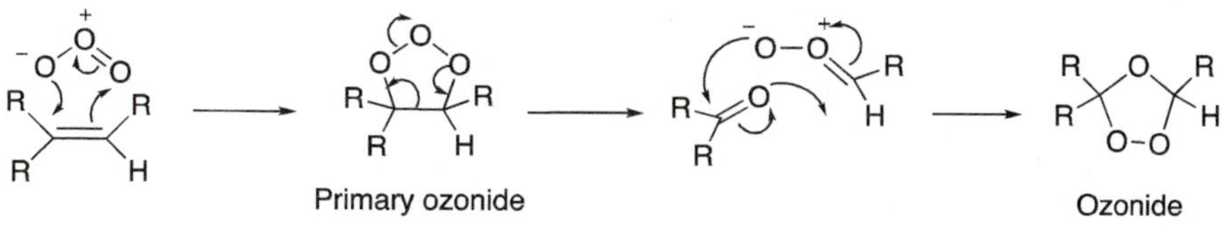

Primary ozonide → Ozonide (11-69)

We will not go into the details of the mechanism for the reaction of the ozonide, but the general reaction is shown in Reaction (11-70).

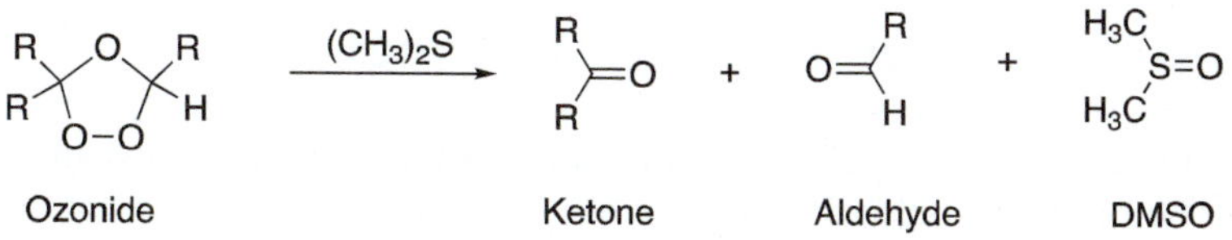

Ozonide $\xrightarrow{(CH_3)_2S}$ Ketone + Aldehyde + DMSO (11-70)

Reaction (11-71) gives an example of an ozonolysis reaction.

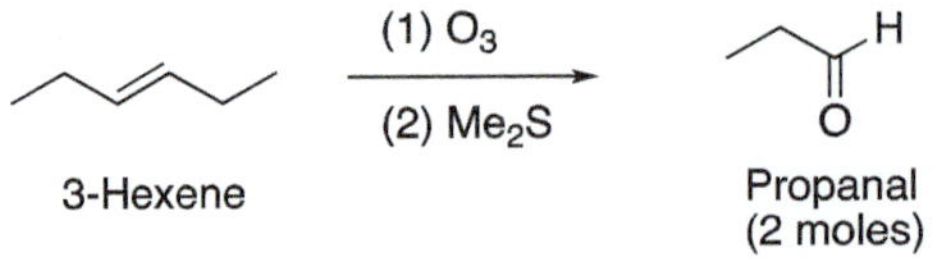

$$\text{3-Hexene} \xrightarrow[\text{(2) Me}_2\text{S}]{\text{(1) O}_3} \text{Propanal (2 moles)} \quad (11\text{-}71)$$

Note that there are two moles of the aldehyde produced since the starting compound is a symmetrical alkene. On the other hand, if the alkene were not symmetrical, two products result, as shown in the examples given in Reactions (11-72) and (11-73).

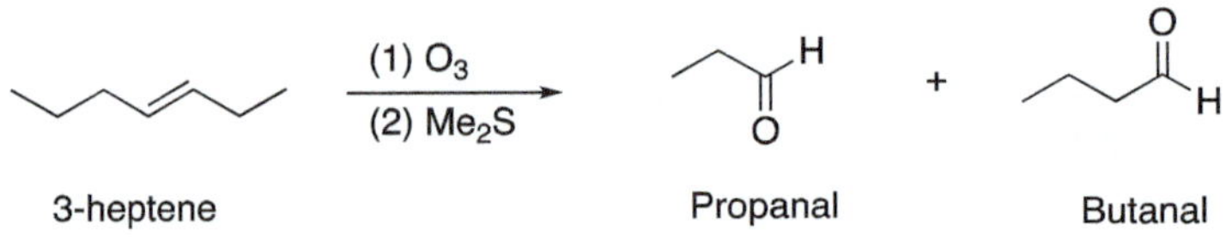

$$\text{3-heptene} \xrightarrow[\text{(2) Me}_2\text{S}]{\text{(1) O}_3} \text{Propanal} + \text{Butanal} \quad (11\text{-}72)$$

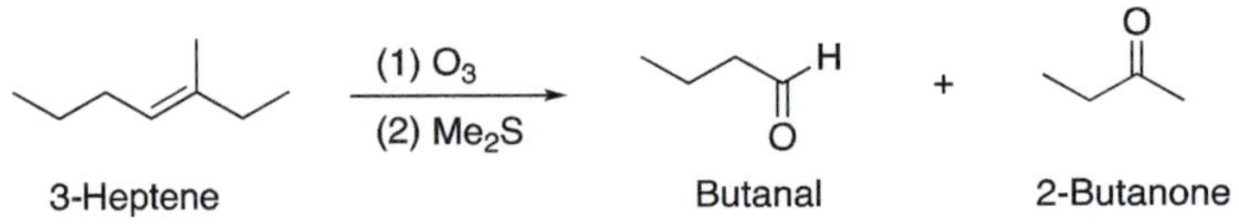

$$\text{3-Heptene} \xrightarrow[\text{(2) Me}_2\text{S}]{\text{(1) O}_3} \text{Butanal} + \text{2-Butanone} \quad (11\text{-}73)$$

If the alkene is part of a ring, a dicarbonyl compound results, as shown in Reaction (11-74).

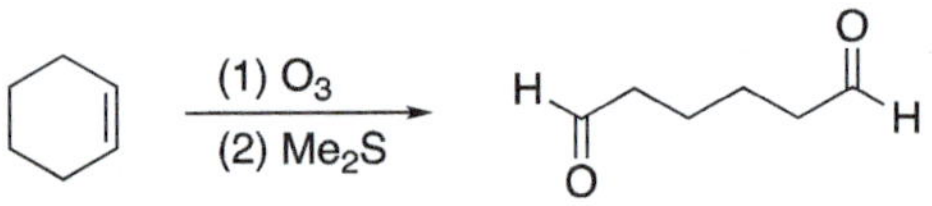

$$\text{cyclohexene} \xrightarrow[\text{(2) Me}_2\text{S}]{\text{(1) O}_3} \text{OHC-(CH}_2)_4\text{-CHO} \quad (11\text{-}74)$$

Based on the expected products of this type of reaction, you should be able to utilize this information to determine a possible starting compound. In fact, this technique is often used to determine the structure of an unknown alkene. Based on the product obtained, one can easily determine the original reactant. The next problem is designed to test your ability to make this determination.

Problem 11.13

i) An unknown compound is treated by ozone, followed by Me$_2$S using the reaction shown below, what is the structure of the reactant?

$$? \xrightarrow[\text{(2) Me}_2\text{S}]{\text{(1) O}_3} \text{(ketone-aldehyde product)}$$

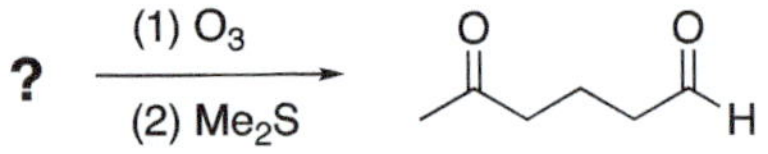

ii) An unknown compound is treated by ozone, followed by Me$_2$S using the reaction shown below, what is the structure of the reactant?

$$? \xrightarrow[\text{(2) Me}_2\text{S}]{\text{(1) O}_3} \text{(ketone)} + \text{(aldehyde)}$$

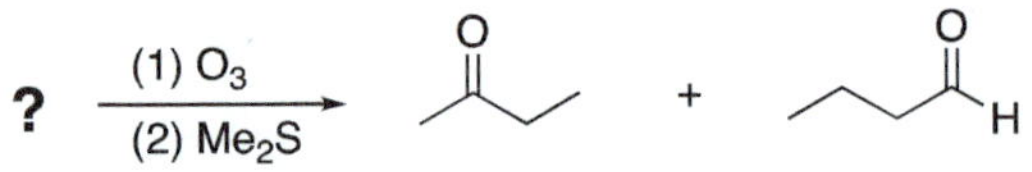

11.6 Applications of Oxidation Reactions of Alkenes

In this section, we apply our knowledge of oxidation reactions to determine the structures of unknown starting compound by thinking in reverse, this process requires analytical thinking! The first type of reaction that we will examine is the use of KMNO$_4$ for the oxidation of alkenes.

Reaction (11-75) shows the oxidation of an unknown starting compound, and it was demonstrated from a separate experiment that the unknown compound discolored bromine in the presence of carbon tetrachloride as solvent.

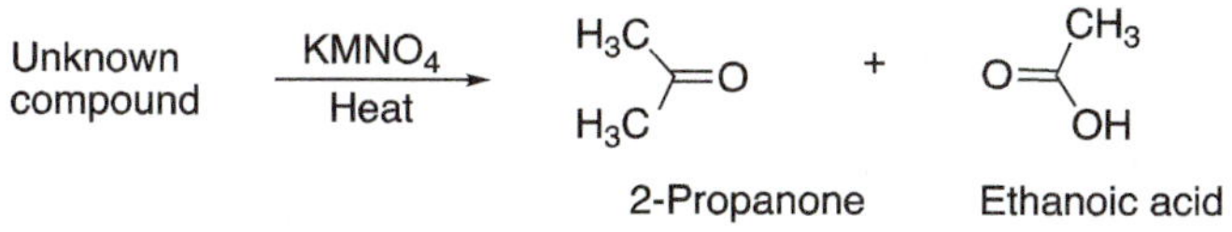

$$\text{Unknown compound} \xrightarrow[\text{Heat}]{KMnO_4} \underset{\text{2-Propanone}}{(CH_3)_2C{=}O} + \underset{\text{Ethanoic acid}}{CH_3COOH} \quad (11\text{-}75)$$

In order to determine the structure of the unknown starting reactant, we will have to make some very important observations before a conclusion can be made about the structure of this unknown compound. The first is that that it discolors bromine solution. You will recall from Chapter 8 that the reddish color of a solution of bromine solution becomes colorless when added to alkenes due to the addition reaction that occurs. Thus, this information implies that the unknown compound is an alkene. The next observation is that the reagent that is used for reaction in Reaction (11-75) is a strong oxidizing reagent in the presence of heat, which would cleave a carbon–carbon double bond to produce the products shown. Based on the structures of the products, the conclusion is that the unknown compound has the structure shown in Reaction (11-76), which also shows the carbon–carbon double bond that must be cleaved to produce the products shown.

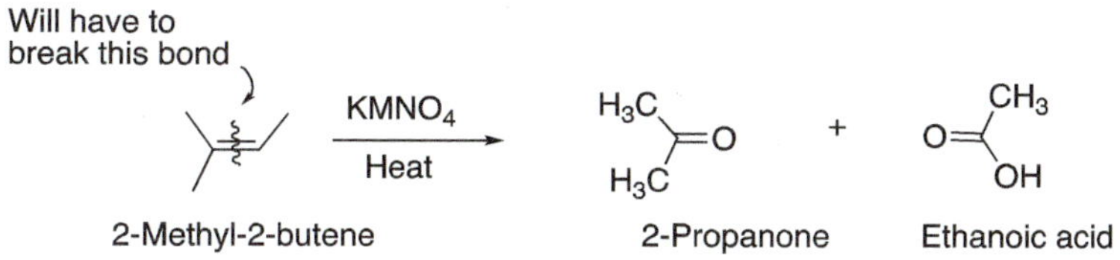

$$\underset{\text{2-Methyl-2-butene}}{\text{(Will have to break this bond)}} \xrightarrow[\text{Heat}]{KMnO_4} \underset{\text{2-Propanone}}{(CH_3)_2C{=}O} + \underset{\text{Ethanoic acid}}{CH_3COOH} \quad (11\text{-}76)$$

Problem 11.14

For each of the following pairs of compounds, determine the reactants based on the reaction with KMnO$_4$ at elevated temperatures.

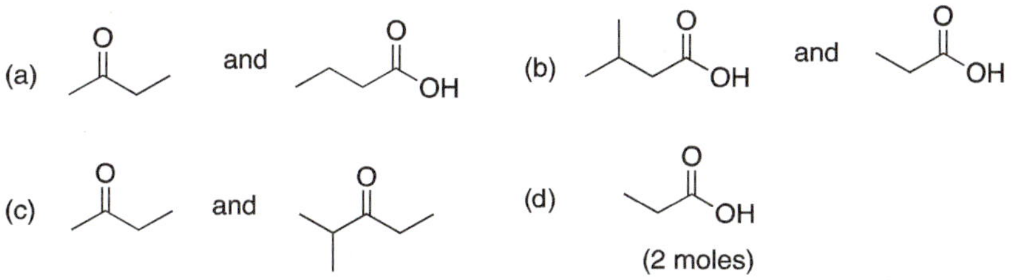

(a) [ketone] and [carboxylic acid with OH]

(b) [ketone with OH] and [carboxylic acid with OH]

(c) [ketone] and [ketone]

(d) [carboxylic acid with OH] (2 moles)

Another example of the application of analytical thinking to determine the structure of the reactant of an oxidation reaction is shown in Reaction (11-77). For this reaction, an unknown compound gave only one product and it is difunctional and was shown from a separate experiment that the unknown compound is an alkene.

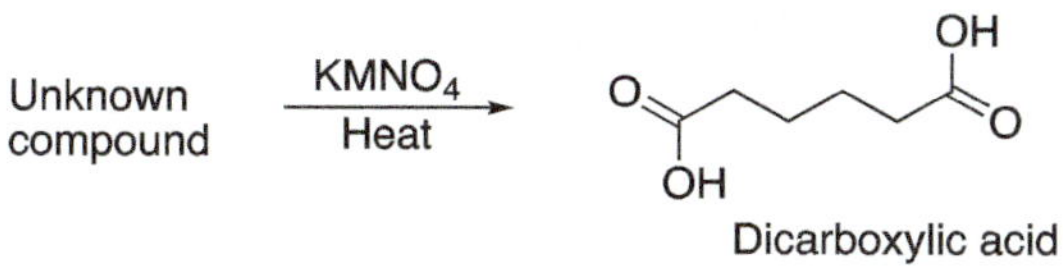

$$\text{Unknown compound} \xrightarrow[\text{Heat}]{KMnO_4} \text{Dicarboxylic acid} \quad (11\text{-}77)$$

Utilizing this information, we can conclude that since the only product obtained is a difunctional carboxylic acid, the unknown reactant is more than likely cyclic and is oxidized to the dialdehyde and further oxidized to the dicarboxylic acid as shown in Reaction (11-78).

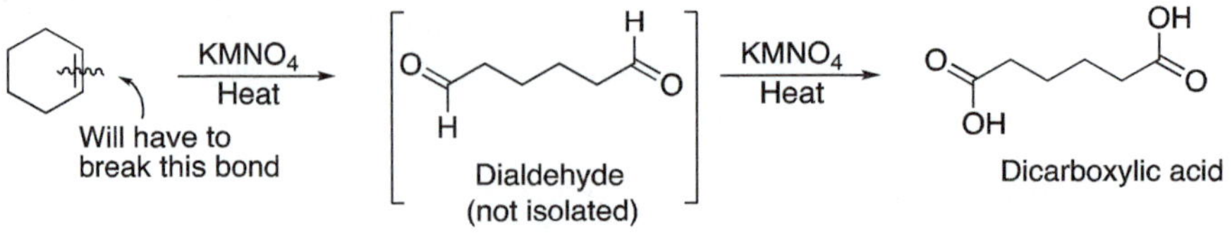

11 Oxidation Reactions in Organic Chemistry

$$\text{(cyclohexene)} \xrightarrow[\text{Heat}]{\text{KMNO}_4} [\text{Dialdehyde (not isolated)}] \xrightarrow[\text{Heat}]{\text{KMNO}_4} \text{Dicarboxylic acid} \quad (11\text{-}78)$$

Will have to break this bond

Again, since there is only one product obtained, it is obvious that the reactant is a cyclic alkene and that it has six carbons. A good conclusion is that it has to be cyclohexene.

Problem 11.15

The following difunctional molecules were obtained from separate reactions carried out in KMnO$_4$ at elevated temperatures. Determine the structure of the starting alkenes.

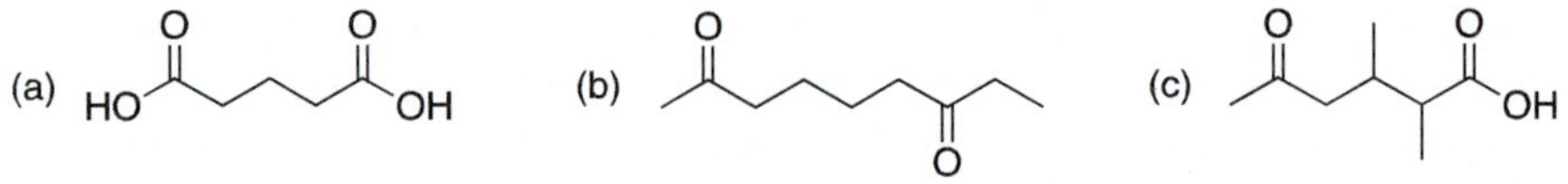

(a) HOOC–CH$_2$–CH$_2$–COOH   (b) CH$_3$–CO–CH$_2$–CH$_2$–CH$_2$–CO–CH$_2$–CH$_3$ (c) CH$_3$–CO–CH$_2$–CH(CH$_3$)–COOH

The same analysis can be carried out for ozonolysis to determine the structure of unknown compounds, but for these reactions, aldehydes can be isolated, as shown for the reactions shown in Reaction (11-79).

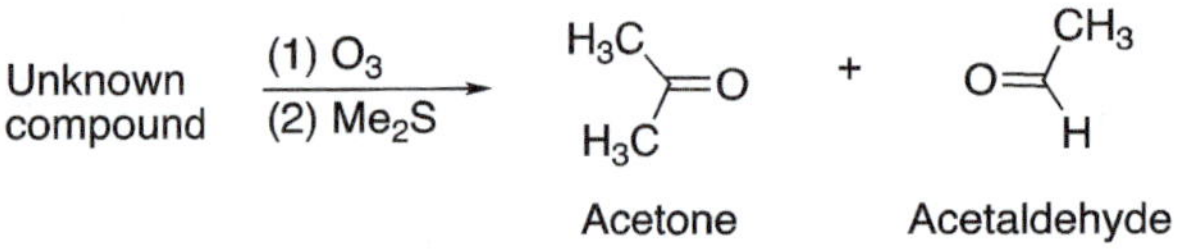

$$\text{Unknown compound} \xrightarrow[\text{(2) Me}_2\text{S}]{\text{(1) O}_3} \underset{\text{Acetone}}{(CH_3)_2C{=}O} + \underset{\text{Acetaldehyde}}{CH_3CH{=}O} \quad (11\text{-}79)$$

Using a similar approach as outlined for the use of KMnO$_4$, the conclusion of the unknown reactant is shown in the reaction given in Reaction (11-80).

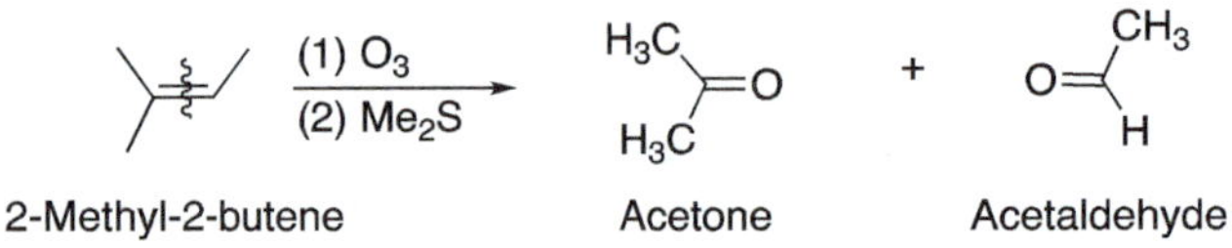

$$\underset{\text{2-Methyl-2-butene}}{(CH_3)_2C{=}CHCH_3} \xrightarrow[\text{(2) Me}_2\text{S}]{\text{(1) O}_3} \underset{\text{Acetone}}{(CH_3)_2C{=}O} + \underset{\text{Acetaldehyde}}{CH_3CH{=}O} \quad (11\text{-}80)$$

Another example is shown in Reaction (11-81), but in this case only one production is obtained and it is a dialdehyde.

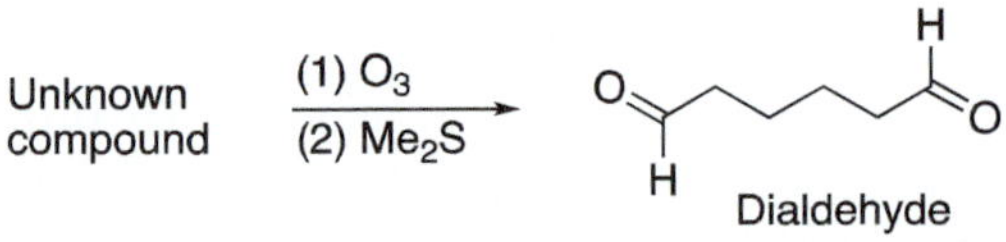

$$\text{Unknown compound} \xrightarrow[\text{(2) Me}_2\text{S}]{\text{(1) O}_3} \text{Dialdehyde} \quad (11\text{-}81)$$

Since there is only one product obtained, it is obvious that the reactant is a cyclic alkene and that it has six carbons. A good conclusion is that the unknown reactant is cyclohexene, as shown in Reaction (11-82).

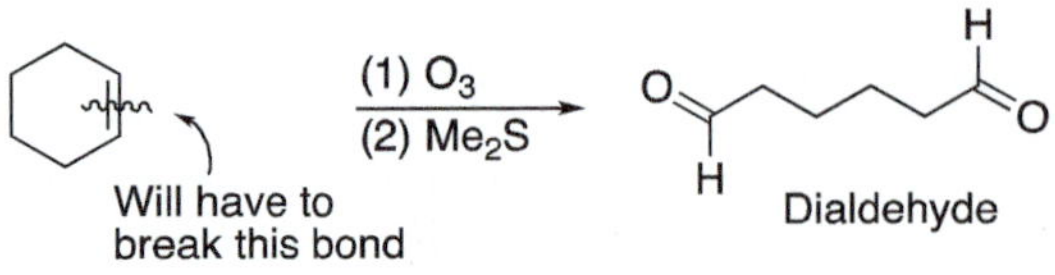

$$\text{(cyclohexene)} \xrightarrow[\text{(2) Me}_2\text{S}]{\text{(1) O}_3} \text{Dialdehyde} \quad (11\text{-}82)$$

Will have to break this bond

Problem 11.16

The following products were obtained from separate ozonolysis reactions of unknown compounds. Determine the structure of the starting alkene for each unknown.

(a) CH₃CH₂-C(=O)-CH₂CH₃ and O=CH-CH₂-C(=O)-H

(b) CH₃-C(=O)-CH₂-CH(=O) and CH₃-CH₂-C(=O)-H

(c) CH₃CH₂-C(=O)-CH₂CH₃ and (CH₃)₂CH-C(=O)-CH₂CH₃

(d) CH₃CH₂-C(=O)-H (2 moles)

(e) H-C(=O)-CH(CH₃)-CH₂-C(=O)-H

(f) CH₃-C(=O)-CH₂CH₂CH₂-C(=O)-CH₂CH₃

11.7 Oxidation of Alkynes

Like alkenes, alkynes can be oxidized, but as you can imagine, the oxidizing reagent has to be a very strong oxidizing agent since the carbon–carbon triple bond is a very strong covalent bond. As mentioned earlier, KMnO₄ is a strong oxidizing agent, and it can be used to oxidize alkynes by adding two moles of —OH groups across the triple bond as shown in Reaction (11-83).

$$R-C\equiv C-R \xrightarrow{KMnO_4} R-\underset{\underset{OH}{|}}{\overset{\overset{OH}{|}}{C}}-\underset{\underset{OH}{|}}{\overset{\overset{OH}{|}}{C}}-R \tag{11-83}$$

The product of Reaction (11-84) is not very stable and, in the presence of an acid, protonation of one of the OH groups occurs and converts it to a good leaving group and the result is the formation of a carbonyl functionality as shown in Reaction (11-84).

$$R-\underset{\underset{OH}{|}}{\overset{\overset{OH}{|}}{C}}-\underset{\underset{OH}{|}}{\overset{\overset{OH}{|}}{C}}-R \xrightarrow{H_3O^+} R-\underset{\underset{:OH}{|}}{\overset{\overset{H_2O^+}{|}}{C}}-\underset{\underset{OH}{|}}{\overset{\overset{OH}{|}}{C}}-R \xrightarrow{-H_3O^+} R-\underset{\underset{O}{\parallel}}{C}-\underset{\underset{OH}{|}}{\overset{\overset{OH}{|}}{C}}-R \tag{11-84}$$

As you can imagine, a similar reaction will take place at the other carbon, which contains two OH groups, as shown in Reaction (11-85).

$$R-\underset{\underset{O}{\parallel}}{C}-\underset{\underset{OH}{|}}{\overset{\overset{OH}{|}}{C}}-R \xrightarrow{H_3O^+} R-\underset{\underset{O}{\parallel}}{C}-\underset{\underset{:OH}{|}}{\overset{\overset{H_2O^+}{|}}{C}}-R \xrightarrow[-H^+]{-H_2O} R-\underset{\underset{O}{\parallel}}{C}-\underset{\underset{O}{\parallel}}{C}-R \tag{11-85}$$

Diketone

The overall reaction of an internal alkyne in the presence of an oxidizing reagent, such as KMnO₄, results in a diketone as shown in the example in Reaction (11-86).

$$CH_3CH_2-C\equiv C-CH_2CH_3 \xrightarrow{KMnO_4, H_2O} CH_3CH_2-C(=O)-C(=O)-CH_2CH_3 \tag{11-86}$$

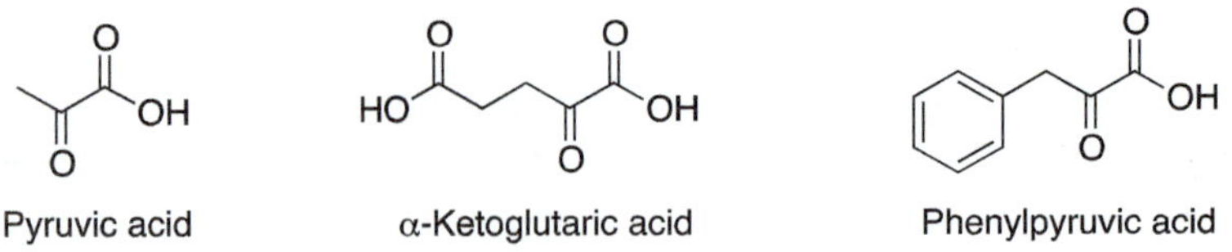

Pyruvic acid   **α-Ketoglutaric acid**   **Phenylpyruvic acid**

Figure 11.1 Important biological α-keto acids.

Reactions involving terminal alkynes result in α-keto acids as shown in Reactions (11-87) and (11-88).

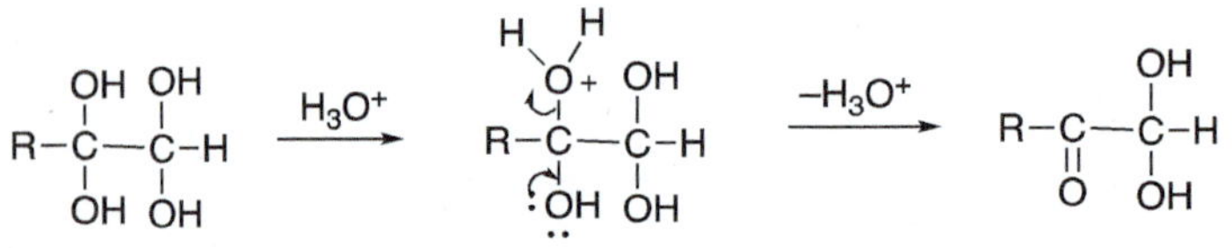

$$\text{R-C(OH)(OH)-C(OH)(OH)-H} \xrightarrow{H_3O^+} \text{R-C(OH_2^+)(OH)-C(OH)(OH)-H} \xrightarrow{-H_3O^+} \text{R-C(=O)(OH)-C(OH)-H} \quad (11\text{-}87)$$

In the next step of the mechanism, one of the OH groups is protonated and converts it to a good leaving group to generate an α-keto acid as shown in Reaction (11-88).

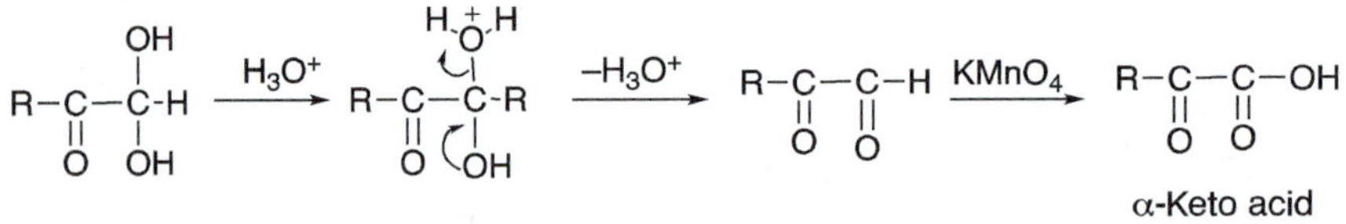

$$\text{R-C(=O)-C(OH)(OH)-H} \xrightarrow{H_3O^+} \text{R-C(=O)-C(OH_2^+)(OH)-R} \xrightarrow{-H_3O^+} \text{R-C(=O)-C(=O)-H} \xrightarrow{KMnO_4} \text{R-C(=O)-C(=O)-OH} \quad (11\text{-}88)$$

α-Keto acid

One of the most important keto acids in biology is pyruvic acid, which is important for glycolysis. Common keto acids are shown in Figure 11.1.

As you might expect, under more severe conditions of oxidation, alkynes can be oxidized to give the cleaved products, as shown in the reaction given in Reaction (11-89). The mechanism for such cleavage is similar to the mechanism for the cleavage of the double bonds of alkenes, which was described earlier.

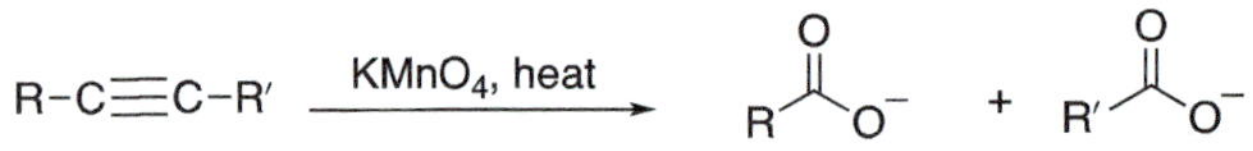

$$\text{R-C≡C-R'} \xrightarrow{KMnO_4, \text{ heat}} \text{R-CO-O}^- + \text{R'-CO-O}^- \quad (11\text{-}89)$$

Problem 11.17

Give the products for the following reactions.

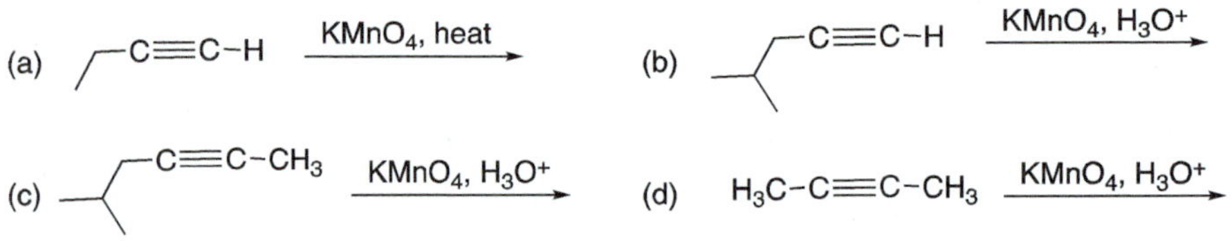

(a) R-C≡C-H $\xrightarrow{KMnO_4, \text{ heat}}$

(b) R-C≡C-H $\xrightarrow{KMnO_4, H_3O^+}$

(c) R-C≡C-CH$_3$ $\xrightarrow{KMnO_4, H_3O^+}$

(d) H$_3$C-C≡C-CH$_3$ $\xrightarrow{KMnO_4, H_3O^+}$

11.8 Oxidation of Aromatic Compounds

Aromatic compounds are cyclic compounds with a specific number of conjugated double bonds; these compounds will be covered in more detail in Chapter 17. Aromatic compounds are exceptionally stable and very resistant to many reaction conditions. In fact, some aromatic compounds such as benzene and toluene are used as solvents. In the presence of an oxidizing agent, they can be oxidized as shown in Reaction (11-90).

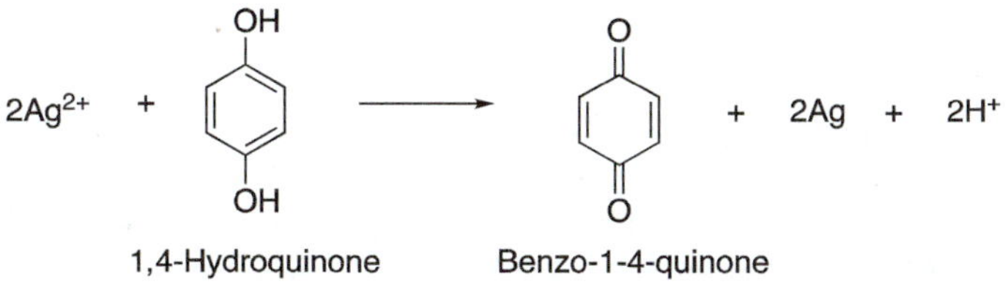

$$2Ag^{2+} + \text{1,4-Hydroquinone} \longrightarrow \text{Benzo-1-4-quinone} + 2Ag + 2H^+ \qquad (11\text{-}90)$$

There are many important compounds in chemistry and biology in which the oxidized form of the quinone moiety is transformed into the reduced form. A classic example is that of coenzyme Q, also known as ubiquinone, structure shown below. Upon reduction, ubiquinone is transformed into the reduced form, ubiquinol as shown in Reaction (11-91).

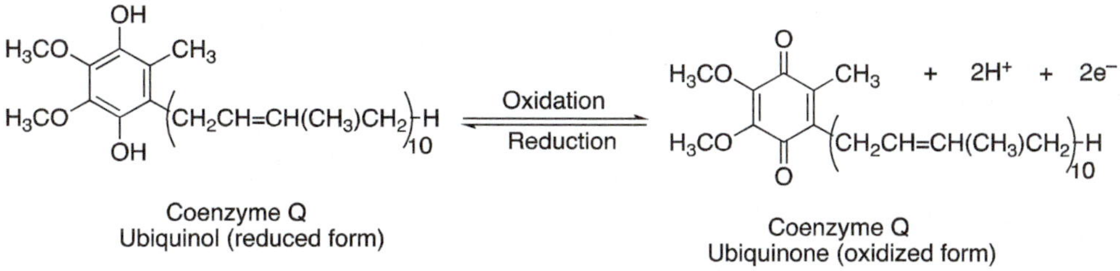

$$\text{Coenzyme Q Ubiquinol (reduced form)} \xrightleftharpoons[\text{Reduction}]{\text{Oxidation}} \text{Coenzyme Q Ubiquinone (oxidized form)} + 2H^+ + 2e^- \qquad (11\text{-}91)$$

Vitamin K_2 is another important aromatic system known as quinone, its structure is shown below, note carefully the oxidized quinone moiety.

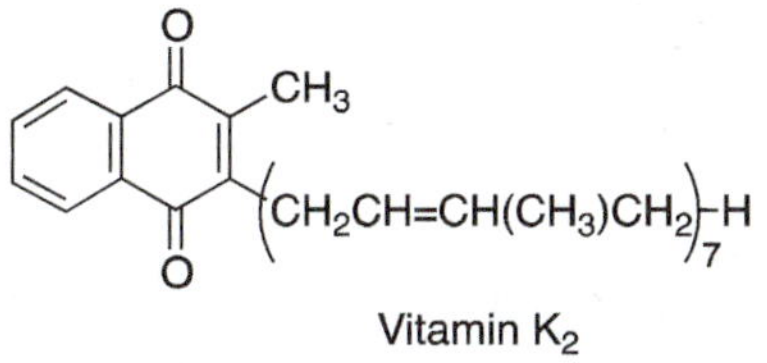

Vitamin K_2

Vitamin K_2 is essential for the synthesis of a blood-clotting agent prothrombin. A deficiency in Vitamin K_2 typically results in a fatal illness in which the blood is slow to clot. As a treatment for this problem, a modified form of Vitamin K_2, menadione, can be used. Menadione can be made from the oxidation of 2-methylnapthalene, as shown in Reaction (11-92).

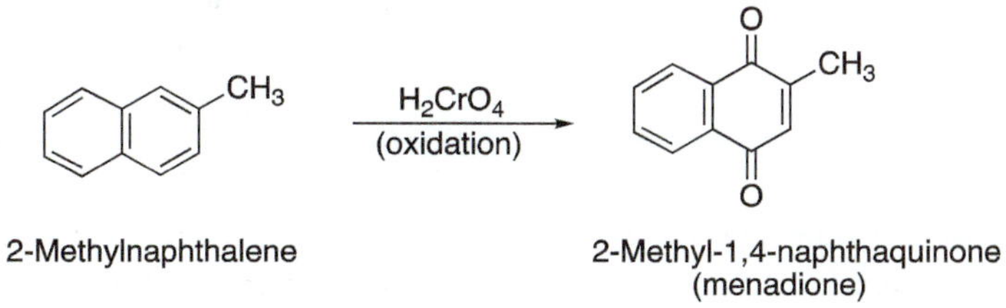

$$\text{2-Methylnaphthalene} \xrightarrow[\text{(oxidation)}]{H_2CrO_4} \text{2-Methyl-1,4-naphthaquinone (menadione)} \qquad (11\text{-}92)$$

11.9 Autooxidation of Ethers and Alkenes

Extreme care must be exercised in the lab when using diethyl ether as a solvent due to the possibility of auto oxidation of ether, which results in an explosive product. Reaction (11-93) gives the products on oxidation with oxygen in the presence of light; the peroxides produced from these reactions are extremely explosive.

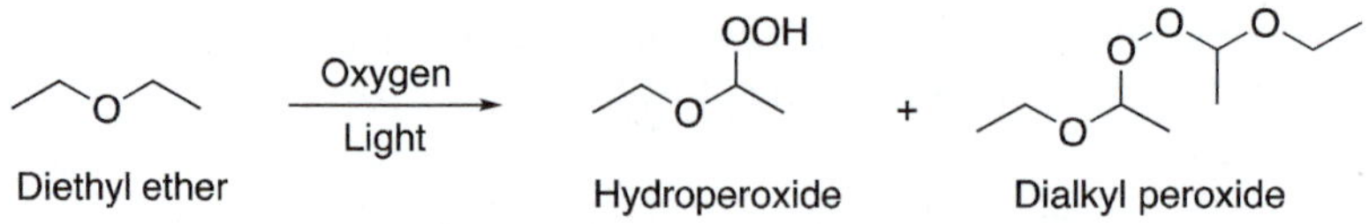

$$\text{Diethyl ether} \xrightarrow[\text{Light}]{\text{Oxygen}} \text{Hydroperoxide} + \text{Dialkyl peroxide} \tag{11-93}$$

Alkenes can undergo a similar oxidation with oxygen in the presence of light. Cooking oil, which contains unsaturated systems, in the presence of oxygen will undergo autooxidation and this oxidation explains the observation that if a bottle of cooking oil remains unopened for a long period of time, when it is eventually opened, air rushes in to replace oxygen used up due to the oxidation process. The oxygen present in the unopened bottle will be used up due to oxidation of the alkenes present in the oil; and Reaction (11-94) shows an oxidation reaction that is possible.

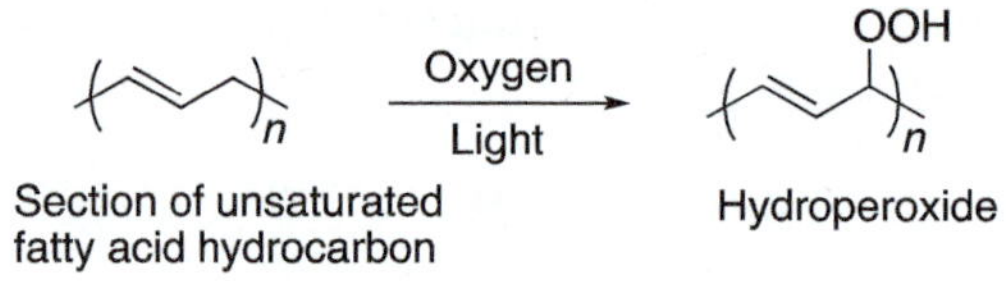

$$\text{Section of unsaturated fatty acid hydrocarbon} \xrightarrow[\text{Light}]{\text{Oxygen}} \text{Hydroperoxide} \tag{11-94}$$

The mechanism for the oxidation of alkenes by oxygen in the presence of light is shown below.

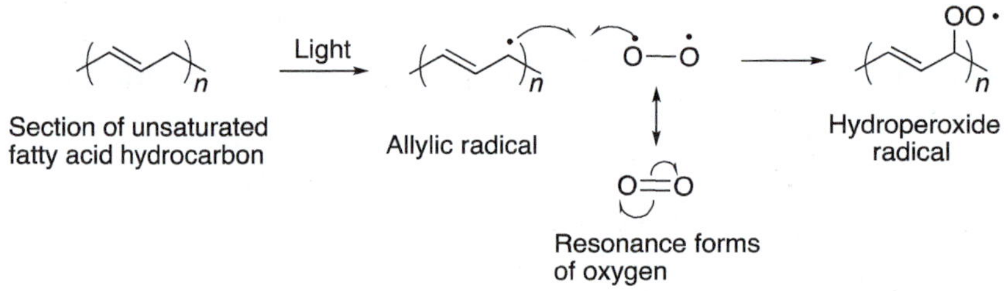

$$\text{Section of unsaturated fatty acid hydrocarbon} \xrightarrow{\text{Light}} \text{Allylic radical} \xrightarrow{\text{O-O}} \text{Hydroperoxide radical} \tag{11-95}$$

Resonance forms of oxygen

Once the radical is formed, it will react with oxygen to form a hydroperoxide radical, which abstracts a hydrogen atom from another molecule of the alkene to produce a new radical.

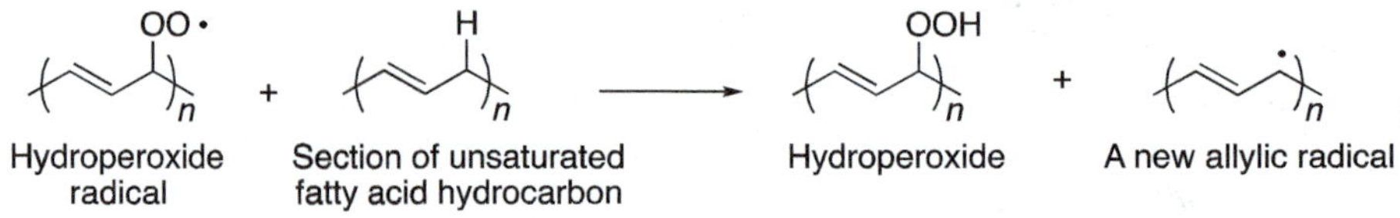

$$\text{Hydroperoxide radical} + \text{Section of unsaturated fatty acid hydrocarbon} \longrightarrow \text{Hydroperoxide} + \text{A new allylic radical} \tag{11-96}$$

11.10 Applications of Oxidation Reactions to Synthesis

We have covered a number of different reaction types and now we are ready to make strategic choices about appropriate combinations of reaction types and specific reactions that can be utilized to synthesize target molecules. First, let us try and carry out the following transformation shown in Reaction (11-97).

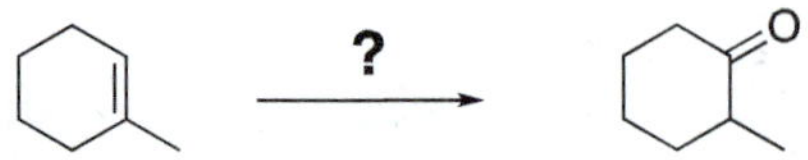

$$\text{cyclohexene} \xrightarrow{?} \text{2-cyclohexenone} \tag{11-97}$$

As pointed out in previous chapters, one of the best approaches to determine the appropriate type of reactions and reaction conditions to carry out a particular transformation is that it is best to carry out a complete analysis of the target molecule focusing on the functional group and deciding the best reaction for its synthesis. The functional group in the target molecule is a ketone and as we have discussed in this chapter, a ketone can be synthesized by an oxidation reaction of an alcohol. Thus, an appropriate choice of reaction would be to convert an alcohol to a ketone, as shown in Reaction (11-98).

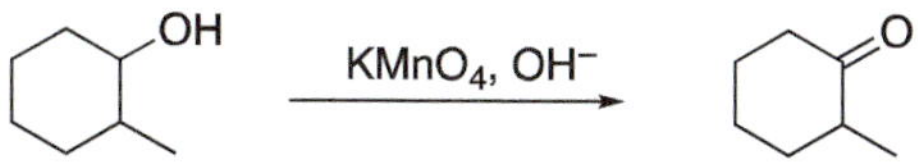

$$\text{cyclohexanol with methyl} \xrightarrow{KMnO_4, OH^-} \text{cyclohexanone with methyl} \tag{11-98}$$

Of course, there are other possible oxidizing regents that could have been used, including $Na_2Cr_2O_7/H_2SO_4$, PCC, and so on. Now that we have figured out the last step in this transformation, let us carry out a similar analysis to determine how best to synthesize the alcohol that is needed to make the target molecule. Reflecting on the reactions that we have studied, one type that comes to mind is an addition reaction involving akenes, more specifically, hydration of alkenes, but we will have to be careful since hydration can take place in one of two ways: Markovnikov or anti-Markovnikov addition. By looking at the structure of the starting material, the addition is an anti-Markovnikov addition, and the appropriate reagents are shown in Reaction (11-99).

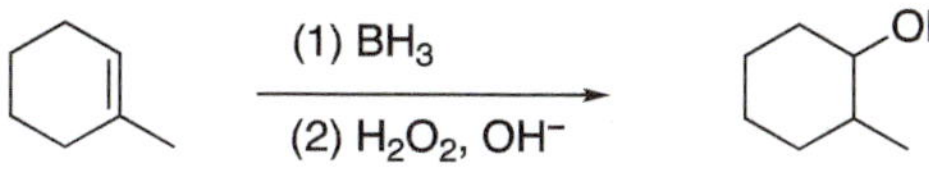

$$\text{methylcyclohexene} \xrightarrow[\text{(2) } H_2O_2, OH^-]{\text{(1) } BH_3} \text{cyclohexanol with methyl} \tag{11-99}$$

Thus, the overall sequence of reactions for the transformation is shown in Reaction (11-100).

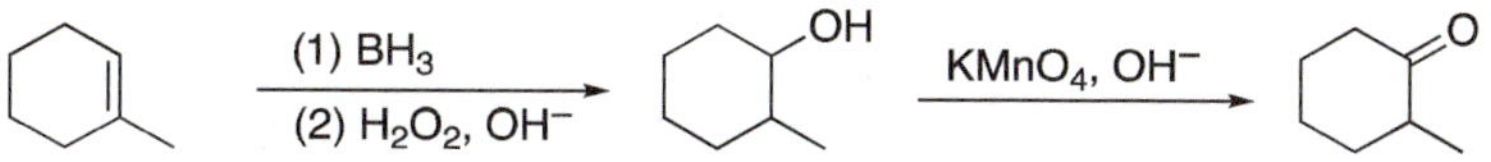

$$\text{methylcyclohexene} \xrightarrow[\text{(2) } H_2O_2, OH^-]{\text{(1) } BH_3} \text{cyclohexanol} \xrightarrow{KMnO_4, OH^-} \text{cyclohexanone} \tag{11-100}$$

Note that an alternate way of writing this sequence of reactions is shown in Reaction (11-101), where numbers indicate the separate reaction steps in the sequence of reactions.

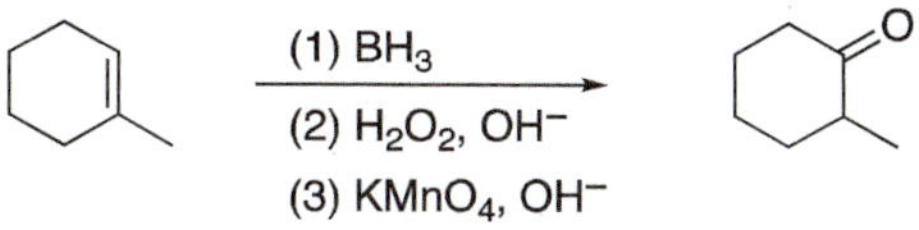

$$\text{methylcyclohexene} \xrightarrow[\substack{\text{(2) } H_2O_2, OH^- \\ \text{(3) } KMnO_4, OH^-}]{\text{(1) } BH_3} \text{cyclohexanone} \tag{11-101}$$

Problem 11.18

i) One of the constituents of turpentine is α-pinene, formula $C_{10}H_{16}$. Based on the reaction scheme shown below, give the structures of compounds **A–D**; show stereochemistry where appropriate.

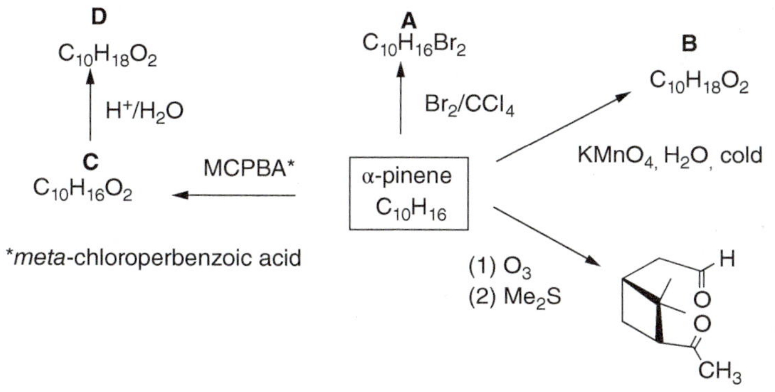

D $C_{10}H_{18}O_2$ ← H^+/H_2O — **C** $C_{10}H_{16}O_2$ ← MCPBA* — α-pinene $C_{10}H_{16}$ — Br_2/CCl_4 → **A** $C_{10}H_{16}Br_2$

α-pinene → $KMnO_4, H_2O$, cold → **B** $C_{10}H_{18}O_2$

α-pinene → (1) O_3 (2) Me_2S → [bicyclic dialdehyde/ketone structure with H, O, O, CH_3]

*meta-chloroperbenzoic acid

ii) Show how to carry out the following transformations. Clearly show each step in your synthesis.

(a)  isobutylene → 2-methylpropanal

(b)  cyclohexene → trans-2-methylcyclohexanol

End of Chapter Problems

11.19 Write balanced equations to show the combustion (oxidation) of each of the following alkanes to yield carbon dioxide and water.

a) C_5H_{12} b) octane c) cyclopentane d) C_7H_{16}

11.20 Complete the following reactions by supplying the missing reactant or major organic product.

(a) 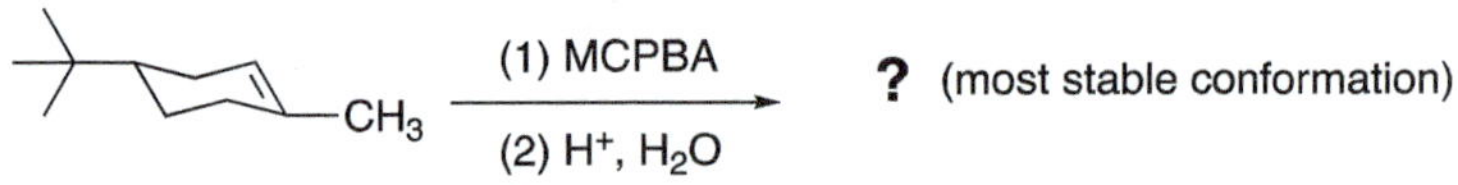 4-methyl-1-tert-butylcyclohexene $\xrightarrow{\text{(1) MCPBA} \quad \text{(2) H}^+, \text{H}_2\text{O}}$? (most stable conformation)

(b) 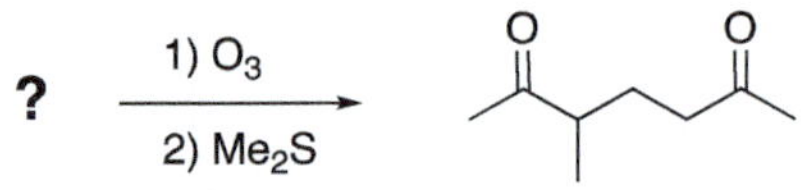 ? $\xrightarrow{\text{1) O}_3 \quad \text{2) Me}_2\text{S}}$ 2,6-dimethyl-3,5-heptanedione-like diketone

(c) 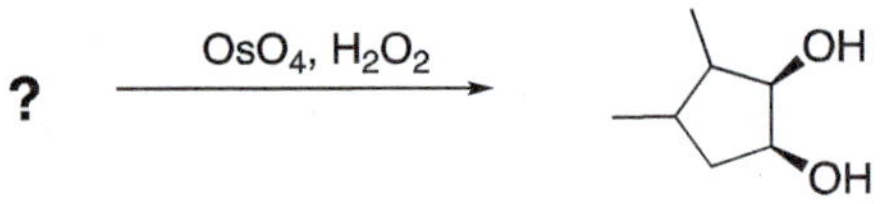 ? $\xrightarrow{\text{OsO}_4, \text{H}_2\text{O}_2}$ cis-diol cyclopentane derivative

11.21 Complete the reactions shown below, by giving the structures of the major organic products.

(a) 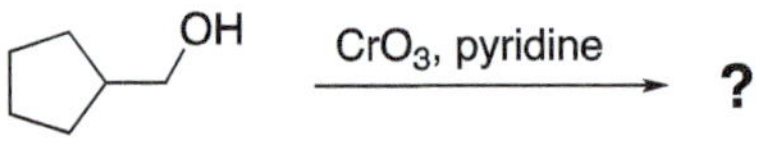 cyclopentyl methanol $\xrightarrow{\text{CrO}_3, \text{ pyridine}}$?   (b) 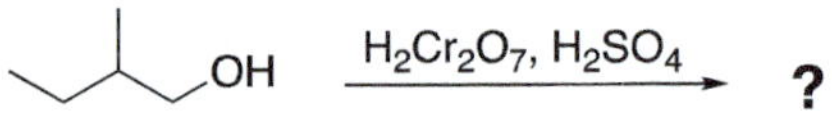 isobutanol $\xrightarrow{\text{H}_2\text{Cr}_2\text{O}_7, \text{H}_2\text{SO}_4}$?

11.22 Give the oxidized organic product that results from the reaction of each of the following with PCC.

(a) 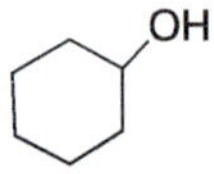 cyclohexylmethanol (b) 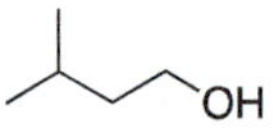 1-propanol derivative (c) 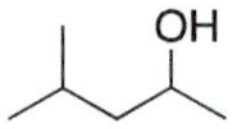 2-methyl-1-propanol (d)  1-methylcyclohexanol

11.23 Give the oxidized organic product for each of the reactions shown below.

(a) 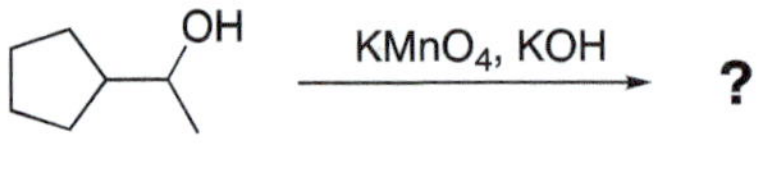 cyclopentyl secondary alcohol $\xrightarrow{\text{KMnO}_4, \text{KOH}}$?   (b) 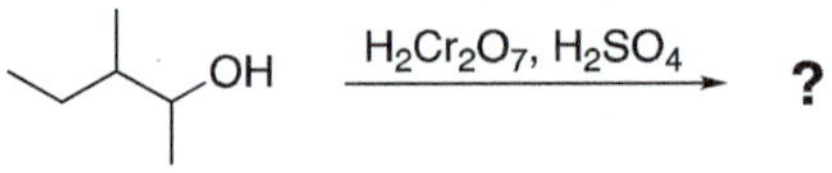 tert-butyl alcohol $\xrightarrow{\text{H}_2\text{Cr}_2\text{O}_7, \text{H}_2\text{SO}_4}$?

11.24 Give the alcohol and an appropriate oxidizing agent for the synthesis of the following aldehydes and ketones.

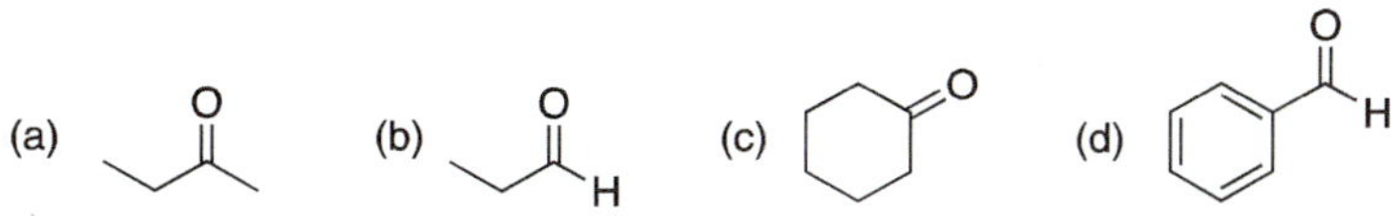

(a), (b), (c), (d)

11.25 Separate ozonolysis [O3 followed by (CH$_3$)$_2$S] of four (4) different compounds, **A, B, C,** and **D**, was carried out. The products obtained upon ozonolysis of each compound are shown below. Give the structures of the lettered starting compounds.

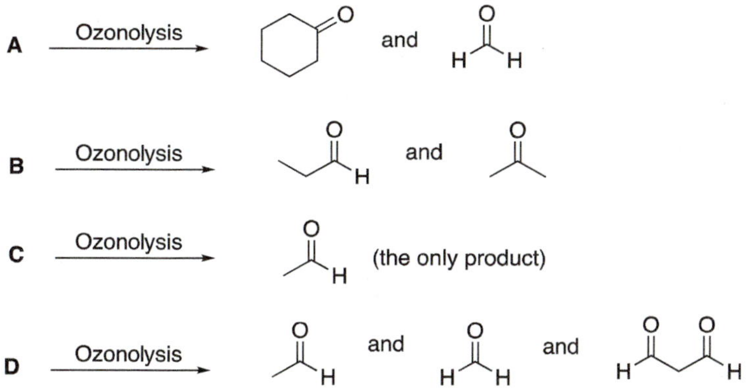

11.26 East Indian sandalwood oil contains a hydrocarbon given the name santene (C$_9$H$_{14}$). Ozonolysis of santene gives compound X, as shown below. The same product was also obtained after the reaction of santene with KMnO$_4$ at elevated temperatures. What is the structure of santene?

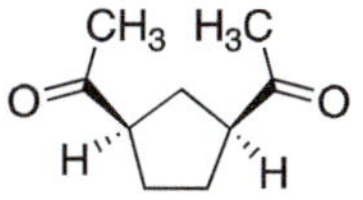

Compound X

11.27 Ambuic acid, which has been isolated from a fungus that is associated with some forest plants, has been shown to have anti-inflammatory action, its structure is shown below. Give the product of the reaction of ambuic acid under acid hydrolysis, include stereo-chemistry where appropriate.

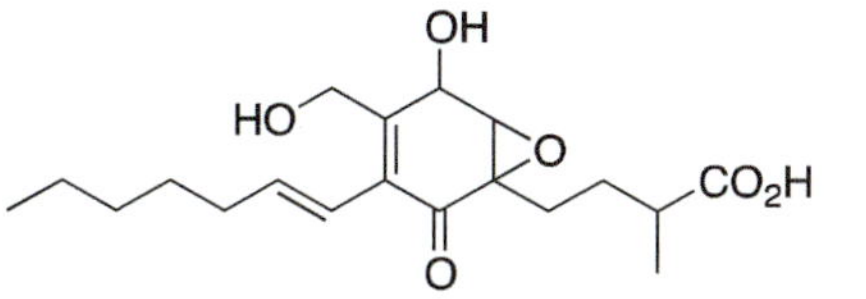

Ambuic acid

11.28 Carissone, an extract from asteraceae, can be synthesized from the starting compound shown below. Show how to carry out the following transformation (hint: consider using an oxidation reaction and a reduction reaction; also, don't forget the use of protecting groups).

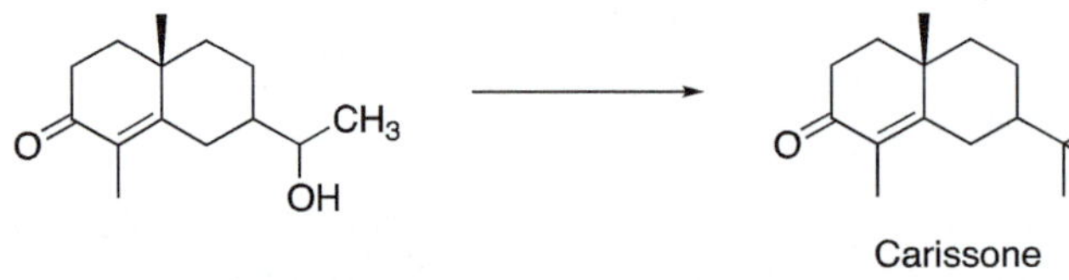

11.29 An unknown compound, **W** (C_7H_{14}), reacts readily with bromine in CCl_4 resulting in a colorless solution ($C_7H_{14}Br_2$). When **W** reacts with BH_3, followed by H_2O_2 in the presence of NaOH, a new compound, **X** ($C_7H_{16}O$) results, which when treated with $KMnO_4$ gives a carboxylic acid, **Y** ($C_7H_{14}O_2$). When **W** was subjected to ozonolysis, followed by the reaction with Me_2S, compound **Z** ($C_6H_{12}O$) is formed along with formaldehyde. It was shown that the oxidation of 3-hexanol using $KMnO_4$ also gave compound **Z**. What are the structures of **W, X, Y,** and **Z**?

11.30 One of the constituents of turpentine is α-pinene, formula $C_{10}H_{16}$. Based on the reaction scheme shown below, give the structures of the lettered compounds.

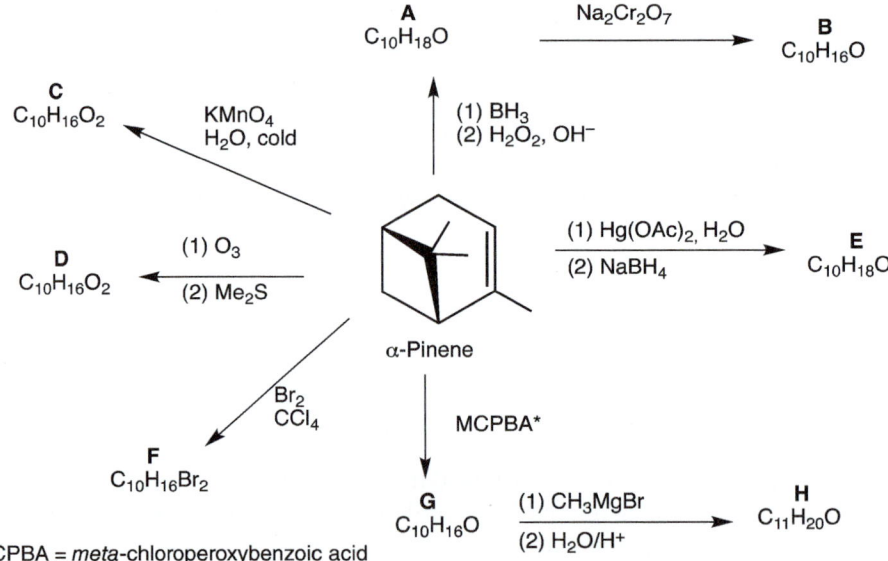

*MCPBA = meta-chloroperoxybenzoic acid

11.31 Show how to carry out the following transformations, include solvents and reaction conditions where appropriate.

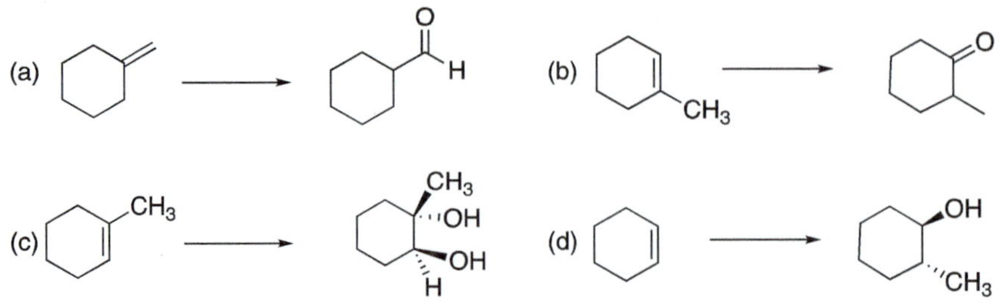

11.32 An unknown compound decolorizes bromine in carbon tetrachloride, and it undergoes catalytic reduction using hydrogen in the presence of a catalyst to give decalin. When the same unknown compound is treated with potassium permanganate at elevated temperature, it gives cyclohexane-1,2-dicarboxylic acid and oxalic acid (which was further oxidized, but not shown in reaction scheme below). Propose a structure for the unknown compound.

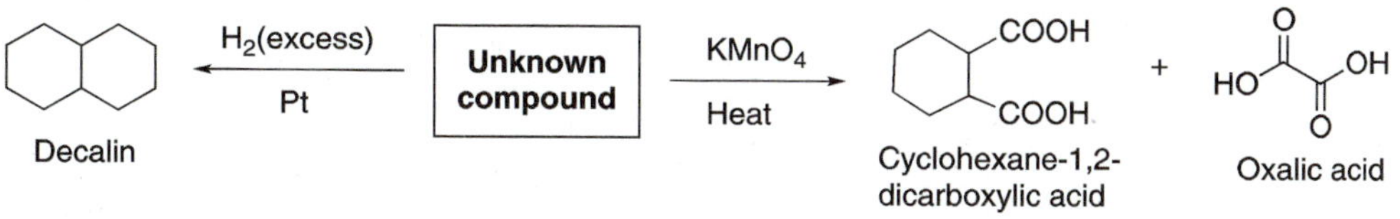

11.33 An unknown compound, **A**, reacts with an excess of hydrogen in the presence of platinum as a catalyst to give compound **B**. When **A** is subjected to ozonolysis, the products **C**, **D**, and **E** are formed. What is the structure of the unknown alkene, **A**?

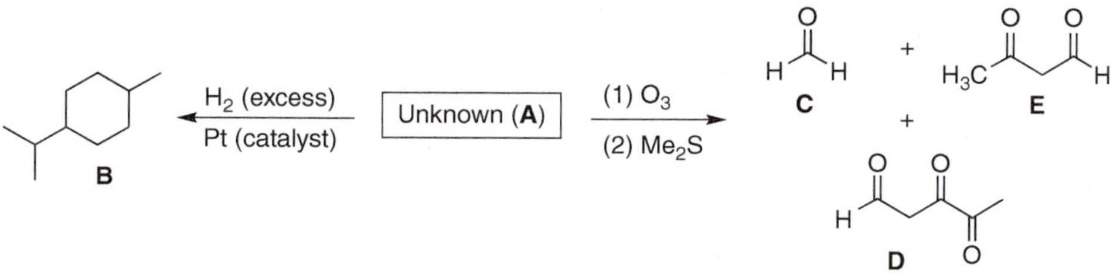

12

Elimination Reactions of Organic Chemistry

12.1 Introduction

Elimination reactions are another very important category of reactions of organic chemistry. As the name suggests, these are reactions in which atoms (or groups of atoms) are removed from a reactant molecule to form a new product. These reactions are the opposite of addition reactions that were covered in Chapter 8. Typically, the atoms or groups of atoms are eliminated from two adjacent atoms and, as a result, are referred to as β-elimination, since β is the second letter of the Greek alphabet. Another term that is used for this type of elimination is a 1,2-elimination since the atoms or group of atoms that leave are originally on carbons 1 and 2 of the reactant. The product of an elimination reaction is an unsaturated functionality (alkene or alkyne). Typically, the atoms that are removed are a proton and a good leaving group (a weak conjugate base) with its pair of bonding electrons, as shown in Reaction (12-1). Since ions of halogens (especially bromine and iodine) are weak conjugate bases, they are typically good leaving groups for these types of reactions. Remember that a proton is really a hydrogen atom without its electron, a positively charged electrophile; and a leaving group is one that has a pair of electrons, a Lewis base.

$$\underset{\text{Potential leaving group}}{\underset{\overset{\overset{\text{Potential proton}}{\frown}}{\underset{X}{\overset{H}{-C-C-}}}}{}} \xrightarrow{\text{Elimination of HX}} \underset{\text{Alkene}}{\text{>C=C<}} + \underset{\text{Proton}}{H^+} + \underset{\text{Leaving group}}{X^-} \quad (12\text{-}1)$$

Of course, one could imagine different ways (mechanisms) in which the elimination of the atoms (H^+ and X^-) can take place to give the alkene product. In this chapter, we will examine possibilities (mechanisms) for elimination reactions and apply that knowledge to design the synthesis of different target molecules.

12.2 Mechanisms of Elimination Reactions

One of the mechanisms that will be examined in this chapter involves the proton and the leaving group departing at the same time, also called a concerted reaction. For a reaction to proceed through this type of mechanism, typically a base is introduced in the reaction mixture to assist in the abstraction of hydrogen as a proton (i.e. hydrogen without its bonding electrons) to form the carbon–carbon double bond. This mechanism is referred to as bimolecular elimination (E2).

Organic Chemistry: Concepts and Applications, First Edition. Allan D. Headley.
© 2020 John Wiley & Sons, Inc. Published 2020 by John Wiley & Sons, Inc.
Companion website: www.wiley.com/go/Headley_OrganicChemistry

Another mechanism by which the reaction can take place involves two steps. In the first step, the leaving group departs taking its bonding electrons, and in a different step, the proton is abstracted and its bonding electrons form the carbon–carbon double bond. This mechanism is referred to as unimolecular elimination (E1). Another possibility that will not be examined in this chapter involves the proton being abstracted first to form a conjugate base, then the electrons of this intermediate fold in to form a double bond while the leaving group leaves. This mechanism is called a unimolecular elimination, conjugate base (E1cB) mechanism. The first two mechanisms will be discussed in details in the next sections of this chapter.

12.2.1 Elimination Bimolecular (E2) Reaction Mechanism

If the reaction takes place in a manner so that the proton (H^+) and leaving group leave at the same time to form the carbon–carbon double bond in a single step, the mechanism is described as bimolecular (E2). This type of mechanism is described as bimolecular because two molecules are involved in this very important step of the reaction. A base accomplishes the removal of the proton at the same time as the departure of the leaving group with its bonding pair of electrons; the electrons from the C—H bond forms the double bond in the product, as shown in Figure 12.1.

Since this reaction occurs in a single step, there is only one transition state and no intermediates for this type of reaction. The dotted lines shown in the transition state represent partial bonds that are being formed and broken as the reaction proceeds from reactants to products. The energy profile for this type of single-step reaction is shown in Figure 12.2. In this mechanism, the reactants proceed directly to the products through the transition state and then to products.

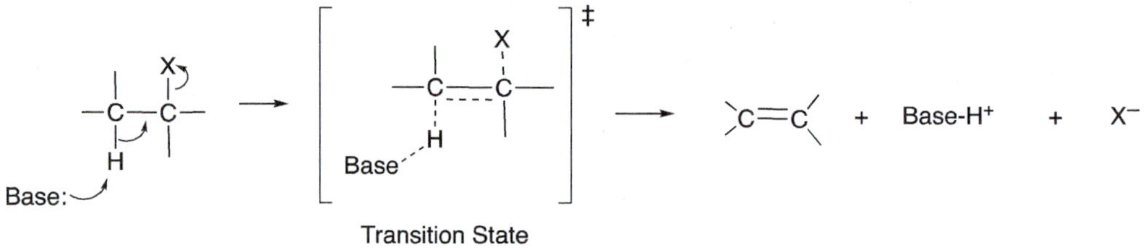

Figure 12.1 Mechanism for a β-elimination-bimolecular (E2) elimination reaction in which as Lewis base is used to abstract a proton from the β-carbon adjacent to the carbon that has the leaving group.

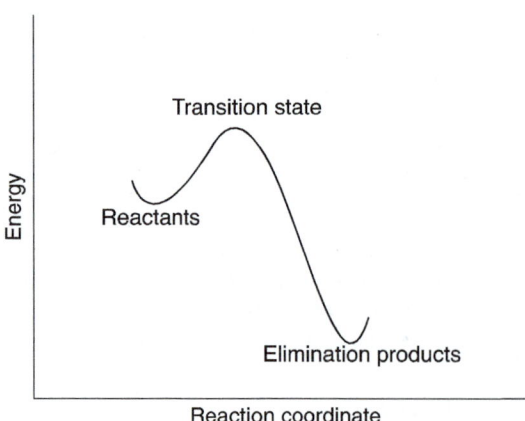

Figure 12.2 Energy profile of a β-elimination reaction. Note that this is a bimolecular single-step reaction (E2).

Shown below are two examples of β-elimination Reactions (12-2) and (12-3). Note that heat is typically applied for elimination reactions, and we will see later that this is one way to ensure that an elimination reaction takes place, compared to another type of reaction, substitution reaction, which will be covered in Chapter 15.

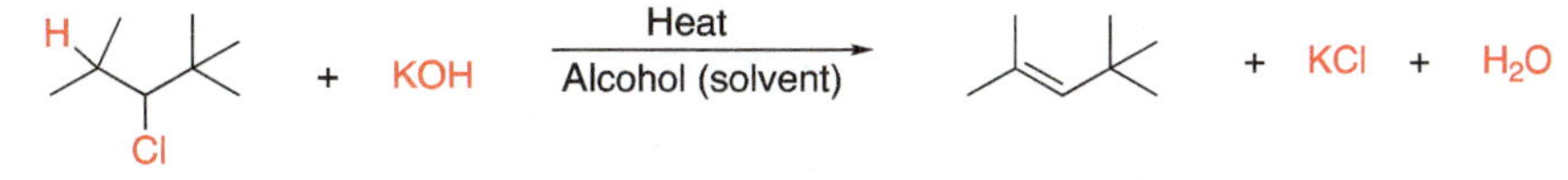

$$\text{3-Chloro-2,2,4-trimethylpentane} + \text{KOH} \xrightarrow[\text{Alcohol (solvent)}]{\text{Heat}} \text{2,4,4-Trimethyl-2-pentene} + \text{KCl} + \text{H}_2\text{O} \quad (12\text{-}2)$$

For the reaction shown in Reaction (12-2), the base, potassium hydroxide, abstracts the proton shown and as the leaving group, the chloride anion, leaves, the double bond is formed. Later in the chapter, we will examine other types of bases that can be used for these reactions. Note that the product shown is the only possible elimination product since there is not another hydrogen adjacent to the leaving chloride anion. The same is true for the reaction shown in 12-3. Later in the chapter, we will see that there are other possible elimination products.

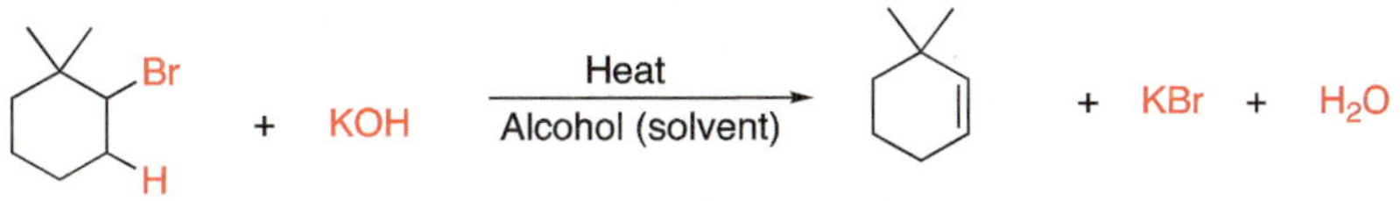

$$\text{2-Bromo-1,1-dimethyl cyclohexane} + \text{KOH} \xrightarrow[\text{Alcohol (solvent)}]{\text{Heat}} \text{3,3-Dimethyl cyclohexene} + \text{KBr} + \text{H}_2\text{O} \quad (12\text{-}3)$$

Another important feature of this mechanism for the elimination of a proton and a leaving group as shown in Figure 12.1 is that of the geometry; the hydrogen and the leaving group must be in the *trans* arrangement. Since the leaving group and hydrogen abstracted are typically opposite to each other, this type of elimination is also called a *trans*-elimination. For the elimination reaction shown in Reaction (12-4), note that the elimination takes place across the carbons that have the leaving group and the hydrogen that are in the *trans* arrangement. The other hydrogen that is adjacent to the leaving group is *cis* to the leaving group, hence elimination will not occur in that position.

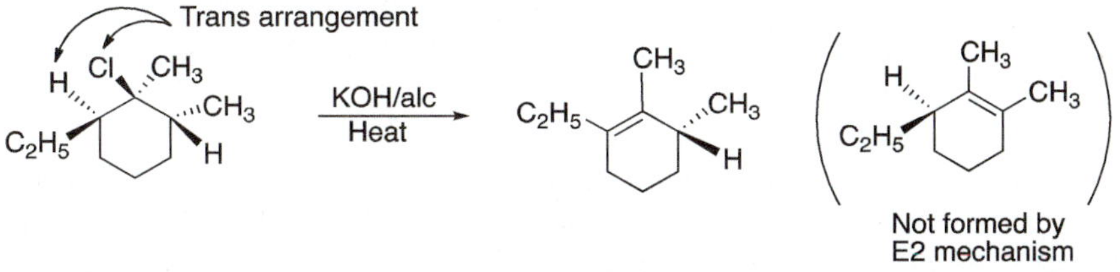

Trans arrangement

$$\xrightarrow[\text{Heat}]{\text{KOH/alc}} \quad + \quad (\text{Not formed by E2 mechanism}) \quad (12\text{-}4)$$

Note that if the stereochemistry of the leaving group and the hydrogen is not designated by the usual dashed-wedge or another method to show stereochemistry in the reactant, the elimination product could be from either side of the leaving group, as shown in Reaction (12-5).

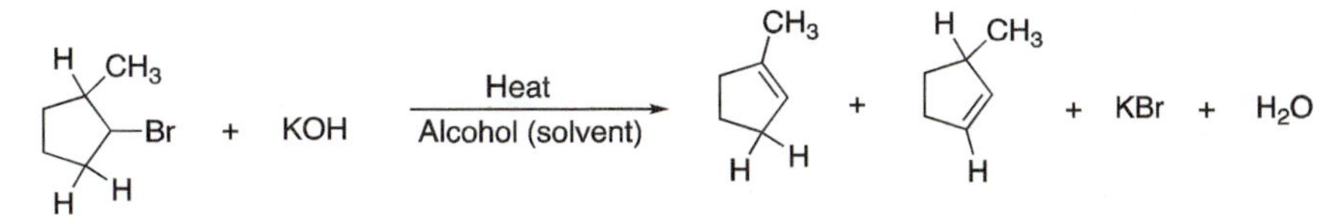

$$\text{1-Bromo-2-methyl cyclopentane} + \text{KOH} \xrightarrow[\text{Alcohol (solvent)}]{\text{Heat}} \text{1-Methyl cyclopentene} + \text{3-Methyl cyclopentene} + \text{KBr} + \text{H}_2\text{O} \quad (12\text{-}5)$$

Note for alkyl halides that have a hydrogen on either of the carbons that are adjacent to the carbon that has the leaving group, there are two possible products, in which the double bonds are in different locations from the leaving group as shown in Reaction (12-5).

Problem 12.1

Give the elimination product for the reaction of KOH/alcohol at elevated temperature with each of the following alkyl halides. Note for some, it is possible to have more than one product, in which the double bonds are in different locations from the leaving group. Give all possible products.

(a) [1-bromoethylbenzene] (b) [3-chloro-1-butene] (c) [bromocyclohexane] (d) [2-bromo-2,3-dimethylbutane]

As shown in Reaction (12-5) and Problem 12.1, it is possible to abstract a proton from either carbon that is adjacent to the leaving group, which results in two different alkenes. Experimental conditions can be designed so that one alkene is the major product and the other is the minor product.

The outcome of getting one alkene as the major alkene product is dependent on two aspects: the size of the base and second the number and size of the groups surrounding the proton to be abstracted. As shown in Reaction (12-6), if there is steric bulk around the hydrogen to be extracted as a proton, it is extremely difficult for the base to accomplish that task.

$$R\text{-}\underset{H}{\overset{R}{C}}\text{-}\underset{H}{\overset{X}{C}}\text{-}\underset{H}{\overset{H}{C}}\text{-}H \quad \longrightarrow \quad \underset{R}{\overset{R}{>}}C=C\underset{CH_3}{\overset{}{<}} + \text{Base-H}^+ + X^- \quad (12\text{-}6)$$

(Bulky R groups; Small base)

This does not mean that the abstraction will not take place, however, but it would require a base small enough (hard base) to accomplish the abstraction. A small base such as sodium hydroxide (NaOH) can accomplish this task. You may ask at this point, why wouldn't the base abstract the adjacent proton instead since it is more accessible? The challenge is that the product that results from the abstraction of the proton from the carbon that has more steric bulk results in a more stable alkene. We have demonstrated in Chapter 4 that alkenes that have more alkyl groups surrounding the double bond are more stable than alkenes that have less alkyl groups surrounding the double bond.

On the other hand, if a very bulky base, such as lithium diisopropylamide (LDA) or potassium *tert*-butoxide, is used for the reaction, either of these bulky bases will not be able to abstract the proton from the carbon that has a lot of surrounding steric bulk. You will recall that we discussed these types of bases in the chapter on acids and bases (Chapter 7). As a result, it will abstract a proton from the carbon that has less steric bulk, i.e. from the carbon that has more hydrogen atoms owing to the very small size of the hydrogen atom. This concept is illustrated in Reaction (12-7).

$$R\text{-}\underset{H}{\overset{R}{C}}\text{-}\underset{H}{\overset{X}{C}}\text{-}\underset{}{\overset{H}{C}}\text{-}H \quad \longrightarrow \quad R\text{-}\underset{H}{\overset{R}{C}}\text{-}C=CH_2 + \text{Base-H}^+ + X^- \quad (12\text{-}7)$$

(Bulky R groups; small hydrogens; Bulky Base)

Note in this case, the product formed is the least stable since the double bond has less alkyl groups surrounding the double bond, compared to the alkene formed in the example shown in Reaction (12-6). Thus, it is possible for a reactant to form either of the two different major alkene products depending on the reaction conditions. The less substituted alkene product is often referred to as the Hofmann product and the more substituted alkene product as the Zaytsev product. This concept is summarized in Reaction (12-8).

$$\text{CH}_2\text{=cyclohexane} \xleftarrow{\text{Bulky base, LDA, heat}} \text{1-Chloro-1-methyl cyclohexane} \xrightarrow{\text{Small base, KOH/H}_2\text{O, }\Delta} \text{methylcyclohexene} \tag{12-8}$$

Hofman product (less substituted alkene) — 1-Chloro-1-methyl cyclohexane — Zaytsez product More substituted alkene

Problem 12.2

Give the major organic products for the following elimination reactions. Pay special attention to the size of the base used; KOH is a small base, whereas LDA is a bulky base.

(a) cyclopentyl-Br, KOH, H$_2$O, heat

(b) 2-chloro-3-methylbutane, KOH, H$_2$O, heat

(c) cyclopentyl-Br, LDA, heat

(d) 2-chloro-3-methylbutane, LDA, heat

Another approach that can be used to obtain the less stable alkene is instead of using a bulky base is to have the leaving group be a bulky leaving group. The outcome again depends on the ability of the base to access the acidic hydrogen for the elimination to occur. Reaction (12-9) shows the elimination reaction in which the bulky trimethyl ammonium ion is the leaving group; hence, the use of a small base, such as the hydroxide anion, can be used to carry out the elimination to give the least stable alkene as shown in Reaction (12-9).

Bulky leaving group: $(CH_3)_3N^+$, I^-

$$R-\underset{H}{\underset{|}{C}}-\underset{N(CH_3)_3^+}{\underset{|}{C}}-\underset{H}{\underset{|}{C}}-H \longrightarrow R-\underset{H}{\underset{|}{C}}-\overset{R}{\underset{|}{C}}=\overset{H}{\underset{|}{C}}-H + \text{Base-H}^+ + (CH_3)_3N \tag{12-9}$$

Small base — Hofmann product

This type of reactant gives the product with the less substituted double bond, and is often referred to as the Hoffman elimination, an example is shown in Reaction (12-10).

$$\text{cyclopentyl-N(CH}_3\text{)}_3^+ \text{ I}^- \xrightarrow{\text{NaOH, heat}} \text{cyclopentene} + \text{H}_2\text{O} + \text{N(CH}_3\text{)}_3 + \text{NaI} \tag{12-10}$$

Hoffman product (Less substituted alkene)

Problem 12.3

Give the Hofmann elimination product for each of the following compounds.

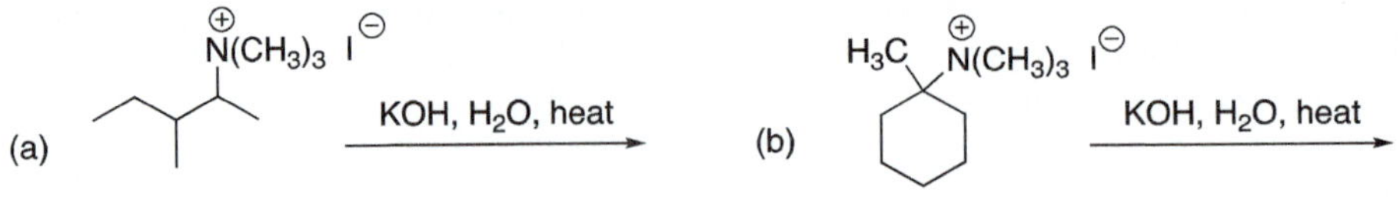

(a) [structure: isobutyl group with $\overset{+}{N}(CH_3)_3$, I^-] → KOH, H_2O, heat

(b) [structure: cyclohexane with H_3C and $\overset{+}{N}(CH_3)_3$, I^-] → KOH, H_2O, heat

As you can imagine, it is possible to have an intramolecular elimination where the base and the leaving group are on the same molecule. This type of reaction is shown in Reaction (12-11).

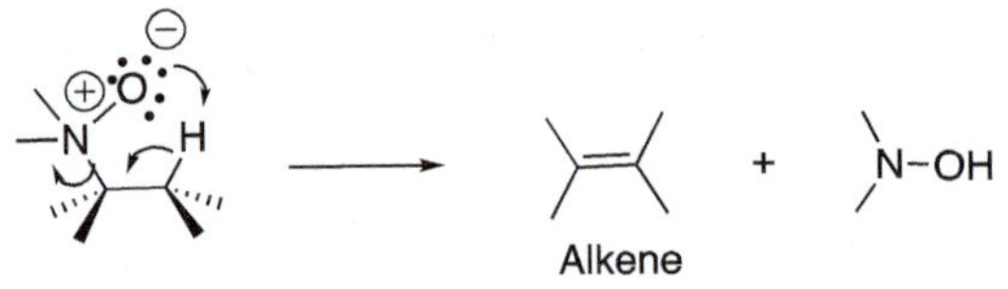

$$\text{[N-oxide structure]} \longrightarrow \text{Alkene} + \text{N–OH} \tag{12-11}$$

The N-oxide can be made from an oxidation reaction of an amine as shown in Reaction (12-12).

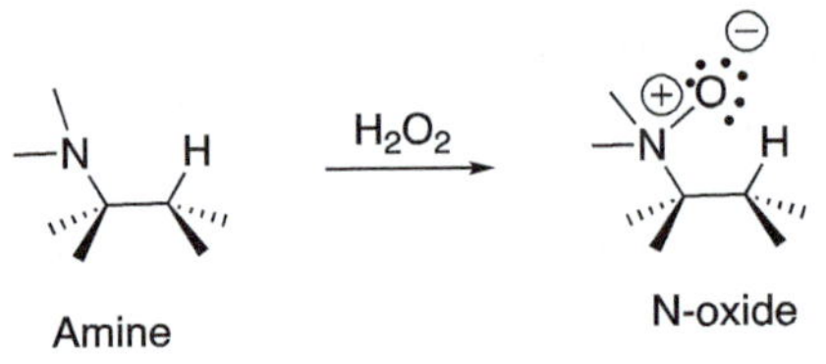

$$\text{Amine} \xrightarrow{H_2O_2} \text{N-oxide} \tag{12-12}$$

12.2.2 Elimination Unimolecular (E1) Reaction Mechanism

As mentioned earlier, there is an alternate mechanism by which elimination reactions can occur. During the course of an elimination reaction, the leaving group may leave with no influence from the base or another molecule. You will recall from Chapter 7 that very stable conjugate bases are good leaving groups. Thus, for a reaction to proceed by this mechanism, the leaving groups are typically very good leaving groups (stable conjugate bases), as illustrated in Reaction (12-13).

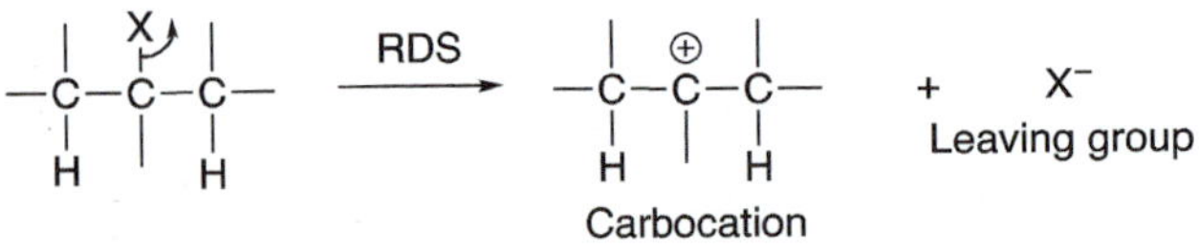

$$\text{[C-C-C with X leaving group]} \xrightarrow{\text{RDS}} \text{Carbocation} + X^- \text{ Leaving group} \tag{12-13}$$

In a second step, the base removes a proton to form the alkene, which is shown in Reaction (12-14).

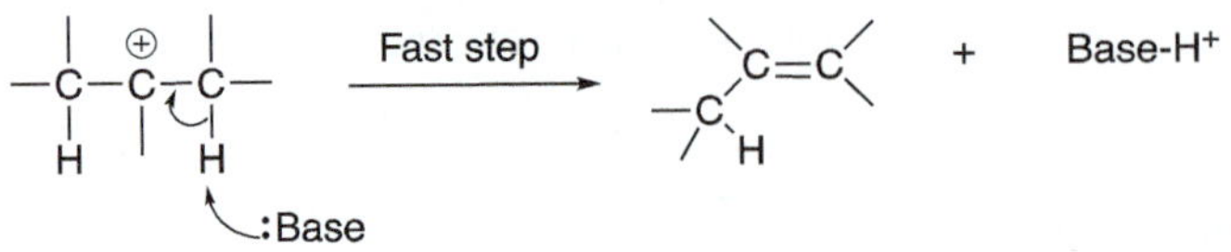

$$\text{[Carbocation + :Base]} \xrightarrow{\text{Fast step}} \text{C=C} + \text{Base-H}^+ \tag{12-14}$$

Note that if there is a hydrogen on the adjacent carbon of the carbocation, the double bond can be formed to the adjacent carbon as shown in Reaction (12-15).

Figure 12.3 Energy profile of E1 reaction. Note that first step is the slowest step (the RDS) of the reaction mechanism (higher activation barrier (E_a), compared to the second step, which is fast.

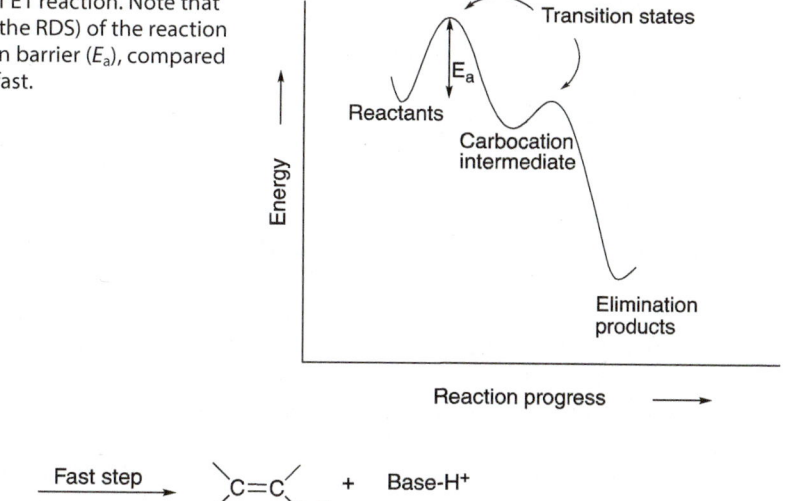

$$\underset{\text{Base}}{\overset{}{-\underset{H}{\overset{|}{C}}-\underset{|}{\overset{\oplus}{C}}-\underset{H}{\overset{|}{C}}-}} \xrightarrow{\text{Fast step}} \underset{}{\overset{}{>C=C<}}\underset{H}{\overset{}{\underset{|}{C}}} + \text{Base-H}^+ \tag{12-15}$$

The formation of the carbocation in the first step of the reaction mechanism is the slowest and most important step of the reaction mechanism since it determines the overall rate of the reaction. Hence, this step has a special name, the rate-determining step (RDS), or rate-limiting step. Since only one reactant is involved in the most important step of this reaction mechanism, this mechanism is classified as a unimolecular elimination reaction (E1). The two steps of the reaction mechanism given in Reactions (12-14) and (12-15) can be represented by a reaction profile showing the two steps of the overall reaction, in which the first step has a higher activation barrier and hence slower, compared to the last step, which is fast. The reaction profile for this two-step reaction given in Figure 12.3 is different from the one-step reaction shown for the bimolecular E2 reaction shown earlier Figure 12.2.

12.2.3 Elimination Unimolecular – Conjugate Base (E1cB) Reaction Mechanism

A third possibility for elimination reactions occurs if the leaving group is a poor leaving group (relatively strong conjugate base), such as a hydroxyl or alkoxide group. For such reactions, which are carried out typically in a very basic medium, the first step involves the base abstracting a relatively acidic proton from the reactant to form a stable conjugate base intermediate as shown in Reaction (12-16).

$$\tag{12-16}$$

Resonance stabilized conjugate base

In a second and concerted step, the lone pair of electrons of the conjugate base folds in to form the double bond at the same time that the leaving group leaves as shown in Reaction (12-17).

$$\tag{12-17}$$

Since the leaving group is typically poor, these reactions are carried out at elevated temperatures. A specific example of a reaction that proceeds via an E1cB is shown in Reaction (12-18).

$$\text{PhCH(OH)CH}_2\text{COCH}_3 + \text{NaOH} \xrightarrow{\text{Heat}} \text{PhCH=CHCOCH}_3 \quad (12\text{-}18)$$

In this course, however, we will concentrate mostly on elimination reactions that proceed via E2 and E1 mechanisms and not much on E1cB reactions.

12.3 Elimination of Hydrogen and Halide (Dehydrohalogenation)

In the next sections of this chapter, we will determine how to apply the concepts of these two reaction mechanisms in order to achieve desired elimination products for different reactions. The first mechanism that we will examine is the E2 mechanism, in which a proton and a halogen anion are eliminated from a molecule in a basic medium to produce the elimination product. Reaction (12-19) gives a specific reaction that proceeds via the E2 mechanism.

$$\underset{\text{2-Bromo-2-methylbutane}}{\text{H}_3\text{C}-\underset{\underset{\text{Br}}{|}}{\overset{\underset{\text{CH}_3}{|}}{\text{C}}}-\text{CH}_2\cdot\text{CH}_3} \xrightarrow[\text{CH}_3\text{CH}_2\text{OK}]{\text{CH}_3\text{CH}_2\text{OH}} \underset{\substack{\text{2-Methyl-2-butene}\\ \text{(major product)}}}{\text{H}_3\text{C}-\overset{\underset{\text{CH}_3}{|}}{\text{C}}=\text{CH}-\text{CH}_3} + \underset{\substack{\text{2-Methyl-1-butene}\\ \text{(minor product)}}}{\text{H}_2\text{C}=\overset{\underset{\text{CH}_3}{|}}{\text{C}}-\text{CH}_2\cdot\text{CH}_3} \quad (12\text{-}19)$$

We have seen from Reaction (12-5) that there are two possible products for these elimination reactions. Consider the reaction given in Reaction (12-19), since this reaction is carried out in a strong basic medium, an alkoxide anion, the mechanism is more than likely an E2. The base is relatively small and will abstract a proton to result in the more stable alkene as the major product. Note that since the carbon–carbon bond that is bonded to the leaving group can rotate, at some point making an acidic hydrogen trans to the leaving group, to give the E2 product. The E2 mechanism for the Reaction (12-19) is shown in Reaction (12-20).

$$\text{H}_3\text{C}-\underset{\underset{\text{CH}_3\ \ \text{H}}{|}}{\overset{\underset{\text{Br}}{|}}{\text{C}}}-\text{CH}-\text{CH}_3 \quad \longrightarrow \quad \left[\text{H}_3\text{C}-\underset{\underset{\text{CH}_3\ \ \text{H}\text{-}\text{-}\text{-}\ddot{:}\text{OEt}}{|}}{\overset{\underset{\text{Br}}{|}}{\text{C}}}\text{-}\text{-}\text{-}\text{CH}-\text{CH}_3\right]^{\ddagger} \quad \longrightarrow \quad \underset{\text{Major product}}{\text{H}_3\text{C}-\overset{\underset{\text{CH}_3}{|}}{\text{C}}=\text{CH}-\text{CH}_3} \quad (12\text{-}20)$$
$$+ \ \text{EtOH} + \text{Br}^-$$
$$\text{Transition state}$$

The minor product will be obtained via the mechanism given in Reaction (12-21). In this mechanism, the base abstracts a proton from one of the methyl groups adjacent to the bromide leaving group.

$$\underset{\underset{\ddot{:}\text{OEt}}{}}{\text{H}-\overset{\underset{\text{H}}{|}}{\underset{\underset{\text{H}}{|}}{\text{C}}}-\overset{\underset{\text{Br}}{|}}{\underset{\underset{\text{CH}_3}{|}}{\text{C}}}-\text{CH}-\text{CH}_3} \quad \longrightarrow \quad \left[\text{H}-\overset{\underset{\text{H}}{|}}{\underset{\underset{\text{H}\ \ \ddot{:}\text{OEt}}{}}{\text{C}}}\text{-}\text{-}\text{-}\overset{\underset{\text{Br}}{|}}{\underset{\underset{\text{CH}_3\ \ \text{H}}{|}}{\text{C}}}-\text{CH}-\text{CH}_3\right]^{\ddagger} \quad \longrightarrow \quad \underset{\text{Minor product}}{\text{H}_2\text{C}=\overset{\underset{\text{CH}_3}{|}}{\text{C}}-\text{CH}_2\text{-}\text{CH}_3} \quad (12\text{-}21)$$
$$+ \ \text{EtOH} + \text{Br}^-$$
$$\text{Transition state}$$

12.3 Elimination of Hydrogen and Halide (Dehydrohalogenation)

If the same reaction were carried out in the absence of a strong base and just a solvent such as ethanol, the loss of the acidic hydrogen is much more difficult. As a result, the carbocation will be formed since the bromide is a good leaving group as shown in Reaction (12-22).

$$\underset{\substack{|\\CH_3\ H}}{H_3C-\overset{Br}{\underset{|}{C}}-CH-CH_3} \xrightarrow[\text{EtOH (solvent)}]{\text{RDS}} \underset{\substack{|\\CH_3\ H}}{H_3C-\overset{\oplus}{\underset{|}{C}}-CH-CH_3} + Br^- \qquad (12\text{-}22)$$

Carbocation intermediate

If the leaving group is an extremely good leaving group, such as the bromide or an iodide anion, this first step is very likely. In the presence of the solvent, which is only slightly basic, abstraction of the proton occurs to give the final alkene product, as shown in Reaction (12-23). Note that the solvent abstracts the proton to give the more stable alkene product, which is the alkene with more alkyl groups bonded to the alkene carbons.

$$H_3C-\overset{\oplus}{\underset{\substack{|\\CH_3}}{C}}-\underset{H}{\overset{|}{CH}}-CH_3 \xrightarrow{\text{Fast}} H_3C-\underset{\substack{|\\CH_3}}{C}=CH-CH_3 + \overset{H}{\underset{H\cdots Et}{\overset{|}{\overset{\oplus}{O}}}} \qquad (12\text{-}23)$$

You may be wondering why the Br⁻ does not abstract the proton to form the alkene. In this case, the Br⁻ is a much weaker base than that of C_2H_5OH. Remember that the pK_a of HBr is −9.0; whereas, the pK_a of protonated alcohols are around 0.0 (the pK_a of H_3O^+ is −1.7). Thus, the major and minor elimination products of this E1 mechanism result from the intermediate carbocation that is formed in the RDS, and Reaction (12-24) shows the final steps to give both possible products.

$$(12\text{-}24)$$

Acidic hydrogens: $H_3C-\underset{\substack{|\\CH_3}}{\overset{+}{C}}-CH_2-CH_3$

EtOH → $H_3C-\underset{\substack{|\\CH_3}}{C}=CH-CH_3$ + $\overset{H}{\underset{H\cdots Et}{\overset{|}{\overset{\oplus}{O}}}}$

Tri-substituted more stable alkene (Zaitsev product)

EtOH → $H_2C=\underset{\substack{|\\CH_3}}{C}-CH_2-CH_3$ + $\overset{H}{\underset{H\cdots Et}{\overset{|}{\overset{\oplus}{O}}}}$

Di-substituted less stable alkene (Hofmann product)

As pointed out earlier, the alkene that has the most alkyl groups bonded to the carbons of the double bond is the most stable alkene, also known as the Zaitsev product, compared to the alkene with the least alkyl groups bonded to the carbons of the double bond, this product is known as the Hofmann product. The elimination reaction of another molecule, 1-bromo-1-methylcyclohexane, is shown in Reaction (12-25).

$$\text{1-Bromo-1-methylcyclohexane} \xrightarrow[\text{Heat}]{C_2H_5OH} \text{1-Methylcyclohexene} + HBr \qquad (12\text{-}25)$$

12 Elimination Reactions of Organic Chemistry

For this reaction, more than likely the mechanism is E1. The steps for the E1 mechanism are shown in Reactions (12-26) through (12-28). In the first step of the mechanism, the very good leaving group, Br⁻, leaves to form a carbocation and a bromide anion.

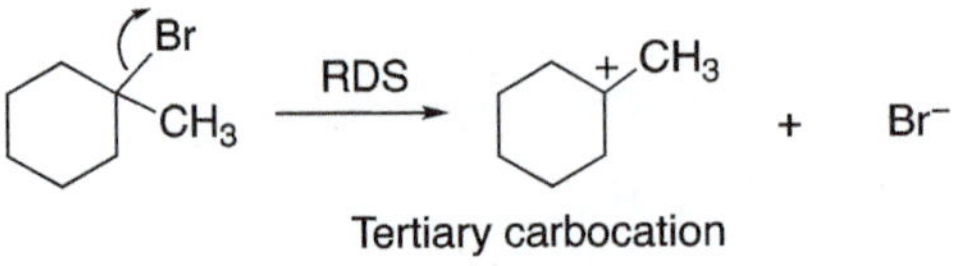

$$\text{(cyclohexane with Br and CH}_3\text{)} \xrightarrow{\text{RDS}} \text{(tertiary carbocation with CH}_3\text{)} + Br^- \quad (12\text{-}26)$$

Tertiary carbocation

Owing to the positive character of the carbocation, the electrons of the bonds adjacent to the charge will be attracted to that carbon. As a result of this shift in electron density, the hydrogens on the adjacent carbons are relatively acidic, and the pair of electrons of the C—H bond will serve to neutralize the carbocation by forming another bond, a pi (π) bond, and the result is the formation of a double bond as shown in Reaction (12-27).

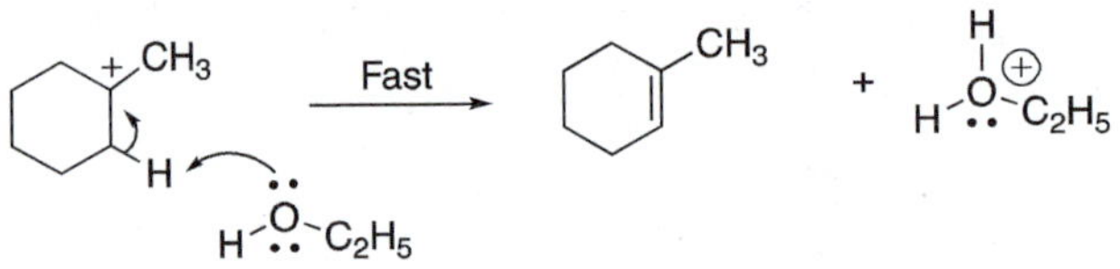

$$(12\text{-}27)$$

In the final step of the mechanism, the bromide anion abstracts a proton from the protonated alcohol to form two very stable and neutral compounds as shown in Reaction (12-28).

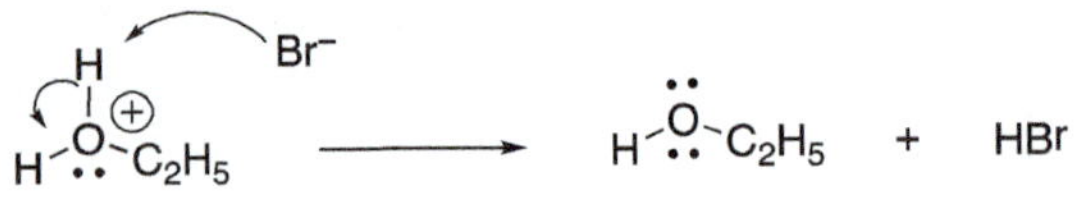

$$(12\text{-}28)$$

By increasing the base strength of an elimination reaction, the percentage of elimination product by way of E2 mechanism is increased, compared to the E1 pathway. For example, when the reaction below is carried out in ethanol, compared to ethanol in the presence of potassium ethoxide, there is a higher percentage of the elimination product that proceeds via the E2 mechanism, compared to the E1 mechanism, as shown in Reaction (12-29).

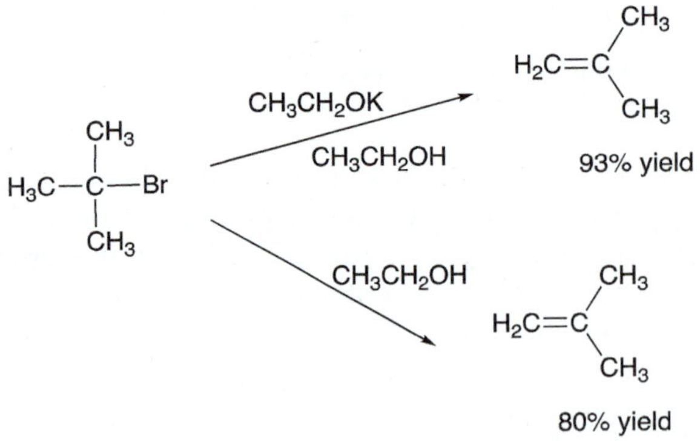

$$H_3C-\underset{\underset{CH_3}{|}}{\overset{\overset{CH_3}{|}}{C}}-Br \quad \xrightarrow[CH_3CH_2OH]{CH_3CH_2OK} \quad H_2C=C\underset{CH_3}{\overset{CH_3}{\diagup}} \quad 93\% \text{ yield}$$

$$\xrightarrow{CH_3CH_2OH} \quad H_2C=C\underset{CH_3}{\overset{CH_3}{\diagup}} \quad 80\% \text{ yield}$$

$$(12\text{-}29)$$

Thus, by altering the reaction conditions, an elimination reaction can be made to proceed predominantly by either of the mechanisms described in this section, E2 or E1.

12.4 Elimination of Water (Dehydration)

Problem 12.4

Give the major elimination products for each of the following reactions.

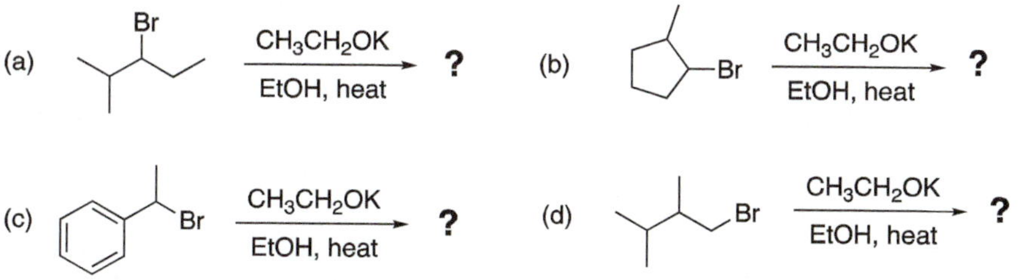

(a) 3-bromo-2-methylpentane + CH₃CH₂OK / EtOH, heat → ?

(b) bromocyclopentane (with methyl substituent) + CH₃CH₂OK / EtOH, heat → ?

(c) (1-bromoethyl)benzene + CH₃CH₂OK / EtOH, heat → ?

(d) 1-bromo-2,3-dimethylbutane + CH₃CH₂OK / EtOH, heat → ?

12.4 Elimination of Water (Dehydration)

In the presence of an acid, alcohols are protonated, and the very poor leaving —OH group is converted to an extremely good leaving group, H_2O. Once water leaves with its bonding electrons, a carbocation is formed, hence the likely mechanism is an E1 mechanism as shown in the example given in Reaction (12-30).

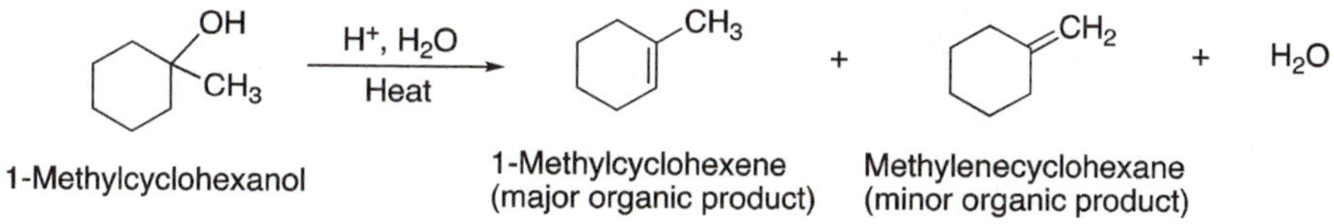

$$\text{1-Methylcyclohexanol} \xrightarrow{H^+, H_2O, \text{Heat}} \text{1-Methylcyclohexene (major organic product)} + \text{Methylenecyclohexane (minor organic product)} + H_2O \tag{12-30}$$

12.4.1 Dehydration Products

Now that we have established that the most likely mechanism is E1, a close examination of the E1 reaction mechanism will give a rationale for the formation of the different dehydration products shown in Reaction (12-30). In the first step of the mechanism, the alcohol OH group is protonated, and it is converted to a good leaving group, H_2O. In the next step, a cation is formed by the elimination of water as shown in Reaction (12-31).

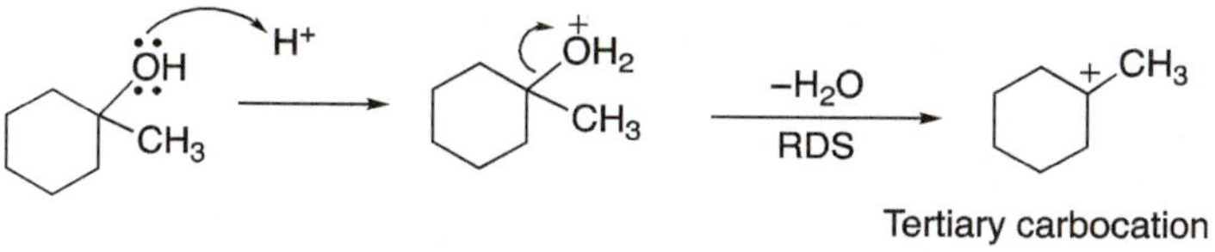

$$\text{ROH} \xrightarrow{H^+} \text{ROH}_2^+ \xrightarrow{-H_2O, \text{RDS}} \text{Tertiary carbocation} \tag{12-31}$$

In the next step of the E1 mechanism, water abstracts a proton from the adjacent carbon of the carbocation to form the alkene products as shown in Reactions (12-32a) and (12-32b).

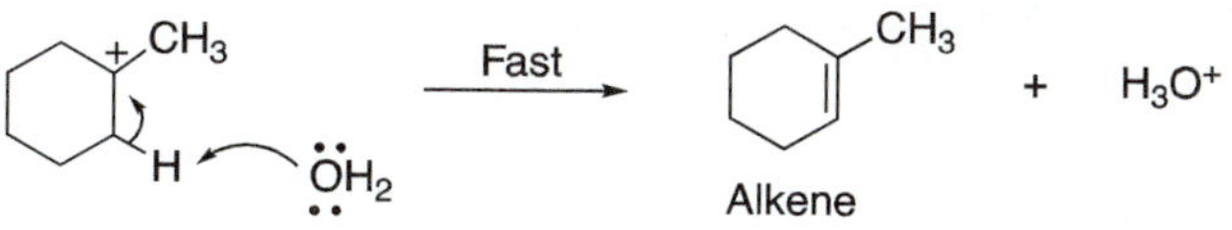

$$\text{carbocation} + OH_2 \xrightarrow{\text{Fast}} \text{Alkene} + H_3O^+ \tag{12-32a}$$

$$\text{carbocation} + OH_2 \xrightarrow{\text{Fast}} \text{Alkene} + H_3O^+ \tag{12-32b}$$

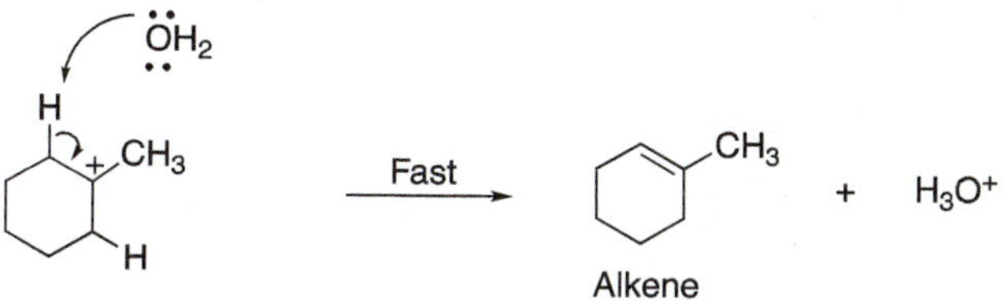

Note that the alkenes formed from Reactions (12-32a) and (12-32b) are the same. Abstraction however from the proton from the methyl group results in a different alkene product, as shown in Reaction (12-33).

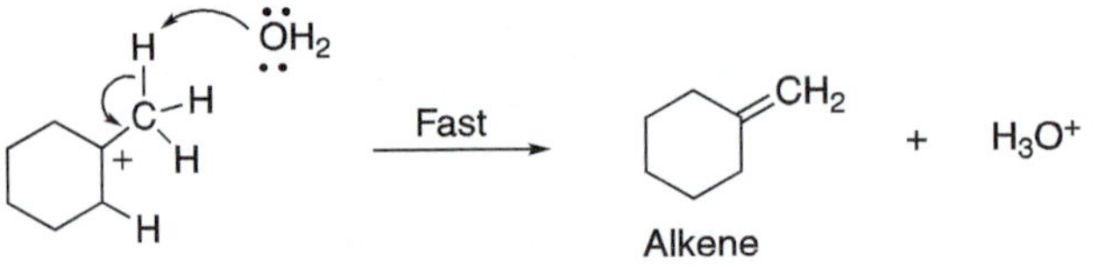

$$\text{[cyclohexyl cation with methyl and H}_2\text{O abstracting methyl H]} \xrightarrow{\text{Fast}} \text{methylenecyclohexane (Alkene)} + H_3O^+ \tag{12-33}$$

As pointed out earlier, the more stable alkene product is the one that has the most alkyl substitution about the carbon–carbon double bond; Thus, the alkene product shown in Reactions (12-32a) and (12-32b) is the major product and Reaction (12-34) gives a summary of the elimination product distribution.

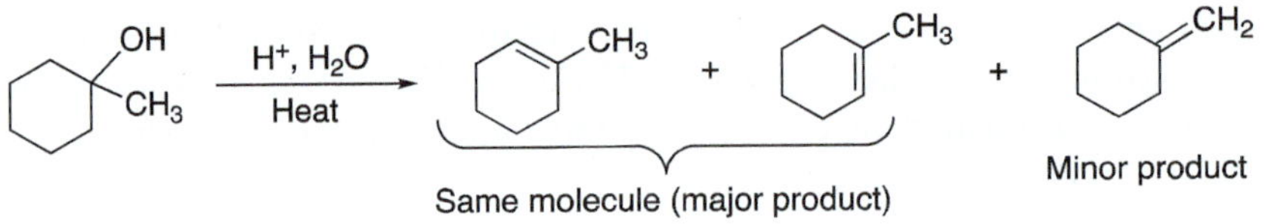

$$\text{1-methylcyclohexanol} \xrightarrow[\text{Heat}]{H^+, H_2O} \underbrace{\text{1-methylcyclohexene} + \text{1-methylcyclohexene}}_{\text{Same molecule (major product)}} + \underbrace{\text{methylenecyclohexane}}_{\text{Minor product}} \tag{12-34}$$

Let us examine a slightly more complex alcohol, in which it is possible to produce even more possible elimination products. Shown in Reaction (12-35) is the first step in the dehydration reaction of 1,2-dimethylcyclohexanol to form a carbocation.

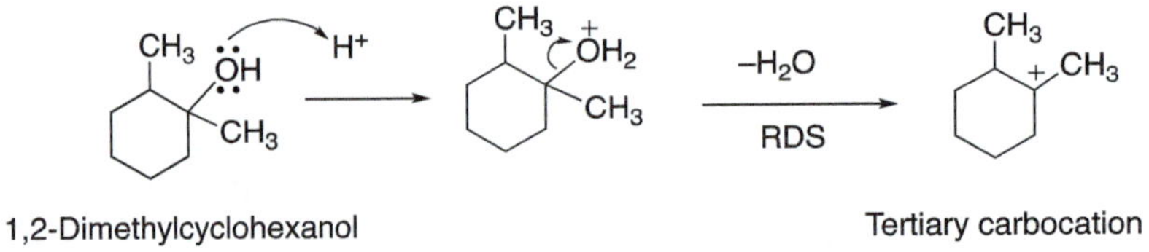

$$\underset{\text{1,2-Dimethylcyclohexanol}}{\text{[structure]}} \xrightarrow{H^+} \text{[protonated]} \xrightarrow[\text{RDS}]{-H_2O} \underset{\text{Tertiary carbocation}}{\text{[structure]}} \tag{12-35}$$

From this carbocation, the elimination of either of the hydrogens of the three adjacent carbons results in three different products as shown in Reactions (12-36)–(12-38).

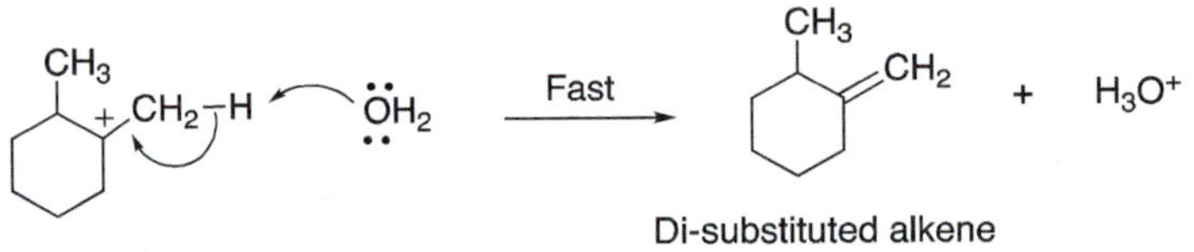

$$\text{[carbocation + OH}_2\text{ abstracting CH}_2\text{-H]} \xrightarrow{\text{Fast}} \underset{\text{Di-substituted alkene}}{\text{[alkene with CH}_3\text{ and =CH}_2\text{]}} + H_3O^+ \tag{12-36}$$

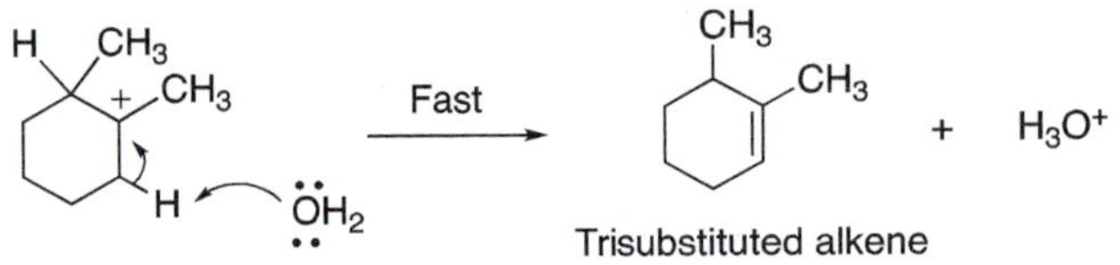

$$\text{[carbocation + OH}_2\text{ abstracting H]} \xrightarrow{\text{Fast}} \underset{\text{Trisubstituted alkene}}{\text{[alkene]}} + H_3O^+ \tag{12-37}$$

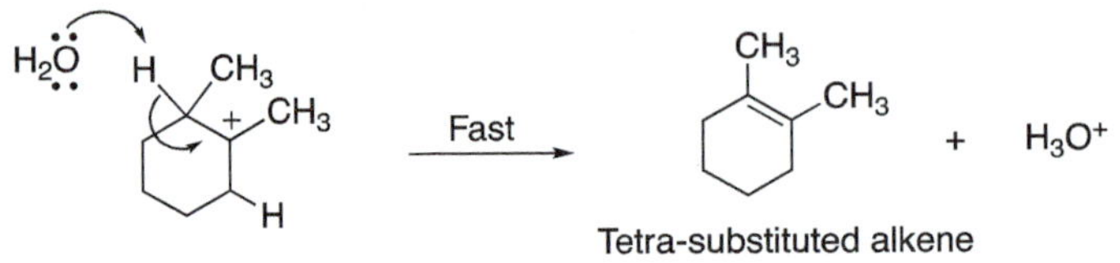

$$\text{[carbocation + H}_2\text{O abstracting H]} \xrightarrow{\text{Fast}} \underset{\text{Tetra-substituted alkene}}{\text{[alkene]}} + H_3O^+ \tag{12-38}$$

12.4 Elimination of Water (Dehydration)

Note that there are three alkenes possible from the initially formed carbocation, the disubstituted, trisubstituted, and tetrasubstituted alkenes. We have seen from the study of the stability of alkenes that the more substituted alkenes are the more stable alkenes. Hence, the more stable alkene will be the major organic product for the above reaction (Reaction 12-38). Another example is shown in Reaction (12-31).

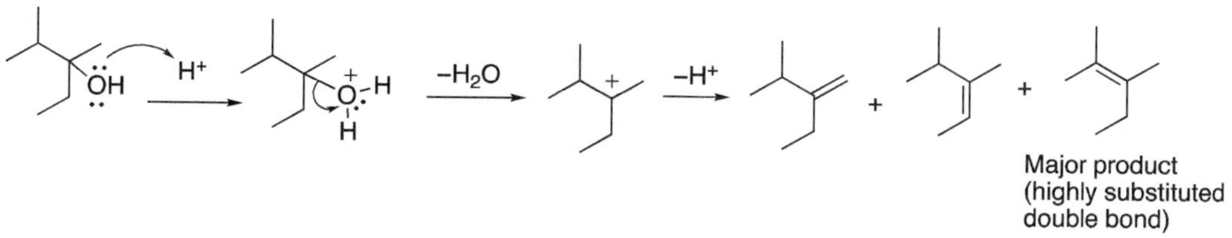

Major product
(highly substituted
double bond)

(12-39)

At this point, students should be able to predict the major product, based on knowledge of the type of mechanism and a rationale for predicting the major product based on the mechanism. Problem 12.5 is designed to get students to think analytically through this process.

Problem 12.5

Give the major organic product for the following reactions.

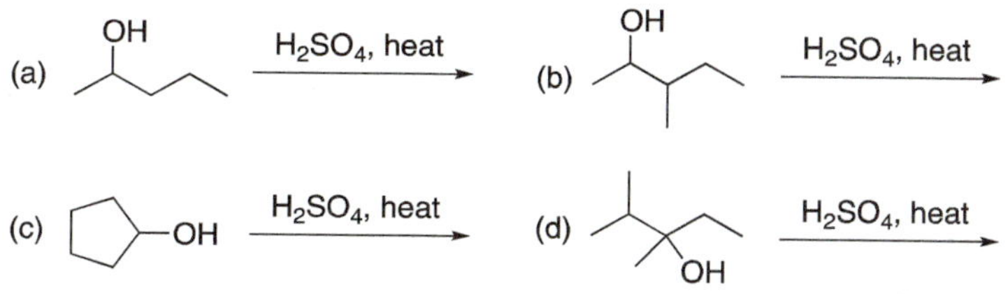

(a) H_2SO_4, heat

(b) H_2SO_4, heat

(c) H_2SO_4, heat

(d) H_2SO_4, heat

12.4.2 Carbocation Rearrangement

Carbocations once formed will rearrange to form a more stable carbocation if possible. Since carbocations are formed for the E1 mechanism, unexpected elimination products are sometimes observed. An example is shown in Reaction (12-40), where an unexpected product is obtained due to rearrangement of a cation intermediate.

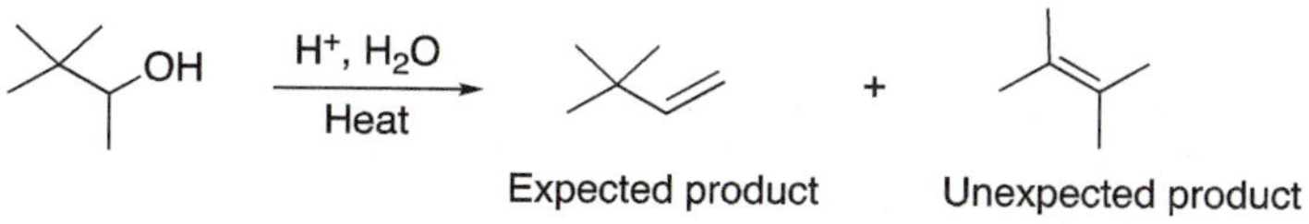

H^+, H_2O
Heat

Expected product   Unexpected product

(12-40)

Since this is an E1 mechanism and involves the formation of a carbocation, the expected product is formed by an abstraction of a proton from the adjacent methyl group as shown in Reaction (12-41).

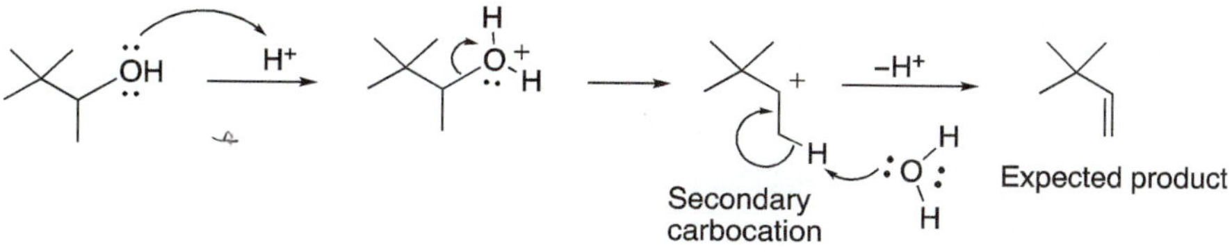

Secondary carbocation   Expected product

(12-41)

As we have seen before, it is possible for a secondary carbocation to become a more stable tertiary carbocation by migration of an adjacent methide group to form a more stable tertiary carbocation as shown in Reaction (12-42).

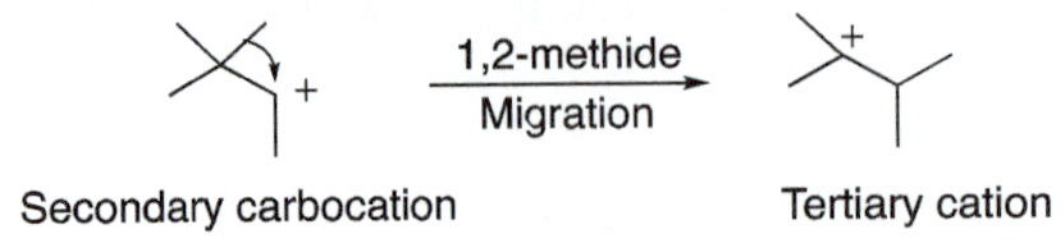

$$\text{Secondary carbocation} \xrightarrow{\text{1,2-methide Migration}} \text{Tertiary cation} \tag{12-42}$$

In the last step of the mechanism, deprotonation of this tertiary carbocation occurs to form the alkene as shown in Reaction (12-43).

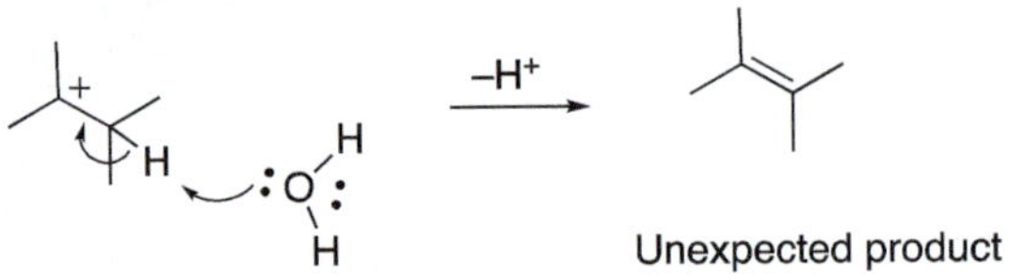

$$\xrightarrow{-H^+} \text{Unexpected product} \tag{12-43}$$

Problem 12.6

Provide a step-by-step mechanism to explain the formation of the unexpected products shown for the following reactions.

(a) 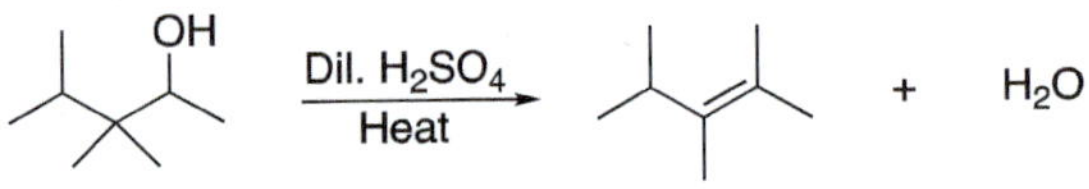 $\text{(alcohol)} \xrightarrow[\text{Heat}]{\text{Dil. } H_2SO_4} \text{(alkene)} + H_2O$

(b) 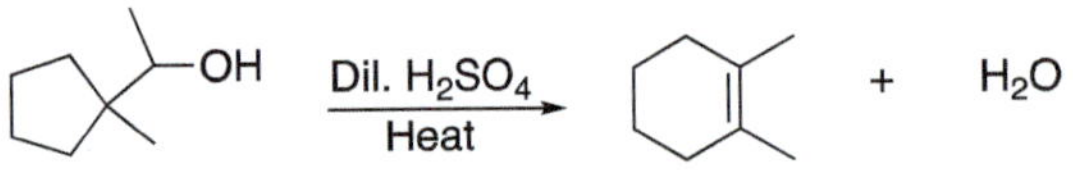 $\text{(cyclopentyl alcohol)} \xrightarrow[\text{Heat}]{\text{Dil. } H_2SO_4} \text{(cyclohexene)} + H_2O$

12.4.3 Pinacol Rearrangement

Let us now consider a diol, in which the alcohol functionalities are adjacent to each other. In the presence of an acid, the protonation of one OH functionality will occur and then water leaves to form a carbocation, as shown in Reaction (12-44).

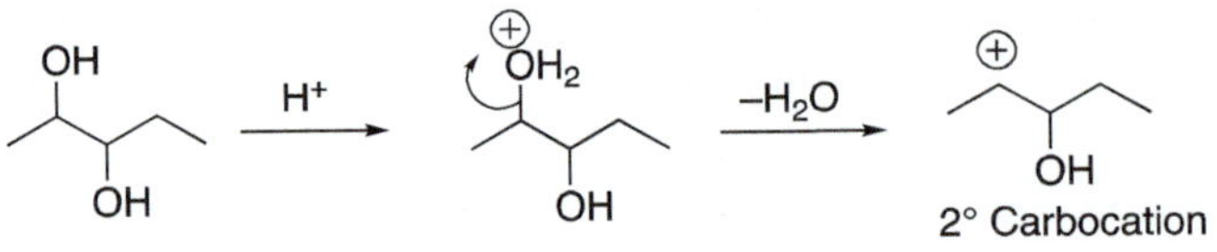

$$\xrightarrow{H^+} \xrightarrow{-H_2O} \text{2° Carbocation} \tag{12-44}$$

The secondary carbocation formed in Reaction (12-44) is fairly stable since it is a secondary carbocation, but if there were an alkyl migration from the adjacent carbon, a carbocation with greater stability would result since the new carbocation would be in resonance with the electrons of the adjacent oxygen atom as shown in Reaction (12-45). Deprotonation of the carbocation intermediate results in a final product, which is an aldehyde.

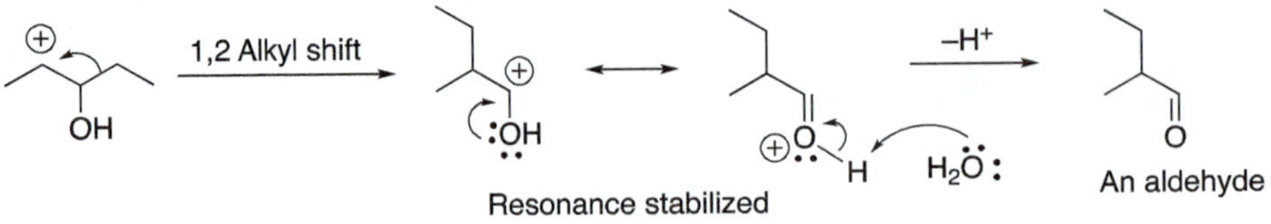

$$\xrightarrow{\text{1,2 Alkyl shift}} \text{Resonance stabilized} \xrightarrow{-H^+} \text{An aldehyde} \tag{12-45}$$

Dehydration of diols of this type is known as the pinacol rearrangement. Note that the pinacol rearrangement occurs for 1,2-diols. As you can imagine, there are many possible products depending on the type of diol and which OH group is protonated. The protonation of the other —OH group for the reaction given in Reaction (12-46) results in the same product as shown in Reactions (12-46) and (12-47).

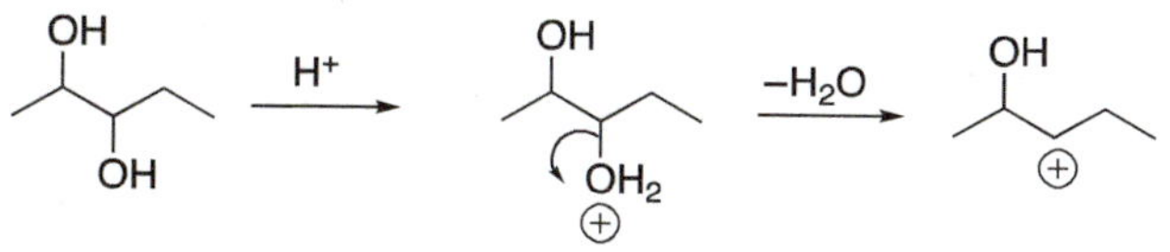

$$\text{(diol)} \xrightarrow{H^+} \text{(protonated)} \xrightarrow{-H_2O} \text{(carbocation)} \quad (12\text{-}46)$$

Even though this carbocation is different from the one generated in Reaction (12-26), the migration of the methyl group generates the same resonance-stabilized carbocation, which gives the same aldehyde.

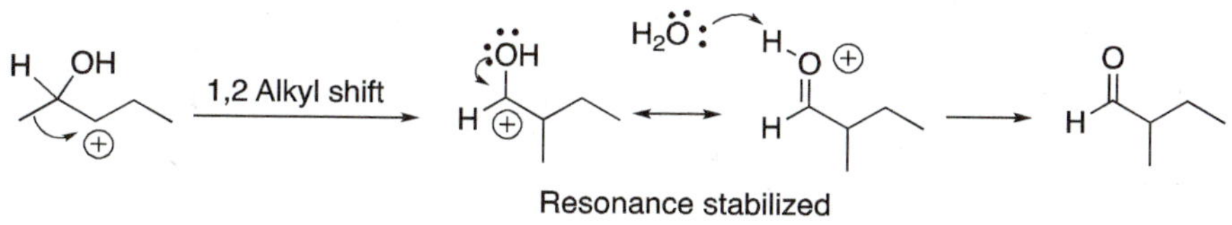

$$\xrightarrow{1,2 \text{ Alkyl shift}} \underbrace{\longleftrightarrow}_{\text{Resonance stabilized}} \longrightarrow \text{aldehyde} \quad (12\text{-}47)$$

Thus, for this pinacol rearrangement, the overall reaction is shown in Reaction (12-48).

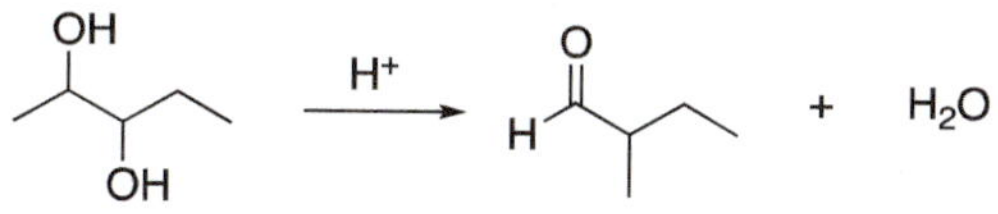

$$\text{diol} \xrightarrow{H^+} \text{aldehyde} + H_2O \quad (12\text{-}48)$$

Problem 12.7

Predict the pinacol rearranged products of the following reactions.

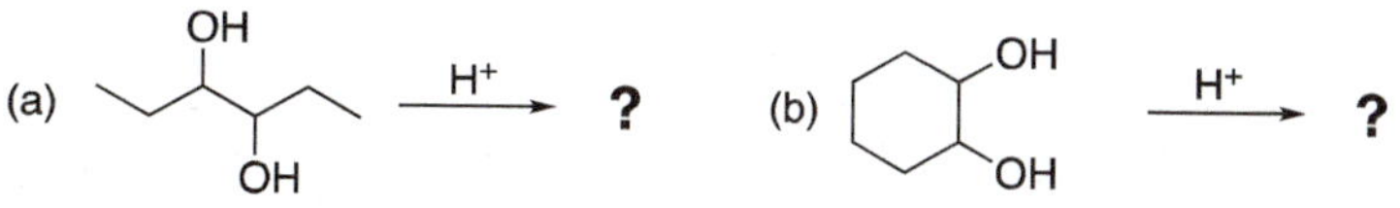

(a) [diol structure] $\xrightarrow{H^+}$?   (b) [cyclohexane-1,2-diol] $\xrightarrow{H^+}$?

12.5 Applications of Elimination Reactions to Synthesis

Now we need to concentrate on the application of elimination reactions to synthesize specific target molecules that have double bonds. The same approach should be used as discussed in previous sections; that is, identify the functional group in the target molecule, in this case the double bond, and determine the best type of reaction and needed molecule that can be used to synthesize the target molecule that contains the double bond functionality. This concept is illustrated below.

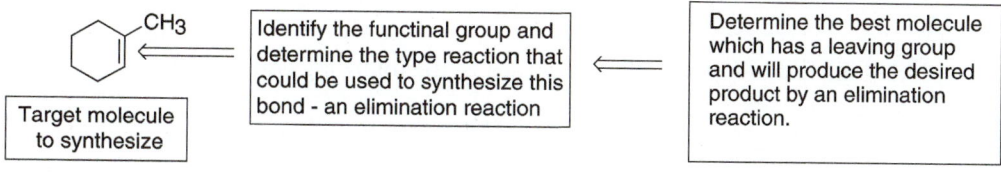

The best approach in solving synthesis problems is to work backward by looking carefully at the target molecule and making some important observations. For example, identify the functional group present and then determine how best to make that functional group based on the reactions that we have discussed. The strategy that can be used is that an appropriate starting material to make the alkene functionality would be an alkyl halide, with of course the same carbon skeleton. An elimination reaction involving an alkyl halide that should produce the desired product is shown in Reaction (12-49).

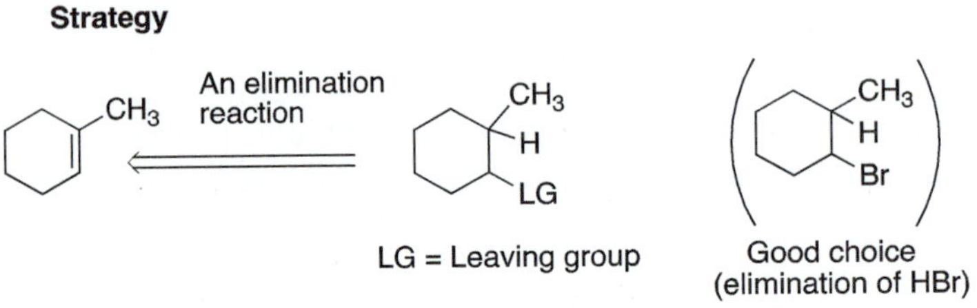

Strategy

methylcyclohexene ⇐(An elimination reaction)= methylcyclohexane with CH₃, H, LG (= methylcyclohexane with CH₃, H, Br — Good choice (elimination of HBr))   (12-49)

LG = Leaving group

You will readily recognize that there is another possibility as shown below in Reaction 12-50.

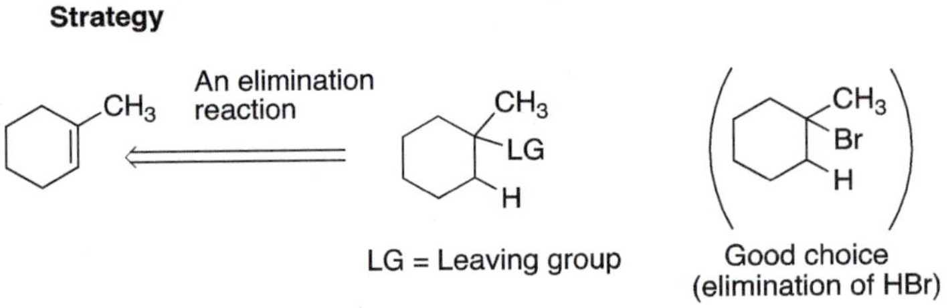

Strategy

methylcyclohexene ⇐(An elimination reaction)= methylcyclohexane with CH₃, LG, H (= methylcyclohexane with CH₃, Br, H — Good choice (elimination of HBr))   (12-50)

LG = Leaving group

To determine the possible reactions that can be used, appropriate reagents and reactions conditions for an elimination reaction are needed, as shown for the reactions discussed in section 12.3 and given in Reactions (12-51) and (12-52) (note that the reactions are shown in reverse).

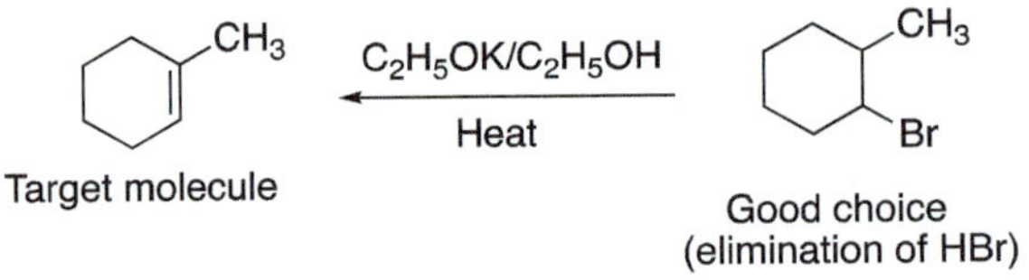

Target molecule (methylcyclohexene) ⇐[C₂H₅OK/C₂H₅OH, Heat]= methylcyclohexane-CH₃ with Br (Good choice, elimination of HBr)   (12-51)

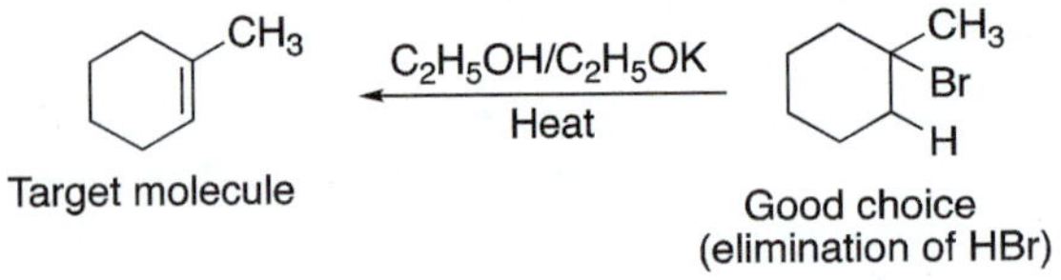

Target molecule (methylcyclohexene) ⇐[C₂H₅OH/C₂H₅OK, Heat]= methylcyclohexane-CH₃ with Br, H (Good choice, elimination of HBr)   (12-52)

We will see in Chapter 14 that one of these reactions is a better choice, compared to the other based on the ease in the synthesis of the starting compound.

Problem 12.8

Starting with any alkyl halide, show how to synthesize the following compounds.

(a) 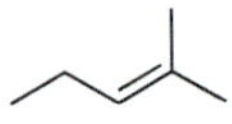 2-methyl-2-butene   (b)  cyclopentene

12.5 Applications of Elimination Reactions to Synthesis

Let us now turn our attention to the combination of reactions studied so far for the synthesis of a target molecule starting from a specific starting compound (multistep synthesis), as shown in Reaction (12-53).

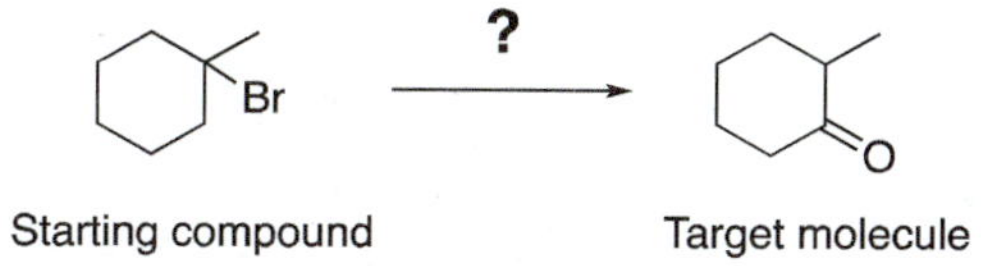

$$\text{Starting compound (cyclohexyl bromide with methyl)} \xrightarrow{?} \text{Target molecule (2-methylcyclohexanone)} \tag{12-53}$$

Using the strategy described earlier in this section shows that an oxidation of an alcohol could be used to synthesize the target molecule, which means that the intermediate molecule would have to be an alkene. We learned in this chapter that elimination reaction of alkyl halides results in the formation of alkenes. Thus, the strategy outlined below could be used for this transformation.

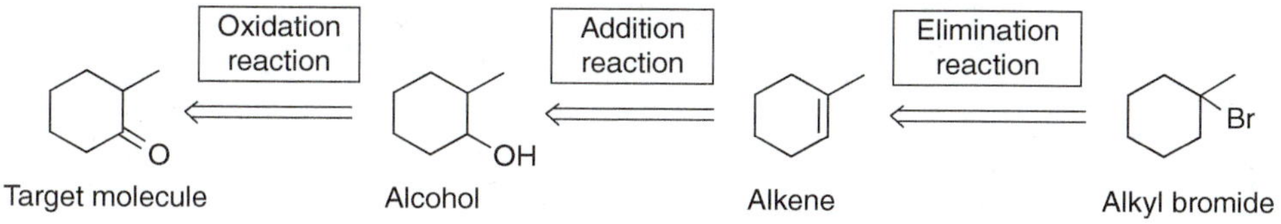

Target molecule ⇐[Oxidation reaction]= Alcohol ⇐[Addition reaction]= Alkene ⇐[Elimination reaction]= Alkyl bromide

Once the strategy is worked out, we now need to select appropriate reactions and reaction conditions to accomplish the multistep synthesis, a possibility is outlined in Reaction (12-54).

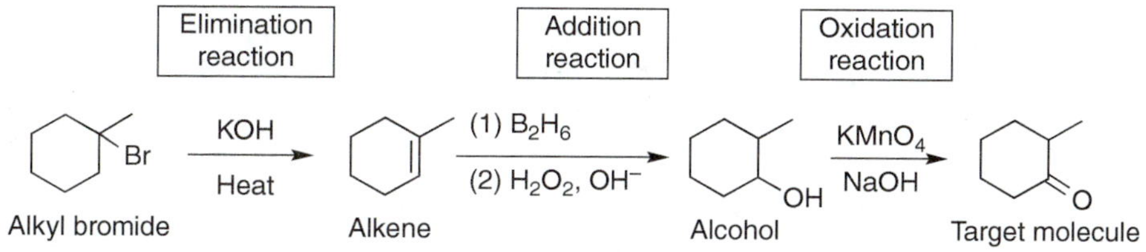

Alkyl bromide $\xrightarrow[\text{Heat}]{\text{KOH}}$ [Elimination reaction] Alkene $\xrightarrow[\text{(2) H}_2\text{O}_2,\ \text{OH}^-]{\text{(1) B}_2\text{H}_6}$ [Addition reaction] Alcohol $\xrightarrow[\text{NaOH}]{\text{KMnO}_4}$ [Oxidation reaction] Target molecule (12-54)

Problem 12.9

Using a multi-step approach, show how to carry out the following transformations.

(a) 2-bromo-3-methylbutane → 3-bromo-2-methylbutane

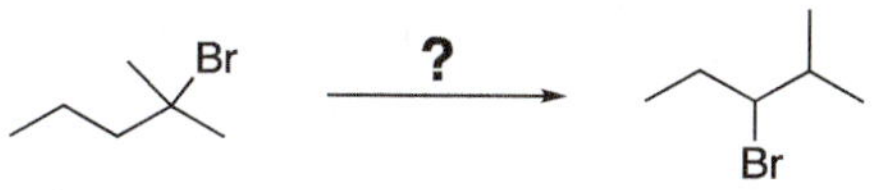

(b) 3-methyl-2-butanone → 2-methyl-2-butanol

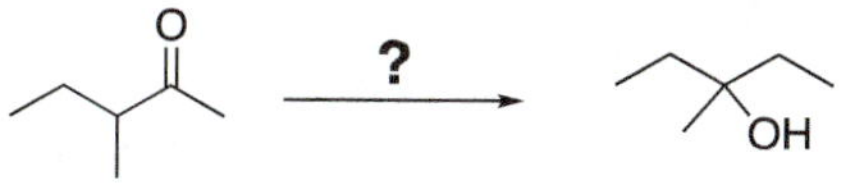

(c) 2-bromo-2-methylbutane → 3-methyl-2-butanone

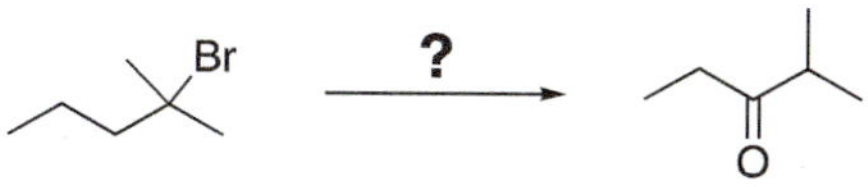

End of Chapter Problems

12.10 Give the major organic products of each of the following reactions.

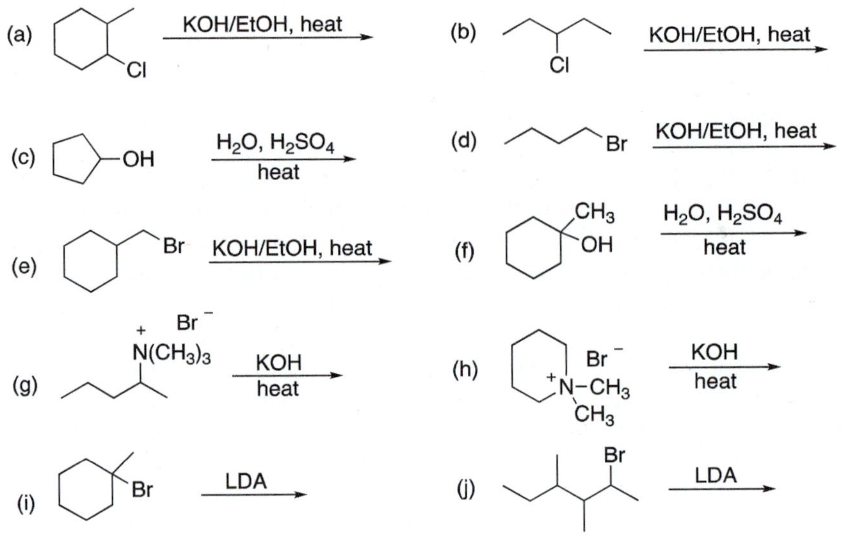

(a) [cyclohexane with methyl and Cl substituents] → KOH/EtOH, heat

(b) [alkyl chloride] → KOH/EtOH, heat

(c) [cyclopentanol] → H_2O, H_2SO_4, heat

(d) [alkyl bromide] → KOH/EtOH, heat

(e) [cyclohexylmethyl bromide] → KOH/EtOH, heat

(f) [cyclohexanol with CH₃ and OH] → H_2O, H_2SO_4, heat

(g) [alkyl ammonium with $N(CH_3)_3^+$ and Br^-] → KOH, heat

(h) [piperidine derivative with $^+N-CH_3$, CH_3, Br^-] → KOH, heat

(i) [cyclohexyl bromide with methyl] → LDA

(j) [alkyl bromide] → LDA

12.11 Give the major organic products for the reaction of 3-methyl-2-pentene with each of the following reagents

a) (i) $Hg(OAc)_2/H_2O$ ii) $NaBH_4$
b) (i) BH_3 ii) H_2O_2, OH^-
c) $KMnO_4$, room temp.
d) Br_2/H_2O
e) i) O_3 ii) Me_2S
f) $KMnO_4$, heat

12.12 What is the best reagent(s) to carry out the following transformations?

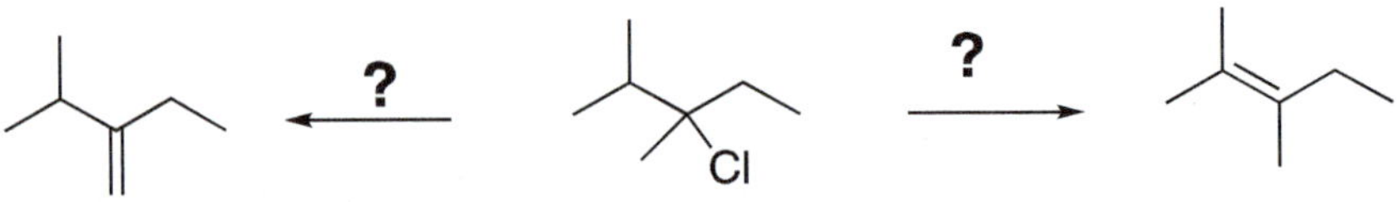

[alkene] ← ? [alkyl chloride] → ? [alkene]

12.13 Using arrow-pushing formulism to indicate electron movement, propose a reasonable mechanism to explain the reaction shown below.

[tertiary alcohol] → H^+ → [alkene] + H_2O

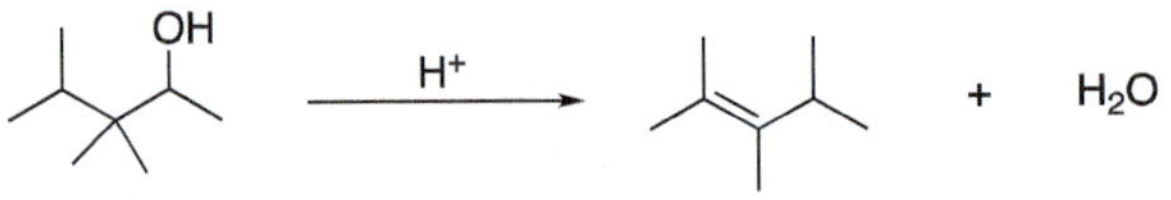

12.14 Using arrow-pushing formulism to indicate electron movement, propose a reasonable mechanism to explain the reaction shown below

[unsaturated alcohol] → H^+ → [methylcyclohexene] + H_2O

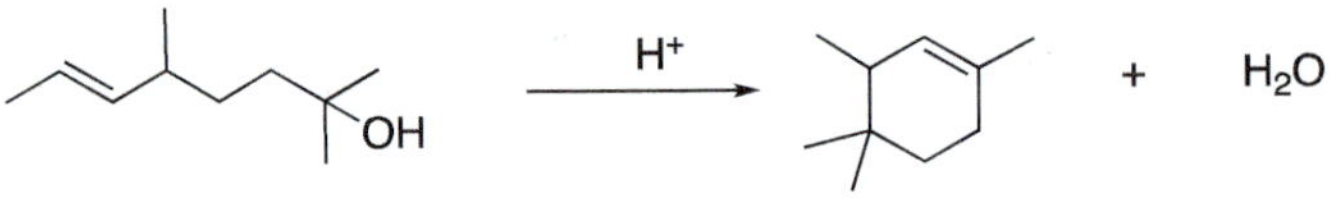

End of Chapter Problems | 327

12.15 Using a multistep approach, show how to carry out the following transformations.

(i) 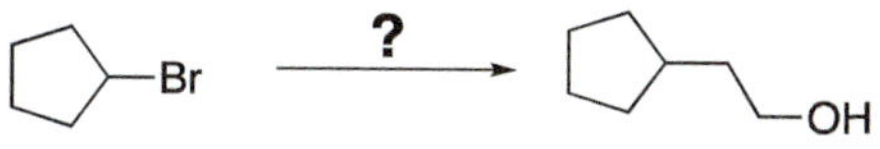 cyclopentyl-Br →?→ cyclopentyl-CH$_2$CH$_2$-OH

(ii) 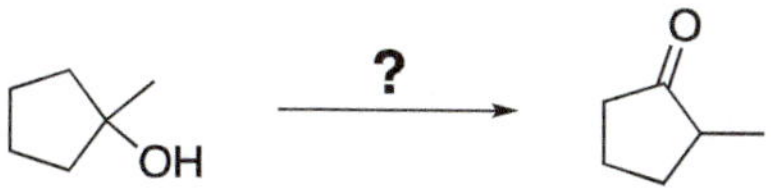 1-methylcyclopentan-1-ol →?→ 2-methylcyclopentan-1-one

12.16 Predict the pinacol rearranged product for the following reaction and propose a mechanism to explain your answer.

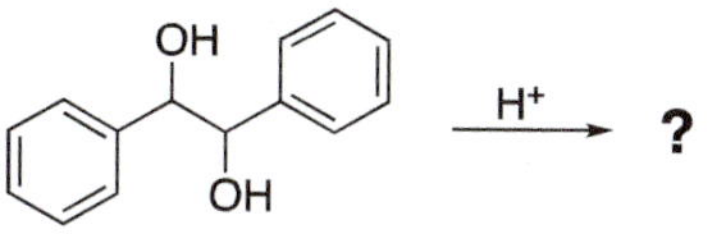

1,2-diphenyl-1,2-ethanediol (both OH groups shown) →H⁺→ ?

12.17 Briefly explain the outcome of the reactions shown below.

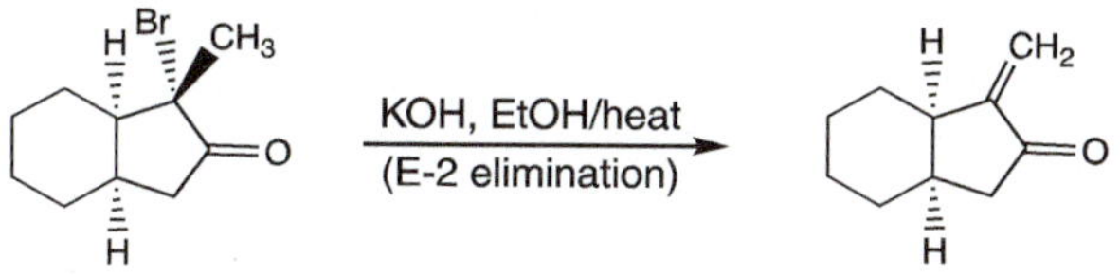

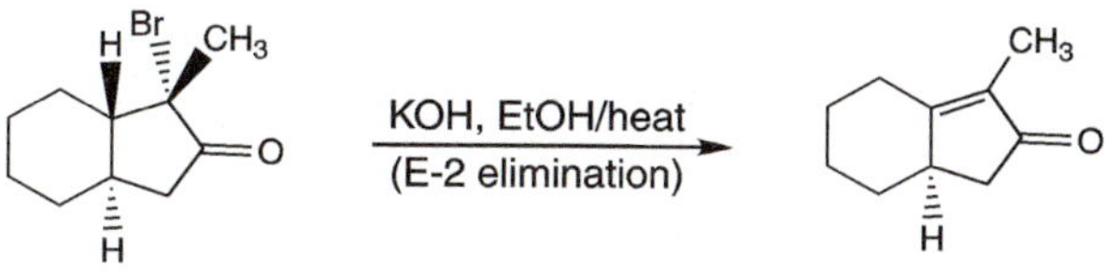

bicyclic ketone with Br and CH$_3$ (trans) →KOH, EtOH/heat (E-2 elimination)→ bicyclic ketone with =CH$_2$ (exocyclic)

bicyclic ketone with Br and CH$_3$ (cis) →KOH, EtOH/heat (E-2 elimination)→ bicyclic ketone with CH$_3$ on endocyclic alkene

12.18 Complete the following reactions by giving the structures of the major organic product. Indicate stereochemistry where appropriate.

(a) 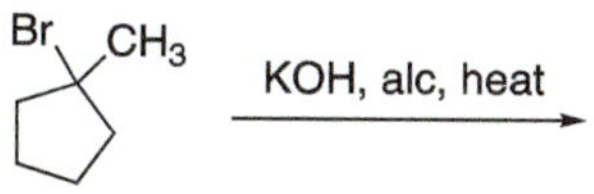 1-bromo-1-methylcyclopentane →KOH, alc, heat→

(b) 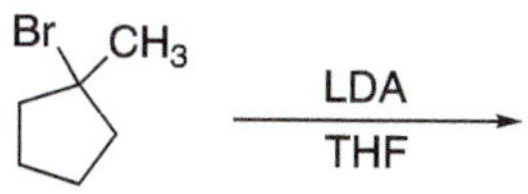 1-bromo-1-methylcyclopentane →LDA / THF→

(c) 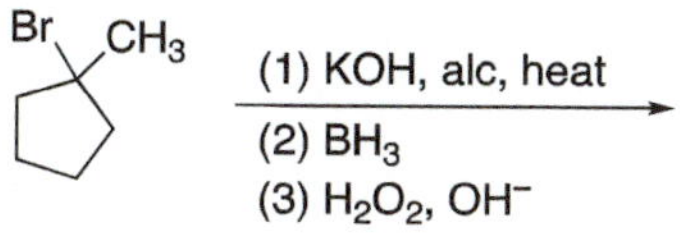 1-bromo-1-methylcyclopentane →(1) KOH, alc, heat (2) BH$_3$ (3) H$_2$O$_2$, OH⁻→

(d) 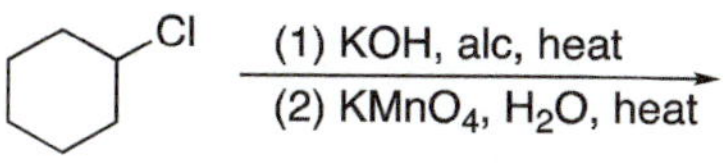 chlorocyclohexane →(1) KOH, alc, heat (2) KMnO$_4$, H$_2$O, heat→

(e) 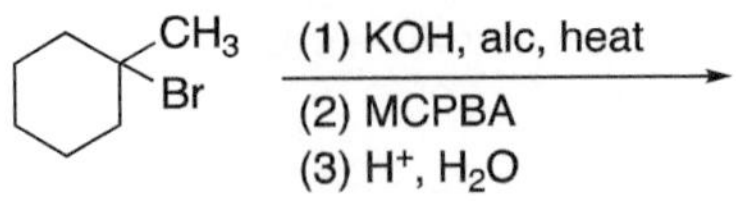 1-bromo-1-methylcyclohexane
(1) KOH, alc, heat
(2) MCPBA
(3) H⁺, H₂O

(f) 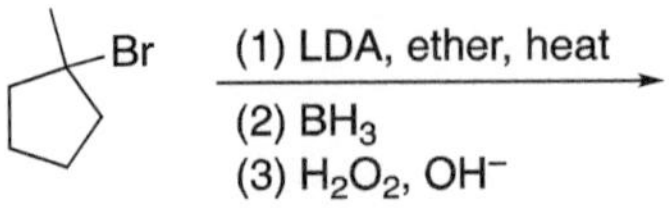 1-bromo-1-methylcyclopentane
(1) LDA, ether, heat
(2) BH₃
(3) H₂O₂, OH⁻

(g) 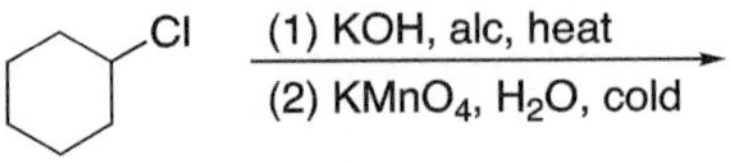 chlorocyclohexane
(1) KOH, alc, heat
(2) KMnO₄, H₂O, cold

(h) 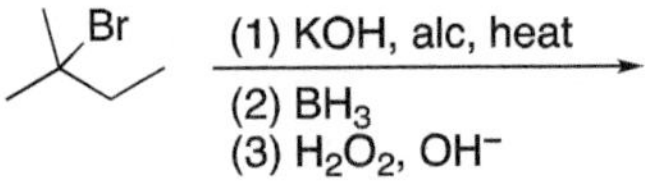 2-bromo-2-methylbutane (tert-amyl bromide) — drawn as (CH₃)₂C(Br)CH(CH₃) structure
(1) KOH, alc, heat
(2) BH₃
(3) H₂O₂, OH⁻

(i) 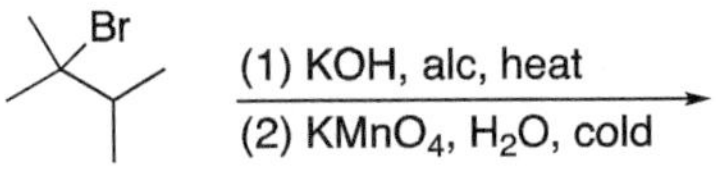 2-bromo-3-methylbutane
(1) KOH, alc, heat
(2) KMnO₄, H₂O, cold

(j) 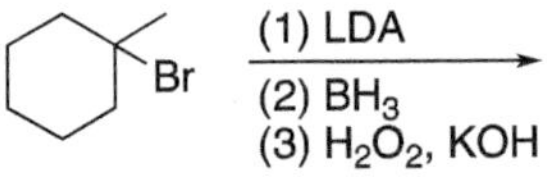 1-bromo-1-methylcyclohexane
(1) LDA
(2) BH₃
(3) H₂O₂, KOH

(k) 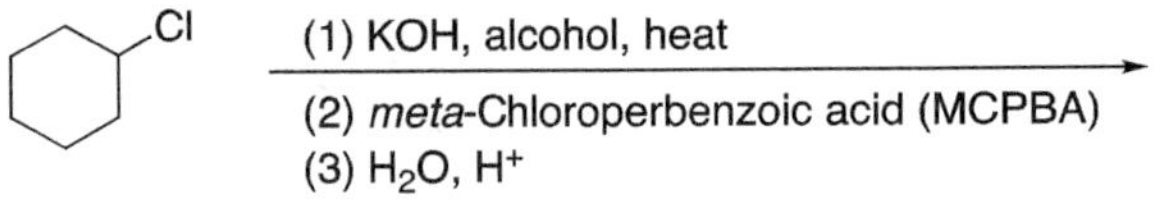 chlorocyclohexane
(1) KOH, alcohol, heat
(2) *meta*-Chloroperbenzoic acid (MCPBA)
(3) H₂O, H⁺

(l) 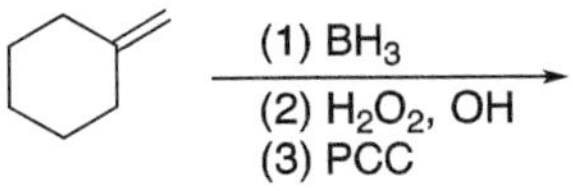 methylenecyclohexane
(1) BH₃
(2) H₂O₂, OH
(3) PCC

(m) 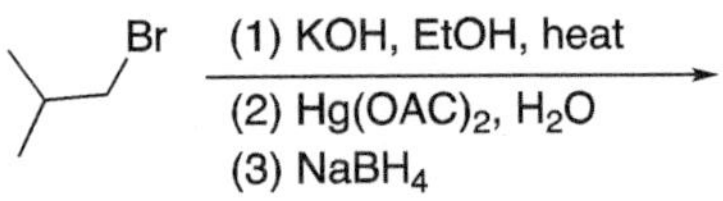 1-bromo-2-methylpropane (isobutyl bromide)
(1) KOH, EtOH, heat
(2) Hg(OAC)₂, H₂O
(3) NaBH₄

(n) 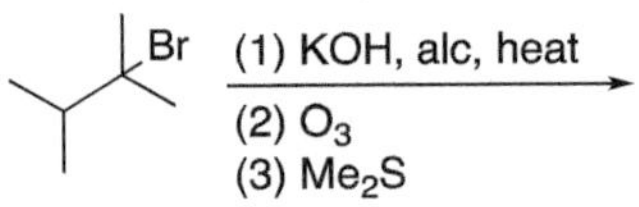 2-bromo-2,3-dimethylbutane
(1) KOH, alc, heat
(2) O₃
(3) Me₂S

(o) 1-Ethyl-1-cyclohexanol
(1) H₂SO₄, heat
(2) BH₃
(3) H₂O₂, KOH

(p) 1-Bromo-1-methylcyclopentane $\xrightarrow{\text{(1) KOH, EtOH, heat}}_{\text{(2) KMnO}_4\text{, H}_2\text{O, room temp.}}$

(q) [structure: bicyclic compound with CH₃, Br, H, H, and =O substituents] $\xrightarrow[\text{(E-2 elimination)}]{\text{KOH, EtOH/heat}}$

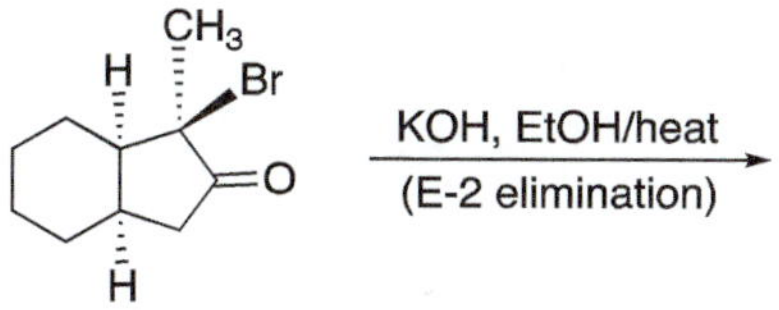

(r) 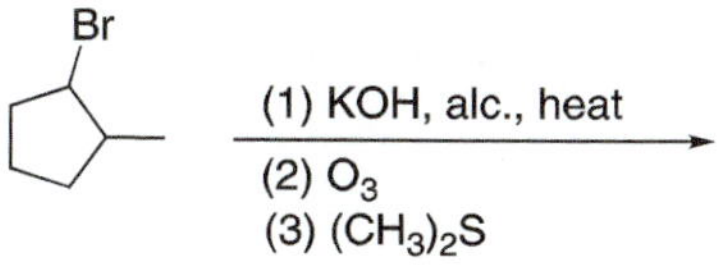 [structure: 1-bromo-2-methylcyclopentane] $\xrightarrow{\begin{array}{l}\text{(1) KOH, alc., heat}\\\text{(2) O}_3\\\text{(3) (CH}_3\text{)}_2\text{S}\end{array}}$

(s) [cyclohexene] $\xrightarrow{\begin{array}{l}\text{(1) MCPBA/CH}_2\text{Cl}_2\\\text{(2) CH}_3\text{MgCl}\\\text{(3) H}^+\text{/H}_2\text{O}\end{array}}$

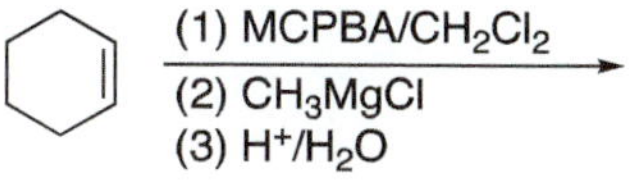

(t) 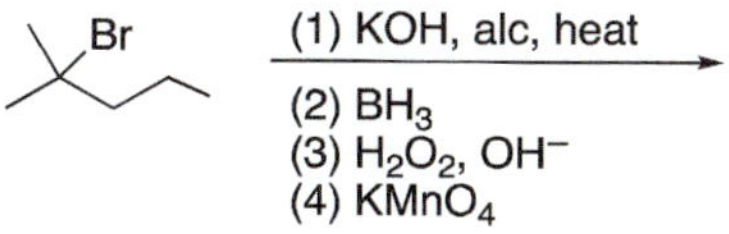 [structure: 2-bromo-3-methylbutane] $\xrightarrow{\begin{array}{l}\text{(1) KOH, alc, heat}\\\text{(2) BH}_3\\\text{(3) H}_2\text{O}_2\text{, OH}^-\\\text{(4) KMnO}_4\end{array}}$

12.19 Treatment of compound **A** ($C_7H_{13}Br$) with sodium ethoxide (a base) in ethanol at 60 °C converts **A** into **B** (C_7H_{12}). Ozonolysis of **B** gives **C** (shown below) as the only product. Deduce the structures of **A** and **B**.

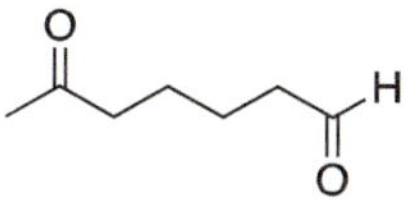

[structure of Compound C: a diketone/keto-aldehyde]

Compound **C**

12.20 Give the structures of compounds **A** through **D** for the reaction scheme shown below.

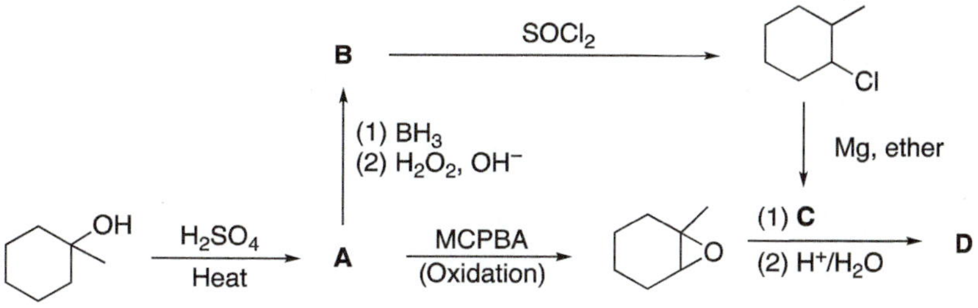

[Reaction scheme: cyclohexanol $\xrightarrow[\text{Heat}]{\text{H}_2\text{SO}_4}$ **A** $\xrightarrow[\text{(Oxidation)}]{\text{MCPBA}}$ [epoxide] $\xrightarrow[\text{(2) H}^+/\text{H}_2\text{O}]{\text{(1) C}}$ **D**

A $\xrightarrow{\begin{array}{l}\text{(1) BH}_3\\\text{(2) H}_2\text{O}_2\text{, OH}^-\end{array}}$ **B** $\xrightarrow{\text{SOCl}_2}$ [chlorocyclohexane with methyl] $\xrightarrow{\text{Mg, ether}}$ **C**]

12.21 Give the structural formulas for the lettered compounds of the following sequence of reactions.

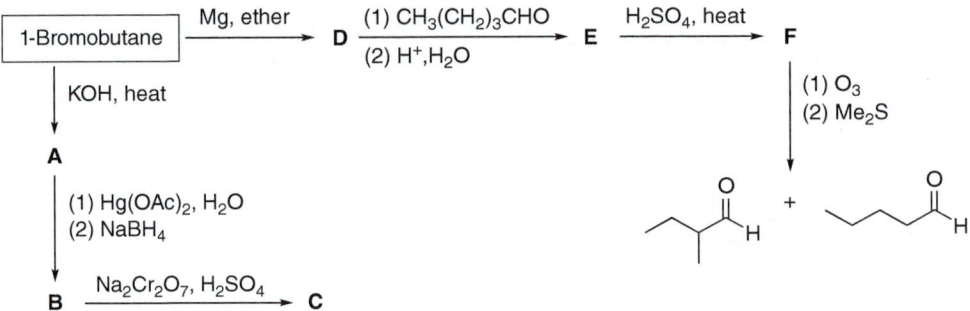

12.22 Give the structures for the lettered compounds in the reaction scheme below.

13

Spectroscopy Revisited, A More Detailed Examination

13.1 Introduction

In Section 3.16 of Chapter 3, the topic of spectroscopy, specifically infrared (IR) spectroscopy, was introduced in the context of functional groups and how this type spectroscopy can be used to assist in the identification of specific functional groups of different organic molecules. In this Chapter, IR spectroscopy, along with additional types of spectroscopy, will be examined in greater details. There are literally millions of different types of organic compounds, and everyday new organic compounds are being isolated from natural sources. Many of these compounds are described as natural products. Most have healing and medicinal benefits and they have been reproduced in large quantities by various industries, mainly for pharmaceutical uses. There are other organic compounds that are synthesized in various research labs. Thus, an integral aspect of an organic chemist's work is the utilization of spectroscopy in the determination of the actual structures of newly discovered and synthesized compounds.

The types of spectroscopy that will be covered in this chapter include ultraviolet, IR, mass spectrometry, and nuclear magnetic resonance (NMR) spectroscopy. Spectroscopy involves the interaction of molecules with different forms of energy. Of interest to chemists, and especially to organic chemists, is the nature of the interaction of molecules when they come into contact with energy. Once energy of a specified frequency interacts with a molecule, the molecule becomes excited and undergoes different types of transitions depending on the frequency of the energy. When some molecules are exposed to energy in the form of light, weak bonds are broken to form radicals as the case of peroxides and halogens. On the other hand, when water molecules, for example are excited by microwave energy, they rotate and, as a result, heat is generated, which is used to heat our food in the microwave oven. Based on the nature of the interaction of molecules with energy of different frequencies, important information can be gained about molecules and used to assist in structure determination. The forms of energy that will be used in spectroscopy encountered in this course include visible light, ultraviolet light, IR, and radio waves.

13.2 The Electromagnetic Spectrum

Energy can be described as massless photons traveling from one point to another at the speed of light, and the travel can be described in the form of waves. In this section, a detailed review of the electromagnetic spectrum will be carried out to ensure that students appreciate the amount of energy transmitted at the various regions of the electromagnetic spectrum. Energy travels at the speed of light and characteristics, such as wavelength and frequencies, can be

Organic Chemistry: Concepts and Applications, First Edition. Allan D. Headley.
© 2020 John Wiley & Sons, Inc. Published 2020 by John Wiley & Sons, Inc.
Companion website: www.wiley.com/go/Headley_OrganicChemistry

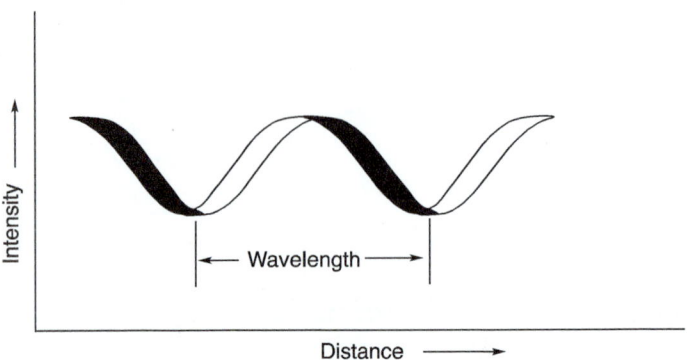

Figure 13.1 Representation of energy in the form of a wave.

THE ELECTROMAGNETIC SPRECTRUM

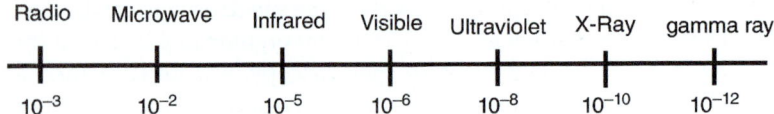

Figure 13.2 Relationship between the wavelength and the different regions of the electromagnetic spectrum.

used to describe the energy at various regions of the electromagnetic spectrum. Wavelength (λ) is the distance from one peak (or valley) of a wave to the next as shown in Figure 13.1. This distance is typically reported in centimeters (cm).

The relationship between wavelength and the various regions of the electromagnetic spectrum is shown in Figure 13.2.

There is a mathematical equation that relates the wavelength to the energy of the different regions of the electromagnetic spectrum and is given in Eq. (13-1).

$$\Delta E = hc / \lambda \tag{13-1}$$

In Eq. (13-1), ΔE is the relative energy, h is a constant (Planck's constant), c is the velocity of light, and λ is the wave length; note that the energy is inversely proportional to wavelength (λ). That is, if the wave length is short, there is a lot of energy. Table 13.1 gives the relative energies and the wavelength ranges for the various regions of the electromagnetic spectrum as well as descriptions of different regions and their effects on molecules.

Based on this information, gamma rays are the most dangerous. Another characteristic of energy is the frequency, which is the number of cycles that passes a specific point in a given time, typically in a second and is usually represented by the Greek letter ν and given in hertz (Hz). Equation (13-2) gives the relationship between frequency and energy.

$$\Delta E = h\nu \tag{13-2}$$

From the relationship in Eq. (13-2), note that energy is directly proportional to the frequency.

The wave number ($\tilde{\nu}$) is another variable that is used to represent energy and the units are in reciprocal centimeter, cm^{-1}. Equation (13-3) gives the relationships with energy and various variables, note that the energy is directly proportional to the wave number and inversely proportional to the wavelength.

$$\Delta E = h\nu = hc / \lambda = hc\tilde{\nu} \tag{13-3}$$

Table 13.1 Relationship between different regions of the electromagnetic spectrum, their energies and effects on molecules.

Wavelength, λ (cm)	Region of EM	Energy (kJ mol^{-1})	Effect of molecule
10^{-9}	Gamma (γ) rays	10^7	Ionization
10^{-7}	X-rays	10^5	Ionization
10^{-7}	Vacuum ultraviolet	10^3	Ionization
10^{-}	Near ultraviolet	10^3	Electronic transition
10^{-4}	Visible	10^2	Electronic transition
10^{-3}	Infrared	10	Molecular vibrations
10^{-1}	Microwave	10^{-2}	Molecular rotations
10^2–10^4	Radio waves	10^{-4}–10^{-6}	Nuclear spin transition

13.2.1 Types of Spectroscopy Used in Organic Chemistry

There are different types of spectroscopy that are used for the determination of the presence of different functional groups and the structure of compounds. Spectroscopy involves the interpretation of the various ways that molecules interact with different forms of energy that are described in Table 13.1. The energy used in organic spectroscopy is from different regions of the electromagnetic spectrum, which include visible light, ultraviolet light, microwave, IR, and radio waves. Since each molecule interacts differently in the presence of different types of energy, the nature of these interactions will give an idea of the structure of unknown compounds. Once energy of a specified frequency is exposed to molecules, an interaction occurs that causes the molecules to be in an excited state owing to the energy absorbed. Once this energy is absorbed, there are different outcomes. Thus, based on how molecules interact with energy of the various regions of the electromagnetic spectrum, that information can be used to assist in determining the identity of unknown molecules and identify the presence of different functional groups.

> **DID YOU KNOW?**
>
> When water molecules are excited by energy in the microwave region of the electromagnetic spectrum. Energy from the microwave oven causes polar water molecules to rotate and, as a result generate heat, which is used to heat our food in the microwave oven. You know from experience that if water is heated in a Styrofoam cup using the microwave oven, the cup is not heated, just the water. For the package of energy that is used in microwave ovens, water molecules are excited and not Styrofoam for example.
>
>

13.3 UV-Vis Spectroscopy and Conjugated Systems

Let us start our discussion of UV-Vis spectroscopy by first examining the effects of energy on molecular orbitals of ethylene. You will recall from Chapter 1 that the two carbons of ethylene are both sp^2 hybridized, which means that each carbon contains an unhybridized p orbital, which is perpendicular (orthogonal) to the plane of the atoms of the ethylene molecule. Each p orbital has one electron and these electrons form two molecular pi (π) orbitals, one lower in energy than the other. The representation is shown in Figure 13.3.

ΔG in Figure 13.3 is the energy required to promote one electron from the lower molecular orbital, also known as the highest occupied molecular orbital (HOMO), to the higher antibonding orbital, which is also known as the lowest unoccupied molecular orbital (LUMO). There is an exact amount of energy that is required for this electron promotion that can be expressed in terms of wavelength by the relationship that we saw earlier in the chapter. For this single isolated double bond, the exact wavelength (λ) is 165 nm as shown in Figure 13.4.

You will recall that this wavelength is in the region of the UV-Vis region of the electromagnetic spectrum. The instrument that is used to supply energy in this region of the electromagnetic spectrum is called a UV-Vis spectrophotometer. By scanning the UV-Vis region, a signal is obtained if a sample has a double bond that corresponds to this specific wavelength. Thus, if a molecule has an isolated double bond, you would expect a signal in the region of 165 nm of the UV-Vis spectrum. This type of information can be used to confirm that an unknown compound has a double bond if it has a signal in this region of the UV-Vis spectrum.

Let us now consider another system where there is another double bond next to an ethylene double bond, as in 1,3-butadiene. The pi (π) molecular orbitals are shown in Figure 13.5.

The energy gap ΔG between the HOMO and LUMO is smaller for 1,3-butadiene, compared to ethylene. This type of shift to a longer wavelength (less energy) is referred to a *bathochromic shift*. 1,3-Butadiene has two double bonds that are in conjugation with each other. The energy that is required for the promotion of an electron from the HOMO to the LUMO is in the

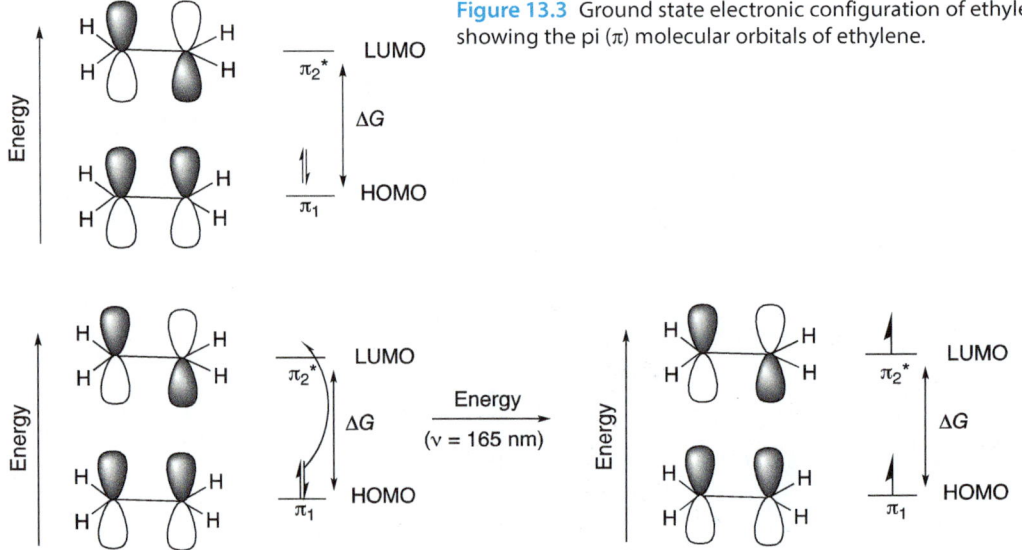

Figure 13.3 Ground state electronic configuration of ethylene showing the pi (π) molecular orbitals of ethylene.

Figure 13.4 Ground state electronic configuration and excited state of ethylene showing the pi (π) molecular orbitals of ethylene after specific energy is absorbed.

Figure 13.5 pi (π) orbitals of 1,3-butadiene.

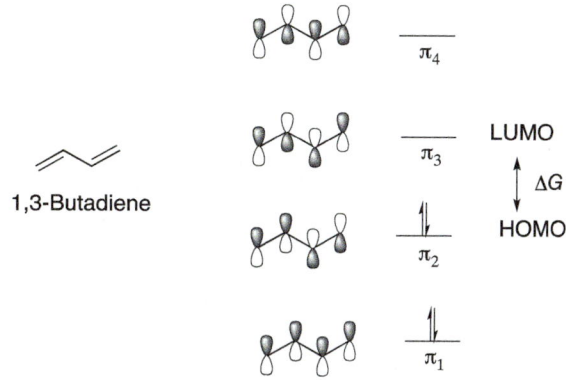

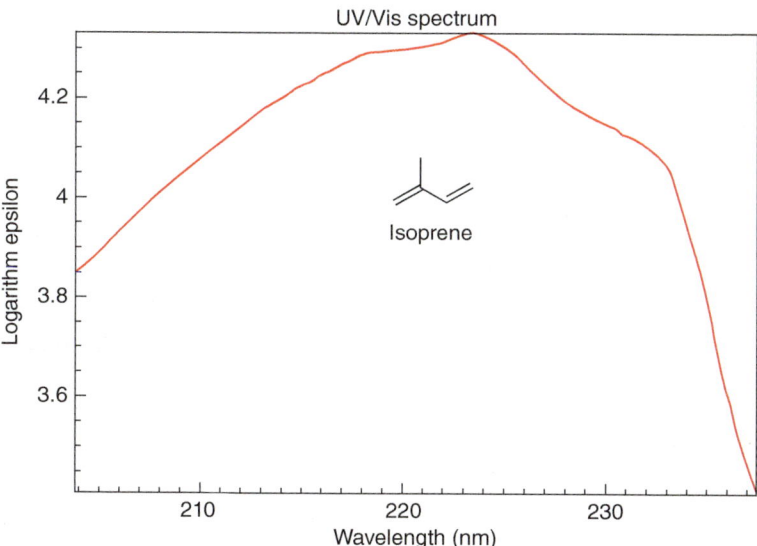

Figure 13.6 UV-Vis spectrum for isoprene, which has two double bonds in conjugation, λ_{max} is 217 nm. *Source:* with permission from NIST.

UV-Vis range of the electromagnetic spectrum and the wavelength is 217 nm. The UV-Vis spectrum of isoprene, which has two double bonds in conjugation, is shown in Figure 13.6.

Notice that the energy requirement is now approximately 217 nm, compared to 165 nm of ethylene. If there were another double bond in conjugation as in 1,3,5-triene, the energy requirement for promotion would be less as reflected by the wavelength of approximately 258 nm. The human eye sees only the colors in the visible region of the electromagnetic spectrum. Thus, if there is a compound that absorbs strongly in this region, that color is absorbed in the electronic transition and what we see is the rainbow colors, minus that color. A large percentage of organic compounds contain various numbers of conjugated double bonds. β-Carotene, which is a highly colored compound, has 11 double bonds in conjugation; β-carotene is a precursor of vitamin A, which is needed for good health and vision. β-Carotene, which structure is shown below, absorbs at 454 nm (blue region), so its absorbance is in the visible region of the electromagnetic spectrum. The human eye detects the β-carotene in carrots as orange since that energy at that wavelength is absorbed in the transition and the human eye sees the remainder of colors.

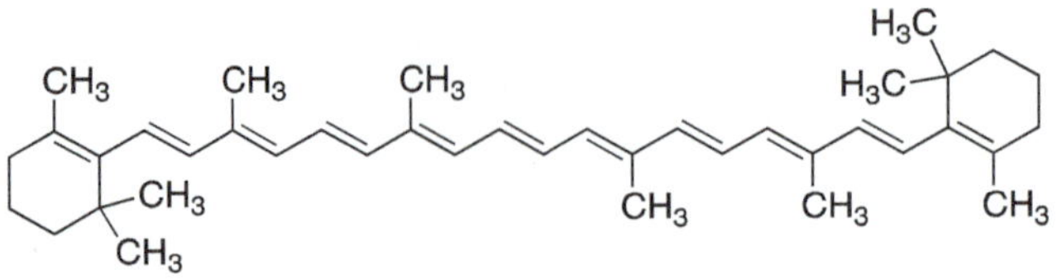

β-Carotene

Another compound with extended conjugation is lycopene, which is a bright red carotene. Carotenoid pigment and other phytochemicals are found in tomatoes and other red fruits and vegetables. The λ_{max} for lycopene is 474 nm. As you will conclude from these spectra, the energy requirement for the transition in the UV-Vis is lower for extended conjugation.

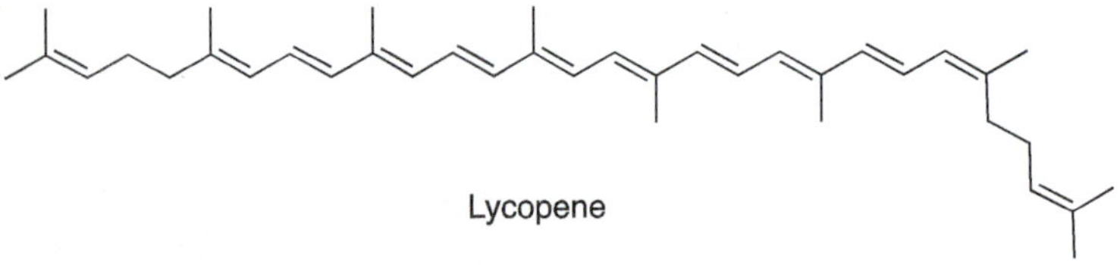

Lycopene

DID YOU KNOW?

Some of the constituents of fruits and vegetables have photochemicals that have different numbers of double bonds that are conjugated and give rise to their different colors. The bright colors of carrots and tomatoes are due to the presence of highly conjugated systems, β-carotene and lycopene, respectively. The human eye sees only the colors in the visible region of the electromagnetic spectrum and if there is a compound that absorbs strongly in this region, that color is absorbed in the electronic transition and what we see is the rainbow colors, minus that color. β-Carotene absorbs at 454 nm, and lycopene absorbs at 471 nm in the region of the electromagnetic spectrum.

Conjugation may also involve unshared electrons on an atom adjacent to double bonds. Curcumin and Ruhemann's purple both have extended conjugations that involve a pair of non-bonding electrons of the oxygen atoms.

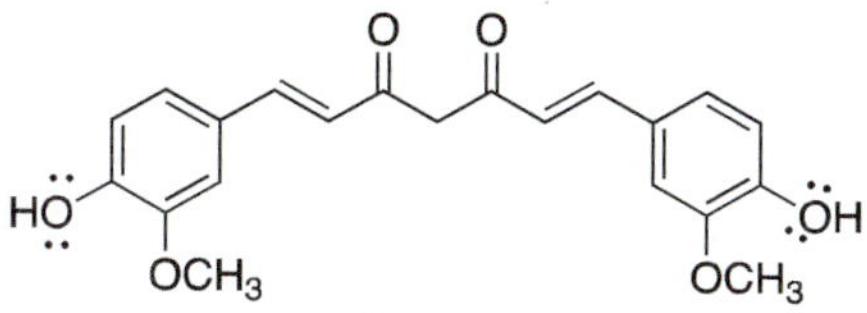

Curcumin

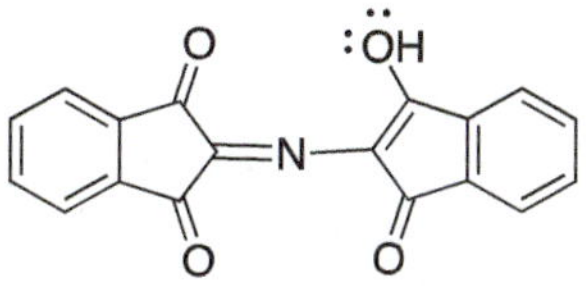

Ruhemann's Purple

Curcumin is a compound that is found in turmeric, which is a well-known spice. Ruhemann's purple is produced in the detection of fingerprints as we will see in more detail in Chapter 20 on the section of peptides and amino acids. As you can imagine, there are ways to calculate the λ_{max} for different conjugated systems based on the number of double bonds in conjugation and the number and types of substitutions about the double bonds. The Wood–Fieser and Fieser–Kuhn rules can be used to get good estimates of λ_{max}, but the details will not be discussed in this chapter, but are handled in more advanced spectroscopy courses. We should be able to estimate, however, the region of absorption for molecules with extended conjugated double bonds.

Problem 13.1

i) The structure of vitamin E is shown below; would you expect it to have a similar UV-Vis spectrum as β-carotene? Explain your answer.

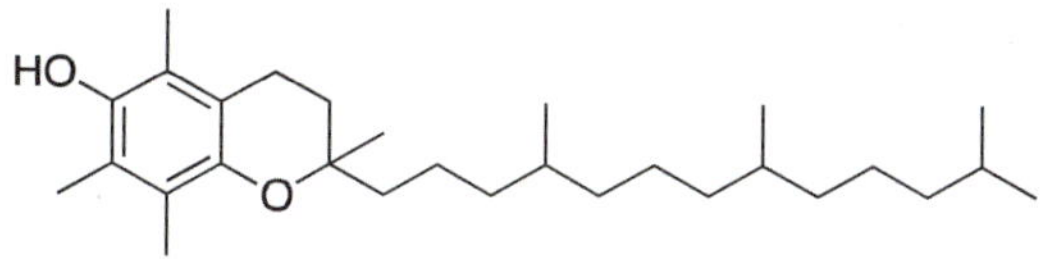

Vitamin E

ii) Based on your knowledge of the UV-Vis spectra of ethylene and 1,3-butadiene, estimate if the maximum absorption of the molecules shown below is greater than or less than that of ethylene.

a) 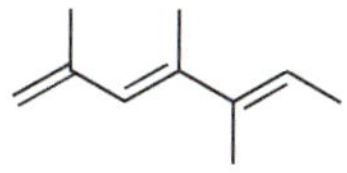   b)

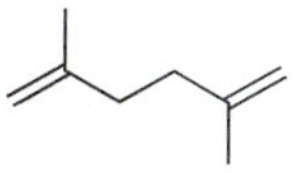

13.4 Infrared Spectroscopy

Energy of the IR region of the electromagnetic spectrum causes molecular vibrations within the molecules. The bonds in a molecule can be described as springs, which hold the atoms of the molecule together and that there is a constant vibration of the bonds similar to the vibration of a spring. As a result, the energy required to vibrate a strong bond in a molecule is less than the energy required to vibrate a weaker bond. The energy needed for bond vibrations come from the IR region of the electromagnetic spectrum. Thus, a scan of a molecule in the IR region of the EM spectrum gives an indication of the strengths of the different bonds, and hence the type of bonds and functional groups that are present in a molecule.

The bond spring-like vibrations described above are one form of molecular vibration, molecular vibrations can take many forms. Most organic molecules have many atoms, and hence a number of different bonds resulting in different modes of possible molecular vibrations. For a particular molecule, the number of possible molecular vibrations can be calculated. A molecule with N atoms has 3N − 6 normal vibrational modes, and if the molecule is linear, there are 3N − 5 modes since there is a rotation about the linear axis, which cannot be observed. Thus, rotation about a linear axis of a linear molecule, such as CO_2, cannot be observed. As a result, the number of possible vibrations for CO_2 is 3(3) − 5 = 4, these four vibrational modes for CO_2, along with the energy required for each vibrational, are illustrated in Figure 13.7.

The IR spectrum of carbon dioxide is shown in Figure 13.8. It should be pointed out that vibrations that do not result in a net dipole change do not show in IR spectra. As a result, there are only three IR signals shown for CO_2 since the vibration at $1340\,\text{cm}^{-1}$ as shown in Figure 13.7 will not give rise to an IR signal.

Water is triatomic molecule and is nonlinear; hence, water has 3(3) − 6 = 3 molecular vibrations. The IR spectrum of water was given in Figure 3.4 of Chapter 3. You will recall that there is a noticeable broad band around $3500\,\text{cm}^{-1}$, which represents two vibrational modes of water and another band around $1600\,\text{cm}^{-1}$, which represents the third vibrational mode of water.

Table 13.2 gives the regions of the IR spectrum where the different functional groups occur in the IR region of the electromagnetic spectrum.

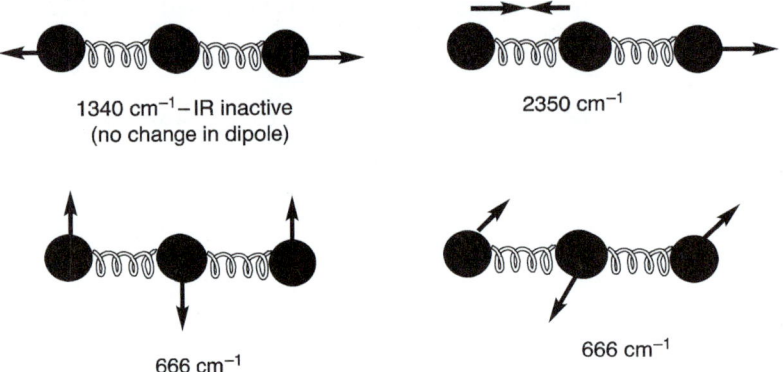

Figure 13.7 Vibrational modes and IR frequencies of carbon dioxide, which is a linear molecule.

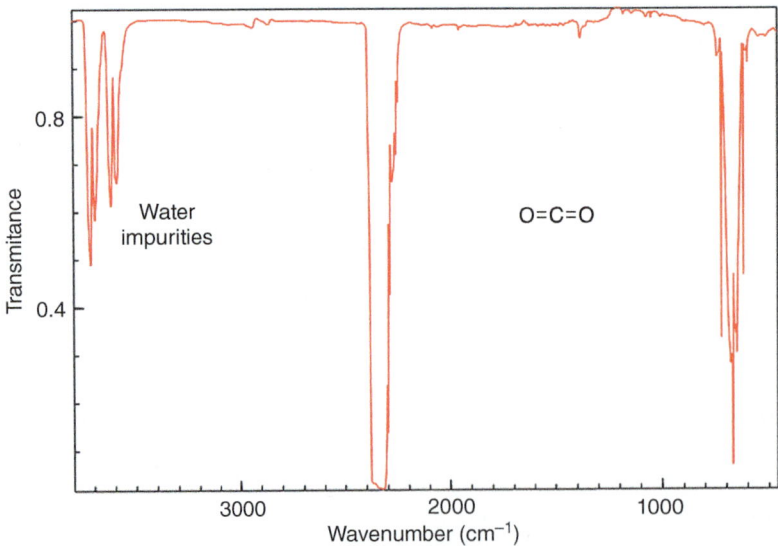

Figure 13.8 IR spectrum of carbon dioxide showing the two vibrational modes. *Source:* with permission from NIST.

Table 13.2 IR frequencies of various functional groups.

Bond	Functional group	Type vibration	Frequency (cm^{-1})	Intensity
C—H	Alkane	Stretch	3000–2850	s
	CH$_3$	Bend	1450 and 1375	m
	–CH$_2$–	Bend	1465	m
	Alkenes	Stretch	3100–3000	m
		Out-of-plane bend		
	Aromatics	Stretch	3150–3050	s
		Out-of-plane bend	900–690	s
		Stretch		w
C≡C	Alkyne	Stretch	~3300	s
C=O	Aldehyde		1740–1720	s
	Ketone		1725–1705	s
	Carboxylic acid		1725–1700	s
	Ester		1750–1730	s
	Amide		1670–1640	s
	Anhydride		1810–1740	s
	Acid chloride		1800	s
C—O	Alcohol, ethers, esters, carboxylic acids, anhydrides		1300–1000	s
O—H	Alcohols, phenols		3650–3600 (free) 3650–3600 (H-bonded)	m
	Carboxylic acid		3400–2400	m
N—H	Primary and secondary amines and amides	Stretch	2500–3100	m
		Bend	1640–1550	m–s
C—N	Amines		1350–1000	m–s
C=N	Imines and oximes		1690–1640	m–s
C≡N	Nitriles		2260–2240	m
X=C=Y	Allenes, ketenes, isocyanates, isothiocyanates		2270–1950	m–s
N=O	Nitro (–NO$_2$)		1550–1350	s
S—H	Thiols		2250	w
S=O	Sulfoxides		1050	s
	Sulfones, sulfonyl chlorides, sulfates, sulfonamides		1375–1300	s
C—X	X = fluorine		1400–1000	s
	X = chlorine		800–600	s
	X = bromine, iodine		<667	s

s, strong; m, medium; w, weak.

The information given in Table 13.2 can be used to determine different types of functional groups present in organic molecules. For example, formaldehyde has one functional group (a carbonyl), and based on the information gained from Table 13.2, there should be an IR absorption band around $1700\,cm^{-1}$. The IR spectrum of formaldehyde is shown in Figure 13.9.

As shown in Figure 13.9, the carbonyl functional vibration is shown at $1717\,cm^{-1}$. The other signal that is prominent occurs around $3000\,cm^{-1}$ and represents the C–H vibrations.

The IR spectrum of formic acid is shown in Figure 13.10, note that there are a few prominent bands below $1000\,cm^{-1}$. There is the expected carbonyl frequency around $1700\,cm^{-1}$ similar to the carbon–oxygen double bond in the previous molecule, formaldehyde; there is also a band around $2900\,cm^{-1}$, which indicates the presence of C–H vibrations. There is an obvious broad band that occurs around $3400–3100\,cm^{-1}$, which indicates the presence of the O–H frequency. Thus, based on the information from the IR spectrum, one conclusion is that the molecule probably has a carboxylic acid functionality. Of course, extreme caution must be exercised since the molecule could be a difunctional molecule with a ketone or aldehyde and an alcohol. We will see later in the chapter, that another type of spectroscopy can be used to gain more specific information about the nature of the functional group that is actually present in the molecule.

As we have demonstrated in the previous section, the correlation of the wavenumbers for the bands of the IR spectrum of an unknown compound with the values given in the table for functional groups can give important information about the type of functional groups that are present in an unknown molecule. This approach can be used for the analysis of larger and more complex molecules. Shown in Figure 13.11 is an IR spectrum for benzoic acid, which has more functional groups, than the ones mentioned earlier.

Note the prominent broad band around $3300–3100\,cm^{-1}$, which is an indication of the presence of an OH group. There is another prominent band around $1700\,cm^{-1}$, which is an indication of a carbon–oxygen double bond. There are also some bands below $1000\,cm^{-1}$. These bands represent carbon–carbon and carbon–hydrogen bond vibrations of the benzene ring.

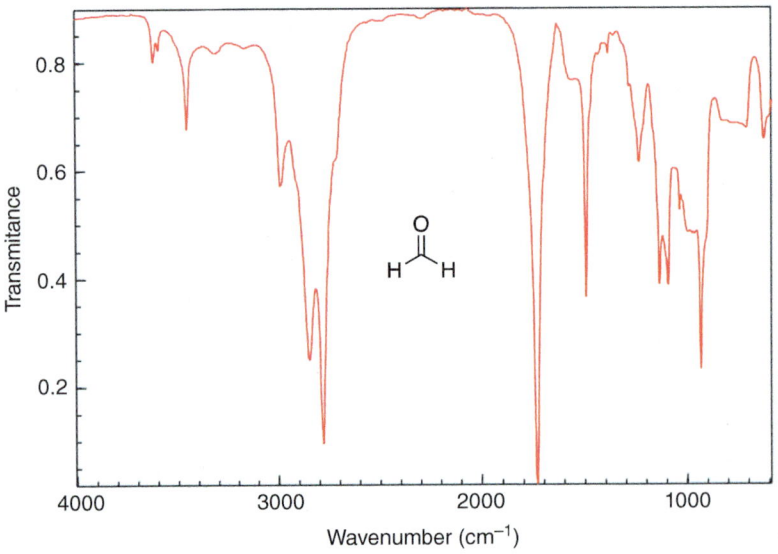

Figure 13.9 IR spectrum of formaldehyde (HCOH). *Source:* with permission from NIST.

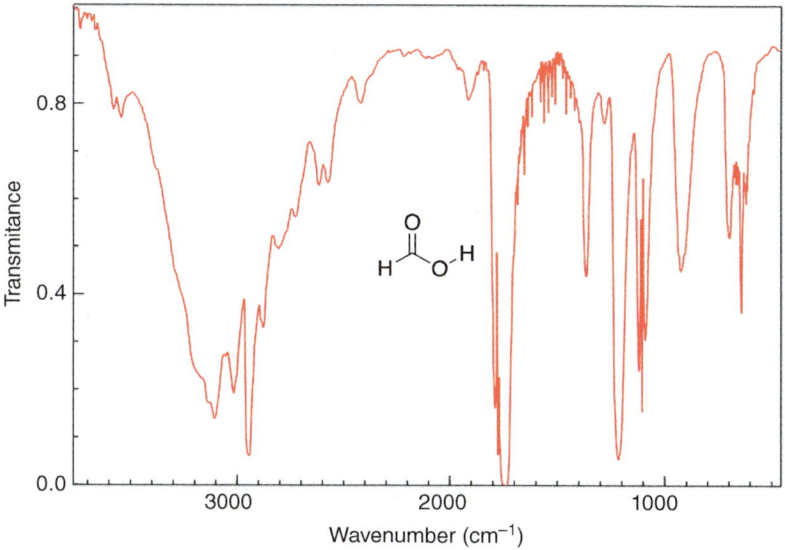

Figure 13.10 IR spectrum of formic acid (HCO₂H), showing the O—H and carboxylic acid carbonyl double bond.

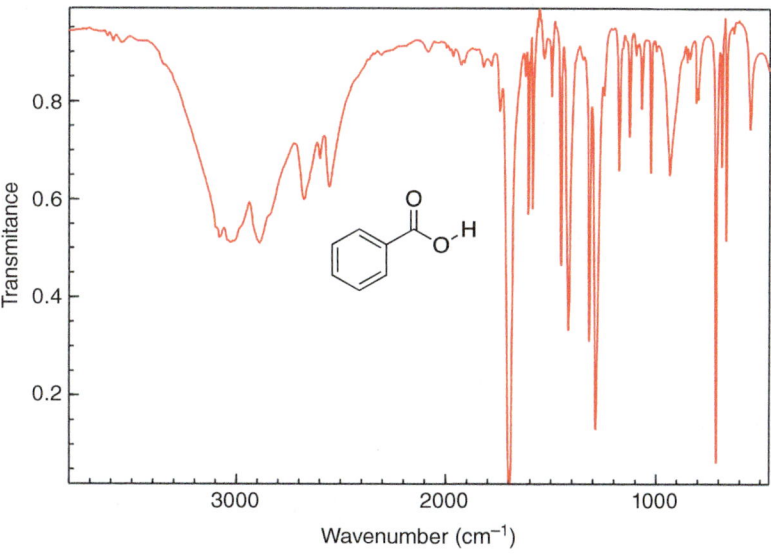

Figure 13.11 IR spectrum of benzoic acid, showing the OH and carboxylic acid carbonyl frequencies.

Figure 13.12 shows the IR spectrum of benzylamine. Note carefully the frequency, which occurs in the region of $3200 \, \text{cm}^{-1}$, this band appears with a slight overlap of two bands, which is indicative of a primary amine. Note also that since there is not a carbonyl present in this compound, there is not a band around $1700 \, \text{cm}^{-1}$. The bands above $1000 \, \text{cm}^{-1}$ represent the carbon–carbon and carbon–hydrogen vibration of the benzene ring.

Figure 13.13 shows the IR spectrum of benzyl alcohol. Note carefully the frequency, which occurs in the region of $3200 \, \text{cm}^{-1}$, is slightly different for the benzylamine. For benzyl alcohol,

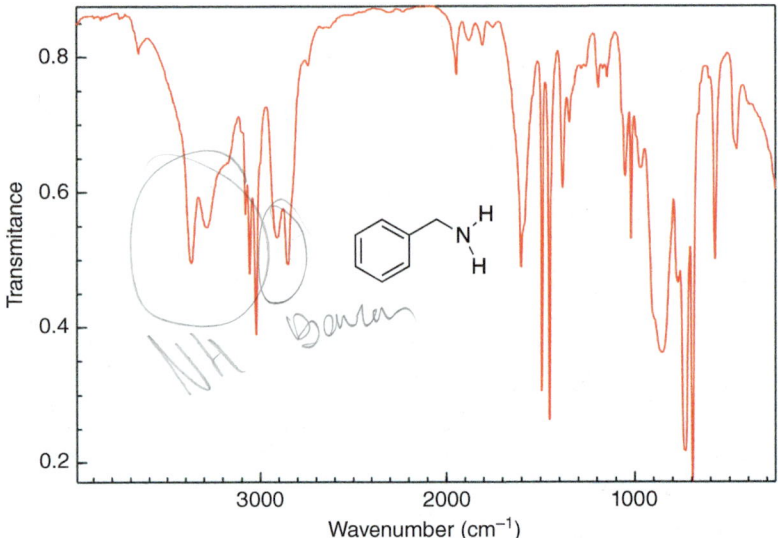

Figure 13.12 IR spectrum of benzylamine, showing the N–H frequencies. Note that there is no frequency in the region of 1710 cm^{-1} since this molecule does not have a carbonyl group. *Source:* with permission from NIST.

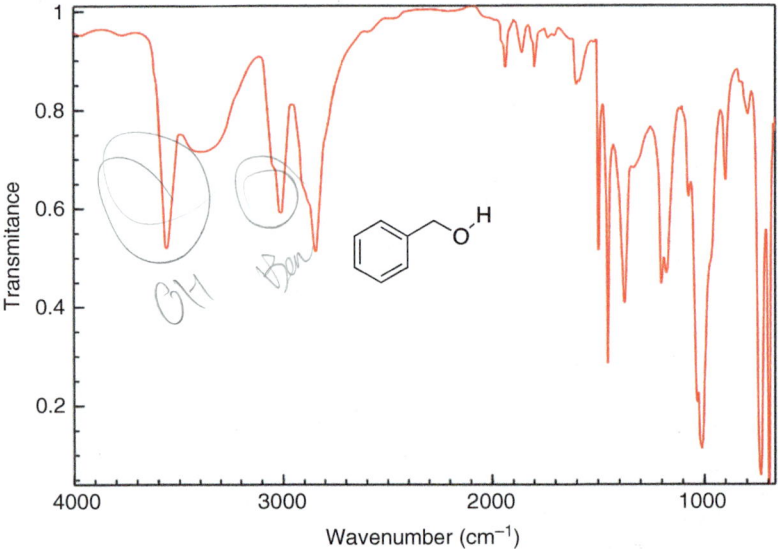

Figure 13.13 IR spectrum of benzyl alcohol, showing the O–H frequency. Note that there is no frequency in the region of 1710 cm^{-1} since this molecule does not have a carbonyl group. *Source:* with permission from NIST.

this band is broader and represents the OH stretch. Note also that since there is no carbonyl present in this compound, there is not a band around 1700 cm^{-1}. The bands above 1000 cm^{-1} represent the carbon–carbon and carbon–hydrogen vibration of the benzene ring. The slight difference between the spectra of benzylamine and benzyl alcohol can be used to identify these compounds.

Problem 13.2

An unknown compound is acyclic and has a molecular formula of $C_5H_{10}O$ and a prominent frequency in the IR spectrum occurs around $1710\,cm^{-1}$. Give a likely structure of this unknown compound.

13.5 Mass Spectrometry

Mass spectrometry does not involve the use of energy from the electromagnetic spectrum to excite molecules and is a destructive method that comes about from the bombardment of electrons into a molecule. The result is that the molecule is destroyed, but in a consistent and often predictable manner so that the information can be used to assist in the determination of the structure of an unknown compound. When electrons collide into a molecule, the electrons typically knock out bonding electrons to create a charged species and based on the charge, the relative mass of each ion can be determined, hence giving an idea of the molecular mass of an unknown compound. Shown below is an example that involves the electronic bombardment of a simple molecule, methane, as shown in Reaction (13-4).

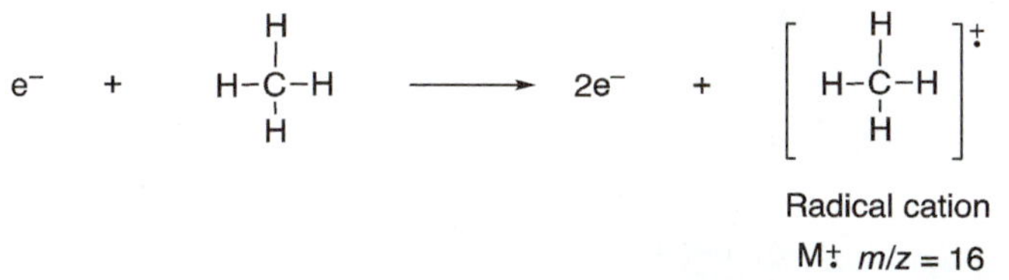

$$e^- + CH_4 \longrightarrow 2e^- + [CH_4]^{+\cdot} \quad (13\text{-}4)$$

Radical cation
M⁺˙ m/z = 16

The radical cation can be considered as the molecule, but with one of its bonding electrons removed, which results in a radical (since it lacks just one electron). Also, since it lacks an electron, it is also considered positive (cation); hence, called a radical cation. This species that results from the loss of only an election is called the molecular ion. The molecular ion is not stable and will undergo fragmentation to form other radicals and cations. Another example, which involves the bombardment of electron into ethane, is shown in Reaction (13-5).

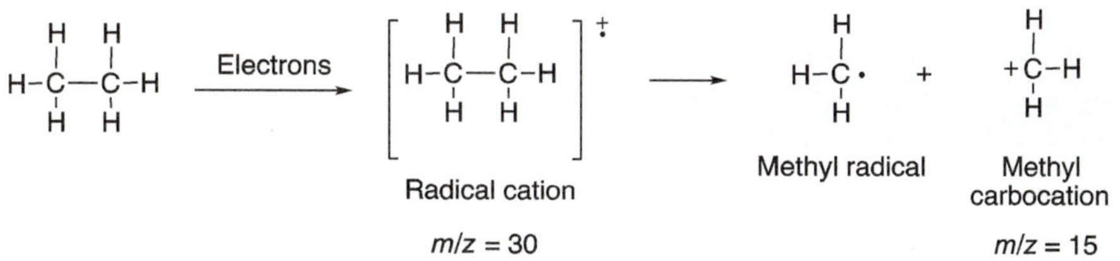

$$CH_3CH_3 \xrightarrow{\text{Electrons}} [CH_3CH_3]^{+\cdot} \longrightarrow CH_3\cdot + {}^+CH_3 \quad (13\text{-}5)$$

Radical cation Methyl radical Methyl carbocation
m/z = 30 m/z = 15

The mass spectrometer is designed to detect the relative mass-to-charge ratio of the different species generated from the bombardment of molecules with electrons. The schematics showing the components of the mass spectrometer is shown in Figure 13.14.

As shown in the schematics of the instrument, the sample is vaporized typically under reduced pressure as shown at the left of the diagram. The sample is then ionized by a beam of electrons, which typically is generated by heating an element, which is connected to electrodes. The ions are then separated based on their mass-to-charge ratio. The mass-to-charge of the

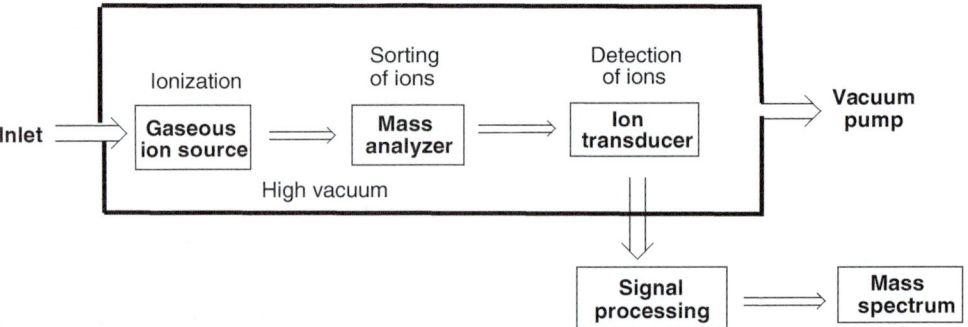

Figure 13.14 Schematics showing the components of the electron impact mass spectrometer.

Figure 13.15 Electrospray ionization mass spectrometer.

ions are determined and recorded as a spectrum. Another type instrument uses electrospray ionization, instead of electron impact. Electrospray ionization is a non-destructive way of getting the molecules into the gas-phase. This method is commonly used to analyze organic molecules especially those with acidic and basic functional groups because they naturally form charged molecules that can be measured by MS as intact molecules. An example of one such instrument is shown in Figure 13.15.

Thus, for the mass spectrum of ethane, one would see essentially the signals for the molecular ion peak ($m/z = 30$) and another peak at $m/z = 15$. There is also another signal at M-1 for the

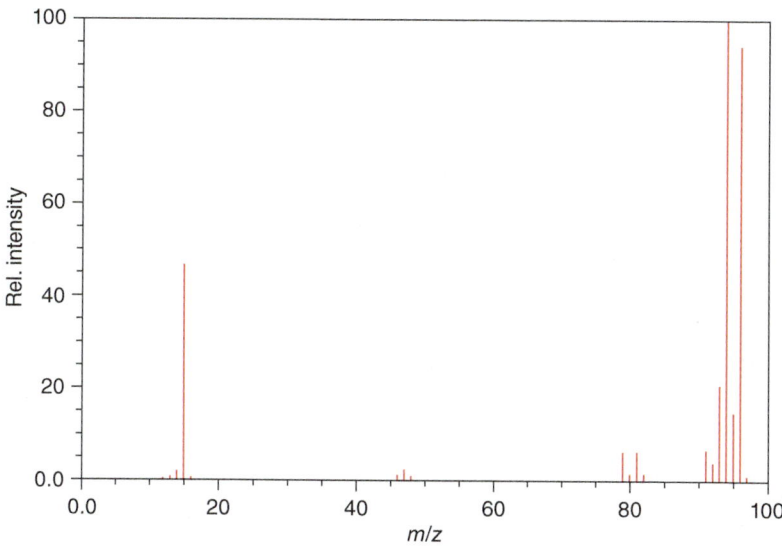

Figure 13.16 Electron ionization mass spectrum of methylbromide (CH₃Br). *Source:* with permission from NIST.

loss of a hydrogen atom from the molecular ion fragment, another fragmentation pattern is shown in Reaction (13-6).

$$e^- + \text{Ethane} \longrightarrow [\text{Radical cation}]^{\bullet +} \longrightarrow \text{Ethyl cation} + \text{H}^\bullet \quad (13\text{-}6)$$

The mass spectrum for methylbromide (CH₃Br) is shown in Figure 13.16. We know from the periodic table that the molecular weight of methylbromide is 95 g mol⁻¹, which means that the molecular ion peak should be $m/z = 95$. There are two other peaks and much larger peaks, however, at $m/z = 94$ and $m/z = 96$. These peaks come about due to the presence of two isotopes of bromine which represent an isotope of ^{81}Br, which exists in natural abundance 49.31%, compared to ^{79}Br, which is 50.69%, these percentages that they occur in natural abundance are also reflected in the intensities of these peaks. Note that there is a peak at $m/z = 15$, which represents the methyl cation or radical. There is also another pair of signals around $m/z = 80$, which reflects the isotopes of bromine.

Figure 13.17 shows the mass spectrum for 2,4-dimethylpentane. Note that there is a very small peak at $m/z = 86$, which is the one mass unit greater than the major peaks. This peak is called the $M + 1$ peak and represents a mixture of isotopes of carbon. You will remember from general chemistry that the natural abundance of the isotopes of carbon are ^{12}C, ^{13}C, and ^{14}C. The percentage of ^{14}C is extremely small and the $M + 1$ comes mainly from the ^{13}C that is present naturally in any sample of organic compound. As we saw in the above example, compounds that have a greater natural abundance of different isotopes, such as bromine, the ratio of those signals reflects the percentage that exists in natural abundance.

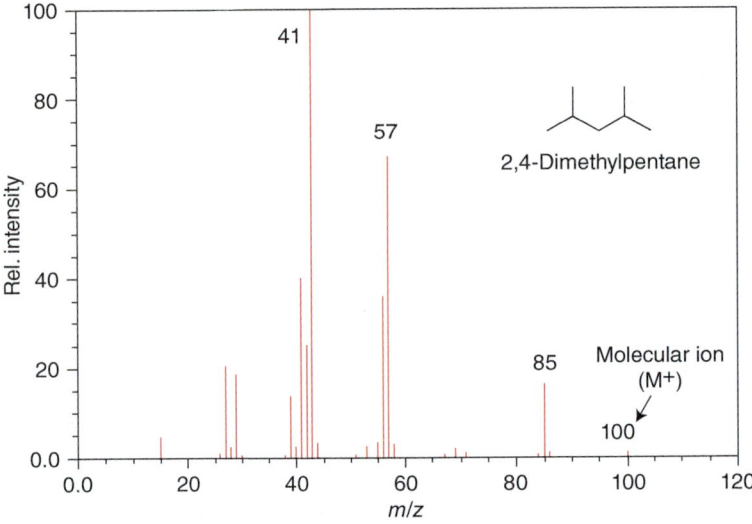

Figure 13.17 Electron ionization mass spectrum of 2,4-dimethylpentane. *Source:* with permission from NIST.

Problem 13.3

The mass spectrum and IR spectrum for an unknown compound are shown below, what is the likely structure of this unknown compound? The molecular formula of the unknown compound is C_3H_6O.

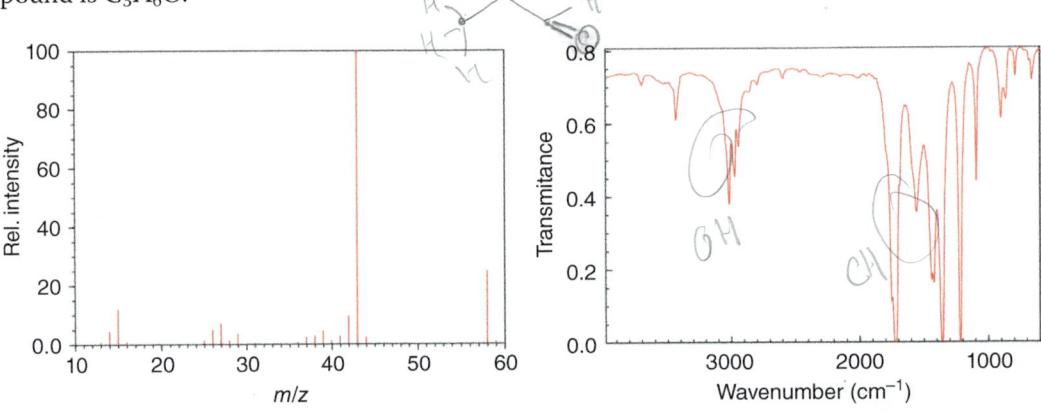

13.6 Nuclear Magnetic Resonance (NMR) Spectroscopy

NMR spectroscopy is one of the most important types of spectroscopy that is used to assist in the determination of the structure of organic compounds. This technique is nondestructive to the compound and gives very detailed information about the structure of an unknown compound. In order to determine the complete structure of an unknown, a combination of the various types of spectroscopy is typically used. For NMR spectroscopy, the connectivity of the different atoms in a molecule can be determined; these atoms include H, C, N, F, and P. The most commonly used type in NMR spectroscopy is the ^1H or proton, hence known as proton NMR spectroscopy, or just simply ^1H NMR. Another type of NMR spectroscopy that is commonly used to examine the structure of compounds is ^{13}C NMR spectroscopy. In this section, both ^1H NMR and ^{13}C NMR spectroscopies will be covered.

13.6.1 Theory of Nuclear Magnetic Resonance Spectroscopy

A nucleus with an odd number of electrons (odd atomic number) has a nuclear spin that can be observed in the NMR spectrometer. As mentioned earlier, the most popular ones are the following: H, C, N, F, and P. The simplest nucleus of course is the hydrogen (the proton) NMR. A hydrogen atom with one electron can be visualized as a rotating sphere of positive charge. This rotation generates a magnetic field (magnetic moment), which looks like a small bar magnet (with two poles) as illustrated in Figure 13.18.

Notice that these tiny bar representations of the protons have random orientations. When these small bar magnets (the protons) are the presence of an external magnet (H_0), they are aligned with the external magnetic field as illustrated in Figure 13.19.

There are two possible modes of alignments (orientations) of the tiny magnetic bars with the external magnetic field, H_0, as shown in Figure 13.20.

There is one orientation, in which the magnets are aligned and it is the more favored orientation. Hence it is lower in energy, compared to the other orientation, which is higher in energy. A specific amount of energy is required for the transition from the more favored lower energy alignment to the excited state for each proton (here represented by tiny bar magnets). The

Figure 13.18 Representation of spinning protons, which give rise to tiny magnetic bar-like magnets.

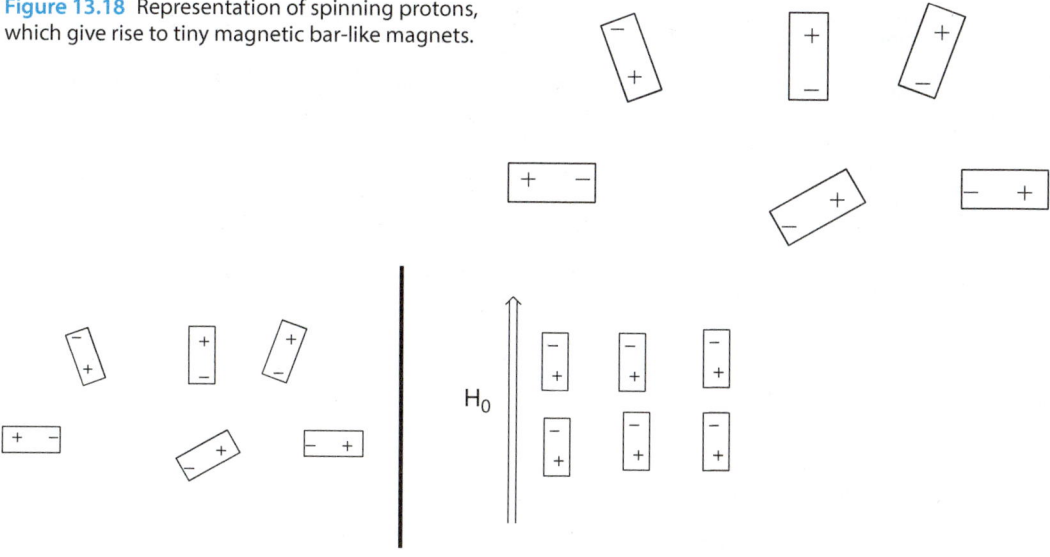

Figure 13.19 Nonaligned nuclear spins in the absence of an external magnetic field (left) and aligned nuclear spins in the presence of an external magnetic field, H_0.

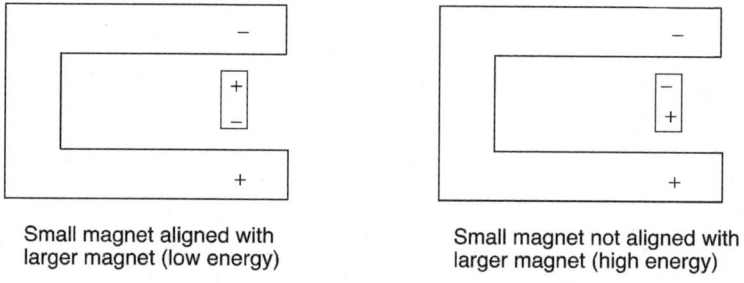

Small magnet aligned with larger magnet (low energy)

Small magnet not aligned with larger magnet (high energy)

Figure 13.20 Representation of two possible orientations of a small bar magnet in the presence of a larger external magnet field.

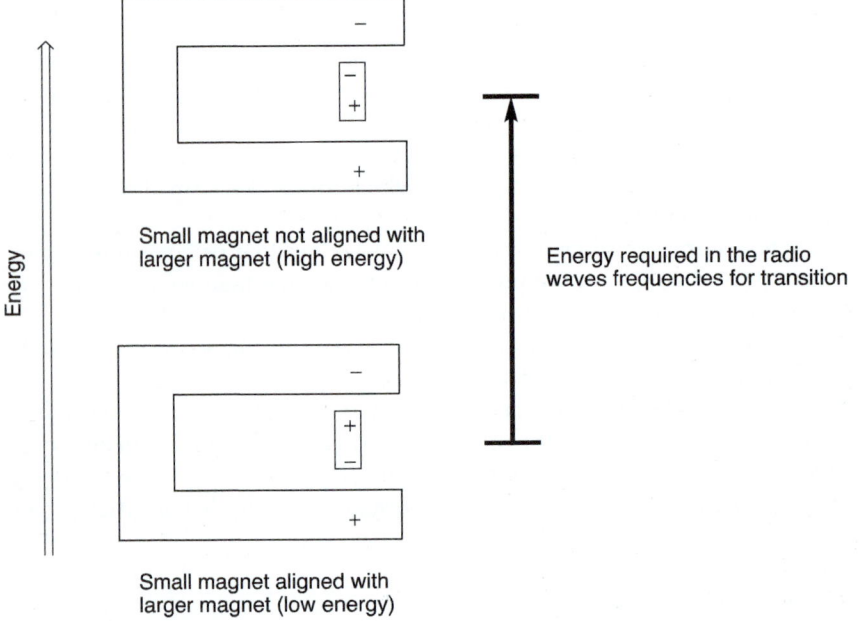

Figure 13.21 Illustration of energy requirement for transition of various orientations of protons (represented as tiny bar magnets) in the presence of a larger magnetic field.

energy required for this transition is in the *radio frequency* range of the electromagnetic spectrum. This concept is illustrated in Figure 13.21.

13.6.2 The NMR Spectrometer

The major component of an NMR instrument is a very strong magnet; also required is a radio frequency transmitter, a method to detect signals, and a recorder (printer). The schematics of the NMR instrument are shown in Figure 13.22.

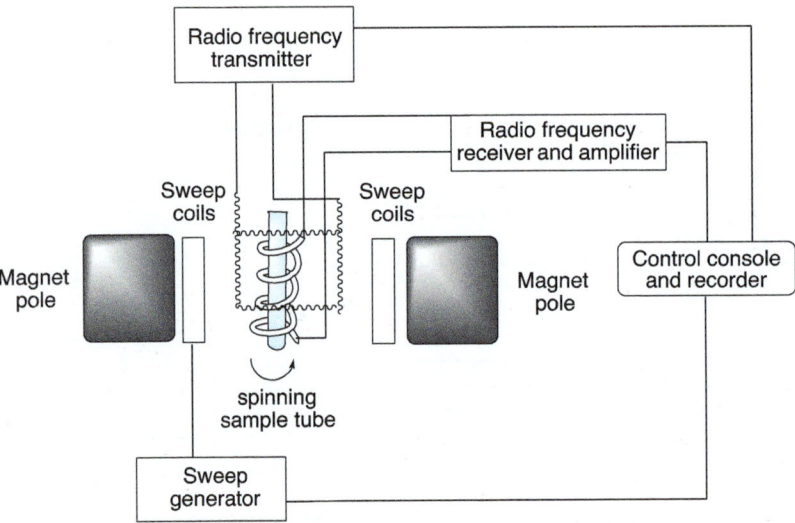

Figure 13.22 Schematics of the nuclear magnetic resonance (NMR) spectrometer.

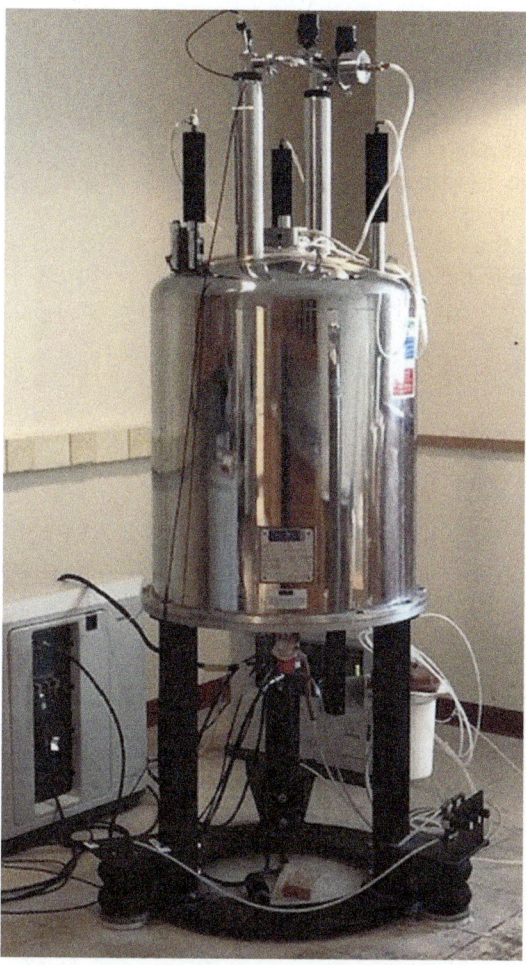

Figure 13.23 300 MHz nuclear magnetic resonance (NMR) instrument.

The sample is inserted within the field of a very strong magnet, typically an electromagnetic, which uses a combination of liquid helium and liquid nitrogen to keep it cool. Once the sample is in the magnetic field, energy is supplied to the sample in the form of radio frequency. The range of radio frequency is typically scanned and signals are detected based on the environment of the different protons. A picture of an NMR instrument is shown in Figure 13.23.

13.6.3 Magnetic Shielding

Imagine that each bar magnet is a proton on an organic molecule that has a number of other atoms and that this molecule is placed in the presence of an external magnet, H_0. The surrounding atoms of that proton, which also have electrons, will produce small electric fields, which will shield the proton on the molecule from the external field. As a result, the proton under consideration is considered to be more *shielded* from the effect of the external field, H_0. The opposite possibility also exists where the surrounding environment of a proton is less shielded from the effect of the external magnet. Protons that are more shielded are excited at

DID YOU KNOW?

The very powerful electromagnet required for NMR spectroscopy has to be kept extremely cold. This is accomplished by using liquid nitrogen, which has a temperature of −196 °C. Liquid nitrogen is stored and transported under pressure. In order to achieve an even lower temperature, liquid helium, which is extremely expensive, but is colder than liquid nitrogen is also used to keep the magnet cool, its temperature is −270 °C. For efficiency, the liquid helium keeps the magnet cool, and the liquid nitrogen keeps the liquid helium from evaporating too fast.

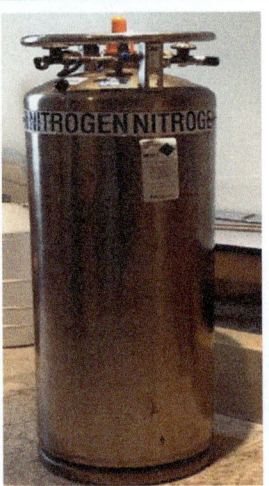

smaller frequencies, compared to protons that are less shielded, which are excited at larger frequencies. Remember that the larger the frequency, the greater the energy. Methanol, which has two different sets of protons, is shown in Figure 13.24.

Since the environments of each proton on a molecule are different, it is possible for NMR spectroscopy to detect different types of hydrogens in a molecule based on the specific frequency or energy required for excitation in the presence of an external magnetic field. The question now becomes how to distinguish different hydrogens that are in different environments. For the molecule given in Figure 13.24, the three hydrogens of the methyl carbon of methanol are all in the same environment, and hence the energy required to excite these three hydrogens is the same. On the other hand, the hydrogen of the alcohol functionality is in a different environment, and hence the energy required to excite that proton is different from the energy required to excite the three methyl hydrogens. Thus, it is expected in the NMR spectrum that there should be two signals for methanol, one for the excitation of the three methyl hydrogens and another for the excitation of the hydrogen bonded to the oxygen.

13.6.4 The Chemical Shift, the Scale of the NMR Spectroscopy

As mentioned in the previous section, there are different environments that protons of organic molecules exist in and, as a result, require different energies for each transition. In order to standardize the energy requirements for the protons in the many different environments of

Figure 13.24 Methanol showing the different types of protons in different environments.

an organic molecule, a specific compound is needed. That specific compound is tetramethylsaline (TMS), and the structure of TMS is shown below.

```
      H
      |
  H—Si—H
      |
      H
```

Tetramethylsalane (TMS)

Note that all the protons are in the same environment (environmentally equivalent). and in the NMR instrument. These protons will appear as a single signal and are assigned a zero on a ppm scale. Using TMS as a reference signal, the NMR spectrum of methanol, which was discussed earlier in Figure 13.24, is shown in Figure 13.25. As pointed out earlier, methanol has two environmentally different types of protons, and hence the NMR spectrum should show two different signals, as shown in NMR the spectrum for methanol in Figure 13.25.

The different signals in a NMR spectrum are due to the different environments that protons exist in a molecule, such as methanol. The terms upfield and downfield are often used to describe the relative locations of signals in the NMR spectrum. A signal that is upfield is to the right of another signal that is downfield. Thus, the signal at 3.35 ppm is considered to be upfield to the signal that is at 4.80 ppm. Another way of stating the same observation is that the signal that is at 4.80 is downfield, relative to the signal at 3.35 ppm. There is a chemical shift table, which gives estimates of different types of hydrogens in molecules, based on the surrounding environments. These values and environments are shown in Table 13.3.

Thus, it is possible to use the values shown in this chemical shift table to approximate the environment of each proton on a molecule and assist in the determination of its structure.

13.6.5 Significance of Different Signals and Area Under Each Signal

In principle, each type of hydrogen will give a single signal in the NMR spectrum. For example, *tert*-butyl methyl ether will give only two different signals: one for the three protons of the methyl group and one for the nine protons of the *tert*-butyl group, as shown in Figure 13.26. As expected, these signals will have different chemical shifts, due to shielding.

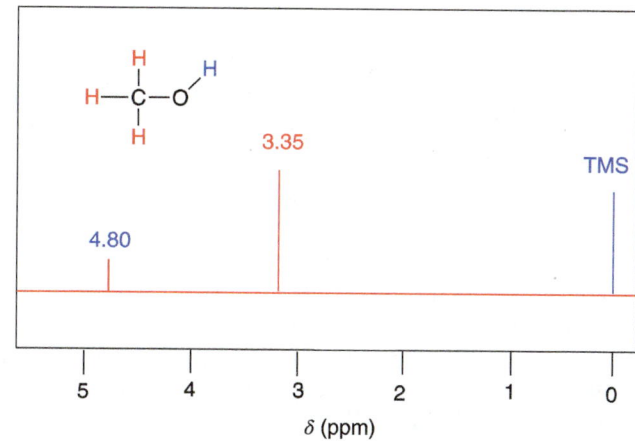

Figure 13.25 NMR spectrum of methanol, showing TMS as zero on the ppm or δ scale.

Table 13.3 Chemical shifts of different protons based on different environments.

Type of proton	Type of compound	Chemical shift range (ppm)
R—CH$_3$	1° aliphatic	0.9
R$_2$CH$_2$	2° aliphatic	1.3
R$_3$CH	3° aliphatic	1.5
—C=C—H	Vinylic	5.5–7.5
—C≡C—H	Acetylenic	2–3
Ph—H	Aromatic	6–8.5
Ph—C—H	Benzylic	2.2–3
—C=C—CH$_3$	Allylic	1.7
R—COOCH	Ester	3.3–4
HC—COOR	Ester	2–2.2
R—COH	Aldehyde	9–10
R—COCH	Ketone	2–2.7
R—COOH	Carboxylic acid	10–13.2
Ph—OH	Phenolic	4–12

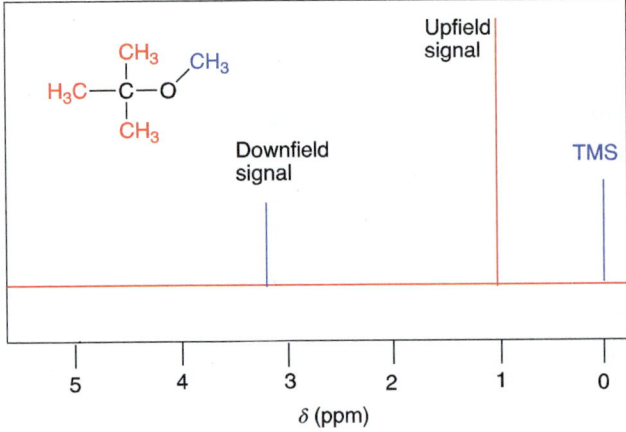

Figure 13.26 Proposed NMR spectrum of *tert*-butyl methyl ether.

The intensities for the signals shown in Figure 13.27 are not the same, one signal is larger than the other signal. The area under each signal corresponds to the number of protons that give rise to that signal. The instrument has an application, which integrates the relative areas under the peaks and the integrations may not necessarily correspond to exact number of hydrogens in a molecule. The actual NMR spectrum of tert-butyl methyl ether is shown in Figure 13.27.

Figure 13.27 300 MHz ¹H NMR spectrum of *tert*-butyl methyl ether in CDCl₃ as solvent. Chemical shifts values are 3.212 ppm (downfield) and 1.189 ppm (upfield). *Source:* with permission from AIST.

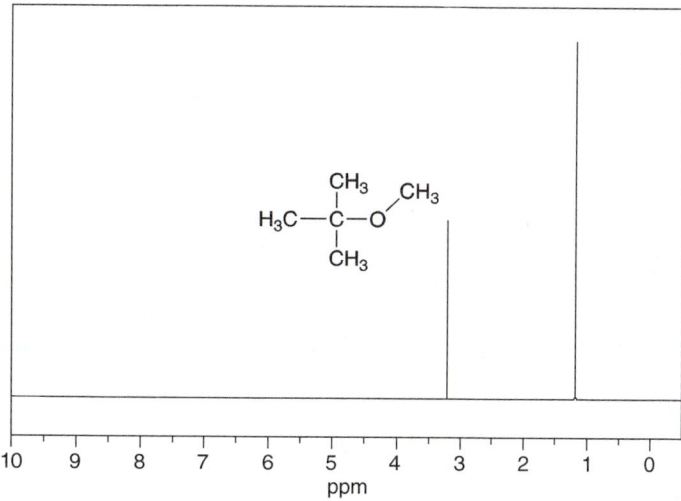

Note that the signal that is upfield (1.0 ppm) is three times larger than that of the signal at 3.1 ppm. The upfield signal has three times as many protons giving rise to that signal, compared to the signal downfield (3.1 ppm). One conclusion is that the signal upfield comes from nine hydrogens and the signal downfield comes from three hydrogens. This approach to determine the relative number of protons per signal in the use of NMR spectra to analyze the structures of molecules will be very important in trying to deduce the structure of unknown compounds.

13.6.6 Splitting of Signals

For two adjacent protons in a molecule that are in different environments, the signals for these two protons will not appear as two single signals, but as split signals. The splitting of the signals typically follows a $N + 1$ rule, where N is the number of protons on the adjacent carbon. Consider the molecule, 1,1,2-trichloroethane, which has three protons, two that are equivalent (or in the same environment) and they are labeled as H_b. The other proton is in a different environment, and it is labeled as H_a (structure shown below).

$$\text{Cl}-\underset{\underset{\text{Cl}}{|}}{\overset{\overset{H_a}{|}}{C}}-\underset{\underset{H_b}{|}}{\overset{\overset{H_b}{|}}{C}}-\text{Cl}$$

1,1,2-Trichloroethane

Since these two types of protons (H_a and H_b) are in different environments, it is expected that there should be two signals: a smaller signal downfield for H_a since that carbon is bonded to two electronegative chlorine atoms. The other signal, which has two protons (H_b), should be upfield to the signal for H_a. Since these two types of protons are adjacent to each other, they will not appear as two simple signals, but will appear as split signals. Since there are two hydrogens (H_b) on the carbon that has one chlorine atom, the signal on the adjacent carbon will follow the $N + 1$ rule and appear as a triplet $(2 + 1 = 3)$. On the other hand, there is one hydrogen (H_a) on the carbon that has the two chlorine atoms, hence the signal of the proton on the adjacent carbon will appear as a doublet $(1 + 1 = 2)$. For 1,1,2-trichloroethane, the splitting of signals is illustrated below.

Based on this information, the splitting pattern is illustrated in the proposed NMR spectrum shown in Figure 13.28.

Shown in Figure 13.29 is an actual NMR spectrum for 1,1,2-trichloroethane.

Let us look at the different isomers of butanol to better understanding splitting patterns observed in ^{1}H NMR spectra. The first spectrum to be examined is *tert*-butanol (2-methyl-2-propanol) along with the signal assignments, which are shown in Figure 13.30.

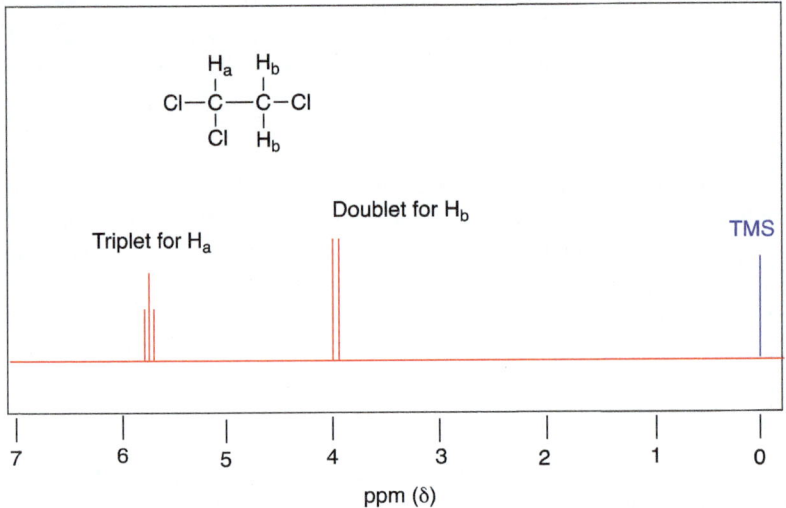

Figure 13.28 Proposed NMR spectrum for 1,1,2-trichloroethane showing the different splitting pattern of the signals.

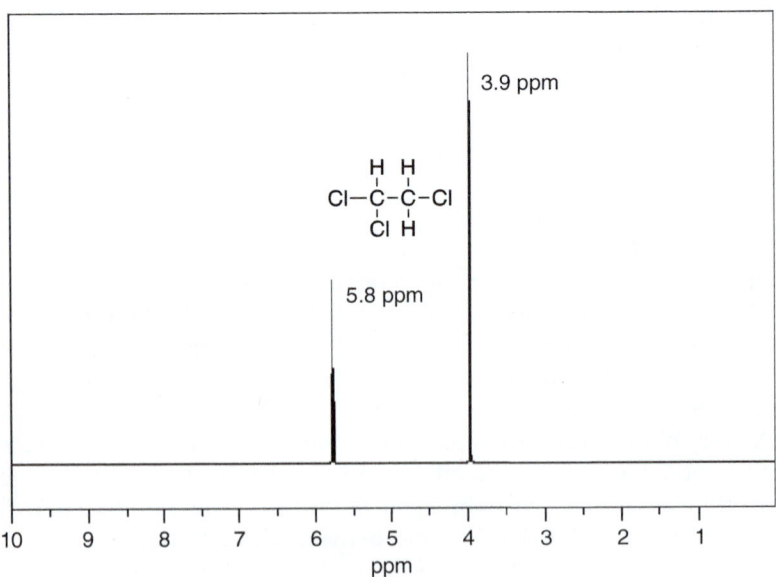

Figure 13.29 Actual 300 MHz ^{1}H NMR spectrum for 1,1,2-trichloroethane showing the different signals. *Source:* with permission from AIST.

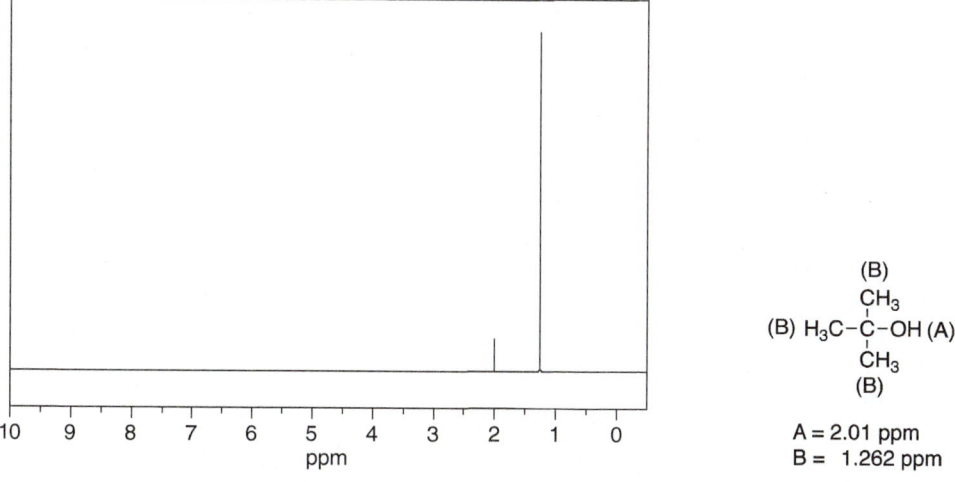

Figure 13.30 300 MHz ^{1}H NMR spectrum for 2-methyl-2-propanol in CDCl$_3$. *Source:* with permission from AIST.

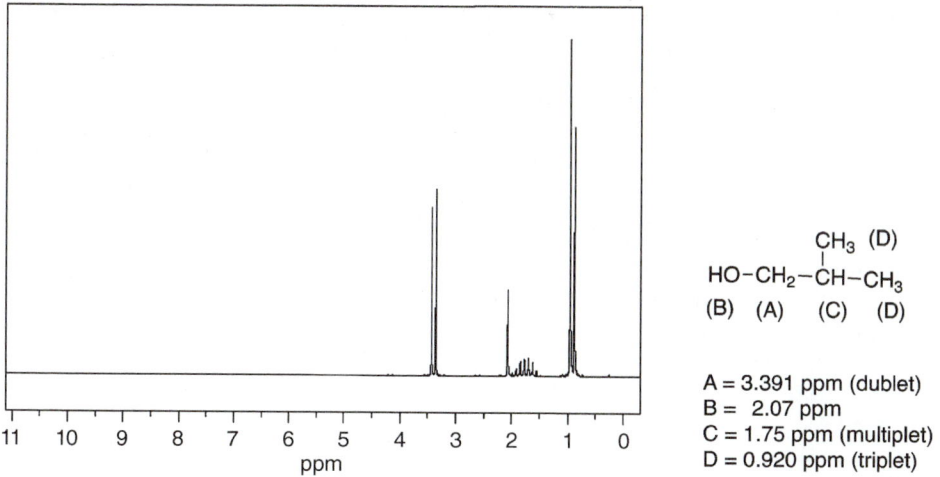

Figure 13.31 300 MHz ^{1}H NMR spectrum for 2-methyl-1-propanol in CDCl$_3$. *Source:* with permission from AIST.

This spectrum is not surprising since all nine hydrogens of the *tert*-butyl group are in the same environment and the O–H proton is in a different environment; note that the OH signal is downfield since the hydrogen is bonded to an electronegative oxygen.

Shown in Figure 13.31 is the NMR spectrum of 2-methyl-1-propanol along with the NMR assignments.

Note that this spectrum is a bit more complex, compared to that given in Figure 13.30 for *tert*-butanol. For 2-methyl-1-propanol, the two methyl groups (labeled as D) are equivalent and appear as one signal upfield at 0.920 ppm. Since these protons are both bonded to an adjacent carbon that has one hydrogen, that signal appears as a doublet based on the formula given earlier. Note that the proton labeled as (C) appears as a multiplet since the adjacent carbons have a total of 8 hydrogens. The hydrogens labeled as (A) appear as a doublet since it is bonded to an adjacent carbon that has two hydrogens. Typically, hydrogens on adjacent heteroatoms do not split the signal of protons that are bonded to carbons. As a result, the hydrogens labeled as (A) appear as a doublet and not a multiplet. Also, these hydrogens are the furthest downfield since the carbon is bonded to the electronegative OH group. The hydrogen bonded to the OH group appears as a singlet at 2.07 ppm.

The next spectrum to be examined is that for 2-butanol and it is given in Figure 13.32. Note that this spectrum is even a bit more complex than that of *tert*-butanol or 2-methyl-1-propanol. The signal that stands out is the multiplet at 3.79 ppm for the proton labeled (B). For this proton, there are three hydrogens on the adjacent carbon to the right and two hydrogens on the adjacent carbon to the left. This hydrogen is bonded to a carbon that is bonded to an electronegative OH group. As a result, this signal is the most downfield signal, and it appears as a multiplet owing to the number hydrogens on the adjacent carbons. The hydroxyl proton (E) is also downfield since it is bonded to an electronegative oxygen. The other signals are all upfield and in the same approximate region, but you may be able to recognize a triplet upfield at 0.93 ppm, which represents the hydrogens labeled as (D), and they appear as a triplet since they are bonded to an adjacent carbon that has two hydrogens.

The last isomer of butanol to be examined is that of 1-butanol, and the spectrum is shown in Figure 13.33, along with the NMR assignments. For the spectrum of 1-butanol, the signal at 3.626 ppm is easily recognized since it is downfield and is a triplet. This signal is downfield since it is bonded to a carbon that is bonded to the electronegative OH group. This signal is a triplet and belongs to the hydrogens labeled (A) since they are bonded to an adjacent carbon that has two hydrogens. Another recognizable signal is the most upfield signal, which is a triplet and belong to protons labeled (E). This signal appears as a triplet since the hydrogens are bonded to an adjacent carbon that has two hydrogens. It is expected that the signal for the protons labeled as (D) should appear as a multiplet, likewise the protons labeled as (C). The multiplet in the region of 1.53–1.39 ppm reflects these protons.

Even though it is possible to determine the structures of the isomers of butanol as discussed in the previous paragraph, if these compounds were unknown compounds, the presence of the OH functional group could have been determined from IR spectroscopy and the molecule weight from mass spectrometry.

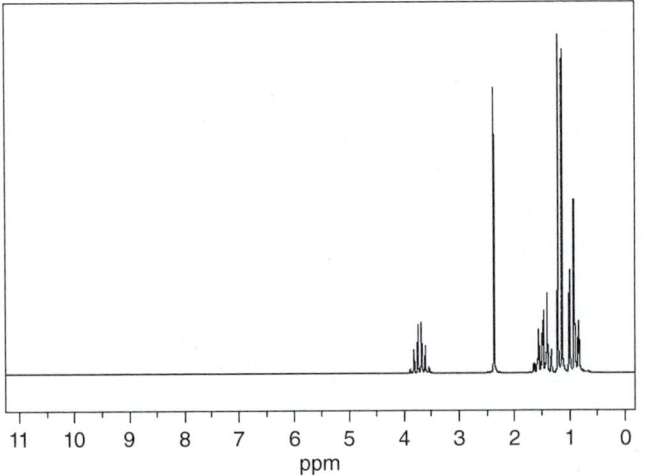

Figure 13.32 300 MHz ^{1}H NMR spectrum for 2-butanol in CDCl$_3$. *Source:* with permission from AIST.

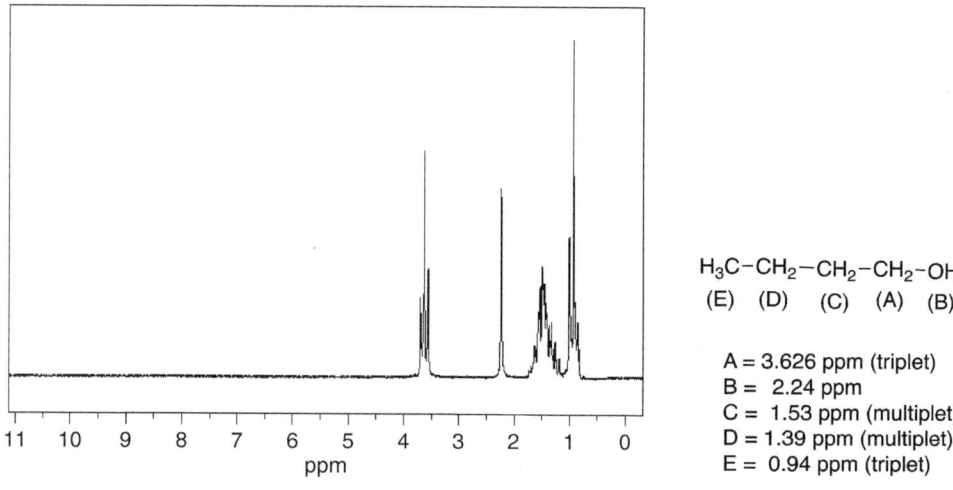

Figure 13.33 300 MHz ^{1}H NMR spectrum for 1-butanol in CDCl$_3$. *Source:* with permission from AIST.

Problem 13.4

An unknown compound showed a molecular weight ion peak at 88.1 *m/z*. The ^{1}H NMR and IR spectra are given below. Give a likely structure of this unknown compound.

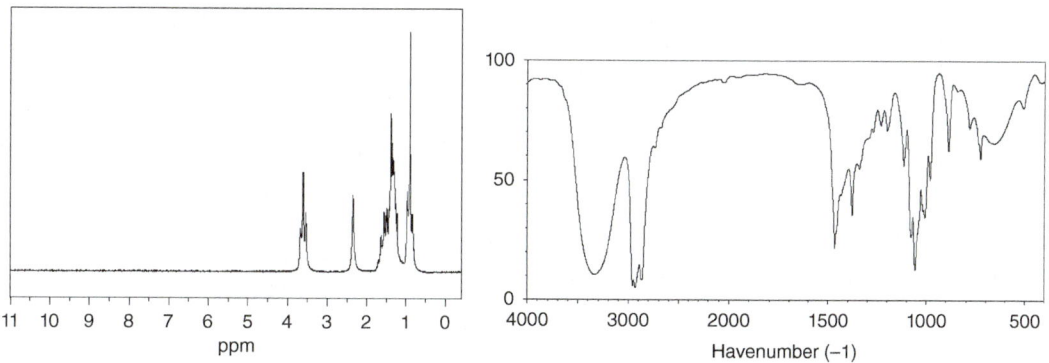

A large number of organic compounds contain the benzene ring in their structures, so it becomes important to be able to recognize this functionality in molecules. The spectrum for benzene is shown in Figure 13.34.

Note that since all hydrogens of benzene are equivalent, there is only one signal that appears at 7.339 ppm. For substituted benzene, however, the hydrogens of the benzene ring are no longer equivalent. The ^{1}H NMR spectrum of nitrobenzene is shown in Figure 13.35.

Note that since all hydrogens of the benzene ring of nitrobenzene are not equivalent, hence there are different signals with different chemical shifts. The protons labeled H$_a$ are equivalent since the molecule is symmetrical. Note also that this signal is furthest downfield since these hydrogens are closest to the electronegative –NO$_2$ group. This signal appears as a doublet since the adjacent carbon has one hydrogen. The protons labeled H$_b$ are equivalent due to the symmetry in the molecule and they appear as a multiplet since the adjacent carbons have one hydrogen each. The proton H$_c$ appears as a multiplet for the same reason of having adjacent carbons with one hydrogen on each carbon.

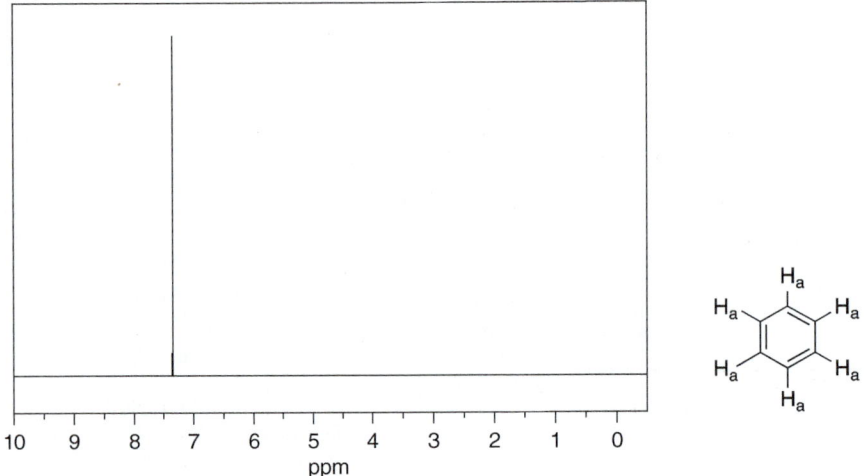

Figure 13.34 300 MHz ^{1}H NMR spectrum of benzene in CDCl$_3$. *Source:* with permission from AIST.

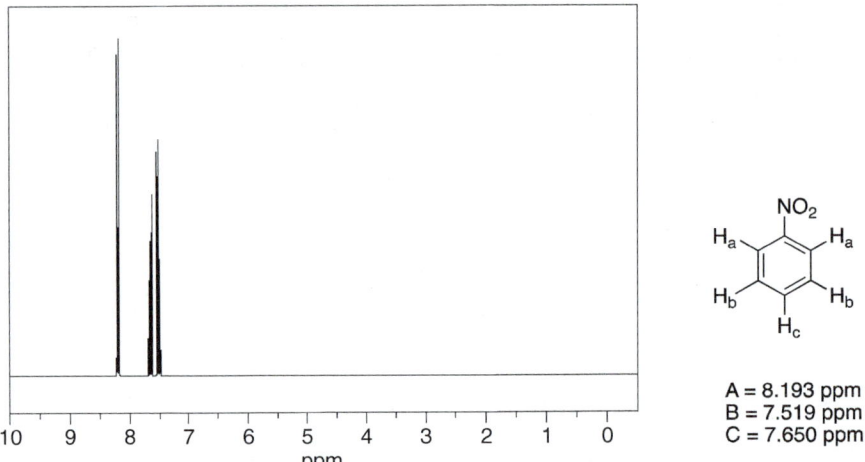

Figure 13.35 300 MHz ^{1}H NMR spectrum of nitrobenzene in CDCl$_3$. *Source:* with permission from AIST.

The ^{1}H NMR spectrum for 1,4-dinotrobenzene is given in Figure 13.36. Note that all four of the hydrogens of 1,4-dinotrobenzene are equivalent, and as result, they appear as a single signal. Owing to the electronegative nitro groups on the benzene ring of 1,4-dinitrobenzene, the chemical shift of this single signal is further downfield, compared to that of benzene; 8.49 ppm for 1,4-dinitrobenzene, compared to 7.339 ppm for benzene.

The ^{1}H NMR spectrum of 1,2-dinitrobenzene, however, is different from 1,4-dinitrobenzene as shown in Figure 13.37.

For this molecule, there are two different types of hydrogens as shown in Figure 13.37. As a result, there are two signals. Since H$_a$ protons are closer to the electronegative nitro group and that signal is downfield, compared to the signal for the protons of H$_b$, which are further from the electronegative nitro groups.

The spectrum of 1,3-dinitrobenzene is shown in Figure 13.38. Note that the signal for H$_a$ is furthest downfield and appears as a singlet since there are no hydrogens on the adjacent carbons and it is next to two nitro groups. The protons shown as H$_b$ are equivalent and appear

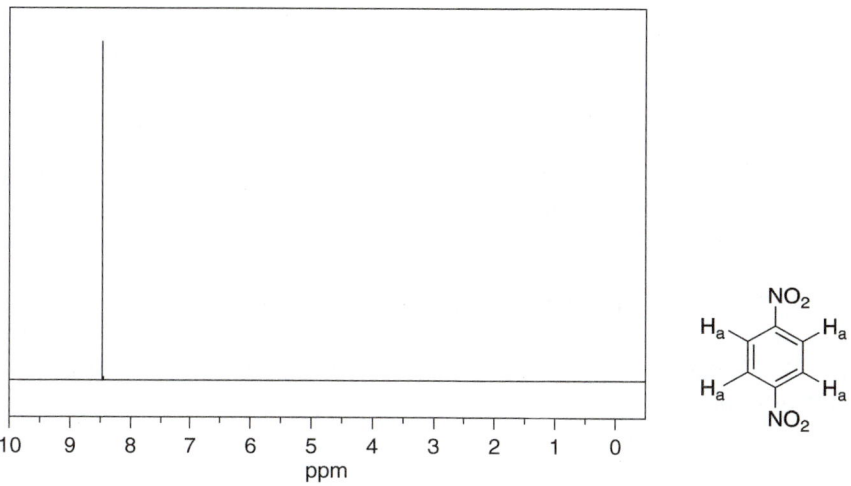

Figure 13.36 300 MHz ^{1}H NMR spectrum of 1,4-dinitrobenzene in CDCl$_3$. *Source:* with permission from AIST.

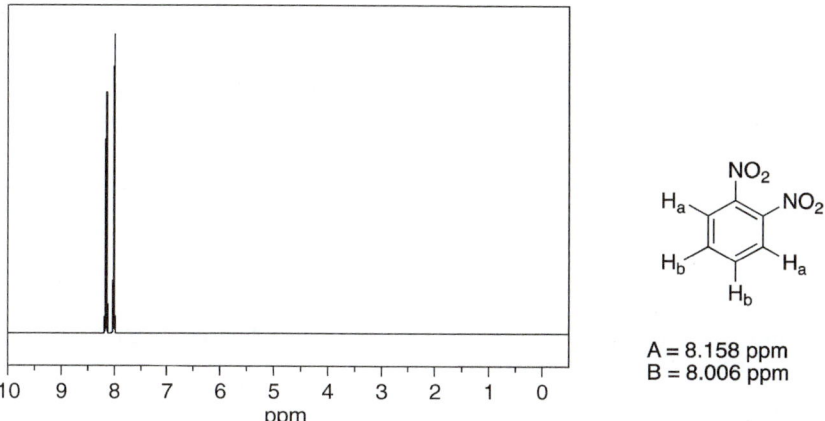

Figure 13.37 300 MHz ^{1}H NMR spectrum of 1,2-dinitrobenzene in acetone. *Source:* with permission from AIST.

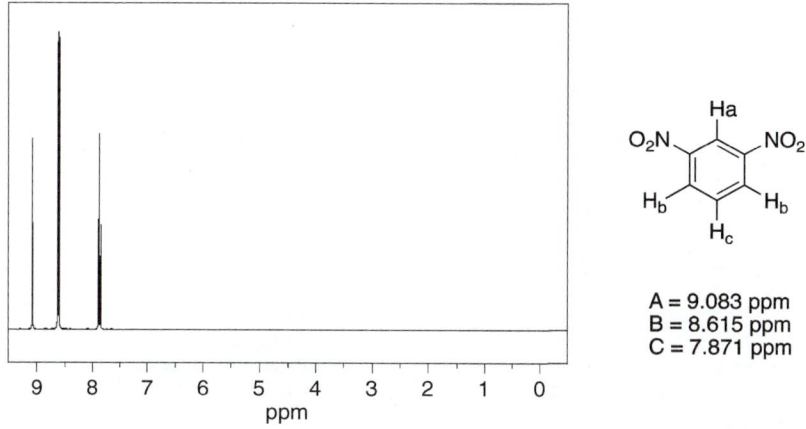

Figure 13.38 300 MHz ^{1}H NMR spectrum of 1,3-dinitrobenzene in acetone. *Source:* with permission from AIST.

as one signal but shows as a doublet since they are adjacent to a carbon that has one proton, H_c. H_c appears as a multiplet, also known as doublet of doublet, since it is next to carbons that have one hydrogen each.

With this information about substituted benzenes, you should be able to identify molecules that have the benzene ring in the structures. In addition, you should be able to identify the type substitution on the ring; that is, *ortho*, *meta*, or *para* substitution. Let us now examine a more complex molecule. The ^{1}H NMR of ethyl benzoate is shown in Figure 13.39, along with the assignments.

The aromatic protons are readily noticeable, being downfield and around 7 ppm. Also note the splitting pattern, which is similar to that of nitrobenzene, which is monosubstituted. The next observation is the splitting pattern of the other two signals at 4.36 and 1.38 ppm. They are quartet and triplet, respectively. This splitting pattern suggests a CH_2 next to a CH_3 group. Note also that the downfield signal would indicate that it is adjacent to an electronegative atom and from the table, the best match is next to an oxygen.

The spectrum for phenyl propionate is given in Figure 13.40.

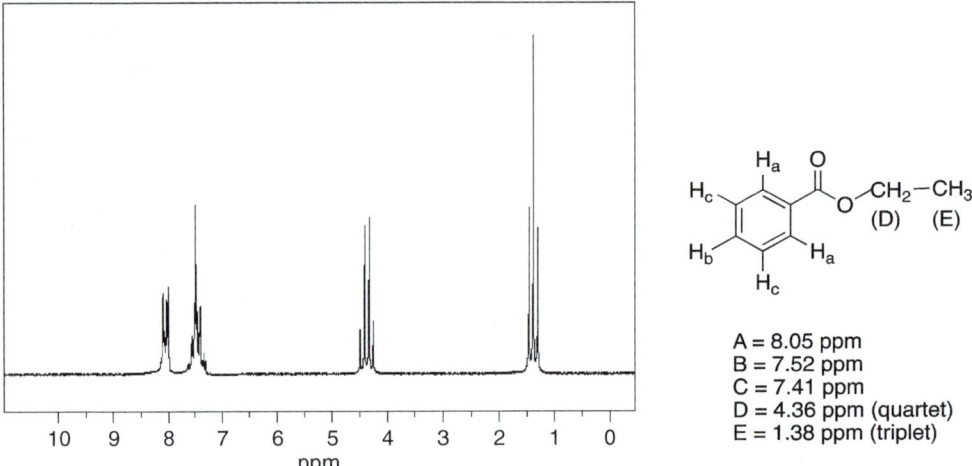

Figure 13.39 300 MHz ^{1}H NMR spectrum of ethyl benzoate in CDCl$_3$. *Source:* with permission from AIST.

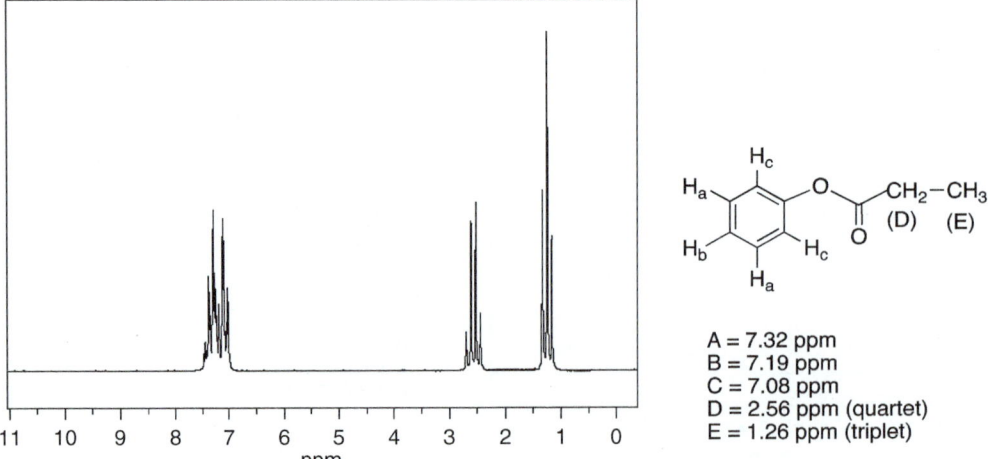

Figure 13.40 300 MHz ^{1}H NMR spectrum of phenyl propionate in CDCl$_3$. *Source:* with permission from AIST.

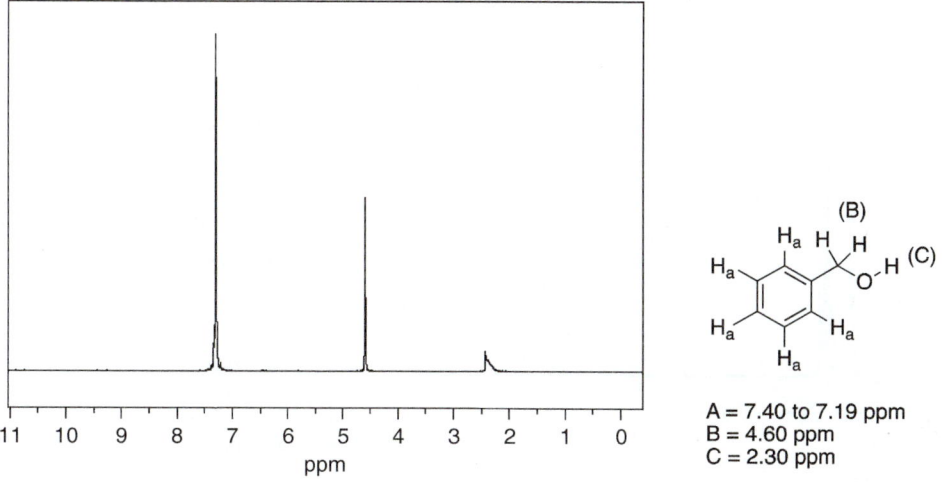

Figure 13.41 300 MHz ^{1}H NMR spectrum of benzyl alcohol in CDCl$_3$. *Source:* with permission from AIST.

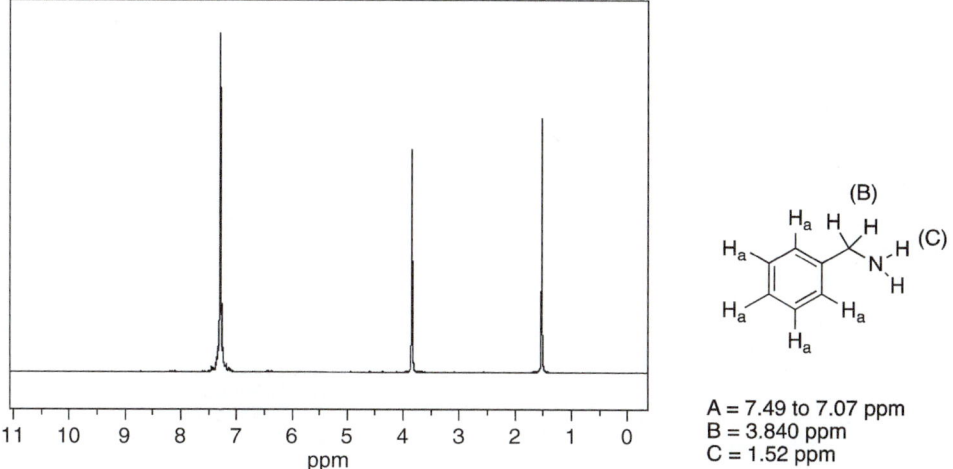

Figure 13.42 300 MHz ^{1}H NMR spectrum of benzyl amine in CDCl$_3$. *Source:* with permission from AIST.

This spectrum is very similar to that for ethyl benzoate, which is given in Figure 13.39, except that the chemical shifts for the upfield signals are different. The signal for the CH$_2$, which appears as a quartet, is further upfield compared to that of ethyl benzoate. Since the CH$_2$ is bonded to a carbonyl carbon and not the electronegative oxygen as in ethyl benzoate, the signal appears further upfield. As a result, it is possible for the chemical shifts to give evidence of the connectivity of atoms in the structure of an unknown compound.

Shown in Figures 13.41 and 13.42 are the ^{1}H NMR spectra of benzyl alcohol and benzylamine, respectively. As expected, there are three signals for the three different types of protons of these compounds. Note that the chemical shifts for the two signals for the upfield signals are slightly different for each compound however. Due to the presence of the more electronegative oxygen, compared to nitrogen, the signals are more downfield for benzyl alcohol, compared to benzylamine.

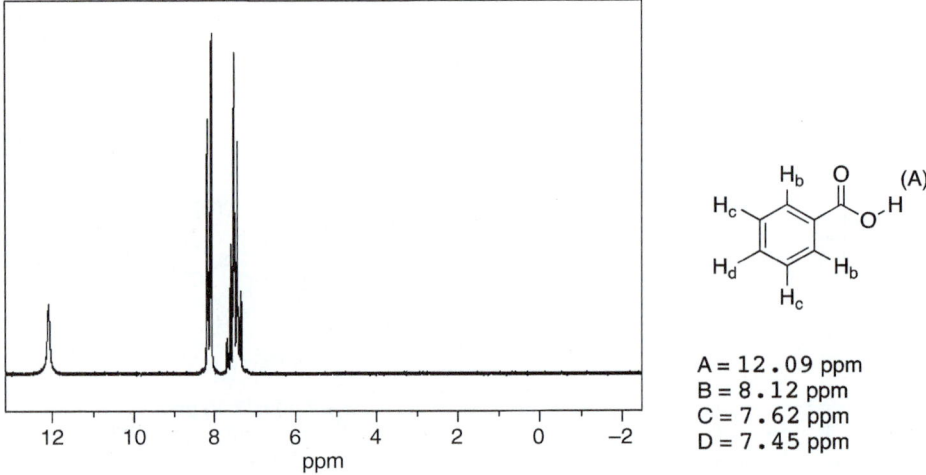

Figure 13.43 300 MHz ^{1}H NMR spectrum of benzoic acid in CDCl$_3$. *Source:* with permission from AIST.

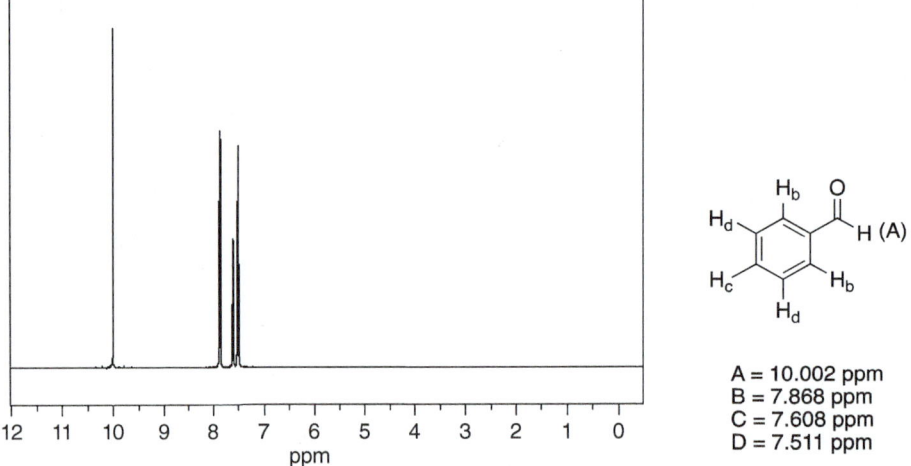

Figure 13.44 300 MHz ^{1}H NMR spectrum of benzaldehyde acid in CDCl$_3$. *Source:* with permission from AIST.

The last set of spectra to be examined are those of benzoic acid and benzaldehyde, which are shown in Figures 13.43 and 13.44, respectively.

The spectra for these two compounds are very similar, except the downfield signal. For benzoic acid, it appears at 12.09 ppm, whereas for benzaldehyde, it appears at 10.02 ppm. As expected, since the proton in benzoic acid is bonded to the electronegative oxygen, it appears further downfield, compared to the proton that is bonded to the carbonyl carbon of benzaldehyde, which appears further upfield. Again, these two examples demonstrate that the NMR can be used to gain the atom connectivity in different unknown molecules.

Problem 13.5

An unknown compound showed a molecular weight ion peak at 134.2 *m/z*. The ^{1}H NMR and IR spectra are given below. Give a likely structure of this unknown compound.

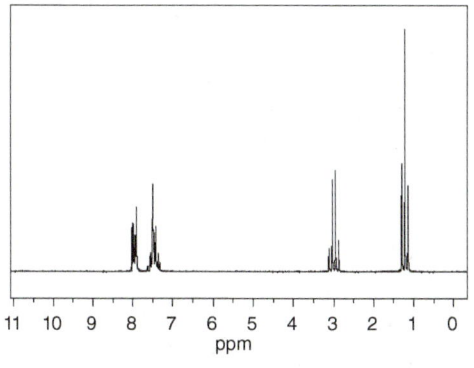
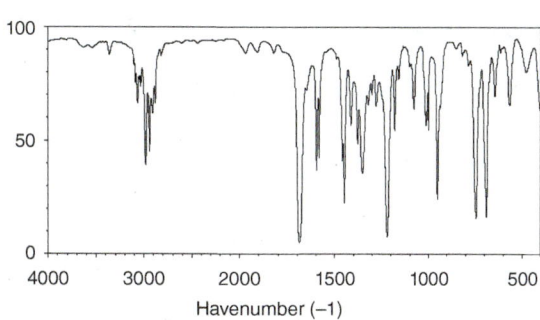

13.6.7 Carbon-13 NMR (^{13}C NMR)

The carbon-13 isotope is another nucleus that has an odd number of electrons and hence has a spin number of ½, which makes it NMR active as described earlier for the ^{1}H. ^{13}C spectra give details about the carbon skeleton of an unknown compound. The natural abundance of ^{13}C is 1.1% of the total carbon atom (^{12}C, ^{13}C, and ^{14}C). The natural abundance of ^{12}C is >98% of the carbon atoms in any molecule. Of the isotopes of carbon, only the ^{13}C gives rise to an NMR signal. Owing to the extremely low abundance of natural occurring ^{13}C NMR in any sample, a high concentration of sample is typically required to gain good ^{13}C NMR spectra, but owing to the increased sensitivity of the modern instruments, good ^{13}C NMR spectra can be obtained even at relatively low sample concentrations.

13.6.8 Carbon-13 Chemical Shifts and Coupling

The same concept as that described in the previous section for the ^{1}H NMR spectroscopy applies the C-13 NMR spectra. The environment around a carbon affects the magnetic environment about each carbon of a molecule. The chemical shift values are different, however, for ^{13}C NMR, compared to ^{1}H NMR. For the excitation of protons, different energies are required, compared to the excitation of carbons. Examples of chemical shift values are shown in Table 13.4.

Splitting does not occur as discussed in the proton NMR spectroscopy since the probability of two C-13 atoms appearing next to each other in a molecule is very low. As a result, the signals for the C-13 NMR appear as singlets and reflect the environment of each carbon. The C-13 NMR spectra give information about the different types of carbons that are present in an organic molecule. Shown in Figure 13.45 is the ^{13}C NMR of benzene.

Since all the carbons of benzene are equivalent, they all appear as a single signal at 128.36 ppm in the C-13 NMR spectrum, which matches the expected chemical shift from Table 13.4. On the other hand, the C-13 spectrum for nitrobenzene is shown in Figure 13.46, along with the assignments.

Since nitrobenzene is a symmetrical molecule, there are four different signals for the four different carbons of nitrobenzene; note that two types of carbons are equivalent. The chemical shift for the carbon that is bonded to the nitro group is furthest downfield as expected based on the electronegativity of the nitro group. The spectrum of 1,3-dinitrobenzene is shown in Figure 13.47.

Table 13.4 Carbon-13 chemical shift values based on different environments of different carbons.

Type of carbon	Type of compound	Chemical shift range (ppm)
R—CH$_3$	Aliphatic	8–35
R$_2$—CH	Aliphatic	15–50
R$_3$—CH	Aliphatic	20–60
R$_4$—C	Aliphatic	30–40
R—CONR$_2$	Amide	165–175
R—CO$_2$R	Ester	165–175
R—CO$_2$H	Carboxylic acid	175–185
R—COH	Aldehyde	190–200
R—COR	Ketone	205–220
—C≡C—	Alkyne	65–85
—C=C—	Alkene	100–150
Ph—C	Phenyl	110–170
Ph—C (benzylic)	Benzylic	30–90

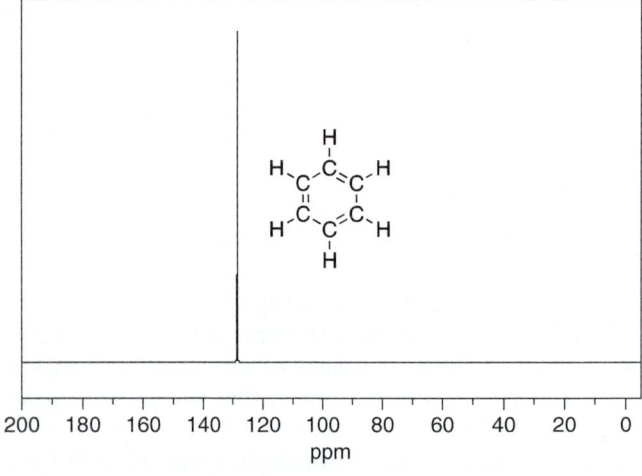

Figure 13.45 300 MHz ^{13}C NMR spectrum of benzene in CDCl$_3$. *Source:* with permission from AIST.

Since 1,3-dinitrobenzene is a symmetrical molecule, it shows four different signals, which correspond to the four different carbons of 1,3-dinitrobenzene.

The ^{13}C NMR spectrum of a slightly more complex molecule, benzyl alcohol, is shown in Figure 13.48.

An obvious signal of the spectrum appears upfield at 64.70 ppm. The chemical shift of this carbon corresponds to the chemical shift of the benzyl carbon as shown in Table 13.4. Two molecules that are similar in structure to benzyl alcohol, but have different functional groups, the carboxylic acid and the aldehyde functionalities are shown in Figures 13.49 and 13.50.

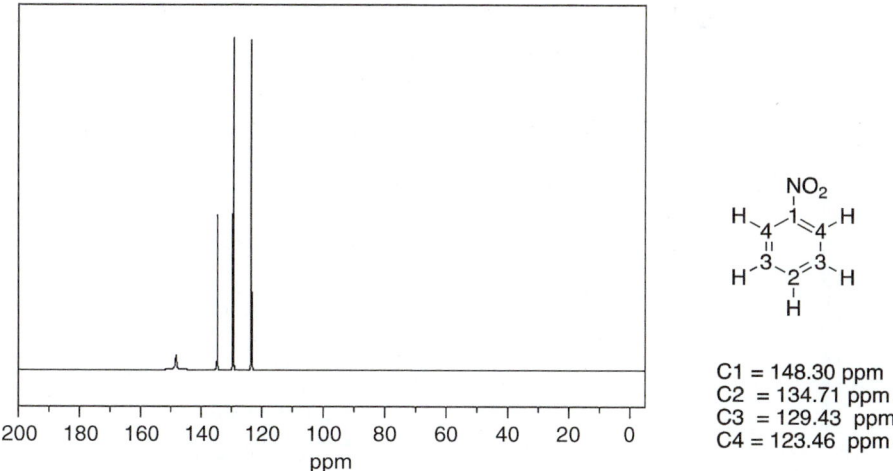

Figure 13.46 300 MHz ^{13}C NMR spectrum of nitrobenzene in CDCl$_3$. *Source:* with permission from AIST.

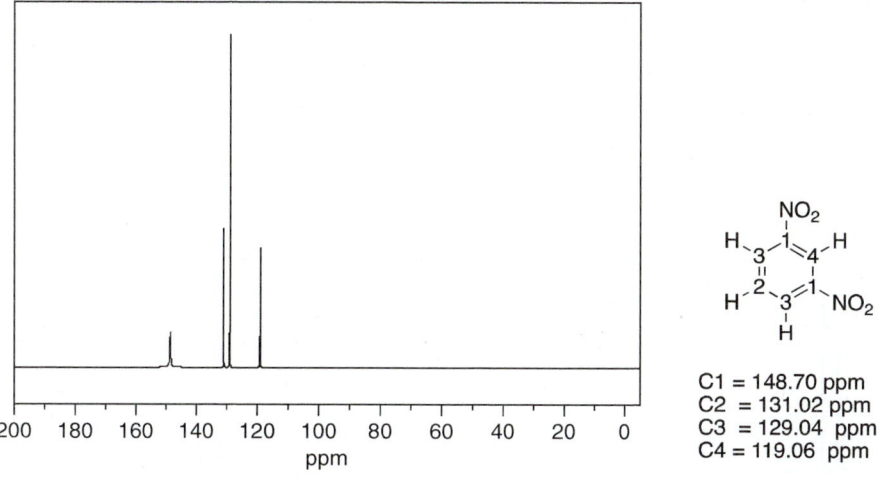

Figure 13.47 300 MHz ^{13}C NMR spectrum of 1,3-dinitrobenzene in CDCl$_3$. *Source:* with permission from AIST.

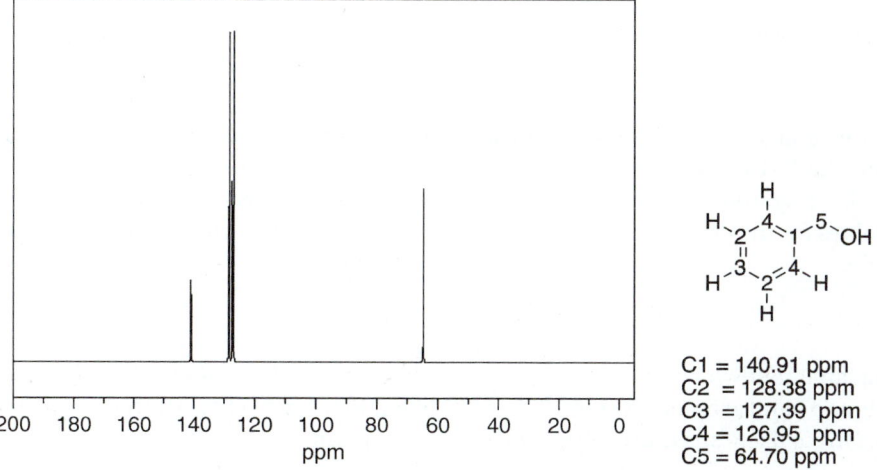

Figure 13.48 300 MHz ^{13}C NMR spectrum of benzyl alcohol in CDCl$_3$. *Source:* with permission from AIST.

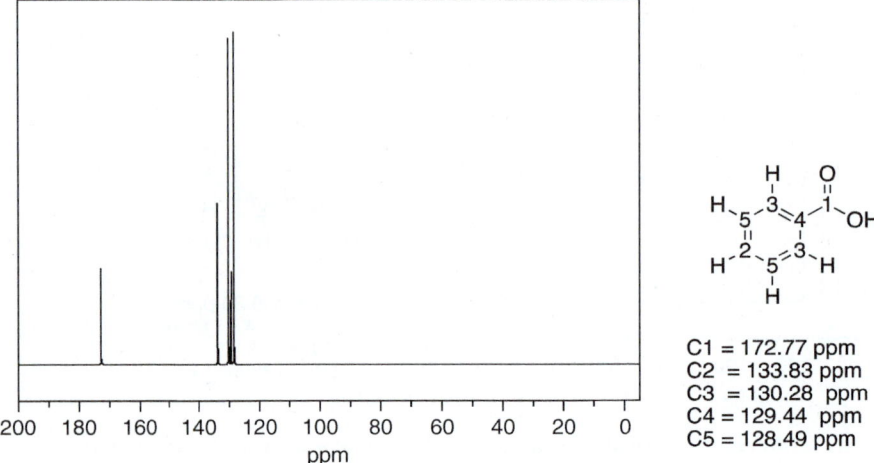

Figure 13.49 300 MHz ^{13}C NMR spectrum of benzoic acid in CDCl$_3$. *Source:* with permission from AIST.

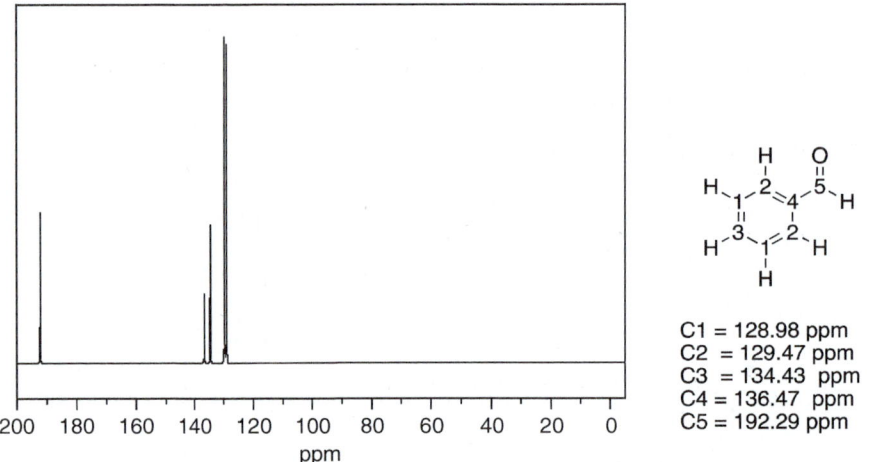

Figure 13.50 300 MHz ^{13}C NMR spectrum of benzaldehyde in CDCl$_3$. *Source:* with permission from AIST.

Both spectra are similar except for the chemical shift of downfield signals are different. As you will see from the C-13 NMR table, the signal for the carbonyl carbon of the aldehyde is further downfield, compared to that of the carboxylic acid. Thus, based on the difference in chemical shift values for the downfield signal, it is possible to distinguish between benzoic acid and benzaldehyde.

Problem 13.6

It was shown that an unknown compound has a molecular formula of C$_7$H$_8$O, and the IR spectrum showed a broad band around 3400 cm^{-1}. The ^{13}C NMR (left) and ^{1}H NMR (right) spectra are given below. Give a likely structure of this unknown compound.

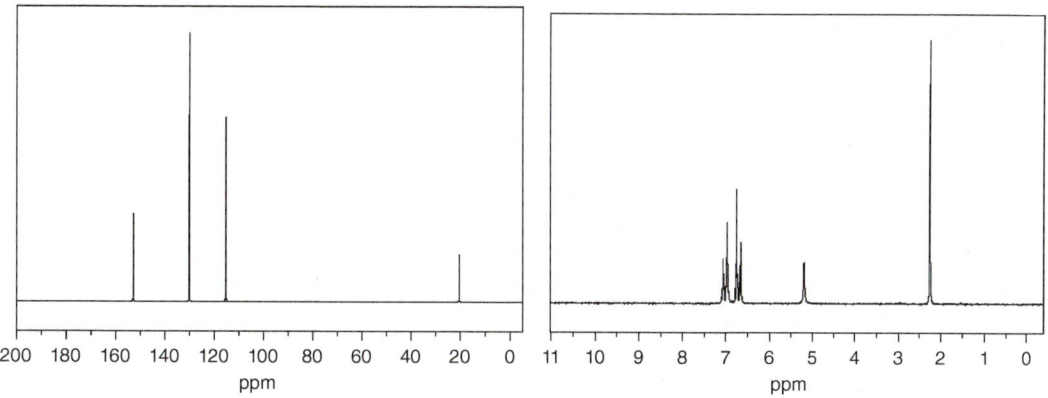

End of Chapter Problems

13.7 The following reaction was carried out, determine how would you use spectroscopy to determine if there was complete conversion to product?

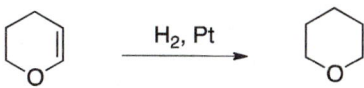

13.8 An unknown compound showed a molecular weight ion peak at 134.2 *m/z*. The ^{1}H NMR and IR spectra are given below. Give a likely structure of this unknown compound.

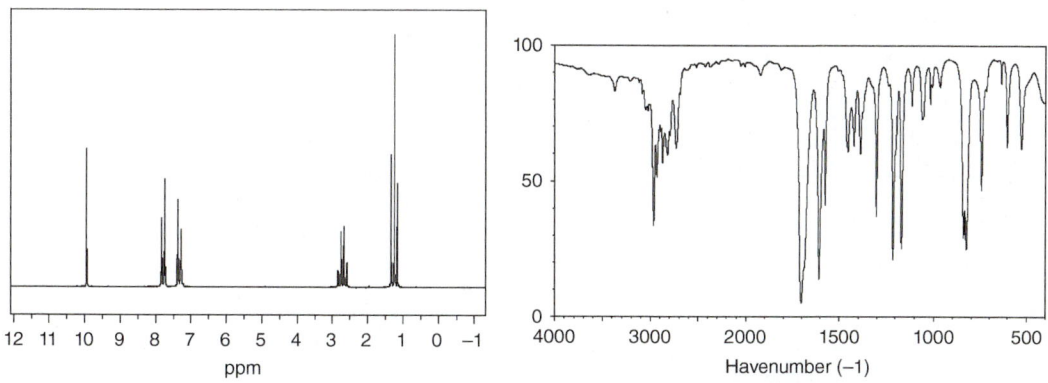

13.9 A tertiary alcohol (**A**) is treated with the strong base, sodium hydride, to form a salt, which then reacts with methyl bromide to form a new compound **B**. The mass spectrum for compound **A** has a *M* + 1 peak at *m/z* = 117 and other peaks at *m/z* = 87 and *m/z* = 73. Propose structures for compounds **A** and **B**.

13.10 The hydration of 2-pentene gave an unknown alcohol. Give a brief description of how you could use spectroscopy to determine the structure of the hydration product.

13.11 The NMR of *N,N*-dimethylacetamide (structure shown below) consists of three singlets at δ = 2.1, δ = 2.93, and δ = 3.03. If the temperature should increase, the NMR spectrum of this compound changes slightly resulting in just two signals: δ = 2.1 and δ = 2.98 with the signals at δ = 2.93 and δ = 3.03 coalesce into the signal at 2.98 with twice the intensity. Explain this observation.

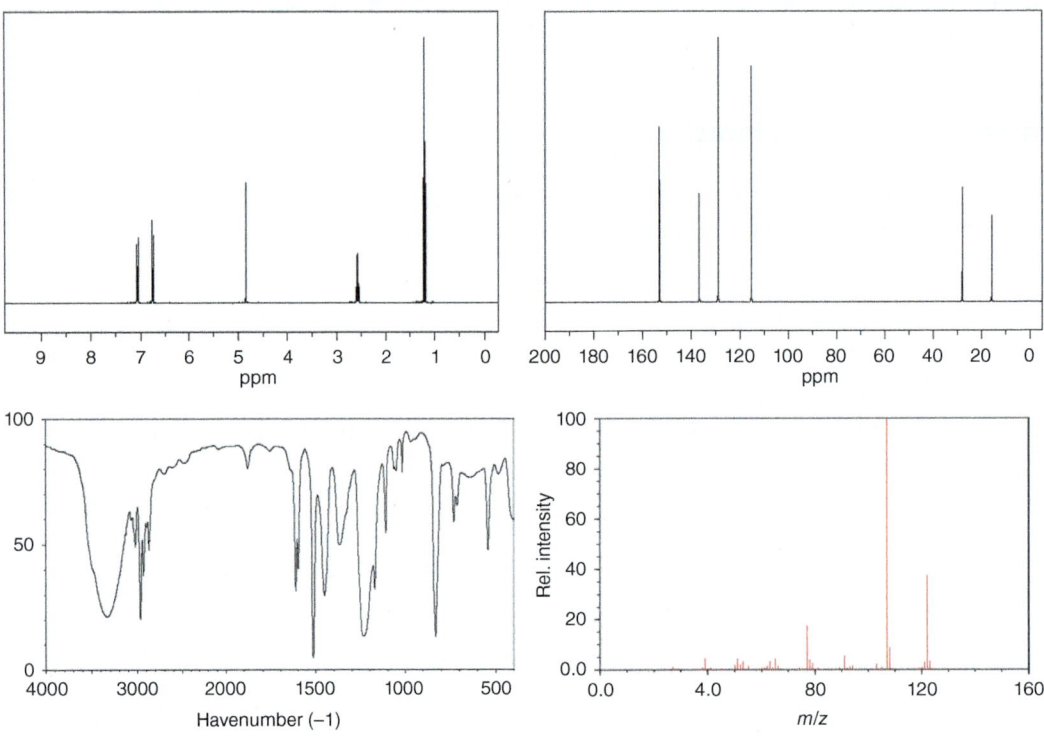

N,N-Dimethylacetamide

13.12 The spectra for an unknown compound are given below. Based on the spectra provided, determine the structure of the unknown compound.

14

Free Radical Substitution Reactions Involving Alkanes

14.1 Introduction

Saturated hydrocarbons, also called alkanes, are fairly unreactive, but they do undergo some reaction types mentioned in Chapter 6. We have seen that most alkanes readily undergo oxidation under specific reaction conditions to liberate heat, along with water and carbon dioxide. In this chapter, we will examine another reaction type that alkanes undergo, and that is substitution reactions. Substitution reactions involving different types of alkanes with chlorine or bromine in the presence of energy in the form of heat or light to form much more reactive compounds, alkyl halides will be examined. Alkyl halides are used frequently for the synthesis of larger and other useful compounds, which will be covered in this chapter. We have already covered some of the reactions of alky halides, which include elimination reactions. Alkyl halides are alkanes in which one or more of the hydrogens of the alkanes has been substituted for one or more halogens. The usual representation of alkyl halides is R-X, where R represents the alkyl group, and X represents the halogen. Some alkyl halides are used as solvents in organic and industrial labs. Such halogenated solvents are immiscible with water and are denser than water (Figure 14.1).

Chloroform was one of the first compounds discovered to induce general anesthesia. The use of this compound opened the possibility to perform surgery on patients who became unconscious when vapors of chloroform were inhaled. It was later discovered that chloroform is toxic and carcinogenic, so its use for this purpose was stopped. Several compounds, such as N_2O and $BrClCHCF_3$, have been used instead. Over the years, less toxic alkyl halides, such as ethyl chloride haloethane and isoflurane, were used.

Ethyl chloride   Haloethane   Isoflurane

Chlorofluorocarbons (CFCs) are very volatile compounds, and they were used as the propellant for aerosols in deodorants, paints, hair spray, and other spray products. They are also called Freon and are used as refrigerants in air conditioners, refrigerators, and freezers. Since these CFCs are inert compounds, they were considered ideal volatile propellants until in the 1970s, it was discovered that they ultimately find their way to the stratosphere where they deplete the ozone (O_3) layer. Even though CFCs are fairly inert, they deplete the ozone layer by reacting with ozone in the presence of ultraviolet radiation. The ozone layer helps to shield us from harmful ultraviolet radiation. The ozone hole, a term used to describe this depletion, is

Organic Chemistry: Concepts and Applications, First Edition. Allan D. Headley.
© 2020 John Wiley & Sons, Inc. Published 2020 by John Wiley & Sons, Inc.
Companion website: www.wiley.com/go/Headley_OrganicChemistry

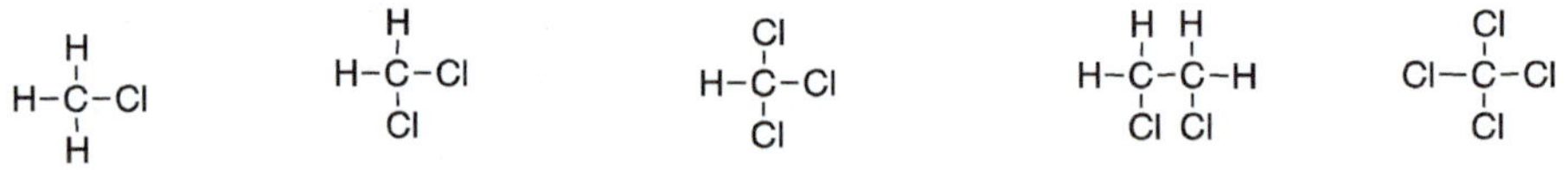

14 Free Radical Substitution Reactions Involving Alkanes

Chloromethane CH_3Cl **Dichloromethane** CH_2Cl_2 **Trichloromethane** $CHCl_3$ (also known as chloroform and was used as an anesthethic) **1, 2-Dichloroethane** $ClCH_2CH_2Cl$ **Tetrachloromethane** CCl_4

Figure 14.1 Common alkyl halides that are used as common organic solvents.

constantly getting larger over time. It is only recently that it was reported that the depletion of the ozone hole is not as severe as earlier years.

Dichlorodifluoromethane (Freon 12) CF_2Cl_2

Trichlorofluoromethane (Freon 11) $CFCl_3$

Alkyl chlorides have been used also as insecticides, three examples are shown below.

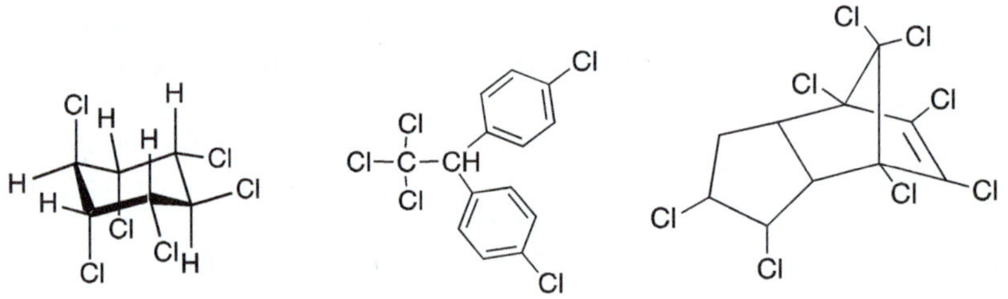

Lindane, an insecticide   DDT, insecticide   Chlordane, insecticide

Lindane is an insecticide and is also used for the pharmaceutical treatment of scabies and lice. DDT was also used in World War II to control various diseases, including malaria, typhus, and bubonic plague. DDT was a very common pesticide, but its use was banned in the early 1970s due to negative environmental consequences associated with its use. Chlordane was used widely in the United States as a pesticide for a wide variety of agricultural products, but due to environmental concerns and health risks to humans, its use was banned in the early 1980s. Polybrominated diphenylethers are used as flame retardants; they have been used in the production of various other products, including plastics, foams, and textiles.

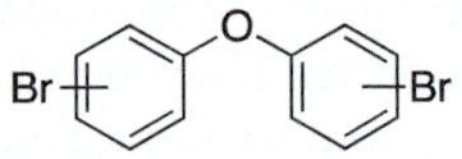

Polybrominated diphenylether

A very important use of alkyl halides, however, is that they are the starting materials for the synthesis of a lot of other useful organic compounds, including various polymers. In this chapter, we will learn how to synthesize alkyl halides from saturated hydrocarbons, and in other chapters, we will examine additional reactions that alkyl halides undergo.

14.2 Types of Alkanes and Alkyl Halides

14.2.1 Classifications of Hydrocarbons

As pointed out in Chapter 4, hydrocarbons, and especially alkanes, are the source for many organic compounds, including alkyl halides. There are literally thousands of different alkanes and they can be categorized into four common categories based on the number of hydrogens and carbons that are bonded to a specific carbon, as shown below (Figure 14.2).

In the first example, the methyl carbon, and hence methyl hydrogen, is derived from methane since there is only one carbon atom involved. The primary carbon is bonded to one alkyl group (R) and hence the hydrogens bonded to that carbon are described as primary hydrogens; the symbol that is used to represent a primary carbon and primary hydrogens is 1°. The secondary carbon is bonded to two alkyl groups and hence the hydrogens that are bonded to that carbon are called secondary hydrogens the symbol that is used to represent a secondary carbon and secondary hydrogens is 2°. The tertiary carbon is bonded to three R groups and has only one hydrogen bonded to that carbon; the hydrogen that is bonded to this carbon is classified as a tertiary hydrogen; the symbol that is used to represent a tertiary carbon and tertiary hydrogens is 3°.

Let us now apply that classification to propane, which has different types of hydrogens and carbons. Propane has two types of hydrogens, primary and secondary hydrogens, as shown in Figure 14.3.

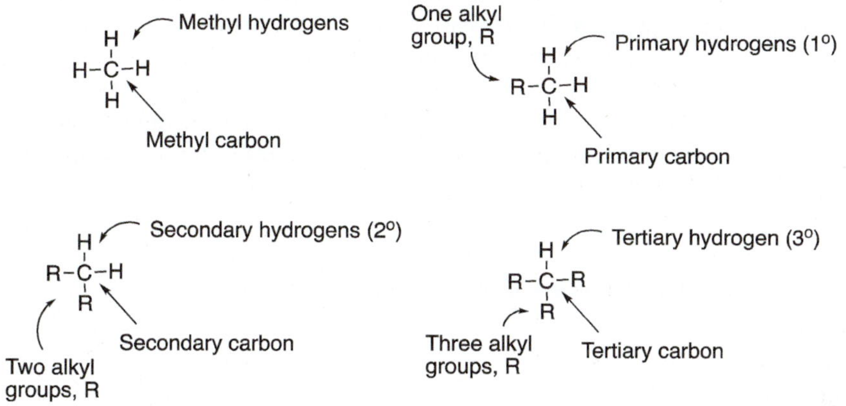

Figure 14.2 Types of carbons and hydrogens of alkanes.

Figure 14.3 Types of carbons and hydrogens of propane.

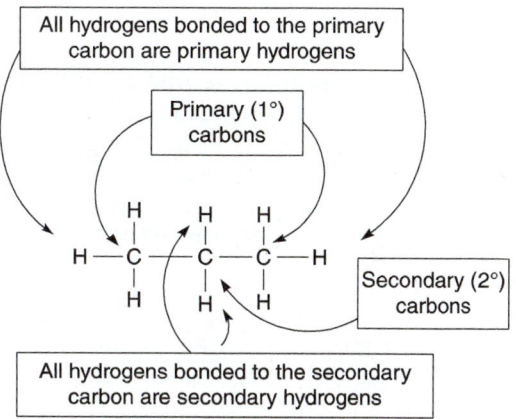

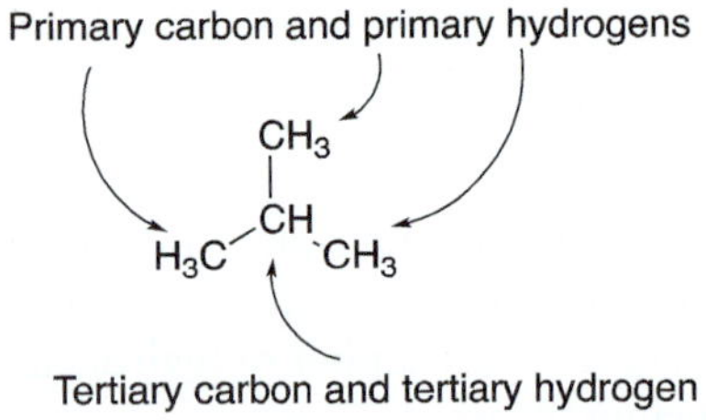

14 Free Radical Substitution Reactions Involving Alkanes

Figure 14.4 Types of hydrogens bonded to 2-methylpropane (isobutane).

Primary carbon and primary hydrogens — CH$_3$ groups of H$_3$C–CH–CH$_3$

Tertiary carbon and tertiary hydrogen — central CH

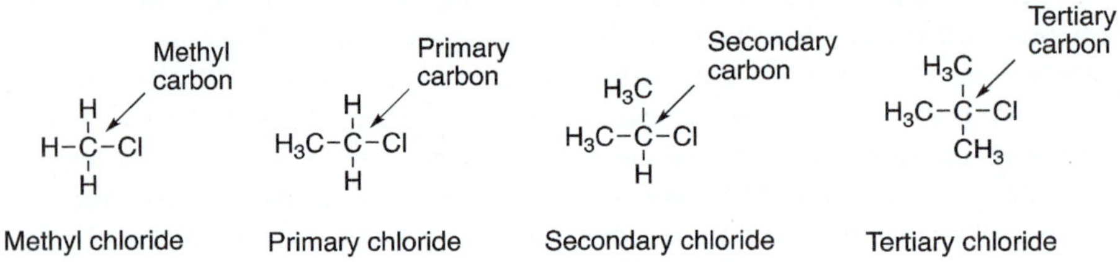

Figure 14.5 Examples of the classification of different types of alkyl chlorides.

- Methyl carbon: H–C(H)(H)–Cl — Methyl chloride
- Primary carbon: H$_3$C–C(H)(H)–Cl — Primary chloride
- Secondary carbon: H$_3$C–C(H)(CH$_3$... shown as H$_3$C)–Cl with H — Secondary chloride
- Tertiary carbon: H$_3$C–C(CH$_3$)(CH$_3$)–Cl — Tertiary chloride

Figure 14.4 gives another example showing the different types of carbons and hydrogens of 2-methylpropane (isobutane).

This same system of classification can be used to classify alkyl halides. The alkyl halides that will be encountered in this chapter can be classified as: methyl, primary, secondary, or tertiary halides. A methyl halide is an alkyl halide in which the halogen is bonded to a methyl carbon; a primary halide is an alkyl halide in which the halogen is bonded to a primary carbon; a secondary halide is an alkyl halide in which the halogen is bonded to a secondary carbon; and a tertiary halide is one in which the halogen is bonded to a tertiary carbon. Examples of the types of alkyl halides are shown in Figure 14.5.

Problem 14.1

i) Draw the expanded structure of the following alkanes and classify each hydrogen as 1°, 2°, or 3°.

   a) 2-Methylbutane
   b) 2,2-Dimethylpentane
   c) Methylcyclopentane

ii) Classify the following alkyl halides as methyl, 1°, 2°, or 3°.

   (a) 2-bromo structure (b) Cl on branched structure (c) Br on branched (d) Br on cyclopentane

In addition to the primary, secondary, and tertiary classifications described above, there are other commonly used classifications, which are shown below.

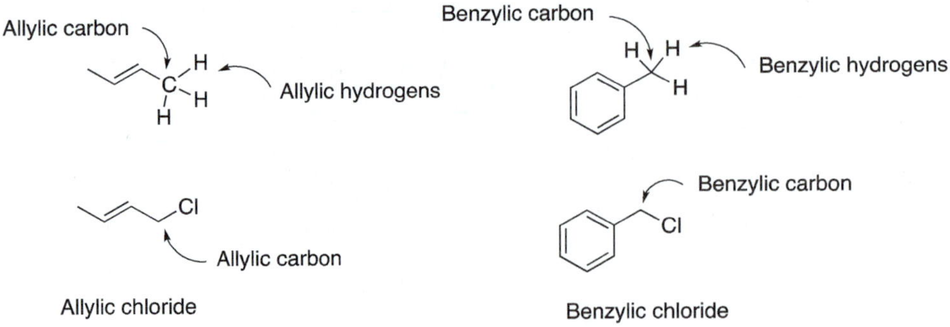

- Allylic carbon / Allylic hydrogens (C=C–CH with H's)
- Benzylic carbon / Benzylic hydrogens (phenyl–CH with H's)

Allylic chloride (CH$_2$=CH–CH$_2$–Cl) — Allylic carbon

Benzylic chloride (Ph–CH$_2$–Cl) — Benzylic carbon

Table 14.1 Bond dissociation energies for different types of hydrocarbons.

Bond	Energy (kcal mol^{-1})	Type of C—H bond
CH$_3$—H	105	Methyl
CH$_3$CH$_2$—H	98	Primary (1°)
(CH$_3$)$_2$CH—H	95	Secondary (2°)
(CH$_3$)$_3$C—H	93	Tertiary (3°)
C$_6$H$_5$CH$_2$—H	90	Benzylic
CH$_2$=CHCH$_2$—H	89	Allylic

A major difference between the allylic and benzylic system of classification is the carbon adjacent to the carbon being classified; an allylic carbon is adjacent to a carbon–carbon double bond, and a benzylic is adjacent to a benzene ring.

14.2.2 Bond Dissociation Energies of Hydrocarbons

The bond strengths of the C—H covalent bond for the different types of hydrocarbons discussed in the previous section are different. Some require more energy to break in a homolytic manner, compared to others that require less energy. Table 14.1 gives the bond dissociation energies for different types of C—H bonds.

Note that it requires more energy to break a methyl C—H bond, compared to a primary C—H bond. Likewise, it is easier to break an allylic or benzylic C—H bond, compared to a secondary or tertiary C—H bond. This concept will be very important when we start our analysis of reactions of alkanes. Now that we have examined the covalent bond, let us look at some of the properties of this type of bond. The typical covalent bond length is about 0.74 Å to about 2.0 Å. Short covalent bonds are typically stronger than longer covalent bonds. Triple bonds are short and strong, while single bonds are longer and weaker than triple bonds. It is possible to supply enough energy to break covalent bonds, especially a single bond so that the cleavage results in an equal distribution of the bonding electrons between the two atoms of that bond. The energy needed to break covalent bonds is called the ***bond dissociation energy***. There are two possible ways in which a single covalent bond can be broken. Since there are two electrons in a single covalent bond, it is possible to break such a bond so that each atom of the bond gets one electron each. This type of bond cleavage is called a ***homolytic*** cleavage. The species that are produced in this case are called ***radicals***. For the homolytic cleavage of hydrogen molecule, the amount of energy that must be supplied to break one mole of the hydrogen–hydrogen bond of the hydrogen molecule is 104 kcal, i.e. the bond dissociation energy is 104 kcal mol^{-1} as shown in Reaction (14-1).

$$H-H \longrightarrow H\cdot + H\cdot \quad \Delta H = +104 \text{ kcal mol}^{-1} \quad (14\text{-}1)$$

Note that a single-barbed arrow is used to represent the movement of one electron for the bond breakage and that the value of the dissociation energy is positive since energy must be supplied in order to break the bond. Thus, the formation of one mole of hydrogen gas from the atoms will liberate the same quantity of energy as shown in Reaction (14-2).

$$H\cdot + \cdot H \longrightarrow H-H \quad \Delta H = -104 \text{ kcal mol}^{-1} \quad (14\text{-}2)$$

The same analysis can be applied to the bonds of hydrocarbons, as shown in Reactions (14-3) and (14-4).

$$\text{CH}_2=\text{CH}-\text{CH}_3 \longrightarrow \text{CH}_2=\text{CH}-\text{CH}_2\cdot + \text{H}\cdot \qquad \Delta H = +89 \text{ kcal mol}^{-1} \qquad (14\text{-}3)$$

Allylic radical

$$(\text{H}_3\text{C})_2\text{CH}-\text{H} \longrightarrow (\text{H}_3\text{C})_2\text{CH}\cdot + \text{H}\cdot \qquad \Delta H = +95 \text{ kcal mol}^{-1} \qquad (14\text{-}4)$$

Secondary radical

The homolytic cleavage of the allylic C—H bond results in a allylic radical and a hydrogen atom, whereas the homolytic cleavage of the secondary C—H bond results in a secondary radical and a hydrogen atom. Based on the bond dissociation energy values, it requires more energy to cleave the secondary C—H bond, compared to the allylic C—H bond. One conclusion that can be drawn from this observation is that the allylic radical is more stable than the secondary radical since reactions tend to proceed to give the more stable product or intermediate which would require the least amount of energy. Figure 14.6 gives an illustration of this concept.

14.2.3 Structure and Stability of Radicals

Since radicals contain a single electron in the p orbital, it would much prefer to have another electron to have a full orbital. If there are electrons surrounding the radical, this will provide some degree of stability to the radical. Obviously, the more electrons in the vicinity of the radical, the more stable will be the radical. We have seen a similar situation when we examined carbocation. More alkyl groups around a radical provide stability to the radicals. Hence, similarly the stability trend for radicals is shown below.

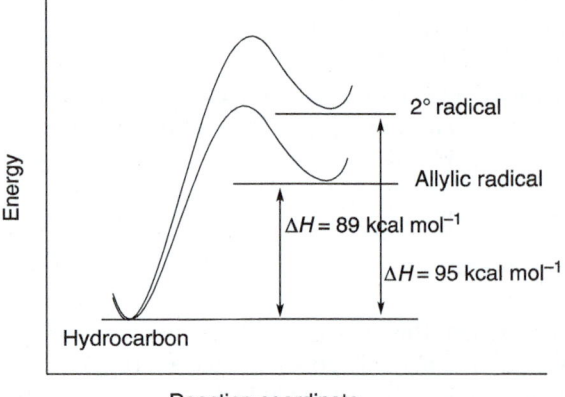

Figure 14.6 Relative energies showing the difference in energies for different types of radicals.

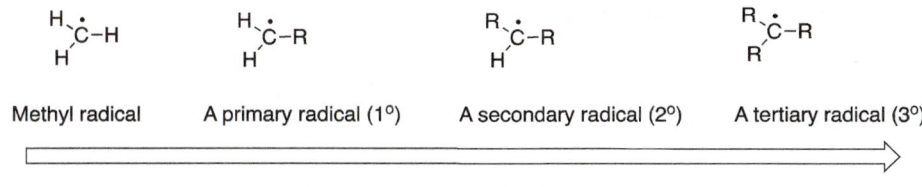

Methyl radical   A primary radical (1°)   A secondary radical (2°)   A tertiary radical (3°)

→ Increasing stability of radicals

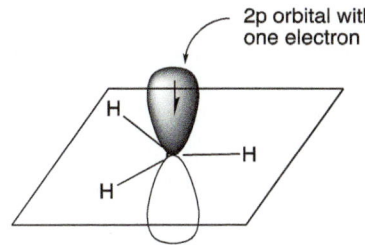

Methyl radical with no electron releasing alkyl group, R, to stabilize the radical

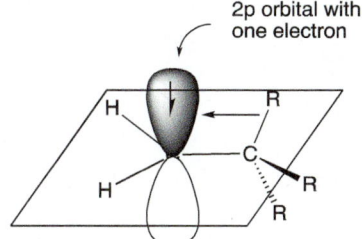

Primary radical with one electron releasing alkyl group, R, to stabilize the radical

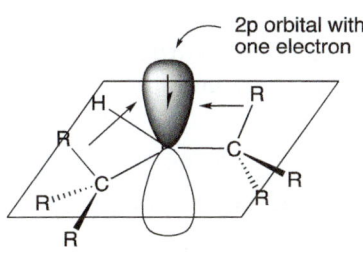

Secondary radical with two electron releasing alkyl groups, R, to stabilize the radical

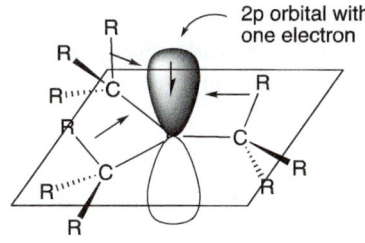

Tertiary radical with three electron releasing alkyl groups, R, to stabilize the radical

Problem 14.2

Of the following pairs of radicals, determine which is more stable?

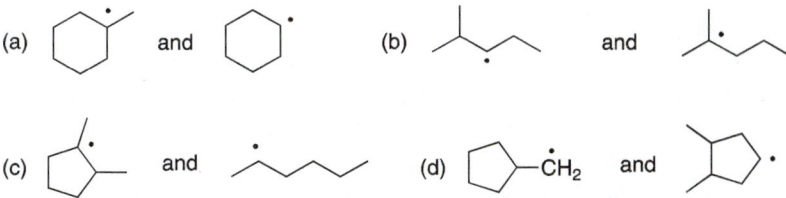

A radical that is adjacent to a carbon–carbon double bond is called an allylic radical as we have seen in the previous section. Allylic radicals are stable due to a flow of electrons, which are located in the adjacent orbitals into the singly occupied p orbital of the radical. As a result of

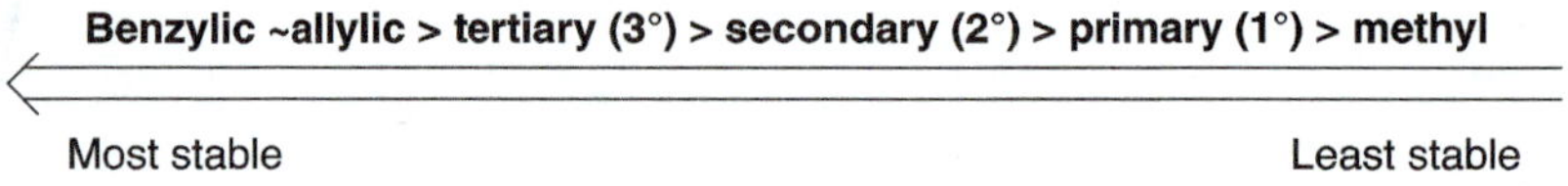

Benzylic ~allylic > tertiary (3°) > secondary (2°) > primary (1°) > methyl

Most stable   Least stable

Figure 14.7 Order of stability of radicals.

such electron flow, an allylic radical is more stable than a tertiary radical; a similar case exists for benzylic radicals. This concept of electron flow in the stabilization of allylic and benzylic radicals can be illustrated by resonance structures as shown below.

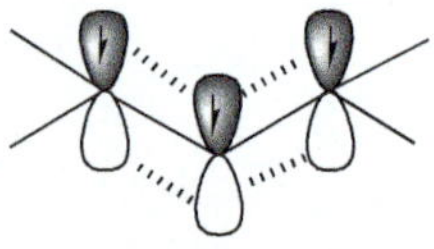

Allylic radical showing conjugatd electrons    Resonance structures for the allylic radical

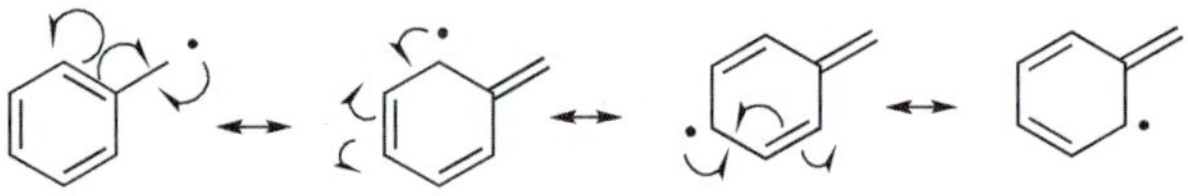

Resonce-stabilized benzylic radical

Thus, the relative magnitudes of the bond dissociation energies can be used to determine the relative stabilities of different types of radicals, and the order is shown in Figure 14.7.

Problem 14.3

Of the following pairs of radicals, determine which of the following radicals is more stable?

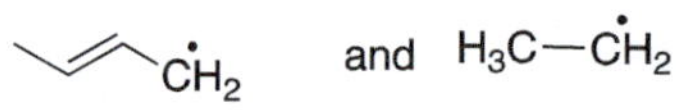

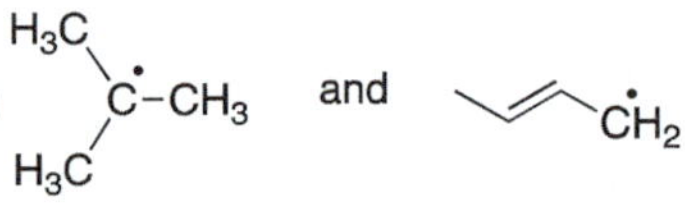

(a) $(H_3C)_2\dot{C}H$ and $H_3C-\dot{C}H_2$ (b) $CH_2=CH-\dot{C}H_2$ and $H_3C-\dot{C}H_2$

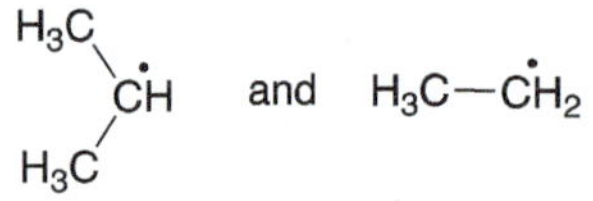

(c) $(H_3C)_3\dot{C}-CH_3$ and $CH_2=CH-\dot{C}H_2$ (d) $PhCH\dot{C}H$ and $(H_3C)_2\dot{C}-CH_3$

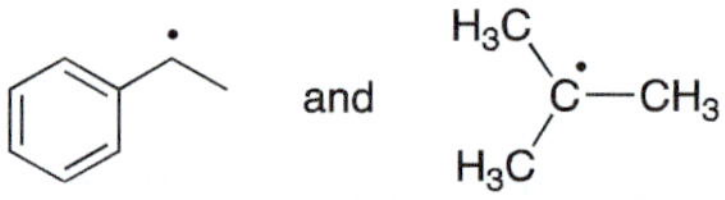

14.3 Chlorination of Alkanes

Alkanes are fairly inert compounds; some alkanes are used as solvents to provide inert media for different reactions that we will see in later chapters. In addition to combustion, alkanes undergo another type of reaction that is very important to organic chemists, and that is the reaction with halogens, specifically chlorine or bromine in the presence of energy in the form of heat or light. As the name suggests, this reaction involves the reaction of alkanes and bromine or chlorine and is called bromination or chlorination of alkanes, respectively. These reactions are performed in the presence of light or heat and are described as substitution reactions since a hydrogen or more than one hydrogen atoms of an alkane reactant are substituted for a halogen or more than one halogen in the product. Most of the reactions that will be encountered in this chapter involve the substitution of one hydrogen in the alkane for a halogen in the product.

The products produced upon the chlorination of alkanes are alkyl halides, most are important industrial raw material for the synthesis of numerous other chemicals. For example, chloromethane was used once as a refrigerant, but discontinued owing to its flammability and toxicity. Today, it is widely used as a chemical intermediate for the production of different compounds, including polymers. Other chloroalkanes are widely used as a solvent in the research labs and in the production of rubber and in the petroleum refining industry. The chlorination of methane is shown in Reaction (14-5).

$$CH_4 + Cl-Cl \xrightarrow{\text{Heat or light}} CH_3Cl + H-Cl \qquad (14\text{-}5)$$

To be able to predict the products of these and other similar reactions, a thorough understanding of how the reaction occurs is essential. Before we examine the reaction mechanism for the chlorination of alkanes, let us examine the process of bond reorganization for the reaction shown in Reaction (14-5). It should be obvious that this reaction is a substitution reaction in which a Cl—Cl bond has to be broken and the chlorine atoms form two new bonds, a new H—Cl bond and a new C—Cl bond in the CH_3Cl product. This also means that a C—H bond had to be broken in the reactant molecule in order for these new bonds to be formed in the products.

14.3.1 Mechanism for the Chlorination of Methane

Now that we have an idea of what has to occur for a chlorination substitution reaction to take place, let us now try to put that concept in the form of a reaction mechanism. To better understand how the reaction shown in 14-5 occurs, imagine a reaction vessel, which contains methane and chlorine. These two compounds will exist in this vessel until they are exposed to heat or light. Under either of these reaction conditions, the energy will break the weakest bond of either of these two reactant molecules. The bond dissociation energy for the C—H bond of methane is 105 kcal mol^{-1}, while the bond dissociation energy for the Cl—Cl bond is 58 kcal mol^{-1} as shown in Reactions (14-6) and (14-7).

$$CH_4 + 105 \text{ kcal mol}^{-1} \longrightarrow CH_3\cdot + H\cdot \qquad (14\text{-}6)$$

$$Cl-Cl + 58 \text{ kcal mol}^{-1} \longrightarrow 2\,Cl\cdot \qquad (14\text{-}7)$$

Stronger bonds, such as the carbon–hydrogen bonds of methane, are not readily broken in the presence of heat or light. Thus, by introducing energy in the reaction vessel, which contains chlorine and methane, a homolytic cleavage of the Cl—Cl bond will occur, rather than the C—H bond as shown in Reaction (14-8). Note again that a single-barbed arrow is used to indicate the movement of one electron and to demonstrate how bond cleavage occurs.

$$Cl\frown Cl \xrightarrow[\text{Homolytic cleavage}]{\text{Energy}} \underset{\text{Chlorine radicals}}{Cl\cdot \;+\; Cl\cdot} \qquad (14\text{-}8)$$

This initial step of this reaction mechanism is also known as the ***initiation step***. In this step, free radicals (chlorine radicals) are generated. A free radical is an atom or group of atoms that has an unpaired electron present. Free radicals are intermediates that are typically generated throughout the course of some reaction and most are fairly reactive. Once a fairly reactive radical intermediate is generated, it will react in order to gain another odd electron, usually to form a covalent bond. Since the chlorine radical is a very reactive intermediate, it will react with methane in an effort to become neutral. This reaction can be accomplished by the abstraction of a hydrogen atom (a hydrogen radical) to form HCl. However, a newly formed radical is also formed in this process as shown below, in this case, a methyl radical as shown in Reaction (14-9).

$$CH_4 + \cdot Cl \longrightarrow \cdot CH_3 + HCl \qquad (14\text{-}9)$$

This organic-free radical intermediate that is formed is also very reactive, and this methyl radical will react with unreacted chlorine molecules by the abstraction of a chlorine atom (a chlorine radical) to form CH_3Cl and a chlorine atom, as shown in Reaction (14-10).

$$\cdot CH_3 + Cl\text{-}Cl \longrightarrow CH_3Cl + Cl\cdot \qquad (14\text{-}10)$$

The steps of the reaction mechanism in which radicals are generated are called the ***propagation steps***. As you can see the intermediate methyl radical eventually goes on to form one of the final products as shown in Reaction (14-10). Another way in which radicals can gain another electron to form a covalent bond is for two radicals to combine to form a neutral molecule. If a methyl radical and a chlorine radical come into contact with each other, they will form a neutral product, chloromethane, as shown in Reaction (14-11).

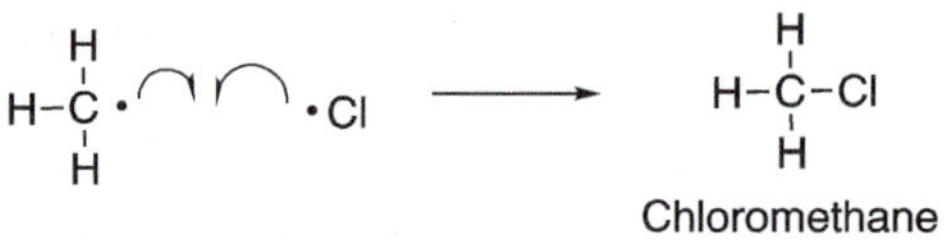

$$\cdot CH_3 + \cdot Cl \longrightarrow CH_3Cl \qquad (14\text{-}11)$$
Chloromethane

This step of the mechanism is called a ***termination step*** since radicals are converted into neutral products. As you can imagine, this is not the only termination step possible since there are other radicals that are present in the reaction vessel. Another termination reaction that is possible is shown in Reaction (14-12).

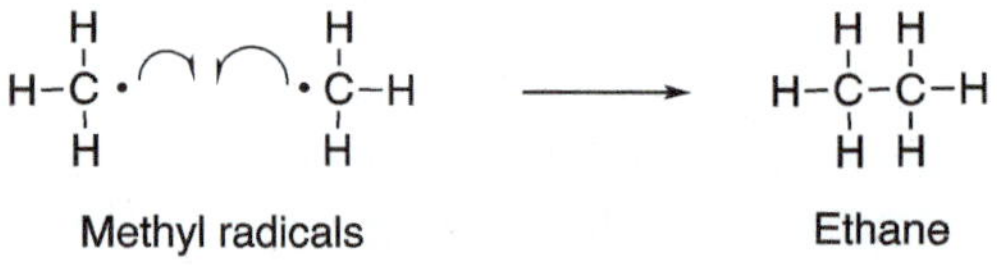

$$\cdot CH_3 + \cdot CH_3 \longrightarrow CH_3\text{-}CH_3 \qquad (14\text{-}12)$$
Methyl radicals   Ethane

As you can imagine from the mechanism described above, there are other possible products. For example, the chloromethyl radical could be generated during a propagation step of the reaction mechanism, as shown in Reaction (14-13).

$$CH_3Cl + \cdot Cl \longrightarrow \cdot CH_2Cl + HCl \qquad (14\text{-}13)$$

It is possible for this chloromethane radical to react with another chlorine atom or even with another chloromethyl radical to produce different products as shown in Reactions (14-14) and (14-15).

$$\text{Cl-CH}_2\text{H} \cdot \curvearrowright \cdot \text{Cl} \longrightarrow \text{Cl-CH}_2\text{-Cl} \quad \text{(Dichloromethane)} \tag{14-14}$$

$$\text{Cl-CH}_2\text{H} \cdot \curvearrowright \cdot \text{CH}_2\text{-Cl} \longrightarrow \text{Cl-CH}_2\text{-CH}_2\text{-Cl} \quad \text{(1,2-Dichloroethane)} \tag{14-15}$$

Depending on the reaction conditions, it is possible to target one of these products as the major organic product. A high concentration of chlorine will result in various multi-chlorination products. On the other hand, a high concentration of methane will result in chloromethane as the major organic product. For our discussion, we will assume that these substitution reactions give the mono-chlorinated product as the major organic product.

14.3.2 Chlorination of Other Alkanes

Let us now look at the chlorination of ethane, the reaction is shown in Reaction (14-16).

$$\text{CH}_3\text{-CH}_3 + \text{Cl-Cl} \xrightarrow{\text{Heat or light}} \text{CH}_3\text{-CH}_2\text{-Cl} + \text{H-Cl} \tag{14-16}$$

All six C—H bonds of ethane are equivalent, and they are all primary hydrogens. Thus, the mono-chlorination product shown in Reaction (14-16) is not surprising based on the mechanism described earlier. The chlorination of propane, however, gives two possible mono-chlorination products as shown in Reaction (14-17).

$$\text{Propane} \xrightarrow{\text{Cl}_2,\ \text{heat}} \text{2-Chloropropane (55\%)} + \text{1-Chloropropane (45\%)} \tag{14-17}$$

Note that there is a mixture of mono-chlorinated products, and they are formed in almost equal amounts. Thus, if one wanted to synthesize a large percentage of 2-chloropropane, only about 55% of the desired product would be obtained; the other product would be 1-chloropropane, an unwanted product. Let us take a closer look at the structures of methane, ethane and propane. All the C—H bonds of methane are equivalent and classified as methyl hydrogens. All the C—H bonds of ethane are equivalent and are described as primary hydrogens, but not all the C—H bonds of propane are equivalent. For propane, there are primary and secondary hydrogens and this difference in the type of hydrogen results in different products for the chlorination of propane, compared to methane and ethane. For molecules that have only one type of carbon–hydrogen bonds or hydrogens, only one mono-chlorination product will result, as shown in Reaction (14-18) for the chlorination of cyclohexane and Reaction (14-19) for the chlorination of 2,2-dimethylpropane.

$$\text{Cyclohexane (only 2° hydrogens)} \xrightarrow{\text{Cl}_2,\ \text{heat}} \text{Chlorocyclohexane} + \text{HCl} \tag{14-18}$$

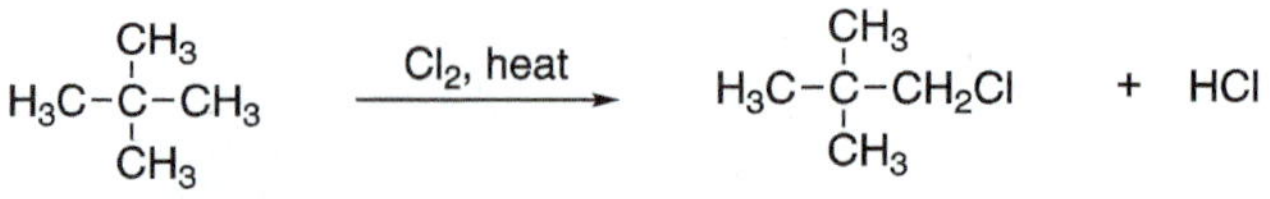

14 Free Radical Substitution Reactions Involving Alkanes

$$\underset{\substack{\text{2,2-Dimethylpropane} \\ \text{(only 1° hydrogens)}}}{\underset{\substack{|\\CH_3}}{\overset{\substack{CH_3\\|}}{H_3C-C-CH_3}}} \xrightarrow{Cl_2,\ heat} \underset{\text{1-Chloro-2,2-dimethylpropane}}{\underset{\substack{|\\CH_3}}{\overset{\substack{CH_3\\|}}{H_3C-C-CH_2Cl}}} + HCl \qquad (14\text{-}19)$$

Thus, it is extremely important to be able to recognize different types of C—H bonds or hydrogens of alkanes in order to properly predict the organic products for these types of free radical substitution reactions.

Problem 14.4

i) Identify the different types of hydrogens of in the following molecules?

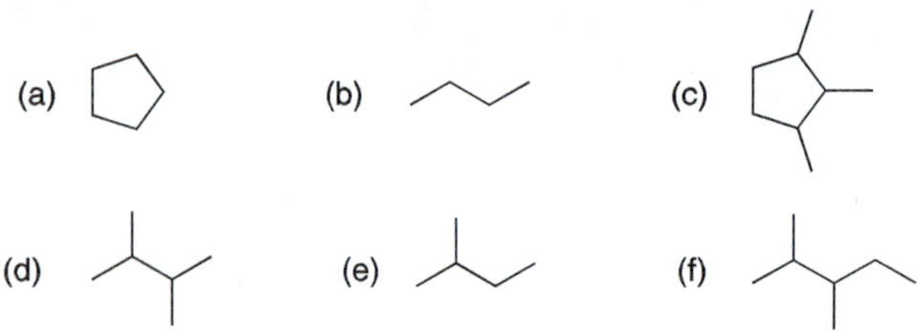

(a) cyclopentane (b) propane (c) methylcyclopentane

(d) neopentane (2,2-dimethylpropane) (e) isobutane (f) 2,2-dimethylbutane

ii) Give the mechanism (initiation, propagation, and termination steps) for the reaction shown below.

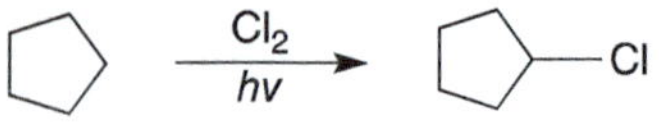

cyclopentane $\xrightarrow{Cl_2,\ h\nu}$ chlorocyclopentane

iii) Determine the chlorination products for the following reactions. Note that some may give more than one chlorination product.

(a) cyclopentane $\xrightarrow{Cl_2,\ heat}$ (b) propane $\xrightarrow{Cl_2,\ heat}$

(c) neopentane $\xrightarrow{Cl_2,\ heat}$ (d) isobutane $\xrightarrow{Cl_2,\ heat}$

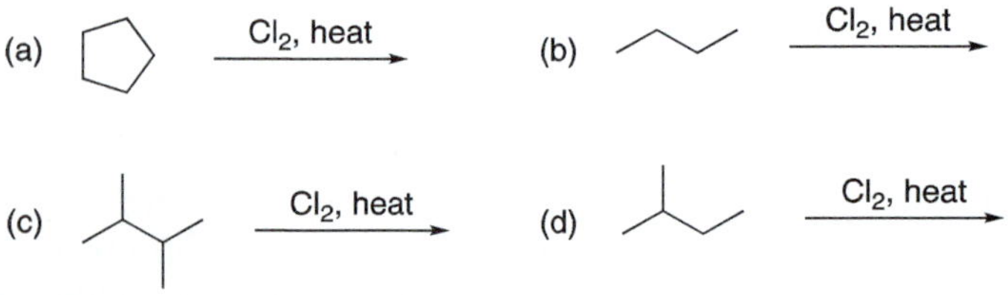

14.4 Bromination of Alkanes

A major observation for the chlorination of alkanes is that there is almost an equal mixture of different mono-chlorination organic products obtained where the alkane has different types of hydrogens. Thus, if we were trying to get a high yield of a particular halogenated product using an alkane as a starting compound, chlorination would not be the ideal reaction to use in order to give a specific halogenated product as the major organic product in high yield.

14.4.1 Bromination of Propane and Other Alkanes

It is known from various experiments that if bromine is used for the same type of free radical substitution reaction instead of chlorine, the outcome is very different, as shown in Reaction (14-20), compared with Reaction (14-17).

$$\text{Propane} \xrightarrow{\text{Br}_2,\ \text{heat}} \text{2-Bromopropane (98\%)} + \text{1-Bromopropane (2\%)} \quad (14\text{-}20)$$

From Reaction (14-20), it is obvious that utilizing bromine gives an extremely high percentage of one halogenated product (2-bromopropane) compared to the other halogenated product (1-bromopropane). You will also notice that compared to the chlorination products, there is a drastic difference in the distribution of the halogenated products. In order to explain this difference, we will have to examine the mechanism for this type of free radical substitution reaction again. In the first step of the mechanism as described in the previous section, there is a homolytic cleavage of the halogen bond due to the energy supplied in the form of heat or light. Let us examine the bond dissociate energies of the halogens, which are shown in Table 14.2.

You will notice that less energy is required to break the Br—Br bond, compared to the Cl—Cl bond. An important consequence of this energy difference is that the bromine radical is more stable and less reactive than the chlorine radical. Let us now look at the next step of the reaction mechanism, the propagation step where the halogen radical abstracts a hydrogen atom from the alkane. You will notice from the dissociation energies given in Tables 14.1 and 14.3 that the energy required to break different C—H bonds varies. More energy is required to break a primary C—H bond, compared to a tertiary C—H bond. This information is presented differently in Table 14.3, which shows the type of radical formed from the cleavage of a different C—H bond.

For alkanes, the tertiary C—H bond is the weakest and hence much easier to break (97 kcal mol^{-1}), compared to the next most difficult, the secondary (99 kcal mol^{-1}) and especially the primary (101 kcal mol^{-1}) and methyl C—H (105 kcal cal^{-1}) bonds. Figure 14.8 gives an illustration of the different types of hydrogens, along with the energy requirements for homolytic cleavage.

Table 14.2 Bond dissociation energies for different halogens.

Bond	Energy (kcal mol^{-1})
F—F	38
Cl—Cl	58
Br—Br	46
I—I	36

Table 14.3 Dissociation bond energies for different C—H bonds.

Hydrocarbon	Radical	Type radical	Bond dissociation energies (kcal mol^{-1})
CH$_3$—H	CH$_3\cdot$	Methyl	105
CH$_3$CH$_2$—H	CH$_3$CH$_2\cdot$	Primary (1°)	101
(CH$_3$)$_2$CH—H	(CH$_3$)$_2$CH$\cdot$	Secondary (2°)	99
(CH$_3$)$_3$C—H	(CH$_3$)$_3$C$\cdot$	Tertiary (3°)	97
C$_6$H$_5$CH$_2$—H	C$_6$H$_5$CH$_2\cdot$	Benzylic	90
CH$_2$=CH—CH$_2$—H	CH$_2$=CH—CH$_2\cdot$	Allylic	89

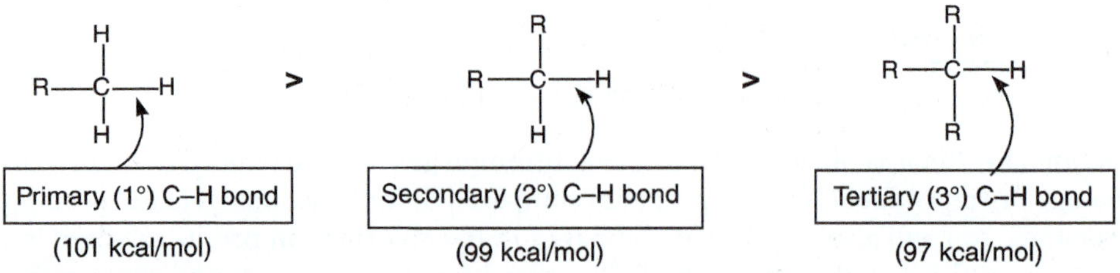

14 Free Radical Substitution Reactions Involving Alkanes

Figure 14.8 Relative strengths of the different types of C–H bonds, the strongest is the primary C–H bond and the weakest is the tertiary C–H bond.

Primary (1°) C–H bond (101 kcal/mol) > Secondary (2°) C–H bond (99 kcal/mol) > Tertiary (3°) C–H bond (97 kcal/mol)

We can utilize the mechanism that was developed to rationalize the different outcomes for these free radical substitution reactions. Let us examine the mechanism for the bromination of propane. The initiation step is shown in Reaction (14-21).

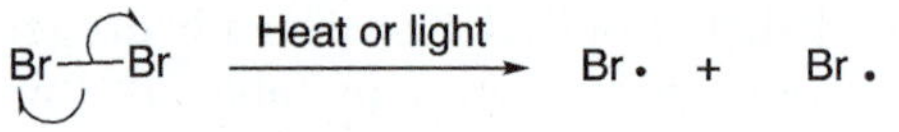

$$Br-Br \xrightarrow{\text{Heat or light}} Br\cdot + Br\cdot \qquad (14\text{-}21)$$

In the next step of the reaction mechanism, the bromine radical, which is fairly stable compared to the chlorine atom, will abstract a hydrogen from propane. For propane, however, there are two types of hydrogens as shown below.

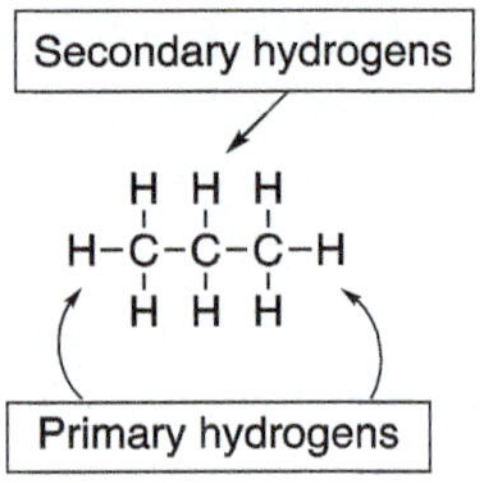

Bromine atom will abstract the hydrogen from the C–H bond that is weakest or will generate a stable secondary radical as shown in Reaction (14-22).

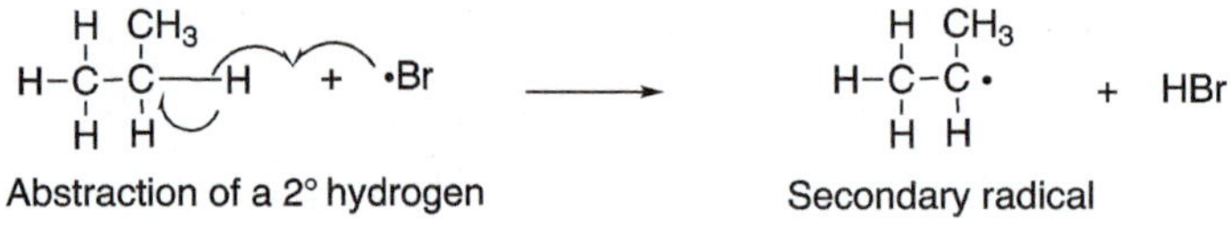

$$\text{Abstraction of a 2° hydrogen} \longrightarrow \text{Secondary radical} + HBr \qquad (14\text{-}22)$$

Bromine atom will also abstract the hydrogen from the C–H bond, but with much more difficulty to generate a primary radical as shown in Reaction (14-23).

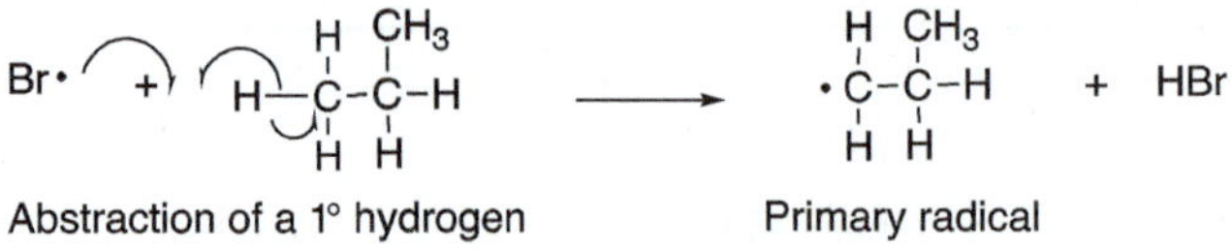

$$\text{Abstraction of a 1° hydrogen} \longrightarrow \text{Primary radical} + HBr \qquad (14\text{-}23)$$

Since the secondary radical is more stable, compared to the primary radical, the major product will result from the addition of bromine radical in a termination step to form the major product, as shown in 14-24.

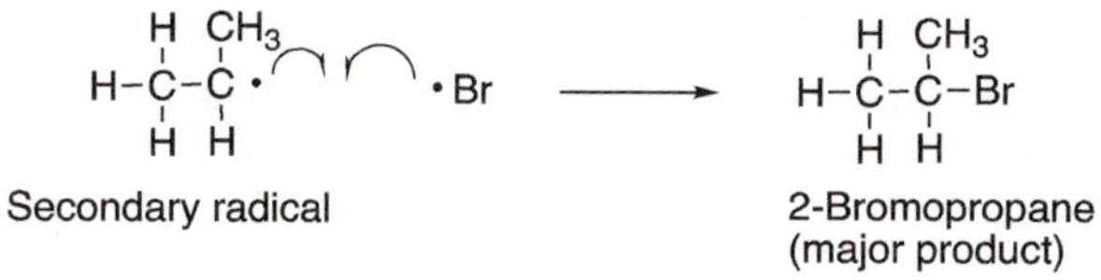

$$\underset{\text{Secondary radical}}{\overset{\begin{array}{cc}H & CH_3\end{array}}{H-\underset{\underset{H}{|}}{\overset{\overset{|}{}}{C}}-\underset{\underset{H}{|}}{\overset{\overset{|}{}}{C}}\cdot}} \;\cdot Br \longrightarrow \underset{\substack{\text{2-Bromopropane}\\\text{(major product)}}}{\overset{\begin{array}{cc}H & CH_3\end{array}}{H-\underset{\underset{H}{|}}{\overset{\overset{|}{}}{C}}-\underset{\underset{H}{|}}{\overset{\overset{|}{}}{C}}-Br}} \qquad (14\text{-}24)$$

On the other hand, the less stable primary radical will coupled with a bromine atom to form the minor product, as shown in Reaction (14-25).

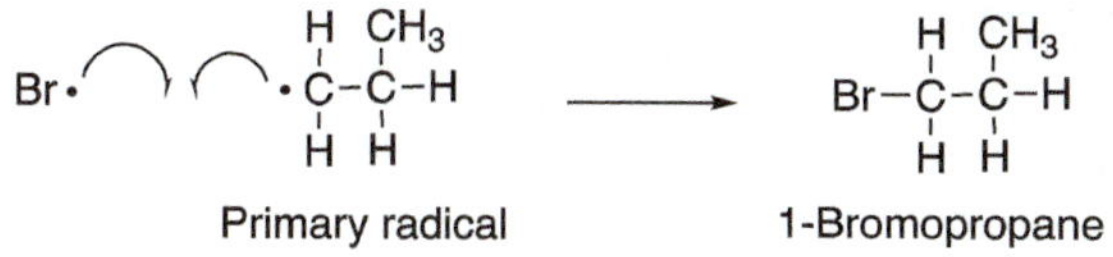

$$Br\cdot \quad \underset{\text{Primary radical}}{\cdot \overset{\begin{array}{cc}H & CH_3\end{array}}{\underset{\underset{H}{|}}{\overset{\overset{|}{}}{C}}-\underset{\underset{H}{|}}{\overset{\overset{|}{}}{C}}-H}} \longrightarrow \underset{\text{1-Bromopropane}}{\overset{\begin{array}{cc}H & CH_3\end{array}}{Br-\underset{\underset{H}{|}}{\overset{\overset{|}{}}{C}}-\underset{\underset{H}{|}}{\overset{\overset{|}{}}{C}}-H}} \qquad (14\text{-}25)$$

Thus, for the substitution reaction of bromine with propane, the major substitution product will result from the abstraction of the secondary C—H bond, compared to abstraction of the other stronger primary C—H bonds.

With this knowledge, you are now able to predict the major substitution bromination reaction product involving just about any hydrocarbon. You will first have to analyze the different C—H bonds of the starting hydrocarbon. An analysis of 2-methylpropane reveals that it has a tertiary carbon–hydrogen, and a tertiary C—H bond is weaker than either a secondary or primary C—H bond. Thus, the major organic product for the reaction of 2-methylpropane with bromine in the presence of heat or light is 2-bromo-2-methylpropane, and the minor product is 1-bromo-2-methylpropane as shown in Reaction (14-26).

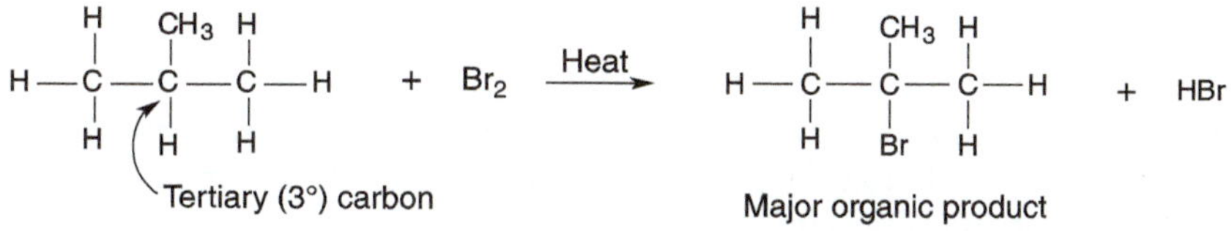

$$\text{H}-\underset{\underset{\text{H}}{|}}{\overset{\overset{\text{H}}{|}}{\text{C}}}-\underset{\underset{\text{H}}{|}}{\overset{\overset{\text{CH}_3}{|}}{\text{C}}}-\underset{\underset{\text{H}}{|}}{\overset{\overset{\text{H}}{|}}{\text{C}}}-\text{H} \; + \; \text{Br}_2 \; \xrightarrow{\text{Heat}} \; \underset{\text{Major organic product}}{\text{H}-\underset{\underset{\text{H}}{|}}{\overset{\overset{\text{H}}{|}}{\text{C}}}-\underset{\underset{\text{Br}}{|}}{\overset{\overset{\text{CH}_3}{|}}{\text{C}}}-\underset{\underset{\text{H}}{|}}{\overset{\overset{\text{H}}{|}}{\text{C}}}-\text{H}} \; + \; \text{HBr} \qquad (14\text{-}26)$$

Tertiary (3°) carbon

On the other hand, the chlorination of 2-methylpropane will give a mixture of substitution products. As a result, bromine is more selective in a substitution reaction for the weaker C—H bond, compared to chlorine for the halogenation of hydrocarbons.

Problem 14.5

i) Predict the major organic product for the reaction of methylcyclohexane with bromine in the presence of heat (hint: find the weakest C—H bond, i.e. the tertiary C—H bond and that will be the point of substitution).

ii) Complete the following reactions by giving the major organic products.

(a) [decalin with methyl group] $\xrightarrow{\text{Br}_2,\text{ heat}}$?   (b) [isopentane] $\xrightarrow{\text{Br}_2,\text{ heat}}$?

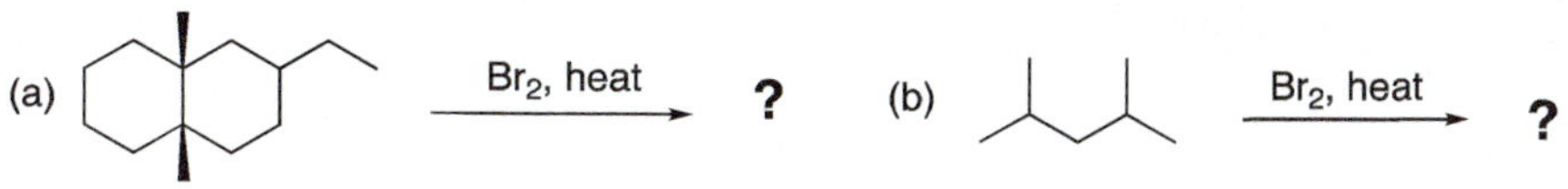

iii) Predict the major organic products for the bromination of each of the following molecules.

(a) [methylcyclopentane]   (b) CH_3CH_3 (c) [2,3-dimethylbutane]   (d) [2-methylbutane]

Based on the results of various experiments, it was observed that the bromination of benzylic and allylic compounds occurs almost exclusively at one position, the benzylic or allylic positions, as shown in Reactions (14-27) and (14-28).

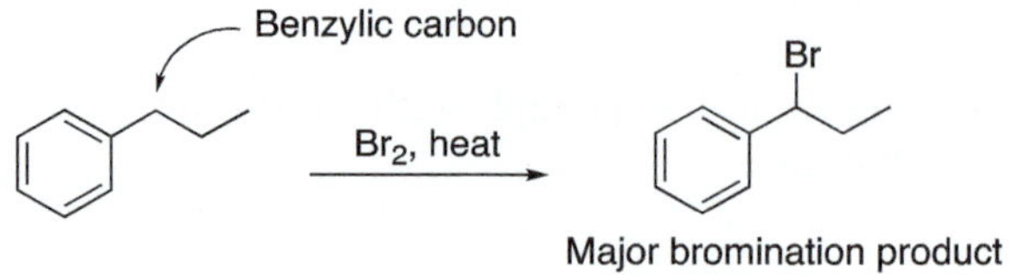

$$\text{PhCH}_2\text{CH}_2\text{CH}_3 \xrightarrow{\text{Br}_2,\ \text{heat}} \text{PhCHBrCH}_2\text{CH}_3 \qquad (14\text{-}27)$$

(Benzylic carbon indicated; Major bromination product)

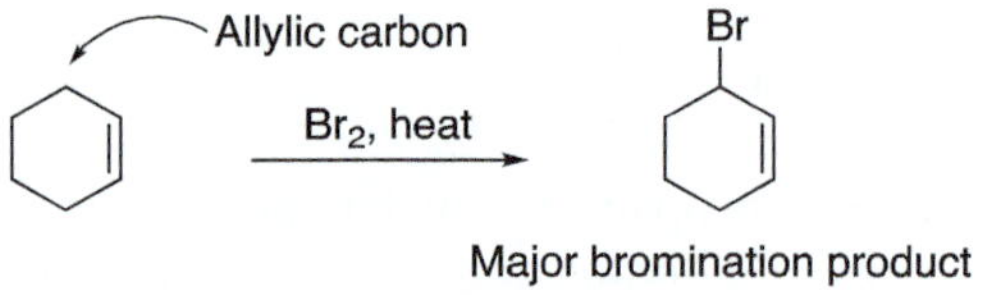

$$\text{cyclohexene} \xrightarrow{\text{Br}_2,\ \text{heat}} \text{3-bromocyclohexene} \qquad (14\text{-}28)$$

(Allylic carbon indicated; Major bromination product)

As shown in Table 14.3, the energy required for homolytic cleavage of the benzylic and allylic C—H bonds are 90 and 89 kcal mol^{-1}, respectively. As a result, these bonds are weaker than those of a tertiary or secondary C—H bond, and hence the benzylic and allylic substitution products will be the major products as shown in Reactions (14-27) and (14-28), respectively. At this point, you may be wondering what about the other halogens? The fluorination of methane is an explosive reaction, whereas iodination of alkanes is unfavorable since the iodine radical is extremely stable and hence unreactive.

Problem 14.6

Give a step-by-step mechanism to explain the unexpected product formed for the reaction shown below. That is, give the initiation, propagation, and termination steps for the reaction. Label the different steps of your mechanism.

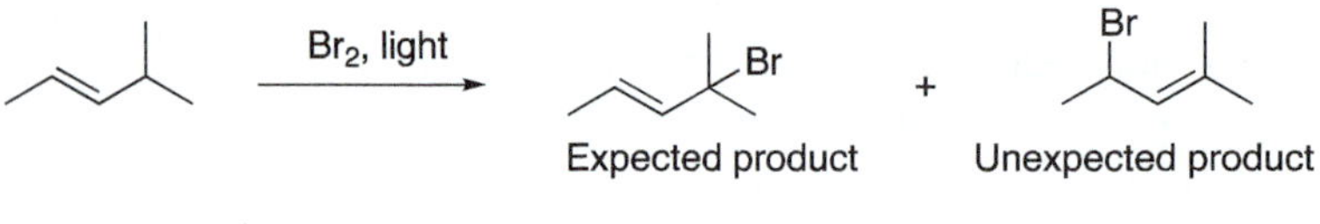

$$\text{(CH}_3\text{)}_2\text{C=CHCH}_3 \xrightarrow{\text{Br}_2,\ \text{light}} \text{Expected product} + \text{Unexpected product}$$

Allylic radicals are readily brominated using a compound called N-bromosuccinamide (NBS), as shown in Reaction (14-29).

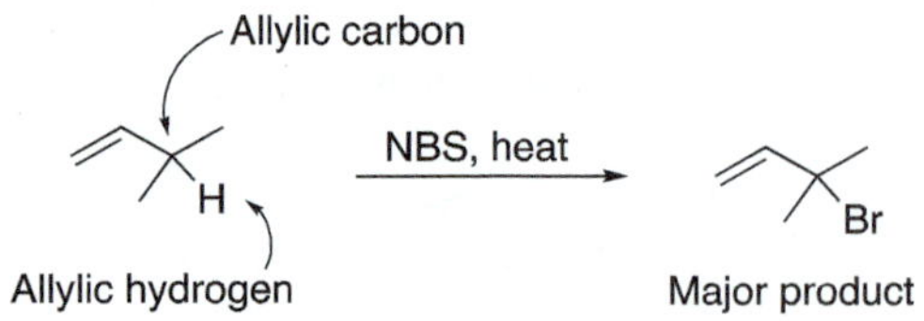

$$\xrightarrow{\text{NBS, heat}} \qquad (14\text{-}29)$$

(Allylic carbon and allylic hydrogen indicated; Major product)

Note that by using NBS, bromine substitution takes place at the allyl position and there is not an addition reaction across the double bond as you might expect since bromine molecule is produced in this reaction. NBS provides a low concentration of bromine so that substitution takes place to produce the more stable allylic radical and bromine does not add across the double bond. In the first step of the reaction using NBS, the N—Br bond, which is a weak bond, is broken in the presence of heat or light, as shown in Reaction (14-30).

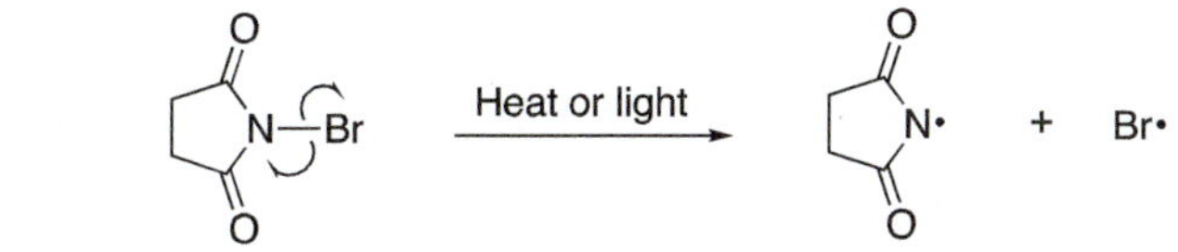

14.4 Bromination of Alkanes

$$\text{N-Bromosuccinamide (NBS)} \xrightarrow{\text{Heat or light}} \text{N}\cdot + \text{Br}\cdot \qquad (14\text{-}30)$$

Once the bromine radicals are produced, it abstracts a hydrogen atom from the allylic position of the reactant to produce the stable allylic radical, as shown in Reaction (14-31).

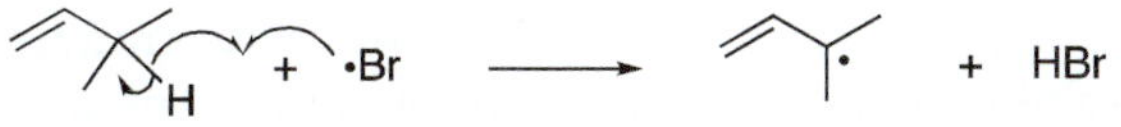

$$\text{allyl-H} + \cdot\text{Br} \longrightarrow \text{allyl}\cdot + \text{HBr} \qquad (14\text{-}31)$$

Since this reaction also generates a very strong acid, HBr, it reacts with more NBS as shown in Reaction (14-32).

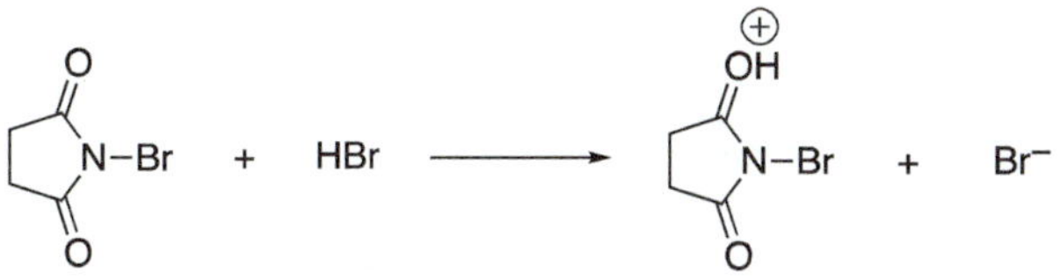

$$\text{NBS} + \text{HBr} \longrightarrow \text{protonated NBS} + \text{Br}^- \qquad (14\text{-}32)$$

This next step involves a substitution reaction of the bromide anion to form bromine as shown in Reaction (14-33) to form bromine and succinamide anion.

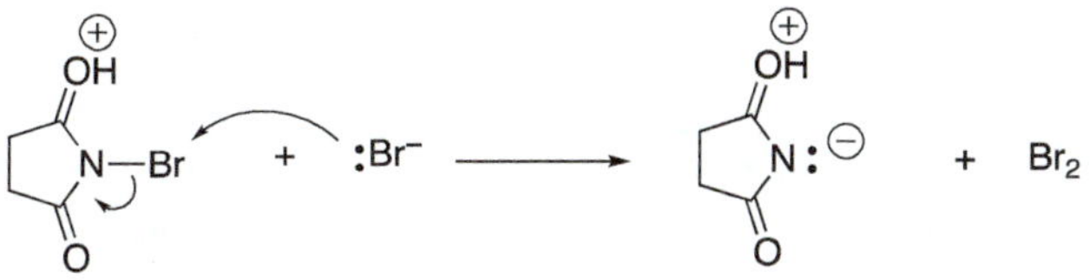

$$\text{protonated NBS} + :\text{Br}^- \longrightarrow \text{succinamide anion} + \text{Br}_2 \qquad (14\text{-}33)$$

The next step of this reaction involves a proton transfer to generate succinamide as shown in Reaction (14-34).

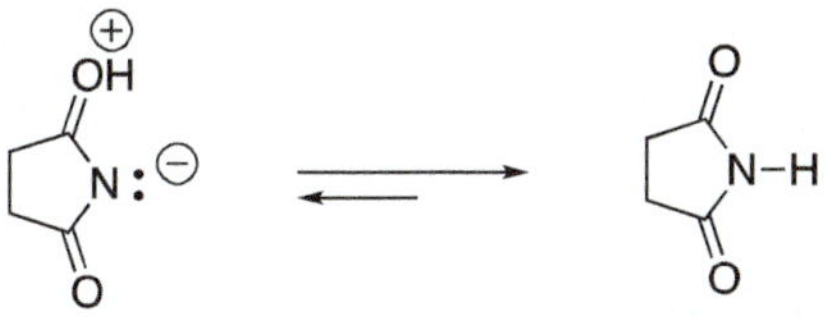

$$\text{succinamide anion} \rightleftharpoons \text{Succinamide} \qquad (14\text{-}34)$$

Once there is a constant and low concentration of bromine as generated as shown in Reaction (14-33), it will react as we have seen in the previous sections to give the brominated substitution product and another bromine radical.

$$\text{allyl}\cdot + \text{Br}\text{—}\text{Br} \longrightarrow \text{allyl-Br} + \text{Br}\cdot \qquad (14\text{-}35)$$

In the last steps of the reaction, radicals react to form neutral molecules as shown in Reactions (14-36) and (14-37).

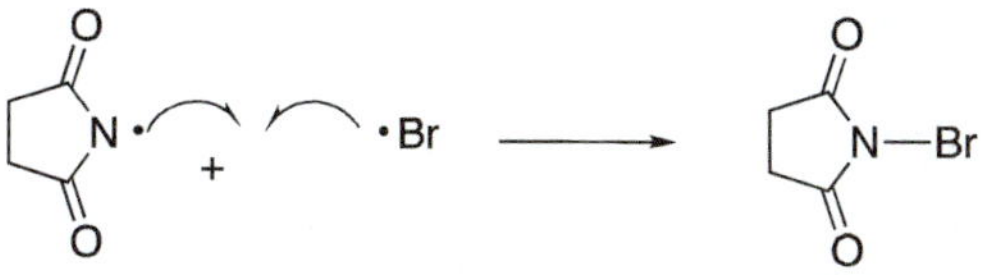

$$\text{N}\cdot + \cdot\text{Br} \longrightarrow \text{N—Br} \qquad (14\text{-}36)$$

$$\text{allyl}\cdot + \cdot\text{Br} \longrightarrow \text{allyl-Br} \qquad (14\text{-}37)$$

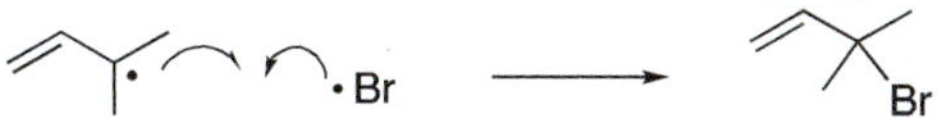

Problem 14.7

Predict the major organic product for the bromination of each of the following molecules in the presence of NBS and heat.

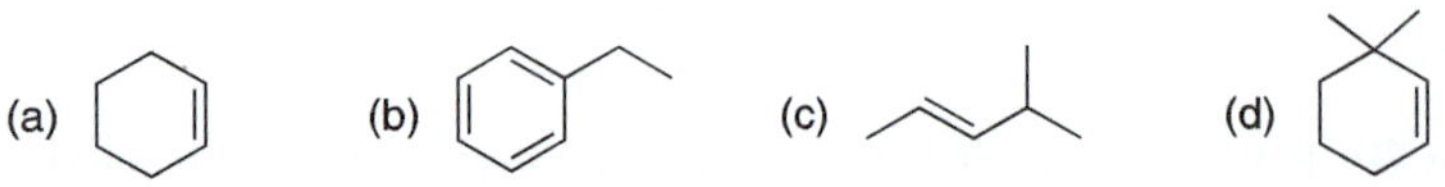

(a) cyclohexene (b) ethylbenzene (c) 2-methyl-2-butene (d) 1,1-dimethylcyclohexene

A consequence of a reaction that proceeds through a mechanism that has a radical as an intermediate is that racemic products are produced if there is a chiral carbon in the product. Since the intermediate radical is planar, the addition of the bromine radical in the termination step will take place equally from opposite sides of the radical resulting in a racemic mixture for molecules that have a chiral center at the point of substitution as shown in Reaction (14-38).

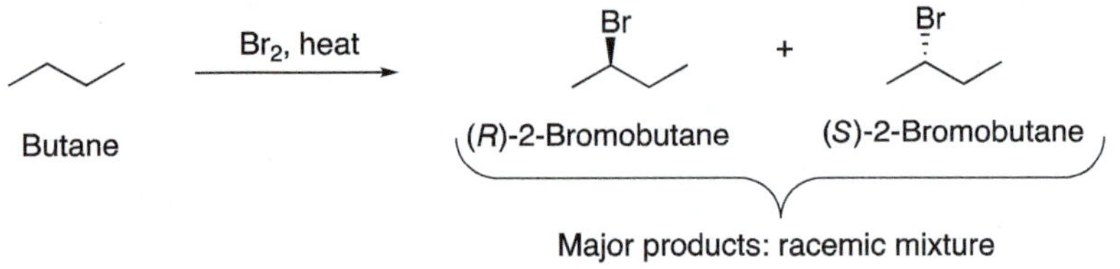

$$\text{Butane} \xrightarrow{Br_2,\ heat} (R)\text{-2-Bromobutane} + (S)\text{-2-Bromobutane} \quad (14\text{-}38)$$

Major products: racemic mixture

The same is true for a reaction in which the reactant is chiral, the stereochemistry is lost in the product, as shown below in Reaction (14-39).

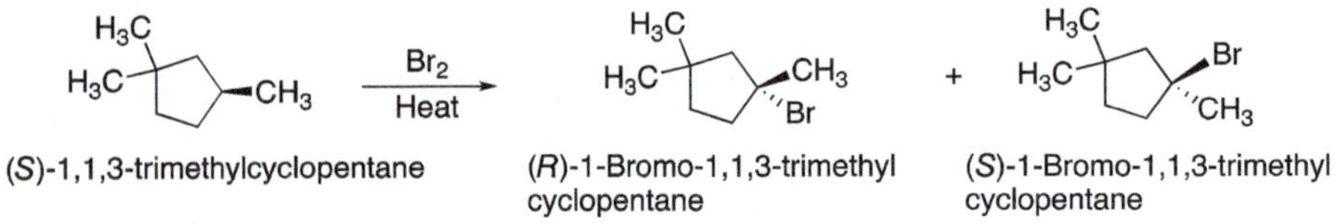

$$(S)\text{-1,1,3-trimethylcyclopentane} \xrightarrow{Br_2,\ Heat} (R)\text{-1-Bromo-1,1,3-trimethyl cyclopentane} + (S)\text{-1-Bromo-1,1,3-trimethyl cyclopentane} \quad (14\text{-}39)$$

For the halogenation of alkanes, bromine is more selective and there is regioselectivity for these reactions. For the bromination of compounds that have allylic or benzylic hydrogens, the use of NBS serves to maintain a low concentration of bromine and bromine radicals to reduce bromination of the double bond and other reactions that will occur in the presence of a high concentration of bromine.

14.5 Applications of Free Radical Substitution Reactions

As pointed out in previous chapters, one of the major goals of chemists is the synthesis of specific target organic compounds by using different types of reactions that we have studied and of course making strategic selections of specific reactions. For example, we have seen that it is much better to use bromination since it is more selective, compared to chlorination if the starting compound has different types of hydrogens. On the other hand, if all the hydrogens are the same, the chlorination is a more economic choice, compared to bromination; bromine is costlier, compared to chlorine. Let us look how best to carry out the following transformation shown in Reaction (14-40) where you need to determine the appropriate reactions and reaction conditions to convert methylcyclohexane to methyl-1-cyclohexene.

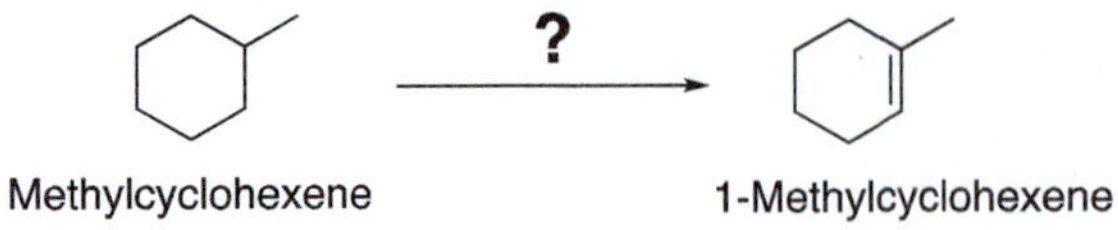

$$\text{Methylcyclohexene} \xrightarrow{?} \text{1-Methylcyclohexene} \quad (14\text{-}40)$$

It is best to focus on the target molecule, in this case methyl-1-cyclohexene, and determine the type of reaction that can be used for its synthesis. Analysis of the target molecule reveals that the functional group contained is an alkene functionality, and we learned from Chapter 11 that elimination reactions are used to synthesize alkenes. Thus, appropriate elimination reactions for its synthesis are shown below.

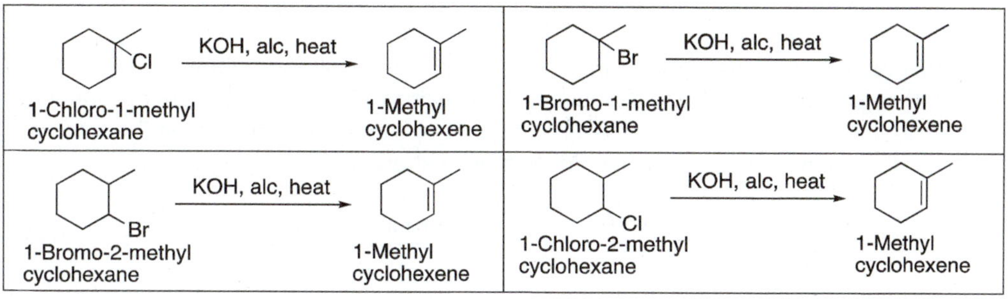

1-Chloro-1-methyl cyclohexane → (KOH, alc, heat) → 1-Methyl cyclohexene	1-Bromo-1-methyl cyclohexane → (KOH, alc, heat) → 1-Methyl cyclohexene
1-Bromo-2-methyl cyclohexane → (KOH, alc, heat) → 1-Methyl cyclohexene	1-Chloro-2-methyl cyclohexane → (KOH, alc, heat) → 1-Methyl cyclohexene

Any of these four reactions will give the desired product, but now comes the question of the appropriate selection! The challenge now comes in the synthesis of one of these alkyl halides in adequate amounts to be used in this step of the elimination reaction. The synthesis of the first reactant (1-chloro-1-methylcyclohexane) from methylcyclohexane using chlorine is not efficient, since it is not very selective and will give a low yield of this desired tertiary substitution. For the second reaction, 1-bromo-1-methylcyclohexane can be easily made via a bromination reaction to give a high yield of 1-bromo-1-methylcyclohexane since bromination is selective for a tertiary C—H substitution. For the third reaction, the substitution required to make 2-bromo-1-methylcyclohexane would be extremely difficult since the reaction would much prefer substitution at the tertiary C—H bond; the same is true for the fourth synthesis of 1-chloro-2-methylcyclohexane. Based on this information, the best elimination reaction to use for the synthesis would be the second reaction in which the highly selective bromination reaction is used for its synthesis as shown in Reaction (14-41).

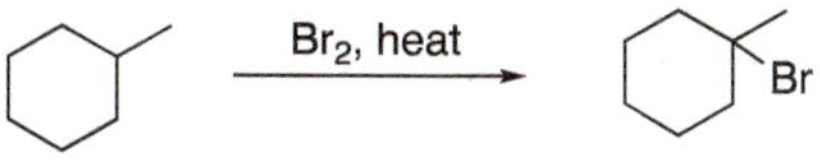

methylcyclohexane —Br₂, heat→ 1-bromo-1-methylcyclohexane (14-41)

Thus, the overall sequence of reactions for the transformation given in Reaction (14-40) is shown in Reaction (14-42).

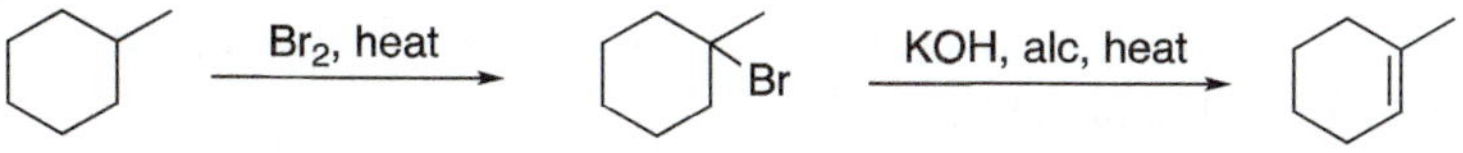

methylcyclohexane —Br₂, heat→ 1-bromo-1-methylcyclohexane —KOH, alc, heat→ 1-methylcyclohexene (14-42)

The same is true for the transformation shown in Reaction (14-43).

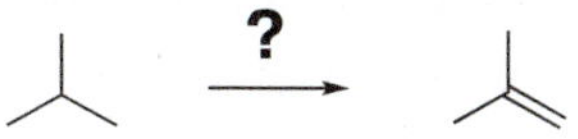

isobutane —?→ isobutylene (14-43)

The last reaction in a sequence of reactions to produce the alkene would be an elimination reaction, and the best reactant for that reaction would be 2-bromo-2-methylpropane as shown in Reaction (14-44).

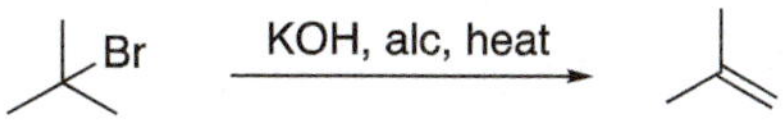

2-bromo-2-methylpropane —KOH, alc, heat→ isobutylene (14-44)

14 Free Radical Substitution Reactions Involving Alkanes

For the above target molecule, the functional group is a bromide and most specific, a tertiary bromide. We have just examined a specific type of reaction, which can perform this task of free radical halogenation of hydrocarbons. However, the appropriate hydrocarbon must be determined, in this case, one that after bromination will give the desired product is shown in Reaction (14-45).

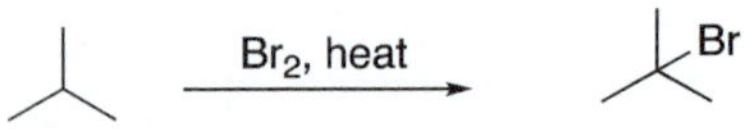

$$\text{isobutane} \xrightarrow{\text{Br}_2,\ \text{heat}} \text{tert-butyl bromide} \tag{14-45}$$

For most organic reactions, only the major organic product is shown. Thus, for the above reaction, even though 1-bromo-2-methylpropane and hydrobromic acid are the other products, they are not written in the reaction. Hence, we do not have to worry about balancing the reactions.

The overall sequence of reactions would then be as shown in Reaction (14-46).

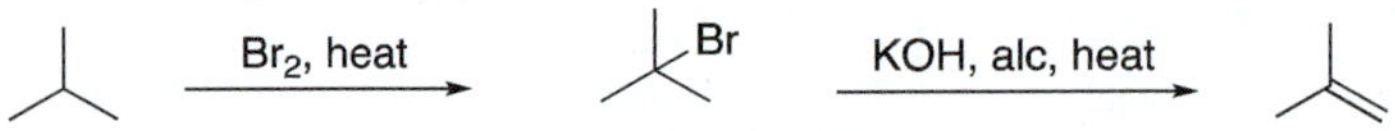

$$\text{isobutane} \xrightarrow{\text{Br}_2,\ \text{heat}} \text{tert-butyl bromide} \xrightarrow{\text{KOH, alc, heat}} \text{isobutylene} \tag{14-46}$$

Problem 14.8

Starting with any saturated hydrocarbon, show how to synthesize cyclopentene.

14.6 Free Radical Inhibitors

For reactions that proceed via a free radical mechanism, if a stable-free radical were introduced, the reaction will stop owing to the coupling of the introduced stable radical with the radical intermediates of the reaction. Stable radicals if used in this sense are called inhibitors or **antioxidants**, and they are frequently used in the food industry as preservatives. Butylated hydroxytoluene (BHT) is one such example.

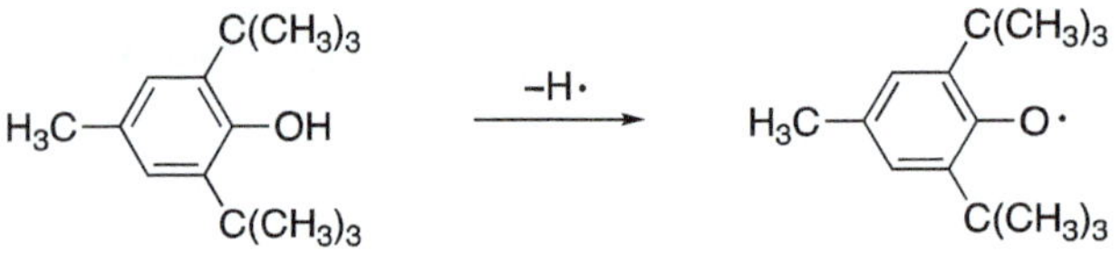

Butylated hydroxytoluene (BHT) $\xrightarrow{-\text{H}\cdot}$ Resonance stabilized radical

Problem 14.9

Using single-barbed arrows to indicate electron movements, give two resonance structures of the radical of butylated hydroxytolune (BHT).

Another radical that is of biological importance is the radical derived from Vitamin E, also known as α-tocopherol radical, its structure is shown below. Due to the stability of this α-tocopherol radical and its ability to act as a radical scavenger, it plays an important role in protecting the organism against oxidative damage.

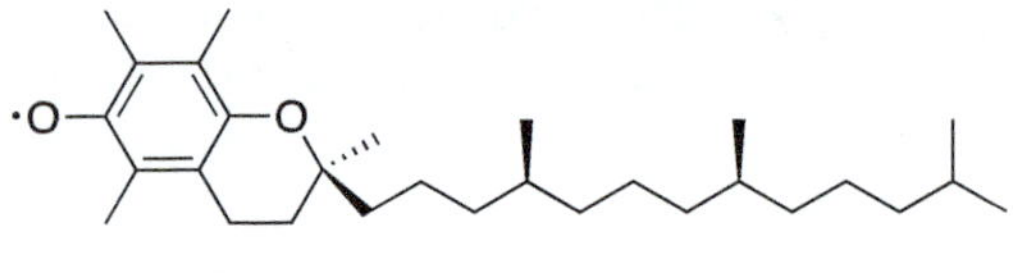

Radical derived from α-tocopherol (vitamin E)

14.7 Environmental Impact of Organohalides and Free Radicals

Ozone is very important to our existence in that it protects us from the harmful radiation of the sun. It exists in equilibrium with oxygen and oxygen radical as shown below. This means that ozone is constantly produced and destroyed in the atmosphere.

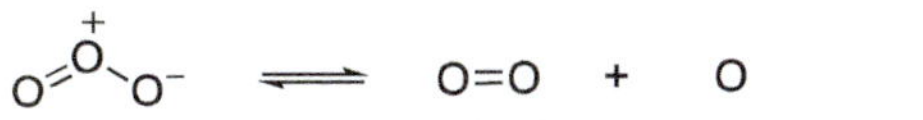

$$O=\overset{+}{O}-O^{-} \rightleftharpoons O=O + O$$
Ozone　　　　　　　Oxygen　Oxygen atom

This delicate equilibrium is very important in that the forward reaction is initiated by light as shown below.

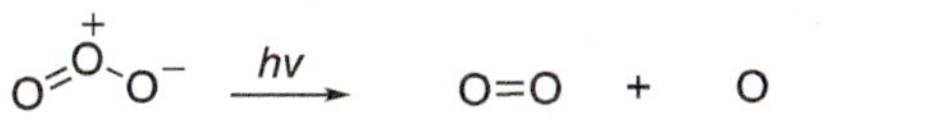

$$O=\overset{+}{O}-O^{-} \xrightarrow{h\nu} O=O + O$$
Ozone　　　　　　　　Oxygen　Oxygen atom

The reverse is spontaneous and produces heat as shown below.

$$O=O + O \longrightarrow O=\overset{+}{O}-O^{-} + heat$$
Oxygen　Oxygen atom　　　　　Ozone

A major benefit of this process is that ultraviolet light from the sun is essentially converted to heat through this process. A major problem encountered in the use of organohalides is that they eventually deplete the ozone in the atmosphere. Organic halides have been used as refrigerants for air conditioners and refrigerators. They have been used also as propellants for various sprays. Some of the commonly used ones are shown below.

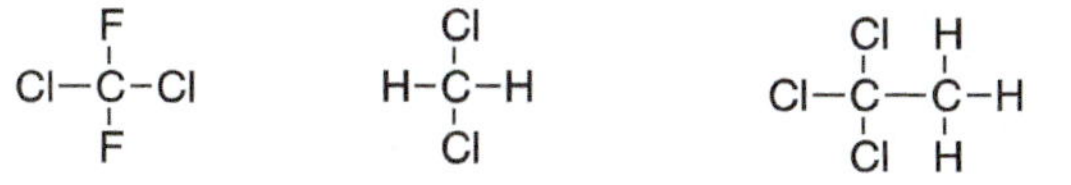

```
    F              Cl            Cl  H
    |              |             |   |
Cl–C–Cl        H–C–H         Cl–C–C–H
    |              |             |   |
    F              Cl            Cl  H

Freon-12       Dichloromethane   1,1,1-Trichloroethane
refrigerant    common solvent    propellant for aerosols
```

Several metric tons of these organohalides, also called CFCs, are released into the atmosphere yearly. Unfortunately, these compounds pose various challenges for the environment. For example, Freon-12, in the presence of light, breaks apart to give chlorine atoms, as shown below.

$$\underset{F}{\overset{F}{Cl-C-Cl}} \xrightarrow{UV-Light} \underset{F}{\overset{F}{Cl-C\cdot}} + Cl\cdot$$

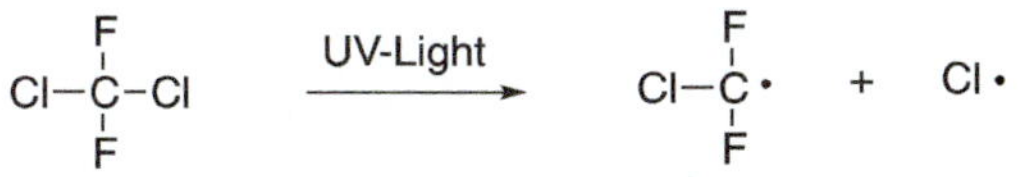

There is a reaction between chlorine atoms and ozone as shown below.

$$Cl\cdot + O=\overset{+}{O}-O^{-} \longrightarrow ClO\cdot + O=O$$
Chlorine atom　　Ozone　　　　　　　　　　　Oxygen

As you can see in this propagation step, more radicals are produced, in addition to oxygen, but most important, ozone has been consumed in this process. There are many possible propagation steps, which eventually generate more chlorine atoms as shown below.

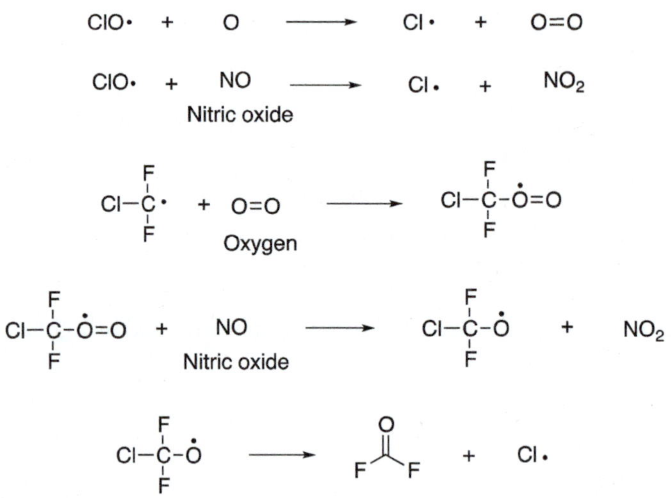

$$\text{ClO} \cdot + \text{O} \longrightarrow \text{Cl} \cdot + \text{O}=\text{O}$$

$$\text{ClO} \cdot + \underset{\text{Nitric oxide}}{\text{NO}} \longrightarrow \text{Cl} \cdot + \text{NO}_2$$

[Cl–CF$_2$• + O=O (Oxygen) → Cl–CF$_2$–O–O•]

[Cl–CF$_2$–O–O• + NO (Nitric oxide) → Cl–CF$_2$–O• + NO$_2$]

[Cl–CF$_2$–O• → F–C(=O)–F + Cl•]

These are all propagation steps that generate chlorine radicals, and chlorine radicals are very destructive to the ozone layer.

Hydroflurorcarbons (HFCs) were used to replace CFC. These compounds were thought to be better than CFC since they contain a weaker C—H bond, compared to a C—Cl bond and, as a result, are more easily destroyed in the atmosphere, where hydroxide radicals are present, and these HFCs could be destroyed before they reach the stratosphere where most of the ozone exists. The rational is that these HFCs do not generate the destructive chlorine radicals.

Some of the propagation steps in which the hydroxide radicals are used are shown below.

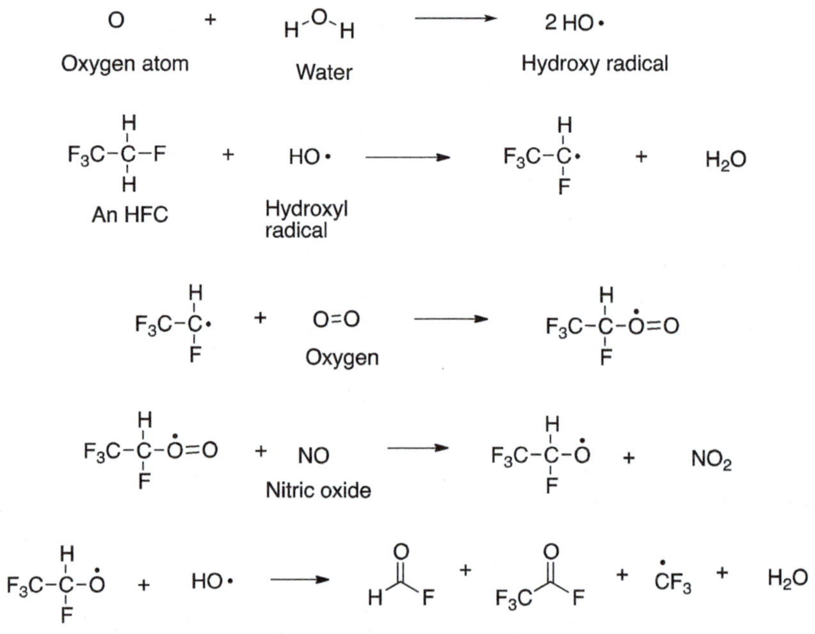

$$\underset{\text{Oxygen atom}}{\text{O}} + \underset{\text{Water}}{\text{H–O–H}} \longrightarrow 2\,\text{HO}\cdot \quad \text{Hydroxy radical}$$

[F$_3$C–CH$_2$F (An HFC) + HO• (Hydroxyl radical) → F$_3$C–CHF• + H$_2$O]

[F$_3$C–CHF• + O=O (Oxygen) → F$_3$C–CHF–O–O•]

[F$_3$C–CHF–O–O• + NO (Nitric oxide) → F$_3$C–CHF–O• + NO$_2$]

[F$_3$C–CHF–O• + HO• → H–C(=O)–F + F$_3$C–C(=O)–F + CF$_3$• + H$_2$O]

Before too long after the introduction of these compounds, they came under much scrutiny, and recently, negotiators from more than 170 countries reached a legally binding accord in Kigali, Rwanda, to drastically reduce the use of HFCs.

Problem 14.10

The propagation step of a free radical reaction is shown below. By using the curved arrow formulism, show how all the products given are formed.

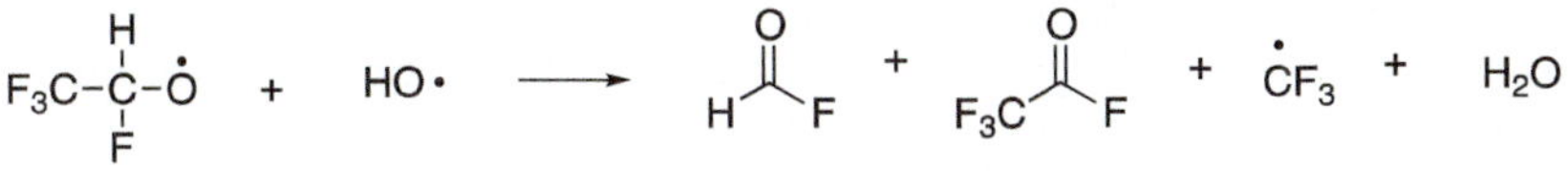

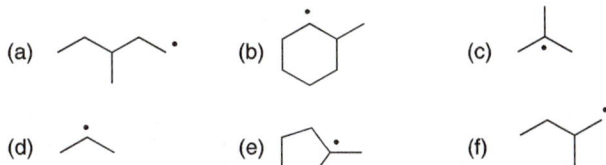

End of Chapter Problems

14.11 Classify each of the following radicals as primary, secondary, or tertiary.

14.12 Give the major organic product for the monobromination of each of the following compounds?

(a) cyclohexane (b) methylcyclohexane (c) 2-methylpropane (d) 1,1-dimethylcyclopentane

14.13 For the chlorination of cyclopentane, cyclopentylcyclopentane was isolated as one of the products. Explain by an appropriate mechanism how this product was formed.

14.14 Give the structure of the reactant or major organic product needed to complete the following reactions.

(a) ? — Br$_2$, heat → Bromomethane   (b) 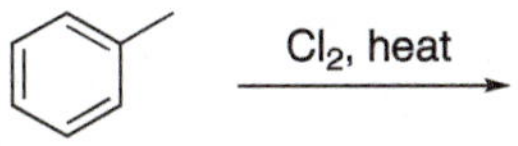 benzene — Cl$_2$, heat → ?

(c) 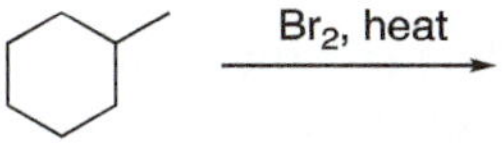 cyclohexane — Br$_2$, heat → ?   (d) ? — Cl$_2$, heat → chlorocyclohexane

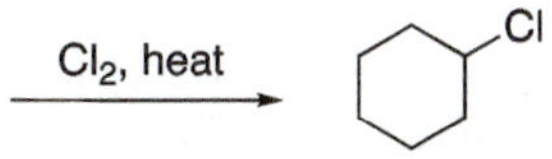

14.15 For each of the following compounds, predict the major product of free radical bromination. Remember that bromination is highly selective.
   a) Cyclohexane
   b) 2,2,3-Trimethylbutane
   c) 2,3-Dimethylbutane
   d) 1,2-Dimethylcyclohexane

14.16 Starting with 2-methylbutane, show how you would synthesize 2-methyl-2-butene. Indicate <u>clearly</u> all reagents and reaction conditions used in the reaction.

14.17 Two hydrocarbons, **A** and **B**, each has the formula C_5H_{10}. In a reaction with Cl_2, **A** gave one monochlorinated product, C_5H_9Cl. **B**, however, gave four different monochlorinated products, each being C_5H_9Cl. What are the structures of **A** and **B**?

14.18 A hydrocarbon with the formula C_5H_{12} reacts with Cl_2 in the presence of light gave a mixture of four structural isomers, each with the formula $C_5H_{11}Cl$. What is the structure of the starting hydrocarbon, C_5H_{12}?

14.19 Give the structure of a cyclic unsaturated hydrocarbon of C_5H_{10} which when it reacts with chlorine in the presence of light gives three different mono-chlorinated products. Give the structure of each of the product produced.

14.20 2-Methylbutane has different types of hydrogens, i.e. primary, secondary, and tertiary. Give complete structures for each of the different mono chlorination substitution products that would result from the reaction of 2-methylbutane with chlorine in the presence of light.

14.21 Give complete structures and IUPAC names for each of the different mono chlorination substitution products that would result from the reaction of methylcyclopentane with chlorine in the presence of light.

14.22 For the molecule shown below, there is a homolytic cleavage of the O—Cl bond in the presence of heat to produce two radicals, which eventually yield the products shown. Give a mechanistic explanation for the formation of the products shown.

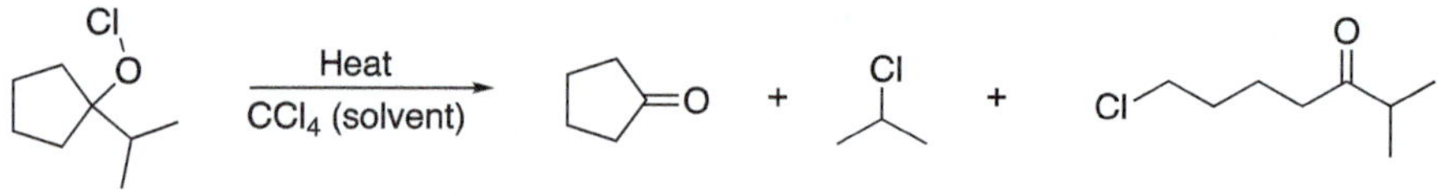

14.23 For the reaction shown below, two isomers were obtained: the expected free radical substitution product and an unexpected free radical substitution product. Using appropriate arrows to indicate electron movement, give a step-by-step description of a mechanism to explain the products shown for this free radical substitution reaction, i.e. initiation, propagation, and termination steps.

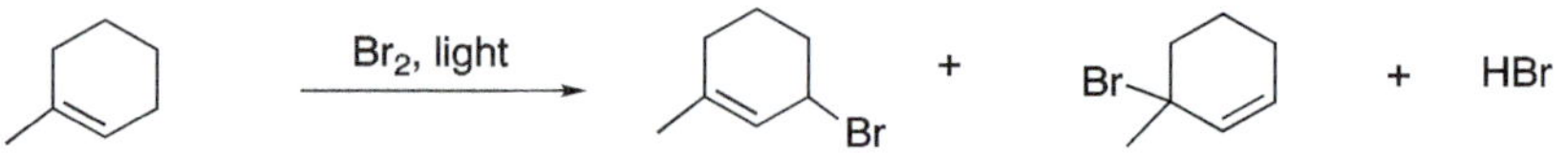

14.24 In an attempt to synthesize compound **C** (shown in the reaction below), the following sequence of reactions was carried out. The actual product isolated, however, was not the desired product, **C**, but an unexpected isomer, which is shown in the box. Give structures for the lettered compound in the sequence of reactions and give a mechanistic explanation for the formation of this unexpected isomer that was formed in the final step of the reaction sequence.

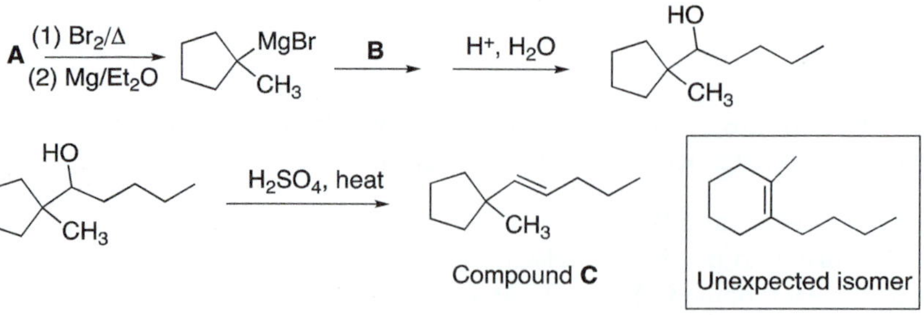

15

Nucleophilic Substitution Reactions at sp³ Carbons

15.1 Introduction

Before we actually start looking at the substitution reactions of this chapter, let us review some of the important concepts and terminologies that we will be using throughout this chapter and the rest of our course in organic chemistry. An important concept to remember is that the basis for most reactions that will be encountered in this chapter and most of the future chapters involves the attraction of a species that has a pair of electrons or is a negative species to another species that has a positive or partially positive charge. You will recall that there are different terminologies to describe the species involved and we will review the meaning of these terminologies and concepts in the next few sections of this chapter then we will apply those concepts as we analyze various substitution reactions at sp³ carbons.

15.2 The Electrophile

The substitution of an electronegative atom, such as chlorine, for hydrogen of an alkane introduces a polar covalent bond into the molecule. The carbon–halogen bond is a polar covalent bond since halogens are more electronegative than carbon, the bonding electrons are attracted closer to the halogen (X) than to the carbon. This concept is illustrated in the figure below.

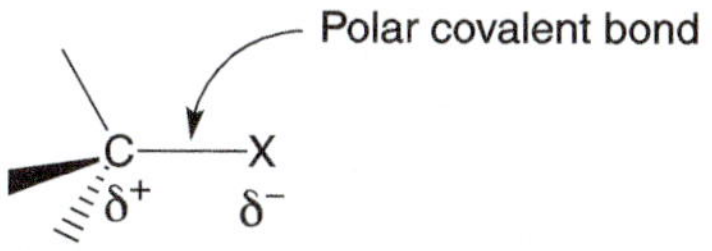

The result is that the carbon has a partial positive charge (δ^+) and of course the halogen, which is more electronegative, has a partial negative charge (δ^-). Since an electrophile is described as an electron-loving species, in the above example, the carbon bonded to the electronegative halogen is an electrophilic atom and the molecule is classified as an electrophile, as shown below.

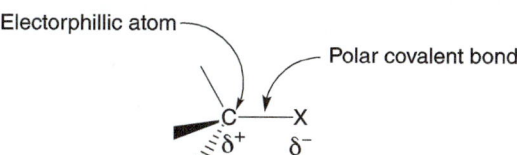

Representation of an electrophilic molecule

Organic Chemistry: Concepts and Applications, First Edition. Allan D. Headley.
© 2020 John Wiley & Sons, Inc. Published 2020 by John Wiley & Sons, Inc.
Companion website: www.wiley.com/go/Headley_OrganicChemistry

Some electrophiles are fully positive, as is the case of a proton, which is a cation whereas others are neutral molecules, but carry a partially positive atom, as in the example above. It is important to be able to easily recognize electrophiles in organic chemistry. Examples of molecules that have electrophilic atoms are shown below, the partial positive charge (δ^+) indicates the electrophilic atoms.

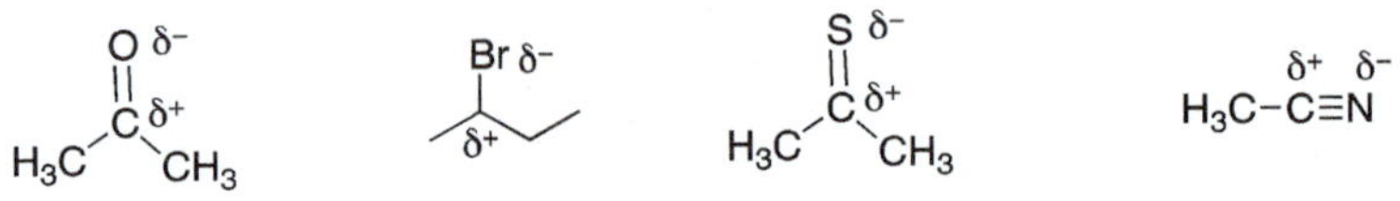

Problem 15.1

Using the symbol (δ^+), identify the electrophilic atom in each of the following molecules.

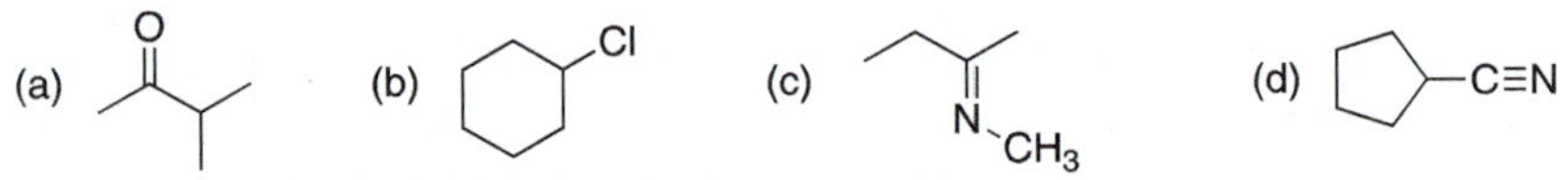

15.3 The Leaving Group

We have already pointed out that the bonding electrons for the carbon–halogen bond are not equally shared, but they are closer to the highly electronegative halogen. The halogens that we will encounter in this course are mostly chlorine, bromine, and iodine. As a result, if the carbon halogen bond were to be broken, it would break heterolytically with the halogen taking the bonding electrons. The atom or a group of atoms that takes the electrons when a bond breaks heterolytically is called a leaving group, as illustrated in Reaction 15-1 where chlorine is classified as the leaving group.

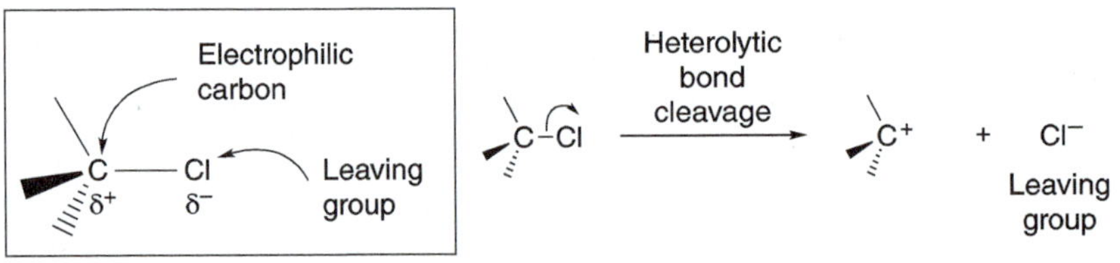

$$(15\text{-}1)$$

In this chapter, we will encounter not only halogens as leaving groups but also many other types of leaving groups, and we will need a method to determine the leaving group and their ability to leave in a heterolytic cleavage of a covalent bond. This information can be gained from our knowledge of acid–base reactions that was presented in Chapter 7. One important concept covered is the relationship between conjugate acids and bases and specifically that strong acids result in weak conjugate bases in an acid–-base reaction. For example, hydrobromic acid has a pK_a of −9.0, which makes it a very strong acid, and as a result, its conjugate base in the form of the bromide anion (Br$^−$) is a weak base since it is large and polarizable and hence can accommodate the negative charge. Weak conjugate bases are good leaving groups owing to the ability to stabilize the negative charge. You may recall that the negative charge can be stabilized not only by polarizability effect but also inductive and resonance effects. Thus, groups such as the acetate anion are good leaving groups. You will recall from Chapter 7 on acids and bases that the conjugate base of acetic acid, the acetate anion is resonance stabilized, making it a fairly stable base, hence fairly good leaving group as shown in Reaction 15-2.

15.3 The Leaving Group

$$\text{(Reaction showing anhydride cleavage to carbocation and resonance-stabilized acetate anion)} \quad (15\text{-}2)$$

Carbocation   Resonance-stabilized acetate anion

15.3.1 Converting Amines to Good Leaving Groups

The R_2N groups are extremely poor leaving groups. You will recall that the pK_a value of R_2NH is approximately 38. These groups can be converted to better leaving groups if protonated in the presence of an acid, which will result in neutral molecules after the leaving group has taken the bonding electrons, as shown in Reaction (15-3).

$$\text{Amine} + H^+ \xrightarrow{\text{Acid-base reaction}} \text{Protonated amine} \longrightarrow \text{Carbocation} + \text{An amine, (a good leaving group)} \quad (15\text{-}3)$$

An alternate method of converting an amine into a good leaving group is to carry out a substitution reaction to convert the amine to an ammonium ion, which is a good leaving group as shown in Reaction 15-4. You will recall that the pK_a of the ammonium salt is 9.4.

$$\text{Amine} + CH_3I \xrightarrow{\text{Substitution reaction}} \text{Ammonium ion salt} \xrightarrow{\text{Heterolytic bond cleavage}} \text{Carbocation} + \text{An amine, (a good leaving group)} \quad (15\text{-}4)$$

15.3.2 Converting the OH of Alcohols to a Good Leaving Group in an Acidic Medium

The —OH group is an extremely poor leaving group. You will recall that the pK_a value of water is approximately 15. On the other hand, the pK_a of H_3O^+ is −1.7. The —OH group of an alcohol can be converted to a better leaving group if protonated, which will result in a neutral molecule after the leaving group has taken the bonding electrons, as shown in Reaction (15-5).

$$\text{Alcohol} + H^+ \xrightarrow{\text{Acid-base reaction}} \text{Protonated alcohol} \xrightarrow{\text{Heterolytic bond cleavage}} \text{Carbocation} + \text{Water, an extremely good leaving group} \quad (15\text{-}5)$$

Problem 15.2

i) Using the pK_a table in Chapter 7, determine which of the following is the weakest base, and hence the best leaving group.
   a) Cl^-; Br^-; F^-; I^-
   b) $CH_3CO_2^-$; CH_3O^-; $C_6H_5O^-$
   c) SCN^-; H_2O; NH_3; OH^-

ii) Of the following pairs of ions, which is the better leaving group?
   a) I^- and Cl^-
   b) CH_3O^- and $CH_3CO_2^-$
   c) OH^- and H_2O

15.3.3 Converting the OH of Alcohols to a Good Leaving Group Using Phosphorous Tribromide

As mentioned above, the —OH group of alcohols is a poor leaving group, but can be converted to a much better leaving group like a bromide in the presence of phosphorous tribromide, as shown below in Reaction 15-6.

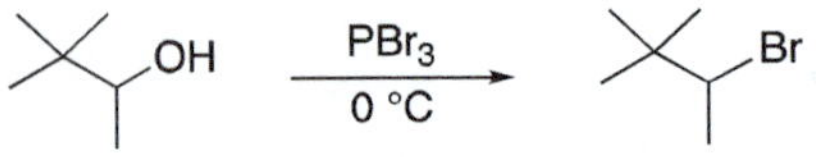

$$\text{(CH}_3\text{)}_3\text{C-OH} \xrightarrow[0\,°C]{PBr_3} \text{(CH}_3\text{)}_3\text{C-Br} \tag{15-6}$$

The mechanism for the reaction using 1-propanol with PBr_3 is shown in Reaction 15-7.

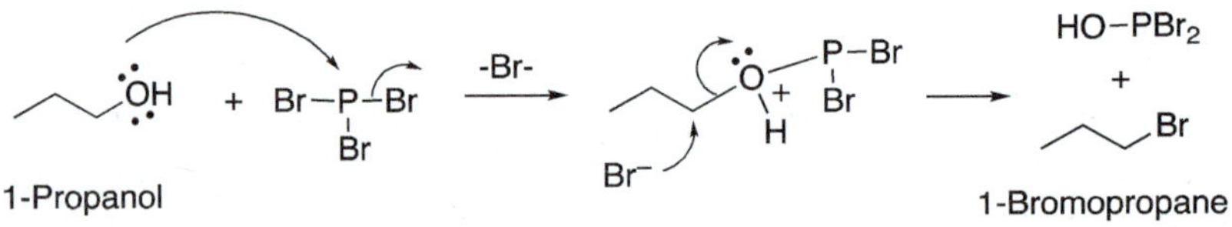

$$\underset{\text{1-Propanol}}{CH_3CH_2CH_2\ddot{O}H} + Br-P(Br)-Br \xrightarrow{-Br^-} CH_3CH_2CH_2-\overset{+}{O}(H)-PBr_2 \xrightarrow{Br^-} \underset{\text{1-Bromopropane}}{CH_3CH_2CH_2-Br} + HO-PBr_2 \tag{15-7}$$

In this reaction, the nucleophilic hydroxyl group attacks the highly electrophilic phosphorous displacing a bromide anion. Since the bromide anion is nucleophilic, it will displace the leaving group as shown in the mechanism.

15.3.4 Converting the OH of Alcohols to a Good Leaving Group Using Thionyl Chloride

Thionyl chloride is a highly reactive compound, which will convert the —OH of alcohols to alkyl chlorides, the chloride anion is a much better leaving group than the —OH group. An example of this conversion is shown in Reaction 15-8. These reactions will be discussed in more detail in the next chapter.

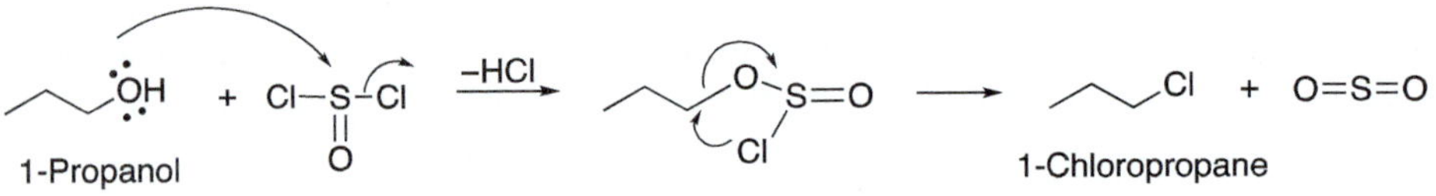

$$\underset{\text{1-Propanol}}{CH_3CH_2CH_2\ddot{O}H} + Cl-\underset{O}{\overset{O}{\underset{\|}{S}}}-Cl \xrightarrow{-HCl} CH_3CH_2CH_2-O-S(=O)-Cl \longrightarrow \underset{\text{1-Chloropropane}}{CH_3CH_2CH_2-Cl} + O=S=O \tag{15-8}$$

15.3.5 Converting the OH of Alcohols to a Good Leaving Group Using Sulfonyl Chlorides

Another method that is widely used to convert the —OH of alcohols to a good leaving group is given in the example shown in Reaction 15-9, where 4-toluenesulfonyl chloride (TsCl) reacts with butanol to convert the OH to an extremely good leaving group (-OTS).

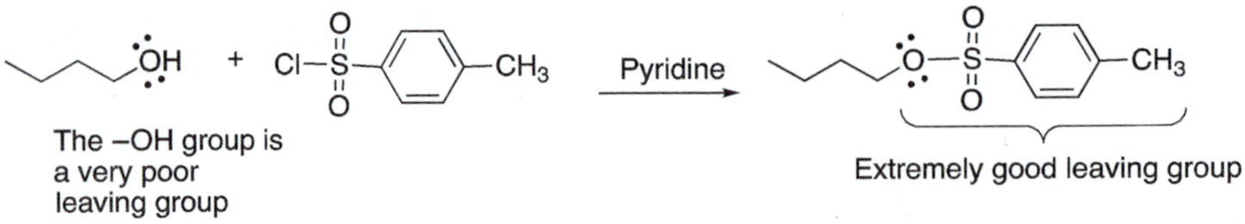

$$\underset{\substack{\text{The —OH group is} \\ \text{a very poor} \\ \text{leaving group}}}{R-\ddot{O}H} + Cl-\underset{O}{\overset{O}{\underset{\|}{S}}}-\text{C}_6\text{H}_4-CH_3 \xrightarrow{\text{Pyridine}} \underset{\text{Extremely good leaving group}}{R-O-\underset{O}{\overset{O}{\underset{\|}{S}}}-\text{C}_6\text{H}_4-CH_3} \tag{15-9}$$

As you can imagine, one of the driving forces for this reaction is that the nucleophilic hydroxide group reacts readily with the electrophilic sulfur displacing the chloride anion. The acidity of the conjugate acid of the leaving group, 4-toluenesulfonic acid, is −2.8 making the conjugate base shown in Reaction 15-9 an extremely good leaving group.

15.4 The Nucleophile

A nucleophile is defined as a species that has at least one unshared pair of electrons, which you will recall from Chapter 7 on acid and bases, is the same definition as a Lewis base. Nu: or Nu:⁻ is the abbreviation used for a nucleophile, which means a nucleus loving species, remember that a nucleus is positive owing to the presence of protons. Nucleophiles can be neutral or can have a negative formal charge, such as the hydroxide ion, which has a formal charge of −1. Other examples of nucleophiles include H_2O, $R_3N:$, I^-, $CN:^-$, and H_2S. For nucleophiles, typically there is only one atom, which is the nucleophilic atom and that is the atom with the unshared pair of electrons, examples are shown below.

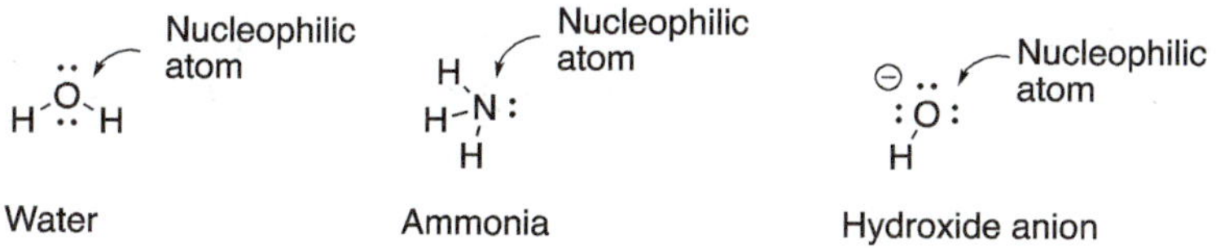

Water   Ammonia   Hydroxide anion

Problem 15.3

Which of the following molecules are electrophiles and which are nucleophiles? For each electrophile identified, indicate the electrophilic atom and for the nucleophile, identify the atom with the unshared pair of electrons.

(a) $CH(H_3C)(H_3C)(H)Cl$ (b) $K^+\ :C\equiv N$ (c) $\ddot{N}H_3$ (d) cyclohexyl-Br   (e) H_2S

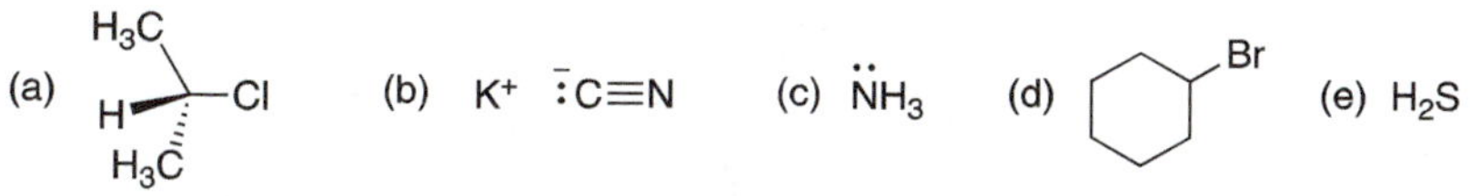

15.5 Nucleophilic Substitution Reactions

As the name suggest and as mentioned in Chapter 6, the nucleophilic substitution reaction involves a nucleophile reacting with an electrophile in which the leaving group in the reactant is substituted by the nucleophile to form a product. The general feature of a nucleophilic substitution reaction is shown in Reaction (15-10).

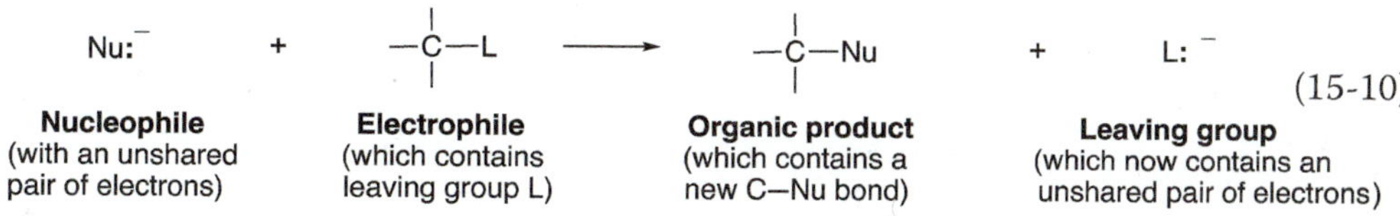

$$Nu:^- \ + \ \overset{|}{\underset{|}{C}}-L \ \longrightarrow \ \overset{|}{\underset{|}{C}}-Nu \ + \ L:^- \quad (15\text{-}10)$$

Nucleophile (with an unshared pair of electrons)   **Electrophile** (which contains leaving group L)   **Organic product** (which contains a new C—Nu bond)   **Leaving group** (which now contains an unshared pair of electrons)

An example of a nucleophilic substitution reaction is shown in Reaction (15-11), which involves the reaction of chloroethane and hydroxide anion.

$$H_3C-\underset{H}{\overset{H}{C}}-\ddot{\underset{..}{Cl}}: \;+\; ^{\ominus}:\ddot{O}H \;\longrightarrow\; H_3C-\underset{H}{\overset{H}{C}}-\ddot{O}H \;+\; :\ddot{\underset{..}{Cl}}:^{\ominus} \tag{15-11}$$

1-Chloroethane (electrophile)   Hydroxide anion (nucleophile)   Ethanol (organic product)   Chloride anion (leaving group)

You should be able to identify some important species of Reaction (15-11); the first is the electrophile in the reaction. After you have identified the electrophile, you should be able to identify the electrophilic atom. Next, you should be able to identify the nucleophile. Once the nucleophile has been identified, you should be able to identify the nucleophilic atom and the unshared pair of electrons that are used to form the new bond in the product. The next is the leaving group that is typically bonded to the electrophilic atom of the electrophile, in this case, chlorine is the leaving group since it is substituted for an OH in the product and it is a stable anion (weak base). You should recognize that a major driving force for these substitution reactions to occur is that owing to the partial positive character of the electrophilic carbon of electrophiles, an attraction of the unshared pair of negative electrons and the nucleophile occurs and in the process the leaving group leaves. Thus, a reaction between a nucleophile and an electrophile is favorable because these species have opposite charges (or partial charges) and opposite charges (or partial charges) always attract each other. In the process of the reaction, the leaving group leaves because it can accommodate a formal negative charge (or a neutral molecule in some cases, as we have seen with H_2O and R_3N). Shown in Reactions 15-12 and 15-13 are examples of substitution reactions with different leaving groups and nucleophiles.

$$\text{\textasciitilde}\text{Cl (Electrophile)} \;+\; Na^+ SH^- \text{ (Nucleophile)} \;\longrightarrow\; \text{\textasciitilde}SH \;+\; NaCl \tag{15-12}$$

$$\text{\textasciitilde}\text{Cl (Electrophile)} \;+\; Na^+ OH^- \text{ (Nucleophile)} \;\longrightarrow\; \text{\textasciitilde}OH \;+\; NaCl \tag{15-13}$$

The leaving group can give information about a substitution reaction; a reaction that contains an electrophile with a good leaving group is a better electrophile for a substitution reaction, compared to another reaction in which the electrophile has a poor leaving group. Consider the reactions shown in Reactions (15-14) and (15-15). For Reaction 15-14, the leaving group is the chloride anion, which is not as good a leaving group as the iodide anion shown in Reaction (15-15).

$$HO^- + CH_3-CH_2-Cl \longrightarrow CH_3-CH_2-OH + Cl^- \tag{15-14}$$

$$HO^- + CH_3-CH_2-I \longrightarrow CH_3-CH_2-OH + I^- \tag{15-15}$$

Note that if the nucleophile is a salt, the cation is ionically bonded to the leaving group to form a salt in the product as shown in Reaction (15-16).

$$\text{cyclopentyl-Br} + NaOH \longrightarrow \text{cyclopentyl-OH} + NaBr \tag{15-16}$$

On the other hand, if the nucleophile is bonded to a proton, such as H₂O or NH₃, the proton is bonded to the leaving group in the product, as shown in Reaction 15-17.

$$\text{cyclopentyl-Br} + H_2O \longrightarrow \text{cyclopentyl-OH} + HBr \qquad (15\text{-}17)$$

Problem 15.4

Which of the two reactions shown below is the most favorable substitution reaction? Explain your answer.

(a)  (CH₃)₂CH–I + K⁺CN⁻ ⟶ (CH₃)₂CH–CN + K⁺I⁻

(b)  (CH₃)₂CH–Cl + K⁺CN⁻ ⟶ (CH₃)₂CH–CN + K⁺Cl⁻

Once the electrophile, nucleophile, and leaving group have been identified, it is easy to predict the products of substitution reactions, and Problem 15.5 is designed to assist students recognize these features of substitution reactions and to apply that knowledge to predict the products of substitution reactions.

Problem 15.5

Give the products (both organic and inorganic) for the following reactions. Hint: be sure to identify the electrophilic and nucleophilic atoms before predicting the products.

(a)  cyclohexyl-Br + NH₃ ⟶ ? + ?

(b) 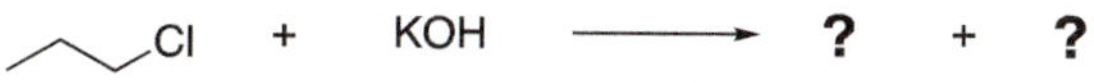 CH₃CH₂CH₂Cl + KOH ⟶ ? + ?

(c)  CH₃CH₂CH₂Cl + H₂S ⟶ ? + ?

(d)  cyclopentyl-I + KCN ⟶ ? + ?

15.5.1 Mechanisms of Nucleophilic Substitution Reactions

In this section, we will examine possible routes for substitution reactions. One possibility is that the nucleophile attacks the electrophile from the rear and pushes out the leaving group and in the process a new carbon–nucleophile bond is formed. Another possibility by which the reaction can take place is for the leaving group to depart first, without the assistance of the nucleophile, to form a fully positive species, also known as a carbocation, and the leaving group. In a second separate step, the nucleophile attacks the positive carbocation to form a carbon–nucleophile bond of the product. Note that either of these two routes gives the same substitution products. These different possible routes are referred to as mechanisms and they have specific names, bimolecular nucleophilic substitution (S_N2) is the term that is used to describe the first mechanism; and unimolecular nucleophilic substitution (S_N1) is the term used to describe the second mechanism. In the next sections, we will go into these mechanisms in more detail.

15.6 Bimolecular Substitution Reaction Mechanism (S_N2 Mechanism)

For a nucleophilic substitution reaction to take place via a S_N2 mechanism, the electrons of the nucleophile must get close enough to the partial positive electrophilic carbon of the electrophile to actually "push" out the leaving group, and in the process a new organic product is formed as shown in Reaction (15-18).

$$\text{Nu:}^- + \quad \underset{\text{Electrophile}}{\overset{}{\text{C}-X}} \longrightarrow \underset{\text{Organic product}}{\text{Nu}-\text{C}} + \underset{\text{Leaving group}}{:X^-} \tag{15-18}$$

If this general substitution reaction were visualized in slow motion, one would observe that as the nucleophile approaches the electrophile, a partial bond is formed between the nucleophilic atom and the electrophilic atom; at same time, the bond between the atom of the leaving group and the electrophilic carbon is partially broken. As the reaction proceeds further, the bond to the electrophile and the nucleophile is fully formed; and also, the bond to the leaving group and the electrophile is fully broken to form the products. Reaction (15-19) gives an illustration of this process.

$$\underset{\text{Nucleophile}}{\text{HO:}^-} + \underset{\text{Electrophile}}{\text{H}_3\text{C}-\text{Br:}} \longrightarrow \underset{\text{Transition state}}{\left[\text{HO}\cdots\overset{\text{CH}_3}{\underset{\text{H H}}{\text{C}}}\cdots\text{Br}\right]^{\ddagger}} \longrightarrow \underset{\text{Organic product}}{\text{HO}-\text{CH}_3} + \underset{\text{Leaving group}}{:\text{Br:}^-} \tag{15-19}$$

with labels "Bond being formed" and "Bond being broken" on the transition state.

This type of reaction is described as bimolecular because two molecules, the nucleophile and the electrophile, are involved in this very important step of the mechanism for the formation of the products; this is called the rate-determining step (RDS). As a result, this mechanism is described as a S_N2 mechanism, substitution nucleophilic bimolecular, and Figure 15.1 gives the energy profile for the mechanism of this type of substitution reaction.

15.6.1 The Electrophile of S_N2 Reactions

For a reaction to proceed via this pathway, or mechanism, it is obvious that there must be enough space for the nucleophile to approach the electrophilic carbon. A methyl halide, which has three very small hydrogen atoms bonded to the electrophilic carbon, has enough space to accommodate the nucleophile and, as a result, a substitution reaction involving a methyl halide proceeds very favorably by this mechanism. On the other hand, a tertiary halide, which has

Figure 15.1 Energy profile of S_N2 reaction of a nucleophile with an electrophile.

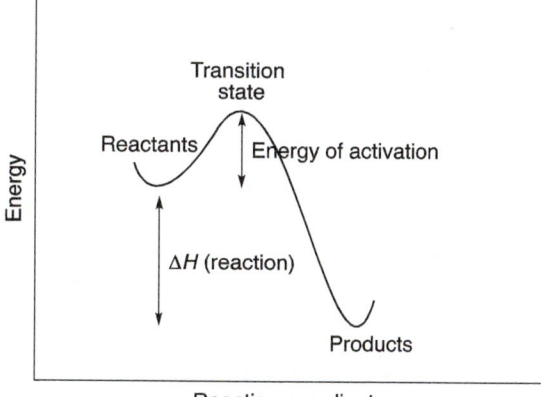

Figure 15.2 Energy profile of S_N2 reaction where E_{a1} represents the reaction with an electrophile that is not crowded, such as methyl halide, and E_{a2} represents the reaction with a more crowded electrophile.

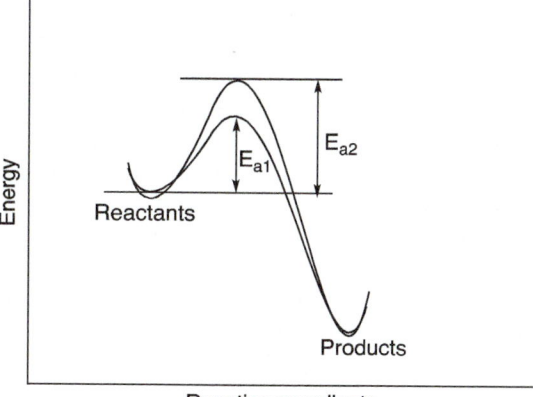

three much larger alkyl groups bonded to its electrophilic carbon, does not have enough space for the nucleophile to approach the electrophilic carbon of the electrophile. Hence, a substitution reaction involving a tertiary halide will proceed extremely slowly to the point that it is not considered to proceed by this mechanism (S_N2 mechanism) as illustrated below.

An alternate way of understanding this concept is that the energy of activation for a substitution reaction involving a methyl halide is less than the energy of activation for the reaction involving a more crowded alkyl halide as illustrated in Figure 15.2.

Thus, the order of reactivity of electrophiles for substitution reactions that proceed by a S_N2 mechanism is methyl > 1° > 2° > 3° (the reactivity of the 3° is so slow that it is considered unreactive by this mechanism).

Order of reactivity of electrophiles for substitution reactions that proceed by a S_N2 mechanism: **methyl > 1° > 2° >>> 3°**

Problem 15.6

Which of the following pairs of molecules would react the fastest by a S_N2 reaction mechanism?

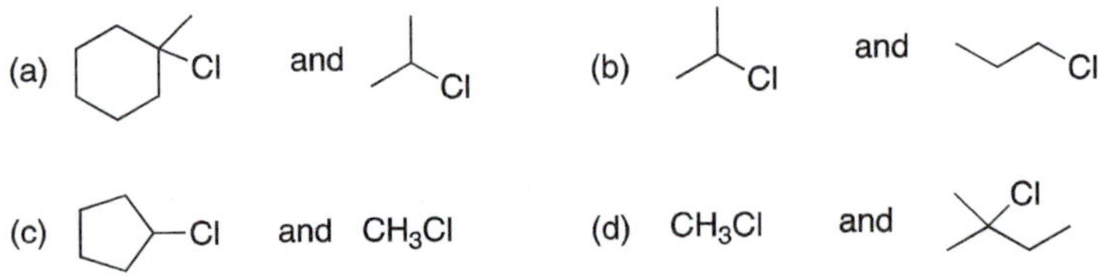

(a) cyclohexyl-Cl and (CH₃)₂CH-Cl

(b) (CH₃)₂CH-Cl and CH₃CH₂CH₂-Cl

(c) cyclopentyl-Cl and CH₃Cl

(d) CH₃Cl and (CH₃)₃C-CH₂Cl

15.6.2 The Nucleophile of S_N2 Reactions

For a substitution reaction to proceed by this S_N2 mechanism, it should now be obvious that the nucleophile must be small enough in order to gain access to the electrophilic carbon. The term nucleophilicity is often used to describe the ability of a nucleophile to gain access to the electrophilic carbon of an electrophile. The nucleophilicity trend is shown in Figure 15.3. This trend is based on the effective size of the nucleophile in a particular solvent.

Thus, the most favorable substitution reactions are those in which there is a good nucleophile (small) and a methyl or primary halide. As pointed out earlier, nucleophiles are also Lewis bases, which contain at least one unshared pair of electrons. As a result, a nucleophile for an S_N2 reaction cannot be a strong base since as you might expect, an acid–base reaction, which leads to an elimination reaction will take place instead of the desired substitution reaction. Large strong bases, such as lithium diisopropylamide (LDA) or potassium *tert*-butoxide, that were discussed in Chapter 7 and shown in Figure 15.4 are typically never used for substitution reactions (Figure 15.10).

You will recall from Chapter 12 that strong bases are required for elimination reactions, specifically E2 reactions, and the nucleophiles described in this chapter can also serve as bases for elimination reactions. Thus, there is always a competition between substitution reactions vs. elimination reactions when these reactions are carried out. Thus, it is very important that the appropriate reaction conditions are in place to ensure the desired type of reaction and eventually the desired organic product. To ensure substitution reactions, a small weak base is preferred. On the other hand, to ensure elimination reaction, strong bases are required, often at elevated temperatures. Strong bulky bases such as potassium *tert*-butoxide or LDA are often used to essentially guarantee elimination reactions. As you might have guessed, bulky electrophile in the presence of even a small strong base will undergo substantial elimination reaction. Problem 15.7 is designed to have students combine the knowledge gained from Chapter 12 on elimination with the information discussed thus far so to determine possible products.

$$CN^- > I^- > OR^- > OH^- > Br^- > Cl^- > ROH > H_2O$$

Figure 15.3 Nucleophilicity trend for nucleophiles of S_N2 reactions.

Potassium *tert*-butoxide: $H_3C-C(CH_3)_2-\ddot{O}:^- K^+$

Lithium diisopropylamide (LDA): $(iPr)_2\ddot{N}^- Li^+$

Figure 15.4 These bulky strong Lewis bases do not serve as good nucleophiles; instead, they serve primarily as strong bases for acid–base reactions.

Problem 15.7

Give the substitution and elimination products for each of the reactions shown below. Indicate which product (elimination or substitution) do you think would be the major product.

(a) 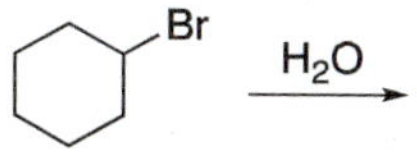 cyclohexyl-Br + H₂O →

(b) 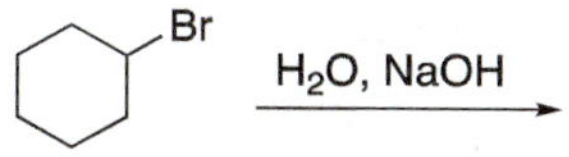 cyclohexyl-Br + H₂O, NaOH →

15.6.3 The Solvents of S$_N$2 Reactions

As you might expect, these reactions are carried out in a solvent and the nature of the solvent can influence the outcomes of substitution reactions. Nonpolar, aprotic solvents are typically the best solvents for substitution reactions that proceed via this mechanism pathway. The fluoride anion is a very small nucleophile owing to fluorine being the most electronegative atom of the periodic table. Hence, it is a good nucleophile for S$_N$2 reactions, but its ability to act as a good nucleophile depends of the solvent. In polar protic solvents, such as alcohols, it is highly solvated, which increases its effective size, and does not make it a good nucleophile for S$_N$2 reactions in these types of solvents, as illustrated in Figure 15.5.

On the other hand, in a nonpolar aprotic solvent, such as acetone, the fluoride is not highly solvated, and hence, it is effectively a small and effective nucleophile. Table 15.1 gives a description of commonly used solvents of organic chemistry.

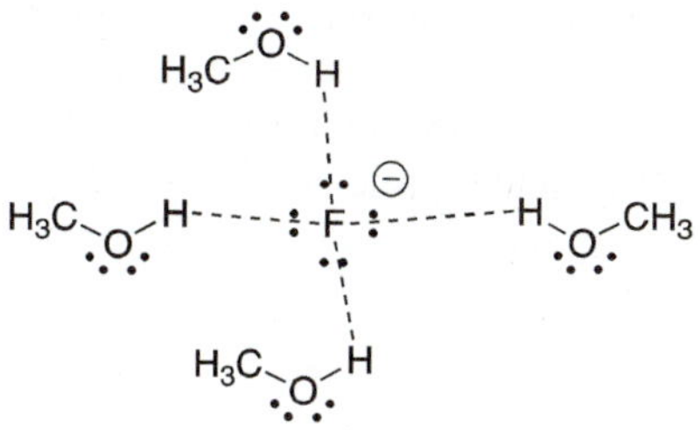

Figure 15.5 Solvated fluoride anion with a protic solvent, methanol.

Table 15.1 Common protic and aprotic organic solvents.

Solvent	Type of solvent	Structure
Water	Protic-polar	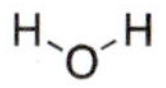 H–O–H
Acetone	Nonprotic-polar	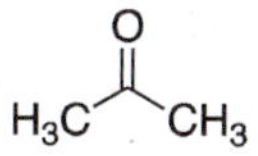 H₃C–C(=O)–CH₃
Methanol	Protic-polar	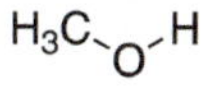 H₃C–O–H
Ethanol	Protic-polar	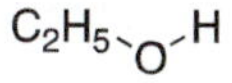 C₂H₅–O–H
Dimethyl sulfoxide (DMSO)	Nonprotic-polar	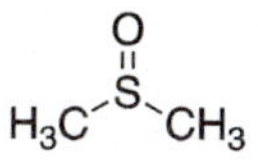 H₃C–S(=O)–CH₃
Diethyl ether	Nonprotic	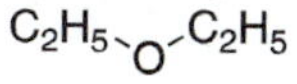 C₂H₅–O–C₂H₅
Tetrahydrofuran (THF)	Nonprotic-polar	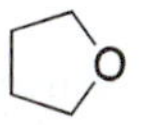 (cyclic ether)
Carbon tetrachloride	Nonprotic-polar	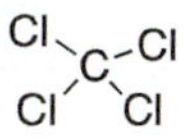 CCl₄

Problem 15.8

i) Of the following pairs of nucleophiles, which would be a better nucleophile in an aprotic polar solvent, such as dimethyl sulfoxide (DMSO)?

(a) Na$^+$ I$^-$ and Na$^+$ Cl$^-$ (b) Na$^+$ OH$^-$ and Na$^+$ SH$^-$

ii) Select different solvents that could be used to carry out the following S$_N$2 reactions.

(a) 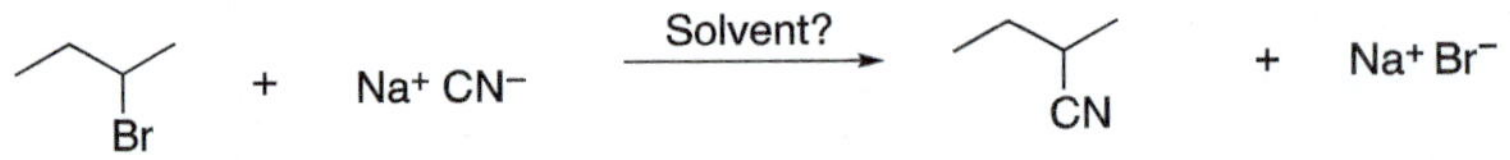 (CH$_3$)$_2$CHCH$_3$-Br + Na$^+$ CN$^-$ $\xrightarrow{\text{Solvent?}}$ (CH$_3$)$_2$CHCH$_3$-CN + Na$^+$ Br$^-$

(b) 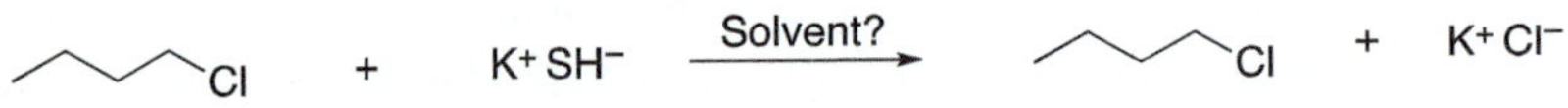 CH$_3$CH$_2$CH$_2$-Cl + K$^+$ SH$^-$ $\xrightarrow{\text{Solvent?}}$ CH$_3$CH$_2$CH$_2$-Cl + K$^+$ Cl$^-$

15.6.4 Stereochemistry of the Products of S$_N$2 Reactions

One important outcome of substitution reactions that proceed via the S$_N$2 mechanism is if the electrophilic carbon is stereogenic, the product has the opposite configuration compared to the starting material. A look at Reaction (15-20) explains why an inversion takes place. If the nucleophile approaches from the opposite side of the leaving group, then an inversion of configuration takes place to form the product.

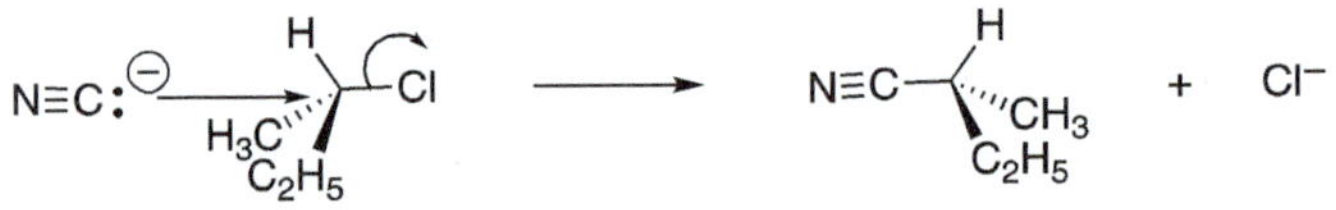

$$N\equiv C:^{\ominus} + \underset{H_3C\,\, C_2H_5}{\overset{H}{\underset{|}{C}}}-Cl \longrightarrow N\equiv C-\underset{C_2H_5}{\overset{H}{\underset{|}{C}}}''''CH_3 + Cl^- \qquad (15\text{-}20)$$

Problem 15.9

For the following reactions, give the major organic product and assign absolute configuration of the chiral carbon in both reactant and product.

(a) 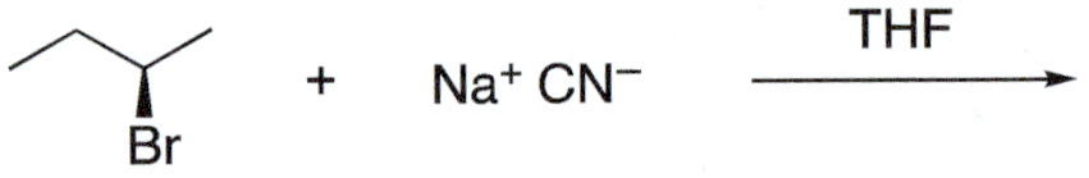 (CH$_3$)$_2$CH-CH(Br)-CH$_3$ + Na$^+$ CN$^-$ $\xrightarrow{\text{THF}}$

(b) 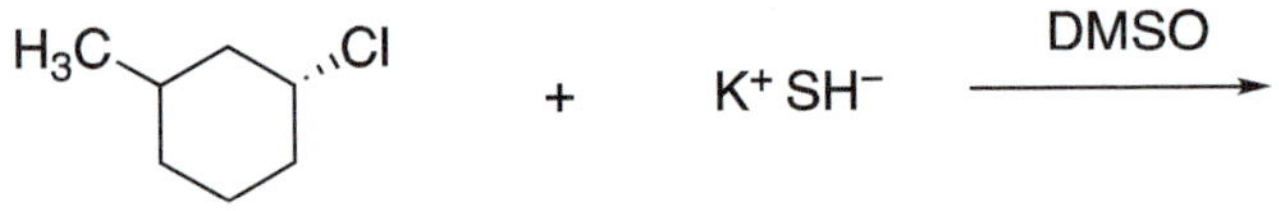 [trans-1-methyl-2-chlorocyclohexane] + K$^+$ SH$^-$ $\xrightarrow{\text{DMSO}}$

(c) 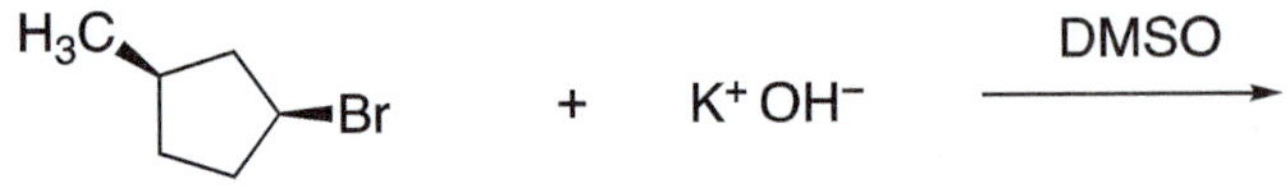 [trans-1-methyl-2-bromocyclopentane] + K$^+$ OH$^-$ $\xrightarrow{\text{DMSO}}$

Thus, this process can be used to synthesize a particular enantiomer by taking advantage of the stereospecificity of S$_N$2 reactions and a reaction that was covered in Section 15-3-5 in which an OH group was converted to a good leaving group using TsCl. This concept is shown in Reaction (15-21).

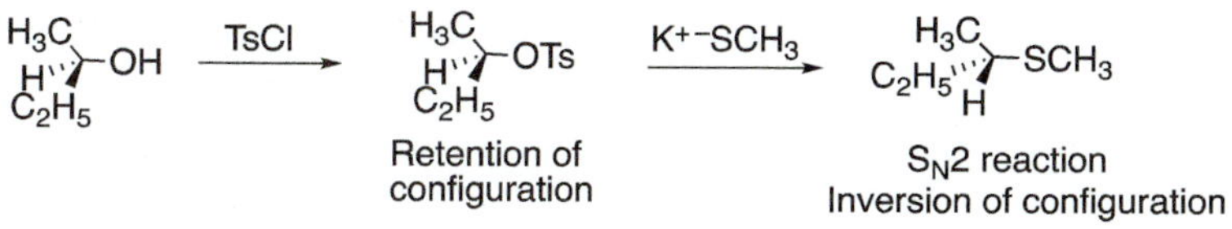

15.6 Bimolecular Substitution Reaction Mechanism (S$_N$2 Mechanism)

$$\underset{C_2H_5}{\overset{H_3C}{\underset{H}{\rightarrow}}}OH \xrightarrow{TsCl} \underset{\underset{\text{Retention of configuration}}{C_2H_5}}{\overset{H_3C}{\underset{H}{\rightarrow}}}OTs \xrightarrow{K^+ {}^-SCH_3} \underset{\underset{\text{Inversion of configuration}}{C_2H_5}}{\overset{H_3C}{\underset{H}{\rightarrow}}}SCH_3 \quad \text{S}_N\text{2 reaction} \qquad (15\text{-}21)$$

Problem 15.10

Show how to carry out the following transformations, carefully note the stereochemistry in both the reactant and product.

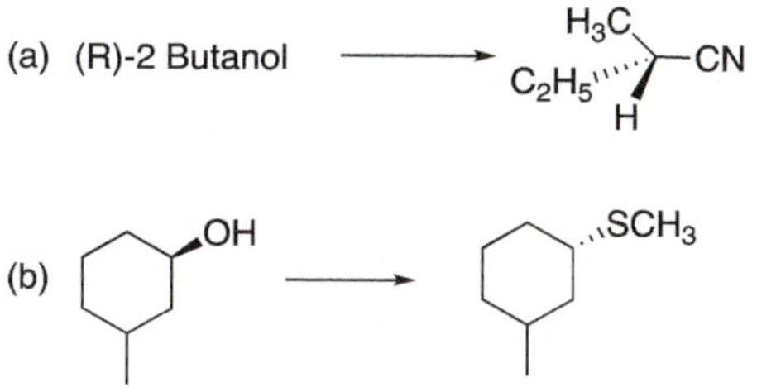

(a) (R)-2 Butanol $\longrightarrow$ $\underset{H}{\overset{H_3C}{\underset{C_2H_5}{\rightarrow}}}CN$

(b) [cyclohexane with OH and methyl] $\longrightarrow$ [cyclohexane with SCH$_3$ and methyl]

In analyzing S$_N$2 reactions to predict the organic products, the following factors should be considered.

1) *Electrophile*: CH$_3$ > 1° > 2° ≫ 3°
2) *Stereochemistry*: stereospecific reaction in which inversion results. (Remember that a stereospecific reaction converts a reactant with a specific stereochemistry into another as a product.)
3) *Nucleophile*: small and weak base.
4) *Leaving group*: need leaving groups that are weak stable conjugate bases (good leaving group)
5) *Solvent*: must be nonionizing, which are typically nonprotic solvents. Good solvents include acetone and dimethyl sulfoxide (DMSO).
6) *Mechanism*: concerted in that the attack of the nucleophile and the leaving of the leaving group occur in a single step.

15.6.5 Intramolecular S$_N$2 Reactions

Note that it is possible for the nucleophile and the electrophile to be on the same molecule. In this case, the reaction is described as an intramolecular substitution reaction as shown in Reaction (15-22), the mechanism is shown below.

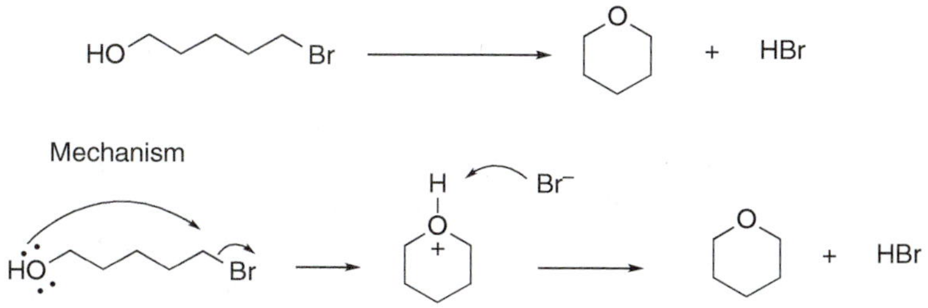

$$HO\text{-}(CH_2)_n\text{-}Br \longrightarrow \text{[tetrahydropyran]} + HBr \qquad (15\text{-}22)$$

Mechanism

A similar intramolecular substitution S_N2 reaction can be used to synthesize oxiranes as shown in Reaction (15-23).

$$\text{alkoxide with L} \longrightarrow \text{Oxirane} + \text{L}^- \qquad (15\text{-}23)$$

For these reactions, a weak base, such as sodium hydroxide, can be used to shift the equilibrium in the direction of the formation of an alkoxide ion for the reaction to occur as shown in Reaction (15-24). As expected, there are other organic products including the elimination products and substitution products.

$$\text{halohydrin} \xrightarrow{\text{NaOH}} \text{alkoxide} \longrightarrow \text{Oxirane} + \text{L}^- \qquad (15\text{-}24)$$

We have already discussed in Chapter 8 that halohydrin can be formed from the addition reaction of water and bromine to alkenes. Using an alkene as a starting compound for the synthesis of an epoxide by a combination of these reactions is illustrated in Reaction (15-25).

$$\text{alkene} \xrightarrow[\text{DMSO}]{\text{Br}_2,\ \text{H}_2\text{O}} \text{halohydrin} \xrightarrow{\text{NaOH}} \text{epoxide} \qquad (15\text{-}25)$$

15.7 Unimolecular Substitution Reaction Mechanism (S_N1 Mechanism)

An alternate mechanism for substitution reactions that was mentioned in the previous section involves the leaving group breaking away from the electrophile and taking its bonding electrons in an independent step to form a carbocation and the leaving group, as shown in Reaction (15-26).

$$\underset{\text{Electrophile}}{\text{C—L}} \xrightarrow{\text{RDS}} \underset{\text{Carbocation}}{\text{C}^+} + \underset{\text{Leaving group}}{:\text{L}^-} \qquad (15\text{-}26)$$

In a second step, the nucleophile attacks the carbocation formed in the initial step to produce the product as shown in Reaction (15-27).

$$\underset{\text{Carbocation}}{\text{C}^+} + \underset{\text{Nucleophile}}{:\text{Nu}^-} \xrightarrow{\text{Fast}} \underset{\text{Substitution product}}{\text{C—Nu}} \qquad (15\text{-}27)$$

This mechanism is called an S_N1 since in the first step and most important step of the reaction mechanism (also called the rate-determining step, RDS) only one molecule (the electrophile) is involved as shown in Reaction (15-26). The overall substitution reaction, which is the same using the mechanism discussed in the previous section, is shown in Reaction (15-28).

$$\underset{\text{Electrophile}}{\overset{|}{\underset{|}{C}}-L} \quad + \quad \underset{\text{Nucleophile}}{:Nu^-} \quad \longrightarrow \quad \underset{\text{Substitution product}}{\overset{|}{\underset{|}{C}}-Nu} \quad + \quad \underset{\text{Leaving group}}{:L^-} \qquad (15\text{-}28)$$

This type of reaction mechanism is described as unimolecular because as mentioned earlier, only one molecule, the electrophile, is involved in a very important step, the RDS, of the mechanism in the formation of the products. This concept of a two-step mechanism can be illustrated using an energy profile diagram as shown in Figure 15.6.

Since the RDS involves just one molecule, the mechanism for substitution reactions that proceed through this mechanism is called an S_N1 – nucleophilic substitution *unimolecular* mechanism. The RDS is sometimes referred to as the slow step. As shown in Figure 15.6, the energy of activation for the first step of the reaction is greater than the energy of activation for the second step. Hence, the second step of the reaction is also referred to as the fast step. In the next section, we will examine substitution reactions that occur by this mechanism and the factors that influence substitution reactions that proceed via this mechanism.

15.7.1 The Nucleophile and Solvents of S_N1 Reactions

In this section, we will examine the factors that influence substitution reactions that proceed via the S_N1 mechanism. The first step of this reaction mechanism involved the heterolytic breakage of the bond of the leaving group and the electrophile to form a carbocation and the leaving group. Since a charged carbocation is formed, dipolar protic solvents, such as water or ethanol, are typically used to promote this very important step in the reaction mechanism. Thus, polar protic solvents shown in Table 15.1 favor this reaction mechanism. In the second step of the reaction mechanism, the nucleophile attacks the carbocation to form the product. For reactions that proceed by this mechanism, typically the solvent is also the nucleophile; hence, reactions that proceed by this mechanism are also called **solvolysis** reactions. Since the nucleophile reacts in the second step of the mechanism, which is not the RDS, the size and nature of the nucleophile is not as important as the case for the S_N2 mechanism. Thus, the rate of S_N1 reactions is not dictated by the nature or concentration of the nucleophile as we saw in the S_N2 reaction mechanism.

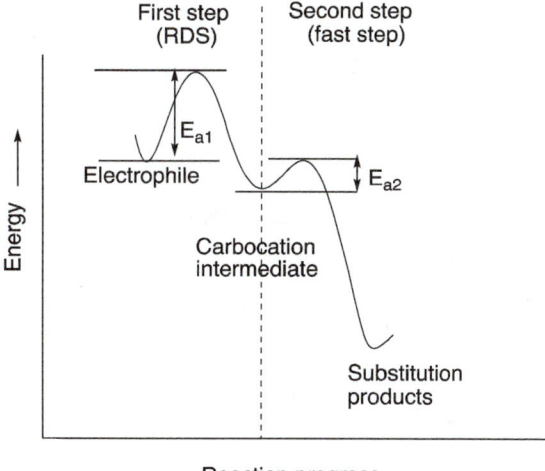

Figure 15.6 Energy profile for a typical S_N1 reaction, which takes place in two steps giving a carbocation as an intermediate and then the product. Note that E_{a1} for the first step (the RDS) is greater than E_{a2} for the second and fast step (fast step).

Problem 15.11

i) Predict the products for the following S_N1 reactions.

(a) cyclopentyl-Br + H_2O ⟶

(b) (CH$_3$)$_2$CHCH(Br)CH$_3$ (2-bromo-3-methylbutane) + CH_3OH ⟶

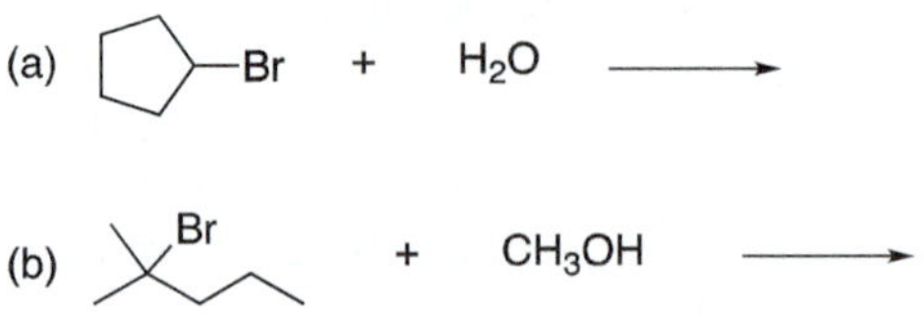

ii) Give the structure of an appropriate electrophile to complete the following S_N1 reactions.

(a) ? + H_2O ⟶ 2-methylpentan-2-ol type product with OH

(b) ? + CH_3OH ⟶ cyclohexyl-OCH_3

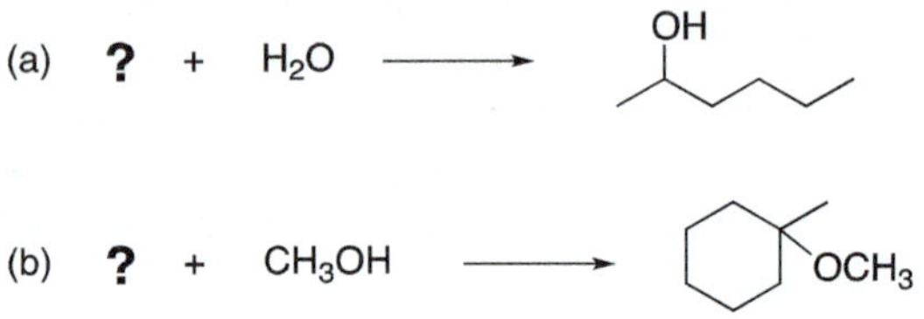

15.7.2 Stereochemistry of the Products of S_N1 Reactions

For electrophiles, such as alkyl halides, that have stereogenic centers bonded to the leaving group, there is an equal mixture of both enantiomers in the product. To explain this outcome, compared to the outcome of the S_N2 mechanism, we will have to look in more detail at the S_N1 mechanism. The carbocation that is formed during the S_N1 mechanism is sp^2 hybridized and flat with a vacant p orbital perpendicular to the plane of the three bonds bonded to the carbocation. As a result, an attack of the nucleophile, in the second step of the mechanism to form a new bond with the vacant p orbital, can take place from either side of the carbocation. That is, there is an equal probability of an attack of the nucleophile at the top side or from the bottom side of the vacant p orbital of the carbocation to form the product. The result is a racemic mixture as illustrated in Figure 15.7.

As a result, it is very difficult to convert an optically active alcohol to an optically active bromide using an S_N1 substitution reaction owing to the formation of a carbocation intermediate as the reaction proceeds to form products. This concept is illustrated in Reaction (15-29).

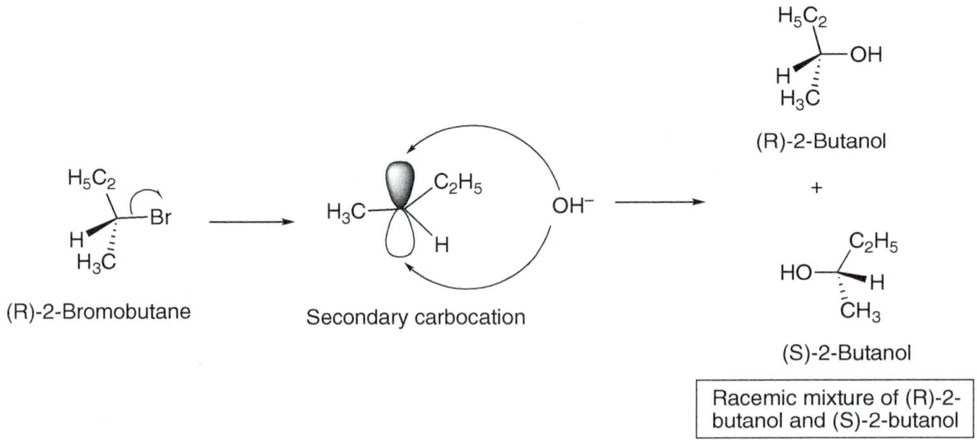

Figure 15.7 The attack of a flat carbocation by a nucleophile to give a pair of enantiomers.

(R)-2-Bromobutane ⟶ Secondary carbocation + OH^- ⟶ (R)-2-Butanol + (S)-2-Butanol

Racemic mixture of (R)-2-butanol and (S)-2-butanol

15.7 Unimolecular Substitution Reaction Mechanism (S_N1 Mechanism)

$$\underset{\substack{\text{Stereogenic}\\\text{electrophile}}}{\overset{H_3C}{\underset{C_2H_5}{\overset{H}{\diagdown}}}\!\!-\!OH} + HBr \longrightarrow \underbrace{\underset{C_2H_5}{\overset{H_3C}{\underset{}{\overset{H}{\diagdown}}}\!\!-\!Br} + \underset{H}{\overset{H_3C}{\underset{C_2H_5}{\overset{}{\diagdown}}}\!\!-\!Br}}_{\text{Racemic mixture since } S_N1 \text{ mechanism}} + H_2O \qquad (15\text{-}29)$$

Problem 15.12

Using dashed-wedge or Fischer representation, give the structures of the racemic organic products of the S_N1 reactions of CH_3OH with each of the following.

(a) $C_2H_5\!\!-\!\!\underset{Cl}{\overset{H}{|}}\!\!-\!\!CH_3$ (b) $C_2H_5\!\!-\!\!\underset{CH_3}{\overset{Br}{|}}\!\!-\!\!C_3H_7$ (c) $\underset{H_3C}{\overset{H_5C_2}{\diagdown}}\!\!\underset{H}{\overset{}{\diagup}}\!\!-\!Cl$

15.7.3 The Electrophile of S_N1 Reactions

As mentioned earlier, the leaving groups for reactions that proceed via the S_N1 mechanism are typically very good leaving groups. An exceptionally good leaving group is protonated water, which is created upon the protonation of an alcohol, to form OH_2^+. Protonated alcohols are produced in the presence of an acidic solution. The mechanism that is typically encountered is the S_N1 mechanism if the alcohols are secondary or tertiary. The substitution reaction of alcohols and hydrogen chloride gives an alkyl chloride as the product. These reactions can be used to determine the type of an unknown alcohol. Owing to the stability of tertiary carbocations, tertiary alcohols react very fast with hydrogen chloride to produce the alkyl chloride product. Consider the substitution reaction involving a tertiary alcohol as shown in Reaction (15-30).

$$\underset{\text{Tertiary alcohol}}{R\!\!-\!\!\underset{R}{\overset{OH}{\overset{|}{\underset{|}{C}}}}\!\!-\!\!R} \xrightarrow{HCl} R\!\!-\!\!\underset{R}{\overset{Cl}{\overset{|}{\underset{|}{C}}}}\!\!-\!\!R + H_2O \qquad (15\text{-}30)$$

In the first step of the mechanism, the strong acid protonates the alcohol to form protonated water, which is a very good leaving group, as shown in Reaction (15-31).

$$\underset{\text{Tertiary alcohol}}{R\!\!-\!\!\underset{R}{\overset{R}{\overset{|}{\underset{|}{C}}}}\!\!-\!\ddot{O}H} + H\!\!-\!\!Cl \longrightarrow R\!\!-\!\!\underset{R}{\overset{R}{\overset{|}{\underset{|}{C}}}}\!\!-\!\overset{+}{O}H_2 + Cl^- \qquad (15\text{-}31)$$

To form the carbocation ion, the extremely good H_2O leaving group leaves to create the carbocation in a slow step of the mechanism as shown in Reaction (15-32).

$$R\!\!-\!\!\underset{R}{\overset{R}{\overset{|}{\underset{|}{C}}}}\!\!-\!\overset{+}{O}H_2 \xrightarrow{RDS} \underset{\substack{\text{Tertiary}\\\text{carbocation}}}{R\!\!-\!\!\underset{R}{\overset{R}{\overset{|}{\underset{|}{C^+}}}}} + \underset{\text{Leaving group}}{H_2O} \qquad (15\text{-}32)$$

In the final step of the mechanism, the nucleophilic chloride anion reacts with the carbocation to form the final product as shown in Reaction (15-33).

$$R-\overset{R}{\underset{R}{C}}{}^{+} + :\ddot{\underset{..}{Cl}}:^{-} \xrightarrow{\text{Fast}} R-\overset{R}{\underset{R}{C}}-\ddot{\underset{..}{Cl}}: \tag{15-33}$$

Tertiary carbocation — Racemic tertiary alkyl chloride if R groups are different

A similar analysis can be carried out for a secondary alcohol, but in these reactions, a secondary carbocation is formed, which is not as stable as a tertiary carbocation. Hence, the reaction rate is a bit slower than that of a tertiary alcohol. In the first step of the mechanism, the acid protonates the secondary alcohol to form protonated water, which leaves to create a secondary carbocation in the RDS, as shown in Reaction (15-34).

$$R-\overset{R}{\underset{H}{C}}-\ddot{O}H + H-Cl \xrightarrow{-Cl^-} R-\overset{R}{\underset{H}{C}}-\overset{+}{O}H_2 \xrightarrow[\text{RDS}]{-H_2O} R-\overset{R}{\underset{H}{C}}{}^{+} \tag{15-34}$$

Secondary alcohol — Secondary carbocation

In the final step of the mechanism, the nucleophilic chloride anion reacts with the carbocation to form the final product as shown in Reaction (15-35).

$$R-\overset{R}{\underset{H}{C}}{}^{+} + :\ddot{\underset{..}{Cl}}:^{-} \xrightarrow{\text{Fast}} R-\overset{R}{\underset{H}{C}}-\ddot{\underset{..}{Cl}}: \tag{15-35}$$

Secondary carbocation — Secondary alkyl chloride

On the other hand, primary alcohols react very slowly with hydrogen chloride to form the corresponding alkyl chloride. In fact, the reaction is so slow that a catalyst, $ZnCl_2$ is used to increase the rate of the reaction to an observable rate as shown in Reaction (15-36).

$$R-CH_2-OH \xrightarrow[ZnCl_2]{HCl} R-CH_2-Cl \tag{15-36}$$

Primary alcohol

Since the rates of the reaction of different types of alcohols with hydrogen chloride are different, even in the presence of the catalyst, $ZnCl_2$, the difference in rates can be used to determine the type of alcohol present. This test is known as the **Lucas Test**. The rate of the reaction of a secondary alcohol with hydrogen chloride is faster than that of primary alcohols, even in the presence of the catalyst, $ZnCl_2$, but slower than that of tertiary alcohols.

Problem 15.13

Which of the following pairs of molecules would react the fastest using a Lucas test?

(a) cyclohexanol with OH and (CH₃)₃C—OH

(b) (CH₃)₂CH—OH and CH₃CH₂CH₂—OH

(c) cyclopentanol—OH and CH₃OH

(d) CH₃OH and (CH₃)₃C—CH(OH)CH₃

15.7 Unimolecular Substitution Reaction Mechanism (S$_N$1 Mechanism)

Reactions that proceed via a S$_N$1 mechanism are not the best to be used for the synthesis of a target molecule. In addition to giving racemic mixtures, rearrangement of the carbocation typically occurs giving rise to unexpected products. Reaction (15-37) gives an example of a S$_N$1 reaction that gives an unexpected product.

$$\text{(CH}_3\text{)}_2\text{CHCH(Br)CH}_3 \xrightarrow{\text{CH}_3\text{OH}} \text{Expected product (OCH}_3\text{)} + \text{Unexpected product (OCH}_3\text{)} \tag{15-37}$$

The explanation of the expected product is pretty straightforward as shown in the mechanism in Reaction (15-38).

$$\text{R-Br} \xrightarrow{-\text{Br}^-} \text{Secondary carbocation} \xrightarrow{\text{HOCH}_3} \text{R-O}^+\text{(H)CH}_3 \xrightarrow{-\text{H}^+} \text{Expected product (OCH}_3\text{)} \tag{15-38}$$

In order to fully understand and explain the appearance of the unexpected product, we will have to examine the S$_N$1 mechanism closely and concentrate on the carbocation intermediate that is formed during the RDS. As we know, the stability of carbocations varies and depends on the number of alkyl groups that are bonded to the carbon of the carbocation. Since carbocations are positively charged species, any effect that will delocalize or stabilize this positive charge will stabilize the carbocation. Alkyl groups are considered to be electron releasing and, as a result compared to hydrogens, will stabilize carbocations. Since tertiary carbocations have three alkyl groups bonded to the carbocationic carbon, this type of carbocation is more stable than one that has one carbon and two hydrogens, i.e. secondary or primary carbocations. As we have seen from previous chapters, the relative stability of carbocations is: Tertiary (3°) > Secondary (2°) > Primary (1°) > Methyl (CH$_3$). Carbocations often rearrange to form more stable carbocations. For the reaction given above, the initial carbocation that is formed during the course of the reaction is a secondary carbocation, which will undergo a rearrangement to form a more stable tertiary carbocation as shown in Reaction (15-39).

$$\text{Secondary carbocation} \xrightarrow[\text{Migration}]{1,2\text{-Methide}} \text{Tertiary carbocation} \tag{15-39}$$

Even though the secondary carbocation is fairly stable, but the tertiary carbocation is more stable and its stability is the driving force for the rearrangement. As shown in Reaction (15-39), there is a migration of the methyl group, with its bonding electrons from the adjacent carbon, to form the tertiary carbocation. This migration is also called a methide migration and specifically 1,2 methide migration since the methide anion migrates across two carbons. The tertiary carbocation, once formed, will react with the nucleophile to form the observed major product as shown in Reaction (15-40).

$$\text{Tertiary cation} \xrightarrow{\text{HOCH}_3} \text{R-O}^+\text{(H)CH}_3 \xrightarrow{-\text{H}^+} \text{Unexpected product (OCH}_3\text{)} \tag{15-40}$$

Note the hydrogen that is bonded to the oxygen nucleophilic atom becomes acidic once the nucleophile bonds to the carbocation. This hydrogen is abstracted as a proton in the final step to form the product shown. Thus, the overall reaction mechanism to form the unexpected product is shown in Reaction (15-41).

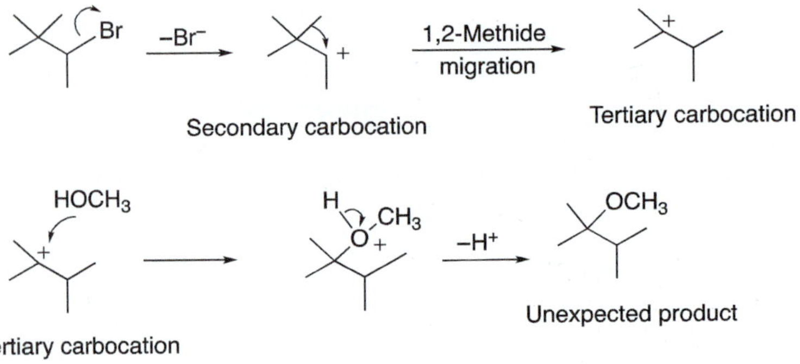

(15-41)

Problem 15.14

Using arrows to show electron movement, propose a mechanism to explain the products of the following reaction.

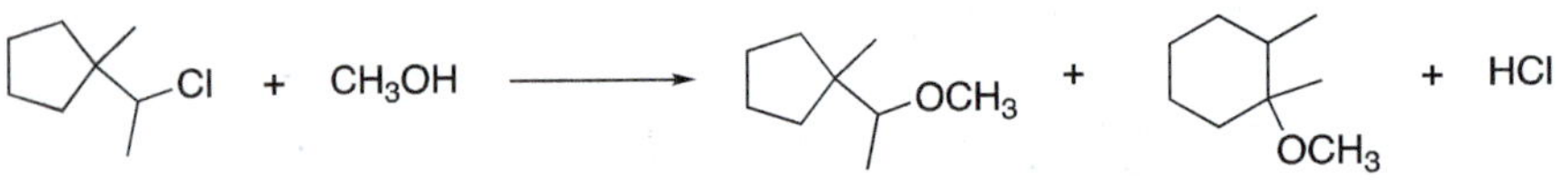

There are some electrophiles that result in very stable carbocations after the leaving group leaves. The allylic carbocation and the benzylic carbocation are examples of very stable carbocations and they are more stable than even the tertiary carbocation due to the delocalization of the electrons, as shown in Reactions (15-42) and (15-43).

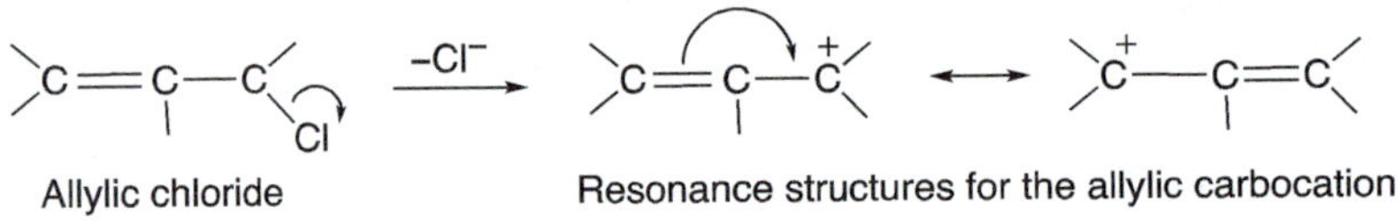

(15-42)

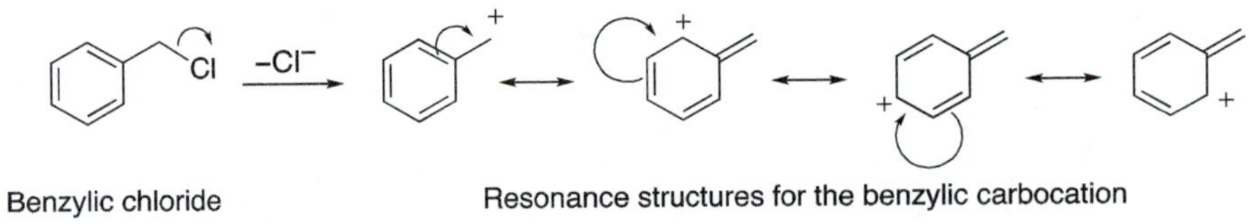

Allylic chloride   Resonance structures for the allylic carbocation

(15-43)

Benzylic chloride   Resonance structures for the benzylic carbocation

These carbocations are stable owing to the delocalization of the positive charge over the adjacent pi (π) network. Even though rearrangement does not occur with these very stable carbocations; sometimes, unexpected products are obtained as shown in Reaction (15-44).

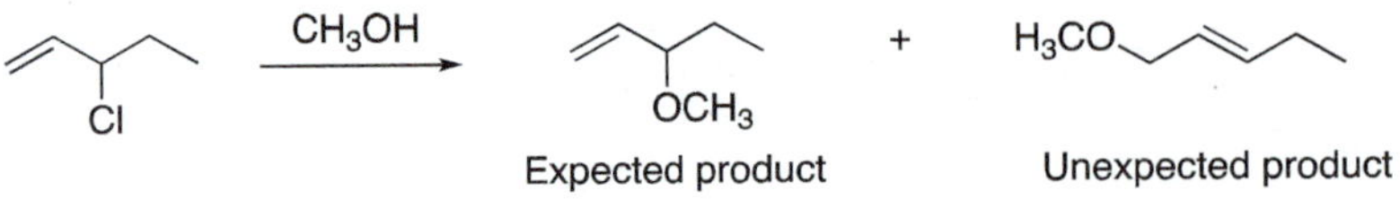

(15-44)

15.7 Unimolecular Substitution Reaction Mechanism (S_N1 Mechanism)

The formation of the unexpected can be explained by the reaction mechanism shown below.

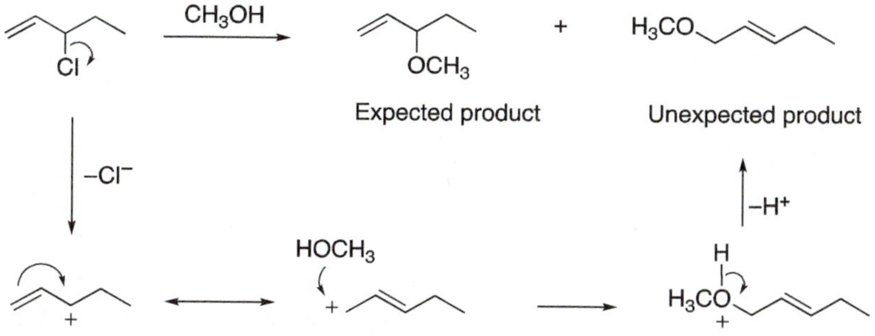

[Reaction scheme: allylic chloride + CH_3OH → expected product (with OCH_3 at original position) + unexpected product (H_3CO at rearranged position). Loss of Cl^- gives allylic carbocation with two resonance structures (note: the second resonance structure is not a primary carbocation); attack by $HOCH_3$ followed by loss of H^+ gives the unexpected product.]

Another similar example is shown in Reaction (15-45) where there are expected and unexpected products observed.

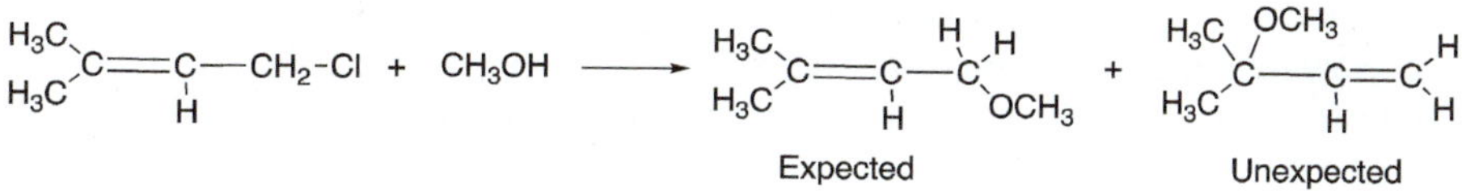

$$(CH_3)_2C=CH-CH_2-Cl + CH_3OH \longrightarrow (CH_3)_2C=CH-CH(OCH_3)(H) + (CH_3)_2C(OCH_3)-CH=CH_2 \quad (15\text{-}45)$$

Expected   Unexpected

The mechanism to explain the unexpected product is shown in Reaction (15-46).

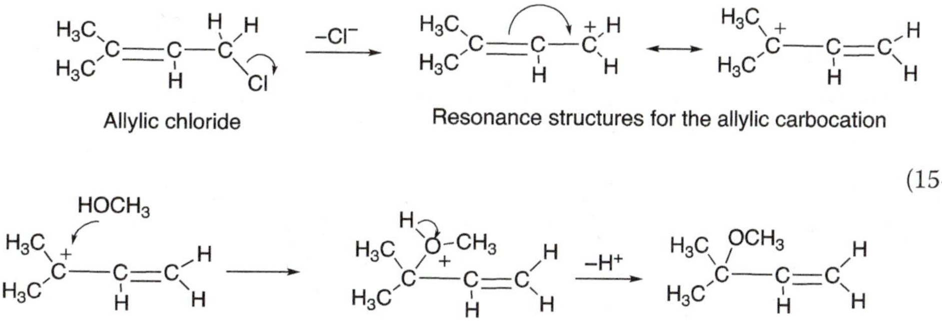

[Mechanism: Allylic chloride loses Cl^- to form allylic carbocation with two resonance structures; addition of $HOCH_3$, then loss of H^+, gives the unexpected product.]

(15-46)

A similar unexpected product does not appear for reactions in which the benzylic carbocation is formed. The six pi (π) electrons of the ring form an exceptionally stable system that we will study in Chapter 17 and unexpected products where that cyclic six-member ring system is disturbed tend not to occur as shown in Reaction (15-47).

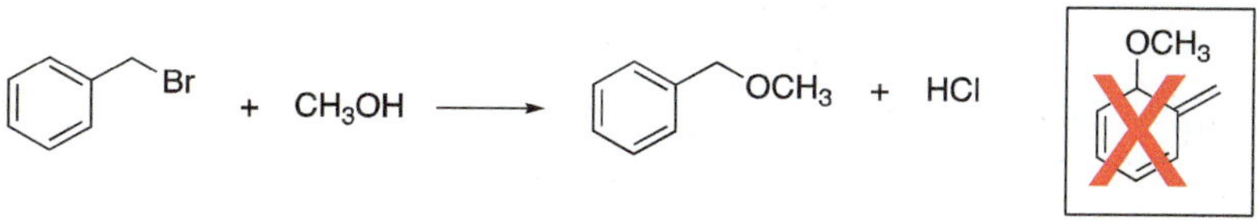

$$PhCH_2Br + CH_3OH \longrightarrow PhCH_2OCH_3 + HCl \quad (15\text{-}47)$$

[Struck-through structure showing OCH_3 substituted on the ring — not formed.]

For S_N1 reactions, these types of halides react very fast owing to the quick formation of the very stable intermediate carbocation that is initially formed in the RDS. These intermediate carbocations are resonance-stabilized carbocations.

Problem 15.15

Give the products (expected and unexpected) for the reaction of the alkyl halides shown below with each of the following nucleophiles: (a) H₂O and (b) CH₃OH.

(a) [cyclohexenyl-CH₂Cl]   (b) [bicyclic alkene with Br]

Summary: For S_N1 reactions, the following factors should be considered.

1) *Electrophile*: The reactivity of the electrophile in S_N1 reactions is: Allylic ~ benzylic > 3° > 2° > 1° >> CH₃
2) *Stereochemistry*: An equal mixture of stereoisomers (racemic mixture) results if starting substrate is optically active
3) *Nucleophile*: The size of the nucleophile does not matter since attack of nucleophile occurs after the RDS and typically the solvent is the nucleophile.
4) *Leaving group*: For these reactions, the leaving groups are typically extremely good leaving groups, which are also weak stable conjugate bases.
5) *Solvent*: Solvents must be ionizing, which are typically protic solvents. Good solvents include water and alcohols, which are also typically the nucleophiles and the reactions are also called solvolysis.
6) *Mechanism*: Stepwise mechanism, in which the first step is the RDS and involves the formation of the carbocation, the second step is fast and involves the attack of the nucleophile on the carbocation intermediate.

15.8 Applications of Nucleophilic Substitution Reactions – Synthesis

As pointed out in previous chapters, organic synthesis is a very important aspect of organic chemistry and this process typically involves forming new covalent bonds in a target product molecule. Since a single covalent bond has two electrons, there are two possibilities of making such a bond using nucleophilic substitution reactions, as illustrated in Figure 15.8.

Of course, based on your basic knowledge of chemistry, the nucleophilic reactant must be the fragment (or atom) that can best accommodate the pair of electrons and can be neutral or an anion, i.e. it must be fairly electronegative. The electrophilic fragment (or atom) is usually the partially positive carbon of a polar covalent bond to carbon, i.e. an electrophilic carbon

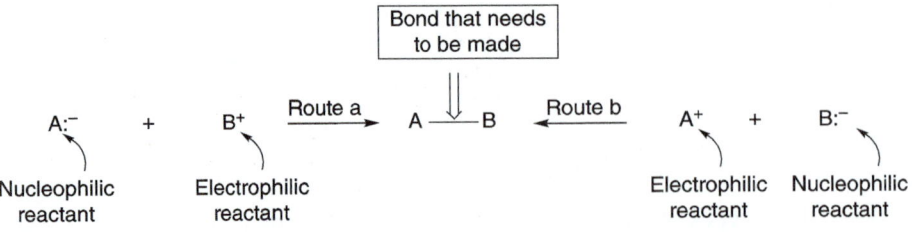

Figure 15.8 Two different ways that a single covalent bond can be synthesized.

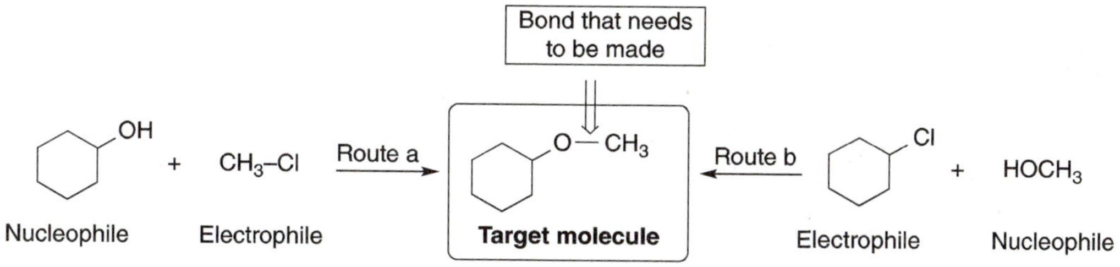

15.8 Applications of Nucleophilic Substitution Reactions – Synthesis

Figure 15.9 Two different ways of synthesizing the bond of the ether functionality.

bonded to an electronegative atom such as chlorine. In synthesis, the bonds of the functional groups of the target molecule are identified, and types of reactions are chosen that can be used to make the various functional groups of the target molecule. Thus, care must be exercised in deciding which electrophile, nucleophile, and the type of reactions to be used.

15.8.1 Synthesis of Ethers

The possible ways of synthesizing an ether functionality is illustrated in Figure 15.9.

Note that you must first identify the functional group of your target molecule and then determine the bond that needs to be made that will result in that functional group. After careful analysis and application of the knowledge that we have learned thus far, route *b* would appears to be slightly better since it involves the formation of a secondary carbocation and route *a* involves the formation of a methyl carbocation which is not stable. On the other hand, the synthesis via route *a* could be accomplished if the conditions were right for a S_N2 mechanism. As you will see, there are sometimes different possibilities for the synthesis of a target molecule. You will have to think critically of possibilities and then make a creative decision on the best route based on your knowledge of reactions.

As we have seen from the above example for the synthesis of the ether target molecule, there are two ways of making the ether bond. Alexander Williamson developed the Williamson reaction for the synthesis of ethers in 1850 and it involves an S_N2 substitution reaction of an alkoxide anion and alkyl halide. As a result, the electrophile for the Williamson synthesis must be either primary or secondary and the nucleophile is made from a corresponding alcohol. An example of the Williamson synthesis of an ether is shown in Reaction (15-48).

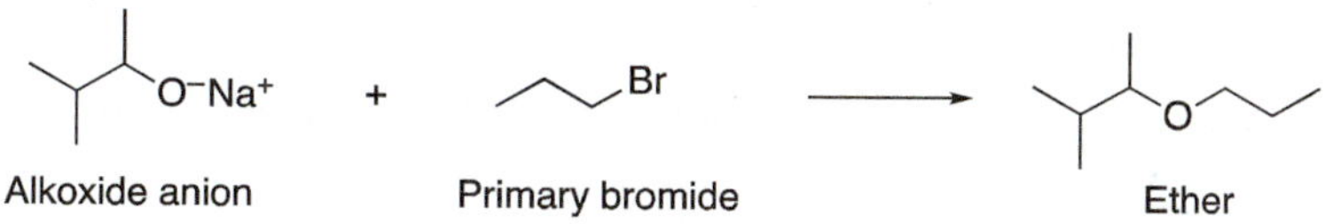

(15-48)

Problem 15.16

Utilize the Williamson synthesis to make the following compounds.

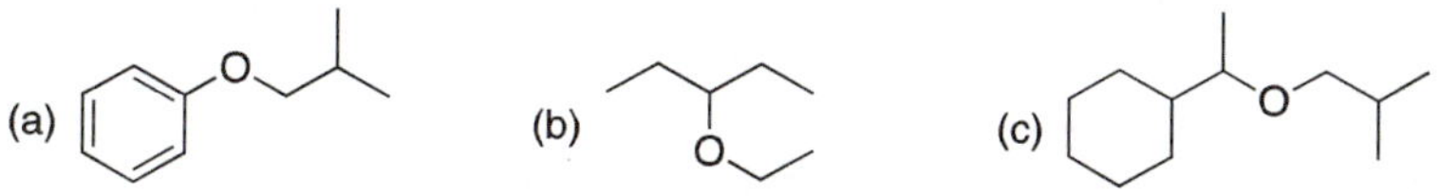

15.8.2 Synthesis of Nitriles

One of the most common methods of synthesizing nitriles involves a substitution reaction of the leaving group of an alkyl halide with a nitrile group. The synthesis of this bond is typically accomplished via a S_N2 substitution reaction in a fairly polar nonprotic solvent as shown in Reaction (15-49).

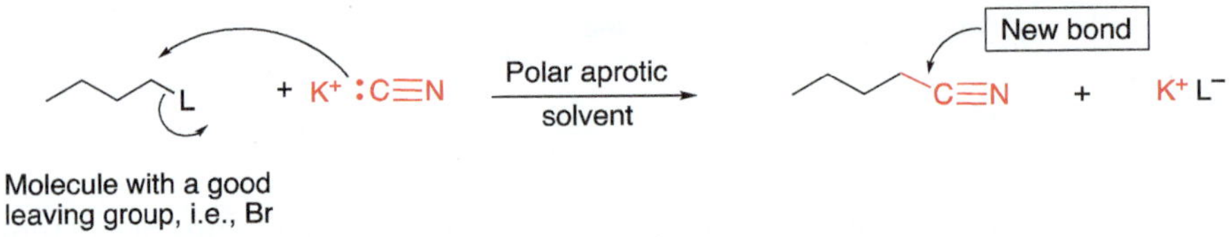

$$\text{Molecule with a good leaving group, i.e., Br} + K^+ \text{:} C \equiv N \xrightarrow{\text{Polar aprotic solvent}} \text{R-}C \equiv N \text{ (New bond)} + K^+ L^- \quad (15\text{-}49)$$

As we have seen, alkyl halides make good electrophiles and hence must be methyl, primary, or secondary, but never tertiary, if used for this type of synthesis. The two starting compounds that can be used for the synthesis of 3-methylpentanenitrile are shown in Reaction (15-50).

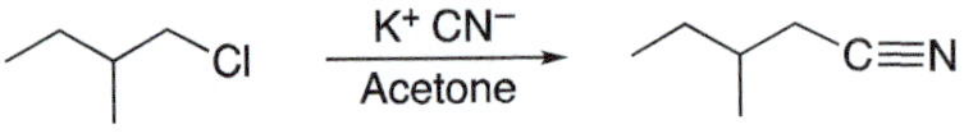

$$\text{R-Cl} \xrightarrow{K^+ CN^-, \text{ Acetone}} \text{R-}C \equiv N \quad (15\text{-}50)$$

As pointed out, the reaction mechanism is an S_N2 mechanism and, as a result, the electrophile must be methyl, primary or rarely secondary, but never tertiary.

Problem 15.17

Show how to synthesize the nitriles shown below in which substitution reactions are used.

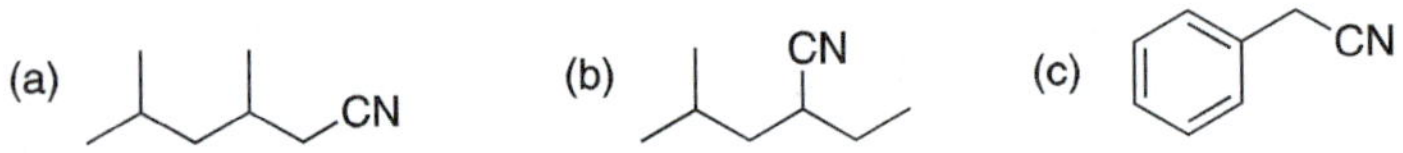

(a) [structure with CN] (b) [structure with CN] (c) [benzyl CN structure]

15.8.3 Synthesis of Silyl Ethers

An important category of compounds in organic chemistry are silyl ethers, an example of a silyl is shown below.

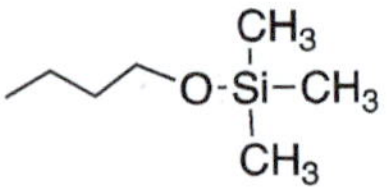

Butyltetramethylsilyl ether

Its synthesis is shown below, and it involves the substitution of alkoxide group for a chloride anion leaving group, but most important, the reaction center is not a carbon as seen for most of the reactions thus far, its center is a silicon atom. As you know, silicon is just below carbon on the periodic table and hence its reactivity is very similar to that of carbon. Hence, the products of the reaction shown in Reaction (15-51) are not surprising and should be predictable based on utilizing the knowledge gained thus far.

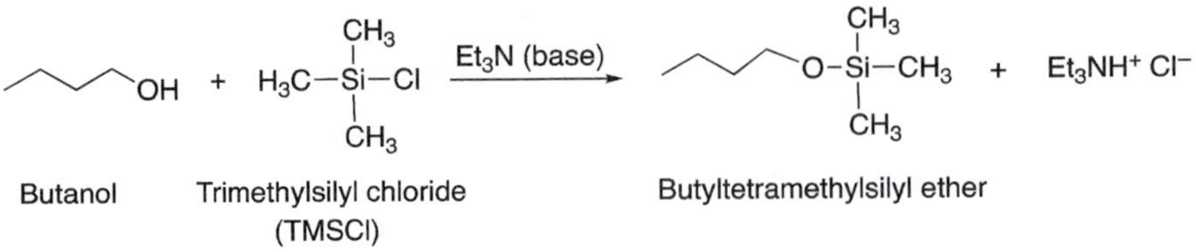

$$\text{Butanol} + \text{Trimethylsilyl chloride (TMSCl)} \xrightarrow{\text{Et}_3\text{N (base)}} \text{Butyltetramethylsilyl ether} + \text{Et}_3\text{NH}^+ \text{Cl}^- \quad (15\text{-}51)$$

Problem 15.18

Give major organic products or starting alcohols to complete the reactions below.

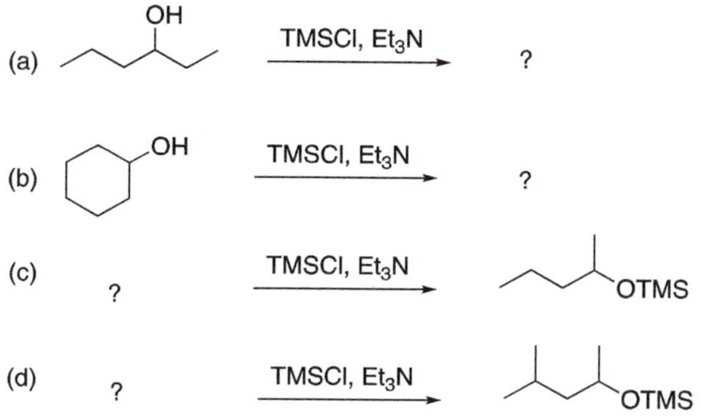

(a) CH₃CH₂CH(OH)CH(CH₃)CH₃ (2-methyl-3-pentanol-like structure with OH on carbon) + TMSCl, Et₃N → ?

(b) cyclohexyl-CH₂OH + TMSCl, Et₃N → ?

(c) ? + TMSCl, Et₃N → CH₃CH₂CH₂CH₂-OTMS (structure with OTMS)

(d) ? + TMSCl, Et₃N → (CH₃)₂CHCH(CH₃)-OTMS

The TMS group can be used as a protecting group so that another reaction can take place at another functionality of a difunctional molecule. We have seen protecting groups before in the use of ethylene glycol to protect aldehydes and ketones when carrying out a reaction elsewhere in a molecule. In this case, the TMS can be used to protect an alcohol so that another reaction can take place elsewhere in a molecule. Consider the difunctional molecule, 5-bromopentanol, which contains a good leaving group in the form of a bromine and a nucleophilic alcohol functionality. Thus, a likely reaction for this molecule is the intramolecular substitution reaction shown in 15-52.

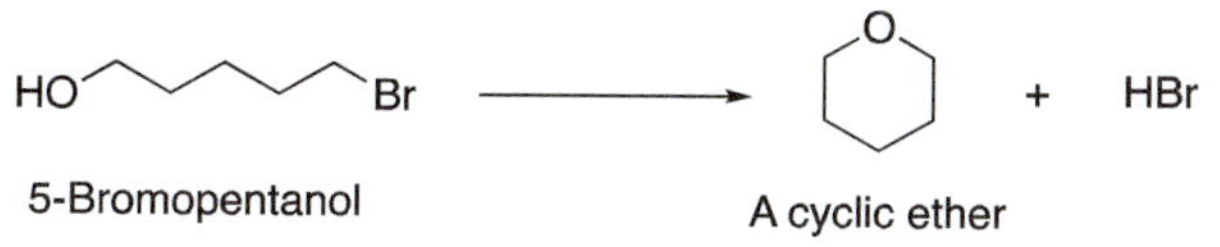

HO–(CH₂)₅–Br ⟶ (cyclic tetrahydropyran) + HBr   (15-52)

5-Bromopentanol   A cyclic ether

Another possible reaction that this molecule could be involved in is with another molecule depending on its concentration in the medium, as shown in Reaction (15-53).

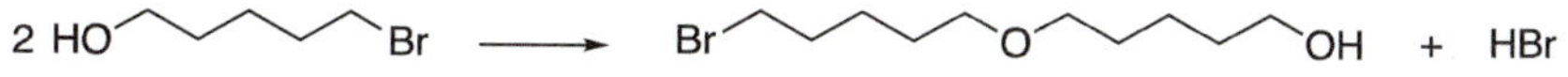

2 HO–(CH₂)₅–Br ⟶ Br–(CH₂)₅–O–(CH₂)₅–OH + HBr   (15-53)

Thus, if we were trying to carry out the reaction shown in 15-54, the result will be a mixture of unwanted products.

HO–(CH₂)₅–Br + K⁺ ⁻C≡N ⟶ HO–(CH₂)₅–C≡N + HBr   (15-54)

If you wanted to produce only the cyanide as the product and not the products like those obtained from Reactions (15-52) and (15-53), protection of the —OH functionality is required so that the required reaction can take place at the other functionality. This strategy of protecting an alcohol functionality and then its removal to produce a target molecule by a multi-step synthesis is shown below.

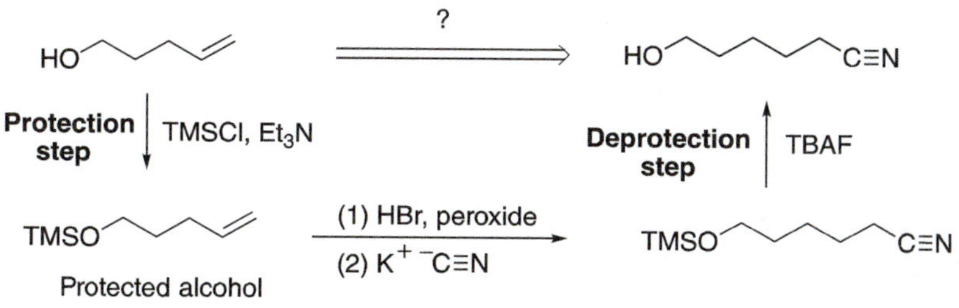

HO–CH₂CH=CH₂ ═══?═══> HO–(CH₂)₄–C≡N

Protection step | TMSCl, Et₃N ↓   Deprotection step ↑ TBAF

TMSO–CH₂CH=CH₂   —(1) HBr, peroxide→   TMSO–(CH₂)₄–C≡N
                  (2) K⁺ ⁻C≡N
Protected alcohol

The protected alcohol is not a good nucleophile and hence will not react with the bromide functionality. As a result, the added reagent, the cyanide nucleophile, will displace the leaving group (the bromide anion) to form the product shown at the bottom of the scheme. To obtain the desired product, the TMS protection group must be removed and tetrabutyl ammonium fluoride (TBAF) is used to accomplish that task. The explanation of this reaction is shown in Reaction (15-55).

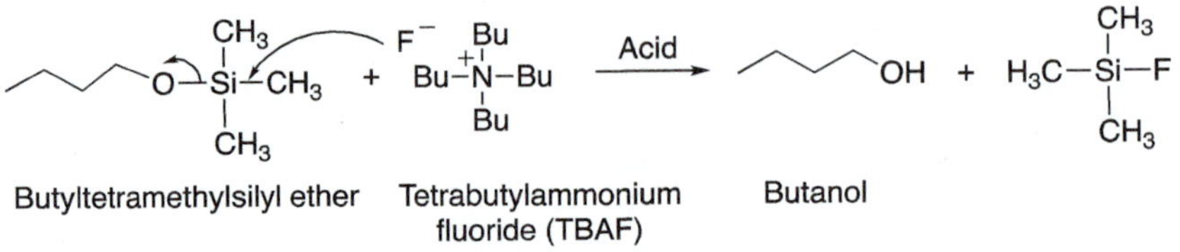

$$\text{Butyltetramethylsilyl ether} + \text{Tetrabutylammonium fluoride (TBAF)} \xrightarrow{\text{Acid}} \text{Butanol} + H_3C-\underset{\underset{CH_3}{|}}{\overset{\overset{CH_3}{|}}{Si}}-F \quad (15\text{-}55)$$

For the above reaction, the fluoride anion, being a good nucleophile, attacks the Si displacing the RO group, which is eventually protonated by the solvent or an acid in the reaction medium.

Problem 15.19

Show how to carry out the following transformations.

(a) [structure: HO-cyclohexyl-CH₂Cl → HO-cyclohexyl-CH₂-S-propyl]

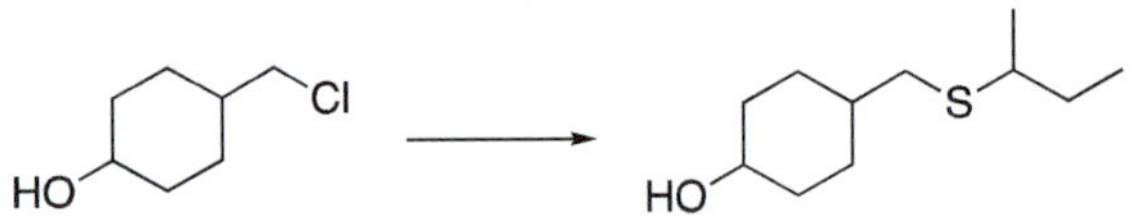

(b) [structure: OH-CH(CH₃)-CH₂-CH₂-Br → OH-CH(CH₃)-CH₂-CH₂-CN]

15.8.4 Synthesis of Alkynes

Alkynes that have the ability to form anions will react just as other nucleophiles described thus far. An example of the nucleophilic substitution reaction is shown in Reaction (15-56).

$$R-C\equiv C{:}^- \, Na^+ + R'-X \longrightarrow R-C\equiv C-R' + Na^+ \, X^{:-} \quad (15\text{-}56)$$

These reactions are S_N2 reactions and care must be exercised in the selection of an appropriate solvent. The acetylide anion can be made easily by an appropriate acid–base reaction to remove the acidic proton of a terminal alkyne. Since the acetylide anion is a base, a protic solvent, such as water cannot be used as a solvent. Nonprotic solvents, such as acetone or DMSO, are typically good solvents for the synthesis of substituted alkynes by this route. An example for the synthesis of substituted alkyne is shown in Reaction (15-57).

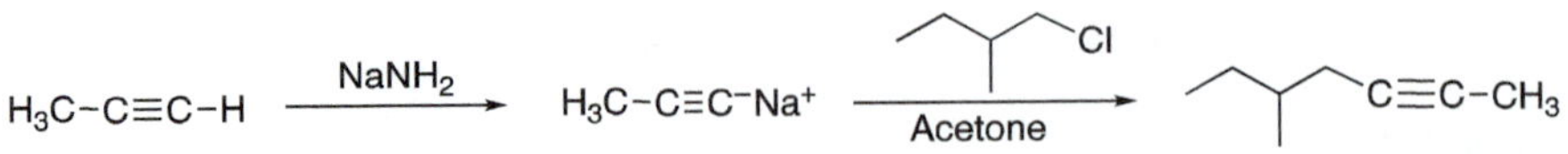

$$H_3C-C\equiv C-H \xrightarrow{NaNH_2} H_3C-C\equiv C^- Na^+ \xrightarrow[\text{Acetone}]{\text{CH}_3\text{CH}(Cl)\text{CH}_2\text{CH}_3} \text{product with } C\equiv C-CH_3 \quad (15\text{-}57)$$

Problem 15.20

Starting with any terminal alkyne, show how to synthesize the following molecules.

(a)  [structure] (b) 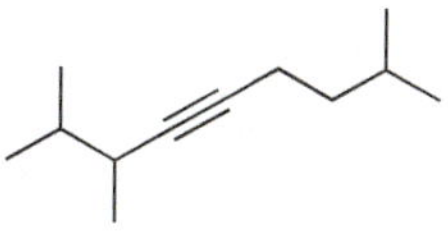 [structure] (c) 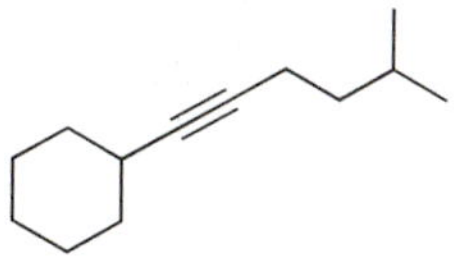 [structure]

15.8.5 Synthesis of α-Substituted Carbonyl Compounds

As we have seen in Chapter 7 on acids and bases, enolates are bases that are resonance stabilized and hence they are also of synthetic importance since they can also serve as nucleophiles for substitution reactions. Enolates react in a similar manner as any of the nucleophiles mentioned above, as shown in Reaction (15-58) in which there is a reaction with an enolate with an alkyl chloride to create a larger molecule with a new carbon–carbon bond in the α-position of the initial carbonyl compound.

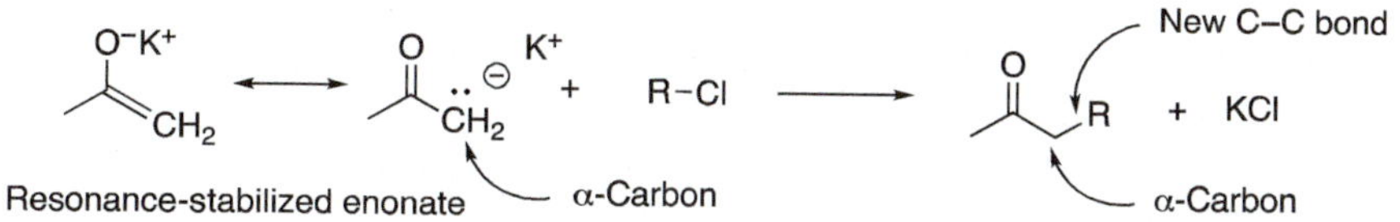

$$\text{Resonance-stabilized enonate} \quad + \quad R\text{-Cl} \quad \longrightarrow \quad \text{product} \quad + \quad KCl \quad (15\text{-}58)$$

A specific application of this type of reaction in which a carbonyl compound is transformed into a larger α-substituted compound is shown below in Reaction (15-59).

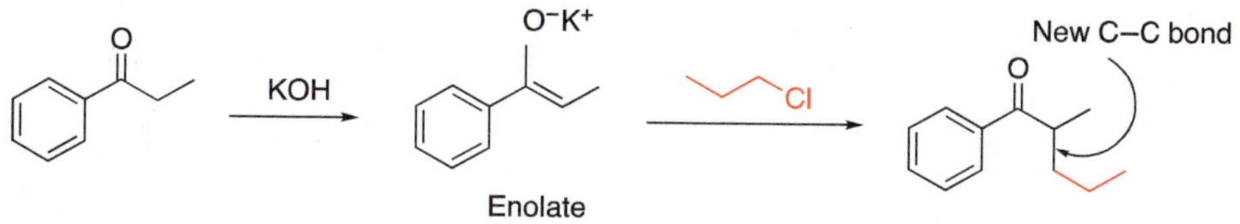

$$\text{(15-59)}$$

Problem 15.21

Starting with any carbonyl compound, show how to synthesize the following compounds. Consider making the bond in red.

(a) 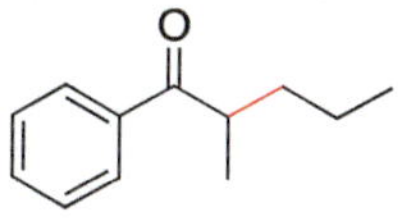 [structure] (b) 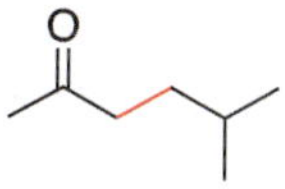 [structure] (c) 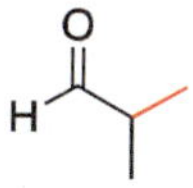 [structure]

The stability of enolates can be increased if the negative charge of the conjugate enolate is delocalized over a larger number of atoms. Diethyl malonate (malonic ester) is one such example and it is a fairly acidic molecule. These compounds are of importance in the synthesis of substituted acetic acids. Shown in Reaction (15-60) is an example of a reaction in which malonic ester is used as a nucleophile after deprotonation to synthesize a larger molecule.

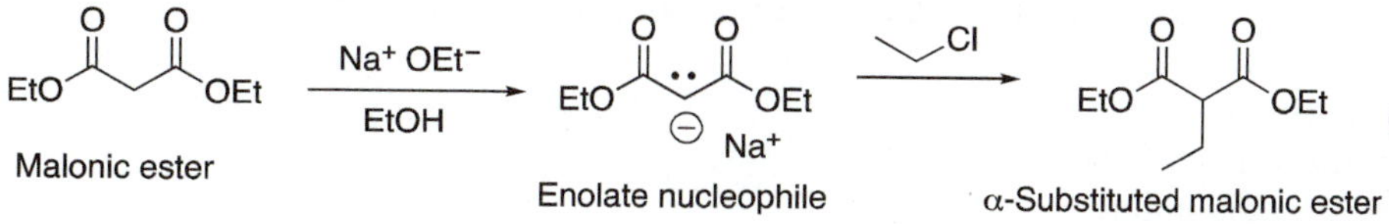

$$\text{Malonic ester} \xrightarrow{\text{Na}^+ \text{OEt}^-, \text{EtOH}} \text{Enolate nucleophile} \xrightarrow{\text{Cl}} \alpha\text{-Substituted malonic ester} \quad (15\text{-}60)$$

15 Nucleophilic Substitution Reactions at sp³ Carbons

Note that the α-substituted malonic ester that is produced by this reaction has an α-hydrogen that can be abstracted to form a new nucleophile, which can undergo a similar reaction as shown in Reaction (15-61) to make an even larger and more complex molecule.

$$\text{EtO-CO-CH(Et)-CO-OEt} \xrightarrow{\text{Na}^+ \text{OEt}^-, \text{EtOH}} \text{EtO-CO-C}^-(\text{Et})\text{-CO-OEt} \xrightarrow{\text{CH}_3\text{Cl}} \text{EtO-CO-C(Et)(CH}_3\text{)-CO-OEt} \quad (15\text{-}61)$$

In later chapters, we will learn further transformations that these molecules can undergo to eventually synthesize molecules as the target molecule shown in Reaction (15-62).

$$\text{EtO-CO-C(Et)(CH}_3\text{)-CO-OEt} \xrightarrow{?} \xrightarrow{?} \text{HO-CO-CH(CH}_3\text{)-CH}_2\text{CH}_3 \quad (15\text{-}62)$$

Problem 15.22

Show how to carry out the following transformation, clearly show all reagents in your reaction sequence

EtO-CO-CH₂-CO-OEt (Malonic ester) $\xrightarrow{?}$ EtO-CO-CH(CH₂Ph)-CO-OEt

In summary, when designing a synthetic route for a target molecule, you must first identify the functional group in the target molecule and determine the bond of the functional group that must be made. Once you have made that identification, you should next devise a route to make that bond using an appropriate electrophile and also an appropriate nucleophile. Finally, review your synthesis to make sure that you have made a strategic choice. That is, if you want the reaction to proceed via an S_N1 route, then you must make sure of appropriate solvent and type of electrophile. If you want the reaction to proceed via an S_N2 route based on perhaps a required stereochemistry, then you will have to make sure that the solvent is nonionizing, the nucleophile is small and not very basic and the electrophile is primary or secondary, but never tertiary.

End of Chapter Problems

15.23 Classify each of the following alkyl halides as methyl, primary, secondary, or tertiary.

(a) isobutyl-Br   (b) n-propyl-Cl   (c) 1-chloro-1-methylcyclobutane   (d) cyclopentyl-Cl

15.24 Using no more than five carbons, give the structures and IUPAC names of the following type of molecules.
  a) Primary halide
  b) Secondary halide
  c) Tertiary halide

15.25 Give the structure and IUPAC names for all isomers that have the formula $C_5H_{11}Cl$. Also, classify each as primary, secondary, or tertiary.

15.26 For each over-all substitution reaction below, identify the electrophilic atom in the organic substrate, the nucleophile and its nucleophilic atom, and the leaving group.
a) $CH_3CH_2Br + Na^+ \ ^-:SC_2H_5 \rightarrow (CH_3CH_2)_2S + Na^+ \ Br^-$
b) $C_6H_5CH_2O_2CH_3 + Na^+ \ ^-:CN \rightarrow C_6H_5CH_2CN + CH_3CO_2^- \ Na^+$
c) $CH_3CH_2OH + HI_{(excess)} \rightarrow CH_3CH_2I + H_2O$
d) $CH_3CH_2Br + :NH_2CH_3 \rightarrow CH_3CH_2N^+HCH_3 \ Br^-$
e) $CH_3CH_2I + :P(C_2H_5)_2 \rightarrow (CH_3CH_2)_3P^+ + I^-$

15.27 Of the following pairs of alkyl halides, which is the most reactive by an S_N2 mechanism?
a) CH_3Cl and CH_3CH_2Cl
b) CH_3Cl and CH_3Br
c) $CH_3CH_2CH_2Cl$ and $(CH_3)_2CHCl$
d) $(CH_3)_3CCl$ and $(CH_3)_2CHCl$

15.28 Of the following pairs of alkyl halides, which is the most reactive by an S_N1 mechanism?
a) CH_3Cl and $CH_3CH(CH_3)Cl$
b) $CH_3CH(CH_3)Cl$ and $CH_3CH(CH_3)Br$
c) $CH_3CH_2CH_2Cl$ and $CH_3CH(CH_3)Cl$
d) $(CH_3)_3CCl$ and $(CH_3)_2CHCl$

15.29 $(CH_3)_3CCH(Br)CH_2CH_3$ undergoes $S_N1/E1$ reactions in aqueous solvent, give the structures of the products that are formed.

15.30 What is the principal organic product of the E2 reaction of (1R, 2S)-1-iodo-1,2-diphenylbutane in the presence of hydroxide ions (HO^-)?

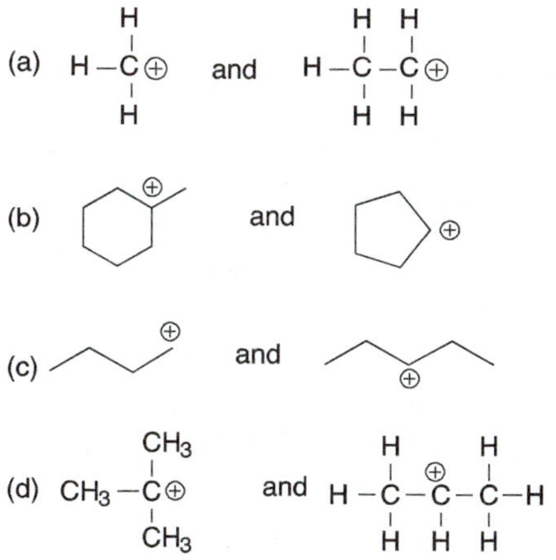

15.31 Of the following pairs of carbocation, which is the most stable?

15.32 Show how to carry out the transformation shown below. Clearly indicate different steps in your synthesis, solvents, reaction conditions and any other reagents used for your synthesis.

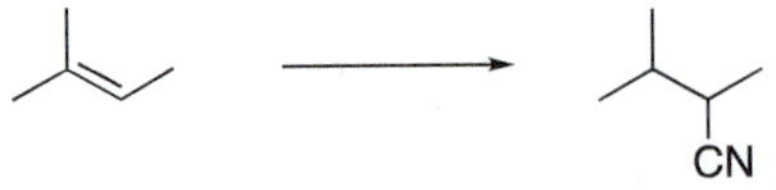

15.33 In an aqueous solution, I⁻ catalyzes the S_N2 reaction shown below.

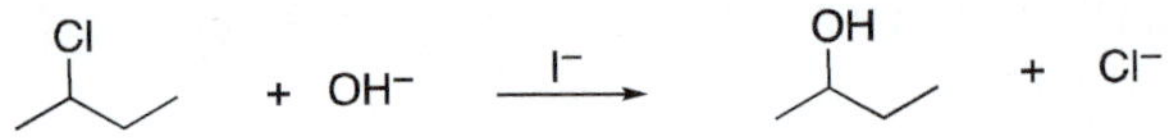

The catalysis depends upon I⁻ being a better nucleophile than OH⁻ and a better leaving group than Cl⁻.
a) Give a two-step sequence for the I⁻ catalysis reaction.
b) Starting with (R)-2-chlorobutane, show the stereochemistry of both steps of your mechanism.

15.34 Suggest an alkyl halide and appropriate nucleophile that would react to form the products shown.

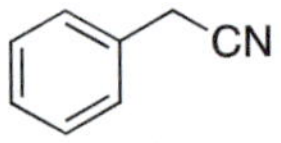

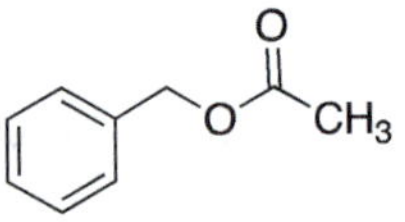

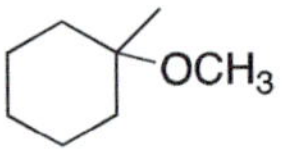

15.35 For the following reaction, explain (with the aid of an appropriate mechanism) the formation of each of the products shown.

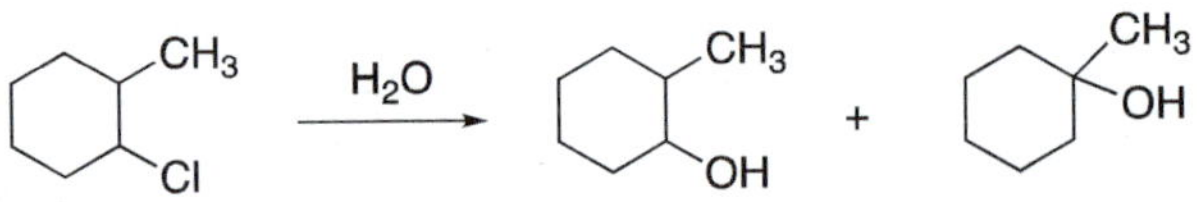

15.36 Predict the *major* product for each of the following reactions (include stereochemistry where appropriate).

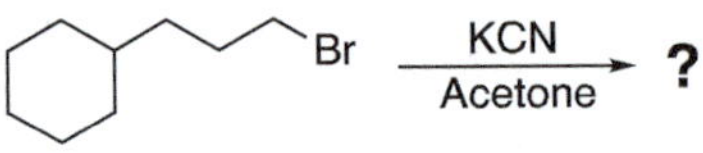

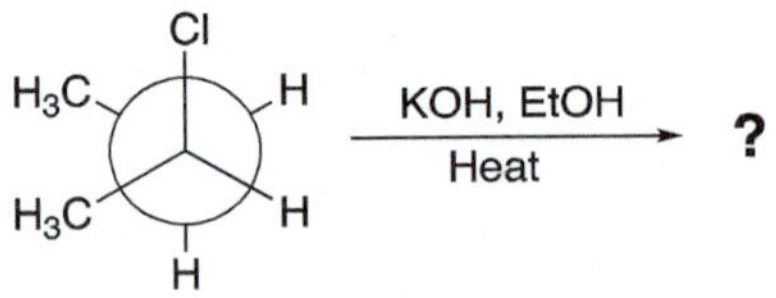

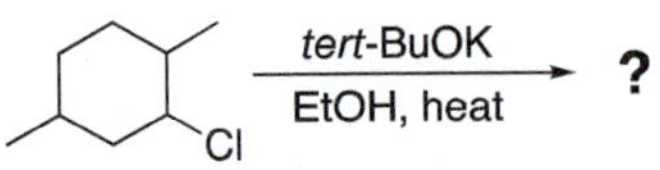

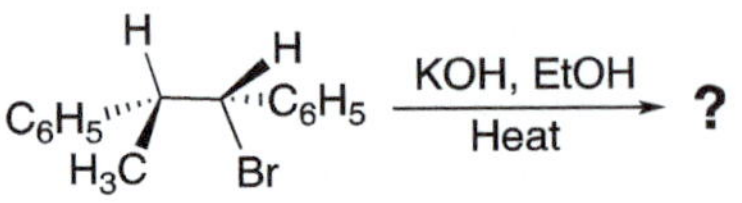

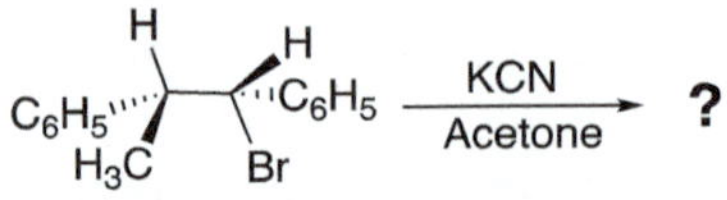

(g) 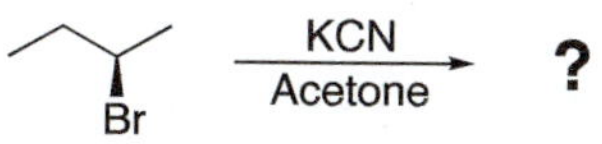 [sec-butyl bromide] —KCN/Acetone→ ?   (h) 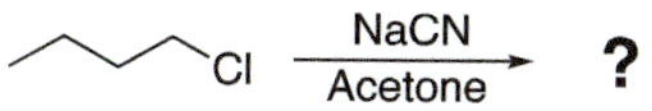 [1-chloropropane] —NaCN/Acetone→ ?

(i) 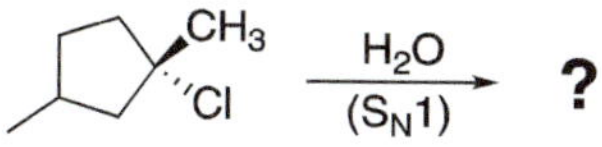 [cis-1-chloro-3-methylcyclopentane with Cl wedge] —H₂O (S$_N$1)→ ?   (j) 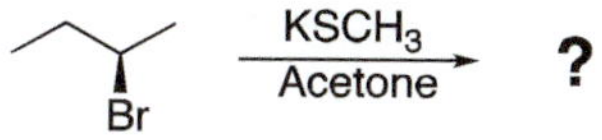 [isopropyl bromide] —KSCH₃/Acetone→ ?

15.37 Complete the following reactions by giving the *structures* of the reactant, missing reagents above the arrow or *major organic product*. Note: some transformations may require more than one step and these steps must be indicated clearly. Indicate stereochemistry where appropriate.

(i) 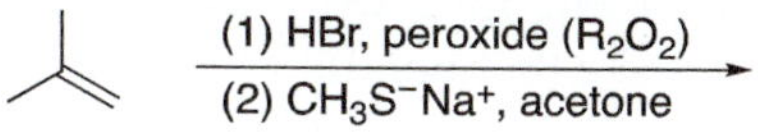 [isobutylene] $\xrightarrow{\text{(1) HBr, peroxide (R}_2\text{O}_2\text{)}}{\text{(2) CH}_3\text{S}^-\text{Na}^+\text{, acetone}}$

(ii) 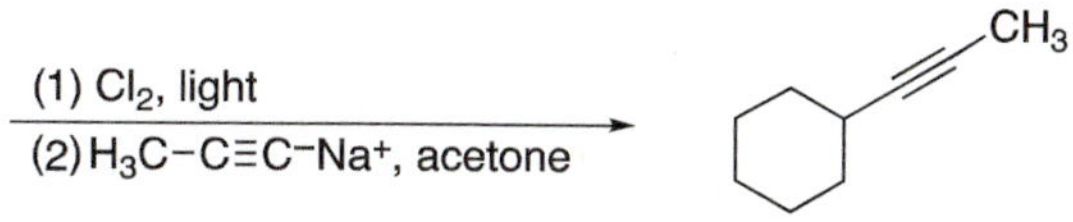 [ethylcyclohexane] $\xrightarrow{\text{(1) Cl}_2\text{, light}}{\text{(2) H}_3\text{C-C≡C-Na}^+\text{, acetone}}$ [cyclohexyl-C≡C-CH₃]

(iii) 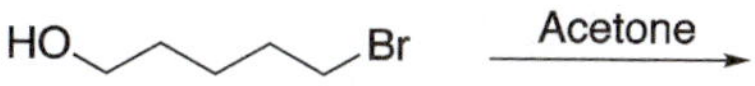 HO~~~~~Br —Acetone→
(hint: consider an intramolecular S$_N$2 substitution reaction)

(iv) 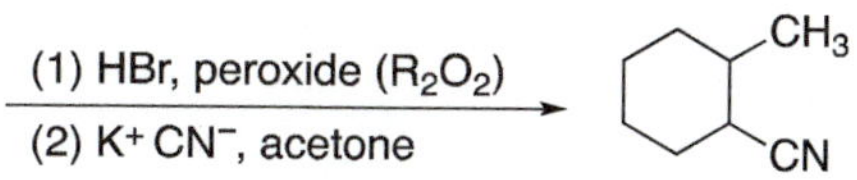 $\xrightarrow{\text{(1) HBr, peroxide (R}_2\text{O}_2\text{)}}{\text{(2) K}^+\text{ CN}^-\text{, acetone}}$ [methylcyclohexane with CN]

(v) 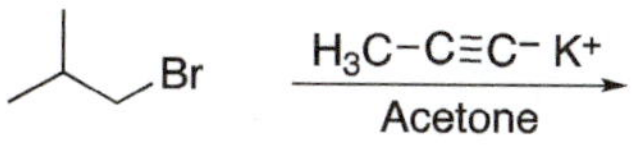 [isobutyl bromide] —H₃C-C≡C-K⁺ / Acetone→

(vi) 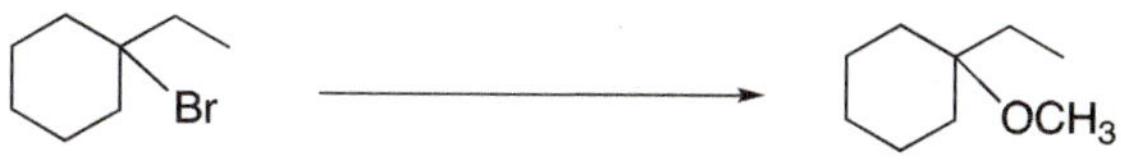 [1-bromo-1-ethylcyclohexane] ——→ [1-ethyl-1-methoxycyclohexane]

(vii) 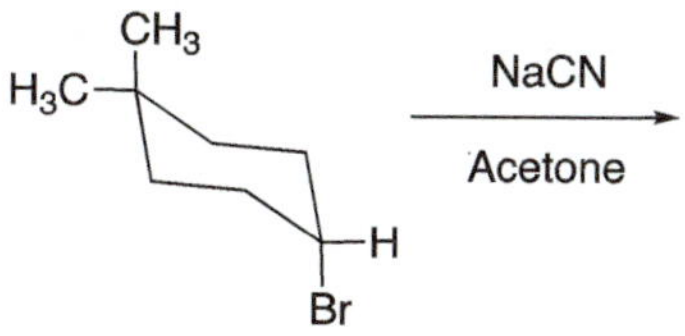 [trans-1,1-dimethyl...cyclohexane with axial Br and H] —NaCN / Acetone→

(viii) ——NaI, acetone——→ [Fischer-like: I up, H₃C back wedge, H, CH₂CH₃; central C with I on top]

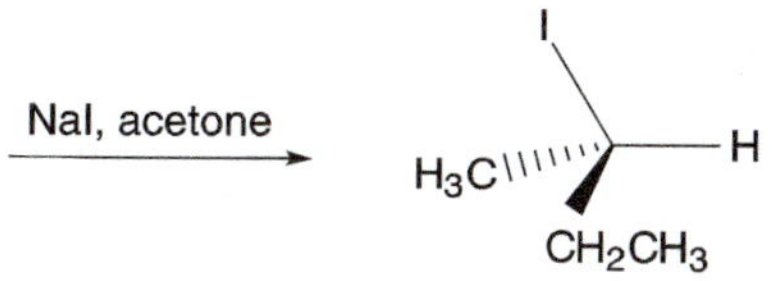

(ix) (S)-2-Bromopentane —KCN/Acetone→

15.38 Rearrangement of carbocation intermediates often occurs for S_N1 reactions. With this information in mind, explain by appropriate *mechanisms* (i.e., step-by-step descriptions and curved arrows to indicate electron movements) the formation of the organic products shown for the following reaction.

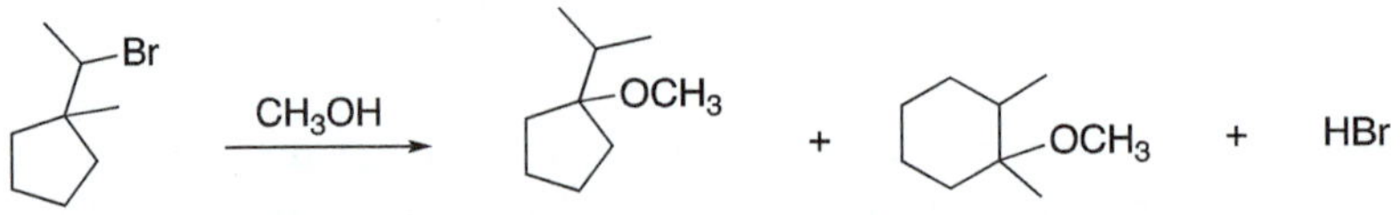

[cyclopentane with CH(Br)CH(CH₃)₂ substituent] $\xrightarrow{CH_3OH}$ [cyclopentane with CH(OCH₃)CH(CH₃)₂] + [cyclohexane with OCH₃ and CH₃] + HBr

15.39 Show how to carry out the following transformations. For each step in your synthesis, *clearly* show all reagents and reaction conditions. (Hint: identify the bond of the functional group in the product that must be made, then work backward to identify an appropriate molecule and reaction type to make that bond).

(a) Any saturated hydrocarbon $\longrightarrow$ [2-methyl-2-butanol structure with H₃C and OH]

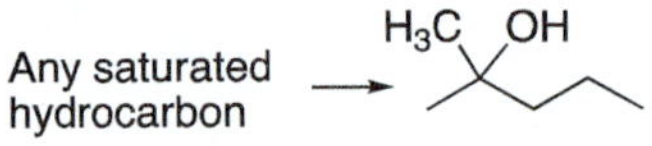

Hint: Consider using a free radical substitution reaction in your synthesis.

(b) [methylcyclohexane] $\longrightarrow$ [1-methyl-1-methoxycyclohexane]

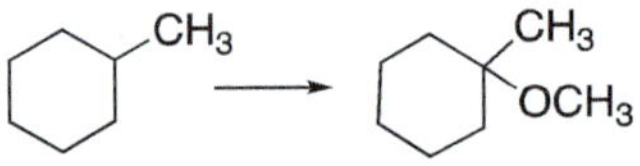

Hint: Consider using a free radical substitution reaction in your synthesis.

(c) Butane $\longrightarrow$ [(CH₃)₂CH–C≡CCH₃]

Hint: Consider using a S_N2 reaction in your synthesis.

(d) Any saturated hydrocarbon $\longrightarrow$ [branched structure with SCH₃]

Hint: Consider adding a leaving group to an alkene in an anti-Markovnikov addition.

16

Nucleophilic Substitution Reactions at Acyl Carbons

16.1 Introduction

In the previous chapter, we examined substitution reactions that involve the substitution at sp^3 carbons. In this chapter, we will examine substitution reactions that take place at sp^2 hybridized carbons, and specifically acyl carbons. You will recall that the acyl group is one that has a carbon–oxygen double bond. For substitution reactions that take place at sp^2 hybridized carbons, one atom (or group of atoms) bonded to the acyl carbon in the reactant is substituted for another atom (or group at atoms) in the product. Let us examine the structure of compounds that have sp^2 carbons and will undergo these types of substitution reactions. These reactions typically occur at sp^2 carbons that are bonded to an electronegative oxygen atom with a double bond in the form of a carbonyl carbon and also contain a good leaving group. In the last sections of the chapter, we will examine substitution reactions that take place with molecules that have other atoms, such as a sulfur, instead of a carbon that is bonded to an oxygen. Shown in Reaction (16-1) is the generic substitution reaction for molecules capable of undergoing an acyl substitution.

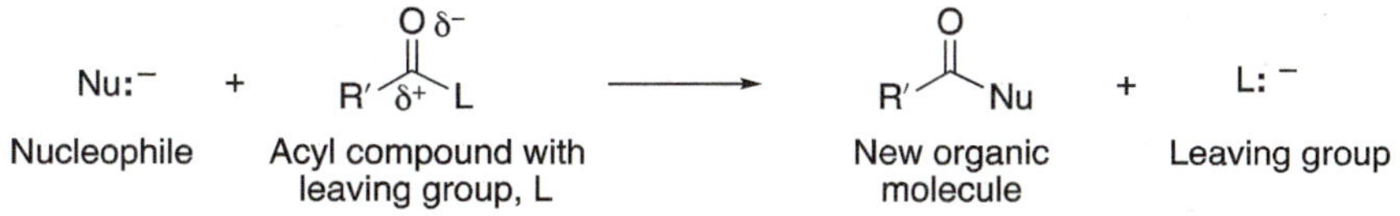

$$\text{Nu:}^- \ + \ \underset{\substack{\text{Acyl compound with} \\ \text{leaving group, L}}}{\overset{\overset{\displaystyle O\ \delta-}{\|}}{R'\overset{\delta+}{-}L}} \ \longrightarrow \ \underset{\substack{\text{New organic} \\ \text{molecule}}}{\overset{\overset{\displaystyle O}{\|}}{R'-Nu}} \ + \ \underset{\text{Leaving group}}{L:^-} \quad (16\text{-}1)$$

Note that the acyl carbon is double bonded to a very electronegative oxygen atom, which makes the covalent bond a polar covalent bond with the bonding electrons more closely associated with the electronegative oxygen, rather than the carbon. As a result, the carbon is partially positive, making it electrophilic and can react with a nucleophile. As we will see in this chapter, the leaving groups are similar to the ones that we have seen before; likewise, the nucleophiles are much the same as those covered in the previous chapter.

Problem 16.1

Analyze the following reactions carefully and predict the substitution products. (Note that if there is an acidic medium for the reaction, it could convert a poor leaving group, such as $-NH_2$ and $-OH$ to a much better leaving group, such as $-NH_3^+$ $-OH_2^+$, respectively).

(a) Na⁺ OH⁻ + CH₃C(O)Br ⟶ ?

(b) H₂O + C₆H₅C(O)NH₂ —HCl→ ?

(c) CH₃OH + C₆H₁₁C(O)Cl ⟶ ?

(d) CH₃NH₂ + CH₃C(O)Cl ⟶ ?

16.2 Mechanism for Acyl Substitution

For the general reaction given in Reaction (16-1), the substitution takes place at the acyl carbon and the leaving group, L, is replaced with the nucleophile, Nu. These types of reactions occur in two steps, in which the first step involves the attack of the nucleophile on the electrophilic carbon of the electrophile, converting it to a sp^3 carbon, which is also called a *tetrahedral intermediate*, as shown in Reaction (16-2).

$$\text{Nu:}^- + \underset{\text{Electrophile}}{\text{R-C(=O)-L}} \longrightarrow \underset{\text{Tetrahedral intermediate}}{\text{R-C(O}^-\text{)(Nu)-L}} \quad (16\text{-}2)$$

Since the tetrahedral intermediate contains a leaving group and there are lone pairs of electrons on the negatively charged oxygen, a favorable double bond will re-form with the carbon and the oxygen. Simultaneously, the leaving group leaves with its bonding electrons as shown in Reaction (16-3).

$$\underset{\text{Tetrahedral intermediate}}{\text{R-C(O}^-\text{)(Nu)-L}} \longrightarrow \underset{\text{Substitution product}}{\text{R-C(=O)-Nu}} + \underset{\text{Leaving group}}{\text{L:}^-} \quad (16\text{-}3)$$

From the above general mechanism, the nucleophile first uses its unshared pair of electrons to bond to the electrophilic carbon of the acyl compound to form a tetrahedral intermediate. This intermediate is unstable since it contains a leaving group and has the possibility of reforming a very stable carbon–oxygen double bond, the overall reaction mechanism is shown in Reaction (16-4).

$$\text{Nu:}^- + \underset{\text{Electrophile}}{\text{R-C(=O)-L}} \longrightarrow \underset{\text{Tetrahedral intermediate}}{\text{R-C(O}^-\text{)(Nu)-L}} \longrightarrow \underset{\text{Substitution product}}{\text{R-C(=O)-Nu}} + \underset{\text{Leaving group}}{\text{L:}^-} \quad (16\text{-}4)$$

Before we actually start our examination of these substitution reactions, let us first refresh our understanding of leaving groups, electrophile and the nucleophile. The next sections examine the reaction of various acyl compounds with different nucleophiles.

16.2.1 The Leaving Group of Acyl Substitution Reactions

As we have pointed out in the previous chapter, one of the major factors for a substitution reaction is the ability of the leaving group to leave. For these reactions, the leaving group departs from the tetrahedral intermediate after the electrons from the oxygen return to re-form a carbon–oxygen double bond. The ability of a leaving group to leave is reflected by its ability to stabilize the negative charge after leaving. Weak conjugate bases are good leaving groups, while strong bases are poor leaving groups. Based on the pK_a values shown in Table 7.1 of Chapter 7, the leaving group ability of a specific atom or group of atoms can be determined. Recall that strong acids, which have more negative pK_a values result in conjugate bases that are weak conjugate bases, hence good leaving groups. For acyl substitution reactions that will be covered in this chapter, the groups that are bonded to the acyl carbon and are potential leaving groups as shown below.

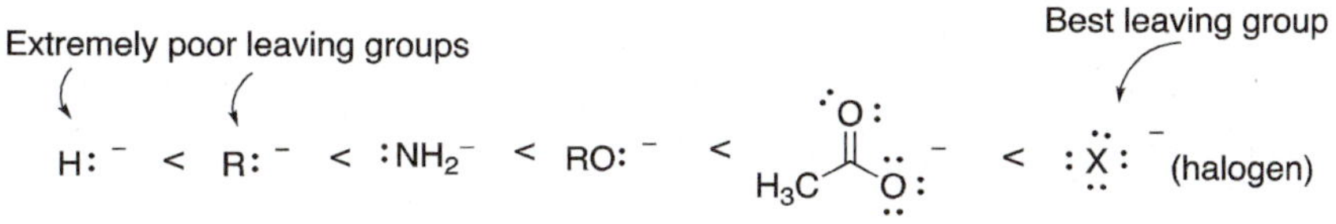

Extremely poor leaving groups

Best leaving group

$$H:^- \; < \; R:^- \; < \; :NH_2^- \; < \; RO:^- \; < \; H_3C-C(=O)-O:^- \; < \; :\ddot{X}:^- \text{ (halogen)}$$

This trend implies that some of the most reactive acyl compounds are the ones in which the carbonyl carbon is bonded to a halogen and the least reactive are acyl compounds that have hydrogen or alkyl groups bonded to the carbonyl carbon. In this chapter, we will examine compounds that have the following leaving groups: $-Cl^-, -Br^-, -OR^-, -OH^-, -O_2CR^-, -NH_3^+, -OH_2^+$. Note that the last two leaving groups of the list are protonated leaving groups. As we have discussed in Chapter 7, the H_2O^+ is a weaker base than OH^- and the same is true for NH_3^+, compared to NH_2.

16.2.2 Reactivity of Electrophiles of Acyl Substitution Reactions

As you can imagine, the size of the R group bonded to the electrophilic carbon that contains the leaving group will affect the stability of the tetrahedral intermediate since the ideal bond angles about a sp^3 carbon are 109.5°. As a result, large alkyl R groups that are bonded to the carbonyl carbon of the acyl functionality will serve to destabilize the tetrahedral intermediate and small R groups will offer more stability to the tetrahedral intermediate. This factor will dictate the reactivity of the acyl compound. For example, acyl compounds where R is hydrogen is more reactive than an acyl compound in which the R group is a *tert*-butyl $-C(CH_3)_3$ group. A trend for the reactivity of different acyl compounds is shown below.

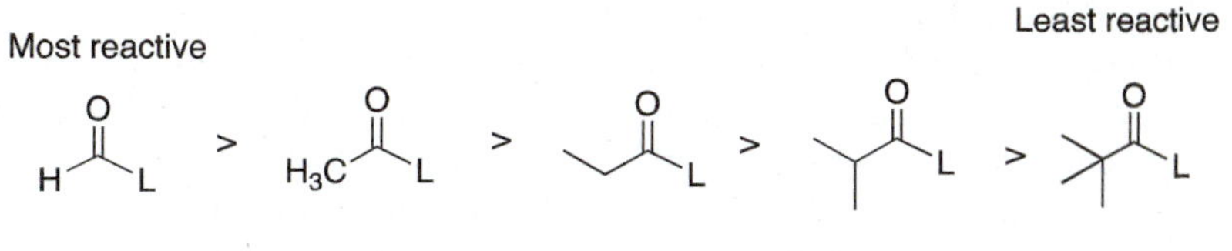

Most reactive   **Least reactive**

H–C(=O)–L > H$_3$C–C(=O)–L > CH$_3$CH$_2$–C(=O)–L > (CH$_3$)$_2$CH–C(=O)–L > (CH$_3$)$_3$C–C(=O)–L

Reactivity of acyl compounds toward substitution reactions

Another factor that affects the reactivity of acyl compounds toward a substitution reaction is the electronegativity of the group bonded to the acyl carbon. A very electronegative atom or group makes the acyl carbon more electrophilic since it is already bonded to the electronegative oxygen. As a result, acyl compounds that contain an electronegative atom or groups bonded to the acyl carbon are more reactive than acyl compounds that do not have an electronegative atom or group bonded to the acyl carbon. Examples of different acyl compounds that demonstrate this concept are shown below.

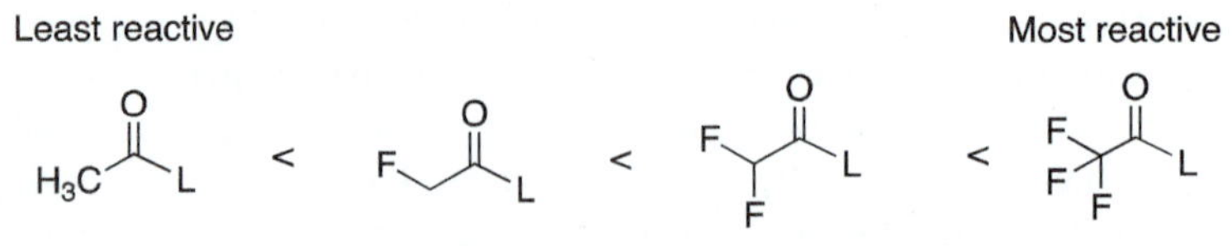

Least reactive   Most reactive

$$H_3C-C(=O)-L \; < \; F-CH_2-C(=O)-L \; < \; F_2CH-C(=O)-L \; < \; F_3C-C(=O)-L$$

Reactivity of acyl compounds toward substitution reactions

Problem 16.2

Of the following pairs of acyl compounds, which is more reactive toward substitution reactions?

(a) $H_3C-C(=O)-Cl$ and $F_3C-C(=O)-Cl$

(b) $F_3C-C(=O)-Cl$ and $F_3C-C(=O)-OCH_3$

(c) $(CH_3)_2CH-C(=O)-Cl$ and $H_3C-C(=O)-Cl$

(d) $H_3C-C(=O)-Cl$ and $(CH_3)_2CH-C(=O)-OCH_3$

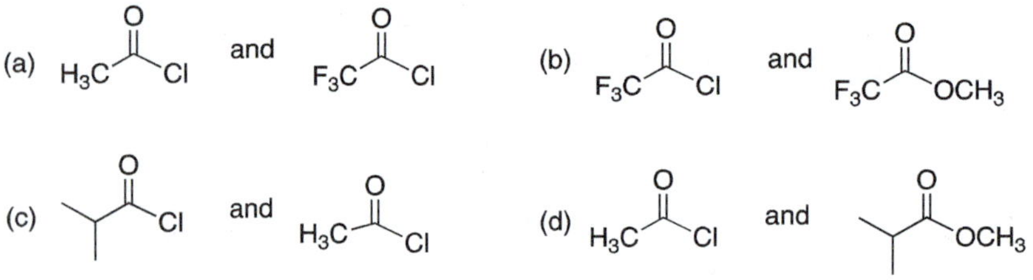

16.2.3 Nucleophiles of Acyl Substitution Reactions

As we have seen from the previous chapter, a nucleophile is defined as a species that has at least one unshared pair of electrons, which as you will recall is the same definition as a Lewis base. Nucleophiles are nucleus-loving species. As we have seen, nucleophiles can be neutral or can have a formal negative charge, and those that we will be examining in this chapter include the following: OH^-, H_2O; NH_2^-, NH_3; RO^-, and ROH (alcohols), which all have at least one lone-pair of electrons.

16.3 Substitution Reactions Involving Acid Chlorides

As we have seen above, alkanoyl chlorides (also known as acid chlorides) are one of the most reactive acyl compounds, and as a result, we will start our discussion of substitution reactions by looking at the substitution reactions of acid chlorides with various nucleophiles. Acid chlorides are very reactive molecules owing to the presence of a very good leaving group, the chloride anion (Cl^-). Acid chlorides react readily with nucleophiles to form various products, including carboxylic acids, esters, and amides. The general mechanism for the reaction of an acid chloride with a nucleophile is shown in Reaction (16-5).

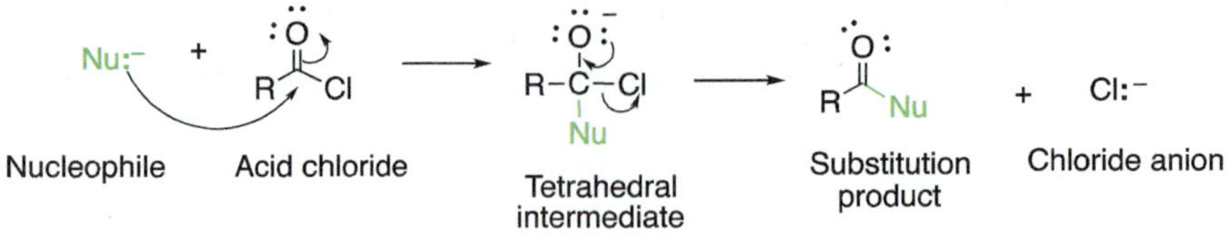

$$Nu:^- + R-C(=O)-Cl \longrightarrow R-C(O^-)(Nu)-Cl \longrightarrow R-C(=O)-Nu + Cl:^- \quad (16\text{-}5)$$

Nucleophile   Acid chloride   Tetrahedral   Substitution   Chloride anion
                                  intermediate   product

16.3 Substitution Reactions Involving Acid Chlorides

The first step of the reaction mechanism is an addition reaction and the second step is an elimination reaction of the leaving group. The various nucleophiles that will be examined in this section include: H_2O, NH_3, ROH, ROOH, $LiAlH_4$ (hydride ion) and organometallic reagents. The products of the reaction of these nucleophiles with acid chlorides give molecules that have different functionalities.

16.3.1 Substitution Reactions Involving Acid Chlorides and Water

Acid chlorides are very reactive and react readily with water. Owing to the moisture on our bodies, acid chlorides will cause severe irritation when exposed to these compounds. Special precautions must be taken when working with acid chlorides in the laboratory. Acid chlorides, if left open in the lab, will react with moisture in the atmosphere to form the corresponding carboxylic acid. An example of the reaction of ethanoyl chloride and water is shown in Reaction (16-6), these reactions are also referred to as hydrolysis reactions of acid chlorides.

$$H_3C-COCl + H_2O \longrightarrow H_3C-COOH + HCl \qquad (16\text{-}6)$$

Ethanoyl chloride → Ethanoic acid

As we have pointed out earlier, the size of the alkyl group bonded to the carbonyl carbon will dictate the reactivity of the acid chloride, and acid chlorides that have large groups react slower than those with smaller groups. Large groups destabilize the tetrahedral intermediate compared to smaller groups. Also, electronegative groups bonded to the carbonyl carbon will increase the reactivity of these types of compounds. The mechanism for the hydrolysis of ethanoyl chloride is shown in Reaction (16-7).

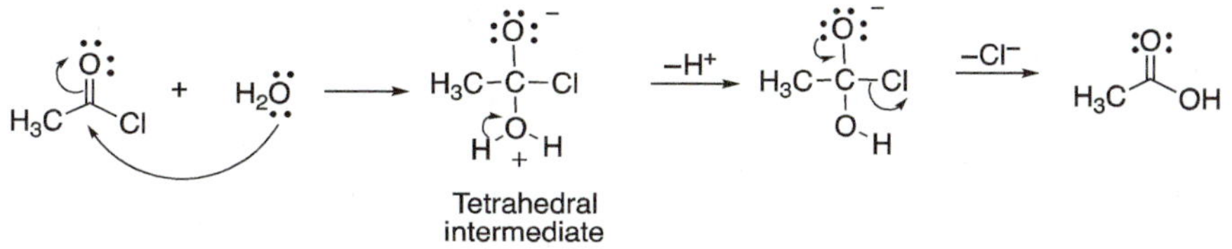

$$(16\text{-}7)$$

Tetrahedral intermediate

Problem 16.3

i) Give the products of the following reactions.

(a) (4-methylpentanoyl chloride) + H_2O → ?

(b) (cyclopentanecarbonyl chloride) + H_2O → ?

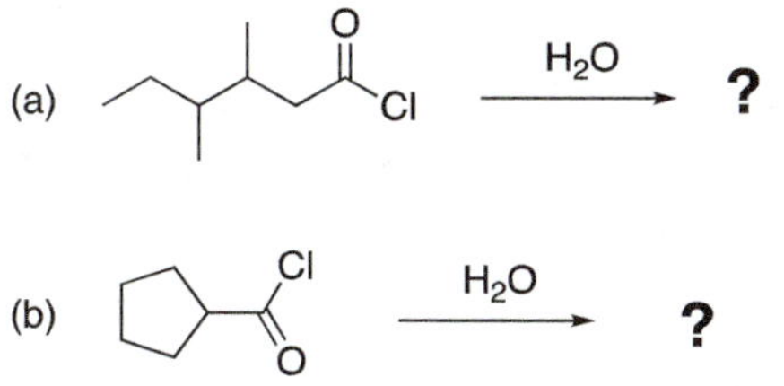

ii) Give the structure of the acid chloride reactant needed to complete the following reactions.

(a) ? + H_2O → (3-methylbutanoic acid)

(b) ? + H_2O → (benzoic acid)

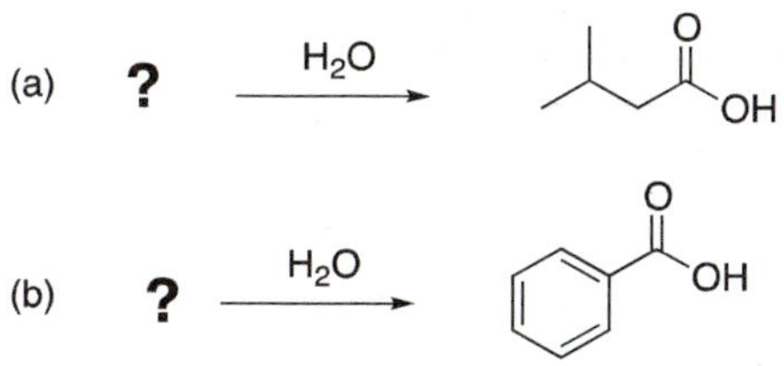

iii) Arrange the following acid chlorides in order of their reactivity (consider carefully the size of the alkyl groups).

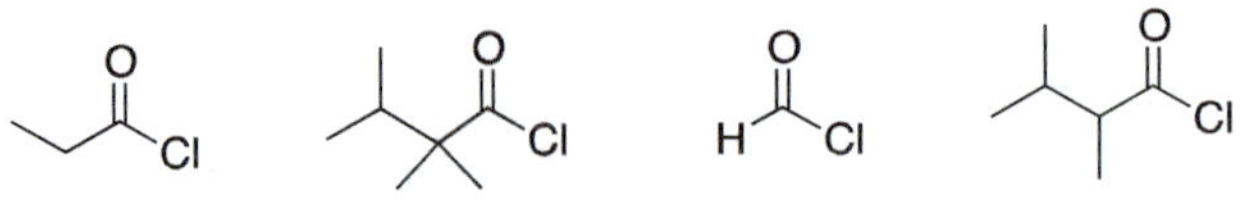

16.3.2 Substitution Reactions Involving Acid Chlorides and Alcohols

The reaction of ethanoyl chloride and methanol is shown in Reaction (16-8). Note that in this case, the product is not a carboxylic as we saw in the reaction with water, but a different functional group, an ester.

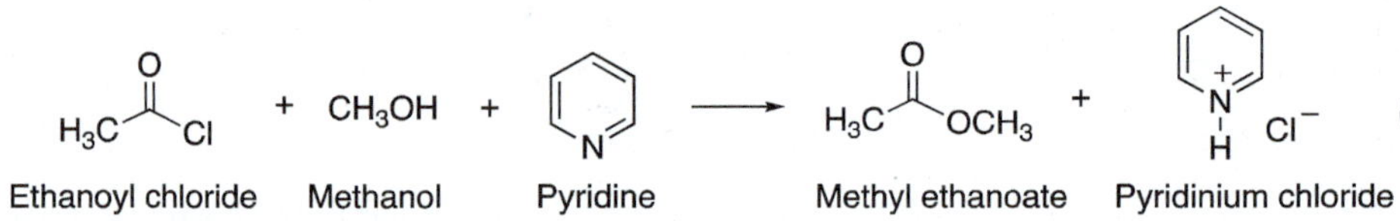

$$H_3C-COCl + CH_3OH + \text{Pyridine} \longrightarrow H_3C-COOCH_3 + \text{Pyridinium chloride} \quad (16\text{-}8)$$

Ethanoyl chloride   Methanol   Pyridine   Methyl ethanoate   Pyridinium chloride

The hydrochloric acid that is liberated is usually scavenged by a mild base, which produces a salt that is precipitated out from the reaction solution and helps drive the reaction to the product. Pyridine is typically used as the base for these reactions. As you can imagine, the mechanism for these reactions is similar to that of hydrolysis of acid chlorides, except these reactions involve alcohols. In the first step of the mechanism, the alcohol attacks the carbonyl carbon to form the tetrahedral intermediate. Pyridine, being a base, abstracts the proton from the protonated alcohol to form protonated pyridine, as shown in Reaction (16-9)

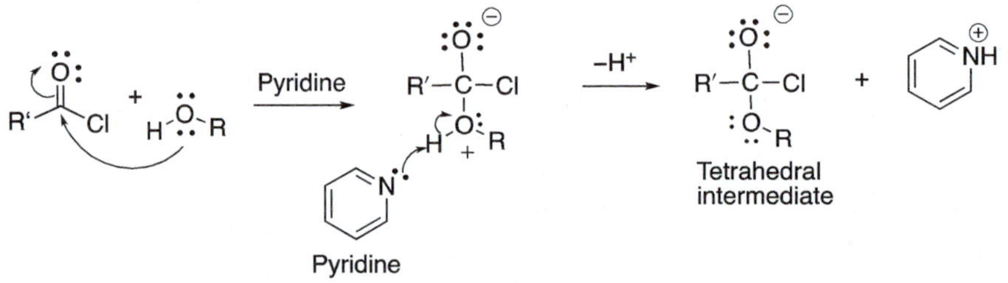

$$(16\text{-}9)$$

Tetrahedral intermediate

In the final step of the mechanism, the oxygen–carbon double bond forms while the chloride anion leaves, as shown in Reaction (16-10).

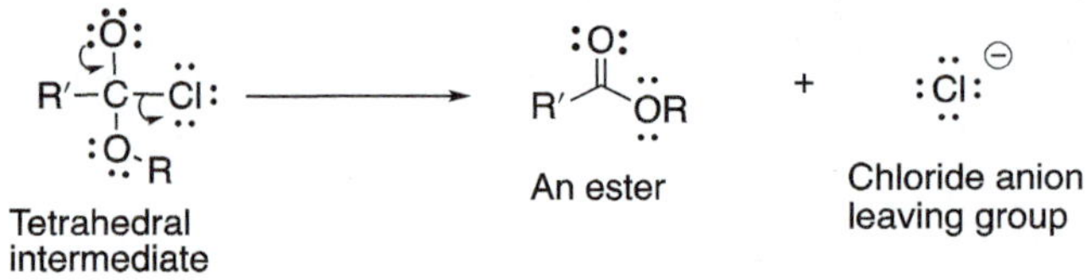

$$(16\text{-}10)$$

Tetrahedral intermediate   An ester   Chloride anion leaving group

Problem 16.4

i) Give the major organic products of the following reactions.

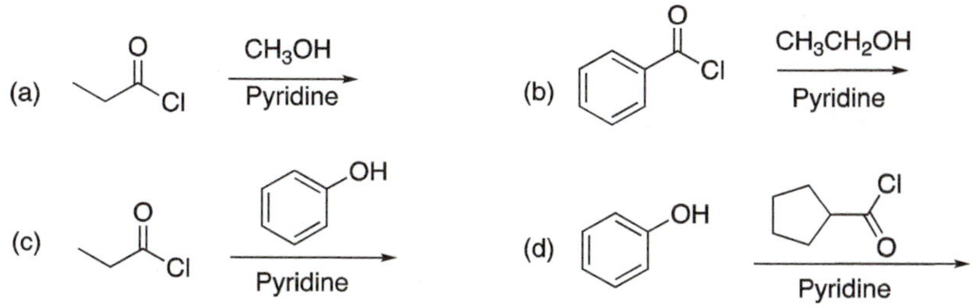

(a) CH₃COCl + CH₃OH → (Pyridine)

(b) C₆H₅COCl + CH₃CH₂OH → (Pyridine)

(c) CH₃COCl + phenol (PhOH) → (Pyridine)

(d) PhOH + cyclopentanecarbonyl chloride → (Pyridine)

ii) Give a reasonable mechanism for the following reaction.

Phenol + Propanoyl chloride + pyridine → Ester (phenyl propanoate) + pyridinium chloride

As expected, acid chlorides that have bulky R groups are less reactive than acid chlorides with less bulky R groups toward the reaction with alcohols, and acid chlorides that have electronegative groups bonded to the carbonyl carbon are more reactive, compared to acid chlorides that have less electronegative groups bonded to the carbonyl carbon.

16.3.3 Substitution Reactions Involving Acid Chlorides and Ammonia and Amines

Amines and ammonia are Lewis bases and also nucleophilic, and they react with acid chlorides to form amides. The type of amide produced as products depends on the type of amine used as reactant. The products of the reaction of ammonia and acid chlorides are primary amides; primary amines react with acid chlorides to form secondary amides; secondary amines react with acid chlorides to form tertiary amides. Examples of these reactions are shown in Reactions (16-11) through (16-13). The HCl that is produced in the reaction reacts with excess amines or ammonia to form the corresponding ammonium chlorides.

Benzoyl chloride + 2 NH_3 → Benzamide (a primary amide) + NH_4Cl (16-11)

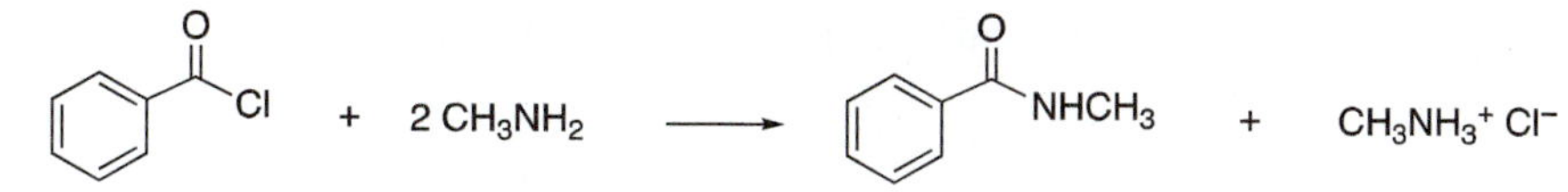

Benzoyl chloride + 2 CH_3NH_2 → N-Methylbenzamide (a secondary amide) + $CH_3NH_3^+ Cl^-$ (16-12)

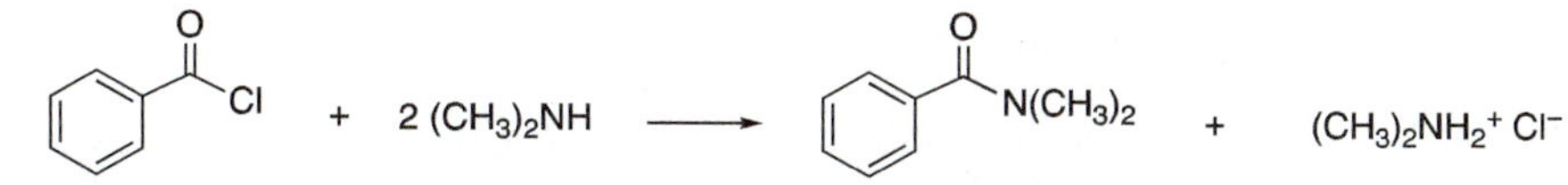

Benzoyl chloride + 2 $(CH_3)_2NH$ → N,N-Dimethylbenzamide (a tertiary amide) + $(CH_3)_2NH_2^+ Cl^-$ (16-13)

The amide is a very important functional group, especially in biology. The amide bond as we will see in Chapter 20 bonds amino acids to form peptides, which are the building blocks of protein.

Problem 16.5

i) Give the organic products for the reactions shown below.

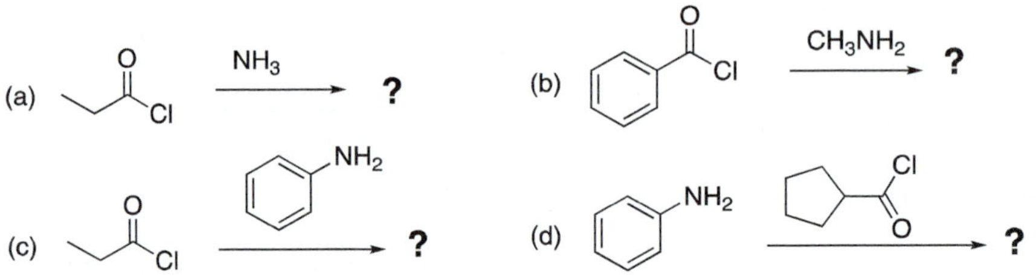

(a) propanoyl chloride + NH_3 → ?

(b) benzoyl chloride + CH_3NH_2 → ?

(c) propanoyl chloride + aniline ($C_6H_5NH_2$) → ?

(d) aniline + cyclopentanecarbonyl chloride → ?

ii) Provide an appropriate acid chloride and amine reactant molecules to complete the following reactions.

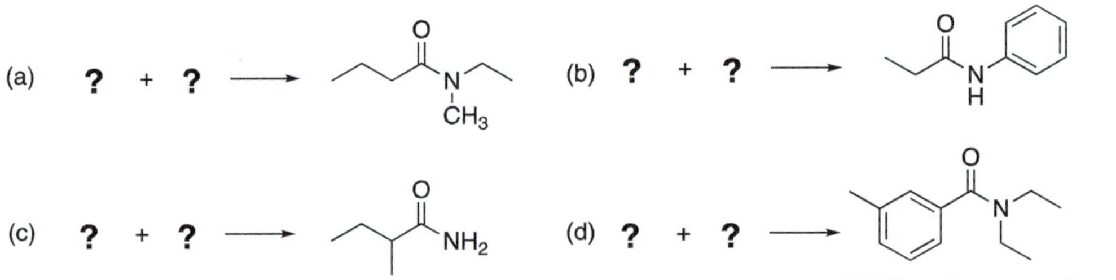

(a) ? + ? → N-ethyl-N-methylbutanamide

(b) ? + ? → N-phenylpropanamide

(c) ? + ? → 2-methylpropanamide (isobutyramide)

(d) ? + ? → DEET, an insect repellent

16.3.4 Substitution Reactions Involving Acid Chlorides and Carboxylate Salts

The conjugate bases of carboxylic acids are carboxylic acid salts and they are nucleophilic, even though they are not very good nucleophiles, but they react with acid chlorides to form the new functional group, *anhydride*. The general reaction is given in Reaction (16-14).

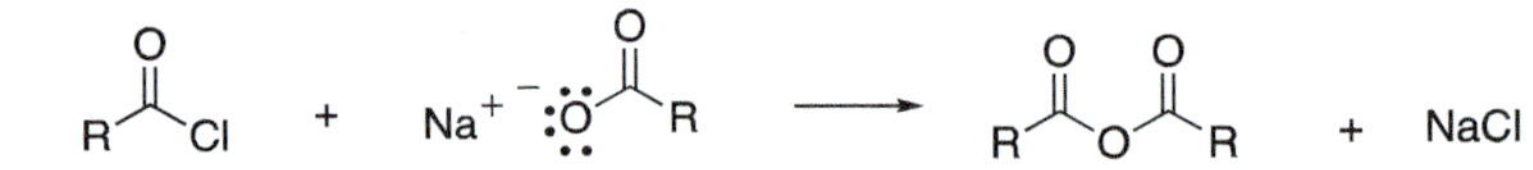

$$R\text{-COCl} + Na^+ \;{}^-\!\!:\!O\text{-CO-}R \longrightarrow R\text{-CO-O-CO-}R + NaCl \qquad (16\text{-}14)$$

Acid chloride   Sodium alkanoate   Anhydride

If the R group of the acid chloride is the same as that of the nucleophile, then the result as shown above is a symmetrical anhydride. A specific reaction is shown in the example given Reaction (16-15).

Benzoyl chloride + Sodium benzoate → Benzoyl anhydride + NaCl   (16-15)

If, on the other hand, the R groups are different, nonsymmetrical anhydrides are the products. As we will see when we look at the applications of these compounds at the end of the chapter, these compounds are most useful if they are symmetrical.

Problem 16.6

Give the major organic product for each of the following reactions.

(a) 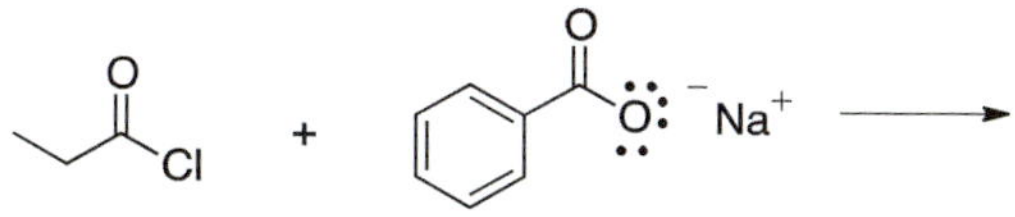 CH₃C(O)Cl + PhC(O)O⁻ Na⁺ ⟶

(b) 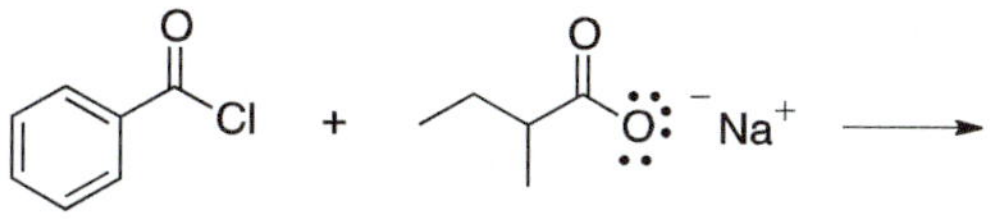 PhC(O)Cl + (CH₃)₂CHC(O)O⁻ Na⁺ ⟶

16.3.5 Substitution Reactions Involving Acid Chlorides and Soft Organometallic Reagents

The reaction of soft organometallic reagents, such as an organocuprate, results in a ketone, as shown in the example in Reaction (16-16).

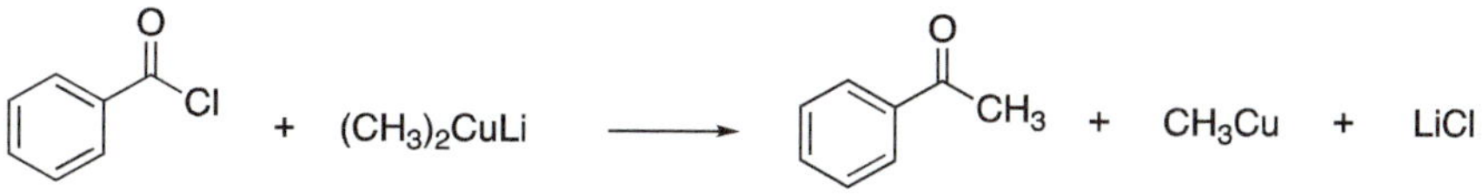

PhC(O)Cl + (CH₃)₂CuLi ⟶ PhC(O)CH₃ + CH₃Cu + LiCl   (16-16)

Problem 16.7

i) Give the major organic product for each of the following reactions.

(a) 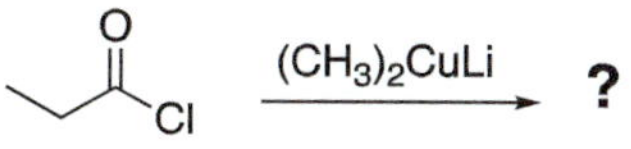 CH₃C(O)Cl $\xrightarrow{(CH_3)_2CuLi}$?

(b) 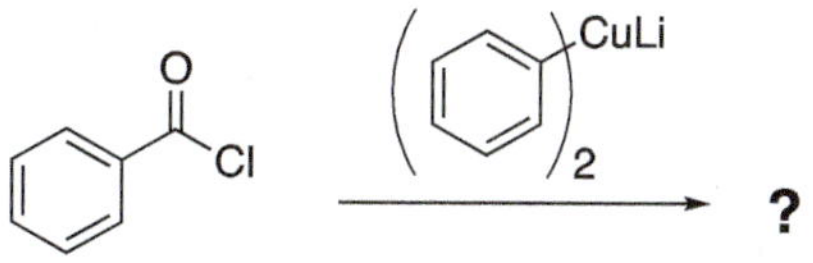 PhC(O)Cl $\xrightarrow{(Ph)_2CuLi}$?

ii) For the target ketone molecules shown below, there are two different acid chlorides and two different organocuprates that can be used in separate syntheses to give the desired product. Give the structures of the acid chloride and organocuprate reactants that could be used to synthesize these target molecules.

(a) 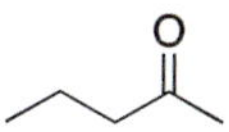 CH₃CH₂CH₂C(O)CH₂CH₃   (b) 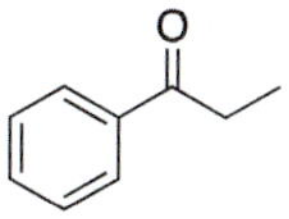 PhC(O)CH₂CH₃

16.3.6 Substitution Reactions of Acid Chlorides with Hard Organometallic Reagents

On the other hand, in the reaction of acid chlorides with much harder nucleophiles, in this case, also known as very strong reducing agents, the reaction will first give the acyl substitution reaction to give the corresponding ketone, as shown in Reaction (16-17).

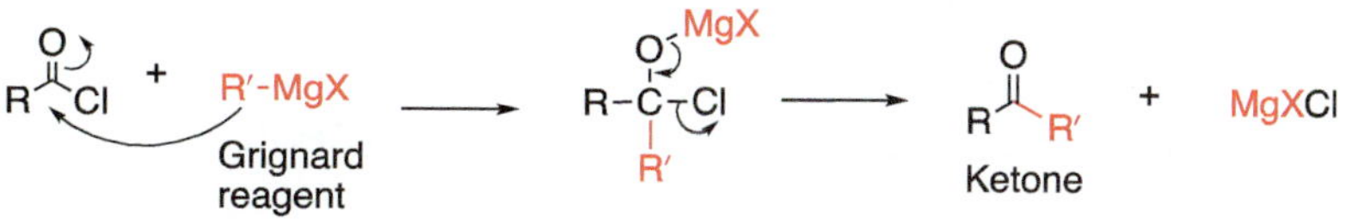

RC(O)Cl + R'−MgX (Grignard reagent) ⟶ R−C(OMgX)(R')−Cl ⟶ RC(O)R' (Ketone) + MgXCl   (16-17)

You will recall from Chapter 11, in which we examined reduction of carbonyl compounds, that ketones readily react with Grignard reagents to give an alcohol salt as shown in Reaction (16-18).

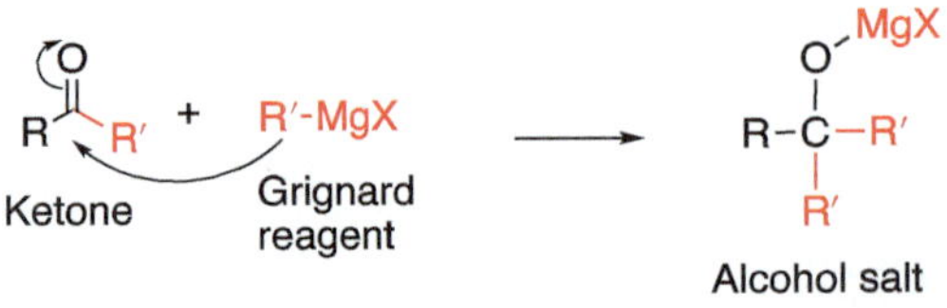

$$\underset{\text{Ketone}}{R\overset{\overset{\displaystyle O}{\|}}{-}R'} + \underset{\substack{\text{Grignard} \\ \text{reagent}}}{R'\text{-MgX}} \longrightarrow \underset{\text{Alcohol salt}}{R-\underset{\underset{\displaystyle R'}{|}}{\overset{\overset{\displaystyle O^-\text{MgX}}{|}}{C}}-R'} \tag{16-18}$$

The salts are readily hydrolyzed in the presence of an acidic medium to give alcohols. In this case, tertiary alcohols result since two alkyl groups from the Grignard reagent are added to the carbon that contains the alcohol functional group, as shown in Reaction (16-19).

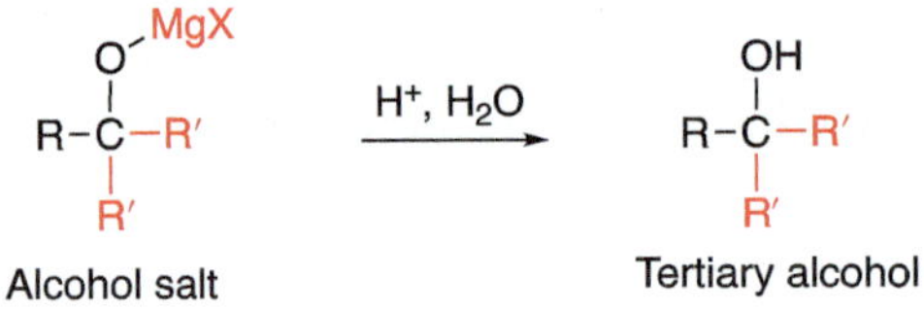

$$\underset{\text{Alcohol salt}}{R-\underset{\underset{\displaystyle R'}{|}}{\overset{\overset{\displaystyle O^-\text{MgX}}{|}}{C}}-R'} \xrightarrow{H^+, H_2O} \underset{\text{Tertiary alcohol}}{R-\underset{\underset{\displaystyle R'}{|}}{\overset{\overset{\displaystyle OH}{|}}{C}}-R'} \tag{16-19}$$

An example of the overall reaction is shown in Reaction (16-20).

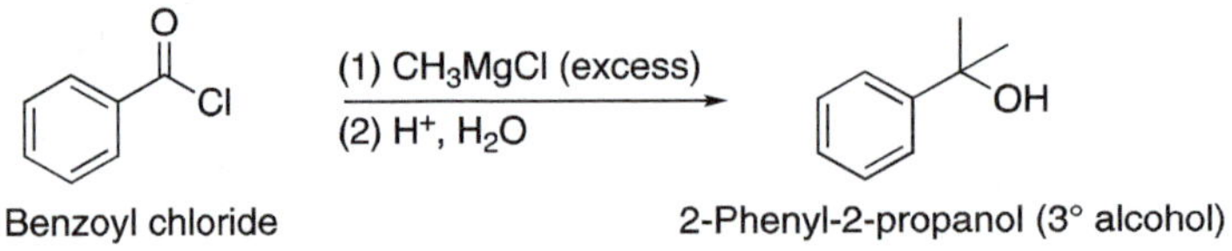

$$\underset{\text{Benzoyl chloride}}{\text{PhCOCl}} \xrightarrow[\text{(2) } H^+, H_2O]{\text{(1) } CH_3MgCl \text{ (excess)}} \underset{\text{2-Phenyl-2-propanol (3° alcohol)}}{\text{PhC(CH}_3)_2\text{OH}} \tag{16-20}$$

Problem 16.8

Give the major organic products of the following reactions.

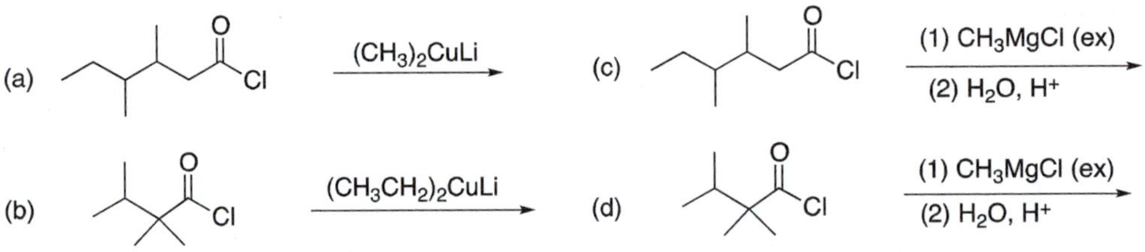

(a) (CH$_3$)$_2$CHCH$_2$CH(CH$_3$)COCl $\xrightarrow{(CH_3)_2CuLi}$

(b) (CH$_3$)$_3$CCOCl $\xrightarrow{(CH_3CH_2)_2CuLi}$

(c) (CH$_3$)$_2$CHCH$_2$CH(CH$_3$)COCl $\xrightarrow[\text{(2) } H_2O, H^+]{\text{(1) } CH_3MgCl \text{ (ex)}}$

(d) (CH$_3$)$_3$CCOCl $\xrightarrow[\text{(2) } H_2O, H^+]{\text{(1) } CH_3MgCl \text{ (ex)}}$

16.3.7 Substitution Reactions of Acid Chlorides with Soft Metal Hydrides Reagents

The reaction of acid chlorides with soft metal hydrides results in aldehydes. It is possible to deliver just one mole of a hydride ion if a soft bulky reducing agent is used, such as lithium tri-*tert*-butoxyaluminum hydride, an example in which this reagent is used is given in Reaction (16-21).

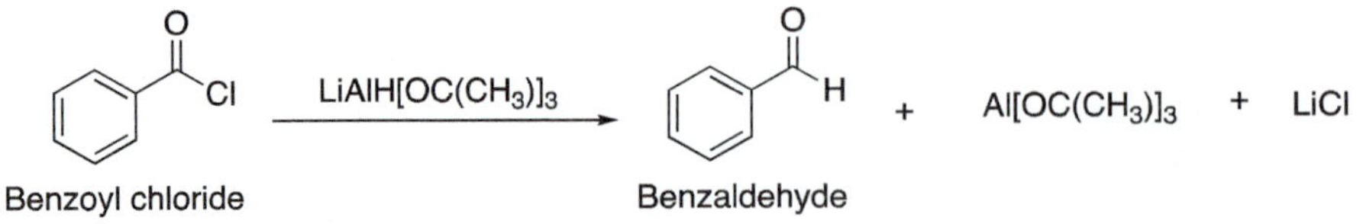

$$\underset{\text{Benzoyl chloride}}{\text{PhCOCl}} \xrightarrow{\text{LiAlH[OC(CH}_3)_3]_3} \underset{\text{Benzaldehyde}}{\text{PhCHO}} + \text{Al[OC(CH}_3)_3]_3 + \text{LiCl} \tag{16-21}$$

16.3.8 Substitution Reactions of Acid Chlorides with Hard Metal Hydrides Reagents

On the other hand, in the reaction of acid chlorides with much harder nucleophiles, in this case also known as a very strong metal hydride reducing agents, the reaction will first give an addition of the hydride anion to the acyl carbon, followed by elimination of the chloride anion to give the corresponding aldehyde, as shown in Reaction (16-22).

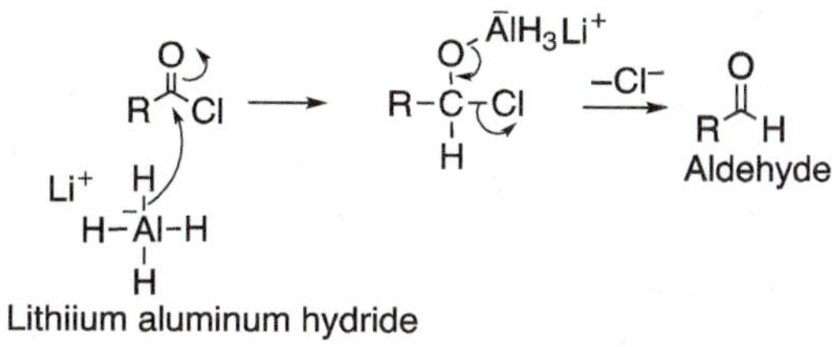

$$\text{R-C(=O)-Cl} \xrightarrow{\text{LiAlH}_4} \text{R-C(H)(O}^-\text{AlH}_3\text{Li}^+\text{)-Cl} \xrightarrow{-\text{Cl}^-} \text{R-CHO (Aldehyde)} \tag{16-22}$$

In the presence of the strong reducing agent, lithium aluminum hydride (LiAlH$_4$), the initial aldehyde that is produced readily reacts with more LiAlH$_4$ agent to give an alcohol salt as shown in Reaction (16-23).

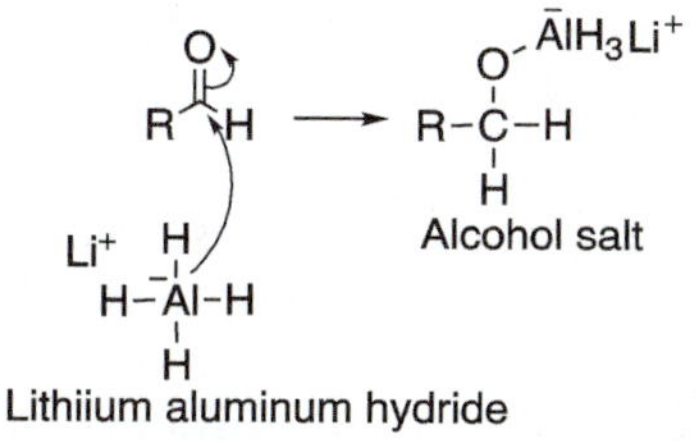

$$\text{R-CHO} \xrightarrow{\text{LiAlH}_4} \text{R-CH}_2\text{-O}^-\text{AlH}_3\text{Li}^+ \quad \text{(Alcohol salt)} \tag{16-23}$$

As you can imagine, the salts are readily hydrolyzed in the presence of an acidic medium to give alcohols. In this case, primary alcohols result since two hydrogen atoms are added to the carbon that contains the alcohol functional group, as shown in Reaction (16-24).

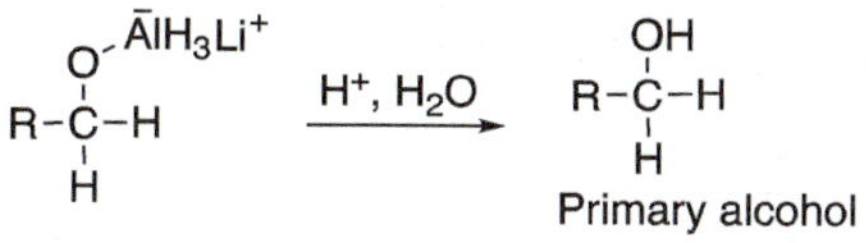

$$\text{R-CH}_2\text{-O}^-\text{AlH}_3\text{Li}^+ \xrightarrow{\text{H}^+, \text{H}_2\text{O}} \text{R-CH}_2\text{-OH} \quad \text{(Primary alcohol)} \tag{16-24}$$

An example of the overall reaction is shown in Reaction (16-25).

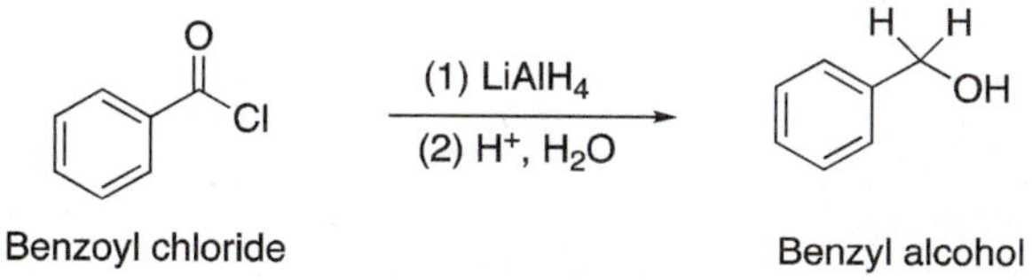

$$\text{PhCOCl (Benzoyl chloride)} \xrightarrow[\text{(2) H}^+, \text{H}_2\text{O}]{\text{(1) LiAlH}_4} \text{PhCH}_2\text{OH (Benzyl alcohol)} \tag{16-25}$$

More examples are shown in Reactions (16-26) and (16-27).

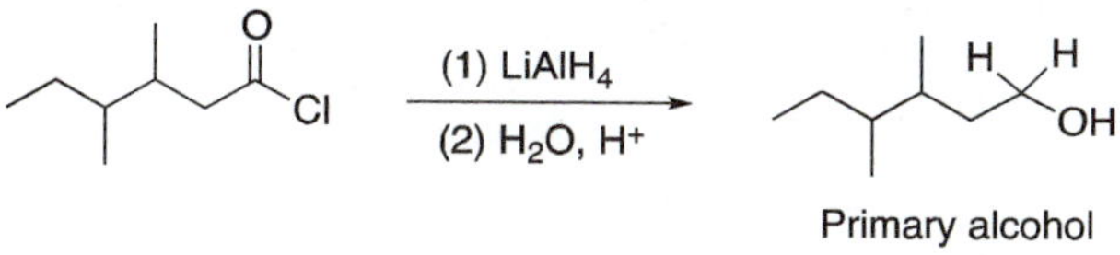

$$(\text{CH}_3)_2\text{CHCH}_2\text{CH}(\text{CH}_3)\text{COCl} \xrightarrow[\text{(2) H}_2\text{O, H}^+]{\text{(1) LiAlH}_4} \text{Primary alcohol} \tag{16-26}$$

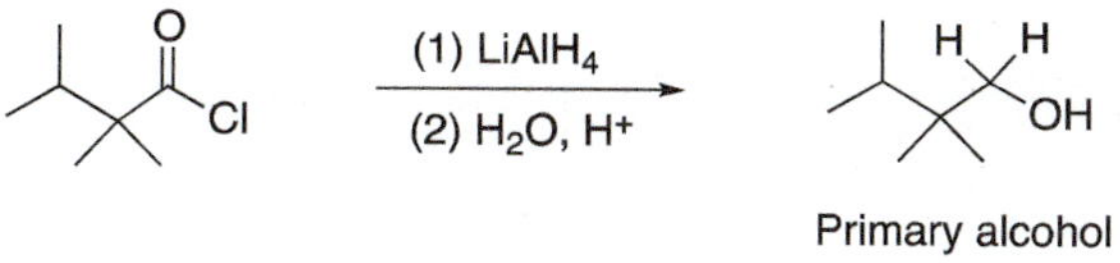

$$(\text{CH}_3)_3\text{CCOCl} \xrightarrow[\text{(2) H}_2\text{O, H}^+]{\text{(1) LiAlH}_4} \text{Primary alcohol} \tag{16-27}$$

Problem 16.9

Give the major organic products of the following reactions.

(a) [structure: 3-methylpentanoyl chloride] (1) LiAlH₄ (2) H₂O, H⁺ → ? (c) [structure: 3-methylpentanoyl chloride] LiAlH[OC(CH₃)]₃ → ?

(b) [structure: 3,3-dimethylbutanoyl chloride] (1) LiAlH₄ (2) H₂O, H⁺ → ? (d) [structure: 3,3-dimethylbutanoyl chloride] LiAlH[OC(CH₃)]₃ → ?

Since acid chlorides can be transformed to a wide range of other functional groups, acid chlorides are very important starting compounds in the synthesis of many organic compounds. A summary of some of the major reactions that acid chlorides undergo is shown in Figure 16.1.

16.4 Substitution Reactions Involving Anhydrides

There are two important features of anhydrides that lead to the type reactions observed for these molecules. First, the carbonyl double bond is a polarized bond. Thus, the carbon will accept a nucleophile. Second, anhydrides have a very good leaving group in the form of the carboxylate anion. Thus, similar substitution reactions will take place at the carbonyl carbon of anhydrides of the reactions of acid chlorides. A major difference, however, between the reactions of acid chlorides and anhydrides is that the leaving group for anhydrides is not as good a

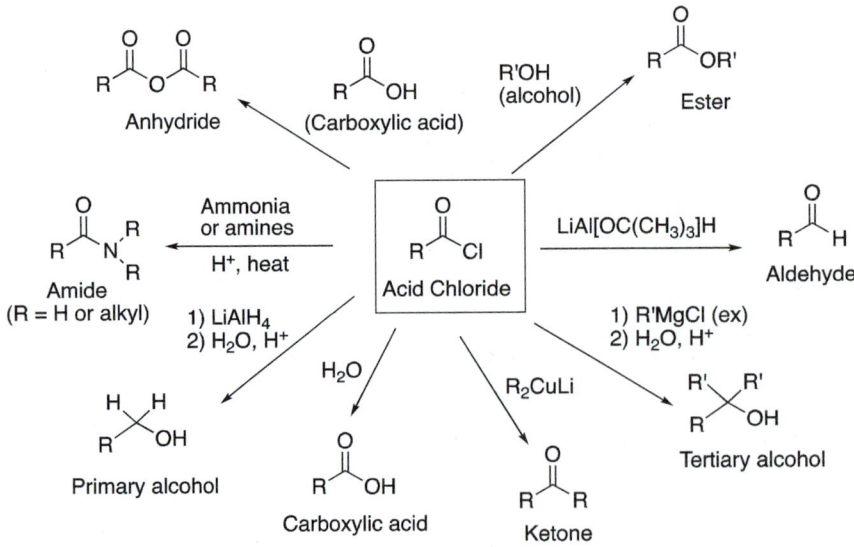

Figure 16.1 Summary of the reactions of acid chlorides.

leaving group as the chloride anion. You will recall from Chapter 7 on acids and bases that the pK_a of carboxylic acids is approximately 4.7, whereas the pK_a for HCl is −7.0. This difference in pK_a values indicates that the chloride is a much better leaving group, compared to the carboxylate anion. The general reaction of anhydrides with a nucleophile is shown in Reaction (16-28).

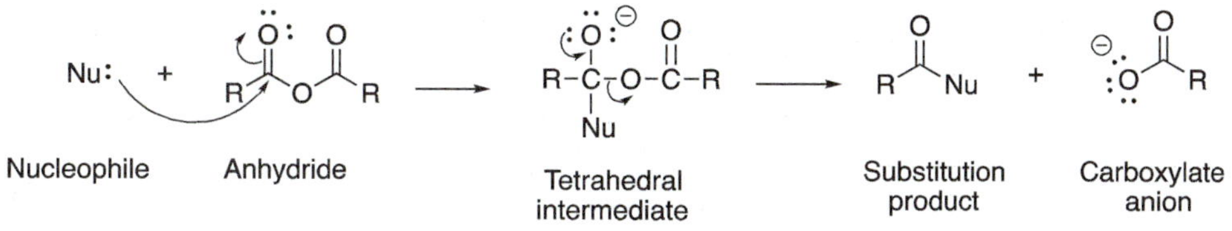

$$\text{Nu:} + \underset{\text{Anhydride}}{R\overset{\overset{\overset{..}{\overset{..}{O:}}}{\|}}{C}-O-\overset{\overset{O}{\|}}{C}-R} \longrightarrow \underset{\text{Tetrahedral intermediate}}{R-\underset{\underset{Nu}{|}}{\overset{\overset{\overset{..}{\overset{..}{O:}^{\ominus}}}{|}}{C}}-O-\overset{\overset{O}{\|}}{C}-R} \longrightarrow \underset{\text{Substitution product}}{R\overset{\overset{O}{\|}}{C}Nu} + \underset{\text{Carboxylate anion}}{{}^{\ominus}:\overset{..}{O}-\overset{\overset{O}{\|}}{C}-R}$$

(16-28)

The first step of the reaction mechanism is the addition of the nucleophile to one the carbonyl carbons of the anhydride. The stability of the resulting intermediate determines the overall reactivity of anhydrides. The second step is the elimination of the carboxylate anion, RCOO⁻. Thus, the overall reaction is a nucleophilic substitution reaction at a carbonyl carbon in which a RCOO⁻ anion is substituted by Nu⁻ to form the product. Note that since both carbonyl carbons of the anhydride are electrophilic, there is a probability that the nucleophile will attack either carbonyl carbon to give a mixture of organic products if the R groups are different. Thus, if this type of reaction is used as a synthetic reagent, the anhydrides should be symmetric anhydrides. Owing to the mixture of organic products that result from the reaction of unsymmetrical anhydrides with a nucleophile, most of the anhydrides that are used in the laboratories are symmetrical anhydrides.

16.4.1 Substitution Reactions of Anhydrides with Water

The reaction of anhydrides with water is very similar to that of the hydrolysis of acid chlorides, except as pointed out earlier the leaving group for the anhydride is the carboxylate anion and in the presence of an acidic medium, another mole of carboxylic acid is formed as shown in Reaction (16-29).

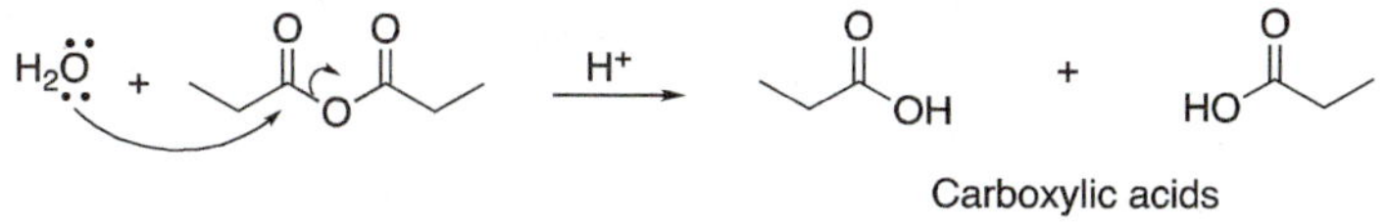

(16-29)

Carboxylic acids

As you can imagine, the size of the alkyl group bonded to the carbonyl carbon dictates the reactivity of anhydrides. Anhydrides with large groups react slower than anhydrides with smaller groups bonded to the carbonyl carbon. Examples of the relative reactivity of two anhydrides are shown below.

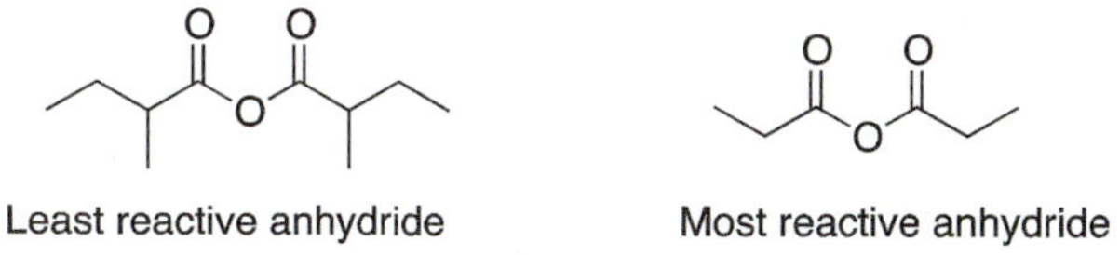

Least reactive anhydride   Most reactive anhydride

Problem 16.10

Give the major organic products of the following reactions.

(a) 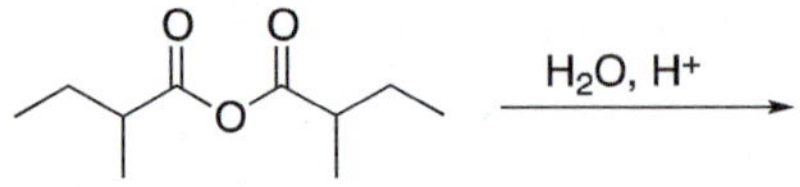 (isobutyric anhydride) $\xrightarrow{H_2O, H^+}$

(b) 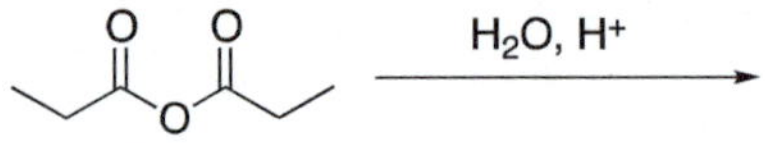 (propionic anhydride) $\xrightarrow{H_2O, H^+}$

16.4.2 Substitution Reactions of Anhydrides with Alcohols

The reaction of acetic anhydride with methanol is shown in Reaction (16-30). You will notice that based on your knowledge of the reactions of acid chlorides with alcohols, the outcome of the reaction of anhydrides with alcohols is not surprising.

$$CH_3OH \;+\; \text{An anhydride} \;\longrightarrow\; \text{Ester} \;+\; \text{Carboxylic acid} \tag{16-30}$$

In the first step of the reaction mechanism, the nucleophilic alcohol reacts with the electrophilic carbonyl center to form a tetrahedral intermediate, which is followed by a proton transfer as shown in Reaction (16-31).

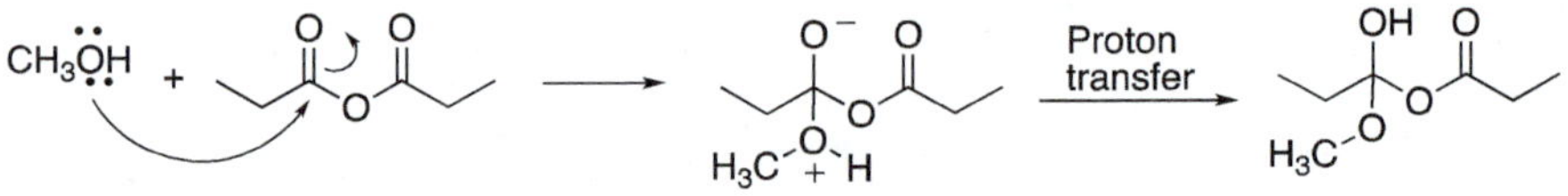

$$CH_3\ddot{O}H \;+\; \text{anhydride} \;\longrightarrow\; \text{tetrahedral intermediate} \;\xrightarrow{\text{Proton transfer}}\; \text{protonated intermediate} \tag{16-31}$$

In the next step of the mechanism, the tetrahedral intermediate forms a carbon–oxygen double bond as the leaving carboxylate anion leaves as shown in Reaction (16-32).

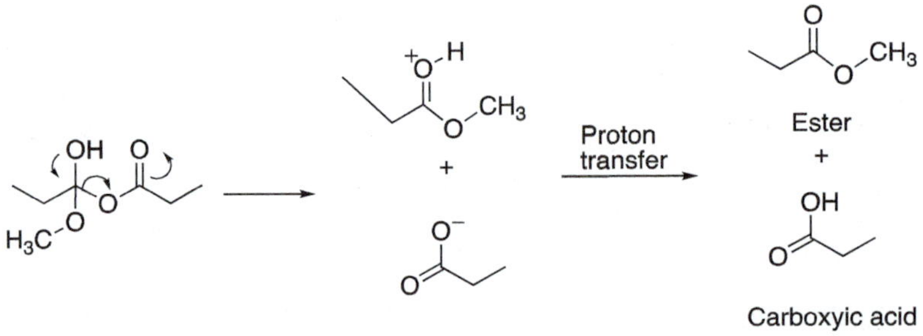

$$\text{intermediate} \;\longrightarrow\; \text{protonated ester + carboxylate} \;\xrightarrow{\text{Proton transfer}}\; \text{Ester} + \text{Carboxylic acid} \tag{16-32}$$

The products of these reactions are esters and carboxylic acids.

Problem 16.11

Give the products of the following reactions.

(a) 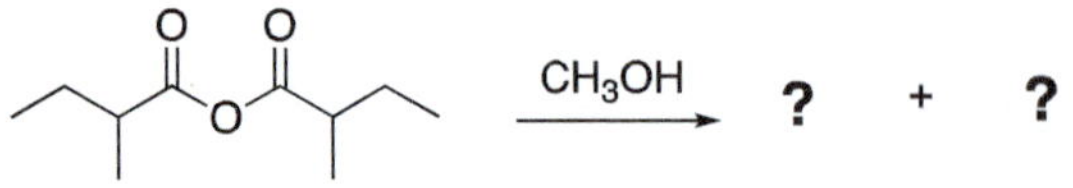 (isobutyric anhydride) $\xrightarrow{CH_3OH}$? + ?

(b) 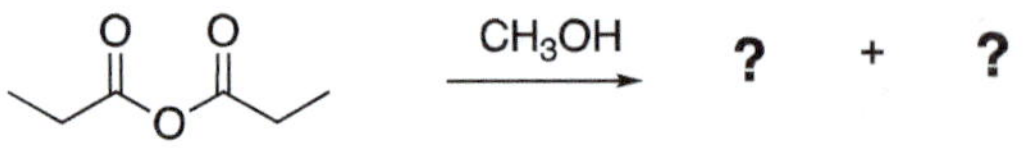 (propionic anhydride) $\xrightarrow{CH_3OH}$? + ?

16.4.3 Substitution Reactions of Anhydrides with Ammonia and Amines

Since ammonia and amines are nucleophilic molecules, they will react with anhydrides to form primary amides, secondary amides, and tertiary amides, and of course, carboxylic acids. The carboxylic acid that is produced in the reaction, reacts with excess ammonia or amines to form the corresponding ammonium salt. The mechanism for these types of reactions is similar to that shown in the previous section, except that the nucleophile in this case is the nitrogen of ammonia or amine. The first step of the mechanism involves the bonding of the nucleophilic nitrogen to the electrophilic carbonyl carbon to form a tetrahedral intermediate, followed by a proton transfer as shown in Reaction (16-33).

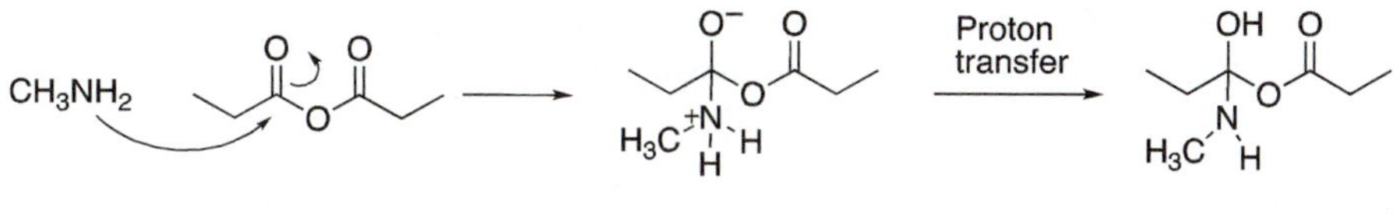

$$(16\text{-}33)$$

In the next step of the mechanism, the carbon–oxygen double bond forms and the carboxylate anion leaves as shown in Reaction (16-34) to form the amide and carboxylic acid products.

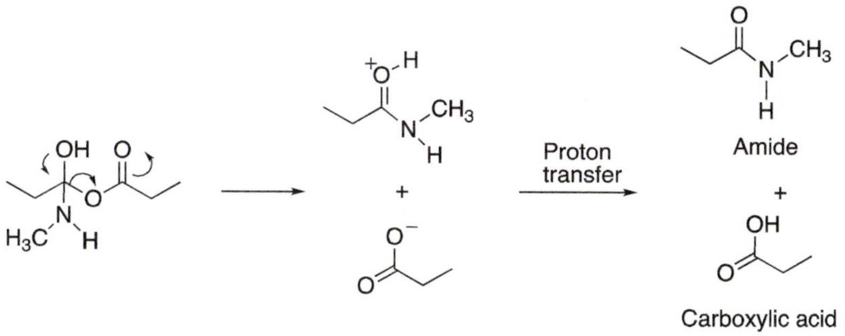

$$(16\text{-}34)$$

Problem 16.12

Give the products for the reaction of each of the molecules shown below with ethanoic anhydride (ethanoic ethanoic anhydride).

a) CH_3NH_2
b) NH_3
c) $(CH_3CH_2)_2NH$
d) $CH_3CH_2NH_2$

16.4.4 Substitution Reactions of Anhydrides with Carboxylate Salts

As you can imagine, the reaction of an anhydride with carboxylate salts gives another anhydride and a carboxylate salt. These reactions are not very useful for the synthesis of new compounds since they generate a nonsymmetric anhydride as shown in in the example given in Reaction (16-35).

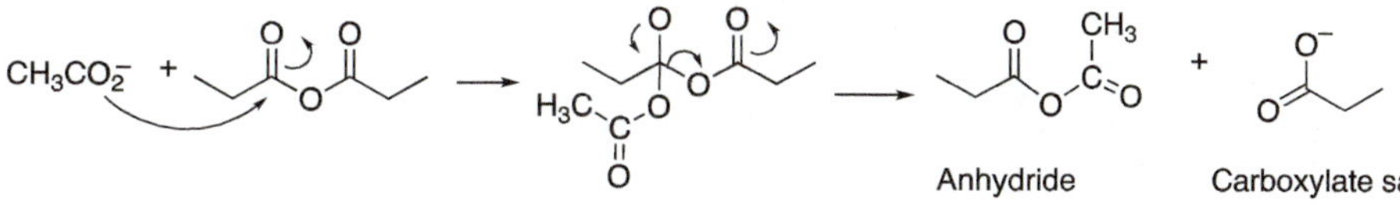

$$(16\text{-}35)$$

Problem 16.13

Give the product for the reaction of each of the molecules shown below with sodium ethanoate.

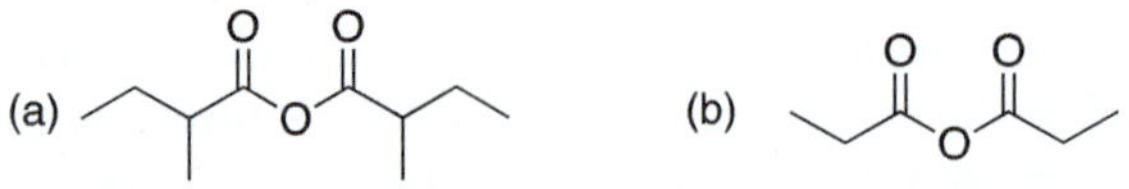

(a) [isobutyric anhydride structure]   (b) [propanoic anhydride structure]

16.4.5 Substitution Reactions of Anhydrides with Soft Organometallic Reagents

A soft reducing agent, such as the organocuprate, will react with anhydrides to give the corresponding ketone and carboxylate salt, as shown in Reaction (16-36).

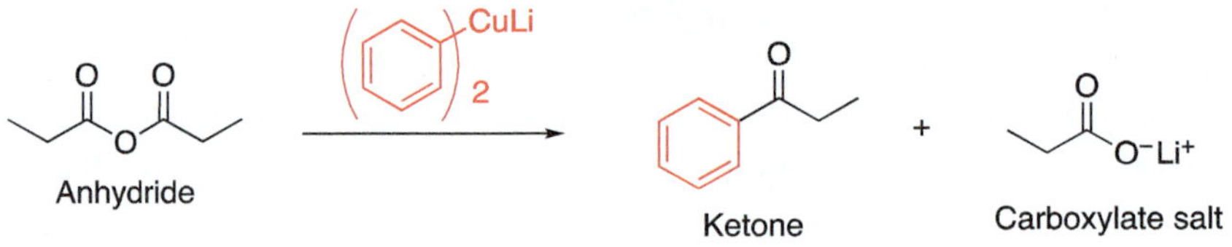

Anhydride + (C₆H₅)₂CuLi → Ketone + Carboxylate salt (O⁻Li⁺)   (16-36)

Problem 16.14

Give the products for the reaction of each of the molecules shown below with diethyl lithium cupurate.

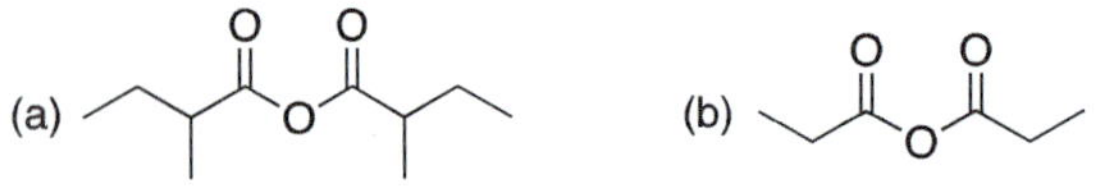

(a) [isobutyric anhydride structure]   (b) [propanoic anhydride structure]

16.4.6 Substitution Reactions of Anhydrides with Hard Organometallic Reagents

As we have pointed out earlier in Chapter 10 when we examined reducing agents, organometallic reagents in addition to being very strong bases and good reducing agents, they are also good nucleophiles. Grignard reagents, for example, react with anhydrides to produce tertiary alcohols as the final product after hydrolysis, as shown in Reaction (16-37).

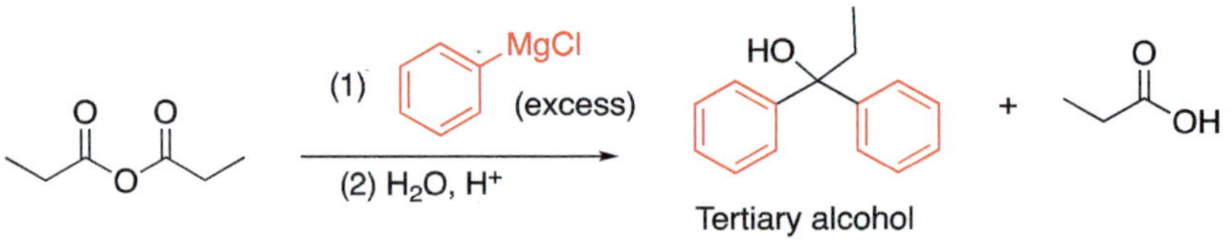

Anhydride (1) C₆H₅MgCl (excess); (2) H₂O, H⁺ → Tertiary alcohol + carboxylic acid (OH)   (16-37)

Problem 16.15

Give the products for the reaction of each of the molecules shown below with phenyl Grignard reagent (C₆H₅MgCl), followed by acid hydrolysis.

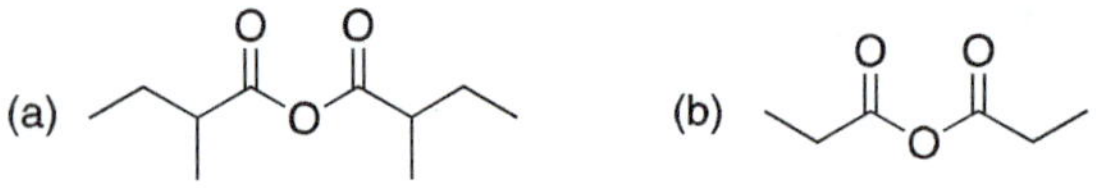

(a) [isobutyric anhydride structure]   (b) [propanoic anhydride structure]

16.4.7 Substitution Reactions of Anhydrides with Soft Metallic Hydrides

The reaction of anhydrides with soft metal hydrides, such as lithium di-*tert*-butoxide hydride, results in aldehydes and carboxylate salts. As you would expect, the reaction will add just one mole of a hydride ion to produce an aldehyde as a final product, as shown in Reaction (16-38).

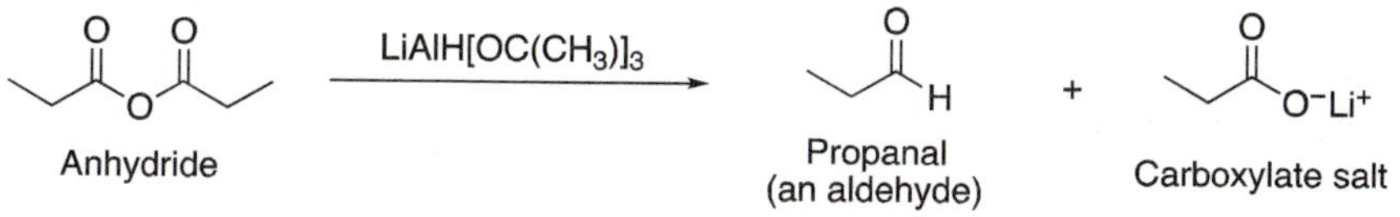

$$\text{Anhydride} \xrightarrow{\text{LiAlH[OC(CH}_3)_3]_3} \text{Propanal (an aldehyde)} + \text{Carboxylate salt} \qquad (16\text{-}38)$$

In a second step, the carboxylate salt can be converted to a carboxylic acid by an acid–base reaction.

Problem 16.16

Give the products for the reaction of each of the molecules shown below with lithium tri-*tert*-butoxyaluminum hydride.

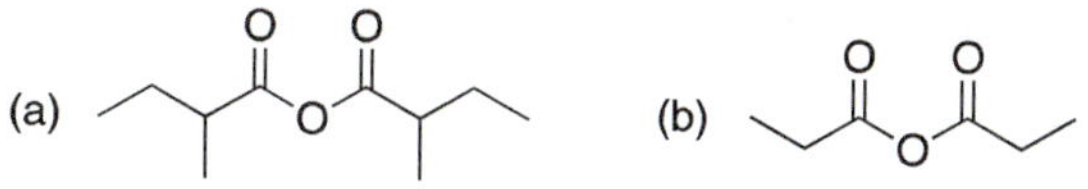

(a) [structure of anhydride] (b) [structure of anhydride]

16.4.8 Substitution Reactions of Anhydrides with Hard Metallic Hydrides

On the other hand, the reaction of anhydrides with a much harder nucleophile, in this case, also known as a very strong reducing agent, results in first an acyl substitution as shown in Reaction (16-39).

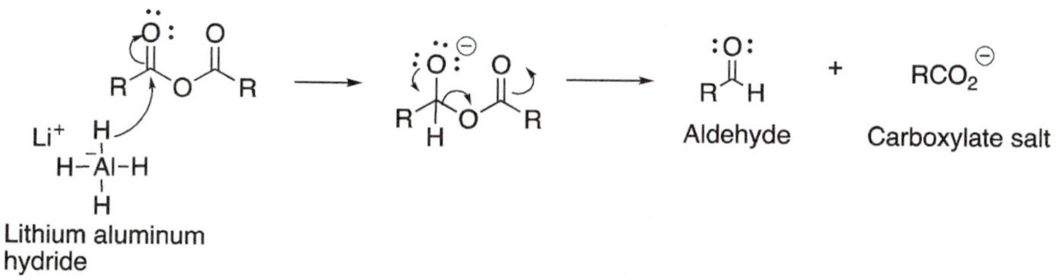

$$\text{[Anhydride + LiAlH}_4\text{]} \longrightarrow \text{[tetrahedral intermediate]} \longrightarrow \text{Aldehyde} + RCO_2^{\ominus} \text{ Carboxylate salt} \qquad (16\text{-}39)$$

Lithium aluminum hydride

The aldehyde that is produced readily accepts another mole of hydride anion from the reducing LiAlH$_4$ agent to give an alcohol salt, which upon hydrolysis gives a primary alcohol as shown in Reaction (16-40).

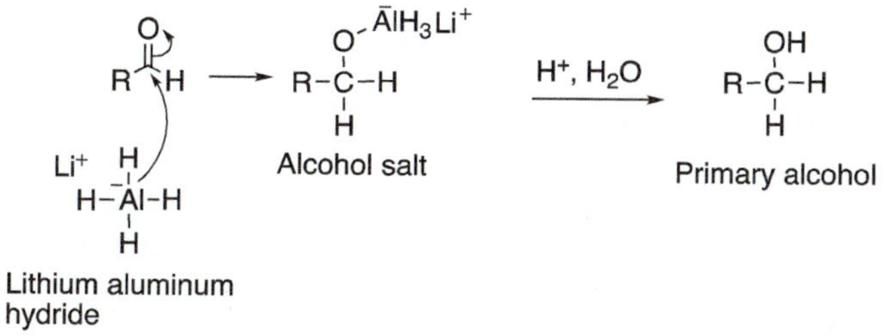

$$\underset{\text{Lithium aluminum hydride}}{R\text{-CHO} + \text{LiAlH}_4} \longrightarrow \underset{\text{Alcohol salt}}{R-\underset{H}{\underset{|}{\overset{|}{C}}}-H \ \ O^-\bar{A}lH_3Li^+} \xrightarrow{H^+, H_2O} \underset{\text{Primary alcohol}}{R-\underset{H}{\underset{|}{\overset{OH}{\underset{|}{C}}}}-H} \qquad (16\text{-}40)$$

As we will see later in the chapter, the carboxylate salt can be reduced in the presence of a strong reducing agent, such as $LiAlH_4$, to give the corresponding alcohol, as shown in Reaction (16-41).

$$\underset{\text{Carboxylate salt}}{RCO_2^-} \xrightarrow[\text{(2) } H_2O,\ H^+]{\text{(1) } LiAlH_4} \underset{\text{Primary alcohol}}{RCH_2OH} \qquad (16\text{-}41)$$

Thus, the final product of a reduction of an anhydride using a strong metal hydride reducing agent, such as $LiAlH_4$, is shown in the examples are shown in Reactions (16-42) and (16-43).

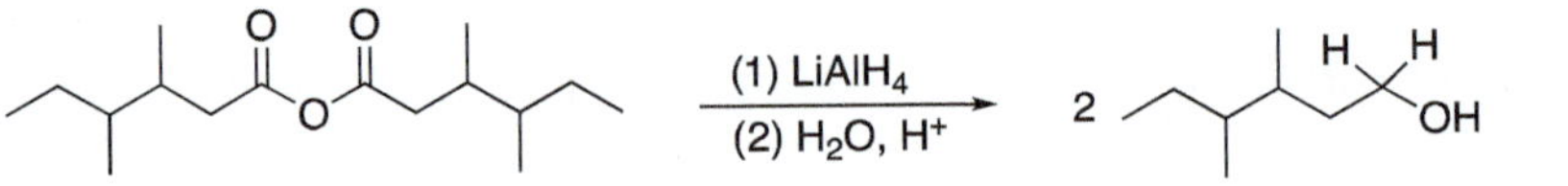

(16-42)

Primary alcohol (2 moles)

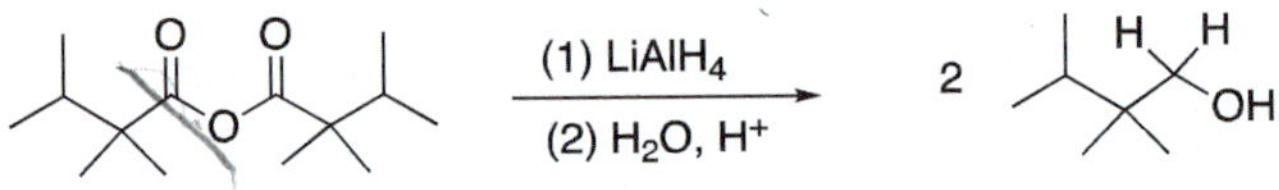

(16-43)

Primary alcohol (2 moles)

Note that these examples use symmetrical anhydrides so that there are no mixed organic products produced, Reactions (16-44) and (16-45) give examples using unsymmetrical anhydrides.

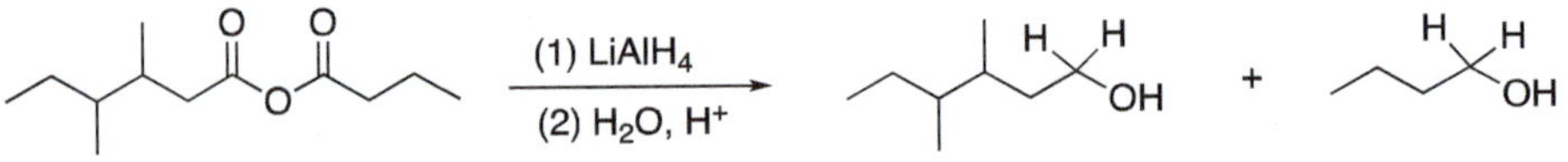

(16-44)

Mixed primary alcohols as products

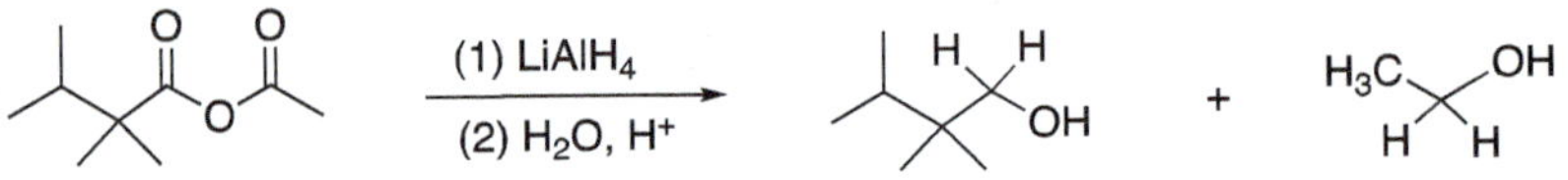

(16-45)

Mixed primary alcohols as products

Problem 16.17

Give the organic products for the reaction of each of the molecules shown below with $LiAlH_4$, followed by acidic work-up.

(a) 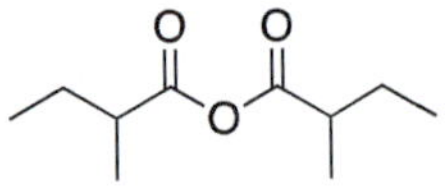 (b) 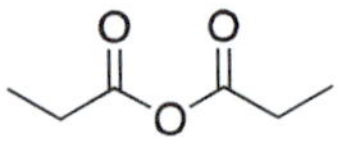

16.5 Substitution Reactions Involving Esters

For substitution reactions involving esters, the leaving group is not as a good leaving group as those that we have seen thus far in this chapter, which are the chloride anion and the carboxylate anion. For reactions involving esters, the leaving group is an alkoxide anion. As we have demonstrated in Chapter 7 when we looked at acids and bases, we saw that the conjugate acid of the alkoxide anion is an alcohol. Alcohols have pK_a values of approximately 20, which means

16.5 Substitution Reactions Involving Esters

that they are very weak acids, and as a result, the conjugate bases, alkoxide anions, are extremely strong bases and hence a very poor leaving groups as shown in Reaction (16-46).

$$\text{Nu:}^- + R-\overset{\overset{\displaystyle :O:}{\|}}{C}-OCH_3 \longrightarrow R-\overset{\overset{\displaystyle :O:^-}{|}}{\underset{\displaystyle Nu}{C}}-OCH_3 \longrightarrow R-\overset{\overset{\displaystyle :O:}{\|}}{C}-Nu + {:}\overset{..}{\underset{..}{O}}CH_3 \quad (16\text{-}46)$$

Alkoxide anion — Very poor leaving group

In an acidic medium, however, the alkoxide ion is converted into an extremely good leaving group, a protonated alcohol. Thus, the hydrolysis of esters is typically acid catalyzed. As you will see from the mechanism, however, the alkoxide leaving group is not first protonated in order to make it a good leaving group. Let us look at the mechanism a bit closer. In the first step of the mechanism, the proton protonates the most basic site of the ester molecule, as shown in Reaction (16-49).

$$\underset{\text{Ester}}{R'-\overset{\overset{\displaystyle :O:}{\|}}{C}-\overset{..}{\underset{..}{O}}R} \xrightarrow{H^+} \underset{\text{Two resonance structures of protonated ester}}{R'-\overset{\overset{\displaystyle \overset{+}{O}-H}{\|}}{C}-OR \longleftrightarrow R'-\overset{\overset{\displaystyle :O-H}{|}}{C}=\overset{+}{O}R} \quad (16\text{-}47)$$

Note that the most basic site of the ester molecule is the lone pair of electrons on the carbonyl oxygen and not the lone pair of electrons on the alkoxide oxygen as demonstrated by the stable resonance structure that results upon protonation at the carbonyl oxygen site. In another step, the nucleophile attacks the carbonyl carbon to form a tetrahedral intermediate as shown in Reaction (16-48).

$$R'-\overset{\overset{\displaystyle O}{\|}}{C}-OR \xrightarrow{H^+} R'-\overset{\overset{\displaystyle \overset{+}{O}-H}{\|}}{C}-OR \longleftrightarrow R'-\overset{\overset{\displaystyle O-H}{|}}{C}=\overset{+}{O}R \xrightarrow{Nu:^-} R'-\overset{\overset{\displaystyle OH}{|}}{\underset{\displaystyle Nu}{C}}-OR \quad (16\text{-}48)$$

Tetrahedral intermediate

As pointed out previously, the stability of the tetrahedral intermediate depends on the size of the R group that is bonded to the carbonyl carbon. Very crowded tetrahedral intermediates have bond angles that are strained and are less than the desired 109.5° and, as a result, less stable than tetrahedral intermediates with smaller and less bulky alkyl groups bonded to the carbonyl carbon. In the presence of an acid, protonation of the OR will occur and as we have seen earlier, this protonation transforms the OR group into a good leaving group as shown in Reaction (16-49).

$$\underset{\substack{\text{Tetrahedral}\\\text{intermediate}}}{R'-\overset{\overset{\displaystyle OH}{|}}{\underset{\displaystyle Nu}{C}}-OR} \xrightarrow{H^+} R-\overset{\overset{\displaystyle \overset{..}{O}H}{|}}{\underset{\displaystyle Nu}{C}}-\overset{+}{O}\overset{H}{\underset{\displaystyle R}{}} \rightleftharpoons R'-\overset{\overset{\displaystyle \overset{+}{O}H}{\|}}{\underset{\displaystyle Nu}{C}} + ROH \quad (16\text{-}49)$$

In a final equilibrium step of the mechanism, the proton is lost to generate the organic product as shown in Reaction (16-50).

$$R'-\overset{\overset{\displaystyle \overset{+}{O}-H}{\|}}{C}-Nu \xrightarrow{-H^+} \underset{\substack{\text{Product with}\\\text{new C–Nu bond}}}{R'-\overset{\overset{\displaystyle :O:}{\|}}{C}-Nu} \quad (16\text{-}50)$$

This step demonstrates that the proton is really a catalyst since it was used in the first step and regenerated in this step. It is consumed in the first step of the mechanism and liberated in the last step of the mechanism.

Thus, even though this leaving group of esters is not a very good leaving group, compared to those of the acid chlorides and anhydrides, under acidic conditions, it is converted to a very good leaving group, ROH.

16.5.1 Substitution Reactions of Esters with Water

The reaction of esters with water is very similar to that of the hydrolysis of the previous acyl compounds that we have examined earlier, except that the leaving group for the ester is the alkoxide anion, which is a poor leaving group. In the presence of a basic aqueous medium and at elevated temperatures, hydrolysis of esters can still occur, despite having a very poor leaving group. The first step of the reaction of methyl propanoate under basic conditions is shown in Reaction (16-51).

$$\ddot{O}H^- + C_2H_5-\overset{O}{\underset{}{C}}-OCH_3 \rightleftharpoons C_2H_5-\overset{O^-}{\underset{O-H}{C}}-OCH_3 \qquad (16\text{-}51)$$

In the next step of the mechanism, the alkoxide anion, which is a very poor leaving group leaves as shown in Reaction (16-52).

$$C_2H_5-\overset{\ddot{O}:}{\underset{O-H}{C}}-OCH_3 \rightleftharpoons C_2H_5-\overset{O}{\underset{}{C}}-OH + OCH_3^- \qquad (16\text{-}52)$$

Very poor leaving group — Alkoxide anion

Since the alkoxide anion is a very strong base, an acid–base reaction will occur involving the carboxylic acid to give the carboxylate salt and an alcohol, as shown in Reaction (16-53).

$$C_2H_5-\overset{O}{\underset{}{C}}-OH + OCH_3^- \rightleftharpoons C_2H_5-\overset{O}{\underset{}{C}}-O^- + CH_3OH \qquad (16\text{-}53)$$

Alkoxide anion   Carboxylate anion

The basic hydrolysis of esters is also called **saponification** and this type of reaction is used in the production of soap. As pointed out earlier, these reactions are carried out at elevated temperatures. If the hydrolysis is carried out using the esters of fatty acid and lye (a base), the carboxylate salt that is produced is used as **soap**. In Chapter 20, a more detailed discussion of the type of esters that are used to produce soaps and how soaps work will be discussed. The saponification of a triglyceride (tri-ester) is shown in Reaction (16-53).

$$\begin{array}{l}CH_2\text{-}O\text{-}\overset{O}{C}\text{-}(CH_2)_{14}CH_3 \\ | \quad O \\ HC\text{-}O\text{-}\overset{O}{C}\text{-}(CH_2)_7CH\text{:}CH\text{-}(CH_2)_7CH_3 \\ | \quad O \\ CH_2\text{-}O\text{-}\overset{O}{C}\text{-}(CH_2)_{16}CH_3\end{array} \xrightarrow[\text{heat}]{\text{NaOH} \atop H_2O} \begin{array}{l}CH_2\text{-}OH \\ | \\ HC\text{-}OH \\ | \\ CH_2\text{-}OH\end{array} \begin{array}{l}Na^+{}^-O\text{-}\overset{O}{C}\text{-}(CH_2)_{14}CH_3 \\ + \\ Na^+{}^-O\text{-}\overset{O}{C}\text{-}(CH_2)_7CH_2\text{-}CH_2(CH_2)_7CH_3 \\ + \\ Na^+{}^-O\text{-}\overset{O}{C}\text{-}(CH_2)_{16}CH_3\end{array} \qquad (16\text{-}54)$$

Problem 16.18

Give the products that result from the basic hydrolysis of the compound shown below.

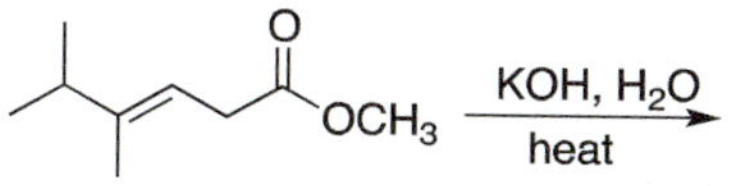

[structure of 4-methylpent-3-enoic acid methyl ester] $\xrightarrow{\text{KOH, H}_2\text{O, heat}}$

Hydrolysis of esters carried out under acidic conditions promotes the departure of a better leaving group as an alcohol. As a result, an acid-catalyzed reaction will occur, the mechanism for the hydrolysis of methyl propanoate is shown in Reactions (16-55) and (16-56).

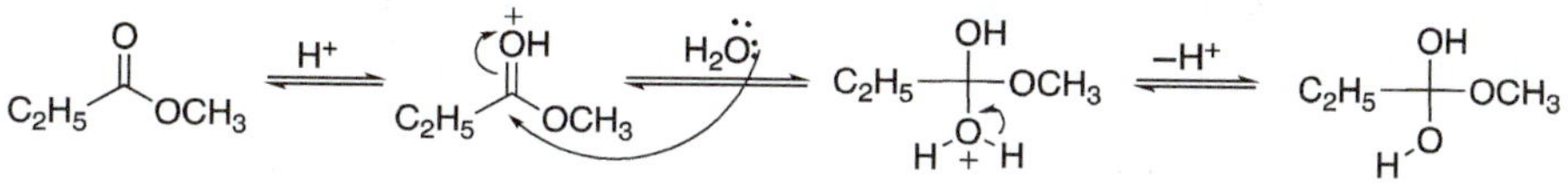

[Mechanism scheme showing protonation of methyl propanoate, addition of water, proton transfers] (16-55)

[Continued mechanism scheme showing loss of methanol to give propanoic acid + HOCH₃] (16-56)

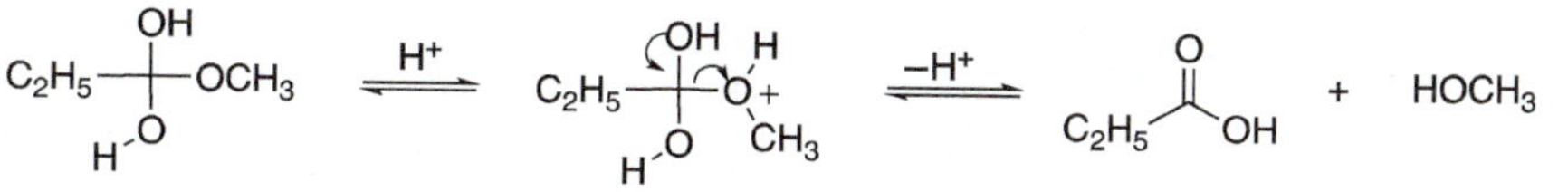

16.5.2 Substitution Reactions of Esters with Alcohols

The reaction of esters with alcohols is also called transesterification since another ester is produced as a product. As you can imagine, the reaction is very similar to that of the reaction of water with esters and an example of a transesterification is shown in Reaction (16-57).

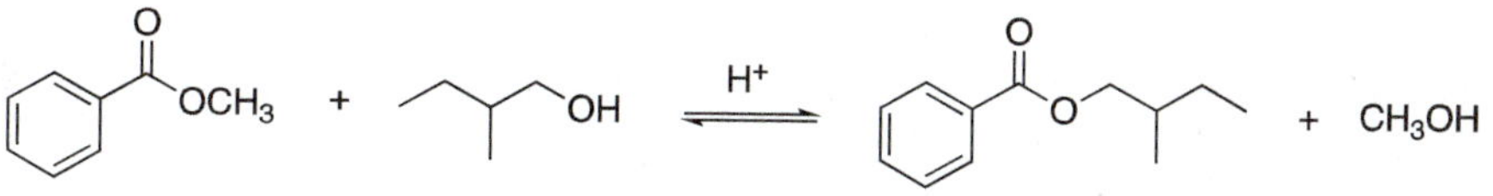

[methyl benzoate] + [isobutanol] $\xrightleftharpoons{\text{H}^+}$ [isobutyl benzoate] + CH_3OH (16-57)

Note that these reactions are shown as equilibrium reactions, and the equilibrium can be manipulated as you have seen in general chemistry via *Le Chatelier's* Principle. By removing the methanol in the product or by having a large concentration of the reactant alcohol, it is possible to shift the equilibrium to the right in order to produce the ester shown in the product. On the other hand, if the concentration of the methanol is high, the equilibrium will favor the reverse direction.

Problem 16.19

Give the organic products of the following transesterification reactions.

(a) [benzyl alcohol] + [methyl 2-methylpropanoate] $\xrightleftharpoons[\text{heat}]{\text{H}^+}$

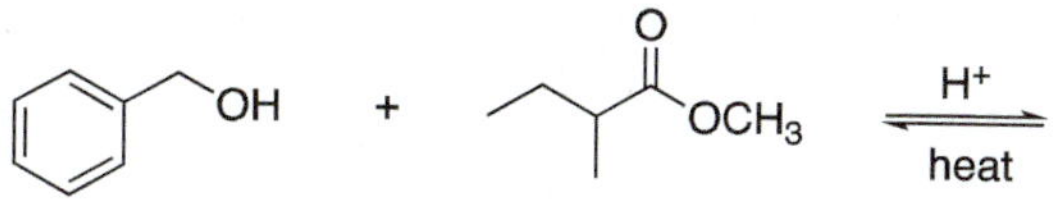

(b) [ethyl benzoate] + [3-methylbutan-2-ol] $\xrightleftharpoons[\text{heat}]{\text{H}^+}$

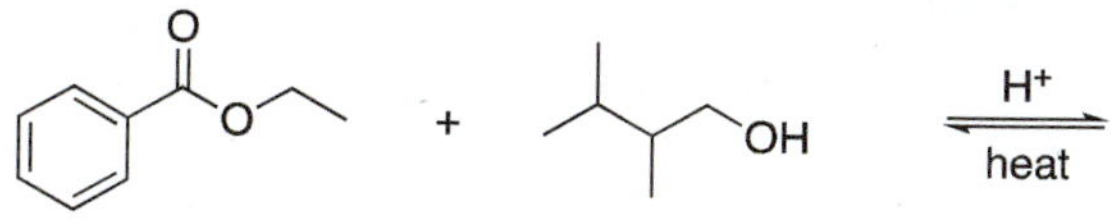

Intramolecular reactions involving difunctional molecules with an alcohol and ester functionalities are possible and produce cyclic esters, also called lactones as products. The general reaction is given in Reaction (16-58).

$$\text{HO}\overset{\text{Alcohol}}{\frown}_n\overset{\text{Ester}}{\underset{\text{O}}{\overset{\text{O}}{\|}}}\text{OR} \xrightarrow{\text{H}^+} \underset{\text{Lactone (cyclic ester)}}{\text{cyclic ester}} + \text{ROH} \qquad (16\text{-}58)$$

Reaction (16-59) gives an example of a transesterification reaction to form a six-member ring cyclic ester (lactone).

$$\text{HO}\frown\frown\overset{\text{O}}{\underset{}{\|}}\text{O}\frown \xrightleftharpoons{\text{H}^+} \underset{\text{A lactone}}{\bigcirc} + \underset{\text{Ethanol}}{\frown\text{OH}} \qquad (16\text{-}59)$$

16.5.3 Substitution Reactions of Esters with Ammonia and Amines

Let us now examine the substitution reaction of esters with ammonia and amines. The mechanism for these reactions is very similar to those of acid chlorides, except as mentioned, an acidic catalyst must be used and these reactions are carried out at elevated temperatures. Since ammonia and amines are nucleophilic molecules, they will react with ester to form the corresponding primary amide (with ammonia); secondary amide (with primary amines); and tertiary amide (with secondary amines). The other product that is produced is an alcohol.

Reaction (16-60) gives the reaction of an ester with ammonia.

$$\underset{\text{Methyl benzoate}}{\text{Ph-COOCH}_3} + \underset{\text{Ammonia}}{\text{NH}_3} \xrightleftharpoons[\text{Heat}]{\text{H}^+} \underset{\substack{\text{Benzamide} \\ \text{(a primary amide)}}}{\text{Ph-CONH}_2} + \underset{\text{Methanol}}{\text{CH}_3\text{OH}} \qquad (16\text{-}60)$$

Reaction (16-61) shows the reaction of an ester with a primary amine, methylamine.

$$\underset{\text{Methyl benzoate}}{\text{Ph-COOCH}_3} + \underset{\text{Methylamine}}{\text{CH}_3\text{NH}_2} \xrightleftharpoons[\text{Heat}]{\text{H}^+} \underset{\substack{N\text{-Methylbenzamide} \\ \text{(a secondary amide)}}}{\text{Ph-CONHCH}_3} + \underset{\text{Methanol}}{\text{CH}_3\text{OH}} \qquad (16\text{-}61)$$

Reaction (16-62) gives the reaction of an ester with a secondary amine, dimethylamine.

$$\underset{\text{Methyl benzoate}}{\text{Ph-COOCH}_3} + \underset{\text{Dimethylamine}}{(\text{CH}_3)_2\text{NH}} \xrightleftharpoons[\text{Heat}]{\text{H}^+} \underset{\substack{N,N\text{-Dimethylbenzamide} \\ \text{(a tertiary amide)}}}{\text{Ph-CON(CH}_3)_2} + \underset{\text{Methanol}}{\text{CH}_3\text{OH}} \qquad (16\text{-}62)$$

Problem 16.20

Give the products of the following reactions.

(a) 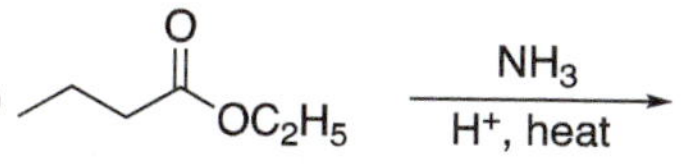 $CH_3CH_2CH_2C(O)OC_2H_5 \xrightarrow{NH_3, \; H^+, \text{ heat}}$

(b) 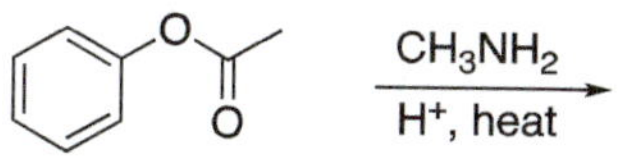 $C_6H_5OC(O)CH_3 \xrightarrow{CH_3NH_2, \; H^+, \text{ heat}}$

16.5.4 Substitution Reactions of Esters with Soft Organometallic Reagents

As you might predict, the reactions of organometallic reagents with esters are similar to those of acid chlorides. Owing to the reactivity of organometallic reagents, a catalyst is not required for these reactions. Substitution reactions involving different organocuprates and esters are shown in Reactions (16-63) and (16-64).

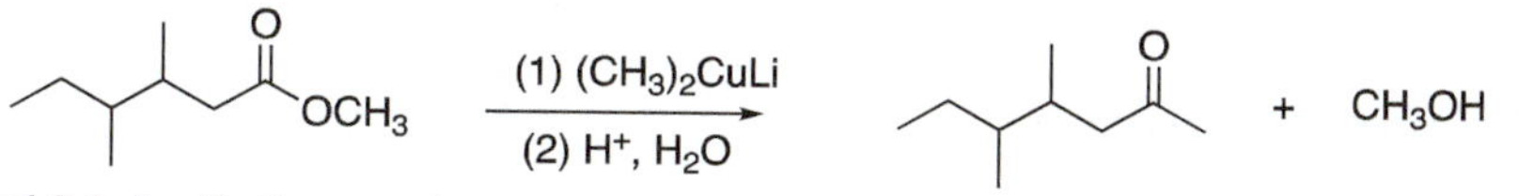

Methyl 3,4-dimethylhexanoate $\xrightarrow[(2) \; H^+, H_2O]{(1) \; (CH_3)_2CuLi}$ 4,5-Dimethyl-2-heptanone + CH_3OH (16-63)

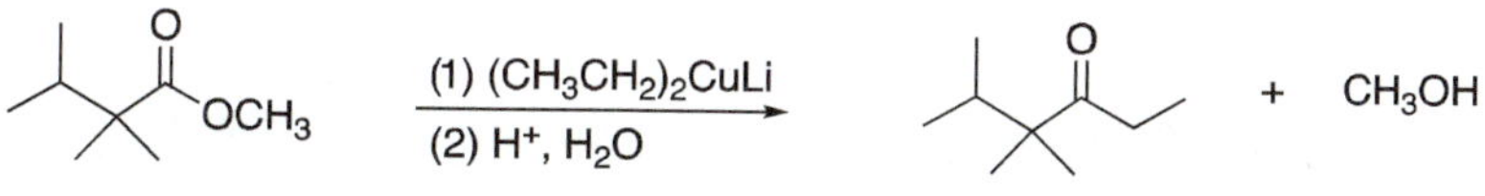

Methyl 2,2,3-trimethylbutanoate $\xrightarrow[(2) \; H^+, H_2O]{(1) \; (CH_3CH_2)_2CuLi}$ 4,4,5-Trimethyl-3-hexanone + CH_3OH (16-64)

16.5.5 Substitution Reactions of Esters with Hard Organometallic Reagents

As mentioned several times earlier, Grignard reagents in addition to being very strong bases are also good reducing agents, and they will react with esters to eventually form the corresponding alcohol after hydrolysis in the presence of excess Grignard reagents. The reactions shown in Reactions (16-65) and (16-66) are examples of Grignard reagents reacting with esters to produce alcohols after hydrolysis.

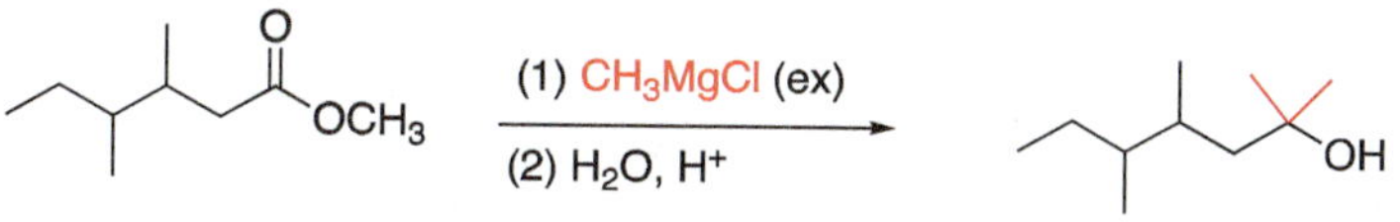

Methyl 3,4-dimethylhexanoate $\xrightarrow[(2) \; H_2O, H^+]{(1) \; CH_3MgCl \; (ex)}$ 2,4,5-Trimethyl-2-heptanol (16-65)

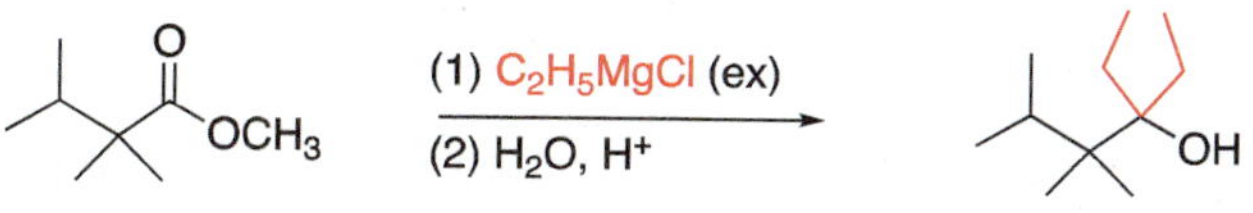

Methyl 2,2,3-trimethylbutanoate $\xrightarrow[(2) \; H_2O, H^+]{(1) \; C_2H_5MgCl \; (ex)}$ 3-Ethyl-4,4,5-trimethyl-3-hexanol (16-66)

Note that one of the outcomes for these reactions is that esters are transformed into tertiary alcohols, which contains two of the same alkyl groups that came from the Grignard reagent. These groups are shown in red in Reactions (16-65) and (16-66).

Problem 16.21

Show how to synthesize each of the following alcohols by using an ester and a Grignard Reagent. Hint: identify the two groups that are the same, they are from the Grignard reagent.

(a) 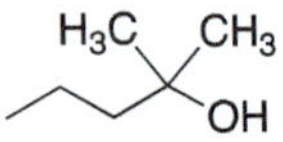 H₃C-CH(CH₃)-CH₂-C(CH₃)₂-OH... [structure: 2-methyl with OH and two CH₃ groups]

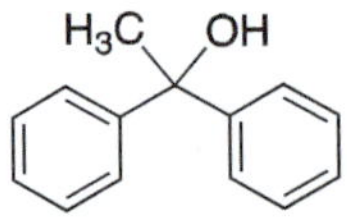

(b) diphenyl(methyl)methanol — Ph₂C(OH)(CH₃)

(c) 1-phenyl-1-propanol-type: PhC(OH)(C₂H₅)(C₂H₅)

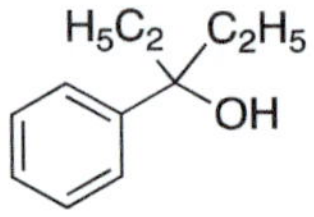

16.5.6 Substitution Reactions of Esters with Soft and Hard Metallic Hydrides

The reactions of esters with metal hydrides, such as tri-*tert*-butoxyaluminum hydride, result in aldehydes. Reactions shown in (16-67) and (16-68) are examples of the reaction of esters with tri-*tert*-butoxyaluminum hydride to produce aldehydes.

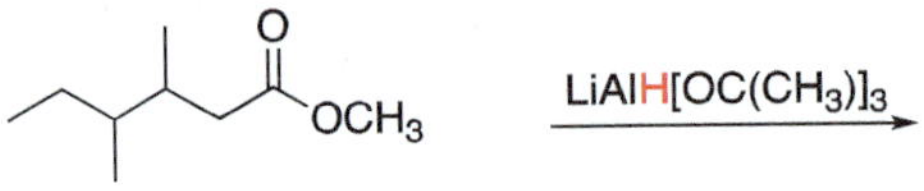

Methyl 3,4-dimethylhexanoate $\xrightarrow{\text{LiAlH[OC(CH}_3)_3]_3}$ 3,4-Dimethylhexanal (16-67)

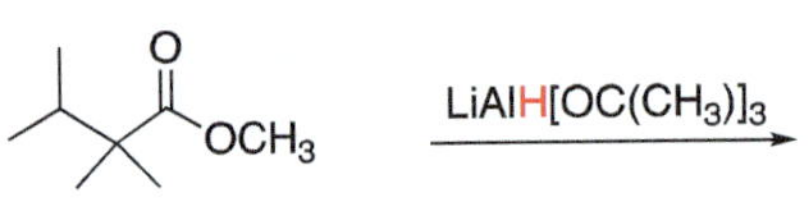

Methyl 2,2,3-trimethylbutanoate $\xrightarrow{\text{LiAlH[OC(CH}_3)_3]_3}$ 2,2,3-Trimethylbutanal (an aldehyde) (16-68)

On the other hand, in the reaction with a much stronger nucleophilic reducing agent, such as LAH, the corresponding alcohol will be the product as shown Reactions (16-69) and (16-70).

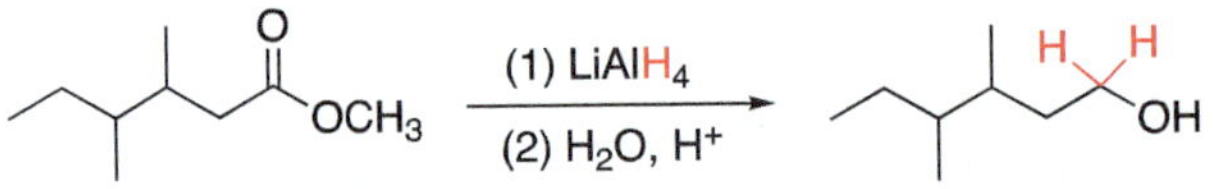

Methyl 3,4-dimethylhexanoate $\xrightarrow[\text{(2) H}_2\text{O, H}^+]{\text{(1) LiAlH}_4}$ 3,4-Dimethylhexanol (a primary alcohol) (16-69)

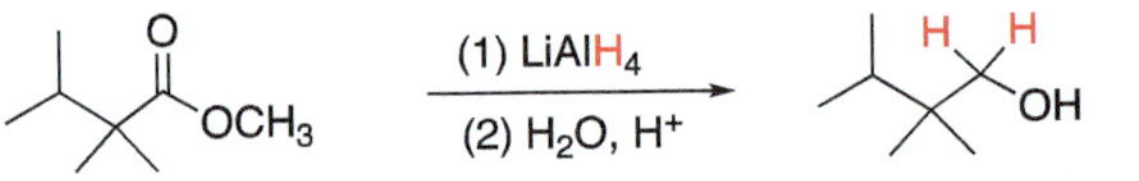

Methyl 2,2,3-trimethylbutanoate $\xrightarrow[\text{(2) H}_2\text{O, H}^+]{\text{(1) LiAlH}_4}$ 2,2,3-Trimethylbutanol (a primary alcohol) (16-70)

Problem 16.22

Show the structures of appropriate esters that can be used as starting compounds for the synthesis of the molecules shown below. Utilize LiAlH₄ as a reducing agent, followed by acidic workup to obtain the final product.

(a) 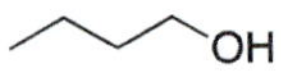 CH₃CH₂CH₂CH₂OH (b) 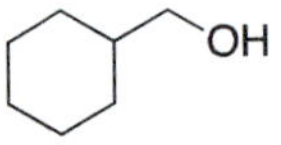 cyclohexyl-CH₂OH (c) 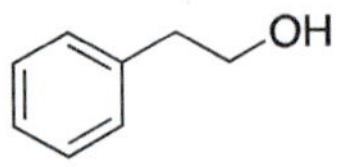 PhCH₂CH₂OH

16.5.7 Substitution Reactions of Esters with Enolates of Esters

Enolates of esters are formed by the deprotonation of an ester that contain at least one α-hydrogen, as shown in Reaction (16-71).

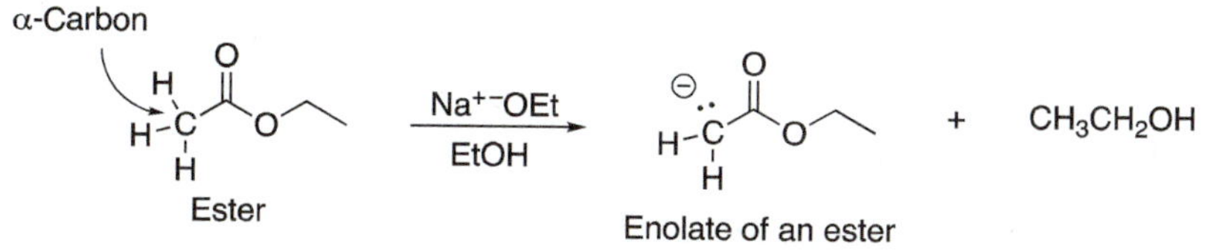

$$\text{Ester} \xrightarrow[\text{EtOH}]{\text{Na}^+ \ ^-\text{OEt}} \text{Enolate of an ester} + CH_3CH_2OH \tag{16-71}$$

The reactions that enolates of esters undergo are very similar to the substitution reactions of enolates that were discussed in Chapter 15, except in this case, the enolate can react with another mole of an ester by an acyl substitution reaction as shown in Reaction (16-72) to produce a β-keto ester.

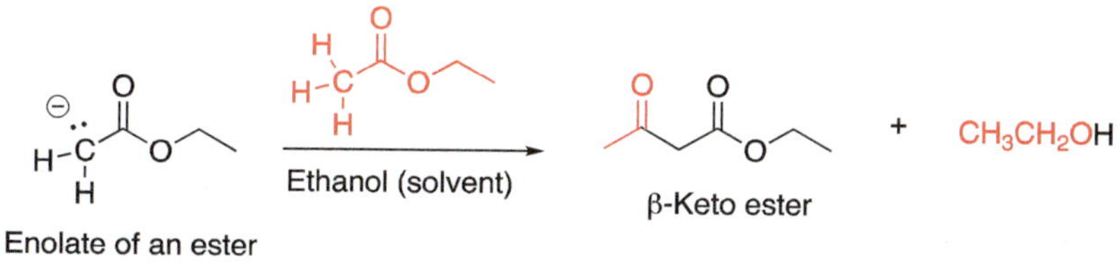

$$\text{Enolate of an ester} \xrightarrow{\text{Ethanol (solvent)}} \text{β-Keto ester} + CH_3CH_2OH \tag{16-72}$$

This type of acyl substitution condensation reaction is known as a Claisen condensation reaction. As you can imagine, it is possible for an intramolecular Claisen reaction to occur as shown in Reaction (16-73).

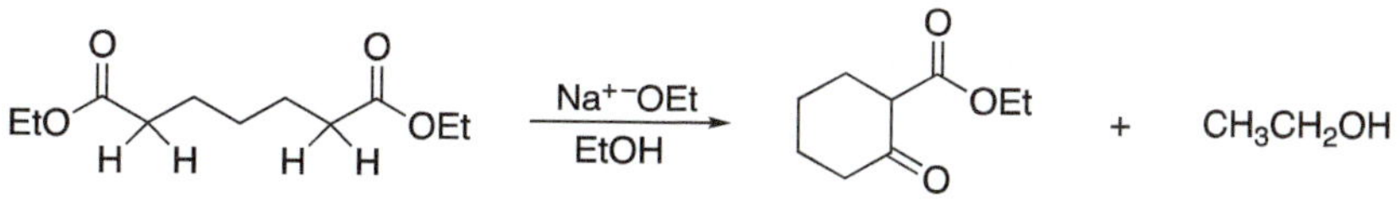

$$\xrightarrow[\text{EtOH}]{\text{Na}^+ \ ^-\text{OEt}} + CH_3CH_2OH \tag{16-73}$$

In the first step of the mechanism, the ethoxide anion base abstracts a proton from the alpha carbon to form an enolate anion, which then attacks the carbonyl carbon of the ester as shown in Reaction (16-74) to form a tetrahedral intermediate.

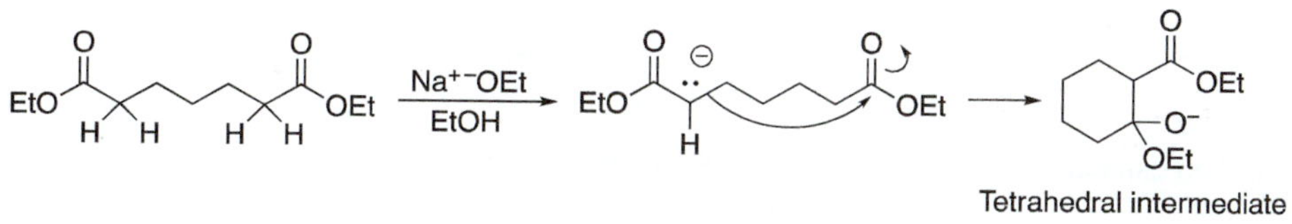

$$\xrightarrow[\text{EtOH}]{\text{Na}^+ \ ^-\text{OEt}} \longrightarrow \text{Tetrahedral intermediate} \tag{16-74}$$

In the next step of the mechanism, the tetrahedral intermediate forms the carbonyl compound with the ethoxide anion being protonated by the solvent to form ethanol as the other product as shown in Reaction (16-75).

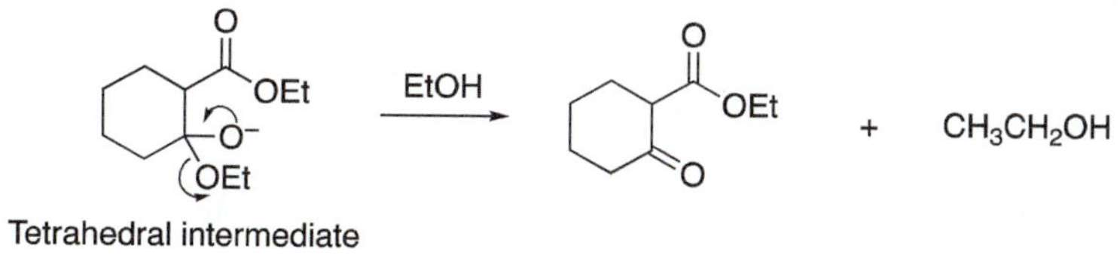

$$\text{Tetrahedral intermediate} \xrightarrow{\text{EtOH}} + CH_3CH_2OH \tag{16-75}$$

This type of intramolecular Claisen condensation is known as the Dieckmann reaction. Another example of the Dieckmann reaction is shown in Reaction (16-76).

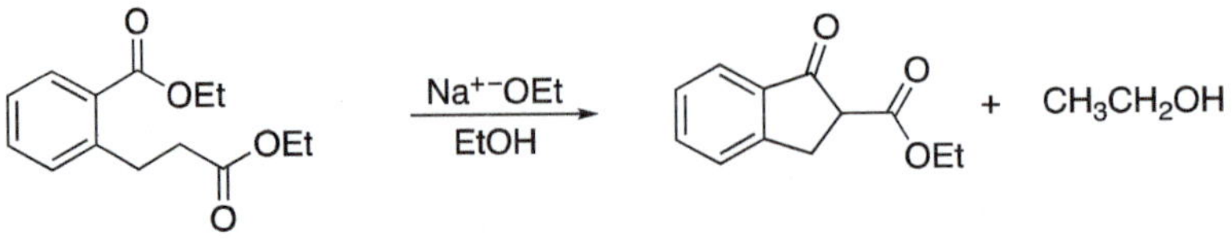

$$\text{(diester)} \xrightarrow[\text{EtOH}]{\text{Na}^+ \text{}^-\text{OEt}} \text{(cyclic β-keto ester)} + CH_3CH_2OH \qquad (16\text{-}76)$$

As you can imagine, it is possible for enolates of esters that have α-hydrogen(s) to react with esters that do not have α-hydrogen(s) as shown in the reaction in Reaction (16-77).

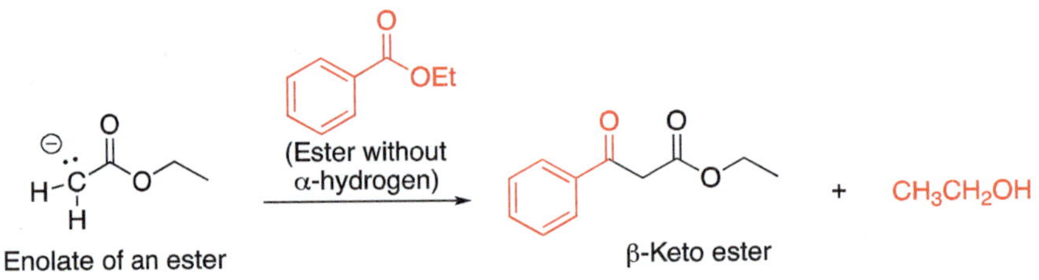

$$\text{Enolate of an ester} \xrightarrow{\text{(Ester without α-hydrogen)}} \text{β-Keto ester} + CH_3CH_2OH \qquad (16\text{-}77)$$

These types of reactions are known as crossed Claisen condensation reactions.

Since the compounds produced from these reactions are esters, they can be hydrolyzed in the presence of acid and water to form carboxylic acids as shown in the example given in Reaction (16-78).

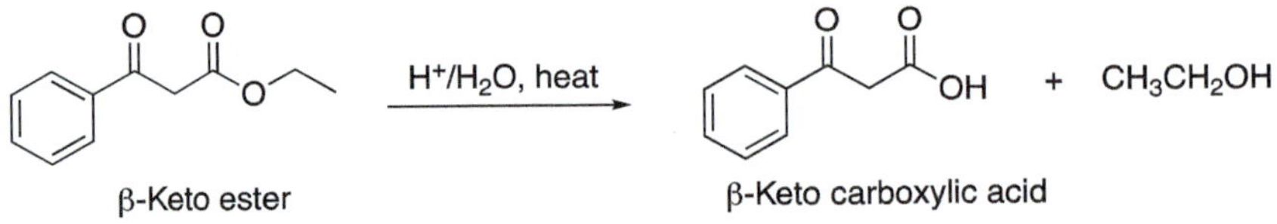

$$\text{β-Keto ester} \xrightarrow{\text{H}^+/\text{H}_2\text{O, heat}} \text{β-Keto carboxylic acid} + CH_3CH_2OH \qquad (16\text{-}78)$$

In the presence of heat, β-keto carboxylic acids readily decompose to give carbon dioxide and a ketone as shown in the example given in Reaction (16-79).

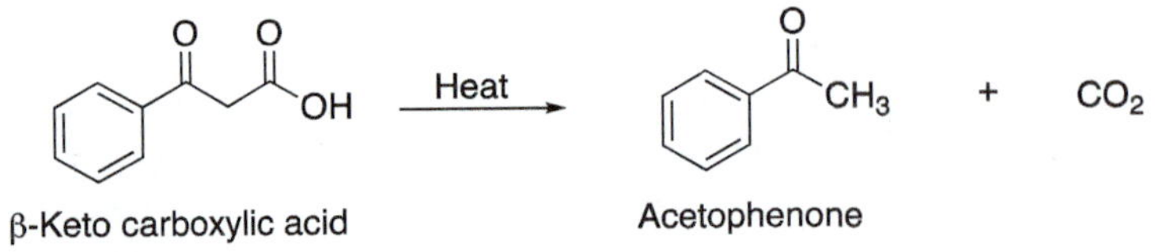

$$\text{β-Keto carboxylic acid} \xrightarrow{\text{Heat}} \text{Acetophenone} + CO_2 \qquad (16\text{-}79)$$

The loss of carbon dioxide from a β-keto carboxylic acid results in the formation of an enol as shown in the mechanism given in Reaction (16-80).

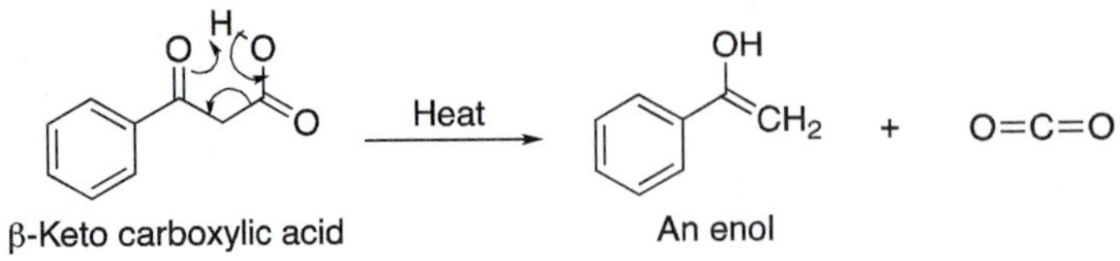

$$\text{β-Keto carboxylic acid} \xrightarrow{\text{Heat}} \text{An enol} + O=C=O \qquad (16\text{-}80)$$

As we have seen previously, enols readily tautomerize to give the more stable keto form, as shown in the example given in Reaction (16-81).

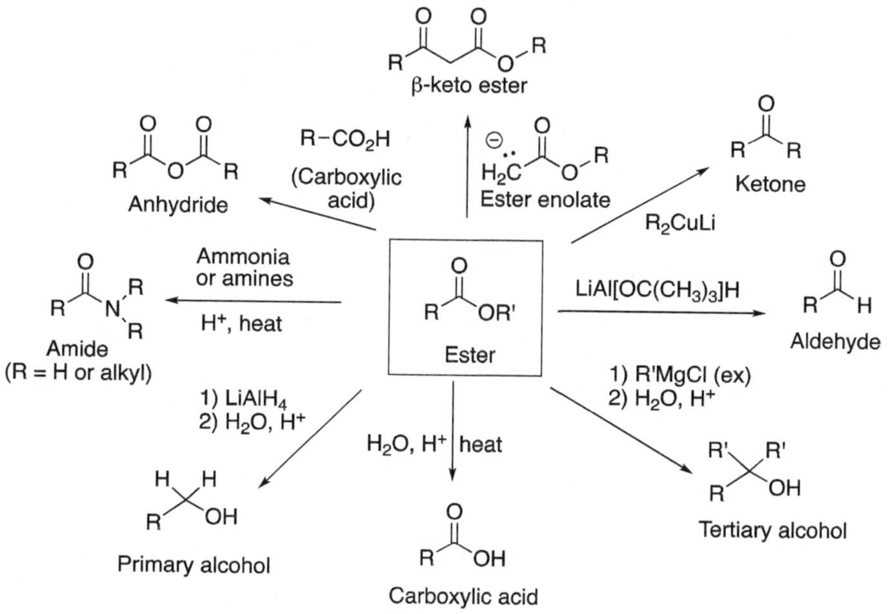

Figure 16.2 Summary of the reactions of esters.

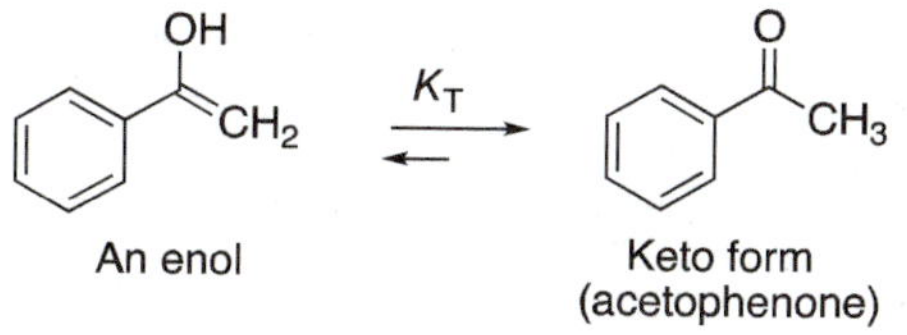

$$\text{An enol} \underset{}{\overset{K_T}{\rightleftharpoons}} \text{Keto form (acetophenone)} \tag{16-81}$$

Shown in Figure 16.2 is a summary of reactions that esters undergo.

16.6 Substitution Reactions Involving Amides

There are two important features of amides that lead to the type of reactions observed for these molecules. They have similar features as those of esters: they have a polarized carbonyl double bond, and they have a potential leaving group. Amides, however, have an extremely poor leaving group in the form of the amide anion. Thus, the general reaction type that these molecules undergo is very similar to those that we have covered but will closely mimic the substitution reactions of esters. The general reaction that these types of molecules undergo is shown in Reaction (16-82).

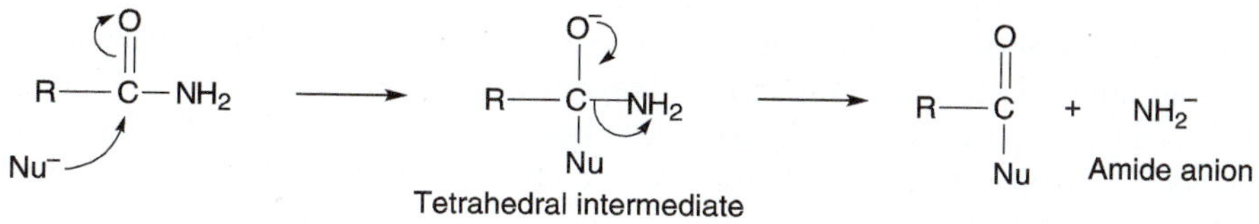

$$R-\underset{Nu^-}{\overset{O}{\overset{\|}{C}}}-NH_2 \longrightarrow R-\underset{Nu}{\overset{O^-}{\underset{|}{C}}}-NH_2 \longrightarrow R-\overset{O}{\overset{\|}{C}}-Nu + NH_2^- \tag{16-82}$$

Tetrahedral intermediate — Amide anion

The first step of the reaction mechanism is an addition reaction to the electrophilic carbon of the carbon–oxygen double bond to form a tetrahedral intermediate. The second step involves the elimination of the very basic and extremely poor leaving group, the amide anion. You will recall from Chapter 7 on acids and bases that the pK_a of the conjugate acid of the amide anion is ammonia, which has a pK_a of 35. This means that the amide anion is a very strong base and an extremely poor leaving group. As a result, these substitution reactions take place under acid catalysis in order to convert the very poor amide anion to protonated amide, which is a much better leaving group. These reactions also take place at elevated temperatures.

The first step of the mechanism for the reaction of amides with a nucleophile in an acidic medium is shown in Reaction (16-83).

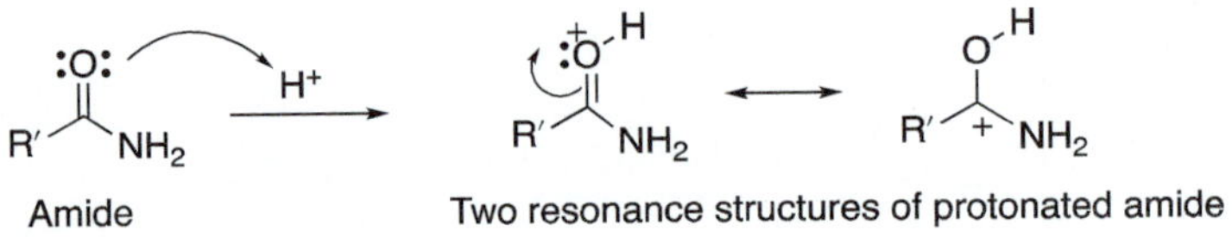

$$\text{Amide} \xrightarrow{H^+} \text{Two resonance structures of protonated amide} \tag{16-83}$$

In the second step as shown in Reaction (16-84), the nucleophile attacks the electrophilic carbon of the carbon–oxygen bond.

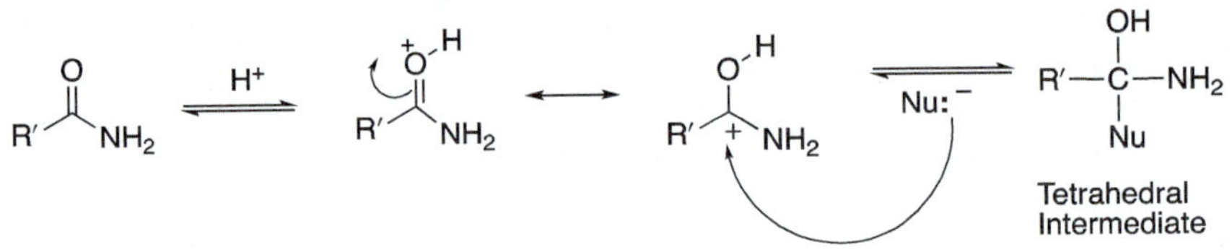

$$\tag{16-84}$$

Since the medium is acidic, protonation of the amide group occurs as shown in Reaction (16-85).

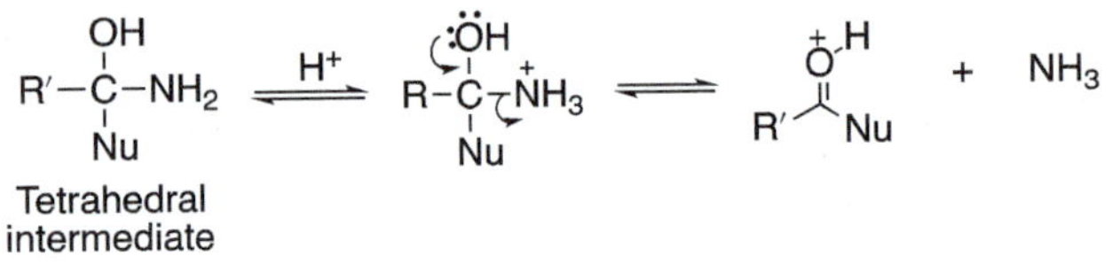

$$\tag{16-85}$$

In the last step of the mechanism, there is an acid–base reaction as given in Reaction (16-86).

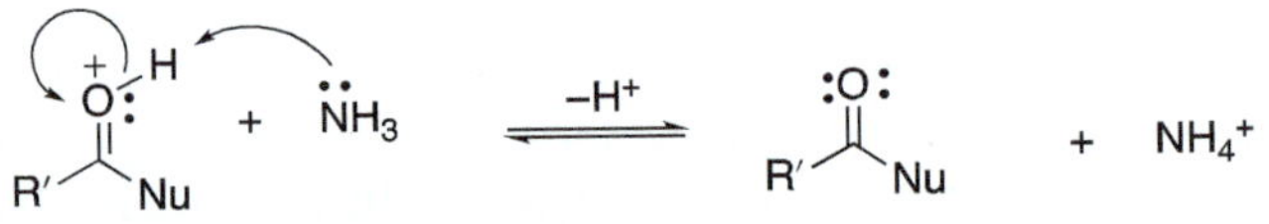

$$\tag{16-86}$$

16.6.1 Substitution Reactions of Amides with Water

It is obvious from the above mechanism that this reaction will not likely take place under basic conditions, but under acidic conditions where the amide anion can be protonated converting it to a better leaving group. Amides are hydrolyzed under acidic conditions to the corresponding carboxylic acid as given in the example in Reaction (16-87) for the hydrolysis of acetamide to produce acetic acid and an ammonium salt.

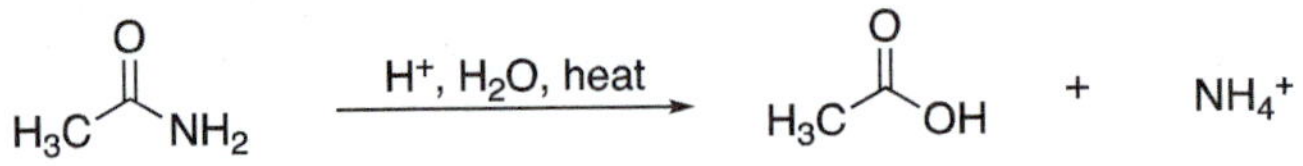

$$H_3C-C(=O)-NH_2 \xrightarrow{H^+, H_2O, \text{heat}} H_3C-C(=O)-OH + NH_4^+ \tag{16-87}$$

Note that these reactions typically occur at elevated temperatures. Another example is given below for the hydrolysis of benzamide in which the mechanism for the hydrolysis is shown below.

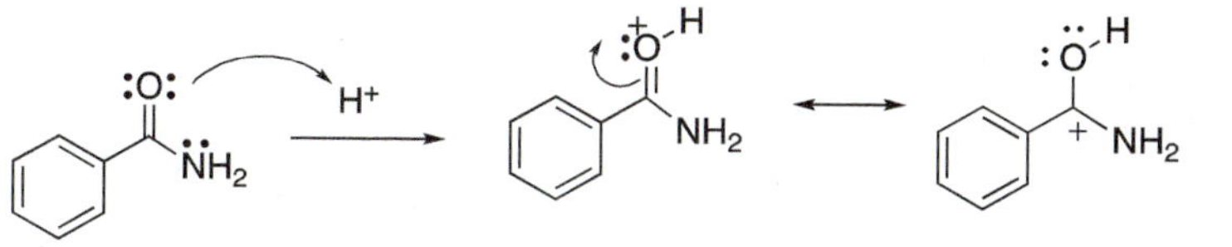

Since this newly formed intermediate is charged, the attack of the nucleophilic water can be accomplished.

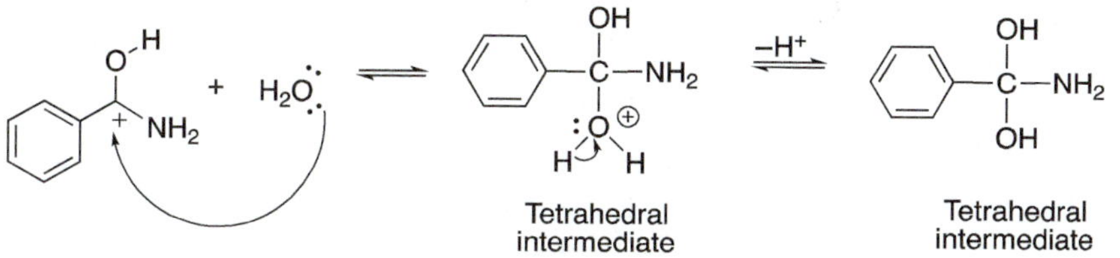

The tetrahedral intermediate is then protonated on the nitrogen to form a new protonated intermediate, which loses ammonia.

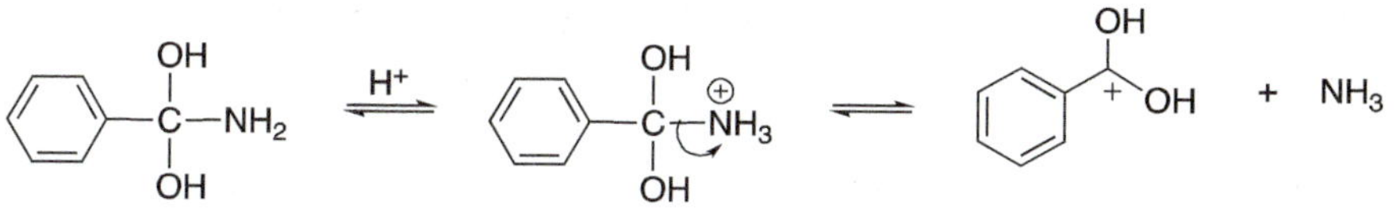

By losing a proton, this intermediate will generate the carboxylic acid.

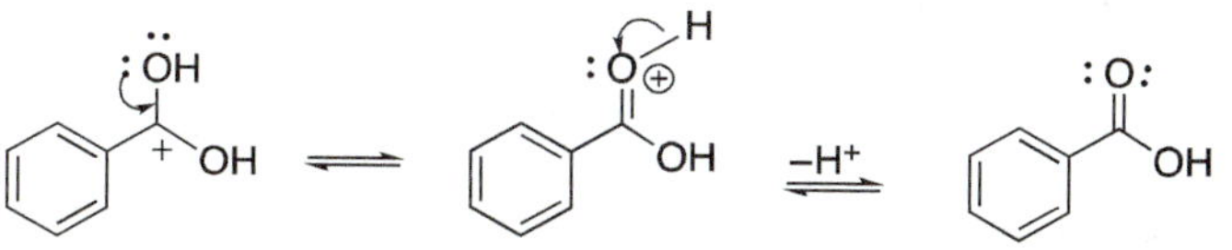

In the last step of the reaction, the ammonia is protonated since the medium is acidic.

$$NH_3 \underset{}{\overset{H^+}{\rightleftharpoons}} NH_4^+$$

Problem 16.23

Give the products that result from the acid hydrolysis of the following amides.

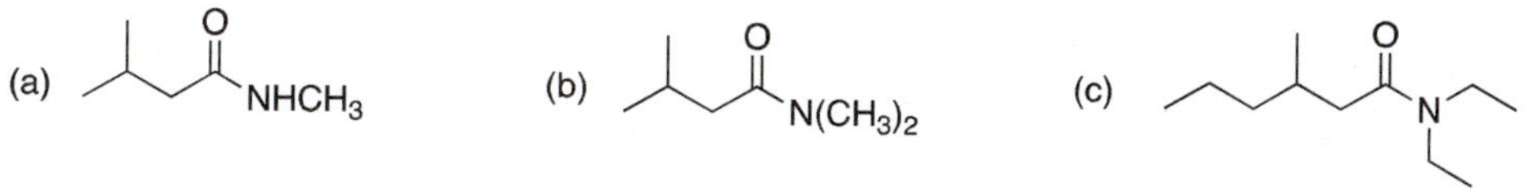

(a) [structure with NHCH$_3$] (b) [structure with N(CH$_3$)$_2$] (c) [structure with N(ethyl)(methyl)]

16.6.2 Substitution Reactions of Amides with Hard Metallic Hydrides

Amides can be reduced to form the corresponding amine. Owing to the polarity of the carbonyl bond of amides, hydride ions can be added to the carbonyl carbon from strong reducing agents such as LiAlH$_4$, as shown in Reaction (16-88).

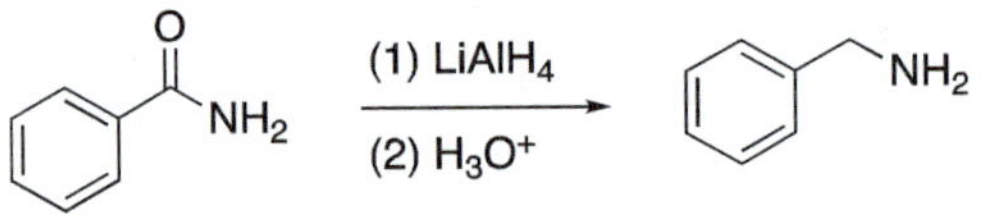

$$\text{PhC(O)NH}_2 \xrightarrow[\text{(2) H}_3\text{O}^+]{\text{(1) LiAlH}_4} \text{PhCH}_2\text{NH}_2 \quad (16\text{-}88)$$

The mechanism for the above reaction involves an attack of the carbonyl carbon by the hydride ion from LiAlH$_4$ as shown in Reaction (16-89).

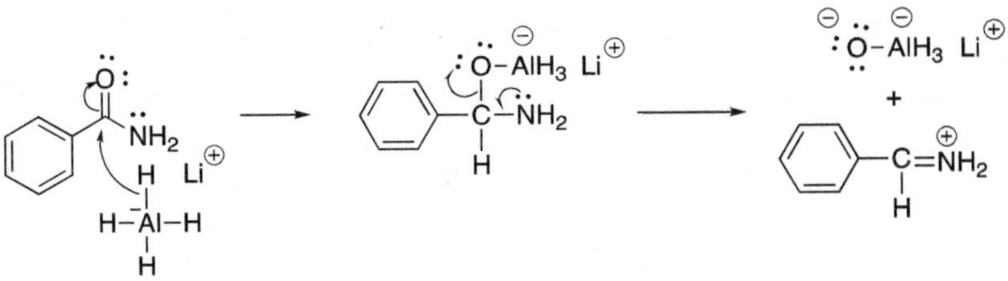

$$(16\text{-}89)$$

You will recall that the NH$_2$ is very basic and hence an extremely poor leaving group. Also, in this reaction using LAH, there is no proton present to convert the NH$_2$ group into a good leaving group. As a result, the aluminum acts as a Lewis acid to assist with the departure of the oxygen as shown in Reaction (16-89). The iminium-type of intermediate is then attacked by another mole of hydride ions as shown in Reaction (16-90) to form the final amine product.

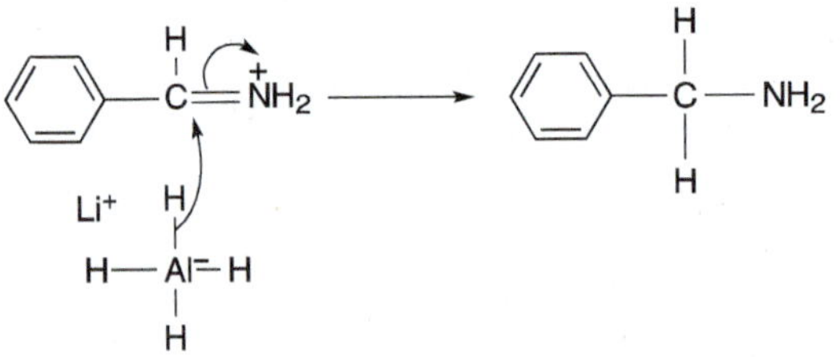

$$(16\text{-}90)$$

Problem 16.24

Give the appropriate starting amide, which upon reduction with LiAlH$_4$ will give the amines shown below.

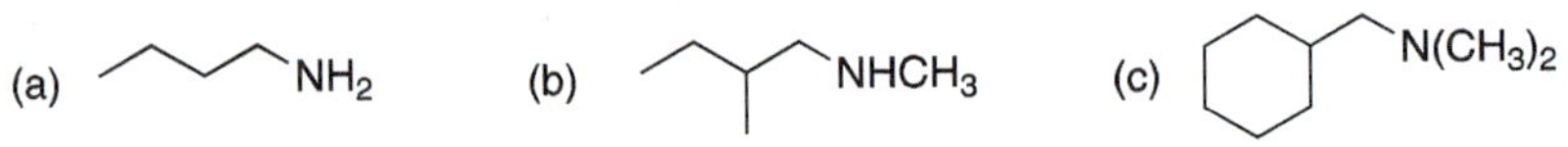

(a) CH$_3$CH$_2$CH$_2$NH$_2$ (b) (CH$_3$)$_2$CHCH$_2$NHCH$_3$ (c) C$_6$H$_{11}$CH$_2$N(CH$_3$)$_2$

16.7 Substitution Reactions Involving Carboxylic Acids

As you can imagine, carboxylic acids are not very reactive since the OH group is a very poor leaving group as shown in Reaction (16-91).

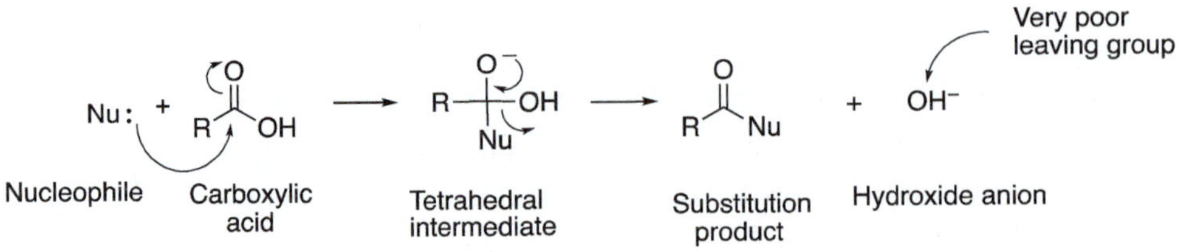

$$\underset{\text{Nucleophile}}{\text{Nu:}^-} + \underset{\text{Carboxylic acid}}{\text{R–C(O)–OH}} \longrightarrow \underset{\text{Tetrahedral intermediate}}{\text{R–C(O}^-\text{)(Nu)–OH}} \longrightarrow \underset{\text{Substitution product}}{\text{R–C(O)–Nu}} + \underset{\text{Hydroxide anion}}{\text{OH}^-}$$

Very poor leaving group

$$(16\text{-}91)$$

16.7 Substitution Reactions Involving Carboxylic Acids

It is possible to carry out a substitution reaction in the presence of acid, which will protonate the —OH group and convert it to a good leaving group as shown in the general reaction given in Reaction (16-92).

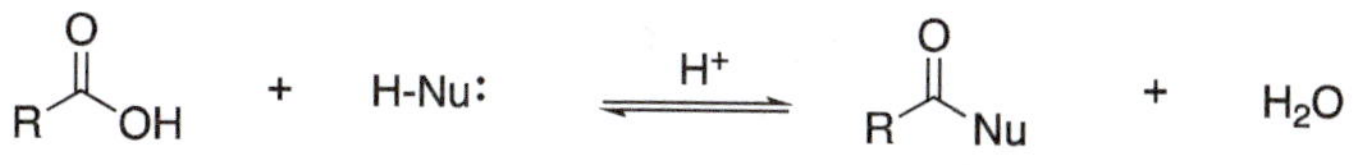

$$R-C(=O)-OH + H-Nu: \xrightleftharpoons{H^+} R-C(=O)-Nu + H_2O \tag{16-92}$$

The mechanism for the acid-catalyzed reaction is shown below. In the first step, the carbonyl oxygen is protonated to give a resonance-stabilized carbocation as shown in Reaction (16-93).

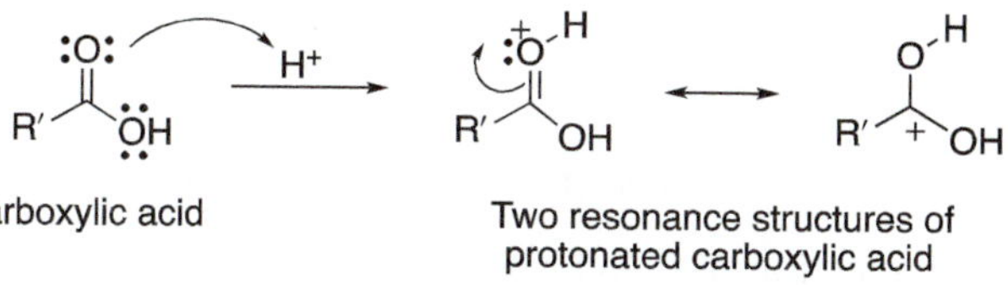

Carboxylic acid → Two resonance structures of protonated carboxylic acid (16-93)

In the next step of the reaction mechanism, the nucleophile attacks the protonated carboxylic acid, as shown in Reaction (16-94), to produce a tetrahedral intermediate.

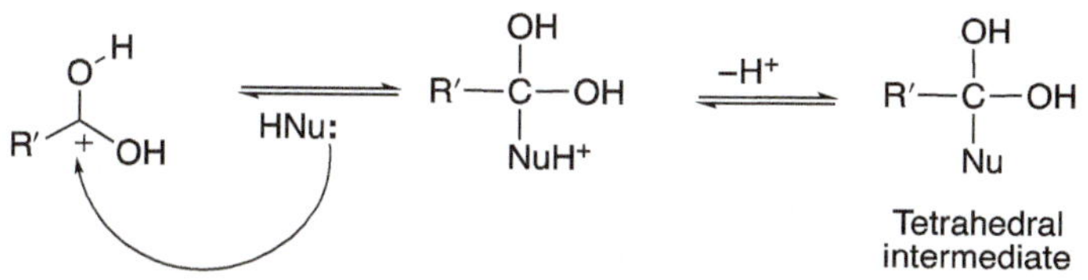

(16-94) Tetrahedral intermediate

In the next step of the mechanism, protonated hydroxide leaves to produce water, as shown in Reaction (16-95) and an acid catalyst is regenerated to give the products.

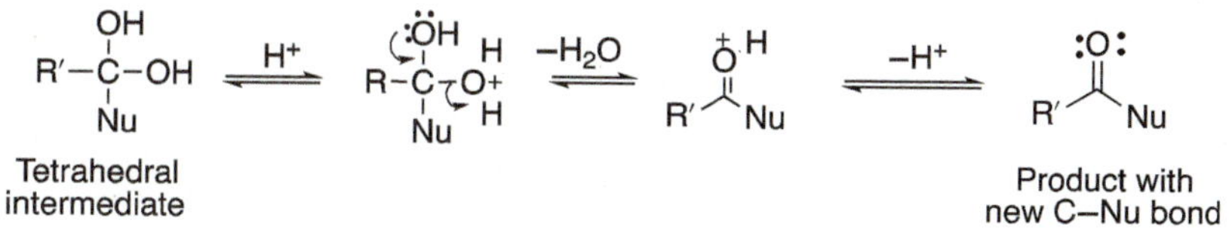

Tetrahedral intermediate → Product with new C–Nu bond (16-95)

16.7.1 Substitution Reactions of Carboxylic Acids with Alcohols

A typical acid-catalyzed reaction of a benzoic acid with ethanol is shown in Reaction (16-96).

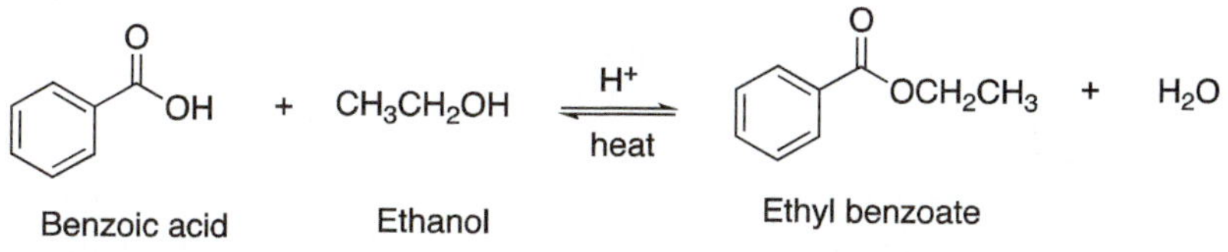

Benzoic acid + Ethanol $\xrightarrow{H^+, \text{heat}}$ Ethyl benzoate + H_2O (16-96)

Note that the product of an acid-catalyzed reaction of a carboxylic acid and an alcohol is a new molecule with a different functional group, an ester.

Problem 16.25

Give esters that result from each of the esterification reactions shown below.

(a) 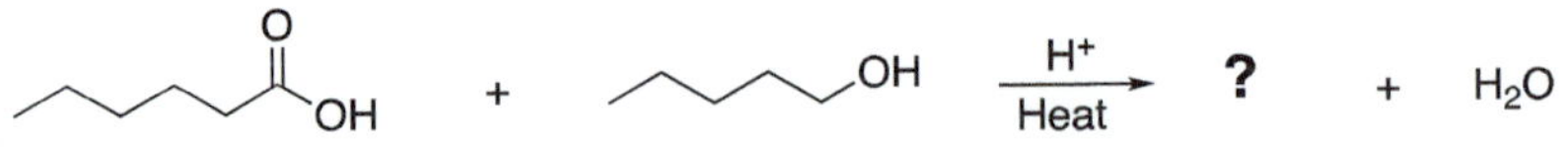 CH₃CH₂CH₂CH₂COOH + CH₃CH₂CH₂CH₂OH $\xrightarrow{\text{H}^+, \text{Heat}}$? + H₂O

(b) 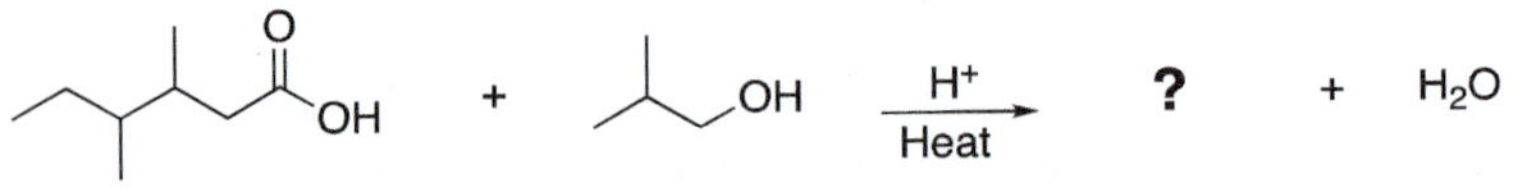 (CH₃)₂CHCH₂CH₂COOH + (CH₃)₂CHOH $\xrightarrow{\text{H}^+, \text{Heat}}$? + H₂O

(c) 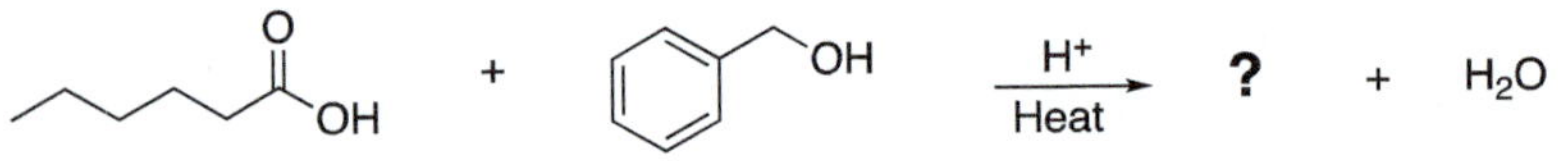 CH₃CH₂CH₂CH₂COOH + C₆H₅CH₂OH $\xrightarrow{\text{H}^+, \text{Heat}}$? + H₂O

16.7.2 Substitution Reactions of Carboxylic Acid with Ammonia and Amines

Let us now examine the substitution reaction of carboxylic acids with ammonia and amines. These reactions are very similar to those that we have seen before, except as mentioned, an acidic catalyst must be used as shown in Reaction (16-97) for the acid-catalyzed reaction of benzoic acid with ammonia.

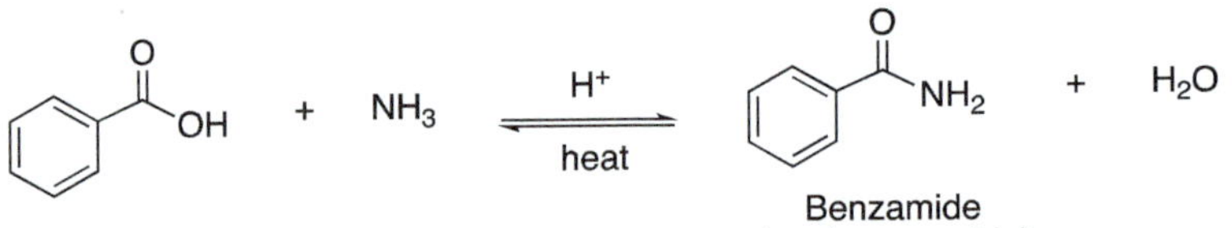

C₆H₅COOH + NH₃ $\xrightleftharpoons[\text{heat}]{\text{H}^+}$ C₆H₅CONH₂ + H₂O   (16-97)

Benzoic acid   Benzamide
                                 (a primary amide)

Reaction (16-98) shows the reaction of benzoic acid with a primary amine, methylamine.

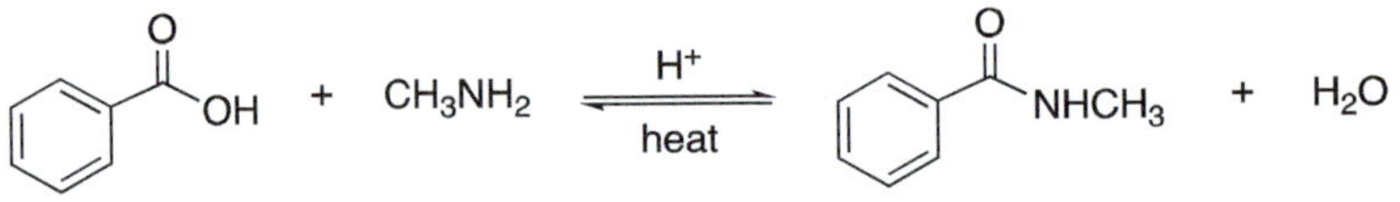

C₆H₅COOH + CH₃NH₂ $\xrightleftharpoons[\text{heat}]{\text{H}^+}$ C₆H₅CONHCH₃ + H₂O   (16-98)

Benzoic acid   Methylamine   N-Methylbenzamide
                                (a secondary amide)

Reaction (16-99) shows the reaction of an ester with a secondary amine, dimethylamine.

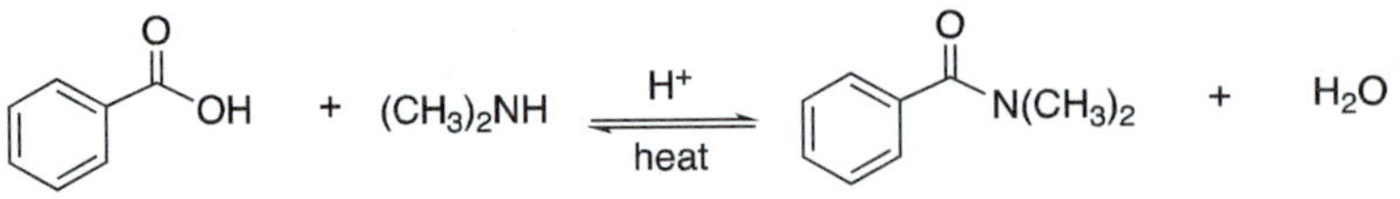

C₆H₅COOH + (CH₃)₂NH $\xrightleftharpoons[\text{heat}]{\text{H}^+}$ C₆H₅CON(CH₃)₂ + H₂O   (16-99)

Benzoic acid   Dimethylamine   N,N-Dimethylbenzamide
                                (a tertiary amide)

Since ammonia and amines are nucleophilic molecules, they will react with carboxylic acids to form the corresponding primary amide (with ammonia); secondary amide (with primary amines); and tertiary amide (with secondary amines).

Problem 16.26

Give the organic products of the following reactions.

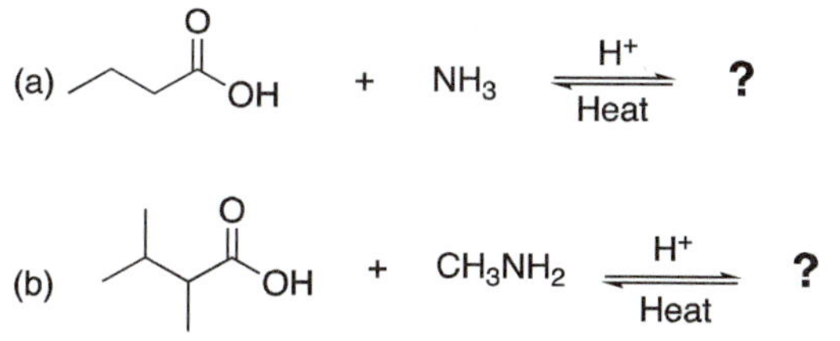

(a) CH₃CH₂C(=O)OH + NH₃ $\xrightleftharpoons[\text{Heat}]{\text{H}^+}$?

(b) (CH₃)₂CHC(=O)OH + CH₃NH₂ $\xrightleftharpoons[\text{Heat}]{\text{H}^+}$?

16.7.3 Substitution Reactions of Carboxylic Acids with Hard Metallic Hydrides

Carboxylic acids can be reduced to form the corresponding alcohol by the supply of electrons and hydrogens, i.e. the hydride ion (H⁻). Owing to the polarity of the carbonyl bond of carboxylic acids, hydride ions can be added, from strong reducing agents, such as LiAlH₄, to the polarized carbonyl bond of carboxylic acids. Reaction (16-100) gives an example of the reduction of a carboxylic acid by LAH.

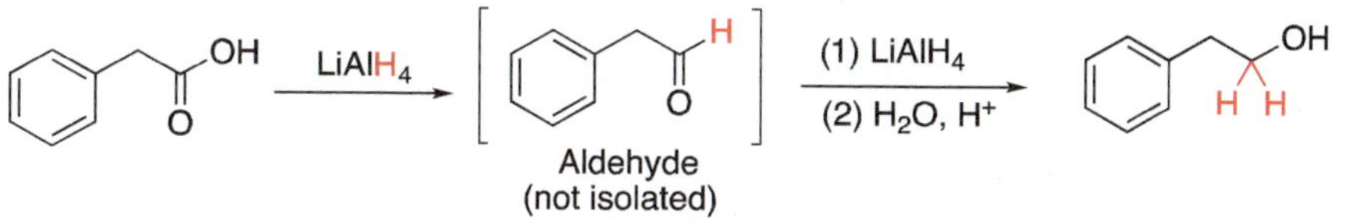

PhCH₂COOH $\xrightarrow{\text{LiAlH}_4}$ [PhCH₂CHO] (Aldehyde, not isolated) $\xrightarrow[(2)\ \text{H}_2\text{O, H}^+]{(1)\ \text{LiAlH}_4}$ PhCH₂CH₂OH (16-100)

Since carboxylic acids are fairly acidic and LiAlH₄ is an extremely strong base, the first step of the reaction is an acid–base reaction as shown in Reaction (16-101).

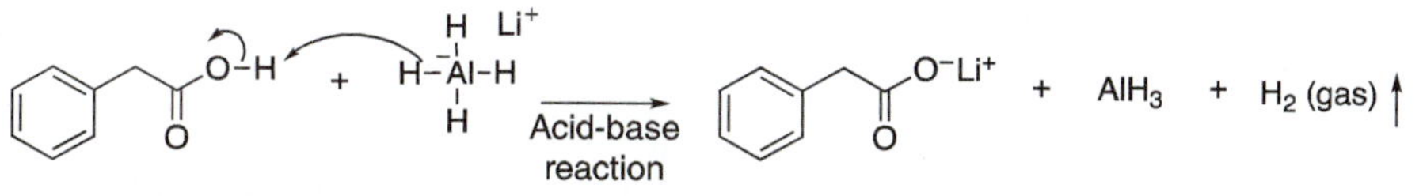

PhCH₂C(=O)O–H + H–AlH₃⁻ Li⁺ $\xrightarrow{\text{Acid-base reaction}}$ PhCH₂C(=O)O⁻Li⁺ + AlH₃ + H₂ (gas)↑ (16-101)

Once deprotonation occurs to form the carboxylate salt, a reduction takes place. In the first step, a hydride ion from LiAlH₄ is delivered to the carbonyl carbon as shown in Reaction (16-102) to form the aldehyde.

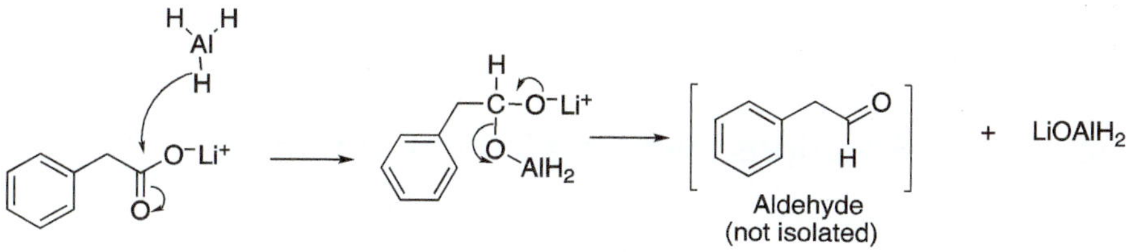

PhCH₂C(=O)O⁻Li⁺ + AlH₄⁻ → PhCH₂CH(O⁻Li⁺)(O–AlH₂) → [PhCH₂CHO] (Aldehyde, not isolated) + LiOAlH₂ (16-102)

In the next step of the mechanism, the aldehyde is reduced by another mole of LiAlH₄ to form the alkoxide salt. In a separate reaction involving an acid–base reaction, the alkoxide anion is neutralized to give the final alcohol product as shown in Reaction (16-103).

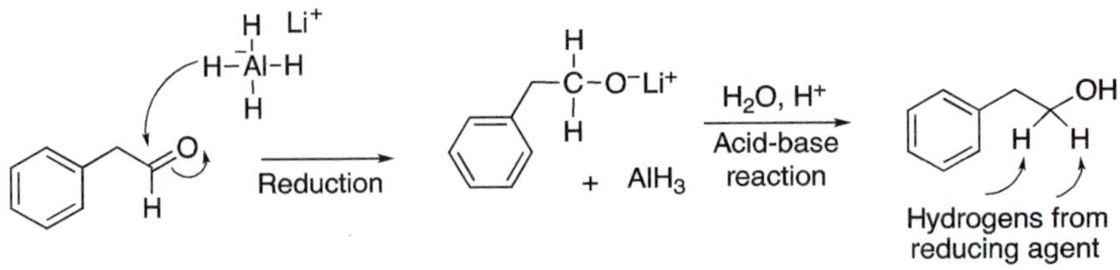

16 Nucleophilic Substitution Reactions at Acyl Carbons

$$\text{PhCH}_2\text{CHO} \xrightarrow[\text{Reduction}]{\text{LiAlH}_4} \text{PhCH}_2\text{CH(H)}-\text{O}^-\text{Li}^+ + \text{AlH}_3 \xrightarrow[\text{Acid-base reaction}]{\text{H}_2\text{O, H}^+} \text{PhCH}_2\text{CH}_2\text{OH} \qquad (16\text{-}103)$$

(Hydrogens from reducing agent)

Problem 16.27

Give the reduction products of the following reactions.

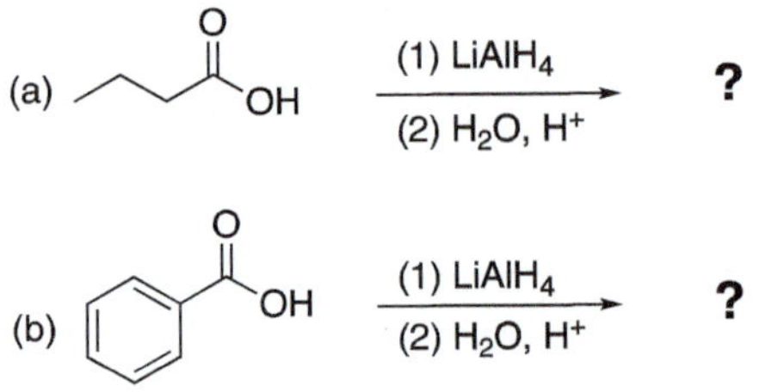

(a) CH$_3$CH$_2$CH$_2$C(O)OH $\xrightarrow{\text{(1) LiAlH}_4 \;\; \text{(2) H}_2\text{O, H}^+}$?

(b) PhC(O)OH $\xrightarrow{\text{(1) LiAlH}_4 \;\; \text{(2) H}_2\text{O, H}^+}$?

16.8 Substitution Reactions Involving Oxalyl Chloride

Oxalyl chloride is a unique organic compound that has adjacent carbonyl groups and chlorines bonded to each carbonyl carbon as shown below. These structural features make oxalyl chloride a very reactive compound, which is primarily used to convert carboxylic acids to acid chlorides. This reaction is a substitution reaction in which the —OH group from the carboxylic acid is substituted with a chlorine from the oxalyl chloride. An example of this reaction is shown in Reaction (16-104).

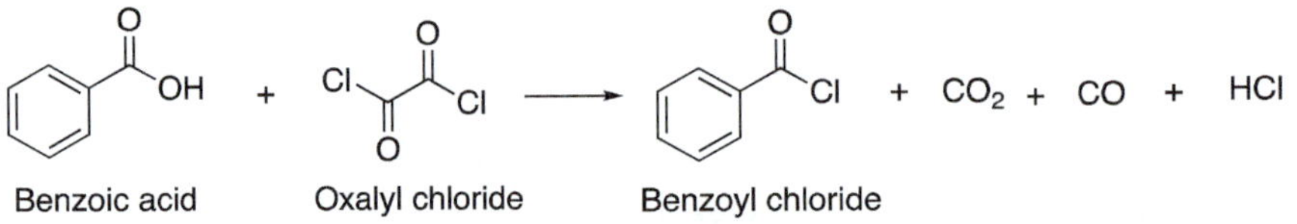

$$\underset{\text{Benzoic acid}}{\text{PhCOOH}} + \underset{\text{Oxalyl chloride}}{\text{ClC(O)C(O)Cl}} \longrightarrow \underset{\text{Benzoyl chloride}}{\text{PhCOCl}} + \text{CO}_2 + \text{CO} + \text{HCl} \qquad (16\text{-}104)$$

A driving force for this reaction to proceed to give the products is that gaseous products are produced, an increase in entropy for the reaction.

16.9 Substitution Reactions Involving Sulfur Containing Compounds

Owing to the similarity of the carbon oxygen double bond with the sulfur oxygen double bond, we will examine substitution reactions involving thionyl chloride and phenyl sulfonyl chloride. Interestingly, these reactions are used to convert the poor leaving —OH group to a better leaving group, the chloride and the tosylate, respectively. Reaction (16-105) gives an example of the reaction of thionyl chloride with an optically active alcohol.

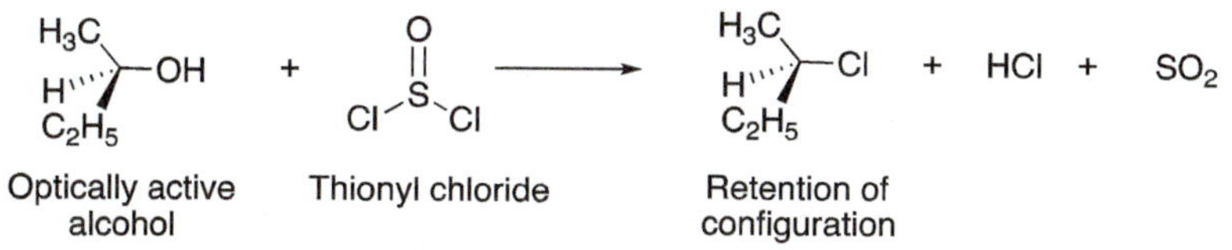

16.9 Substitution Reactions Involving Sulfur Containing Compounds

$$\text{Optically active alcohol} + \text{Thionyl chloride} \longrightarrow \text{Retention of configuration} + HCl + SO_2 \quad (16\text{-}105)$$

These reactions result in retention of configuration if the reactant is an optically active alcohol. The mechanism given below explains this outcome. In the first step, the nucleophilic alcohol attacks the electrophilic sulfur atom and a chloride ion is eventually displaced as shown in Reaction (16-106)

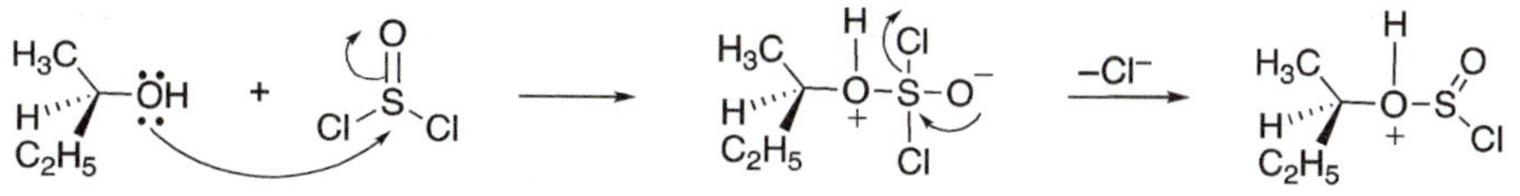

$$(16\text{-}106)$$

In another step, the chloride anion abstracts a proton to form HCl and the resulting intermediate breaks apart liberating sulfur dioxide (SO_2) an excellent leaving group, which leaves as a gas. Simultaneously, the other chloride anion leaves then bonds to the carbon from the same side, hence the retention of configuration as shown in Reaction (16-107).

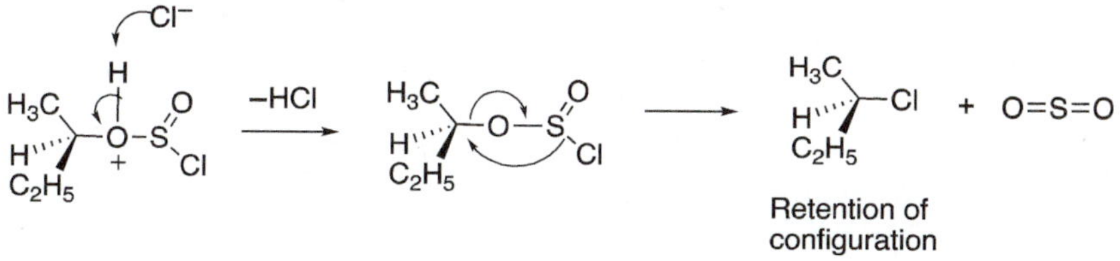

$$(16\text{-}107)$$

Note in the last step of the mechanism, there is an internal S_N2 attack of the chloride anion resulting in a substitution. Since this step involves an S_N2 attack, these reactions work best for primary or secondary alcohols.

Problem 16.28

Give the products of the reaction of each of the compounds shown below with $SOCl_2$, give stereochemistry where appropriate.

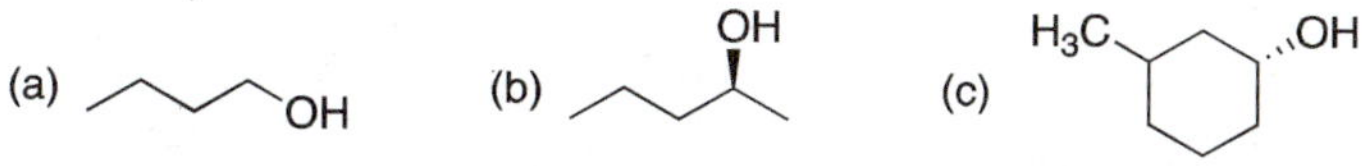

(a), (b), (c)

Another sulfur reagent that can be involved in similar substitution reactions is *p*-toluenesulfonyl chloride, also known as tosyl chloride (TsCl). The general reaction of TsCl with a nucleophile is shown in Reaction (16-108).

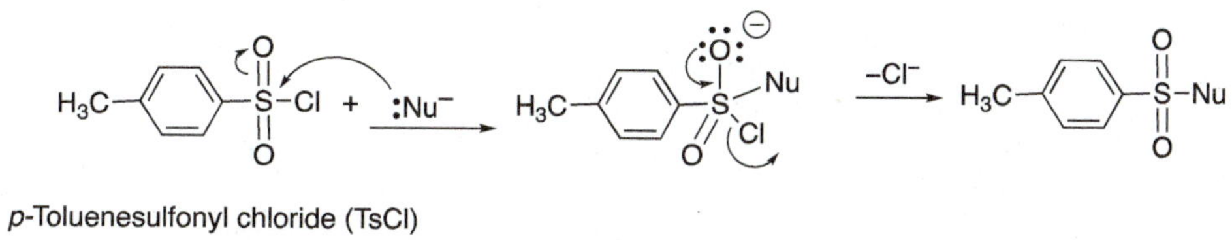

p-Toluenesulfonyl chloride (TsCl)

$$(16\text{-}108)$$

As you can imagine, the sulfur atom is very electrophilic since it is bonded to two electronegative oxygen atoms and also to a chlorine, which is a good leaving group. A specific example of this substitution reaction is given in Reaction (16-109).

$$\text{H}_3\text{C}-\text{C}_6\text{H}_4-\text{S(O)}_2-\text{Cl} + \text{(CH}_3\text{)}_2\text{CHOH} \longrightarrow \text{H}_3\text{C}-\text{C}_6\text{H}_4-\text{S(O)}_2-\text{OCH(CH}_3\text{)}_2 + \text{HCl} \quad (16\text{-}109)$$

An advantage of reactions of this type is that the very poor —OH leaving of alcohols is essentially converted to an extremely good leaving group. Reactions of this type give another option of converting the —OH of alcohols to a good leaving group as demonstrated in the reaction in Reaction (16-110).

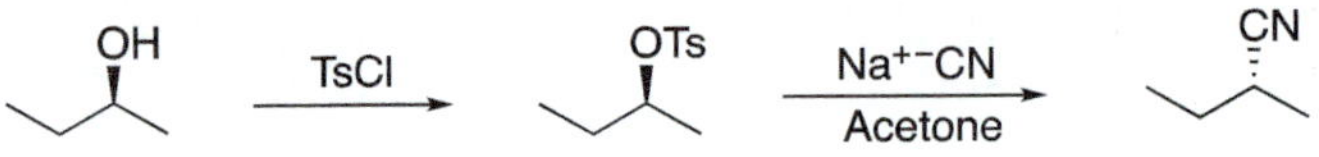

$$\text{R-OH} \xrightarrow{\text{TsCl}} \text{R-OTs} \xrightarrow[\text{Acetone}]{\text{Na}^+\text{-CN}} \text{R-CN} \quad (16\text{-}110)$$

For this above reaction, note that the stereochemistry of the alcohol reactant is retained for the first reaction. Since the second reaction in the given sequence of reactions is S$_N$2 (as indicated by the solvent), there is an inversion of stereochemistry in the product.

16.10 Applications of Acyl Substitution Reactions

16.10.1 Preparation of Esters

As pointed out earlier, for the synthesis of esters, there are two important bonds that can be made to put this type of molecule together. A bond that can be made is the bond to the carbon of the carbonyl carbon–oxygen bond. This bond can be made from a reaction of the following type. Below is shown a specific example to synthesize propyl ethanoate. The leaving group can of course be a chloride anion, in that case, the starting material is an acid chloride.

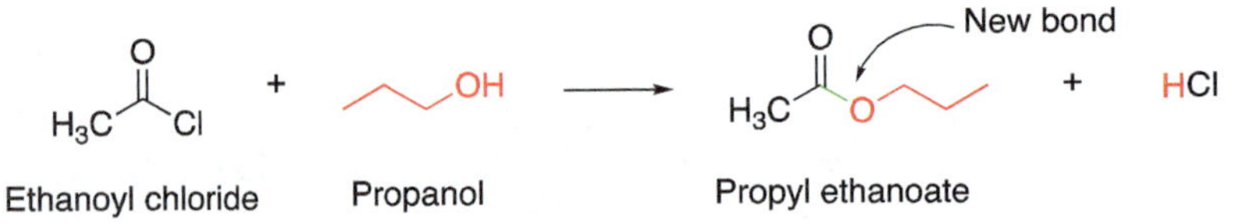

$$\underset{\text{Ethanoyl chloride}}{\text{H}_3\text{C-COCl}} + \underset{\text{Propanol}}{\text{CH}_3\text{CH}_2\text{CH}_2\text{OH}} \longrightarrow \underset{\text{Propyl ethanoate}}{\text{H}_3\text{C-CO-OCH}_2\text{CH}_2\text{CH}_3} + \text{HCl} \quad (16\text{-}111)$$

Another molecule that can be used as a starting material is a carboxylic acid. Since the OH is a very poor leaving group, however, such reactions are carried out in an acidic medium to convert the OH into a much better leaving group, OH$_2$. If you were not given the starting two reactants to synthesize the product, propyl ethanoate, would you be able to think of the two reactants needed to synthesize that molecule? The key is to think of how best to make the bond highlighted. So, let us think backward in terms of identifying what is needed to make a new molecule, and specifically the two fragments needed to make a new bond.

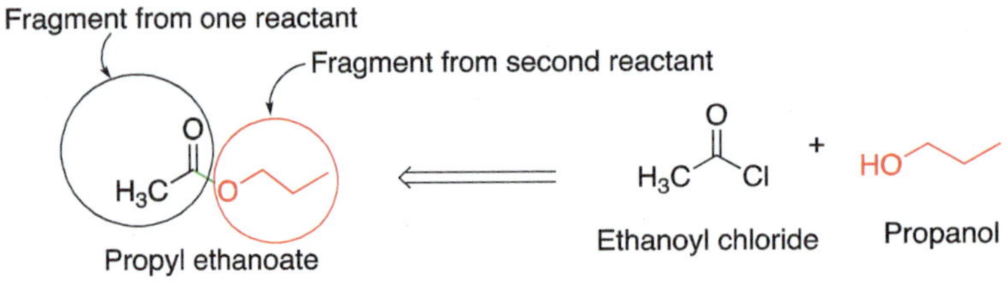

$$\underset{\text{Propyl ethanoate}}{\text{H}_3\text{C-CO-OCH}_2\text{CH}_2\text{CH}_3} \Longleftarrow \underset{\text{Ethanoyl chloride}}{\text{H}_3\text{C-COCl}} + \underset{\text{Propanol}}{\text{HO-CH}_2\text{CH}_2\text{CH}_3} \quad (16\text{-}112)$$

The other method to synthesize the same molecule is shown in Reaction (16-113).

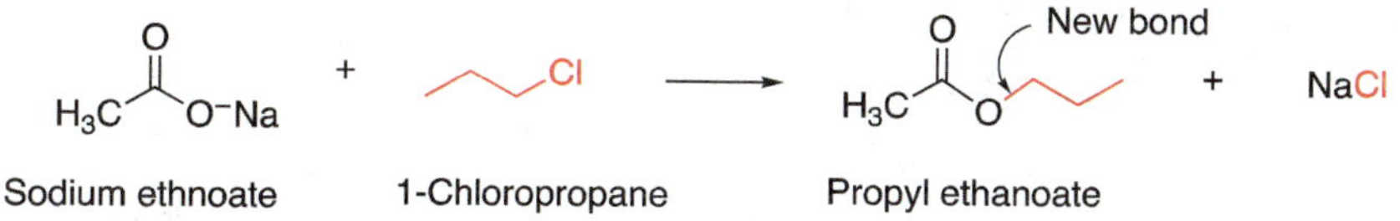

$$\underset{\text{Sodium ethnoate}}{H_3C-C(=O)-O^-Na^+} + \underset{\text{1-Chloropropane}}{CH_2CH_2CH_3-Cl} \longrightarrow \underset{\text{Propyl ethanoate}}{H_3C-C(=O)-O-CH_2CH_2CH_3} \text{ (New bond)} + NaCl \quad (16\text{-}113)$$

Using the same approach as above, once the bond to be made is recognized, the fragments needed as reactants can be identified as shown above. Note that this reaction involves using sodium acetate as the nucleophile for an S_N2 reaction to react with a primary halide to create the target molecule.

Problem 16.29

Using the method outlined above, try to figure out two synthetic routes for the synthesis of the molecules shown below.

(a) 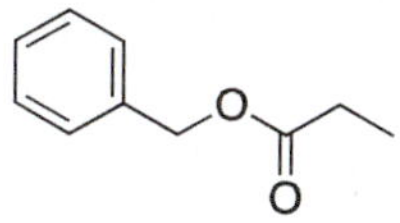 [structure: benzyl propanoate] (b) 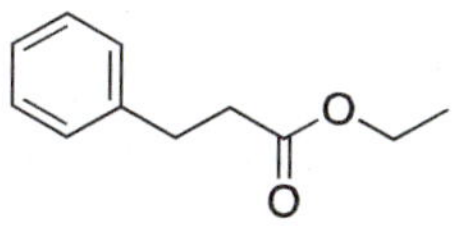 [structure: ethyl 3-phenylpropanoate]

16.10.2 Preparations of Amides

For the synthesis of amides, there is one important bond that should be considered. This bond is the carbonyl carbon–nitrogen bond. This bond can be made from a substitution reaction at a acyl carbon. Amides can be made by synthesizing the carbonyl carbon–nitrogen bond, bond **a** shown below.

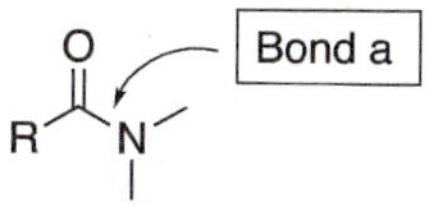

[structure: R-C(=O)-N with "Bond a" label]

We have seen this type of reaction before when we used ammonia or amines to react using a substitution at a carbonyl carbon as shown in Reaction (16-114).

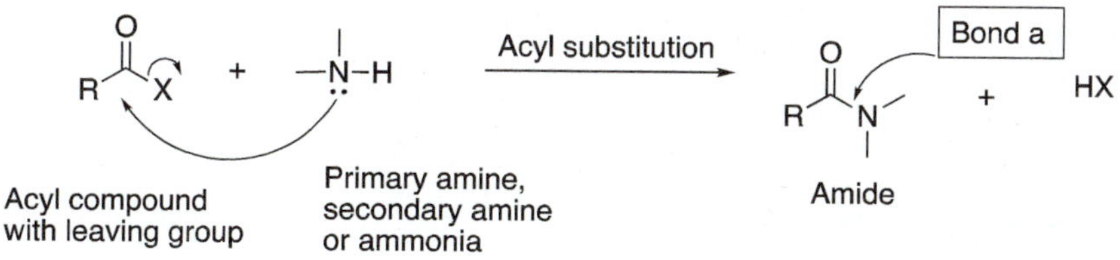

$$\underset{\substack{\text{Acyl compound} \\ \text{with leaving group}}}{R-C(=O)-X} + \underset{\substack{\text{Primary amine,} \\ \text{secondary amine} \\ \text{or ammonia}}}{H-N<} \xrightarrow{\text{Acyl substitution}} \underset{\text{Amide}}{R-C(=O)-N<} \text{ (Bond a)} + HX \quad (16\text{-}114)$$

Since acid chlorides are the most reactive molecules by a substitution reaction at an acyl carbon, they are frequently used to make the carbonyl carbon–nitrogen bond of the amide as shown in Reaction (16-115).

$$H_3C-C(=O)-Cl + R_2NH \longrightarrow H_3C-C(=O)-NR_2 + HCl \quad (16\text{-}115)$$

Cl is used as a good leaving group, but any other good leaving group could be used. Groups such as the carboxylate anion (anhydrides) or alkoxide (esters) can be used.

Consider the synthesis of the *N,N*-dimethylacetamide, shown below.

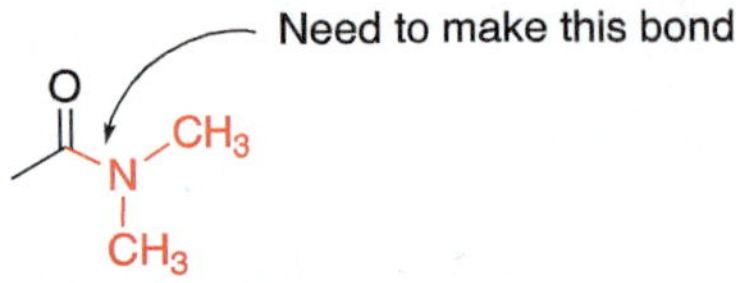

N,N-Dimethylacetamide

Two strategies are shown below, in which an electrophilic anhydride and nucleophilic dimethyl amine are used as starting reagents as illustrated in the reaction scheme given in Reaction (16-116).

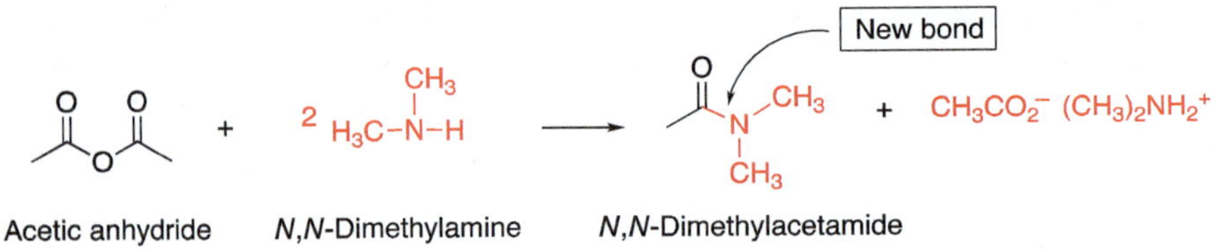

Acetic anhydride + 2 N,N-Dimethylamine → N,N-Dimethylacetamide + $CH_3CO_2^-$ $(CH_3)_2NH_2^+$ (16-116)

It is possible to use a different electrophile and same nucleophile to synthesize the target molecule, as shown in Reaction (16-117)

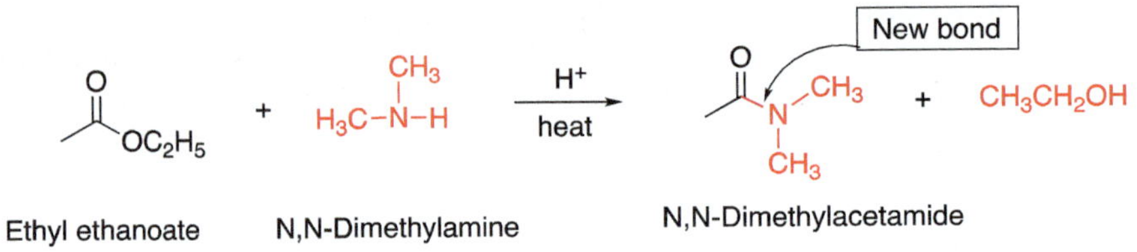

Ethyl ethanoate + N,N-Dimethylamine $\xrightarrow{H^+, \text{heat}}$ N,N-Dimethylacetamide + CH_3CH_2OH (16-117)

Problem 16.30

Give appropriate starting reagents that could be used for the synthesis of the amides shown below.

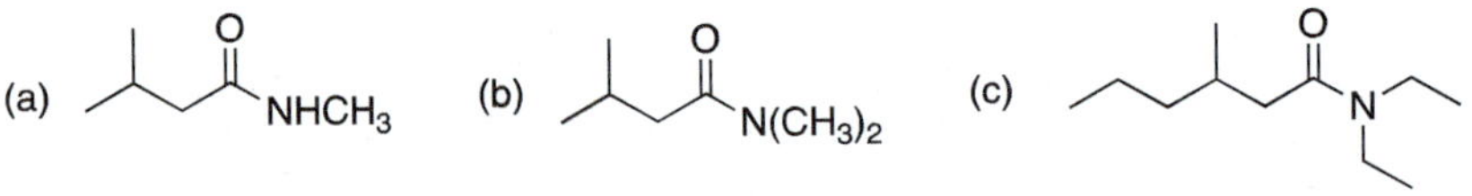

(a), (b), (c)

End of Chapter Problems

16.31 Show how the following compounds can be prepared from benzoyl chloride.

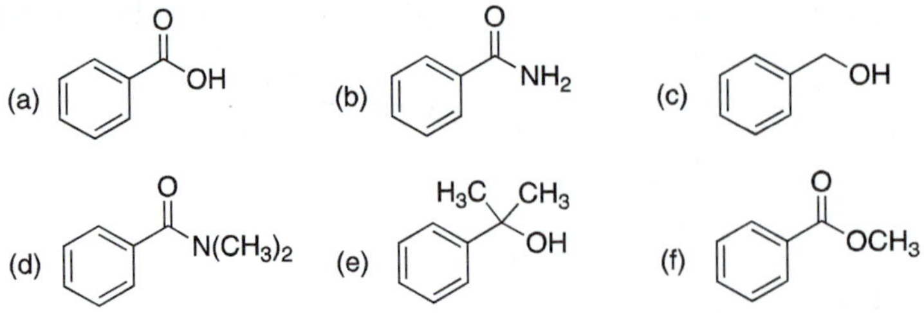

(a), (b), (c), (d), (e), (f)

16.32 Give the structural formulas for the final major organic products for each of the following reactions or sequence of reactions.

(a) Cyclohexanecarboxylic acid $\xrightarrow{\text{(1) SOCl}_2}{\text{(2) NH}_3}$

(b) Ethyl methanoate $\xrightarrow{\text{(1) CH}_3\text{MgCl (ex)}}{\text{(2) H}^+, \text{H}_2\text{O}}$

(c) PhC(O)OCH$_3$ $\xrightarrow{\text{(1) CH}_3\text{MgCl (ex)}}{\text{(2) H}^+, \text{H}_2\text{O}}$

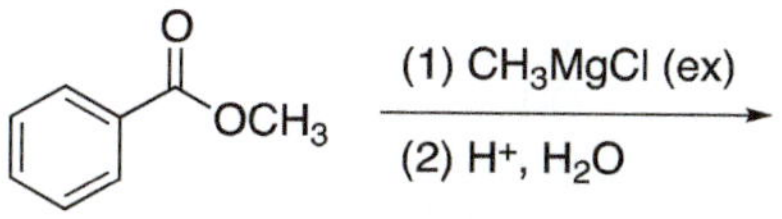

(d) 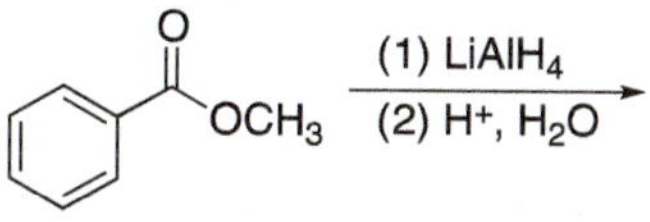 PhC(O)OCH$_3$ $\xrightarrow{\text{(1) LiAlH}_4}{\text{(2) H}^+, \text{H}_2\text{O}}$

(e) 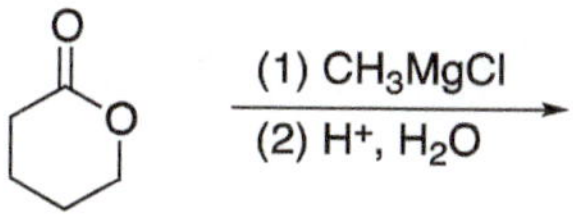 δ-valerolactone $\xrightarrow{\text{(1) CH}_3\text{MgCl}}{\text{(2) H}^+, \text{H}_2\text{O}}$

16.33 Show how to carry out the following transformations. Clearly indicate all reagents and reaction conditions used in your synthesis.

(a) 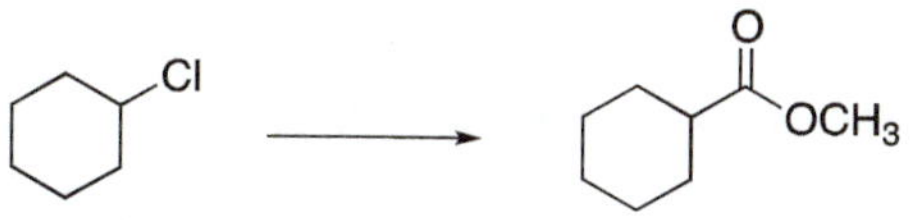 cyclohexyl chloride ⟶ methyl cyclohexanecarboxylate

(b) 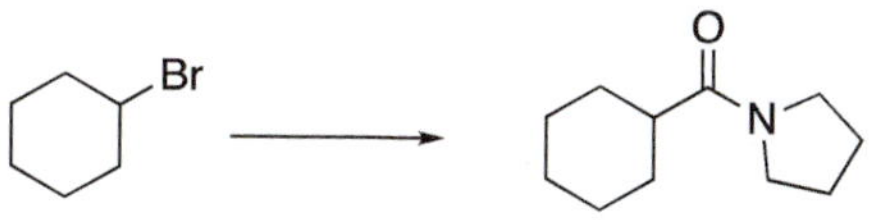 cyclohexyl bromide ⟶ 1-(cyclohexanecarbonyl)pyrrolidine

(c) 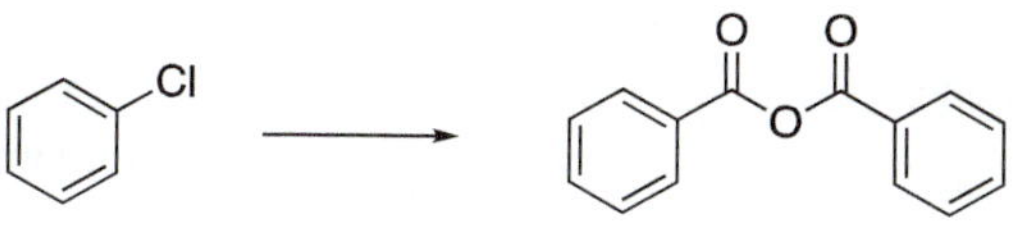 benzoyl chloride ⟶ benzoic anhydride

(d)  cyclohexyl bromide ⟶ cyclohexylmethanol

(e) 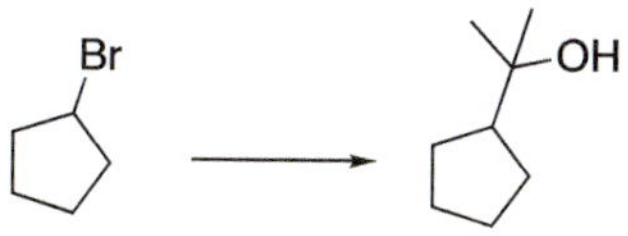 cyclopentyl bromide ⟶ 2-cyclopentylpropan-2-ol

(f) 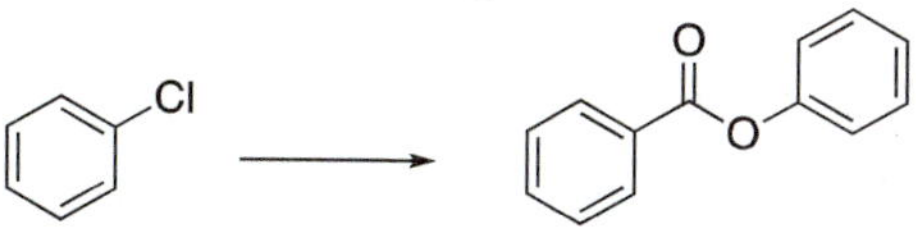 benzoyl chloride ⟶ phenyl benzoate

(g) 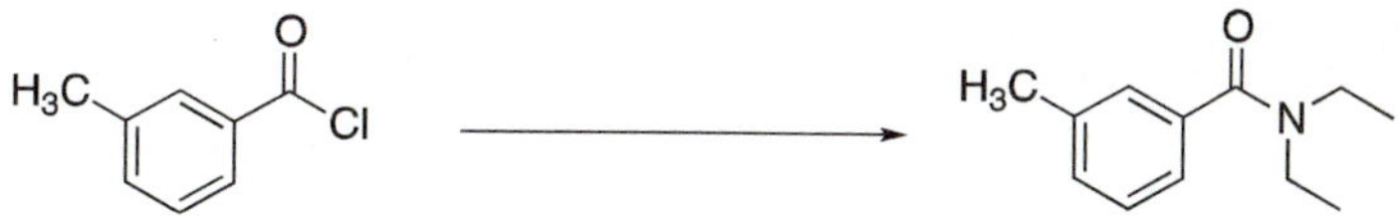 3-methylbenzoyl chloride ⟶ DEET (N,N-diethyl-3-methylbenzamide)

16.34 Use an appropriate mechanism (arrow pushing) to account for the fact that when a carboxylic acid is dissolved in water labeled with ^{18}O, the reaction shown below is observed.

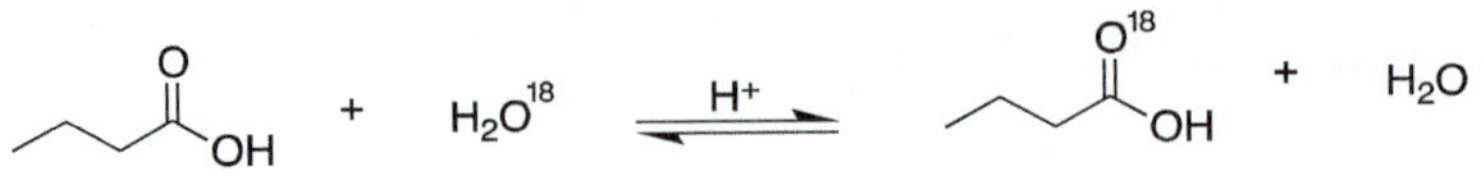

$$\text{CH}_3\text{CH}_2\text{CH}_2\text{COOH} + H_2O^{18} \underset{}{\overset{H^+}{\rightleftharpoons}} \text{CH}_3\text{CH}_2\text{CH}_2\text{CO}^{18}\text{OH} + H_2O$$

16.35 Give the major organic product, which are all lactones, of the following reactions.

(a) 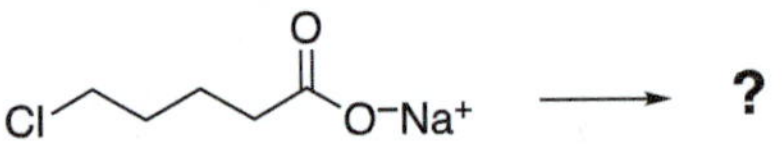 Cl-CH$_2$CH$_2$CH$_2$CH$_2$C(O)O⁻Na⁺ → ?

(b) (CH$_3$)$_2$C(OH)CH$_2$C(O)OCH$_3$ $\xrightarrow{H^+}$?

(c) HOCH(CH$_3$)CH$_2$CH$_2$C(O)Cl → ?

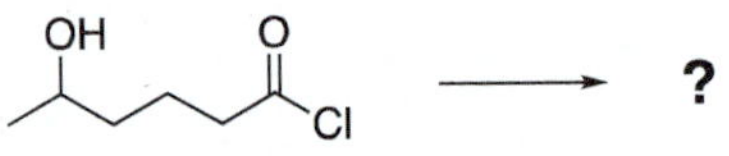

(d) 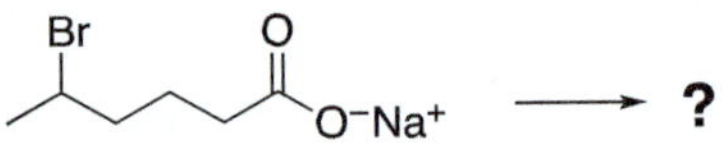 CH$_3$CHBrCH$_2$CH$_2$C(O)O⁻Na⁺ → ?

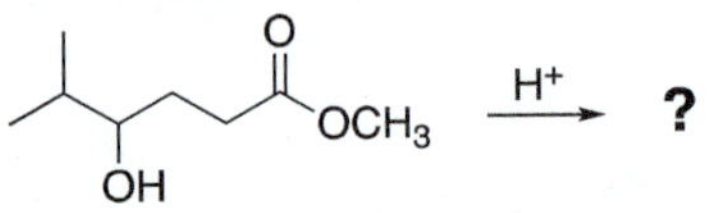

16.36 Determine which of the reactions shown below is more favorable, explain your answer.

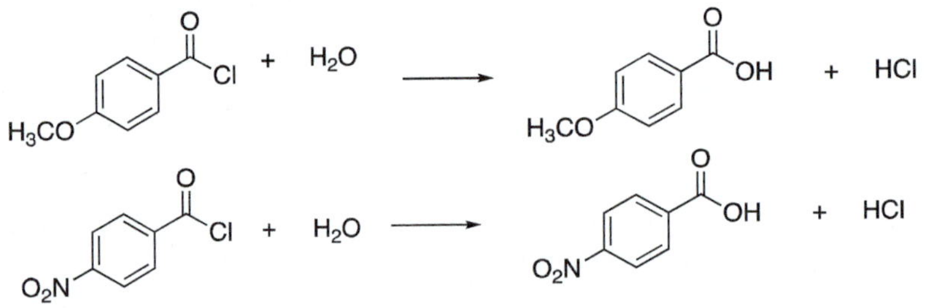

$$\text{H}_3\text{CO-C}_6\text{H}_4\text{-COCl} + H_2O \rightarrow \text{H}_3\text{CO-C}_6\text{H}_4\text{-COOH} + HCl$$

$$\text{O}_2\text{N-C}_6\text{H}_4\text{-COCl} + H_2O \rightarrow \text{O}_2\text{N-C}_6\text{H}_4\text{-COOH} + HCl$$

16.37 Show how to carry out the following transformation. N,N-Diethyl meta-toluamide (DEET) is the main ingredient in insect repellent.

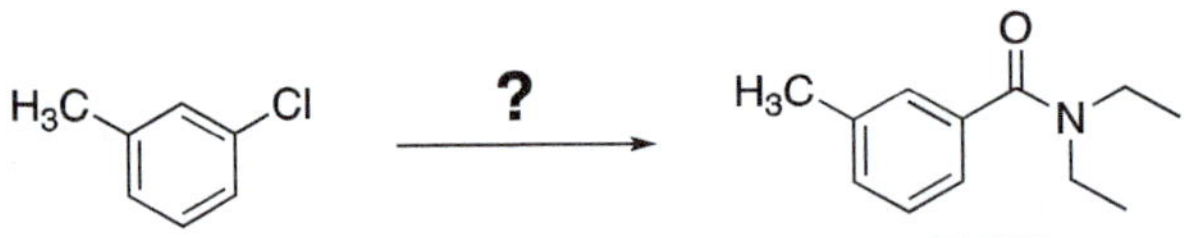

$$\text{H}_3\text{C-C}_6\text{H}_4\text{-Cl} \xrightarrow{?} \text{H}_3\text{C-C}_6\text{H}_4\text{-C(O)N(C}_2\text{H}_5)_2$$

DEET

16.38 Complete the following reactions by giving the structures of the missing reactant, reaction conditions, or major organic product. Indicate stereochemistry where appropriate.

(i) 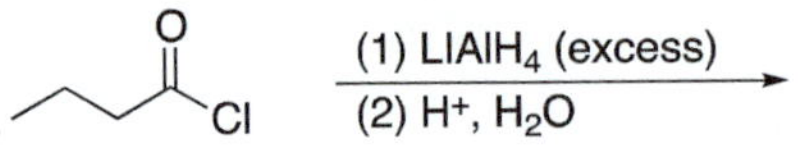 CH$_3$CH$_2$CH$_2$C(O)Cl $\xrightarrow[\text{(2) H}^+, H_2O]{\text{(1) LiAlH}_4 \text{ (excess)}}$

(ii) An acid chloride $\xrightarrow[\text{(2) H}^+, H_2O]{\text{(1) LiAlH}_4 \text{ (excess)}}$ (CH$_3$)$_2$CHCH(CH$_3$)OH

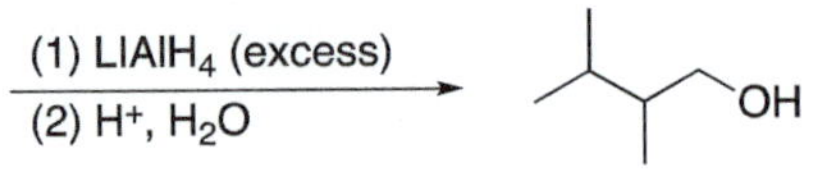

(iii) 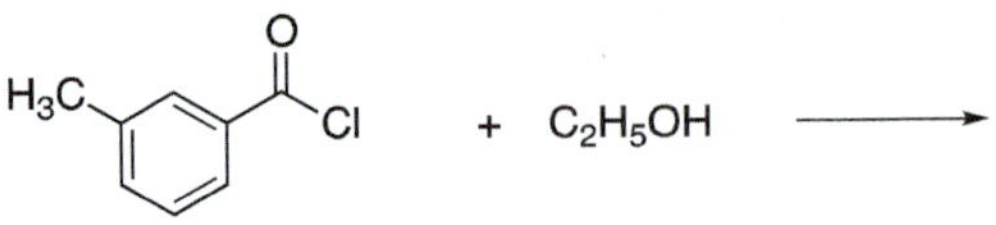 H$_3$C-C$_6$H$_4$-C(O)Cl + C$_2$H$_5$OH →

16.39 Sulfamethoxazole is an antibacterial compound from a group of compounds called sulfa drugs, which are widely used as antibacterial agents, its structure is shown below. Determine the structure of the amine necessary to carry out the transformation for its synthesis. Also determine how to remove the protecting group for the amine functionality that is bonded to the phenyl ring to give the target sulfamethoxazole molecule.

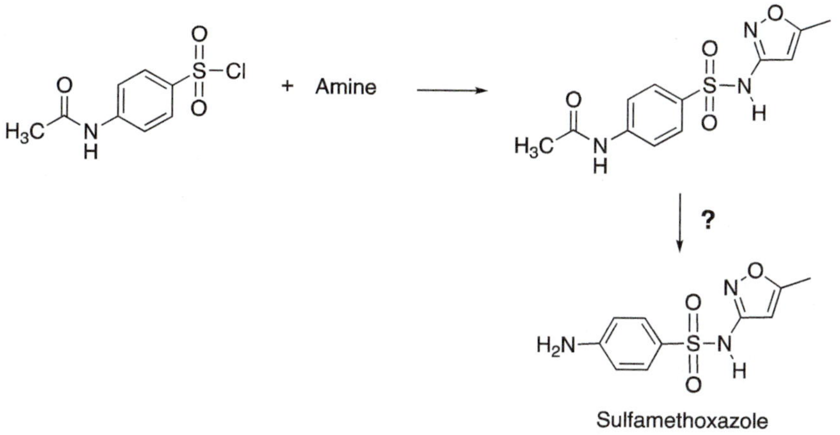

16.40 Using arrow formulism to indicate electron movement, provide a step-by-step mechanism for the reaction shown below.

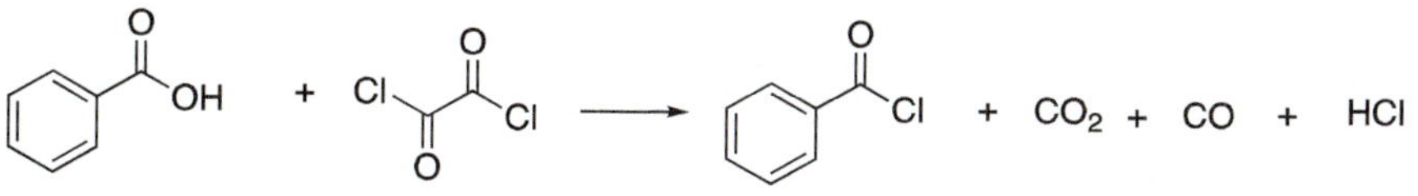

17

Aromaticity and Aromatic Substitution Reactions

17.1 Introduction

Some compounds are often described as having distinct aromas or fragrances. For example, coffee is sometimes described as having a delightful aroma. The same is true for cherries, peaches and almonds; these compounds all have distinctive aromas or fragrances. One of the compounds that accounts for the sweet aroma of cherries and almonds is benzaldehyde. Another example is that cinnamaldehyde accounts for the distinctive fragrance of the flavoring, cinnamon; the same is true for vanillin, which is present in vanilla.

> **DID YOU KNOW?**
>
> The distinctive aromas and fragrances of various fruits and spices are typically due to the presence of aromatic compounds. Cinnamaldehyde accounts for the distinctive fragrance of the flavoring, cinnamon; vanillin is another aromatic compound found in vanilla, and thymol is found in thyme.
>
>

Benzene is another one of these organic compounds with a very distinctive aroma and is responsible for the odor of coal distillate. It was discovered that these organic compounds all have

Organic Chemistry: Concepts and Applications, First Edition. Allan D. Headley.
© 2020 John Wiley & Sons, Inc. Published 2020 by John Wiley & Sons, Inc.
Companion website: www.wiley.com/go/Headley_OrganicChemistry

a special feature, which is described as aromaticity, and as result, these compounds are referred to as aromatic compounds. As you can imagine, there are numerous compounds that can be described as aromatic compounds. There are other aromatic compounds that may not have a distinct characteristic odor. Some include aspirin, morphine, norepinephrine, epinephrine, nicotine, vitamin E, and estrone. It is difficult to study all aromatic compounds individually, so we will carry out our study of aromatic compounds on the most common aromatic compound, benzene, and apply the knowledge gained to other compounds that are similar and aromatic.

Benzene was isolated from compressed illuminating gas in 1825 by Michael Faraday of the Royal Institution. In 1834, Eilhardt Mitscherlich of the University of Berlin synthesized benzene by heating benzoic acid with calcium oxide and he also showed that the molecular formula of benzene is C_6H_6. Today, benzene is obtained primarily from the distillation of petroleum, and it is used in the synthesis of many useful chemicals in our society. Benzene is also a very common solvent that is used in most chemical labs, both industrial and academic. However, it was removed from teaching labs, since it was shown to cause cancer when given to laboratory animals in very large dosage. Benzene is absorbed in the blood stream from inhalation and is converted to phenols and excreted by the liver. Benzene causes skin irritation; it is thought to cause leukemia. If absorbed in the blood, it will increase the number of white blood cells and it also causes damage to bone marrow. However, it is still sometimes used, under very controlled conditions, in industrial and research labs.

Since most of the chemical properties of compounds in this very large category of aromatic compounds are similar, we will study in detail the chemistry of only one of these compounds, benzene. By knowing the chemical and physical properties of benzene, we will be able to predict the properties of other aromatic compounds. In this chapter, we will first define aromaticity in a chemical sense and then apply that definition to other molecules to determine if they are aromatic or not. In the later sections of the chapter, we will examine the reactions of aromatic compounds.

17.2 Structure and Properties of Benzene

It was shown that the chemical formula of benzene is C_6H_6. Based on our knowledge of the number of hydrogens that are bonded to the carbons of saturated hydrocarbons, it is obvious that benzene is not a saturated hydrocarbon. In 1866, Friedrich Kekulé proposed that benzene had a cyclic structure with three alternating double bonds, as shown in Figure 17.1.

If this representation of benzene were correct, there should be another isomer in which the double bonds are in different locations. Such possible isomers of benzene are more obvious for 1,2-dichlorobenzene, which are shown below.

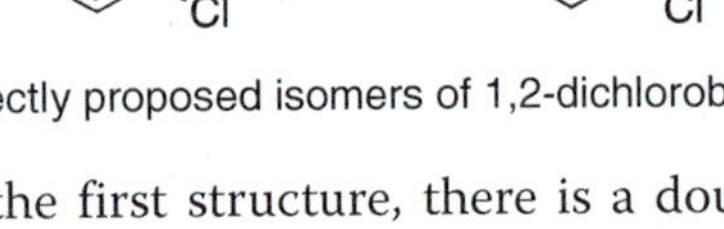

Incorrectly proposed isomers of 1,2-dichlorobenzene

In the first structure, there is a double bond between the carbons that have the chlorine atoms, and in the other structure, there is a single bond between the same two carbons. It was

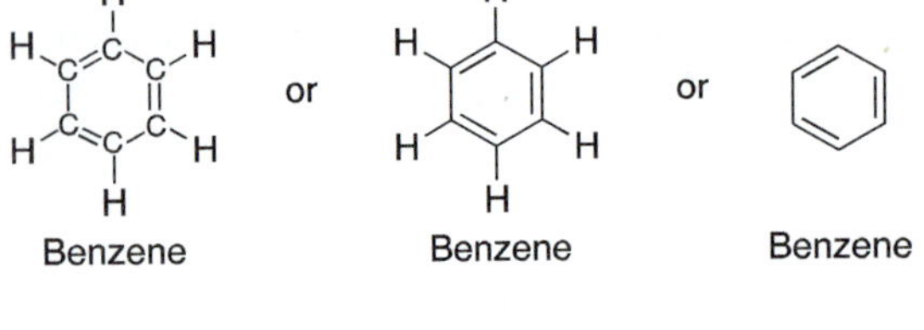

Figure 17.1 Kekulé's structure of benzene.

Benzene   Benzene   Benzene

known then that 1,2-dichlorobenzene has no isomers. To explain this discrepancy, Kekulé incorrectly proposed that a very rapid and undetectable equilibrium exists between both isomers, as shown in Reaction (17-1).

$$\text{(structures of 1,2-dichlorobenzene shown with equilibrium arrows crossed out)} \tag{17-1}$$

It was later shown by X-ray spectroscopy and other experimental analysis that all the carbon–carbon bond lengths of benzene are equal and that they are 1.387 Å. This bond length of 1.387 Å is between the lengths of a single carbon–carbon bond (1.48 Å) and an isolated carbon–carbon double bond (1.34 Å). Thus, the structures proposed by Kekulé for benzene, and shown in Figure 17.1 in which there is a rapid equilibrium between isomers were incorrect.

The model that is presently used to represent benzene is one that describes benzene as a flat molecule and one that has six carbon atoms, which are all sp^2 hybridized, and in each of the unhybridized p orbitals, there is one electron, and each carbon atom is bonded to one hydrogen atom. This representation is shown in Figure 17.2.

The six electrons are delocalized throughout the entire framework of the six carbon atoms, and based on this model, all the carbon–carbon bond lengths are equal. Thus, it is technically incorrect to designate any one carbon–carbon bond of benzene as a double bond or as a single bond. However, it is difficult to represent the carbon–carbon bond as a bond and a half, so throughout the years, it has become acceptable to draw benzene as either of the structures shown in Figure 17.3.

Note that the first two structures are exactly the same as those proposed by Kekulé; however, either one is used in today's scientific literature to represent benzene only because they are less cumbersome representations of the very complex picture of the benzene molecule shown in Figure 17.2. Figure 17.4 shows the structures of some common aromatic compounds.

Other compounds that contain the benzene ring system are shown in Figure 17.5. Norepinephrine is used to treat low blood pressure; aspirin (acetylsalicylic acid) is an analgesic; vitamin E is an antioxidant that occurs in foods such as leafy green vegetables, nuts and seeds; atorvastatin is the active ingredients in the lipid-lowering drug, lipotor; and citalopram is a commonly used antidepressant.

Figure 17.2 Model of benzene showing p orbitals and the nature of the bonding involving the pi (π) electrons and all sp^2 hybridized carbons.

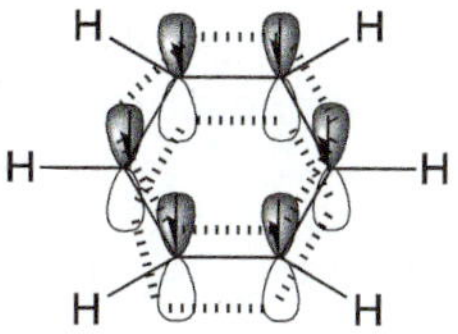

Figure 17.3 Common and acceptable representations of benzene.

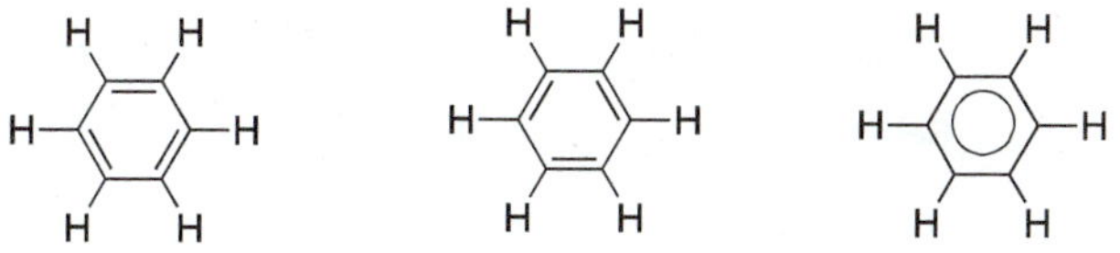

Figure 17.4 Familiar pleasant-smelling aromatic compounds, which all contain the benzene ring system.

Benzaldehyde

Vanillin

Cinnamaldehyde

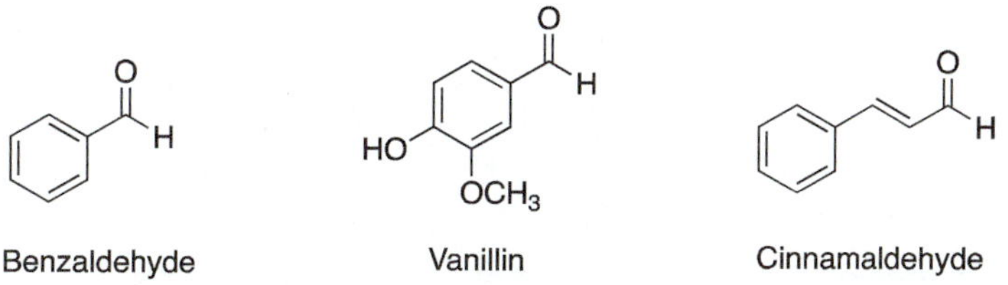

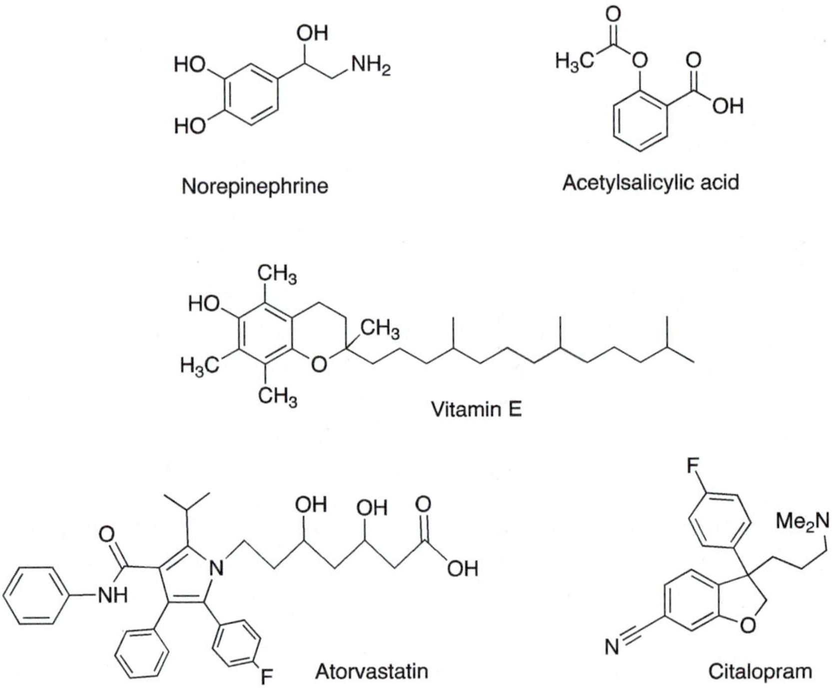

Norepinephrine

Acetylsalicylic acid

Vitamin E

Atorvastatin

Citalopram

Figure 17.5 Commonly used compounds that contain the benzene ring system.

Problem 17.1

Identify the benzene ring aromatic system in the examples given in Figures 17.4 and 17.5.

17.3 Nomenclature of Substituted Benzene

17.3.1 Nomenclature of Monosubstituted Benzenes

The removal of one hydrogen and its substitution by a different atom, or group of atoms, results in a monosubstituted benzene as shown below Reaction (17-2).

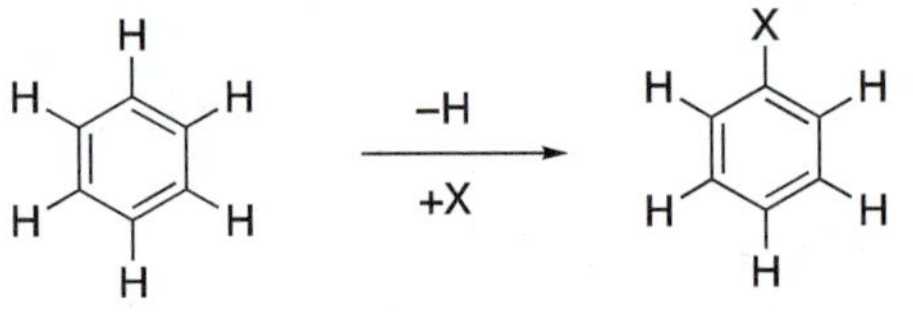

Benzene → Mono-substituted benzene (17-2)

These compounds are named as derivatives of benzene. For example, if X were methyl (CH_3), the name is methylbenzene. Other examples of monosubstituted benzenes are shown below.

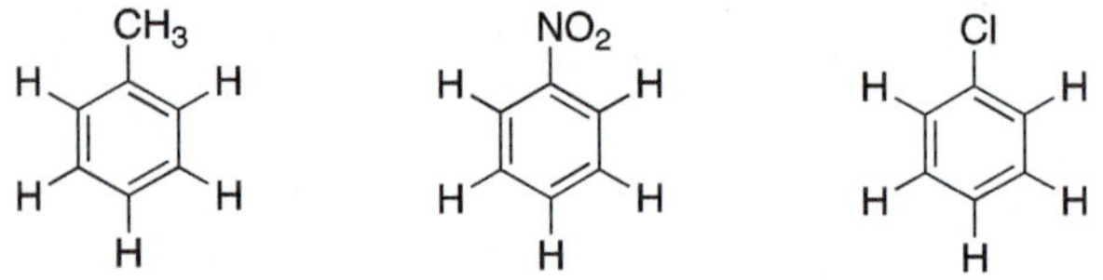

Methylbenzene   Nitrobenzene   Chlorobenzene

17.3 Nomenclature of Substituted Benzene

Table 17.1 Names that are used for some monosubstituted benzene molecules.

X	Common name
CH_3	Toluene
NH_2	Aniline
OH	Phenol
COOH	Benzoic acid
CHO	Benzaldehyde
OCH_3	Anisole

There are other names that are frequently used if specific substituents are bonded to the benzene ring; in fact, these names are used as root names for the IUPAC system of nomenclature of substituted benzene molecules. Table 17.1 shows the most commonly used names based on different monosubstituents.

Problem 17.2

i) Give the structure of each of the following compounds.
   (a) Bromobenzene, (b) hydroxybenzene (phenol), (c) fluorobenzene.
ii) Give the names for the following molecules.

(a) 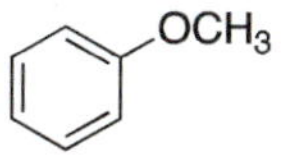 OCH_3-substituted benzene (b) 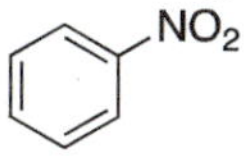 NO_2-substituted benzene (c) 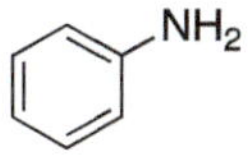 NH_2-substituted benzene (d) 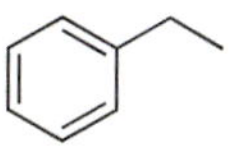 ethyl-substituted benzene

17.3.2 Nomenclature of Di-Substituted Benzenes

If there are two substituents bonded to the benzene ring (sometimes referred to as the phenyl ring), the exact relationship between the substituents on the ring must be specified. There is the possibility of different isomers based on the relationship of the groups on the phenyl ring. Thus, if benzene has two methyl groups bonded to the ring, there is the possibility of three different isomers – different molecules as shown in Figure 17.6.

In naming such disubstituted benzene compounds, as expected, one substituent must take priority and it is assigned #1 and priority is given based on the alphabet. The other substituent gets an assigned number based on its relationship to that group that is in position number 1. Examples are shown in Figure 17.7.

Often times there is an easily recognizable portion of the disubstituted benzene that has a name as shown in Table 17.1. In that case, the name of the disubstituted benzene is based on the root name shown in Table 17.1. Note that the position of the group that is used to determine the name of the compound gets number 1. Examples are shown in Figure 17.8.

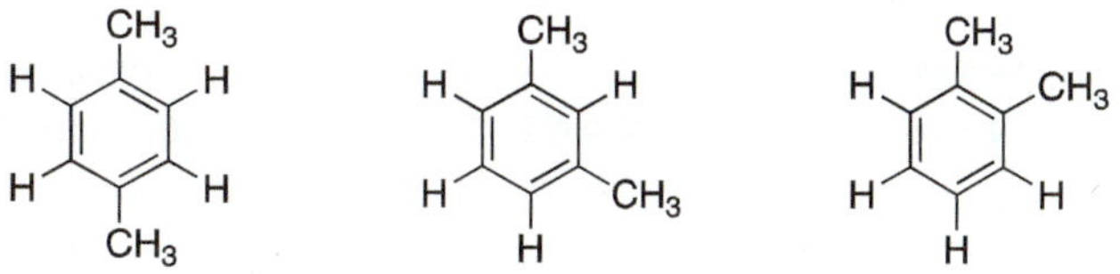

Figure 17.6 Different isomers of dimethylbenzenes.

1,3-Dibromobenzene 1-Bromo-3-nitrobenzene 1-Methyl-2-nitrobenzene 1,4-Dinitrobenzene

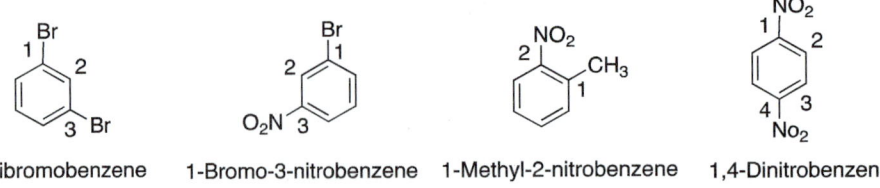

Figure 17.7 Examples of disubstituted benzene.

3-Nitrotoluene 3-Nitroaniline 3-Nitrophenol

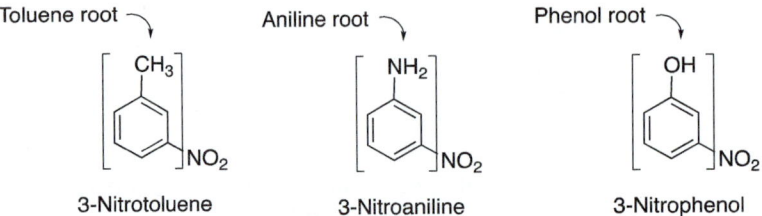

Figure 17.8 Examples of disubstituted benzenes.

Instead of using numbers in the nomenclature of disubstituted compounds, there is another method of naming these molecules. The nomenclature is accomplished by assigning common terms to the numbered relationships mentioned above. The common terms for the relationships are shown in Figure 17.9.

Examples using this system of nomenclature are shown in Figure 17.10.

Problem 17.3

i) Give the structures for each of the following compounds.
   (a) *meta*-Bromobenzaldehyde, (b) *p*-Chlorophenol,
   (c) 2-Fluoroaniline, (d) *o*-Hydroxybenzoic acid, and (e) 4-Nitrophenol.

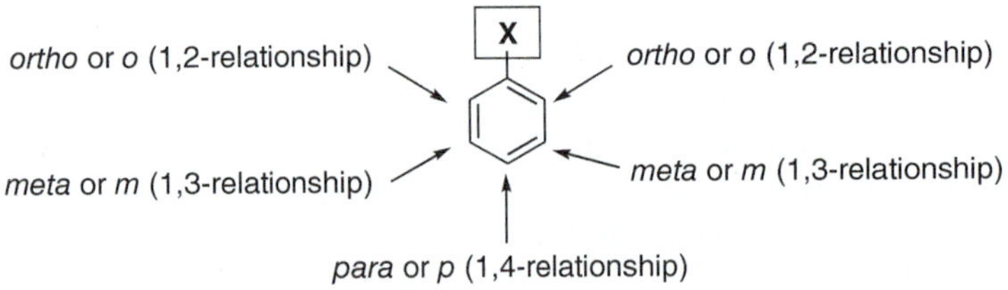

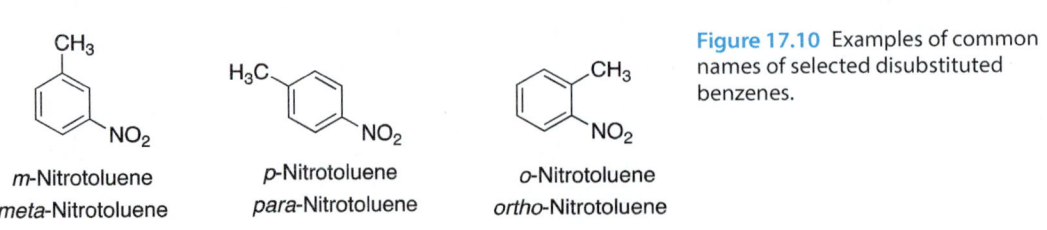

Figure 17.9 Common representations for the numbered relationships used in the naming of disubstituted benzenes.

m-Nitrotoluene *meta*-Nitrotoluene | *p*-Nitrotoluene *para*-Nitrotoluene | *o*-Nitrotoluene *ortho*-Nitrotoluene

Figure 17.10 Examples of common names of selected disubstituted benzenes.

ii) Give the names of the following molecules.

(a) 3-nitrobenzoic acid structure: O₂N — C₆H₄ — C(=O)OH

(b) 2-nitroaniline structure: C₆H₄ with NH₂ and NO₂

(c) 3-chloroaniline structure: C₆H₄ with NH₂ and Cl

(d) 4-bromophenol structure: Br — C₆H₄ — OH

(e) 3-nitrobenzaldehyde structure: O₂N — C₆H₄ — CHO

(f) 3-nitrotoluene structure: O₂N — C₆H₄ — CH₃

17.4 Stability of Benzene

It was mentioned earlier in the chapter that benzene is often used as a solvent in industrial and academic research labs. Solvents are used in reactions to provide an inert medium so that the reactions can take place; that is, they do not normally react with the reactants or the products. Since benzene is a commonly used solvent, the implication is that it is a very stable and an inert molecule. To fully understand why benzene is inert, let us examine the heat liberated upon hydrogenation of benzene compared to similar compounds. For a particular compound, the amount of heat liberated upon hydrogenation can be used to give a very good approximation of the stability, compared to similar compounds. We used this concept when we examined the stability of the different isomers of a particular alkene in Chapter 4.

For different compounds that form the same product by the same type of reaction, the reactant that liberates the most heat is the least stable, as illustrated in Figure 17.11.

The hydrogenation of cyclohexene, cyclohexadiene, cyclohexatriene (not a real molecule), and benzene all give the same product, cyclohexane. The hydrogenation of cyclohexene is an exothermic reaction and 28.6 kcal mol⁻¹ of heat is liberated, as given in Reaction (17-3).

$$\text{Cyclohexene} + H_2 \xrightarrow{\text{Pt (catalyst)}} \text{Cyclohexane} + 28.6 \text{ kcal mol}^{-1} \tag{17-3}$$

Figure 17.11 Energy diagram illustrating the relationship between the stability of reactants that give the same product, via the same type of reaction.

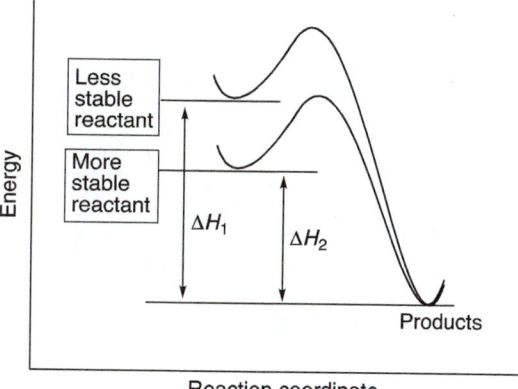

The molecule with two double bonds that are not in conjugation with each other (1,4-cyclohexadiene) liberates approximately twice as much heat, as the hydrogenation of cyclohexene Reaction (17-4).

$$\text{1,4-Cyclohexadiene} + 2\,H_2 \xrightarrow{\text{Pt (catalyst)}} \text{Cyclohexane} + 57.4 \text{ kcal mol}^{-1} \quad (17\text{-}4)$$

The hypothetical molecule with three double bonds in a cyclic six-member ring, cyclohexatriene, which is not a real molecule (one of Kekulé's incorrect structures of benzene) is expected to liberate three times the amount of heat upon hydrogenation as that liberated from hydrogenation of cyclohexene. The hypothetical reaction is shown in Reaction (17-5).

$$\text{Hypothetical 1,3,5-Cyclohexatriene} + 3\,H_2 \longrightarrow \text{Cyclohexane} + 85.8 \text{ kcal/mol (predicted)} \quad (17\text{-}5)$$

Note that 85.8 kcal mol^{-1} of heat is the predicted value for the hypothetical cyclohexatriene molecule, which is three times that for cyclohexene. This prediction is reasonable since the actual heat liberated from 1,4-cyclohexadiene (which has two double bonds) is twice that of cyclohexene. Benzene, the real molecule, liberates only 49.8 kcal mol^{-1} of heat as shown in Reaction (17-6).

$$\text{Benzene (the real molecule)} + 3\,H_2 \xrightarrow{\text{Pt (catalyst)}} \text{Cyclohexane} + 49.8 \text{ kcal mol}^{-1} \quad (17\text{-}6)$$

The energy profile for these reactions is shown in Figure 17.12, which shows benzene to liberate the least amount of heat.

It is obvious from these results that benzene is more stable than expected when compared to the hypothetical cyclohexatriene. Cyclohexatriene is expected to liberate 85.8 kcal mol^{-1} of heat, but the real molecule, benzene, yields only 49.8 kcal mol^{-1} upon hydrogenation. Thus, benzene is 36.0 Kcal/mol more stable than expected (85.8–49.8 kcal mol^{-1}). This energy,

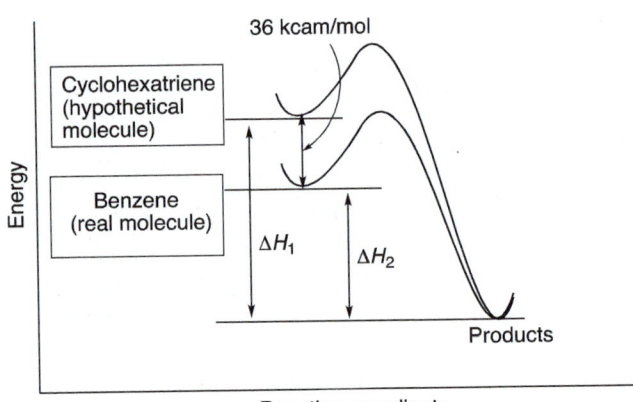

Figure 17.12 Representation of the exothermic reactions of benzene and the hypothetical, 1,3,5 cyclohexatriene.

36.0 kcal mol^{-1}, is also called the **resonance energy** of benzene. Since benzene is much more stable than expected, this observation explains why it can be used as a solvent and is fairly nonreactive.

17.5 Characteristics of Aromatic Compounds

As mentioned earlier, benzene is just one of the many compounds of this very large category of compounds that are called aromatic compounds. Based on our knowledge of benzene, it is possible to compare certain characteristics of benzene with other molecules in order to determine if other molecules are aromatic or not. Let us now carry out a more detailed examination of benzene. An important observation about benzene is that it is cyclic and planar, i.e. flat. Thus, all aromatic compounds should be cyclic and flat. Since all the carbons that make up the aromatic system are sp^2 hybridized, and we know that sp^2 hybridized carbons are trigonal planar, the benzene molecule has to be flat. This concept is very important because the electrons in the unhybridized p orbitals can delocalize about the plane perpendicular to the atoms of the molecule.

Another very important characteristic of benzene and hence other aromatic molecules is the number of electrons that actually are in conjugation with each other. Note that if the electrons are not in conjugation, the system cannot be aromatic. For benzene, there are six electrons, and a very important question should come to mind: is it possible to have aromatic compounds with more or less than six electrons? This question was answered by Erich Hückel, a theoretical physicist, in that he examined numerous compounds in the early 1930s and showed that those molecules that have similar properties as benzene and hence aromatic compounds (neutral molecules as well as ions) all have 4n + 2 pi (π) electrons, where n is an integer, such as 0,1,2,3,4, etc. Thus, for benzene which has 6 pi (π) electrons, n is one. This method of determining if a system is aromatic or not based on the number of electrons of the aromatic system is known as the *Hückel's rule*.

In summary, all aromatic compounds must meet the following criteria in order to be considered aromatic:

1) Must follow the Hückel's Rule (4n + 2) electrons that are in conjugation. These are typically pi (π) electrons, but could include electrons from an adjacent heteroatom
2) The structure must be a cyclic conjugated system
3) Each carbon atom must have at least one unhybridized p orbital
4) There must be overlap of the "p" orbitals so that there is a continuous ring of electrons, i.e. conjugated.
5) The energy of the system must be lower than the hypothetical system in which the electrons are localized and not in resonance.

17.5.1 Carbocyclic Compounds and Ions

In this section, we will apply the observations and criteria listed above to different molecules and ions to determine if they are aromatic or not. Let us apply these requirements to the ions shown in Figure 17.13.

1,3,5 Cycloheptatriene cation has six pi (π) electrons, which are all in conjugation with each other. Note that all the carbons of the ring are sp^2 hybridized and the carbon of the carbocation has no electrons, whereas the other orbitals each contain one electron in each p orbital. Thus, this cation contains six pi (π) electrons, which meets Hückel's rule and also meets the other

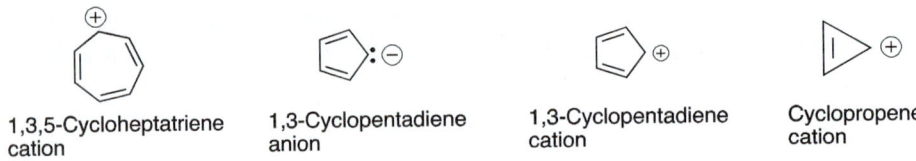

1,3,5-Cycloheptatriene cation 1,3-Cyclopentadiene anion 1,3-Cyclopentadiene cation Cyclopropene cation

Figure 17.13 Carbocyclic-conjugated ions.

characteristics outlined above for aromaticity. Thus, 1,3,5 cycloheptatriene cation is aromatic. The next ion shown in Figure 17.13 is the cyclopentadiene anion. For this anion, the carbons are also all sp^2, and the caboanion carbon contains two electrons in the p orbital. Thus, the total pi (π) electrons are six, which again meet Hückel's rule. As a result, this anion is also aromatic. The next ion shown in Figure 17.13 is the cyclopentadiene cation. For this cation, the carbons are also all sp^2, and the carbocation carbon contains no electrons in the p orbital. Thus, the total pi (π) electrons are four, which does not meet Hückel's rule requirement, since n equals half (½). As a result, this cation is not aromatic. Regarding the last ion is the cyclopropene cation, note that for this ion, the carbocation carbon has no electrons in the p orbital. Hence, the total number of pi (π) electrons is two (from the double bond). Two electrons also satisfy the Hückel's rule since the integer n for the Hückel equation is zero (0).

Problem 17.4

Determine which of the following molecules or ions is(are) aromatic? Briefly explain your answer.

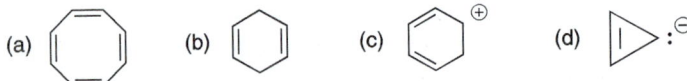

17.5.2 Polycyclic Compounds

In this section, we will examine molecules that have more than one ring fused to form polycyclic molecules. Let us examine the three molecules shown in Figure 17.14 and determine if they are aromatic or not.

Napthalene, which is used as an insecticide and pest repellent, is flat since all the carbons are sp^2 hybridized. It contains ten (10) pi (π) electrons that are in conjugation with each other. Based on the characteristics outlined above for aromaticity, naphthalene is aromatic. The second molecule is anthracene, which is used in the preservation of wood and for coating various materials. A similar analysis reveals that it is also flat and the electrons are in conjugation with each other. Equally important, there are 14 pi (π) electrons, which meets Hückel's rule requirement. Hence, anthracene is aromatic. The last molecule shown is phenanthrene, which is used to make dyes, plastics, and pesticides and also used in the pharmaceutical industry to make various drugs. Even though the three rings of this molecule are fused differently than that of anthracene, it meets the requirements to be aromatic and is an aromatic molecule.

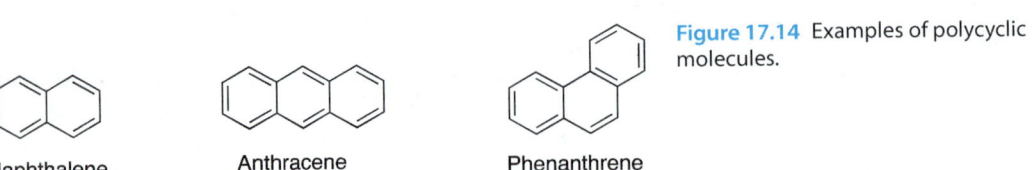

Figure 17.14 Examples of polycyclic molecules.

Naphthalene Anthracene Phenanthrene

Figure 17.15 Examples of heterocyclic compounds.

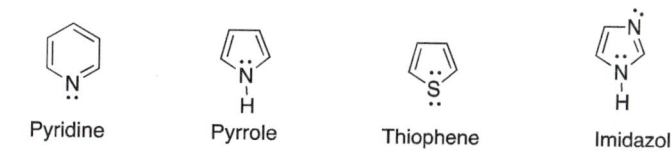

17.5.3 Heterocyclic Compounds

Heterocyclic compounds are compounds that are cyclic and contain at least one heteroatom in the ring. Examples of heterocyclic compounds are shown in Figure 17.15.

The first molecule, pyridine, is used in the synthesis of a host of different products, including food flavorings, paints, dyes, rubber products, and adhesives. Let us concentrate first on the heteroatom in pyridine. First, it is sp^2 hybridized and it has a lone pair of electrons in one of the sigma orbitals and it also contains one electron in the p orbital, which is in conjugation with the electrons of the conjugated system. Note that the pair of electrons in the sigma orbital are orthogonal to the six electrons of the conjugated system, and hence is not considered in the analysis of aromaticity as shown in Figure 17.16.

There are six electrons that are in conjugation, which meet the Hückel's requirement for aromaticity; hence, pyridine is an aromatic molecule. The second molecule shown in Figure 17.15 is pyrrole, which is used as an intermediate in the synthesis of a host of pharmaceutical products and agrochemicals. Initial inspection of the heteroatom, nitrogen, in pyrrole would indicate that it is sp^3 hybridized, with a lone pair of elections in a sp^3 hybridized orbital. Since these electrons would prefer to be in conjugation with the four electrons of the conjugated two double bonds resulting in essentially six electrons that are in conjugation in the cyclic system which would be aromatic. Hence, pyrrole is an aromatic molecule as illustrated in Figure 17.17. Note that in order to have complete conjugation with all six electrons, the heteroatom rehybridizes.

The next molecule shown in Figure 17.15 is thiophene, which is also widely used as an intermediate in the agrochemical and pharmaceutical industries. Initial inspection of the sulfur in thiophene would indicate that it is sp^3 hybridized like that of nitrogen in pyrrole. Thiophene has two nonbonded pairs of electrons and you can imagine, there is only one pair that can be in conjugation with the other four elections of the conjugated two double bonds. In order to have complete conjugation with all six electrons, the heteroatom rehybridizes as shown in Figure 17.18.

Figure 17.16 Model of pyridine showing p orbitals and the nature of the bonding involving the pi (π) electrons; note that the carbons and nitrogen are all sp^2 hybridized.

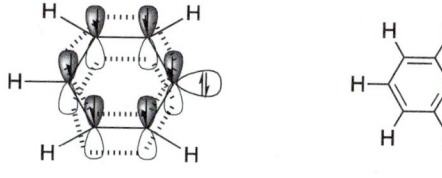

Figure 17.17 Delocalization of electrons about the plane of pyrrole.

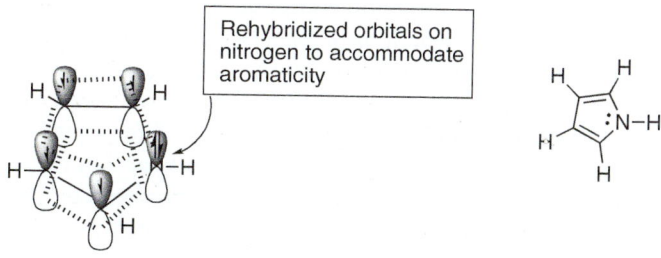

Rehybridized orbitals on nitrogen to accommodate aromaticity

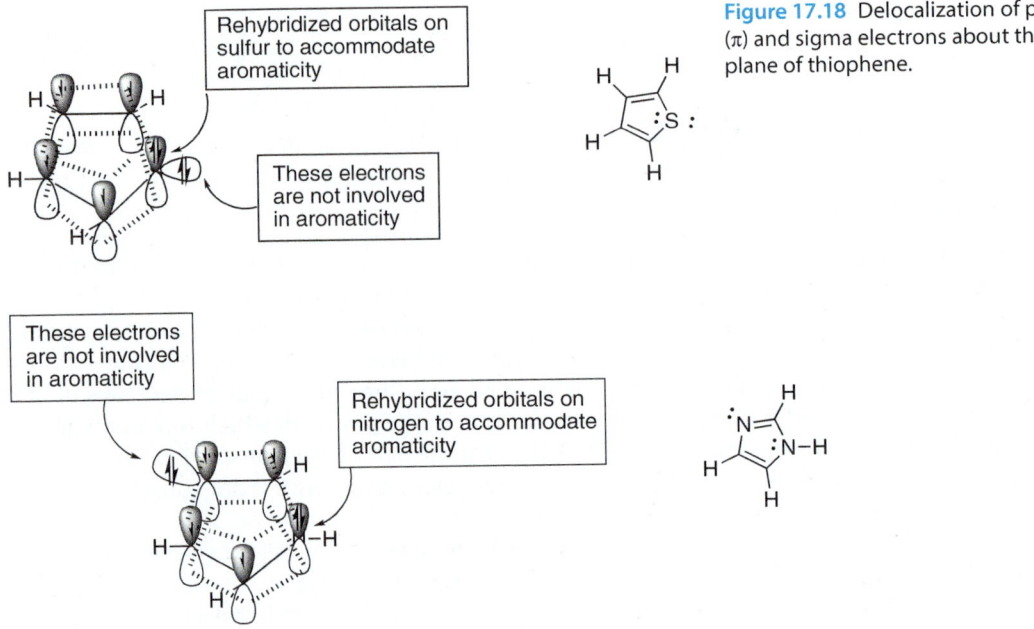

Figure 17.18 Delocalization of pi (π) and sigma electrons about the plane of thiophene.

Figure 17.19 Imidazole, a heterocyclic aromatic molecule.

The last molecule is imidazole, which is used as an important intermediate for the synthesis of various pharmaceuticals and agrichemicals. It has two heteroatoms, which must be analyzed separately. A similar analysis as that carried out for pyrrole shows that in order to have complete conjugation with all six electrons, the heteroatom at the right of moelcule rehybridizes. The other nitrogen atom is sp^2 hybridized similar to that of pyridine. As a result, imidazole is an aromatic molecule as illustrated in Figure 17.19.

Problem 17.5

Determine which of the following molecules is (are) aromatic, note that the unshared electrons are not shown?

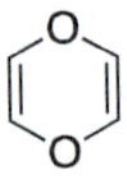

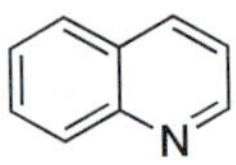

17.6 Electrophilic Aromatic Substitution Reactions of Benzene

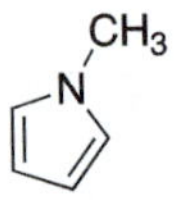

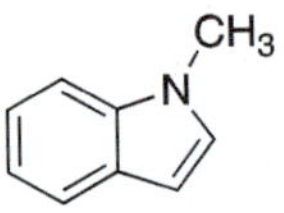

The reactions that aromatic compounds undergo are unique in that they do not undergo the expected addition reactions of alkenes that we covered in Chapter 8. Addition reactions would be a predicted type of reaction owing to the electron density of these compounds. One type of reaction that these molecules undergo is described as **electrophilic aromatic substitution**. For these reactions, an electrophile is substituted for a hydrogen atom of the benzene or aromatic molecule to form the product as shown in Reaction (17-7). The symbol E$^+$ is typically used to represent an electrophile. You will recall that electrophiles are electron-loving species, and another definition is that electrophiles are electron-deficient species.

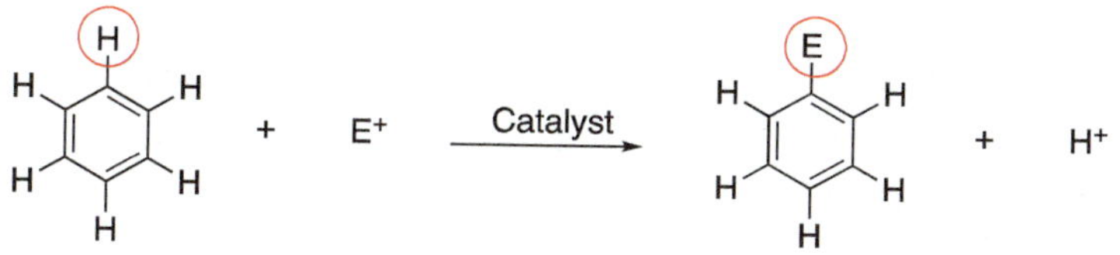

17.6 Electrophilic Aromatic Substitution Reactions of Benzene

$$\text{Benzene} + E^+ \xrightarrow{\text{Catalyst}} \text{Substituted benzene} + H^+ \quad (17\text{-}7)$$

Owing to the exceptional stability of benzene and other aromatic compounds, they are not very reactive and most require a catalyst for this type of reaction to occur. In addition, you will notice that the organic product of these types of reactions is also aromatic. If the reaction that occurred were an addition reaction, as shown in Reaction (17-8), the product would not be aromatic.

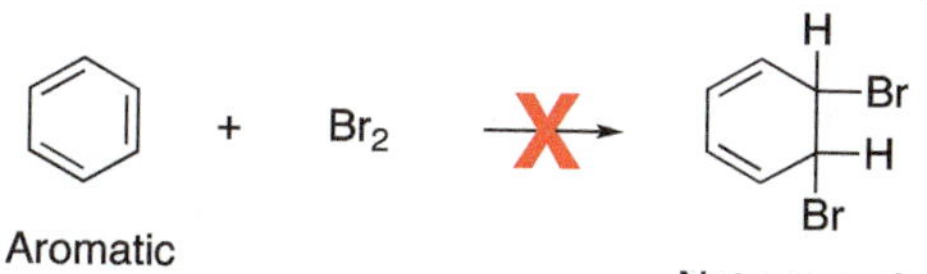

$$\text{Aromatic} + Br_2 \xrightarrow{X} \text{Not aromatic} \quad (17\text{-}8)$$

Since aromatic molecules are more stable than comparable molecules, aromaticity is conserved for most reactions that involve aromatic compounds. Thus, if an addition reaction were to occur for benzene, 36 kcal mol^{-1} would be lost since the aromaticity of the molecule would be destroyed. Instead, a substitution reaction occurs so that the aromaticity is conserved. The general mechanism for this type of substitution reaction is shown in Reaction (17-9).

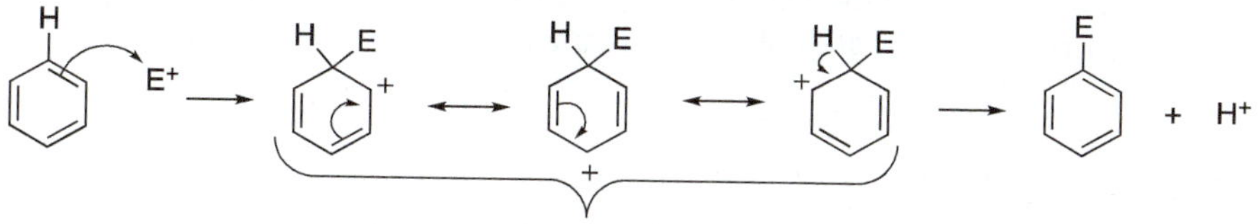

$$\text{(Resonance stabilized intermediate)} \quad (17\text{-}9)$$

Typical electrophiles that will be discussed in this chapter include the nitronium ion (NO_2^+), the halogen cation (Cl^+ and Br^+), carbocation (R^+), and acyl cation ($RC{=}O^+$). Reactions involving these electrophiles will be covered in the next sections.

17.6.1 Substitution Reactions Involving Nitronium Ion

The nitration of benzene is shown in Reaction (17-10). The electrophile for the nitration of benzene is NO_2^+, and it is generated from the mixture of nitric and sulfuric acid.

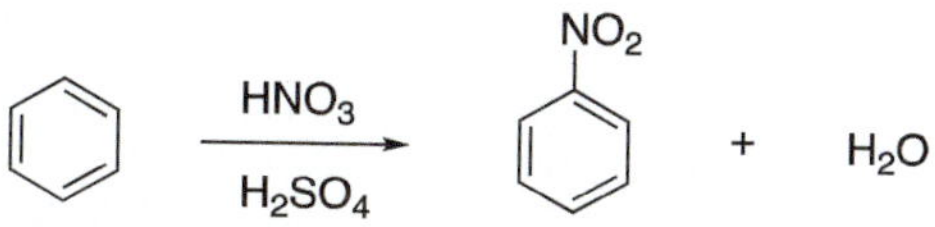

$$\text{Benzene} \xrightarrow[H_2SO_4]{HNO_3} \text{Nitrobenzene} + H_2O \quad (17\text{-}10)$$

Sulfuric acid is a stronger acid than nitric acid; therefore, in a mixture of nitric and sulfuric acids, protonation of nitric acid occurs to give a very good leaving group, OH_2. The elimination of water from protonated nitric acid gives the NO_2^+ electrophile. The mechanism for the formation of NO_2^+ from sulfuric and nitric acids is shown in Reaction (17-11).

$$\text{HO-S(=O)(=O)-OH} + \text{HO-N}^{+}\text{(=O)(O}^{-}\text{)} \rightleftharpoons \text{HO-S(=O)(=O)-O}^{-} + \text{H}_2\text{O-N}^{+}\text{(=O)(O}^{-}\text{)} \tag{17-11}$$

Sulfuric acid + Nitric acid

$$\text{H}_2\text{O-N}^{+}(\text{=O})(\text{O}^{-}) \longrightarrow \text{H}_2\text{O} + \text{O=N=O}^{+} \text{ (Nitronium ion)}$$

In order to account for the formation of nitrobenzene from benzene in the presence of NO_2^+, it is best to understand how this reaction proceeds, i.e. the mechanism. First, since the benzene molecule is electron rich (nucleophilic), the electrons of benzene attack the electrophile to form a charged intermediate, as shown in Reaction (17-12). This step is known as the rate determining step (RDS), also known as the slow step.

$$\text{C}_6\text{H}_6 + \text{O=N=O}^{+} \xrightarrow{\text{RDS}} [\text{Resonance stabilized charged intermediate}] \tag{17-12}$$

Even though the charged intermediate is relatively stable as shown by the three possible resonance structures that can be drawn, this ion can become even more stable (aromatic) if it could rearomatize to become a neutral molecule. Aromatization can be accomplished by the loss of a proton (H^+) as shown in the final step of the mechanism in Reaction (17-13).

$$[\text{C}_6\text{H}_6\text{-NO}_2]^+ \xrightarrow{\text{Fast step}} \text{C}_6\text{H}_5\text{NO}_2 + \text{H}^+ \tag{17-13}$$

Nitrobenzene

The driving force for this fast and favorable step of the reaction mechanism is the formation of the very stable aromatic benzene system.

17.6.2 Substitution Reactions Involving the Halogen Cation

The mechanism for the halogenation of benzene is similar to the nitration of benzene except that the electrophile is the X^+ (halogen) ion, Br^+ or Cl^+. The bromination of benzene is shown in Reaction (17-14).

$$\text{Benzene} \xrightarrow[\text{FeBr}_3 \text{ (catalyst)}]{\text{Br}_2} \text{Bromobenzene} + \text{HBr} \tag{17-14}$$

For this reaction, the electrophile is Br^+ and it is generated from Br_2 and $FeBr_3$. $FeBr_3$ is a Lewis acid, and it has a vacant orbital. In a mixture of bromine and $FeBr_3$, one of the lone pairs of electrons on bromine coordinates with $FeBr_3$ to form a complex, as shown in the first step of the reaction mechanism, which is given in Reaction (17-15).

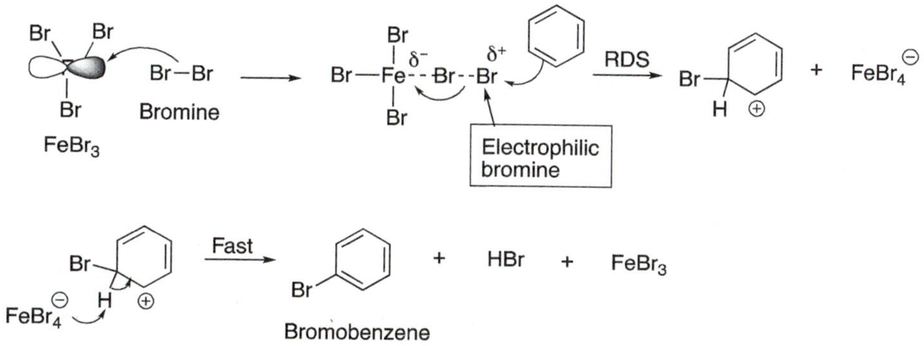

17.6 Electrophilic Aromatic Substitution Reactions of Benzene

$$\text{(17-15)}$$

In this complex, the bromine is polarized so that the electron distribution is closer to the iron, making one of the bromine atoms electrophilic. Thus, a partial positive charge is on one of the bromine atoms, the one farthest from the iron shown in Reaction (17-15). Thus, for the bromination of benzene by Br_2 and $FeBr_3$, the Lewis acid, $FeBr_3$, helps in the generation of the electrophilic Br^+ ion from Br_2. In the second step of the mechanism, the electrons from the benzene ring then react with the electrophilic Br^+ ion to form a charged intermediate, which rearomatizes after the loss of a proton to form the product, bromobenzene in the final step of the mechanism.

17.6.3 Substitution Reactions Involving Carbocations

Benzene can be alkylated in the presence of an alkyl halide and a Lewis acid, as a catalyst, to form an alkylbenzene, as shown in Reaction (17-16).

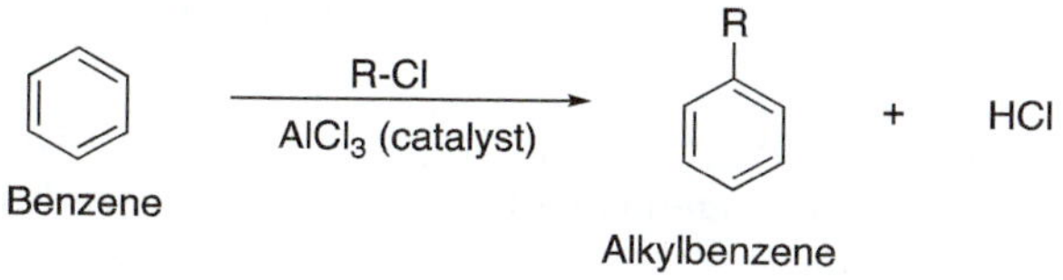

$$\text{(17-16)}$$

For these substitution reactions, the electrophile is a carbocation R^+. This carbocation is usually generated from the breakage of a carbon–halogen bond. In the presence of a catalyst, such as a Lewis acid, typically $AlBr_3$ or $AlCl_3$, a complex is formed between the Lewis acid and the alkyl halide, similar to the complex described in the previous section. In this complex, one pair of the electrons of the halogen occupies the empty orbital of the Lewis acid, and hence the carbon–halogen bond is weakened liberating the alkyl group as electrophilic, as shown in the first step of the reaction mechanism given in Reaction (17-17).

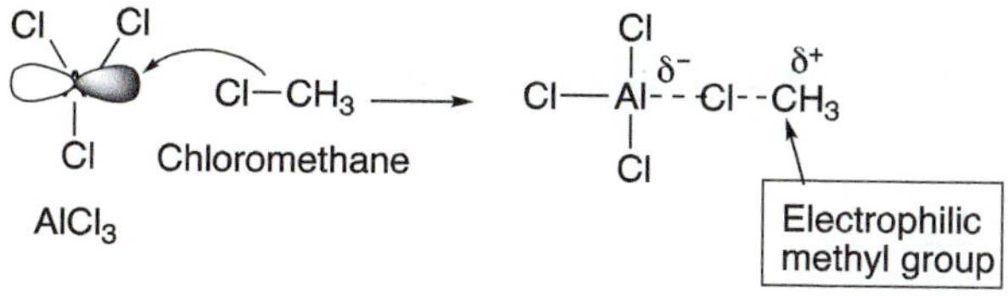

$$\text{(17-17)}$$

In the next step of the reaction mechanism, the electrophilic alkyl group is attacked by the nucleophilic electrons of the benzene ring to form the charged benzenium ion intermediate, which then rearomatizes to form the alkyl benzene, as given Reaction (17-18).

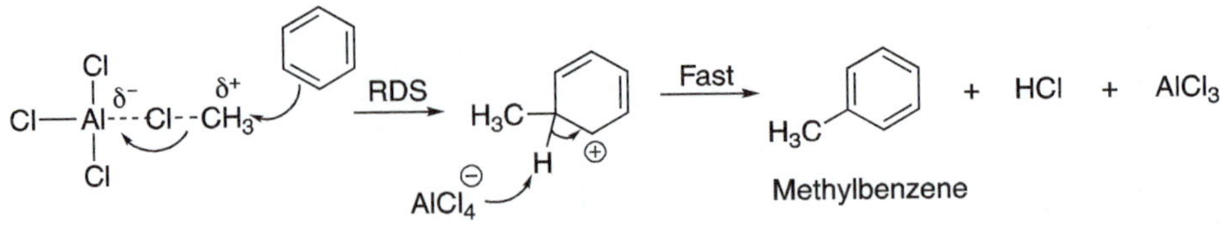

17 Aromaticity and Aromatic Substitution Reactions

$$\text{Cl-Al(Cl)(Cl)} \cdots \text{Cl--CH}_3 + \text{C}_6\text{H}_6 \xrightarrow{\text{RDS}} [\text{H}_3\text{C-C}_6\text{H}_6]^+ \; \text{AlCl}_4^- \xrightarrow{\text{Fast}} \text{H}_3\text{C-C}_6\text{H}_5 + \text{HCl} + \text{AlCl}_3$$

Methylbenzene

(17-18)

Reactions of this type, in which AlCl$_3$ or AlBr$_3$ is used as catalyst in conjunction with alkyl halides to form an alkylbenzenes from benzene are known as the **Friedel Crafts Alkylation**.

For these alkylation reactions, unexpected products are sometimes observed depending on the type of alkyl halide used, as given in Reaction (17-19).

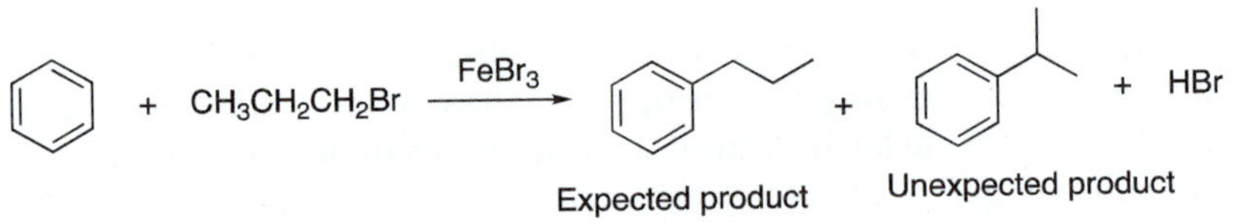

$$\text{C}_6\text{H}_6 + \text{CH}_3\text{CH}_2\text{CH}_2\text{Br} \xrightarrow{\text{FeBr}_3} \text{Ph-CH}_2\text{CH}_2\text{CH}_3 + \text{Ph-CH(CH}_3)_2 + \text{HBr}$$

Expected product   Unexpected product

(17-19)

For the reaction given in Reaction (17-19), the carbocation initially formed is $CH_3CH_2CH_2^+$, which is a primary carbocation. As we have seen previously, primary carbocations are unstable and will rearrange to the more stable secondary carbocation if possible. The rearrangement of the primary propyl cation generated in Reaction (17-19) to form the secondary carbocation occurs by a hydride migration, as shown in Reaction (17-20).

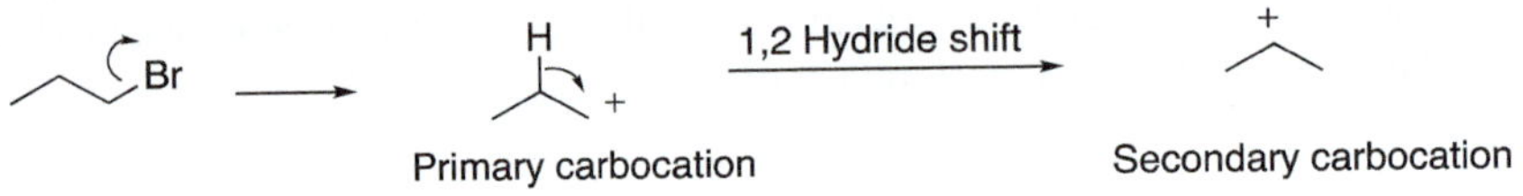

$$\text{CH}_3\text{CH}_2\text{CH}_2\text{Br} \longrightarrow \text{CH}_3\text{CH}_2\text{CH}_2^+ \xrightarrow{\text{1,2 Hydride shift}} (\text{CH}_3)_2\text{CH}^+$$

Primary carbocation   Secondary carbocation

(17-20)

Thus, for the reaction given in Reaction (17-19), the major organic product is shown in Reaction (17-21), which occurs due to the rearrangement of the intermediate carbocation generated throughout the course of the reaction.

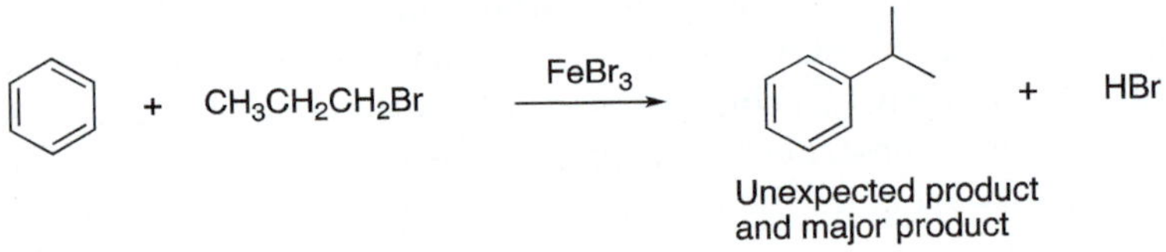

$$\text{C}_6\text{H}_6 + \text{CH}_3\text{CH}_2\text{CH}_2\text{Br} \xrightarrow{\text{FeBr}_3} \text{Ph-CH(CH}_3)_2 + \text{HBr}$$

Unexpected product and major product

(17-21)

Problem 17.6

Give the expected organic products for each of the following reactions.

(a) $\text{C}_6\text{H}_6 + (\text{CH}_3)_2\text{CHCH}_2\text{Br} \xrightarrow{\text{FeBr}_3} ?$

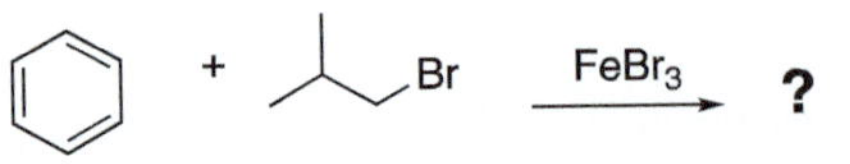

(b) $\text{C}_6\text{H}_6 + (\text{CH}_3)_3\text{CBr} \xrightarrow{\text{FeBr}_3} ?$

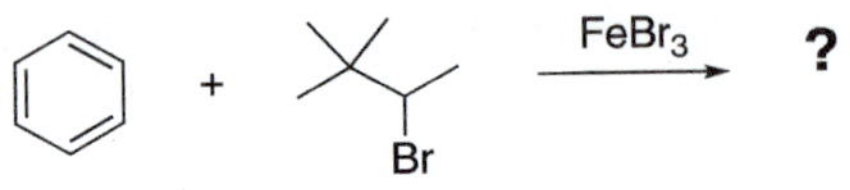

17.6.4 Substitution Reactions Involving Acyl Cations

The acyl cation $^+\overset{\overset{O}{\|}}{C}-R$, which is an electrophile, can be introduced on the aromatic ring of benzene by the electrophilic aromatic substitution reaction shown in Reaction (17-22).

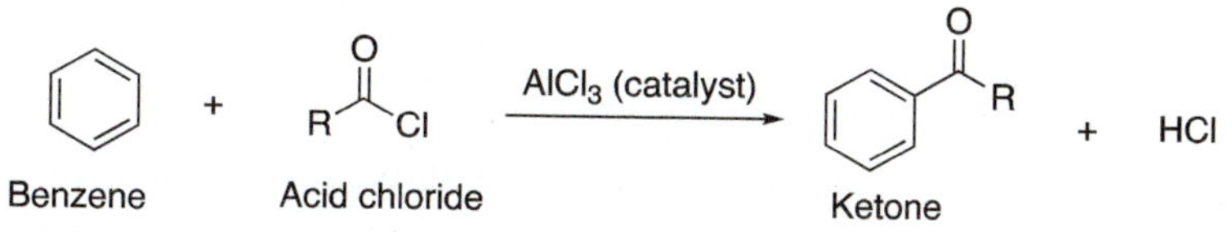

$$\text{Benzene} + \text{Acid chloride} \xrightarrow{\text{AlCl}_3 \text{ (catalyst)}} \text{Ketone} + \text{HCl} \qquad (17\text{-}22)$$

The reaction of acid chlorides with benzene in a presence of the catalyst, AlCl$_3$, is known as the **Friedel–Crafts acylation** and the product are aryl ketones. A specific example of the Friedel–Crafts acylation is given in Reaction (17-23).

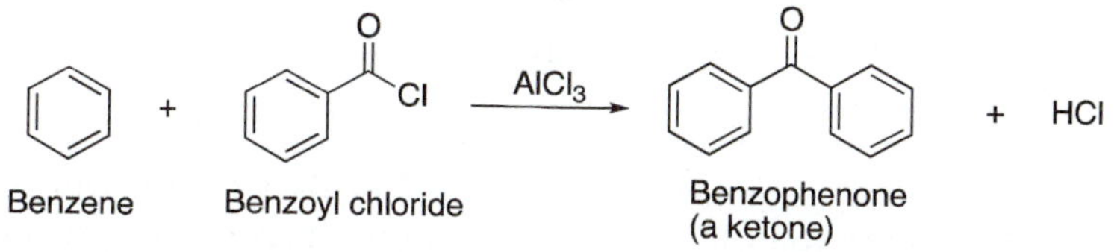

$$\text{Benzene} + \text{Benzoyl chloride} \xrightarrow{\text{AlCl}_3} \text{Benzophenone (a ketone)} + \text{HCl} \qquad (17\text{-}23)$$

As you can imagine, the catalyst, AlCl$_3$ serves to liberate the acyl cation from the acid chloride, as shown in the first step of the mechanism given in Reaction (17-24). Note that owing to the stability of the acyl cation, it does not rearrange like carbocations.

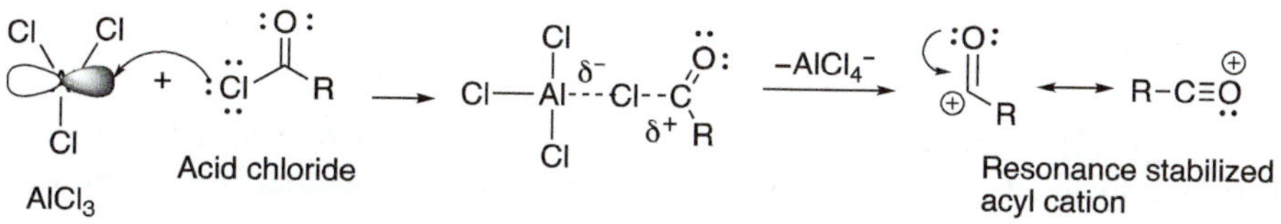

$$(17\text{-}24)$$

In the next step of the mechanism, the acyl cation is attached by the electron-rich benzene ring to form the charged benzenium intermediate, which then rearomatizes to form the product shown in Reaction (17-25) along with the catalyst and HCl.

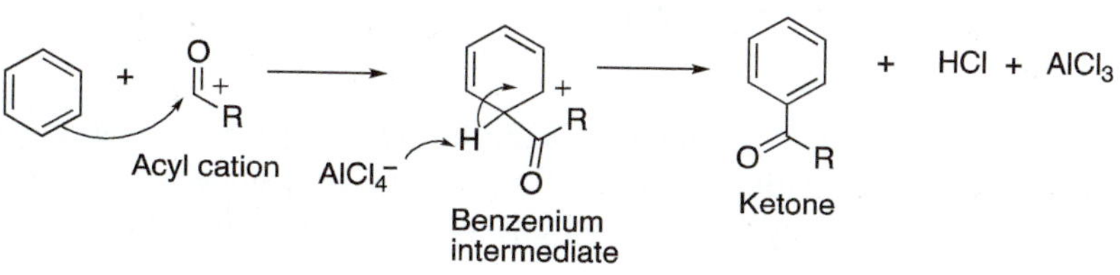

$$(17\text{-}25)$$

Problem 17.7

i) Give the major organic product for each of the following reactions.

(a) [acetyl chloride + benzene] $\xrightarrow{\text{AlCl}_3}$?

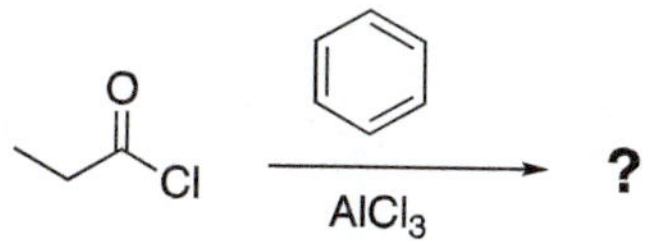

(b) [benzene + cyclopentanecarbonyl chloride] $\xrightarrow{\text{AlCl}_3}$?

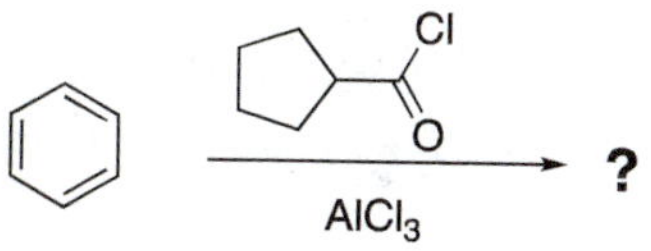

ii) Give the structure of the major organic product of the reaction of benzene with each of the following reagents.
   a) $Br_2/FeBr_3$
   b) $CH_3CH_2Cl/AlCl_3$
   c) $CH_3CH_2COCl/AlCl_3$

17.6.5 Substitution Reactions Involving Sulfonium Ion

Sulfur trioxide is an electrophilic molecule and the electrophilic atom is sulfur, as shown in Reaction (17-26). Based on the structure shown below for sulfur trioxide, the sulfur is electrophilic and for the reaction with a nucleophilic reagent such as benzene, the bond is formed with benzene and sulfur.

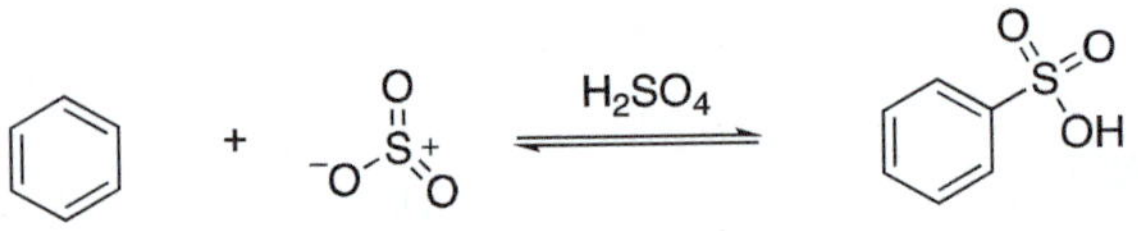

$$\text{benzene} + SO_3 \underset{}{\overset{H_2SO_4}{\rightleftharpoons}} \text{C}_6\text{H}_5\text{SO}_3\text{H} \tag{17-26}$$

This reaction is carried out in "fuming" sulfuric acid, which is 7% SO_3 in H_2SO_4, which generates protonated sulfur trioxide and as you can imagine it is an extremely good electrophile as shown in Reaction (17-27).

$$SO_3 + H_2SO_4 \rightleftharpoons {}^+SO_3H + HSO_4^- \tag{17-27}$$

Sulfur trioxide

In the next step of the reaction mechanism, the nucleophilic benzene attacks the protonated sulfur trioxide to form a benzenium intermediate, which then rearomatizes to form the stable substituted benzene, as shown in Reaction (17-28).

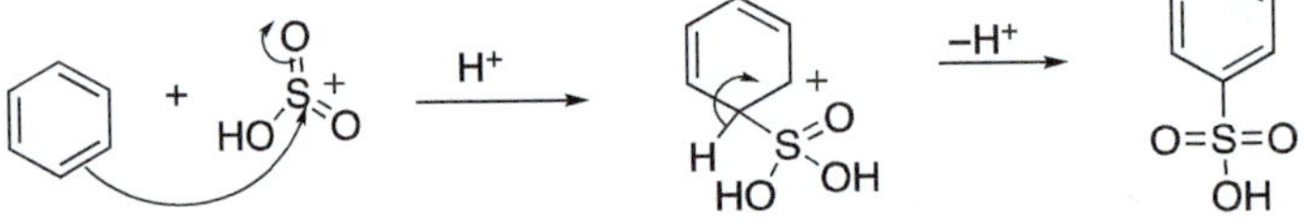

$$\text{benzene} + {}^+SO_3H \xrightarrow{H^+} \text{intermediate} \xrightarrow{-H^+} C_6H_5SO_3H \tag{17-28}$$

Since these reactions are reversible as shown in Reaction (17-26), this substituent can be used as a protecting group for a specific position of the benzene ring since it can be removed by reversing the reaction. This strategy will be used later in the chapter to synthesize specific target compounds.

17.7 Electrophilic Aromatic Substitution Reactions of Substituted Benzene

For substituted benzenes that are involved in electrophilic substitution reactions, as you can imagine, there are three possible products, as shown in Reaction (17-29).

17.7 Electrophilic Aromatic Substitution Reactions of Substituted Benzene

$$\text{Substituted benzene} \xrightarrow{E^+} \text{ortho} + \text{meta} + \text{para} \quad (17\text{-}29)$$

Possible products for electrophilic aromatic substution

Depending on the nature of the substituent, X in Reaction (17-29), the distribution of the products shown can be different. Compared to benzene, some substituted benzenes are more reactive than benzene, and others are less reactive than benzene. Depending on the nature of the substituent on the phenyl ring, it can make substituted benzenes more reactive or less reactive than benzene itself. For example, aniline is much more reactive toward electrophilic reagents than benzene, and the electrophile is always bonded either at the *ortho* or *para* positions of the product, relative to the amino group of aniline for electrophilic aromatic substitution reactions. Reaction (17-30) shows the products of the bromination of aniline.

$$\text{Aniline} \xrightarrow{Br_2} \text{4-Bromoaniline} + \text{2-Bromoaniline} + \text{2,4,6-Tribromoaniline} + HBr \quad (17\text{-}30)$$

Note that aniline is so reactive toward electrophilic aromatic substitution reactions that a catalyst is not needed for the bromination of aniline. On the other hand, if the substituent is a nitro group the product distribution is very different, as shown in Reaction (17-31), only the *meta* substituted product is obtained and a catalyst is needed.

$$PhNO_2 + Br_2 \xrightarrow{FeBr_3} \text{3-bromonitrobenzene} + HBr \quad (17\text{-}31)$$

The question now becomes why are the product distributions different for different substituted benzene and is it possible to predict the outcomes for electrophilic aromatic substitution reaction of other substituted benzenes? We will have to examine the reaction mechanism to answer these questions.

17.7.1 Electron Activators for Electrophilic Aromatic Substitution Reactions

For aniline, the electron density in the aromatic ring is greater than that of benzene owing to the presence of the lone pair of electrons of the nitrogen atom that is adjacent to the phenyl ring. The resonance structures of aniline are shown in Reaction (17-32) and illustrate the increased electron density of the phenyl ring, compared to benzene. Equally important is the location of increased charge density about the phenyl ring for aniline.

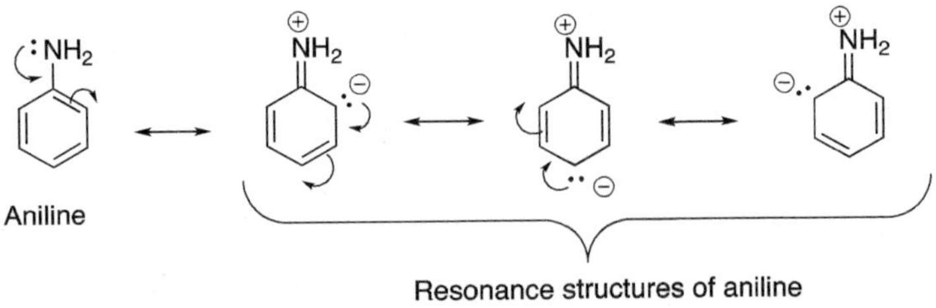

Aniline

$$\text{Aniline} \longleftrightarrow \text{Resonance structures of aniline} \quad (17\text{-}32)$$

Thus, the phenyl ring of aniline is more nucleophilic than benzene. From the resonance structures above, note that the positions that carry a higher electron density are the *ortho* and the *para* positions. Thus, in an electrophilic aromatic substitution reaction, the electrophile will react at these positions. For substituted benzene, if the substituent has a pair of electrons adjacent to the phenyl ring, the ortho and para positions carry a higher electron density, compared to the meta position. As a result, the electrophilic aromatic substitution products involving such molecules will be the *ortho* and *para* substituted products. Reaction (17-33) gives the first step in the reaction mechanism for the reaction of anisole with an electrophile E^+, note that the lone pair of electrons of the OCH_3 group offers an additional stability to the intermediate.

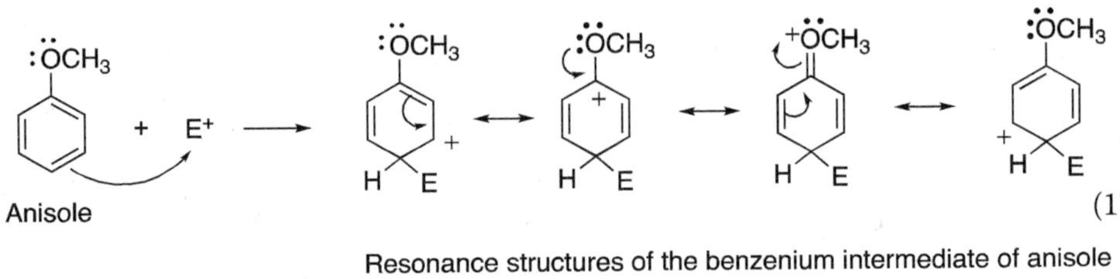

$$\text{Anisole} + E^+ \longrightarrow \text{Resonance structures of the benzenium intermediate of anisole} \quad (17\text{-}33)$$

Reaction (17-34) gives the final step of the mechanism, in which the final product is formed from the charged benzenium intermediate.

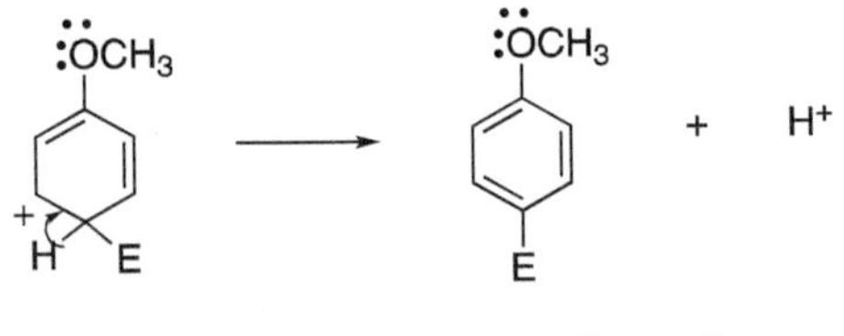

$$\longrightarrow \quad + \quad H^+ \quad (17\text{-}34)$$

A similar reaction will occur in the ortho position as shown in Reaction (17-35).

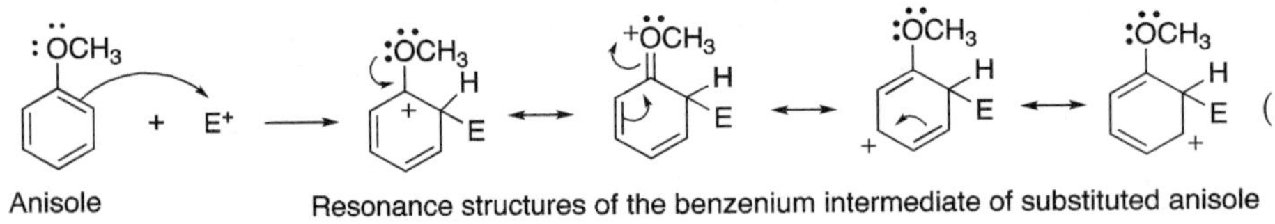

$$\text{Anisole} + E^+ \longrightarrow \text{Resonance structures of the benzenium intermediate of substituted anisole} \quad (17\text{-}35)$$

In the final step of the reaction mechanism, a proton is abstracted for rearomatization to produce the substituted anisole product, as shown in Reaction (17-36).

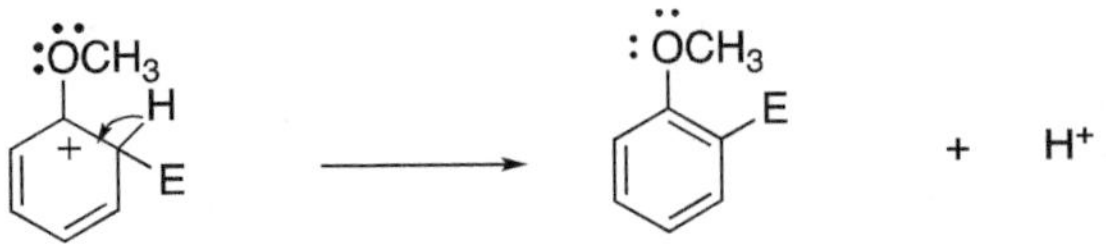

$$\longrightarrow \quad + \quad H^+ \quad (17\text{-}36)$$

It is possible to have a meta substitution as shown in Reaction (17-37), but note that the OCH₃ group is not involved in the stabilization of the benzenium intermediate, compared to the stabilization of those of the para and ortho substitution given above.

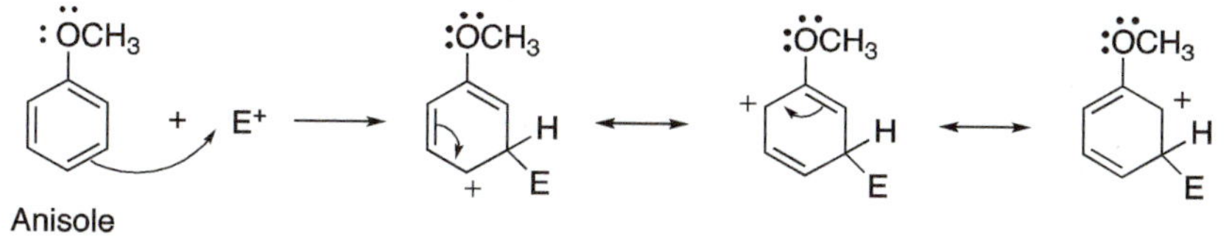

$$\text{Anisole} + E^+ \longrightarrow \cdots \longleftrightarrow \cdots \longleftrightarrow \cdots \quad (17\text{-}37)$$

In the final step of the reaction mechanism, a proton is lost to give the minor substitution product as shown in Reaction (17-38).

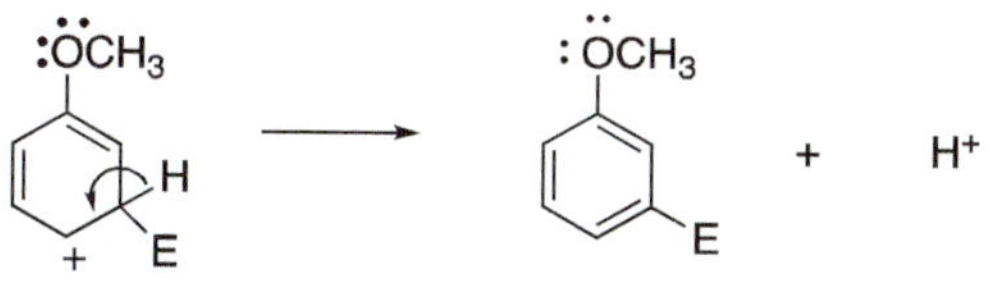

$$\cdots \longrightarrow \cdots + H^+ \quad (17\text{-}38)$$

The reaction between anisole and ethanoyl chloride in the presence of a catalyst is given in Reaction (17-39) where all possible products are shown, but the major product is the para substituted product.

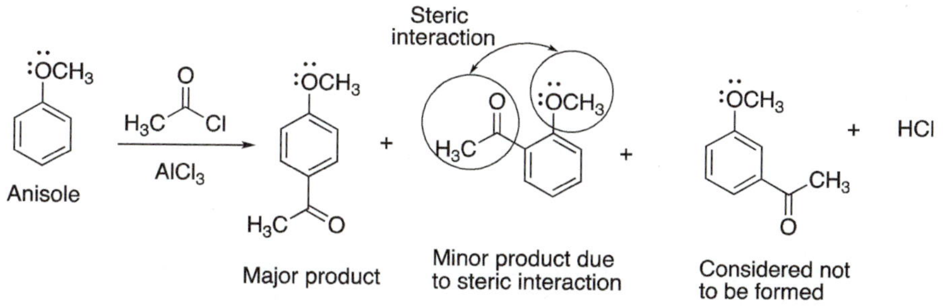

$$\text{Anisole} + H_3C\text{-COCl} \xrightarrow{AlCl_3} \text{Major product} + \text{Minor product due to steric interaction} + \text{Considered not to be formed} + HCl \quad (17\text{-}39)$$

Note that since there is at least one pair of unshared pair of electrons on the oxygen of anisole, this group will activate the ortho and para positions of the benzene ring toward electrophilic substitution. Of the *ortho* and *para* products, typically the *para* product is the major product owing to steric interaction between the substituents in this position of the *ortho* product as shown in Reaction (17-39). Since the methoxy group does not assist in the stabilization of the intermediate that results from a meta substitution, the meta substituted product is not considered to be formed as a reaction product for these reactions.

Since alkyl groups are known to be electron-releasing groups, the major products of electrophilic aromatic substitution reactions of alkyl benzenes will also be the *para* substituted products. Examples of these reactions are shown in Reaction (17-40).

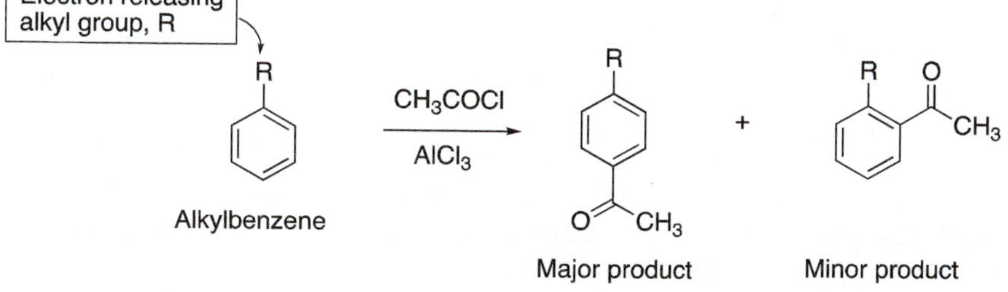

$$\text{Alkylbenzene (R, electron releasing alkyl group)} \xrightarrow[AlCl_3]{CH_3COCl} \text{Major product} + \text{Minor product} \quad (17\text{-}40)$$

Substituents of this type are called **activating substituents or ortho para directors** because they activate the aromatic ring toward electrophilic aromatic substitution reactions (compared to benzene) and they direct the incoming electrophile to the *ortho* and *para* positions.

Problem 17.8

i) Give the major organic product for each of the following reactions.

(a) toluene + CH₃COCl →(AlCl₃) ?

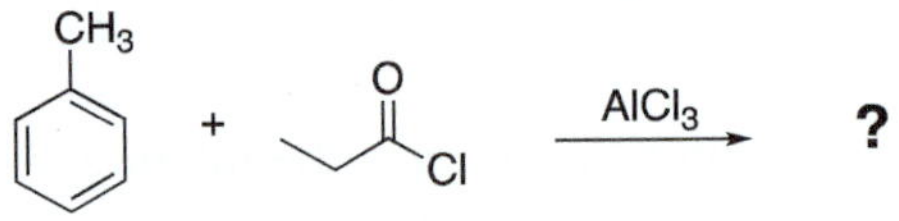

(b) anisole + cyclopentanecarbonyl chloride →(AlCl₃) ?

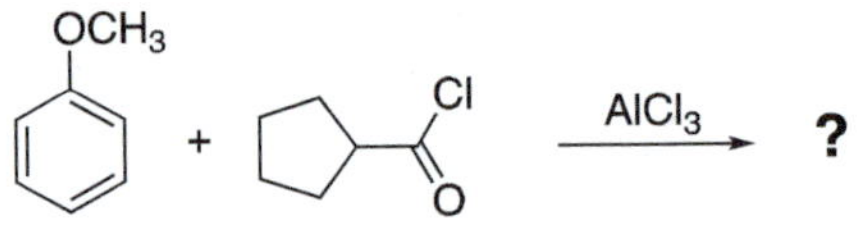

(c) toluene + CH₃CH₂CH₂Cl →(AlCl₃) ?

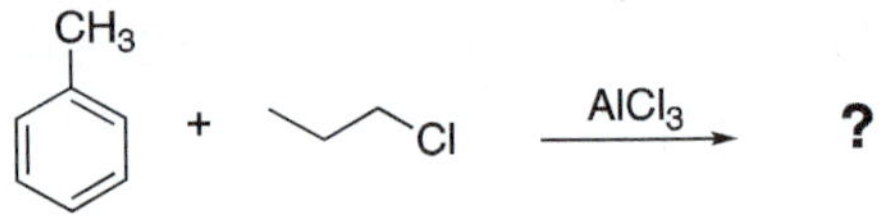

ii) Give the structure of the major organic product of the reaction of anisole with each of the following reagents.
   a) $Br_2/FeBr_3$
   b) $CH_3CH_2Cl/AlCl_3$
   c) $CH_3CH(CH_3)CH_2Cl/AlCl_3$

17.7.2 Electron Deactivators for Electrophilic Aromatic Substitution Reactions

On the other hand, the electron density in the aromatic ring of nitrobenzene is less than that of benzene owing to the presence of an electron withdrawing group, which has a double bond that is adjacent to the phenyl ring of nitrobenzene, as illustrated by the resonance structures of nitrobenzene shown in Reaction (17-41).

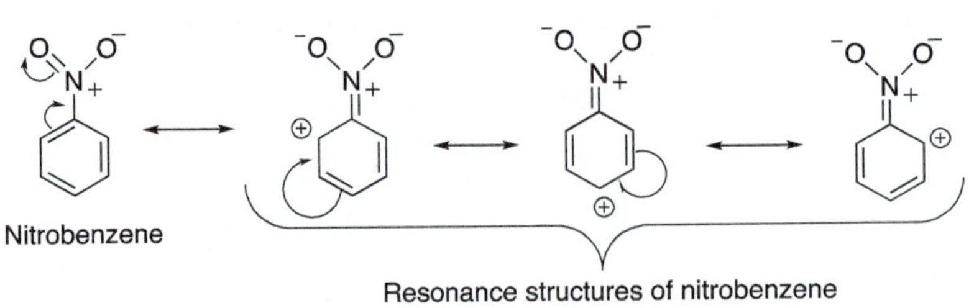

(17-41)

Nitrobenzene

Resonance structures of nitrobenzene

Thus, the phenyl ring of nitrobenzene is less nucleophilic than benzene. From the resonance structures in Reaction (17-41), note that the *ortho* and the *para* positions are positive. Thus, for an electrophilic aromatic substitution reaction, the electrophile will not react at these positions, but instead at the *meta* position. Thus, it appears that if a double bond is in conjugation with the phenyl ring, the electrophilic aromatic substitution product will be the meta-substituted product. This is especially true if the double bond is bonded to an

17.7 Electrophilic Aromatic Substitution Reactions of Substituted Benzene | 489

electron-withdrawing atom. The first step in the reaction between nitrobenzene and an electrophile, E^+ is shown in Reaction (17-42).

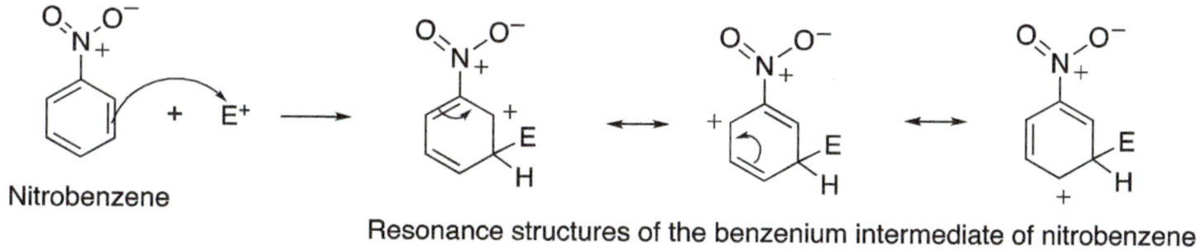

Resonance structures of the benzenium intermediate of nitrobenzene

(17-42)

The product-forming step of the reaction is given in Reaction (17-43).

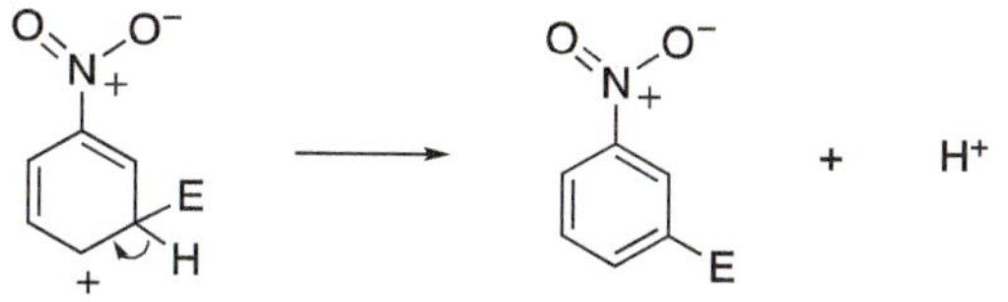

(17-43)

Thus, any substituted benzene that has a double bond that is adjacent to the phenyl ring and especially if the double bond is bonded to an electronegative atom as is the case with the nitro group, the electrophilic aromatic substitution products of such a substituted benzene will be the *meta* product. The reaction of nitrobenzene with ethanoyl chloride is given in Reaction (17-44).

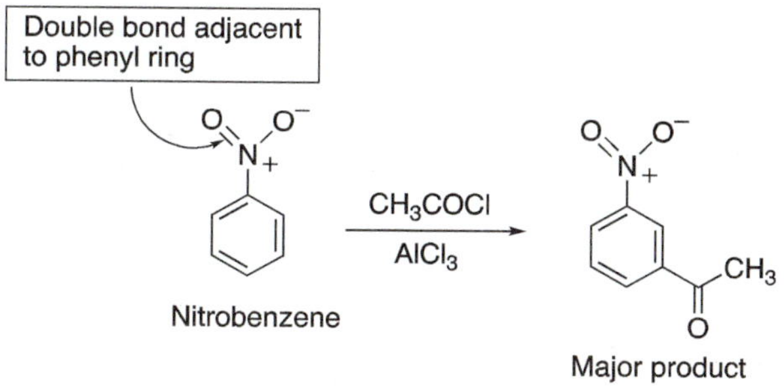

(17-44)

Another example of a reaction of a substituted benzene which has a substituent that has a double bond adjacent to the phenyl ring (acetophenone) with ethanoyl chloride is given in Reaction (17-45).

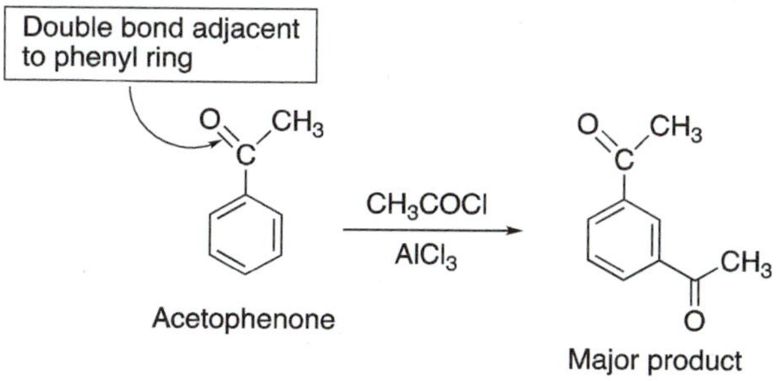

(17-45)

Substituents of this type are called **deactivating substituents or meta directors** because they deactivate the aromatic ring toward aromatic electrophilic substitution reactions (relative to benzene), and they direct incoming electrophiles to the *meta* position.

17 Aromaticity and Aromatic Substitution Reactions

Problem 17.9

i) Give the major organic product for the reactions given below.

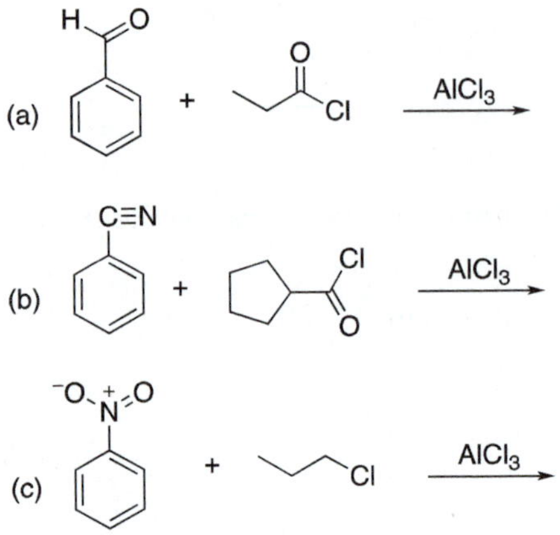

(a) benzaldehyde + propanoyl chloride $\xrightarrow{AlCl_3}$

(b) benzonitrile + cyclopentanecarbonyl chloride $\xrightarrow{AlCl_3}$

(c) nitrobenzene + propyl chloride $\xrightarrow{AlCl_3}$

ii) Give the structure of the major organic product of the reaction of benzaldehyde with each of the following reagents.
 a) $Br_2/FeBr_3$
 b) $CH_3CH_2Cl/AlCl_3$
 c) $CH_3CH(CH_3)CH_2Cl/AlCl_3$

Examples of activating and deactivating groups for electrophilic aromatic substitution reactions are shown in Table 17.2.

17.7.3 Substitution Involving Disubstituted Benzenes

It is possible to have a substitution reaction involving benzene molecules that have two substituents, compared to one that we have discussed in the previous sections. In this case, the effect of each substituent must be examined carefully. Consider the molecule below, it has an activator and a deactivator group present.

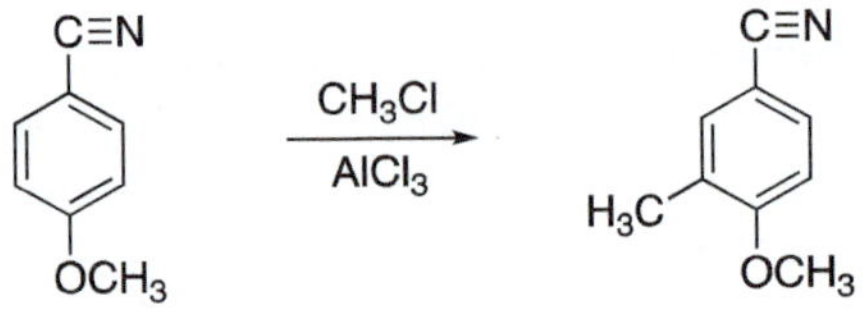

4-methoxybenzonitrile $\xrightarrow[AlCl_3]{CH_3Cl}$ 3-methyl-4-methoxybenzonitrile (17-46)

Table 17.2 Examples of activating and deactivating substituents for electrophilic aromatic substitution reactions.

Activating groups	Deactivating groups
NH_2	COR
OH	CHO
OCH_3	CN
R (alkyl)	NO_2

The reactant has cyano group, which is a deactivator, and hence, a meta director. It also has a methoxy group, which is an activator and also *otho, para* director. The result is that the incoming electrophile will be directed to one position, meta by the cyano and to the ortho as directed by methoxy group. Note that since the para position is blocked, only the ortho position is available for a substitution reaction. The concept can be used in the synthesis of specific compounds, such as the synthetic scheme outlined in Reaction (17-47).

$$\text{H}_3\text{C-C}_6\text{H}_4\text{-SO}_3\text{H} \xrightarrow{\text{Cl}_2, \text{FeCl}_3} \text{(Cl, H}_3\text{C)C}_6\text{H}_3\text{-SO}_3\text{H} \xrightarrow{\text{H}^+} \text{(Cl, H}_3\text{C)C}_6\text{H}_4 + \text{SO}_3 \quad (17\text{-}47)$$

You will recall from the previous section that the substitution reaction involving SO_3 and benzene is a reversible reaction and that the SO_3H substituent is a meta-director. The starting compound of Reaction (17.27) can be made as outlined in Reaction (17-48) from benzene.

$$\text{C}_6\text{H}_6 \xrightarrow{\text{CH}_3\text{Cl}, \text{AlCl}_3} \text{H}_3\text{C-C}_6\text{H}_5 \xrightarrow{\text{-O-S}^+\text{O}_2 / \text{H}_2\text{SO}_4} \text{H}_3\text{C-C}_6\text{H}_4\text{-SO}_3\text{H} \quad (17\text{-}48)$$

Note that for the synthesis, the *o,p*-director goes on first so that it will direct the second electrophile in the para position. Carrying out this reaction by introducing the SO_3H substituent first would result in directing the second substituent in the wrong position and the target molecule would not be achieved. The last step in the sequence of the reaction is shown in Reaction (17-49) where the SO_3H substituent is removed to achieve the target molecule.

$$\text{(Cl, H}_3\text{C)C}_6\text{H}_3\text{-SO}_3\text{H} \xrightleftharpoons{\text{H}^+} \text{(Cl, H}_3\text{C)C}_6\text{H}_4 + \text{SO}_3\text{H}^+ \quad (17\text{-}49)$$

17.8 Applications – Synthesis of Substituted Benzene Compounds

As we have seen at the beginning of the chapter, there are numerous useful compounds that are substituted benzenes. In this section, we will utilize the reactions studied earlier to convert benzene into various substituted benzene compounds. Since benzene is readily available from petroleum, it is a good starting compound from which a host of substituted benzene compounds can be made. In the first example, the oxidation of a CH group directly bonded to a benzene ring (also known as a benzylic CH group) by a strong oxidizing agent is shown in Reaction (17-50).

$$\text{C}_6\text{H}_5\text{-CR}_2\text{H} \xrightarrow{\text{KMnO}_4, \text{heat}} \text{C}_6\text{H}_5\text{-CO}_2\text{H} \quad (17\text{-}50)$$

Note that this oxidation converts a benzylic CH into a carboxylic acid group. Also note that this reaction essentially converts an alkyl *o,p*-director to a meta-director, as shown in the reaction in Reaction (17-51).

$$\text{Benzene} \xrightarrow{\text{CH}_3\text{Cl, FeCl}_3} \text{toluene (}o,p\text{-director)} \xrightarrow{\text{KMnO}_4, \text{ heat}} \text{benzoic acid (meta-director)} \quad (17\text{-}51)$$

This strategy can be used to synthesize slightly different compounds, as illustrated in Reactions (17-52) and (17-53).

$$\text{Benzene} \xrightarrow[\text{FeCl}_3]{\text{CH}_3\text{Cl}} \text{(}o,p\text{-Director)} \xrightarrow[\text{Heat}]{\text{KMnO}_4} \text{(meta-Director)} \xrightarrow[\text{FeCl}_3]{\text{Cl}_2} \text{m-chlorobenzoic acid} \quad (17\text{-}52)$$

$$\text{Benzene} \xrightarrow[\text{FeCl}_3]{\text{CH}_3\text{Cl}} \text{(}o,p\text{-Director)} \xrightarrow[\text{FeCl}_3]{\text{Cl}_2} \text{p-chlorotoluene} \xrightarrow[\text{Heat}]{\text{KMnO}_4} \text{p-chlorobenzoic acid} \quad (17\text{-}53)$$

Even though the same reactions and starting compound are used for both Reactions (17-52) and (17-53), the final products are different; the only difference is the sequence of these reactions. For Reaction (17-52), the methyl group is converted to the carboxylic acid group, which makes it a meta-director for the incoming chlorine. For Reaction (17-53), the methyl group is used as a *o,p*-director for the incoming chlorine.

A similar strategy is illustrated for the sequence of reactions in Reaction (17-54).

$$\text{Benzene} + (\text{CH}_3)_2\text{CHCOCl} \xrightarrow{\text{AlCl}_3} \text{isobutyrophenone (meta-Director)} \xrightarrow[\text{(reduction)}]{\text{Zn(Hg), HCl}} \text{isobutylbenzene (}o,p\text{-Director)}$$

$$\downarrow \text{CH}_3\text{COCl, AlCl}_3 \qquad \qquad \downarrow \text{CH}_3\text{COCl, AlCl}_3$$

$$\text{m-diacyl product} \qquad \qquad \text{p-isobutylacetophenone} \quad (17\text{-}54)$$

Again, the same starting compound, benzene and essentially the same reactions are used in the reaction scheme above, yet the final products are different just based on a slight difference in the sequence of reactions. One of these reaction sequences can used to synthesize an

17.8 Applications – Synthesis of Substituted Benzene Compounds

important compound, ibuprofen, and is outlined in the reaction scheme in Reactions (17-55) and (17-56).

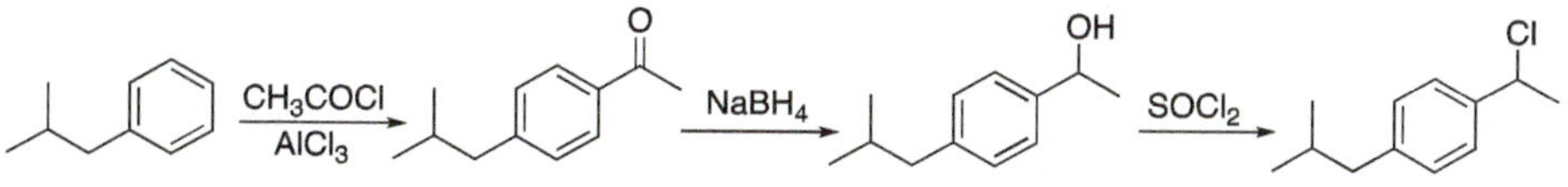

$$\text{isobutylbenzene} \xrightarrow[\text{AlCl}_3]{\text{CH}_3\text{COCl}} \text{(aryl ketone)} \xrightarrow{\text{NaBH}_4} \text{(secondary alcohol)} \xrightarrow{\text{SOCl}_2} \text{(alkyl chloride)} \tag{17-55}$$

Note that since the relationship between the groups in our target molecule, ibuprofen, is para, the reaction selected from Reaction (17-54) is the reaction shown to the extreme top right. For the reaction sequence given in Reaction (17-55), reduction of the carbonyl gives a secondary alcohol, which can be converted to a chloride using thionyl chloride ($SOCl_2$).

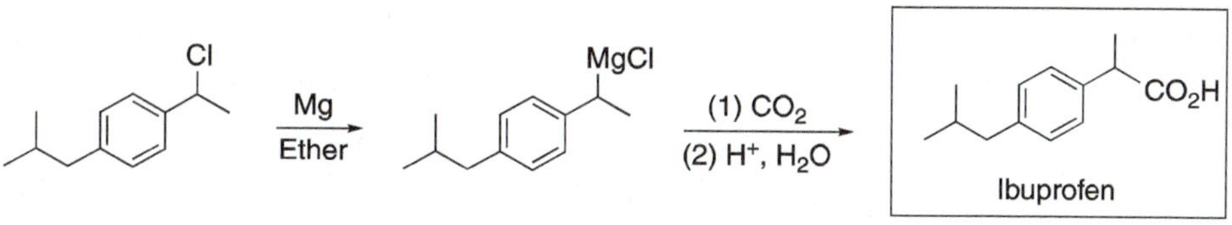

$$\text{(alkyl chloride)} \xrightarrow[\text{Ether}]{\text{Mg}} \text{(Grignard, MgCl)} \xrightarrow[\text{(2) H}^+, \text{H}_2\text{O}]{\text{(1) CO}_2} \boxed{\text{Ibuprofen, CO}_2\text{H}} \tag{17-56}$$

The chlorine functionality can be converted to an organometallic reagent using a Grignard reaction and the Grignard reagent used to reduce carbon dioxide gives the product, ibuprofen, after acidic hydrolysis.

Another example of a similar strategy of introducing a specific group onto a benzene ring and converting it to a different director is shown in the example in Reaction (17-57). The reduction of a nitro group to form an amine is an important reaction, as shown below in Reaction (17-57).

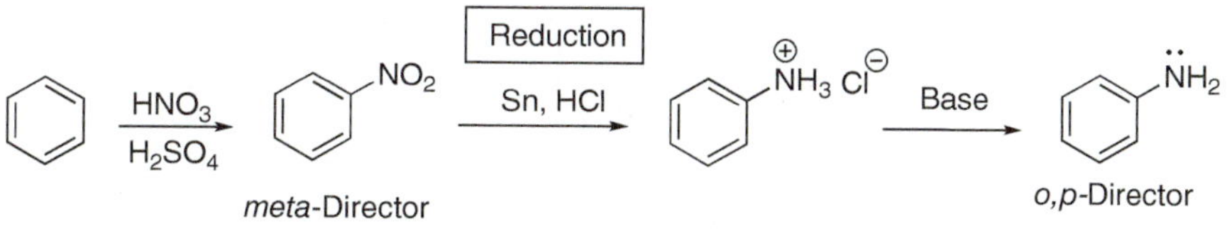

$$\text{benzene} \xrightarrow[\text{H}_2\text{SO}_4]{\text{HNO}_3} \underset{\textit{meta-Director}}{\text{Ph-NO}_2} \xrightarrow[]{\boxed{\text{Reduction}} \atop \text{Sn, HCl}} \text{Ph-NH}_3^+\text{Cl}^- \xrightarrow{\text{Base}} \underset{\textit{o,p-Director}}{\text{Ph-NH}_2} \tag{17-57}$$

Note for this set of reactions, the nitro group, which is a meta-director is converted to a o,p-director. We can take advantage of this in making different substituted benzenes, as shown in Reaction (17-58).

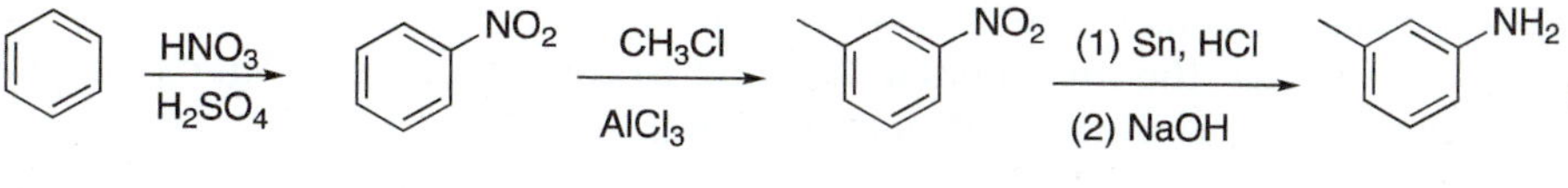

$$\text{benzene} \xrightarrow[\text{H}_2\text{SO}_4]{\text{HNO}_3} \text{Ph-NO}_2 \xrightarrow[\text{AlCl}_3]{\text{CH}_3\text{Cl}} \text{m-CH}_3\text{-Ph-NO}_2 \xrightarrow[\text{(2) NaOH}]{\text{(1) Sn, HCl}} \text{m-CH}_3\text{-Ph-NH}_2 \tag{17-58}$$

Since the nitro group is a *meta*-director, the incoming methyl group will be directed to the *meta* position of the benzene ring. Once the reduction is accomplished, additional reactions can be accomplished as shown in Reaction (17-59).

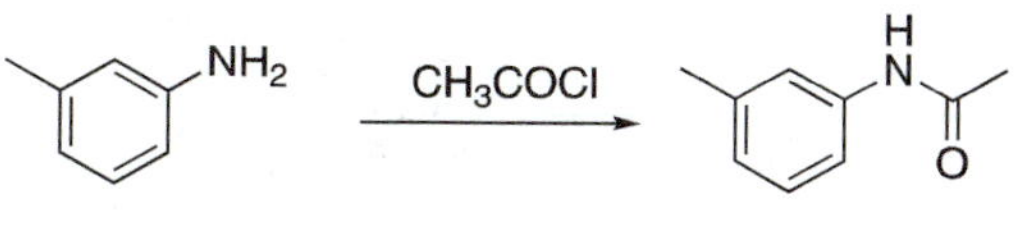

$$\text{m-CH}_3\text{-Ph-NH}_2 \xrightarrow{\text{CH}_3\text{COCl}} \text{m-CH}_3\text{-Ph-NHC(O)CH}_3 \tag{17-59}$$

Problem 17.10

Show how to carry out the following transformation.

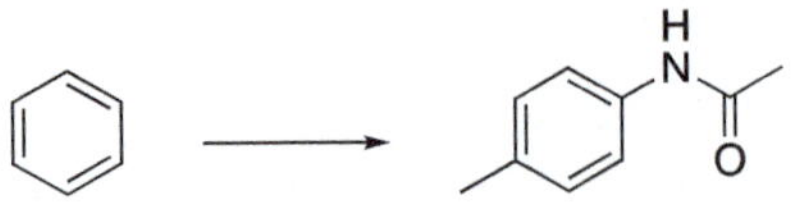

benzene ⟶ 4-methylacetanilide (N-(4-methylphenyl)acetamide)

17.9 Electrophilic Substitution Reactions of Polycyclic Aromatic Compounds

Close examination of naphthalene reveals that there are two different types of hydrogens and they are shown in Reaction (17-60) as H_α and H_β. Hydrogens labeled H_α are close to the point of ring fusion and the other types of hydrogens, H_β, are removed from the point of ring fusion by a carbon. Thus, there are two products possible for the electrophilic substitution reaction as shown in Reaction (17-60).

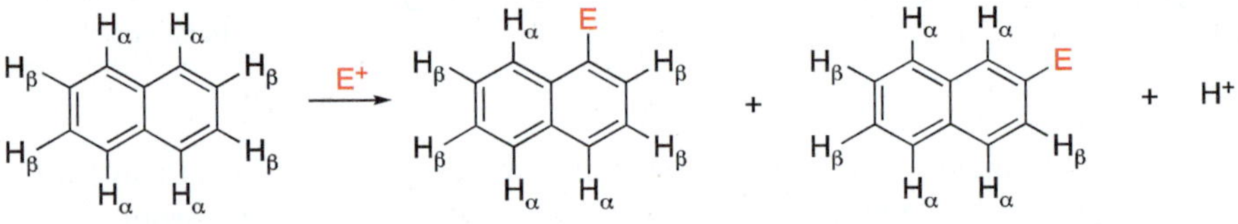

Naphthalene + E^+ ⟶ α-substituted product + β-substituted product + H^+ (17-60)

By varying the reaction conditions for this substitution reaction, it is possible to produce one of these products as the major product compared to the other. If E^+ is bulky, such as the sulfonyl group, there will be steric crowing around the electrophile in the product if the substitution is in the alpha (α) position. On the other hand, if the substitution is in the beta (β) position, there would be less steric crowding. This concept is illustrated in Reaction (17-61).

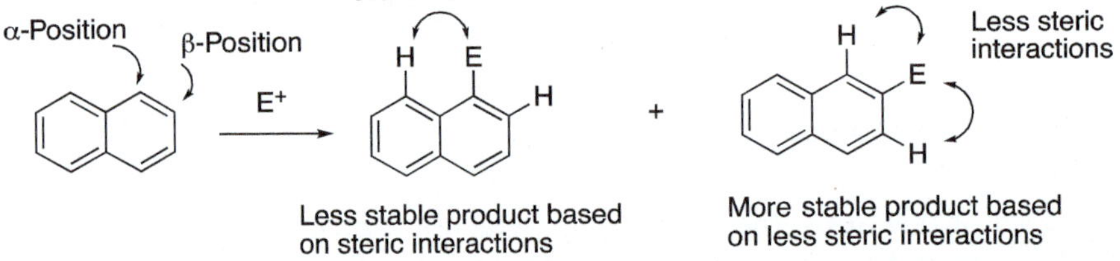

α-Position, β-Position + E^+ ⟶ Less stable product based on steric interactions (steric interaction between H and E) + More stable product based on less steric interactions (Less steric interactions) (17-61)

There is another factor that must be considered in predicting the major and minor products for this type of substitution reaction. A close examination of the reaction mechanism shows that the intermediate leading to the formation of the product in which there is an α-substitution is more stable than the intermediate leading to the formation of the product where there is a β-substitution, as shown in Reaction (17-62).

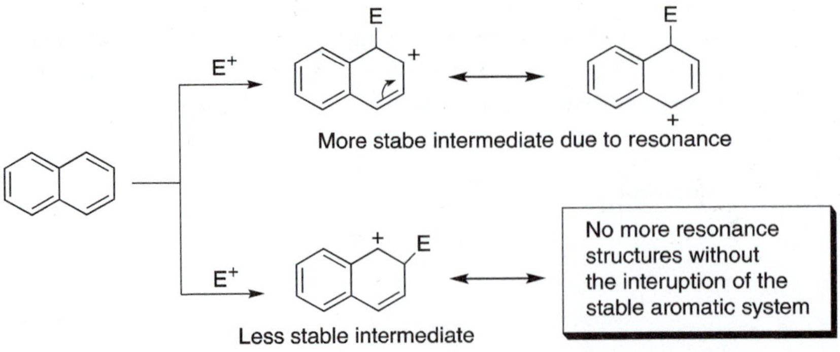

Naphthalene + E^+ ⟶ More stable intermediate due to resonance

Naphthalene + E^+ ⟶ Less stable intermediate (No more resonance structures without the interuption of the stable aromatic system) (17-62)

Thus, the energy of activation leading to the less stable intermediate is greater than that leading to the more stable intermediate. As a result, in order to form the substitution product with the substitution in the β-position, additional energy must be supplied in order to overcome the higher energy of activation. On the other hand, to form the least stable product in which the substitution is in the α-position, a lower temperature is required. Varying the temperature of a reaction to have the reaction go through different pathways is known as thermodynamic and kinetic control of the reaction. This concept is illustrated in sulfonation of naphthalene shown Reactions (17-63) and (17-64).

$$\text{naphthalene} \xrightarrow[80\ °C]{SO_3/H_2SO_4} \text{1-naphthalenesulfonic acid (SO}_3\text{H at α)} \quad (17\text{-}63)$$

$$\text{naphthalene} \xrightarrow[160\ °C]{SO_3/H_2SO_4} \text{2-naphthalenesulfonic acid (SO}_3\text{H at β)} \quad (17\text{-}64)$$

For Reaction (17-63), the temperature is lower than that of the reaction in Reaction (17-64), hence the kinetic product is formed as the major product. On the other hand, for the reaction in Reaction (17-64), which is carried out at a higher temperature, the more stable less sterically hindered product is produced since there is enough energy to overcome a high activation barrier. This is a classic example of kinetic vs. thermodynamic control of reactions and is illustrated in the energy profile diagram shown in Figure 17.20.

For the substitution reaction in which the electrophile is not very bulky, steric stability of the product is not a major factor, but the stability on the resonance stabilized intermediate carbocation is important. Thus, at room temperature, the bromination of naphthalene readily occurs as shown in Reaction (17-65).

$$\text{naphthalene} + Br_2 \xrightarrow{CCl_4} \text{1-bromonaphthalene} + HBr \quad (17\text{-}65)$$

The intermediate formed in the first step of the mechanism is the more stable intermediate as shown in Reaction (17-66).

$$\text{(mechanism with Br–Br, }-Br^-\text{, resonance stabilized intermediate)} \quad (17\text{-}66)$$

Resonance stabilized intermediate

Figure 17.20 Energy profile for the electrophilic substitution of naphthalene under different reaction temperatures (black represents thermodynamic control and red represents kinetic control).

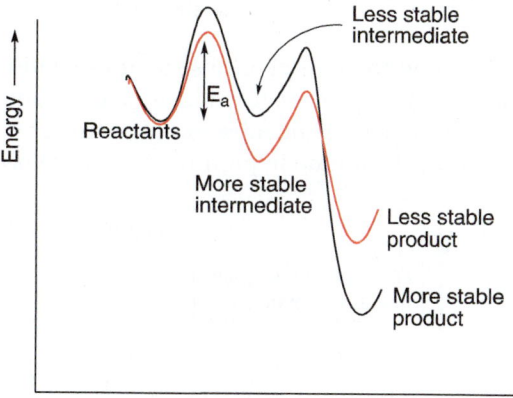

Problem 17.11

i) Give the organic products for the reaction of naphthalene with each of the following reagents.
   a) Cl₂/FeCl₃
   b) Br₂/FeBr₃
   c) CH₃COCl/AlCl₃

ii) Shown below are the structures of two polycyclic aromatic compounds. Give the expected organic product for the reaction of each with Br₂ in carbon tetrachloride as the solvent at an elevated temperature.

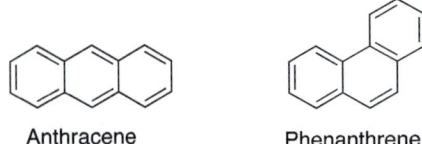

Anthracene   Phenanthrene

17.10 Electrophilic Substitution Reactions of Pyrrole

Pyrrole is a five-member ring aromatic compound, but unlike benzene it has a dipole moment greater than zero. To fully appreciate the difference, an examination of the dipole moment and its direction must be carried out. For pyrrole, the dipole moment is 1.81 debyes, but the direction is toward the aromatic ring as shown in Figure 17.21.

As a result, the aromatic ring is electron rich and will prefer to react with electrophiles and undergo electrophilic aromatic substitution similar to that of benzene. For pyrrole, however, there are two possible positions on the aromatic ring to which the electrophile can bond and the intermediates generated are shown in Reaction (17-67).

(17-67)

Based on these resonance structures, addition of the electrophile to position 2 produces a more stable intermediate, compared to the addition of the electrophile to position 3. Thus, the major product for the nitration of pyrrole, which is given in Reaction (17-68), is the product with the substitution in position #2 and not position #3.

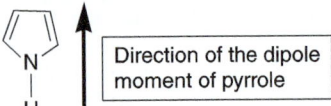

Figure 17.21 Direction of the dipole moment of pyrrole.

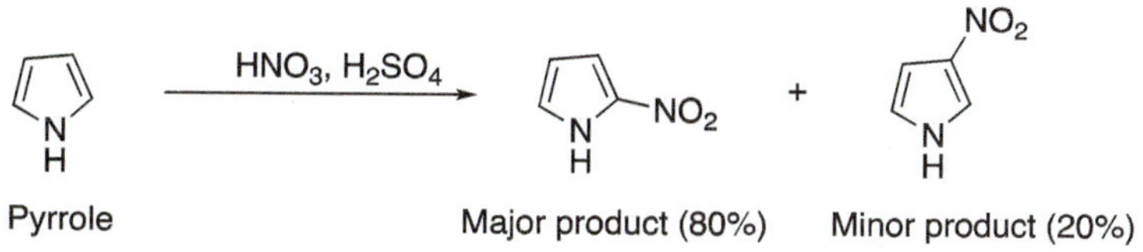

$$\text{Pyrrole} \xrightarrow{HNO_3, H_2SO_4} \text{Major product (80%)} + \text{Minor product (20%)} \qquad (17\text{-}68)$$

Problem 17.12

Give the organic products for the reaction of pyrrole with each of the following reagents.
a) $Br_3/FeBr_3$
b) $CH_3CH_2Cl/AlCl_3$
c) $CH_3CH_2COCl/AlCl_3$

17.11 Electrophilic Substitution Reactions of Pyridine

Even though pyridine looks a lot like benzene, the products of the reaction of pyridine with electrophiles are slightly different from those of benzene, polycyclic aromatic compounds, or pyrrole. To fully understand why the difference, we need to examine an important property of pyridine and that is the dipole moment. Pyridine has a dipole moment of 2.26 debyes and the direction of the dipole is away from the aromatic ring and toward the nitrogen, as shown below.

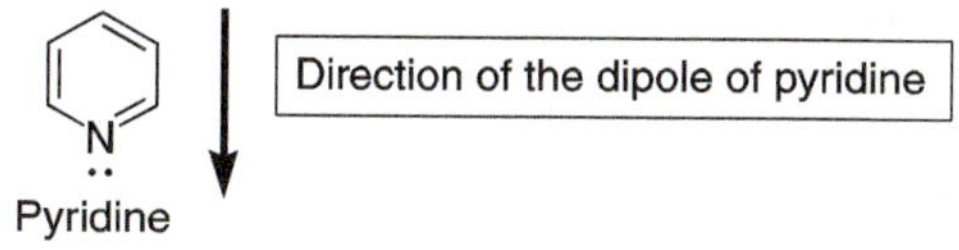

Direction of the dipole of pyridine

Pyridine

Based on this dipole moment, the electron density is toward the nitrogen atom and away from the aromatic ring. Reactions (17-69)–(17-71) give the resonance structures of the intermediates that result from the RDS step of the reaction mechanism after the addition of an electrophile.

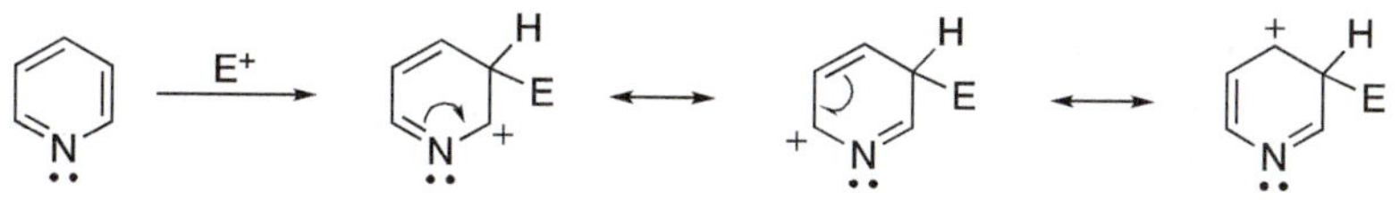

$$(17\text{-}69)$$

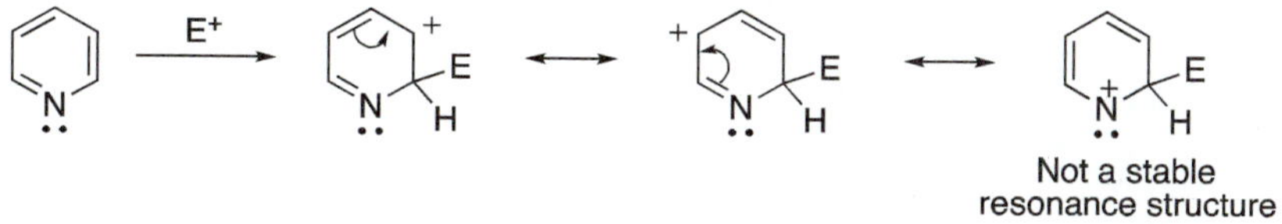

$$(17\text{-}70)$$

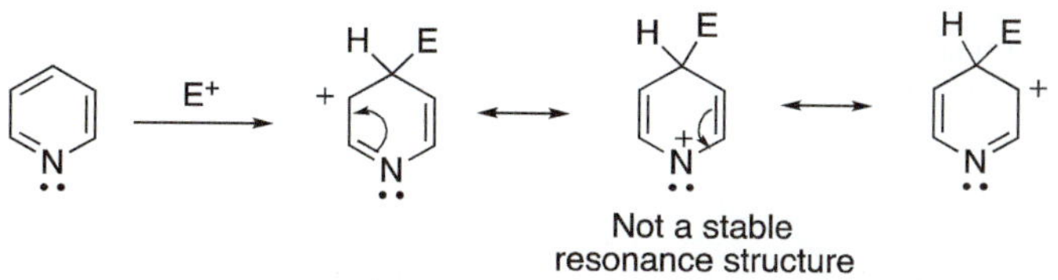

Not a stable resonance structure

$$(17\text{-}71)$$

Not a stable resonance structure

You will notice that for the addition to positions #2 and #4 (Reactions (17-70) and (17-71)), there is a resonance structure in each that is not stable since the positive charge is on the very electronegative nitrogen. Since the pathway that leads to the more stable intermediate is the attraction of the electrophile to the #3 position of the pyridine ring, the major product will be the substitution in the #3 position. The final step in the reaction mechanism to give the major product is shown in Reaction (17-72). Note that the major product shown is typically formed in very low yields since pyridine is less reactive, compared to the other aromatic molecules.

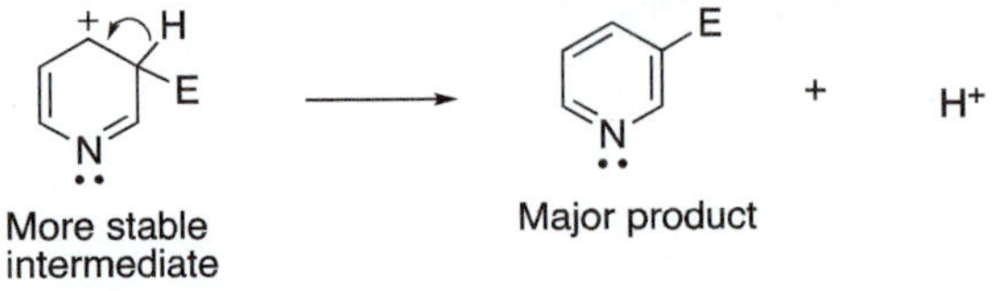

$$\text{More stable intermediate} \longrightarrow \text{Major product} + H^+ \tag{17-72}$$

If the least stable intermediates were to form products, the final steps in the reaction mechanism to give the minor products would be as shown in Reactions (17-73) and (17-74).

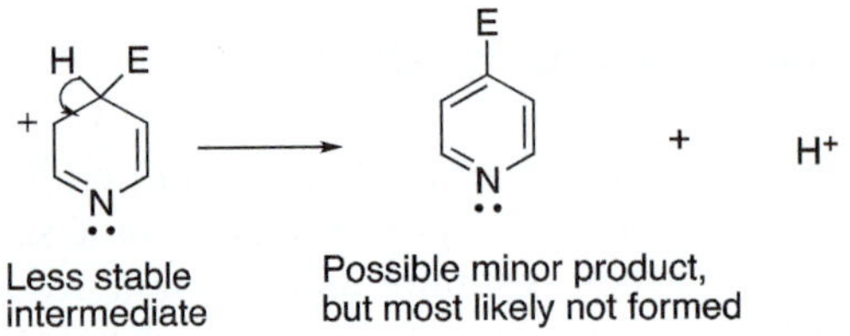

$$\text{Less stable intermediate} \longrightarrow \text{Possible minor product, but most likely not formed} + H^+ \tag{17-73}$$

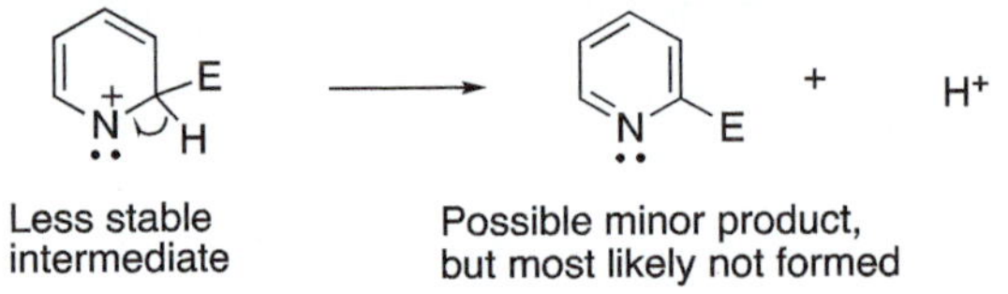

$$\text{Less stable intermediate} \longrightarrow \text{Possible minor product, but most likely not formed} + H^+ \tag{17-74}$$

The overall aromatic substitution reaction is shown in Reaction (17-75).

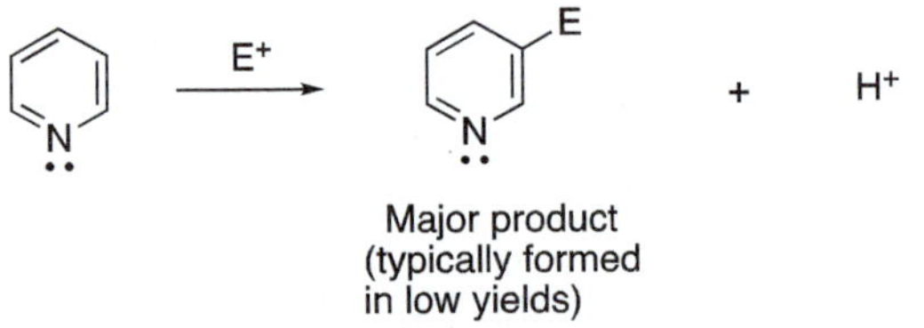

$$\text{Pyridine} \xrightarrow{E^+} \text{Major product (typically formed in low yields)} + H^+ \tag{17-75}$$

An example of the electrophilic substitution involving pyridine and showing only the major organic product is given in Reaction (17-76).

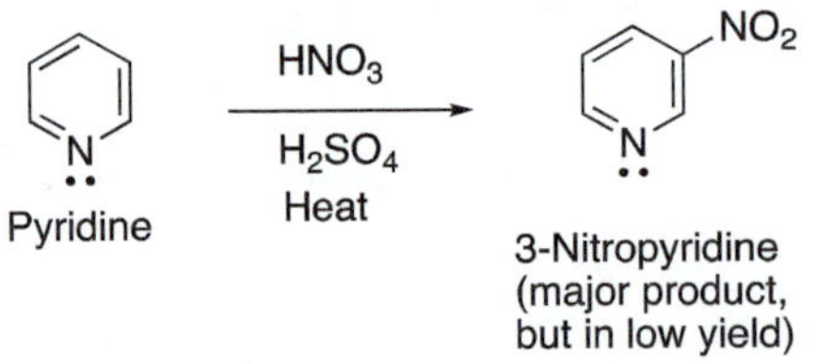

$$\text{Pyridine} \xrightarrow[\text{Heat}]{\text{HNO}_3, \text{H}_2\text{SO}_4} \text{3-Nitropyridine (major product, but in low yield)} \tag{17-76}$$

Problem 17.13

Give the major organic product for the reaction of pyridine with $Br_2/FeBr_3$.

It should be noted that since pyridine has a pair of electrons on the nitrogen that is not involved in the aromaticity, it is nucleophilic and in the presence of an electrophile, the possibility exists for a substitution reaction as shown in the examples below.

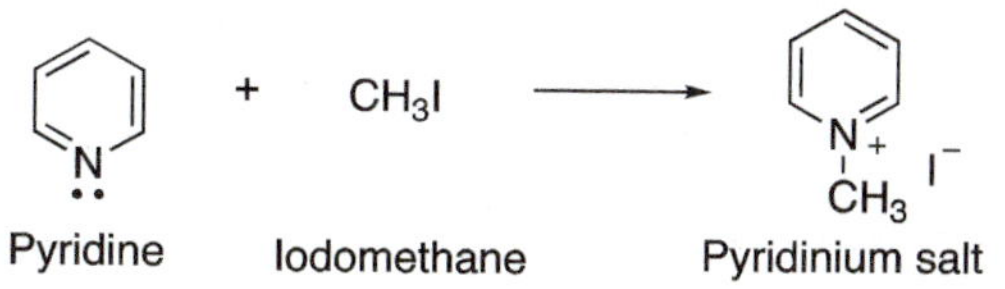

$$\text{Pyridine} + CH_3I \longrightarrow \text{Pyridinium salt} \quad (17\text{-}77)$$

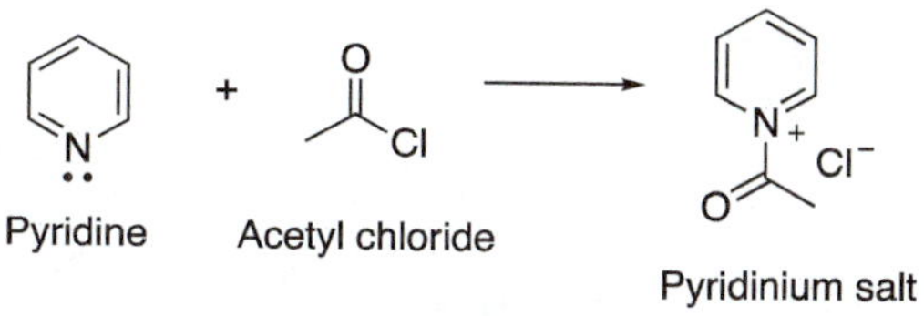

$$\text{Pyridine} + \text{Acetyl chloride} \longrightarrow \text{Pyridinium salt} \quad (17\text{-}78)$$

17.12 Nucleophilic Aromatic Substitution

17.12.1 Nucleophilic Aromatic Substitution Involving Substituted Benzene

We have demonstrated that the benzene ring is nucleophilic owing to the pi (π) electrons that are present in the aromatic ring. The electron density of the aromatic ring can be reduced if electron-withdrawing groups are bonded to the ring. For example, the high acidity of picric acid is due to the reduced electron density of the aromatic ring, which is brought about by the electron withdrawing nitro groups pulling electrons from the aromatic system, as shown in the acid-base equilibrium below.

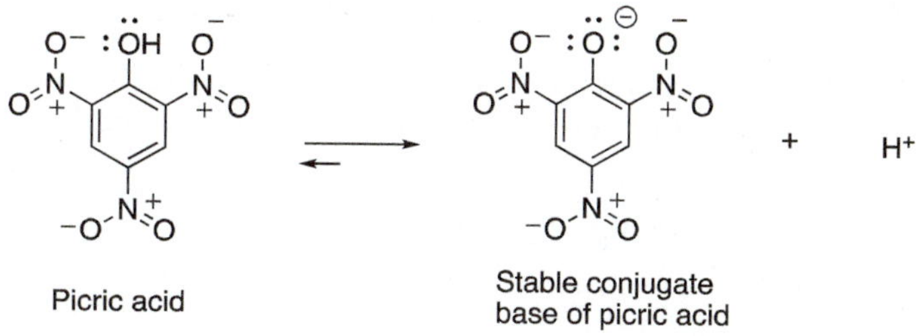

$$\text{Picric acid} \rightleftharpoons \text{Stable conjugate base of picric acid} + H^+$$

As you can imagine, if a good leaving group is bonded to a molecule that has a low electron density an electronic environment is generated for substitution reactions. In this case, the aromatic ring would be electrophilic if a substitution reaction were to occur. This type of reaction would be a nucleophilic aromatic substitution, as shown in the reaction given in Reaction (17-79).

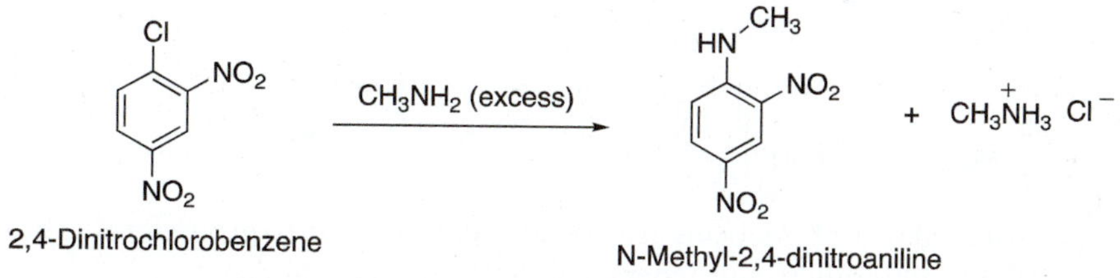

$$\text{2,4-Dinitrochlorobenzene} \xrightarrow{CH_3NH_2 \text{ (excess)}} \text{N-Methyl-2,4-dinitroaniline} + CH_3\overset{+}{N}H_3 \; Cl^- \quad (17\text{-}79)$$

2,4-Dinitrochlorobenzene has a good leaving group in the form of the chloride anion and the electron density of the aromatic ring is reduced owing to the presence of two electron withdrawing nitro groups. In the presence of a nucleophile, such as the methylamine, a reaction will occur and the mechanism is shown below.

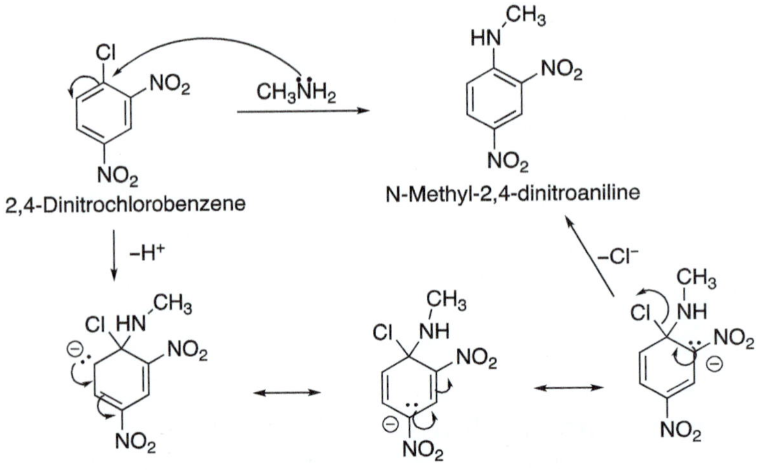

17 Aromaticity and Aromatic Substitution Reactions

[Reaction scheme: 2,4-Dinitrochlorobenzene reacting with CH₃NH₂ to give N-Methyl-2,4-dinitroaniline, showing resonance-stabilized Meisenheimer intermediates with loss of H⁺ and Cl⁻.]

Since the nitro groups are in the *ortho* and *para* positions to the chlorine, the negative charge is stabilized by the nitro groups as shown in the resonance example given in Reaction (17-80).

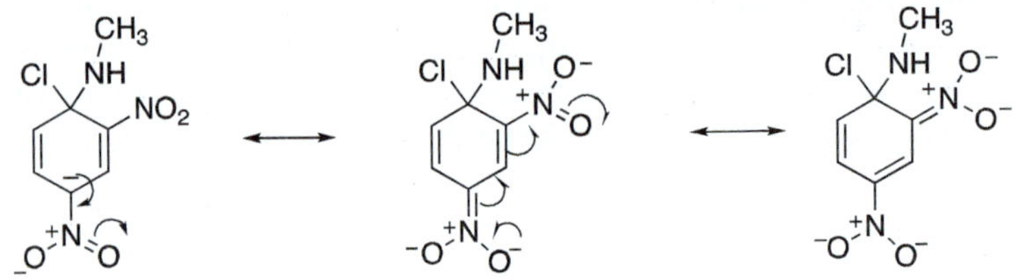

[Resonance structures showing delocalization of negative charge onto ortho and para nitro groups.] (17-80)

Note that the nitro groups in the *ortho* or *para* positions can assist in the stabilization of the negatively changed intermediate better than if it were in the *meta* positions to the leaving group.

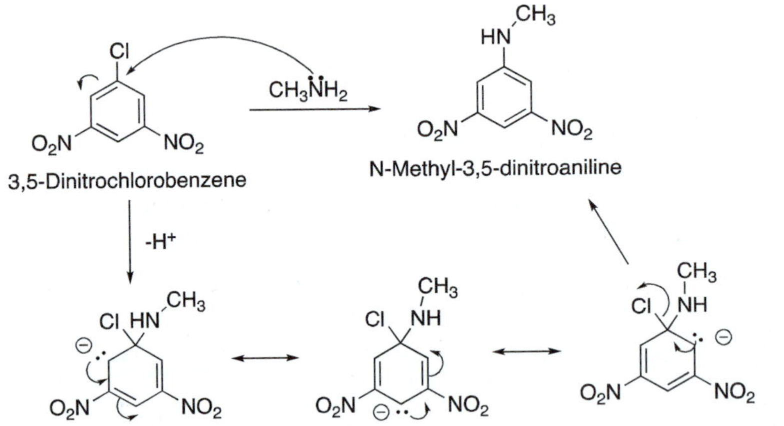

[Reaction scheme: 3,5-Dinitrochlorobenzene reacting with CH₃NH₂ to give N-Methyl-3,5-dinitroaniline, showing intermediates where negative charge is not stabilized by the meta nitro groups.]

Note that the unshared pair of electrons are not adjacent to the nitro group for this molecule, compared to 2,4-dinitrochlorobenzene, where stabilization can occur.

Problem 17.14

Which of the molecules shown below is more reactive toward nucleophilic aromatic substitution? Explain your answer.

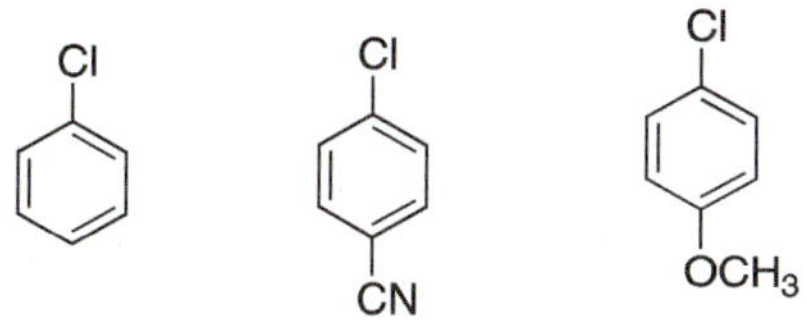

17.12 Nucleophilic Aromatic Substitution

[Structures: chlorobenzene; 4-chlorobenzonitrile (Cl with CN para); 4-chloroanisole (Cl with OCH₃ para)]

It was shown that the reaction given in Reaction (17-81) could occur under extremely severe reaction conditions, such as extremely high temperatures (~300 °C) and using a good nucleophile, such as the hydroxide anion, even though there are no electron withdrawing groups on the benzene ring.

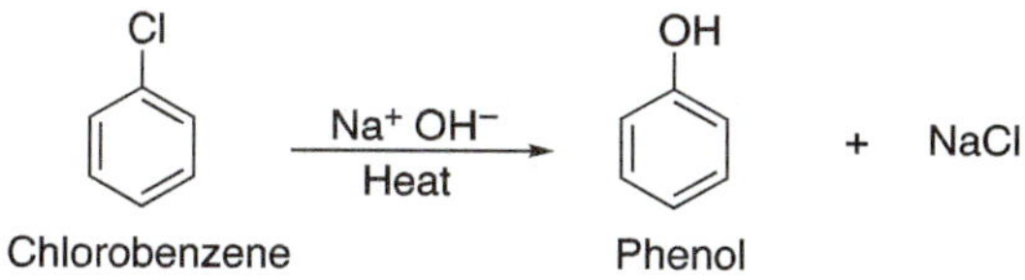

$$\text{Chlorobenzene} \xrightarrow[\text{Heat}]{\text{Na}^+ \text{OH}^-} \text{Phenol} + \text{NaCl} \tag{17-81}$$

Using an even stronger nucleophile, such as the amide anion, the reaction requires less severe reaction conditions as shown in Reaction (17-82).

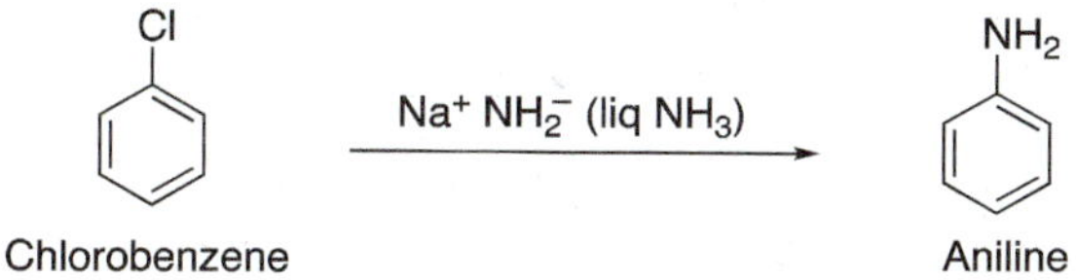

$$\text{Chlorobenzene} \xrightarrow{\text{Na}^+ \text{NH}_2^- \text{(liq NH}_3\text{)}} \text{Aniline} \tag{17-82}$$

These reactions occur since the nucleophiles are extremely good (also very strong bases) and under more severe reaction conditions, compared to the reactions discussed earlier. Experiments that were carried out proved that the reaction pathway might not be as straightforward as the nucleophile attacking the electrophilic carbon that contains the leaving group to accomplish the substitution. It was shown that if labeled chlorobenzene, in which the carbon that contains the leaving group is labeled with C-14, were used as the reactant, there was scrambling to provide a 50 : 50 mixture in the products as shown in the reaction in Reaction (17-74).

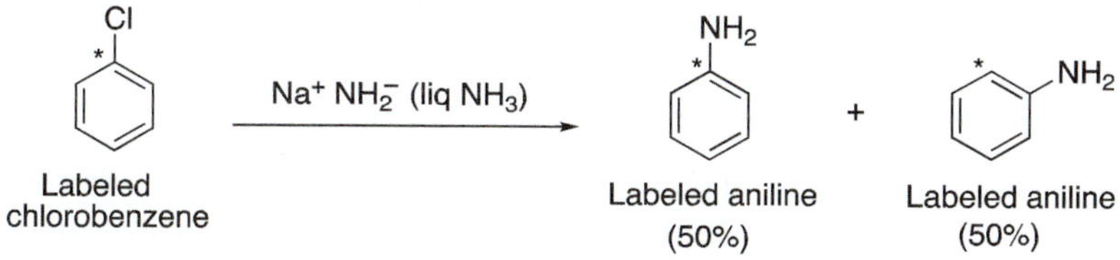

$$\text{Labeled chlorobenzene} \xrightarrow{\text{Na}^+ \text{NH}_2^- \text{(liq NH}_3\text{)}} \text{Labeled aniline (50\%)} + \text{Labeled aniline (50\%)} \tag{17-83}$$

A plausible explanation for this observation is that a symmetrical intermediate was formed throughout the course of the reaction. In the first step of the mechanism, the very strong base abstracts a proton adjacent to the carbon containing the leaving group and forms a triple bond as the chloride leaves as shown it Reaction (17-84).

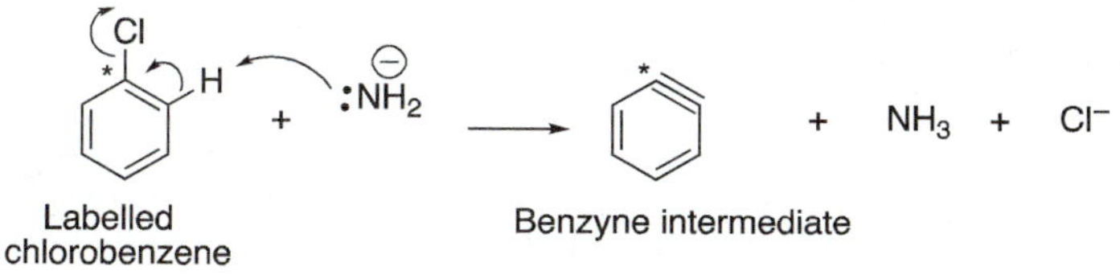

$$\text{Labelled chlorobenzene} + :\text{NH}_2^- \longrightarrow \text{Benzyne intermediate} + \text{NH}_3 + \text{Cl}^- \tag{17-84}$$

The intermediate is called a benzyne intermediate and as you can imagine, this intermediate is very unstable due to the geometry of this highly strained intermediate. In the next step of the mechanism, the benzyne is attacked by the nucleophile. As you can imagine, there is an equal probability of attack coming at either carbon as shown in Reactions (17-85) and (17-86).

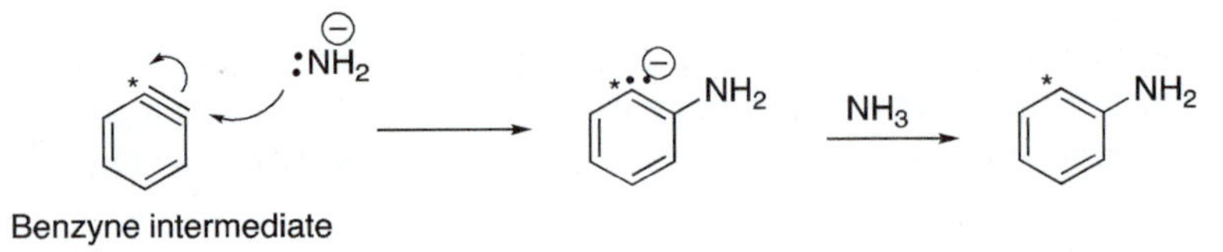

(17-85)

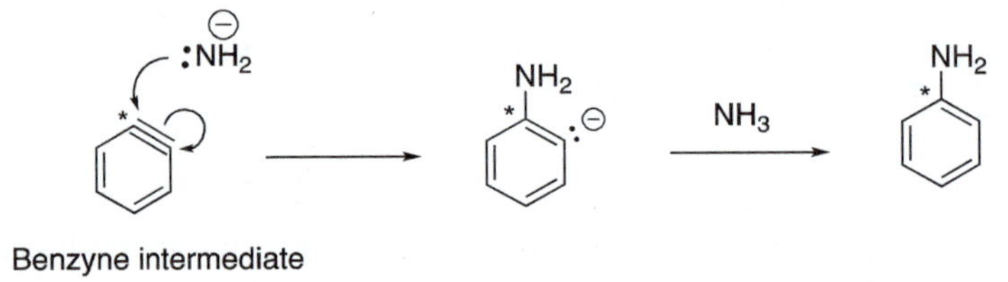

(17-86)

After attack of the nucleophilic amide anion on the benzyne intermediate, it is protonated by the solvent, ammonia, to form the product shown. Even though the benzyne intermediate has never been isolated and actually studied, this is a classic representation of using scientific reasoning to provide a plausible explanation, which does not violate the basic chemistry principles and knowledge, for a very unusual observation.

Problem 17.15

Would you expect scrambling for the reaction shown below, the star shows the point of carbon labeling? Explain your answer.

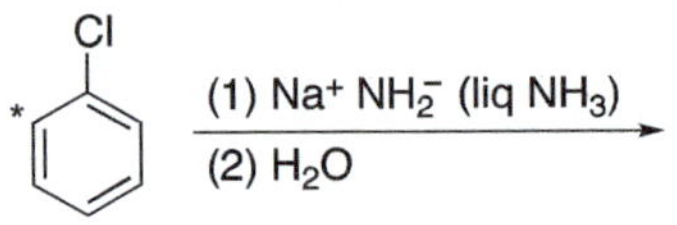

17.12.2 Nucleophilic Aromatic Substitution Involving Substituted Pyridine

As pointed out, pyridine has a dipole moment of 2.26 debyes and the direction of the dipole is away from the aromatic ring and toward the very electronegative nitrogen. This means that the electron density of the pyridine ring is less than that of benzene. Thus, pyridine is more susceptible to nucleophilic attack, compared to benzene. The resonance structures shown in Reaction (17-87) give an idea of the regions of the molecule of lowest electron density, and hence most electrophilic.

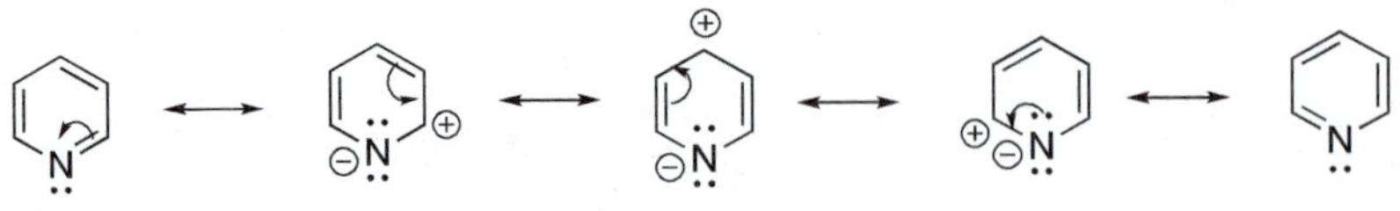

(17-87)

Thus, placing a leaving group in either of the positions that are electrophilic would result in a facile nucleophilic displacement reaction, as illustrated in Reactions (17-88) and (17-89).

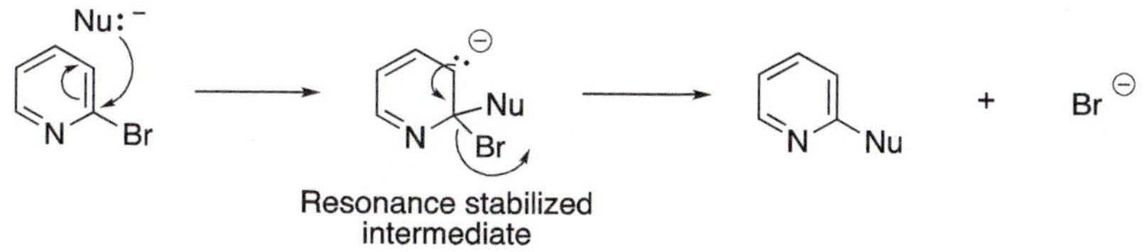

17.12 Nucleophilic Aromatic Substitution

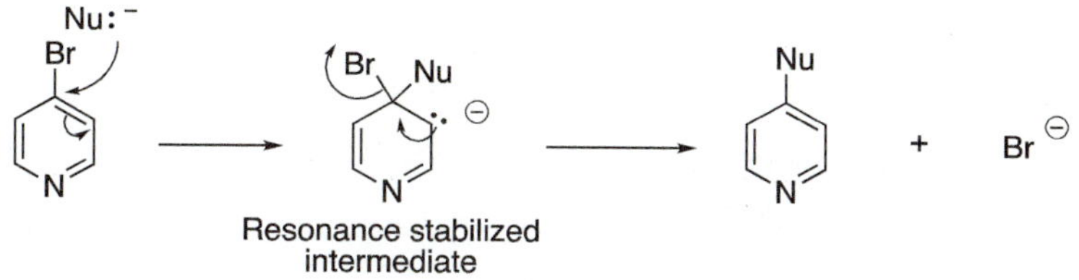

(17-88) [Reaction showing 2-bromopyridine + Nu:⁻ → resonance stabilized intermediate → 2-substituted pyridine + Br⁻]

Resonance stabilized intermediate

(17-89) [Reaction showing 4-bromopyridine + Nu:⁻ → resonance stabilized intermediate → 4-substituted pyridine + Br⁻]

Resonance stabilized intermediate

A specific reaction involving 2-bromopyridine and the NH_2^- nucleophile is shown in Reaction (17-90).

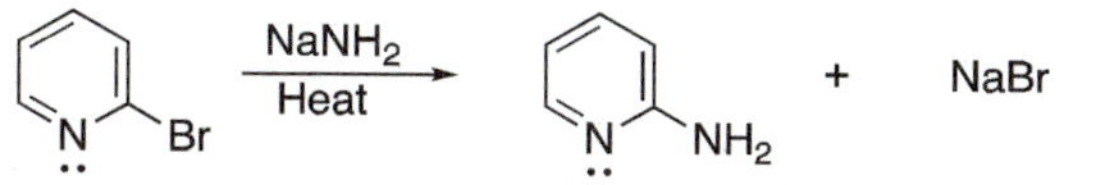

$$\text{2-bromopyridine} \xrightarrow{\text{NaNH}_2, \text{Heat}} \text{2-aminopyridine} + \text{NaBr} \quad (17\text{-}90)$$

Even though the hydride ion is possibly the strongest base, it can be made to be a leaving group if it leaves as hydrogen gas. Of course, the driving force is the formation of the gas, which is entropy favored. In the example given in Reaction (17-91), the nucleophile is the strongly basic nucleophilic NH_2^-, but extreme conditions, such as high pressure and temperature, are required.

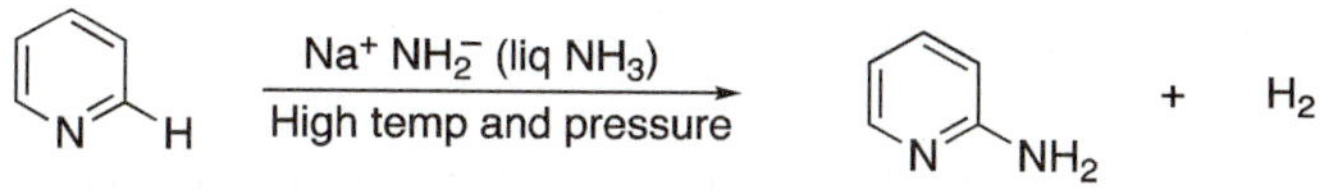

$$\text{pyridine} \xrightarrow[\text{High temp and pressure}]{Na^+ NH_2^- \text{ (liq NH}_3\text{)}} \text{2-aminopyridine} + H_2 \quad (17\text{-}91)$$

As you can imagine, these nucleophilic substitution reactions work best if there is a good leaving group already bonded to the pyridine ring. Another example in which an organometallic reagent is used is given in Reaction (17-92).

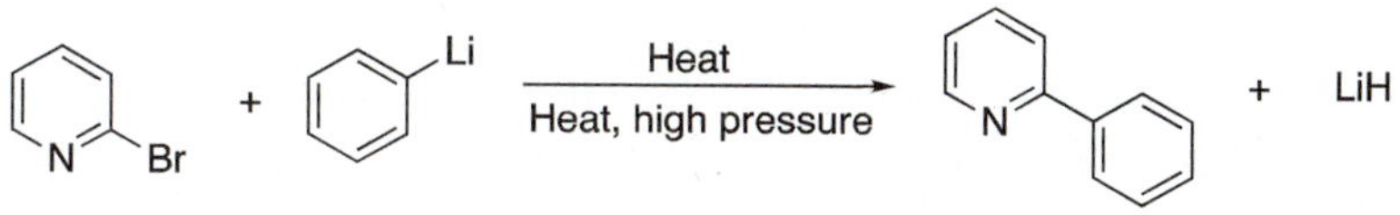

$$\text{2-bromopyridine} + \text{PhLi} \xrightarrow[\text{Heat, high pressure}]{\text{Heat}} \text{2-phenylpyridine} + LiH \quad (17\text{-}92)$$

Problem 17.16

Give the product for each of the following reactions.

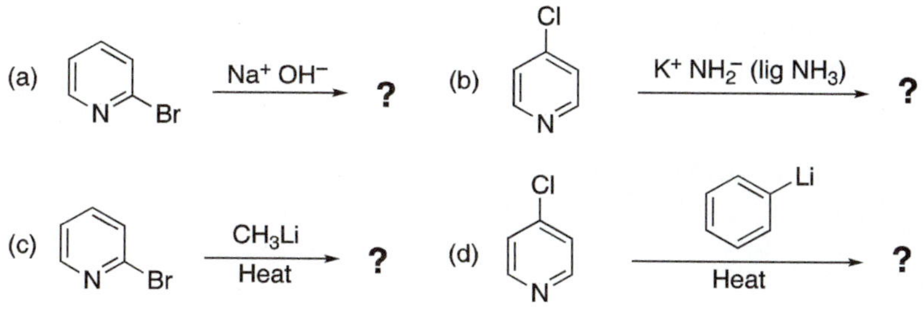

(a) 2-bromopyridine $\xrightarrow{Na^+ OH^-}$?

(b) 4-chloropyridine $\xrightarrow{K^+ NH_2^- \text{ (liq NH}_3\text{)}}$?

(c) 2-bromopyridine $\xrightarrow[\text{Heat}]{CH_3Li}$?

(d) 4-chloropyridine + PhLi $\xrightarrow{\text{Heat}}$?

End of Chapter Problems

17.17 Which of the following molecules is(are) aromatic? Explain your answer.

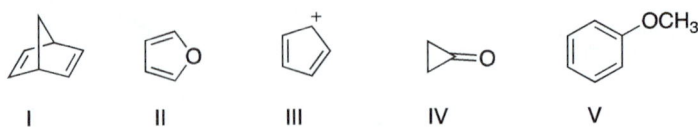

I   II   III   IV   V

17.18 Give IUPAC names for the following compounds.

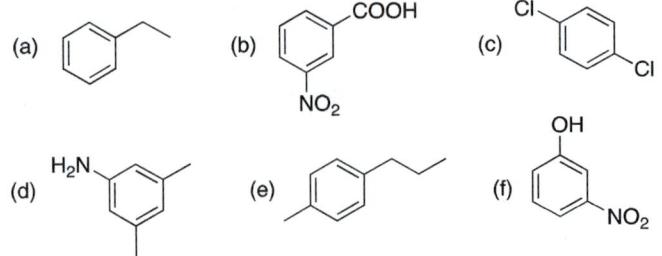

17.19 Give the structure of each of the compounds shown below.

a) 1-Bromo-2-chlorobenzene
b) 1-Iodo-2,3-dimethylbenzene
c) *m*-Chlorophenol
d) 2,3-Dichlorophenol
e) 2,4,6-Trinitrotoluene
f) *p*-Chlorobenzoic acid
g) *o*-Nitrophenol
h) 4-Chloroaniline

17.20 Give the <u>structure</u> of the expected principal organic product when benzene reacts with the specified set of reagents.

a) HNO_3/H_2SO_4
b) $CH_3COCl/AlCl_3$
c) $Cl_2/FeCl_3$
d) SO_3/H_2SO_4

17.21 Complete the following reactions by giving the structures of missing reagents above the arrow or major organic product. Note: some transformations may require more than one step and these steps must be indicated clearly.

(i) 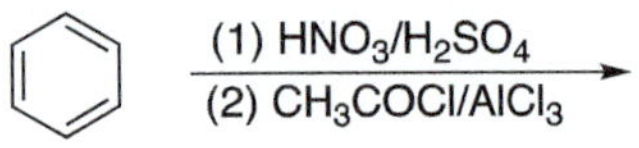 benzene → (1) HNO_3/H_2SO_4 (2) $CH_3COCl/AlCl_3$

(ii) 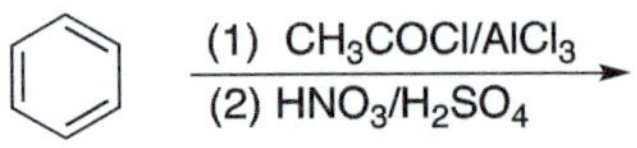 benzene → (1) $CH_3COCl/AlCl_3$ (2) HNO_3/H_2SO_4

(iii) 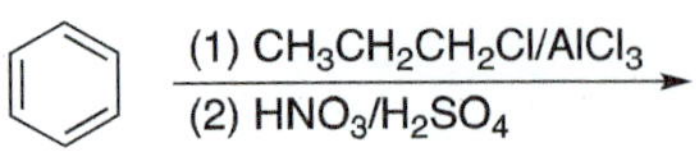 benzene → (1) $CH_3CH_2CH_2Cl/AlCl_3$ (2) HNO_3/H_2SO_4

(iv) 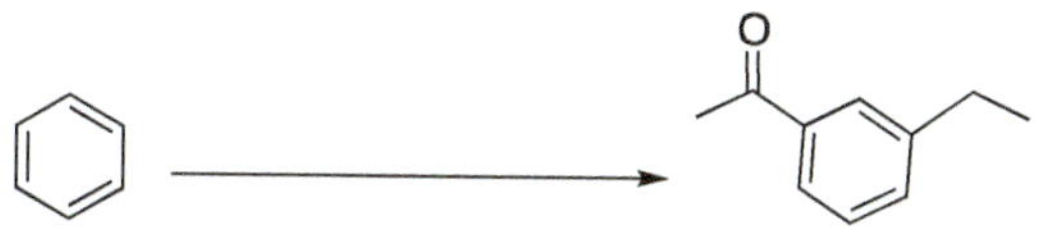 benzene ⟶ 1-(3-ethylphenyl)ethan-1-one

(v) 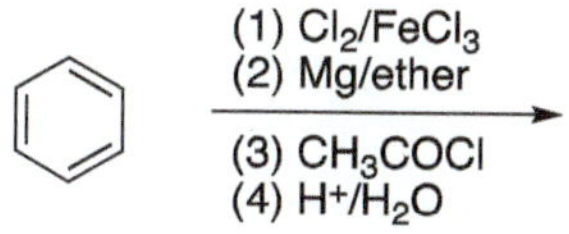 benzene
(1) Cl₂/FeCl₃
(2) Mg/ether
(3) CH₃COCl
(4) H⁺/H₂O

(vi)  benzene ⟶ 1-ethyl-3-nitrobenzene

(vii) 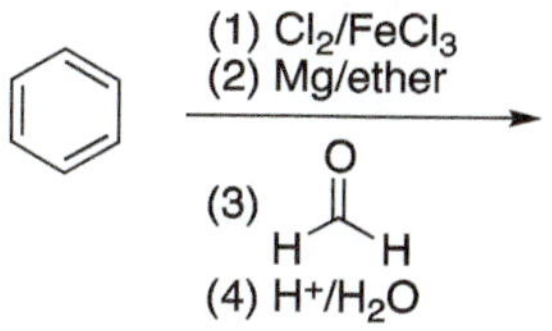 benzene
(1) Cl₂/FeCl₃
(2) Mg/ether
(3) H–C(=O)–H
(4) H⁺/H₂O

17.22 Using arrow-pushing formulism to indicate electron movement, propose a reasonable mechanism to account for the following observation.

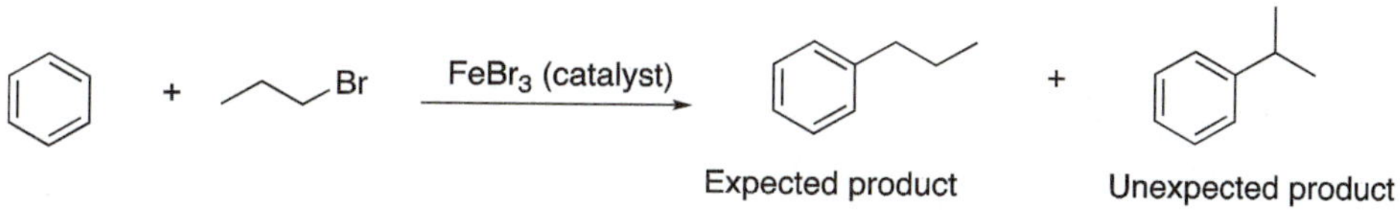

benzene + CH₃CH₂CH₂Br —FeBr₃ (catalyst)→ propylbenzene (Expected product) + isopropylbenzene (Unexpected product)

17.23 Provide the necessary reagents and reaction conditions needed to carry out each lettered transformation (note: it may be necessary to have more than one step to accomplish a particular transformation). For example: (i) C_6H_5MgCl/ether, (ii) H_3O^+.

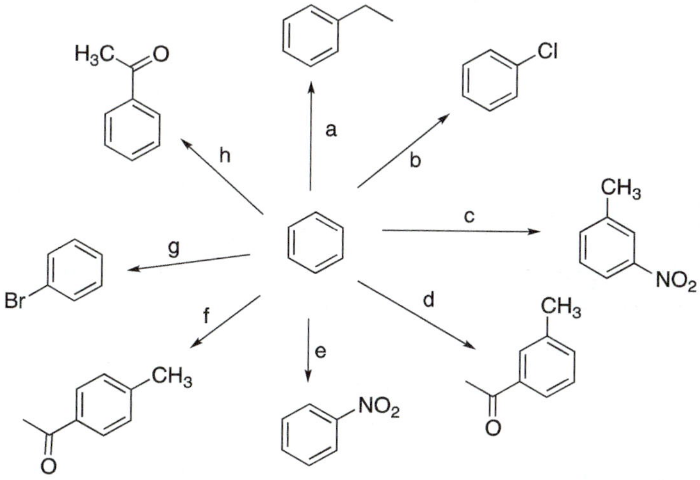

17 Aromaticity and Aromatic Substitution Reactions

17.24 Using arrow-pushing formulism to indicate electron movement, provide a step-by-step mechanism for the following reaction.

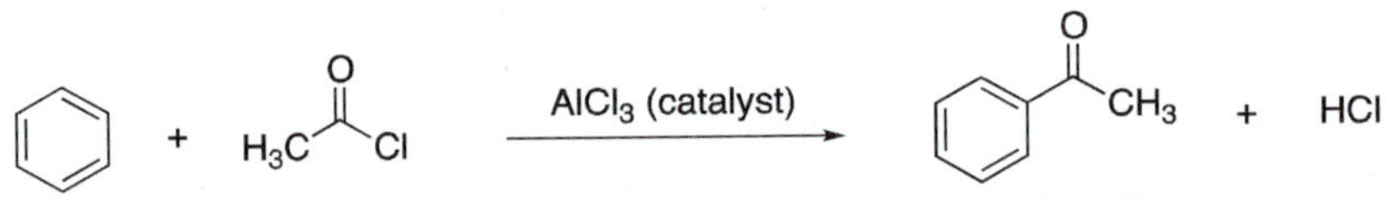

benzene + H₃C−C(=O)−Cl →[AlCl₃ (catalyst)] acetophenone (C₆H₅−C(=O)−CH₃) + HCl

17.25 Indicate clearly, by giving appropriate reaction conditions, how you would carry out the transformations shown below. Note that some transformations may require more than one step.

(a) 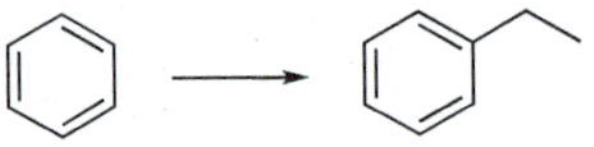 benzene → ethylbenzene

(b)  benzene → chlorobenzene

(c) 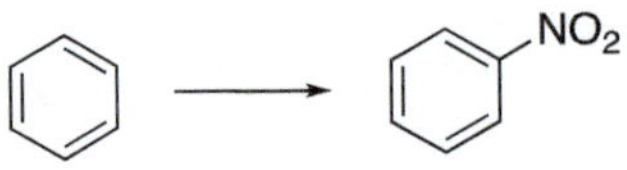 benzene → nitrobenzene

(d) 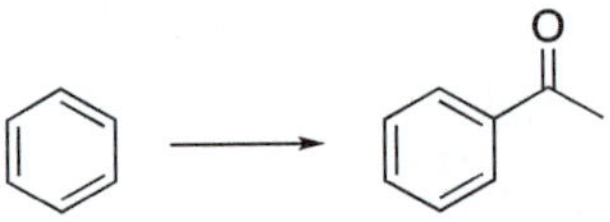 benzene → acetophenone

(e) 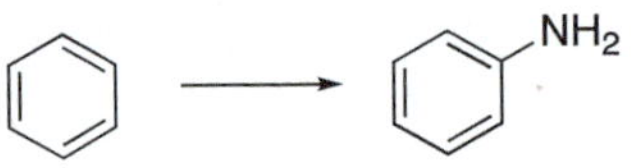 benzene → aniline

(f) 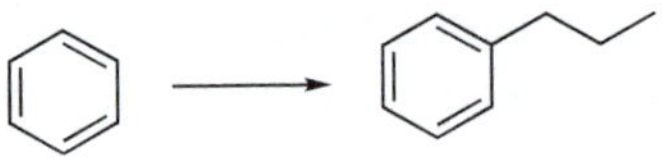 benzene → ethylbenzene

17.26 Provide the reagents and reaction conditions necessary to carry out the transformations shown below.

(a) 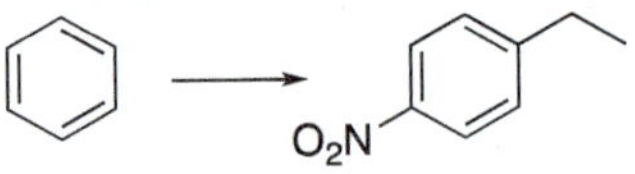 benzene → 4-nitroethylbenzene (O₂N at para to ethyl)

(b)  benzene → 4-chloro-isopropylbenzene

(c) 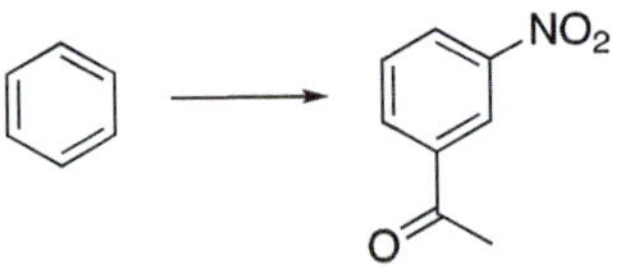 benzene → 3-nitroacetophenone

(d) 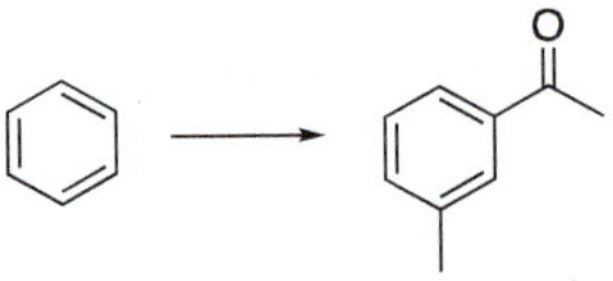 benzene → 3-methylacetophenone

(e) 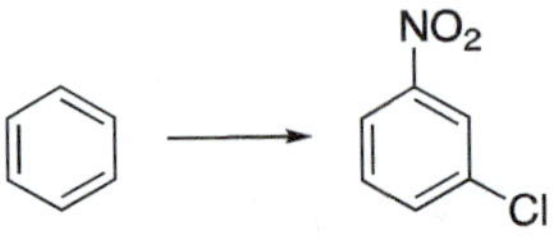 benzene → 3-nitrochlorobenzene

(f) 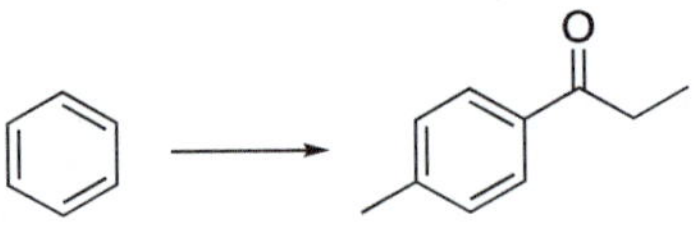 benzene → 4-isopropylpropiophenone

17.27 Give the structures of the expected principal organic products when naphthalene reacts with the specified set of reagents.

a) H₂SO₄/HNO₃
b) CH₃COCl/AlCl₃
c) Cl₂/FeCl₃
d) SO₃/H₂SO₄, 60 °C

17.28 Give the <u>structure</u> of the expected principal organic product when pyrrole reacts with the specified set of reagents.

a) H₂SO₄/HNO₃
b) CH₃COCl/AlCl₃
c) Cl₂/FeCl₃
d) SO₃/H₂SO₄

17.29 Picric acid is one of the strongest organic acids, with a pK_a = 0.3. Draw resonance structures for the molecule shown below and explain its acidity.

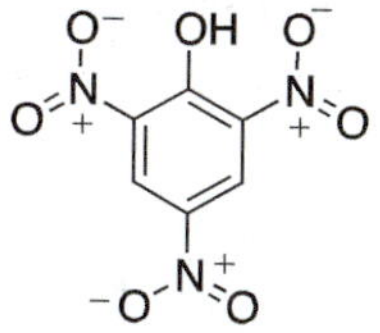

Picric acid

17.30 Using appropriate mechanisms, account for the following observations.

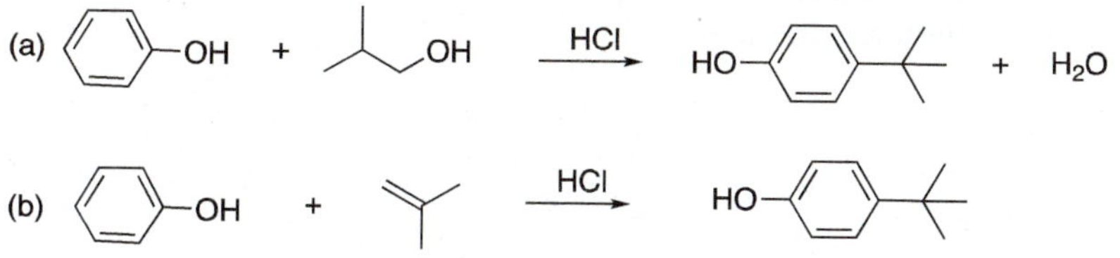

(a) Ph–OH + (CH₃)₂CHCH₂OH $\xrightarrow{HCl}$ HO–C₆H₄–C(CH₃)₃ + H₂O

(b) Ph–OH + (CH₃)₂C=CH₂ $\xrightarrow{HCl}$ HO–C₆H₄–C(CH₃)₃

17.31 Explain the following observation.

PhCF₃ $\xrightarrow{HNO_3, H_2SO_4}$ m-NO₂-C₆H₄-CF₃

PhCH₃ $\xrightarrow{HNO_3, H_2SO_4}$ p-NO₂-C₆H₄-CH₃

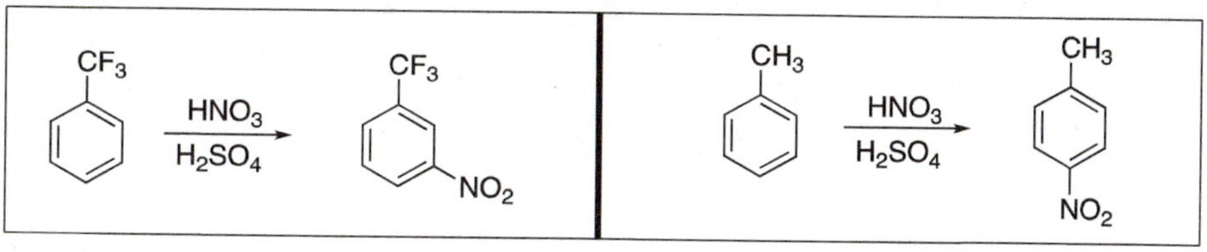

17.32 Give a mechanism to explain the reaction shown below.

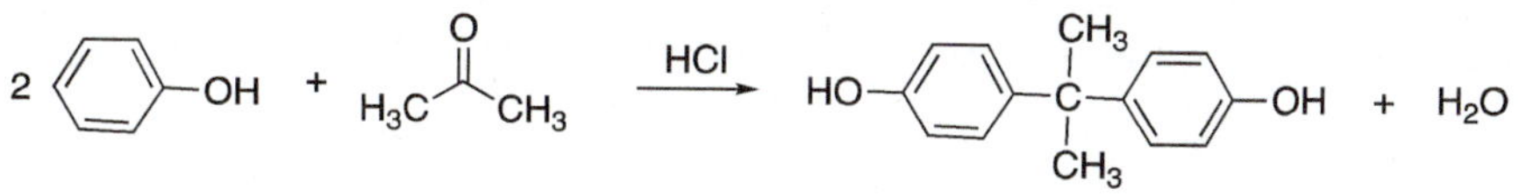

2 Ph–OH + (CH₃)₂C=O $\xrightarrow{HCl}$ HO–C₆H₄–C(CH₃)₂–C₆H₄–OH + H₂O

Bisphenol A

17.33 Give the missing reagents to complete the synthesis of benzocaine hydrochloride, a reagent that is routinely used as an anesthetic.

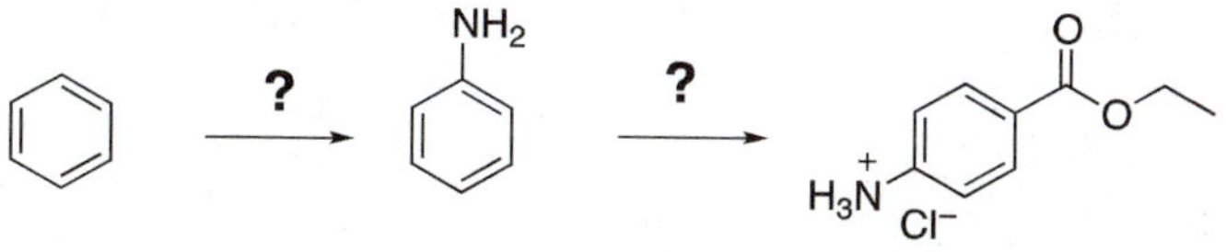

benzene $\xrightarrow{?}$ aniline $\xrightarrow{?}$ Benzocaine hydrochloride

17.34 Provide a mechanism to explain the following reaction, this mechanism is similar to the mechanism for the Friedel–Crafts acylation. (hint: in the first step, consider using a lone pair of electrons on the carbonyl oxygen to bond to the vacant orbital of the Lewis acid catalyst, AlCl₃).

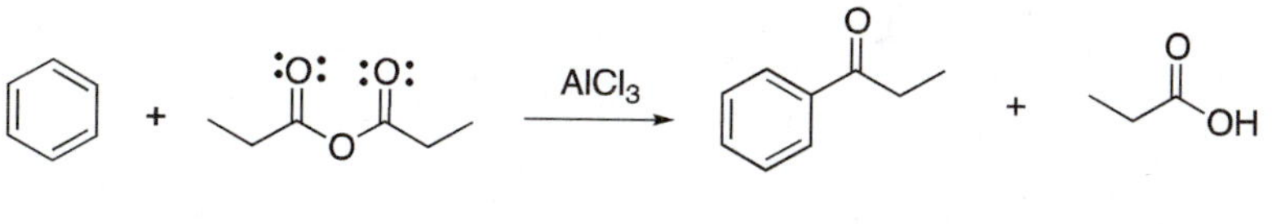

benzene + (CH₃CH₂CO)₂O $\xrightarrow{AlCl_3}$ PhCOCH₂CH₃ + CH₃CH₂COOH

17.35 A possible synthesis for ibuprofen is given in Reactions (17-54) to (17-56). An organic chemistry student proposed that a slight modification to ibuprofen should still make it to be effective and the structure is shown below. Using benzene as your starting compound, show how to synthesize this modified ibuprofen.

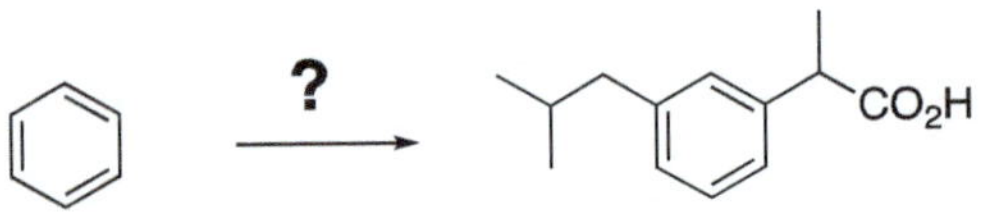

Modified ibuprofen

17.36 It was mentioned that the SO₃H group can be used as a protecting group for a specific position on the benzene ring. Provide reagents and appropriate reaction conditions to carry out each of the lettered transformation.

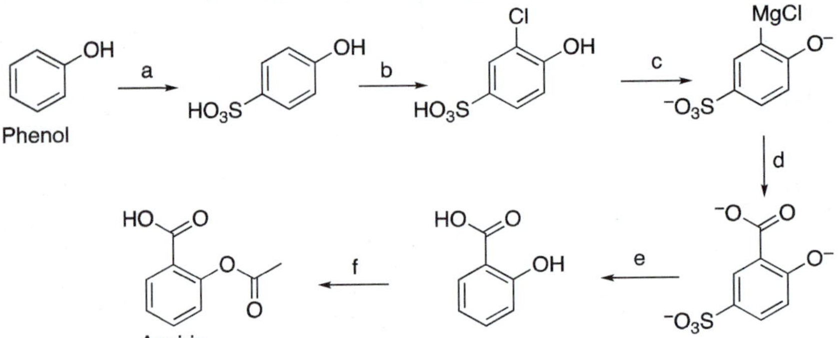

Phenol → a → → b → → c → → d → → e → → f → Aspirin

17.37 Show how to carry out the following transformations. For each step in your synthesis, clearly show all reagents and reaction conditions. (hint: identify the bond of the functional group in the product that must be made, then work backward to identify an appropriate molecule and reaction type to make that bond).

(a)

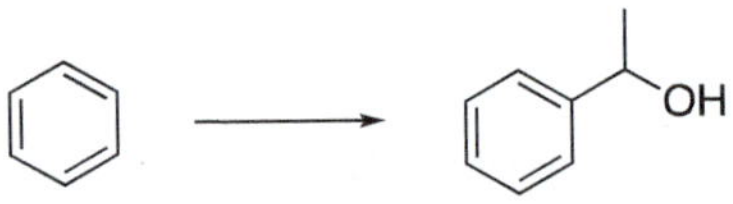

(b)

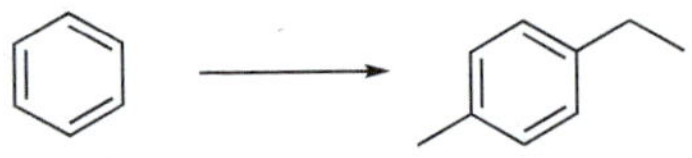

(c)

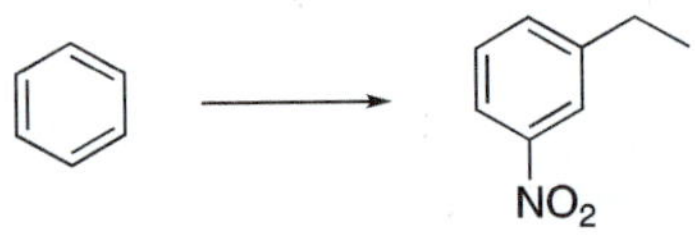

(d)

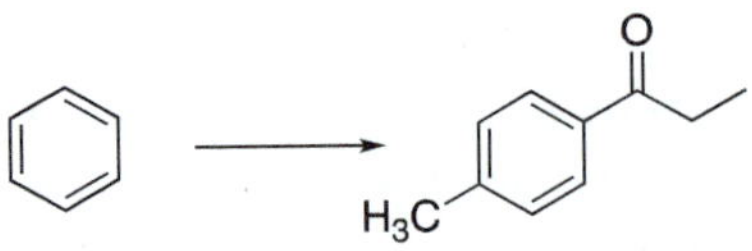

17.38 At temperatures in excess of 160 °C, 2,4,5-trichlorophenol, which is a constituent of the warfare defoliant "Agent Orange," dimerizes to give dioxin. Propose a reasonable mechanism for the reaction shown below.

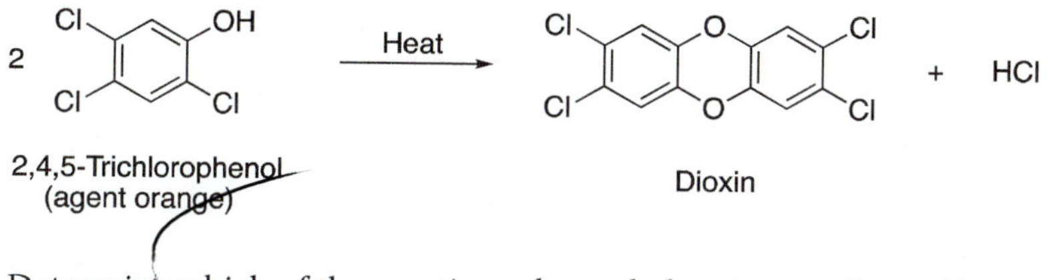

2 (2,4,5-Trichlorophenol (agent orange)) —Heat→ Dioxin + HCl

17.39 Determine which of the reactions shown below is most favorable.

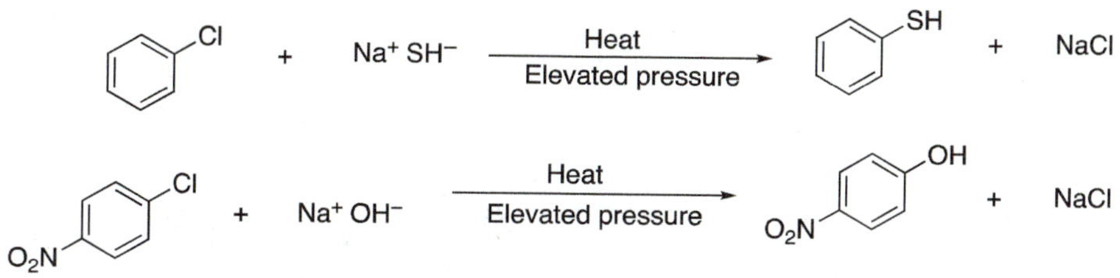

PhCl + Na⁺ SH⁻ —Heat, Elevated pressure→ PhSH + NaCl

4-O₂N-C₆H₄-Cl + Na⁺ OH⁻ —Heat, Elevated pressure→ 4-O₂N-C₆H₄-OH + NaCl

17.40 The ¹H NMR spectrum for cyclopentadiene salt consists of just one signal, whereas the NMR spectrum of cyclopentadiene is more complex. Explain this observation.

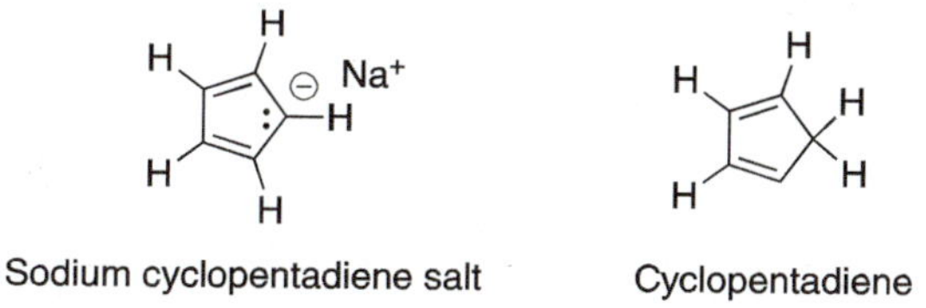

Sodium cyclopentadiene salt   Cyclopentadiene

17.41 Show how to carry out the following transformations. For each step in your synthesis, clearly show all reagents and reaction conditions. (hint: identify the bond of the functional group in the product that must be made, then work backward to identify an appropriate molecule and reaction conditions to make that bond).

(a) 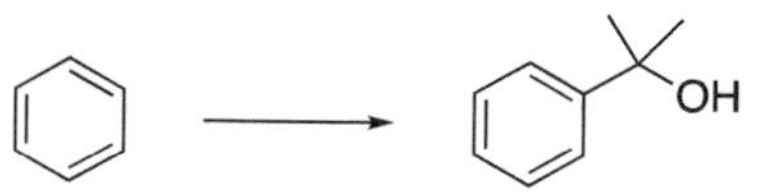 benzene → 2-phenyl-2-propanol

(b) 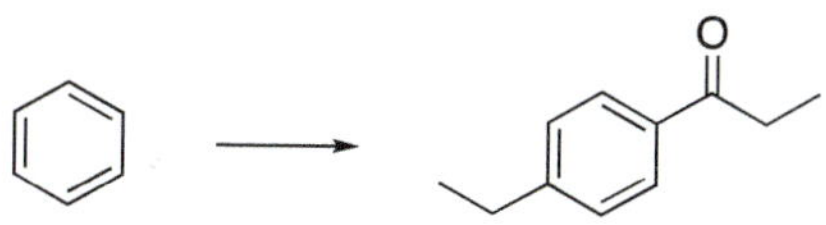 benzene → 1-(4-ethylphenyl)propan-1-one

18

Conjugated Systems and Pericyclic Reactions

18.1 Conjugated Systems

Conjugated systems are found routinely in nature, most are typically highly colored compounds. 2-Methyl-1,3-butadiene, also known as isoprene, is one of the simplest conjugated systems and its structure is shown below. Isoprene is a colorless liquid that is produced and emitted by many types of trees, including the oak and eucalyptus trees. Shown also below are two other examples of simple conjugate compounds.

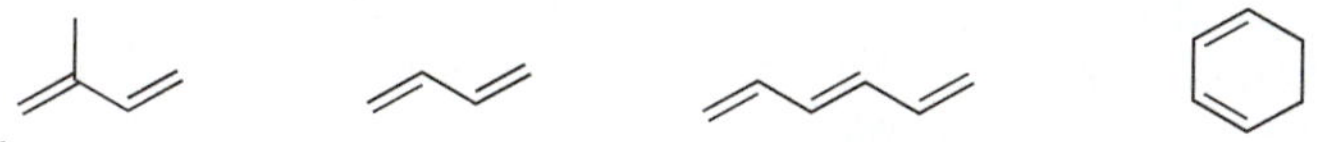

Isoprene   1,3-Butadiene   1,3,5-hexatriene   1,3-Cyclohexadiene

As we have seen on the section on UV–Vis spectroscopy in Chapter 13, a large percentage of organic compounds, especially natural products, contain extended conjugated double bonds. β-Carotene and lycopene, for example, are highly colored compounds due to the presence of extended conjugated double bonds. In this chapter, we will examine the reactions that conjugated systems undergo and apply that knowledge to the synthesis of various other more complex molecules.

18.1.1 Stability of Conjugated Alkenes

In Chapter 4, we briefly looked at the stability of alkenes based on the heat liberated from hydrogenation reactions. We examined the relationship between the amount of heat liberated to the relative stability of different alkenes. Let us revisit the hydrogenation of different isomers of pentene. Consider the hydrogenation of 1-pentene and 2-pentene, which are shown in Reactions (18-1) and (18-2).

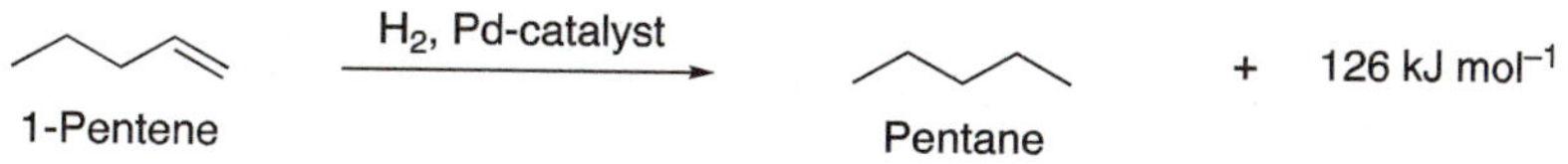

1-Pentene $\xrightarrow{\text{H}_2,\ \text{Pd-catalyst}}$ Pentane + 126 kJ mol^{-1} (18-1)

2-Pentene $\xrightarrow{\text{H}_2,\ \text{Pd-catalyst}}$ Pentane + 114 kJ mol^{-1}  (18-2)

The heat liberated from the hydrogenation of 1-pentene, which has a terminal double bond is 126 kJ mol^{-1} (Reaction (18-1), is more than the heat liberated from the hydrogenation of 2-pentene, which has an internal double bond. An extrapolation of this information to

Organic Chemistry: Concepts and Applications, First Edition. Allan D. Headley.
© 2020 John Wiley & Sons, Inc. Published 2020 by John Wiley & Sons, Inc.
Companion website: www.wiley.com/go/Headley_OrganicChemistry

conjugated systems would imply that the heat liberated from 1,3-pentadiene should be 126 + 114 kJ mol^{-1} (240 kJ mol^{-1}), which would be the sum of the two reactions shown in Reactions (18-1) and (18-2). Reaction (18-3) shows that the heat liberated is less than predicted, actually 225 kJ mol^{-1}.

$$\text{1,3-Pentadiene} \xrightarrow{\text{H}_2,\text{ Pd-catalyst}} \text{Pentane} + 225 \text{ kJ mol}^{-1} \quad (18\text{-}3)$$

The implication is that the conjugated diene is more stable than a molecule that has two isolated double bonds. Another analysis that demonstrates that observation is that the heat liberated from the hydrogenation of 1,4-pentadiene, which has two isolated terminal double bonds, is 252 kJ mol^{-1}, which is the sum of two moles of 1-pentene shown in Reaction (18-1) (126 + 126 kJ mol^{-1}).

$$\text{1,4-Pentadiene} \xrightarrow{\text{H}_2,\text{ Pd-catalyst}} \text{Pentane} + 252 \text{ kJ mol}^{-1} \quad (18\text{-}4)$$

The implication from these observations is that molecules that have conjugated double bonds are more stable than molecules that have isolated double bonds. In the case of pentadienes, 1,3 pentadiene is more stable than 1,4-pentadiene, which has isolated double bonds. Figure 18.1 gives the energy diagram for this concept.

The stability of the conjugated alkene comes from the fact that the electrons of the double bond can freely delocalize across the pi (π) system. We can use resonance to illustrate this concept as shown in resonance structures given in (18-5) and (18-6).

(18-5)

(18-6)

An alternate representation of the electronic distribution is shown below.

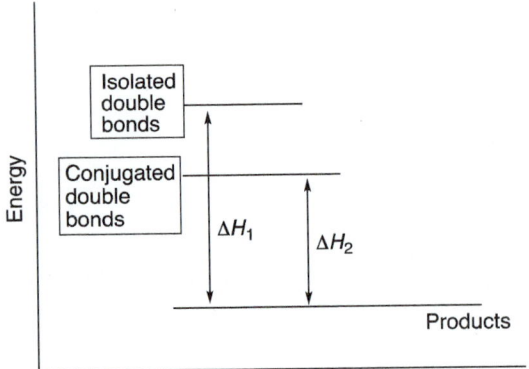

Figure 18.1 Energy diagram of the relative stability of molecules with conjugated double bonds compared to molecules with isolated double bonds.

As we have seen from various addition reactions studied thus far, unexpected products are sometimes produced owing to the possibility of resonance, an example is shown in Reaction (18-7).

$$\text{1,3-Pentadiene} + \text{HBr} \longrightarrow \text{4-Bromo-2-pentene} + \text{1-Bromo-2-pentene} \tag{18-7}$$

The same type of electron delocalization is true for molecules that have atoms with unshared electrons adjacent to a double bond, as shown in the resonance structures given in (18-8).

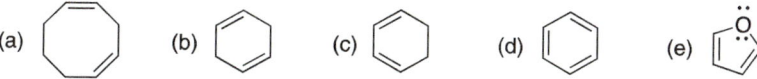
(18-8)

Problem 18.1

i) Determine which of the following molecules or ions have conjugated electrons?

(a) (b) (c) (d) (e)

ii) For structures that are conjugated, use arrows to show electron movement that results in another resonance structure.

18.2 Pericyclic Reactions

The mechanisms for most of the reactions that we have looked at so far involve intermediates and a rate-determining step leading to the formation of one of the intermediates. Some reactions are concerted reactions, that is, no intermediates are formed as the reactants proceed to products by going through only a transition state. We have seen these type reactions for the E2 and $S_N 2$ mechanisms. Pericyclic reactions are another type of concerted reactions, where reactants proceed directly to products by going through a transition state. Pericyclic reactions have the following characteristics: (i) concerted; (ii) cyclic transition state; (iii) energy is usually supplied by heat or light; (iv) solvents have little or no effect on the reaction rate; and (v) they are stereospecific. In this section, we will examine three types of pericyclic reactions: (i) cycloaddition, (ii) electrocyclic and (iii) sigmatropic reactions.

18.2.1 Cycloaddition Reactions

For these reactions, the π electrons from two different molecules react to give a single cyclic product. These reactions involve π electrons, and the number of π electrons involved is important for these reactions. The first pericyclic cycloaddition reaction that we will examine is the reaction of two ethylene molecules to form a cyclobutane.

18.2.1.1 Cycloaddition Reactions [2+2]

The simplest cycloaddition reaction is the reaction of two ethylene functionalities to form cyclobutane. The reactions of ethylene with cis and trans 2-butene are shown in Reactions (18-9) and (18-10).

$$\text{Ethylene} + \text{cis-2-Butene} \xrightarrow{\text{energy}} \text{cis-1,2-Cyclobutane} \quad (18\text{-}9)$$

$$\text{Ethylene} + \text{trans-2-Butene} \xrightarrow{\text{energy}} \text{trans-1,2-Cyclobutane} \quad (18\text{-}10)$$

Note from the reaction, that the methyl groups of *cis*-2-butene remain on the same side to give the product. Hence, the stereochemical outcomes of these reactions are predictable based on the stereochemistry of the reactants. These types of reactions are called stereospecific reactions.

The source of energy needed for these [2+2] reactions is light and not heat, and we will now use the molecular orbital theory to explain this requirement and also to explain the stereochemical outcome of these reactions. The energy that is needed for this reaction to proceed is light at the proper wavelength of the electromagnetic spectrum. The *frontier molecular orbital* method is an approach that is widely used to gain a better understanding of how pericyclic reactions proceed from reactants to give the product. For the analysis of these reactions, only the π orbitals and π electrons of the reactants are considered.

The carbons of ethylene are sp² hybridized; thus, there are two p orbitals associated with each carbon atom and each contains a single electron. In order to form the π bond, these two p atomic orbitals must form molecular orbitals. Since there are two p atomic orbitals, they form two molecular orbitals; one molecular is lower in energy (π_1) and the other orbital (π_2^*) is higher in energy as illustrated in Figure 18.2.

Since the molecular orbital π_1 is the lowest in energy and each orbital can accept two electrons, this molecular orbital contains both p electrons and is called the HOMO, which means *highest occupied molecular orbital*. The other molecular orbital, which is higher in energy, does not contain any electrons and, as a result, is called the LUMO, which means *lowest unoccupied molecular orbital*. For the reaction to proceed, electrons must flow from the HOMO of one reactant to the LUMO of the other reactant. In addition, the orbitals must have the correct phase for interaction of the orbitals to occur and eventually for the reaction to proceed from reactants to product. If the orbitals do not have the correct phase, then the reaction is *symmetry forbidden* (as shown in Figure 18.3), and if they have the correct phase, the reaction is *symmetry allowed* as shown in Figure 18.4.

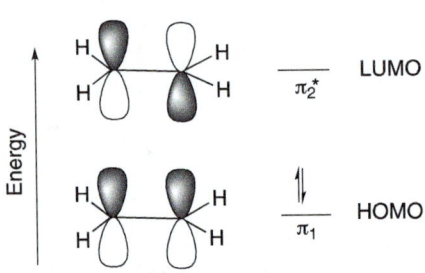

Figure 18.2 Molecular orbitals of ethylene molecule with its p electrons.

Figure 18.3 Illustration of incorrect phase of the orbitals of two ethylene molecules.

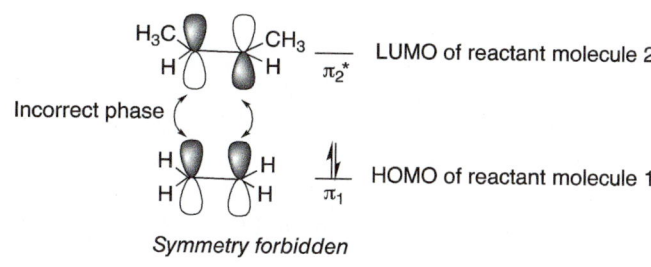

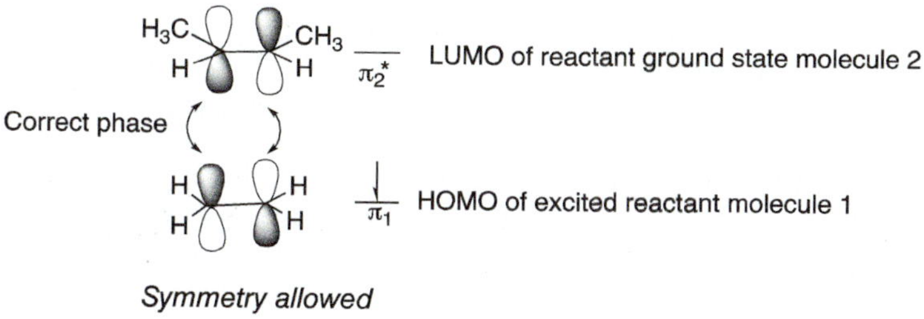

Figure 18.4 Illustration of correct phase of the orbitals of two ethylene molecules for the reaction to occur.

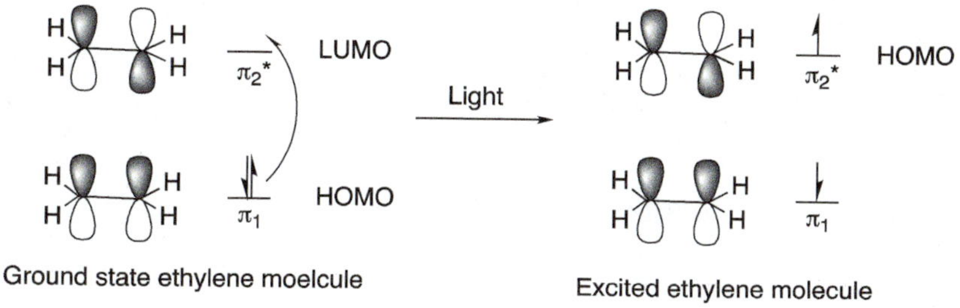

Figure 18.5 Promotion of electron from the HOMO to LUMO for light.

Under conditions without additional energy, such as light, it is obvious that the phase of HOMO of one reactant does not match the phase of LUMO of the other reactant.

In order to meet the symmetry allowed requirement as shown in Figure 18.4, an electron must be promoted from π_1 to π_2^* and this electron promotion can be accomplished by light as shown in Figure 18.5. When this happens, the molecule is described as being in the excited state.

Note that this electron promotion can occur in either the ethylene or 2-butene to form a new HOMO in the excited molecule. As a result, the phase will be correct for an interaction between both orbitals, i.e. the reaction is symmetry allowed and electrons can flow from the HOMO of one reactant to the LUMO of the other reactant to form the new bonds. This reaction will not occur in heat since a specific energy is required to promote the electron from π_1 to π_2^*, which comes from light.

Reaction 18-11 shows the proposed mechanism for the [2 + 2] cycloaddition of cis 2-butene and ethylene.

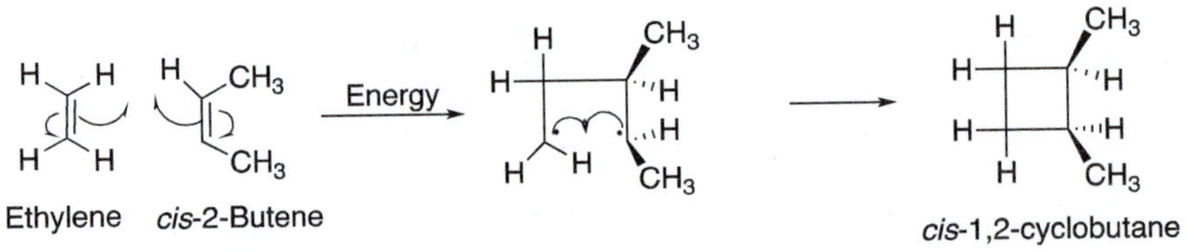

Ethylene + cis-2-Butene →(Energy) → cis-1,2-cyclobutane (18-11)

The mechanism for the trans addition is shown in Reaction 18-12

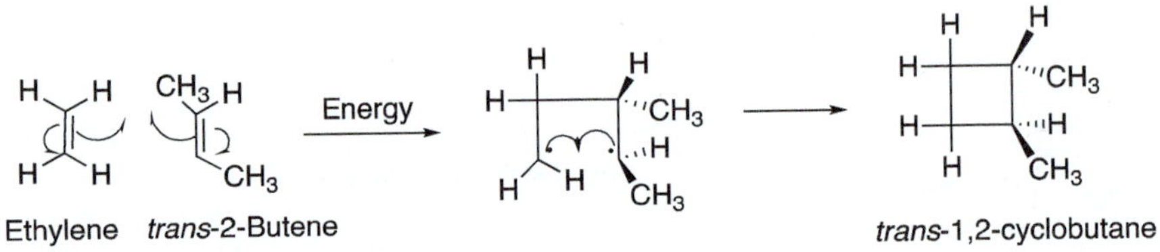

Ethylene + trans-2-Butene →(Energy) → trans-1,2-cyclobutane (18-12)

Based on this information, the reaction conditions can be given more specifically to indicate light and not just energy as given earlier. These reactions, with specific reaction conditions, are given in Reaction (18-13) for the reaction of ethylene with cis butene and Reaction (18-14) for the reaction of ethylene with trans-2-butene.

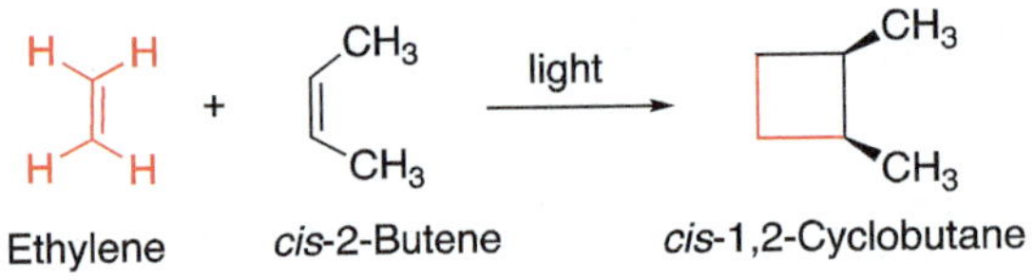

Ethylene + cis-2-Butene →(light) → cis-1,2-Cyclobutane (18-13)

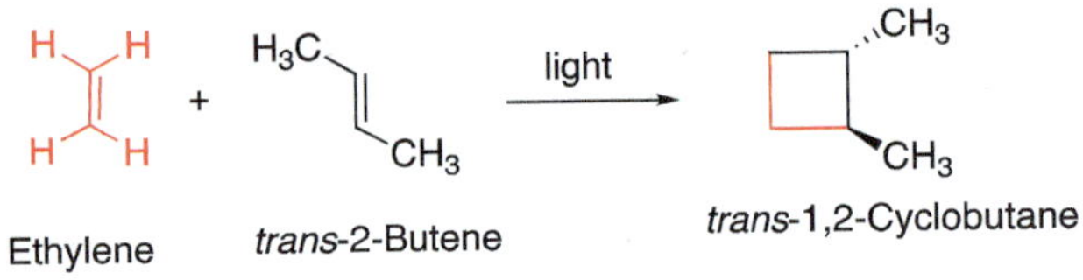

Ethylene + trans-2-Butene →(light) → trans-1,2-Cyclobutane (18-14)

Problem 18.2

Give the product of the following [2+2] cycloaddition reaction.

cyclopentenone + cyclopentene →(Light)

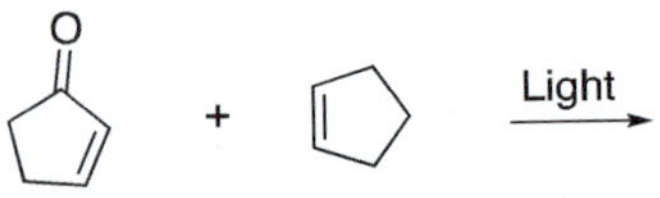

18.2.1.2 Cycloaddition Reactions [4+2]

Let us now consider the reaction involving 1,3-butadine and ethylene as shown in Reaction (18-15).

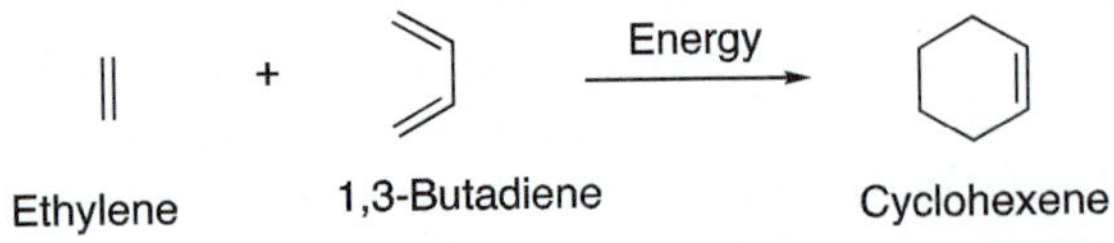

Ethylene + 1,3-Butadiene →(Energy) → Cyclohexene (18-15)

This type of reaction is also known as a [4+2] cycloaddition owing to the fact that there are two π electrons from one reactant and four π electrons from the other reactant. This type of reaction is often referred to as the *Diels Alder reaction*, named after the German chemists, Otto Diels and Kurt Alder, who discovered this reaction. They discovered the reaction in 1938 and received the Nobel Prize for their contribution to the advancement of chemistry in 1950. Their discovery opened up a new arena for synthesizing cyclic compounds that are of extreme value in the synthesis of important cyclic compounds. For this reaction only the π electrons participate as shown in Reaction (18-16) in which the arrow-pushing formulism is used to explain the mechanism.

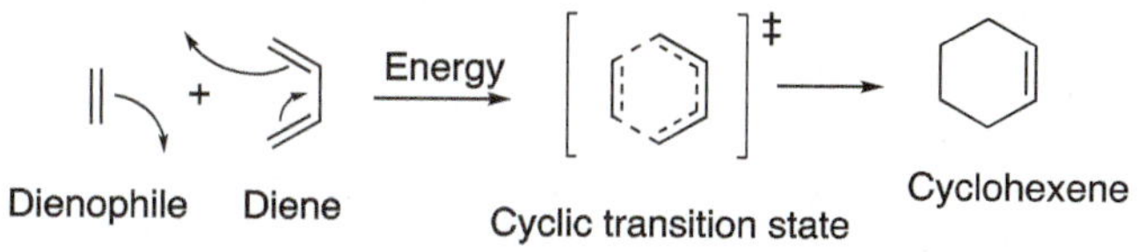

$$\underset{\text{Dienophile}}{\|} + \underset{\text{Diene}}{\diagup\!\!\!\diagdown} \xrightarrow{\text{Energy}} \left[\underset{\text{Cyclic transition state}}{\bigcirc} \right]^{\ddagger} \longrightarrow \underset{\text{Cyclohexene}}{\bigcirc} \qquad (18\text{-}16)$$

For this reaction, the diene is the molecule with two double bonds in conjugation with each other, and the other molecule is called the dienophile, which means diene-loving. We can use the *frontier molecular orbital* method to analyze this reaction to determine if light or heat is required for this reaction. Shown in Figure 18.6 are the molecular orbitals for the diene (1,3-butadiene) and dienophile (ethylene). Shown in Figure 18.7 is the correct phase requirement of the orbitals of the diene (1,3-butadiene) and the dienophile (ethylene) for a reaction to occur.

For these two molecules, the HOMO and the LUMO have the correct symmetry and the electrons will flow from the HOMO from one of the reactants into the LUMO of the other reactant molecule to form the product as shown in Reaction 18-16. As a result, excitation by light is not needed for these [4+2] reactions, but they are typically carried out at elevated temperatures.

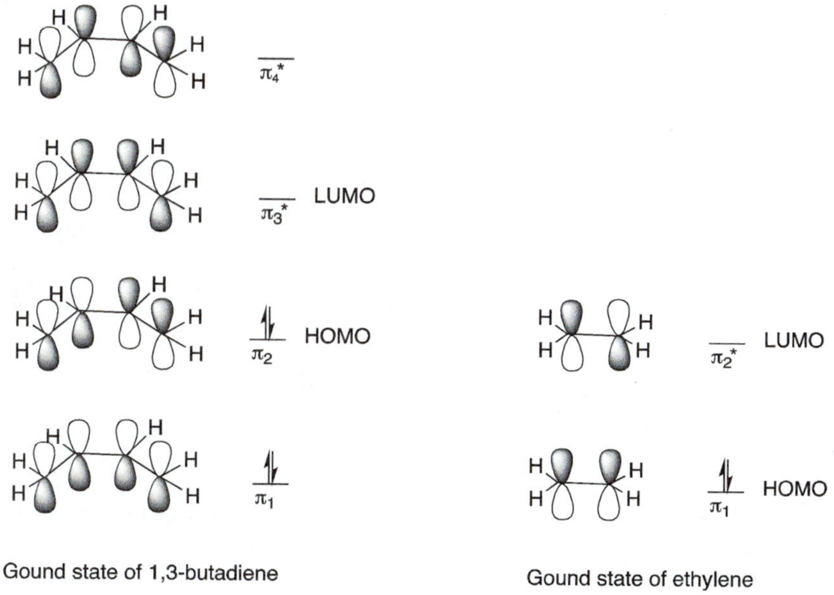

Gound state of 1,3-butadiene   Gound state of ethylene

Figure 18.6 Molecular orbitals of the ground state of 1,3-butadiene and ethylene.

Figure 18.7 Illustration of the correct phase involving the HOMO of 1,3-butadiene and the LUMO of ethylene.

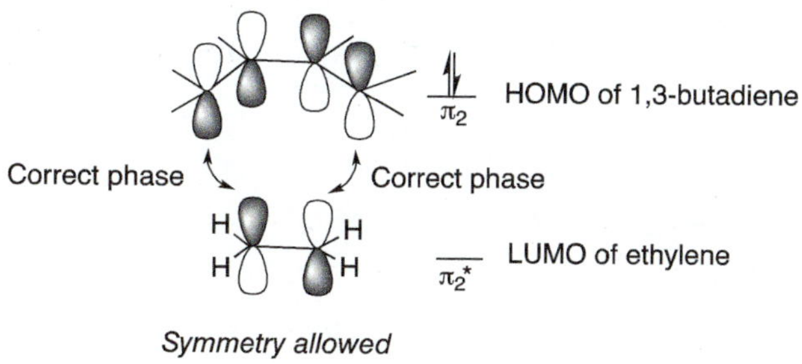

Symmetry allowed

For these reactions, stereochemistry is conserved, as shown in Reactions (18-17) and (18-18).

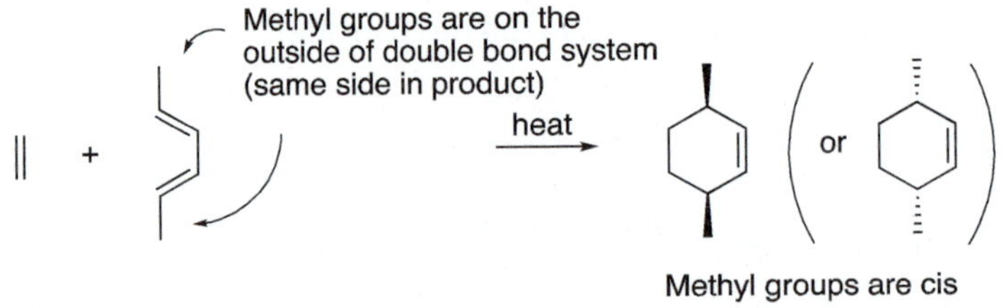
Methyl groups are on the outside of double bond system (same side in product)
heat
or
Methyl groups are cis

(18-17)

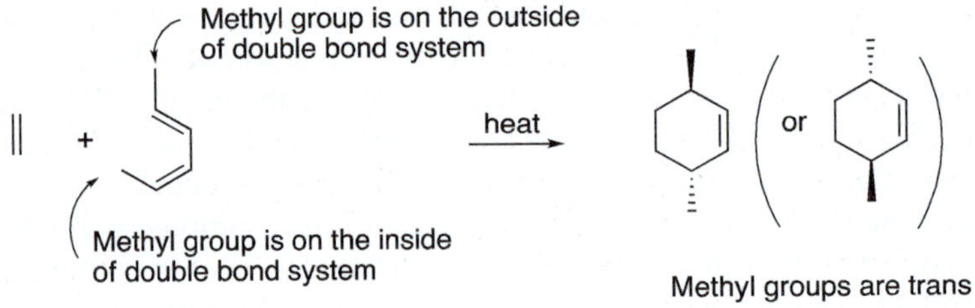
Methyl group is on the outside of double bond system
Methyl group is on the inside of double bond system
heat
or
Methyl groups are trans

(18-18)

For more complex molecules, the possibility exists for different products as shown in Reaction (18-19).

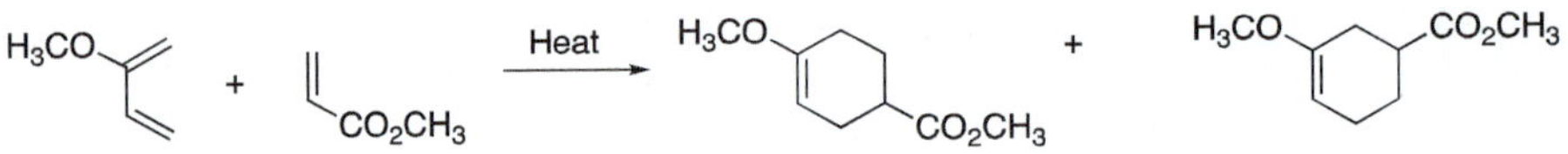
H3CO
CO2CH3
Heat
H3CO
CO2CH3
H3CO
CO2CH3

(18-19)

In order to determine the major product, we will have to examine the mechanism again, specifically the electronic distribution of the reactants leading to the transition states for the different products. Shown below are the resonance structures of the diene and dienophile.

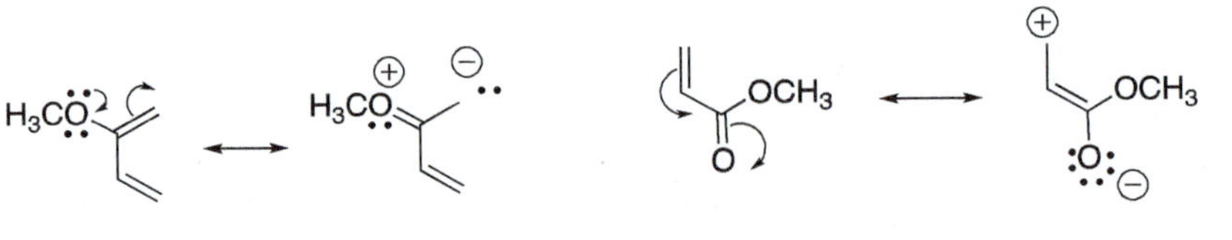
H3CO
H3CO
OCH3
OCH3

Resonance structures of the diene

Resonance structures of the dienophile

It is obvious that the cycloaddition that leads to the 1,4-addition product (Reaction 18-20) is more favorable, compared to the 1,3-addition product shown in Reaction (18-21).

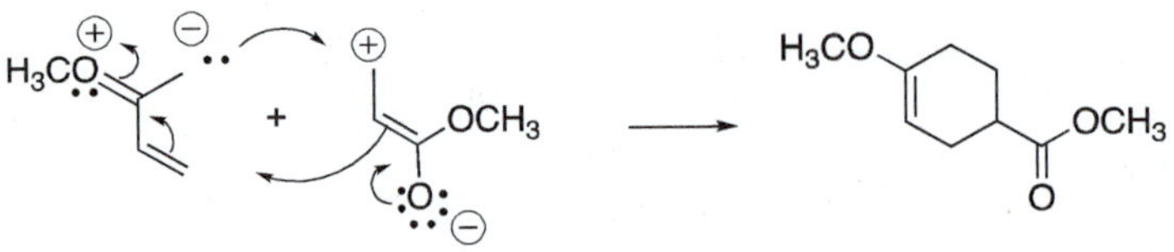
H3CO
OCH3
H3CO
OCH3

(18-20)

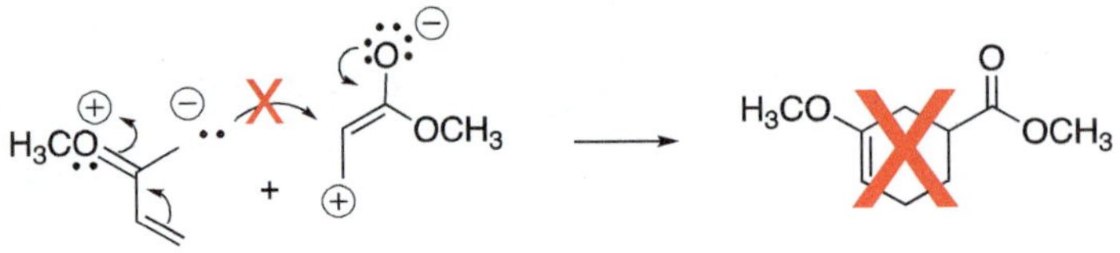
H3CO
OCH3
H3CO
OCH3

(18-21)

Reaction (18-20) is favorable since the electron-releasing methoxy group makes the adjacent carbon nucleophilic as shown in the resonance structure of the diene to the left. Regarding the dienophile, the resonance structure shows that the electrons of the double bond are delocalized into the carbonyl of the ester functionality, making the carbon furthest from the ester functionality electrophilic. This electron distribution occurs since the ester functionality is electron withdrawing. As a result, both reactant molecules in Reaction (18-20) are perfectly oriented for the nucleophilic portion of the diene to react with the electrophilic portion of the dienophile as illustrated in Reaction (18-20). On the other hand, a different orientation of the dienophile as given in Reaction (18-21) shows that the nucleophilic end of the diene and the electrophilic end of the dienophile are not oriented for an effective reaction to occur. As a result, the major product of the reaction shown in Reaction (18-19) is given below in Reaction (18-22). In general, Diels-Alder reactions are favored if electron-donating groups are bonded to the diene and electron-withdrawing groups bonded to the dienophile.

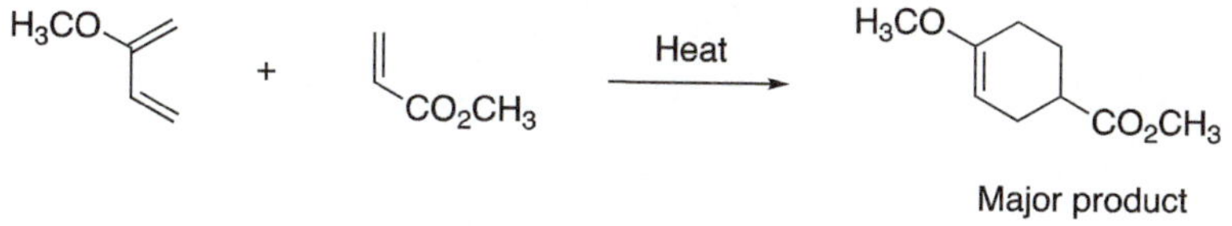

$$H_3CO\text{-diene} + CH_2=CH\text{-}CO_2CH_3 \xrightarrow{\text{Heat}} \text{cyclohexene product (Major product)} \quad (18\text{-}22)$$

Problem 18.3

Give the major organic products of the following [4+2] cycloaddition reaction.

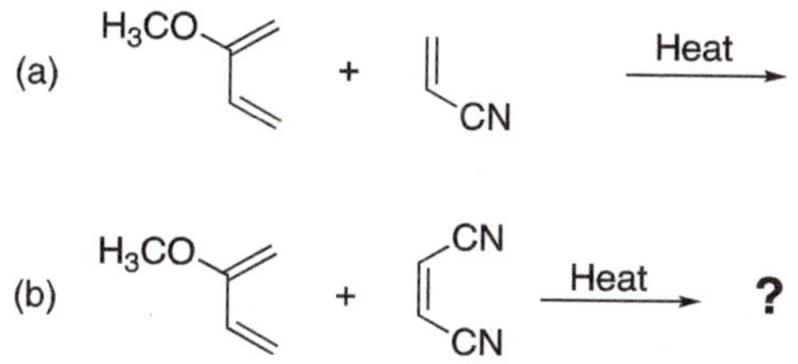

(a) H_3CO-diene $+$ $CH_2=CH\text{-}CN$ $\xrightarrow{\text{Heat}}$?

(b) H_3CO-diene $+$ $NC\text{-}CH=CH\text{-}CN$ $\xrightarrow{\text{Heat}}$?

18.2.2 Electrocyclic Reactions

Another category of pericyclic reactions involves reactions in which the π electronic systems of a conjugated system are rearranged to form cyclic compounds. Examples of the reactions of two isomers of 2,4-hexadiene are shown in Reactions (18-23) and (18-24) in which each reaction is initiated by a different form of energy, heat in one case and light for the other.

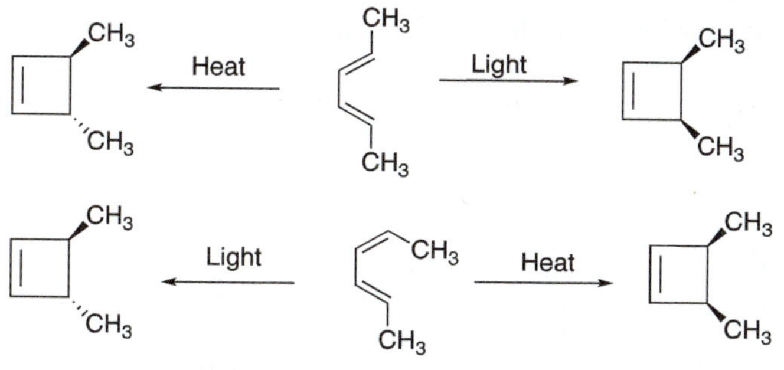

$$\text{cis,cis-cyclobutene (CH}_3\text{, CH}_3\text{)} \xleftarrow{\text{Heat}} \text{2,4-hexadiene} \xrightarrow{\text{Light}} \text{trans,trans-cyclobutene (CH}_3\text{, CH}_3\text{)} \quad (18\text{-}23)$$

$$\text{trans,trans-cyclobutene} \xleftarrow{\text{Light}} \text{2,4-hexadiene isomer} \xrightarrow{\text{Heat}} \text{cis,cis-cyclobutene} \quad (18\text{-}24)$$

Note the stereochemical outcome for each of these reactions is different depending on the type of energy used. To better visualize the electrocyclic ring closure reactions 18-23 and

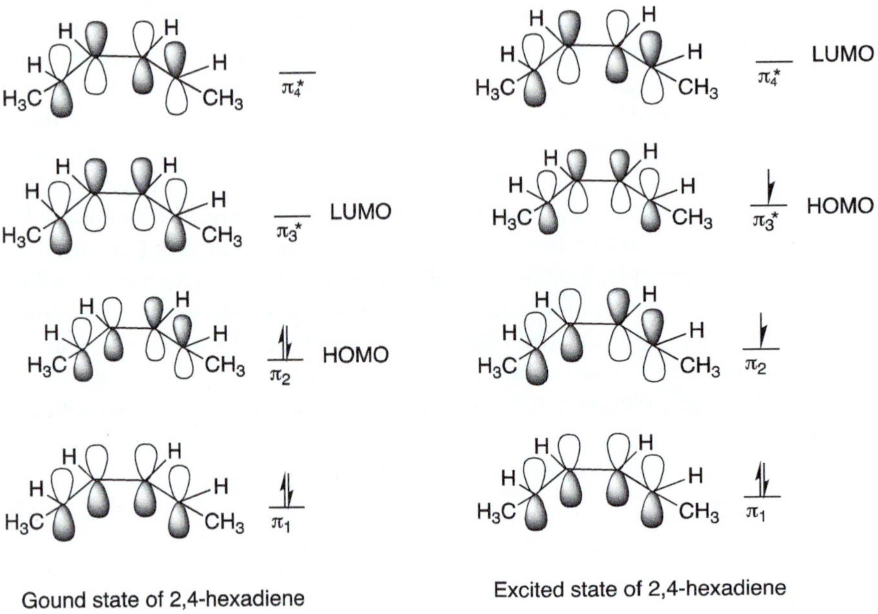

18 Conjugated Systems and Pericyclic Reactions

Ground state of 2,4-hexadiene

Excited state of 2,4-hexadiene

Figure 18.8 Ground and excited states of 2,4-hexadiene.

18-24, the process can be viewed as the formation of a new σ-bond, which is made at the ends of the polyene as shown in Reaction 18-25.

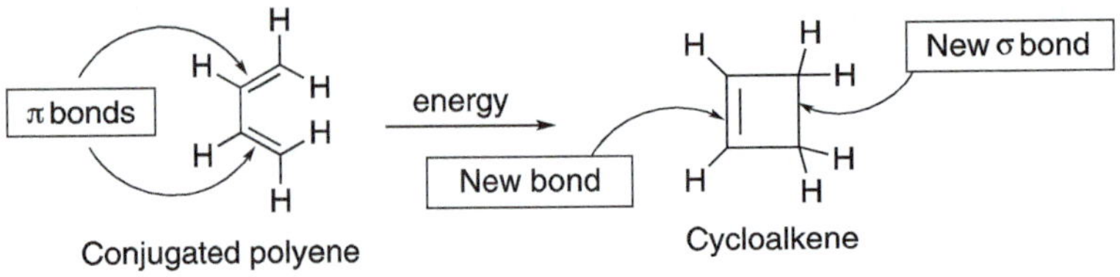

$$\text{Conjugated polyene} \xrightarrow{\text{energy}} \text{Cycloalkene} \qquad (18\text{-}25)$$

The frontier molecular orbital method can be used to explain the outcomes of these reactions. Figure 18.8 gives the molecular orbitals for the HOMO of the ground state and excited state of 2,4-hexadiene. For the ring closure reaction of the ground state molecule, which involves the HOMO, the phase for the orbitals are correct for ring closure with a conrotatory closure as illustrated in Figure 18.9.

Let us now look at the ring closure for the excited molecule, which is shown in Figure 18.10. Note that the stereochemistry of the products is dictated by the different required rotations for these reactions as shown in Figures 18.9 and 18.10.

As you can imagine, the number of π electrons will dictate the type of ring closure and Table 18.1 gives a summary.

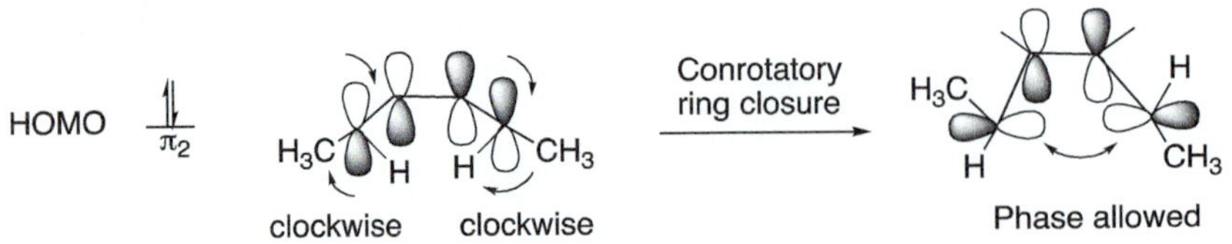

Figure 18.9 Conrotatory ring closure of ground state 1,3-hexadiene to give the products shown in Reactions (18-23) and (18-24).

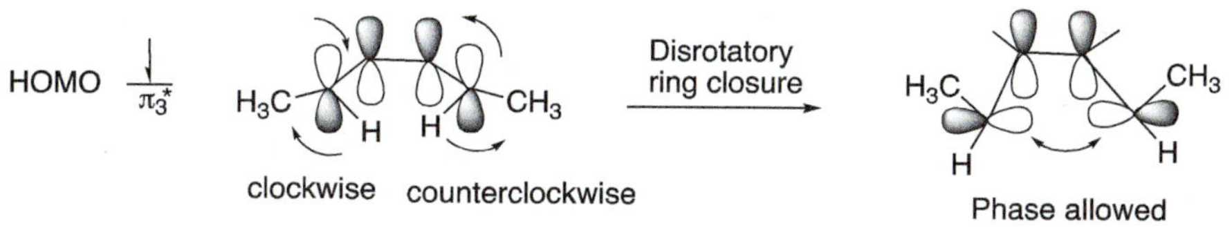

18.2 Pericyclic Reactions

Figure 18.10 Disrotatory ring closure of excited 1,3-hexadiene to give the products shown in Reactions (18-23) and (18-24).

Table 18.1 Relationship between number of pi (π) electrons of conjugated systems and type of ring closure for electrocyclic reactions.

Number of pi (π) electrons	Type energy	Type of ring closure
4	Heat	Conrotatory
4	Light	Disrotatory
6	Heat	Disrotatory
6	Light	Conrotatory

Shown in Reaction (18-26) is the electrocyclic reaction of 2,4,6-octatriene.

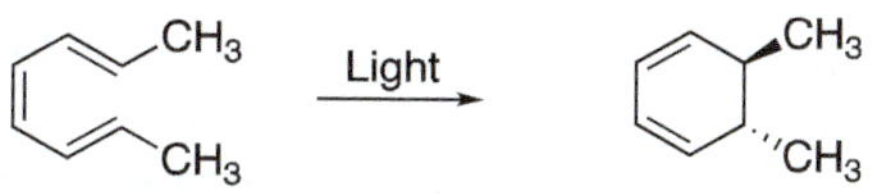

(18-26)

Problem 18.4

Give the product of the following electrocyclic reactions, include appropriate stereochemistry where appropriate.

(a) [structure] Heat → ?   (b) [structure] Light → ?

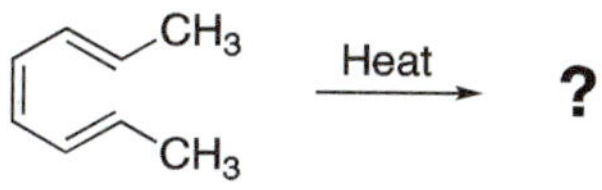

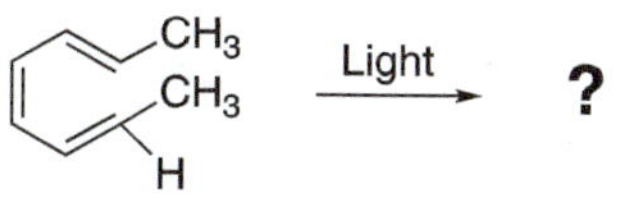

18.2.3 Sigmatropic Reactions

For these reactions, a sigma bond is formed and another sigma bond is broken. In the process, the π bonds are rearranged as shown in Reaction (18-27).

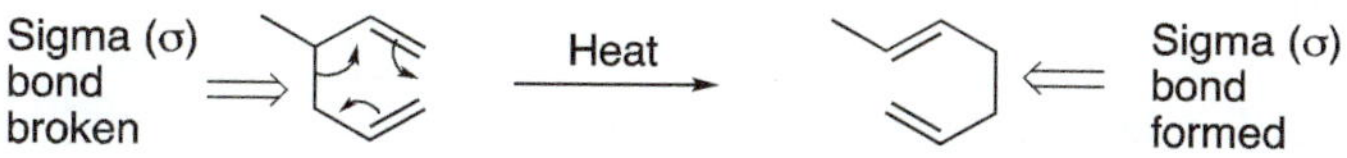

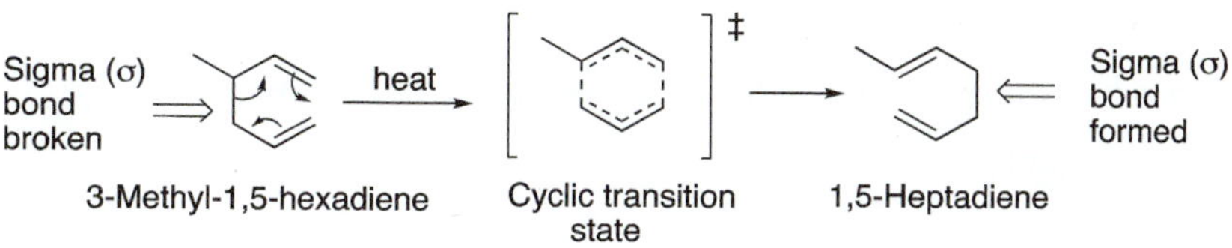

Sigma (σ) bond broken   Heat →   Sigma (σ) bond formed   (18-27)

As you can imagine, the transition states for these reactions are cyclic and shown in Reaction (18-28) is the sigmatropic rearrangement reaction involving 3-methyl-1,5-hexadiene.

Sigma (σ) bond broken   heat   Cyclic transition state   1,5-Heptadiene   Sigma (σ) bond formed   (18-28)

3-Methyl-1,5-hexadiene

Note that for these reactions, the π electrons are not conjugated, but since a sigma bond is involved in the rearrangement and since a cyclic transition state is formed as shown in Reaction (18-28), these reactions are classified as pericyclic reactions. The type of reaction given in Reaction (18-28) is described as a [3,3] sigmatropic rearrangement and is also known as the *Cope rearrangement*. The number three indicates the three atoms that are separated from the sigma bonds to be formed and broken, as illustrated below. A look at the transition state clearly shows the three atoms that are involved in the bond-making and bond-breaking process.

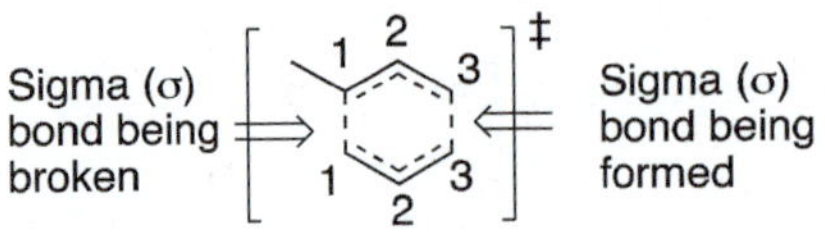

Transition state for a [3,3] sigmatropic rearrangement

Another example of the nomenclature of a cope rearrangement is shown in Reaction (18-29).

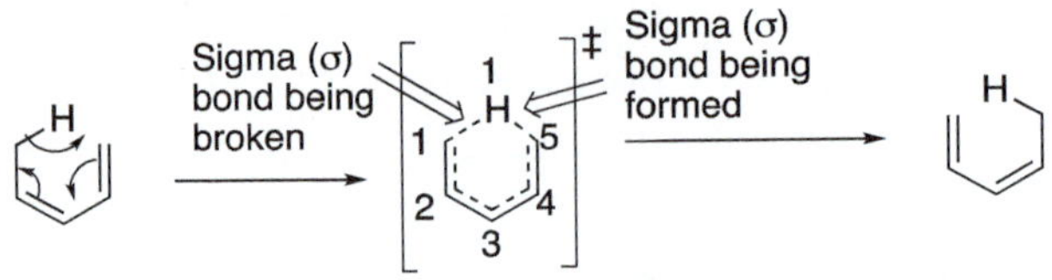

(18-29)

Transition state for a [1,5] sigmatropic rearrangement

Note that for this sigmatropic reaction, there is one atom (the hydrogen) on one side of the sigma bond to be formed and broken and five atoms on the other side, hence the name [1,5] sigmatropic reaction. If the reactants, and hence products, contain an oxygen atom as part of the cyclic rearrangement framework, this type of reaction is known as a *Claisen rearrangement*, as shown in Reaction (18-30).

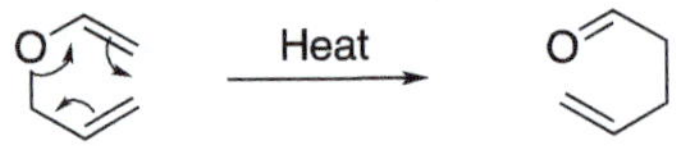

(18-30)

Problem 18.5 Give the product for the following Claisen rearrangements.

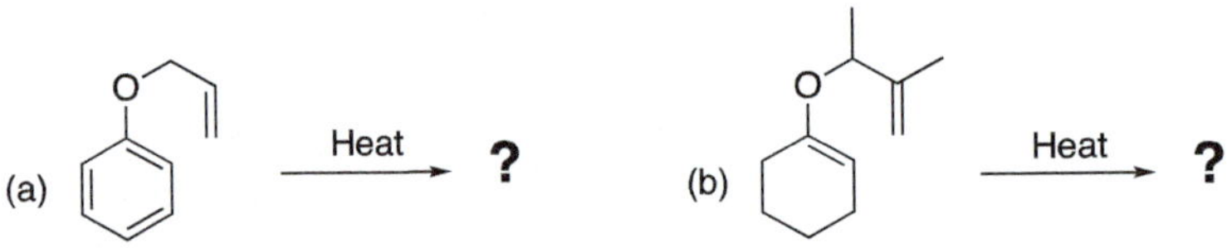

(a) (b)

End of Chapter Problems

18.6 Which of the following molecules is(are) conjugated?

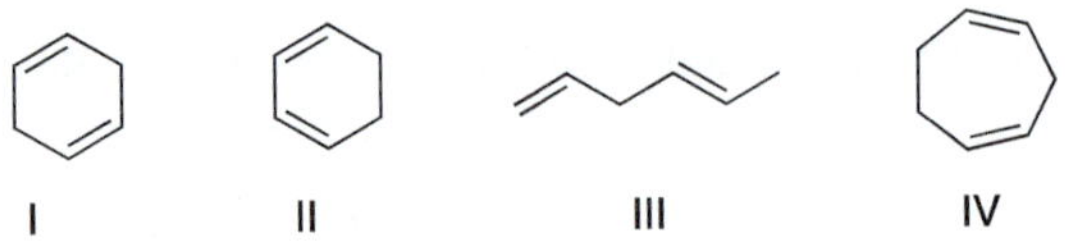

I   II   III   IV

18.7 Predict the product for the Diels–Alder [4+2] reaction given below.

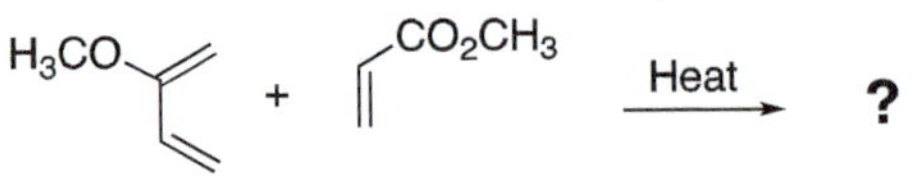

18.8 Complete the following reactions by providing the dienes and dienophiles.

(a) 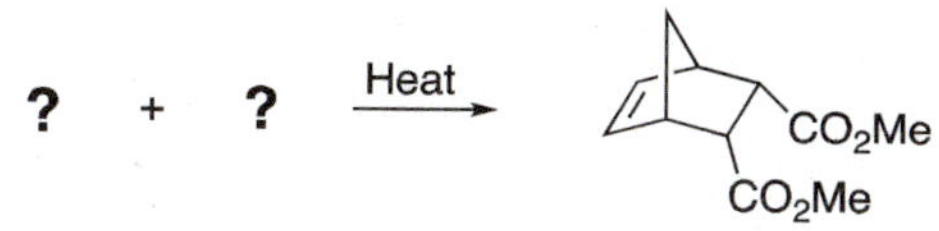 ? + ? $\xrightarrow{\text{Heat}}$ [norbornene with CO$_2$Me, CO$_2$Me]

(b) 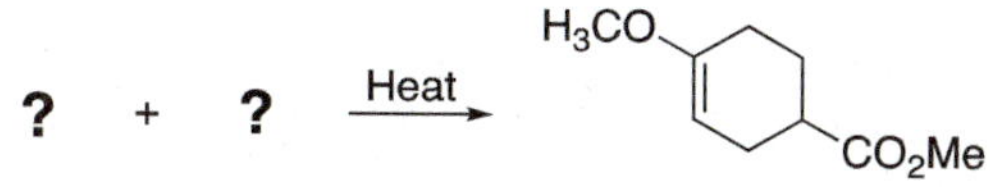 ? + ? $\xrightarrow{\text{Heat}}$ [cyclohexene with H$_3$CO and CO$_2$Me substituents]

(c) 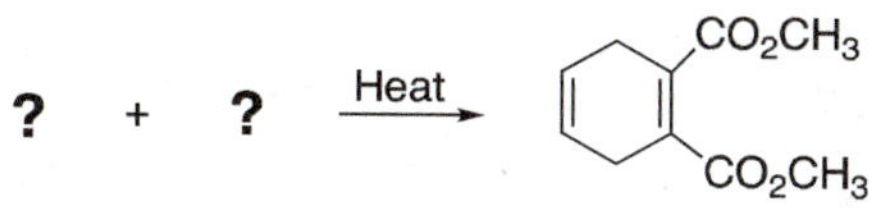 ? + ? $\xrightarrow{\text{Heat}}$ [benzene ring with CO$_2$CH$_3$, CO$_2$CH$_3$]

18.9 Give the structures of compounds **A** and **B** (including stereochemistry where appropriate) in the reaction sequence shown below, the last step involves a pericyclic reaction.

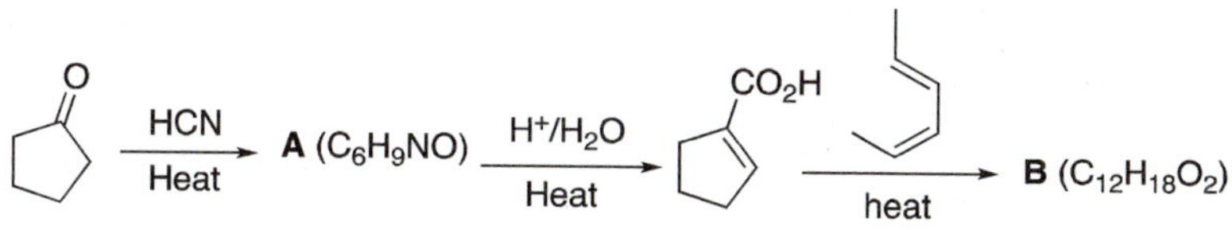

cyclopentanone $\xrightarrow[\text{Heat}]{\text{HCN}}$ **A** (C$_6$H$_9$NO) $\xrightarrow[\text{Heat}]{\text{H}^+/\text{H}_2\text{O}}$ [cyclopentene with CO$_2$H] + [diene] $\xrightarrow{\text{heat}}$ **B** (C$_{12}$H$_{18}$O$_2$)

18.10 For the reaction of diarlysenenoketones with conjugated dienes (shown below), give the major organic product with appropriate stereochemistry.

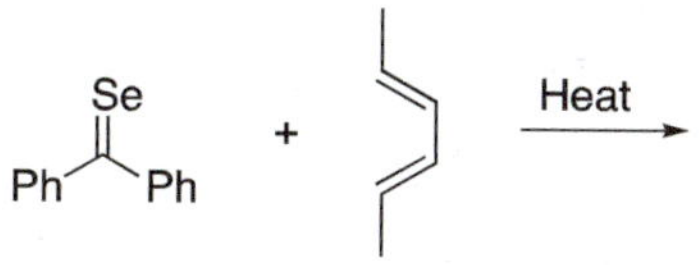

Ph-C(=Se)-Ph + [diene] $\xrightarrow{\text{Heat}}$

18.11 For the following pericyclic reactions, give the major organic products. (hint: consider a sigmatropic rearrangement, i.e. Cope and/or Claisen rearrangements)

(a) 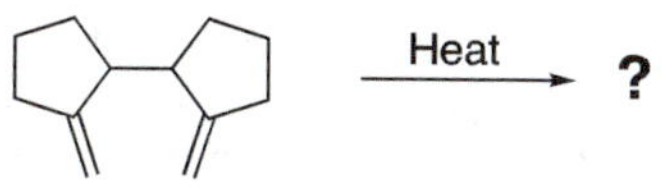 [bicyclic diene structure] $\xrightarrow{\text{Heat}}$?

(b) 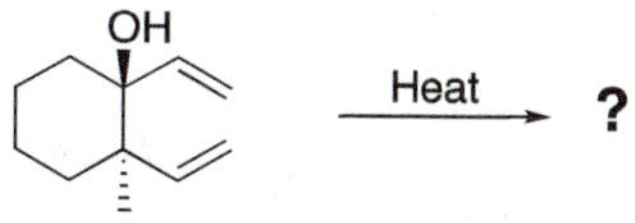 [cyclohexane with OH and vinyl groups] $\xrightarrow{\text{Heat}}$?

18.12 The conversion of the reactant to the product shown below involves a sequence of two pericyclic reactions. The first is an electrocyclic ring opening reaction involving the cyclobutene moiety to form an unstable intermediate **A**. The second reaction involves an intramolecular [4+2] cycloaddition reaction. Give the structure of intermediate **A** and use arrow-pushing formulism to indicate electron movement to explain how the intermediate **A** and product are formed.

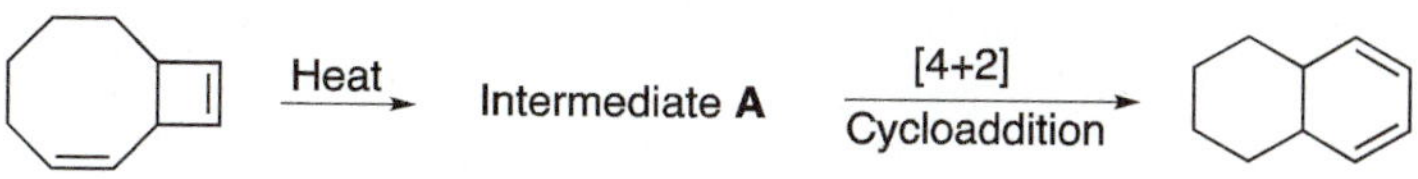

[bicyclic cyclobutene-cyclooctene] $\xrightarrow{\text{Heat}}$ Intermediate **A** $\xrightarrow{[4+2]\text{ Cycloaddition}}$ [decalin diene product]

18.13 For the synthesis of the sex hormone estrone, a sequence of two pericyclic reactions was utilized. The first reaction is an electrocyclic ring opening reaction involving the cyclobutane moiety to form an unstable intermediate **A**; and the second reaction an intramolecular [4+2] cycloaddition reaction. Give the structure of intermediate **A** and explain using arrow-pushing formulism to indicate electron movement to explain how the product, estrone, is formed.

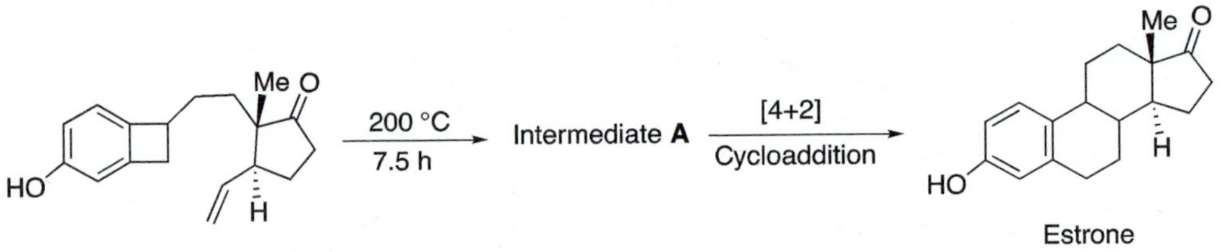

19

Catalytic Carbon–Carbon Coupling Reactions

19.1 Introduction

The use of transition metals in organic chemistry has increased over the years, especially for the synthesis of new carbon–carbon bonds. You will recall that transition metals are in the middle of the periodic table and they have partially filled d and f orbitals. There is a wide variety of groups called ligands that are typically coordinated to transition metals. As a result, transition metals that have ligands coordinated to them are referred to as coordination compounds or coordination transition–metal complexes. The groups that are coordinated to transition metals are sometimes neutral, such as ammonia and triphenylphosphine (PPh_3) or fully charged anion, such as the chloride anion. They are all Lewis bases in that they all have at least one unshared pair of electrons. Examples of different coordinated palladium ion complexes are shown below.

$Pd(PPh_3)_4$

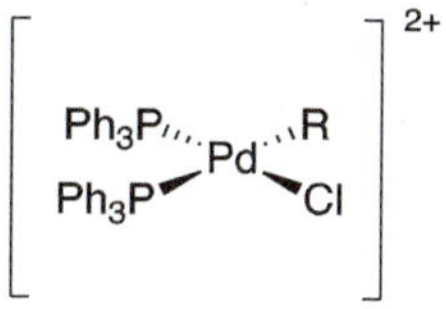

$Pd(PPh_3)_2(R)(Cl)^{2+}$

In this chapter, we will examine the reactions involving various transition–metal complexes and the mechanism by which they catalyze various carbon–carbon bond forming reactions. As we have seen throughout our course in organic chemistry, often times, there is the need to make new carbon–carbon bonds to achieve the goal of synthesizing larger target molecules from smaller molecules. The difficulty encountered in accomplishing such a task is that carbons of a carbon–carbon bonds are of comparable electronegativity, and there are not many options whereby the strategy of using a nucleophile to react with an electrophile to make a new carbon-carbon bonds can be used. In this chapter, we will examine a new strategy of synthesizing new carbon–carbon bonds in which different coordination transition–metal complexes are used as catalysts for these carbon–carbon bond forming reactions.

19.2 Reactions of Transition Metal Complexes

In this section, we will examine the structure and type of reactions that selected coordination transition–metal complexes undergo. Some of these reactions are already mentioned in previous chapters, but in this section, more details will be provided about the mechanism for these reactions.

Organic Chemistry: Concepts and Applications, First Edition. Allan D. Headley.
© 2020 John Wiley & Sons, Inc. Published 2020 by John Wiley & Sons, Inc.
Companion website: www.wiley.com/go/Headley_OrganicChemistry

19.2.1 Oxidative Addition Reactions

In oxidative addition, a metal is inserted into a sigma bond. We have already seen this type reaction in the synthesis of the Grignard reagent, in which magnesium is inserted into the C—X bond to form R-Mg-X, where X is a halogen. For Reaction (19-1), the metal Mg in the reactant is oxidized to Mg^{2+}. As a result, the term oxidative addition is used to describe this type of reaction.

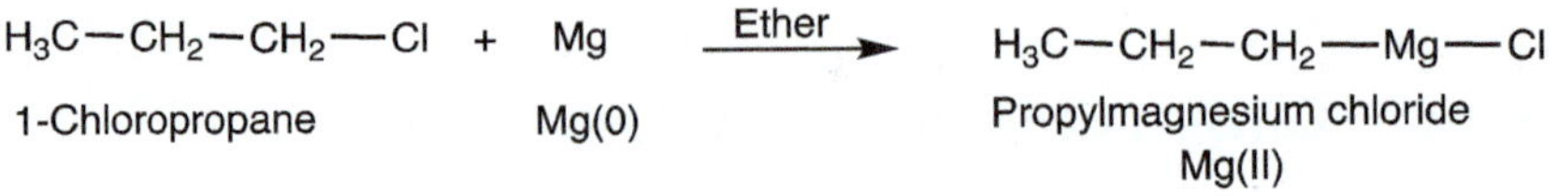

$$H_3C-CH_2-CH_2-Cl \ + \ Mg \ \xrightarrow{Ether} \ H_3C-CH_2-CH_2-Mg-Cl \quad (19\text{-}1)$$

1-Chloropropane   Mg(0)   Propylmagnesium chloride, Mg(II)

As we have mentioned in earlier chapters, Victor Grignard discovered this type of reaction, and the reaction is now known as the Grignard reaction and the organomagnesium halides generated are known as Grignard reagents.

For transition metals, the same type of reaction is possible as shown below in Reaction (19-2), in which Pd is inserted into a Ph—X bond, where Ph is the phenyl group ($-C_6H_5$) and X is a halogen.

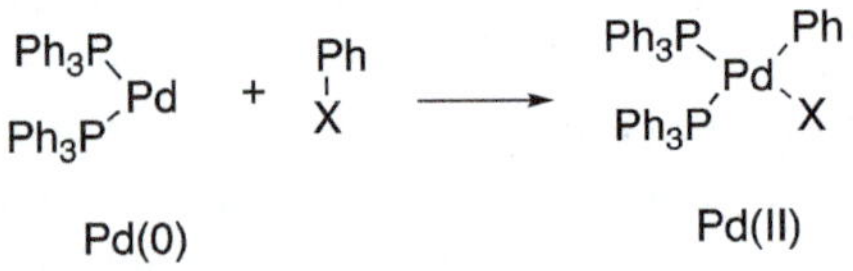

$$\underset{Pd(0)}{Ph_3P-Pd-PPh_3} \ + \ Ph-X \ \longrightarrow \ \underset{Pd(II)}{(Ph_3P)_2Pd(Ph)(X)} \quad (19\text{-}2)$$

Note the change in oxidation state of palladium, it is Pd(0) in the reactant and is oxidized to Pd(II) in the product. The formation of this type of complex is key to the reactions that will be discussed throughout this chapter.

19.2.2 Transmetallation Reactions

Another important reaction that coordinated transition–metal complexes can undergo is transmetallation as shown in Reactions (19-3) through (19-5a).

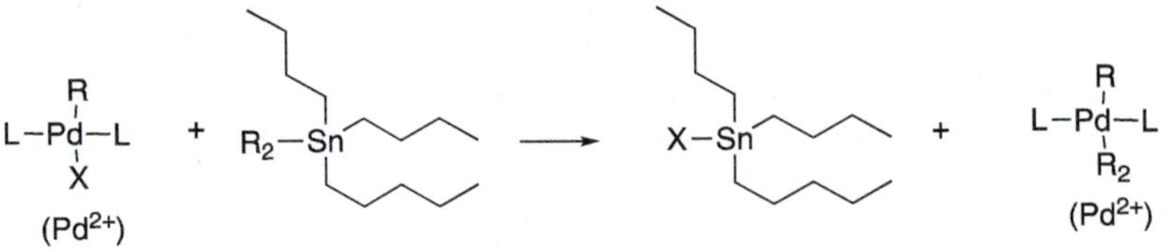

$$\underset{(Pd^{2+})}{L-Pd(R)(X)L} \ + \ R_2-Sn(Bu)_3 \ \longrightarrow \ X-Sn(Bu)_3 \ + \ \underset{(Pd^{2+})}{L-Pd(R)(R_2)L} \quad (19\text{-}3)$$

$$\underset{(Pd^{2+})}{L-Pd(R)(X)L} \ + \ R_2-Zn-X \ \longrightarrow \ X-Zn-X \ + \ \underset{(Pd^{2+})}{L-Pd(R)(R_2)L} \quad (19\text{-}4)$$

$$\underset{(Pd^{2+})}{L-Pd(R)(X)L} \ + \ R_2-B(OR)_3 \ \longrightarrow \ X-B(OR)_3 \ + \ \underset{(Pd^{2+})}{L-Pd(R)(R_2)L} \quad (19\text{-}5a)$$

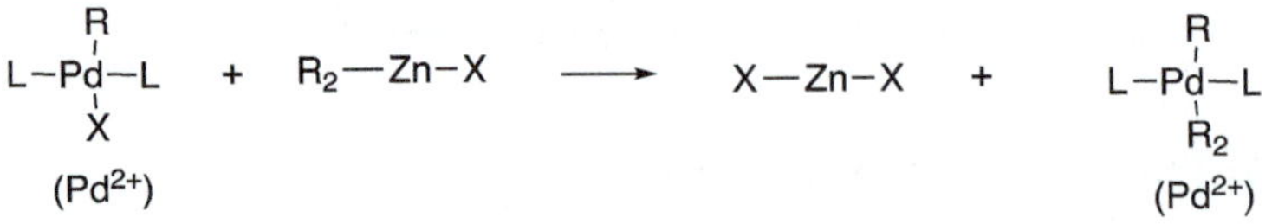

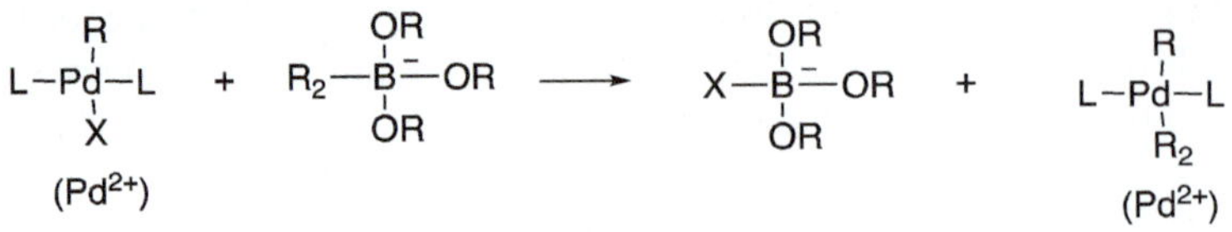

These reactions are very straightforward in that one group is transferred from a coordinated metal ion complex to another. We have seen a similar type of reaction before in the synthesis of organocuprates, also known as the Gillman reagent and given in Reaction (19-5b).

$$2\ R\text{–Li} + CuBr \longrightarrow R\text{–Cu–R}\ ^{\ominus}\ Li^{\oplus} + LiBr \tag{19-5b}$$

19.2.3 Ligand Migratory Insertion Reactions

Another type of a reaction of metal ion complex is the possibility of migratory insertion of a ligand in a bond of the metal ion, as shown below in Reaction (19-6).

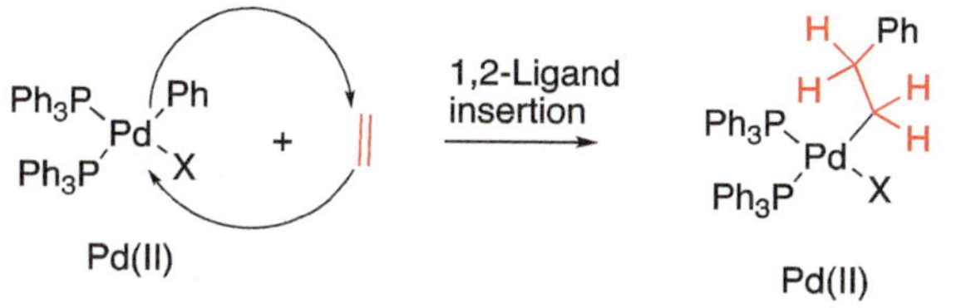

$$(19\text{-}6)$$

These reactions are called 1,2 insertion since the insertion takes place across two carbons of the alkene, and this type of reaction is also key in coupling reactions, as we will see later in the chapter.

19.2.4 β-Elimination Reactions

As the name implies, these reactions involve the migration of a atom or group in the β-position of the metal along with its bonding electrons, to the metal. This reaction is essentially the reverse of ligand insertion as discussed in the previous section. An example is shown below in Reaction (19-7).

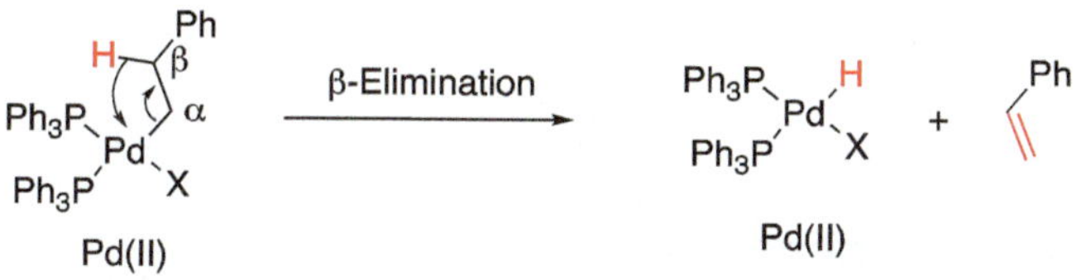

$$(19\text{-}7)$$

This type of elimination is also known as an intramolecular *syn*-elimination.

19.2.5 Reductive Elimination Reactions

As the name implies, this type of reaction is essentially the reverse of oxidative addition. For these reactions, two ligands bonded to the metal leave. An example is given in Reaction (19-8).

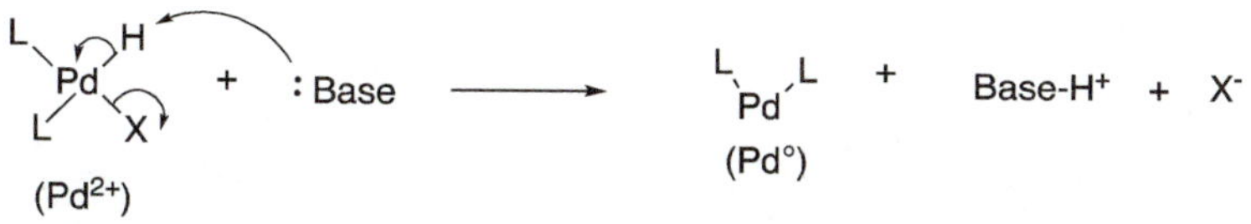

$$(19\text{-}8)$$

For Reaction (19-8), a base is introduced to abstract a proton and the bonding electrons are used to reduce the Pd^{2+} to Pd (0) metal while the other ligand leaves with its bonding electrons. Another example of this type of reductive elimination reaction is given in Reaction (19-9).

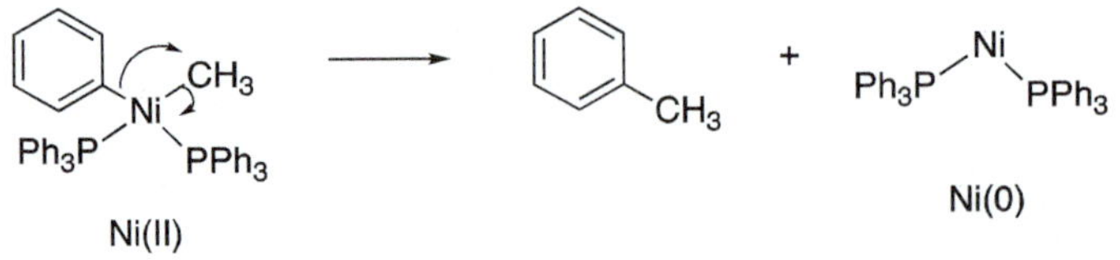

$$\text{(19-9)}$$

For this reaction, which is an intramolecular elimination, the phenyl group migrates to form a new carbon–carbon bond while the Ni is reduced from Ni(II) to Ni(0).

Problem 19.1

For each of the following reaction described above the arrow, give the products.

(a) Ph$_3$P–Pd(Br)–PPh$_3$ with vinyl group + ethylene —1,2-Ligand insertion→

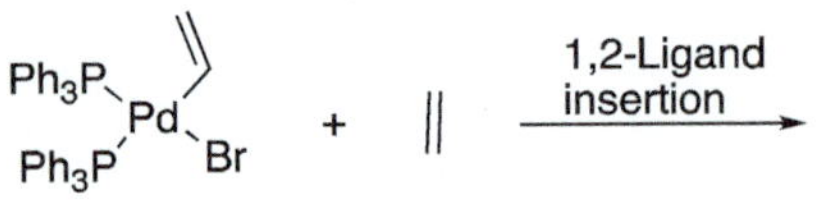

(b) Ph$_3$P–Pd–PPh$_3$ + phenyl iodide —Oxidative addition→

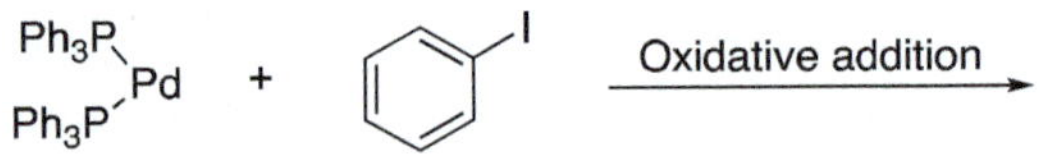

(c) isopropyl–Pd(Br)(PPh$_3$)(PPh$_3$) —Reductive elimination→

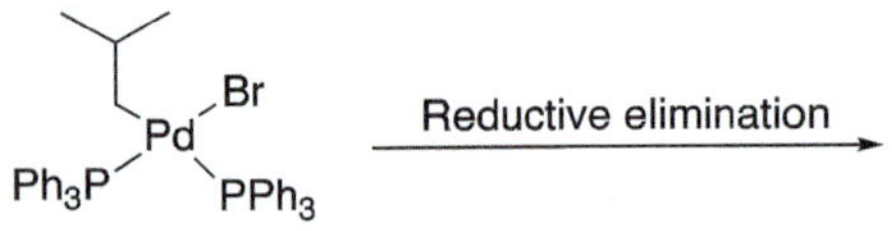

(d) Ph$_3$P–Pd(Br)(PPh$_3$) with allyl group —Reductive elimination→

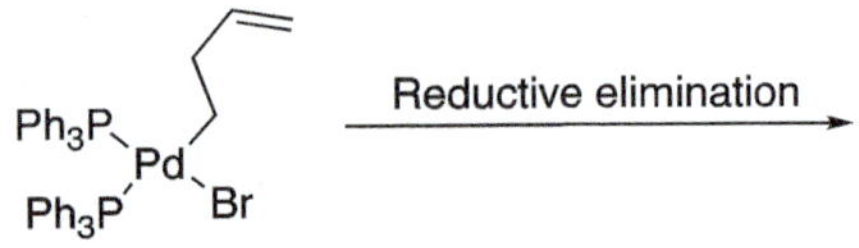

19.3 Palladium-Catalyzed Coupling Reactions

19.3.1 The Heck Reaction

This first coupling reaction that we will examine in which a transition metal is used is the Heck reaction. This reaction is named after American chemist, Richard Frederick Heck, who is known for the discovery of this reaction. This reaction involves the coupling of aryl halides with alkenes in the presence of a base and a palladium catalyst. A general example of the Heck reaction is shown in Reaction (19-10a).

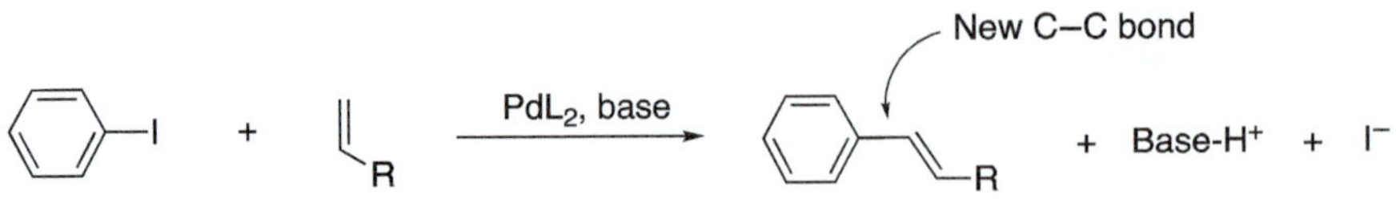

$$\text{Ph–I} + \text{CH}_2=\text{CHR} \xrightarrow{\text{PdL}_2,\text{ base}} \text{Ph–CH=CH–R} + \text{Base-H}^+ + \text{I}^- \quad (19\text{-}10\text{a})$$

New C–C bond

The first step of this reaction involves an *oxidative addition* of the organoiodide and the Pd complex, as shown in Reaction (19-10b).

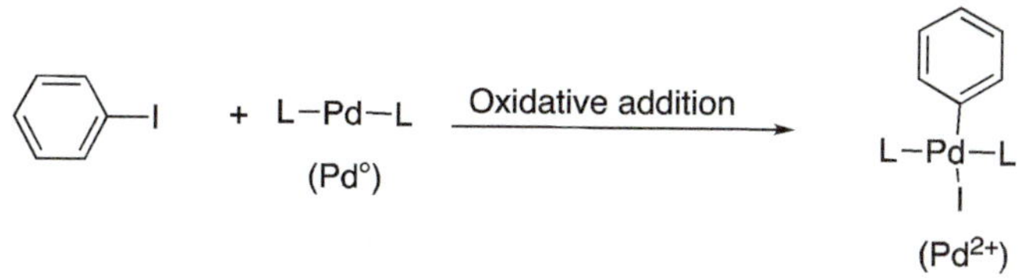

$$\text{Ph–I} + \text{L–Pd–L} \xrightarrow{\text{Oxidative addition}} \underset{(Pd^{2+})}{\text{L–Pd(Ph)(I)–L}} \qquad (19\text{-}10b)$$
$$(Pd^0)$$

Note that the Pd changes oxidation state from 0 to 2^+, hence an oxidation. In the next step of the reaction, the alkene is inserted into the Pd-phenyl bond, as shown in Reaction (19-11).

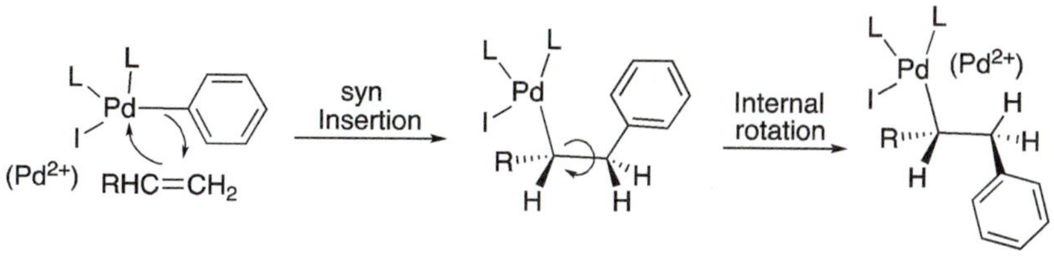

$$\text{(19-11)}$$

The next step involves a β-hydride elimination, as shown in Reaction (19-12). Note that the elimination is a *syn*-elimination in that the hydride and Pd must be coplanar in order for the reaction to take place.

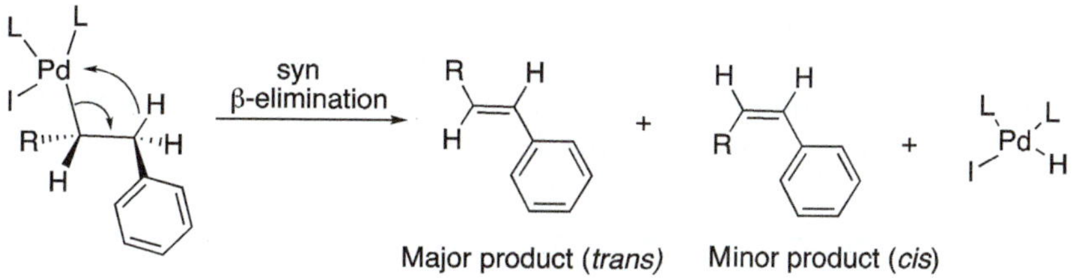

$$\text{(19-12)}$$

Major product (*trans*)   Minor product (*cis*)

The final step of the reaction is a reductive elimination reaction as shown in Reaction (19-13).

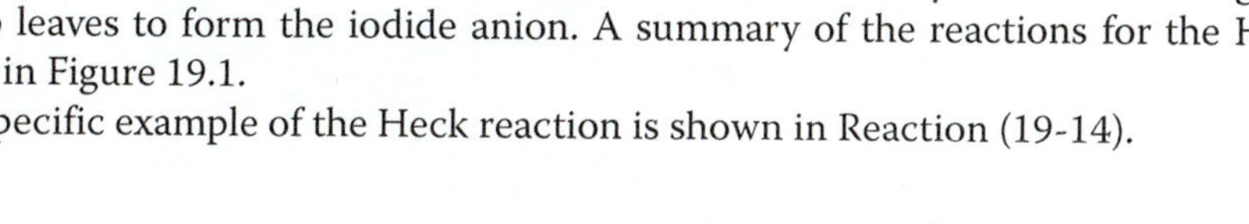

$$\text{L–Pd(H)(I)–L} + \text{:Base} \xrightarrow{\text{Reductive elimination}} PdL_2 + \text{Base-H}^+ + I^- \qquad (19\text{-}13)$$

An amine base is typically introduced, which abstracts a proton, and the bonding electrons are used to reduce the Pd (II) to Pd (0), and simultaneously, the iodine being a good leaving group leaves to form the iodide anion. A summary of the reactions for the Heck reaction is given in Figure 19.1.

A specific example of the Heck reaction is shown in Reaction (19-14).

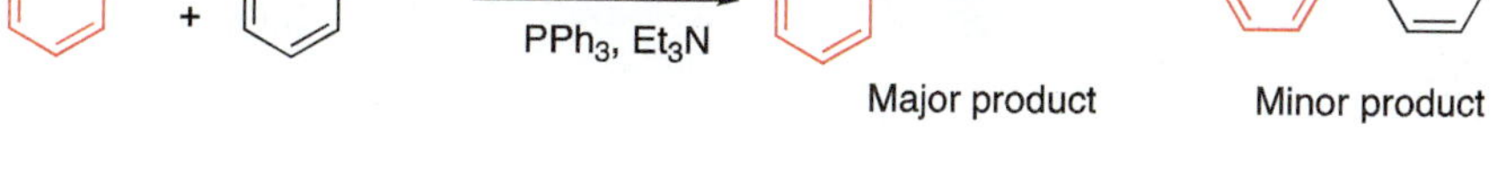

$$\text{PhI} + \text{CH}_2\text{=CHPh} \xrightarrow[\text{PPh}_3,\ \text{Et}_3\text{N}]{\text{Pd(OAc)}_2} \text{trans-stilbene} + \text{cis-stilbene} \qquad (19\text{-}14)$$

Major product   Minor product

19 Catalytic Carbon–Carbon Coupling Reactions

Figure 19.1 A summary of the Heck catalytic coupling reactions.

In this case, the alkene is not symmetrical, and as a result, there are two possible products as shown. The coupling reaction typically occurs at the less substituted end of the alkene and the major product typically has the *trans*-stereochemistry. The base, triethylamine, is used for the reductive elimination reaction. The Heck reaction is a very important reaction, especially in the pharmaceutical industry; a lot of pharmaceutical products or intermediates leading to the final product have the alkene functionality that can be synthesized using this type of reaction. In addition, since only small amounts of catalysts are needed for these reactions, it can be regenerated at the end of the reaction and recycled.

> **DID YOU KNOW?**
>
> The Heck reaction is used in the industrial production of octyl methoxycinnamate, also known as ethylhexyl methoxycinnamate or octinoxate and marketed as trade names Uvinul MC80 and Eusolex 2292. This compound can absorb ultraviolet (UV) light from sunlight and fluorescent sources and is an ingredient in some lip balms and sunscreens.

Octyl methoxycinnamate

19.3 Palladium-Catalyzed Coupling Reactions

Problem 19.2

i) Give the product for the following Heck reactions.

(a) Ph–I + cyclohex-2,5-diene-1,4-dione $\xrightarrow{\text{Pd(OAc)}_2, \text{PPh}_2, \text{NEt}_3}$?

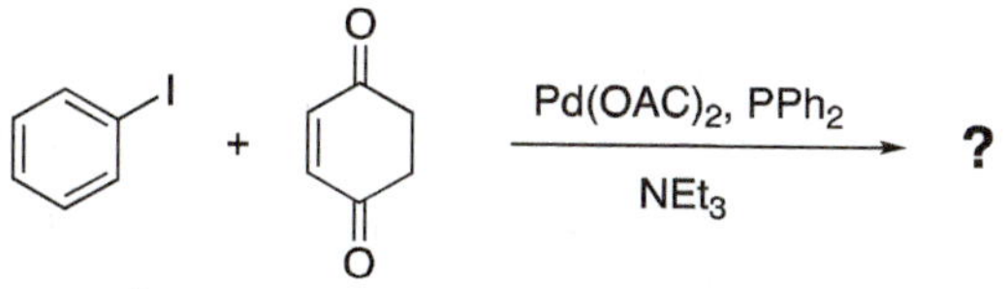

(b) 2-bromo-(but-3-enyl)benzene $\xrightarrow{\text{Pd(OAc)}_2, \text{PPh}_2, \text{NEt}_3}$?

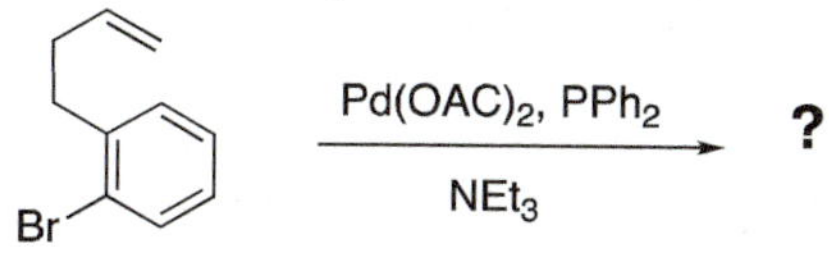

ii) Give the reactants to complete the Heck reactions.

(a) ? + ? $\xrightarrow{\text{Pd(OAc)}_2, \text{PPh}_2, \text{NEt}_3}$ 4-methoxycinnamamide

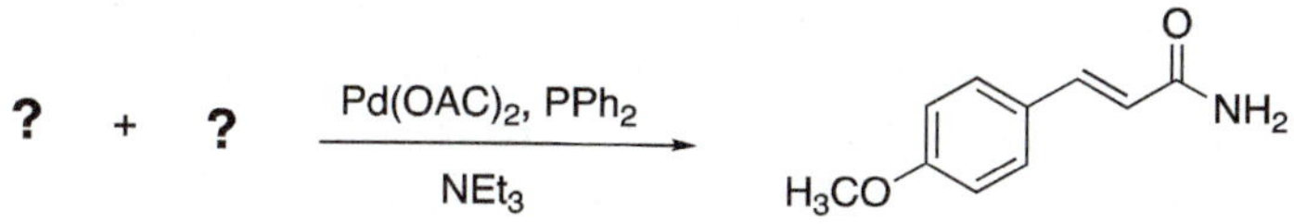

(b) ? + ? $\xrightarrow{\text{Pd(OAc)}_2, \text{PPh}_2, \text{NEt}_3}$ methyl 5-cyclohexylpenta-2,4-dienoate

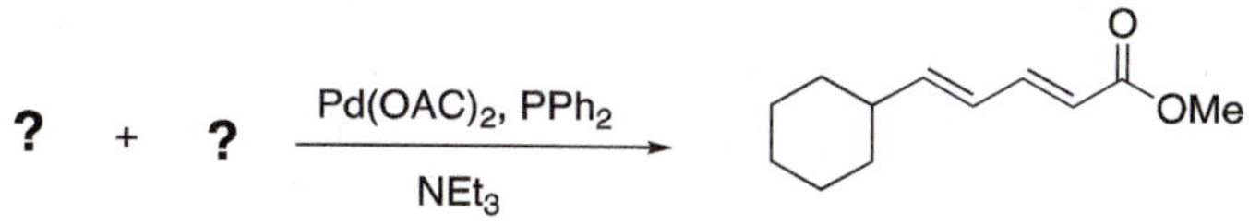

19.3.2 The Suzuki Reaction

The Suzuki reaction is named after Professor Akira Suzuki, a Japanese chemist of Hokkaido University. This reaction is similar to the Heck reaction; but in the Suzuki reaction, organoboranes are used. An example of the Suzuki reaction is given in Reaction (19-15).

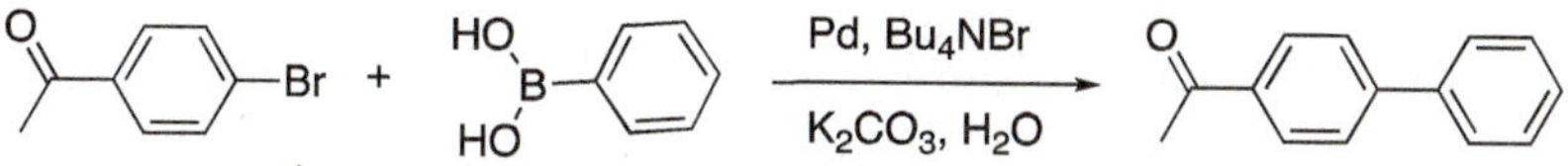

$$\text{4-bromoacetophenone} + \text{PhB(OH)}_2 \xrightarrow{\text{Pd, Bu}_4\text{NBr}, \text{K}_2\text{CO}_3, \text{H}_2\text{O}} \text{4-acetylbiphenyl} \quad (19\text{-}15)$$

Note carefully for this reaction, like the Heck reaction, a carbon–carbon bond is formed at the point of attachment from the C—Br and C-borane of the reactants. The mechanism to explain this reaction is shown below. The first step of the reaction involves an insertion into the C—Br bond as shown in Reaction (19-6) *(oxidative addition)*.

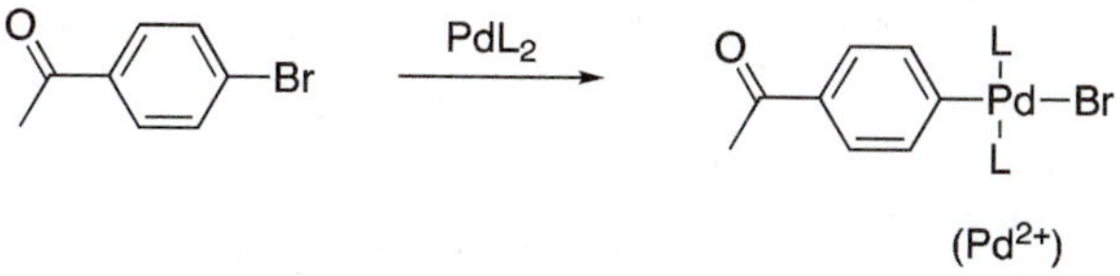

$$\text{Ar-Br} \xrightarrow{\text{PdL}_2} \text{Ar-Pd(L)}_2\text{-Br} \quad (19\text{-}16)$$

(Pd^{2+})

In the next step of the reaction mechanism, boric acid is activated with a base to form a borate complex as shown in Reaction (19-17).

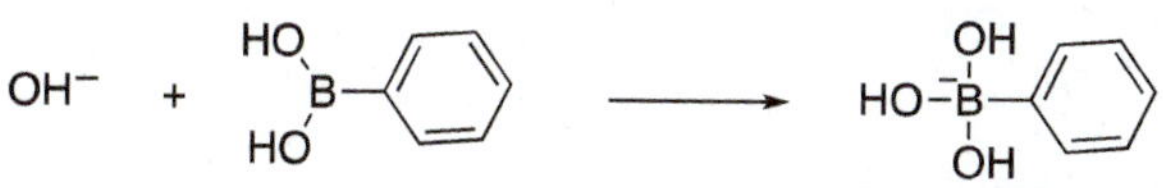

$$\text{OH}^- + \text{PhB(OH)}_2 \longrightarrow \text{PhB(OH)}_3^- \quad (19\text{-}17)$$

The next step is a transmetallation step, as shown in Reaction (19-18).

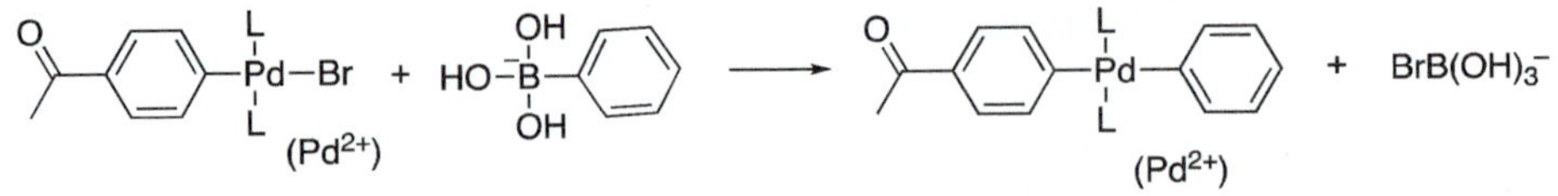

$$\text{[Acetyl-phenyl-Pd(L)}_2\text{-Br]} + \text{HO-B(OH)-phenyl} \longrightarrow \text{[Acetyl-phenyl-Pd(L)}_2\text{-phenyl]} + \text{BrB(OH)}_3^- \quad (19\text{-}18)$$

(Pd²⁺) → (Pd²⁺)

The next step of the mechanism is a *reductive elimination* in which the catalyst is regenerated as shown in Reaction (19-19).

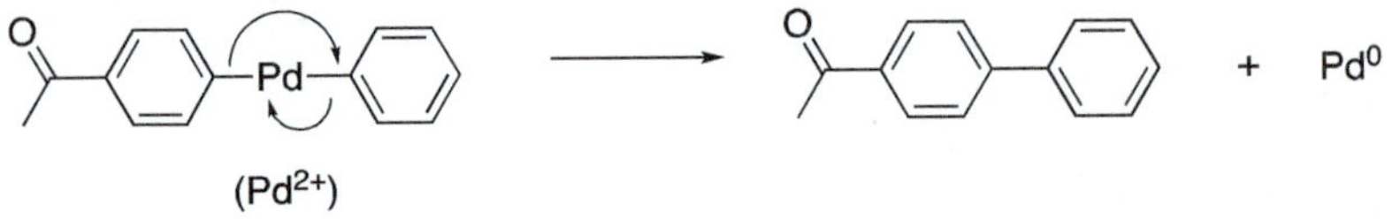

$$\text{[Acetyl-phenyl-Pd-phenyl]} \longrightarrow \text{[Acetyl-biphenyl]} + \text{Pd}^0 \quad (19\text{-}19)$$

(Pd²⁺)

In this step, the catalytic species is regenerated and can be used again. A major advantage of these reactions is that they can be carried out in aqueous media. This is a plus for "Green Chemistry" owing to the advantages of water as a solvent, compared to organic solvents, which are typically toxic and present a disposal problem. Another example of the Suzuki reaction is shown in Reaction (19-20).

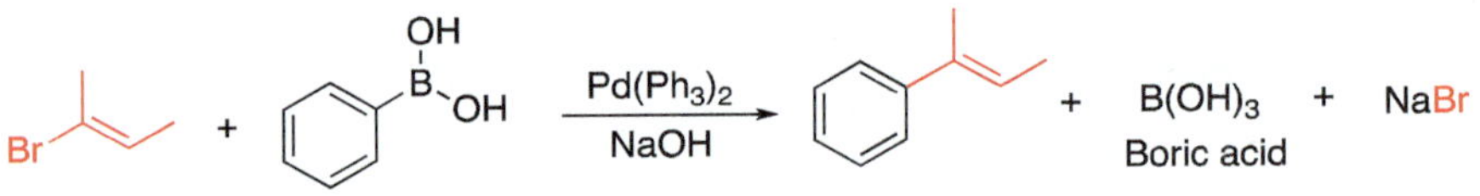

$$\text{Br-vinyl} + \text{phenyl-B(OH)}_2 \xrightarrow[\text{NaOH}]{\text{Pd(Ph}_3)_2} \text{styrene} + \text{B(OH)}_3 + \text{NaBr} \quad (19\text{-}20)$$

Boric acid

Boronate esters can also be used for the Suzuki reaction, and they can be synthesized as outlined in the example shown in Reaction (19-21).

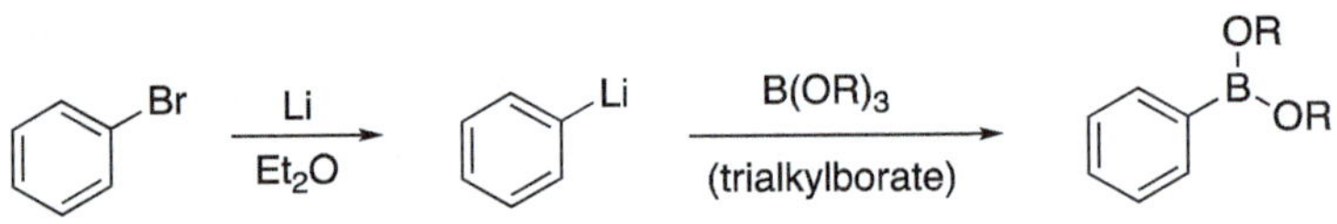

$$\text{Ph-Br} \xrightarrow[\text{Et}_2\text{O}]{\text{Li}} \text{Ph-Li} \xrightarrow{\text{B(OR)}_3} \text{Ph-B(OR)}_2 \quad (19\text{-}21)$$

(trialkylborate)

A summary of the Suzuki reaction is shown in Figure 19.2.

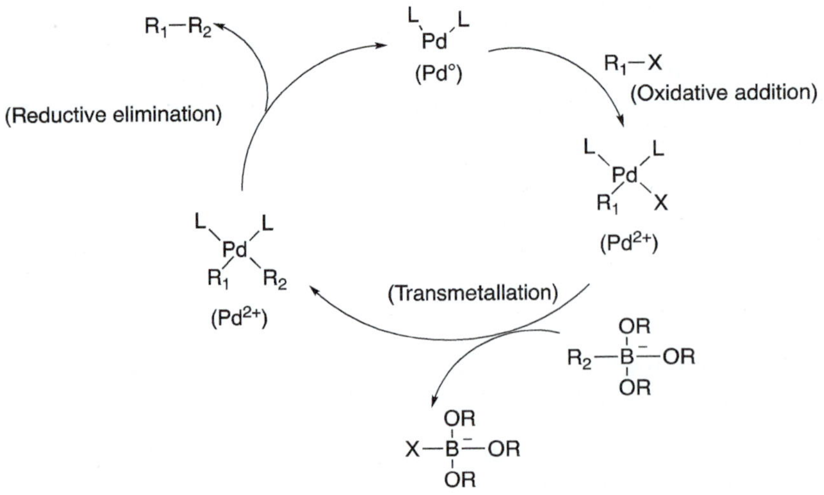

Figure 19.2 A summary of the Suzuki catalytic coupling reactions.

Problem 19.3

i) Give the major organic products to complete the Suzuki reactions.

(a) Ph–B(OMe)(OMe) + CH$_2$=CH–OTf $\xrightarrow{\text{Pd(PPh}_3)_4\text{, NaOH}}$?

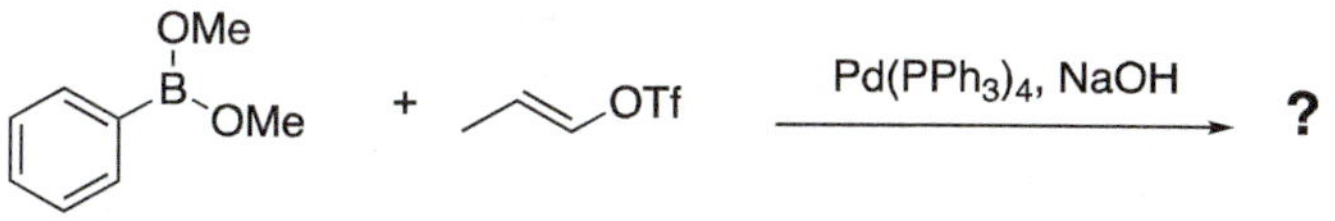

(b) (2,3-dihydro-1H-isoindol-2-yl)boronic vinyl + CH$_2$=CH–OTf $\xrightarrow{\text{Pd(PPh}_3)_4\text{, NaOH}}$?

ii) Give the reactants to complete the Suzuki reactions.

(a) ? + ? $\xrightarrow{\text{PdPPh}_3)_4\text{, NaOH}}$ PhCH=C(CH$_3$)CH$_2$CH$_3$

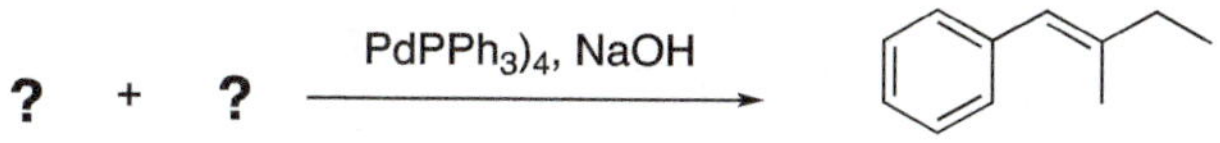

(b) ? + ? $\xrightarrow{\text{PdPPh}_3)_4\text{, NaOH}}$ PhCH=CH–C(O)NH$_2$

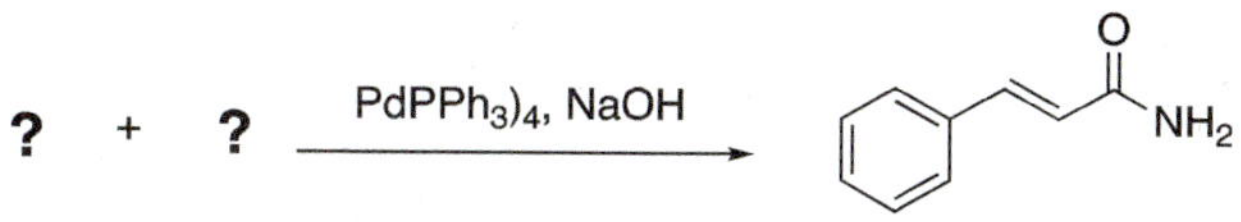

19.3.3 The Stille Coupling Reaction

This reaction, named after John Stille of Colorado State University, is another method of synthesizing new carbon–carbon bonds to make larger molecules from smaller molecules. The general reaction scheme is shown in Reaction (19-22).

$$R_1\text{–SnBu}_3 + R_2\text{-X} \xrightarrow{\text{Pd}} R_1\text{–}R_2 + \text{SnBu}_3X \qquad (19\text{-}22)$$

A specific reaction is shown below. Note that this reaction essentially couples two benzene rings, a very difficult task to accomplish as we have seen using the chemistry described in the earlier chapters of this book.

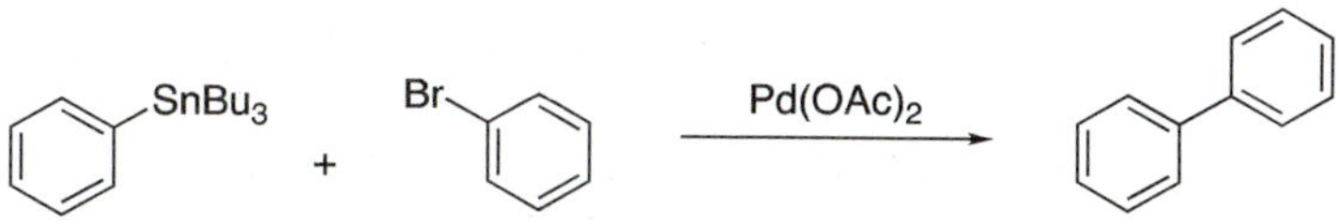

$$\text{Ph–SnBu}_3 + \text{Br–Ph} \xrightarrow{\text{Pd(OAc)}_2} \text{Ph–Ph} \qquad (19\text{-}23)$$

A summary of the Stille reaction is given in Figure 19.3.

Problem 19.4

i) Give the major organic products to complete the Stille coupling reactions.

(a) (3-iodo-6-isopropylcyclohex-2-ene) + CH$_2$=CH–SnBu$_3$ $\xrightarrow{\text{Pd(OAc)}_2}$?

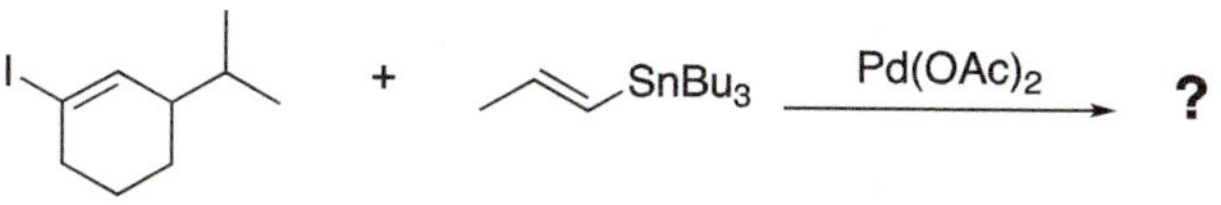

(b) (2-vinyl-2,3-dihydro-1H-indene) + CH$_2$=CH–SnBu$_3$ $\xrightarrow{\text{Pd(OAc)}_2}$?

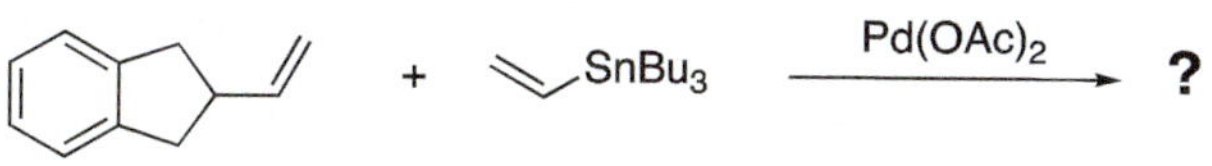

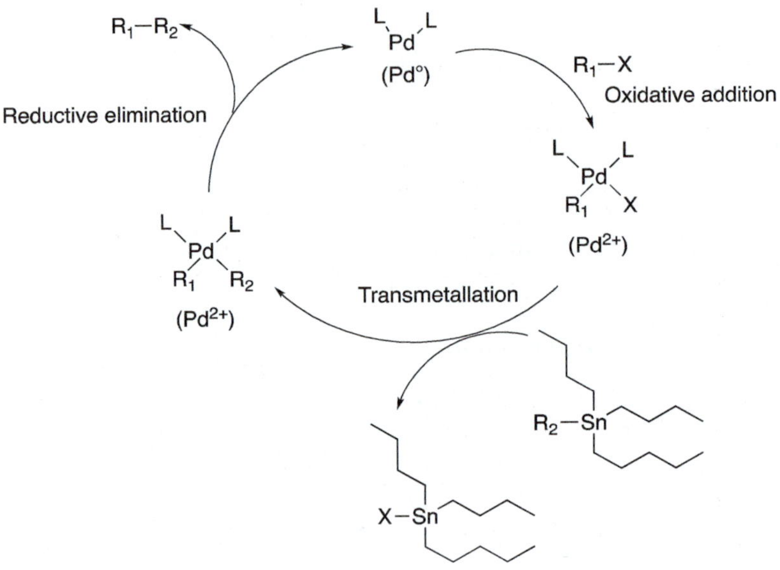

Figure 19.3 A summary of the Stille catalytic coupling reactions.

ii) Give the reactants to complete the Stille coupling reactions.

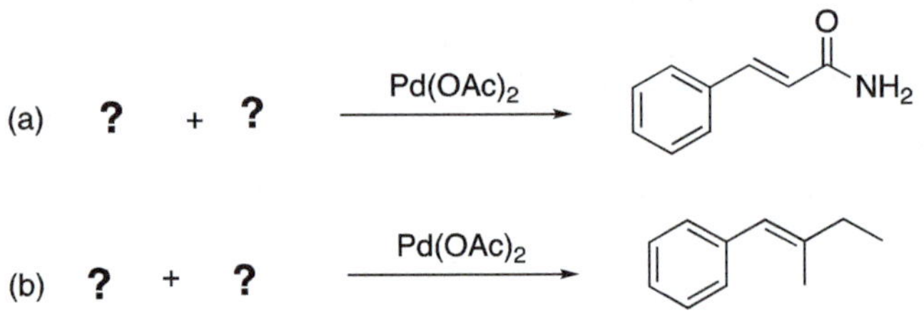

(a) ? + ? $\xrightarrow{Pd(OAc)_2}$ cinnamamide (PhCH=CH-C(O)NH$_2$)

(b) ? + ? $\xrightarrow{Pd(OAc)_2}$ (E)-2-phenyl-2-butene (PhC(CH$_3$)=CHCH$_3$)

19.3.4 The Negishi Coupling Reaction

The last reaction that will be covered is the Negishi coupling reaction. This reaction is named after Ei-ichi Negishi, a Japanese chemist, who carried out most of his research work at Purdue University and shared the Nobel Prize in Chemistry in 2010 with Heck and Suzuki for the discovery of the palladium-catalyzed cross-coupling reactions in organic synthesis. The general reaction is shown in Reaction (19-24).

$$R_1-ZnX + R_2-X \xrightarrow{Pd} R_1-R_2 + ZnX_2 \tag{19-24}$$

A summary of the Negishi palladium-catalyzed cross couplings is given in Figure 19.4.

Problem 19.5

Give the reactants to complete the Negishi cross-coupling reactions.

(a) (cyclohexenyl-CH(Cl)-CH$_3$) + CH$_2$=CH-ZnCl $\xrightarrow{Pd}$

(b) (indanyl-CH$_2$Cl) + CH$_2$=CH-ZnCl $\xrightarrow{Pd}$

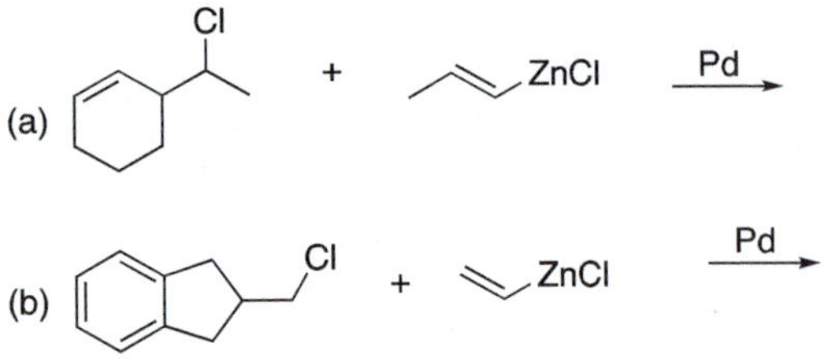

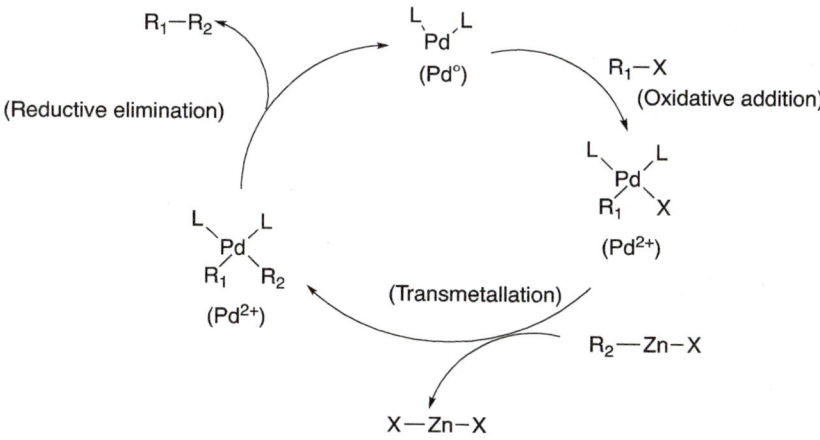

Figure 19.4 A summary of the Negishi catalytic coupling reactions.

End of Chapter Problems

19.6 Provide a mechanism similar to the one shown in Figure 19.1 to explain the following Heck reaction.

PhI + PhCH=CH₂ →[Pd⁰, Et₃N] trans-stilbene

19.7 Provide the products for the following catalytic coupling reactions.

(a) 2-bromo(allyl)benzene →[Pd⁰, Et₃N] ?

(b) (iodo-methylcyclohexenyl with CN) →[Pd⁰, Et₃N] ?

(c) PhI + CH₂=CH–CHO →[Pd⁰, Et₃N] ?

20

Synthetic Polymers and Biopolymers

20.1 Introduction

Polymers can be divided into two broad categories, synthetic polymers and biopolymers. Biopolymers are polymers that are made by living organisms; these include carbohydrates, nucleic acids, proteins, and peptides and will be covered in the last section of this chapter. Synthetic polymers are polymers that are typically made from simple starting compounds and converted into polymers through various chemical reactions. Many chemical industries specialize in the synthesis of different types of polymers. Synthetic polymers are very important compounds in our everyday lives. We just have to look around, and we can see the many applications of polymers. The physical properties of synthetic polymers allow these polymers to be used for different purposes. Some polymers are very hard and durable, and as a result, they are used as desk surfaces, pipes, CDs, etc., whereas some polymers are more flexible and are used as bottles, clothing, and wraps. Polymers, as the name suggests, are many "mers," which means units. Thus, a polymer is made up of many simple units. The process of transforming the simple units of polymers to the actual polymer is called ***polymerization***, as illustrated in Reaction (20-1), where n represents numbers that are typically in the hundreds.

$$n\text{X (Monomer)} \xrightarrow{\text{Polymerization}} -\text{X}-\text{X}-\text{X}-\text{X}-\text{X}-\text{X}-\text{X}- \text{ (Polymer)} \tag{20-1}$$

Some of the most useful polymers are shown in Table 20.1, along with the monomers from which they are made.

In this chapter, we will examine the various reactions that lead to polymerization and the mechanism involved in each type reaction. We will also look at ways to identify polymers and the characterization of polymers.

20.2 Cationic Polymerization of Alkenes

In this section, we will take advantage of the knowledge we gained from studying addition reactions involving alkenes that were covered in Chapter 8. As we saw, different reaction paths can be achieved depending on the type of reagent added to alkenes. For example, in the addition of an electrophile to alkenes, the intermediates that result are carbocations; the addition of a nucleophile to alkenes results in carboanions, and the addition of radicals to alkenes results in radical intermediates. The intermediates that are generated from an initial reaction of a reagent as described above with an alkene are reactive and can react with the initial alkene reactant

Organic Chemistry: Concepts and Applications, First Edition. Allan D. Headley.
© 2020 John Wiley & Sons, Inc. Published 2020 by John Wiley & Sons, Inc.
Companion website: www.wiley.com/go/Headley_OrganicChemistry

Table 20.1 Selected polymers produced from various monomers through polymerization.

Monomer	Polymer	Properties	Uses
$CH_2{=}CH_2$	Polyethylene	Flexible	Plastic containers
$CH_2{=}CH{-}Cl$	Polyvinyl (PVC)	Rigid	Plumbing pipes
$Ph{-}CH{=}CH_2$	Polystyrene	Semi-rigid	Containers and packing materials
$CF_2{=}CF_2$	Polytetrafluoroethylene (Teflon)	Extremely high melting point	Engine gaskets
$CH_2{=}C(CH_3)CO_2CH_3$	Poly(methylmethacrylate)	Flexible, clear	Lenses, windows
$CH_2{=}CH{-}CN$	Polyacrylonitrile	Strong, crystalline	Fibers

alkene to form polymeric compounds. In the next sections, we will examine possible reactions of alkenes using different types of reactants.

20.2.1 Cationic Polymerization of Isobutene

We have seen from previous chapters that alkenes in the presence of a Lewis acid will form carbocations, which are good electrophiles and can be attacked by a nucleophilic alkene to form another carbocation, as shown in Reaction (20-2).

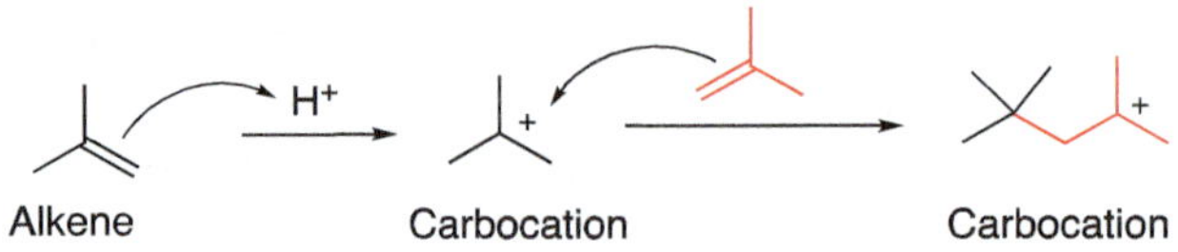

$$\text{Alkene} \xrightarrow{H^+} \text{Carbocation} \longrightarrow \text{Carbocation} \tag{20-2}$$

As you can imagine, in the presence of an excess of an alkene, this process will continue as illustrated in Reaction (20-3).

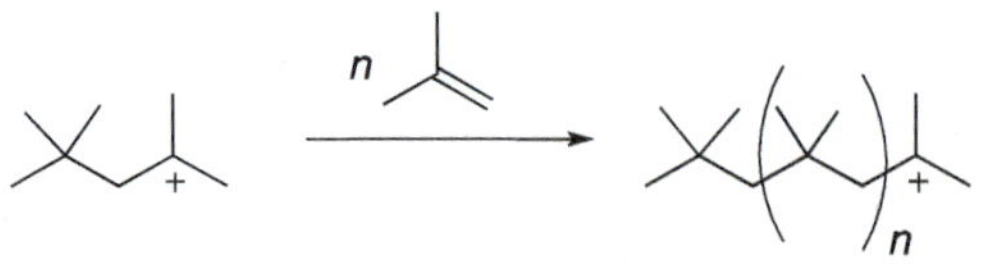

$$\tag{20-3}$$

As we have learned in previous chapters, a carbocation can lose a proton to form an alkene. This polymerization of 2-methylpropene, also known as isobutene, can be terminated by loss of a proton, as shown in Reaction (20-4).

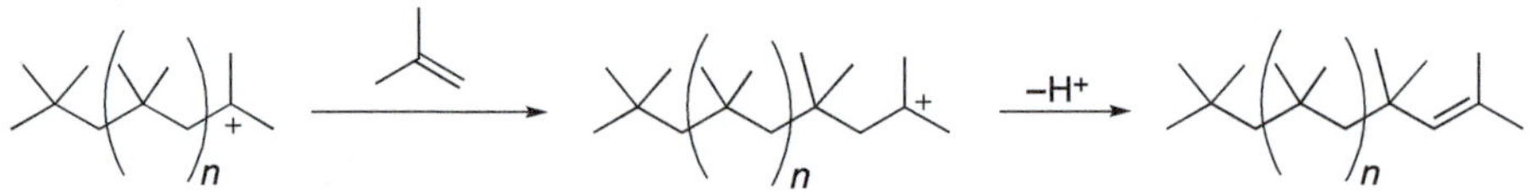

$$\xrightarrow{-H^+} \tag{20-4}$$

Polymerization of isobutene is used in the manufacture of rubber.

20.2.2 Cationic Polymerization of Styrene

Polystyrene is used in a vast majority of household containers and packing materials. The starting material for polystyrene, as the name suggests, is styrene. Note that styrene has a carbon–carbon

20.2 Cationic Polymerization of Alkenes

double bond feature as the alkene used above, and hence the addition can take place in two possible ways to produce two different carbocation intermediates. As shown in Reaction (20-5), one carbocation is more stable than the other.

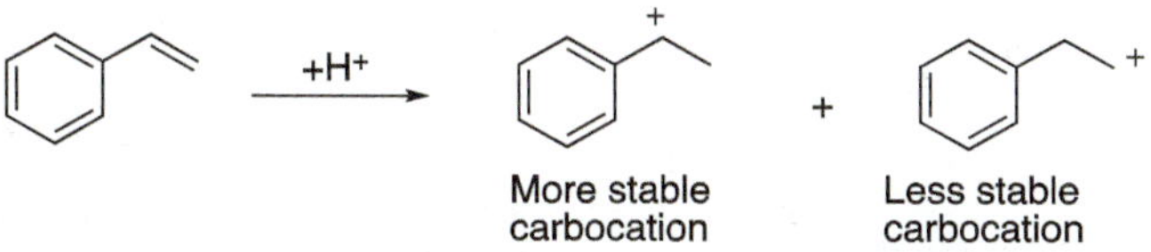

$$\text{styrene} \xrightarrow{+H^+} \text{More stable carbocation} + \text{Less stable carbocation} \tag{20-5}$$

In the next step of the reaction, the carbocation is attacked by another mole of the alkene to generate two different additional carbocations, as shown below in Reaction (20-6). Again, one is more stable than the other carbocation.

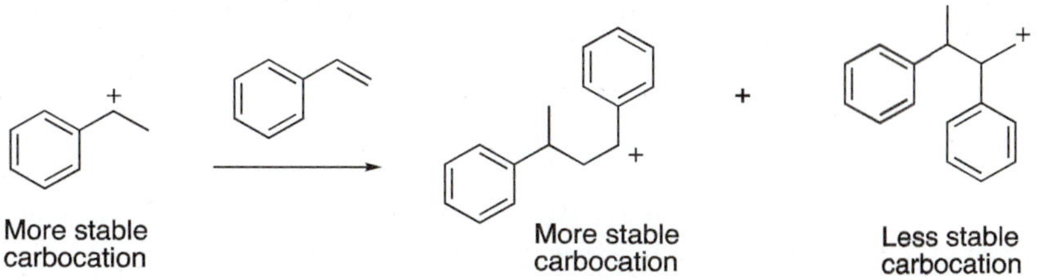

$$\text{More stable carbocation} + \text{styrene} \longrightarrow \text{More stable carbocation} + \text{Less stable carbocation} \tag{20-6}$$

In the above reaction, the more stable carbocation is again the benzylic carbocation. The reaction in which the least stable carbocation reacts with another mole of styrene is shown in Reaction (20-7).

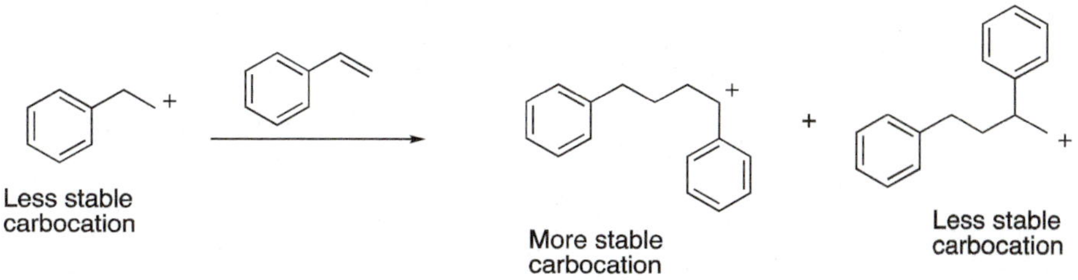

$$\text{Less stable carbocation} + \text{styrene} \longrightarrow \text{More stable carbocation} + \text{Less stable carbocation} \tag{20-7}$$

Note that this coupling results in a different set of carbocations, compared to the coupling reaction given in Reaction 20-6. The coupling of the stable benzylic carbocation with styrene to generate polystyrene is shown below.

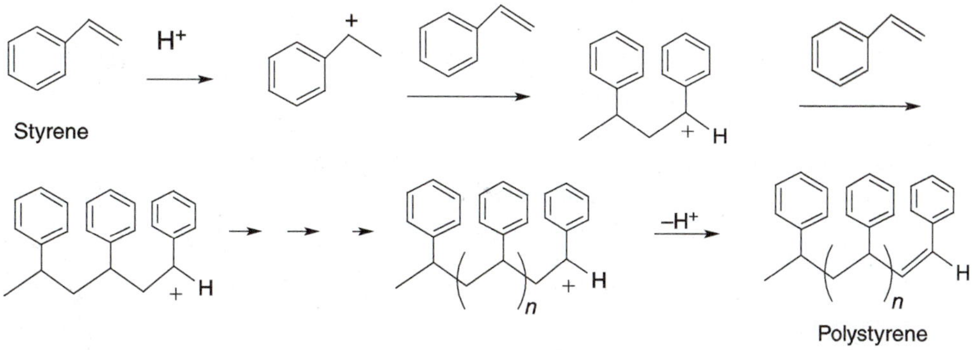

Styrene $\xrightarrow{H^+}$ ... $\xrightarrow{-H^+}$ Polystyrene

This process results in polystyrene as the product, and polystyrene is one of the most useful polymers in everyday life. Polystyrene is a clear, brittle plastic that is used for transparent containers, insulation, and even inexpensive lenses.

Problem 20.1

i) Give the polymer that would result from the polymerization of ethylene ($CH_2=CH_2$), which is also called polyethylene and used as thin plastic wrap.

ii) Give the polymer that would result from the polymerization of tetrafluoro ethylene ($CF_2=CF_2$), which is also known as Teflon.

20.3 Anionic Polymerization of Alkenes

As pointed out earlier, the reaction of alkenes with anions results in carboanionic intermediates. Carboanions can be stabilized through resonance and hence alkenes that have electron-withdrawing groups that can stabilize anions result in alkene reactants that are ideal for the synthesis of polymers via an anionic intermediate pathway.

20.3.1 Anionic Polymerization of Vinylidene Cyanide

A classic example is the polymerization of 1,1-dicyanoethene, also known as vinylidene cyanide in the presence of a nucleophile, such as the hydroxide anion which forms a very strong glue. The reactions are shown below.

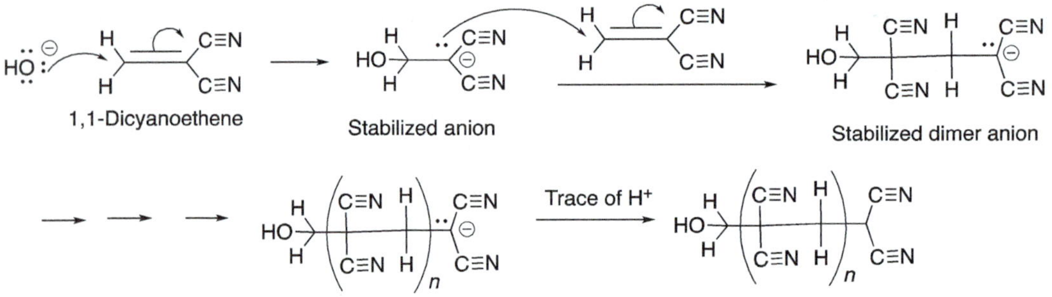

When the tube of this glue is opened, a trace of moisture, in which just a minute quantity of hydroxide anions is present starts the polymerization reaction. After the polymer is formed, a trace of protons terminates the reaction and the result is a solid polymer, which is used to join two surfaces together.

20.4 Free Radical Polymerization of Alkenes

The principle involved in free radical polymerization is one that we have utilized before when we examined free radical halogenation of alkanes. Like free radical halogenation of alkanes, free radical polymerization of alkenes involves three steps: an initiation step, a propagation step, and a termination step. The initiation step is the first step of the reaction mechanism and it is the step that produces radicals. Usually the radicals are generated by the homolytic cleavage of a weak nonpolar covalent bond to produce fairly stable radicals. The compounds that are used to produce such radicals are called radical initiators, and an example of a radical initiator reaction is shown in Reaction (20-8).

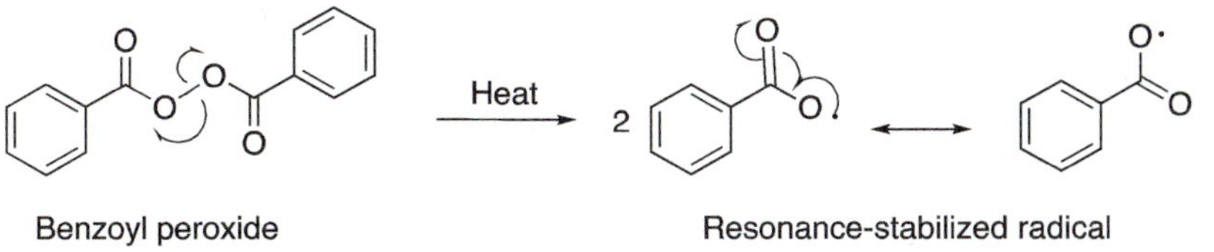

20.4 Free Radical Polymerization of Alkenes

$$\text{Benzoyl peroxide} \xrightarrow{\text{Heat}} 2 \text{ benzoyloxy radical} \longleftrightarrow \text{Resonance-stabilized radical} \qquad (20\text{-}8)$$

For Reaction (20-8), the O—O bond of the benzoyl peroxide is a very weak nonpolar covalent bond, and if this compound is heated, the O—O bond will break in a homolytic manner to produce radicals. As shown, these radicals are fairly stable owing to the resonance gained through conjugation with the adjacent carbonyl group. Owing to the favorable release of carbon dioxide, which is a gas, the initially formed radical often undergoes a further reaction to form CO_2 and another radical, the phenyl radical, as shown in Reaction (20-9).

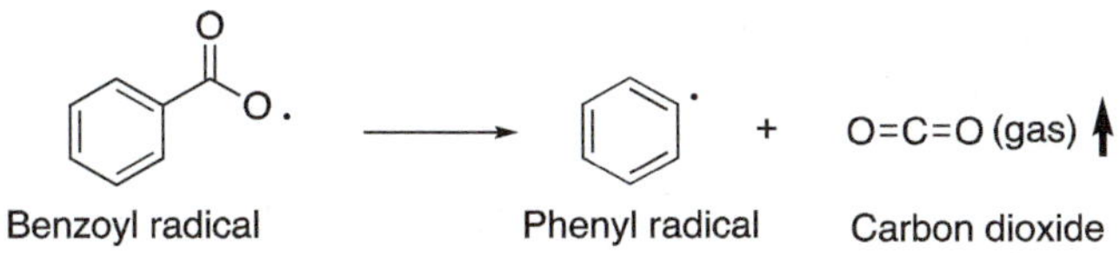

$$\text{Benzoyl radical} \longrightarrow \text{Phenyl radical} + O=C=O \text{ (gas)} \uparrow \qquad (20\text{-}9)$$

These phenyl radicals, as well as the benzoyl radicals, in the presence of an alkene functionality, react to produce more radicals, a *propagation step*. The RO· notation is typically used to represent such radicals.

Problem 20.2

The molecule shown below is often used as a radical initiator. By heating it slightly, a stable radical is produced, in addition to nitrogen gas. Give the structure of the radical produced and demonstrate its stability by drawing a resonance structure.

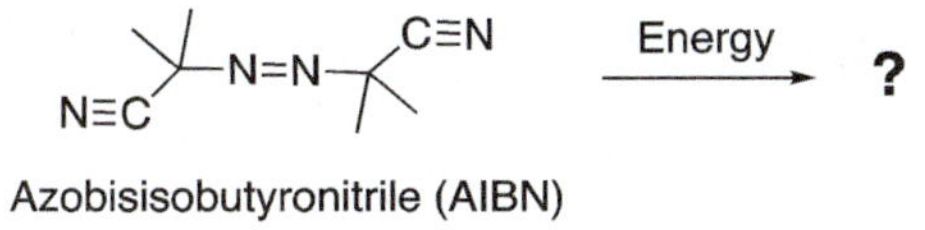

Azobisisobutyronitrile (AIBN) $\xrightarrow{\text{Energy}}$?

20.4.1 Free Radical Polymerization of Isobutylene

For the free radical polymerization of isobutylene, the first step of the reaction involves the generation of radicals using a radical initiator as shown in Reaction (20-10).

$$\text{RO—OR} \xrightarrow{\text{Energy}} 2 \text{ RO·} \qquad (20\text{-}10)$$

Peroxide (radical initiator)

Once radicals are produced, and in the presence of an isobutylene, an addition will occur to generate another radical, which reacts with another mole of the alkene to form a coupled radical, as shown in Reaction (20-11).

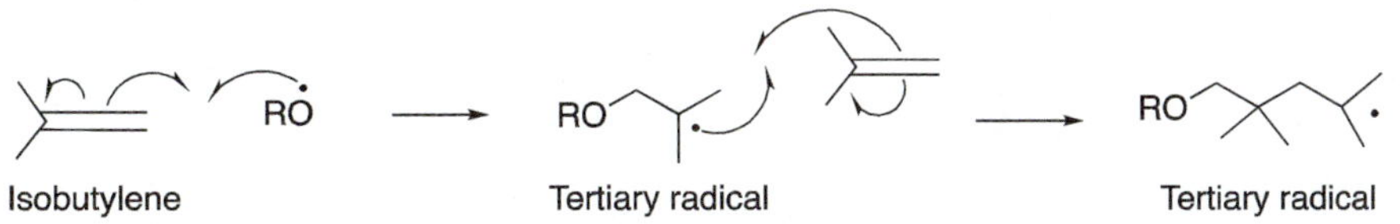

$$\text{Isobutylene} + \text{RO·} \longrightarrow \text{Tertiary radical} \longrightarrow \text{Tertiary radical} \qquad (20\text{-}11)$$

Since new radicals are produced in this step, it is possible for this newly formed radical to react with another molecule of the alkene to produce yet another radical, which has a longer chain. This process will continue until the chain of the radical is fairly long as shown in Reaction (20-12), and n represents an integer, typically in the hundreds.

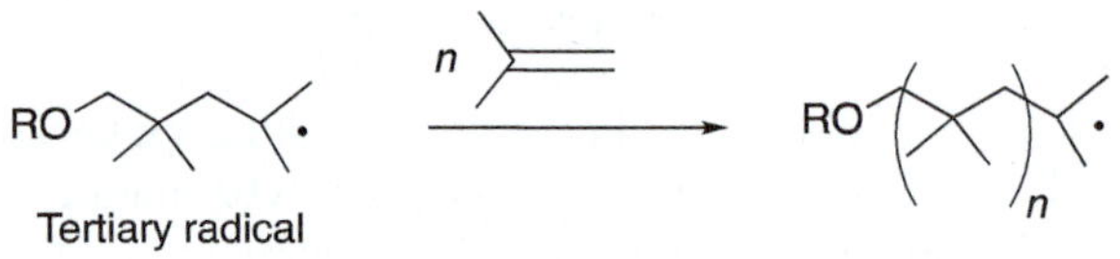

$$\text{RO-(tertiary radical)} \xrightarrow{n \text{ alkene}} \text{RO-(chain)}_n\cdot \qquad (20\text{-}12)$$

Tertiary radical

Polymers are not radical, but they are neutral molecules; thus, a terminating step is needed for this mechanism. Termination can occur in one of two ways, *radical coupling* or *disproportionation*. In a terminating step, which occurs by radical coupling, two radicals come together to form one molecule. Since there are two electrons in a single covalent bond, and each radical has one electron, coupling is a very favorable step. Radicals can form neutral molecules by another way, disproportionation. In this step, a radical abstracts a hydrogen atom (a hydrogen with one electron) from another radical to form two products. Reaction (20-13) shows a termination step where two radicals couple to form a neutral product.

$$\text{RO-(chain)}_n\cdot + \text{RO}\cdot \longrightarrow \text{RO-(chain)}_n\text{-OR} \qquad (20\text{-}13)$$

20.5 Copolymerization of Alkenes

So far, we have examined polymers that have the same type of repeating unit, but it is possible to have polymers that have different types of repeating units. There are different ways in which the different monomers can be arranged in the polymer. Some examples are shown below.

-A-B-A-B-A-B-A-   or   -A-A-A-A-B-B-B-B-A-A-A-A-   or   -A-A-B-B-B-A-B-A-
Alternating polymer   Block polymer   Random polymer

Such type polymers are referred to as **copolymers**.

20.5.1 Cationic Copolymerization

An example of the copolymerization of styrene and isobutylene is shown below.

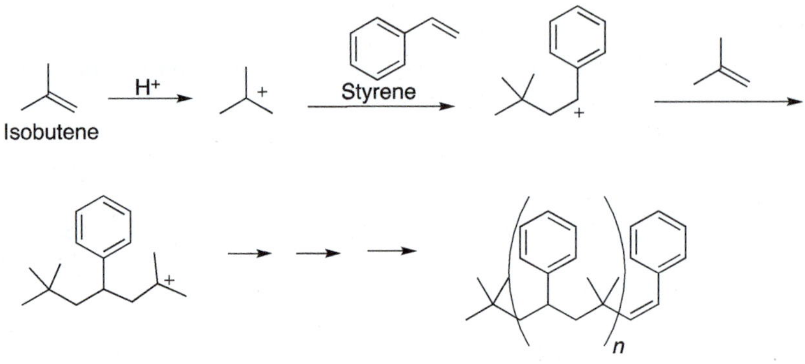

Isobutene $\xrightarrow{H^+}$ (cation) $\xrightarrow{\text{Styrene}}$ (intermediate) $\longrightarrow$

$\longrightarrow \longrightarrow \longrightarrow$ Copolymer with styrene and isobutene

Copolymers have different properties and as a result have different uses. For example, butyl rubber is a copolymer of isobutylene and isoprene that was first produced in 1937 by William Sparks and Robert Thomas of Exxon Corporation. Butyl rubber is impermeable to air and, as a result, is used as liners for airtight containers. Another copolymer is styrene-butadiene rubber, which is a copolymer of styrene and butadiene, which was developed by Goodyear and used in car tires

20.5.2 Epoxy Resin Copolymers

Resins are polymers that are typically solids or highly viscous. Resins that are made by copolymerization of an epoxide with another compound, typically one having two hydroxyl groups. Resins are mainly used as adhesives and coatings. An example of the synthesis of epoxy resin is shown in Reaction (20-14).

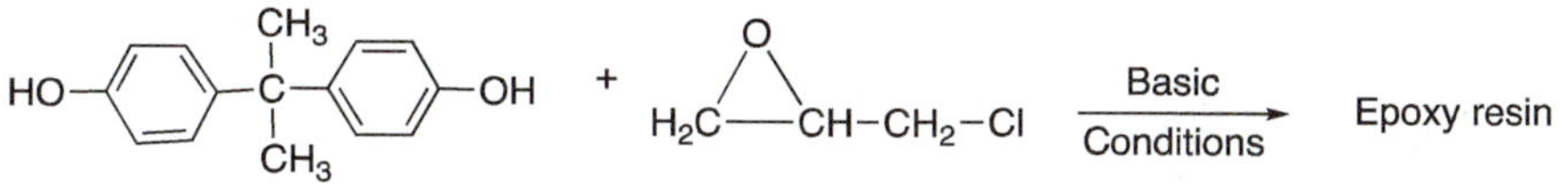

$$\text{HO-}\underset{\underset{CH_3}{|}}{\overset{\overset{CH_3}{|}}{C}}\text{-OH} \;+\; H_2C\overset{O}{\underset{}{-}}CH-CH_2-Cl \;\xrightarrow{\text{Basic Conditions}}\; \text{Epoxy resin} \qquad (20\text{-}14)$$

The reaction mechanism shown below explains the reaction and the formation of the epoxy resin as product.

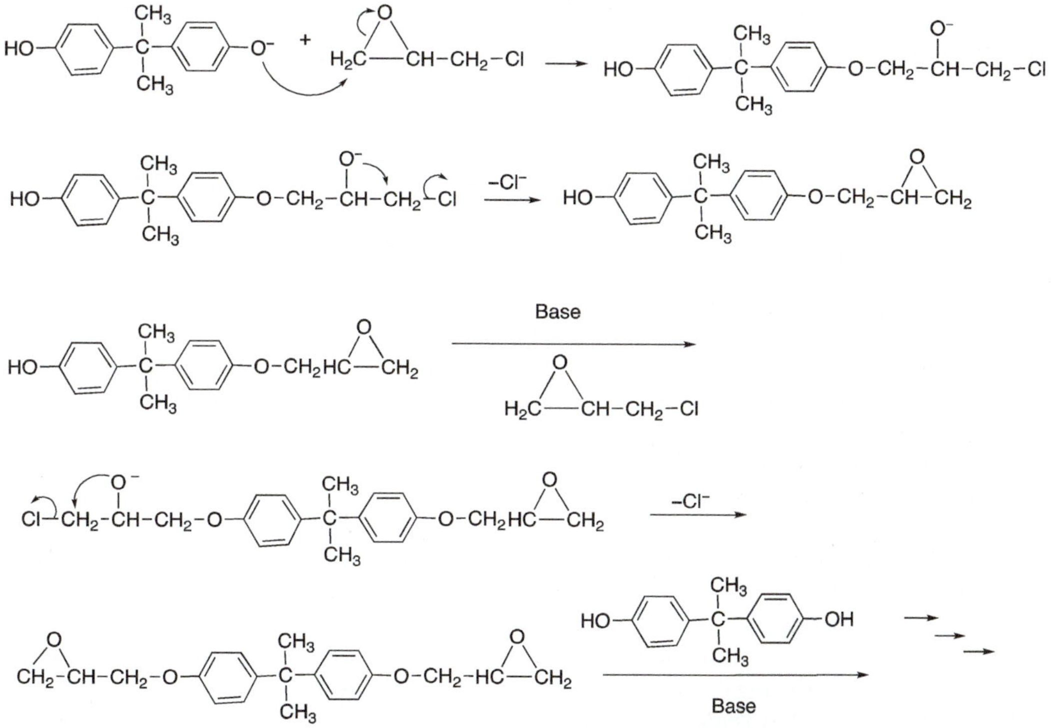

The product that is isolated is the resin. As you will notice, there are many possible molecules that can be used to accomplish a similar reaction leading to slightly modified resins. In industry, various starting compounds are used to make different types of resins.

20.6 Properties of Polymers

Since polymers play an essential role in our everyday lives, it is very important to know about the properties of the different polymers. The properties of polymers are typically divided into

(a)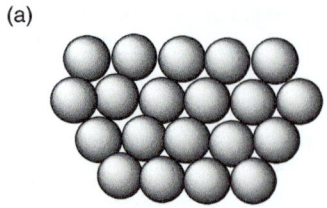
Molecular arrangement of crystalline solid

(b)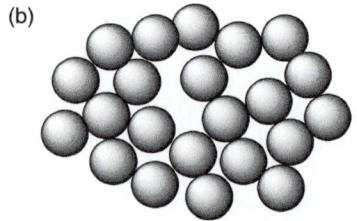
Molecular arrangement of amorphous solid

Figure 20.1 Illustration of the molecular arrangements of crystalline and amorphous solid polymers. (a) Molecular arrangement of crystalline solid. (b) Molecular arrangement of amorphous solid.

two main categories, their solubility in various solvents and properties that they exhibit at elevated temperatures.

20.6.1 Solubility of Polymers

You will recall from general chemistry that the solubility of solutes depends primarily on possible intermolecular interactions between the solute and solvent. If the solute and solvent have similar features to allow for effective intermolecular interactions, the solute will be soluble in the solvent. Thus, polystyrene, for example, which is a non-polar molecule, is more likely to be soluble in toluene than water owing to the similar van der Waals interactions between polystyrene and the organic solvent, toluene. On the other hand, if there are polar or ionic side chains on polymers, they will likely be soluble in water and other polar solvents.

20.6.2 Thermal Properties of Polymers

Polymers are typically solids, and they can be classified into two very broad categories: crystalline and amorphous. You will recall from general chemistry that crystalline solids such as sodium chloride have a regular arrangement of the ions, in this case, sodium cation and chloride anion are arranged in a regular manner. Since polymers consist of molecules and not ions or atoms, the possibility exists for the molecules to have regular orderly arrangements or not. The molecules of crystalline polymers have regular arrangement, and crystalline polymers are typically rigid and incompressible and hold a definite and fixed shape. On the other hand, amorphous solids do not have a regular arrangement of the molecules and, as a result, are not as rigid as crystalline solids. Figure 20.1 gives an illustration of the molecular arrangements of both types of solid polymers.

Crystalline solids have distinct precise melting points, whereas amorphous solids typically melt over a temperature range at which point they become rubbery and viscous. This temperature is typically referred to as glass-transition temperature (T_g).

20.7 Biopolymers

In the previous section, we examined polymers that are made from simple alkenes and other simple molecules to make larger molecules, polymers. In this section, we will examine polymers that are found naturally. These compounds are described as biopolymers and they include peptides, proteins, and carbohydrates. Proteins and peptides are natural polymers, compared to synthetic polymers that were discussed in the previous chapter. Peptides contain repeating units of α-amino acids, and the bond that links the amino acids together is called the peptide bond, or amide bond. Proteins and peptides play very important roles in the structure and function of cells. Carbohydrates are another type of polymer that are found naturally. They are found predominantly in the plant world. They make up more than 50% of the dry weight of the earth's biomass. Chemical energy is stored in the form of carbohydrates. The monomers of carbohydrates have the empirical formula $(CH_2O)_n$, where n represents an integer. The formula for glucose, which is one of the repeating units of carbohydrates, is $(CH_2O)_6$ or $C_6H_{12}O_6$. As a result, carbohydrates are also referred to as hydrates of carbon. A large number of naturally occurring compounds, including cholesterol, fall under a category of compounds called lipids. Lipids are typically not considered to be polymers, but they are fairly large organic molecules that are insoluble in water, but soluble in most organic solvents, such as diethyl ether. Lipids can be divided into three types: (i) triglycerides and waxes; (ii) phospholipids; and (iii) steroids, prostaglandins, and terpenes. Triglycerides, waxes, and phospholipids can be hydrolyzed under acidic or basic conditions to produce smaller molecules, typically carboxylic acids and alcohols. In this chapter, the emphasis will be on a study of the structure, properties, and reactions of the monomers of biopolymers.

20.8 Amino Acids, Monomers of Peptides and Proteins

Amino acids all have the same basic structural features, which are shown below.

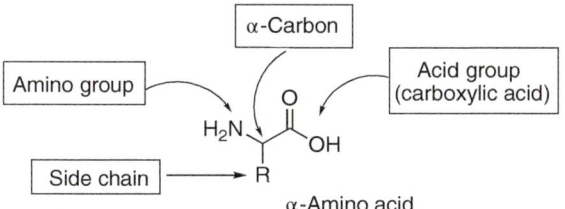

Amino acids with these features are also referred to as α-amino acids because on the α-carbon (the first carbon next to the acid functionality) is an amino functionality. There are 20 amino acids that occur naturally, and they can be classified based on the nature of the side chain. Table 20.2 shows the classification, the structure of the side chains, R and their abbreviations.

With the exception of glycine, all naturally occurring amino acids are optically active, that is, they will all rotate the plane of polarized light; they are also classified as L-configuration. The assignment of L and D for optically active amino acids is based on the system derived from the assignment of L and D for glyceraldehyde. The D and L means that the OH group that is present on the stereogenic carbon is on the right or left, respectively. The (+) and (−) means that these compounds rotate the plane of polarized light to the right or to the left, respectively. The terms dextrorotatory and levorotatory are also used to represent (+) and (−), respectively.

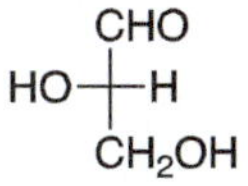

```
    CHO                    CHO
HO─┼─H                 H─┼─OH
   CH₂OH                  CH₂OH
L-(−) Glyceraldehyde   D-(+) Glyceraldehyde
[α]_D = −8.7           [α]_D = +8.7
```

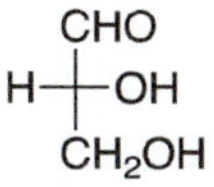

For amino acids, if the amino acid is drawn in the Fischer projection with the carboxylic acid group at the top and the R group is at the bottom, then the L and D assignment is based on the

Table 20.2 Side-chain groups of amino acids that occur naturally.

Classification	R	Amino acid	Abbreviations
Alkyl side chain	—H	Glycine	Gly
	—CH_3	Alanine	Ala
	—$CH(CH_3)_2$	Valine	Val
	—$CH_2CH(CH_3)CH_s$	Leucine	Leu
	—$CH_2CH_2CH(CH_3)$	Isoleucine	Ile
HO side chain	—$CH_2CH(OH)CH_3$	Threonine	Thr
	—CH_2OH	Serine	Ser
Sulfur side chain	—CH_2CH_2SH	Cysteine	Cys
	—$CH_2CH_2SCH_3$	Methionine	Met
Aromatic side chain	p-HO-$C_6H_4CH_2$	Tyrosine	Tyr
	$C_6H_5CH_2$	Phenylalanine	Phe
	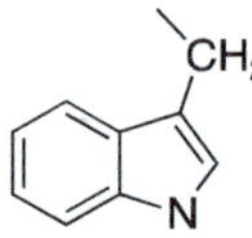 (indole-CH_2)	Tryptophan	Trp
	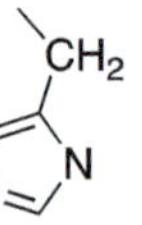 (imidazole-CH_2)	Histidine	His
Acidic side chain	—CH_2COOH	Aspartic acid	Asp
	—CH_2CH_2COOH	Glutamic acid	Glu
Basic side chain	CH_2CONH_2	Asparagine	Asn
	$CH_2CH_2CONH_2$	Glutamine	Glu
	$(CH_2)_4NH_2$	Lysine	Lys
	$CH_2CH_2CH_2NH(NH)NH_2$	Arginine	Arg
Amino group part of ring	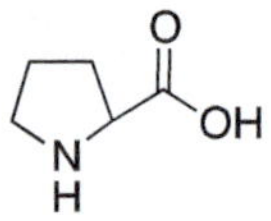 (proline structure)	Proline	Pro

position of the amino group (NH_2), similar to that of the —OH group in glyceraldehyde. If the amino group is to the left, then the assignment is the L; and if the amino group is to the right, the assignment is the D.

```
      COOH              COOH
H₂N──┼──H           H──┼──NH₂
       R                  R
  L-Amino acid       D-Amino acid
```

It should be pointed out that the absolute R and S configurations discussed in Chapter 5 are determined by a different method and hence there is not necessarily a direct correlation between the R and S, and the L and D assignments of these compounds. Shown below is the absolute configuration of the amino acids shown above (the assumption is made that the R group is the third priority).

```
        COOH                COOH
H₂N ──┼── H            H ──┼── NH₂
        R                    R
   S-Amino acid         R-Amino acid
```

Problem 20.3

Determine the absolute configuration (R or S) of the following α-amino acids.

(a), (b), (c) [structures of phenylalanine, alanine, and cysteine with H₂N and COOH groups]

20.9 Acid–Base Properties of Amino Acids

As is obvious from the structure of amino acids, they are both acidic and basic molecules – they contain a carboxylic acid functionality and an amine functionality. In solutions of different pH, amino acids exist in different forms, as shown in Reaction (20-15).

$$H_2N-CH(R)-COO^- \xrightarrow{H^+} H_2N-CH(R)-COOH \rightleftharpoons H_3N^+-CH(R)-COO^- \xrightarrow{H^+} H_3N^+-CH(R)-COOH$$

High pH (Basic solution)   Neutral pH   Low pH (acidic solution)

(20-15)

Problem 20.4

a) Give the structure of (R)-alanine at a pH of 10 (basic solution).
b) Give the structure of valine at a pH of 1 (acidic solution).

Note that in solutions of low or high pH, the actual amino acid carries a net charge, i.e. at low pH, the charge is positive, and at high pH, the net charge is negative. For each amino acid, there is a pH where the net charge will be zero. This pH is close to neutral, but not necessarily pH 7. This pH is also referred to as the pI, the isoelectric pH. As we know from physics, like charges will repel each other and unlike charges will be attracted to each other. If amino acids are placed in a solution of specified pH and electrodes are immersed in the solution, migration of the charged amino acids will occur. At low pH, the protonated amino acid (positively charged amino acid) will migrate toward the negative electrode, and at high pH, the anionic amino acid will migrate to the cathode. This process is known as electrophoresis.

Problem 20.5

To what electrode (anode or cathode) will leucine go in a solution of pH 11.0? Explain your answer.

20.10 Synthesis of α-Amino Acids

20.10.1 Synthesis of α-Amino Acids Using the Strecker Synthesis

A method to synthesize amino acids was developed by a German chemist, Adolph Strecker, and involves an addition reaction to an aldehyde, followed by a hydrolysis reaction. The general synthetic route is shown in Reaction 20-16.

$$O=CH \atop R \quad \xrightarrow[-H_2O]{NH_3/H^+} \quad H_2\overset{+}{N}=CH \atop R \quad \xrightarrow{KCN} \quad H_2N-\underset{R}{\overset{H}{C}}-C\equiv N \quad \xrightarrow[Heat]{H^+, H_2O} \quad H_3\overset{+}{N}-\underset{R}{C}-\underset{OH}{\overset{O}{C}} \quad (20\text{-}16)$$

Aldehyde   Iminium salt   Protonated amino acid

We have seen the type of reactions used in this synthesis throughout our course. The first reaction is an addition–elimination reaction that involves the addition of the nucleophilic ammonia to the electrophilic carbon of an aldehyde, followed by the elimination of water to form an iminium salt. In the next step of the reaction sequence, which is an addition reaction, the nucleophilic nitrile is added to the iminium carbon and the proton is added to the electronegative oxygen to form an amino nitrile. In the last step of the reaction sequence, the nitrile is hydrolyzed under acidic conditions to form the protonated amino acid. Note this sequence of reactions is not stereospecific, and a racemic mixture of the amino acid results.

Problem 20.6

Show how to make the following amino acids using the Strecker synthesis.

(a) phenylalanine structure   (b) alanine structure   (c) valine structure

20.10.2 Synthesis of α-Amino Acids Using Reductive Amination

The synthesis of amino acids using reductive amination involves the reduction of an imine carboxylic acid to form the amino acid, as shown in Reaction (20-17).

$$\underset{R}{\overset{O}{C}}-\underset{OH}{\overset{O}{C}} \quad \xrightarrow[-H_2O]{NH_3} \quad \underset{R}{\overset{HN}{C}}-\underset{OH}{\overset{O}{C}} \quad \xrightarrow{H_2,\ Pt} \quad \underset{R}{\overset{H_2N}{C}}-\underset{OH}{\overset{O}{C}} \quad (20\text{-}17)$$

α-Keto carboxylic acid   α-Imino carboxylic acid   α-Amino acid

The first step of the reaction involves an addition-elimination reaction, in which ammonia adds, with the accompanying loss of water, to the keto functionality of the α-keto carboxylic acid starting compound. α-Keto carboxylic acids, such as pyruvic acids, play an important role in biology. The second step involves a reduction reaction, in which hydrogen gas reduces the imine to the amine.

Problem 20.7

Starting with any α-keto carboxylic acid, show how to make the following amino acids using reductive amination.

(a) phenylalanine structure   (b) alanine structure   (c) valine structure

20.10.3 Synthesis of α-Amino Acids Using Hell Volhard Zelinsky Reaction

This reaction utilizes the reactivity of a hydrogen on the α-carbon of a carboxylic acid. The α-hydrogen is first changed into a good leaving group, in this case, a bromide to form a α-bromocarboxylic acid. The reaction of an α-bromocarboxylic acid with a nucleophilic ammonia results in a substitution reaction in which the ammonia replaces the bromide to result in an α-amino acid as shown in Reaction (20-18).

$$\underset{\substack{\text{Carboxylic acid that} \\ \text{has an α-hydrogen}}}{\text{H-}\underset{\substack{| \\ R}}{\overset{H}{\underset{|}{C}}}\text{-}\overset{O}{\underset{}{C}}\text{-OH}} \quad \xrightarrow[\text{(2) H}_2\text{O}]{\text{(1) Br}_2\text{, PBr}_3} \quad \underset{\text{α-Bromocarboxylic acid}}{\text{Br-}\underset{\substack{| \\ R}}{\overset{H}{\underset{|}{C}}}\text{-}\overset{O}{\underset{}{C}}\text{-OH}} \quad \xrightarrow{\text{NH}_3} \quad \underset{\text{α-Amino acid}}{\text{H}_2\text{N-}\underset{\substack{| \\ R}}{\overset{H}{\underset{|}{C}}}\text{-}\overset{O}{\underset{}{C}}\text{-OH}} \quad (20\text{-}18)$$

Problem 20.8

Show how to make the following amino acids using the Hell–Volhard–Zelinsky Reaction.

(a) phenylalanine (b) alanine (c) valine

20.10.4 Synthesis of α-Amino Acids Using the Gabriel Malolic Ester Synthesis

This synthesis takes advantage of introducing an amino group as a primary amine using phthalimide into a molecule that can be converted into an amino acid through a series of reactions as outlined in Reactions (20-19) through (20-21). The first step of the reaction sequence is a Gabriel substitution reaction in which a phthalimide salt reacts with α-bromo-substituted malonic ester by a substitution reaction to form a phthalimide-substituted ester.

Potassium phthalamide + α-Bromo substituted maloic ester → Phthalimide substituted malonic ester + KBr   (20-19)

In the next step, the phthalimide substituted malonic ester reacts with hydrazine to liberate the amino substituted malonic ester as shown in Reaction (20-20).

Phthalimide substituted malonic ester $\xrightarrow{\text{NH}_2\text{-NH}_2\text{, heat}}$ 2,3-Dihydro phthalazine-1,4-dione + Amino substituted malonic ester   (20-20)

In the last step of the synthesis, the amino substituted malonic ester is hydrolyzed to liberate the amino acid as shown in Reaction (20-21).

$$\underset{}{\text{H}_2\text{N}\underset{\substack{| \\ \text{CO}_2\text{Et}}}{\overset{\text{CO}_2\text{Et}}{\underset{|}{-}}}\text{R}} \xrightarrow[\text{Heat}]{\text{H}^+\text{, H}_2\text{O}} \underset{\text{α-Amino acid}}{\text{H}_3\overset{+}{\text{N}}\text{-CH-}\underset{\substack{| \\ R}}{\overset{O}{\underset{}{C}}}\text{-OH}} + \text{EtOH} + \text{CO}_2 \quad (20\text{-}21)$$

Problem 20.9

Show how to make the following amino acids using the Gabriel malonic ester synthesis reaction.

(a) 2-methylphenylalanine (H₂N-CH(CH₂-C₆H₄-CH₃)-COOH)

(b) cyclopentylglycine (H₂N-CH(cyclopentyl)-COOH)

(c) valine (H₂N-CH(CH(CH₃)₂)-COOH)

20.11 Reactions of α-Amino Acids

Amino acids are bifunctional molecules in that they have both an amine functionality and a carboxylic acid functionality. As we have discussed in Chapter 15, carboxylic acids will undergo substitution reactions at acyl carbon, but the OH is a poor leaving group, so these types of substitution reactions are typically acid catalyzed; nonetheless, a carboxylic acid functionality is a reactive functionality. The amine functionality is basic and hence will undergo an acid–base reaction in addition to a nucleophilic substitution reaction. As we have seen for the reactions that involve polyfunctional molecules, protection of functionalities typically has to take place to prevent unwanted reactions. This strategy is used for most of the reactions of amino acids. Consider the reaction given in Reaction (20-22), in which it is desired to have the nucleophilic Grignard reagent react with the ester functionality of proline ester to produce the tertiary alcohol shown in the box.

Ester of proline + (1) PhMgBr, (2) H⁺, H₂O → Target molecule (tertiary alcohol with two phenyl groups on proline) + side product (N-acylated proline ester) (20-22)

For Reaction (20-22), it is expected that there will be a substitution reaction at the acyl group in which the nucleophilic Grignard substitutes the methoxy (–OCH₃) group to produce the target molecule. The challenge is that another reaction can take place in which the anime nitrogen acts as the nucleophile to give a different molecule as shown. Thus, in order to achieve the goal of the synthesis of the target molecule, the use of a protecting group must be considered to protect the amine functionality.

20.11.1 Protection–Deprotection of the Amino Functionality

The reaction that is typically used to protect the amino functionality is shown in Reaction (20-23).

Proline methyl ester + Di-*tert*-butyldicarbonate (Boc anhydride) → Protected amine (Boc-proline methyl ester) + CO_2 + isobutylene (20-23)

In the first step of the reaction mechanism, the nucleophilic amine attacks the electrophilic carbonyl carbon of di-*tert*-butyldicarbonate as shown in Reaction (20-24).

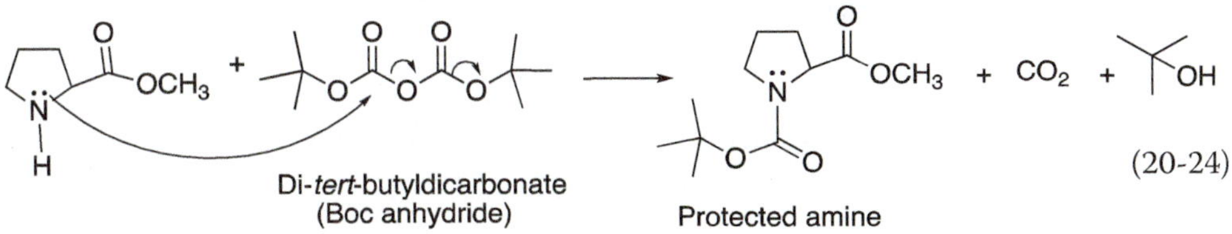

[Reaction 20-24: Proline methyl ester + Di-*tert*-butyldicarbonate (Boc anhydride) → Protected amine + CO_2 + *tert*-butanol]

(20-24)

A driving force for this reaction to occur as you can imagine is the release of carbon dioxide gas. In addition, *tert*-butanol readily loses water to form isobutene, as shown in Reaction (20-25).

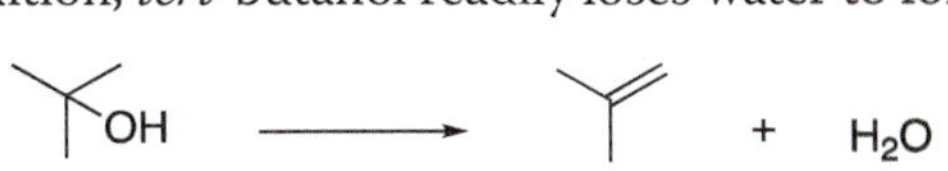

[Reaction 20-25: *tert*-butanol → isobutene + H_2O]

(20-25)

With the protected amino acid ester in hand, the desired reaction can be carried out as shown in Reaction (20-26) to obtain the salt of the desired product.

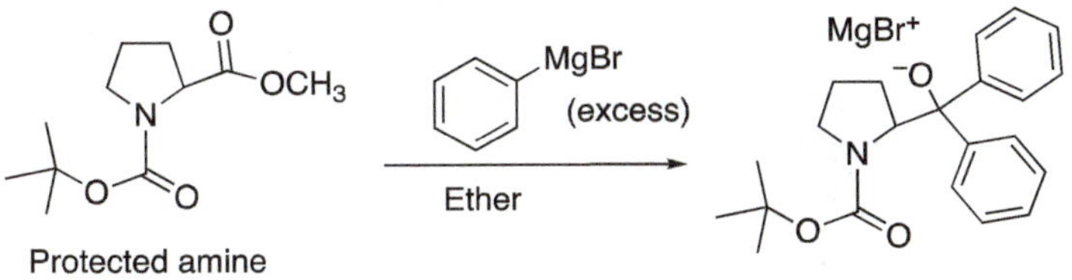

[Reaction 20-26: Protected amine + PhMgBr (excess) in Ether → Grignard adduct with $MgBr^+$ salt]

(20-26)

Deprotection of the amine functionality to give the target molecule can be accomplished as shown in Reaction (20-27).

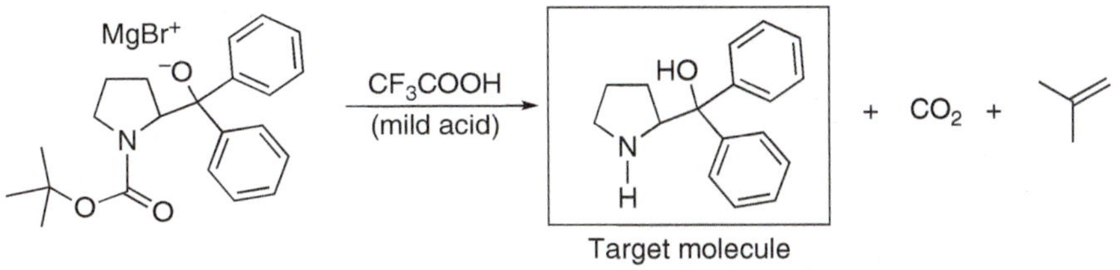

[Reaction 20-27: Grignard salt + CF_3COOH (mild acid) → Target molecule + CO_2 + isobutene]

(20-27)

Since the last step in which the removal of the protecting boc group requires an aqueous acidic solution, it is not necessary to introduce an additional step to convert the Grignard salt to a protected alcohol before removal of the protecting group.

20.11.2 Reactions of the Carboxylic Acid Functionality

Sometimes it will be necessary to protect the carboxylic acid functionality in order to carry out a reaction at the amine functionality of an amino acid. Typical protection includes converting the carboxylic acid group to an ester. Reagents such as thionyl chloride and methanol or trimethylchlorosilane (TMSCl) with methanol have been successfully used and are shown in Reactions (20-28) and (20-29).

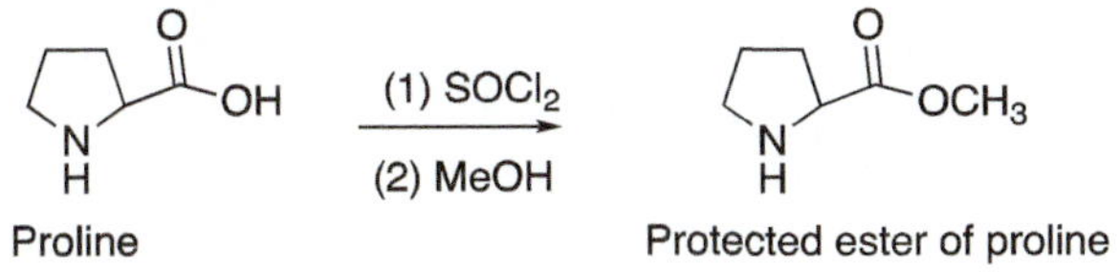

[Reaction 20-28: Proline + (1) $SOCl_2$ (2) MeOH → Protected ester of proline]

(20-28)

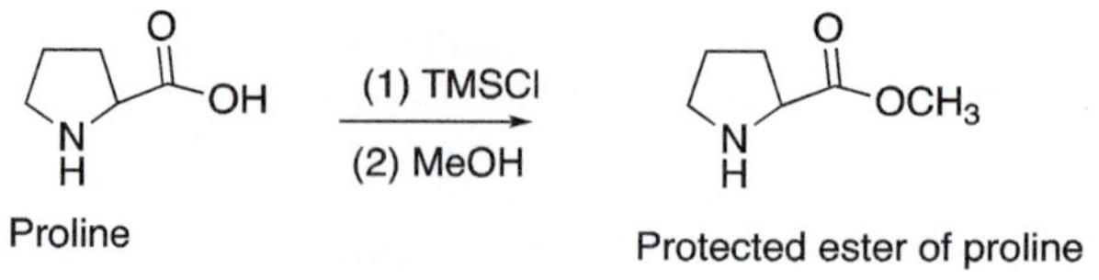

20 Synthetic Polymers and Biopolymers

$$\text{Proline} \xrightarrow[\text{(2) MeOH}]{\text{(1) TMSCl}} \text{Protected ester of proline} \tag{20-29}$$

Deprotection of the ester functionality can take place in the presence of acid hydrolysis of the ester group as shown in Reaction (20-30)

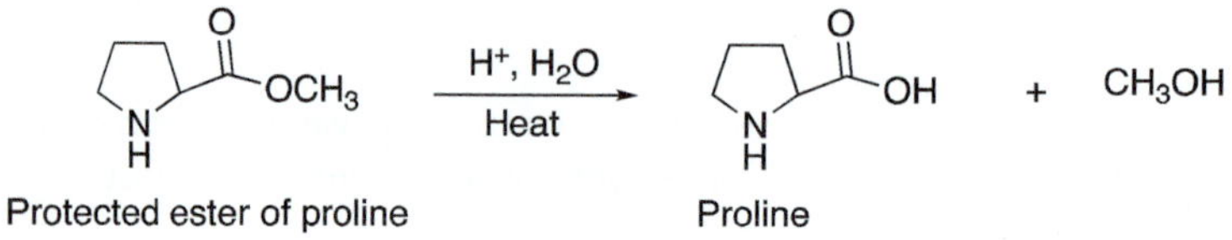

$$\text{Protected ester of proline} \xrightarrow[\text{Heat}]{\text{H}^+, \text{H}_2\text{O}} \text{Proline} + \text{CH}_3\text{OH} \tag{20-30}$$

Problem 20.10

Show how to carry out the following transformation.

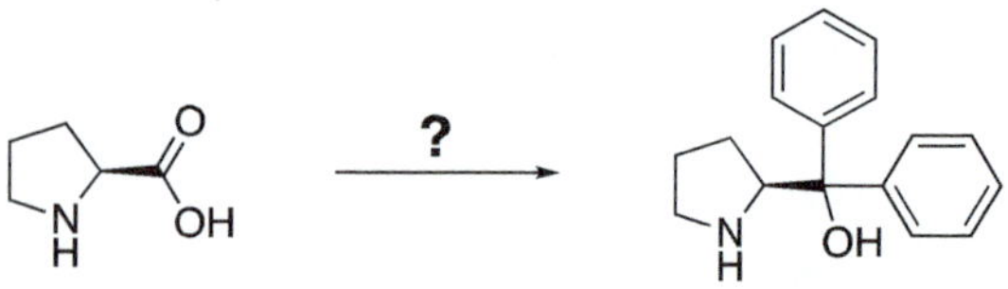

20.11.3 Reaction of α-Amino Acids to Form Dipeptides

Amino acids are bonded together by a special bond, the peptide bond, to form peptides and proteins. The general representation of the peptide bond is shown below.

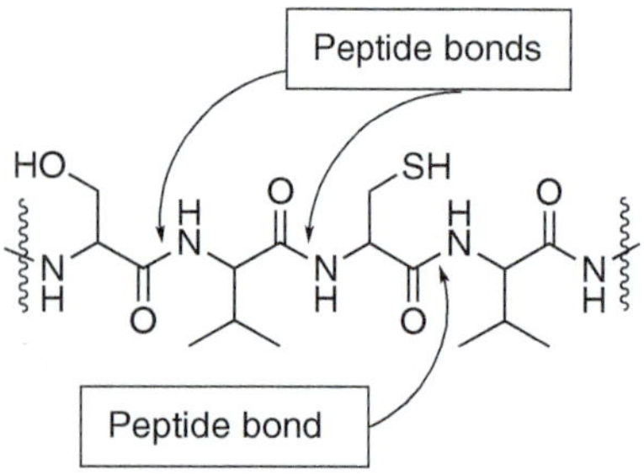

A short section of a peptide

If the desire is to synthesize a specific dipeptide by mixing two amino acids, no reaction will occur. Amino acids do not automatically react with each other to form peptides. Furthermore, if two amino acids were to react to form a dipeptide, there are two possible dipeptides, as shown in Reaction (20-31).

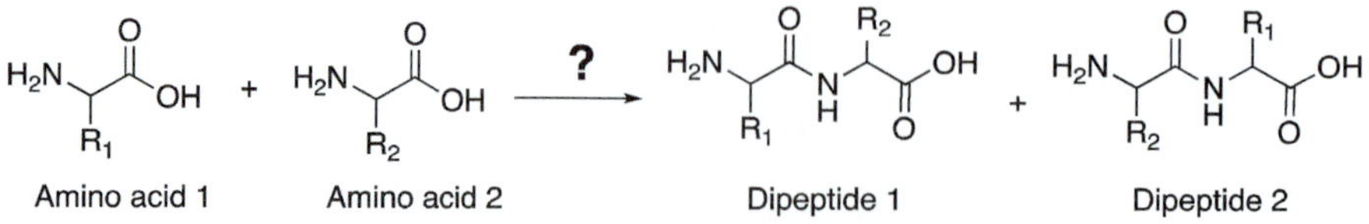

$$\text{Amino acid 1} + \text{Amino acid 2} \xrightarrow{?} \text{Dipeptide 1} + \text{Dipeptide 2} \tag{20-31}$$

20.11 Reactions of α-Amino Acids

Thus, there are a number of considerations that must be kept in mind when synthesizing peptides from amino acids. First, how to get the desired amino and carboxylic acid functionalities to react with each other to get the desired peptide. Another consideration is the type of reactions that should be considered to achieve the target peptide. As we have seen from the previous section, protecting groups must be employed in order to achieve desired reactions. Let us look at a strategy for the synthesis of dipeptide 1 in Reaction (20-32). First, the protection of the amino functionality of amino acid 1 must be accomplished. For amino acid 2, protection of the carboxylic acid functionality must also occur as illustrated in Reaction (20-32) where the circles indicate desired protection of the functional groups shown.

$$\text{Protected amino acid 1} + \text{Protected amino acid 2} \xrightarrow{?} \text{Protected dipeptide} \tag{20-32}$$

Once the coupling of both protected amino acids occurs to form a new peptide bond and the protected dipeptide is formed, the protections can be removed to liberate the desired dipeptide.

Let us first concentrate on the protection of the amino functionality and the protection group that is typically used is the boc protection, and this reaction is shown Reaction (20-33).

$$\text{Amino aicd} \xrightarrow{\text{Boc anhydride}} \text{Boc protected amino acid} \tag{20-33}$$

Next, the protection of the carboxylic acid functionality of the other amino acid is required, as shown in Reaction (20-34).

$$\text{Amino acid 2} \xrightarrow[\text{(2) CH}_3\text{OH}]{\text{(1) SOCl}_2} \text{Protected amino acid 2} \tag{20-34}$$

The next step in the sequence of reactions for peptide synthesis is the coupling of the two protected amino acids. For the coupling of the protected amino acids, a special compound dicyclohexylcarbodimide (DCC) is used to assist with the coupling and the formation of the peptide bond. First, the boc-protected amino acid is mixed with DCC to form a complex as shown in Reaction (20-35).

$$\text{Boc protected amino acid} + \text{DCC} \longrightarrow \tag{20-35}$$

In the next step of the reaction sequence, the protected amino acid 2 is added, which reacts with the complex formed in the reaction above as shown in Reaction (20-36).

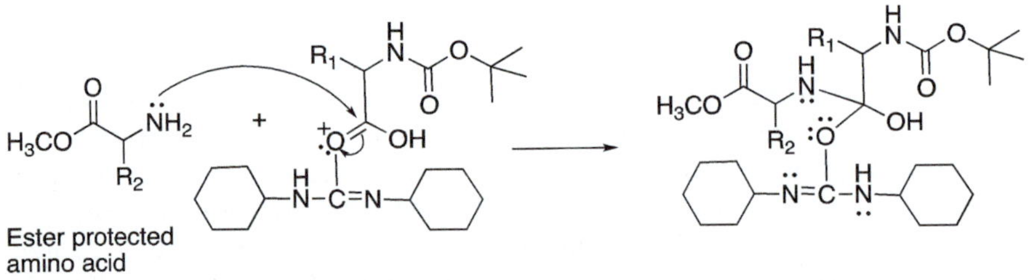

In the next step of the reaction, the protected dipeptide is liberated, along with 1,3-dicyclohexylurea as shown in Reaction (20-37)

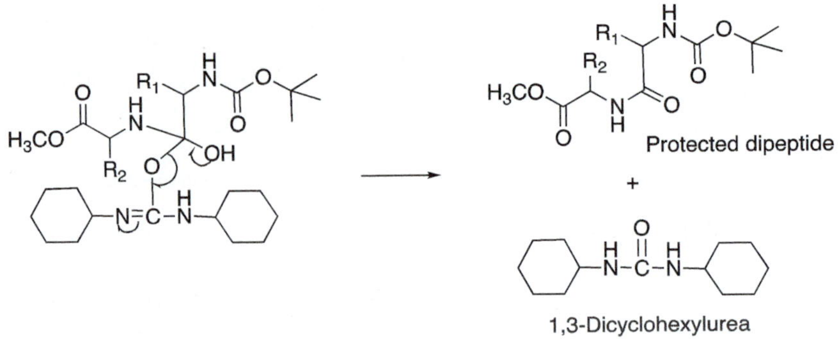

$$\text{(20-37)}$$

The last step in this sequence of reactions is the deprotection of the amine and carboxylic acid functionalities, which is shown in Reaction (20-38).

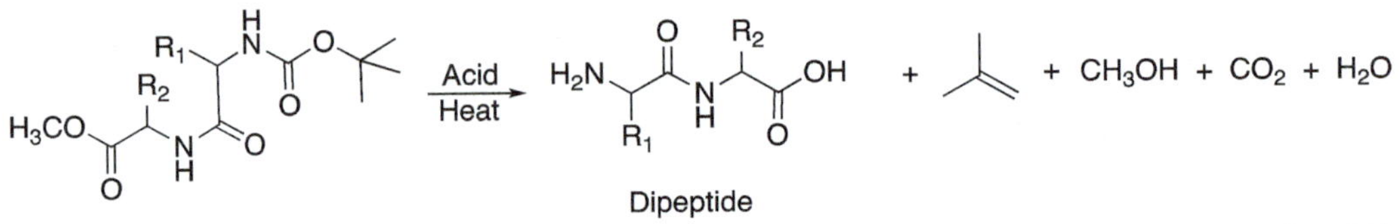

$$\text{(20-38)}$$

Note that in order to synthesize peptide 2 as shown in Reaction (20-31), the protection of the starting amino acids would have to be the reverse.

Problem 20.11

Show how to synthesize the dipeptide shown below.

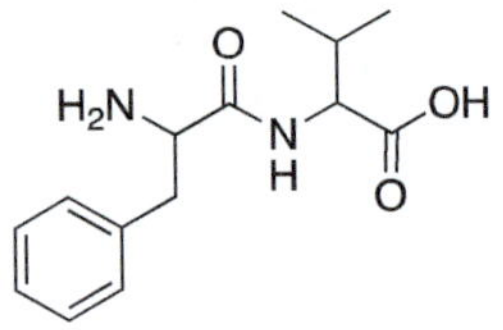

20.11.4 Reaction of α-Amino Acids With Ninhydrin

If an unknown substance is mixed with ninhydrin and a violet color appears, which is also known as Ruhemann's purple, this observation is a strong indication that the unknown substance could be an amino acid owing to the reaction of amino acids with ninhydrin, which is shown in Reaction (20-39).

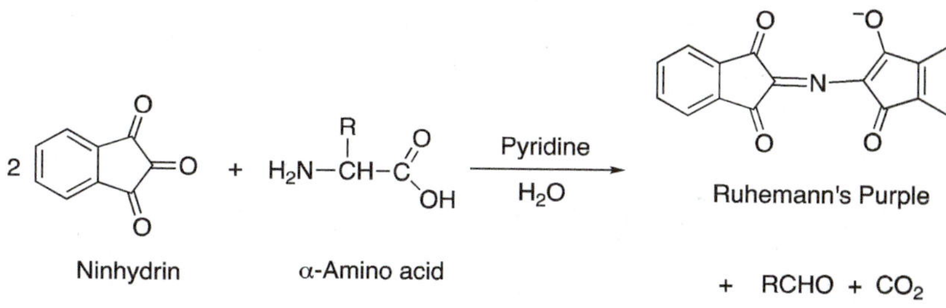

20.11 Reactions of α-Amino Acids

$$2 \text{ Ninhydrin} + \text{H}_2\text{N-CHR-COOH} \xrightarrow[\text{H}_2\text{O}]{\text{Pyridine}} \text{Ruhemann's Purple} + \text{RCHO} + \text{CO}_2 \quad (20\text{-}39)$$

This reaction is typically used to detect the presence of amino acids and is sometimes used in the detection of fingerprints since traces of amino acids are typically present with fingerprints. The first step in the mechanism for the reaction of ninhydrin and an α-amino acid is shown Reaction (20-40).

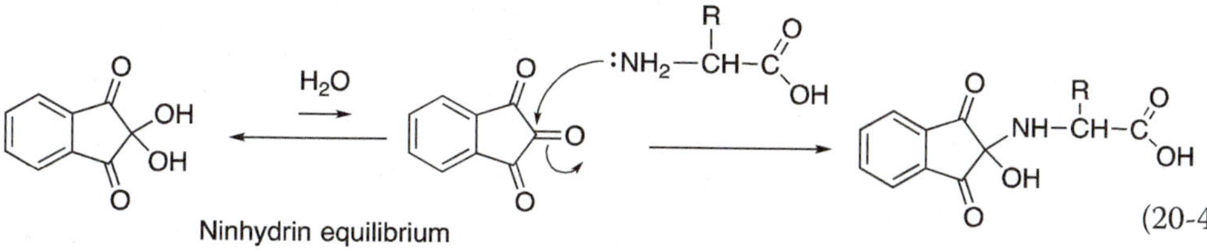

$$\text{Ninhydrin equilibrium} \quad (20\text{-}40)$$

In an aqueous solution, ninhydrin exists as shown in Reaction (20-40) as an equilibrium. Even though the equilibrium lies to the left, a minute amount of the keto form reacts with the nucleophilic amine functionality of amino acids. In the next step of the mechanism, the elimination of water occurs to form the imine, as shown in Reaction (20-41).

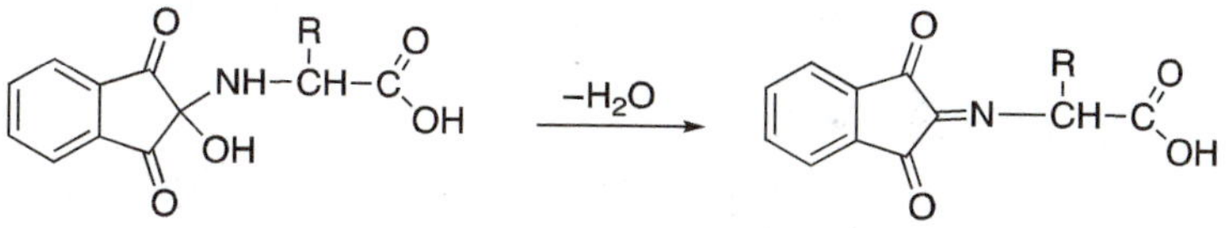

$$(20\text{-}41)$$

In the next step, carbon dioxide is lost and an equilibrium is established as shown in Reaction (20-42)

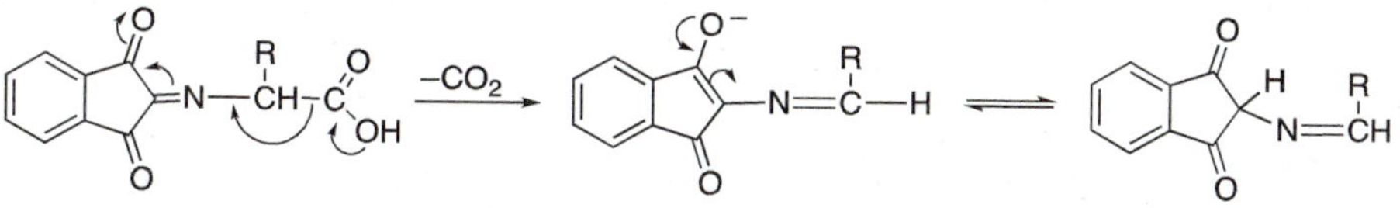

$$(20\text{-}42)$$

In the next step of the reaction, water adds to the electrophilic carbon that is bonded to the nitrogen, followed by the loss of RCHO (an aldehyde) as shown in Reaction (20-43).

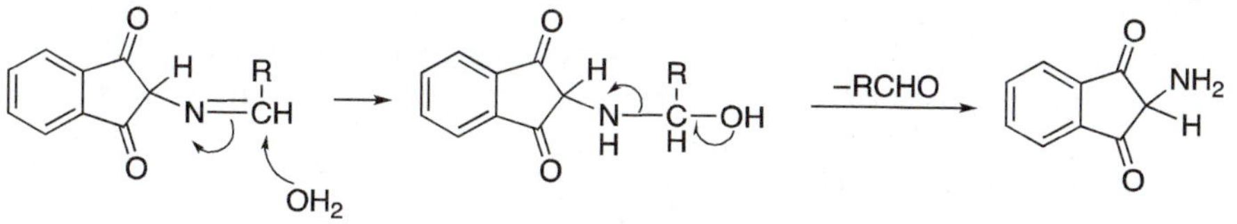

$$(20\text{-}43)$$

In the next step of the reaction mechanism, the nucleophilic nitrogen attacks another mole of ninhydrin as shown in Reaction (20-44).

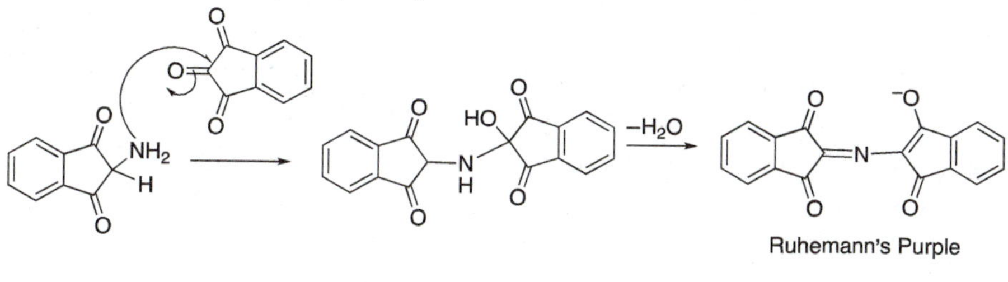

$$(20\text{-}44)$$

Ruhemann's Purple

20.12 Primary Structure and Properties of Peptides

Amino acids are bonded to each other via peptide bonds (also called amide bonds) to form peptides. Shown below is a portion of a peptide that contains four amino acids, in which the peptide bonds are indicated.

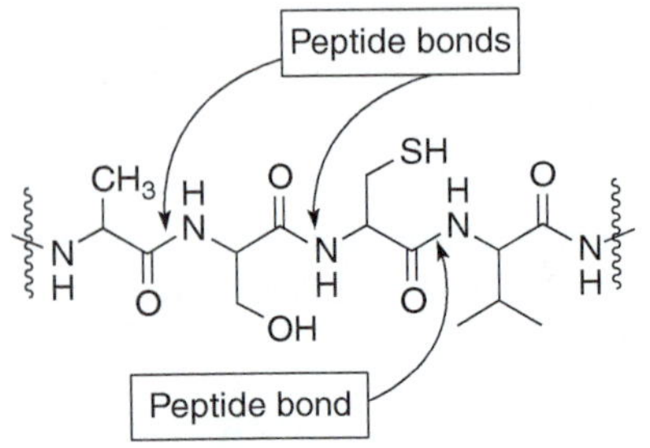

Often times, the three-letter abbreviation is used to indicate different amino acids of the peptide; the dash between the amino acids indicates the peptide bond. Thus, Ala-Cys-Gly represents a tripeptide containing the amino acids: alanine, cysteine, and glycine.

Problem 20.12

With the aid of Table 20.2, give the structure of the tripeptide: Ala-Cys-Gly.

The backbone of proteins consists of repeating units of amino acids. To determine the individual acid residues, two tasks must be accomplished: (i) identification of the amino acids that are present and (ii) identification of the sequence of the amino acids of the peptide.

20.12.1 Identification of Amino Acids of Peptides

To determine the type amino acids that are present in a peptide, hydrolysis of the amide bonds can be accomplished. Reaction (20-45) shows the hydrolysis of a dipeptide.

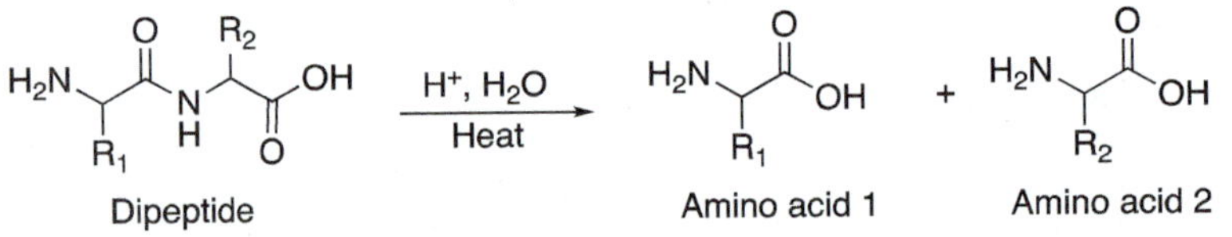

(20-45)

After the hydrolysis, the individual amino acids must be separated and identified. The mixture of amino acids can be separated by column chromatography. Once separated, the identification of the amino acids of the above dipeptide, for example, can be accomplished by spectroscopy as discussed in Chapter 13. The other challenge now becomes the determination of the sequence of the amino of the original peptide and will be covered in the next section.

20.12.2 Identification of the Amino Acid Sequence

Based on the two amino acids determined from the hydrolysis reaction of the peptide shown above, there are two possible peptides that could give the results shown for a hydrolysis reaction.

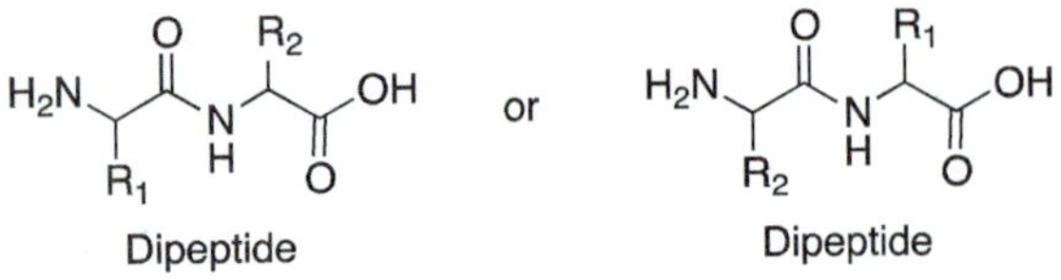

20.12 Primary Structure and Properties of Peptides

The terminal amino acids of a peptide can be determined since they react differently due to the presence of the different functional groups, the amine and carboxylic acid functionalities. The N-terminal has an amino group, while the C-terminal has a carboxylic acid functionality. Frederick Sanger devised a method to detect the N-terminal amino acid of peptides. The reaction with peptide 1 is shown in Reaction (20-46).

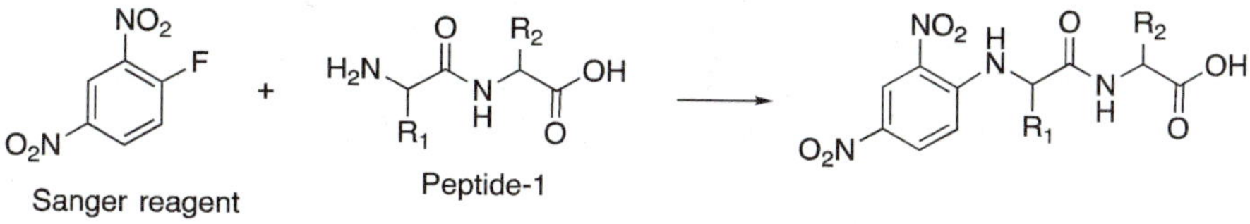

$$\text{Sanger reagent} + \text{Peptide-1} \longrightarrow \text{product} \quad (20\text{-}46)$$

The hydrolysis of the product of Reaction (20-46) is shown below in Reaction (20-47).

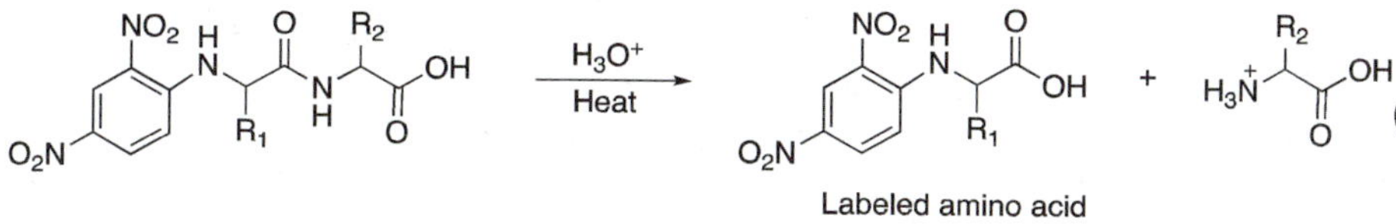

$$\xrightarrow{H_3O^+,\ \text{Heat}} \text{Labeled amino acid} + \text{amino acid} \quad (20\text{-}47)$$

Whereas the reaction of the Sanger reagent with peptide 2 gives a different product, as shown in Reaction (20-48).

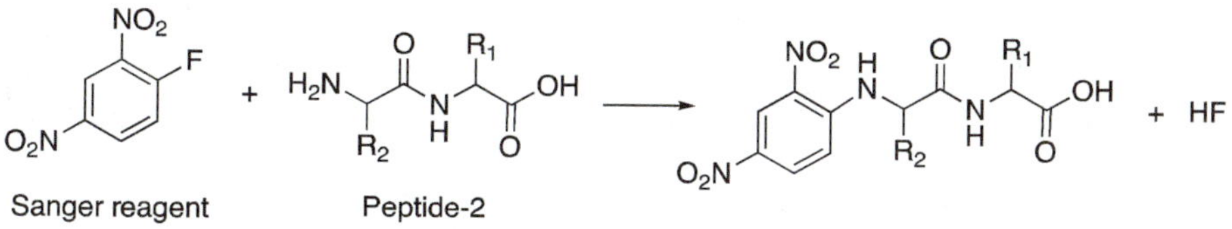

$$\text{Sanger reagent} + \text{Peptide-2} \longrightarrow \text{product} + HF \quad (20\text{-}48)$$

The hydrolysis of the product of Reaction (20-48) is shown in Reaction (20-49).

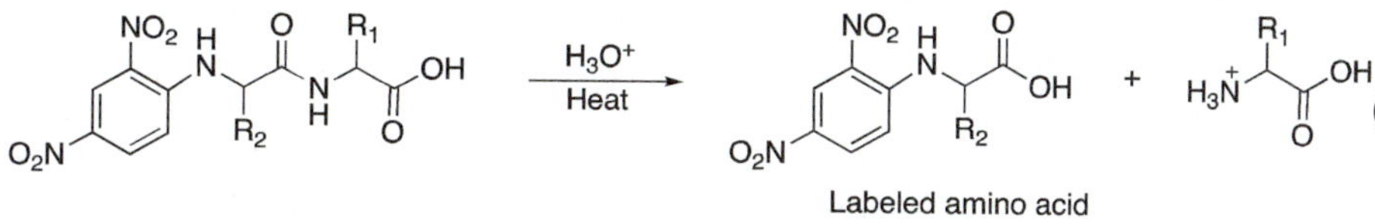

$$\xrightarrow{H_3O^+,\ \text{Heat}} \text{Labeled amino acid} + \text{amino acid} \quad (20\text{-}49)$$

The resulting hydrolysis of the peptide that has the labeled dinitrobenzene will give an indication of the N-terminal amino acid.

The Edman's reagent is another reagent that is often used to determine the N-terminal amino acid of a peptide. This reaction is used to determine the terminal amino end of a peptide and was discovered by Pehr Victor Edman, a Swedish biochemist, in 1952. The reaction is shown in Reaction (20-50).

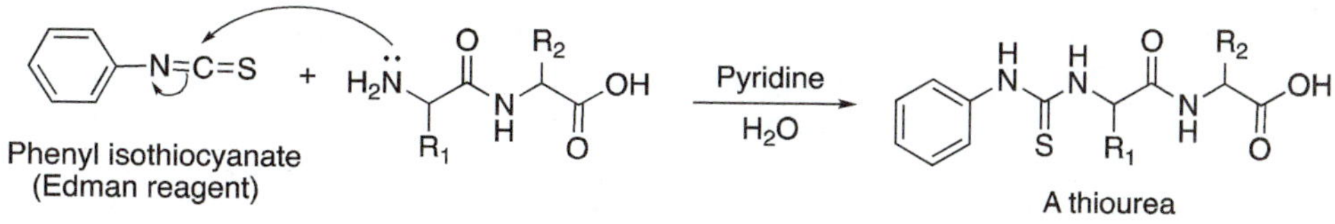

$$\text{Phenyl isothiocyanate (Edman reagent)} + \text{peptide} \xrightarrow{\text{Pyridine},\ H_2O} \text{A thiourea} \quad (20\text{-}50)$$

Note that the N-terminal of the dipeptide has a thiourea label attached. The reaction of the product thiourea with trifluoroacetic acid (TFA) and heat results in hydrolysis of the amide bond giving rise to a thiazolinone of the N-terminal amino acid and the other amino acid of the dipeptide as shown in Reaction (20-51).

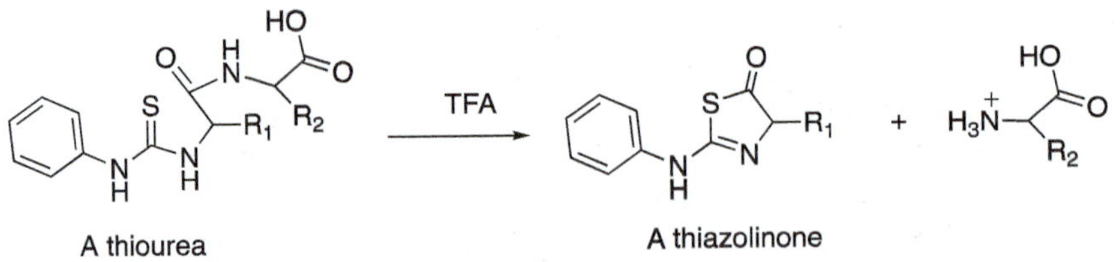

[Reaction scheme 20-51: A thiourea → (TFA) → A thiazolinone + H₃N⁺-CHR₂-COOH]

(20-51)

The thiazolinone is then hydrolyzed in acidic aqueous solution and undergoes a rearrangement to give a phenylthiahydantoin (PTH) as shown in Reaction (20-52).

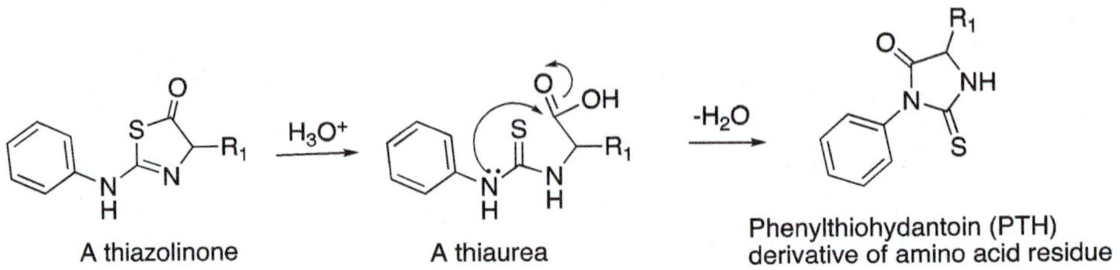

[Reaction scheme 20-52: A thiazolinone → (H₃O⁺) → A thiaurea → (-H₂O) → Phenylthiohydantoin (PTH) derivative of amino acid residue]

(20-52)

Identification of the PHT will indicate the structure of the N-terminal amino acid.

Problem 20.13

Give the structures of the products of the reaction of the tri-peptide from Problem 20.12 with the Edman's Reagent, followed by hydrolysis.

After the identification of the N-terminal amino acid, the other task is to determine the amino acid sequence of the peptide. There are certain chemicals and enzymes that will cleave specific peptide bonds depending on the amino acids that make up the peptide bond. For example, trypsin, which is an intestinal digestive enzyme, specifically hydrolyses polypeptides at the carbonyl end of arginine and lysine. With this method, a segment of a peptide that has these adjacent amino acid residues can be determined. There are other enzymes that will perform similar cleavages, but at peptide bonds that have different amino acid residues. With this information, scientists can determine the actual sequence and type of amino acid in each peptide, no matter how many amino acids that are present in the peptide.

20.13 Secondary Structure of Proteins

Now that we have covered the primary structure of proteins, in this section, we will briefly examine the intermolecular and intramolecular interactions of peptides in proteins. Shown below is the type of hydrogen bond that exists between strands of peptides. These intermolecular interactions make up the secondary structure of proteins. As shown, there are strands of peptides that are hydrogen bonded together to form sheets, hence the term β-sheets are also used to describe this structure of peptides and proteins.

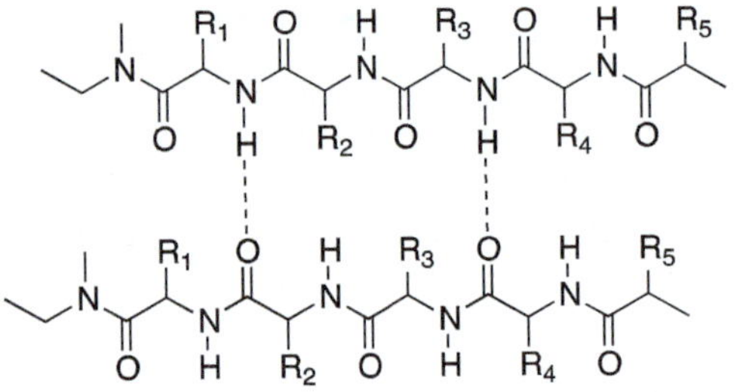

[Structural diagram showing two parallel peptide strands with hydrogen bonding between them, with side chains labeled R_1, R_2, R_3, R_4, R_5]

It is possible to have hydrogen bonds within the same strand of a peptide. Such structures are typically referred to as α-helix.

20.14 Monosaccharides, Monomers of Carbohydrates

Carbohydrates are another type of biopolymers. The monomers of carbohydrates are called monosaccharides, and they are the building blocks of more complex sugars such as disaccharides and polysaccharides. Monosaccharides have different names, and the type of reactions that monosaccharides undergo can be determined by the name of the monosaccharide. If the name of a monosaccharide ends with "OSE," then that monosaccharide is a reducing sugar. That is, present on the molecule is a functionality that has a carbonyl group or an OH group, which can be oxidized and hence serve as a reducing agent. An aldose has an aldehyde present in the molecule, and a ketose has a ketone present. The number of carbons present in a monosaccharide is represented within the type name. For example, if the monosaccharide has six carbons, it is called an aldohexose. Shown below are examples of aldohexoses.

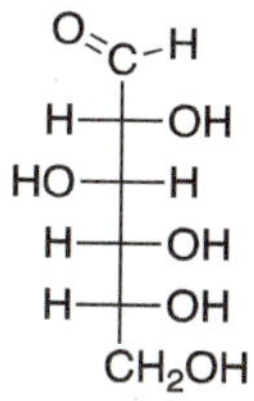

Glucose 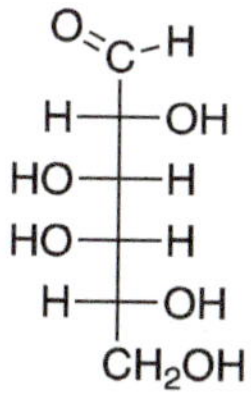   Galactose 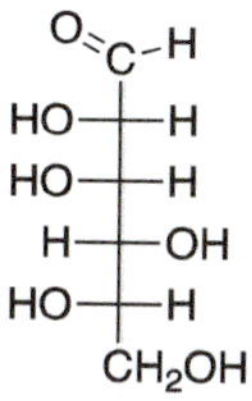   Mannose

From the above monosaccharides, you will notice that they all have more than one stereogenic center, and hence several stereoisomers. It was observed that natural occurring monosaccharides have the same configuration as D-(+) glyceraldehyde, which was discussed in Section 20.8.

Since monosaccharides have more than one stereogenic center, monosaccharides with the stereogenic center farthest from carbon 1 (or the carbon of the aldehyde functionality) is used to assign a D or L configuration. For a particular monosaccharide, if the OH group that is furthest from the aldehyde is on the right, the monosaccharide is assigned a D configuration. On the other hand, if the OH group is on the left, the monosaccharide is assigned the L configuration as shown in the examples below.

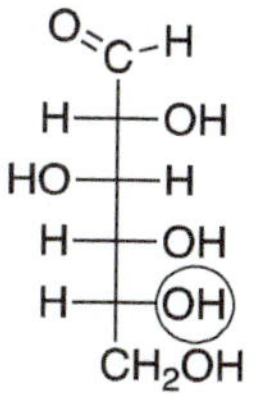

D-Glucose 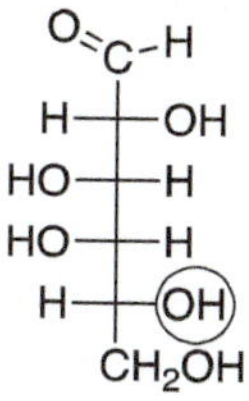   D-Galactose 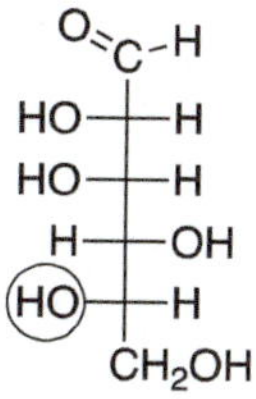   L-Mannose

Epimers are diastereomers that differ in configuration at only one of the chiral carbon atoms. Thus, D-glucose and D-galactose are epimers since at carbon 4 and only at that carbon, there is a difference in stereochemistry.

Problem 20.14

What are the structures of L-glucose, L-galactose, and D-mannose?

20.15 Reactions of Monosaccharides

20.15.1 Hemiacetal Formation Involving Monosaccharides

The reactions that monosaccharides undergo are the same as those of molecules that have the same functionalities, alcohols and aldehyde and were covered in earlier chapters. Hemiacetal formation reactions involve an addition reaction of alcohols to an aldehyde to form hemiacetals, as shown in Reaction (20-53).

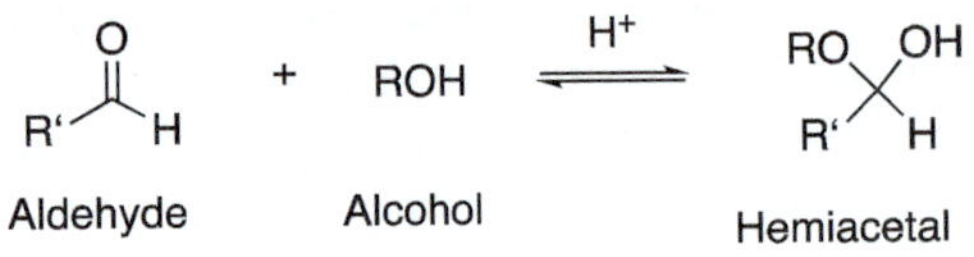

$$\text{Aldehyde} + \text{ROH} \underset{}{\overset{H^+}{\rightleftharpoons}} \text{Hemiacetal} \tag{20-53}$$

Monosaccharides contain both functionalities on the same molecule, and as a result, the possibility exists for intramolecular reactions. Cyclic hemiacetal formation reactions are one of the most important reactions of monosaccharides. An example of an intramolecular hemiacetal formation reaction is given in Reaction (20-54).

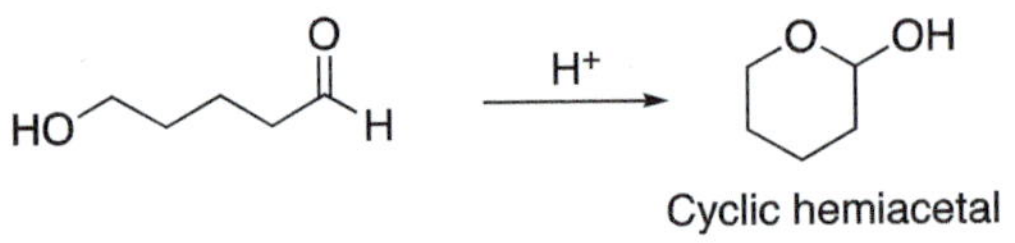

$$\tag{20-54}$$

Problem 20.15

Give the cyclic hemiacetal or hemiketal for each of the compounds shown below.

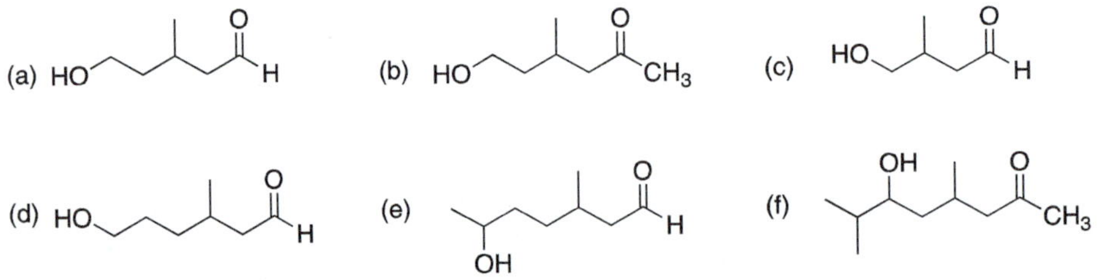

(a), (b), (c), (d), (e), (f)

A similar intramolecular reaction occurs with glucose as shown in Reaction (20-55).

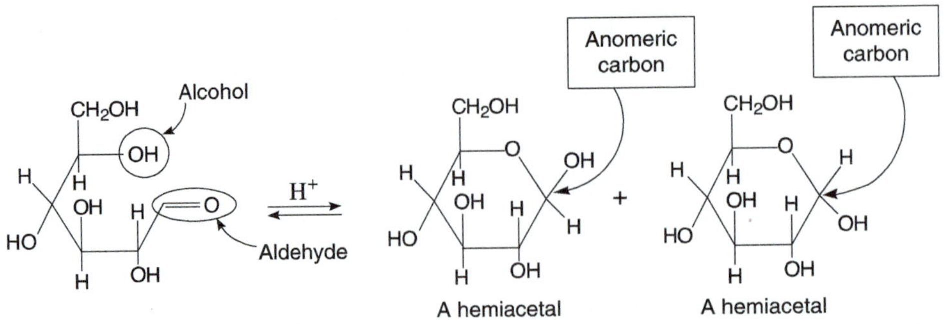

$$\tag{20-55}$$

Note that the two hemiacetal products are stereoisomers and are known as anomers. These two stereoisomers differ from each other in the orientation of the OH; the OH group can be in the equatorial position (up) or in the axial position (down) of this six-member cyclic compound. Anomers are diastereomers that differ in configuration around only one atom, the anomeric atom. The reaction given in Reaction (20-55) can be represented differently in an aqueous solution and is shown in Reaction (20-56) to better illustrate the equatorial and axial groups.

(20-56)

The name of these cyclic hemiacetals is derived from pyran due to the similarity of the structure of both these compounds; both structures are six-member ring compounds, which contain an oxygen atom as part of the ring.

Pyran

The cyclic hemiacetal of glucose is called a glucopyranose. The carbon at which the reaction takes place is called the ***anomeric carbon*** as shown in Reaction (20-55). Both anomers have a root name of D-glucose since the starting compound is D-glucose. The Greek letters α and β are used to differentiate these anomers; if the OH group is down or axial, the α notation is used, and if the OH group is up or equatorial, the β notation is used. Since the hemiacetal reactions are equilibrium reactions, all three species exist in equilibrium in solution; however, one dominates in concentration at equilibrium. The chair conformer of glucopyranose that has the OH group in the equatorial position is the more stable conformation and hence the one that is in abundance, compared to the other isomers, as shown in Reaction (20-57). The equatorial position has more room to accommodate the fairly large OH group, and hence this epimer is more stable, compared to the other epimer, which has the OH in the axial position.

α-D-Glucopyranose (<40%)   D-Glucose (open chain) (~0%)   β-D-Glucopyranose (>60%)

(20-57)

Problem 20.16

Give the most stable product for the hemiacetal formation of galactose.

You may be wondering if there are other possible cyclic hemiacetals that could be formed? The answer is yes; however, the formation of seven-member ring hemiacetal is more difficult, compared to six-member ring hemiacetals. The five-member ring hemiacetal is however fairly stable and easier to form than the seven-member ring due to the proximity of the two reacting groups. The reaction for the formation of the five-member ring hemiacetal is shown in Reaction (20-58).

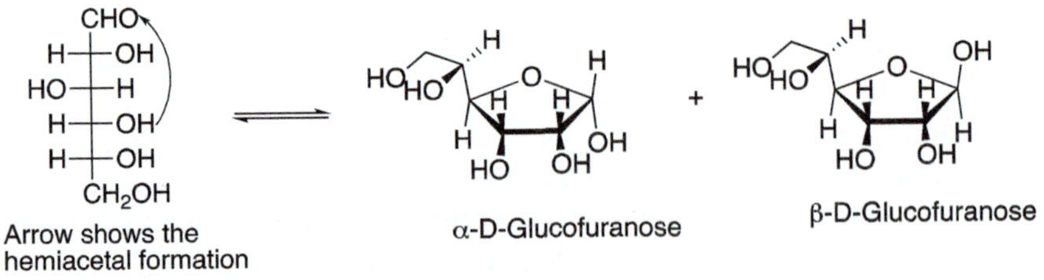

20 Synthetic Polymers and Biopolymers

$$\text{Glucose (Fischer)} \rightleftharpoons \alpha\text{-D-Glucofuranose} + \beta\text{-D-Glucofuranose} \quad (20\text{-}58)$$

Arrow shows the hemiacetal formation

The root name of these five-member ring hemiacetals is derived from furan, which is a five-member cyclic compound which contains four carbons and one oxygen.

Furan

A similar system as that described for the nomenclature of different six-member ring hemiacitals (anomers) of glucose can be used to name the different epimers of the five-member ring hemiacetals of glucose. There are two possible anomers; in this case, however, the anomers are called glucofuranose and specifically α-D-glucofuranose and β-D-glucofura-nose, depending on the orientation of the –OH group as shown in Reaction (20-58).

20.15.2 Base-catalyzed Epimerization of Monosaccharides

Since the α-hydrogen of monosaccharides is acidic, in the presence of a base, that proton can be abstracted to form an enolate anion, as shown below in Reaction (20-59).

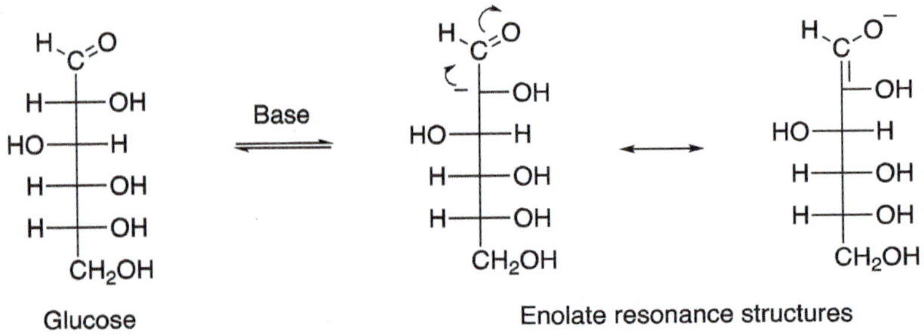

$$\text{Glucose} \xrightarrow{\text{Base}} \text{Enolate resonance structures} \quad (20\text{-}59)$$

Since the enolate anion formed is sp^2 and flat, it can acquire another proton for the solvent, this proton can add to either side to create the opposite configuration of the original conformer as shown in Reaction (20-60).

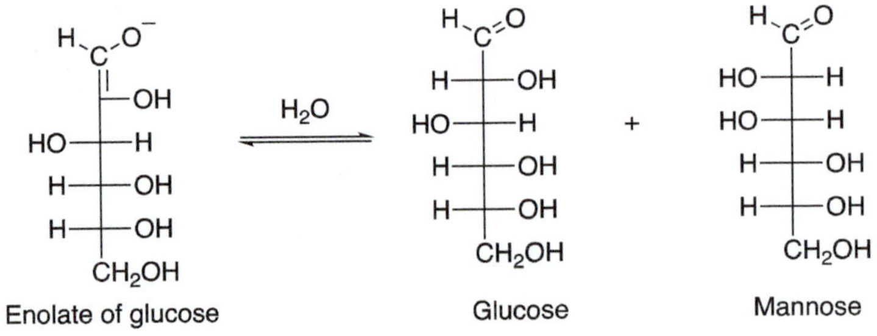

$$\text{Enolate of glucose} \xrightarrow{H_2O} \text{Glucose} + \text{Mannose} \quad (20\text{-}60)$$

As shown in the reaction, there is a mixture of glucose and mannose.

20.15.3 Enediol Rearrangement of Monosaccharides

Another reaction that is possible once the enoate is formed by the removal of the α-hydrogen of glucose is a possible rearrangement to form the keto monosaccharide as shown in Reaction (20-61).

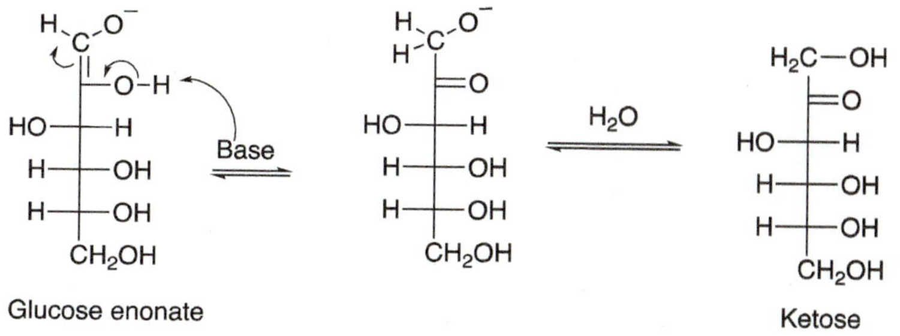

$$\text{Glucose enonate} \xrightleftharpoons{\text{Base}} \xrightleftharpoons{H_2O} \text{Ketose} \tag{20-61}$$

20.15.4 Oxidation of Monosaccharides with Silver Ions

Since monosaccharides have an aldehyde functionality, and as we saw in Chapter 11, aldehydes are easily oxidized to carboxylic acids or carboxylate salts, depending on the reagent used. If the monosaccharide is shown in the hemiacetal form, it is not immediately obvious that it can be oxidized, but in solution, there is an equilibrium between the open chain aldehyde and the hemiacetal forms, and it is the aldehyde form that is oxidized. The Tollen's reagent is often used as a mild oxidizing reagent to test the presence of a monosaccharide of an unknown sample. The presence of a silver mirror is an indication of the presence of the aldehyde functionality as shown in Reaction (20-62).

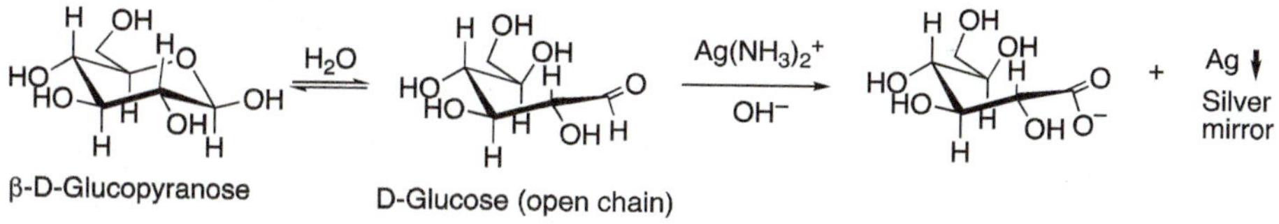

$$\beta\text{-D-Glucopyranose} \xrightleftharpoons{H_2O} \text{D-Glucose (open chain)} \xrightarrow[\text{OH}^-]{Ag(NH_3)_2^+} + \text{Ag} \downarrow \text{Silver mirror} \tag{20-62}$$

Other reagents that are frequently used for the oxidation of monosaccharides include the **Benedict's Reagent** and the **Fehlings Reagent**. Both reagents contain Cu^{2+}, which is blue colored, and its reaction with a monosaccharide gives Cu_2O, which is a red solution. The main difference between the two reagents is that the Benedict's solution has a citric acid complex, whereas the Fehling's reagent is made with a tartaric acid complex, but a color change is an indication of the presence of a monosaccharide that was oxidized.

20.15.5 Oxidation of Monosaccharides with Nitric Acid

Another important oxidation reaction is shown below, in which the terminal hydroxyl group of glucose is oxidized to a carboxylic acid using nitric acid to form glucaric acid as shown in Reaction (20-63).

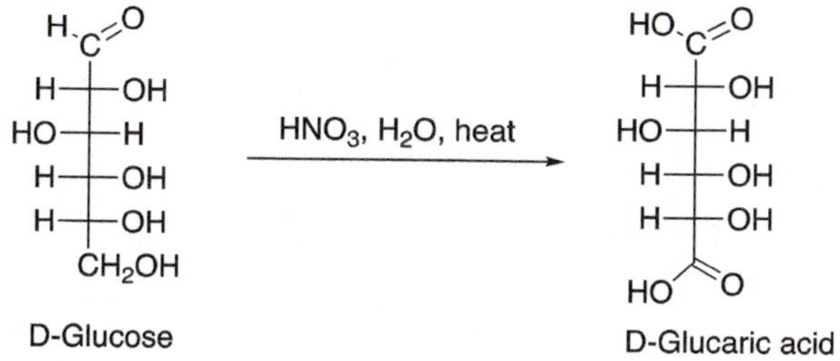

$$\text{D-Glucose} \xrightarrow{HNO_3, H_2O, \text{heat}} \text{D-Glucaric acid} \tag{20-63}$$

D-Glucaric acid, also known as saccharic acid, is a good chelating agent; its salt is used in dishwasher detergents to assist in the chelation of calcium and magnesium ions, which makes detergents more efficient in hard water. As you can imagine, the primary alcohol is much easier to oxidize, compared to the other alcohols, which are secondary.

20.15.6 Oxidation of Monosaccharides with Periodic Acid

As we have seen from Chapter 11 on oxidation reactions, periodic acid is used to oxidize and cleave the carbon–carbon single bonds of compounds that have 1,2-diols functionalities. Examples are given in Reactions (20-64) and (20-65).

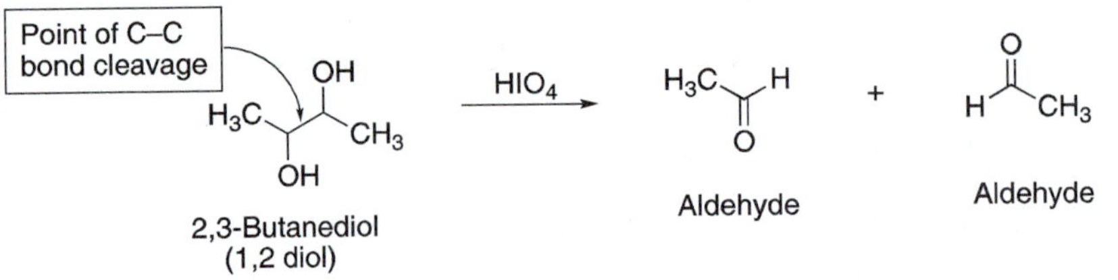

$$\text{2,3-Butanediol (1,2 diol)} \xrightarrow{HIO_4} \text{H}_3\text{C–CHO (Aldehyde)} + \text{H}_3\text{C–CHO (Aldehyde)} \quad (20\text{-}64)$$

Point of C–C bond cleavage shown between the two CHOH carbons.

$$\text{HOCH}_2\text{–(CHOH)}_4\text{–CH}_2\text{OH} \xrightarrow{HIO_4} 2\,\text{HCHO (Formaldehyde)} + 4\,\text{HCOOH (Formic acid)} \quad (20\text{-}65)$$

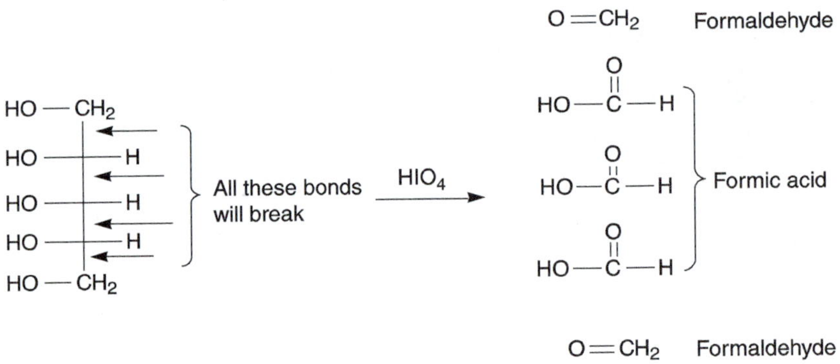

All these bonds will break.

Note the products that are produced from each carbon that contained the hydroxyl functional group. This type of oxidation occurs with not only alcohols that are adjacent to each other but also molecules that contain the two oxidizable functionalities in a 1,2-relationship as shown in Reaction (20-66).

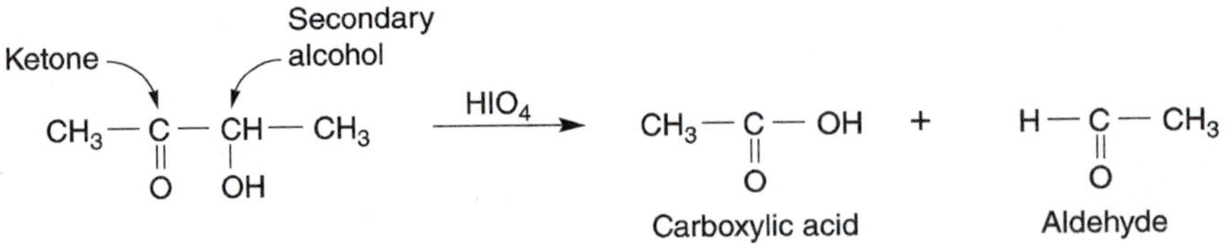

$$\underset{\text{Ketone}}{\text{CH}_3-\underset{\underset{O}{\|}}{C}}-\underset{\underset{\text{Secondary alcohol}}{\text{OH}}}{\text{CH}}-\text{CH}_3 \xrightarrow{HIO_4} \underset{\text{Carboxylic acid}}{\text{CH}_3-\text{COOH}} + \underset{\text{Aldehyde}}{\text{H–CO–CH}_3} \quad (20\text{-}66)$$

Problem 20.17

i) Give the products for the periodic oxidation reaction shown below and predict how many moles of products are formed.

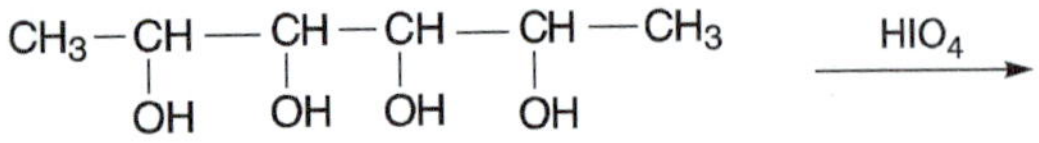

$$\text{CH}_3-\underset{\text{OH}}{\text{CH}}-\underset{\text{OH}}{\text{CH}}-\underset{\text{OH}}{\text{CH}}-\underset{\text{OH}}{\text{CH}}-\text{CH}_3 \xrightarrow{HIO_4}$$

ii) Give the structure of an unknown monosaccharide, if after oxidation, 4 mol of formic acid and 2 mol of formaldehyde are produced.

20.15.7 Reduction of Monosaccharides

The reduction of glucose gives a very important compound that is used as a sugar substitute and the reaction is given in Reaction (20-67).

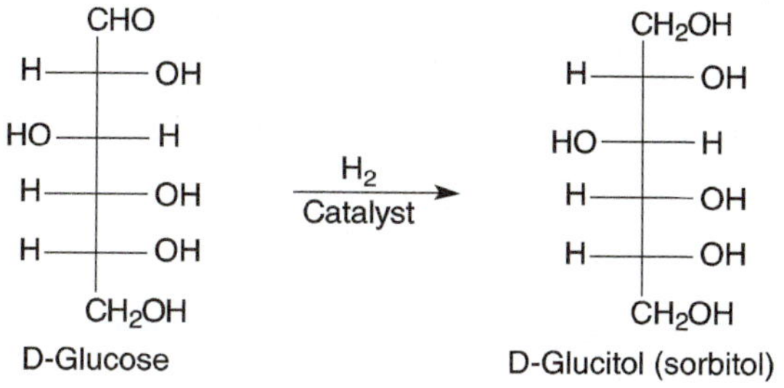

$$\text{D-Glucose} \xrightarrow{H_2, \text{Catalyst}} \text{D-Glucitol (sorbitol)} \tag{20-67}$$

20.15.8 Ester Formation of Monosaccharides

Each monosaccharide has several alcohol functionalities, and as we saw in Chapter 7, alcohols react with carboxylic acid derivatives, such as acid chlorides or acid anhydrides, to form esters as shown in Reaction (20-68), Ac = CH_3CO_2 (acetate).

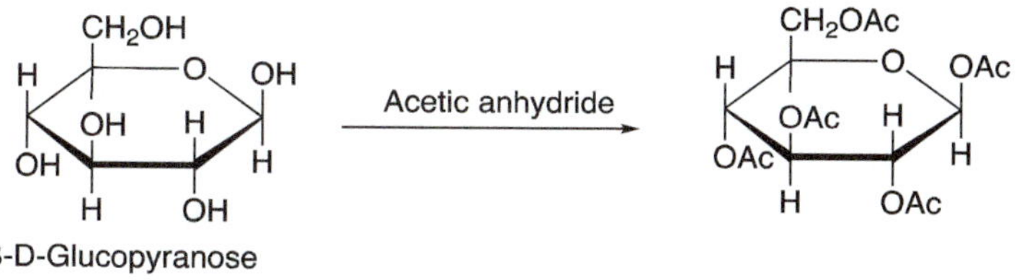

$$\beta\text{-D-Glucopyranose} \xrightarrow{\text{Acetic anhydride}} \text{(peracetylated product)} \tag{20-68}$$

20.15.9 Ether Formation of Monosaccharides

The alcohol groups of monosaccharides can be alkylated with alkyl halides using a substitution reaction that was used for the synthesis of ethers. This type of reaction you will recall involves the use of a base to deprotonate the hydrogen of the alcohol functionality, followed by a substitution reaction using an electrophilic alkyl halide. This reaction involving glucopyranose is shown in Reaction (20-69).

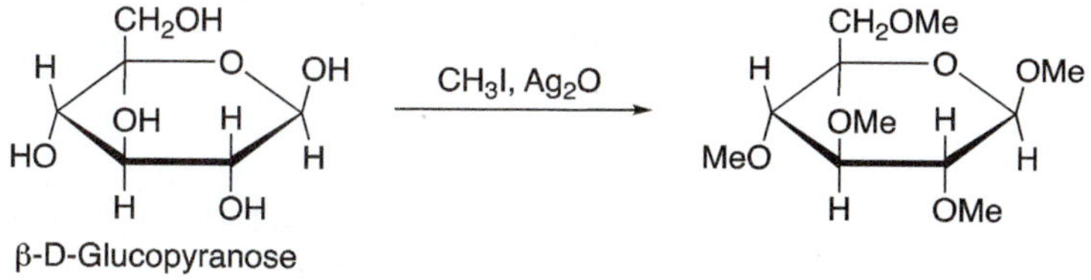

$$\beta\text{-D-Glucopyranose} \xrightarrow{CH_3I, Ag_2O} \text{(permethylated product)} \tag{20-69}$$

20.15.10 Intermolecular Acetal Formation Involving Monosaccharides

As we have seen in Chapter 12, the reaction of an aldehyde with 2 mol of an alcohol gives an acetal after first forming a hemiacetal as shown in Reaction (20-70).

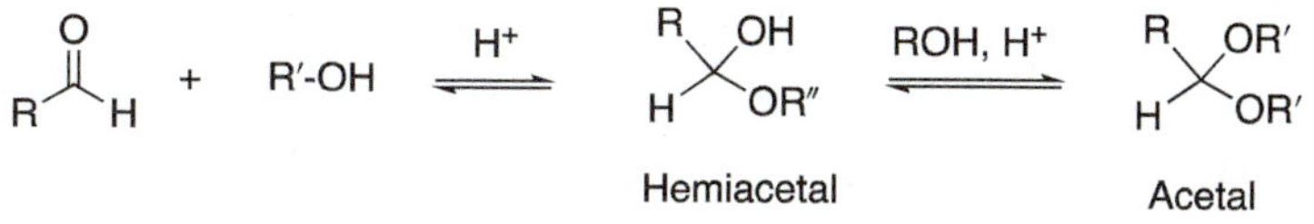

$$RCHO + R'\text{-OH} \xrightleftharpoons{H^+} \underset{\text{Hemiacetal}}{R\text{-CH(OH)(OR'')}} \xrightleftharpoons{ROH, H^+} \underset{\text{Acetal}}{R\text{-CH(OR')}_2} \tag{20-70}$$

The same type of reaction can occur with difunctional molecules. First, an intramolecular reaction would yield the cyclic hemiacetal, followed by the reaction with a mole of an alcohol that would yield an acetal, as shown in the example given in Reaction (20-71).

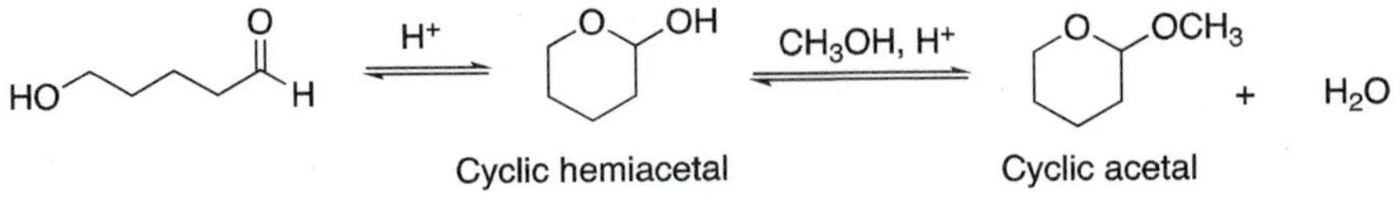

$$\text{HO-CH}_2\text{-CH}_2\text{-CH}_2\text{-CH}_2\text{-CHO} \xrightleftharpoons{H^+} \text{Cyclic hemiacetal} \xrightleftharpoons{CH_3OH, H^+} \text{Cyclic acetal} + H_2O \quad (20\text{-}71)$$

Problem 20.18

Give the acetal that will result from the reaction of methanol with each of the following compounds.

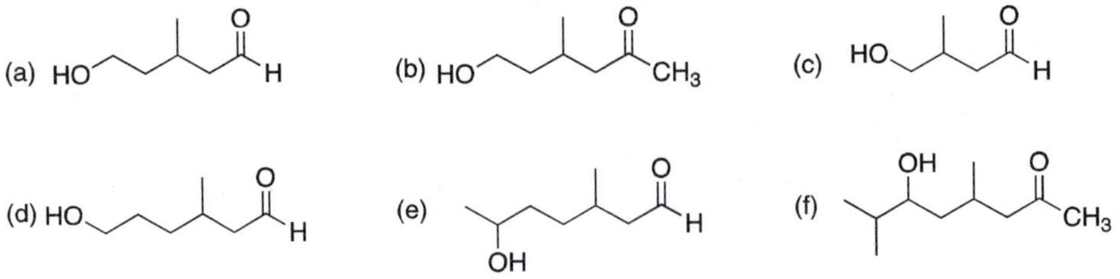

(a), (b), (c), (d), (e), (f)

A similar reaction occurs with monosaccharides, a glycopyranoside results as shown in Reaction (20-72).

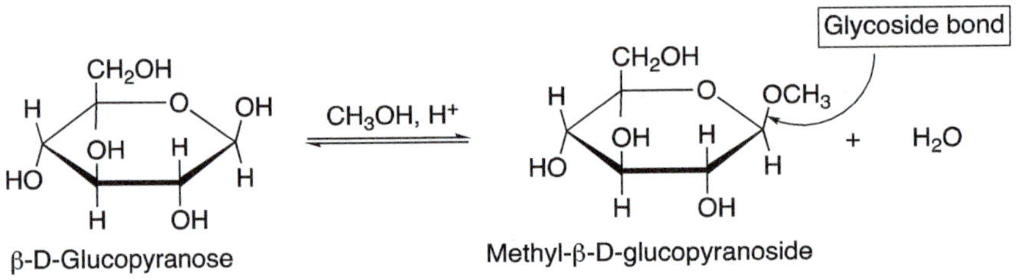

$$\beta\text{-D-Glucopyranose} \xrightleftharpoons{CH_3OH, H^+} \text{Methyl-}\beta\text{-D-glucopyranoside} + H_2O \quad (20\text{-}72)$$

Glycoside bond

Based on the mechanism for the acetal formation, an equal mixture of epimer will be the product as shown below in Reaction (20-73).

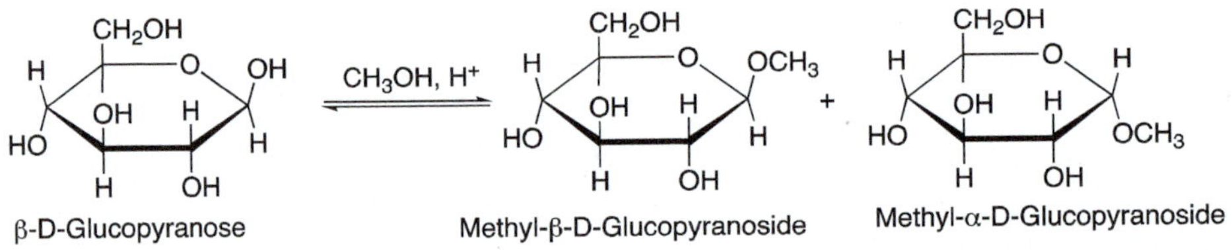

$$\beta\text{-D-Glucopyranose} \xrightleftharpoons{CH_3OH, H^+} \text{Methyl-}\beta\text{-D-Glucopyranoside} + \text{Methyl-}\alpha\text{-D-Glucopyranoside}$$

(20-73)

We will see in the next section that glycoside bonds are the bonds of polysaccharides that are formed from the monosaccharides.

20.16 Disaccharides and Polysaccharides

The glycoside bond is the bond that bonds two or more monosaccharides to form polysaccharides. Glycoside bonds are typically formed at the anomeric carbon of a monosaccharide and the hydroxyl group of another monosaccharide. Shown below are three commonly known disaccharides, maltose, lactose, and sucrose.

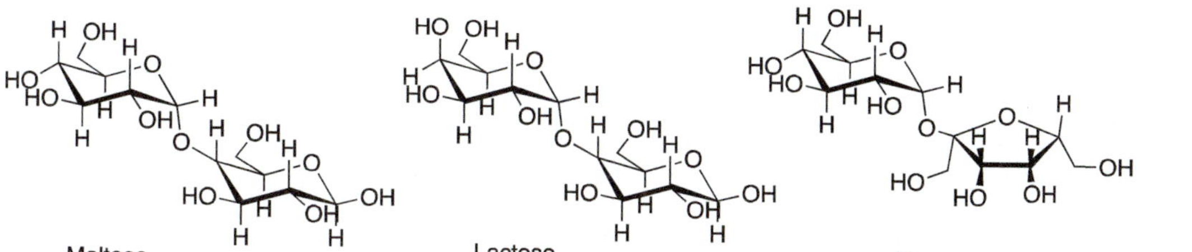

Maltose   Lactose   Sucrose

Maltose is a disaccharide, which is obtained from the partial hydrolysis of starch. Complete hydrolysis of maltose gives only one monosaccharide, D-glucose, which implies that there are two glucose monosaccharides in maltose. Lactose is a disaccharide consisting of galactose and glucose. Lactose makes up about 2–8% of milk, but lactose cannot be absorbed by some individuals, and there is a buildup of this disaccharide in the blood and urine of these individuals, which is called galactosemia (or lactose intolerant). On the other hand, sucrose contains no galactose (just glucose and fructose). Sucrose is often added to candy to give it a smooth and creamy texture. In the manufacture of candy, sucrose will disrupt the slow cooling crystal formation process of the crystallization of candy and the result is smooth and creamy candies.

20.17 N-Glycosides and Amino Sugars

In a comparable reaction to the glycoside formation described in the previous section, amines can react with monosaccharides to form N-glycoside bonds as shown in Reaction (20-74).

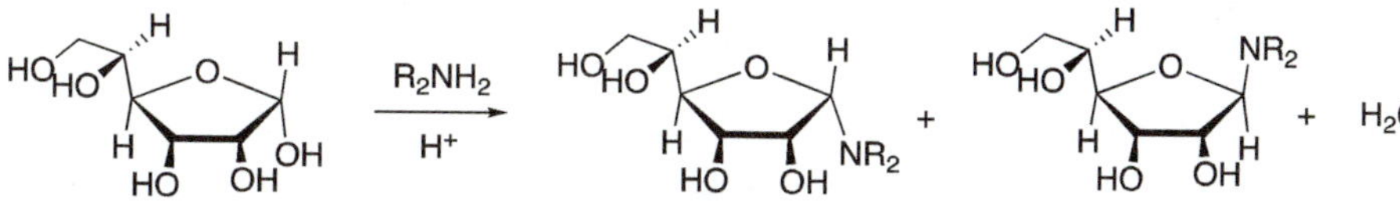

$$\alpha\text{-D-Glucofuranose} \xrightarrow{R_2NH_2,\ H^+} \alpha\text{-D-N-Glycoside} + \beta\text{-D-N-Glycoside} + H_2O \quad (20\text{-}74)$$

The products of these reactions are very important in biology in that they serve as building blocks for RNA and DNA, examples of these molecules are shown below.

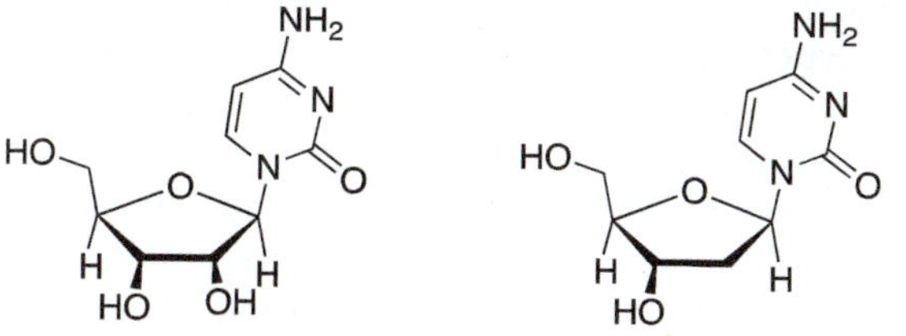

A Ribonucleoside   A Deoxyribonucleoside

By changing the amine bases bonded through the N-glycoside bond, the different bases of DNA are formed and they are shown below.

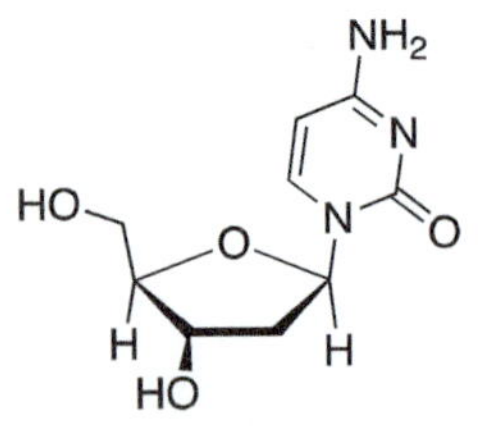

Deoxycytidine

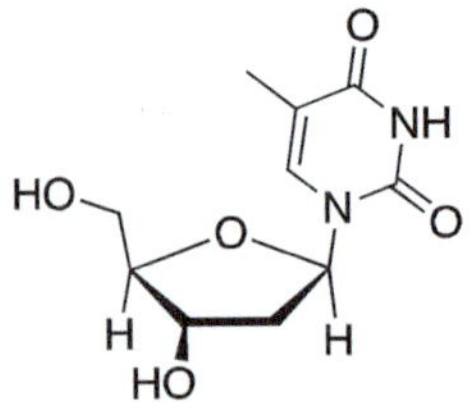

Deoxythymidine

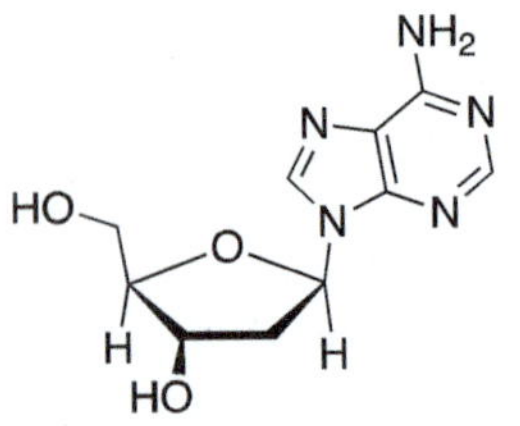

Deoxyadenosine

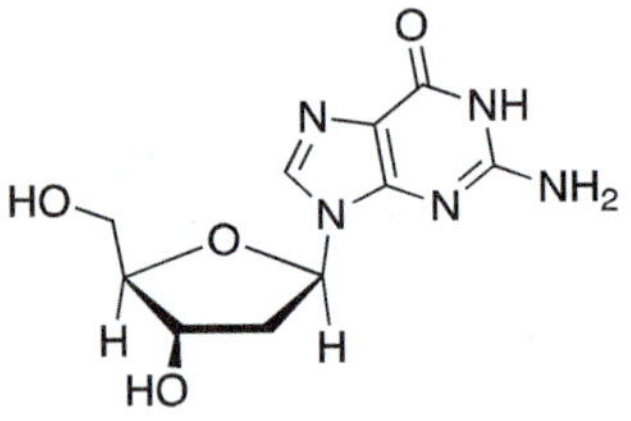

Deoxyguanosine

Replacing one of the hydroxyl groups of a monosaccharide with an anime results in a new type of molecule. For example, the substitution of the OH group on the α-carbon of D-glucose results in glucosamine. Glucosamine is used to stimulate the formation and repair of cartilage found naturally in the body that cushions joint, its structure is shown below.

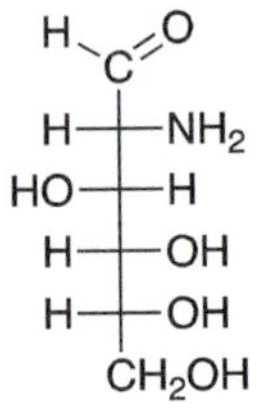

D-Glucosamine

20.18 Lipids

Lipids are not polymers, but most are large naturally occurring compounds. A large number of naturally occurring compounds, including cholesterol, fall under a category of compounds called lipids. Lipids are typically insoluble in water, but soluble in most organic solvents, such as diethyl ether. Lipids can be divided into three types: (i) triglycerides and waxes; (ii) phospholipids; and (iii) steroids, prostaglandins, and terpenes (Figure 20.2).

Triglycerides, waxes, and phospholipids can be hydrolyzed under acidic or basic conditions to produce smaller molecules, typically carboxylic acids and alcohols. Phospholipids make up our cells' membrane and are critical to the various functions of the cell. Steroids, prostaglandins, and terpenes are classified based on structural features of these types of molecules and the special functions that they perform (Figure 20.3).

Steroids are compounds used to treat a wide variety of inflammation and to reduce the activity of the immune system. Prostaglandins are hormone-like compounds that participate in a wide range of functions of the body. Functions such as contraction and relaxation of smooth muscles and dilation and constriction of blood vessels are affected by prostaglandins. Terpenes

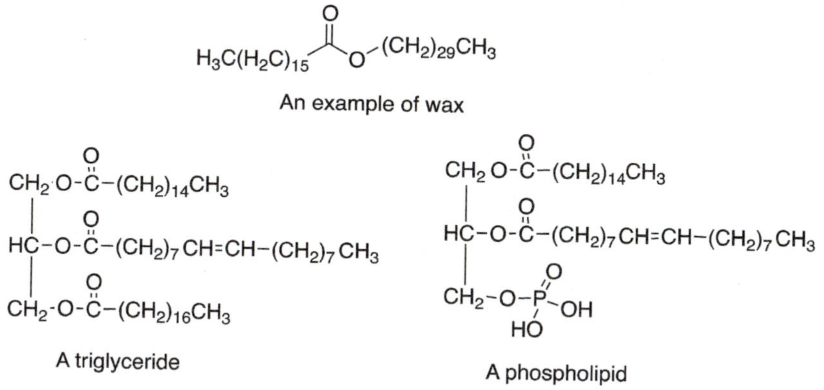

Figure 20.2 Examples of a wax, a triglyceride, and a phospholipid.

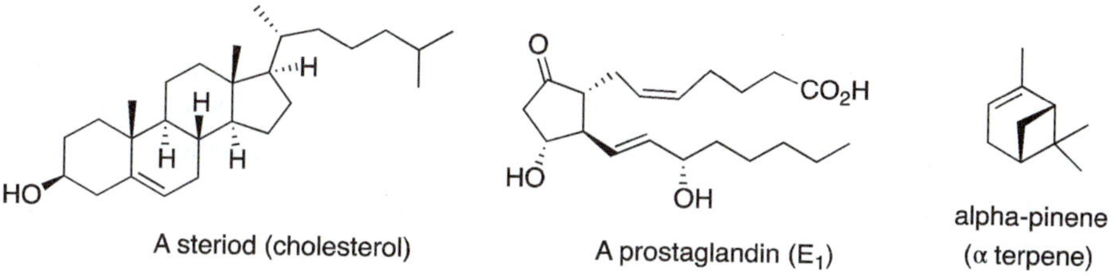

Figure 20.3 Examples of a steroid, a prostaglandin, and a terpene.

are natural products with unique fragrances; they are widely used in the chemical and pharmaceutical industries for a variety of products, such as cosmetics. Alpha pinene is one of the most commonly known terpenes and is used for its anti-inflammatory properties.

20.19 Properties and Reactions of Waxes

Wax can be divided into three categories: animal-, plant-, and petroleum-derived waxes; the most common animal wax is beeswax. Plants secrete wax through their leaves as a way to control evaporation. Petroleum-derived waxes come from petroleum; we are all familiar with candles, which is a form of wax. Since waxes are really esters, they can undergo hydrolysis reaction that we have seen before. An example of a base hydrolysis of an ester is shown in Reaction (20-75).

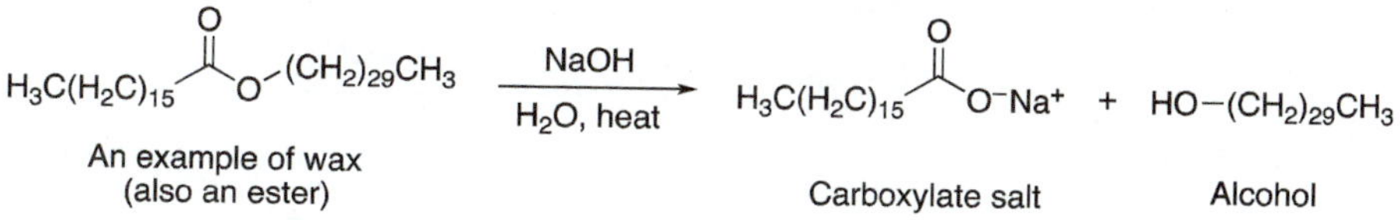

$$H_3C(H_2C)_{15}\text{-COO-}(CH_2)_{29}CH_3 \xrightarrow{\text{NaOH}, H_2O, \text{heat}} H_3C(H_2C)_{15}\text{-COO}^-Na^+ + HO\text{-}(CH_2)_{29}CH_3 \quad (20\text{-}75)$$

An example of wax (also an ester) → Carboxylate salt + Alcohol

20.20 Properties and Reactions of Triglycerides

Triglycerides gained their name from glycerol, also known as glycerin, the basic structure from which they were derived.

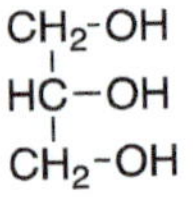

CH$_2$-OH
HC-OH
CH$_2$-OH
Glycerol

Triglycerides are triesters of glycerol and examples are shown below.

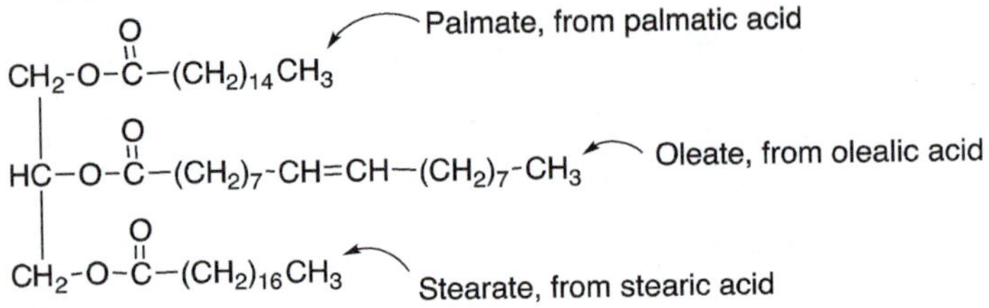

CH$_2$-O-C(=O)-(CH$_2$)$_{14}$CH$_3$ — Palmate, from palmatic acid
HC-O-C(=O)-(CH$_2$)$_7$-CH=CH-(CH$_2$)$_7$-CH$_3$ — Oleate, from olealic acid
CH$_2$-O-C(=O)-(CH$_2$)$_{16}$CH$_3$ — Stearate, from stearic acid

Animal fats are typically solids at room temperature, and plant oils are typically liquids at room temperature; they are forms of triglycerides. One explanation for this difference in states between animal fats and plant oils at room temperature is that animal fats are typically saturated esters and hence more structured in their molecular orientation resulting in a higher melting point and therefore often solids at room temperature. On the other hand, plant oils contain unsaturated portions and hence cannot attain a structured arrangement so the result is lower melting points and are therefore liquids at room temperature.

20.20.1 Saponification (Hydrolysis) of Triglycerides

Since triglycerides are also esters, they will undergo hydrolysis, similar to those that we have seen earlier for esters. Shown in Reaction (20-76) is the basic hydrolysis of a triglyceraldehyde.

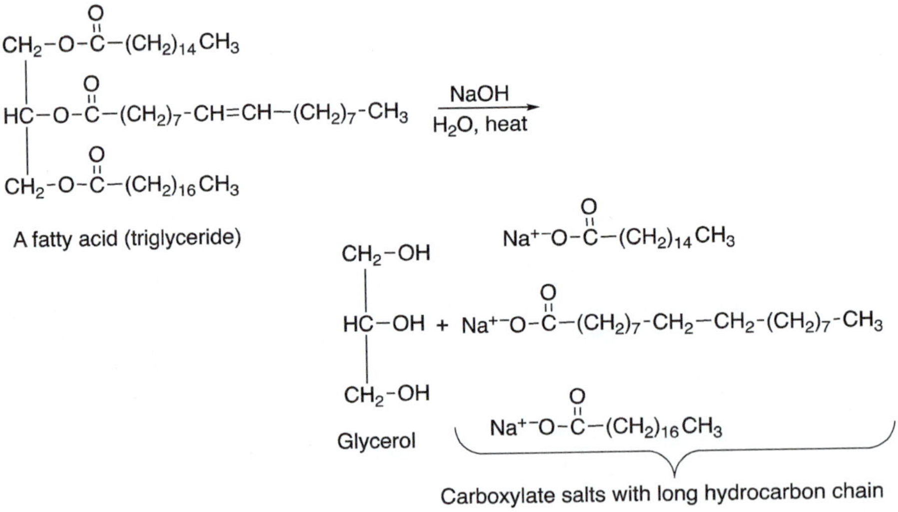

$$\text{A fatty acid (triglyceride)} \xrightarrow[\text{H}_2\text{O, heat}]{\text{NaOH}} \text{Glycerol} + \text{Carboxylate salts with long hydrocarbon chain} \quad (20\text{-}76)$$

The hydrolysis of animal fats under basic conditions is sometimes described as saponification since one of the products of the reaction is soap. As shown in the reaction above, the products are salts of carboxylic acids that have a long hydrocarbon tail, sometimes referred to as hydrophobic since not soluble in water. The carboxylate salt end is polar and hence hydrophilic, since water soluble. Carboxylate salts with long hydrocarbon tails are used as soaps since they have hydrophobic and hydrophilic properties and as shown in Figure 20.4 and can surround grease particles and extract them during the washing process.

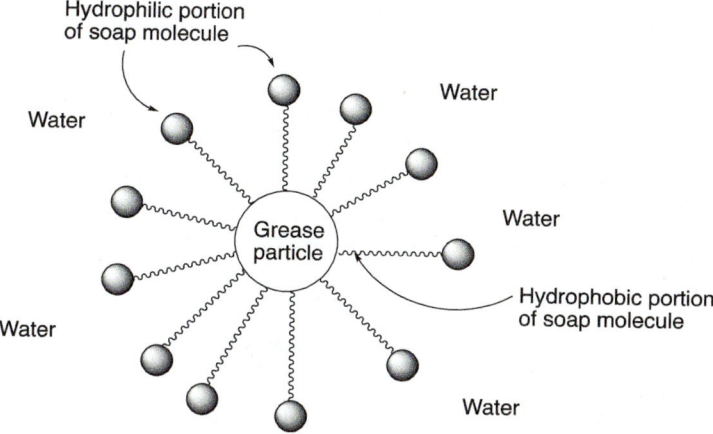

Figure 20.4 Representation of soap molecules in water surrounding grease particles.

20.20.2 Reduction of Triglycerides

Reactions of these molecules are the same as that of compounds that we have seen before. Shown below is the reduction (hydrogenation) of the alkene functionality of the triglyceride shown in Reaction (20-77).

$$\begin{array}{c}\text{CH}_2\text{-O-CO-(CH}_2)_{14}\text{CH}_3 \\ | \\ \text{HC-O-CO-(CH}_2)_7\text{-CH=CH-(CH}_2)_7\text{-CH}_3 \\ | \\ \text{CH}_2\text{-O-CO-(CH}_2)_{16}\text{CH}_3\end{array} \xrightarrow{\text{H}_2,\ \text{Pt}} \begin{array}{c}\text{CH}_2\text{-O-CO-(CH}_2)_{14}\text{CH}_3 \\ | \\ \text{HC-O-CO-(CH}_2)_7\text{-CH}_2\text{-CH}_2\text{-(CH}_2)_7\text{-CH}_3 \\ | \\ \text{CH}_2\text{-O-CO-(CH}_2)_{16}\text{CH}_3\end{array}$$

(20-77)

These reductions are frequently carried out in industry, and the products are sold under different names, such as shortenings, Crisco, and margarine. Some are partially hydrogenated oils. For trans-fatty acids, the process of hydrogenation of plant oil results in saturated fatty acids.

20.20.3 Transesterification of Triglycerides

Biodiesel is made by the process of transesterification of triglycerides or fatty acids as shown in the reaction below in Reaction (20-78).

$$\begin{array}{c}\text{CH}_2\text{-O-CO-(CH}_2)_{14}\text{-CH}_3 \\ | \\ \text{HC-O-CO-(CH}_2)_7\text{CH=CH-(CH}_2)_7\text{CH}_3 \\ | \\ \text{CH}_2\text{-O-CO-(CH}_2)_{16}\text{-CH}_3 \\ \text{Triglyceride}\end{array} \xrightarrow[\text{Heat}]{\text{MeOH, OH}^-} \begin{array}{c}\text{CH}_2\text{OH} \\ | \\ \text{HC-OH} \\ | \\ \text{CH}_2\text{-OH} \\ \text{Glycerol}\end{array} + \begin{array}{c}\text{H}_3\text{CO-CO-(CH}_2)_{14}\text{-CH}_3 \\ + \\ \text{H}_3\text{CO-CO-(CH}_2)_7\text{CH}_2\text{-CH}_2\text{(CH}_2)_7\text{CH}_3 \\ + \\ \text{H}_3\text{CO-CO-(CH}_2)_{16}\text{CH}_3 \\ \text{Methyl esters (biofuel)}\end{array}$$

(20-78)

20.21 Properties and Reactions of Phospholipids

An example of the basic structure of phospholipids is shown in Figure 20.5, and as you can imagine, there are different variations that can be made on the side chains that will result in different phospholipids. Phospholipids are compounds of biological importance. For example, lecithin is used in treating memory disorders, such as dementia and Alzheimer's disease; the structure of lecithin is shown in Figure 20.5.

The reactions that phospholipids undergo are similar to that of esters, specifically hydrolysis of the ester functionality, as shown below.

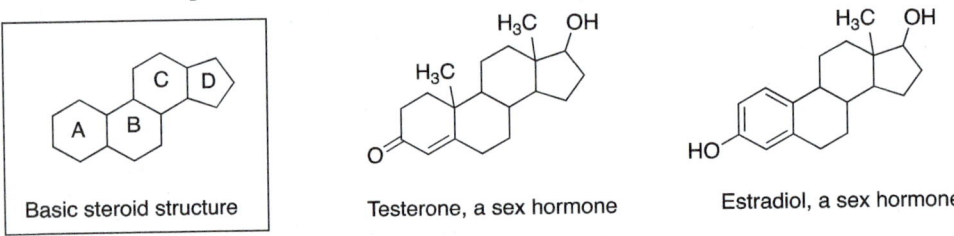

20.22 Structure and Properties of Steroids, Prostaglandins, and Terpenes

Steroids are compounds that have the basic structure shown below.

Basic steroid structure

Testerone, a sex hormone

Estradiol, a sex hormone

Progesterone, a female hormone

Cholesterol

Figure 20.5 Lecithin, a phospholipids, which is used for the treatment of various memory disorders.

A lecithin

There are three six-member rings that are fused and a five-member ring. These rings are often described as A, B, C, and D rings. For example, steroids have the A, B, C, and D rings and all have specific biological functions.

Prostaglandins are compounds that have a wide range of hormone-like effects on the body. For example, prostaglandins are involved in the contraction and relaxation of smooth muscle, the dilation and constriction of blood vessels, the control of blood pressure, and the modulation of inflammation. Thromboxane is a type of prostaglandin that stimulates constriction and clotting of platelets, and prostaglandin A_2 (PGA$_2$) is another prostaglandin that has been shown to have some antiviral/antitumor activity.

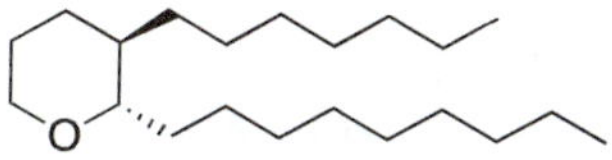

Thromboxane

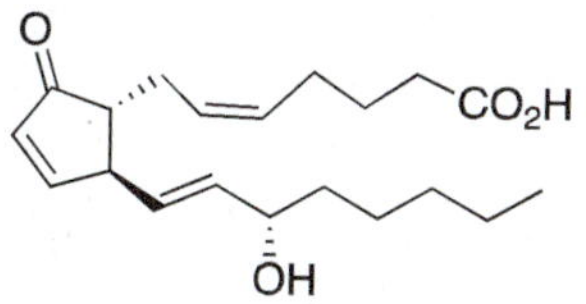

Prostaglandin A_2

Terpenes are typically found as constituents of essential oils, they are mostly hydrocarbons and have a chemical structure of repeated isoprene units. Terpenes are secreted by many plants in order to offer protection from predators. Terpenes can be classified as monoterpenes, which are terpenes with 10 carbon atoms and two isoprene units; sesquiterpenes have 15 carbon atoms and three isoprene units; diterpenes have 20 carbon atoms and four isoprene units; triterpenes have 30 carbon atoms and six isoprene units; and tetraterpenes have 40 carbon atoms and eight isoprene units. Terpenes can also be classified based on the number of cyclic rings that are contained in their structures, i.e. acyclic, monocyclic, and bicyclic. Examples of terpenes are shown below.

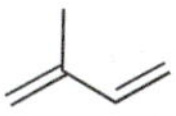

Isoprene 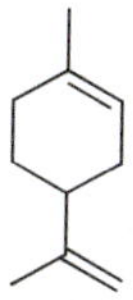   Limonene 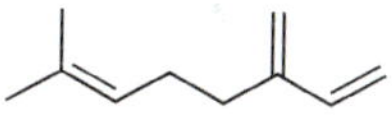   β-Myrcene 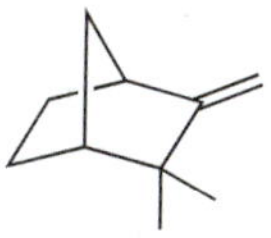   Camphene    α-Pinene

Limonene has a lemon-orange smell and is found in the peels of citrus fruits and in other plants. Camphene is a colorless crystal with a camphor-like odor and is used as a food additive. Myrcene is used in the perfume industry, and α-pinene has a variety of uses, including healing properties.

End of Chapter Problems

20.19 Give the polymer that would result from the polymerization of each of the following alkenes.

(i) 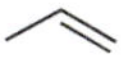   (ii) 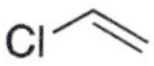 Cl   (iii) 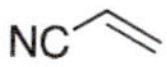 NC

20.20 Give the starting monomer that is used for the synthesis of the polymers; isoprene is made from a monomer and Buna-N is a copolymer, made from two different monomers.

(a) 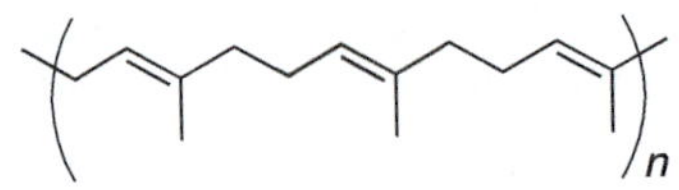 Isoprene rubber

(b) 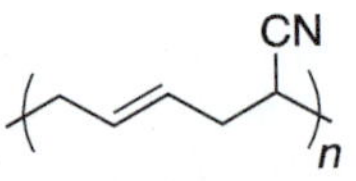 Buna-N, a synthetic rubber

20.21 Give the mechanism for the radical polymerization of vinyl chloride. Clearly identify the three steps of the mechanism, initiation, propagation, and termination.

20.22 The polymer that results from the copolymerization of hexamethylenediamine and adipic acid is Nylon-66, named since each monomer has six carbons. Give the structure of Nylon-66.

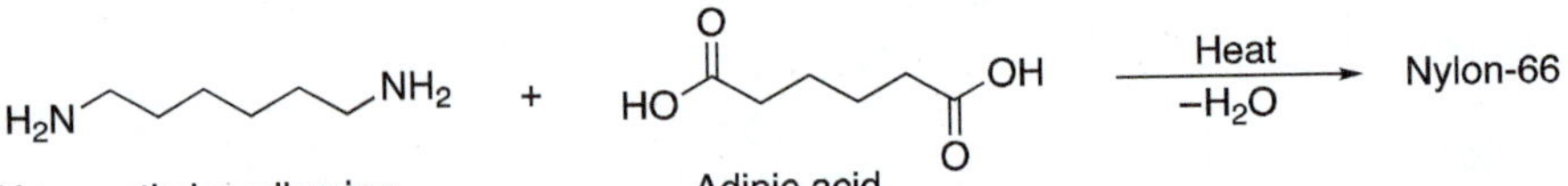

Hexamethylenediamine + Adipic acid $\xrightarrow[-H_2O]{\text{Heat}}$ Nylon-66

20.23 Draw the chair conformation of the predominant diastereomer (epimer) formed in solution when D-mannose cyclizes to form a hemiacetal.

20.24 Fructose is another monosaccharide and is also known as fruit sugar and sometimes added to candy to make them soft or amorphous. The open-chain structure of fructose is shown below, give the five-member ring monosaccharide.

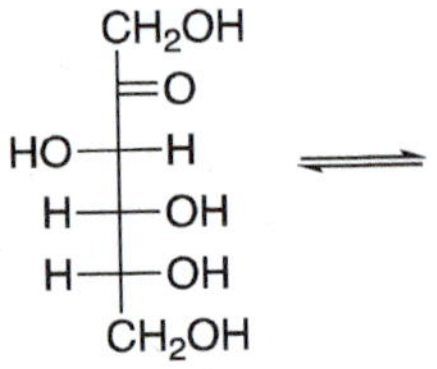

D-Fructose

20.25 List the products of the treatment of D-fructose with an excess of periodic acid.

20.26 Two compounds (**A** and **B**) were subjected to oxidation with HIO_4. The following products were obtained from each reaction, respectively. What are the structures of compounds **A** and **B**?

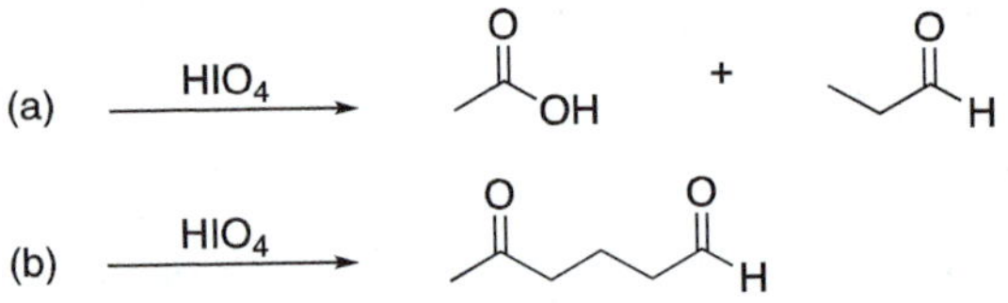

(a) $\xrightarrow{HIO_4}$ acetic acid + acetaldehyde

(b) $\xrightarrow{HIO_4}$ (the only product)

20.27 Give the structure of valine in a solution of pH = 2.

20.28 Give the structure of phenylalanine in a solution of pH = 12.

20.29 Give the structure of a tripeptide that contains valine, alanine, and phenylalanine.

20.30 Give the structure of the products of structure of your tripeptide in question 20.29 (above) with Sanger's reagent, followed by the hydrolysis of the Sanger product with 6.0 M HCl at 110 °C.

Index

a

acetals and ketals, as protection groups   234–235
acetone   69
acetylene   54–55
  carbon-carbon triple bond   26–27
acetylide anion   259
acetylides, reduction using   259–260
acid-base properties of amino acids   547
acid-base reactions   145–148, 165 *see also* acids; bases
  applications of   176–180
acid-catalyzed dehydration   155
acid chlorides, reactions   428–429
  IUPAC nomenclature of   82–83
  nomenclature of   82–83
  with alcohols   430–431
  with ammonia and amines   431–432
  with carboxylate salts   432–433
  with hard metal hydrides reagents   435–436
  with hard organometallic reagents   433–434
  with soft metal hydrides reagents   434
  with soft organometallic reagents   433
  with water   429–430
acidic medium, alcohols conversion to   395–396
acidic workup   151
acidity equilibrium constant (K_a)   167
acidity of compounds, factors affecting
  electronegativity   171
  hybridized orbitals, types of   171–172
  inductive effect   175
  polarizability/atom size   174
  resonance structures   172–174

acids   146–147
  Lewis   148, 165–166, 184, 205, 481
  pK_a values for   168, 170
  relative strengths of   166–167, 169–170
  weak   165
activating substituents   488
acyclic saturated hydrocarbon compounds   41, 42
acyl carbons, nucleophilic substitution reactions   425–426
  acid chlorides   428–436
  acyl substitution, mechanism for   426–428
  acyl substitution reactions, applications of   460–462
  amides   451–454
  anhydrides   436–442
  carboxylic acids   454–458
  esters   442–451
  oxalyl chloride   458
  sulfur containing compounds   458–460
acyl cation, substitution reactions involving   483–484
acyl chlorides   82
acyl substitution reactions   426
  applications of   460–462
  electrophiles of   427–428
  leaving group of   427
  nucleophiles of   428
addition reactions   149
alcohol functionality   64
alcohols and aldehydes, oxidation of   279–280
  chromic acid (H_2CrO_4)   281–282
  Dess-Martin oxidation   284
  nitrous acid   286–287

Organic Chemistry: Concepts and Applications, First Edition. Allan D. Headley.
© 2020 John Wiley & Sons, Inc. Published 2020 by John Wiley & Sons, Inc.
Companion website: www.wiley.com/go/Headley_OrganicChemistry

alcohols and aldehydes, oxidation of (cont'd)
   periodic acid   287–288
   potassium permanganate
      ($KMnO_4$)   280–281
   pyridinium chlorochromate   285
   silver ions   286
   Swern oxidation   283–284
alcohols, substitution reactions involving   63
   acid chlorides   430–431
   anhydrides   438
   carbonyl compounds   230–235
   carboxylic acids   455–456
   esters   445–446
   types of   65
aldehydes   279–280
   nomenclature of   70–71
   structure and properties of   69
aliphatic hydrocarbons *see* saturated
      hydrocarbons
alkanals   70
alkanamide   89
alkanenitrile   90
alkanes   369 *see also* saturated hydrocarbons
   bromination of   380–386
   chlorination of   376–380
   conformational isomers of   104–108
   stability of   119–121
alkanes and alkyl halides
   bond dissociation energies of
      hydrocarbons   373–374
   hydrocarbons, classifications of   371–373
   structure and stability of radicals   374–376
alkanethiols   68
alkanoates   80
alkanoic acids   75
alkanoic alkonoic anhydrides   84
alkanols   66
alkanones   71
alkanoyl chlorides   82
alkene geometric stereoisomers, IUPAC
      nomenclature of   116–117
alkenes
   addition reactions involvement in   183–185
   anionic polymerization of   540
   autooxidation of   301–302
   carbenes, addition to   207–209
   copolymerization of   542–543
   epoxidation of   288–291
   free radical polymerization of   540–542

   halogens and water to   198–199
   halogens to   196–198
   hydrogen halide to (*see* hydrohalogenation of
      alkenes)
   oxidation reactions of   296–299
   ozonolysis of   295–296
   reduction of   268–272
   stability of   121–122
   water addition to (*see* hydration of alkenes)
alkenes with bond cleavage, oxidation
      of   293
   $KMnO_4$ at elevated temperatures
      293–295
   ozonolysis   295–296
alkenes without bond cleavage, oxidation
      of   288
   epoxidation of alkenes   288–291
   osmium tetroxide (OsO_4)   292–293
   potassium permanganate
      ($KMnO_4$)   291–292
alkoxide anion   444
alkoxide salt   255, 267
alkyl chlorides   370, 371
alkyl groups   187, 194
alkyl halides   369, 370
alkylsulfonium salt   283
alkynes   209
   bromine to alkynes   209–210
   hydrogen halide to alkynes   210–211
   IUPAC nomenclature of   60–61
   oxidation of   299–300
   reduction of   268–272
   stability of   122–123
   synthesis of   418–419
   water, addition to   211–213
allylic carbocation   412
allylic radical   375
amide bond   432
amides   88, 451–452
   IUPAC nomenclature of   89–90
   nomenclature of   89–90
   preparation of   461–462
   primary, secondary and tertiary   89
   reactions with hard metallic hydrides   453–454
   reactions with water   452–453
amines, substitution reactions involving   84
   acid chlorides   431–432
   anhydrides   439
   carbonyl compounds   240–241

carboxylic acids 456–457
  esters 446–447
  IUPAC nomenclature of 86
amino acids 545–547, 556
  acid-base properties of 547
  α-amino acids 550
  carboxylic acid functionality 551–552
  dipeptides 552–554
  protection-deprotection of amino functionality 550–551
  reaction with ninhydrin 554–555
  synthesis of 547–550
amino functionality, protection-deprotection of 550–551
amino sugars 567–568
ammonia (NH_3), substitution reactions involving 85
  acid chlorides 431–432
  anhydrides 439
  carboxylic acids 456–457
  esters 446–447
  molecular orbital picture 22
  sp^3 hybridized orbitals 23
ammonium cyanate 1
ammonium salts 85
amphetamine 85
anhydrides 432, 436–437
  IUPAC nomenclature of 84
  nomenclature of 84
  reactions with alcohols 438
  reactions with ammonia and amines 439
  reactions with carboxylate salts 439–440
  reactions with hard metallic hydrides 441–442
  reactions with hard organometallic reagents 440
  reactions with soft metallic hydrides 441
  reactions with soft organometallic reagents 440
  reactions with water 437–438
  structural feature of 83
aniline 87, 485–486
animal fats 570
anionic polymerization of alkenes 540
anomeric carbon 561
anti-bonding molecular orbital 21
anti-Markovnikov addition 303
  to alkenes 192–196, 206
antioxidants 388
aprotic solvents 165

aromatic compounds
  characteristics of 475–478
  oxidation of 300–301
  reduction of 268–272
aromaticity and aromatic substitution reactions 467–468
  aromatic compounds, characteristics of 475–478
  benzene, structure and properties of 468–470
  electrophilic aromatic substitution reactions of benzene 478–484
  nomenclature of substituted benzene 470–473
  nucleophilic aromatic substitution 499–503
  polycyclic aromatic compounds, electrophilic substitution reactions of 494–496
  pyridine, electrophilic substitution reactions of 497–499
  pyrrole, electrophilic substitution reactions of 496–497
  stability of benzene 473–475
  substituted benzene compounds 491–494
  substituted benzene, electrophilic aromatic substitution reactions of 484–491
aromatic protons 360
aromatics 43
arrow-pushing formulism 234
asphyxiants 44
asymmetric molecules 127
atoms 414 see also chemical bonds
  electronic configuration 6–8
  Lewis dot structures 8
  orbitals 4–6
  two-dimensional illustration 6
atorvastatin 130
Aufbau principle 7
autooxidation of ethers and alkenes 301–302
axial hydrogens 109, 110

b

Baker's yeast 139
barbituric acid 45
base-catalyzed epimerization of monosaccharides 562
bases
  Lewis 148, 165–166, 251, 525
  relative strengths of 169–170

basicity of compounds, factors affecting
    electronegativity 171
    hybridized orbitals, types of 171–172
    inductive effect 175
    polarizability/atom size 174
    resonance structures 172–174
bathochromic shift 334
Benedict's Reagent 563
benzene 54–55, 268, 467–468
    electrophilic aromatic substitution reactions of 478–484
    stability of 473–475
    structure and properties of 468–470
benzoic acid 145
    IR spectrum of 340, 341
benzyl alcohol, IR spectrum of 341, 342
benzylamine, IR spectrum of 341, 342
benzylic carbocation 412, 413
bicyclic compounds, IUPAC nomenclature of 52–54
bimolecular substitution reaction mechanism (S_N2 mechanism) 400
    electrophile of 400–402
    intramolecular of 405–406
    nucleophile of 402–403
    solvents of 403–404
    stereochemistry of 404–405
biodiesel 571
biological α-keto acids 300
biopolymers 544–545
Birch reduction 268
boat conformations of cyclohexane 110, 111
boiling points
    of alkanes 32–33
    of 2,2-dimethylpropane 32, 33
    of liquid 32
    of pentane 32, 33
bond angles 23, 27
    of ammonia 22
    of methane 21
    in oxiranes 91
bond dissociation energies of hydrocarbons 373–374
bond dissociation energy 373
bonding electrons 9, 277
bonding features of carbon 28, 29
bond length 21, 24, 26
bond polarities 14
bond spring-like vibrations 337

boron 205
branched alkanes, IUPAC nomenclature of 46–49
branched alkenes, IUPAC nomenclature of 56
branched cyclic alkanes, IUPAC nomenclature of 51–52
breath analyzer test 282
bridged mercury cationic intermediate 203
bromide anion 394
bromination of alkanes 380–386
bromination reaction of alkene 149
bromine 196–197
    addition to alkynes 209–210
bromine radical 193
2-bromobutane, Fischer projection 128, 129
2-bromo-3-chlorobutane, stereoisomers of 134, 135
1-bromo-1-chloroethane 131–133
    Fischer projection of 129
bromonium ion 197, 198
Bronsted-Lowry acid 146
Bronsted-Lowry base 147
bulky trimethyl ammonium ion 313
butanol 356
butylated hydroxytoluene (BHT) 64

c
Cahn-Ingold-Prelog system 131
camphene 573
camphor 53
carbenes, addition to alkenes 207–209
carbinoid 208
carboanions 540
carbocation rearrangement 321–322
carbocations 166, 180, 187, 189, 201, 281, 399, 411
    substitution reactions involving 481–482
carbocyclic compounds 475–476
carbocyclic-conjugated ions 476
carbohydrates, monomers of 559
carbon atom 7
carbon-carbon double bond 294, 295, 309, 375
    of alkenes 184, 193, 199, 203
carbon-carbon multiple bonds 269
carbon-carbon triple bond 28
carbon–halogen bond 394
carbon-lithium bond 253
carbon monoxide, carbon-heteroatom triple bond 27, 28

carbon–nitrogen double bond   240
carbon–nitrogen triple bond   242
carbon-13 NMR (^{13}C NMR)   363
   chemical shifts and coupling   363–367
carbon–nucleophile bond   399
carbon, oxidation state of   276, 277
carbon–oxygen double bond   226, 340, 427
carbon radicals   194
carbonyl carbon–nitrogen bond   461
carbonyl compounds   223–224
   alcohols reaction with   230–235
   amines reactions with   240–241
   enolates reactions with   237–239
   hydrogen cyanide (HCN) to   224–226
   water reactions with   226–229
   ylides reactions with   235–237
carbonyl double bond   436
carbonyl oxygen   260
carboxylate salts
   and acid chlorides, substitution reactions involving   432–433
   anhydrides with, substitution reactions involving   439–440
carboxylic acid functionality   551–552
carboxylic acids   282, 454–455
   nomenclature of   75–78
   purification of   138, 139
   reactions with alcohols   455–456
   reactions with ammonia and amines   456–457
   reactions with hard metallic hydrides   457–458
   structure and properties of   73–75
β-carotene   335
catalyst proton   200
catalytic carbon-carbon coupling reactions   525
   palladium-catalyzed coupling reactions (see palladium-catalyzed coupling reactions)
   transition metal complexes, reactions of (see transition metal complexes, reactions of)
catalytic coupling reactions   158–159
catalytic hydrogenation
   of cis-2-butene and trans-2-butene   122
   of 1-hexyne and 3-hexyne   123
   of 3-methyl-1-butene and 2-methyl-2-butene   121, 122
   reduction using   269–272
cationic copolymerization   542–543
cationic polymerization of alkenes   537–538
   isobutene   538
   styrene   538–540

chair conformations of cyclohexane   110, 111
C–H bonds, dissociation bond energies for   381
chemical bonds
   covalent bonds   9–11
   ionic bonds   9
chemical formulas   18–20
chirality, significance of   129–131
chiral molecules   126
   nomenclature of absolute configuration   131–133
chiral stereoisomers   126–129
chlordane   370
chloride anion   435
chlorination of alkanes   376–377, 379–380
chlorination of methane, mechanism for   377–379
chlorine   196
chlorine functionality   493
chlorofluorocarbons (CFCs)   369
chloroform   369
chloromethane   377
   dipole-dipole attraction   29–30
chlorosulfonium salt   283
chromic acid (H_2CrO_4), oxidation using   281–282
chromium salts   283
cis-addition   205, 206, 270, 271
cis-1,3-dichlorocyclopentane   110
cis-dimethylcyclobutane   109
cis-dimethylcyclopentanes   109, 110
cis-hydrogenation   269
citric acid   145
Claisen condensation reaction   449–450
Claisen rearrangement   522
Clemmensen reduction   260, 262
C=O and C=S containing compounds, reduction of   255
   acetylides   259–260
   hydrogen with a catalyst   261
   metals   260–261
   $NaBH_4$ and $LiAlH_4$, reduction using   255–257
   organometallic reagents   257–259
   Wolff Kishner reduction   261–263
combustion reaction   275, 277
   of alkanes   120, 121
   of octane and 2,2,3,3-tetramethylbutane   120
common alkanes, IUPAC nomenclature of   50

concerted reaction   309
conformational isomers   103, 125
    of alkanes   104–108
    of cycloalkanes   108–114
conjugate bases   169
    of alcohols   178
    deprotonation of methane   176
    of phenol   172, 173
    relative strengths of   166–167
conjugated alkenes, stability of   511–513
conjugated double bonds   512
conjugated systems   334–337, 511
    conjugated alkenes, stability of   511–513
constitutional isomers *see* structural isomers
coordinate covalent bond   28
cope rearrangement   521
copolymerization of alkenes   542–543
copolymers   542–543
core electrons   8
covalent bond
    carbon-carbon double bond   23–25
    carbon-carbon triple bond   26–27
    carbon-heteroatom double bond   25–26
    carbon-heteroatom triple bond   27–28
    single bond to carbon   21–22
    single bond to heteroatoms   22–23
    single bond to hydrogen   20–21
cracking process   43
12-crown-4 ether   92
crown ethers   92
crude oil   276
    principal components   42, 43
    refining   43
curcumin   337
curved-arrow formulism   16, 184
cyanohydrin   226
cyclic alcohols, IUPAC nomenclature of   67
cyclic alkanes, IUPAC nomenclature of   51
cyclic alkenes, IUPAC nomenclature of   58
cyclic carboxylic acids, IUPAC nomenclature
    of   76–78
cyclic compounds, with more than one
    stereogenic center   136–137
cyclic esters, IUPAC nomenclature of   80–81
cyclic ethers   91
cyclic hemiacetals   231
cyclic ketones, IUPAC nomenclature of   72–73
cyclic saturated hydrocarbon compounds   41, 42
cycloaddition reactions   513–519

cycloalkanecarboxylic acid   77
cycloalkanes   41, 42
    conformational isomers of   108–114
cycloalkanol   67
cycloalkanones   72
cycloalkenes   58
cycloalkyl group   77
cyclobutane, conformational isomers of   109
cyclohexane
    chair and boat conformations of   110, 111
    conformational isomers of   110–112
    structure   33, 34
cyclohexatriene   474
cyclopentane, conformational isomers
    of   109–110
cyclopropane, conformational isomers
    of   108–109
cysteine   65

d

d-alanine   126
dashed-wedge representation
    of 2-bromobutane   128, 129
    stereoisomers of 2-bromo-3-chlorobutane
        135, 136
dative covalent bond   28
deactivating substituents   489
decalin   53
degenerate orbitals   8
dehydration
    elimination reactions of   319–323
    products   319–321
dehydrohalogenation   155
    elimination reactions of   316–319
delta negative representations   13
delta positive representations   13
deprotonation of methane   176
Des-Martin periodinane (DMP) reaction   284
Dess-Martin oxidation   284
dextrorotatory *(d)* compounds   134
D-glucaric acid   564
diastereomerism   135
diastereomers   135
diazomethane   209
2,3-dibromobutane, stereoisomers of   135
dichlorodiphenyltrichloroethane (DDT)   3, 370
1,2-dichloroethane
    anti-conformer of   125
    gauche conformer of   125

dichromate salts   282
Dieckmann reaction   450
Diels-Alder reaction   158, 516
dienes   57
diethylether   91
diethylhexyl phthalate (DEHP)   79
diethyl malonate (malonic ester)   419
1,2-difluoroethene   116
difluoroformaldehyde   227
difunctional acid chlorides, IUPAC nomenclature of   83
difunctional alcohols, IUPAC nomenclature of   67
difunctional amines, IUPAC nomenclature of   88
difunctional carboxylic acids, IUPAC nomenclature of   76
difunctional ketones, IUPAC nomenclature of   71–72
dimethylbenzenes, isomers of   471
1,4-dimethylcyclohexane   113, 114
1,2-dimethylcyclopropane   108
1,2-dimethylcyclopropane isomers, symmetrical analysis of   137
dimethylether, boiling point   32
dimethylsulfoxide (DMSO)   199
2,4-dinitrochlorobenzene   499
1,4-dioxane   91
dipeptides   552–554
dipole-dipole intermolecular attractions   29–30
dipole moment   14
disaccharides   566–567
dissolving metals, reduction using aromatic compounds, alkynes, and alkenes   268–269
di-substituted benzene   472, 490–491
  nomenclature of   471–473
disubstituted cyclohexane, conformational isomers of   113–114
2,3-disubstituted oxirane   94
diyne   60
double bond   12
drugs, structures of   133

e

Edman's reagent   557
electrocyclic reactions   519–521
electromagnetic spectrum   331–333
electron activators for electrophilic aromatic substitution reactions   485–488
electron deactivators for electrophilic aromatic substitution reactions   488–490
electron distribution   519
electronegative groups   429
electronegativity   12–13, 171
electronic structure of atoms
   electronic configuration   6–8
   Lewis dot structures   8
   orbitals   4–6
electron impact mass spectrometer   344
electron ionization mass spectrometer   344
electron ionization mass spectrum   346
   of methylbromide   345
electrons   4, 5
electrophiles   184, 188, 224, 393–394
   of acyl substitution reactions   427–428
   of S_N2 reactions   400–402
electrophilic addition products   186
electrophilic aromatic substitution reactions of benzene   478–479
   acyl cation   483–484
   carbocations   481–482
   halogen cation   480–481
   nitronium ion   479–480
   sulfonium ion   484
electrophilic aromatic substitution reactions of substituted benzene   484–485
   disubstituted benzenes   490–491
   electron activators for   485–488
   electron deactivators for   488–490
electrophilic atom   393
electrophilic bromonium   197
electrophilic carbocation   200
electrophilic carbon   224
electrophilic fragment   414
electrophilic substitution reactions
   of polycyclic aromatic compounds   494–496
   of pyridine   497–499
   of pyrrole   496–497
electrophoresis   547
electropositive chromium atom   281
β-elimination-bimolecular (E2) elimination reaction   309
elimination bimolecular (E2) reaction mechanism   310–314

elimination of unimolecular (E1) reaction mechanism   314–315
elimination reactions   154–156, 309
　applications of   323–325
　elimination bimolecular (E2) reaction mechanism   310–314
　elimination of unimolecular (E1) reaction mechanism   314–315
　elimination unimolecular-conjugate base (E1cB) reaction mechanism   315–316
　hydrogen and halide (dehydrohalogenation)   316–319
　of water (dehydration)   319–323
β-elimination reactions   527
elimination unimolecular-conjugate base (E1cB) reaction mechanism   315–316
enantiomerism   126
　determination of   127–128
　significance of   129–131
enantiomers   126, 127, 131, 408
　interaction with polarized light   133, 134
　resolution of   137–139
enediol rearrangement of monosaccharides   563
energy   331–332
enolates   173
　to carbonyl compounds   237–239
　of esters, esters with   449–451
epimers   559
epoxidation of alkenes   288–291
epoxides   91, 92
　reactions of   289–290
epoxy resin copolymers   543
equatorial hydrogens   109, 110
equivalent Lewis dot structures   15, 16
ester formation of monosaccharides   565
esters   78, 180, 442–444
　IUPAC nomenclature of   79–80
　preparation of   460–461
　reactions with alcohols   445–446
　reactions with ammonia and amines   446–447
　reactions with enolates of esters   449–451
　reactions with hard organometallic reagents   447–448
　reactions with soft and hard metallic hydrides   448
　reactions with soft organometallic reagents   447
　reactions with water   444–445
ethanol (CH_3CH_2OH)   33
　boiling point   32
　health disadvantage   63

ether formation of monosaccharides   565
ethers
　autooxidation of   301–302
　cyclic   91
　nomenclature of   93
　structure   91
　synthesis of   415
ethylene (C_2H_4)   23, 54–55
　pi (π) bond of   24–25
　sp^2 sigma bonds and unhybridized p orbital of   24
ethylene glycol   63
exothermic reaction, energy profile for   119

f

fatty acids   115
Fehlings Reagent   563
Fischer projection   128
　of 2-bromobutane   128, 129
　of 2-bromo-3-chlorobutane   134, 135
flat carbocation   190
2-fluorobutanoic acid   175
formal charges   14–15
formaldehyde   69, 256
　carbon-heteroatom double bond   25, 26
　IR spectrum of   340
formic acid   73, 74
　IR spectrum of   340, 341
formoterol   64
fractional distillation   43
free radical inhibitors   388
free radical polymerization of alkenes   540–542
free radical substitution reactions, involving alkanes   369–370
　alkanes and alkyl halides, types of (see alkanes and alkyl halides)
　applications of   386–388
　bromination of alkanes   380–386
　chlorination of alkanes (see chlorination of alkanes)
　free radical inhibitors   388
　organohalides and free radicals, environmental impact of   389–391
Friedel-Crafts acylation   483
Friedel-Crafts alkylation   482
frontier molecular orbital method   514, 517, 520
fumaric acid   115
functional groups
　defined   39
　of organic molecules   39–41

g

Gabriel Malolic Ester synthesis, α-amino acids using 549–550
geometric isomers 103, 125
   of 2-butene 115
   definition 114
   of 1,2-difluoroethene 115
glucosamine 568
glucose 33, 34
glyceraldehyde, specific rotation values 134
glycoside bond 566
"Green Chemistry," 532
Grignard reaction 177
Grignard reagents 177, 253, 257, 434, 440, 447, 493, 526, 550

h

halogenation of alkenes 196–198
halogen cation, substitution reactions involving 480–481
halogens 394
   to alkenes 196–198
   bond dissociation energies for 381
halohydrin formation (halogens and water to alkenes) 198–199
hard and soft acids and bases (HSAB) 166
hard Lewis acids 166
hard Lewis bases 166
hard metal hydrides reagents, acid chlorides with 435–436
hard metallic hydrides, substitution reactions involving
   amides with 453–454
   anhydrides with 441–442
   carboxylic acids with 457–458
hard organometallic reagents, substitution reactions involving
   acid chlorides with 433–434
   anhydrides with 440
   esters with 447–448
Heck catalytic coupling reaction 530
Heck reaction 158–159, 528–531
Hell Volhard Zelinsky reaction, α-amino acids using 548–549
hemiacetal formation involving monosaccharides 560–562
hemiacetals 230, 231
hemiketals 232, 233
heteroatomic groups, IUPAC nomenclature of 49–50

heteroatoms 1
heterocyclic compounds 477–478
hexachlorophene 64
highest occupied molecular orbital (HOMO) 334, 515, 517, 520
high-pressure liquid chromatography (HPLC) technique 138
Hoffman elimination 313–314
Hofmann product 313, 317
homolytic cleavage 373, 374
Hückel rule 475, 476
Hund's rule 8
hybridized orbitals, types of 171–172
hydrate 226, 227
hydration, carbonyl compounds toward 227–229
hydration of alkenes 199–202
   hydroboration-oxidation 204–207
   oxymercuration-demercuration 203–204
hydrazone 262
hydride anion 178, 201, 252, 255, 435
hydroboration-oxidation, hydration by 204–207
hydrocarbons 41
   bond dissociation energies of 373–374
   classifications of 371–373
   oxidation states for 277, 278
   saturated (see saturated hydrocarbons)
hydrochloric acid (HCl) 146, 430
hydrochlorination 149
hydroflurorcarbons (HFCs) 390
hydrogen and halide (dehydrohalogenation), elimination reactions of 316–319
hydrogenation 152
   of 1-pentene 511
hydrogen atom 6, 7, 252
hydrogen bases 178
hydrogen bond, intermolecular 30–31
hydrogen cyanide (HCN)
   carbon-heteroatom triple bond 27, 28
   to carbonyl compounds 224–226
hydrogen halide to alkenes
   anti-Markovnikov addition to alkenes 192–196
   major addition product, prediction of 187–190
   Markovnikov Rule 190–191
   stereochemistry of addition reaction products, prediction of 190
   symmetrical alkenes, addition reactions to 185–186

hydrogen halide to alkenes (cont'd)
    unexpected hydrohalogenation
        products   191–192
    unsymmetrical alkenes, addition reactions
        to   186–187
hydrogen halide to alkynes   210–211
hydrogen molecule (H_2)
    molecular orbital diagram   20, 21
    sigma (σ) orbital   20
    sigma star orbital (σ^*)   20
hydrogen peroxide   288
hydrogen with a catalyst, reduction using   254
    C=O and C=S containing compounds   261
    imines   265–266
hydrohalogenation of alkenes
    anti-Markovnikov addition to
        alkenes   192–196
    major addition product, prediction
        of   187–190
    Markovnikov's Rule   190–191
    stereochemistry of addition reaction
        products, prediction of   190
    symmetrical alkenes, addition reactions
        to   185–186
    unexpected hydrohalogenation
        products   191–192
    unsymmetrical alkenes, addition reactions
        to   186–187
hydrolysis   244
    of esters   443, 445
    of triglycerides   570–571
hydroxide anion   262
hydroxyl proton   356
hypothetical cyclohexatriene molecule   474
hypothetical electrophilic addition reaction,
    energy profile for   185

i

imines
    addition reactions involvement in   241–242
    reduction of   263–266
iminium-type of intermediate   454
inductive effect   175
infrared (IR) spectroscopy   95–98, 337–343
inhibitors   388
initiation step   377
intermolecular acetal formation involving
    monosaccharides   565–566
intermolecular attractions

dipole-dipole   29–30
    intermolecular hydrogen bond   30–31
    London forces   31
intermolecular hydrogen bond   30–31
intermolecular London force attractions   31
International Union of Pure and Applied
    Chemists (IUPAC)   45
intramolecular S_N2 reactions   405–406
ions   475–476
IR frequencies of functional groups   339
isoamyl acetate   79
isobutene, cationic polymerization of   538
isobutylene, free radical polymerization of   541–542
isomers
    definition   103
    of dimethylbenzenes   471
isoprene   57, 511
isopropanol   63, 65
IUPAC nomenclature
    of acid chlorides   82–83
    of alkene geometric stereoisomers   116–117
    of alkynes   60–61
    of amides   89–90
    of amines   86
    of anhydrides   84
    of bicyclic compounds   52–54
    of branched alkanes   46–49
    of branched alkenes   56
    of branched cyclic alkanes   51–52
    of common alkanes   50
    of cyclic alcohols   67
    of cyclic alkanes   51
    of cyclic alkenes   58
    of cyclic carboxylic acids   76–78
    of cyclic esters   80–81
    of cyclic ketones   72–73
    of difunctional acid chlorides   83
    of difunctional alcohols   67
    of difunctional amines   88
    of difunctional carboxylic acids   76
    of difunctional ketones   71–72
    of esters   79–80
    of heteroatomic groups   49–50
    of nitriles   90
    of polyenes   57
    of straight chain alkanes   45, 46
    of substituted benzenes   58–60
    of substituted phenols   68
    of thiols   68

j

Jones reagent 279, 282

k

ketals 232
   and acetals, as protection groups 234–235
β-keto carboxylic acids 450
keto-enol tautomerization 211, 239
ketones 256, 257
   nomenclature of 71–73
   structure and properties of 69

l

lactones *see* cyclic esters
lactose 567
L-alanine 126
leaving group, alcohols conversion to 394–395
   acidic medium 395–396
   of acyl substitution reactions, mechanism for 427
   amines conversion to 395
   phosphorous tribromide 396
   sulfonyl chlorides 396–397
   thionyl chloride 396
Le Chatelier's Principle 445
levorotatory (l) compounds 134
Lewis acids 148, 165–166, 184, 205, 481
Lewis bases 148, 165–166, 251, 525
Lewis dot structures
   of acetylene 26
   of atoms 8, 10
   of carbon dioxide molecule 10, 11
   of covalent molecule 10
   of formaldehyde 25
   of methane 10
   of methyleneimine 25
   of nitric acid 14
ligand migratory insertion reactions 527
limonene 573
lindane 370
Lindlar's catalyst 272
line-angle representation of molecules 18–20
lipids 568–569
lithium alkoxide 151
lithium aluminium hydride (LiAlH$_4$) 150, 151, 435, 453–454
London forces 31, 32
lowest unoccupied molecular orbital (LUMO) 334, 514, 515, 517
Lucas Test 410

m

magnetic shielding 349–350
major resonance contributor 17
maleic acid 115
malonic ester 419
maltose 567
Markovnikov addition product 202, 204, 303
Markovnikov's Rule 190–191
mass spectroscopy 343–346
mass spectrum of ethane 344
mechanisms of elimination reactions 309–310
   elimination bimolecular (E2) reaction 310–314
   elimination of unimolecular (E1) reaction 314–315
   elimination unimolecular-conjugate base (E1cB) reaction 315–316
menadione 301
mercaptans *see* thiols
meso compounds 136
meta-chloroperoxy acid (MCPBA) 289
meta directors 489
metal hydrides
   C=O and C=S containing compounds 260–261
   reducing agents 252–253
metamphetamine 85
methane
   chlorination of 377–379
   molecular orbital diagram 22
   oxidation state of 276, 277
   sp^3 orbitals 21
   tetrahedral arrangement 22
methanol (CH_3OH) 33, 350
   infrared spectrum of 95, 96
   properties and structure 63
methanol molecules, hydrogen bonds 30
methide migration 411
methoxyethane 93
methylbromide, electron ionization mass spectrum of 345
2-methylbutanoic acid, separating enantiomeric mixtures of 138
methyl carbocation 187
methyl carbon 371
methyl cyanoacylate 90
methylcyclohexane
   chair conformers of 112
   with hydrochloric acid, addition reaction of 189

methyleneimine, carbon-heteroatom double bond   25, 26
methyl Grignard   257
methyl group bonds   287
methyl halide   372, 400
methyl hydrogen   371
methyl radical   194
methyl tert-butyl ether (MTBE)   91
Michael addition reaction   239
minor resonance contributor   17
molecular chirality and biological action   130–131
molecular orbital (MO) theory   20
molecules
  asymmetric   127
  chiral and achiral   128
  line-angle representations of   18–20
  with oxygen, oxidation states for   277, 278
  shapes of   12
monomers
  of carbohydrates   559
  of peptides and proteins   545–547
monosaccharides   559
  base-catalyzed epimerization of monosaccharides   562
  enediol rearrangement of monosaccharides   563
  ester formation of monosaccharides   565
  ether formation of monosaccharides   565
  hemiacetal formation involving monosaccharides   560–562
  intermolecular acetal formation involving monosaccharides   565–566
  with nitric acid, oxidation of   563–564
  with periodic acid, oxidation of   564
  reduction of monosaccharides   565
  with silver ions, oxidation of   563
monosubstituted benzene molecules   471
monosubstituted benzenes, nomenclature of   470–471
monosubstituted cyclohexane, conformational isomers of   112

n

$NaBH_4$ and $LiAlH_4$, reduction using C=O and C=S containing compounds   255–257
  imines   263–265
$Na_2Cr_2O_7$ oxidizing agent   154
natural gas   43
N-bromosuccinamide (NBS)   384
Negishi coupling reaction   534–535
net molecular polarity   14
neutralization reaction   151
N-glycosides   567–568
ninhydrin, reactions of α-amino acids with   554–555
nitric acid   166
  resonance structures   16
nitrile rubber   90
nitriles
  addition reactions involvement in   240–244
  IUPAC nomenclature of   90
  nomenclature of   90
  synthesis of   416
nitrogen bases   177
nitronium ion, substitution reactions involving   479–480
nitrous acid, oxidation using   286–287
NMR spectrometer   348–349
NMR spectrum of 2-methyl-1-propanol   355, 356
N,N-diethyl-3-methylbenzamide (DEET)   3
N,N-dimethylacetamide   462
nomenclature of substituted benzene   470–473
nonaligned nuclear spins   347
nonbonding electrons   9
nonionic organic compound   257
nonpolar covalent bonds   13
nonsuperimposable mirror images see enantiomers
nonsymmetric anhydride   439
N-oxide   314
nuclear magnetic resonance (NMR) spectroscopy   346
  carbon-13 chemical shifts and coupling   363–367
  carbon-13 NMR (^{13}CNMR)   363
  chemical shift, scale of   350–351
  magnetic shielding   349–350
  NMR spectrometer   348–349
  significance of different signals   351–353
  splitting of signals   353–363
  theory of   347–348
nucleophile   184, 224, 251, 397
  of acyl substitution reactions   428
  of S_N2 reactions   402–403

nucleophilic alkene attacks   198
nucleophilic aromatic substitution
    substituted benzene   499–502
    substituted pyridine   502–503
nucleophilic attack   227, 243
nucleophilic bromide anion   197
nucleophilic chloride anion   188
nucleophilic substitution reactions   397–400
nucleophilic substitution reactions, at acyl carbons   425–426
    acid chlorides, substitution reactions involving   428–436
    acyl substitution, mechanism for   426–428
    acyl substitution reactions, applications of   460–462
    amides, substitution reactions involving   451–454
    anhydrides, substitution reactions involving   436–442
    carboxylic acids, substitution reactions involving   454–458
    esters, substitution reactions involving   442–451
    oxalyl chloride, substitution reactions involving   458
    sulfur containing compounds, substitution reactions involving   458–460
nucleophilic substitution reactions, at sp^3 carbons   393
    applications of   414–420
    bimolecular substitution reaction mechanism (S_N2 mechanism)   400–406
    electrophile   393–394
    leaving group   394–397
    nucleophile   397
    nucleophilic substitution reactions   397–400
    unimolecular substitution reaction mechanism (S_N1 mechanism)   406–414
nucleophilic water attacks   198
nucleophilic water bonds   201

O

octet rule   10
—OH group   199, 395, 454–455, 458, 460
optically active compounds   134
organic acid   145, 146
organic compounds   1, 2
organic synthesis   3, 159
organocuperates   253
organohalides and free radicals, environmental impact of   389–391
organolithium reducing agents   253
organomagnesium   177
organometallic compounds   253–254
organometallic reagents, reaction with C=O and C=S containing compounds   257–259
ortho para directors   488
2-oxacyclohexanone   81
oxalic acid   145
oxalyl chloride, substitution reactions involving   458
oxidation   275–279
    of alcohols and aldehydes (see alcohols and aldehydes, oxidation of)
    of alkenes with bond cleavage (see alkenes with bond cleavage, oxidation of)
    of alkenes without bond cleavage (see alkenes without bond cleavage, oxidation of)
    of alkynes   299–300
    of aromatic compounds   300–301
    autooxidation of ethers and alkenes   301–302
oxidation of monosaccharides
    with nitric acid   563–564
    with periodic acid   564
    with silver ions   563
oxidation reactions   153–154, 275, 276
    of alkenes   296–299
    applications of   302–304
oxidation state
    of carbon and methane   276, 277
    for hydrocarbons   277, 278
    for molecules with oxygen   277, 278
oxidative addition   528, 531
oxidative addition reactions   526
oxidizing agents   153, 275, 279
oxiranes   91
    nomenclature of   93–94
    reduction of   266–267
oxymercuration-demercuration, hydration by   203–204
ozone hole   369
ozone layer   369
ozonolysis of alkenes   295–296

p

palladium catalyst   159
palladium-catalyzed coupling reactions *see also* palladium-catalyzed coupling reactions
   Heck reaction   528–531
   Negishi coupling reaction   534–535
   Stille coupling reaction   533–534
   Suzuki reaction   531–533
Pauli's exclusion principle   7
pentene   56
peptides
   monomers of   545–547
   primary structure and properties of   556–558
peracids   288
percent enantiomeric (%ee) excess   139
pericyclic reactions   158, 513
   cycloaddition reactions   513–519
   electrocyclic reactions   519–521
   sigmatropic reactions   521–522
periodic acid, oxidation using   287–288
peroxide anion   206
peroxyacids   288
petroleum   276
phenobarbital   45
phenolic compounds   63–64
phenyl (Ph, C_6H_5) groups   236
o-phenylphenol   64
phenyl propionate   360
phosphine oxide   236
phospholipids   568
   properties and reactions of   572
phosphorous tribromide, alcohols conversion to   396
phosphorous ylide   236
   synthesis of   236–237
physical properties of compound   31–32
pi (π) electrons of conjugated systems   521
pinacol rearrangement   322–323
Planck's constant   332
polar covalent bonds   13, 393
polarizability/atom size   174
polar-protic solvents   33
polybrominated diphenylethers   370
polycyclic aromatic compounds, electrophilic substitution reactions of   494–496
polycyclic compounds   476–477
polyenes, IUPAC nomenclature of   57
polymerization   537
polymers   537
   properties of   543–544
polysaccharides   566–567
polystyrene   538
p orbitals   5–6
potassium methoxide   178
potassium permanganate ($KMnO_4$)
   at elevated temperatures, oxidation of alkenes   293–295
   oxidation using   280–281
potassium *tert*-butoxide   178–179
primary alcohol   65
primary amines   85
primary carbocation   187, 189
primary carbon   371
principal quantum numbers   5
proof, defined   63
propagation steps   378
propane   371
   bromination of   380–386
2-propanone, infrared spectrum of   98
propylene glycol   63
prostaglandins   568, 573
   structure and properties of   572–573
protection-deprotection of amino functionality   550–551
proteins
   monomers of   545–547
   secondary structure of   558–559
p-toluenesulfonyl chloride   459
pyranose ring   92
pyridine   430
   electrophilic substitution reactions of   497–499
pyridinium chlorochromate, oxidation using   285
pyrrole, electrophilic substitution reactions of   496–497

q

quinone   301

r

racemic mixture   134
radical cation   343
radical coupling   542
radical propagation step   193
radicals   193, 373
   structure and stability of   374–376

radio frequency   348
rate-determining step (RDS)   315, 400, 480
reaction mechanism   183, 185
reddish bromine solution   196
reducing agents   252
   dissolving metals   254
   hydrogen, in the presence of a catalyst   254
   metal hydrides   252–253
   organometallic compounds   253–254
reduction
   of monosaccharides   565
   of triglycerides   571
reduction reactions   150–152, 251–252
   aromatic compounds, alkynes, and alkenes   268–272
   C=O and C=S containing compounds (*see* C=O and C=S containing compounds, reduction of)
   imines   263–266
   oxiranes   266–267
   reducing agents (*see* reducing agents)
reductive amination   266
   α-amino acids using   548
reductive elimination   532
reductive elimination reactions   527–528
refining process   43
regiospecific reactions   191, 196, 199, 206, 224, 242
relative stabilities of radicals   194
resins   543
resonance energy of benzene   475
resonance structures   15–17, 172–174
ring closure for electrocyclic reactions   520–521
Robinson annulation   239
rules, IUPAC name determination   46–48

S

saccharic acid   564
saponification (hydrolysis)   444
   of triglycerides   570–571
saturated fatty acids   74, 115
saturated hydrocarbons   369
   classification of the carbons   44
   defined   41
   uses and properties of   43
Schiff base   240
secondary alcohol   65
secondary amines   85
secondary carbocation   187, 191, 192
secondary carbon   371
secondary structure of proteins   558–559
short covalent bonds   373
sigmatropic reactions   521–522
silver ions, oxidation using   286
silyl ethers, synthesis of   416–418
Simmons–Smith reaction   208
simvastatin   130
single-barbed arrow   373
single bond   12
single covalent bond   414
singlet carbenes   207
soap   444
sodium amide ($NaNH_2$)   172
sodium borohydride ($NaBH_4$)   150, 151
sodium hydroxide   147
soft and hard metallic hydrides, reactions with esters   448
soft bulky reducing agent   434
soft Lewis acids   166
soft Lewis bases   166
soft metal hydrides reagents, acid chlorides with   434
soft metallic hydrides, anhydrides with   441
soft organometallic reagents, substitution reactions involving
  and acid chlorides   433
  anhydrides with   440
  esters with   447
soft reducing agent   440
solubility of polymers   544
solvents of S_N2 reactions   403–404
solvolysis reactions   407
s orbitals   5
specific rotation [α], 134
spectroscopic methods   94
  infrared spectroscopy   95–98
spectroscopy   331
  electromagnetic spectrum   331–333
  infrared spectroscopy   337–343
  mass spectroscopy   343–346
  nuclear magnetic resonance (NMR) spectroscopy (*see* nuclear magnetic resonance (NMR) spectroscopy)
  types of   333
  UV-Vis spectroscopy and conjugated systems   334–337
spiranes   52

spiratanes 52
spiro compounds 52
splitting of signals 353–363
stability
  of alkanes 119–121
  of alkenes 121–122
  of alkynes 122–123
  of benzene 473–475
  of carbocations 187, 188, 194
  of radicals 194, 195
stereochemistry 125
  of addition reaction products, prediction of 190
  of S_N2 reactions 404–405
stereogenic carbon 126, 190
  compounds with more than one stereogenic carbon 134–137
stereogenic compounds, properties of 133–134
stereogenic molecules 126
  dashed-wedge representation for 128
stereoisomers 103
stereospecific reactions 198, 514
steric interactions 205
steroids 568, 572
  structure and properties of 572–573
Stille coupling reaction 533–534
straight chain alkanes, IUPAC nomenclature of 45, 46
Strecker synthesis, α-amino acids using 547–548
strong bases 165
structural isomers 103
  butane 103
styrene, cationic polymerization of 538–540
substituted benzene 499–502
  electrophilic aromatic substitution reactions of 484–491
  nomenclature of 470–473
substituted benzene compounds 491–494
substituted benzenes, IUPAC nomenclature of 58–60
α-substituted carbonyl compounds 419–420
substituted phenols, IUPAC nomenclature of 68
substituted pyridine 502–503
substitution reactions 156–157, 369
  acid chlorides 428–436
  acyl substitution 426–428
  amides 451–454
  anhydrides 436–442
  1-bromobutane and sodium iodide 157
  carboxylic acids 454–458
  esters 442–451
sucrose 567
sulfonium ion, substitution reactions involving 484
sulfonyl chlorides, alcohols conversion to 396–397
sulfur containing compounds, substitution reactions involving 458–460
sulfuric acid 479
superimposable molecules 126
Suzuki catalytic coupling reaction 533
Suzuki reaction 531–533
Swern oxidation 283–284
symmetrical alkenes, addition reactions to 185–186
symmetrical molecule 127
symmetry allowed 514
symmetry forbidden 514
synthesis of α-amino acids 547–550
synthetic organic chemistry 213
synthetic polymers and biopolymers 537, 544–545
  acid-base properties of amino acids 547
  amino acids, monomers of peptides and proteins 545–547
  α-amino acids, reactions of 550–555
  α-amino acids, synthesis of 547–550
  anionic polymerization of alkenes 540
  cationic polymerization of alkenes 537–540
  copolymerization of alkenes 542–543
  disaccharides and polysaccharides 566–567
  free radical polymerization of alkenes 540–542
  lipids 568–569
  monosaccharides, monomers of carbohydrates 559–566
  N-glycosides and amino sugars 567–568
  peptides, primary structure and properties of 556–558
  phospholipids, properties and reactions of 572
  properties of 543–544
  proteins, secondary structure of 558–559

steroids, prostaglandins, and terpenes   572–573
triglycerides, properties and reactions of   569–571
waxes, properties and reactions of   569

t

tartaric acid, molecular chirality   130
taxol   3
termination step   378
terpenes   568–569, 573
  structure and properties of   572–573
tert-butyl group   351
tertiary alcohol   65
tertiary amines   85
tertiary carbocation   187, 411
tertiary carbon   371
tertiary radical   194
tetrabutyl ammonium fluoride (TBAF)   418
tetrahedral geometry   12
tetrahedral intermediate   426, 453
tetrahydrafuran (THF)   91
tetramethylsaline (TMS)   351
thalidomide   130–131
thermal properties of polymers   544
thiazolinone   558
thiols   65
  IUPAC nomenclature of   68
thionyl chloride, alcohols conversion to   396
300 MHz ^{1}C NMR spectrum of benzaldehyde in CDCl$_3$ 366
300 MHz ^{1}C NMR spectrum of benzoic acid in CDCl$_3$ 366
300 MHz ^{1}C NMR spectrum of benzyl alcohol in CDCl$_3$ 365
300 MHz ^{1}C NMR spectrum of 1,3-dinitrobenzene in CDCl$_3$ 365
300 MHz ^{1}C NMR spectrum of nitrobenzene in CDCl$_3$ 365
300 MHz ^{1}H NMR spectrum for 1-butanol in CDCl$_3$ 357
300 MHz ^{1}H NMR spectrum of benzaldehyde acid in CDCl$_3$ 362
300 MHz ^{1}H NMR spectrum of benzene in CDCl$_3$ 358, 364
300 MHz ^{1}H NMR spectrum of benzoic acid in CDCl$_3$ 362
300 MHz ^{1}H NMR spectrum of benzyl alcohol in CDCl$_3$ 361
300 MHz ^{1}H NMR spectrum of benzyl amine in CDCl$_3$ 361
300 MHz ^{1}H NMR spectrum of 1,2-dinitrobenzene in acetone   359
300 MHz ^{1}H NMR spectrum of 1,3-dinitrobenzene in acetone   359
300 MHz ^{1}H NMR spectrum of 1,4-dinitrobenzene in CDCl$_3$ 358
300 MHz ^{1}H NMR spectrum of ethyl benzoate in CDCl$_3$ 360
300 MHz ^{1}H NMR spectrum of nitrobenzene in CDCl$_3$ 358
300 MHz ^{1}H NMR spectrum of phenyl propionate in CDCl$_3$ 360
300 MHz nuclear magnetic resonance (NMR) instrument   349
thromboxane   573
α-thujene   53
thymol   64
TMS group   417
α-tocopherol radical   388
Tollen's reagent   563
Tollens test for aldehydes   286
tosyl chloride (TsCl)   459
trans-1,3-dichlorocyclopentane   110
trans-dimethylcyclobutane   109
trans-dimethylcyclopentanes   109, 110
trans-diols   289–290
trans-elimination   311
trans-4,5-epoxy-(E)-2-decenal   93
transesterification   445
  of triglycerides   571
transition metal complexes, reactions of   525
  β-elimination reactions   527
  ligand migratory insertion reactions   527
  oxidative addition reactions   526
  palladium-catalyzed coupling reactions (see palladium-catalyzed coupling reactions)
  reductive elimination reactions   527–528
  transmetallation reactions   526–527
transmetallation reactions   526–527
trans product   197
trans-stereochemistry   530
triglycerides, properties and reactions of   569–571

triple bonds 12, 373
tri-*tert*-butoxyaluminum hydride 448

u
ubiquinone 301
unexpected hydrohalogenation products 191–192
unimolecular substitution reaction mechanism (S_N1 mechanism) 406–414
unsaturated fatty acids 74–75, 115
unsaturated hydrocarbons 54–55
- acetylene 54, 55
- benzene 54, 55
- ethylene 54, 55
- examples of 54

unshared electrons 9
unsymmetrical alkenes, addition reactions to 186–187
unsymmetrical anhydrides 442
urea synthesis 1
UV-Vis spectrophotometer 334
UV-Vis spectroscopy 334–337
UV-Vis spectrum for isoprene 335

v
valence electrons 8–10
valence shell electron pair repulsion (VSEPR) theory 12, 268–269
Van der Waals attraction 31
vanillin 64
vibrational modes 338
vinylidene cyanide, anionic polymerization of 540
Vitamin K_2 301

w
water
and acid chlorides, substitution reactions involving 429–430
to alkenes (hydration of alkenes) (*see* water to alkenes (hydration of alkenes))
to alkynes 211–213
amides with, substitution reactions involving 452–453
anhydrides with, substitution reactions involving 437–438
to carbonyl compounds 226–229
esters with, substitution reactions involving 444–445
infrared spectrum of 95
water (dehydration), elimination reactions involving
- carbocation rearrangement 321–322
- dehydration products 319–321
- pinacol rearrangement 322–323

water to alkenes (hydration of alkenes) 199–202
- hydroboration-oxidation 204–207
- oxymercuration-demercuration 203–204

waxes, properties and reactions of 569
weak acids 165
Wittig reaction 236, 245
Wolff Kishner reduction 261–263
Wood–Fieser and Fieser–Kuhn rules 337

x
X-ray spectroscopy 469

y
ylides, addition to carbonyl compounds 235–237

z
Zaitsev product 313, 317